P9-DHA-995

GENETICS

SECOND EDITION

MONROE W. STRICKBERGER

The University of Missouri—St. Louis

MACMILLAN PUBLISHING CO., INC.
NEW YORK
COLLIER MACMILLAN PUBLISHERS
LONDON

Printed in the United States of America

Macmillan Publishing Co., Inc.
866 Third Avenue, New York, New York 10022

Collier Macmillan Canada, Ltd.

Library of Congress Cataloging in Publication Data
Strickberger, Monroe W
Genetics.

Includes bibliographies and indexes.
1. Genetics. I. Title.
QH430.S87 1976 575.1 74-33103
ISBN 0-02-418090-4

Printing: 5678 Year: 89012

To Zelda, Neal, and Paul

PREFACE

The purposes of this book are to provide basic theoretical information about genetics, the study of heredity, and to present some of the experiments and reasoning through which this information has been achieved.

Although genetics is a modern science, it has grown even more rapidly in recent years and its branches now extend to almost all fields of biology. In fact, the diverse specializations of genetics enable it to occupy at present a unique central position among the biological sciences in that it ties together different disciplines that involve form, function, and change. This unifying "core" quality of genetics stems from the many levels in which genetic phenomena operate, from the molecules of cells, through developmental stages of individuals, to populations of organisms. Compartmentalization at each of these levels is common and even unavoidable since research work proceeds separately in each area of genetics. To picture genetics, however, only in terms of its many separate constituent parts leads easily to an unbalanced and fragmented view that does not do justice to the unity of the science as a whole.

Where does the unity of genetics lie? Initially the recognition of heredity began with the simple observation made among different organisms that "like begets like." In the course of history, however, the simplicity of this observation was replaced by many complex questions: From what source does this wonderful correspondence between generations arise? How is the knowledge transmitted that determines biological development? What factors account for similarities between generations and what for differences? What is inherited and what is not? What hereditary factors do members of a species have in common and in what factors do they differ? Why and how do new species of organisms arise? How can we control heredity?

A theme shared by all these questions is a concern with the materials and modes of inheritance. Since genetics is the science that seeks to answer these questions, it can be broadly defined as the study of biological material transmitted between generations of organisms. More exactly, this science encompasses studies of the kind of material transmitted, the manner in which this is accomplished, and the effect of this material on an individual organism and on generations of organisms. If we call this hereditary material *genetic material,* we can mark out the following areas of study.

1. What and where is genetic material?
2. How is it formed, transmitted, and changed?
3. How is it organized and how does it function?
4. What happens to it among groups of organisms as time passes?

The fundamental unifying theme of genetics is thus a material one which can be studied at many different levels of existence. The order of these questions and levels does not at all imply a rank of importance, since each aspect is only one facet of genetics, although some problems may assume more interest at particular historical times. It is with this over-all view in mind that the book is organized.

Within this framework a historical approach to genetics has been presented in many places for a number of reasons. First, in the swift progress of modern genetics, many aspects of our present understanding will rapidly be changed by future discoveries. Rather than be the study of a static set of axioms, a true presentation of genetics should include a sense of its continuity and progress. Second, such an approach provides many opportunities for the development of ideas from the simple to the complex and thereby facilitates learning. Third, the people who have contributed to a science and the relationship between their contributions are an important and interesting aspect of the science and help to encourage student interest.

For teaching purposes, the instructor can select sections of the text according to his training and inclination, but it is suggested that material be used from each of the basic subdivisions of the book with special emphasis on chapters included in the sections on transmission and arrangement of genetic material. A thorough understanding of basic genetic methodology and recombinational principles is extremely important in enabling students to forge ahead intelligently in areas of their own interest. One suggested program for a single-semester genetics course meeting three times weekly is to include Chapters 1 through 7, 9 through 12, 16 and 17, and selected sections of Chapter 8 (e.g., chi-square) and of Chapters 19 through 31. A one-semester course that is oriented toward evolutionary problems can follow a similar program but include material in Chapters 14 and 15, Chapters 32 through 36, and omit material in Chapters 19 and 20, and 24 through 30.

M. W. S.

ACKNOWLEDGMENTS

In the face of the large volume of genetic literature and the many important developments and changes that are continually taking place in genetics, it would be an illusion for an author to maintain, or for a reader to believe, that a textbook in this field can be written without errors or misinterpretations. For the first edition of this book I was therefore fortunate in having extensive sections of the manuscript reviewed by L. Van Valen and C. J. Wills. Various chapters were read and criticized by the late Th. Dobzhansky, N. E. Melechen, H. D. Stalker, and the late R. Rolfe. Exercise problems were offered to me by S. W. Brown, M. M. Green, H. D. Stalker, and the late J. A. Jenkins.

The present edition was undertaken about five years after the first was completed and incorporates a large number of changes and additions. Through the mediation of the Biology Editor for the publisher, Charles E. Stewart, Jr., extensive portions of this material were reviewed by J. Antonovics, J. Boynton, N. W. Gillham, G. B. Johnson, J. H. Postlethwait, E. Simon, and F. W. Stahl. I am also grateful to H. N. Arst, Jr., M. Ashburner, D. J. Cove, A. Derby, G. A. Dover, R. T. O. Kemp, D. W. MacDonald, P. Oliver, H. E. Schaffer, J. M. Thoday, H. L. K. Whitehouse, and A. S. Wiener, who criticized individual chapters, and to many colleagues and students who were kind enough to point out errors and ambiguities in the first edition.

Credits for the original sources of illustrations are given in the figure legends, and the full citations are usually in the references at the end of each chapter. The references also include sources used for the text material as well as selected further articles or books that may be relevant. Citations for authors mentioned in the problems are usually not included in the references, but will mostly be found in the *Answer Manual for Genetics, Second Edition,* published separately by the Macmillan Company. I take responsibility for errors that remain, and I would be grateful if they are brought to my attention in a constructive fashion.

Again, I owe special thanks to my wife, Zelda, for considerable typing and many corrections. A large amount of additional typing for this edition was carefully done by Mrs. Robyn Salvato. Kenneth Johnson and Patricia Comens helped greatly with indexing.

CONTENTS

PART I
IDENTIFICATION OF GENETIC MATERIAL

PART II
TRANSMISSION AND DISTRIBUTION OF GENETIC MATERIAL

PART V
FUNCTION OF GENETIC MATERIAL

PART VI
COURSE OF GENETIC MATERIAL IN POPULATIONS

PART

I

IDENTIFICATION OF GENETIC MATERIAL

Nothing from nothing ever yet was born.

LUCRETIUS
On the Nature of Things

1

HISTORY OF THE PROBLEM

The modern science of genetics originated when Gregor Mendel discovered that hereditary characteristics are determined by elementary units transmitted between generations in uniform predictable fashion. Each such unit, which can be called a genetic unit, or *gene*, is a substance that must satisfy at least two essential requirements: (1) that it is inherited between generations in such fashion that each descendant has a physical copy of this material, and (2) that it provide information to its carriers in respect to structure, function, and other biological attributes. Perhaps as a consequence of this double aspect of the gene, there have been two important historical approaches toward genetic phenomena: one toward identifying its physical substance, the genetic material, and the other toward discovering the manner and ways by which its manifestations, the biological characters, are inherited. Until the twentieth century these lines of investigation were usually separate, because so little was known about both transmitted substances and transmitted characters.

THE CONTINUITY OF LIFE

The first aspect of the problem, to determine the exact form of genetic material, was not an easy task. For a long period of time it was common to think that biological materials did not necessarily have to be transmitted between generations. At least until the middle of the eighteenth century, practically all biologists believed that many organisms, especially small "primitive" ones, could arise spontaneously from various combinations of decaying matter. The spontaneous appearance of flies from refuse, the observation by Leeuwenhoek (1632–1723) of small infusoria "arising" from apparently clear infusions of hay, and many other observations seemed to support the idea that life could arise without the direct transfer of matter from immediate ancestors.

It took a series of experiments begun by two Italian biologists, Redi (1621–1697) and Spallanzani (1729–1799), to seriously question this doctrine of "spontaneous generation." By excluding adult flies from laying eggs on meat, Redi showed that fly larvae would not develop. Spallanzani found that if one boiled sealed flasks of organic matter long enough, the usual tiny "animalcules" observed by Leeuwenhoek would no longer appear spontaneously. Since Spallanzani's sealed boiled flasks now contained heated or "tainted" air, the question remained of whether unheated air would, nevertheless, generate new organisms.

The seventeenth and eighteenth centuries also marked the beginnings of systematic studies or the classification of biological organisms into separate and distinct species. According to Linnaeus (1707–1778), the founder of systematics, there was a "fixity of species," so that organisms of one species could only give rise to organisms of the same type. As the science of classification grew, this view became generally accepted among biologists. At the same time, however, the spontaneous origin of living creatures from organic matter implied a nonfixity of species and this proved quite difficult to reconcile with the Linnaean concept.

Controversy on this subject raged until the nineteenth century, when the idea of spontaneous generation was finally put to rest by Pasteur (1822–1895) and by Tyndall (1820–1893), who showed that the putrefaction of organic matter only occurred under conditions that permitted solid particles to enter a nutrient culture. These solid particles were soon identified as the microbes whose multiplication leads to the fermentation of organic cultures. By common agreement these findings led to the view that, at least for the present, the birth of new organisms arises only through the continuity of life.

PREFORMATIONISM AND EPIGENESIS

The concept of a continuous transfer of living material left important questions to be answered in respect to the specific physical parts transferred by a parental organism and the manner in which this is accomplished. Originally Aristotle had proposed that an organism formed through sexual reproduction receives the "substance" of the female egg and a contribution of "form" by the male seminal fluid. The combined effect produced by these two factors in creating a new organism did not necessarily involve any material transfer between them but occurred through a mystical influence of the male semen, called by Harvey (1578–1657) the *aura seminalis.* Later, during the seventeenth and eighteenth centuries, after the discovery of eggs and sperm and of pollen and ova, the idea was advanced by many biologists that one of the sex cells, or gametes, either sperm or egg, contained within itself the entire organism in perfect miniature form (*preformationism*); for this miniature creature to unfold into its preformed adult proportions only proper nourishment was necessary. There were, of course, many difficulties in accepting this hypothesis, not the least being that such perfect miniature creatures, although imagined (Fig. 1–1), were never truly observed. Nevertheless the concept that an organism develops from a minuscule piece of transmitted matter, "preformed" though it may be, was an important step forward as compared to the idea of spontaneous generation.

When Wolff (1738–1794) showed that different adult structures of both plants and animals develop from uniform embryonic tissues that betray no inkling of their ultimate fate, preformationism was replaced by the more modern, although not necessarily novel, idea of *epigenesis.* The epigenetic view proposed that many new factors, such as tissues and organs, appeared during the development of an organism which were not present in its original formation. Wolff believed these organs arose completely *de novo* through mysterious vital forces, while his famous successor, von Baer (1792–1876), proposed the more accepted view

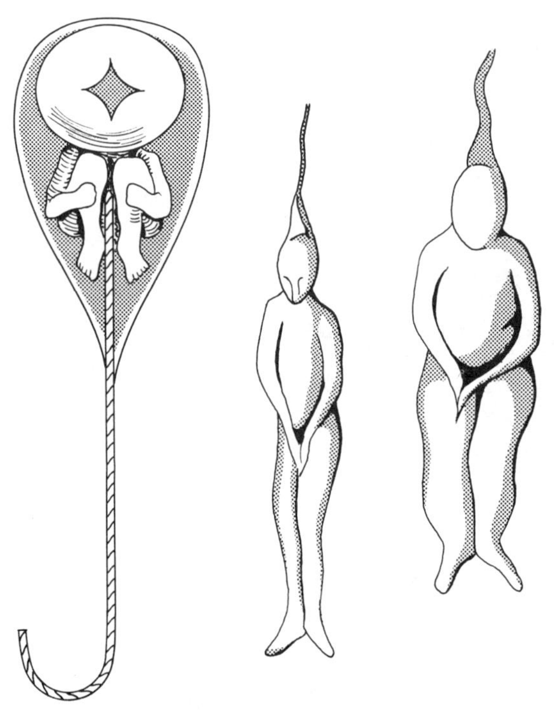

Figure 1–1

"Homunculi" presumed to be seen in human sperm. (After Singer, from Hartsoeker and Delenpatius.)

that they arose through a gradual transformation of increasingly specialized tissue.

PANGENESIS AND THE INHERITANCE OF ACQUIRED CHARACTERS

It is easy to see that with the acceptance of the theory of epigenesis the problem of finding the genetic material of an organism was put back to something invisible and, many believed, something mystical, within the original embryonic cell. Charles Darwin (1809–1882), the founder of modern evolutionary theory, and certainly no mystic, believed (as did many other biologists) that very small, exact, but invisible copies of each body organ and component (*gemmules*) were transported by the bloodstream to the sex organs and there assembled into the gametes. Upon fertilization gemmules of the opposite sex were added and all these miniature elements then separated out to different parts of the body during development to constitute a mixture of maternal and paternal organs and tissues (Fig. 1–2a).

To those who believed in evolution this doctrine of *pangenesis* provided an even further attraction by explaining how heritable changes could occur that might lead to the origin of new species. According to pangenesis, for example, the excess use or disuse of an organ would alter its gemmules and consequently lead to a changed inheritance in the descendants. This theory, named "the inheritance of acquired characters," had a long previous history and even has some few adherents today (Lysenko and his remaining followers in the Soviet Union).

Lamarck (1744–1829), the foremost eighteenth-century popularizer and exponent of this theory, tried to explain the extraordinary ability of these small hereditary agents to respond directly to the environment by assuming that each had a spiritual consciouslike property that could absorb and interpret messages from the outside. Despite the mysticism and many other difficulties involved in the theories of pangenesis and inheritance of acquired characters, many biologists felt it necessary to accept these theories as the only existent reasonable explanations of heredity.

By the end of the nineteenth century a number of important discoveries had been made which fortunately set the stage for a more precise material characterization of the source of heredity. Many details of cell structure and cell division were already known through the researches of a long line of biologists, among them notably Schleiden (1804–1881), Schwann (1810–1882), Nägeli (1817–1891), Virchow (1821–1902), Flemming (1843–1915), and Bütschli (1848–1920). The cell *nucleus*, named and described by Robert Brown (1773–1858) in 1833, was shown by O. Hertwig (1849–1922) to be directly involved in sea urchin fertilization through the union of the sperm and egg nuclei. Similarly Strasburger (1844–1912) had shown such union to occur for plants, and had invented the terms *nucleoplasm* and *cytoplasm* to refer to the protoplasmic material in the nucleus and its surrounding cell body, respectively. The dark-staining nuclear threads,

(a) Pangenesis theory

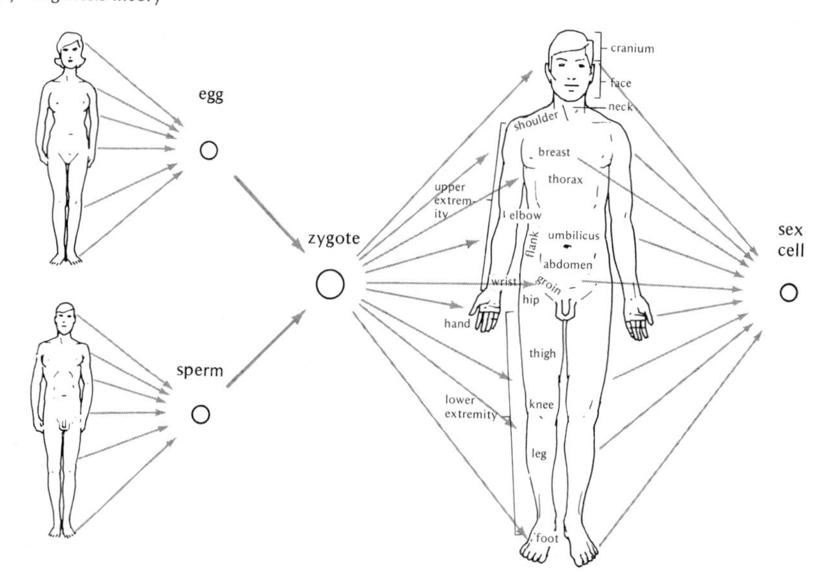

(b) Germplasm theory

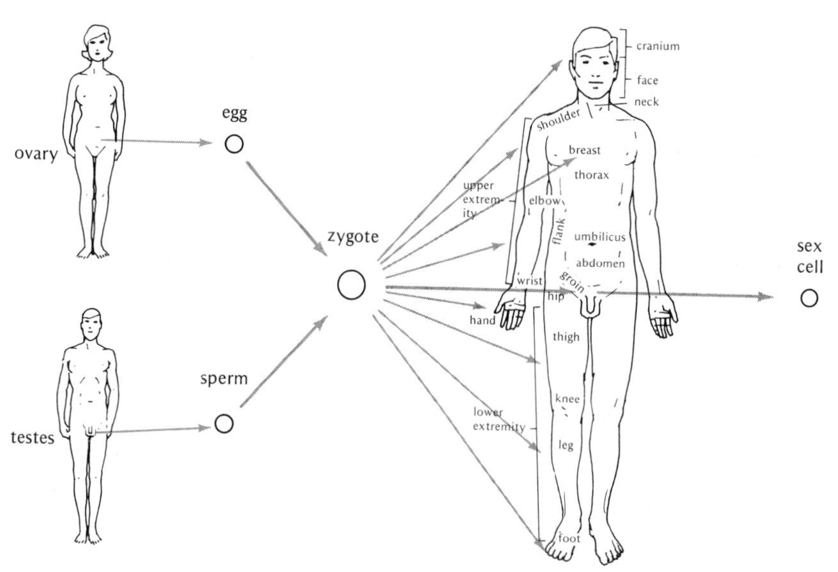

Figure 1–2

Comparison between (a) pangenesis and (b) germplasm theories in the formation of a human. In pangenesis all structures and organs throughout the body contribute copies of themselves to a sex cell. In the germplasm theory the plans for the entire body are contributed only by the sex organs.

named "chromatin," which had been shown first by Schneider (ca. 1873) and later by Flemming to divide longitudinally during cell division, also passed in equal portions to the two daughter cells according to van Beneden (1845–1910). The total number of these individual threads, or *chromosomes*, remained constant in all cells of an organism except during gamete formation. In gametes the number of chromosomes was reduced but was then later restored when the nuclei of the gametes fused during fertilization to form the first embryonic cell.

Thus the cellular link between parents and offspring that occurred through gametes was also accompanied by chromosomal links between each generation. The chromosomes of the offspring, however, were not merely quantitatively equal to the number of parental chromosomes but seemed to be qualitatively equal as well. That is, since the splitting of the parental chromosomes occurred longitudinally, any differences existing along the length of the chromosome were also passed on to offspring.

The close of the century had also fairly well established, through the efforts of Weismann (1834–1914), that pangenesis could not be verified. Weismann showed that even after 22 generations of mice had been denuded of their tails, newborn mice still managed to inherit the complete tail structure. The theory of pangenesis was therefore replaced by Weismann with the "germplasm" theory, which proposed that multicellular organisms give rise to two types of tissue: *somatoplasm* and *germplasm*. The somatoplasm consisted of tissues that were essential for the functioning of the organism but that lacked the property of entering into sexual reproduction. Changes that occurred in somatic tissues (e.g., mouse tails) were thus not passed on in heredity. Germplasm, on the other hand, was set aside for reproductive purposes, and any changes occurring within it could lead to changed inheritance. According to this view, there was a continuity of germplasm between all descendant generations which accounted for the many biological similarities that were inherited (Fig. 1–2b).

By the early twentieth century, most of the morphological features of the cell had been observed under the light microscope, as well as the general features involved in the cell-division processes of *mitosis* (somatic, or body-cell, division) and *meiosis* (germ-cell, or gamete, formation). Work by cytologists such as Boveri (1862–1915), Henking (ca. 1891), Montgomery (1873–1912), and others had shown that such cell divisions led to an accurate partitioning and separation of the nuclear chromosomes. In somatic mitosis, the number of chromosomes in each daughter cell remained identical to that of the parental cell. In meiosis, gametes were formed containing exactly half the parental number, which then combined with a gamete of the opposite sex to produce an individual with a full chromosomal complement.

It was only one step further to hypothesize that the constancy of chromosomes in a species and their precise partitioning in inheritance was a reflection of the similar constancy and inheritance pattern of more easily observable biological characters. That step was rapidly taken when a series of fundamental discoveries by the Austrian monk Gregor Mendel (1822–1884) was brought to light in 1900.

EXPERIMENTS WITH HYBRIDS

Mendel's important experiments will be discussed in detail in Chapters 6 and 7. For the present we can mention that they were an extension and development of the efforts of a long line of biologists who had been studying the effects of unifying (*hybridizing*) two differently appearing parental stocks. It had long been thought that heredity was a "blending" process and that offspring were essentially a "dilution" of the different parental characteristics. Blending inheritance thus helped to explain the observation that children were at times intermediate to both parents in respect to measurable characters such as size. On the other hand, there were also the frequent observations that children could resemble either one parent or the other, and that specific characteristics appeared "undiluted" for many generations. Observations of this latter kind extended to numerous characters in plants and animals and led to speculative explanations ranging from different quantitative contributions by each parent to differences in wind direction at the time of fertilization.

One of the first biologists to perform actual experiments comparing similarities and dissimilarities of plant hybrids and their parents was Kölreuter (1733–1806). He made the important observation that although hybrids between species may show a uniform appearance, their fertile offspring will usually produce considerable diversity. Similar observations were noted throughout the 1800s by various botanists, including Gärtner (1772–1850), Naudin (1815–1899), and Darwin. None of them, however, with perhaps the exception of the bee-hybridizer Dzierzon (1811–1906), paid much attention to the numerical ratio in which the different parental characters appeared in the offspring.

It was Mendel's exceptional contribution to demonstrate that the appearance of different characters in heredity followed specific laws that could be determined merely by counting the diverse kinds of offspring produced from any particular set of crosses. More specifically, Mendel showed that the hybrid between two parental types of garden peas, which had differed in a single character, would produce two types of gametes in equal numbers. Each type of gamete was, in respect to the observed character, an unchanged descendant of one of the original parental gametes. Evidently, for these experiments there was no blending or dilution of inheritance, and inherited characteristics were determined by discrete units which remained unchanged in the presence of other such units in the hybrid. Although published in 1866, Mendel's findings unfortunately lay unrecognized until 1900. When these fundamental experiments were finally discovered, they appeared to fit quite well with the particulate nature of individual chromosomes that separated during cell division. As a consequence, the identification of chromosomes as carriers of genetic material was rapidly put forth by Sutton (1876–1916), Boveri, and others (Chapter 7).

Thus the beginning of the twentieth century saw the basic groundwork of genetics embodied in the idea that an actual hereditary material existed, that it was of a particulate nature, and that its behavior in transmission from one generation to another could be predicted. The next four chapters describe in greater detail some of the fundamental cellular and reproductive mechanisms involved in heredity as well as the important discoveries that led to a chemical and physical identification of the genetic material.

REFERENCES

Hughes, A., 1959. *A History of Cytology*. Abelard-Schuman, New York.

Nordenskiöld, E., 1928. *The History of Biology*. Tudor, New York.

Roberts, H. F., 1929. *Plant Hybridization Before Mendel*. Princeton Univ. Press, Princeton. (Reprinted 1965 by Hafner Publishing Co., New York.)

Singer, C., 1959. *A History of Biology*. Abelard-Schuman, New York.

Sirks, M. J., and C. Zirkle, 1964. *The Evolution of Biology*. Ronald Press, New York.

Stubbe, H., 1972. *History of Genetics*. Massachusetts Institute Tech. Press, Cambridge. (Translated by T. R. W. Waters from the second German edition, 1965, with later additions.)

Taylor, G. R., 1963. *The Science of Life*. McGraw-Hill, New York.

Zirkle, C., 1946. The early history of the idea of the inheritance of acquired characters and of pangenesis. *Trans. Amer. Phil. Soc.*, **35**, 91–151.

The following are collections of original research papers in various fields of genetics. Numerous original contributions cited throughout this book can be found in one or more of these collections. For example, the collection of Gabriel and Fogel contains English translations of papers by Redi, Spallanzani, Pasteur, Flemming, van Beneden, Weismann, and other historical figures in biology.

Abou-Sabé, M. A. (ed.), 1973. *Microbial Genetics.* Dowden, Hutchinson and Ross, Stroudsburg, Pa.

Adelberg, E. A. (ed.), 1966. *Papers on Bacterial Genetics*, 2nd ed. Little, Brown, Boston.

Boyer, S. H., IV (ed.), 1963. *Papers on Human Genetics.* Prentice-Hall, Englewood Cliffs, N.J.

Gabriel, M. L., and S. Fogel (eds.), 1955. *Great Experiments in Biology.* Prentice-Hall, Englewood Cliffs, N.J.

Levine, L. (ed.), 1971. *Papers on Genetics: A Book of Readings.* C. V. Mosby, St. Louis.

Moore, J. A. (ed.), 1972. *Readings in Heredity and Development.* Oxford Univ. Press, New York.

Peters, J. A. (ed.), 1959. *Classic Papers in Genetics.* Prentice-Hall, Englewood Cliffs, N.J.

Raacke, I. D. (ed.), 1971. *Molecular Biology of DNA and RNA.* C. V. Mosby, St. Louis.

Selected Papers in Biochemistry. This is a series of twelve volumes published originally by the University of Tokyo Press during 1971–72. They contain a large number of important English language research papers in biochemistry. Many of these papers pertain to genetics, and the titles of the volumes range from *Bacterial Genetics and Temperate Phage* to *Coding, Protein Synthesis,* and *Allosteric Regulation.* They are edited by J. Tomizawa, S. Nishimura, Y. Kaziro, and others, and are distributed through University Park Press, Baltimore.

Stent, G. S. (ed.), 1965. *Papers on Bacterial Viruses*, 2nd ed. Little, Brown, Boston.

Taylor, J. H. (ed.), 1965. *Selected Papers on Molecular Genetics.* Academic Press, New York.

Voeller, B. R. (ed.), 1968. *The Chromosome Theory of Inheritance, Classic Papers in Development and Heredity.* Appleton-Century-Crofts, New York.

Zubay, G. L., and J. Marmur (eds.), 1973. *Papers in Biochemical Genetics*, 2nd ed. Holt, Rinehart and Winston, New York.

Collections of articles from *Scientific American* on various topics in genetics can be found in:

Hanawalt, P. C., and R. H. Haynes (eds.), 1973. *The Chemical Basis of Life.* W. H. Freeman, San Francisco.

Haynes, R. H., and P. C. Hanawalt (eds.), 1968. *The Molecular Basis of Life.* W. H. Freeman, San Francisco.

Srb, A. M., R. D. Owen, and R. S. Edgar (eds.), 1970. *Facets of Genetics.* W. H. Freeman, San Francisco.

Extensive glossaries of the terms used in genetics can be found in:

King, R. C., 1972. *A Dictionary of Genetics*, 2nd ed. Oxford Univ. Press, New York.

Rieger, R., A. Michaelis and M. M. Green, 1968. *A Glossary of Genetics and Cytogenetics*, 3rd ed. Springer-Verlag, New York.

A bibliography citing hundreds of books and publications containing material or glossaries on genetic nomenclature, symbols, and terminology has been collected in the following source. It also includes a listing of journals that publish reports on genetics and cytology with information as to editor, publisher, publication frequency, and general content.

Boyes, J. W., Y. Nishimura, and B. C. Boyes, 1973. *References to Nomenclature and Other Publications in Genetics and Cytology.* International Genetics Federation, Montreal.

2
CELLULAR DIVISION AND CHROMOSOMES

At present all forms of life consist of some protoplasmic material with a protective and absorptive boundary between itself and its environment. Single-celled organisms possess only one such boundary membrane for each individual, whereas multicellular organisms generally contain many individual units, each with its own membrane. Despite this difference, however, the organization of each such unit, or cell, has some important similarities in all organisms. Cell structure usually consists of two distinct areas which, in living cells, are in constant motion: *cytoplasm*, the major portion of the protoplasmic substance contained in the cell membrane; and, within the cytoplasm, a dark-staining body, the *nucleus*.

In bacteria and blue-green algae, called *procaryotes* ("before the nucleus"), the nuclear material is not separated from the cytoplasm by a discrete membrane. In the cells of the more complex *eucaryotes* ("true nucleus"), which include the majority of living species and multicellular organisms, a nuclear membrane separates the genetic material from the cytoplasm which is then further subdivided by other distinct membranous structures. In viruses, organisms that must utilize the cellular activity of their host, the only membrane present is the viral envelope enclosing primarily the viral genetic material.

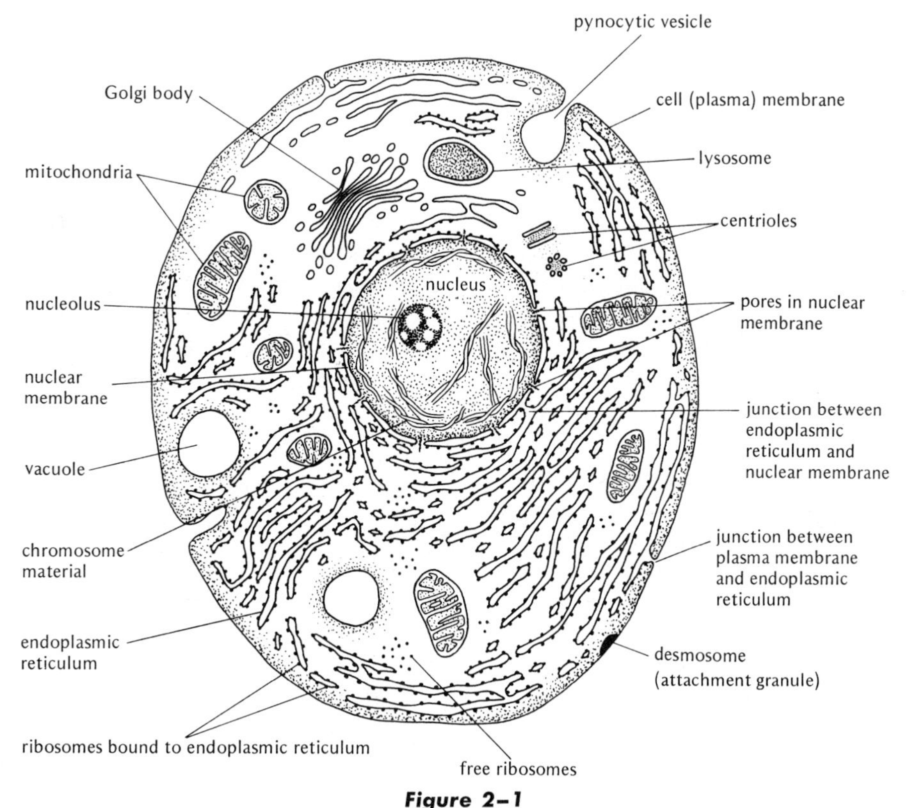

Figure 2–1

Diagrammatic representation of an animal cell showing cross sections through various important cell organelles. Membranous connections have been noted between the cell membrane, nucleus, and other constituents. (Modified after Wolfe.)

CYTOPLASM

Within the cytoplasmic matrix lie a number of organelles concerned with active cell function whose presence or size may vary between different organisms and between different tissues. Generally the following organelles are observed under the electron microscope in eucaryotes (Fig. 2-1): The *mitochondria* are small bodies, usually with distinct shelflike internal layers, whose primary function appear to be as providers of cellular energy through respiration and oxidation. The *Golgi* apparatus consists of netlike staining bodies commonly found in cells engaged in secretion. The *ergastoplasm*, or *endoplasmic reticulum*, is a cytoplasmic double-walled membrane usually folded in layers that appear to be connected with the cell membranes. In certain areas the endoplasmic reticulum is associated with many small particles, called *ribosomes.* Ribosomes are also present in procaryotes such as bacteria, although these organisms lack mitochondria, Golgi bodies, and endoplasmic reticulum as they exist in higher forms. The ribosomes, as will be discussed further, play an important role in the synthesis of long-chained molecules known as *proteins.*

Proteins are among the most characteristic features of living organisms, and various aspects of their function, structure, and synthesis will be presented more fully in Chapters 26 through 30. For the present we can mention that they are composed of subunits, called amino acids, that are linked together by strong covalent chemical bonds known as peptide linkages. These linkages may extend to hundreds or thousands of amino acids arranged in a linear *polypeptide* chain. Each amino acid within this chain may be one of 20 different kinds. Thus two protein chains, or polypeptides, may have the same length but still be considerably different from each other because they bear different sequences of amino acids. For example, if we consider only three kinds of amino acids—glycine, arginine, and lysine—one polypeptide may begin with the sequence glycine-glycine-arginine . . . , while the other begins with the sequence lysine-arginine-glycine. . . . It is the linear sequence of amino acids in a polypeptide chain that determines the way the chain will fold up in the cell, and thus specifies the shape and conformation of the molecule that results. Each kind of protein usually has a unique function associated with its particular composition and structure. There are proteins that transport molecules such as oxygen (e.g., hemoglobin), function as structural building blocks of the organism (e.g., collagen), and help in performing chemical reactions (enzymes).

In the cytoplasm many proteins are aggregated with other chemical compounds into various membranes, granules and fibers, some of which have been previously mentioned. For the most part these inclusions, as most nonnuclear organelles, arise anew each generation through nuclear influence. Occasional cytoplasmic inclusions are found, however, that appear to be descended from previous such inclusions in the parental cell.

Many cytologists, although not all, believe that an important cytoplasmic organelle necessary for cell division in many cells, the *centrosome* and its accompanying *centriole*, appears to duplicate itself and show continuous inheritance between cell generations. In plants the *chloroplast*, a plastid containing chlorophyll and serving as the photosynthetic factory for the transformation of light into usable plant energy, also seems to arise from preexisting chloroplasts. Such self-replicating inclusions may therefore be considered as cytoplasmic agents that contain their own self-reproducing genetic material, *plasmagenes*. Other cytoplasmic forms of inheritance are discussed in Chapter 13.

NUCLEUS

Because of its importance as the primary director of cellular activity and inheritance, the nucleus provides our main focus for identifying the genetic material. Under the electron microscope the eucaryotic nucleus is surrounded by a double membrane that appears to be in active contact with the endoplasmic reticulum and the cell membrane. Under the light microscope staining by various chemical dyes reveals a dark network, called *chromatin*, which, during the process of cellular division, becomes organized as distinct bodies, *chromosomes*. Some areas of the chromosomes stain very darkly (*heterochromatin*), while others stain relatively lightly (*euchromatin*). In addition to chromosomes there are usually one or more rounded bodies, *nucleoli*, attached to specific chromosome regions called *nucleolar organizers*.

CELL DIVISION

A primary problem, undoubtedly faced in the very dawn of life, concerned the physical growth of primitive organisms. Without cell division, growth occurs through an increase in volume and an enlargement of the outer surface membrane. However, as the surface of a spherical organism increases by the square of its radius (r), its volume increases proportionately greater by the cube of this number; i.e., an increase in size (r) produces a relatively smaller increase in surface area (r^2) than in volume (r^3). Quite rapidly, therefore, the inner constituents of such an expanding organism have proportionately less surface area from which to obtain food, oxygen, and the various metabolic necessities, as well as less surface from which to secrete their metabolic wastes and products. In the absence of division death would quickly ensue, both as a result of these causes and because of the many physical stresses and accidents that could rupture such an unwieldy membrane. Some form of cell division must, therefore, have been a primary necessity for the maintenance of life.

MITOSIS

Cell division, however, is not a simple answer to these problems unless assurance is provided that the essential cell constituents are properly distributed to the daughter cells. It is easy to see that an incomplete distribution to cells and their descendants of something as vital as the genetic material, the carrier of biological information, would seriously endanger their future. One successful answer to this problem was the evolution of a *mitotic* mechanism that produces an even cellular division of essential hereditary components.

In procaryotes such as bacteria, partitioning of the nuclear material—that is, the bacterial chromosome—occurs by utilizing the attachment of the chromosome and its newly formed replicate to a section of the bacterial cell membrane. As the bacterium elongates during cell division, one portion of the membrane carrying one chromosome separates from another portion carrying its replicate. When sufficient separation between the two chromosomes has been achieved, the membrane between them invaginates and cleaves, so that each of the chromosomes are now in separate daughter cells. Procaryotic replication mechanisms will be discussed further in later chapters.

In eucaryotes cell division is more complex because more than one chromosome is usually present. Manipulation of eucaryotic chromosomes occurs through *microtubules*, which are structurally similar to the fibers of cilia and flagella. Most often these microtubular fibers are organized into spindle-shaped bodies that appear to be generated by the centrioles, which are, in turn, similar in many respects to the *basal bodies* involved in flagella formation. The eucaryotic modes of cell division undoubtedly evolved early in the history of these organisms, since these processes are shared by most eucaryotes. The remainder of this chapter will concern itself with the eucaryotic processes of mitosis and meiosis. In general, although individual variations of the mitotic process and structures exist, the usual stages of mitosis are as shown in Fig. 2-2, and can be described as follows.

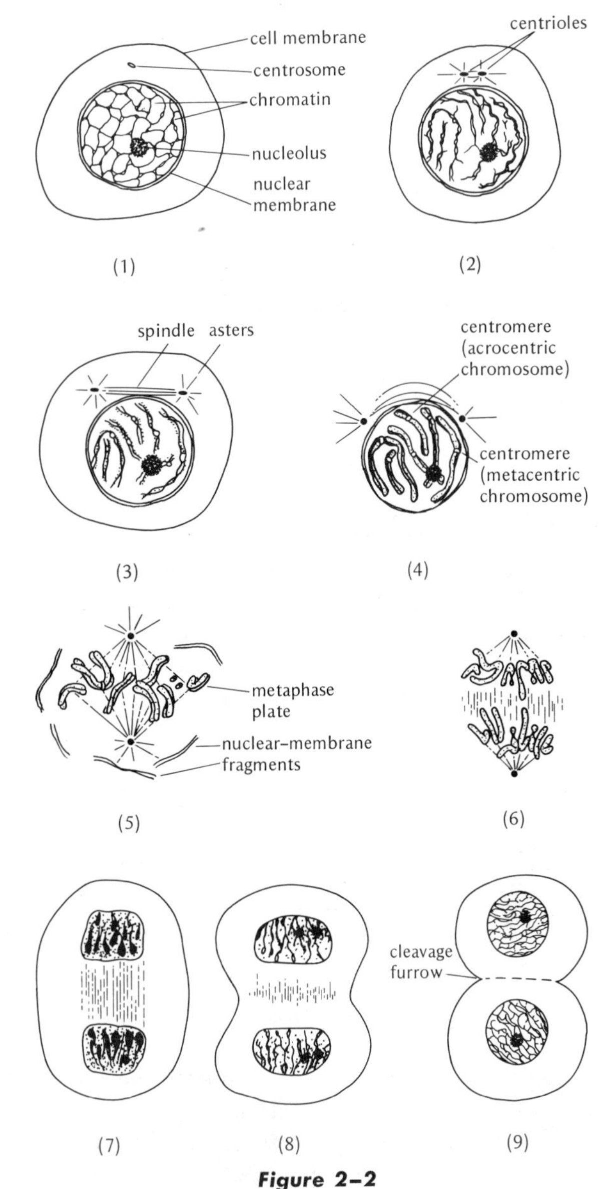

Figure 2-2

Various stages of mitosis in a somatic cell. (1) Interphase; (2, 3, and 4) early, medium, and late prophase stages showing increased condensation of chromosomes; (5) metaphase; (6) anaphase; (7, 8, and 9) various telophase stages. (After Sharp.)

TABLE 2–1
Duration of mitotic stages (in minutes)

	Temp., °C	Prophase	Metaphase	Anaphase	Telophase
mouse (spleen)	38	21	13	5	4
chicken (mesenchyme cells in tissue culture)	39	30–60	2–10	3–7	2–10
frog (fibroblasts in culture)	20–24	32	20–29	— 6–11	—
grasshopper (neuroblasts)	38	102	13	9	57
sea urchin embryo	12	19	17	12	18
onion (root tip)	20	71	6.5	2.4	3.8
pea (root tip)	20	78	14.4	4.2	13.2

INTERPHASE

The interphase period between successive cell divisions consists of processes associated with growth and preparation for mitosis. In many cases the specific chemicals that constitute the newly synthesized chromosomes, as well as the proteins that will soon give rise to the mitotic spindle, are found during this period. As compared to the period of active mitosis, which may range from about 10 minutes to a few hours, the interphase or "intermitotic" period of most cells is usually many times longer, although it may be considerably shortened in some of the rapidly dividing embryonic tissues shown in Table 2–1.

PROPHASE

In prophase, the first stage of mitosis, the preparations for cell division continue, but at a more active pace than during interphase. Chemical constituents of the new chromosomes are now synthesized in those cases where they have not been formed during interphase. More conspicuously prophase includes the coiling of the chromosomes, or their "condensation," to the point where they are now visible as threadlike structures. As a rule each of these mitotic prophase chromosomes appears longitudinally split into two duplicates, each of which is called a *chromatid* as long as it remains connected to its "sister" chromatid. Further prophase events are the splitting of the centrosome and the movement of each half to opposite sides of the nucleus, the synthesis of the mitotic apparatus, the disappearance of the nucleolus, and the beginning of the breakdown of the nuclear membrane.

METAPHASE

Once the new chromosome material has been synthesized and the chromosomes have coiled and condensed, they begin a succession of active movements accompanied by the complete breakdown of the nuclear membrane. These movements are based on the attachment of each chromosome through a specific point along its length, the *centromere*, to a double-poled spindle-shaped structure, the *spindle*. The centromere is usually characterized by a localized constriction that occupies a constant position for any particular chromosome. Chromosomes whose centromere is approximately midway between each end, thereby forming two equal chromosome "arms," are described as *metacentric* (Fig. 2–3a). Chromosomes with a more terminally placed centromere, forming unequal chromosome "arms," are called *acrocentric* (Fig. 2–3b). It is assumed by some investigators that some chromosome material always exists on both sides of the centromere, even in those instances when the centromere appears to be at the very tip of the chromosome (*telocentric*, Fig. 2–3c).

In animal cells the spindle to which the chromosome centromeres attach is usually formed be-

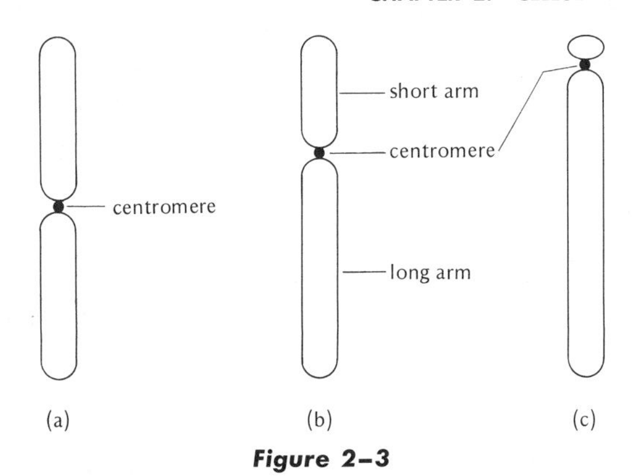

Figure 2–3

Diagrammatic representation of three types of chromosome based on centromere position: (a) metacentric; (b) acrocentric; (c) telocentric.

tween the two centrioles that were formerly together on one side of the nucleus. As these centrioles separate and move to new positions on opposite sides of the nucleus, they appear to radiate distinctive lines (*astral rays*) forming a network between them of continuous microtubular *spindle fibers*. In many plants the centrosomes and their accompanying centrioles are missing, although spindles are nevertheless present. Whether plant or animal, each chromosome is now in duplicated condition as a result of the previous synthesis, and the two sister chromatids remain attached to each other in regions immediately adjacent to their centromeres. One of the most distinctive events in cell division now occurs—the movement and arrangement of all chromosomes on a flat plane (*metaphase plate*) midway between the two poles of the spindle. These movements, and those that follow in the stages discussed below, have been precisely described by Bajer, Molé-Bajer, Nicklas, and others, but the kinetic forces responsible for them are not yet known.

ANAPHASE

The shortest of the mitotic phases, *anaphase*, occurs when the mutual attachment between the two sister chromatids ceases. At that point the centromere of each chromatid on the metaphase plate separates from its sister centromere and the two chromatids move toward opposite poles of the spindle. The appearance of this event is that of an active "repulsion" between sister centromeres, each dragging its chromatid along with it.

TELOPHASE

During the *telophase* period each of the two polar groups of chromatids, which may now be called *daughter chromosomes*, undergoes a reversion to the more extended and swollen interphase state. The nuclear membrane is reestablished and the nucleoli reformed. Division of the cytoplasmic portion of the cell (*cytokinesis*) is also completed during this period, giving rise to a *cell plate* in plant cells and the indented *cell furrow* in animal cells.

MEIOSIS AND HOMOLOGOUS PAIRING

The exact replication and splitting of each chromosome into two identical parts and their subsequent separation into two cells would not ordinarily lead to any change in chromosome number between the parent and daughter cells. The mitotic division of four chromosomes, for example, could only be expected to give rise to daughter cells with a similar number. In organisms whose cells are always formed by asexual means (absence of conjugation between sex cell nuclei), the number of chromosomes should therefore remain constant between generations. In sexually reproducing organisms, however, where a *zygote* (embryonic cell) is formed by fertilization between male and female gametes, i.e., sperm and egg or pollen and ova, the embryonic cells would have double the chromosomes of each parent if no reduction in number occurred during sex-cell formation. The finding that, barring very unusual incidents, the chromosome number remains constant between the generations of a species, indicates that such continuous doubling does not occur. Indeed sexual organisms with continuous doubling of chromosome number would rapidly achieve large unbal-

TABLE 2–2

Chromosome numbers found in different species of animals and plants*

Common Name	Species	Diploid No.
ANIMALS		
man	*Homo sapiens*	46
Rhesus monkey	*Macaca mulatta*	42
cattle	*Bos taurus*	60
dog	*Canis familiaris*	78
cat	*Felis domesticus*	38
horse	*Equus calibus*	64
donkey	*Equus asinus*	62
house mouse	*Mus musculus*	40
rat	*Rattus norvegicus*	42
golden hamster	*Mesocricetus auratus*	44
guinea pig	*Cavia cobaya*	64
rabbit	*Oryctolagus cuniculus*	44
pigeon	*Columbia livia*	ca. 80
chicken	*Gallus domesticus*	ca. 78
alligator	*Alligator mississipiensis*	32
toad	*Bufo americanus*	22
frog	*Rana pipiens*	26
carp	*Cyprinus carpio*	104
starfish	*Asterias forbesi*	36
silkworm	*Bombyx mori*	56
red ant	*Formica sanguinea*	48
house fly	*Musca domestica*	12
fruit fly	*Drosophila melanogaster*	8
mosquito	*Culex pipiens*	6
cockroach	*Blatta germanica*	23♂, 24♀
grasshopper	*Melanoplus differentialis*	24
honeybee	*Apis mellifera*	32
flatworm	*Planaria torva*	16
freshwater hydra	*Hydra vulgaria attenuata*	32
nematode	*Caenorhabditis elegans*	11♂, 12♀

Common Name	Species	Diploid No.
PLANTS		
yeast	*Saccharomyces cerevisiae*	ca. 18
green algae	*Acetabularia mediterranea*	ca. 20
garden onion	*Allium cepa*	16
barley	*Hordeum vulgare*	14
rice	*Oryza sativa*	24
spiderwort	*Tradescantia virginiana*	24
wheat	*Triticum aestivum*	42
corn	*Zea mays*	20
snapdragon	*Antirrhinum majus*	16
squash	*Cucurbita pepo*	40
upland cotton	*Gossypium hirsutum*	52
tomato	*Lycopersicon esculentum*	24
tobacco	*Nicotiana tabacum*	48
evening primrose	*Oenothera biennis*	14
kidney bean	*Phaseolus vulgaris*	22
white oak	*Quercus alba*	24
pine	*Pinus species*	24
garden pea	*Pisum sativum*	14
potato	*Solanum tuberosum*	48
white clover	*Trifolium repens*	32
broad bean	*Vicia faba*	12
		Haploid No.
slime mold	*Dictyostelium discoideum*	7
mold (fungus)	*Aspergillus nidulans*	8
pink bread mold	*Neurospora crassa*	7
penicillin mold	*Penicillium species*	4
green algae	*Chlamydomonas reinhardi*	16

* In most diploid organisms both male and female sexes possess the full diploid number of chromosomes. In a few organisms, such as bees (see also Chapter 12), the sexes differ in the number of sets of chromosomes, the females being diploid and the males haploid.

anced cells with tremendously unwieldy nuclei and insufficient cytoplasm.

At some unknown point in the history of life, sexually reproducing organisms evolved a mechanism that enabled them to regularly reduce the number of chromosomes in each gamete to half the usual number. If, for example, four were the regular *diploid* chromosome number in somatic cells, then the reduced or *haploid* number in the sex cells would be two. *Meiosis* is simply the process by which the chromosomes are separated during the formation of sex cells and their numbers reduced from the diploid to the haploid condition. *Fertilization* then marks the event in which two haploid nuclei join to reform a diploid cell. As shown in Table 2–2, a large number of animal species, as well as higher plants, are ordinarily composed of diploid cells except for their gametes. Others, such as the mold *Neurospora*, are haploid for most of their life cycle but then, through fertilization of two haploid sex cells, produce a diploid zygote that undergoes meiosis to form again a haploid stage (see Chapter 3). The number, as well as the size and shape of the chromosomes of a species, is usually constant and is called its *karyotype*.

In diploid cells prior to meiosis each individual chromosome (e.g., *A*) usually has a pairing mate, or *homologue* (e.g., *A′*), so that a parental karyotype of four chromosomes (diploid number) would consist of two *homologous pairs*, e.g., *AA′* and *BB′*. Meiosis in such an organism would then produce haploid gametes each one containing two individual chromosomes, one from each pair (*A* or *A′*, together with *B* or *B′*, e.g., *AB*, *A′B*, *AB′*, or *A′B′*). Upon fertilization by another gamete, a diploid zygote is formed with two homologous chromosome pairs which, in many higher organisms, now divides mitotically to produce a diploid adult. In this fashion each parent contributes one of a homologous pair of chromosomes to an offspring.

The advantage of homologous pairing derives from differences in biological function between chromosomes. For example, if each pair, *AA′* and *BB′*, affects somewhat different characteristics of the individual, then it will be of advantage for an organism to have a member of each of the two pairs represented. Meiosis ensures the presence of one *A* (or *A′*) and one *B* (or *B′*) chromosome in a gamete by dividing the two chromosomes of each homologous pair into separate gametes. If there were no homologous pairing, it is easy to conceive that the chromosomes would undergo a random reduction producing gametes with varying numbers of chromosomes. And even if the haploid number of chromosomes occurred in each such gamete, many zygotes might still be formed lacking at least one essential chromosome (see Fig. 2–4).

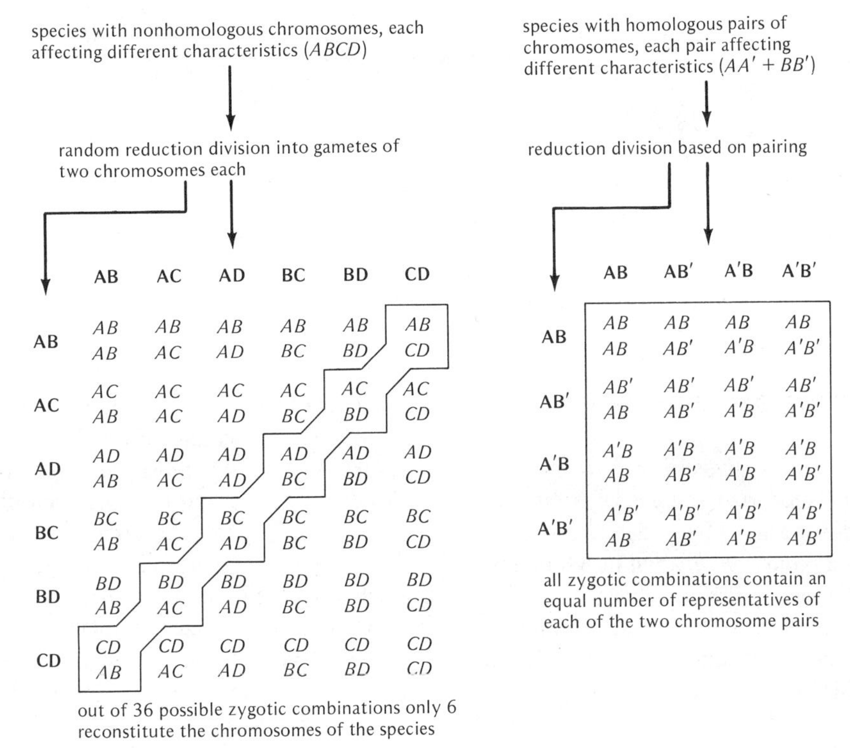

Figure 2–4

Comparison of the distribution of paired and unpaired chromosomes in a reduction division. In both cases a species has only four chromosomes, but in one case (left side) these chromosomes assort randomly to opposite poles, and in the other case (right side) homologous chromosomes pair before separating.

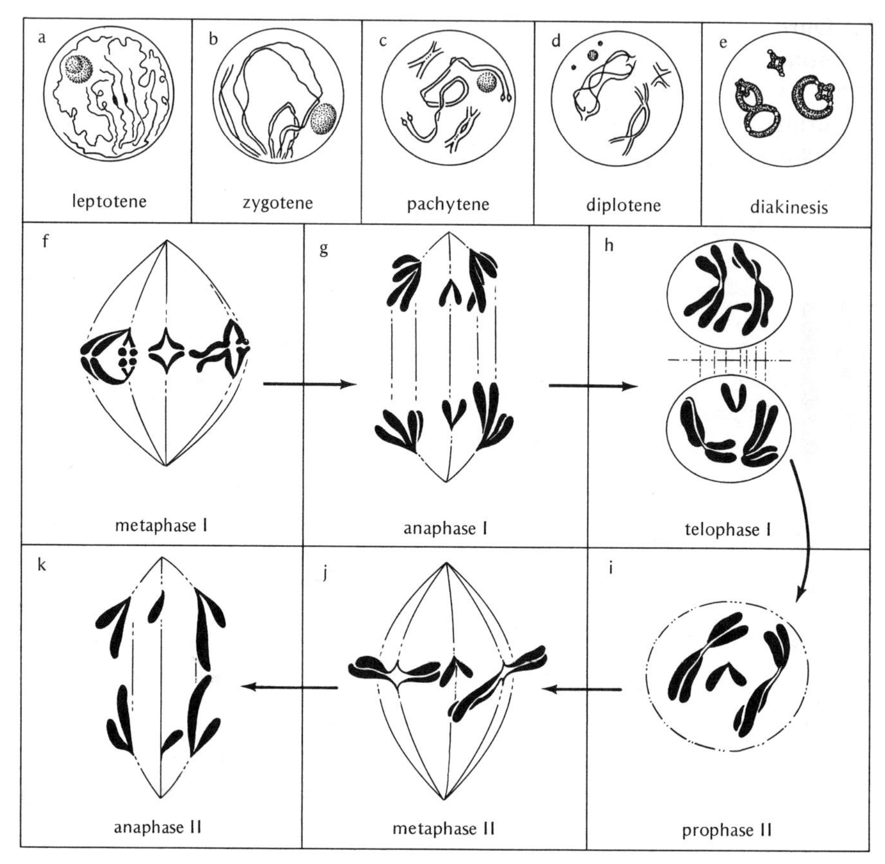

Figure 2–5

Principal stages of meiosis. (From Wolfe, after Lewis and John.)

FIRST MEIOTIC PROPHASE

As a rule the meiotic divisions follow a standard scheme during which two successive divisions of the chromosomes occur (Fig. 2–5). The first division represents a *reduction division* in which members of homologous pairs of chromosomes are separated into daughter cells without duplication; i.e., their numbers are reduced to half. Although variations of this have been discovered, one of the significant and common facts of meiosis is the initial pairing and subsequent separation of homologous chromosomes. From all indications this initial pairing occurs during the fairly lengthy first meiotic prophase, which has been subdivided into the following five stages.

LEPTOTENE. The leptotene is the first of the meiotic stages which differs from the previous interphase. The chromosomes first appear as long, slender threads with many beadlike structures (*chromomeres*) along their length. In some plants the chromosomes are clumped to one side of the nucleus (*synizesis*). In some animals (many insects) they appear polarized with their ends drawn together toward that portion of the nuclear membrane close to the centriole. Although considerable biochemical evidence indicates the replication of chromosome material during this period, it is usually difficult to observe any morphological duplication. Meiotic prophase chromosomes generally appear as single and individual structures until the pachytene stage.

ZYGOTENE. During the zygotene stage homologous chromosomes appear to attract each other and enter into a very close zipperlike pairing (*synapsis*). This pairing is highly specific and occurs between all homologous chromosome sections even if present on different nonhomologous chromosomes. For example, if a piece of chromosome material, e.g., *a*, has been shifted from one of the *A* chromosomes to one of the *B* pair (*translocation*, Chapter 22), the *a* part of that particular *B* chro-

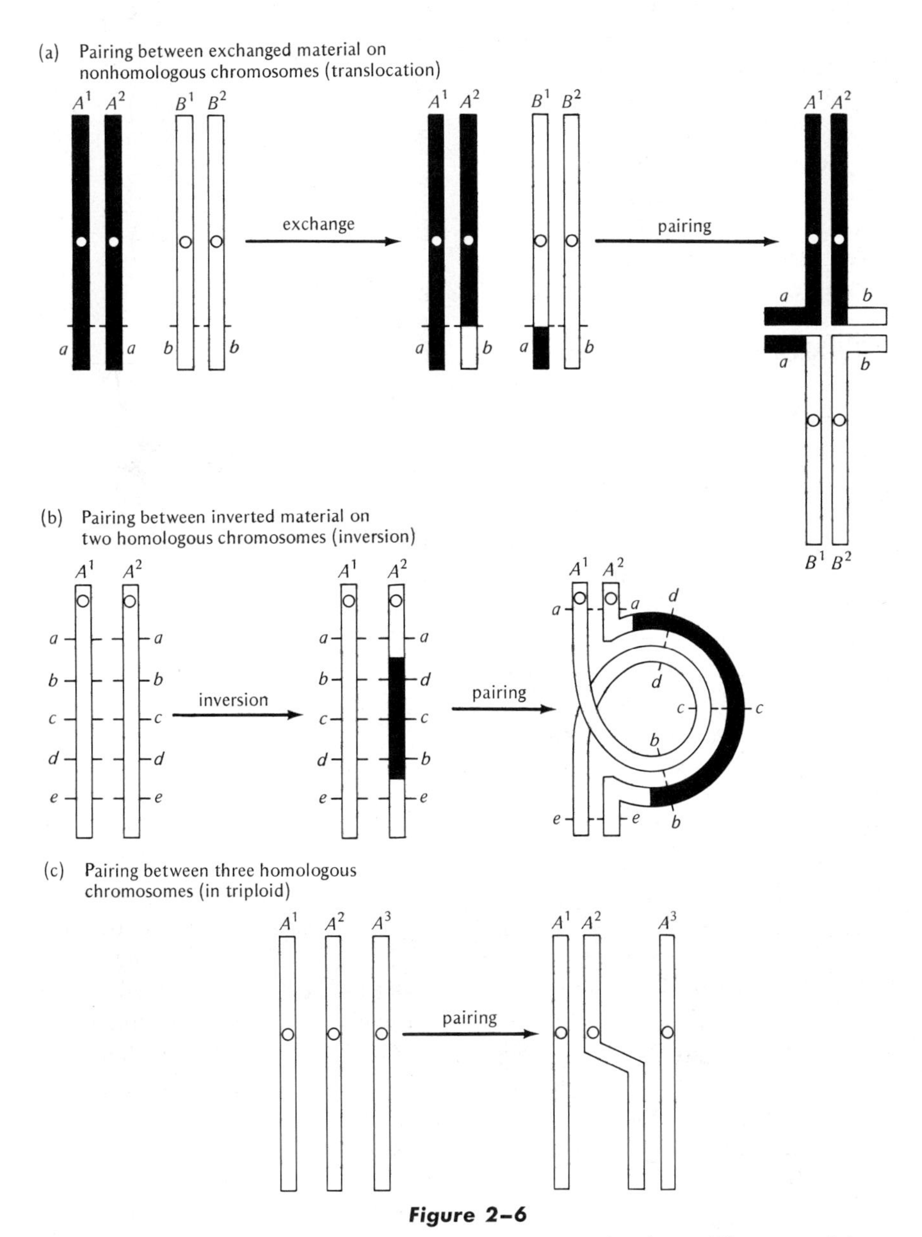

Figure 2–6

Pairing between homologous sections of chromosomes under three different conditions. Although each chromosome is composed of two chromatids, the chromatid structures have been omitted for diagrammatic simplicity. (See, however, Figure 2–7.)

mosome will be attracted to an *A* chromosome still containing the *a* material (Fig. 2–6a). Similarly, if the order of chromosome material is inverted in one member of a chromosome pair, e.g., *a–d–c–b–e* instead of *a–b–c–d–e*, a loop will be formed between the two chromosomes so that tight pairing is preserved (Fig. 2–6b). In almost all cases pairing or synapsis is confined to two homologous chromosome areas at a time. In *triploid* organisms, having three members for each homologous group rather than two, pairing nevertheless occurs only between two chromosomes in any one region (Fig. 2–6c). Recently, however, Comings and Okada have reported that triple pairing of homologous chromosomes can be observed in triploid chickens.

PACHYTENE. The pachytene is a stage of progressive shortening and coiling of the chromosomes that occurs once zygotene pairing has been completed. At pachytene the two sister chromatids of a homologous chromosome are associated with the two sister chromatids of their homologous partner. This group of four chromatids is known as a *bivalent*, or *tetrad*,* and a series of exchanges of genetic material can occur or has already occurred between nonsister homologous chromatids (Fig. 2–7). Such exchanges can be detected by special means (Chapter 16) and signify the genetic mechanism of *crossing over*, or *recombination*. Pachytene is also the stage in which a structure called the *synaptinemal complex* can be observed between synapsed chromosomes through electron microscopy. It appears as a ribbonlike group of three longitudinal components organized in two dense lateral elements and a thin central element composed primarily of proteins (Fig. 2–8). The synaptinemal complex may function to pull chromosomes together helping them to pair more precisely and efficiently. Moses and others have suggested that this structure may be correlated with the occurrence of genetic crossing over (see p. 347, Chapter 17). Under the light microscope a somewhat different but easily observable physical counterpart to genetic crossing over, the *chiasmata*, can be noted during the following stages.

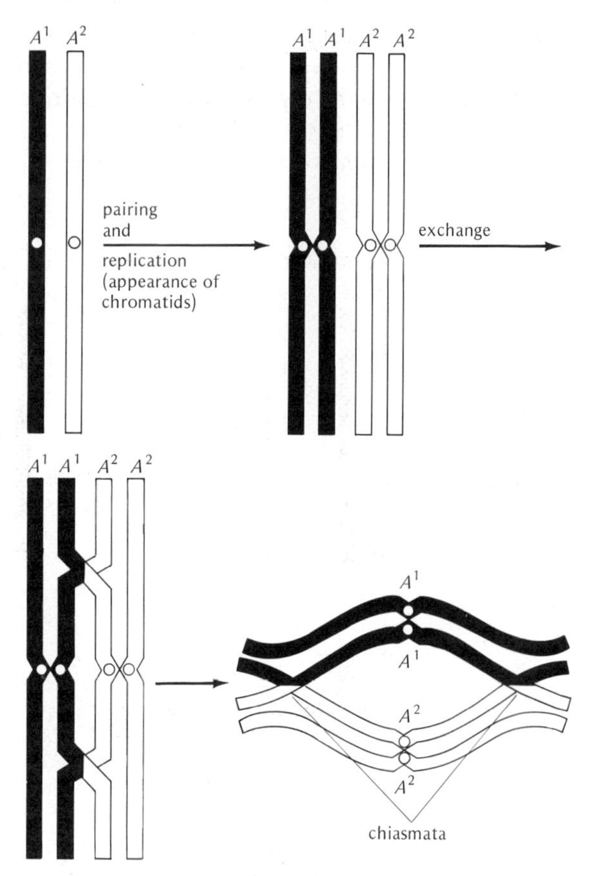

Figure 2–7

Pairing between two homologous metacentric chromosomes, A^1 and A^2, and subsequent exchange of chromosome material in each arm leading to the formation of a bivalent with two chiasmata.

DIPLOTENE. At the diplotene stage each chromosome now acts as though it were repulsing its closely paired homologue, especially near the centromere. Distinctly visible separations then occur between homologous chromosomes except for specific regions where an actual physical crossing over appears to have taken place between homologous chromatids. These crossed areas or chiasmata (singular: *chiasma*), are X-shaped attachments between the chromosomes and seem to be the only remaining force holding each bivalent

** Tetrad* is also used in genetics in a somewhat different context: to describe the four haploid cellular products of a single meiosis in terms of their genetic constitution (see p. 134).

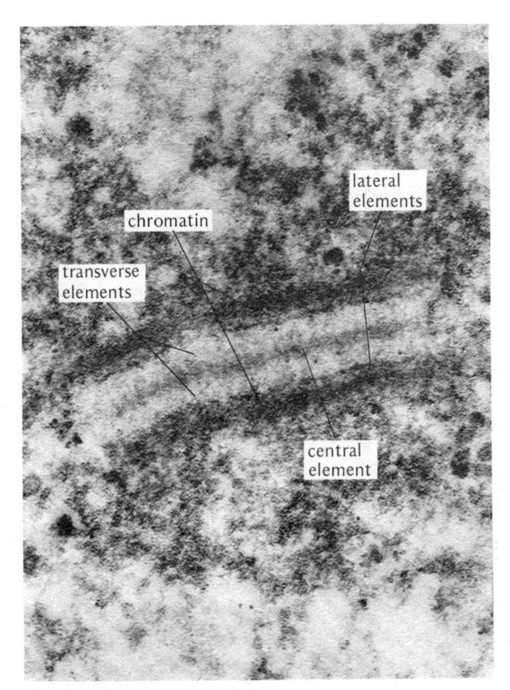

Figure 2–8

Electron micrograph of a synaptinemal complex observed during prophase of the first meiotic division in the rooster, Gallus domesticus. *(Courtesy of M. J. Moses.)*

together until metaphase. In many organisms their position and number seem to be constant for a particular chromosome.

DIAKINESIS. At diakinesis, coiling and contraction of the chromosomes continue until they are thick, heavy-staining bodies. In this process the bivalents usually migrate close to the nuclear membrane and become evenly distributed. The nucleolus either disappears or detaches from its associated chromosome. During the latter part of this stage, or the early part of metaphase, the nuclear membrane dissolves and the bivalents attach themselves by their centromeres to the rapidly formed spindle.

FIRST METAPHASE

In metaphase the chromosomes reach their most condensed state and appear relatively smooth in outline. The chiasmata that had first appeared during diplotene have now moved toward the ends of each chromosome (*terminalization*), leaving only the single terminal attachment between the formerly paired arms of homologous chromsomes (Fig. 2–9). These remaining chiasmata prevent the separation of homologous chromosomes which now lie on each side of the equatorial plate of the spindle stretched by their respective centromeres toward opposite poles.

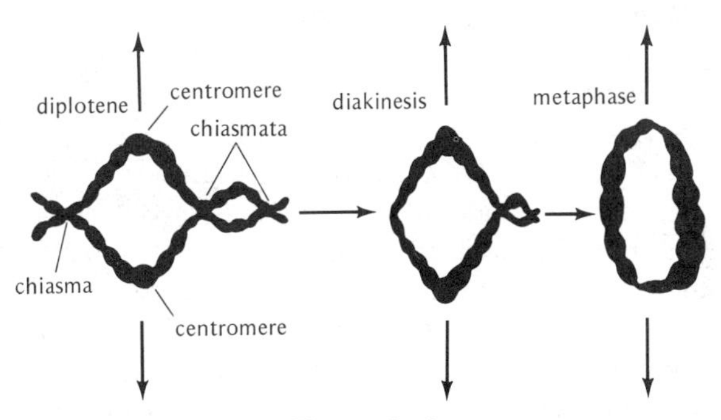

Figure 2–9

Terminalization of chiasmata in a bivalent during meiosis showing the reduction from three chiasmata in diplotene to only two terminal chiasmata in metaphase. The arrows above and below the bivalent point to the direction of separation between the homologues. (After Darlington.)

FIRST ANAPHASE

Although previously duplicated along its entire length, each chromosome still maintains only a single functional centromere for both of its sister chromatids. The separation, or "disjunction," of one homologous chromosome from another in anaphase toward opposite poles therefore results in this single centromere (which may already be structurally split) dragging both chromatids (*dyad*) along with it. The chiasmata slip off the ends of the chromosome as they are pulled apart, and the poleward-moving chromatids are now bound together at only one point, the centromere. Since two chromatids compose each dyad, their appearance depends on the position of the centromere: a double V if the chromosome is either metacentric or acrocentric, and a single V if telocentric.

Our previous view of the first meiotic division as a reduction division that separates two homologous chromosomes into daughter cells must now be modified in accordance with the exchange of chromosomal material manifested by chiasmata. That is, if no exchange occurs between homologous chromosomes in meiosis, their separation is purely *reductional* because a chromosome in one daughter cell does not contain any material from its homologue in the other cell. However, if genetic exchange occurs, each separating dyad in anaphase carries part of its homologue, resulting in an equal (*equational*) division of the exchanged chromosome material to both daughter cells. If we consider that each of a homologous pair of chromosomes is contributed by a different parent, chiasma exchange in meiosis will redistribute chromosome material from both parents to the daughter cells. Gametes formed from such daughter cells may, therefore, carry a wide mixture of chromosome material, part from one parent and part from another.

In addition, since it is a matter of chance which of the parental homologous chromosomes are separated to each daughter cell, the more pairs of chromosomes an organism has, the greater the chances that a gamete will contain material from both parents and the less the chances that all its chromosomes will be from one parent. For example, an organism with two homologous pairs may have received chromosomes A and B from one parent and their homologues A' and B' from the other. This individual, having the constitution $AA'BB'$, can form four different haploid gametes in equal proportions, AB, AB', $A'B$, $A'B'$, of which two are parental. On the other hand, the meiotic division of cells with four pairs of chromosomes (e.g., the fruit fly, *Drosophila melanogaster*) will yield, on the average, only one out of eight gametes with all chromosomes from either of the two parents. In general, for n pairs of homologous chromosomes, only $1/2^{n-1}$ gametes will contain a parental combination. In humans, with 23 homologous pairs, the chances for a meiotic daughter cell to have all chromosomes from one parent is therefore only one out of 4,194,304 (2^{22}) times. In the remainder of occasions, gametes will have a mixed proportion of chromosomes from both parents. Thus the random distribution of parental chromosomes, as well as the crossing over between them, leads to a most important source of differences or variability among the gametes of sexual-reproducing organisms.

TELOPHASE AND INTERPHASE

These stages vary considerably between organisms. In most cases once the dyads reach one of the spindle poles, a nuclear membrane is formed around them, and the chromosomes pass into a short interphase before the second meiotic division begins. In the plant *Trillium* the anaphase group of dyads enters immediately into the second meiotic division, skipping telophase and interphase. Usually the sequence of events is so rapid that the interphase chromosomes are not as physically extended as in mitosis, nor is there sufficient time to form a single large nucleolus. Mechanical division of the cell (cytokinesis) may occur during this stage (in corn) or may be postponed until simultaneous formation of four daughter cells at the end of the second meiotic division.

SECOND MEIOTIC DIVISION

The chromosomes enter the prophase of the second meiotic division as dyads or two sister chromatids connected together in their centromere region. As soon as these connected centromeres divide, each chromatid (*monad*) separates from its sister and moves to the opposite pole in the second anaphase. The second telophase and cytokinesis follow rapidly, giving rise to four haploid cells from each initial diploid cell that entered meiosis. In a cell consisting of a diploid number of four

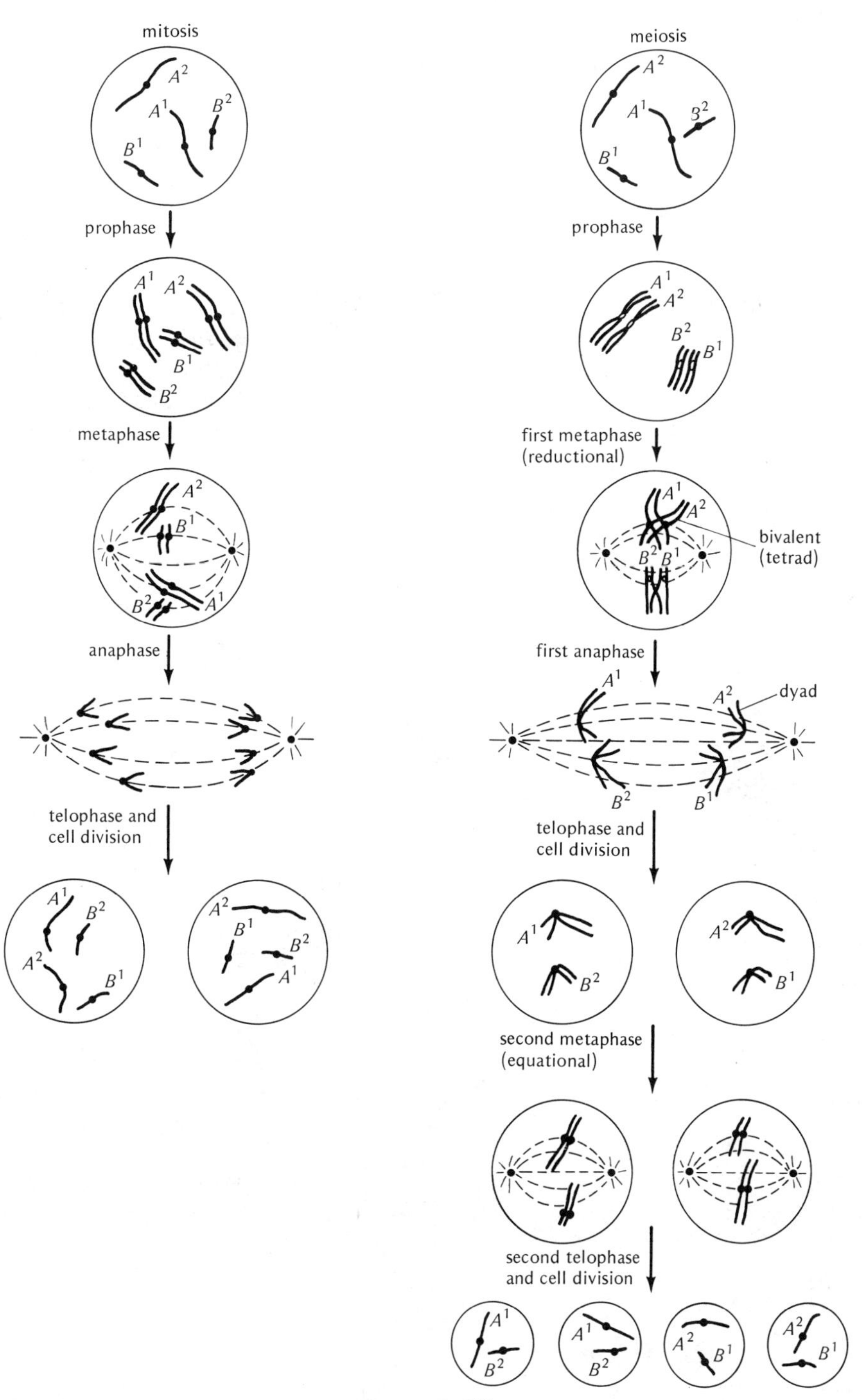

Figure 2–10

Diagrammatic comparison of mitosis and meiosis. The primary differences between these two forms of cell division are (1) homologous pairing in meiosis and its absence in mitosis, and (2) separation between duplicated centromeres in mitosis and the absence of such separation in meiosis until the second meiotic division.

chromosomes, or two pairs of homologues, the meiotic events leading to a reduction in chromosome number can be summarized as follows:

First meiotic division:
- metaphase I: two bivalents or tetrads (four chromosomes or eight chromatids) on the metaphase spindle
- anaphase I: two dyads (two chromosomes or four chromatids) pass to each pole

Second meiotic division:
- metaphase II: two dyads on the spindle in each daughter cell
- anaphase II: two monads (two chromatids) pass to each pole

A diagrammatic comparison between mitosis and meiosis for such an organism is presented in Fig. 2–10. Note that the crucial difference lies in the pairing of homologues in meiotic prophase and the absence of centromere separation in meiosis until the second anaphase.

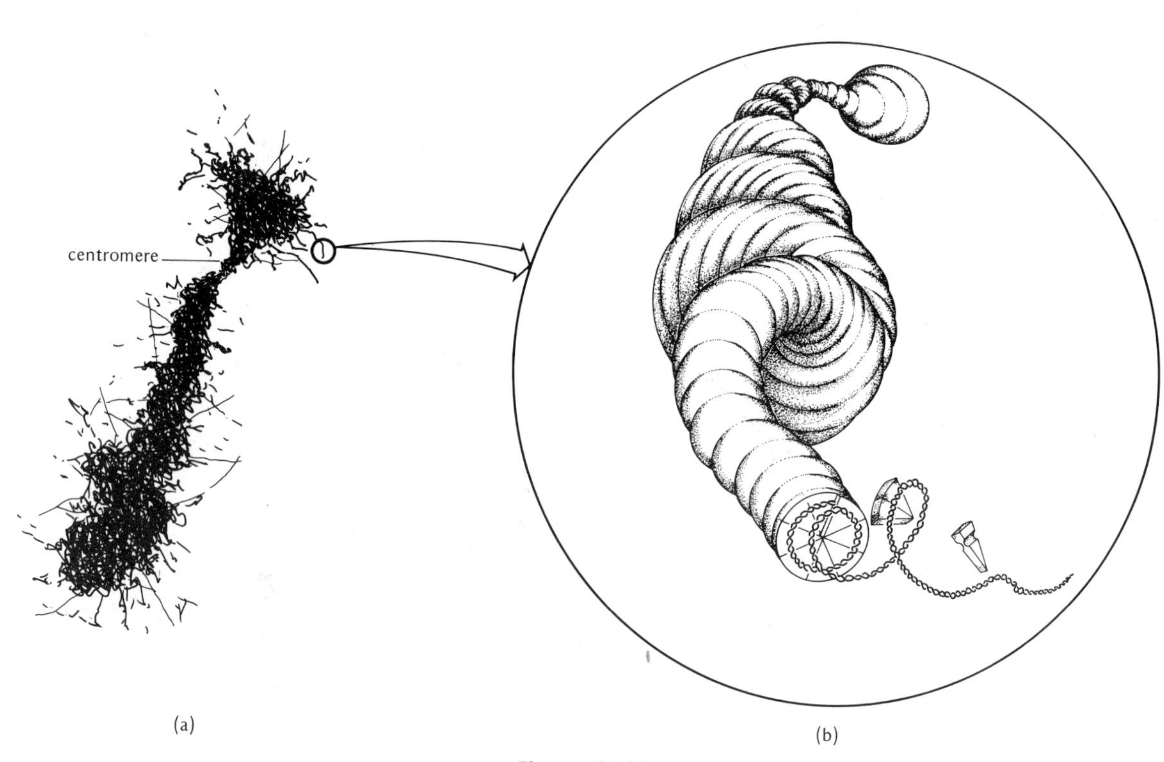

Figure 2–11

(a) Copy of an electron micrograph (21,750x) of a human chromosome at metaphase. The fibers (or fiber?) of which it is composed are obviously heavily coiled, but diminish in number or folding at the narrow centromere region. (b) Structure proposed by DuPraw to explain the form taken by a chromosomal fiber. The thin helical thread at the bottom of the figure represents the genetic material (nucleic acid, see Chapter 4) which is supercoiled into fibers about 100 angstroms in diameter, and held in place by the schematically drawn pie-shaped chromosomal protein material. These thin fibers are then further coiled and packed into 200 angstrom diameter fibers which may then coil and fold still further to form the more visible chromosomal strands observed under the light microscope. (From DuPraw.)

CHROMOSOMES

Although any pair of homologous chromosomes usually maintains constant size and shape at a given stage of the cell cycle, there may be considerable differences between nonhomologous pairs and between the chromosomes of different species. The condensed chromosomes observed in the cell divison of eucaryotes may be as short as 1/4 micron (1 micron = 1 micrometer = .001 millimeter) in fungi and birds, or as long as 30 microns in *Trillium* plants. As a rule, however, most metaphase chromosomes fall within a range represented by about 3.5 microns in *Drosophila*, 5 microns in man, and 8 to 10 microns in corn.

On the electron microscope level the chromosomes appear to be constituted of fibers that may range in thickness from 100 angstroms (1 angstrom (Å) = .0001 micron = .0000001 millimeter) to about 500 angstroms depending upon the treatment to which they have been subjected. Most often the main fibrous element observed is about 250 angstroms in diameter. According to Du Praw and others, the chromatid of a chromosome consists of a single coiled fiber that has become heavily looped and folded during cell division (Fig. 2-11), whereas Ris and others suggest an association of at least two or more of these fibers per chromosome. As yet, the detailed structural organization of the chromosome on this level is not known, although a number of experiments have attempted to distinguish between the two alternatives (see Chapter 5).

MITOTIC CHROMOSOMES

In gross structure (Fig. 2-12) the mitotic chromosome is usually a rodlike body with one constriction at the centromere, called the *primary constriction*. In some chromosomes further constrictions can be seen which may include pinching off a small chromosomal section called the *satellite*. These *secondary constrictions* are often associated with regions where the nucleolus is formed or attached (*nucleolar organizers*). When a mitotic chromosome is stretched out, as in prophase, the knoblike *chromomere* regions along its length often show distinct sizes and occupy specific positions thereby giving each nonhomologous chromosome a morphologically distinct appearance.

Occasionally, extra chromosomes are noted in some individuals of a species. Usually, these *supernumerary*, *accessory*, or "B" chromosomes have no obvious function and can be gained or lost without serious disadvantage. About 600 species of flowering plants and more than 100 animal species are now known to carry such supernumerary chromosomes (Jones). Similarly some chromosomes or portions of chromosomes may be discarded during formation of the somatic tissues (*chromosome dimi-*

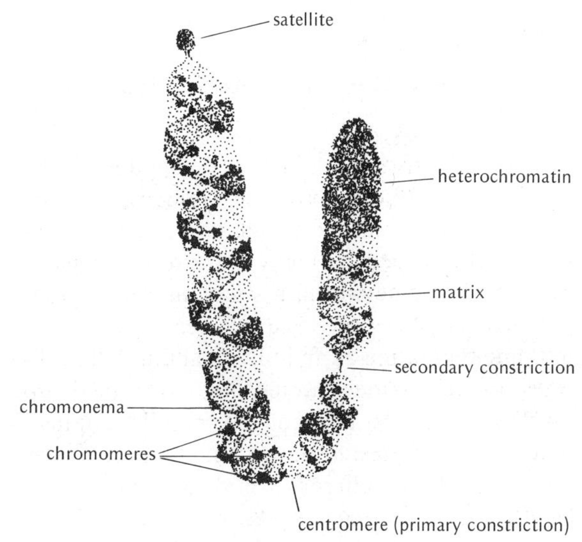

Figure 2-12

A generalized mitotic chromosome with structures often described under the light microscope. Each thin chromosome strand (chromonema) is believed to undergo extensive coiling in certain areas to produce the densely visible chromomeres. (After Ris, from Heitz.)

nution), while the sexual-reproductive tissues maintain a complete chromosome complement. In fact the observation of this phenomenon in some organisms such as *Ascaris megalocephala* (now called *Parascaris equorum*) was used to support that part of Weismann's theory (p. 7) which suggested that the germplasm contained all the hereditary determinants, whereas the somatoplasm lost everything except material necessary for its particular function.

For the most part, however, the number of chromosomes is constant for any plant or animal species, and the difference in function between tissues seems to arise from the functioning of different chromosomes at specific times and at specific places (Chapter 30) rather than from their presence or absence. Most commonly, chromosome numbers range between 12 and 50, i.e., 6 to 25 homologous pairs in the diploid condition. At the extremes, one pair has been found in certain strains of the horse nematode. *Ascaris megalocephala*, and 510 pairs in the fern, *Ophioglossum petiolatum.*

POLYTENE CHROMOSOMES

In certain tissues of insects belonging to the order *Diptera* (flies, mosquitoes, midges, etc.) the cell nuclei have reached a high degree of enlargement accompanied by many extra replications of each chromosome within a single nucleus (endopolyploidy). However, instead of each new chromosome separating as an individual unit, all replicates of the same chromosome are lined up together in parallel fashion. This parallel duplication, or *polyteny*, results in very thick chromosomes that magnify any differences in density along their length (e.g., chromomeres). As shown in Fig. 2-13, such density differences produce a precise "banding" pattern that serves to identify any particular chromosome and its various sections. The numbers of bands vary between different species but are constant for the members of any particular species. In *Drosophila melanogaster* about 5000 bands can be recognized.

In polyteny the two homologous chromosomes of each diploid pair are also often lined up side by side (*somatic pairing*), so that if the total diploid number of chromosomes is eight (four pairs), only four very thick and long chromosomes appear. It is easy to see that if there were differences in banding order between two chromosomes of a pair, these differences would be reflected by loops and other abnormal topological formations between the homologous chromosomes (see, for example, Fig. 2-6). Also of special interest is the change in appearance of certain bands in different tissues and at different times, either condensing or expanding to form "puffs." The utilization of polytene chromosomes in genetic research begun by Painter in 1933 has proved to be of considerable value, and will be discussed in later chapters.

LAMPBRUSH CHROMOSOMES

The oocytes of some vertebrates with large yolky eggs expand greatly during their growth period, forming correspondingly large nuclei at these stages. In some amphibia the meiotic prophase chromosomes of such nuclei can reach about 1000 microns in length, with long lateral loops giving a hairy "lampbrush" appearance (Fig. 2-14a).

Each pair of loops arises from single chromomeres placed at short intervals along the very thin and probably double-stranded chromosome (Fig. 2-14b). As in the puffs of the salivary chromosomes, the loops seem to be elaborating materials that are to be used by the cell (Fig. 2-14c, d). Toward the end of meiotic prophase the loops begin to disappear and the chromosomes contract, so that at metaphase the bivalents are of the usual small size.

THE CONTINUITY OF CHROMOSOMES

Upon first observation chromosomes do not appear to maintain continuity as integral unchanging structures from one cell to another. After dividing

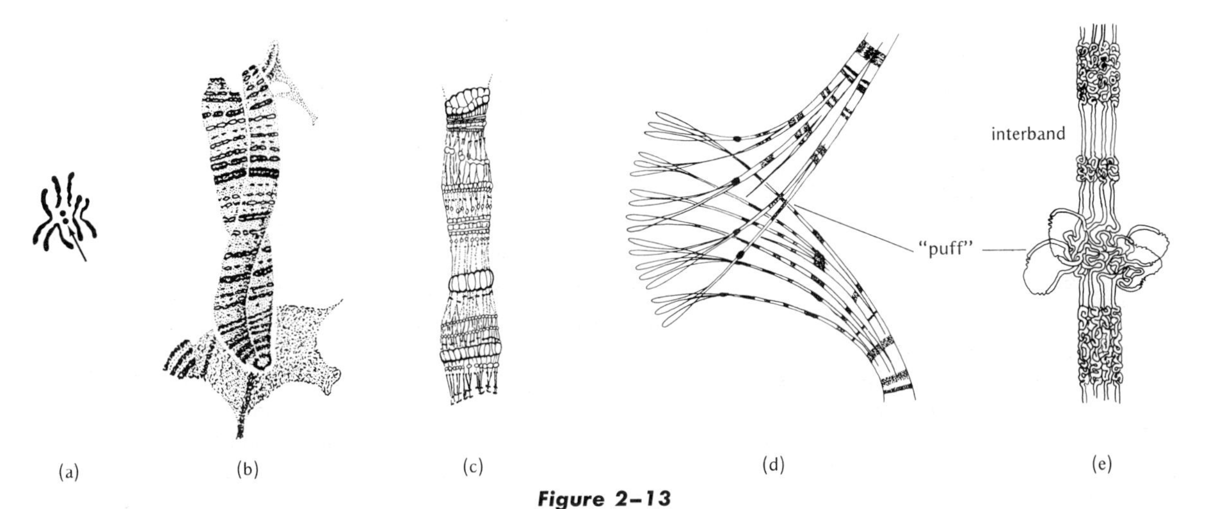

Figure 2–13

Comparison of mitotic (a) and salivary chromosomes (b, c, d, e). (a) Chromosome constitution (greatly enlarged) of a Drosophila melanogaster *female. Arrow points to chromosome 4. (b) The same chromosome in the larval salivary glands. (c) Enlarged section of salivary chromosome in* Simulium. *(d) Structure of puffed region in* Chironomus. *(e) Diagram visualizing the salivary bands as tight coils of chromatid material (chromomeres) and the interband spaces as uncoiled chromatid lengths. Puffing is shown to be caused by expansion of the same chromomere in many paired chromatids. (a and b after Bridges, 1935; c after Painter and Griffen; d after Beermann; e after DuPraw.)*

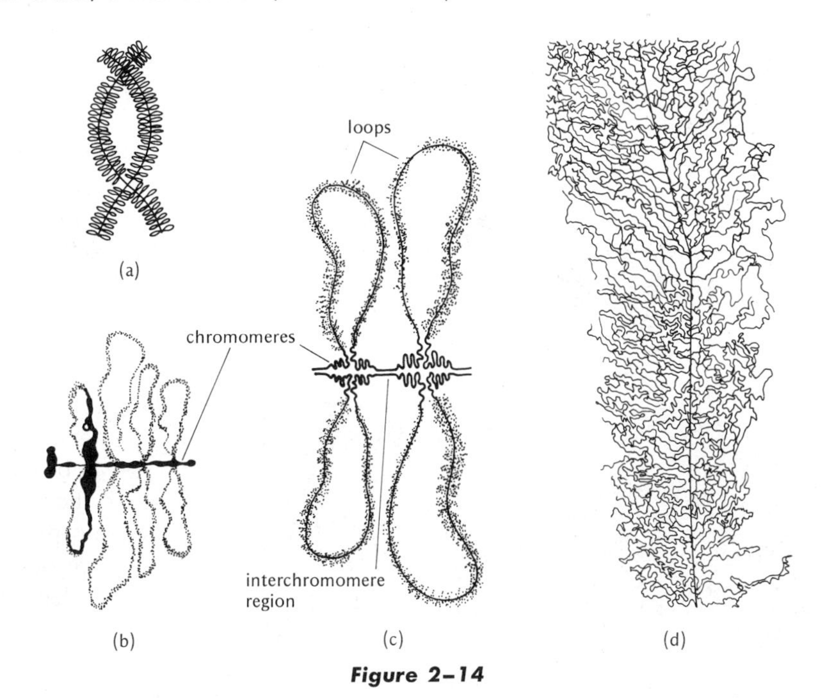

Figure 2–14

Lampbrush chromosomes in the newt, Triturus viridescens. *(a) Paired homologous chromosomes with two chiasmata. (b) Variation between loops of a single chromosome. (c) Enlarged view of two single loops and proposed chromomere structure. (d) Copy of electron micrograph of a section of a chromosome loop showing fibrils emerging from the loop axis. (a, b, and c after Gall; d after Miller and Hamkalo.)*

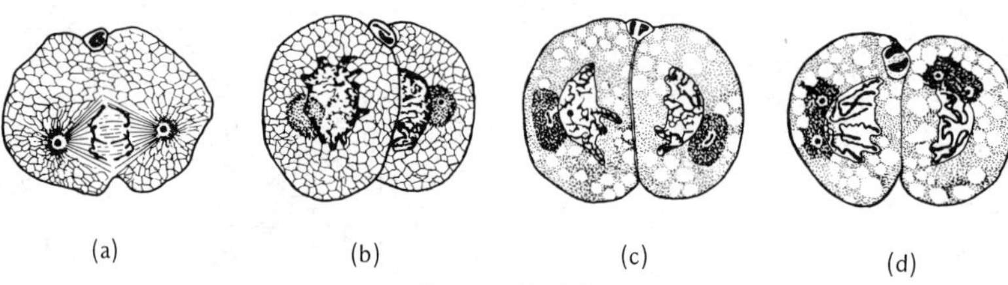

Figure 2–15

Cleavage divisions in eggs of Ascaris megalocephala. *(a) Anaphase of first cleavage division. (b) Two-cell stage, showing chromosome lobes in nuclear membrane. (c) Early prophase of next cleavage division (beginning of four-cell stage) showing chromosome ends appearing in lobes. (d) Late-prophase chromosomes now occupy the same position as they did in the preceding telophase. (After Wilson, from Boveri.)*

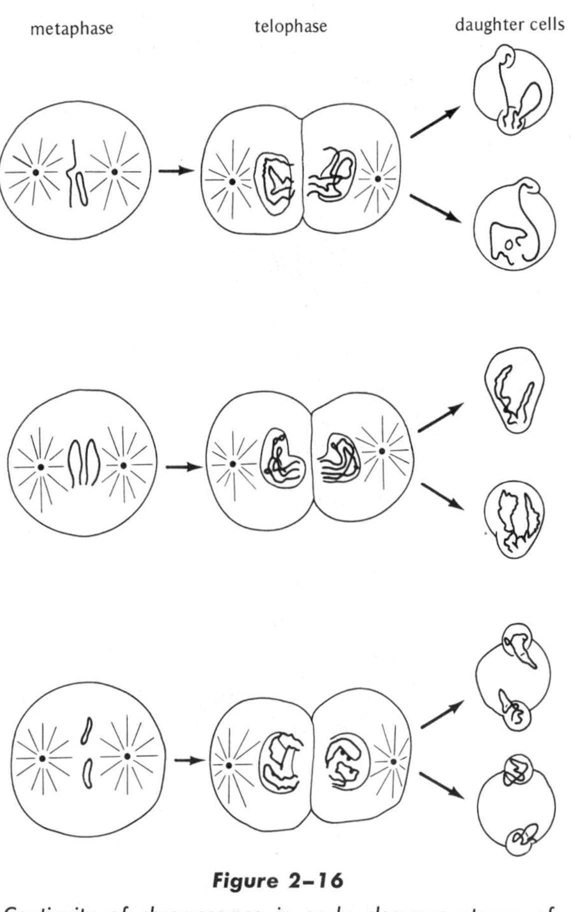

Figure 2–16

Continuity of chromosomes in early cleavage stages of Ascaris megalocephala univalens. *(After Wilson, from Boveri.)*

and separating, they enter the interphase period, during which they are either relatively invisible or, at most, appear as a network of intertwined long thin threads. To say that a particular chromosome at metaphase of one cellular division is identical to that at the metaphase of a later cellular division demands some form of evidence that chromosomes do not lose their identity during the interphase period.

This evidence was first provided by Boveri in experiments with the eggs of the nematode *Ascaris megalocephala*. Cell division in these eggs has the unique property that the telophase chromosomes are not completely withdrawn into a rounded nucleus but project into the cytoplasm so that the nuclear membrane forms lobes around them. When prophase of the next cell division occurs, the chromosomes now present in these lobes can be shown to be the same as those in the previous telophase. In Fig. 2–15 note that there are two pairs of chromosomes in each cell (variety *bivalens*) and therefore eight free ends. In the two daughter cells these ends are placed almost exactly opposite each other, as though they were mirror images. In varieties of *Ascaris* containing only one pair of chromosomes (*univalens*), the continuity between chromosomes is even more clearly expressed because the configuration of chromosomes in the daughter cells in prophase reflects the previous metaphase configuration of the chromosomes in the parental cell (Fig. 2–16).

PROBLEMS

2-1. In an animal with only one pair of homologous telocentric chromosomes, a stage in one of the cell divisions appears as follows:

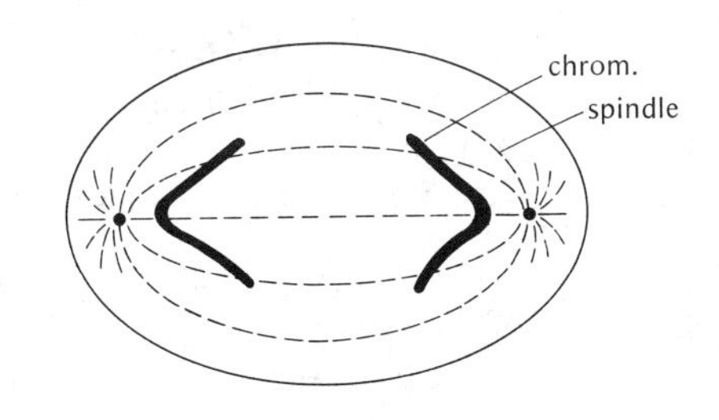

(a) Identify which stage in mitosis or meiosis this is. (b) After the above cell division is completed, two cells will result. Draw the chromosome products that will appear when these two cells divide to form four daughter cells.

2-2. An organism has two homologous pairs of chromosomes, one pair metacentric, and the other pair telocentric. (a) Draw the metaphase plate of the first meiotic division. (b) Draw the metaphase plate of a second meiotic division.

2-3. A cell pictured as follows was observed under the microscope.

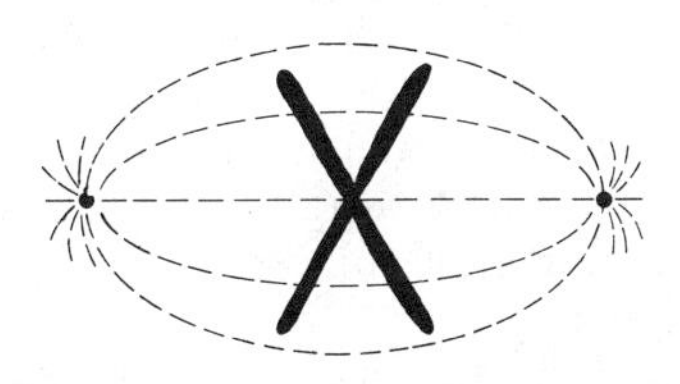

At what stages of mitosis and/or meiosis could this configuration have occurred?

2-4. In a zygote that begins with a complement of two homologous chromosome pairs, *A* and *a*, and *B* and *b*, which of the following chromosome complements would you expect to find in its somatic cells during growth: *AaBB*? *AABb*? *AaBb*? *AABB*? *aabb*? More than one of these combinations?

2-5. If the individual in the previous question becomes an adult, which of the following combinations of chromosomes would you expect to find in the gametes: (a) *Aa*, *AA*, *aa*, *Bb*, *BB*, *bb*? (b) *Aa*, *Bb*? (c) *A*, *a*, *B*, *b*? (d) *AB*, *Ab*, *aB*, *ab*? (e) *Aa*, *Ab*, *aB*, *Bb*?

2-6. The cells of a particular male contain one pair of homologous chromosomes, e.g., *AA′*, and one additional chromosome without a pairing mate, e.g., *B*. What is the chromosome constitution of each of the four gametes produced by one complete meiotic division?

2-7. If a plant with four homologous pairs of chromosomes, *AA*, *BB*, *CC*, and *DD*, is self-fertilized, which of the following chromosome combinations would you expect to find in its offspring: *AB*? *CD*? *ABCD*? *AABB*? *CCDD*? *AABBCC*? *AABBCCDD*? *AAAABBBBCCCCDDDD*?

2-8. A diploid male organism has two homologous telocentric pairs of identifiable chromosomes, *A* and *B* from the maternal parent and *A′* and *B′* from the paternal parent. (a) Draw the anaphase of mitosis in this organism. (b) Draw one possible arrangement of these chromosomes at anaphase of the first meiotic division. (c) On the basis of your drawing in (b) draw the appearance of these chromosomes at anaphase of the second meiotic division. (d) If these same chromosomes were involved in meiosis in a female, would the kinds of egg nuclei produced be different from the sperm nuclei? Would their frequencies be different?

2-9. The broad bean, *Vicia faba*, has a diploid number of 12 chromosomes in its somatic cells, consisting of 6 homologous pairs (that is, 6 maternal and 6 paternal chromosomes). A student stated that only one-fourth of the gametes produced by meiosis in this plant will have all of its chromosomes from either maternal or paternal origin. Explain whether you think the student is correct.

2-10. Explain why you would expect genetic differences between cells to arise from meiosis and not from mitosis.

2-11. In a particular corn plant, one member of a pair of homologous chromosomes (tenth-chromosome pair) has a knob and the other member does not. A second pair of homologous chromosomes (sixth) also shows a difference between the two members—one bears a terminal satellite and the other member lacks a satellite. Draw the possible kinds of gametes produced by meiosis.

2-12. By what mode of cell division would you expect sperm cells to be formed in the haploid male bee?

2-13. Would heredity through cytoplasmic particles affect the degree of resemblance between an offspring and a particular parent?

REFERENCES

ALTMAN, P. L., and D. S. DITTMER (eds.), 1972. *Biology Data Book*, Vol. 1, Fed. Amer. Soc. Exp. Biol., Washington, D.C. (Includes tables of chromosome numbers in plants and animals, and cell division frequencies.)

BAJER, A. S., and J. MOLÉ-BAJER, 1972. *Spindle Dynamics and Chromosome Movements*. Academic Press, New York.

BEERMANN, W., 1956. Nuclear differetiation and functional morphology of chromosomes. *Cold Sp. Harb. Symp.*, **21,** 217–230.

BRIDGES, C. B. 1935. Salivary chromosome maps. *Jour. Hered.*, **26,** 60–64.

CALLAN, H. G. 1963. The nature of lampbrush chromosomes. *Intern. Rev. Cytol.*, **15,** 1–34.

COMINGS, D. E., and T. A. OKADA, 1971. Triple chromosome pairing in triploid chickens. *Nature*, **231,** 119–121.

DARLINGTON, C. D., 1965. *Cytology*. Churchill, London. (Includes a reprint of Darlington's 1937 edition of *Recent Advances in Cytology*.)

DUPRAW, E. J., 1970. *DNA and Chromosomes*. Holt, Rinehart and Winston, New York.

GALL, J. G., 1956. On the submicroscopic structure of chromosomes. *Brookhaven Symp. Biol.*, **8,** 17–32.

JONES, R. N., 1975. B-chromosome systems in flowering plants and animal species. *Intern. Rev. Cytol.*, **40,** 1–100.

MAZIA, D., 1961. Mitosis and the physiology of cell division. In *The Cell*, Vol. III, J. Brachet and A. E. Mirsky (eds). Academic Press, New York, pp. 77–412.

MILLER, O. L., Jr., and B. A. HAMKALO, 1972. Visualization of RNA synthesis on chromosomes. *Intern. Rev. Cytol.*, **33,** 1–25.

MOSES, M. J., 1968. Synaptinemal complex. *Ann. Rev. Genet.*, **2,** 363–412.

NICKLAS, R. B., 1971. Mitosis. *Adv. in Cell Biol.*, **2,** 225–297.

PAINTER, T. S., and A. B. GRIFFEN, 1937. The structure and development of the salivary chromosomes of *Simulium*. *Genetics*, **22,** 612–633.

RHOADES, M. M., 1961. Meiosis. In *The Cell*, Vol. III, J. Brachet and A. E. Mirsky (eds). Academic Press, New York, pp. 1–75.

RIS, H., 1957. Chromosome structure. In *The Chemical Basis of Heredity*, W. D. McElroy and B. Glass (eds.). Johns Hopkins Press, Baltimore, pp. 23–69.

RIS, H., and D. F. KUBAI, 1970. Chromosome structure. *Ann. Rev. Genet.* **4,** 263–294.

Scientific American, 1961. The living cell. Vol. 205, No. 3, September issue.

SHARP, L. W., 1934. *Introduction to Cytology*. McGraw-Hill, New York.

WESTERGAARD, M., and D. VON WETTSTEIN, 1972. The synaptinemal complex. *Ann. Rev. Genet.*, **6,** 71–110.

WILSON, E. B., 1925. *The Cell in Development and Heredity*, 3rd ed. Macmillan, Inc., New York.

WOLFE, S. L., 1972. *Biology of the Cell*. Wadsworth Publ. Co., Belmont, Calif.

3
REPRODUCTIVE CYCLES

We have seen that the gametes produced by meiosis carry a mixture of parental chromosomal material that has been formed both through crossing over between homologues and through redistribution of the different parental chromosomes (p. 22). In effect, a pool of chromosomal material is created in the sex organs of an individual through the contributions of both parents from which, in turn, the gametes may draw various combinations. Since a species consists of many individuals acting as parents, each generation represents a reshuffling of chromosomal material on a very broad scale, permitting the origin of an immense variety of new combinations.

What particular advantage does chromosomal variability offer that makes sexual reproduction such a universal phenomenon? Although the details of such a discussion must be postponed (Chapter 34), it is sufficient for the moment to point out that as a rule the greater the variability of a species, the greater the chances that it will survive in time. All the biological and environmental factors upon which the success of an organism depends are not usually constant from place to place or time to time. Since changes in chromosome material will result, as we will see, in changed biological characteristics, a greater variety of chromosomal materials leads to increased chances that a species will survive under the varied conditions in which it must live. Thus, at some time or other in their life cycles, most living organisms take advantage of the opportunity to exchange and reshuffle their genetic material. The life-cycle descriptions that follow cover a variety of organisms of current interest in genetic studies and are of value for discussions in later chapters.

VIRUSES

With the invention of the light microscope in the seventeenth century and its perfection during the following two centuries, the investigation of the smaller forms of life proceeded at a rapid pace. By the end of the nineteenth century many unicellular organisms and even bacteria had been observed and described, and it seemed as though the most minute forms of life had been discovered. However, in 1892 a new type of organism appeared on the biological horizon through the discoveries of Beijerinck and of Iwanowski that the agent causing mosaic lesions on tobacco plants was small enough to pass through porcelain filters that stopped ordinary bacteria. Within the following few decades many plants, animals, and even bacteria were found to harbor such relatively invisible disease-causing agents.

The structure and life cycle of these miniscule agents, called *viruses* (or *bacteriophage* in bacterial infection), remained unknown until fairly recent times, and important advances in their study still continue. At present, through electron-microscope studies, many different kinds of viruses are known (Fig. 3–1), ranging in shape from round (mumps) to rodlike (tobacco mosaic) to polyhedral with attached tail ("T-even" bacteriophages). Sizes range from the vaccinia virus, as large as a small bacterium, to the minute Japanese B encephalitis, no larger than some protein molecules. In general the structure of a virus is that of a protein coat surrounding the viral genetic material, which is present in the form of a single chromosome.

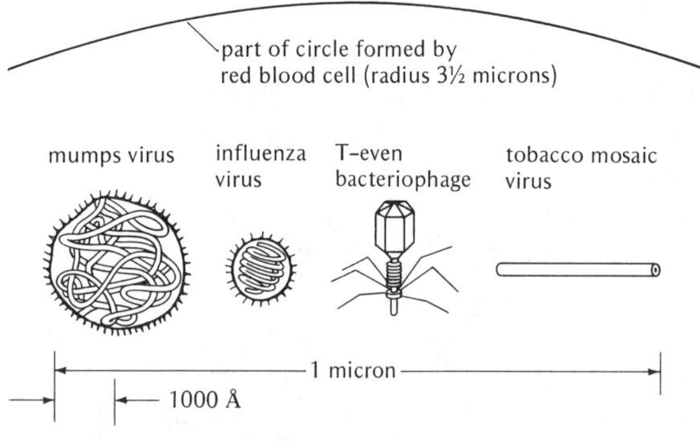

Figure 3–1

Different viruses and their relative size compared to a segment of a human red blood cell. One angstrom (Å) = 1/10,000 micron. (Adapted from Scientific American.)

Most often the viral surface makes firm contact (*adsorption*) with the host's cell membrane and reacts with it, enabling either the entire viral particle or its chromosome to enter at that spot. Once inside the cell the viral chromosome begins a process in which the host's entire metabolic machinery becomes devoted to forming new viral particles. In extensively studied infection of the intestinal bacteria, *Escherichia coli*, by the T2 bacteriophage, the viral chromosome multiples in the host within a few minutes after entry, rapidly forming a number of generations of viral chromosomes. This period is known as the *eclipse period*, and the virus is considered to be *vegetative* because it is reproducing but is not yet infective to other bacteria. At the same time as the viral chromosome is replicating, the host cytoplasm has become converted to the manufacture of the protein coats that will soon surround individual viral chromosomes. Within about 30 minutes of the time of entry, the host's cell membrane dissolves (*lysis*), and approximately 200 to 300 new infective viral particles emerge (Fig. 3–2).

Some viruses, such as bacteriophage lambda (λ), do not kill all the bacteria they infect. Some of the bacterial survivors of λ infections, called *lysogenic* bacteria, carry the virus in an inactive form known as the *prophage*. Bacteriophages that can be carried in such passive fashion are known as *temperate* phages, in contrast to the *virulent* phages which always cause destruction of the bacterial cell upon infection. The prophage attaches itself to a particular spot on the bacterial chromosome and seems to replicate itself only once with each division of the bacterium. When a specific change in the environment occurs (*induction*), some of the lysogenic bacteria become "sensitive," and the viral chromosome dissociates itself from the bacterial chromosome. The virus then replicates itself independently of its host, in sufficient numbers to lyse the bacterium.

Although viruses are seemingly asexual, Del-

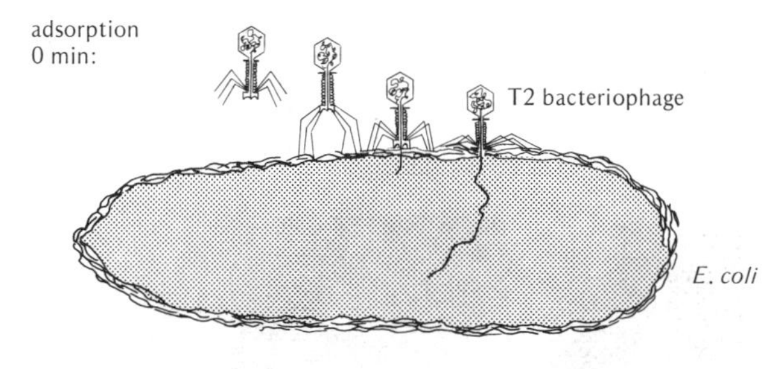

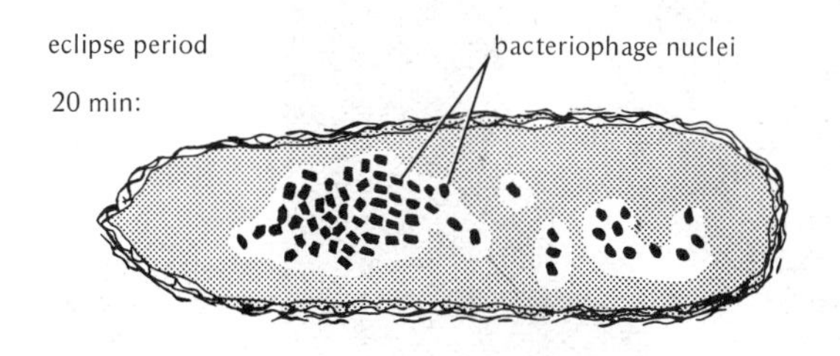

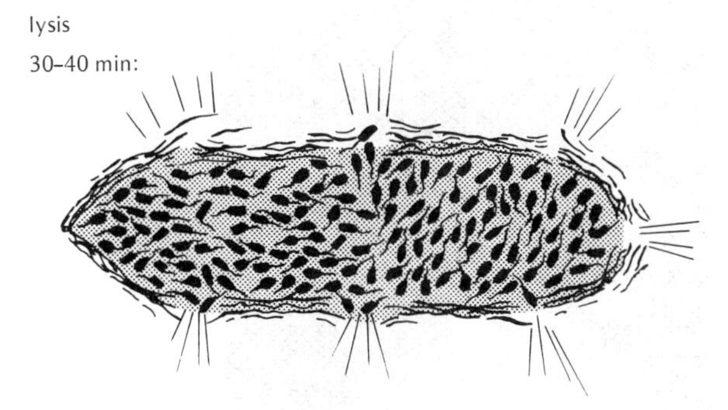

Figure 3–2

Cycle of bacteriophage infection and reproduction.

bruck and Bailey, and Hershey have discovered that when different strains of a particular type of bacteriophage are introduced into a bacterial culture, new combinations of the original viral traits appear among their progeny. By excluding other possible causes, the conclusion was drawn that some form of exchange of genetic material had occurred between the two parental types of virus. Additional findings also suggest that "mating" is not necessarily limited to two bacteriophages, but that if three different bacteriophages enter a host cell, new combinations can occur between all three types of genetic material! These matters will be discussed in detail in Chapter 20.

BACTERIA

Because of their simple structure and very short generation time (about 20 minutes in some cases), bacteria have become among the most popular organisms used in genetic experiments. Microscopically small and single-celled, they have varied shapes and sizes (Fig. 3–3) and divide, as a rule, by simple fission. According to this asexual division process, a single bacterium gives rise to a *clone* of descendants all genetically related to each other through their common ancestor. However, not all clones are identical, although they share a common ancestor. When accidental changes in the genetic constitution (*mutation*) of a bacterium occur, the clone to which it gives rise will, of course, differ to that extent from other clones. Those clones or strains of bacteria that have lost

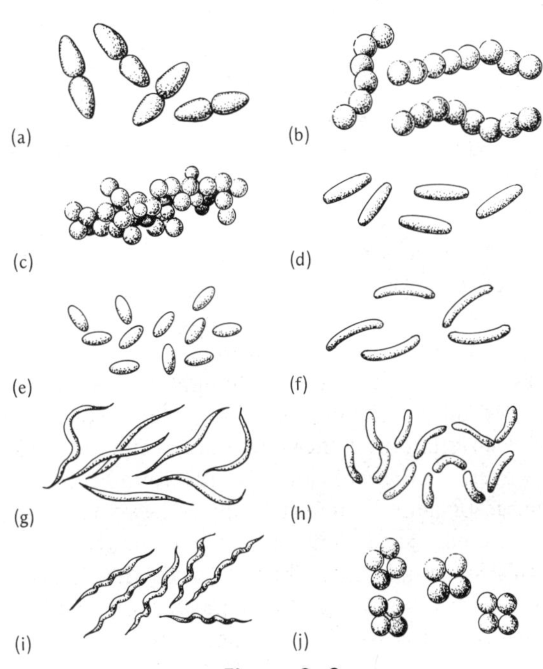

Figure 3–3

Different bacterial forms: (a) diplococci (pairs); (b) streptococci (chains); (c) staphylococci (clusters); (d) bacilli; (e) coccobacilli; (f) fusiform bacilli; (g) filamentous forms; (h) vibrios; (i) spirilla; (j) sarcinae. (After Davis, Dulbecco, Eisen, Ginsberg, and Wood.)

nutritional substance	symbols	
	auxotroph	prototroph
Threonine	T^-	T^+
Leucine	L^-	L^+
Thiamine	B_1^-	B_1^+
Biotin	B^-	B^+
Phenylalanine	Ph^-	Ph^+
Cystine	C^-	C^+

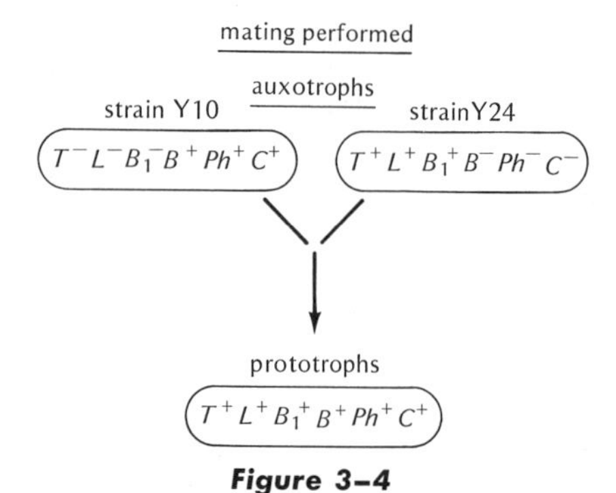

Figure 3–4

Results of mating between E. coli *strains Y10 and Y24 according to the data and symbols used by Lederberg and Tatum.*

the ability to produce certain nutritional substances (usually amino acids) are called *auxotrophs*, and those that can normally produce these constituents are called *prototrophs*.

For a long time it was thought that bacteria divided only asexually and that genetic exchange between different clones could not occur. In 1946, however, Lederberg and Tatum found an exchange of genetic material between two types of *Escherichia coli*, a common bacterial inhabitant of man's intestinal tract. They observed that if each type were auxotrophic for a particular chemical constituent for which the other type was prototrophic, a combined culture gave rise to occasional bacteria uniting the characteristics of both types (Fig. 3–4).

The experimental errors that might have been responsible for such results, such as contamination, new mutation, and other factors, were excluded through various experiments, and the origin of this new type of bacteria was ascribed to a direct exchange of material in cell-to-cell contact. In recent

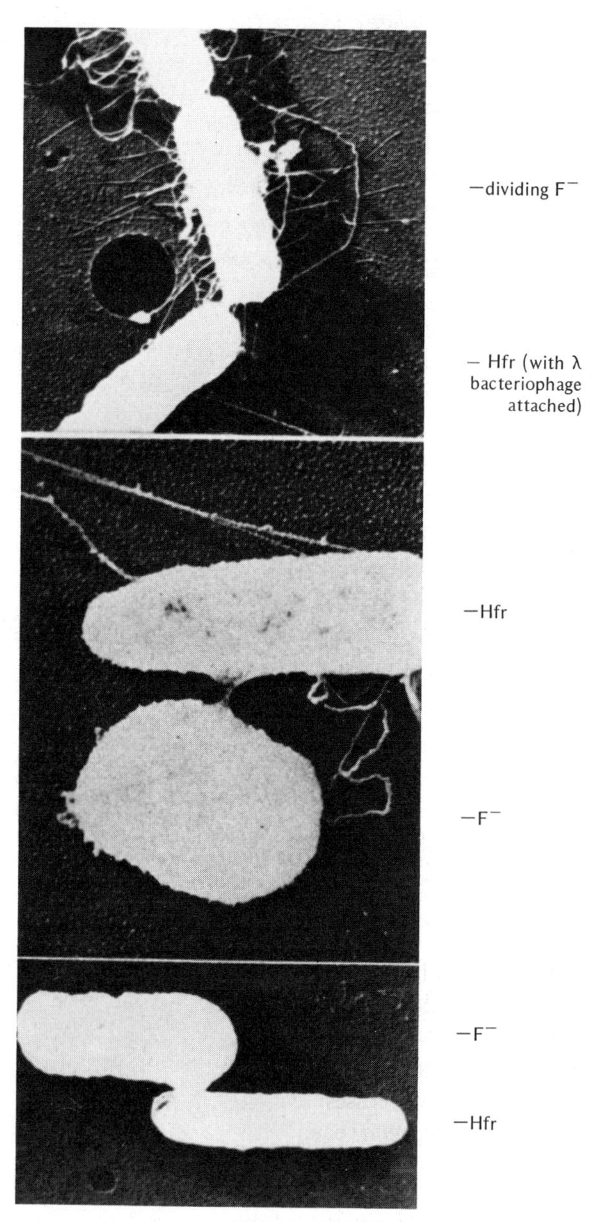

Figure 3–5

Electron micrographs of mating E. coli *cells (Hfr × F^-). (From T. F. Anderson, in E. Wollman, F. Jacob, and W. Hayes, 1956. Conjugation and genetic recombination in* Escherichia coli *K-12. Cold Sp. Harb. Symp.,* **21**, *152.)*

years various microbial geneticists have clarified how this physical exchange takes place.

In brief, there are two mating types of bacteria: one that is the *donor* of genetic material, and one that is the *recipient*. The particular factor determining the mating type of bacteria is called F, and the possessors of F (donors) are of two types, designated F^+ and Hfr (high frequency of recombination), while the recipients, lacking F, are designated F^-. F is of considerable interest in itself and will be discussed later (Chapter 19). At present, it can be pointed out that F has been found to be a piece of genetic material that may (Hfr strains) or may not (F^+ strains) be directly attached to the bacterial chromosome.

When two mating types of bacteria are mixed together (Hfr $\times$ F^-, or $F^+ \times F^-$), chance collisions between them may lead to effective sexual pairing. Transfer of genetic material then occurs, and some electron micrographs show an actual bridge of cellular material between the two different types (Fig. 3–5) through which the chromosome of the donor penetrates the recipient. Once the recipient cell receives either all or part of the donor chromosome, an exchange of genetic material may take place as discussed in Chapter 19, and the recipient may then incorporate some of the inherited characteristics of the donor.

PARAMECIA

In contrast to viruses and bacteria, the reproductive cycle of the eucaryotic single-celled slipper-shaped animal, *Paramecium*, is more easily observable and apparently more complex. As a result of the work of many biologists, including Sonneborn and others, there are at present three reproductive processes known in the organism *Paramecium aurelia*, each accompanied by different behavior of its nuclear complement. This unicellular animal maintains three individual nuclei, of which two are small *micronuclei* containing the diploid number of chromosomes, and the third is a large *macronucleus* with a manyfold multiplication in chromosome number (*polyploid*).

In the ordinary vegetative part of the cycle the animal divides by binary fission, with the micronucleus reproducing mitotically and the macronucleus constricting in the middle and separating into two halves. In suitable cultures *Paramecium* may divide about every 5 hours. However, if two different mating types are brought together under the proper conditions, an initial clumping of many individuals occurs, followed by pairing between single animals and *conjugation* at the oral region (Fig. 3–6). The micronuclei in each animal then undergo meiosis, forming four haploid nuclei from each micronucleus, or a total of eight haploid nuclei per cell. Seven of these are eventually destroyed by the cytoplasm, while the remaining micronucleus enters a "protected" area (*paroral cone*) where it undergoes a single mitotic division to form two gametic nuclei. One of these gametes retreats back into the cell (female), while the other (male) goes forth across the cytoplasmic bridge into its conjugal mate and fuses with the female nucleus.

The two *conjugant Paramecia*, now diploid once more, separate as *exconjugants*. Their macronuclei, having previously broken into a number of fragments, continue to disintegrate until the fragments are no longer visible. Each exconjugant's zygotic nucleus then undergoes two mitoses to form four micronuclei, two of which expand greatly to form individual macronuclei. At the next binary fission each of these macronuclei enters separately into a daughter cell, and the original complement of two micronuclei and one macronucleus per organism is restored. This final division occurs about 18 hours after the beginning of conjugation. Since the macronucleus disintegrates during meiosis and is reformed from the micronuclei, it has been suspected that it plays little role in heredity. Instances, however, are known where some of its small fragments may regenerate, and, under those circumstances, it can have a unique influence on the developing organism.

A third major form of cellular behavior is *autogamy*, in which nuclei follow the exact division pattern of sexual conjugation without the presence of a partner (Fig. 3–7). In this process meiosis of the two micronuclei occurs as before, seven haploid nuclei are destroyed, and one haploid nucleus

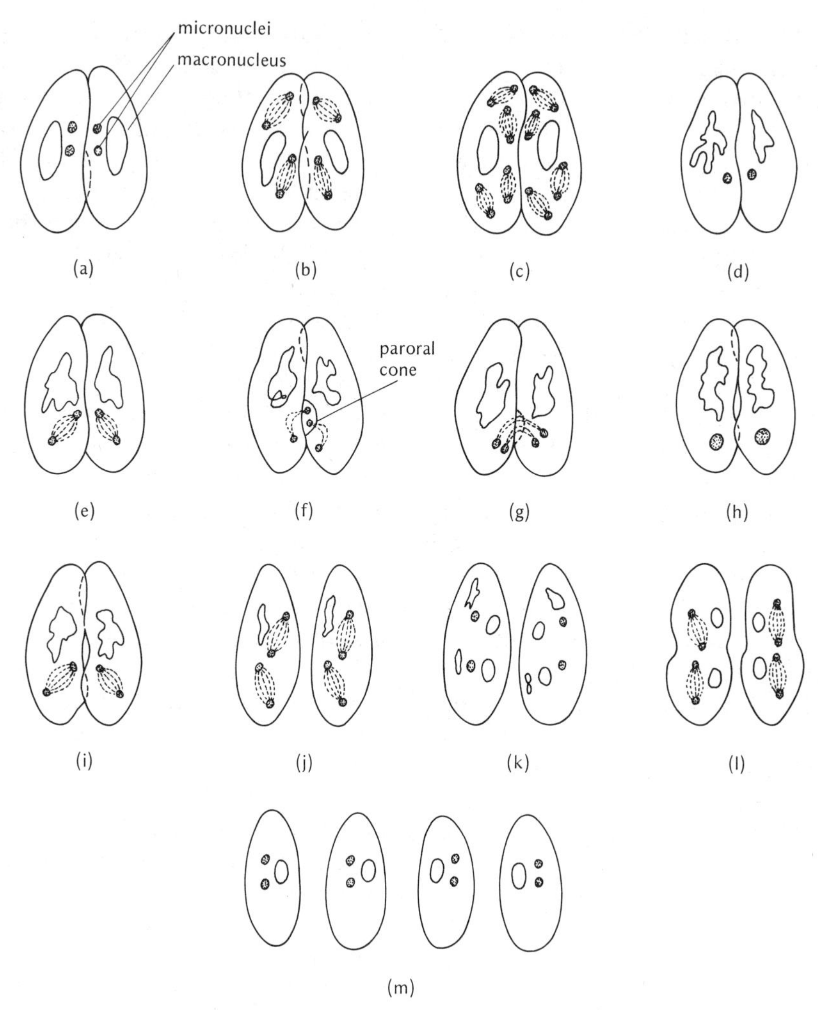

Figure 3–6

Stages in sexual reproduction in Paramecium aurelia. *(a–c) Conjugation followed by meiosis of micronuclei. (d) Only one haploid product in each cell remains. (e) Mitosis of each haploid product occurs. (f and g) Exchange of haploid products and mutual cross-fertilization produce diploid product (h) in each cell. (i and j) Mitosis of diploid-cell nucleus produces four products in each cell, two of which become macronuclei and two micronuclei (k). (l and m) Further division of the micronucleus and cellular fission produce four cells from the original two. (After Wichterman, from Sonneborn.)*

divides mitotically to form two gametes. However, instead of mating with conjugant nuclei, these two gametic nuclei fertilize each other, forming a new micronucleus. Since the new micronucleus owes its origin to the duplication of a single haploid nucleus, all pairs of homologous chromosomes in such animals are now exactly identical. This severely reduces the original variability between homologous chromosomes, a fact that has proved useful in some genetic studies.

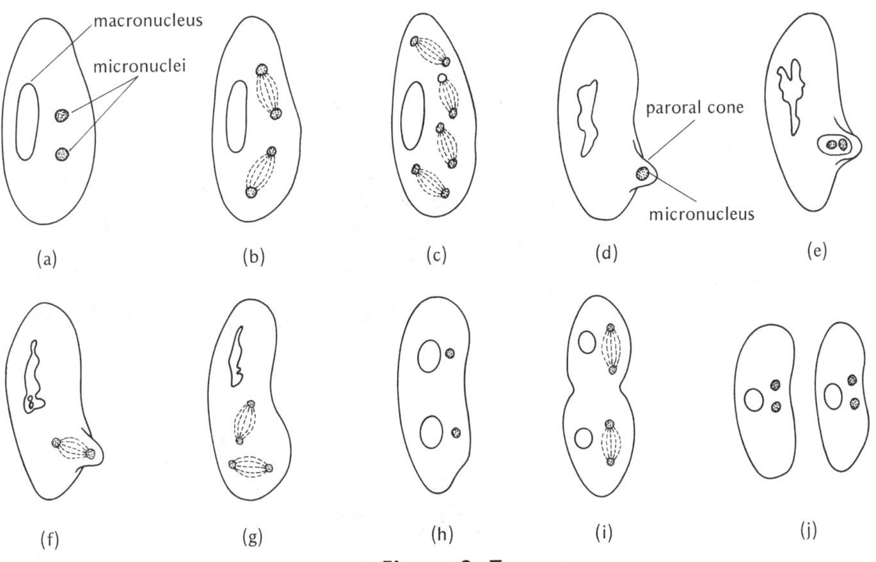

Figure 3–7

Stages in autogamy in Paramecium. (a–c) Meiosis of micronuclei leading to persistence of only one haploid product. (d) Haploid product divides mitotically and then each of these two nuclei unite to form diploid product (e). (f-i) Mitotic division of diploid product leads to production of new micro- and macronuclei. (j) Cell undergoes fission to produce two cells. (After Wichterman, from Diller.)

CHLAMYDOMONAS

In the plant world a simpler unicellular counterpart to *Paramecium aurelia* is the haploid green alga *Chlamydomonas reinhardi* (Fig. 3-8a). It has two motile flagella, a small distinct haploid nucleus, several mitochondria, and a single green chloroplast that occupies about 50 percent of the volume of the cell. The chloroplast enables *Chlamydomonas* to grow phototrophically in light utilizing CO_2 as its carbon source, but it can also grow heterotrophically on an acetate medium in the absence of light.

Two morphologically indistinguishable mating types are noted, mt^+ and mt^-, which can differentiate into gametes in the absence of nitrogen. The gametes then fuse in pairs of opposite mating type (*syngamy*) to form a diploid zygote that encapsulates and undergoes meiosis to yield, in due course, either four or eight haploid *zoospores*, one-half mt^+ and one-half mt^- (Fig. 3–8b). Clones of chromosomally identical descendants are then produced by each haploid cell until the appropriate conditions arise for gamete formation. Generation time for an asexual vegetative division is about 6 hours, and the sexual cycle takes somewhat less than 2 weeks.

NEUROSPORA

In one of the common bread molds *Neurospora crassa* the normal vegetative phase is a branching haploid organism with intertwined filaments, or *hyphae*, forming a spongy pad, the *mycelium*. Each hypha is segmented and each segment usually contains more than one haploid nucleus (*multinucleate*). Occasionally hyphae from one mycelium fuse with those of another while their nuclei remain individually separate although mixed in the

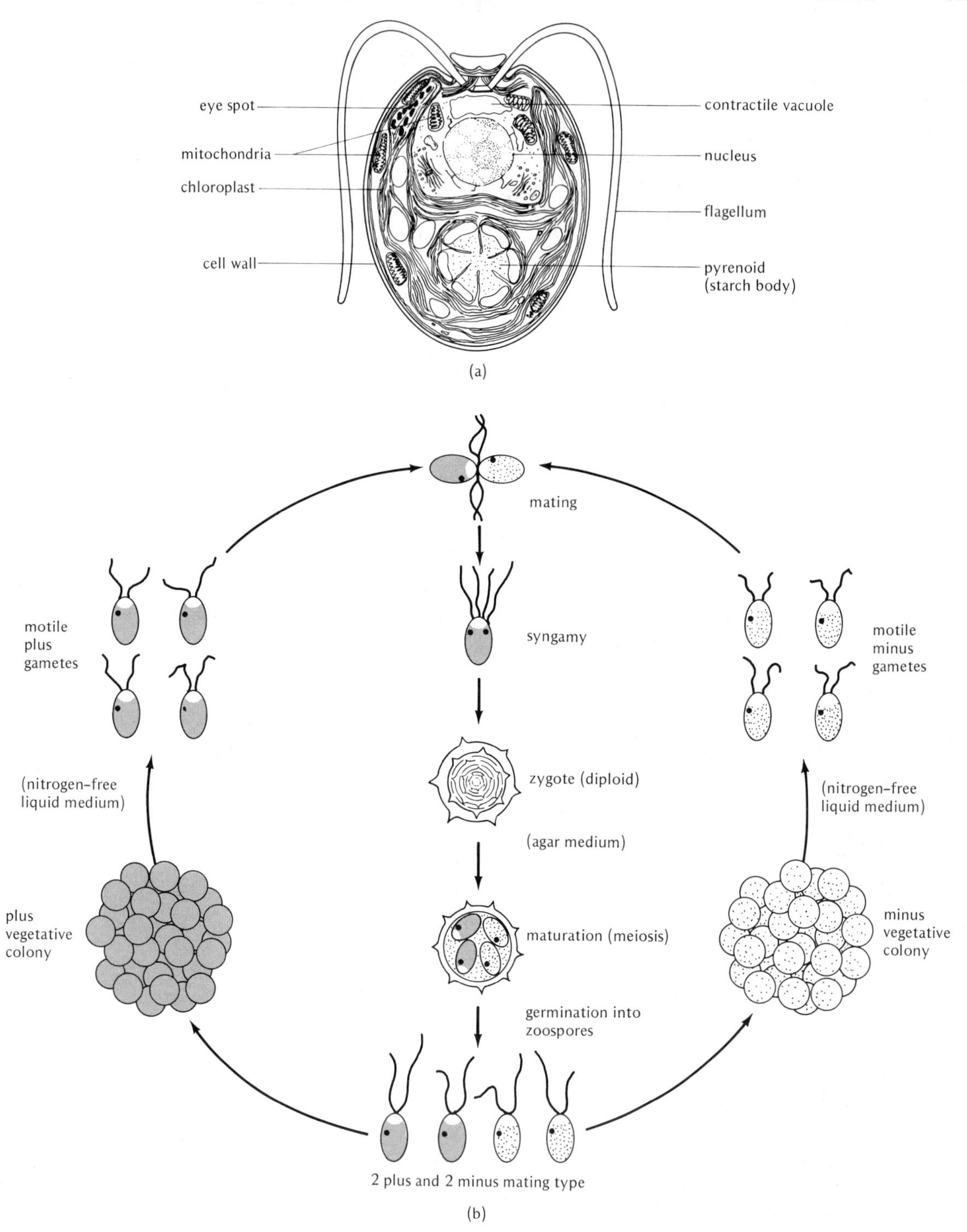

Figure 3–8

(a) Diagrammatic representation of Chlamydomonas reinhardi *according to electron micrograph studies. (After R. Sager, 1972.* Cytoplasmic Genes and Organelles. *Academic Press, New York.) (b) Life cycle of* Chlamydomonas.

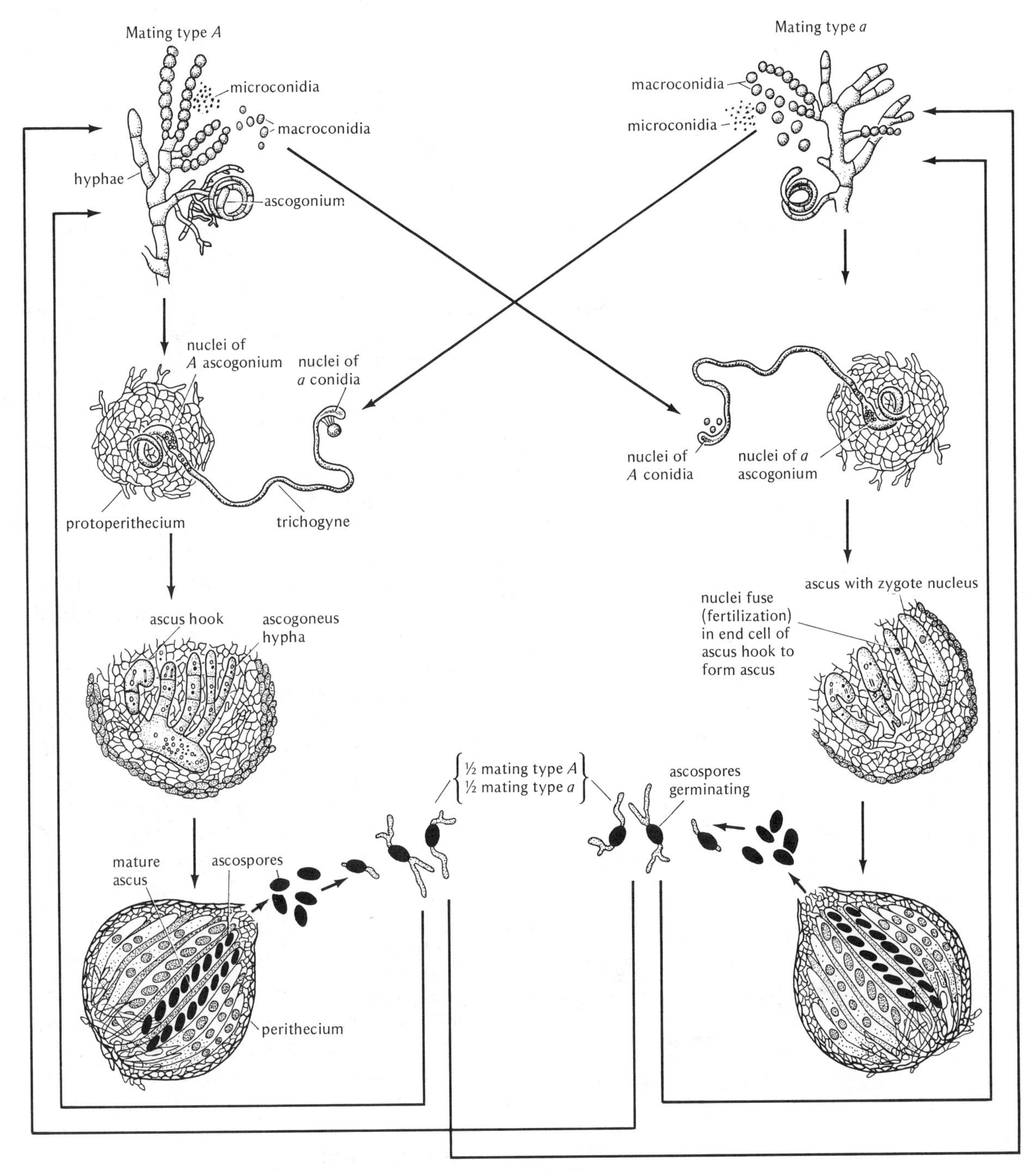

Figure 3–9

Sexual reproduction in Neurospora. *(Modified after Begg.)*

same cytoplasm (*heterocaryon*). Asexual reproduction of new individuals can occur in two ways: either by repeated mitosis of nuclei, leading to growth and fragmentation of hyphae; or by the mitotic formation of special haploid spores, *conidia*, that can form new mycelia. As shown in Fig. 3–9, the mycelia also produce immature sexual fruiting bodies (*protoperithecia*) containing haploid maternal nuclei and special mating filaments, called *trichogynes*, that extend outside the fruiting body. When a hyphal fragment or conidium from one mating type contacts a receptive trichogyne of the opposite mating type, it moves rapidly into the fruiting body, dividing mitotically as it goes along, and fertilizes a number of maternal nuclei. Diploid sexual zygotes are thus formed that are each enclosed in an oval *ascus sac* (plural: *asci*). A mature fruiting body (*perithecium*) can contain as many as 300 asci.

The zygote in each ascus immediately divides meiotically to yield four haploid cells, which then undergo a single mitosis to furnish a total of eight haploid cells per ascus. Each of these eight cells is called an *ascospore*, and they are arranged in a particular linear order in the ascus serving to identify the linear order of meiotic division products. This correlation arises from the linear separation of each of the first meiotic division products, which then divide to form two haploid cells that are also linearly separated from each other in the ascus (Fig. 3–10). The final mitosis only duplicates every haploid cell but does not change their meiotic relationship. Thus the first meiotic division gives rise to separate halves of the ascus, while the second meiotic division gives rise to separate quarters in each half. The genetic exchanges that occurred between chromosomes in any step of meiosis are easily determined by dissecting out and isolating each ascospore according to its linear order and then observing its characteristics or that of the culture it produces (p. 319). The interval between which an ascospore serves as the founder of a new mycelium, and the time when the mycelium produces new ascospores is about 2 weeks.

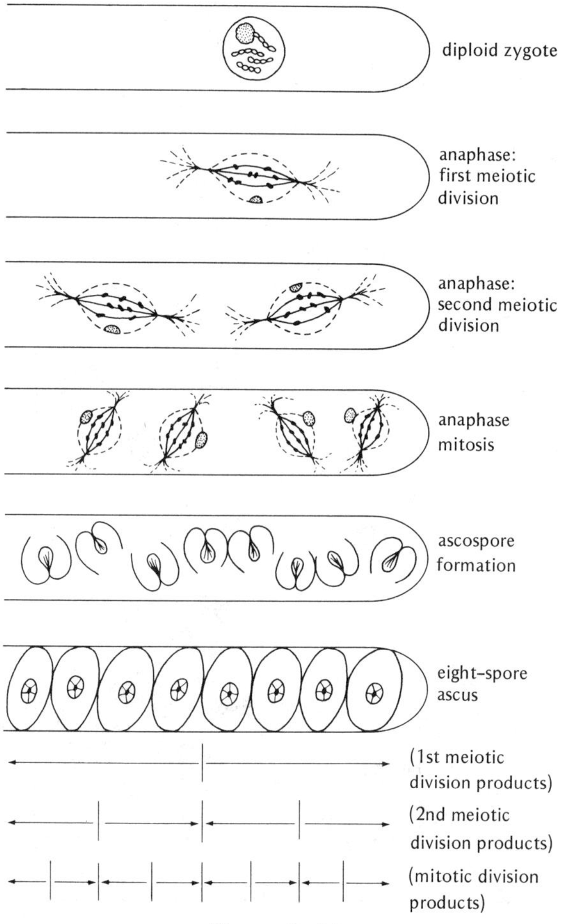

Figure 3–10

Ascospore formation in Neurospora, showing the linear separation between the meiotic and mitotic products. (After Olive.)

YEASTS

Although yeasts do not produce true branching mycelial growths, they are usually included among the fungi that produce asci and ascospores (*Ascomycetes*). However, in contrast to fungi such as *Neurospora*, yeasts may exist as either haploid or diploid unicellular organisms that have the distinction of multiplying by "budding." In *Saccharomyces cerevisiae* (baker's and brewer's yeast), for example, most cultures consist of diploid oval-shaped cells that grow by extruding small cytoplasmic buds from the cell wall. When the cell

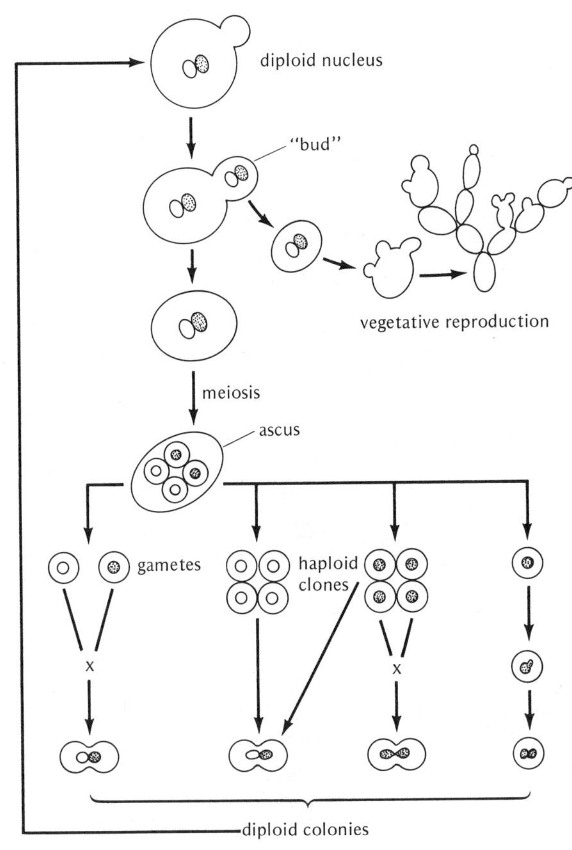

Figure 3–11

Yeast life cycle showing vegetative reproduction (upper part of figure) and formation of meiotic products. Diploidization can occur through the four ways shown at the bottom. (After Ephrussi.)

nucleus divides, the bud receives one of the daughter nuclei and eventually separates as a distinct cell. This kind of vegetative reproduction can continue until unfavorable conditions occur which cause the cell wall to thicken and form an *ascus*. In this process the diploid cell undergoes two meiotic divisions, forming four spherical haploid products known as *ascospores*. When favorable conditions return, the haploid ascospores swell in size and can then form either gametes or colonies of spherical haploid vegetative cells. Both events usually lead to formation of diploid zygotes through fusion between two of the products of the haploid ascospores (Fig. 3–11). Occasionally the nucleus of an individual haploid ascospore can also duplicate itself without cytoplasmic division and then combine with its product to form a diploid cell (direct diploidization).

As in *Neurospora*, some yeasts have mating types which do not permit indiscriminate conjugation between haploid ascospores or their products; only opposite mating types (i.e., a and α) will produce diploid zygotes. Single ascospores, and their clones isolated from such yeasts, will therefore maintain the haploid condition until brought into the presence of cells from the opposite mating type.

CORN (*ZEA MAYS*)

In most lower plants the haploid condition is the rule for the major part of the life cycle, with different haploid mating types producing the gametes that combine to form the diploid zygote. As one progresses from the algae and fungi to the mosses and ferns there is a gradual lengthening of the diploid phase. From its early beginnings as just a simple zygote that immediately undergoes meiosis to form the more permanent haploid stage, the doubled chromosome constitution of the zygote is propagated through mitosis and becomes, in itself, a major part of the life cycle. This diploid phase is called the *sporophyte* because some diploid cells (*sporocytes*) undergo meiosis to form haploid spores (*sporogenesis*). These spores are not gametes but multiply mitotically to yield the *gametophytes* (haploid male and female stages) which, in turn, form the gametes. When the flowering plants (angiosperms), such as corn (*Zea mays*), are reached in this progression, the haploid gametophytes are so reduced in size as to be microscopic.

In corn the two types of gametophytes are represented by small microspores in the stamens (tassels on top of the plant) for the male, and by large megaspores in the pistils (ears of the plant) for the female. In the stamens single diploid *microspore mother cells* (pollen mother cell, or PMC) divide meiotically to yield four haploid microspores, each becoming encapsulated as a *pollen grain* (the male gametophyte). The haploid pollen nucleus then

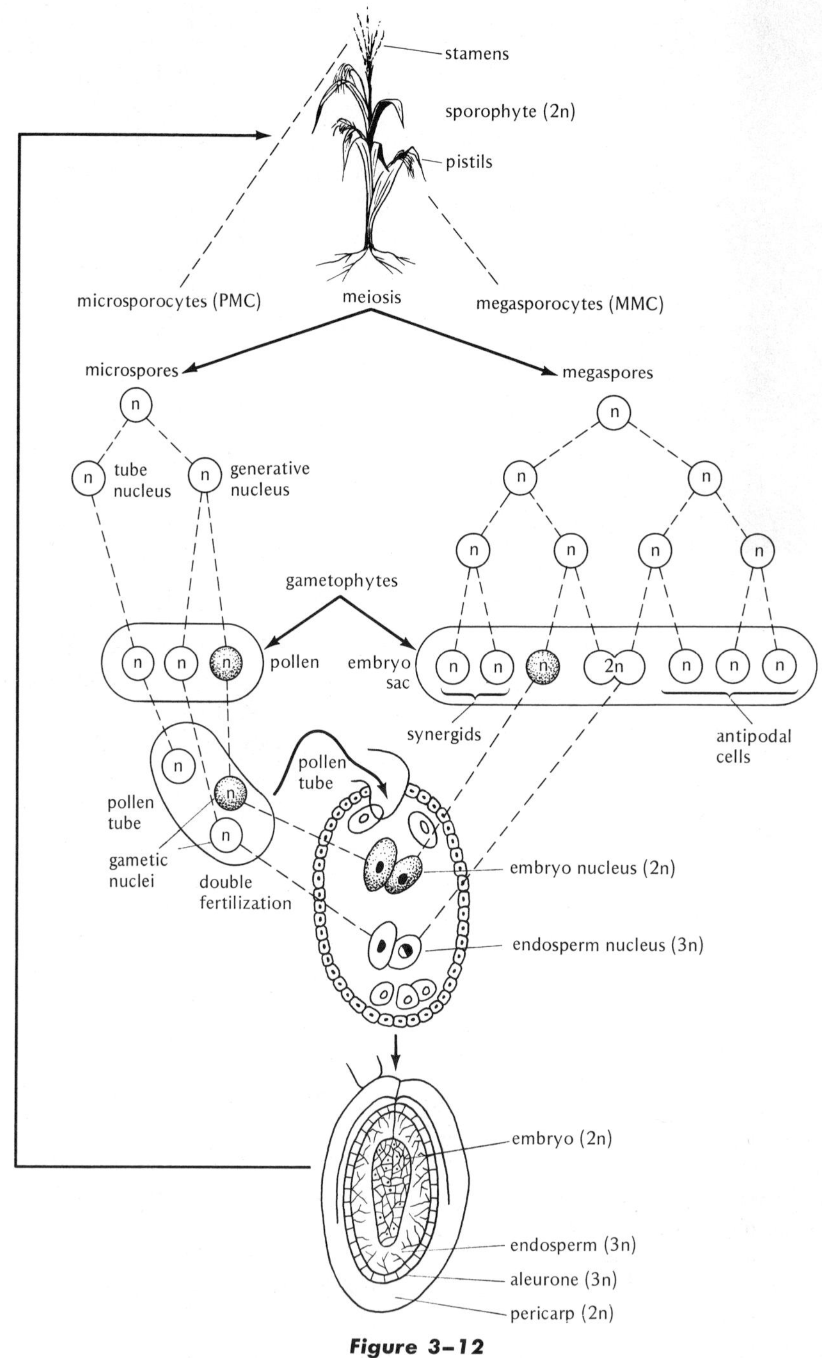

Figure 3–12

Life cycle of corn.

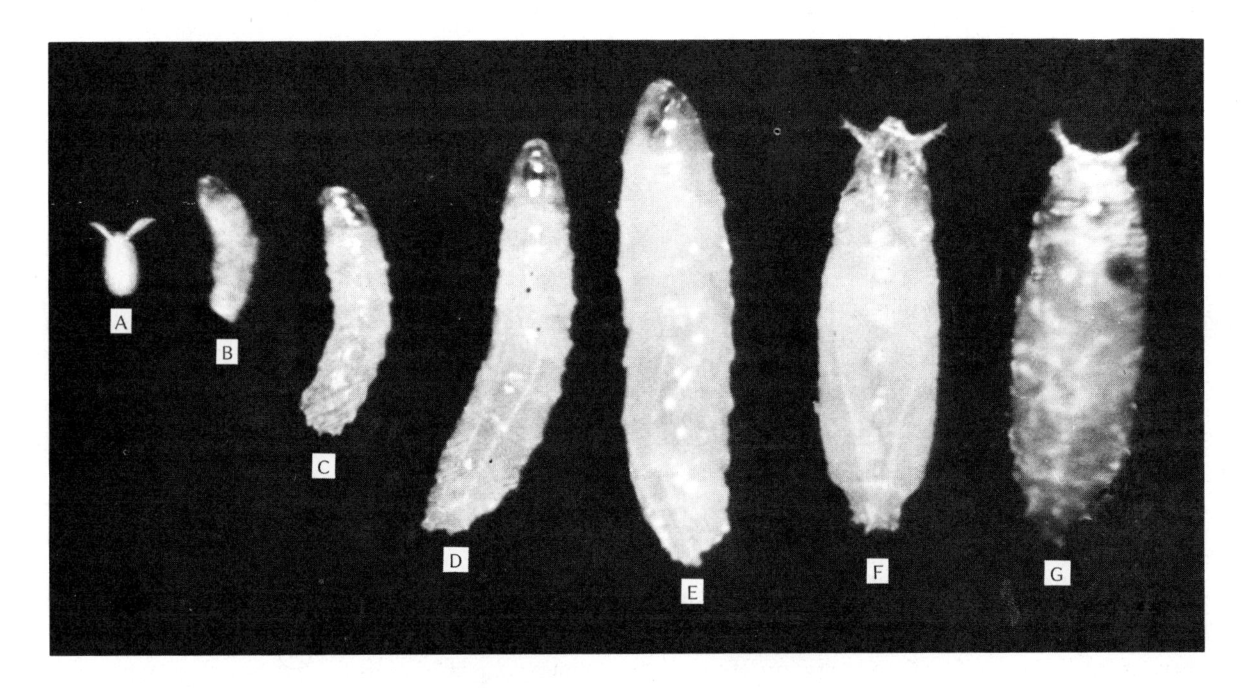

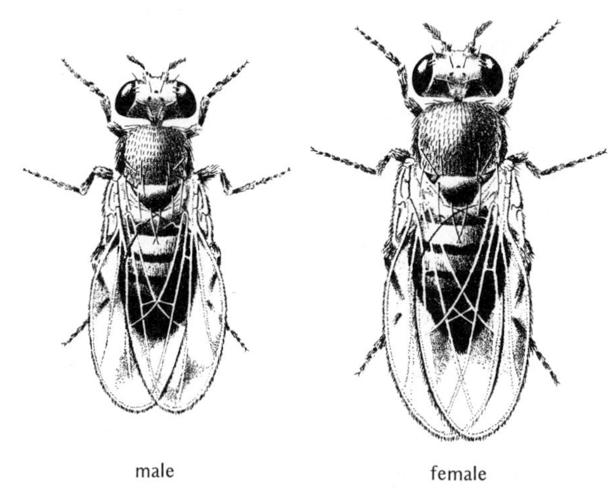

Figure 3–13

Stages in the life cycle of Drosophila melanogaster. *(A) Egg stage, approximately 0.5 mm in length. When fertilization occurs, the egg undergoes a period of rapid cell division and development, producing structures mainly to be used for the coming larval stage. At normal temperatures (25°C) this developmental embryonic stage lasts for about one day, at which point a larva hatches out of the former egg membrane. (B–C) Early larvae, with their anterior ends facing the top of the page. The larvae are motile forms that feed and grow very rapidly. At each of the main larval growth stages, the larva "molts" by discarding its skin or cuticle. It then lays down a new expanded cuticle which it proceeds to fill by rapid feeding. (D–E) Late larval stages showing the relative sizes reached within about five days from the time of fertilization. (F) Prepupa stage reached when the larva stops feeding, ceases motion, and its cuticle begins to harden. (G) Pupa. A hard protective case is formed around the larva and the organism now undergoes metamorphosis for an approximate four-day period. During this period larval structures are replaced by the development of new adult structures, many of which have their origin in small discs of tissue (imaginal discs, see Chapter 30) that were present as far back as the embryonic stage. On the ninth day after fertilization an adult fly emerges from the pupal case. The adult male and female forms are shown below the preadult forms. (From Strickberger)* divides mitotically to form a *tube* nucleus and a *generative* nucleus, the latter then dividing once more to form the two male *gametic* nuclei (Fig. 3–12).

A similar succession of events occurs for each *megaspore mother cell* (MMC) in the pistils, except that only one of the four haploid *macrospore* nuclei becomes the functional occupant of the *embryo sac* (the female gametophyte). The nucleus of this cell then divides mitotically into two daughter nuclei, which divide twice more, forming a total of eight haploid nuclei, four at each end of the embryo sac. A single nucleus from each end group of four then unites at the center to form the diploid *endosperm nucleus*. Of the remaining six nuclei in the embryo sac, the group of three farthest away from the pollen-tube point of entry (*micropyle*) are called the *antipodal* cells, while the other group differentiates into a single female gametic nucleus and two *synergids*.

In the process of fertilization the pollen grain makes contact with the *stigma* (silk) of the pistil and then germinates into a long *pollen tube* carrying the two male sperm nuclei to the embryo sac. One male gamete fertilizes the female gametic nucleus to form the diploid zygote, and the other male gamete combines with the diploid endosperm nucleus to form the *triploid* (diploid + haploid) tissue that will nourish the embryo. Thus this double fertilization forms two types of tissue: *embryonic* and *endosperm*. It is of interest and importance to geneticists that these two tissues differ genetically only in the number of sets of chromosomes from the same female haploid nucleus. Generation time for corn, from seed to sexually mature plant, is approximately 6 months.

DROSOPHILA

Starting with the introduction of the fruit fly, *Drosophila melanogaster*, to genetics about 1906 by Castle and somewhat later by Morgan, its experimental use has led to the discovery of many important genetic laws. In the natural state it is a small diploid fly approximately 2 mm (1/12 inch) in length that arises through the usual insect life cycle progressing from egg to larva to pupa to adult (Fig. 3–13). At 25°C this cycle takes about 10 days, with a manyfold increase in size occurring during the larval stages. After pupal metamorphosis the adults that emerge do not undergo further cell division except in their reproductive organs.

Differing from plants, animals generally produce their haploid gametes by meiosis directly from diploid tissue rather than through the intervention of a haploid gametophyte generation.

In the male reproductive organ, the testes, *spermatogonial* cells divide mitotically, forming a group of diploid *primary spermatocytes* (Fig. 3–14). Each spermatocyte then divides meiotically, forming two *secondary spermatocytes* in the first reduction division and yielding a total of four haploid sperm cells after the second division. In the female, *oogonial* cells in the ovary produce the diploid nuclei for the relatively large egg cells (*oocytes*). Upon penetration of the egg by a male sperm, the egg nucleus undergoes meiosis to yield four haploid nuclei. Only one of these egg nuclei

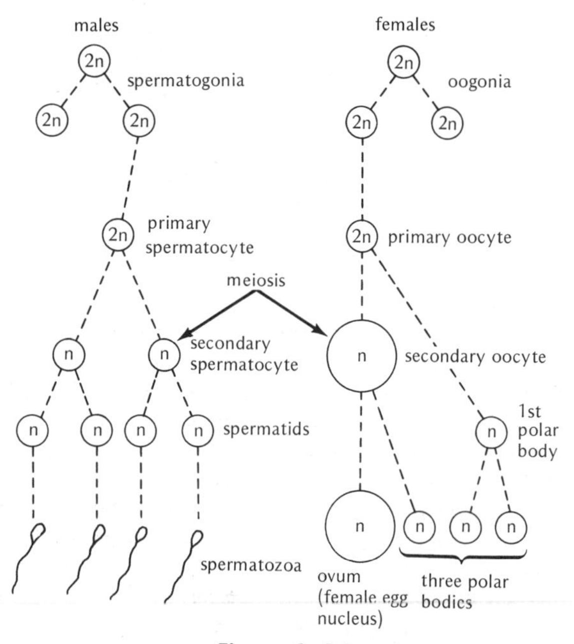

Figure 3–14
Gametogenesis in higher animals.

then combines with the male gamete to form the diploid zygote; the remaining three nuclei (*polar bodies*) degenerate. The diploid life cycle then continues through mitosis except for the cell divisions involved in gamete formation.

REPRODUCTION IN MAMMALS

In respect to cell division the reproductive cycles of mice and men are no different in principle than in *Drosophila*. As in most animals the haploid phase of mammals is reduced to a single-cell stage, the sperm or the egg. The sperm are formed in the male testes in *seminiferous tubules*, and the eggs in *Graffian follicles* of the female ovary. For our purposes the description given in Fig. 3–14 is quite adequate to account for the formation of mammalian gametes in the spermatocytes and oocytes. There is, however, one unique feature in the life cycle of mammals, and that is the maternal nourishment of the mammalian embryo until a late stage of development. In comparison, insect embryos, for example, are "independent" immediately after fertilization. This lengthened relationship between mammalian parent and offspring can lead to important interactions between them (p. 171 ff.).

PROBLEMS

3–1. In a zygote formed by crossing two strains of *Neurospora*, one carrying a chromosome with centromere *A* and the other carrying a homologous chromosome with centromere *a*, assume that the resultant haploid spores can be identified as to their centromere composition, whether *A* or *a*. Using these letters, how would you distinguish which of the two meiotic divisions in the zygote is reductional for the centromere?

3–2. In the following five *Neurospora* asci, each haploid ascospore is labeled to mark the presence of a particular parental chromosomal centromere, *A* or *a*, depending on which of the two parents contributed it. Which of these 5 arrangements would you ordinarily expect to find?

(1) Ⓐ ⓐ ⓐ Ⓐ ⓐ ⓐ Ⓐ Ⓐ
(2) Ⓐ Ⓐ ⓐ ⓐ ⓐ Ⓐ ⓐ Ⓐ
(3) ⓐ ⓐ ⓐ ⓐ Ⓐ Ⓐ Ⓐ Ⓐ
(4) Ⓐ ⓐ ⓐ Ⓐ Ⓐ ⓐ ⓐ Ⓐ
(5) ⓐ Ⓐ Ⓐ ⓐ ⓐ Ⓐ Ⓐ ⓐ

3–3. A diploid organism has two pairs of chromosomes. (a) Would you expect the number of centromeres in each somatic cell to be equally divided between those of maternal and paternal origin? (b) Would you expect equal numbers of maternal and paternal centromeres in all the gametic cells of this diploid? (c) List the possible kinds of centromere combinations in the gametes of an organism that is diploid for two pairs of chromosomes, with centromeres *A* and *A′* in one pair and centromeres *B* and *B′* in the other pair.

3–4. Explain whether the elaborate spindle mechanism found in cellular division in higher organisms would be necessary in lower organisms, such as bacteria, with only a single chromosome.

3–5. A cell has eight pairs of homologous chromosomes. (a) How many bivalents are formed during the first meiotic division? (b) How many functional centromeres are in each of the two daughter cells resulting from the first meiotic division? (c) How many functional centromeres are in a zygote that has been formed immediately after the union of two second meiotic division products?

3–6. Some species may reproduce sexually at certain times and asexually at other times. From which of these modes of reproduction would you expect greater differences between an offspring and its parent? Why?

3–7. The endosperm of corn is triploid, containing three homologues for each chromosome. (a) Explain whether you would expect a difference in the number of chromosomes per cell in different endosperm cells. (b) If a triploid cell became a microsporocyte, explain whether you would expect a difference in the number of chromosomes present in different pollen cells.

3–8. In plants such as corn: (a) How many zygote-producing gametes will result from five microspore mother cells? (b) From five megaspore mother cells? (c) From five pollen cells? (d) From five embryo sacs?

3-9. In mice how many gametes result from: (a) Five primary spermatocytes? (b) Five secondary spermatocytes? (c) Five primary oocytes? (d) Five secondary oocytes?

3-10. In man, with 23 pairs of homologous chromosomes in somatic tissue, how many chromosomes would you expect to find in: (a) Primary spermatocytes? (b) Secondary spermatocytes? (c) Sperm? (d) Spermatogonia? (e) Polar bodies?

3-11. Findings have indicated that a human female is born with all of her primary oocytes already formed, and these oocytes are at the prophase stage of the first meiotic division. (a) If a human female were to produce 400 eggs during her lifetime, from how many primary oocytes do these eggs arise? (b) From how many secondary oocytes? (c) A maximum of how many polar bodies would you expect to find in such a female? (d) In observing one of these primary oocytes, about how many bivalents would you expect to find?

3-12. A horse has a diploid number of 64 chromosomes and a donkey has a diploid number of 62 chromosomes. (a) How many chromosomes would you expect to find in the male (mule) and female (hinny) hybrids between the horse and donkey? (b) Assuming that there is little or no meiotic pairing between the horse and donkey chromosomes, explain whether you would expect to find the horse-donkey hybrids fertile or sterile.

REFERENCES

BEALE, G. H., 1954. *The Genetics of Paramecium aurelia.* Cambridge Univ. Press, Cambridge.

BEGG, C. M. M., 1959. *An Introduction to Genetics.* English Universities Press, London.

DAVIS, B. D., R. DULBECCO, H. N. EISEN, H. S. GINSBERG, and W. B. WOOD, JR., 1973. *Microbiology*, 2nd ed. Harper & Row, New York.

DEMEREC, M. (ed.), 1950. *The Biology of Drosophila.* John Wiley, New York.

EPHRUSSI, B., 1953. *Nucleo-Cytoplasmic Relations in Micro-Organisms.* Oxford Univ. Press, London.

HORNE, R. W., 1963. The structure of viruses. *Sci. American*, **208,** January issue, pp. 48–56.

JACOB, F., and E. L. WOLLMAN, 1961. *Sexuality and the Genetics of Bacteria.* Academic Press, New York.

KIESSELBACH, T. A., 1949. *The Structure and Reproduction of Corn.* Univ. Nebraska Coll. Agric. Exp. Station Res. Bull. No. 161.

KNIGHT, C. A., 1974. *Molecular Virology.* McGraw-Hill, New York.

LEDERBERG, J. and E. L. TATUM, 1946. Gene recombination in *E. coli. Nature*, **158,** 558. (Reprinted in the collections of Levine and of Peters; see References, Chapter 1.)

OLIVE, L. S., 1965. Nuclear behavior during meiosis. In *The Fungi*, Vol. 1, G. C. Ainsworth and A. S. Sussman (eds.). Academic Press, New York, pp. 143–161.

STRICKBERGER, M. W., 1962. *Experiments in Genetics with Drosophila.* John Wiley, New York.

WICHTERMAN, R., 1953. *The Biology of Paramecium.* Blakiston, New York.

WOLLMAN, E. L., F. JACOB, and W. HAYES, 1956. Conjugation and genetic recombination in *Escherichia coli* K-12. *Cold Sp. Harb. Symp.*, **21,** 141–162.

4
NUCLEIC ACIDS

The events of cellular division and reproductive cycles point strongly to chromosomes as the major carriers of genetic material for the following reasons:

1. They duplicate precisely and divide equally in mitosis, furnishing each cell with a full complement of chromosomes.
2. Their behavior in meiosis accords with our expectations of heredity—that it is due to contributions from both parents.
3. Their random mixing and crossing over during meiosis provides an important source for the observed variability between individuals.
4. In addition there is considerable evidence that chromosomes (Chapter 12) and chromosome aberrations (Chapters 21 and 22) can be associated with the inheritance of specific characteristics.

Therefore, if questions about heredity are posed in terms of isolating and identifying the genetic material, it would be of value to ask these questions in terms of chromosomes themselves. Of what are they composed? What evidence is there that the material they carry is the genetic material?

NUCLEIN

Historically the search for the chemical identity of genetic material began about a century ago. In 1871, soon after Charles Darwin's major work on the theory of evolution and Gregor Mendel's founding experiments in genetics, Friedrich Miescher (1844–1895) published a method for separating cell nuclei from the cytoplasm. From these cell nuclei he then extracted an acid material, *nuclein*, that proved to contain an unusually large amount of phosphorus, and, in contrast to proteins, contained no sulfur. Miescher's conclusion was that this substance was unique and "not comparable with any other group known at present."

In later years the nuclein, now named *nucleic acid*, was found to associate with various proteins in combinations called *nucleoproteins*. Of the protein portion, two kinds occurred in significant amounts: *protamine* in fish sperm, and *histone* in other nuclei. Although the histones appeared relatively complex, protamine had a simple structure consisting mostly of linked groups of the amino acid *arginine* (Fig. 4–1). Could either or both of these proteins be the long-sought-for genetic material?

Since each organism undergoes many different biological activities, the diversity within an organism as well as the diversity between organisms demands that genetic material be capable of carrying many complex messages. The various proteins that determine biological activities in an organism (p. 12 and Chapter 26) are usually long chemical chains consisting of many amino acid units, of which about 20 different kinds are known. Although fish are certainly not the most complex of creatures, it seemed difficult to accept the idea that the genetic material of a fish is only a repeating protein chain, protamine, made up primarily of a single amino acid. In what fashion could such a simple chain determine the synthesis of other proteins having many different kinds of amino acids? It also seemed strange that the genetic material might be a histone until sperm formation and then turn into a protamine. The nucleic acid portion, on the other hand, was a constant feature of all cells and contained, as we shall see, much more variability than protamine.

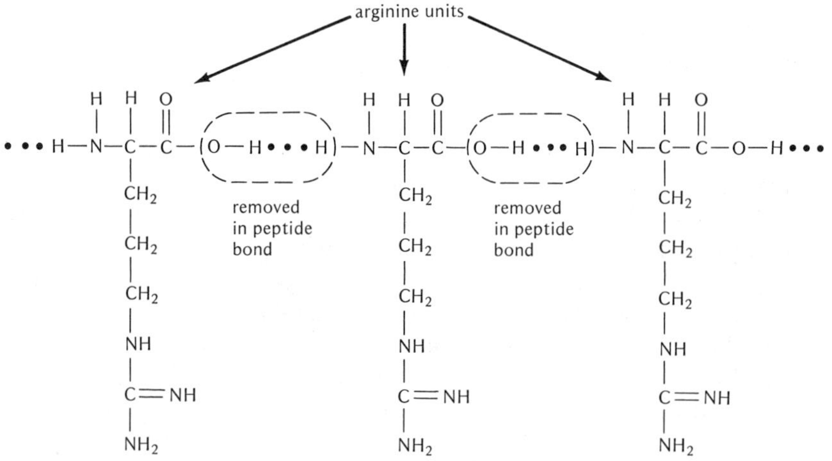

Figure 4–1

Basic protamine structure showing the formation of peptide bonds and the linkages between arginine amino acids. When similar chemical units are linked together in repeated fashion, the multiunit molecule they form is called a polymer. *Since amino acids are used in the present case, the polymer is a* polypeptide *or* protein.

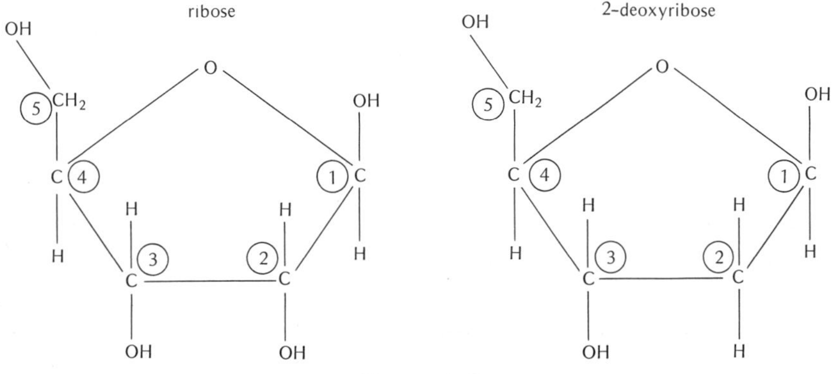

Figure 4–2
Structural formulas for ribose and deoxyribose sugars.

NUCLEOTIDES

When the nucleic acid was separated from the protein, several biochemists, especially Levene (1869–1940), showed that it could be broken down further into smaller sections called *nucleotides.* Each nucleotide, in turn, consisted of a sugar, a phosphate group, and a nitrogen-containing portion. The sugar contained five carbons and had been named *ribose,* or lacking one oxygen atom, *deoxyribose* (Fig. 4–2). Although these are the two main sugars found in nucleic acids, any particular nucleic acid does not ordinarily contain both of these sugars at the same time. As a consequence there are two kinds of nucleic acid: *ribonucleic acid* (RNA), which is found commonly in the cytoplasm, and *deoxyribonucleic acid* (DNA), found with few exceptions only in the nucleus.

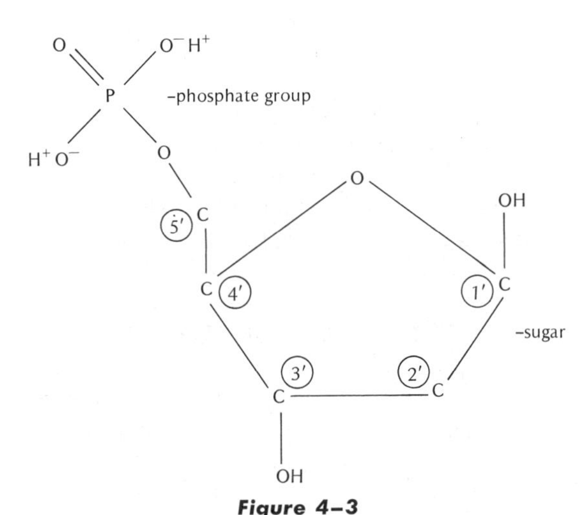

Figure 4–3
Linkage between a phosphate and deoxyribose group yielding a sugar phosphate, or phosphorylated sugar. (Individual H atoms omitted.)

The phosphate group of each nucleotide is attached to the sugar at its number 5 carbon position often designated as the 5′ ("5-prime") position (Fig. 4–3). The term "acid" appended to these chemicals was derived from the common representation of the phosphate group as having a dissociable hydrogen ion (PO_4H^+). Because of the loss of an electron, the H^+ ion has a positive charge, enabling it to combine with molecules that can accept such charges. Actually such an acid would be much too corrosive for biological systems, and a metal ion (sodium, calcium, or magnesium) is usually bound to the phosphate group.

Aside from the sugar and phosphate groups which were constant components of all nucleotides in a nucleic acid, a more variable nitrogen-containing group was also present, associated with each sugar at its number 1 carbon position. The nitrogen unit contained either one or two carbon-nitrogen rings and could function as a *base* (hydrogen-ion acceptor), in contrast to the acidic pyrimidines, one-ring bases:

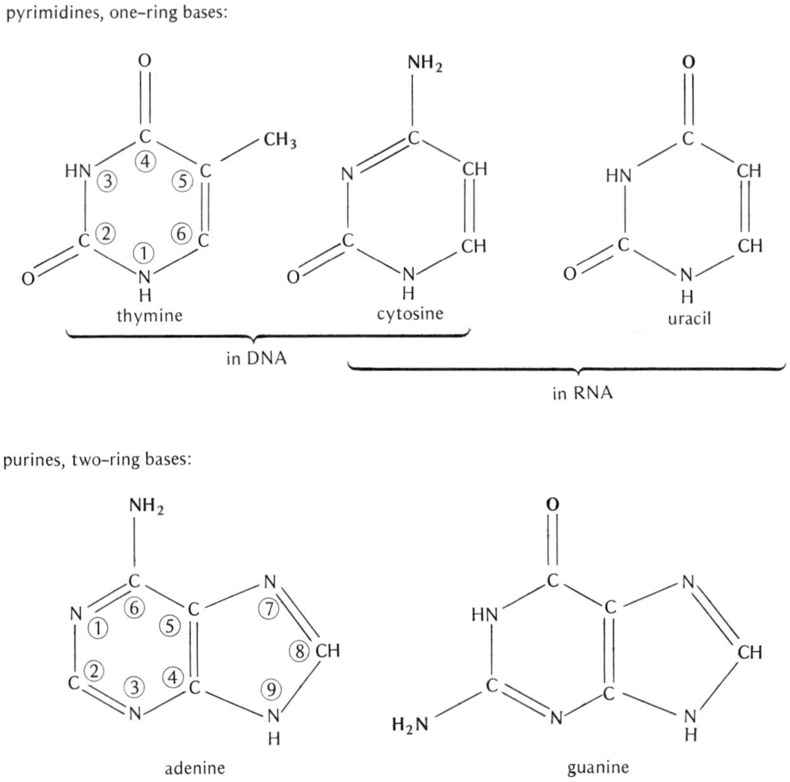

thymine

cytosine

uracil

in DNA

in RNA

purines, two-ring bases:

adenine

guanine

Figure 4–4

Structural formulas of the common bases in nucleic acids. Numbering of the ring positions follows the Chemical Abstracts system.

nature of the phosphate group. Bases containing one carbon-nitrogen ring were the *pyrimidines*, and the two-ring bases were the *purines* (Fig. 4–4). In DNA the two main pyrimidines found were *cytosine* and *thymine*, while RNA carried either cytosine or *uracil*. The difference between thymine and uracil seems to be negligible (the presence of a methyl group, CH_3, at the number 5 position of thymine) but is nevertheless significant, so that thymine is not ordinarily found in RNA nor uracil in DNA. The two main purines, *adenine* and *guanine*, are found in both DNA and RNA.

In addition to the difference in ring structures, the bases also show the presence of an amino group (NH_2) at the number 4 position of one pyrimidine (cytosine) and the number 6 position of a purine (adenine), while a keto group (C=O) occurs at these positions in the other bases (thymine, uracil, guanine). DNA, and RNA, can therefore also be described as containing two kinds of amino bases and two kinds of keto bases, evenly divided between pyrimidines and purines.

THE NUCLEOTIDE CHAIN

Since each different base distinguishes the particular nucleotide carrying it, there are usually four different *deoxyribonucleotides* in DNA and four *ribonucleotides* in RNA. Thus the combination of phosphate-deoxyribose-adenine yields a nucleotide that can be called deoxyadenylic acid or deoxyadenosine 5′ phosphoric acid. If we include the other three major nucleotides in DNA, there are

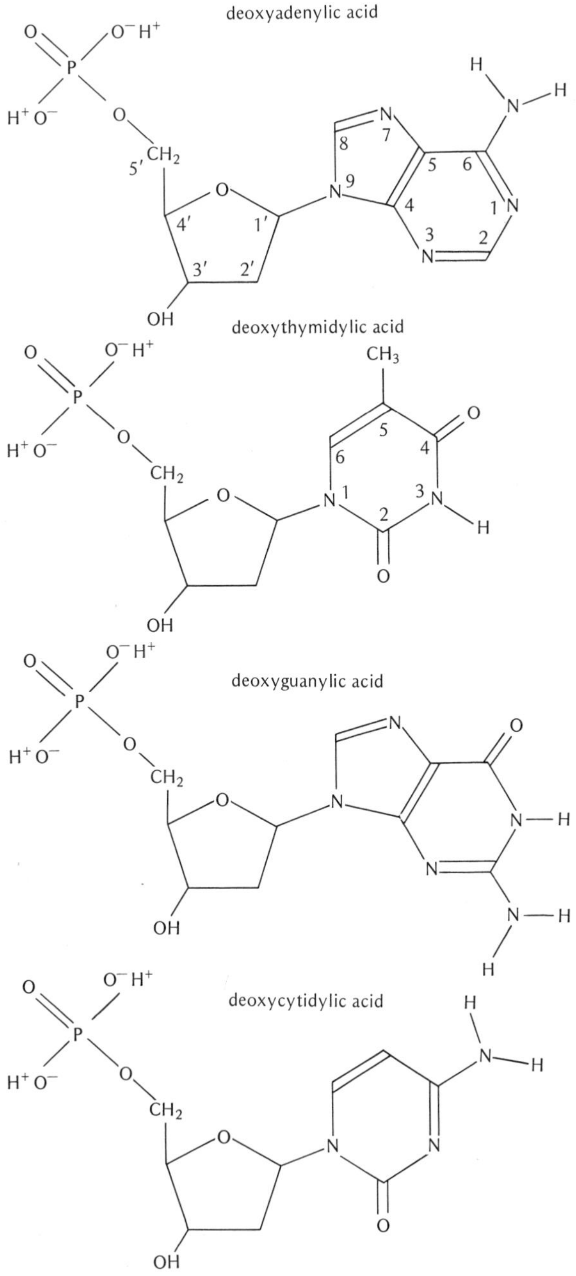

deoxyadenylic acid

deoxythymidylic acid

deoxyguanylic acid

deoxycytidylic acid

Figure 4–5
The four major nucleotides in DNA.

also deoxycytidylic acid, deoxyguanylic acid, and deoxythymidylic acid (Fig. 4–5). The combination of a base and a sugar without the phosphate group is called a *nucleoside*. Restricting ourselves to the bases mentioned above there are thus four *deoxyribonucleosides* and four *ribonucleosides*.*

When Levene discovered the deoxyribonucleotides, he also noted they could be connected together through phosphate-sugar linkages (Fig. 4–6). He proposed that each of the deoxyribonucleotides was present in equal amounts, and connected together in chains in which each of the four different nucleotides was regularly repeated in a *tetranucleotide* sequence (e.g., AGCT AGCT · · ·).

As with protamine, however, the simplicity of such a molecule did not seem able to account for the complexity of biological inheritance, and relatively little attention was paid to DNA until the 1940s. In that decade Chargaff and other biochemists showed that not all the nucleotide bases were present in equal amounts and that the ratio of the different bases changed between different species (see Table 4–1, p. 54). These experiments, in conjunction with improved techniques for isolating long molecular strands, suggested that DNA was not a simple molecule made up of the same repeated tetranucleotide sequence but that it was a very long connected chain consisting of hundreds or thousands of the four different nucleotides linked in varied sequences.

The length and potential variability of this newly envisioned DNA molecule certainly fulfilled the important prerequisite that genetic material must contain all the different biological "messages." Since there was a choice of four nucleotides at each position, long-chained DNA molecules could attain a great degree of variability by simply varying their nucleotide sequence. For example, there are four possible alternatives for one nucleotide location on DNA because there are four different bases: adenine, cytosine, guanine, and thymine (A, C, G, T). If, let us say, each alternative represents a different biological "message," then four messages can be determined or "coded" at a single position. If a sequence of two nucleotides

*A nucleotide can therefore also be described as a phosphorylated nucleoside, e.g., adenylic acid = adenosine monophosphate = AMP; deoxyadenylic acid = deoxyadenosine monophosphate = dAMP.

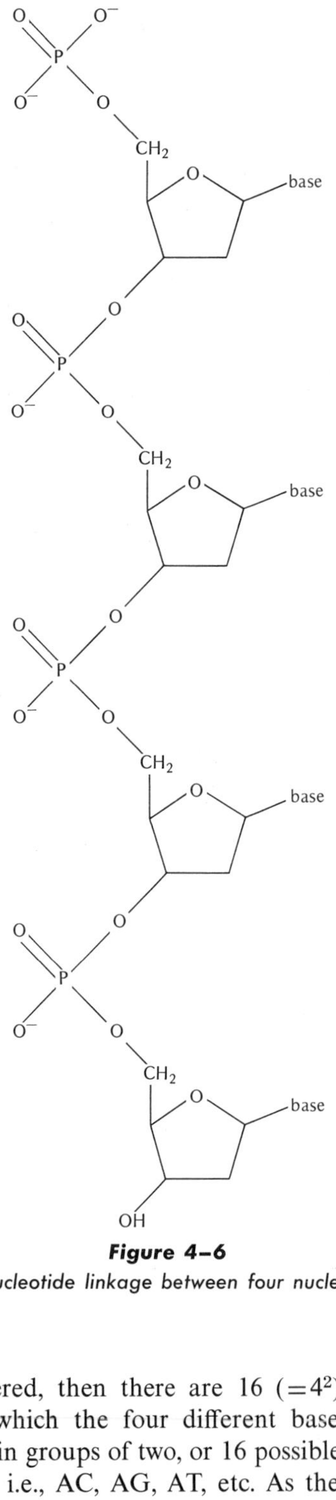

Figure 4–6
Tetranucleotide linkage between four nucleotides.

is considered, then there are 16 ($=4^2$) possible ways in which the four different bases can be arranged in groups of two, or 16 possible different messages, i.e., AC, AG, AT, etc. As the chain of nucleotides increases, the number of different possible messages increases exponentially (4^n, where n is the number of nucleotides in the molecule), and a DNA chain of 10 nucleotides could be constructed in more than a million (4^{10}) different ways. Thus, if a biological product were determined by a sequence of 10 nucleotides, and a total of 10,000 different biological products are needed by an organism, this could easily be accomplished by a total length of $10 \times 10{,}000$, or 100,000, nucleotides.

In reality many more than 10 nucleotides are needed to code a biological product, since these are usually proteins of considerable length and complexity. At the same time, however, the number of nucleotides in most organisms is undoubtedly greater than 100,000 (Fig. 4–7). In man, for instance, the haploid set of chromosomes contains DNA molecules with a total length of more than three billion nucleotides. That amount of nucleotides could produce 100,000 different biological products even if each product is of such length and complexity as to be determined by more than 30,000 nucleotides! Since it is likely that organisms produce less than 100,000 biological products, and that each product is coded in a length much less than 30,000 nucleotides (see Chapter 27), we can consider that the amount of DNA is more than sufficient for the "coding" purposes needed.

THE DOUBLE HELIX

The organization of a DNA chain as a linear sequence of nucleotides (*primary structure*) was only one aspect of its molecular architecture. The next difficult question concerned the existence of higher levels of organization. Did an extended chain of nucleotides possess a particular shape or *secondary structure*? Were these chains always single; were they randomly placed alongside each other in groups of two, three, or more; or was there a regular relationship between them? In the 1940s a number of findings already indicated that the DNA molecule was regularly organized. Chargaff and others had shown that as a rule the numbers of purine bases (A + G) equaled the numbers of

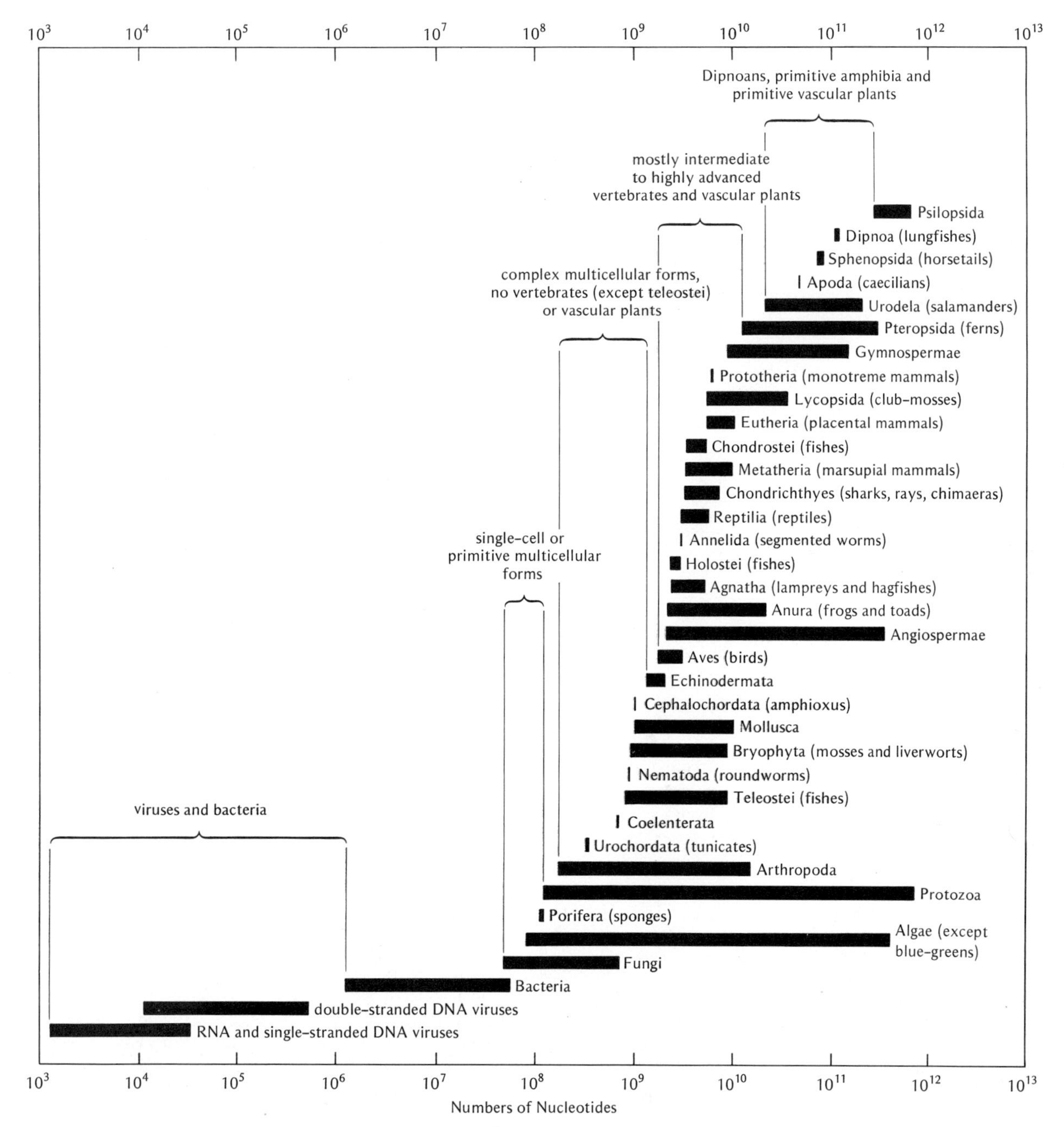

Figure 4–7

Comparison of the numbers of nucleotides in the genetic material of different types of organisms. Each bar in the figure represents the range of nucleotide numbers found among the species sampled in a designated group. The nucleotide values given are largely derived from estimates of the weight of nucleic acid in the haploid complement of an organism according to the formula: 1 gram of nucleic acid $= 2.0 \times 10^{21}$ *nucleotides. Thus the DNA in the haploid complement of human chromosomes* (3.2×10^{-12} *gram) provides man with* $(2.0 \times 10^{21}) \times (3.2 \times 10^{-12}) = 6.4 \times 10^9$ *nucleotides. In terms of double-stranded DNA molecules this therefore represents a value of half that length, or of about* 3.2×10^9 *nucleotide pairs. (Nucleotide lengths may then be converted into physical lengths by the formula* 2.9×10^6 *nucleotide pairs of double helix DNA = 1 millimeter. Thus man has a DNA length per haploid cell of* $3.2 \times 10^9/2.9 \times 10^6$ *or* $1.1 \times 10^3 = 1.1$ *meters.) Note that these data show a general tendency for the increase of amount of DNA in organisms with increasing complexity, although some exceptions exist (e.g., salamanders). As shall be discussed in Chapter 5, some or most of the "extra" DNA is "repetitive" and may serve functions different from that of nonrepetitive DNA. (From Sparrow, Price, and Underbrink.)*

TABLE 4–1

Nucleic acid contents and base compositions in a sampling of organisms

	Haploid DNA content	Purines		Pyrimidines			
	(10^{-12} gram)*	Adenine	Guanine	Cytosine	Thymine	Uracil	$\frac{A + G}{C + T(U)}$
man	3.2	31.0	19.1	18.4	31.5		1.00
cattle	2.8	28.7	22.2	22.0	27.2		1.03
rat	3.1	29.7	20.2	20.9	29.1		1.00
mouse	2.5	29.1	21.1	21.1	29.0		1.00
guinea pig	2.9	29.8	21.7	20.4	28.6		1.05
chicken	1.3	28.0	22.0	21.6	28.4		1.00
frog (*Rana pipiens*)	6.5	26.3	23.5	23.8	26.4		.99
carp	1.6	30.8	18.5	18.9	31.7		.97
sea urchin (*Arbacia punctulata*)	.79	29.9	20.9	19.1	30.1		1.03
fruit fly (*Drosophila melanogaster*)	.18	27.3	22.5	22.5	27.6		.99
protozoan (*Tetrahymena pyriformis*)	13.4	35.4	14.5	14.7	35.4		1.00
corn	7.5	25.6	24.5	24.6	25.3		1.00
tobacco	1.2	29.7	19.8	20.0	30.4		.98
bread mold (*Neurospora crassa*)	.046	23.0	27.1	26.6	23.3		1.00
yeast (*Saccharomyces cerevisiae*)	.018	31.3	18.7	17.1	32.9		1.00
bacteria:							
Staphylococcus aureus	.007	31.0	18.5	19.2	31.2		.98
Neisseria meningitidis	.0016	24.6	25.5	25.0	24.0		1.00
Diplococcus pneumoniae	.002	30.3	21.6	18.7	29.5		1.08
Salmonella typhimurium	.013	22.9	27.1	27.0	23.0		1.00
Escherichia coli	.0047	24.6	25.5	25.6	24.3		1.00
Bacillus subtilus	.003	28.4	21.0	21.6	29.0		.98
viruses:							
herpes simplex	.00011	13.8	37.7	35.6	12.8		1.06
bacteriophage T2	.0002	32.6	18.2	16.6†	32.6		1.03
bacteriophage λ	.000055	26.0	23.8	24.3	25.8		.99
bacteriophage φX174‡	.0000026	24.7	24.1	18.5	32.7		.95
bacteriophage fd	.0000022	24.4	19.9	21.7	34.1		.79
reovirus	.000017	28.0	22.0	22.0		28.0	1.00
tobacco mosaic	.0000033	29.3	25.8	18.1		26.8	1.23
influenza	.0000034	23.0	20.0	24.5		32.5	.75
bacteriophage f2	.0000017	22.2	25.9	26.8		25.1	.92
bacteriophage Qβ	.000002	22.1	23.7	24.7		29.1	.85

Derived mostly from data collected by H. S. Shapiro in Sober 1970.

* Molecular weights, if desired, can be calculated according to the formula 1.67×10^{-24} grams = 1 dalton. Thus the DNA in a haploid cell of man has a molecular weight of $(3.2 \times 10^{-12})/(1.67 \times 10^{-24}) = 1.91 \times 10^{12}$ daltons.

† 5-Hydroxymethylcytosine.

‡ Note that for this virus and some others listed there is no equivalence in base ratios between A–T and G–C. As will be explained below (p. 58), these are single-stranded viruses.

pyrimidine bases (T + C); that there was an equivalence between bases carrying amino groups at the 6 or 4 positions (A + C) and bases carrying keto groups at these positions (T + G); and that the ratios of adenine to thymine and guanine to cytosine were close to 1 in the various eucaryotic

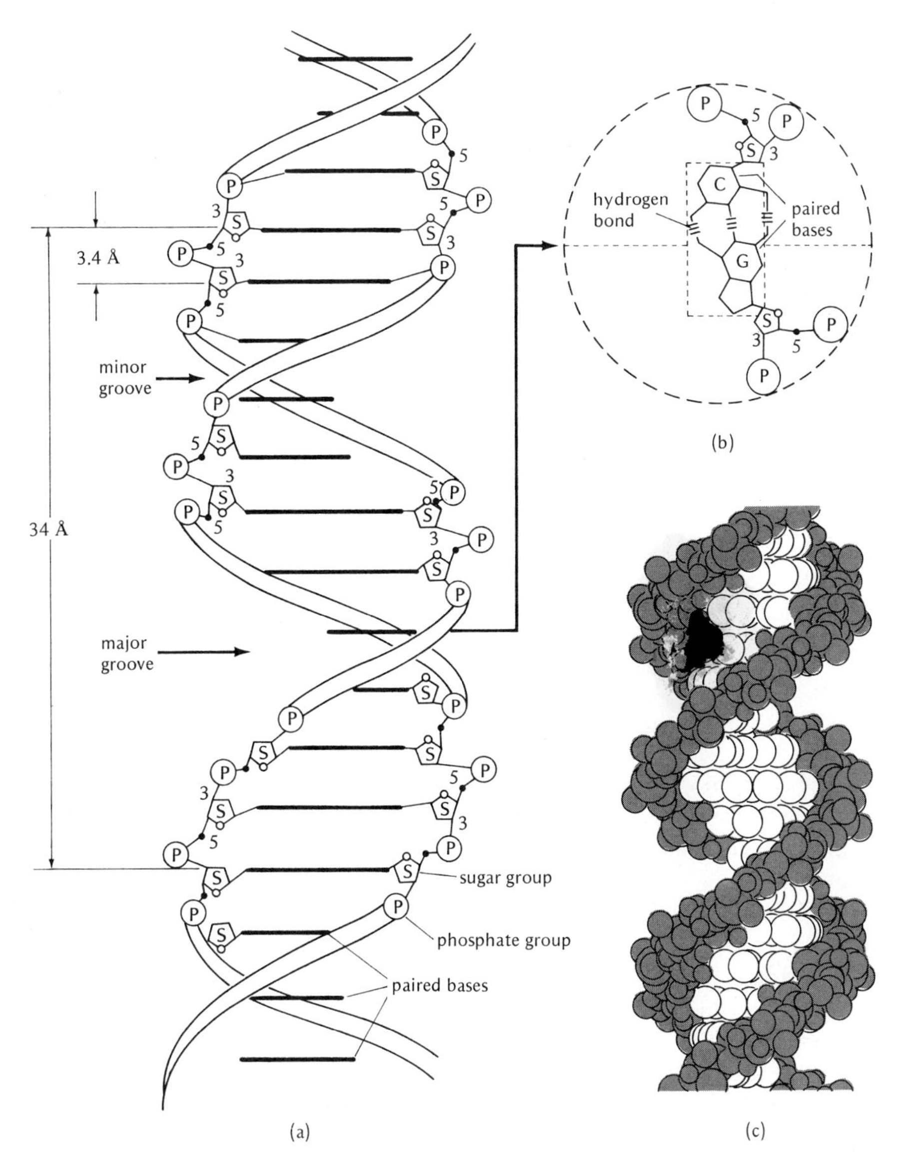

Figure 4–8

The Watson-Crick model of the DNA molecule. (a) Diagram of a double helix with the phosphate-sugar backbone of each strand shown as a long ribbon. Bridging the gap between the two strands are parallel rows of paired nucleotide bases "stacked" at regular 3.4 Angstrom intervals. Some of the phosphates (P) and sugars (S) are drawn individually to indicate the opposite orientations of the sugar groups in each strand. (b) Cross-sectional view of DNA double helix showing two paired nucleotide bases (C = cytosine, G = guanine) extending into the central part of the molecule. The phosphate-sugar groups do not actually lie in the same flat cross-sectional plane as the paired bases, but are merely shown in their approximate relative positions. (After W. Etkin, 1973. A representation of the structure of DNA. Bioscience, ***23****:652–653.) (c) Representation of the molecular structure of the DNA section shown in (a) using space-filling models of individual atoms. (After M. Feughelman et al., 1955. Molecular structure of deoxyribose nucleic acid and nucleoprotein.* Nature, ***175****:834–838.)*

species investigated (Table 4–1). These consistencies were considered by some workers to signify a symmetrical relationship between the nucleotide bases. By the early 1950s, x-ray studies on DNA by Wilkins, Franklin, and others indicated a well-organized, multiple-stranded fiber about 22 angstroms in diameter that was also characterized by the presence of groups spaced 3.4 angstroms apart along the fiber and a repeating unit every 34 angstroms.

Taking into account the facts known at the time, Watson and Crick in 1953 proposed a "double helix" structure for DNA which quickly gained wide acceptance (Fig. 4–8). According to Watson and Crick, the DNA molecule is two-stranded and coiled like a rope (plectonemic), so that only by permitting the ends to revolve freely can the two paired *complementary* strands be separated. The coiling is helical, in the fashion of a circular staircase that always maintains the same diameter and the same width of the steps, with a connecting railing on either side.

To carry the staircase analogy further, we can consider that each complementary strand of DNA is only one-half the circular staircase, either one side or the other, consisting of approximately one-half the width of each step with a railing connecting each of these half-steps. The railing, or "backbone," of a strand is composed of the phosphate-sugar linkages, which are continuously repeated without change. The half-steps of a strand that extend to meet the half-steps of the complementary strand are single purine or pyrimidine bases. Each step is thus a pair of bases, termed *base pair*, or *complementary base pair*, between the two complementary strands of DNA. Dimensionally the steps are 3.4 angstroms apart, and each is turned 36° from the preceding one, so that a complete turn of 360° involves 10 stairs, or 10 base pairs, and is 34 angstroms long.

By proposing these dimensions and limiting the diameter of the double helix to 20 angstroms, Watson and Crick found that only when each base pair consisted of a combination of one purine and one pyrimidine could they fit together in the width of each stair (about 11 angstroms; see Fig. 4–9). The fit was determined by *hydrogen bonding*, which consists of the ability of a single hydrogen atom (positive charge) to be shared between an oxygen atom (slight negative charge) and a nitrogen atom (slight negative charge) or between two nitrogen atoms opposite each other on the complementary bases.

Although such hydrogen bonds are weaker than the usual chemical bonds, the fact that so many of them occur along the length of the DNA double helix, at least two for each base pair, gives a high degree of stability and rigidity to the molecule. The hydrogen bonding is also responsible for the exclusive fitting between the 6-amino purine (adenine) and the 4-keto pyrimidine (thymine), and between the 4-amino pyrimidine (cytosine) and the 6-keto purine (guanine). In other words, the bases of double-stranded DNA should give quantitative relationships in which $A = T$ and $G = C$, or $A + G = C + T$, or $A + G/C + T = 1$. (Note, however, that $A + T$ need not necessarily equal $G + C$.) Other base-pair combinations, such as adenine and cytosine, would lead to the presence of two hydrogen atoms at one bonding position, and none at the others. Also, for the base pairs to fit properly, the sugars of one strand must all be pointed in a direction opposite to the sugars of its complementary strand. Therefore, in contrast to a regular staircase, the railings of each DNA strand are expected to be inclined (polarized) in opposite directions. (Further evidence for this will be presented in Chapter 5.)

This DNA structure offers a ready explanation of how a molecule could form perfect copies of itself as it increased in number. In order to replicate, the helically coiled DNA has merely to separate its two strands, hydrogen-bond available nucleotides in the surrounding medium with the exposed bases of each strand, and then connect these nucleotides into new strands with the aid of appropriate enzymes (Fig. 4–10). Since only adenine pairs with thymine, a strand can hold only thymine at its adenine positions and vice versa. The same relationship is, of course, true for cytosine and guanine. Any separate strand, therefore, can form an exact duplicate of the complementary strand from which it has separated. Here, then, is a molecule that can duplicate itself exactly and

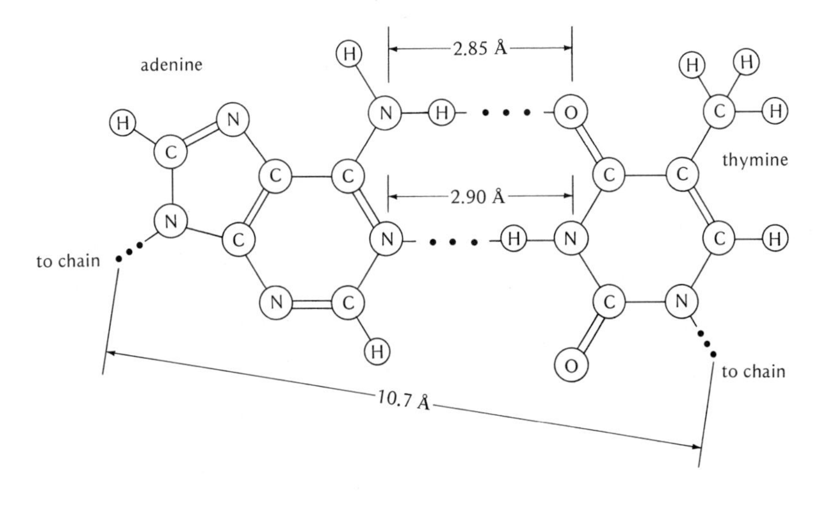

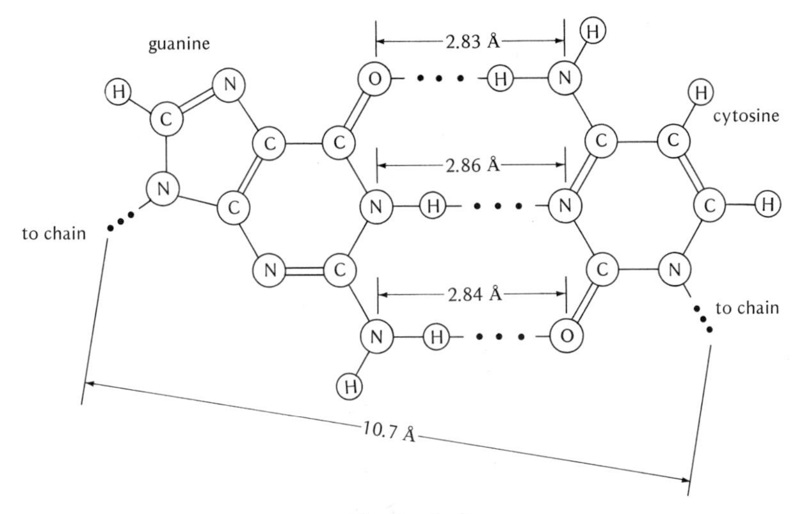

Figure 4–9

Hydrogen-bond pairing and molecular distances between adenine and thymine, and guanine and cytosine. (After Pauling and Corey.)

thereby transmit its structure equally to all cellular division products.

The Watson-Crick model also had the advantage of explaining in a simple fashion how changes in heredity (*mutations*) could happen. That is, occasional errors in base pairing could occur and a different nucleotide would then become substituted during replication for the one usually present at that position. Thus, for example, if a sequence A T C G A A ...... was, by copying error, changed to A T *T* G A A ......, all descendant molecules (except for further mutations) would now bear this new sequence. Such errors are probably very rare under normal conditions, occurring at a frequency of probably no more than one per million or even one per billion replications for any particular nucleotide. The fact that conditions may not always be optimal, and that DNA molecules

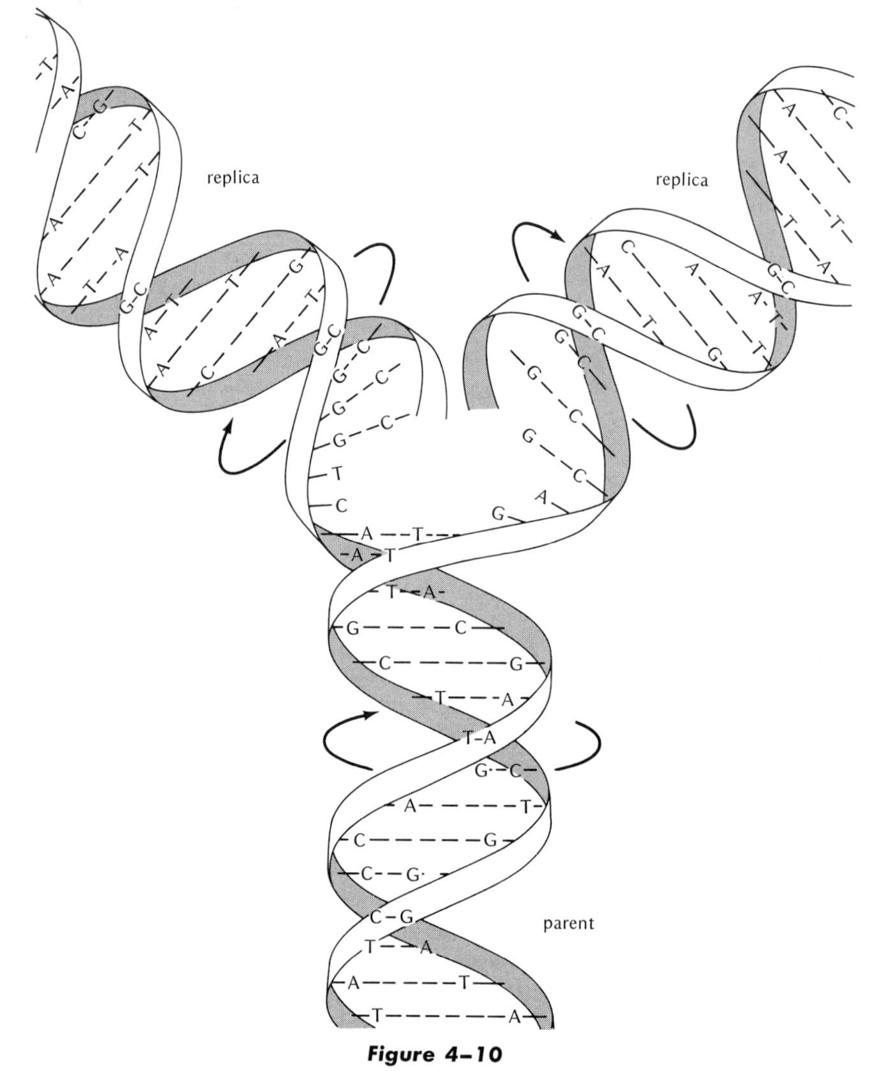

Figure 4–10

Replication of DNA based on unwinding of the DNA molecule and complementary pairing of cellular nucleotides with the exposed bases of the parental strands. (After G. S. Stent, Molecular Genetics: An Introductory Narrative. W. H. Freeman and Company, San Francisco, 1971.)

have many nucleotide positions at which such errors can occur, makes it likely that a long DNA chain may possess one or more nucleotide changes at each replication. Some of the causes for mutation and the kinds of nucleotide errors produced will be discussed further in Chapter 24.

In those instances where equality between the complementary base ratios is lacking ($A \neq T$, $G \neq C$), e.g., bacteriophage ϕX174 (Table 4–1), a single-stranded noncomplementary structure is indicated. Such DNA strands, called "plus," can, however, produce complementary "minus" strands yielding a plus-minus "replicative form" (p. 90). Further plus strands are then produced from minus-strand templates by regular base pairing. Thus, although single-stranded, the mode of replication of such DNA is still expected to be based on exact complementary pairing.

PRESENCE OF DNA IN CHROMOSOMES

The tremendous variability possible in the nucleotide order of DNA, as well as the ability of this molecule to replicate itself perfectly, has been of great theoretical value in identifying DNA as the genetic material. Despite these advantages, however, questions of a more descriptive nature remain to be answered. Where exactly is DNA located and in what amounts? What is the relationship of DNA and other nuclear components, specifically chromosomes?

CHEMICAL COMPOSITION

Through biochemical methods chromosome material from calf thymus glands has been isolated and shown by Mirsky and Ris to consist of two fractions, major (90 percent) and minor (10 percent). The major fraction is extracted with salt solution and consists of 45 percent DNA and a remainder that is a histone-type protein. The histones are believed to surround the DNA double helices and to be intimately involved in aiding or controlling DNA function (Chapter 30). (Histones have not been found in bacterial cells, and other proteins are apparently used for these purposes.)

The minor or "residual" chromosome fraction of mammalian cells has a very small percentage of DNA (1.5 to 2.6 percent), a much larger percentage of RNA (7.5 to 14 percent), and a high percentage of heavy protein (about 80 percent). As will be discussed later in greater detail, the RNA seems to serve a special function in transmitting information for protein synthesis. In some tissues that are actively involved in protein manufacture, such as liver, the residual chromosome fraction increases to about 50 percent. Sperm nuclei, on the other hand, show very little net synthesis of proteins and have a residual chromosome fraction close to 3 percent.

The chemical isolation of DNA from nuclei has enabled various quantitative studies to be made in respect to the amounts of DNA in different tissues. One important result of many such experiments is the relative "constancy" of DNA amounts in practically all diploid tissues of an organism and of a species (Table 4–2). In haploid tissues, e.g., sperm nuclei, the amount of DNA is halved, as would be expected of the genetic material of meiotic products. Furthermore, under conditions of starvation which diminish the amount of most cellular components, the amount of DNA remains unchanged, as does the chromosome number. Although not definitive proof, these facts have served as strong support for the genetic role of DNA.

THE FEULGEN REACTION

In 1912 Feulgen made the discovery that if DNA were treated with warm acid (*hydrolysis*), a reddish-purplish staining reaction would occur upon the addition of a chemical known as Schiff's reagent. This reaction, named the *Feulgen reaction*, has been found to be specific for DNA, and neither RNA nor any other cell substance will give the precise DNA type of staining response. When the reaction is carried out in cells, only the chromosomes show the Feulgen reaction, and all other areas remain relatively colorless. If DNA is re-

TABLE 4–2

Average DNA content per nucleus (10^{-12} gram) for different tissues in a variety of organisms

	Tissue			
ORGANISM	Sperm	Erythrocyte	Liver	Kidney
human	3.25		10.36*	8.6
cattle	3.42		7.05	6.63
chicken	1.26	2.58	2.65	2.28
toad	3.70	7.33		
carp	1.64	3.49	3.33	
shad	.91	1.97	2.01	

From Mirsky and Osawa, 1961.

* This value includes many liver cells possessing four chromosome complements per nucleus instead of the diploid number. If only diploid liver cells are scored by individual examination (cytophotometry, see below), the DNA values are the same as for other diploid tissues.

moved by appropriate enzymes, such as deoxyribonuclease, or through extraction with trichloroacetic acid, there is no visible Feulgen reaction.

The specificity of the Feulgen reaction for chromosomes and DNA has enabled a number of quantitative studies to be made by measuring the amount of light transmitted through Feulgen-stained nuclei (cytophotometry). In support of the chemical studies cited previously, cytophotometric experiments have also shown a constancy of the amount of DNA for all nuclei of an organism except for polyploid tissues with multiple chromosome complements (i.e., certain liver nuclei) or haploid tissues (i.e., sperm). Studies by Swift, Alfert, and others have also indicated that the amount of Feulgen-stained DNA doubles during the interphase period of cell growth, and is then equally distributed at anaphase to the two daughter nuclei. This, again, is the type of behavior that would be expected for genetic material.

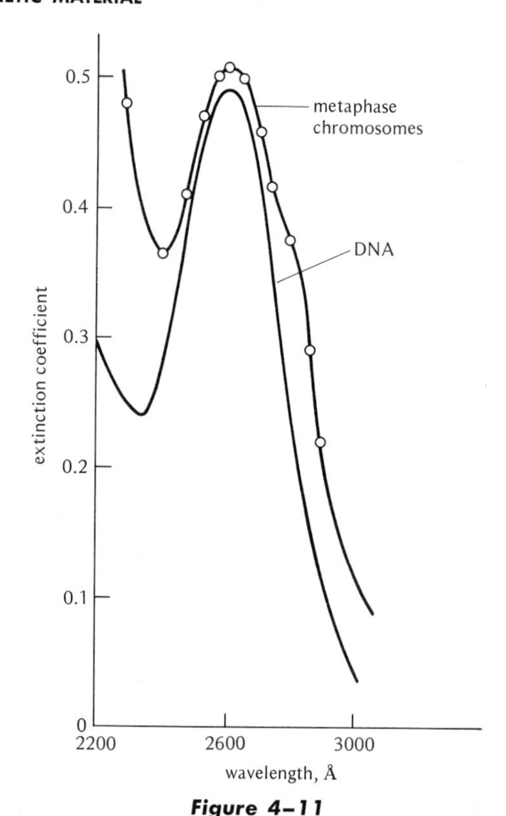

Figure 4–11

Degree of ultraviolet absorption by metaphase chromosomes and DNA at different wavelengths. (After Brachet, from Caspersson.)

ULTRAVIOLET ABSORPTION

The presence of DNA in chromosomes can also be observed without chemical isolation and without staining. Because of the ring structure in their bases, nucleic acids absorb light in the ultraviolet region at a wavelength of about 2600 angstroms.* Ultraviolet microscopes which can refract such wavelengths have therefore been used to study the presence or absence of nucleic acids in various cell structures. Basic work in this field by Caspersson has clearly demonstrated the correspondence between ultraviolet absorption in chromosomes and in DNA (Fig. 4–11). In giant salivary polytene chromosomes (p. 26), the major ultraviolet absorption occurs in the dark band areas; the lighter "interband" spaces show very little absorption. This supports results obtained with Feulgen staining, and indicates that most of the DNA is localized within the stained bands and very little in the interbands.

The specific ultraviolet absorption of nucleic acids at the 2600-angstrom wavelength led to one of the first indications that DNA might, in fact, be the genetic material. In 1941 Hollaender and Emmons found that fungi irradiated at this wavelength developed more hereditary changes (mutations) than when irradiated at any other wavelength. It was, of course, argued that DNA might still not be the genetic material, since the energy absorbed by nucleic acids at this wavelength could presumably be passed on to the true genetic material, whose absorption wavelength was undetermined. However, although not conclusive, these findings, as well as similar observations in other organisms, such as bacteria, corn (Fig. 4–12), and *Drosophila*, intensified the interest of geneticists in nucleic acids.

*Absorption maxima in angstroms for individual DNA bases are: adenine 2625, guanine 2490, cytosine 2760, and thymine 2645.

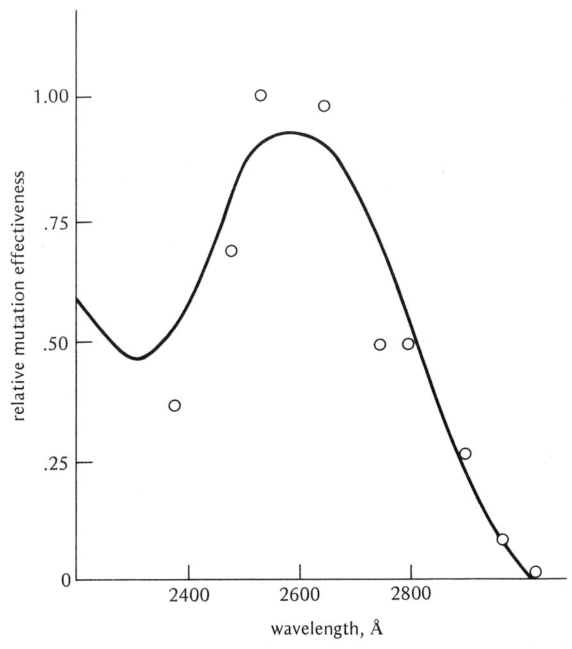

Figure 4–12

Degree of ultraviolet absorption of DNA at different wavelengths (solid line) and relative efficiency of these wavelengths in inducing mutation in corn pollen (open circles). (From Stadler and Uber.)

DNA IN LAMPBRUSH CHROMOSOMES

The greatly extended threadlike loops of lampbrush chromosomes (p. 26) have offered an unusual opportunity to examine the detailed structure of a single chromosome thread. In the electron microscope each single loop appears to consist of two parallel strands of about 100 to 150 angstroms in diameter. Evidence that each strand contains DNA has been derived from experiments of Callan, Gall, and others in which various portions of the strand are digested away. Enzymes that remove protein (proteases) and RNA (ribonucleases) will not affect the integrity of a loop. However, an enzyme that affects DNA (deoxyribonuclease) quickly fragments the loop and its connections. It has therefore been proposed that, at least in the lampbrush chromosome, the DNA molecule is a long continuous structure.

AUTORADIOGRAPHY

The radioactive isotope of hydrogen, known as *tritium*, can be incorporated into various sub-

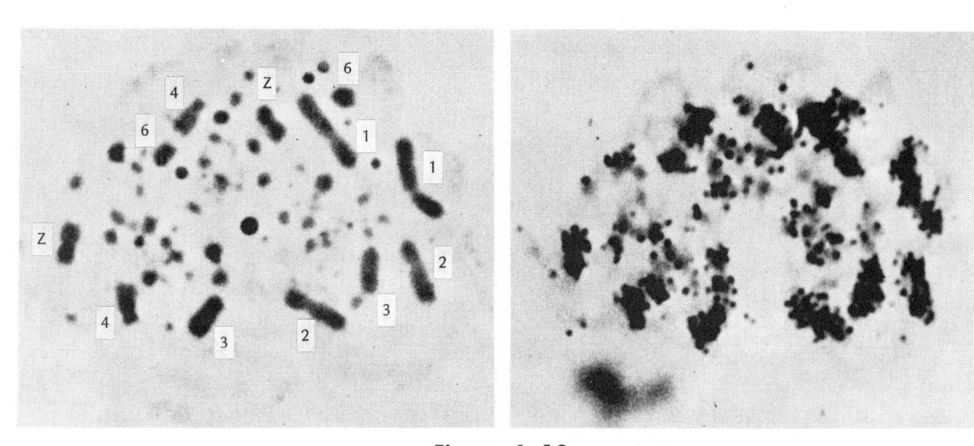

Figure 4–13

Bone-marrow cell of the domestic fowl that has incorporated radioactive thymidine. On the left are some of the chromosomes, and on the right is the appearance of the same cell when covered with film sensitive to radioactivity. Note that the radioactively exposed film grains are restricted to the chromosome areas. (From T. C. Hsu, W. Schmid, and E. Stubblefield, 1964. DNA replication sequences in higher animals. In The Role of Chromosomes in Development, *M. Locke (ed.), Academic Press, New York, p. 89.)*

stances. If it is used as a substitute for hydrogen atoms in thymidine, the deoxyribonucleoside of thymine, then it can be directly incorporated as part of the thymine nucleotide in DNA molecules. The extent to which thymidine incorporation occurs can then be measured by the photographic method known as *autoradiography*. The procedure involves growing the cells in the tritiated thymidine solution and then covering the cells with sensitized photographic film. As the incorporated tritiated thymidine ejects its radioactive particles ("decays"), dark exposed spots are formed on the film immediately above. Autoradiography performed under these conditions by Howard, Pelc, Taylor, and others has generally shown that only the chromosomes appear to be radioactively "labeled" (Fig. 4-13).

Since the incorporation of tritiated thymidine increases with the synthesis of new DNA molecules, further information can be derived as to the time of DNA duplication in dividing cells. Generally, for both mitosis and meiosis, DNA duplication occurs during the interphase period before cell division begins. Evidence from the effect of x-rays on chromosome breakage (Chapter 24) has also indicated that these particular synthesizing periods coincide with the time that chromosomes appear to develop "double" structures.

As a result of all these investigations, the evidence for the generally localized presence of DNA in chromosomes appears quite conclusive. Nevertheless, small amounts of DNA can also be found in such extrachromosomal structures as plastids, mitochondria, and *kappa* particles of *Paramecia* (Chapter 13). In such instances, DNA probably functions as genetic material in a fashion similar to its function in the chromosome.

BIOLOGICAL SIGNIFICANCE OF DNA

There are four general requirements that genetic material can be expected to fulfill:

1. That it replicate itself accurately during cell growth and duplication.
2. That its structure be sufficiently stable so that heritable changes (mutations) occur only very rarely.
3. That it have the potentiality to carry all kinds of necessary biological information.
4. That it transmit this information to the cell.

Although the last point has not yet been discussed (see Chapter 27), DNA appears to fulfill these requirements completely and is certainly a most suitable candidate for the role of genetic material. However, essential as the above qualities may be, at least one important question remains: What is the evidence definitely identifying DNA as *the* genetic material?

The answer to this question requires the isolation of DNA in some biologically effective form in which it is firmly associated with the heredity of a particular biological characteristic. Only in recent decades has this type of isolation been accomplished, and the experiments that follow mark true turning points in the history of genetics.

TRANSFORMATION

In *Diplococcus pneumoniae* bacteria the thin capsule enclosing the cell wall consists of substances called *polysaccharides*, which are of specific types. Type II capsules, for example, elicit the formation of antibodies in the bloodstream of rabbits that are different from antibodies formed with type III capsules. These capsule types are distinct properties of bacterial strains, and bacteria of one kind do not appear in pure cultures of another kind.

On the other hand, the capsules themselves are subject to variability in respect to their presence or absence. That is, any particular type of *D. pneumoniae* may occasionally give rise to bacteria that

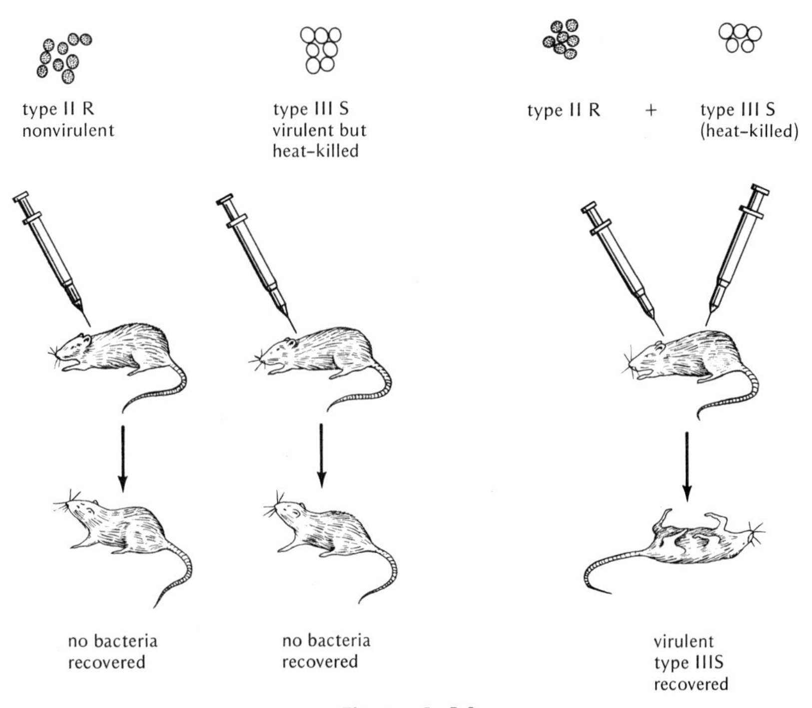

Figure 4–14
Transformation experiment of Griffith.

do not have capsules at all (mutation). Such non-capsulated bacteria have a *rough* (R) appearance when grown on culture plates, and are relatively harmless in contrast to the *smooth* (S) appearance and virulence of encapsulated bacteria. When cultured separately, both R and S colonies will transmit their respective characters to future generations except for rare mutations. Whatever the kind of bacteria, however, they are all sensitive to heat, and if the temperature is raised sufficiently high, the bacteria are *heat-killed* and can no longer divide.

In one of the seminal sets of experiments in modern genetics, Griffith, in 1928, showed that heat-killed bacteria of one type could have a hereditary influence on bacteria of another type. In one experiment he injected heat-killed bacteria of type IIIS into mice carrying type IIR and obtained virulent live cultures which were of the type IIIS variety (Fig. 4–14). Since the injection of heat-killed type IIIS bacteria by itself does not result in any live bacterial cultures, nor does type II mutate into type III, a change or *transformation* of type IIR into type IIIS must have occurred through the transfer of some active substance. Work in the following 15 years showed that Griffith's results could be duplicated by mixing different types of heat-killed and live bacterial strains in mice (*in vivo*) as well as by mixing them in test tubes (*in vitro*).

The search for the specific agent responsible for transformation, the *transforming principle*, continued until 1944. In that year Avery, MacLeod, and McCarty showed that the transforming principle consists entirely of DNA. Their experiment was to extract the DNA from heat-killed cells of type IIIS, and mix this DNA extract directly with *in vitro* cultures of type IIR. A serum was added whose antibodies react with the R cells and cause them to precipitate to the bottom. When transformation occurs, type IIIS cells, not being precipitated, now grow diffusely throughout the medium (Fig. 4–15).

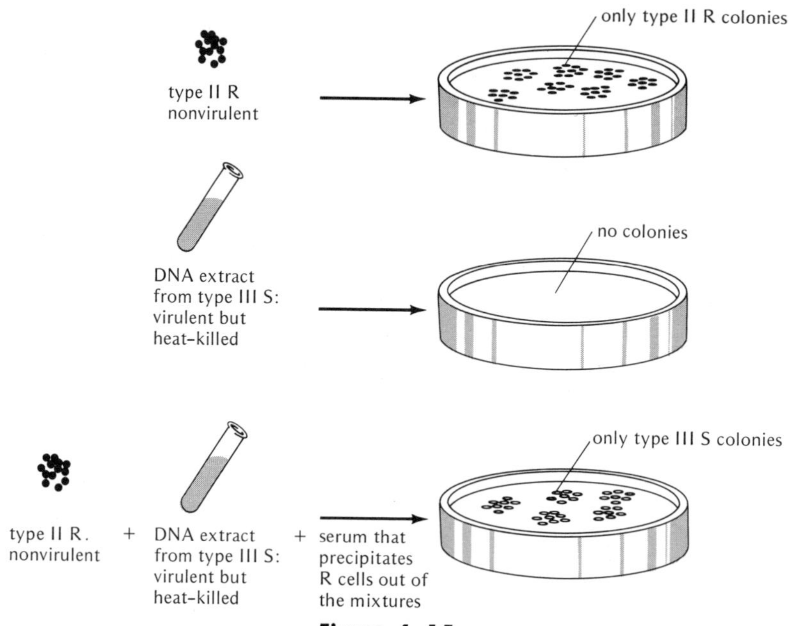

Figure 4–15

Transformation experiment of Avery, MacLeod, and McCarty.

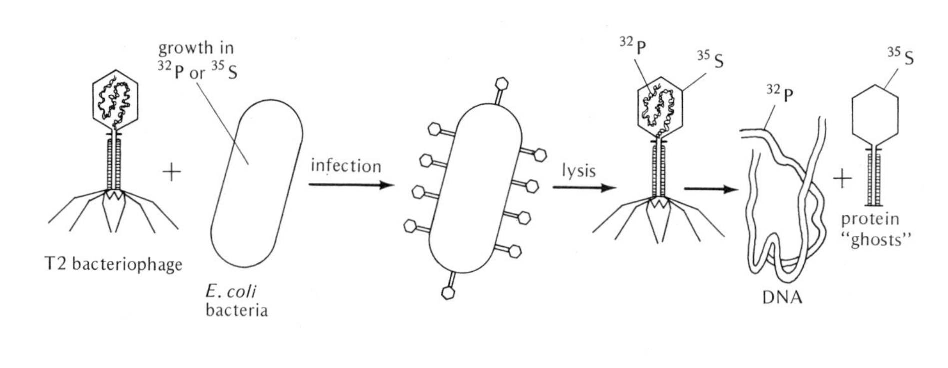

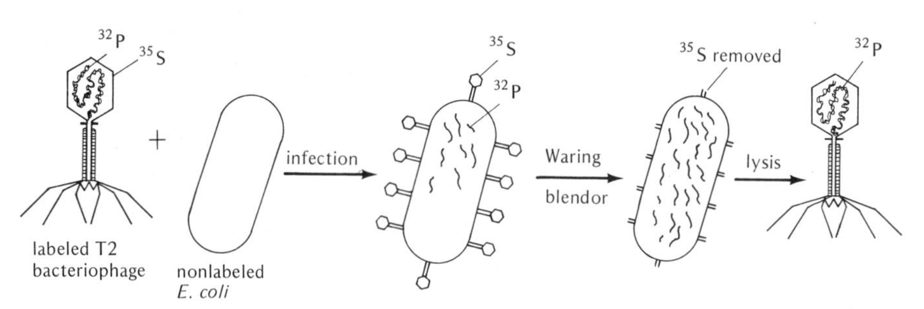

Figure 4–16

Schematic presentation of the experiments of Hershey and Chase.

TABLE 4–3

Some of the genetic characteristics reported to be transformed in bacteria and eucaryotes

	Species
BACTERIA:	
capsule types	
I, II, III, VI, VIII, XIV	*Diplococcus pneumoniae*
a, b, c, d, e, f	*Hemophilus influenzae*
I, II	*Neisseria meningitidis*
filamentous growth	*D. pneumoniae*
drug resistances	
penicillin, streptomycin, sulfanilimide, erythromycin	*D. pneumoniae*
streptomycin, cathomycin	*H. influenzae*
enzyme capacities	
mannitol utilization, salicin fermentation, maltase, lactic acid oxidase	*D. pneumoniae*
indole, anthranilic acid and nicotinic acid utilization, sucrase, β galactosidase	*Bacillus subtilis*
arabinose utilization	*Escherichia coli*
amino acid requirements	
leucine, histidine, arginine threonine, proline	*E. coli*
EUCARYOTES:	
eye color	*Drosophila melanogaster* *Ephestia kuhniella* *Bombyx mori*
melanin synthesis	mouse (*mus musculus*)
cell enzymes	human tissue culture

Most of the bacterial data are after Hotchkiss and after Ravin. *E. coli* listings derive from S. D. Cosloy and M. Oishi, 1973, *Proc. Nat. Acad. Sci.*, **70:**84–87. The eucaryotic transformation data come from various sources (e.g., A. S. Fox, S. B. Yoon, and W. M. Gelbart, 1971, *Proc. Nat. Acad. Sci.*, **68:**342–346; S. Nawa et al., 1971, *Genetics*, **67:**221–234), and some of the mammalian data have been questioned and discussed by Bhargava and Shanmugam. There are also a number of reports on the absorption and transmission of "foreign" procaryotic or eucaryotic DNA by eucaryotic cells (see, for example, Ledoux et al.).

Since 1944 numerous transforming experiments have been done with a variety of bacteria, and some disputed reports of the transformation of eucaryotic cells have also been presented (Table 4–3). In all cases the properties of the transforming principle have been found to be those of DNA. Furthermore, as with DNA, its biological activity is destroyed by a specific enzyme, deoxyribonuclease, but is not destroyed by enzymes that attack RNA or proteins (ribonuclease, trypsin, chymotrypsin). Also, labeling of transforming DNA with radioactive phosphorus results in incorporation of this radioactivity into recipient bacteria in direct proportion to the number of transformants.

The pure DNA nature of the transforming principle in these experiments is, to date, among the strongest evidence that DNA is the genetic material.

VIRAL INFECTION

The enzyme deoxyribonuclease (DNase) does not affect the DNA of bacteriophage ordinarily unless its protein envelope is ruptured (i.e., by *osmotic shock*, rapid dilution from high- to low-salt concentrations). In the presence of DNase such ruptured bacteriophages will yield a solution of small DNA particles in addition to protein "ghosts" representing the ruptured membranes. Through the work of Herriot it was known that if such protein ghosts attach to a bacterial cell, lysis of the cell occurred, but no further bacteriophage was formed. This finding indicated that the DNA played an important role in the production of new viral particles.

A crucial experiment which, probably more than any other, provided the immediate impetus for Watson and Crick to unravel the molecular structure of DNA was performed by Hershey and Chase in 1952. They demonstrated that the production of new viral particles could occur if the protein coats of the virus remained outside the infected bacterial cell and only the viral DNA entered the cell (Fig. 4–16). First, Hershey and Chase showed that if bacteriophage cultures are "labeled" with radioactive phosphorus (^{32}P) or sulfur (^{35}S) and then ruptured in the presence of DNase, the DNA contains all the ^{32}P and the protein ghosts contain all the ^{35}S. They then permitted phages labeled either with ^{32}P or ^{35}S to infect bacteria and shaved off the phage heads from the bacterial cell wall by the action of a Waring Blendor. Since only ^{32}P labeling could be recovered in the newly produced phage, these experiments indicated that DNA was the material transmitted between parent and offspring. Some ^{35}S protein could be shown to have entered the bacterial cell, however, but even this small amount of protein was discarded in the bacterial debris after its destruction (lysis), and was consequently not found in newly produced bacteriophage. Later experiments have shown that some unlabeled protein enters with the infective viral DNA, but this too is a very small amount and may serve the purpose of maintaining a particular physical structure of DNA to permit infection (see, for example, Hirokawa). The conclusions drawn from this experiment are therefore similar to those drawn from the transformation experiments, with the exception that the transforming principle contains almost no detectable protein at all.

Since then, various experiments have indicated that some infective phage DNA can be isolated free of any detectable amounts of protein. For example, the λ phage can be shown to be infective even after the protein envelopes are removed and protein-destroying enzymes are applied (Meyer et al.). Similar infective or "transfective"* activity has been shown by Sekiguchi and his co-workers for single-stranded DNA from φX174 phage. In both cases the purified DNA extracts could attack bacterial protoplasts without cell walls (also called spheroplasts) but not normal bacterial cells. The φX174 DNA could, in fact, attack the protoplasts of bacterial strains that were ordinarily resistant to the complete protein-carrying bacteriophage. Also, in both instances the phage extracts maintained their DNA structure as to single- or double-stranded forms, and their infectivity was destroyed with DNase. Similar results have been obtained by Guthrie, Sinsheimer, and others.

TOBACCO MOSAIC VIRUS INFECTIVITY

Tobacco mosaic virus (TMV) is one of the many viruses using RNA instead of DNA as its sole nucleic acid. Surrounding the RNA coil is a protein that aids in the infection of plant cells and protects the RNA from enzymes such as ribonuclease. The protein is specific for a particular strain, and there are different TMV strains with different protein envelopes. Through biochemical methods, Fraenkel-Conrat and others found that the protein could be separated from the RNA but was not infective by itself. Only when recombined ("reconstituted") with isolated RNA did the viral infectivity return and were new viral particles produced.

The question now asked was whether chemically isolated TMV RNA could be an infective agent on its own and thereby produce new viral particles.

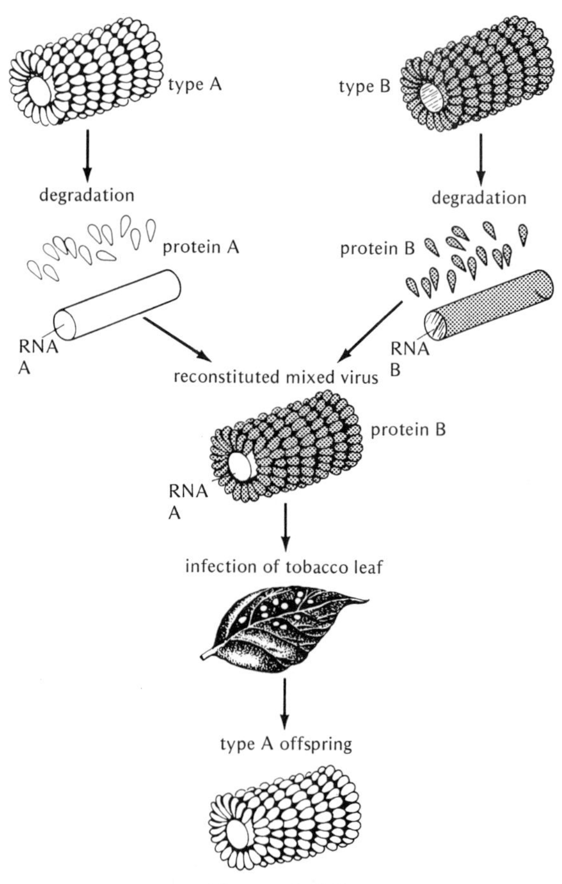

Figure 4–17

Reconstitution of a mixed tobacco mosaic virus and demonstration that the type of offspring produced is determined by the nucleic acid and not by the protein.

*The entry of purified viral nucleic acids into cells, and their subsequent replication or activity, has been called *transfection*.

Experiments by Gierer and Schramm in 1956 showed that this was indeed the case; pure TMV RNA rubbed into tobacco leaves caused the synthesis of new TMV virus and, as expected, such synthesis could be prevented by the ribonuclease enzyme. Furthermore, experiments by Fraenkel-Conrat and Singer showed that reconstituted TMV, containing RNA from one strain and protein from another, always led to the synthesis of viral particles whose protein was determined by the parental RNA rather than the parental protein (Fig. 4–17).

Although this evidence related to the hereditary function of RNA rather than DNA, we should note that it is a nucleic acid rather than protein that is exclusively involved. In DNA-containing organisms, as mentioned above, purified DNA performs a similar hereditary function.

THE CENTRAL DOGMA

These experiments point to the existence of a genetic material, nucleic acid, that is transmitted between generations and found in all living organisms. As described in the following chapter, molecular studies that analyze the replication of nucleic acids support very strongly the structure proposed by Watson and Crick. However, in addition to its own physical replication and transmission, this material also has significant effects on the biological activity of the cell, and therefore on the structural and functional traits of the organism that carries it. In fact an understanding of the exact relationship between genetic material, its cellular products (proteins), and its consequent effects, only began to develop once molecular

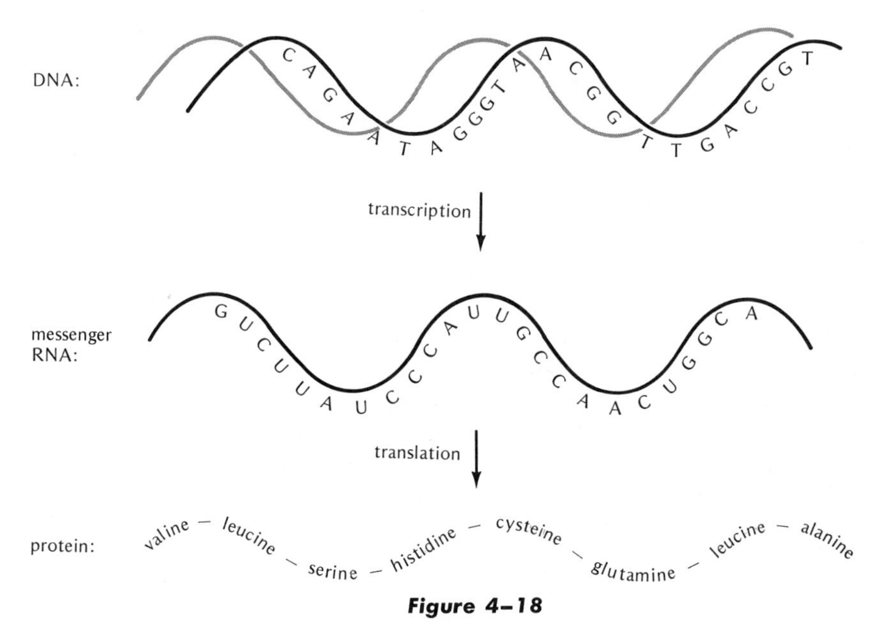

Figure 4–18

A diagrammatic illustration of how information is transferred between the three polymers. One of the two strands of DNA serves as a template upon which a molecule of messenger RNA is transcribed. This messenger then serves, in turn, as a template upon which a molecule of protein is translated. Note that the messenger RNA is exactly complementary to its DNA template, and that a sequence of three nucleotides on the messenger specifies one amino acid (i.e., 24 nucleotides = 8 amino acids).

studies could be combined with a genetic analysis of how biological traits are transmitted. A detailed description of the cellular function of genetic material is therefore postponed (Part V) until the elements of genetic analysis have been discussed.

For the present it may be sufficient to point out that the genetic material affects the production of proteins in the cell through the intermediary of a protein-synthesizing apparatus involving three different kinds of RNA. At the heart of our understanding of the relationship between genetic material and protein is the assumption that the three-dimensional structure of a protein—that is, its form, shape, and subsequent function—is determined essentially by the linear sequence of amino acids of which it is composed. This linear sequence of amino acids is, in turn, determined by the linear sequence of bases in nucleic acid. In brief, the genetic material, through the process of "transcription," produces a molecule of *messenger RNA* which is, base-for-base, a complement to the bases on one of its strands (Fig. 4–18; also p. 100 ff.). Through the mediation of ribosomes, which are themselves composed of *ribosomal RNA* and protein, a sequence of bases in the messenger RNA is then "translated" into a sequence of amino acids. This translation follows the "triplet" rule that a sequence of three messenger RNA bases designates one of 20 different kinds of amino acids.

During this process of translation, no physical material is actually inserted by the messenger RNA into the protein; it is only information that is transferred. That is, the messenger RNA only designates the linear position in which each amino acid is to be placed by special molecules of *transfer RNA* that bring the amino acids to the messenger. With the aid of the ribosome and various enzymes, the amino acids are then connected in sequence through peptide linkages. Thus a polypeptide chain of amino acids is formed in which the precise position of each component has been ultimately designated by the genetic material.

In 1958 Crick proposed that this informational process goes only in the direction of nucleic acid-to-protein but not in reverse. This view, which became known as "The Central Dogma," is diagrammed in Fig. 4–19, and states essentially that

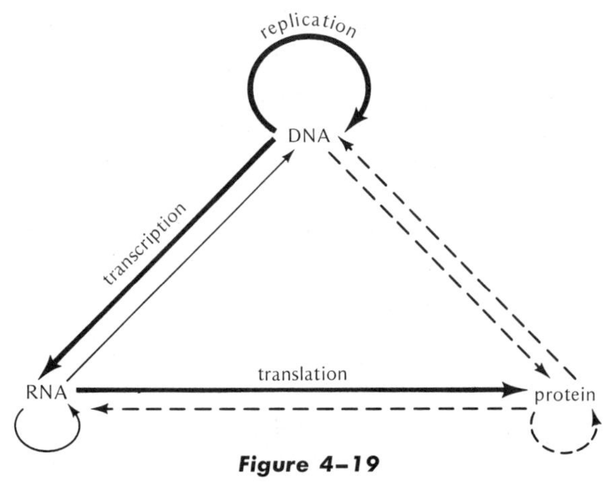

Figure 4–19

Diagram of relationships between the three families of cellular polymers that carry sequence-by-sequence information. As described in the text, such sequential information can be transferred from one polymer to the other through molecular pairing mechanisms unaccompanied by a physical transfer of material. The heavy lines indicate the known directions of transfer of information in cells. That is, DNA can replicate its own information or transfer information to messenger RNA by transcription (Chapter 5), which is then transferred to proteins by translation (Chapters 27 and 28). The thin lines indicate special information transfer mechanisms which have been shown to occur only in cells infected by certain RNA viruses. As shall be discussed in the following chapter, viral RNA can replicate its information by mechanisms similar to DNA replication, but this can also occur through a DNA intermediate formed as a complementary strand to the infecting viral RNA strand. The dashed lines indicate directions of information transfer not yet demonstrated, and believed by many to be nonexistent. (After Crick, 1970.)

once sequential information has passed into protein it cannot get out again. Only nucleic acids, through complementary base pairing, can determine the base sequences of other nucleic acids, and only nucleic acids can determine the amino acid sequences of proteins. This is not to say that proteins cannot *affect* nucleic acids or cause errors in nucleic acid replication (mutations). It only means that a long chain of amino acids in a protein is not translated either into a specific long chain of bases in a nucleic acid or into a similar long-chained sequence of amino acids in another protein.

PROBLEMS

4–1. For each of the following conditions state how long a sequence of nucleotides would be necessary to code for *one* particular "trait." (a) If a cell contained a total of only four different traits. (b) If a cell contained a total of 16 different traits. (c) If a cell contained a total of 64 different traits.

4–2. From your previous answer determine the minimum length of DNA that is necessary to encode one amino acid if there are 20 kinds of amino acid present in the cell.

4–3. Assume that a bacterial cell produced 3000 kinds of protein, each about 100 amino acids long, and that a sequence of three nucleotides is necessary to "code" for each amino acid. How long a molecule of DNA must such an organism possess at the minimum?

4–4. Assume that a primitive living organism discovered on Mars or another planet is found to have a system similar to our own, that is, linear genetic material of one kind ("nucleic acid") is used as a code to specify linear sequences of functional material of another kind ("protein"). In this organism there are only 5 possible kinds of "amino acids" (structural units of the "protein") and only 4 possible kinds of "nucleotides" (structural units of the "nucleic acid"). In toto, this organism produces 100 different "proteins," each 10 "amino acids" long. In order for these 100 proteins to be produced, and no others, what would be the *minimum* number of nucleotides that must be present in the organism's nucleic acid chromosome?

4–5. A single tetranucleotide strand contains the base sequence A–T–C–G. (a) Can you distinguish whether this is DNA or RNA? (b) If this were DNA, what would be the base sequence of its complementary strand? (c) If an RNA molecule were formed complementary to this strand, what would be its base sequence?

4–6. If one kind of biological "message" was coded by the particular base sequence of a tetranucleotide, how many different kinds of messages could be carried by an organism that contained tetranucleotides with all possible base sequences?

4–7. Does each of the two complementary strands of DNA carry the same biological information?

4–8. If the ratio of (A + T) to (G + C) in a particular DNA is 1.00, does this indicate that the DNA is most likely constituted of two complementary strands of DNA, a single strand of DNA, or is more information necessary?

4–9. Explain whether the A + T/G + C ratio in double-stranded DNA is expected to be the same as the A + C/G + T ratio.

4–10. In the table at the foot of this page are base ratios of certain nucleic acid fractions extracted from seven different species. For each one write the type of nucleic acid that is involved (i.e., DNA or RNA) and whether you think it is in the form of a double strand or single strand.

4–11. A single-stranded DNA phage has an A/T base ratio of 0.33, a G/C ratio of 2.0, and an $\frac{A + T}{G + C}$ ratio of 1.33. (a) What is the $\frac{A + G}{T + C}$ ratio in this molecule? (b) If this single-stranded molecule forms a complementary strand, what are these four ratios in the complementary strand? (c) What are these four ratios in both the original and complementary strands together?

4–12. Answer (a), (b), and (c), of the previous question if the original single-stranded DNA phage had an A/T base ratio of 0.50, a G/C ratio of 0.33 and an $\frac{A + T}{G + C}$ ratio of 0.75.

4–13. If an *E. coli* auxotroph A could only grow on a medium containing thymine, and an auxotroph B could only grow on a medium containing leucine, how would you test whether DNA from A could transform B?

Species	Adenine	Guanine	Thymine	Cytosine	Uracil	$\frac{A+T}{G+C}$	$\frac{A+G}{T+C}$
(a)	23	27	23	27	—	—	—
(b)	27	23	27	23	—	—	—
(c)	23	23	27	27	—	—	—
(d)	23	27	—	27	23	—	—
(e)	23	27	—	23	27	—	—
(f)	—	—	—	—	—	1.00	.67
(g)	—	—	—	—	—	.67	1.00

4-14. If the two live bacterial strains in the previous question were mixed together, how would you test whether sexual conjugation had occurred between them?

4-15. If some white fibrous material were extracted from a cell, how would you distinguish whether it is DNA, RNA, or protein?

4-16. In an experiment testing the transforming ability of DNA taken from a streptomycin-resistant strain of pneumococcus on a streptomycin-sensitive strain, Hotchkiss observed that it takes .000000001 grams (10^{-9}) of donor DNA to transform 10,000 streptomycin-sensitive cells. Assuming that each "haploid" pneumococcus cell has 2×10^{-15} grams of DNA (see Table 4-1), what is the proportion of transformed recipient cells to donor cells, or "the efficiency of transformation?"

4-17. Explain why you agree or disagree with the following statements: (a) Organisms with the same base ratios are alike. (b) DNA cannot be the genetic material because it needs the help of proteins, such as enzymes, in order to reproduce.

REFERENCES

AVERY, D. T., C. M. MACLEOD, and M. MCCARTY, 1944. Studies on the chemical nature of the substance inducing transformation of pneumococcal types. Induction of transformation by a deoxyribonucleic acid fraction isolated from pneumococcus type III. *Jour. Exp. Med.,* **79,** 137-158. Reprinted in the collections of Abou-Sabé, of Peters, and of Taylor; see References, Chapter 1.)

BHARGAVA, P. M., and G. SHANMUGAM, 1971. Uptake of nonviral nucleic acids by mammalian cells. *Progr. Nuc. Acid Res. and Mol. Biol.*, **II,** 103-192.

BRACHET, J., 1957. *Biochemical Cytology*. Academic Press, New York.

CALLAN, H. G., and H. C. MACGREGOR, 1958. Action of deoxyribonuclease on lampbrush chromosomes. *Nature*, **181,** 1479-1480.

CHARGAFF, E., 1950. Chemical specificity of nucleic acids and mechanism of their enzymatic degradation. *Experientia*, **6,** 201-209. (Reprinted in the collection of Taylor; see References, Chapter 1.)

CRICK, F. H. C., 1958. On protein synthesis. *Symp. Soc. Exp. Biol.*, **12,** 138-163.

———, 1970. Central dogma of molecular biology. *Nature*, **227,** 561-563.

DAVIDSON, J. N., 1972. *The Biochemistry of the Nucleic Acids*, 7th ed. Chapman and Hall, London.

FRAENKEL-CONRAT, H., and B. SINGER, 1957. Virus reconstitution: combination of protein and nucleic acid from different strains. *Biochim. et Biophys. Acta*, **24,** 540-548. (Reprinted in the collections of Taylor and of Levine; see References, Chapter 1.)

FRANKLIN, R. E., and R. GOSLING, 1953. Molecular configuration of sodium thymonucleate. *Nature*, **171,** 740-741.

FRUTON, J. S., 1972. *Molecules and Life*. Wiley-Interscience, New York. (See Part 3: "From Nuclein to the Double Helix.")

GIERER, A., and G. SCHRAMM, 1956. Infectivity of ribonucleic acid from tobacco mosaic virus. *Nature*, **177,** 702-703. (Reprinted in the collection of Abou-Sabé; see References, Chapter 1.)

GOODGAL, S. H., and R. M. HERRIOT, 1957. Studies on transformation of *Hemophilus influenzae*. In *The Chemical Basis of Heredity*, W. D. McElroy and B. Glass (eds.). Johns Hopkins Press, Baltimore, pp. 336-340.

GRIFFITH, F., 1928. The significance of pneumococcal types. *Jour. Hygiene*, **27,** 113-159.

GUTHRIE, G. D., and R. L. SINSHEIMER, 1960. Infection of protoplasts of *Escherichia coli* by subviral particles of bacteriophage ϕX174. *Jour. Mol. Biol.*, **2,** 297-305.

HERSHEY, A. D., and M. CHASE, 1952. Independent functions of viral protein and nucleic acid in growth of bacteriophage. *Jour. Gen. Physiol.,* **36,** 39-56. (Reprinted in the collections of Abou-Sabé, of Stent, and of Taylor; see References, Chapter 1.)

HIROKAWA, H., 1972. Transfecting deoxyribonucleic acid of *Bacillus* bacteriophage ϕ29 that is protease sensitive. *Proc. Nat. Acad. Sci.,* **69,** 1555-1559.

HOLLAENDER, A., and C. W. EMMONS, 1941. Wavelength dependence of mutation production in the ultraviolet with special emphasis on fungi. *Cold. Sp. Harb. Symp.*, **9,** 179-185. (Reprinted in *Selected Papers in Biochemistry*, Vol. 4; see References, Chapter 1.)

HOTCHKISS, R. D., 1955. The biological role of the deoxypentose nucleic acids. In *Nucleic Acids*, Vol. II, E. Chargaff and J. N. Davidson (eds.). Academic Press, New York, pp. 435-473.

LEDOUX, L., R. HUART, and M. JACOBS, 1974. DNA-mediated genetic correction of thiamineless *Arabidopsis thaliana. Nature*, **249,** 17–21.

LERMAN, L. S., and L. J. TOLMACH, 1957. Genetic transformation I: Cellular incorporation of DNA accompanying transformation in *Pneumococcus. Biochim. et Biophys. Acta*, **26,** 68–82.

LEVENE, P. A., and L. W. BASS, 1931. *Nucleic Acids.* Chemical Catalog Co., New York.

MEYER, F., R. P. MACKAL, M. TAO, and E. A. EVANS, JR., 1961. Infectious deoxyribonucleic acid from λ bacteriophage. *Jour. Biol. Chem.*, **236,** 1141–1143.

MIESCHER, F., 1871. On the chemical composition of pus cells. *Hoppe-Seyler's Med.-Chem. Untersuch.*, **4,** 441–460. (Reprinted in the collection of Gabriel and Fogel; see References, Chapter 1.)

MIRSKY, A. E., and S. OSAWA, 1961. The interphase nucleus. In *The Cell*, Vol. II, J. Brachet and A. E. Mirsky (eds.). Academic Press, New York, pp. 677–770.

OLBY, R., 1974. *The Path to the Double Helix.* Macmillan, London. (A thorough historical study of the various scientific threads leading to the Watson-Crick discovery, including interesting accounts of many of the personal interactions involved, and the laboratories, institutions, and meetings in which they took place.)

PAULING, L., and R. B. COREY, 1956. Specific hydrogen-bond formation between pyrimidines and purines in deoxyribonucleic acids. *Arch. Biochem. Biophys.*, **65,** 164–181.

RAVIN, A. W., 1961. The genetics of transformation. *Adv. in Genet.* **10,** 62–163.

SEKIGUCHI, M., A. TAKETO, and Y. TAKAGI, 1960. An infective deoxyribonucleic acid from bacteriophage φX174, *Biochim. et Biophys. Acta.* **45,** 199–200.

SINSHEIMER, R. L., 1959. A single-stranded deoxyribonucleic acid from bacteriophage φX174, *Jour. Mol. Biol.*, **1,** 43–53.

SOBER, H. A., 1970. *Handbook of Biochemistry*, 2nd ed. Chemical Rubber Co., Cleveland.

SPARROW, A. H., H. J. PRICE, and A. G. UNDERBRINK, 1972. A survey of DNA content per cell and per chromosome of prokaryotic and eukaryotic organisms: some evolutionary considerations. *Brookhaven Symp. Biol.*, **23,** 451–493.

STADLER, L. J., and F. M. UBER, 1942. Genetic effects of ultraviolet radiation in maize: IV. Comparison of monochromatic radiations. *Genetics*, **27,** 84–118.

TAYLOR, J. H., 1956. Autoradiography at the cellular level. In *Physical Techniques in Biological Research*, G. Oster and A. W. Pollister (eds.). Academic Press, New York, pp. 545–576.

WATSON, J. D., 1968. *The Double Helix.* Atheneum, New York. (A revealing personal account of events, motivations, and attitudes involved in the 1953 discovery of DNA structure.)

WATSON, J. D., and F. C. CRICK, 1953a. Molecular structure of nucleic acids. A structure for deoxyribose nucleic acids. *Nature*, **171,** 737–738. (Reprinted in the collections of Peters and of Taylor; see References, Chapter 1.)

———, 1953b. Genetical implications of the structure of deoxyribose nucleic acid. *Nature*, **171,** 964. (Reprinted in the collections of Adelberg and of Taylor; see References, Chapter 1.)

WILKINS, M. H. F., 1963. The molecular configuration of nucleic acids. *Science*, **140,** 941–950.

WILKINS, M. H. F., A. R. STOKES, and H. R. WILSON, 1953. Molecular structure of desoxypentose nucleic acids. *Nature*, **171,** 738–740.

5

REPLICATION AND SYNTHESIS OF NUCLEIC ACIDS

The mechanism for the synthesis of DNA proposed by Watson and Crick (p. 56) offered the important advantage of explaining how new DNA molecules could be exact replicates of the old. According to Watson and Crick, each single strand is a *template* or mold for its complement, and a new helix has one old strand and one that is newly synthesized. This type of replication is called *semiconservative*, in contrast to the *conservative* type, in which two new strands are synthesized in the form of a double helix while the old double helix remains unchanged (Fig. 5-1). Conservative replication could presumably occur were the double helix not to unwind but nevertheless to replicate itself in some unknown fashion (perhaps within the coils of the helix). A third type of replication, *dispersive*, would be possible if the double helical strands were to break down along their length into small pieces. After each individual piece replicates, it would then be randomly reconnected with newly synthesized pieces to form a patchwork single string of "dispersed" old and new pieces.

To decide between these different forms of replication, a number of experiments have been performed utilizing techniques of marking, or "labeling," DNA with specific chemicals that could be detected in later duplications. When formed in the presence of a "label," both strands of newly synthesized DNA double helixes should be labeled in conservative division, while the old double helixes remain unlabeled. In semiconservative division,

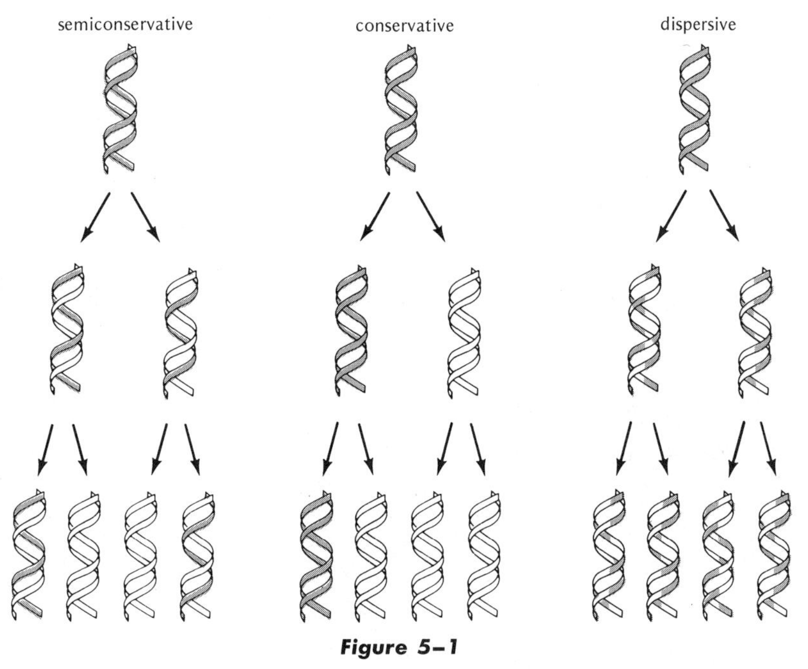

Figure 5–1
Three modes of replication of DNA molecule.

the replicated double helixes carry one old and one new strand and all are therefore labeled. On the other hand, if fully labeled DNA helixes are grown on unlabeled media, the new helixes formed will be unlabeled in conservative replication, and partially labeled, or "hybrid," in semiconservative replication. As can be seen in the following experiment by Meselson and Stahl, further replications on unlabeled media will also show the distinction between semiconservative and dispersive division.

Meselson and Stahl labeled the DNA of *Escherichia coli* bacteria with "heavy" nitrogen, ^{15}N (the normal isotope is ^{14}N), by growing them on an ^{15}N-containing medium that was the sole source of nitrogen for many generations. The DNA extracts of such cells gave a characteristic ultraviolet-light pattern showing a dark band near one end of a tube that had been spun at high speeds in an ultracentrifuge (Fig. 5–2). When these labeled ^{15}N cells were grown on nonlabeled media, the DNA was extracted and shown to consist of a "hybrid" DNA carrying both ^{14}N and ^{15}N at the same time; i.e., the DNA had not replicated in the separately labeled and unlabeled conservative form.

Since conservative replication seemed excluded, the remaining question for Meselson and Stahl was to distinguish between semiconservative and dispersive replication. In answer to this, the next generation of growth on unlabeled media showed that unlabeled DNA was now formed in amounts equal to the partially labeled hybrid DNA. Furthermore, although some hybrid DNA still remained, additional generations of growth on unlabeled media gave a relative increase in the amount of unlabeled DNA. Since unlabeled DNA was always formed despite the presence of labeled hybrid DNA, duplication evidently did not involve random dispersive labeling of all newly formed DNA. The dual $^{15}N/^{14}N$ composition of the "hybrid DNA" could also be demonstrated by heating this band of DNA to about 100°C and then reanalyzing it in the ultracentrifuge. Heating, under appropriate conditions, will break the hydrogen bonds between the two complementary DNA strands "denaturing" them to produce separate

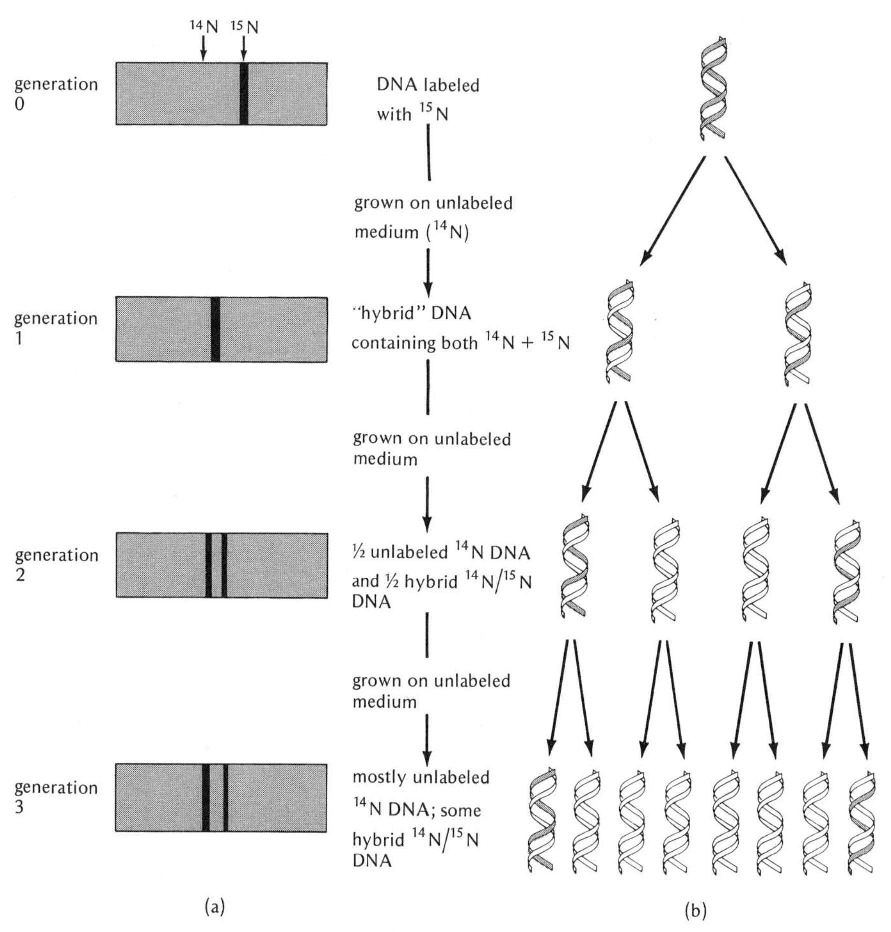

Figure 5–2

Results (left side) and interpretation (right side) of Meselson-Stahl experiment. In the density-gradient centrifugation procedure performed in this and many other experiments, DNA is isolated from cells and is then mixed with a cesium chloride solution and placed in an ultracentrifuge (up to 60,000 revolutions per minute). Rapid spinning for as long as 2 days causes the cesium chloride to be distributed in a density gradient along the ultracentrifuge cell with the least dense layers in that part of the cell that is near the center of the rotor and the most dense layers peripherally. DNA will then settle to that level of the cesium chloride gradient which corresponds to its own density. Since DNA absorbs ultraviolet light, the position of DNA can be detected along the density gradient by an optical system which illuminates the ultracentrifuge cell and registers the position of the DNA bands on photographic plates.

single-stranded DNA.* Under these conditions the hybrid DNA was now found to form two separate bands in the ultracentrifuge density gradient, one heavy (^{15}N) and one light (^{14}N).

The conclusion has therefore been drawn that DNA replication is semiconservative, although it was questioned by Cavalieri and others whether these findings referred specifically to the replication of single DNA double helixes or whether more than one double helix is involved. If, for example, two old and two new double helixes are involved in each replication, it is conceivable that the replication of a single DNA molecule may still be conservative. That is, each of the old double helixes makes a new double helix in conservative fashion, and one old and one new double helix then combines to form an apparent semiconservatively replicated DNA structure.

A more serious objection against semiconservative replication, however, was raised with respect to the amount of time it would take for such replication to occur in a very long molecule. The DNA of bacteriophage T2, for example, is a double-stranded molecule of 130,000,000 molecular weight or about 200,000 nucleotide pairs. Since there are 10 base pairs to each rotation of the double helix and 34 angstroms between rotations, such a molecule would have 20,000 rotations and would be $20{,}000 \times 34$ angstroms, or about 70 microns, in length (Fig. 5–3). Evidently such a long molecule would have to be folded over a number of times to fit into cells or nuclei (the bacterial cell infected by T2 virus is about 2 microns long, with an even smaller nuclear area). This raises two questions: (1) To replicate, must such a molecule first unwind 20,000 times in this folded condition? (2) Can it accomplish this unwinding in about 2 minutes without breaking? When we consider that the DNA of cellular organisms is even longer than that of viruses, the problem of unwinding and replication appears profound. For example, *E. coli* DNA is about 500 times longer than the cell in which it replicates, and the DNA in a single human cell is over two yards long! It is obvious that replication must, in some way, proceed before the parental DNA molecule is completely unwound.

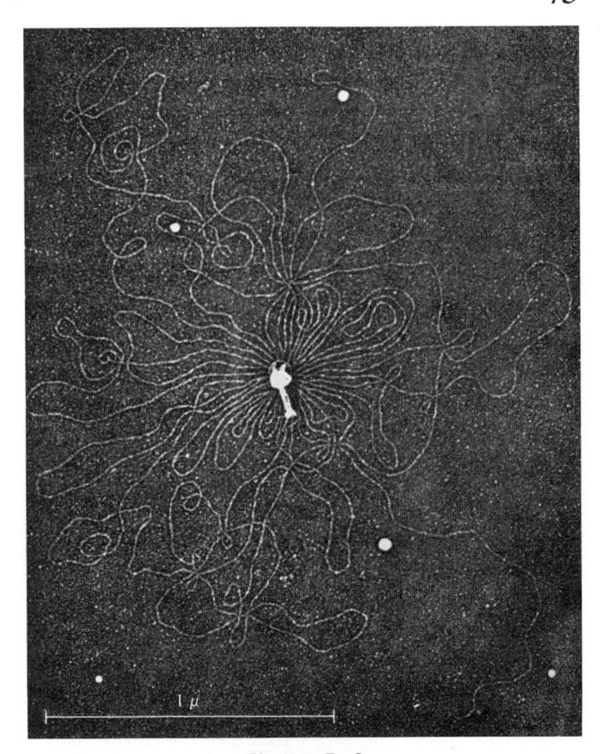

Figure 5–3

DNA molecule extruded by T2 bacteriophage. (From Kleinschmidt, Lang, Jacherts, and Zahn.)

Important evidence on this subject was presented by Cairns in experiments with radioactively labeled *E. coli* chromosomes. The *E. coli* chromosome is circular and, as shown in Fig. 5–4, is apparently also a double-stranded structure. Cairns has shown that the two circular component strands separate during replication, with each

---

*The distinction between single- and double-stranded DNA can be made by measuring their absorption of ultraviolet light. DNA in hydrogen-bonded double-helix form absorbs less ultraviolet light because of the regular stacking of its bases (see Fig. 4–8) compared to the random structure of denatured single-stranded DNA. The double-stranded form is called *hypochromic*, and the loose, single-stranded form is called *hyperchromic*. Characteristically, hyperchromicity can be induced by heating, and the temperature at which one half of the increased ultraviolet absorption is observed is called the *melting temperature*. When this is done, DNA molecules with different base compositions often have different melting points; the higher the melting point, the greater the number of hydrogen bonds between the two DNA strands. Melting temperature is therefore associated with the proportion of guanine and cytosine bases, since these have three hydrogen bonds between them, whereas each A–T pair has only two such bonds.

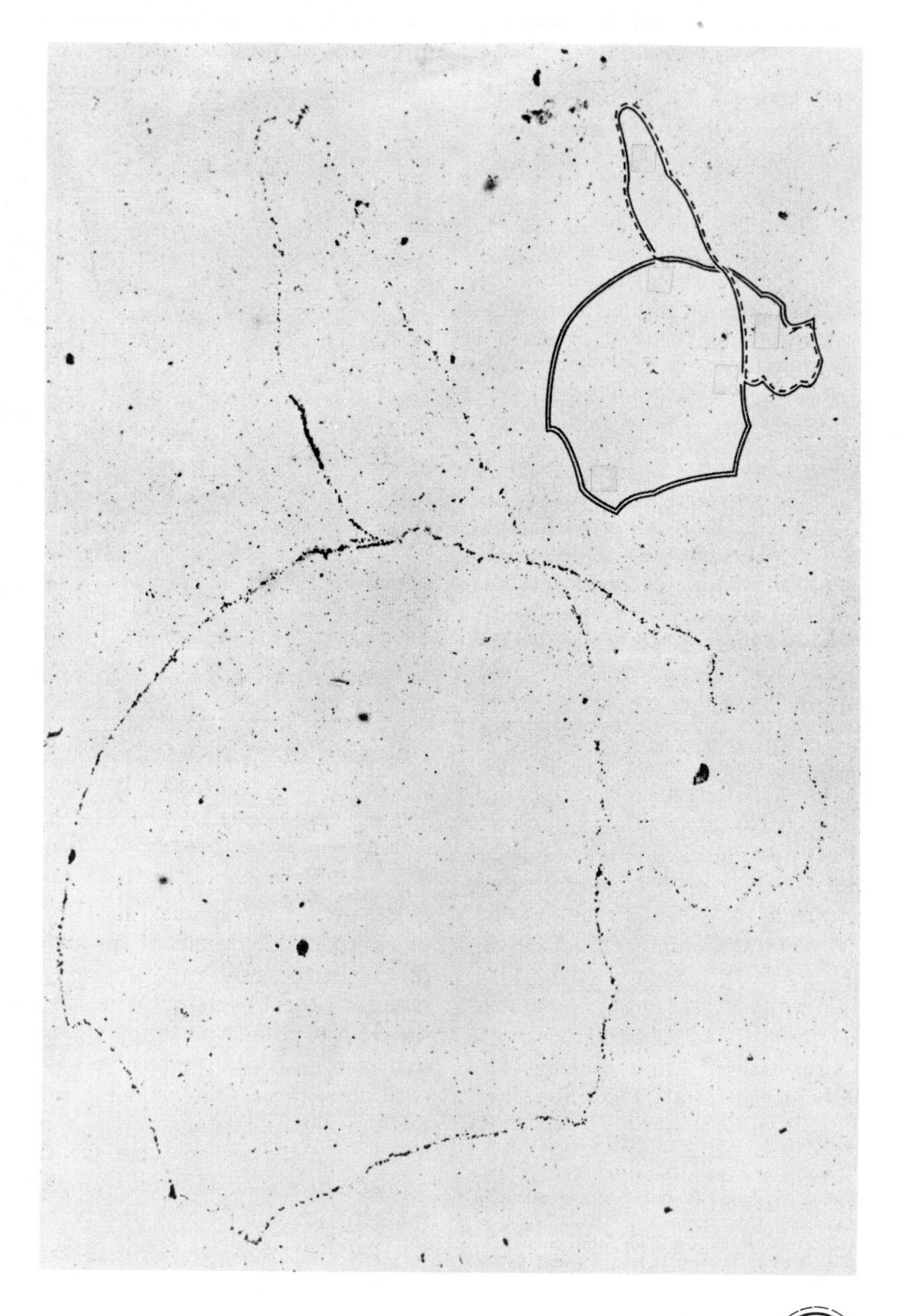

swivel

Figure 5–4

(Opposite page) *Autoradiograph of an* E. coli *chromosome (above) and schematic representation of its mode of duplication (below). (From Cairns, diagram modified.)*

strand duplicating individually. The fact that this duplication produces a Y type of joint indicates that unwinding of the two complementary DNA strands is not completed before replication begins but proceeds simultaneously with the replication. Also, the two-stranded nature of this semiconservative replication indicates that the Cavalieri explanation of multiple double helixes for the observed labeling in the Meselson-Stahl experiment is probably invalid. This conclusion is supported by electron micrographs of procaryote chromosomes (Fig. 5–3) which show almost invariably a single double-stranded DNA molecule. These findings, together with evidence that will be presented in the following sections on the modes of synthesis of DNA, have led to the general acceptance of the semiconservative theory of DNA replication.

BIOLOGICAL REACTIONS

Although DNA replicates in nature inside the cell, the molecular study of this replication is immensely difficult because of the many complex structures and substances that are already present. Without prior knowledge one cannot merely add chemical reagents to a cell and presume that their effect will be confined to DNA alone or that the effect on DNA is not caused by reactions other than those originally planned. Reactions outside the cell, on the other hand, are somewhat easier to control, since the number and types of substances can usually be restricted. Extracellular as well as intracellular reactions, however, have a number of common requirements.

The production or synthesis of any substance demands the presence of the elements or compounds (called *substrates* in cells) that enter into the synthesis, as well as a source of energy enabling the synthesis to occur. The rate of many chemical reactions, however, is slow, and living organisms increase the reaction rate by making extensive use of a class of catalysts called *enzymes.*

Enzymes are proteins that, like all catalysts, help the reaction along but are not themselves used up in the process. Even a small amount of enzyme can have a large effect, although some energy or "fuel" is usually needed. In the cell the fuel for many reactions is a compound called *adenosine triphosphate* (ATP), which can release energy by losing one or two of its three phosphate groups. The cell also maintains a large complement of different enzymes, each type being specific for a particular kind of reaction or synthesis. Some enzymes, for example, can aid in the removal of hydrogen atoms (*dehydrogenases*) and others can aid in the addition of units to a multiunit structure (*polymerases*). A cellular (*in vivo*, in life) reaction therefore usually requires the presence of substrates, enzyme, ATP, and in many cases metallic ions (e.g., magnesium) to activate the enzyme. Many of these substances can also be isolated outside the cell and reactions can be made to occur in a test tube (*in vitro*).

DNA SYNTHESIS

In 1957 Kornberg and co-workers isolated a polymerase enzyme from *Escherichia coli* bacteria that could be used for the *in vitro* synthesis of DNA. The reaction mixture contained the polymerase enzyme, the four different kinds of deoxyribonucleotides in the form of triphosphates (three P groups instead of the usual single P), magnesium ions ($MgCl_2$), and the presence of some previously formed DNA, called *template* DNA. Once all necessary substances were present, DNA production took place in amounts 20 or more times greater than the amount of initial DNA (Fig. 5–5).

Further evidence by Kornberg and others indicated that the enzyme acts by hooking together the free added nucleotide units into a DNA strand. Using different *phosphodiesterase* enzymes that could break the phosphate-sugar ester bonds of the DNA chain, they showed that the new DNA molecule grows by the addition of nucleotides to the

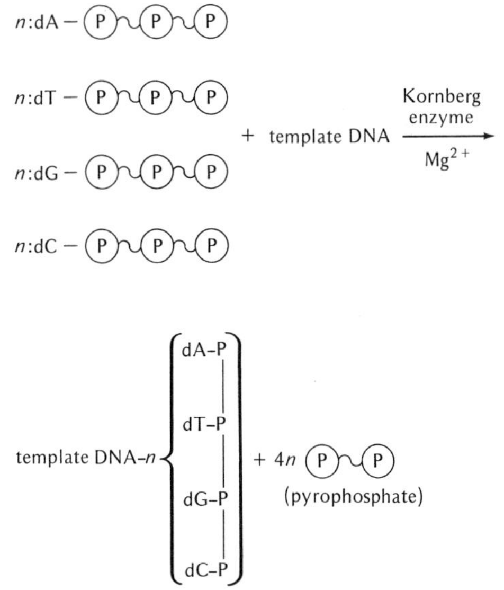

n:dA – (P)∼(P)∼(P)

n:dT – (P)∼(P)∼(P)

n:dG – (P)∼(P)∼(P)

n:dC – (P)∼(P)∼(P)

+ template DNA $\xrightarrow[Mg^{2+}]{\text{Kornberg enzyme}}$

template DNA-n {dA-P | dT-P | dG-P | dC-P} + 4n (P)∼(P) (pyrophosphate)

Figure 5–5

Reaction system for the synthesis of DNA in vitro *using the four different deoxyribonucleotide triphosphates and template DNA.*

hydroxyl group at the number 3 carbon position of the deoxyribose sugar at one end of a chain.

In the experiments performed, a radioactive phosphorus atom (^{32}P) has been incorporated into cytidine deoxyribonucleotides that are entering the DNA chain. Depending on the carbon position of the hydroxyl group on the deoxyribose to which the labeled cytidine nucleotide has been added, different breakage points would be expected to give different radioactive products. For example, if the cytidine nucleotide is added at the number 3 position (left side, Fig. 5–6), the action of *spleen diesterase* (breakage at the 5′ hydroxyl position of the sugar) will cause separation between the ^{32}P label and the cytidine nucleotide. The new connection of the ^{32}P atom will now be to the nucleotide above the cytidine, which may, in turn, be a random sample of any of the four types of bases. In the case of breakdown by *snake venom diesterase* (breakage at the 3′ hydroxyl position), the ^{32}P atom remains connected to the original cytidine base, and the other "labeled" nucleotides are not expected among the breakdown products. On the other hand, if new nucleotides are added at the 5′ sugar position (right side, Fig. 5–6), opposite findings should occur: exclusive cytidine labeling with spleen diesterase, and the absence of ^{32}P-labeled cytidine when snake venom diesterase is used for a limited time.* Actual results showed that ^{32}P-labeled cytidine nucleotide must have been added at the 3′ position, and this appeared to be true for all nucleotides.

By similar methods, Josse, Kaiser, and Kornberg demonstrated that there was a difference in polarity between the complementary strands of the DNA helix, so that the sugars of one strand were oriented in a direction opposite to that of the other. Their technique was to label the four deoxyribonucleotides with ^{32}P and then incorporate each separately into newly synthesized strands of DNA formed by the same "template." As shown previously, a radioactive ^{32}P-labeled nucleotide is incorporated at the 3′ position of the growing DNA strand. Action of the spleen diesterase enzyme on this strand will then separate the ^{32}P atom from the nucleotide which brought it to the growing strand, and will leave this atom attached to the 3′ position of the nucleotide "above" it. For example, one experiment consisted of incubating ^{32}P-labeled cytidylic acid into a DNA-synthesizing reaction mixture containing template DNA as well as unlabeled triphosphates of the other three nucleotides, A, T, G. The newly synthesized DNA product was then broken down by spleen diesterase so that ^{32}P-label was separated from the cytidine and hooked onto the nearest neighboring nucleotide, as shown in the asterisk-marked example given in Fig. 5–6. Through chemical methods these "nearest neighbors" to cytidine can be analyzed and the frequency of each combination determined, i.e., $.18\,\overset{\copyright}{\underset{T}{|}}$, $.190\,\overset{\copyright}{\underset{A}{|}}$, $.267\,\overset{\copyright}{\underset{C}{|}}$, $.362\,\overset{\copyright}{\underset{G}{|}}$. (The circle indicates the introduced labeled nucleotide.) Through the same technique, and using the same DNA template for each synthesis, they analyzed the

* If the enzymes are permitted to completely degrade the DNA, spleen diesterase will act at the 5′ position of the newly added cytidine nucleotide and would be expected to cause the presence of ^{32}P in inorganic phosphate, while ^{32}P would be expected to be absent in inorganic phosphate if snake venom diesterase is used.

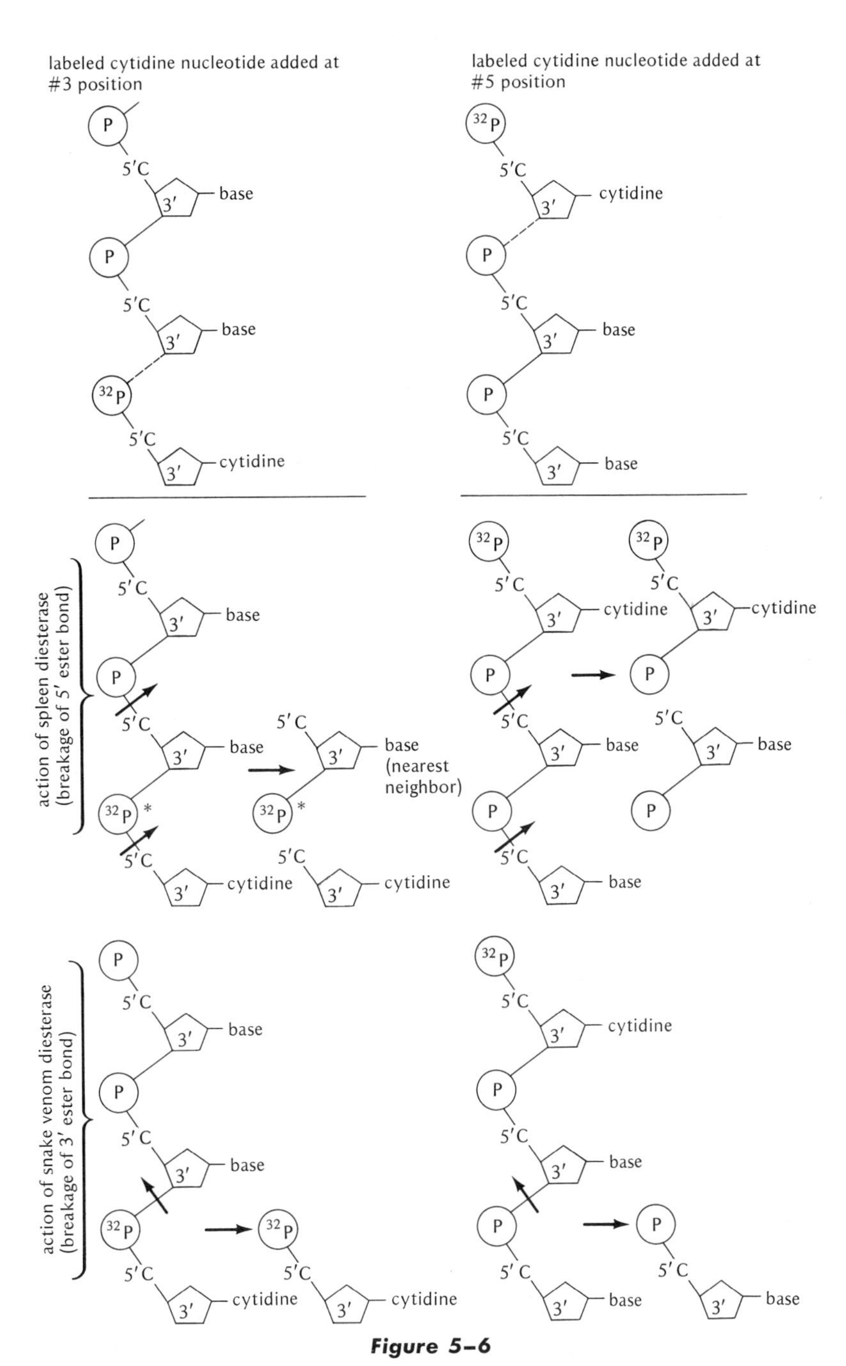

Figure 5–6

Differences in the products formed by spleen and snake-venom diesterase when ^{32}P-labeled nucleotides are added to the DNA chain at the 3′ or 5′ position. In the example on the left-hand side, marked by asterisks, spleen diesterase removes the ^{32}P label that has been brought into the DNA chain by cytidine and causes it to become attached to the base above, called the "nearest neighbor" to cytidine.

nearest neighbors for labeled adenine, thymine, and guanine. The frequencies for nearest neighbors of the labeled nucleotides are shown in Table 5–1. Note that the frequency of $\overset{\text{(T)}}{\underset{\text{T}}{|}}$ is about the same as that for $\overset{\text{(A)}}{\underset{\text{A}}{|}}$, and $\overset{\text{(G)}}{\underset{\text{G}}{|}}$ equals $\overset{\text{(C)}}{\underset{\text{C}}{|}}$. These results mean that the frequency of two consecutive T's along a single DNA strand is accompanied by a similar frequency of two consecutive A's, and

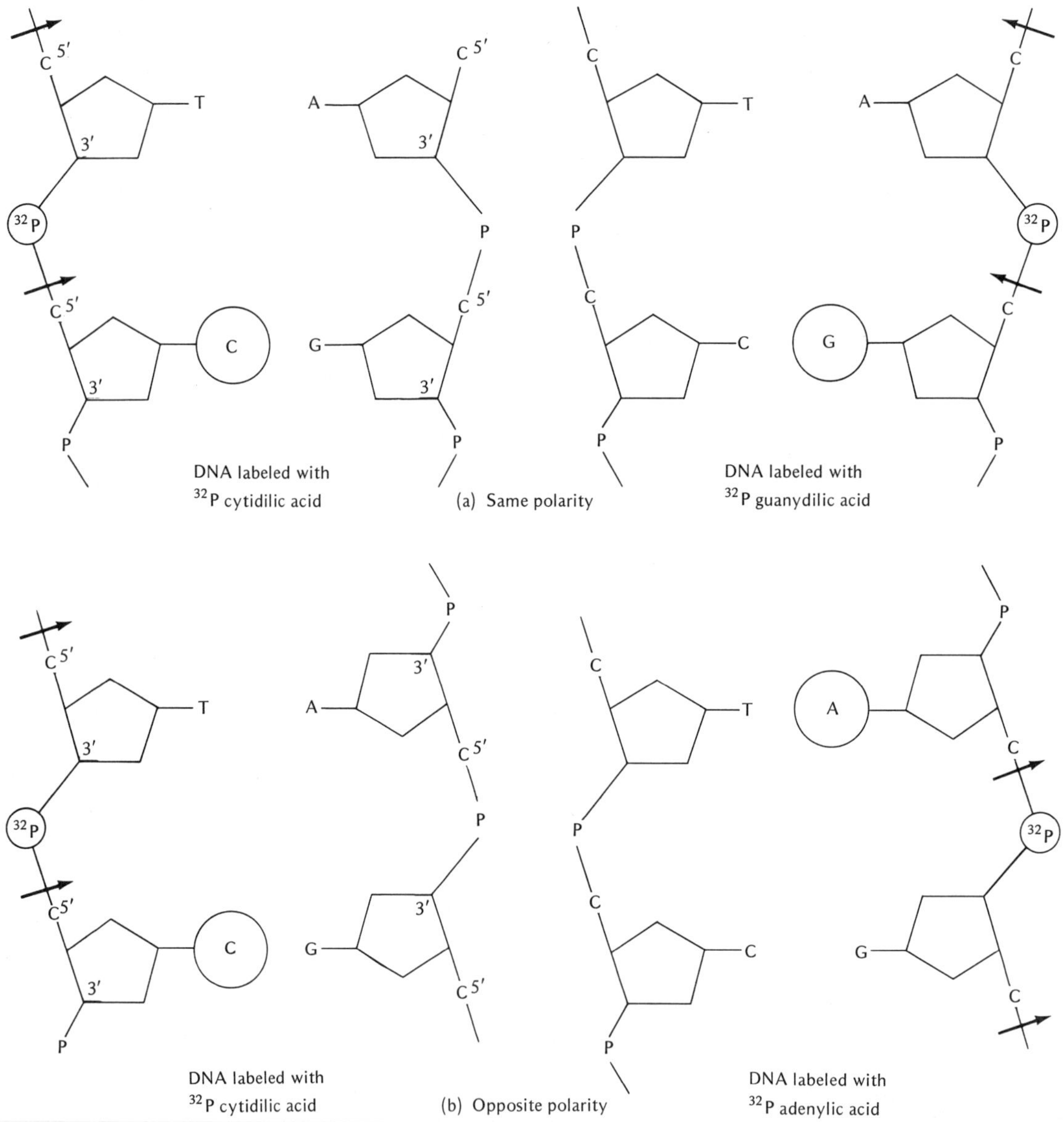

DNA labeled with ^{32}P cytidilic acid

(a) Same polarity

DNA labeled with ^{32}P guanydilic acid

DNA labeled with ^{32}P cytidilic acid

(b) Opposite polarity

DNA labeled with ^{32}P adenylic acid

Figure 5–7

Differences in the products formed by spleen diesterase when the complementary strands of DNA are oriented with the same polarity (above) and opposite polarity (below).

vice versa. Since this also holds true for the relationship between two consecutive C's and G's, these experiments provide important evidence for complementarity between the bases of the two DNA strands.

It was, however, of special interest that the frequencies of certain combinations were approximately equal while others were not. For example, Ⓒ–T equals Ⓐ–G but does not equal A–Ⓖ. The reason for this can be seen in Fig. 5–7, in which the polarity of the deoxyribose sugars determines the origin of the ^{32}P label in a dinucleotide pair. If the sugars in both strands have the same polarity (Fig. 5–7a), it is obvious that the ^{32}P for the complement of Ⓒ–T must originate from the labeled guanidine nucleotide or Ⓖ–A, since the spleen diesterase will break the chain at the number 5 carbon position and attach the radioactive phosphate to the adenylic nucleotide. If, however, the sugars in each strand have opposite polarity (Fig. 5–7b), the ^{32}P for the complement of Ⓒ–T must originate from the labeled adenine nucleotide or Ⓐ–G, since the radioactive ^{32}P now becomes attached to the guanidine nucleotide. Similarly, we would expect equality between Ⓣ–C and A–Ⓖ, Ⓐ–C and T–Ⓖ, etc. The fulfillment of these expectations in accord with opposite polarity is demonstrated in Table 5–1.

TEMPLATE DNA

In the above experiments the role of template DNA seems to have special importance, since, in its presence, DNA synthesis begins immediately, whereas, in its absence, a long lag period of 5 or more hours takes place before any synthesis is noted. In general various findings have pointed to the role of template DNA as a "master copy" upon which the synthesis of new DNA occurs.

The first evidence in this direction has been the correlation between the nucleotide base ratios of the newly synthesized DNA to those of the template DNA (Table 5–2). It would be difficult to explain such a close correlation without the assumption of base-to-base pairing. Second, the nearest-neighbor analysis shows complementary sequences between nearest neighbors in the product and nearest neighbors in the template, e.g., T–T (product) = A–A (template). Third, if one of the deoxyribonucleotides is absent from the substrate,

TABLE 5–1

Nearest-neighbor frequencies of labeled nucleotides (circled) in *Mycobacterium phlei* DNA

^{32}P-labeled Adenine	^{32}P-labeled Thymine	^{32}P-labeled Guanine	^{32}P-labeled Cytosine
T–Ⓐ = .012	T–Ⓣ = .026	T–Ⓖ = .063	T–Ⓒ = .061
A–Ⓐ = .024	A–Ⓣ = .031	A–Ⓖ = .045	A–Ⓒ = .064
C–Ⓐ = .063	C–Ⓣ = .045	C–Ⓖ = .139	C–Ⓒ = .090
G–Ⓐ = .065	G–Ⓣ = .060	G–Ⓖ = .090	G–Ⓒ = .122

Data of Josse, Kaiser, and Kornberg, 1961.

TABLE 5-2
Base ratios (in percent) in template and product DNAs of four sources

		A	T	G	C	$\frac{A+G}{T+C}$	$\frac{A+T}{G+C}$
calf thymus	template	28.9	26.7	22.8	21.6	1.07	1.25
	product	28.7	27.7	21.8	21.8	1.02	1.29
E. coli	template	25.0	24.3	24.5	26.2	.98	.97
	product	26.1	25.1	24.3	24.5	1.02	1.05
Mycobacterium phlei	template	16.2	16.5	33.8	33.5	1.00	.49
	product	16.4	16.2	33.3	34.1	.99	.48
phage T2	template	32.7	33.0	16.8	17.5	.98	1.92
	product	33.2	32.1	17.2	17.5	1.02	1.88

Adapted from Kornberg, 1960.

no product DNA is synthesized. This result would be expected if it were necessary to copy, without gaps and by base pairing, a chain containing many samples of all four nucleotides. Fourth, single strands of template DNA which result from heat-treating normal double-stranded DNA serve as more efficient templates than the double-stranded forms, a result that would also be expected if DNA synthesis occurred through direct base pairing. This improved synthetic efficiency also holds true for the single-stranded DNA found in the bacteriophage ϕX174, since replication can then take place without the necessity of first unwinding two complementary DNA strands. Fifth, *base analogues*, which are identical to the regular nucleotide bases except for minor chemical substitutions (Fig. 5-8), can replace these bases in DNA synthesis. For example, 5-bromocytosine or 5-methylcytosine can be used to replace cytosine deoxyribonucleotides to give a satisfactory chain of DNA. Similarly uracil, 5-bromouracil, or 5-fluorouracil can replace thymine, and hypoxanthine can replace guanine.

The removal of thymine and substitution of its base analogue 5-bromouracil does not affect *in vitro* DNA synthesis, but the substitution of this same analogue for any of the other three bases (adenine, cytosine, or guanine) will prevent DNA synthesis (Table 5-3). Evidently precise hydrogen bonding between the DNA template and added complementary nucleotides places a restriction on which analogues can be substituted for a particular nucleotide. This indicates that the DNA template in single-stranded form probably enters into a pairing relationship with the added free nucleo-

TABLE 5-3
Production of DNA when different analogues are substituted for each of the four different bases in *in vitro* DNA synthesis (*E. coli* polymerase enzyme)*

	Base			
Analogue	Adenine	Thymine	Cytosine	Guanine
uracil	0	54	0	0
5-bromouracil	0	97	0	0
5-fluorouracil	0	32	0	0
5-methylcytosine	0	0	185	0
5-bromocytosine	0	0	118	0
hypoxanthine	0	0	0	25

From Kornberg, 1974.
* Percentages given are those relative to the production of DNA when bases are not substituted.

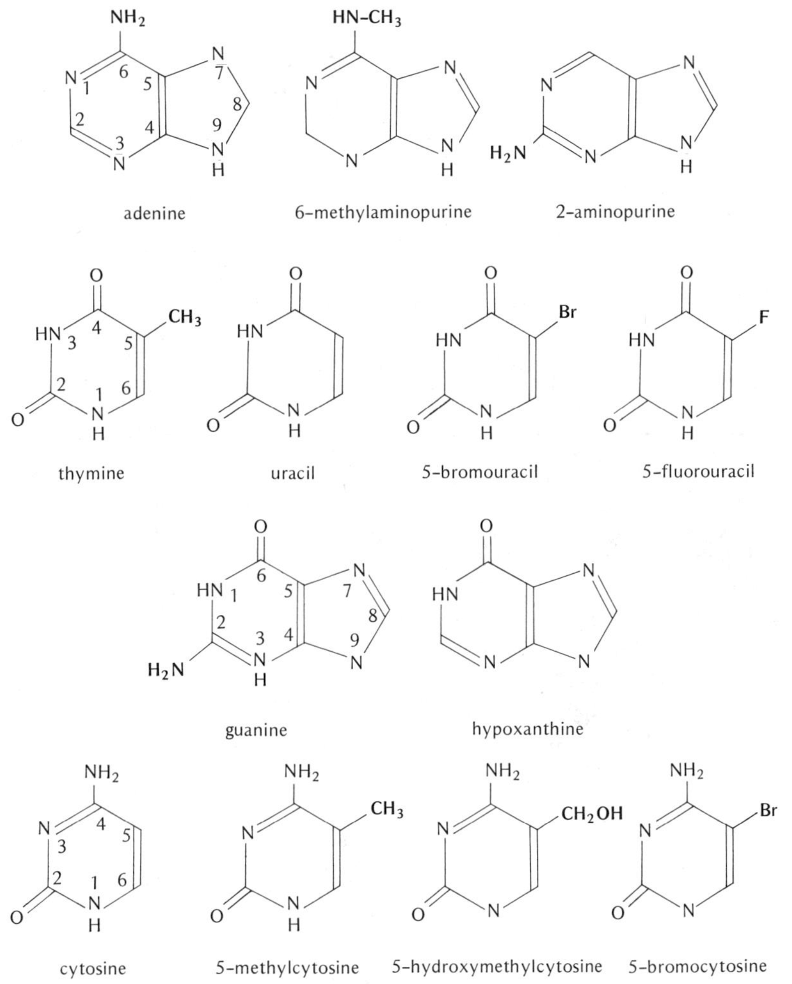

adenine 6-methylaminopurine 2-aminopurine

thymine uracil 5-bromouracil 5-fluorouracil

guanine hypoxanthine

cytosine 5-methylcytosine 5-hydroxymethylcytosine 5-bromocytosine

Figure 5–8

Normal DNA bases and their base analogues.

tides, which are then "zipped" together by the polymerase. Together with other evidence these facts indicate that *in vitro* synthesis occurs through the mechanism of complementary base pairing, and *in vivo* synthesis is probably similar.

The necessity of template DNA for *in vitro* synthesis is not, however, absolute. In the absence of template, and at the end of a long lag period, the polymerase enzyme will nevertheless synthesize polynucleotide chains of DNA that appear to be normal double helixes. In this fashion chains or polymers of synthetic DNAs have been manufactured which have no exact counterpart among living organisms. Two examples are DNA polymers of adenine-thymine ("dAT"), and of adenine-bromouracil ("dA$\overline{\text{BU}}$"). In these synthetic polymers the base sequences along each of the double strands seem to alternate: ATATA . . ., and A$\overline{\text{BU}}$A$\overline{\text{BU}}$A . . ..

Experiments by Wake and Baldwin have indicated that when dA$\overline{\text{BU}}$ is used as a template in a DNA-synthesizing solution containing adenine and thymine nucleotides, hybrid molecules are formed with template dA$\overline{\text{BU}}$ sequences on one

TABLE 5–4

Extent of base pairing between A$\overline{\text{BU}}$:AT, AT:AT, and A$\overline{\text{BU}}$:A$\overline{\text{BU}}$, when adenine and thymine nucleotides are introduced into a solution in which A$\overline{\text{BU}}$:A$\overline{\text{BU}}$ is the template*

Time, minutes	*Extent of newly synthesized material, percent*	*dA$\overline{\text{BU}}$:dAT (hybrid molecules)*	*dAT:dAT*	*dA$\overline{\text{BU}}$:dA$\overline{\text{BU}}$ (template)*
0	0	0	0	100
5	16	28	2	70
9	23	38	4	58
16	37	43	16	41
23	52	49	27	24
33	63	54	36	10
42	68	50	43	7

From Wake and Baldwin, 1962.
* Note that as synthesis proceeds, new A$\overline{\text{BU}}$:A$\overline{\text{BU}}$ template is not formed (because of the absence of $\overline{\text{BU}}$ nucleotides), and the relative frequency of the A$\overline{\text{BU}}$:A$\overline{\text{BU}}$ template gradually decreases.

strand and newly formed dAT sequences on the other. Table 5–4 shows the percentage of these hybrid molecules (dA$\overline{\text{BU}}$:dAT) in the reaction as the synthesis proceeds. Note that as the proportion of hybrid molecules increases, pure dAT:dAT molecules begin to be formed and the dA$\overline{\text{BU}}$:dA$\overline{\text{BU}}$ template appears to be used up. The interpretation of these results is given in Fig. 5–9, which shows how the synthesis of dAT strands would diminish the frequency of dA$\overline{\text{BU}}$:dA$\overline{\text{BU}}$ and increase the frequency of dAT:dAT. These findings therefore support those of the Meselson and Stahl *in vivo* experiments (p. 73) and strongly indicate that *in vitro* replication is semiconservative (formation of hybrid molecules) and nondispersive (formation of pure dAT:dAT molecules).

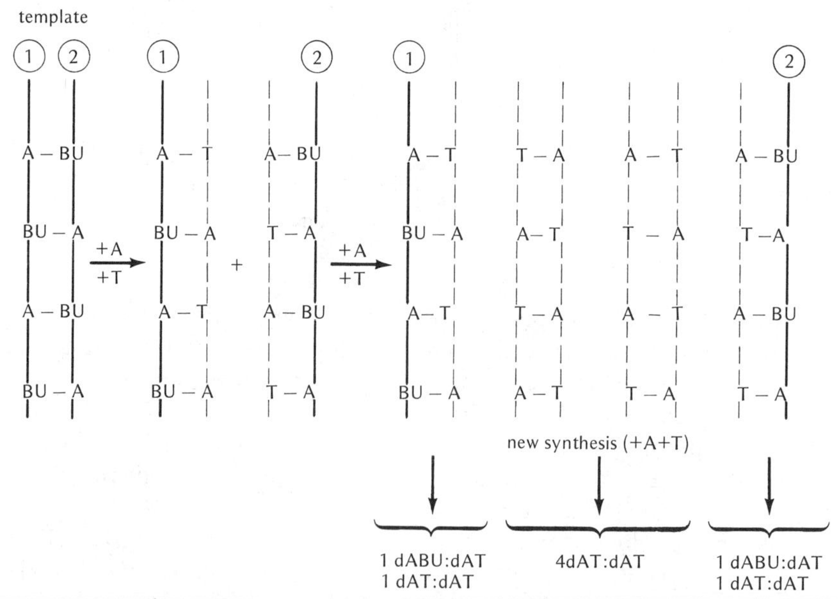

Figure 5–9

Explanation for the reduction in the relative frequency of template molecules and the increase in frequency of dAT:dAT molecules in the experiments of Wake and Baldwin.

DNA POLYMERASES

Central to the issue of how DNA replicates are the enzymes involved. The DNA polymerase which has been discussed up to now is an enzyme about 1000 amino acids long with a diameter of about 65 Å (compared to the 20 Å diameter of the DNA double helix). According to Kornberg and co-workers the enzyme will become associated ("bind") with DNA only if it is single-stranded. When confronted with double-stranded DNA the enzyme will bind only to those regions that may be single-stranded such as a "frayed" unpaired end of a linear DNA duplex, or will bind along the length of the molecule if there are "nicks" in one of the two DNA chains.

Within the enzyme are a number of essential sites that enable it to perform its function (Fig. 5–10). At one site a section of the template DNA is held in place. Another site contains the growing newly synthesized DNA strand complementary to the template, called the *primer*. Right at the tip of the newly growing primer DNA the enzyme bears a site that can match an incoming nucleotide triphosphate to a complementary nucleotide on the DNA template. If the match is successful, that is, if the pairing is A–T or C–G, then the incoming nucleotide is chemically bound by the enzyme to the 3′ position of the primer. Along with these functions there are also sites at which the enzyme has *exonuclease* activity: capability of directionally degrading one of the strands of a DNA double helix into single nucleotides. It appears that the reason that this enzyme can only synthesize DNA in the 5′ to 3′ direction, that is, add new mononucleotides to the 3′ position of the primer, is that the primer terminus site on the enzyme cannot be occupied unless the primer carries at this position a nucleotide that has a 3′ OH terminal group.

The 5′ to 3′ direction of synthesis by the Kornberg enzyme posed serious questions of whether this was the enzyme actually used in normal DNA replication. For if the enzyme can only function along the DNA template in one direction, this means that only one of the parental DNA strands is being continuously replicated while the DNA helix unwinds, whereas the other strand, having opposite polarity, can at best be replicated in discontinuous fashion. As shown in Fig. 5–11, a successive unwinding of template strands A and B leads to synthesis of fragments of 5′ to 3′ DNA sections complementary to that parental strand (B) which is in the 3′ to 5′ orientation. On the other hand, the polymerase on parental strand A can synthesize an unbroken complementary strand no matter how far the DNA double helix unwinds. It seemed inconsistent that the speed and efficiency of replication should differ between the two strands.

Among the further objections raised against complete cellular reliance on the Kornberg enzyme was its relatively slow synthesis of DNA *in vitro* compared to the more rapid rates known to occur *in vivo*: for example, in *E. coli*, there is more than a hundredfold difference in speed between these two types of synthesis. Furthermore, since the enzyme cannot act on double-stranded DNA, either a different enzyme was being used *in vivo* or an additional enzyme was present enabling the double-stranded DNA to unwind. The fact that the enzyme could also function as an exonuclease led to the opinion that its cellular function may really be as a "repair" enzyme; that is, its purpose is to excise loose or defective fragments from one of the two strands of double-stranded DNA and replace these with new complementary strands enabling the restoration of the double helix. Perhaps the most serious question with respect to the *in vivo* function of the Kornberg enzyme was the finding by Cairns and De Lucia of strains of *E. coli* that have less than one percent activity of the Kornberg enzyme, yet still replicate their DNA normally. Since these mutant strains are defective in their ability to repair their DNA when exposed to ultraviolet light, this supports the view that Kornberg's enzyme may function primarily as either an exonuclease or as a repair enzyme, or both.

As a consequence of these findings a search was undertaken for other enzymes capable of polymerizing deoxyribonucleotides. To date, two additional DNA polymerases have been found, named polymerases II and III. In contrast to the Kornberg enzyme (polymerase I), both of these polymerases

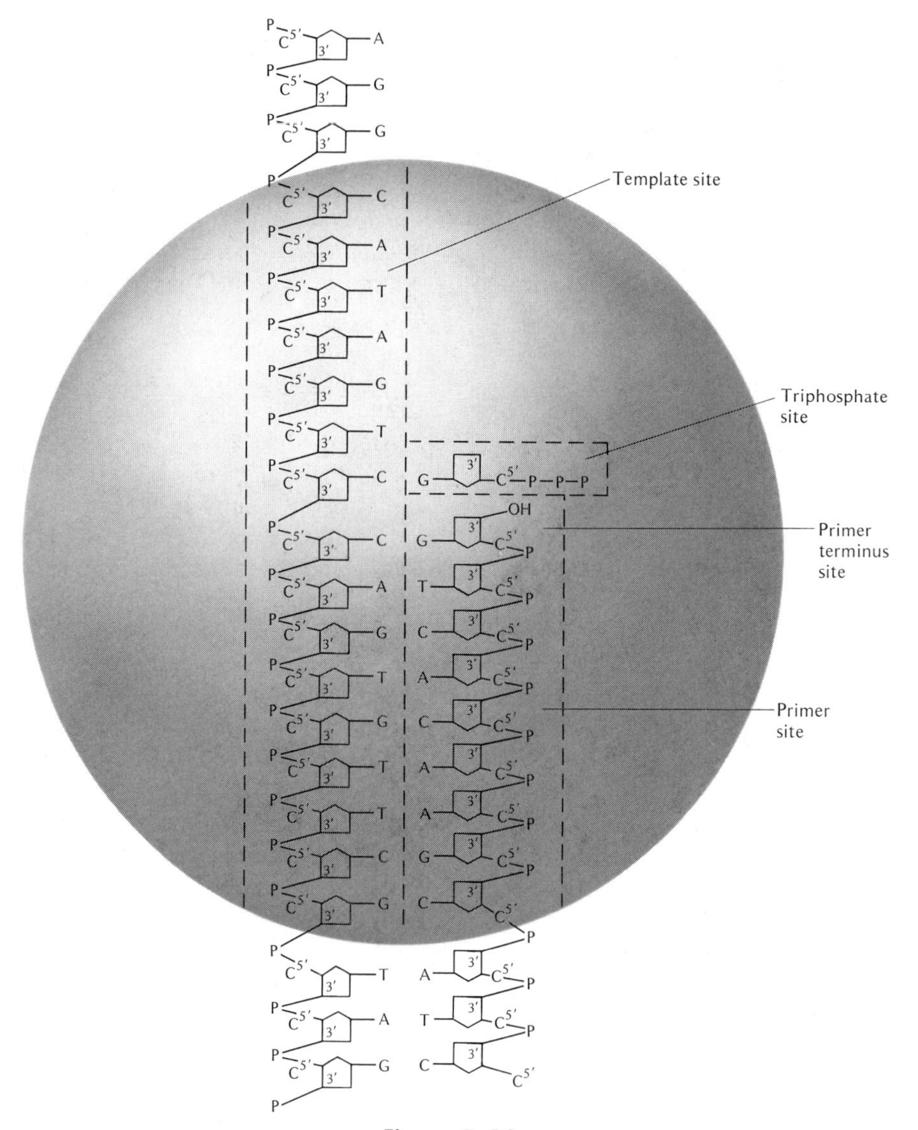

Figure 5–10

Four sites in the active center of DNA polymerase I. As the template strand moves downward on the diagram, complementary deoxyribonucleotide triphosphates which appear at the special triphosphate site are chemically bound to the terminal nucleotide of the primer strand. This reaction removes two phosphate groups from the 5′C position of the entering nucleotide and binds this nucleotide to the oxygen atom at the 3′C position of the primer. The primer strand thus grows from the 5′ to the 3′ direction. As mentioned later in the text (p. 90), it has been suggested that the initial primer sequence in, at least some organisms, is composed of a sequence of RNA nucleotides that pair base-for-base with a special DNA sequence in the template strand at which replication begins. Further DNA nucleotides complementary to the template are then added at the 3′ position of the primer as the template moves along. After DNA synthesis is completed, the RNA section, where present, is removed by exonuclease action and replaced by DNA. Because DNA polymerase I possesses such exonuclease activity, it is believed it may function in this ''repair'' role. (After Kornberg, 1969.)

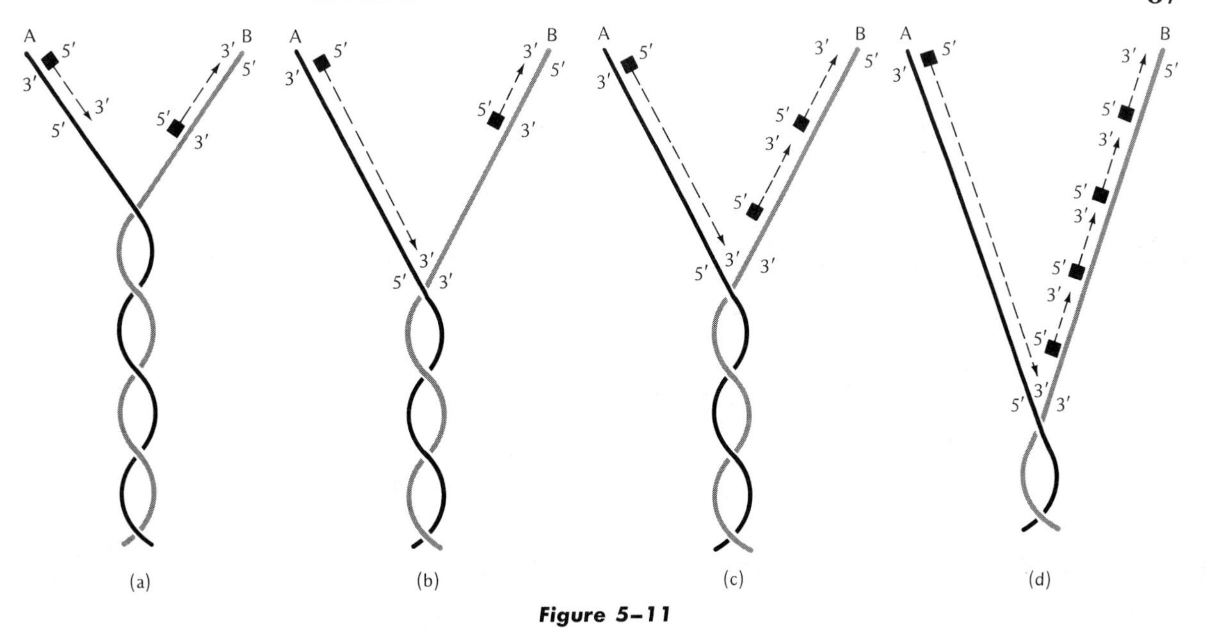

Figure 5–11

Sequence of unwinding and replication at the fork of a DNA double helix caused by the action of polymerases which synthesize DNA strands in the 5′ to 3′ direction. (a) Unwinding of the double helix allows the polymerase to attach at two points indicated by the solid blocks, and to synthesize new complementary strands in the directions indicated by the arrows. (b) As further unwinding occurs, a section of the B strand remains single-stranded, until a polymerase enzyme locates itself near the fork and fills the gap (c). With successive unwinding of the double helix, the polymerase enzyme on strand A can therefore act continuously whereas strand B enzymes would only produce short fragments (d). The initiation points for DNA synthesis shown on this diagram as small solid blocks may, in some cases, be the special RNA primers referred to in previous Figure 5–10, and also on p. 90.

seem to be physically associated with membrane complexes in the cell. It now seems clear that polymerase III is necessary for *in vivo* DNA replication (see Tait and Smith), and that polymerase I may be used to help terminate replication (see Shekman, Weiner and Kornberg).

THE GROWING POINT

Various models for the replication of DNA have been proposed to account for some of the physical and chemical observations. In one model replication begins at a specific initiation point in a closed-circular DNA double helix at which a "nick" has occurred in one of the two strands. The nick, produced by an *endonuclease* enzyme which cuts a single DNA strand at specific points along its length, enables the cut strand to unwind thereby forming two single-strand templates. As replication proceeds, DNA synthesis may go in one direction or both directions from the initiation point (Fig. 5–12), although most evidence now indicates that replication is probably bidirectional. In either case, a forklike Y-joint appears at one side of the separation between the two parental strands and another such Y-joint appears further down the molecule. The circular appearance of the initial parental DNA has therefore changed into a θ (theta)-shaped structure.

It has also been suggested (Gilbert and Dressler) that, since replicating phage DNA is commonly found associated with bacterial membranes, the initial nick in a phage DNA strand causes this strand to open up and become attached to the cell membrane. The remaining strand, still in circular

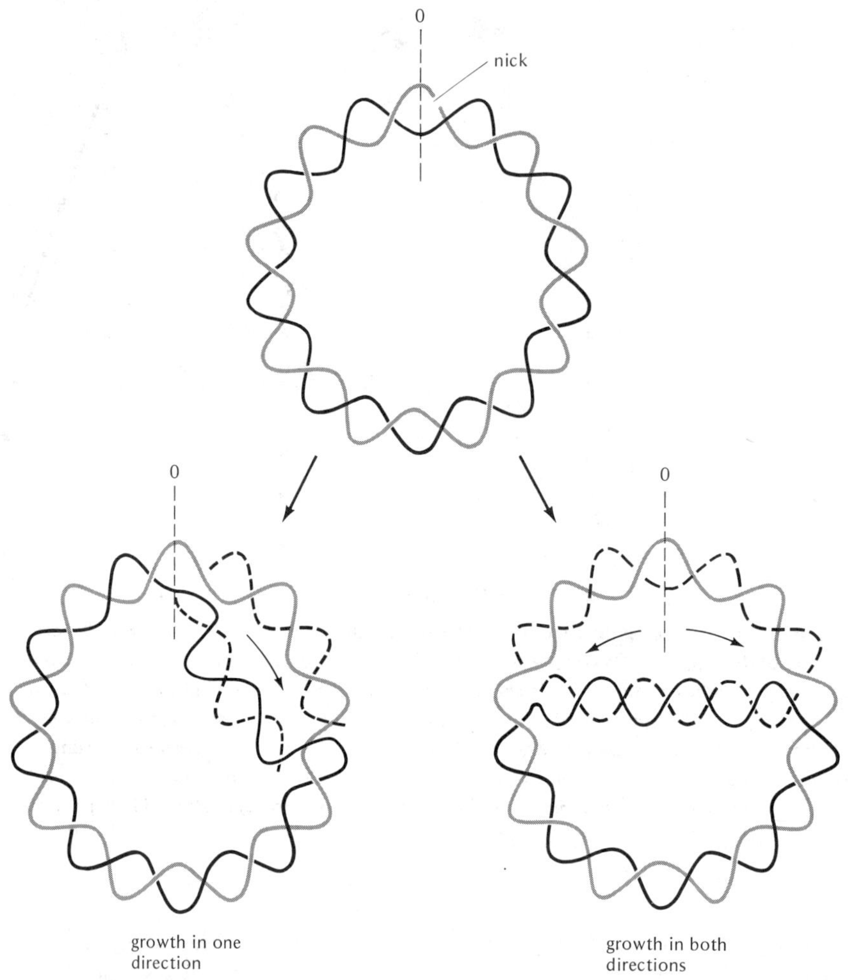

Figure 5–12

Replication model for a circular DNA double helix. 0 represents the initiating point at which replication and unwinding begins via a "nick" produced by an endonuclease enzyme. Whether growth continues in only one direction or in both directions, the molecule in both cases will go through an intermediate phase during which it appears as a θ-shaped structure.

form, then functions as a template to generate a complementary strand as it "rolls" away from the membrane attachment point (Fig. 5–13). By this means a long DNA strand can be generated which remains attached to the membrane and which, in turn, through complementary base pairing, can now generate a new double helix. This *rolling circle* model has been used to explain some observations found in the replication of ϕX174 and, with some modification, the transfer of the *E. coli* sex factor. Graphic support for this replication model has also been obtained from electron micrographs of DNA replication in the African clawed toad *Xenopus laevis*. These micrographs show the presence of rolling circle intermediates in the replication of a fraction of DNA specifically associated with the production of ribosomal RNA (Hourcade et al.). Nevertheless the finding that numerous organisms such as *E. coli*, *Bacillus subtilis*, λ (lambda) virus, polyoma virus, SV40 virus, etc.,

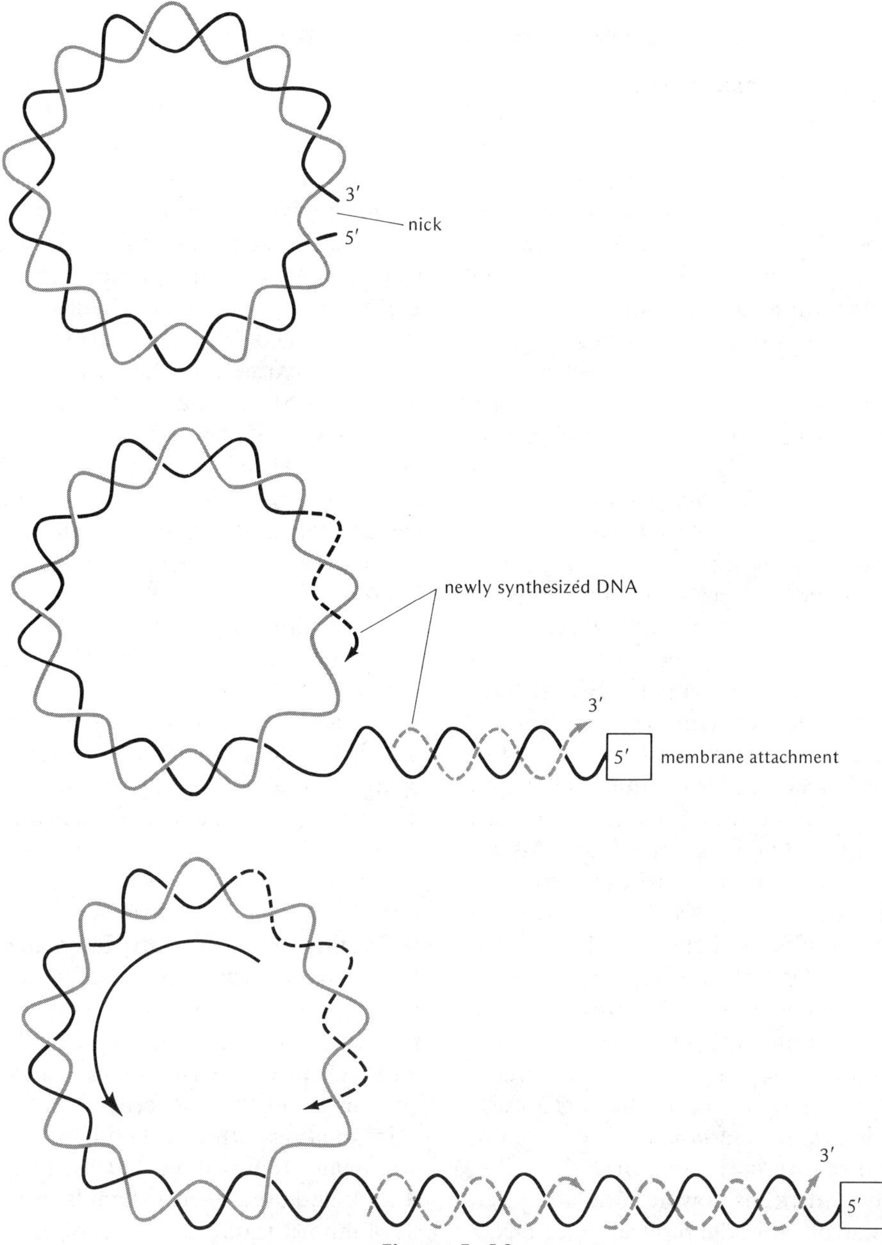

Figure 5–13

Rolling circle replication model of circular DNA double helix. A nick in one of the strands allows the 5′ end to become attached to a membrane while the other end is free to rotate about the unbroken DNA strand. DNA synthesis then begins at the 3′ end of the nicked strand forming a complement to the unbroken parental strand. As synthesis proceeds, the unbroken parental DNA rolls and unwinds, thus extending the template upon which new DNA is synthesized. Meanwhile the "tail" strand attached to the membrane is also lengthened during this process and provides a template upon which complementary DNA can be synthesized. Note that since the complement to the tail strand has its 3′ end toward the membrane, synthesis must proceed in discontinuous fragments. That is, unwinding of the tail strand can act as a template for a section of complementary DNA synthesized in the 5′ to 3′ direction, but further unwinding necessitates repositioning of the polymerase enzyme in order to synthesize a new complementary section. Thus, this model, as that of the Y fork (see Fig. 5–11) must have discontinuous DNA synthesis on at least one strand if it utilizes the known DNA polymerases.

possess observable θ-shaped replicating DNA molecules, is difficult to reconcile with a rolling circle model. Furthermore, bidirectional replication is now also known to occur in eucaryotes such as *Drosophila* (Kriegstein and Hogness) and in the normal chromosomes of *Xenopus* and *Triturus* (p. 94). It is therefore possible that each of these models applies to different organisms, or that some organisms can utilize both of these replication modes for different purposes.

The paradox that known polymerases synthesize new DNA strands only in the 5′ to 3′ direction (see p. 85) has been partially resolved by evidence obtained over the last few years indicating that DNA is synthesized in discontinuous fragments on at least one strand of a replicating double helix. Okazaki and co-workers have shown that tritiated thymine that is permitted to be incorporated into the DNA of *E. coli* during very short intervals ("pulses") can be extracted as part of short fragments of DNA no longer than 1000 to 2000 nucleotides. If incorporation of the label is allowed to proceed for longer periods of time, the tritium is now found as parts of much longer nucleotide chains. Obviously the initial shorter fragments are joined together into longer strands as time continues. A joining enzyme, *DNA ligase*, capable of forming a chemical bond between adjacent single-strand nucleotides that have been separated by a nick, is believed to serve this purpose. Since Okazaki's discovery the discontinuous replication of DNA in short fragments and their subsequent incorporation into long molecules has been demonstrated both *in vitro* and *in vivo*, and in mammalian as well as in procaryotic cells.

Surprisingly evidence is now accumulating that RNA may be involved in the replication of DNA, either as a primer or in some essential mechanical role. In numerous instances so far (bacteriophages ϕX174, M13, P2, as well as the DNA replication of *E. coli* and even chickens), newly synthesized fragments of DNA are found as chemically bonded extensions of even shorter fragments of RNA. The RNA fragments, estimated to be 50–100 nucleotides long, are then removed before the longer DNA fragments are joined together by the ligase enzyme (see, for example, Keller, and Sugino, Hirose, and Okazaki).

IN VITRO SYNTHESIS OF BIOLOGICALLY ACTIVE DNA

The analytical power provided by the many ingenious methods used in unraveling the mechanisms of nucleic acid replication enabled, by 1965, the *in vitro* synthesis of a biologically active nucleic acid. This impressive achievement involved at first the synthesis of the nucleic acid of an RNA virus (p. 105), and was extended within two years to a DNA virus as well. For DNA, Goulian, Kornberg, and Sinsheimer showed that the nucleic acid of the bacteriophage ϕX174 could be replicated in the test tube, and that the *in vitro* product formed was as active and infective as normal viral DNA.

As explained previously (p. 58), ϕX174 is a small virus containing a single circular strand of DNA that infects *E. coli*. In normal infection this "plus" strand acts as a template upon which a complementary "minus" strand is synthesized. These two complementary strands, plus and minus, compose a closed circular double helix called the replicative form, or RF. For about the next 10 minutes the RF molecule generates 15 or more double-stranded daughter helixes via semiconservative replication. After this early interval each RF molecule is replicated asymmetrically, so that only the minus strand serves as a template. New plus strands synthesized this way are then packaged into viral protein envelopes that have meanwhile been made in the host cell.

In their experiment Goulian, Kornberg, and Sinsheimer took tritium-labeled (^{3}H) ϕX174 plus strands and used them as templates for the synthesis of minus strands in a reaction mixture containing Kornberg's DNA polymerase enzyme. In the minus strands, however, the usual thymine was replaced by its heavier analogue 5-bromodeoxyuridine (BU), and radioactive phosphorus (^{32}P) was used to differentiate these strands from the parental tritium-labeled plus strands (Fig. 5–14). Furthermore, in order to allow the minus strands to form a chemically bonded circle, DNA ligase was added.

Upon completion of this stage of the synthesis the double helixes formed were briefly exposed to

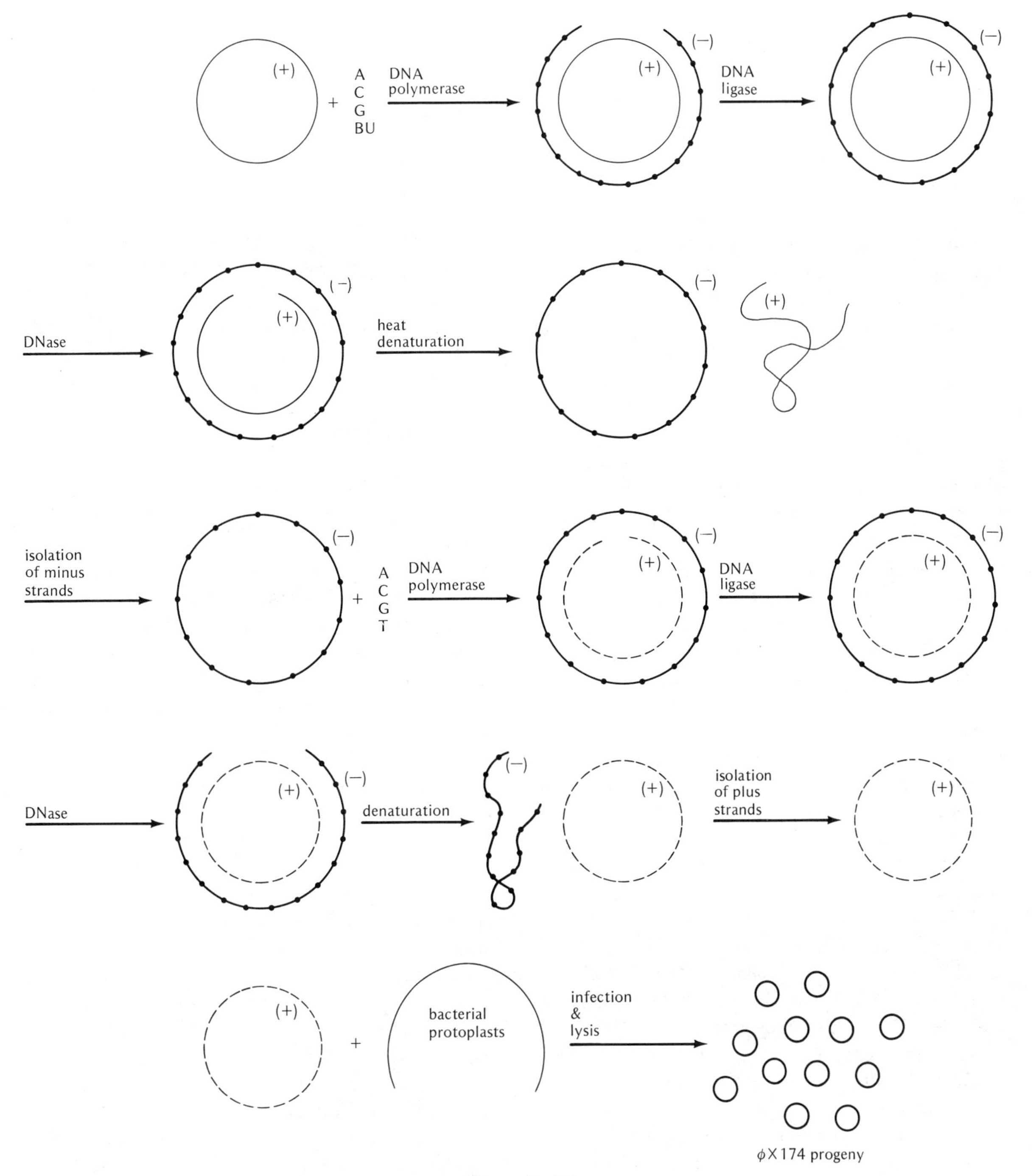

Figure 5–14

Diagrammatic representation of the in vitro synthesis of biologically active DNA from ϕX174, as explained in the text. For simplicity, some products and steps have been omitted. DNase, for example, will nick both plus and minus strands, but these are discarded, and only unnicked strands carrying the desired label are used. (Adapted from Goulian, Kornberg, and Sinsheimer.)

DNase and then denatured by heat to allow isolation through centrifugation of the heavier circular minus strands. These minus strands are characterized by the presence of BU and ^{32}P, and by the fact that they do not contain any of the parental ^{3}H label. Again, a reaction mixture was prepared with DNA polymerase and ligase, but this time the product was a completely synthetic DNA molecule just like the natural replicative form. Treatment with DNase then permitted separation of the newly synthesized plus strands, and these were now shown to be fully infective when exposed to bacterial protoplasts or spheroplasts (see p. 66).* The progeny of such infections were normal ϕX174 viruses.

A further significant achievement has been the chemical synthesis of defined nucleotide sequences undertaken by Khorana and co-workers. The first synthesis of a relatively long DNA sequence of biological importance has been that involved in producing a particular type of transfer RNA in yeast (tRNA^{ala}). This transfer RNA molecule is 77 nucleotides long, and according to the rules of transcription (p. 68, and p. 103) has each of its nucleotides complementary to a nucleotide in a DNA strand. Knowing the sequence of nucleotides in the transfer RNA molecule thus enabled Khorana and his colleagues to synthesize a sequence of DNA nucleotides exactly complementary to it. In effect, such synthesis represents the first creation of a gene (DNA) that specifies a particular cellular product (tRNA).

This synthesis has made use of a number of important techniques. As shown in Fig. 5–15, small sections of DNA were synthesized chemically, nucleotide by nucleotide, so that some nucleotides in one section were complementary to those in another section. Base pairing then enabled a number of these sections to form a double helix. When joining enzyme (ligase) is added, the individual sections become bonded together into long chains, one of which is the exact complement to the nucleotide sequence in the transfer RNA product. Through further development of these methods it is hoped that DNA synthesized *in vitro* will eventually be used to create other specific cellular products.

* In ϕX174 both the plus and minus strands, as well as the replicative form, are capable of infecting spheroplasts, although the plus strand is the normal infective nucleic acid. The DNA synthesized *in vitro* also had these capabilities.

THE CHROMOSOMES OF EUCARYOTES AND DNA REPLICATION

In the early 1950s a number of experiments performed by Swift and others indicated that the replication of DNA in eucaryotic cells, both for mitosis and meiosis, occurs during the interphase period. The period of DNA synthesis was called the "S" (synthetic) period by Howard and Pelc and was separated in time from the previous cell division by a gap called G_1 (Fig. 5–16). After DNA synthesis a further gap called G_2 was found to occur before the next cell division began. The G_1 period showed considerable variability, often ranging from 3 to 4 hours to days, weeks, or months, and appeared to depend upon the kind of cell involved and its physiological condition. As was true of the S period, the G_2 period showed more constancy for a given type of cell; G_2 usually ranging from 2 to 5 hours, and S lasting about 7 to 8 hours. The "trigger" for DNA synthesis and subsequent cell division seemed to occur during the G_1 period, and recent findings support this: specially prepared nuclei exposed to cytoplasm from the G_1 period enter immediately into S phase, whereas exposure to the cytoplasm of other cell stages has no effect (Kumar and Friedman).

To tie together the replication of DNA and the replication of chromosomes in a more precise fashion required the accurate labeling of chromosomes and tracing the exact transmission of this label to their products in cell division. This step was undertaken by Taylor, Woods, and Hughes who labeled dividing root-tip cells of the broad bean *Vicia faba* with radioactive (^{3}H) thymidine, a tracer that is rapidly and specifically incorporated into the newly synthesized DNA of the chromosomes (p. 62). The kind of replication that occurs could then be determined by observing the label-

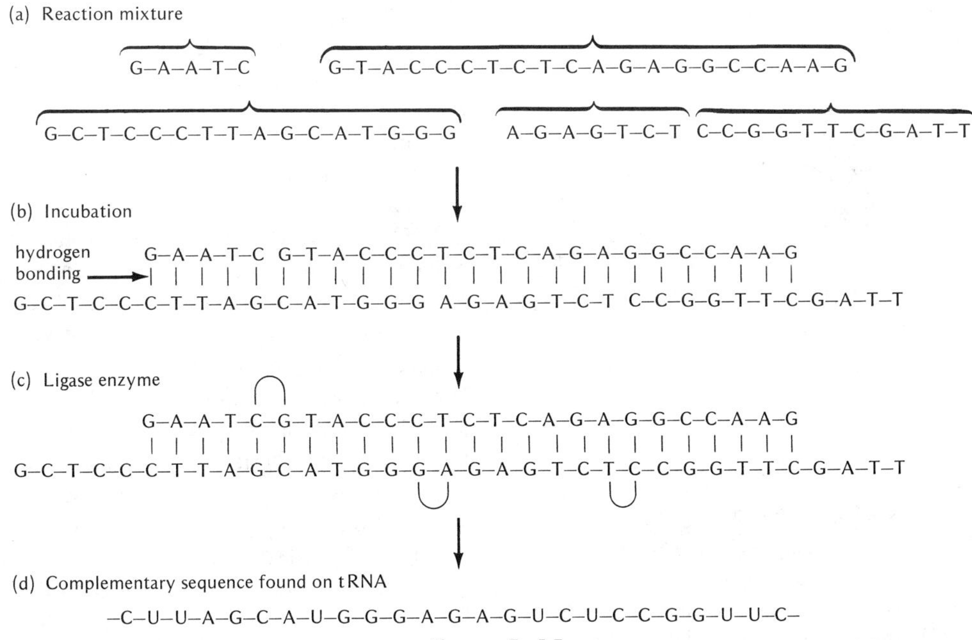

Figure 5–15

Some of the stages in the synthesis of a part of the DNA double helix which is responsible for determining the sequence of nucleotides in a particular type of yeast transfer RNA (tRNA^{ala}). The reaction mixture begins with component segments of DNA (a) which can be annealed by complementary base pairing to form a double-stranded section (b). The joining enzyme, polynucleotide ligase, then chemically bonds together adjacent segments (c). Other parts are synthesized in a similar way and they are then all joined together to form a DNA double helix 77 nucleotides long, of which one strand (upper) is exactly complementary to the nucleotide sequence in the transfer RNA molecule (d). [Actually, as shown later in Fig. 27–8, some of the bases in the transfer RNA molecule are modified in the cell after they are hooked together. It is nevertheless relatively easy to determine the initial unmodified base sequence, and it is this which is given in (d)] (The nucleotide sections in this figure are those given by Khorana et al. for their part "B" of the DNA sequence.)

ing of newly formed chromosomes. In semiconservative replication one would expect that half of each new chromosome is synthesized as a complement to half of the divided parent chromosome. Therefore, in the presence of radioactive substrates, a new chromosome replicating semiconservatively appears as labeled because it consists of half radioactive newly formed chromosome and half parental chromosome. This is precisely what Taylor and his associates found (Fig. 5–17).

In addition, when permitted to grow and divide on unlabeled media, half of the new chromosomes that were formed from such labeled chromosomes were shown to be labeled and the other half were unlabeled. Thus random labeling of chromosomes did not occur, and the mode of duplication was consequently not dispersive. The chromosome, therefore, appears to divide as though it too were a "double" structure like the Watson-Crick DNA helix. Similar labeling experiments, done with the chromosomes of other plants and animals, have also been shown to be interpretable on the basis of semiconservative replication. These experiments have therefore raised the possibility that the semiconservative replication of the eucaryotic chromosome is based essentially on an underlying structure that is a single continuous length of DNA double helix.

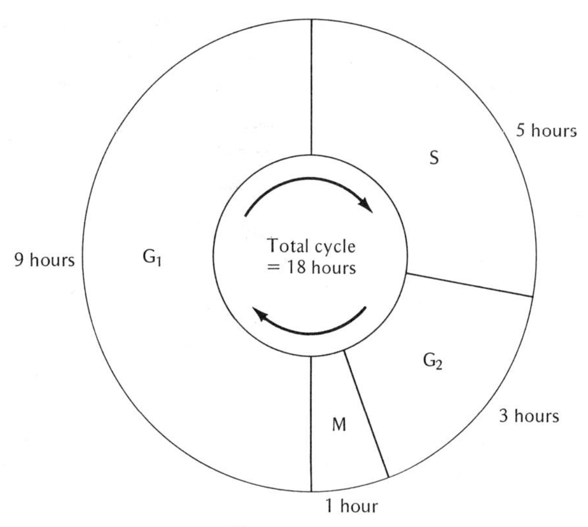

Figure 5–16

Hypothetical cell cycle showing relationship between the period of DNA synthesis (S) and mitosis or meiosis (M). For any given cell type, the G_1 interval (gap between cell division → DNA synthesis) is usually variable, whereas G_2 (gap between DNA synthesis → cell division) and S are fairly constant.

At present this "unineme" view of DNA organization of the chromosome* is supported by various experiments including the following:

1. Gall (1963) showed that the rate at which the enzyme DNase breaks down the thin lateral loops of lampbrush chromosomes indicates a two-stranded structure, i.e., one double helix. In the interchromomere regions of these chromosomes (see Fig. 2–14) breaks occur at a rate showing a four-stranded structure (i.e., two double helixes). Both of these findings make sense if each chromatid in the lampbrush chromosome is only a single double helix, and only when two chromatids are together (interchromomere regions) are there two DNA double helixes.

2. The unineme structure of lampbrush chromosomes is supported by the electron microscope studies of Miller who showed that after protein digestion the remaining thickness of the lateral loops of single chromatids was about equal to the 20 angstrom diameter of a DNA double helix, and the interchromomere widths were, as expected, approximately twice this value (i.e., two chromatids).

3. Callan showed that two amphibian species having a tenfold difference in amount of DNA are probably different in the length of their DNA molecules rather than in the number of DNA molecules. This would certainly be expected if the chromatid were composed of only a single long folded DNA molecule, and chromatids with more DNA had merely longer molecules rather than more side-by-side replicates of the same molecule. The method used by Callan was to compare the number of initiation points for DNA synthesis with the number of replicating units both for *Xenopus* (3×10^{-12} g DNA) and for *Triturus* (30×10^{-12} g DNA) using the fiber autoradiography techniques developed by Huberman and Riggs. This involves radioactive labeling of chromosome material, removal of proteins by enzyme digestion, and the gradual deposition of the DNA fibers on filters. Autoradiographic film is then laid over this material and the film exposed for about four months or more. These studies provide excellent resolution of individual strands and show that DNA molecules begin replication at a number of initiation points along their length. As diagrammed in Fig. 5–18, growth occurs in a bidirectional fashion from these points causing "bubbles" (see also p. 87).

In Callan's experiments the replicating intervals between initiation points was found to be about ten times longer in *Triturus* than in *Xenopus*. (Were the chromosome a multineme structure, the lengthened intervals between initiation points would seem unnecessary, since the increased amount of DNA would be accommodated by more molecules, each with the same replicating intervals.) The replication speed in *Triturus* was, however, more than twice that of *Xenopus*, and replication lasted about four times longer in *Triturus*. Callan interpreted this to mean that *Triturus*, with its increased DNA, has longer initiation intervals

*The "multineme" view is that there are two or more lengths of independent DNA double helixes in a chromosome. This view is supported by microscopic observations of what seem to be multiple strandedness in chromatids (see p. 25). Also, the fact that there are instances of large differences in amounts of DNA between related organisms has been interpreted to mean that some species occasionally possess many new side-by-side repetitions of DNA helixes.

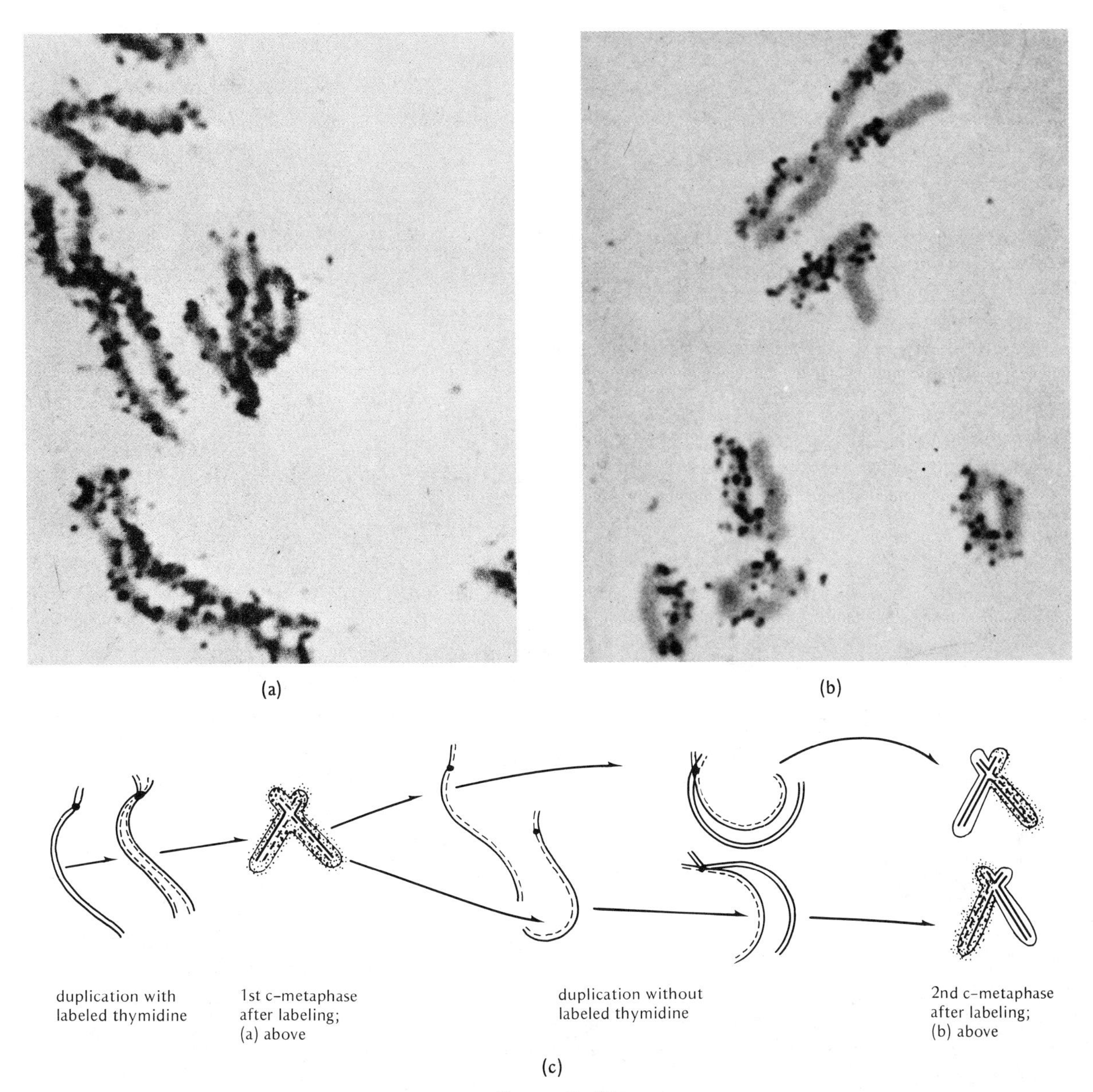

Figure 5–17

(a) Autoradiogram of Vicia faba chromosomes after one replication in radioactively labeled medium. (b) Appearance of such labeled chromosomes after an additional replication in unlabeled medium. (c) Diagrammatic interpretation of the observed events in a and b. Broken lines indicate labeled subunits, solid lines indicate unlabeled subunits. The term "c-metaphase" indicates use of the chemical colchicine to inhibit the formation of spindle fibers and thereby prevent separation of centromeres. Colchicine treatment identifies the two products of a single chromosome replication since they remain attached together at the centromere. (From J. H. Taylor, 1963. The replication and organization of DNA in chromosomes. In Molecular Genetics, Part 1, J. H. Taylor (ed.), Academic Press, New York, pp. 74 and 75.)

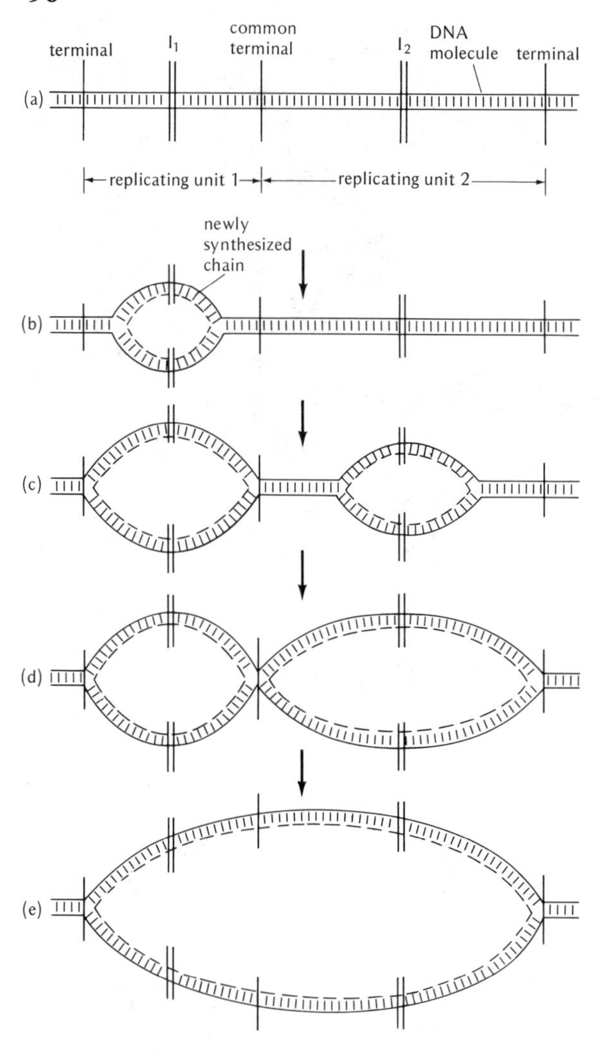

Figure 5–18

Model for DNA replication in two adjacent replicating units in a eucaryotic chromosome as determined by fiber autoradiography. I_1 and I_2 represent respective initiation points for DNA synthesis in units 1 and 2, and are each marked by the appearance of a bubble when replication begins. The termination points for each replicating unit are indicated by a line crossing the DNA molecule at each unit end. (a) Before replication. (b) Replication has begun in unit 1 but not in unit 2. (c) Replication has been completed in unit 1 and begun in unit 2. (d) Replication has been completed in both units. (e) Daughter helixes separated at the common terminal and linked to adjacent helixes. (Adapted from Huberman and Riggs.)

(i.e., longer length) and therefore replicates for a longer period even though at a more rapid rate. If we multiply replication period × rate (4 × 2), this seems sufficient to give *Triturus* about ten times the length of DNA in *Xenopus*, and confirms the observed data. Also, since the number of chromomeres in the meiotic cells of *Triturus* (between 3000 and 4000) is about equal to the calculated total number of initiation points, Callan suggests that the unit of replication may be a single chromomere.

4. Another indication for a folded unineme structure for the eucaryotic chromosome has come from a somewhat unexpected source: the folded chromosome of bacteria. As described previously (see Fig. 5–4), the *E. coli* chromosome is composed of a single circular and continuous DNA double helix about one millimeter long. Since the cellular area occupied by this molecule has a diameter of at most only one micron, the bacterial chromosome must obviously be very stably folded in order to remain within this confined space in the absence of a restraining nuclear membrane. Recently, through the efforts of Stonington and Pettijohn, *E. coli* chromosomes have been extracted in folded condition and shown to contain mostly DNA with some RNA and protein. The structure of these chromosomes, as analyzed by Worcel and Burgi, indicates that they each consist of about 40 to 50 loops held together by RNA. Although the chromosome runs as a continuous molecule throughout all of these loops (Fig. 5–19), each loop is separately tied to RNA and can act independently of others by maintaining different degrees of subsidiary coiling ("supercoils," Fig. 5–19c, d, e). This tertiary loop structure is apparently controlled by the number of enzyme-induced nicks in single strands of the DNA molecule allowing rotation and unwinding for purposes of both replication and transcription. In its fully compacted form, Worcel and Burgi suggest, each loop may have a large number of DNA supercoils, each coil composed of 400 base pairs. The remarkable organization of this folded procaryotic chromosome shows that it is quite possible for the even longer eucaryotic chromosome to maintain a continu-

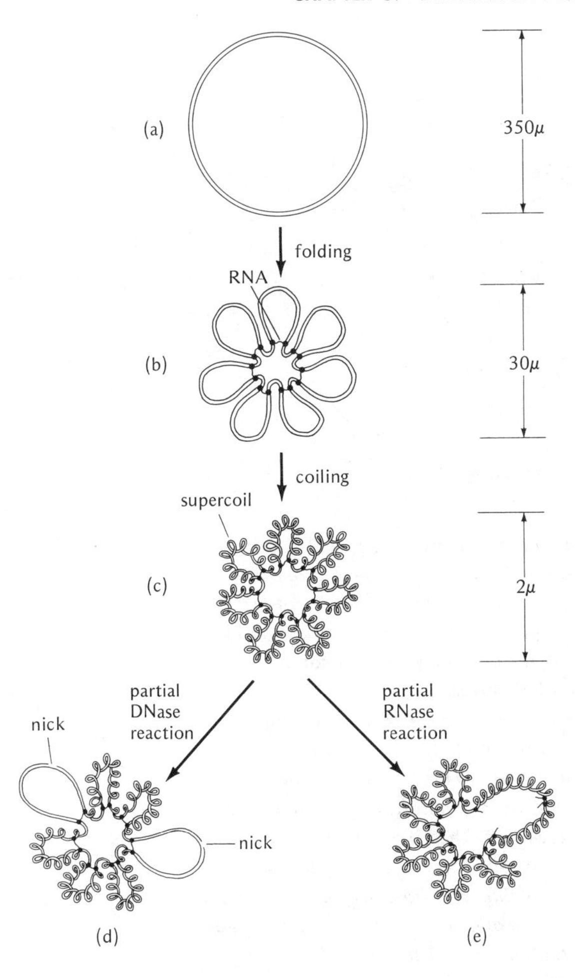

Figure 5–19

A model for the structure of the E. coli *chromosome. (a) Completely unfolded structure of the continuous circular chromosome, similar to that seen in Figure 5–4. (b) The chromosome folded into seven loops attached by RNA connections. (There are actually many more loops, but for diagrammatic simplicity only seven are shown.) (c) The loops undergo subsidiary coiling leading to a supercoiled chromosome. The increase in coiling in the sequence (a) → (c) is accompanied by a reduction in size, until the chromosome reaches a diameter of approximately 2 microns. (d) Endonuclease enzymes can cause a single-strand "nick" in a coil, enabling a supercoiled section to untwist ("relax") without affecting the supercoiling in other loops. (e) A ribonuclease enzyme can attack the RNA connections of two adjacent loops causing them to coalesce without affecting their supercoiling. (After Pettijohn and Hecht.)*

ously long, yet folded, and compact structure (e.g., see Fig. 2–11).

REPEATING DNA SEQUENCES

Soon after the cesium chloride gradient technique began to be used with DNA, it was found that not all the DNA of a particular organism banded uniformly at one position in the density gradient. In mice a small "satellite" band of lighter-weight DNA was found, whereas guinea pigs and cattle also had satellite bands of heavy DNA. There were also considerable differences in the amount of satellite DNA found in related organisms such as 12 percent in the mouse and none in the rat. In the crab *Cancer borealis* about 30 percent of the DNA was in a satellite band consisting almost entirely of a repeating linear sequence ATATAT . . . . . .

The general structure of most satellite DNA remained unknown until a technique was devised by Britten to measure the rate at which the DNA of an organism, sheared to specific sizes and then separated into single strands, reassociates into double strands. Britten found that this reassociation rate is a measure of the complexity of the DNA—that is, DNA consisting of short repeated sequences reassociates more rapidly than DNA of greater complexity with very long "unique" sequences. We can envision the process as one in which a pool of single DNA strands can move freely about, and only those strands that have sections of mutually complementary nucleotide sequences, base-for-base, anneal together to form double helices. Obviously, when there are many strands carrying similar sequences, complementary pairs can "find" and stick to each other more rapidly than when each strand consists of only a relatively rare sequence.

As shown in Fig. 5–20 the most rapidly reassociating nucleic acids are those consisting entirely of complementary strands of single repeating nucleotides (poly U: UUU . . . ; and Poly A: AAA . . .). This finding is certainly expected, since any poly U fragment can complement and anneal

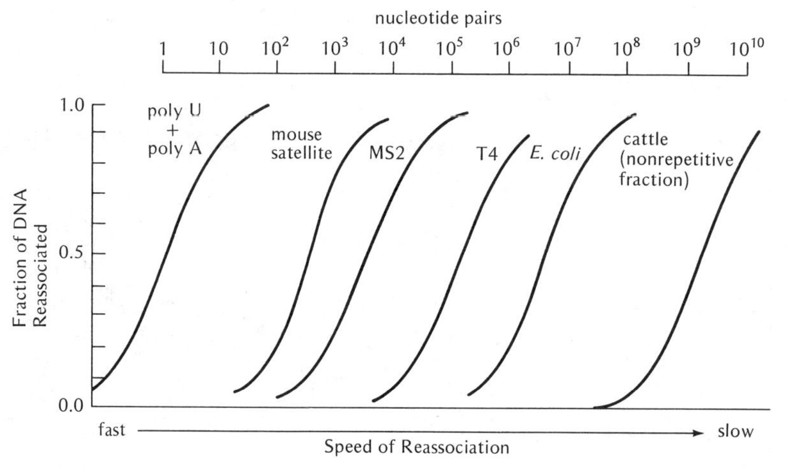

Figure 5–20

Reassociation of double-stranded DNA from various sources and RNA from MS2 virus. The speed of reassociation is measured in terms of the time a given concentration of nucleic acid takes to reassociate at a standard temperature (about 60° C). (In the experiments performed, the unit of measurement is concentration in moles per liter × time in seconds, called Cot.) *Each nucleic acid shows that it contains some fractions that reassociate rapidly and some that reassociate more slowly. Nevertheless each of the nucleic acids in the figure can be characterized as having a rate of reassociation close to the fraction at which one-half of its nucleotides are reassociated (0.5 on the ordinate). Using the known numbers of nucleotide pairs in T4 and* E. coli, *a scale has been constructed above the figure giving the numbers of nucleotide pairs for specific speeds of reassociation. Thus the "repetitive" mouse satellite DNA shows a rate of reassociation equivalent to a sequence that repeats itself every 300 nucleotides, whereas MS2 virus has a unique sequence of about 4000 nucleotides long, and poly-U + poly-A are, of course, restricted to a repeating sequence only one nucleotide long. (Adapted from Britten and Kohne.)*

to any poly A fragment on immediate contact. However, the rate of reassociation slows down for the MS2 and T4 viruses, slows down further still for *E. coli* DNA, and, among the species shown, is relatively slowest for cattle DNA. Since the numbers of nucleotides in some of these DNA's are known (see Fig. 4–7), it became clear that there was a simple linear relationship between the number of nucleotides in the DNA and its reassociation rate (see upper scale Fig. 5–20). Surprisingly the satellite DNA of the mouse reassociated even more rapidly than the DNA of the T4 virus, indicating it was composed of much simpler DNA sequences. If we follow the upper scale in Fig. 5–20, mouse satellite DNA has repeating sequences of about 200 to 300 base pairs long. This finding is not uncommon. European wood mice have satellite DNA's which show even shorter repeating sequences, estimated as between 6 and 20 base pairs long. In some species, such as guinea pigs, there may be more than a million satellite repeats of a small simple sequence, such as CCCTAA/GGGATT, joined together end to end into long molecules.

Through the pioneering efforts of Jones and of Pardue and Gall, information is being obtained for the chromosome locations of some of the highly repetitive satellite DNA. Their procedure involves removing protein and denaturing the DNA of cells into single strands by heat or alkali while they are on a microscope slide. Satellite DNA of the particular organism tested is meanwhile isolated by density-gradient centrifugation and used to synthesize complementary radioactive strands of RNA. The radioactive RNA is then added to the slide so that it can anneal or "hybridize" with the complementary DNA in the cells. Autoradiographs (p. 62) are then made and the localization of the radioactive material in the cells can then be observed. By such means large fractions of satellite DNA are found to be localized primarily in heterochromatic regions of the chromosomes, mainly adjacent to the centromeres. Its precise function, however, remains unknown, although there are suggestions that DNA fibers near the centromere regions may act to hold sister chromatids together.

Not all the repetitive DNA sequences, however, occupy such specific positions, nor is all repetitive DNA of only one kind. In *Xenopus laevis*, for example, somewhat more than half of the DNA is represented by single-copy nonrepetitive sequences and, of the repetitive remainder, 13 percent represents sequences, each of which is repeated 20 times throughout the DNA complement; 67 percent are sequences, each of which is repeated about 1600 times; and the rest consists of "satellite" sequences which are repeated 30,000 times or more. Davidson and co-workers suggest that the nonrepeating single-copy DNA sequences in *Xenopus*, which are about 800 nucleotides long, alternate throughout the chromosomes with repeating sequences about 300 nucleotides long. It is these interspersed repetitive sequences that represent the bulk of repetitive DNA rather than the smaller satellite fraction, and are probably used in the regulation of gene activity (Chapter 30).

RNA

The distinctive features of RNA are, as we have seen, its special ribose sugar, and the substitution of the pyrimidine base, uracil, for thymine. Also, RNA is found throughout the cell, in contrast to DNA, which is concentrated in the nuclei.

In general three types of cellular RNA have been distinguished: nuclear, ribosomal, and soluble, the first in the nucleus and the last two largely in the cytoplasm. Although the ratios may vary between tissues, in some cells the nuclear and soluble portions consist of about 10 percent each of the total RNA, with the remaining 80 percent in the ribosomal portion. Each portion of RNA has been found to represent an essential function in the synthesis of proteins and in the transfer of information from DNA. As has been previously mentioned (p. 68) and will be more thoroughly discussed later (Chapter 27), nuclear RNA is formed as a complementary strand to sections of DNA base sequences. Some of this RNA passes through the nuclear membrane as a "messenger" to the cytoplasmic ribosomes, which contain an RNA-protein mixture of their own ("ribosomal" RNA). Once attached to the ribosomes the messenger RNA serves as a template for the interconnection of different amino acids (see Fig. 27–9) that are carried to this template by the small soluble molecules of "transfer" RNA (70 to 80 nucleotides long).

In RNA the usual absence of equality between the base ratios of guanine to cytosine and adenine to uracil signifies a lack of complementarity between these bases, i.e., the absence of a double helix. As a rule, therefore, RNA appears to main-

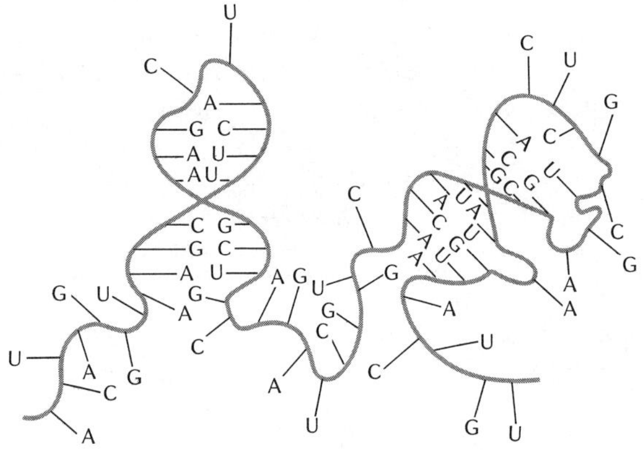

Figure 5–21

Base pairing within a single strand of RNA showing a possible folded structure. (After Fresco and Straus.)

tain a single-stranded structure, although there are forms of RNA that seem to be double-stranded, such as the genetic materials of reo virus (see Table 4–1) and wound tumor virus. However, even when single-stranded, RNA stability seems to be aided by its ability to fold back on itself, so that occasional base-pairing and hydrogen bonding enable some form of paired helical structure (Fig. 5–21; see also Figs. 27–8 and 28–6).

DUPLICATION OF RNA

The similarity in general structure between RNA and DNA, when first noted, suggested a similarity in the duplication mechanism. For some time it was therefore thought that each of these nucleic acids could duplicate independently of each other. That is, just as a strand of DNA could attract free deoxyribonucleotides to form a complementary strand, cellular RNA could do the same with free ribonucleotides. However, except for some RNA viruses which appear to replicate through a double-stranded RNA intermediate, considerable evidence indicates that all or most cellular RNA is of nuclear origin and derives from DNA templates. Some of this evidence is as follows.

CORRESPONDENCE BETWEEN DNA AND RNA BASE RATIOS. If RNA arises directly from DNA, its purine and pyrimidine base composition should be similar to that of the DNA that produced it, since the duplication process seems to occur through the pairing of complementary bases. Of course, some DNA sequences may produce more RNA than other sequences, and this DNA may have a unique base composition as compared to the overall DNA base ratios. For this reason RNA base ratios cannot always be expected to closely reflect the overall DNA base ratios. However, some fractions of RNA may well occur that are formed from a representative sample of DNA.

In the bacteriophage infection of *Escherichia coli*, it is known that upon the entry of viral DNA, the entire synthesis of bacterial host RNA is stopped, and a new type of RNA is produced. Volkin and Astrachan isolated this new RNA, which had been labeled with radioactive phosphorus, ^{32}P, and compared its base composition to that of the bacteriophage DNA. They used two different bacteriophages, T2 and T7, whose DNA base compositions differ, and found that the RNA produced in each case corresponded closely to the DNA of the infecting virus (Table 5–5).

Similar experiments have been done by rapidly labeling ("pulse labeling") the new RNA normally produced in yeast, bacteria, and some other organisms and then demonstrating a correspondence between the base composition of the newly produced RNA and the DNA.

HYBRID MOLECULES. RNA originating from DNA would be expected to have a purine and pyrimidine base sequence which should pair

TABLE 5–5
Base-ratio comparison (in percent) of the DNA of infective bacteriophage and the RNA found after infection

		Adenine	*Thymine (Uracil in RNA)*	*Cytosine*	*Guanine*
T2	DNA	32.5	32.6	16.7*	18.2
	labeled RNA	32.0	32.0	18.0	18.0
T7	DNA	26.0	26.6	23.6	23.8
	labeled RNA	26.7	27.8	23.7	21.8

From Volkin and Astrachan, 1956; Volkin, Astrachan, and Countryman, 1958.
* 5-Hydroxymethylcytosine.

after it is formed, base-for-base, with single strands of its parental DNA. This close pairing, or molecular "hybridization," can occur through a technique by which a mixture of RNA-DNA is heated (formation of single strands) and then gradually cooled (formation of double strands). If the base sequences in this mixture are complementary, close-fitting double strands ("hybrids") are formed which protect the RNA from the degenerative action of the ribonuclease enzyme.

So far, all RNA fractions, messenger, ribosomal, and transfer, have been shown to successfully hybridize with DNA of the same species. On the other hand, since many hereditary differences occur between species, interspecific hybrid DNA-RNA combinations would not be expected, and this is generally found to be the case. As an exception, the transfer RNA of bacteria has been found to hybridize to some extent with the DNA of other closely related species and can, furthermore, function in the protein synthesis of other species. This partial "universality" of transfer RNA may be due to factors which will be discussed later (Chapter 28).

LOCATION AND AMOUNT OF RNA SYNTHESIS. When RNA of animal cells in tissue culture is rapidly labeled with tritiated uridine (uracil replacing thymine), tracing of the label by autoradiography shows that it appears first in the RNA of the nucleus (see Prescott). Only then, after a short interval, does the newly synthesized RNA move to the cytoplasm. The reverse movement of RNA from cytoplasm to nucleus is never observed. Furthermore, if cellular RNA arises from DNA templates, the amount of RNA produced should be proportional in some way to the amount of DNA present in a cell. This relationship was shown by Allfrey and Mirsky to hold true for calf thymus cells; as DNA amounts are diminished by the action of DNase, the production of RNA is correspondingly decreased.

In summary, all these results point to the DNA control of cellular RNA synthesis. The following experiments support this conclusion and have helped to explain some of the mechanisms involved.

RNA TRANSCRIPTION

Two years before Kornberg's isolation of DNA polymerase, Grunberg-Manago and Ochoa isolated an enzyme specifically involved in RNA metabolism. This enzyme, *RNA phosphorylase*, could link together different ribonucleotides into a long RNA chain or break such a chain into smaller sections. In contrast to DNA polymerase, this enzyme could engage in polymerizing activity in the absence of a template, and RNA synthesis proceeds easily no matter what proportions of the four ribonucleotides are used, as long as they are diphosphates rather than mono or triphosphates:

$$n\text{X}\text{—}\textcircled{\text{P}} \sim \textcircled{\text{P}} \xrightleftharpoons[\text{Mg}^{2+}]{\text{phosphorylase}} \underset{\text{(RNA chain)}}{n(\text{X—P})} + n\text{P}$$

where P represents phosphate groups and X represents one or more varieties of the different ribonucleotide bases.

If only one kind of ribonucleotide is provided, RNA chains can be obtained consisting wholly of a single repeating base, e.g., adenine ribonucleotide (polyadenylic acid, "poly-A") or uracil chains (polyuridylic acid, "poly-U"). As expected, a combination of poly-A and poly-U seems to form double helixes as though complementary base pairing occurred. If four kinds of ribonucleotides (A, G, U, C) are present, these bases are polymerized randomly into an RNA molecule whose base composition reflects the initial proportion of each ribonucleotide. The occurrence of this random incorporation even in the presence of a specific DNA indicates that this particular enzyme is not responsible for DNA-directed RNA synthesis. *In vivo*, therefore, the function of RNA phosphorylase is probably not that of synthesizing RNA molecules, but of breaking down RNA molecules that are no longer necessary and of obtaining ribonucleotides for use in other reactions.

In 1959 and 1960 an enzyme that is probably much more important for the synthesis of RNA in the cell was found by Weiss, by Hurwitz, and by Stevens. This enzyme, *RNA polymerase*, only functions in the presence of DNA as a template and hooks together ribonucleotides that have been

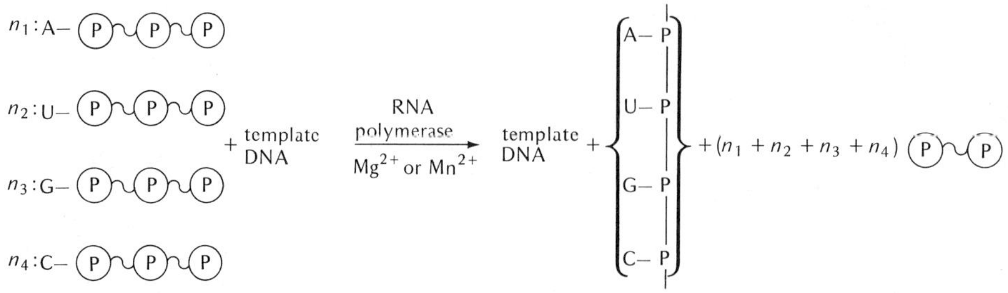

n_1: A–(P)~(P)~(P)
n_2: U–(P)~(P)~(P)
n_3: G–(P)~(P)~(P)
n_4: C–(P)~(P)~(P)
+ template DNA
RNA polymerase
Mg^{2+} or Mn^{2+}
template DNA +
A–P
U–P
G–P
C–P
$+ (n_1 + n_2 + n_3 + n_4)$ (P)~(P)

Figure 5–22

Reaction polymerizing the four kinds of ribonucleotides into an RNA chain using RNA polymerase and template DNA.

added in the form of triphosphates (Fig. 5–22). In these reactions the template DNA is not affected, but remains fully biologically active. For example, when DNA from *D. pneumoniae* is used as template in such an RNA polymerase reaction, it can be reextracted and shown to have retained its full transforming ability. From our previous discussion of RNA *in vivo* synthesis we may then easily conclude that the template DNA acts as the form upon which complementary RNA is made, or, in the terminology used, RNA is *transcribed* from DNA templates.

In one important experiment Hurwitz and his collaborators tested this transcriptional relationship by analyzing RNA base ratios produced by different forms of bacteriophage φX174 in an RNA polymerase system. As shown previously (Table 4–1, and pp. 90–92), φX174 DNA exists in single-stranded condition and the nucleotide base ratios show no evidence of complementary base-pairing except in the replicative form. When such single-stranded DNA is incubated according to Kornberg's method (p. 77), double-stranded DNA is formed; the adenine proportion is now equal in ratio to thymine, and the proportion of guanine is equal to that of cytosine. Since the difference between single-stranded and double-stranded φX174 DNA is reflected in the nucleotide base ratios, they asked the question whether there would be a corresponding difference in the base ratios of RNA synthesized by these two forms of DNA.

Their findings showed that the base composition of the resultant two RNAs corresponded in each case to the DNA base ratios (Table 5–6). Evidently the RNA prepared in this fashion is copied directly from DNA. Furthermore, the fact that double-stranded DNA affects RNA base ratios suggests that RNA *in vitro* is copied from both strands of DNA, and this view is supported by experiments using double-stranded DNA of other organisms.

According to Chamberlin and Berg, the single-

TABLE 5–6

Quantitative measurements of incorporation of ribonucleotides into RNA compared to those expected

DNA Added		Incorporated Ribonucleotides			
		Adenine	Uracil	Cytosine	Guanine
single-stranded φX174	observed	1.02	.79	.72	.62
	expected	1.02	.77	.70	.58
double-stranded φX174	observed	1.16	1.10	.87	.86
	expected	1.16	1.16	.84	.84

After Hurwitz, Furth, and Kahan, 1962.

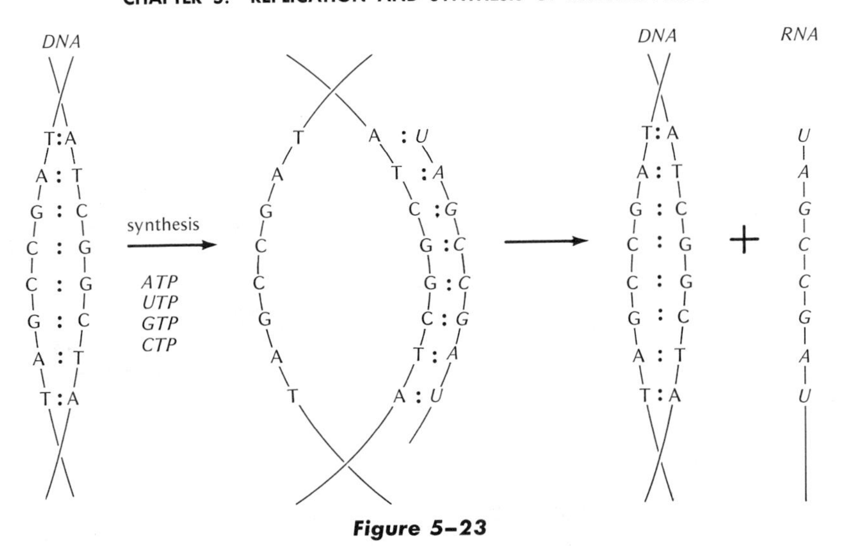

Figure 5–23

Model for the transcription of the nucleotide sequence from one strand of a DNA double helix into an RNA strand. (After Chamberlin.)

stranded φX174 forms a complementary structure with its product RNA which can be isolated as a hybrid DNA/RNA molecule. RNA production in this case therefore occurs through the expected base-pairing mechanism, and must follow the same nucleotide sequence as its DNA template, but in complementary fashion. Double-stranded φX174, on the other hand, does not produce hybrid DNA/RNA molecules. Nevertheless, its production of RNA is presumed to occur through splitting of the hydrogen bonding between the two DNA strands, and consequent base pairing between RNA nucleotides and the exposed DNA nucleotides in the split region (Fig. 5–23).

When experiments are done *in vivo*, however, important evidence exists that RNA is copied from only one strand of DNA called the "sense" strand. One such experiment, done by Marmur and his associates, used bacteriophage SP8 in which the two strands in its DNA double helix differ from each other both in base composition and in molecular weight. That is, heat treatment to break the hydrogen bonding between complementary strands of this DNA produces two distinct kinds of strands that can be separated in a cesium chloride density gradient, one "light" and one "heavy." When SP8 was permitted to infect its bacterial host, *B. subtilis*, and the RNA produced by this infection labeled with tritiated uridine, this RNA was found to hybridize only with the "heavy" DNA strand of SP8, never with the "light."

A reason for the difference between one- and two-strand transcription of DNA may arise from the state of the DNA molecule. According to Hayashi, Hayashi, and Spiegelman, the RNA produced by double-stranded φX174 DNA can be copied *in vitro* from one or both strands depending upon whether the DNA is in circular form or not. If the DNA is extracted in circular form, only one DNA strand is copied; if the DNA circle is disrupted, both DNA strands are copied.

In either case RNA synthesis in DNA organisms is undoubtedly directed by DNA. Further evidence for this exists in the fact that the nearest-neighbor frequencies of RNA are those expected if it were complementary to its DNA templates. Also, the antibiotic actinomycin D has been shown by Reich and others to have little effect on DNA synthesis but a remarkable inhibitory effect on RNA production. According to various studies, actinomycin D seems to bind directly with DNA and prevent RNA polymerase from functioning at

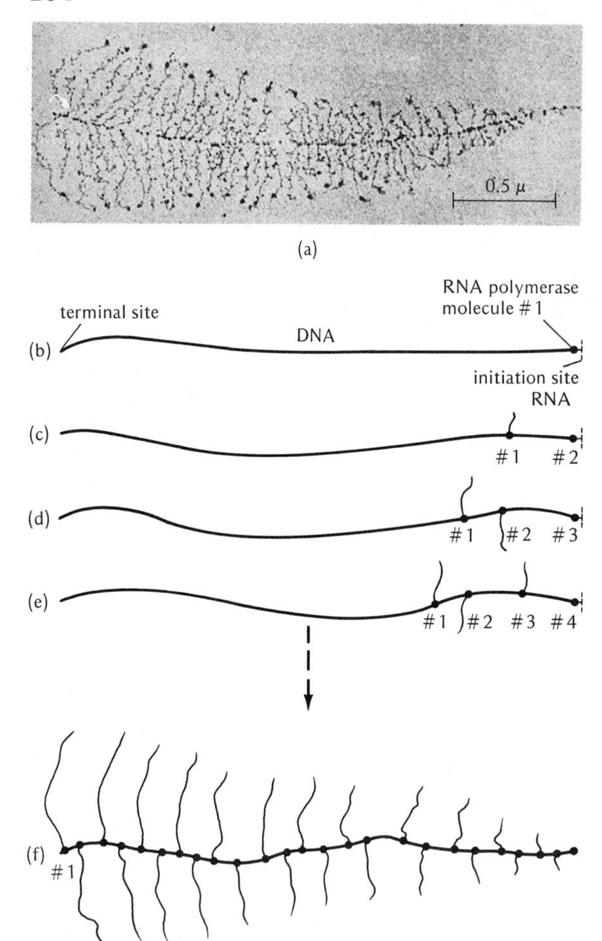

Figure 5–24

(a) Electron micrograph of a DNA sequence in the process of transcription in a Triturus viridescens *oocyte (see also Figure 2–14d). The DNA fiber is difficult to visualize in this reproduction but runs from left to right along the center of the photograph, and the transcribed RNA molecules radiate outward from the DNA axis. At each intersection of an RNA strand and the DNA fiber is a dense protein granule which is most likely the RNA polymerase enzyme. (From Miller, Beatty, Hamkalo, and Thomas.) Figures (b) to (f) show diagramatically the sequence of events believed to lead to the observations made in the electron micrograph above. (b) An RNA polymerase molecule (#1) becomes attached to an initiation site on a DNA sequence awaiting transcription. (c) The polymerase molecule moves along the DNA sequence transcribing it into an RNA sequence. After polymerase #1 has moved a short interval, polymerase #2 becomes attached to the initiation site and also begins transcription. In (d) and (e) further polymerase enzymes (#3, #4) attach to the DNA molecule, and in (f) transcription of the sequence is almost completed for polymerase #1. The RNA molecule associated with #1 therefore represents a sequence complementary to almost the entire length of one of the DNA strands. Upon completion of transcription, the polymerase enzyme is released along with the transcribed RNA strand. Miller and co-workers have also taken similar micrographs of transcription in procaryotes (see Figure 27–12) except that procaryotes appear to have transcription and translation taking place simultaneously. (That is, the ribosomes involved in translating RNA sequences into proteins attach themselves to the messenger RNA molecule as soon as the transcription process begins.)*

these sites. Its action not only prevents the transcription of messenger RNA but also prevents the synthesis of ribosome and transfer RNAs, thus indicating that the latter are also transcribed from DNA templates. The drug rifampicin has a similar inhibitory effect on transcription, binding with RNA polymerase so as to prevent the enzyme from functioning.

Perhaps the most graphic evidence for the DNA-RNA transcription process is a series of electron micrographs taken by Miller and coworkers. For both procaryotes and eucaryotes these photographs show strands of RNA arising along a DNA axis (Fig. 5–24). Also shown are initiation sites where the RNA polymerase becomes attached and transcription begins, as well as terminal sites at which RNA synthesis is apparently completed. Recently factors produced by viruses have been isolated that are involved in the recognition of both of these sites by the RNA polymerase enzyme.

One factor, called *sigma* (σ), is a protein chain that combines with five other "core" protein chains in the polymerase enzyme to form a "holoenzyme" which can then recognize specific transcription initiation sites.* Soon after the initiation of tran-

* The core component of RNA polymerase can transcribe DNA without sigma but it does not recognize specific starting sites on the DNA template (see "promoter" site, Chapter 29), nor does it seem to discriminate between the two DNA strands in a double helix as to which is to be used as a template. Addition of the sigma factor restricts action of the RNA polymerase enzyme to particular sites and to a particular strand.

scription, the sigma factor is released from the enzyme and can then be reused by another enzyme which, in turn, is then also ready to begin transcription. In the case of the T4 bacteriophage the σ factor produced by the virus combines with the host (*E. coli*) RNA polymerase enzyme, so that only viral DNA is transcribed. One of the T4 σ factors is produced very early in infection, transcribing only the earliest functioning genetic material while a different σ factor was found to be involved in the transcription of later-acting genetic material. Some viruses, such as T3 and T7, make their own RNA polymerase enzymes which recognize primarily their own specific initiation sites for transcription.

Since there are undoubtedly other factors or enzyme modifications which allow RNA polymerase to recognize specific transcription initiation sites it is not surprising that a termination factor called *rho* (ρ) has also been isolated. Rho apparently recognizes specific signals on the DNA template during the transcription process and causes RNA synthesis to terminate. Once transcription is terminated, rho dissociates from the polymerase enzyme and is then free to attach to another enzyme that is engaged in transcription.

Interestingly the direction of growth of the RNA strand during transcription is identical to the $5' \rightarrow 3'$ direction of growth of the DNA strand during DNA synthesis (see Fig. 5–10). That is, transcribed RNA grows by the addition of ribonucleotides at the 3′ end, while the 5′ end remains unchanged. Evidence for this view has been obtained in various experiments by observing the disposition of radioactive label used to mark the 5′ or 3′ ends of growing RNA strands. For example, Bremer and co-workers performed an *in vitro* experiment using phage T4 template DNA, *E. coli* DNA-dependent RNA polymerase, and different nucleoside triphosphates necessary for RNA synthesis including adenosine triphosphates bearing a tritium label. After RNA synthesis began, an excess of unlabeled adenine was added in order to dilute out the labeled adenosine nucleotides that were entering the RNA strand. Breakdown of such transcribed RNA molecules by alkaline hydrolysis then showed that the radioactive adenine at the 5′ end (recognized as a tetraphosphate, pppAp) was not affected by the dilution, whereas the dilution caused a reduction in the number of adenosine nucleosides at the 3′ end (no phosphate groups attached). Thus the 5′ end of the RNA molecule must have been synthesized first.

REPLICATION OF RNA VIRUSES

The existence of RNA viruses such as tobacco mosaic viruses, and RNA bacteriophages such as f2, MS2, and Qβ, requires the replication of these nucleic acids in a foreign cytoplasm through a mechanism that does not involve the host DNA as a directing template. At present at least two experiments suggest that the synthesis of many kinds of viral RNA (polio virus, influenza virus, double-stranded RNA phages, most single-stranded RNA phages) is primed directly by the introduced RNA rather then by DNA. First, Reich and co-workers showed that actinomycin D would inhibit 90 percent of DNA-primed RNA synthesis without interfering with the production of an RNA virus. Second, Baltimore and Franklin have shown that the increase in RNA produced by infection of a cell with an RNA virus is associated with a specific cytoplasmic fraction of the cell rather than with the DNA in the nucleus.

The enzyme responsible for such RNA-primed RNA synthesis appears to be unique for each particular RNA virus investigated and has been named *RNA-dependent RNA polymerase* or *replicase*. One form of this enzyme produced by the Qβ bacteriophage was isolated by Spiegelman and co-workers and shown to function in the presence of an RNA template, Mg^{2+} ions, and samples of the four triphosphate ribonucleotides, ATP, UTP, GTP, and CTP, to produce an exact replicate of the template. When the template RNA is derived from the RNA bacteriophage Qβ that ordinarily infects *E. coli*, the resultant RNA produced by this *in vitro* system can enter bacterial protoplasts (see p. 66) and produce fully mature virus particles. These experiments, along with those later performed on a DNA virus (see p. 90), have demonstrated that biologically active nucleic acids can be synthesized *in vitro*.

The direct use of RNA templates by RNA replicase is, however, only one method for the replication of RNA virus. Recently Temin and Mizutani as well as Baltimore discovered the existence of an enzyme, *reverse transcriptase*, in certain single-stranded tumor-producing RNA viruses (Rous sarcoma virus, Rauscher mouse leukemia virus) which can synthesize DNA from an RNA template. Because of this activity the enzyme can also be called *RNA-directed DNA polymerase*. If viral RNA is used as a template by this enzyme, a complementary DNA strand is made which can then become integrated into the host cell nuclear DNA and replicate along with it. Although the details are not yet known, it is believed that when a cellular signal occurs, this section of "viral" DNA can now synthesize viral RNA particles through the mediation of RNA polymerase. The DNA thus effectively functions as a "minus" template for the synthesis of "plus" RNA strands. Evidence for this view is the fact that DNA/RNA hybrids can be detected in which the DNA is complementary to the infectious viral RNA. Once this complementary strand of DNA is synthesized, it then replicates to form double-stranded DNA whose existence can also be demonstrated.

These findings help to explain the fact that the "transformation" of cells into a cancerous form caused by such cancer-producing (oncogenic) RNA viruses can be passed on, like any other characteristic, to each succeeding cell generation without the necessary presence of the RNA virus itself. Temin's suggestion in 1964 that an RNA tumor virus could exist in transformed cells as a "provirus" in DNA form has therefore been thoroughly supported. It is, however, not known at present whether the transformation of host cells occurs because of a special cancer-producing gene carried by the virus ("oncogene") or whether the presence of tumor "proviral" DNA causes transformation through a more complex interaction. In either case there is little question that RNA tumor viruses may be involved or responsible for many kinds of cancer. By using reverse transcriptase to produce radioactively labeled DNA synthesized from RNA viral templates, Spiegelman and coworkers have recently claimed that RNA specific to tumor viruses can be detected in human cancer tissues such as leukemia, breast cancer, sarcomas, and lymphomas. In support of this claim is the demonstration that tumor-specific RNA molecules in mice are probably associated with reverse transcriptase enzymes and localized in particles that are about the size of tumor viruses.

To date, reverse transcriptase is believed to be used by all RNA tumor viruses and also a few nontumor-producing viruses (visna virus, foamy virus). This enzyme may also function to generate short fragments of RNA used as primers in DNA replication (see p. 90).

PROBLEMS

5-1. For the Meselson-Stahl results shown in Fig. 5-2, draw the density gradient bands that would have been expected in generations 1 and 3, if: (a) Replication were conservative. (b) Replication were dispersive.

5-2. The autoradiograph of the *E. coli* chromosome in Figure 5-4 indicates a length of 1100 microns. (a) How many nucleotide base pairs are there in this chromosome? (b) If *E. coli* chromosome replication takes 40 minutes, what is the number of complete revolutions per minute this DNA chain must undergo in order to replicate?

5-3. Explain why it would or would not matter, with respect to the coding properties of RNA, which strand of a DNA double helix it was copied from.

5-4. A DNA double helix of the composition:

A–T
T–A
C–G
G–C

produces RNA which has a base ratio 25%A : 25%U : 25%C : 25%G. Can you distinguish whether the RNA is formed from one or both strands of this DNA?

5-5. The following data gathered by Hurwitz and collaborators show base compositions of two double-stranded DNA sources and their RNA products in *in vitro* experiments:

Species	DNA Base Ratio	RNA Base Ratios	
	$\frac{A+T}{G+C}$	$\frac{A+U}{G+C}$	$\frac{A+G}{U+C}$
B. subtilis	1.36	1.30	1.08
E. coli	1.00	.96	.90

(a) Can you distinguish from these data whether the RNA of these species is copied from single- or double-stranded DNA? (b) Explain how you can tell whether the RNA itself is single- or double-stranded in each of these species.

5-6. Assume that two types of nucleotides, A and B, can be synthetically produced, each different from those commonly used: Type A is incapable of attaching further nucleotides at its 3′ end, and type B is incapable of further attachment of nucleotides at its 5′ end. If either A or B can be incorporated into a DNA chain, which of these will be most effective in preventing further DNA synthesis?

5-7. The DNA isolated from two different species have the same base ratios. Does this mean they have the same nearest neighbor frequencies?

5-8. Would organisms with similar nearest neighbor frequencies have similar base ratios? Would they look alike?

5-9. In a particular experiment, a certain proportion of the nearest neighbors to a phosphorus-labeled cytidine deoxyribonucleotide (Ⓒ) was found to be adenine, that is, Ⓒ–A (Ⓒ above A). (a) On the basis of the particular frequency found for this nearest-neighbor combination, which of the following dinucleotide combinations would you expect to have the same frequency as Ⓒ–A if the two complementary DNA strands have opposite polarity: Ⓐ–C, Ⓖ–T, Ⓖ–A, Ⓣ–G, Ⓣ–A? (b) If they have the same polarity?

5-10. If a complete single-stranded molecule of DNA bears a total of only 6 nucleotides in the sequence, adenine-guanine-guanine-thymine-guanine-cytosine, as shown in the drawing at the top of the next column. (a) What would be the types and frequencies of guanine's nearest neighbors if all guanine nucleotides are labeled with ^{32}P as shown and the DNA is degraded by calf spleen diesterase? (b) Would these frequencies change if this DNA molecule had a complementary strand whose guanine nucleotide(s) were also labeled with ^{32}P?

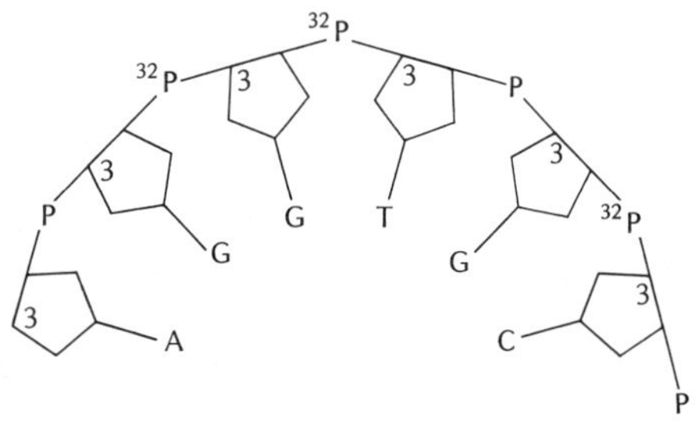

5-11. Using the nearest neighbor frequencies given in Table 5-1 what is the approximate base composition of the DNA in *Mycobacterium phlei*?

5-12. Opposite polarity of the two DNA strands in a double helix, initially demonstrated by Josse, Kaiser, and Kornberg (p. 81), was also confirmed by Chargaff and co-workers on experiments on the DNA of calf thymus. Through hydrolysis, they broke down the DNA into small fragments and then analyzed the frequency of sequences which were only two nucleotides long ("dinucleotides" or "doublets"). By convention such doublets are symbolized according to the position of the phosphate group: a phosphate attached to the 3′ carbon of a nucleotide is placed to the right of the base symbol, and a phosphate attached to the 5′ carbon is placed to the left of the symbol. Thus ApG designates a doublet of structure A3′p5′G. (a) Based on opposite polarity, which doublets would you expect to find equal in frequency to ApG? (b) To GpA?

5-13. Using nearest-neighbor analysis, Mills, Kramer, and Spiegelman obtained the following data in respect to a particular single-strand fragment of the Qβ genotype (RNA), which contains four guanine nucleotides and one each of adenine and cytosine: (1) When ^{32}P-labeled guanine nucleotides are used in the synthesis of this sequence, subsequent hydrolysis yields only guanine nucleotides (nearest neighbors) but no others. (2) When ^{32}P-labeled adenine is used, only guanine nucleotides are found to be labeled on subsequent degradation of the molecule, but no others. (3) When ^{32}P-labeled cytosine is used, only the adenine nucleotide is found to be its nearest neighbor. On the basis of these data, what is the exact sequence of the six nucleotides in this fragment?

REFERENCES

ALLFREY, V. G., and A. E. MIRSKY, 1962. Evidence for the complete DNA-dependence of RNA synthesis in isolated thymus nuclei. *Proc. Nat. Acad. Sci.*, **48,** 1590–1596.

BALTIMORE, D., 1970. RNA-dependent DNA polymerase in virions of RNA tumor viruses. *Nature*, **226,** 1209–1211.

BALTIMORE, D., and R. M. FRANKLIN, 1962. Preliminary data on a virus-specific enzyme system responsible for the synthesis of viral RNA. *Biochem. Biophys. Res. Commun.*, **9,** 388–392.

BESSMAN, M. J., I. R. LEHMAN, J. ADLER, S. B. ZIMMERMAN, E. S. SIMMS, and A. KORNBERG, 1958. Enzymatic synthesis of deoxyribonucleic acid. III. The incorporation of pyrimidine and purine analogues into deoxyribonucleic acid. *Proc. Nat. Acad. Sci.*, **44,** 633–640.

BREMER, H., M. W. KONRAD, K. GAINES, and G. S. STENT, 1965. Direction of chain growth in enzymic RNA synthesis. *Jour. Mol. Biol.*, **13,** 540–553. (Reprinted in *Selected Papers in Biochemistry*, Vol. 5; see References, Chapter 1.)

BRITTEN, R. J., and D. E. KOHNE, 1968. Repeated sequences in DNA. *Science*, **161,** 529–540.

BURGESS, R. P., A. A. TRAVERS, J. J. DUNN, and E. K. F. BAUTZ, 1969. Factor stimulating transcription by RNA polymerase. *Nature*, **221,** 43–47. (Reprinted in the collection of Zubay and Marmur; see References, Chapter 1.)

CAIRNS, J., 1963. The chromosome of *Escherichia coli*. *Cold Sp. Harb. Symp.*, **28,** 43–45. (Reprinted in the collections of Adelberg and of Levine, and in *Selected Papers in Biochemistry*, Vol. 1; see References, Chapter 1.)

CALLAN, H. G., 1972. Replication of DNA in the chromosomes of eukaryotes. *Proc. Roy. Soc. Lond.* (B), **181,** 19–41.

CAVALIERI, L. F., and B. H. ROSENBERG, 1963. Nucleic acids and information transfer. *Progr. Nuc. Acid. Res.*, **2,** 1–18.

CHAMBERLIN, M. J., 1965. Comparative properties of DNA, RNA, and hybrid homopolymer pairs. *Fed. Proc.*, **24,** 1446–1457.

CHAMBERLIN, M. J., and P. BERG, 1963. Studies on DNA directed RNA polymerase; formation of DNA-RNA complexes with single-stranded φX174 DNA as template. *Cold Sp. Harb. Symp.*, **28,** 67–75.

Cold Spring Harbor Symposia on Quantitative Biology, Vol. 35, 1971. *Transcription of Genetic Material.* Cold Spring Harbor Laboratory, Cold Spring Harbor, New York.

DAVIDSON, E. H., B. R. HOUGH, C. S. AMENSON, and R. J. BRITTEN, 1973. General interspersion of repetitive with nonrepetitive sequence elements in the DNA of *Xenopus*, *Jour. Mol. Biol.*, **77,** 1–23.

DELUCIA, P., and J. CAIRNS, 1969. Isolation of an *E. coli* strain with a mutation affecting DNA polymerase. *Nature*, **224,** 1164–1166. (Reprinted in the collection of Zubay and Marmur; see References, Chapter 1.)

FRESCO, J. R., and D. B. STRAUS, 1962. Biosynthetic polynucleotides: Models of biological templates. *Amer. Sci.*, **50,** 158–179.

GALL, J. G., 1963. Kinetics of deoxyribonuclease action on chromosomes. *Nature*, **198,** 36–38.

GILBERT, W., and D. DRESSLER, 1968. DNA replication: the rolling circle model. *Cold Sp. Harb. Symp.*, **33,** 473–484.

GOULIAN, M., A. KORNBERG, and R. L. SINSHEIMER, 1967. Enzymatic synthesis of DNA. XXIV. Synthesis of infectious phage φX174 DNA. *Proc. Nat. Acad. Sci.*, **58,** 2321–2328. (Reprinted in the collection of Zubay and Marmur and in *Selected Papers in Biochemistry*, Vol. 2; see References, Chapter 1.)

GRUNBERG-MANAGO, M., and S. OCHOA, 1955. Enzymatic synthesis and breakdown of polynucleotides; polynucleotide phosphorylase. *Jour. Amer. Chem. Soc.*, **77,** 3165–3166.

GULATI, S. C., R. AXEL, and S. SPIEGELMAN, 1972. Detection of RNA-instructed DNA polymerase and high molecular weight RNA in malignant tissue. *Proc. Nat. Acad. Sci.*, **69,** 2020–2024.

HAYASHI, M., M. N. HAYASHI, and S. SPIEGELMAN, 1964. DNA circularity and the mechanism of strand selection in the generation of genetic messages. *Proc. Nat. Acad. Sci.*, **51,** 351–359.

HOURCADE, D., D. DRESSLER, and J. WOLFSON, 1973. The amplification of ribosomal RNA genes involves a rolling circle intermediate. *Proc. Nat. Acad. Sci.*, **70,** 2926–2930.

HOWARD, A., and S. PELC, 1953. Synthesis of DNA in normal and irradiated cells and its relation to chromosome breakage. *Heredity*, **6,** (Suppl.) 261–273.

HUBERMAN, J. A., and D. A. RIGGS, 1968. On the mechanism of DNA replication in mammalian chromosomes. *Jour. Mol. Biol.*, **32,** 327–341.

HURWITZ, J., J. J. FURTH, and F. M. KAHAN, 1962. Biosynthesis of ribonucleic acid and the role of deoxyribonucleic acid. In *Basic Problems in Neoplastic Disease*, A. Gellhorn and E. Hirschberg (eds.). Columbia Univ. Press, New York, pp. 35–61.

INGRAM, V. M., 1972. *Biosynthesis of Macromolecules*, 2nd ed. W. A. Benjamin, Menlo Park, Calif.

JONES, K. W., 1970. Chromosomal and nuclear location of mouse satellite DNA in individual cells. *Nature*, **225,** 912–915.

JOSSE, J., A. D. KAISER, and A. KORNBERG, 1961. Enzymatic synthesis of deoxyribonucleic acid. VIII. Frequencies of nearest neighbor base sequences in deoxyribonucleic acid. *Jour. Biol. Chem.*, **236,** 864–875.

KELLER, W., 1972. RNA primed synthesis in vitro. *Proc. Nat. Acad. Sci.*, **69,** 1560–1564.

KHORANA, H. G., and 13 co-authors, 1972. Studies on polynucleotides. CIII. Total synthesis of the structural gene for an alanine transfer ribonucleic acid from yeast. *Jour. Mol. Biol.*, **72,** 209–217.

KLEINSCHMIDT, A. K., D. LANG, D. JACHERTS, and R. K. ZAHN, 1962. Darstellung und längenmessungen des gesamten desoxyribosenucleinsäure-inhaltes von T_2-bakteriophagen. *Biochim. Biophys. Acta*, **61,** 857–864.

KORNBERG, A., 1960. Biologic synthesis of deoxyribonucleic acid. *Science*, **131,** 1503–1508.

———, 1969. Active center of DNA polymerase. *Science*, **163,** 1410–1418. (Reprinted in the collection of Zubay and Marmur; see References, Chapter 1.)

———, 1974. *DNA Synthesis*. W. H. Freeman, San Francisco.

KRIEGSTEIN, H. J., and D. S. HOGNESS, 1974. Mechanism of DNA replication in *Drosophila* chromosomes: Structure of replication forks and evidence for bidirectionality. *Proc. Nat. Acad. Sci.*, **71,** 135–139.

KUMAR, K. V., and D. F. FRIEDMAN, 1972. Initiation of DNA synthesis in HeLa cell-free system. *Nature New Biol.*, **239,** 74–76.

LEWIN, B., 1974. *Gene Expression*: Vol. 1 *Bacterial Genomes*, Vol. 2. *Eucaryotic Chromosomes*. Wiley—Interscience, London.

MARMUR, J., C. M. GREENSPAN, E. PALECEK, F. M. KAHAN, J. LEVINE, and M. MANDEL, 1963. Specificity of the complementary RNA formed by *B. subtilis* infected with bacteriophage SP8. *Cold Sp. Harb. Symp.*, **28,** 191–199. (Reprinted in *Selected Papers in Biochemistry*, Vol. 5; see References, Chapter 1.)

MESELSON, M., and F. W. STAHL, 1958. The replication of DNA in *Escherichia coli*. *Proc. Nat. Acad. Sci.*, **44,** 671–682. (Reprinted in the collections of Taylor and in *Selected Papers in Biochemistry*, Vol. 1; see References, Chapter 1.)

MILLER, O. L., Jr., 1965. Fine structure of lampbrush chromosomes. *Nat. Can. Inst. Monogr.*, **18,** 79–99.

MILLER, O. L., Jr., B. R. BEATTY, B. A. HAMKALO, and C. A. THOMAS, Jr., 1970. Electron microscopic visualization of transcription. *Cold Sp. Harb. Symp.*, **35,** 505–512. (Reprinted in the collection of Abou-Sabé; see References, Chapter 1.)

OKAZAKI, R., T. OKAZAKI, K. SAKABE, K. SUGIMOTO, and A. SUGINO, 1968. Mechanism of DNA chain-growth. I. Possible discontinuity and unusual secondary structure of newly synthesized chains. *Proc. Nat. Acad. Sci.*, **59,** 598–605. (Reprinted in the collection of Zubay and Marmur and in *Selected Papers in Biochemistry*, Vol. 1; see References, Chapter 1.)

PARDUE, M. L., and J. G. GALL, 1970. Chromosomal localization of mouse satellite DNA. *Science*, **168,** 1356–1358.

PETTIJOHN, D. E., and R. HECHT, 1973. RNA molecules bound to the folded bacterial genome stabilize DNA folds and segregate domains of supercoiling. *Cold Spring Harb. Symp.*, **38,** 31–41.

PRESCOTT, D. M., 1964. Cellular sites of RNA synthesis. *Progr. Nuc. Acid Res.* **3,** 35–57.

REICH, E., R. M. FRANKLIN, A. J. SHATKIN, and E. L. TATUM, 1962. Action of actinomycin D on animal cells and viruses. *Proc. Nat. Acad. Sci.*, **48,** 1238–1245.

SHEKMAN, R., A. WEINER, and A. KORNBERG, 1974. Multienzyme systems of DNA replication. *Science*, **186,** 987–993.

SPIEGELMAN, S., I. HARUNA, I. B. HOLLAND, G. BOUDREAU, and D. MILLS, 1965. The synthesis of a self-propagating and infectious nucleic acid with a purified enzyme. *Proc. Nat. Acad. Sci.*, **54,** 919–927.

STEVENS, A., 1960. Incorporation of the adenine ribonucleotide into RNA by cell fractions from *E. coli B. Biochem. Biophys. Res. Commun.*, **3,** 92–96.

STONINGTON, O. G., and D. E. PETTIJOHN, 1971. The folded genome of *Escherichia coli* isolated in a protein-DNA-RNA complex. *Proc. Nat. Acad. Sci.*, **68,** 6–9.

SUEOKA, N., 1960. Mitotic replication of deoxyribonucleic acid in *Chlamydomonas reinhardi*. *Proc. Nat. Acad. Sci.*, **46,** 83–91.

SUGINO, A., S. HIROSE, and R. OKAZAKI, 1972. RNA linked nascent DNA fragments in *E. coli*. *Proc. Nat. Acad. Sci.*, **69,** 1863–1867.

TAIT, R. C., and D. W. SMITH, 1974. Roles for *E. coli* DNA polymerases I, II, and III in DNA replication. *Nature*, **249,** 116–119.

Taylor, J. H., 1969. The structure and duplication of chromosomes. In *Genetic Organization*, Vol. I, E. W. Caspari and A. W. Ravin (eds.). Academic Press, New York, pp. 163–221.

Taylor, J. H., P. S. Woods, and W. L. Hughes, 1957. The organization and duplication of chromosomes as revealed by autoradiographic studies using tritium-labeled thymidine. *Proc. Nat. Acad. Sci.*, **48,** 122–128.

Temin, H. M., and S. Mizutani, 1970. RNA-dependent DNA polymerase in virions of Rous sarcoma virus. *Nature*, **226,** 1211–1213.

Volkin, E., and L. Astrachan, 1956. Phosphorus incorporation in *Escherichia coli* ribonucleic acid after infection with bacteriophage T2. *Virology*, **2,** 149–161. (Reprinted in the collection of Taylor; see References, Chapter 1.)

Volkin, E., L. Astrachan, and J. L. Countryman, 1958. Metabolism of RNA phosphorus in *Escherichia coli* infected with bacteriophage T7. *Virology*, **6,** 545–555.

Wake, R. G., and R. L. Baldwin, 1962. Physical studies on the replication of DNA *in vitro*, *Jour. Mol. Biol.*, **5,** 201–216.

Walker, P. M. B., 1971. Repetitive DNA in higher organisms. *Progr. Biophys. Mol. Biol.*, **23,** 145–190.

Weiss, S. B., 1960. Enzymatic incorporation of ribonucleoside triphosphates into the interpolynucleotide linkages of ribonucleic acid. *Proc. Nat. Acad. Sci.*, **46,** 1021–1030.

Wolfson, J., and D. Dressler, 1972. Regions of single-stranded DNA in the growing points of replicating bacteriophage T7 chromosomes. *Proc. Nat. Acad. Sci.*, **69,** 2682–2686.

Worcel, A., and E. Burgi, 1972. On the structure of the folded chromosome of *Escherichia coli*. *Jour. Mol. Biol.*, **71,** 127–147.

PART

TRANSMISSION AND DISTRIBUTION OF GENETIC MATERIAL

It often happens also that the children may appear like a grandfather and reproduce the looks of a great-grandfather, because the parents often conceal in their bodies many primordia mingled in many ways, which fathers hand on to fathers received from their stock; from these Venus brings forth forms with varying lot, and reproduces the countenance, the voice, the hair of their ancestors.

LUCRETIUS
On the Nature of Things

6

MENDELIAN PRINCIPLES: I. SEGREGATION

The information concerning the mechanisms of cell division and the structure of DNA have been of great importance in firmly identifying the genetic material and helping to explain its mode of replication. With a high degree of confidence we can presume that a specific chemical material, DNA, is passed on between generations, and that this material has within itself the potentiality to carry and duplicate many different biological messages.

However, by restricting ourselves only to DNA, our observations become limited to the level of the chemistry of this material, which, interesting as it may be, does not enable us to decipher particular messages or to trace their inheritance. This limitation arises from the fact that an examination of nucleotide sequences, without any further knowledge, does not show any hint of their particular effects, or provide an understanding of how or why they evolved, or enable the prediction of those particular sequences that will appear in different generations and in different individuals. One of the central questions in genetics is, after all, the tracing of biological characteristics: to predict what particular characters will appear in an organism, as well as the occurrence of these characters in subsequent generations. To achieve an understanding of the transmission or modes of inheritance of genetic material we must therefore concern ourselves with obvious biological characteristics that can actually be traced between generations. The analytical methods involved in such determinations have turned out to be among the most powerful of biological techniques. Not only has genetic analysis led to an understanding of the arrangement of genetic determinants (Part

III) and insight into many of their modifications (Part IV) but it lays the essential groundwork for the functional molecular analysis yet to come (Part V). The fundamental step taken along this path was accomplished by Gregor Mendel.

MENDEL

Gregor Mendel was born in 1822 to a family of poor farmers in an area that is now part of Czechoslovakia. As a young man his desire for education was seriously hampered by poverty, and in order to continue his studies he entered the Augustinian monastery at Brünn and was ordained a priest in 1847. A few years after that he was sent to the University of Vienna for training in physics, mathematics, and the natural sciences. Although his performance at the university was not outstanding, his training provided him with many technical and mathematical skills that were of value in performing his later experiments. Upon his return to Brünn in 1854 he became a teacher, and in 1857 began his famous experiments on peas in the monastery garden.

Mendel's experimental use of the garden pea, *Pisum sativum*, was evidently not an accident, but the result of long careful thought. First, pollination could easily be controlled in this plant. Normally the pea plant was self-fertilizing and, consequently, use of one of Mendel's main techniques, "selfing," presented no difficulties. When cross-fertilization between two pea plants was necessary, Mendel had merely to remove the stamens (pollen organs) from one plant and transfer them to another destamenized plant. Second, the pea plant was easy to cultivate, and from one generation to the next took only a single growing season. Third, peas had many sharply defined inherited differences that had been long collected by seedsmen in the form of individual varieties. For his experiments Mendel chose among these varieties seven different "unit characters" to follow in inheritance, ranging from stem size to shape of seed. Each character that he followed had two alternative appearances, or "traits," i.e., tall or short stems, round or wrinkled seeds, etc. (Fig. 6–1).

Although the inheritance of many such differences had been noted by experimenters and plant and animal breeders previously, their observations usually jumbled all kinds of differences together without counting each one separately. The uniqueness of Mendel was that he kept track of each character separately, that he counted the appearance of the different traits for each character among the individuals in every generation, and that he analyzed his numerical results in the form of ratios which expressed underlying laws of heredity.

MENDEL'S EXPERIMENTS

Mendel tested the seven characters individually by crossing a variety carrying a particular trait of a character (e.g., tall) with another variety carrying a different trait of the same character (e.g., short). For simplicity we can consider only two of the characters tested, seed shape and seed color, although the results were strikingly similar for all seven characters.

When he crossed a smooth-seed-shape variety with a wrinkled one, he obtained seeds that were all smooth. Similarly, when he crossed a yellow seed plant to a green seed variety, the seeds produced were all of one type, yellow.* For ease of notation, an initial cross between two varieties is called the parental, or P_1, generation and their offspring, whether in seed form or as plants are called the first filial, or F_1, generation. Succeeding generations descended from this cross are labeled F_2, and so on.

Uniformly, Mendel's crosses between the two different varieties for each character always produced an F_1 that was of only one type. However, when these F_1 plants were permitted to reproduce by self-fertilization, examples of both original varieties now appeared in the F_2. For example, the

* In Mendel's words, "The hybrid forms of the seed-shape and of the albumen [cotyledon color] are developed immediately after the artificial fertilization by the mere influence of the foreign pollen. They can, therefore, be observed even in the first year of experiment, whilst all the other characters [size, etc.] naturally only appear in the following year in such plants as have been raised from the crossed seed."

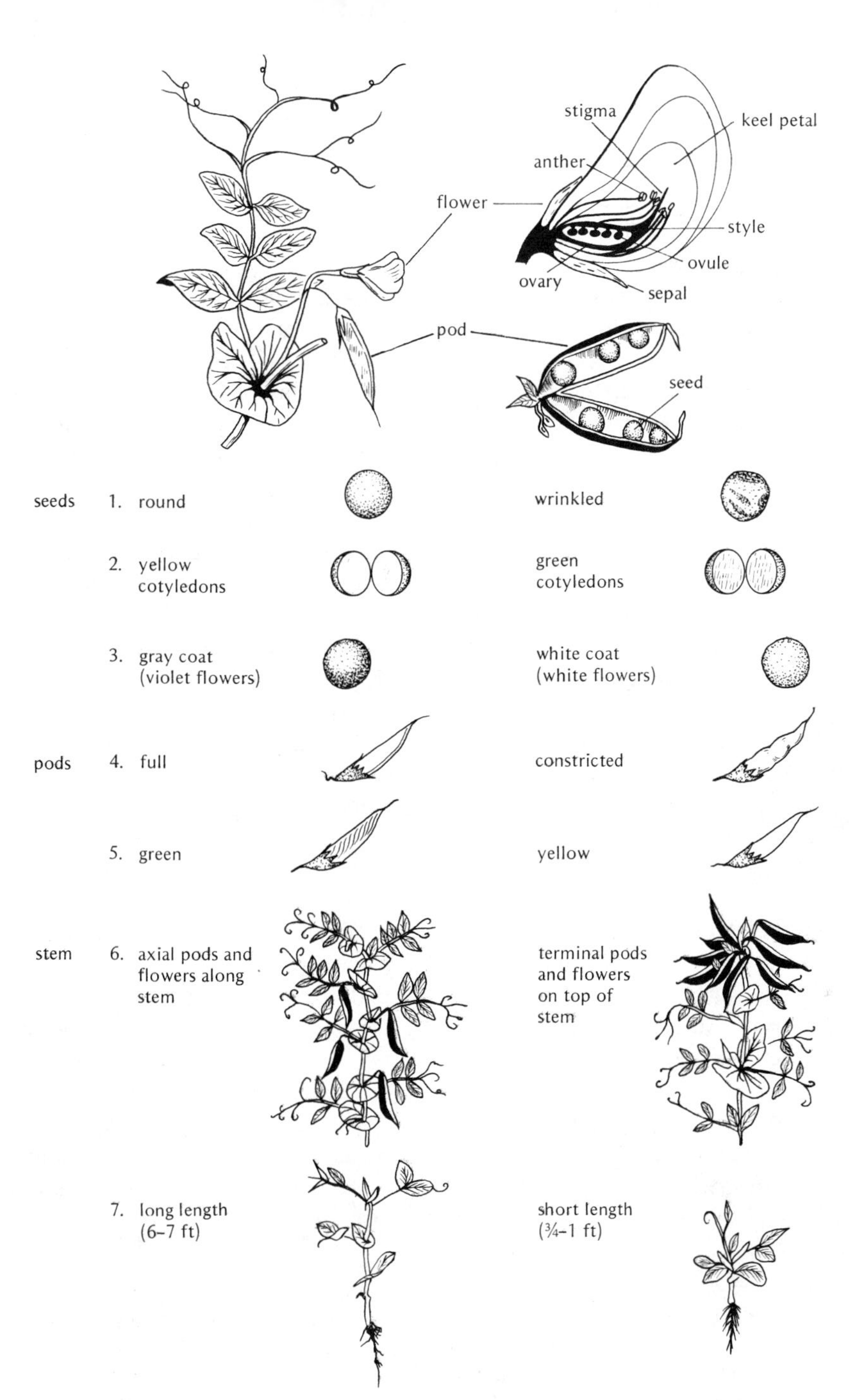

Figure 6–1

Seven characteristics in peas that were observed and scored by Mendel in his published experiments. As in other legumes such as various species of beans (Phaseolus), each flower bears a single ovary, containing up to 10 ovules, which then develops into the seed pod. Although "double fertilization" occurs (see corn, Fig. 3–12), the endosperm is soon resorbed in the pea, and the food reserves of the seed are contained in the two cotyledons ("leaves") of the embryo.

smooth F_1 seeds, i.e., plants from the cross smooth × wrinkled, produced upon self-fertilization an F_2 of 5474 smooth seeds and 1850 wrinkled. Similarly the self-fertilized yellow F_1 produced 6022 yellow and 2001 green seeds in the F_2. These F_2 ratios are in each instance very close to 3:1, and can, therefore, also be written 3/4:1/4, or .75:.25, or 75 percent:25 percent.

For all seven characters tested, the results appeared to fit the following pattern:

1. For any character the F_1 derived from crosses between two different varieties showed only one of the traits and never the other.
2. It did not matter which parent variety provided the pollen and which the ova; "reciprocal crosses" in which each of the two varieties was used to provide male and female parents (e.g., male A × female B, female A × male B) always gave the same results.
3. The trait that had disappeared or been "hidden" in the F_1 reappeared in the F_2, but only in a frequency one-quarter that of the total number.

Mendel called the determining agent responsible for each trait a "factor." From the evidence of the F_1 and F_2, the factor that determines the appearance of a trait could be hidden but not destroyed. This phenomenon, by which one trait appears and the other does not, even though the factors for both are present, is called *dominance.* In Mendel's crosses the factor for smooth seed shape was considered *dominant* over that for the wrinkled, which was considered *recessive.* Symbolically we can represent the factor for smooth by the letter *S*, and use *s* for wrinkled. Similarly, *Y* stands for the dominant yellow seed factor and *y* for the recessive green. (In Figure 6–1, traits that are determined by dominant factors are on the left, and recessive traits are on the right.)

The regular reappearance of various hidden recessive traits was, of course, a notable contribution to hereditary theory which had, for long, considered all traits as becoming "blended" and diluted in hybrid offspring. However, it was still not clear how many factors were involved in the determination of any trait. According to the symbols used, the smooth seed hybrid F_1 contains *S* (since it is smooth), but must also contain *s* (since it produces some wrinkled plants in the F_2). Since *S* is dominant over *s*, Mendel's wrinkled plants must contain *s*, but how many per plant? One? Two? Three? Two *s* and one *S*? Or what?

One way of discriminating between these possibilities was to breed the wrinkled plants and observe their offspring. Upon self-fertilization, Mendel noted that wrinkled plants always gave rise ("bred true") to wrinkled in all generations for which the experiment was carried. Evidently, there were no smooth *S* factors within them. On the other hand, the F_2 plants that appeared smooth did not always breed true; of 565 self-fertilized smooth plants only 193 bred true to smooth while 372 each produced smooth and wrinkled plants in the proportion of 3 smooth:1 wrinkled. In other words, the F_2 smooth plants were of two types, "pure" smooth producers and "hybrid" or mixed smooth-wrinkled producers in the ratio 372 hybrid smooth to 193 pure smooth, or 2/3:1/3.

When the smooth offspring of the F_2 hybrid smooth-wrinkled producers were self-fertilized, they in turn also gave rise to the same two types of plants: one-third pure smooth producers and two-thirds hybrid smooth-wrinkled producers. In summary, therefore, three types of plants could be distinguished in these crosses: pure smooth producers, pure wrinkled producers, and hybrid smooth-wrinkled producers, as follows:

P_1:		smooth × wrinkled ↓		
F_1: (selfed)		smooth (hybrid) ↓		
F_2: (selfed)	smooth: (pure) ↓	smooth: (hybrid) ↘	smooth: (hybrid) ↙	wrinkled (pure) ↓
F_3:	all smooth	3 smooth: 1 wrinkled		all wrinkled

From these results, it is clear that the hybrid smooth plants producing both smooth and wrinkled must contain factors for both. The simplest assumption, therefore is, that a hybrid plant for seed shape contains two factors, *S* and *s*. For the argument to be consistent, the pure smooth and pure wrinkled producers must also contain two factors as well, *SS* and *ss*, respectively. It is therefore clear why the *SS* and *ss* plants breed true; since they are permitted to self-fertilize, only pure lines would be expected from each type. The *Ss* hybrids, on the other hand, do not breed true, since they are each carrying both *S* and *s* factors. If we make the assumption that the *S* and *s* factors can separate or segregate from each other during gamete formation in the hybrid, then some gametes carry *S* and others *s*; i.e., there are two kinds of equally frequent pollen and two kinds of equally frequent ova. The random combinations of these gametes to form zygotes, as seen in Fig. 6–2, will then account for the observed ratios. Thus Mendel demonstrated that a hybrid between two different varieties possesses both types of parental factors, which subsequently separate or segregate in the gametes. This fundamental law is known as the *principle of segregation*, and Mendel's results have since been demonstrated many times and for many hybrid generations (Table 6–1).

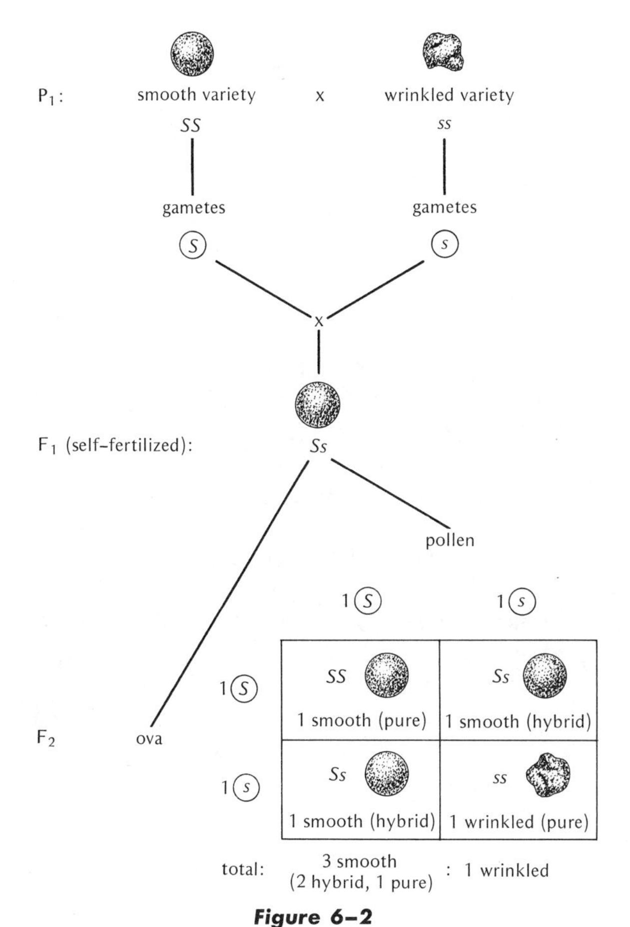

Figure 6–2

Segregation of seed-shape character and explanation for Mendel's observed results in the F_1 and F_2 generations.

TERMINOLOGY

In modern terms, an inherited factor that determines a biological characteristic of an organism is called a *gene*. In diploid organisms such as pea plants, genes exist in pairs, many of which are recognized by their production of a particular biological effect such as seed shape or pod color. The two individual genes in a particular gene pair are known as *alleles*. In some cases these alleles are identical; i.e., a wrinkled plant carries two identical alleles *s* and *s*. In some cases, such as a hybrid smooth plant, the alleles differ, one being the gene or allele for smooth, *S*, and the other being the gene or allele for wrinkled, *s*. The terms allele and gene are thus interchangeable, with the restriction that allele refers only to the genes at a particular gene pair. That is, while the gene for smooth, *S*, is an allele of (or allelic to) that for wrinkled, *s*, it is not an allele of the gene for yellow seed color, *Y*. When a gene pair in an organism contains two identical alleles, e.g., *S* and *S*, the organism is considered *homozygous* for that gene pair and is called a *homozygote*. When two different alleles are present in a single gene pair, e.g., *S* and *s*, the organism is *heterozygous* for that gene pair and is called a *heterozygote*. According to our present use of these terms, the effect of a recessive gene does not appear in a heterozygote that also contains its dominant allele. Recessive characteristics

TABLE 6–1

Segregation for seed color and seed shape in different hybrid generations of garden peas according to various observers

Hybrid Generation	Observer	Seed Shape: No. Round	Seed Shape: No. Wrinkled	Seed Shape: Wrinkled, Percent	Seed Color: No. Yellow	Seed Color: No. Green	Seed Color: Green, Percent
F_2	Mendel	5,474	1,850	25.2	6,022	2,001	24.9
	Correns				1,394	453	24.5
	Tschermak	884	288	24.6	3,580	1,190	24.9
	Bateson	10,793	3,542	24.8	11,903	3,903	24.7
	Hurst	1,335	420	23.9	1,310	445	25.4
	Lock	620	197	24.1	1,438	514	26.2
F_3	Correns				1,012	344	25.5
	Tschermak	2,087	661	24.0	3,000	959	24.2
	Lock	769	259	25.2	3,082	1,008	24.6
F_4	Correns				225	70	23.7
	Lock	2,328	812	25.8	2,400	850	26.1

are therefore only apparent in recessive homozygotes. Of course, an individual that is a recessive homozygote at one gene pair need not necessarily be so at another, e.g., *ssYy*.

The distinction between the effect of the gene and the gene itself must always be kept in mind. The gene for smooth seeds is, of course, not smooth, nor is the gene for yellow seed color yellow. Genes, as we understand them today, are sections of DNA strands and are not miniature replicas of various parts of an organism. The terms *smooth* and *yellow*, or their symbols *S* and *Y*, are merely used as a shorthand notation for the genes responsible for these particular effects.

As an important aid in distinguishing the appearance of an organism from the genetic factors that influence it, the term *phenotype* is used for the former and *genotype* for the latter. In their broad definitions, the phenotype refers to all the manifold biological appearances, including chemical, structural, and behavioral attributes, that we can observe about an organism but excludes its genetic constitution. The genotype defines only the complement of genetic material (or *genome*) that an organism inherits from its parent. Therefore, although the phenotype changes with time as the appearance of the organism changes, the genotype remains relatively constant except for the rare genetic changes known as *mutations*.

TESTING PHENOTYPES

The characteristics of the offspring of Mendel's crosses can be predicted from the genotype of the parents through knowledge of which genes are dominant and which recessive. As we have seen, *ss* individuals are wrinkled, *Ss* and *SS* individuals are smooth. The reverse question then faces us: Can we determine the genotype of an individual by knowing its phenotype? In some cases the appearance of the recessive phenotype, such as wrinkled seeds and green seed color, indicates immediately that the genotype of the plant is homozygous recessive for these factors, i.e., *ss* and *yy*. On the other hand, genotypes causing the dominant phenotype to appear, e.g., smooth, may be either homozygous, *SS*, or heterozygous, *Ss*.

Mendel used two types of tests to distinguish between homozygous and heterozygous dominant phenotypes. The first test has already been described and was to self-fertilize the smooth plants. *SS* genotypes would, of course, only produce smooth offspring, whereas self-fertilized *Ss* plants would produce smooth and wrinkled offspring in a ratio 3:1. Thus the appearance of some wrinkled offspring from self-fertilized smooth plants, or from smooth plants crossed with other smooth plants, is an indication that the smooth parents were heterozygotes.

An additional test performed by Mendel was to *backcross*, or *testcross*, smooth plants to the wrinkled variety. If the smooth plants were homozygous *SS*, the cross, *SS* × *ss*, would produce all *Ss* smooth offspring. On the other hand, if the smooth plants were heterozygotes, the cross would be *Ss* × *ss* and produce 1/2 *Ss* smooth and 1/2 *ss* wrinkled offspring (Fig. 6–3). In the actual test Mendel performed, he backcrossed F_1 hybrid (*Ss*) smooth plants to wrinkled (*ss*) parents and obtained 106 smooth (*Ss*) and 102 wrinkled (*ss*) offspring, or a ratio 1 smooth:1 wrinkled, as expected.

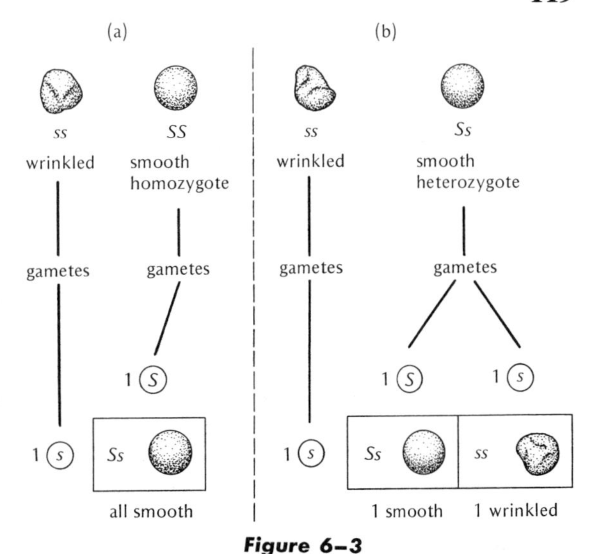

Figure 6–3

Testcross (mating to a homozygous recessive, ss) to determine whether an unknown smooth plant is homozygous (SS) or heterozygous (Ss). (a) The results of the testcross if the smooth parent is homozygous. (b) Results of testcross for a heterozygous parent.

EXAMPLES OF GENE DIFFERENCES AND SEGREGATION

Segregation for single-character differences has been widely documented in almost all organisms that can be closely studied. As an illustration, one trait, widely observed among many animals and plants to be inherited as a single gene difference is *albinism*—the absence or marked reduction of pigment. In humans, many studies have shown the existence of a recessive albino gene which, when in homozygous condition, causes a very light skin in addition to white hair and pink or red eye color (because of the reflection of retinal blood vessels). If we call the normal pigment gene *C* (for colored) and the albino gene *c*, an albino individual can be designated *cc*. The appearance of recessive albinism therefore involves inheritance of one albino gene from each parent. In some cases, of course, one or both parents are also albinos; in others, both parents may appear normally pigmented but carry the albino gene in heterozygous condition.

When sufficient family information is available, the exact mode of inheritance can be traced through the technique of *pedigrees*. For example, Fig. 6–4 illustrates a hypothetical family pedigree for recessive albinism in which the albino individuals are designated by shading the symbols. In such pedigrees, two parents (circles are females, square are males) are connected by a *marriage line* and their offspring (sibs or siblings) are connected to a horizontal *sibship line* below the parents. The offspring are listed in order of birth from left to right with each generation denoted by a roman numeral (I, II, . . .) and the individuals within it by arabic numbers (1, 2, . . .). Twins that have the same connection to the sibship line (II–6 and II–7) are *identical*, or *monozygotic*, and arise from the splitting of a single fertilized zygote. (In some pedigrees identical twins are also connected by a horizontal line.) When the sex is unknown (i.e., II–2) a diamond is used. Twins that have a separate connection to the sibship line (III–3 and III–4) are *fraternal*, or *dizygotic*, and arise from two separately fertilized zygotes. A number that is included within a symbol (II–8 to II–10) refers to sibs not individually listed.

Other inherited characters that can be traced in similar fashion include dominant traits such as

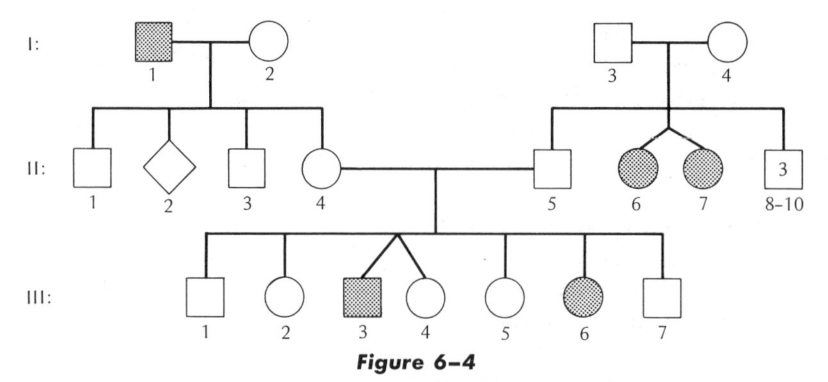

Figure 6–4

Sample pedigree for a recessive trait (albinism) designated by shaded symbols.

chondrodystrophic dwarfism, for which a pedigree is given in Fig. 6–5. Note that in comparing pedigrees of Figs. 6–4 and 6–5 the parents of offspring showing recessive traits may be phenotypically normal (although genotypically heterozygous), whereas at least one of the parents of offspring showing dominant traits is always affected by the trait. Pedigrees of recessive traits are therefore often characterized by the trait appearing in alternate generations, whereas dominant traits of the types shown here do not "skip" generations. That is, parents who do not themselves show evidence of a dominant trait are homozygous for the recessive allele and cannot therefore transmit the dominant allele. Also, the frequency of children that are affected by a recessive trait from a mating of two normal parents is about one-fourth, whereas a dominant trait is usually transmitted by a mating between an affected parent (heterozygote) and a normal parent (recessive homozygote) to about half the offspring.

By convention we usually assume that the genes for a *rare* trait have their origin in the early generations of a pedigree and are not continuously introduced into later generations by matings with individuals outside the pedigree. For example, chondrodystrophy is a relatively rare trait, and its appearance among successive generations in the pedigree in Fig. 6–5 is assumed to be caused by a gene present in the first generation (I–2). If

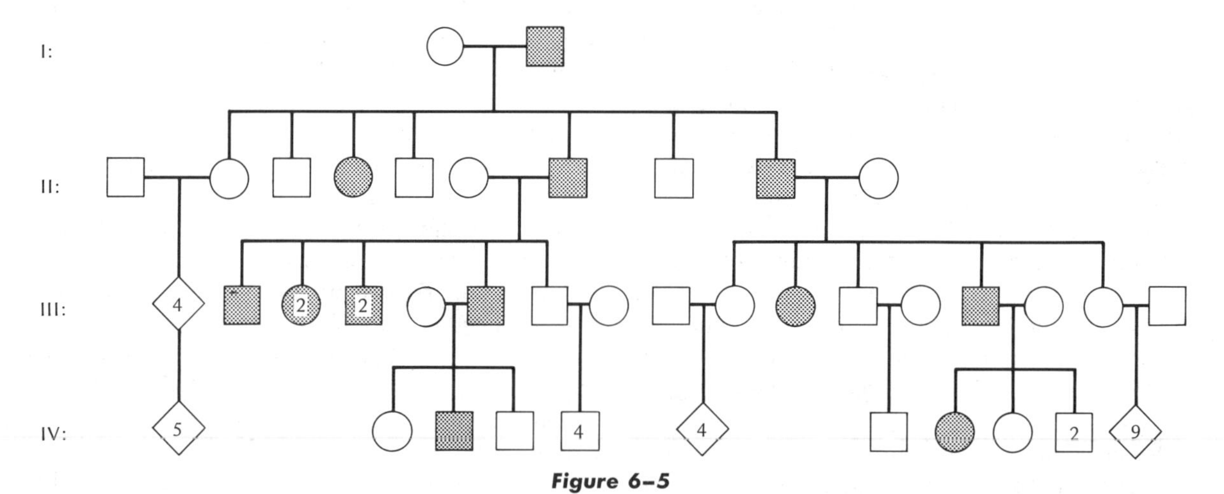

Figure 6–5

Partial pedigree for chondrodystrophic dwarfism (shaded symbols) in a Utah family. (Modified from Stern, after Stephens.)

chondrodystrophy were caused by a recessive gene in this pedigree, we would have to assume that each chrondrodystrophic dwarf who has chondrodystrophic offspring has married a normal-appearing individual who was carrying the gene in heterozygous condition (individuals II–6, II–10, III–10, III–20). In other words, we would be forced to propose that the gene was fairly common to account for the significant frequency of such heterozygotes. Since chondrodystrophy is a quite uncommon trait, the assumption of high frequency of heterozygotes is unwarranted, and this particular pedigree is therefore more in accord with a trait caused by a dominant gene.

THE PURITY AND CONSTANCY OF MENDELIAN FACTORS

Through countless experiments since 1900, mendelian factors, or genes, have been shown to segregate in precise and constant fashion in each generation. Genes, of course, may change accidentally through the rare process of mutation (Chapter 23), but, on the whole, they are not changed by the presence of other genes. Gene *A*, for example, is not expected to change because it finds itself in the presence of a gene *a* in a heterozygote, or in the presence of other genes such as *B*, *C*, etc. According to this fundamental principle, genes are considered "pure," or constant.

One convincing demonstration of this view is the general observation that recessive genes show no modification even when kept for many generations in heterozygous organisms of the dominant phenotype. In *Drosophila*, for example, certain stocks ("*ClB*") have been kept as heterozygotes for both recessive and dominant genes since 1919, or for about 1200 generations. During this long period of time there is no evidence that either the recessive or dominant genes were changed because of their mutual presence in the same organisms. Apparently no "dilution" or blending of genes has occurred.

Another more direct observation of gene purity was in an experiment carried out in guinea pigs by Castle and Phillips in 1909. As in many other mammals, *albino* is a recessive characteristic in guinea pigs, and albino females mated to albino males normally produce only albino offspring. To show that genes are not affected by the genotype or phenotype of the particular body in which they are placed, Castle and Phillips transplanted an ovary from a black guinea pig to an albino female whose ovaries had been removed and then mated this female to an albino male (Fig. 6–6). The resultant offspring were all black, demonstrating that characters acquired in the body of an organism, such as albinism in this case, are not transmitted to the gametes. In the absence of mutation the genotype is therefore constant, no matter in which phenotype it finds itself.

This is not to say that the presence of a particular genotype must always ensure the development of a particular phenotype, and that no variability in development is possible. Many variations of phenotype occur because of environmental differences and other causes (Chapter 10), and it is rare to find that even two identical twins with exactly the same genotypes are exactly alike in all phenotypic respects. Some disease, accident, or difference in upbringing can lead to small or even large

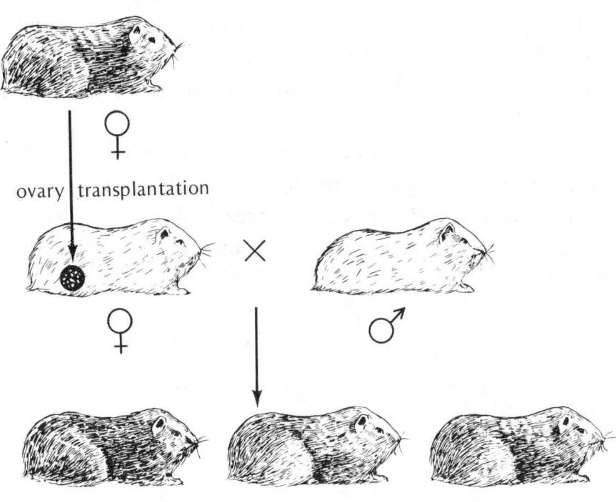

Figure 6–6

Diagram of Castle and Phillips' experiment transplanting an ovary removed from a black guinea pig (above) to a white female whose own ovaries were removed. The lower row shows the observed results when this white female was mated to a white male.

noticeable differences. What the genotype accomplishes is to provide the basic "messages" necessary for the development of an organism and to inscribe limits upon this development. Within these limits, however, the particular form of development depends on factors that are not necessarily constant from one generation to the next or even from one individual to the next. Constancy is therefore a term that we can, for the present, restrict to the stability of genes and their transmission in unchanged condition between cells and between generations.

PROBLEMS

6-1. As Mendel discovered, gray seed color in peas is dominant to white. In the following experiments, parents with known phenotypes but unknown genotypes produced the listed progeny:

Parents	Progeny: Gray	Progeny: White
(a) gray × white	82	78
(b) gray × gray	118	39
(c) white × white	0	50
(d) gray × white	74	0
(e) gray × gray	90	0

Using the letter *G* for the *gray* gene and *g* for *white*, give the most probable genotype of each parent.

6-2. In crosses (b), (d), and (e) of Problem 1, indicate how many of the gray progeny produced by each cross would be expected to produce white progeny when self-fertilized.

6-3. D. F. Jones inbred six corn plants of a special stock and raised approximately 25 seeds from each of them. Some seeds produced green plants and some produced albino white plants, in the following proportions:

Parent	Offspring: White	Offspring: Green
A	5	19
B	9	13
C	4	21
D	4	15
E	8	15
F	4	25
	34	108

Explain the inheritance of the white character.

6-4. Silky feathers in fowl is caused by a gene whose effect is recessive to that for normal feathers. (a) If 96 birds were raised from a cross between individuals that were heterozygous for this gene, how many would be expected to be silky and how many normal? (b) If you had a normal feathered bird, what would be the easiest way to determine whether it is homozygous or heterozygous?

6-5. Groff and Odland found a variety of cucumber whose flowers failed to open when mature. These flowers could nevertheless be pollinated by opening them artificially. The results of their experiments were:

Parents	Phenotypes of Offspring: Open Flowers	Phenotypes of Offspring: Closed Flowers
closed × open	all	none
F_1 (of above) × F_1	145	59
closed × F_1	81	77

Define symbols for the genes involved and indicate the genotypes of: (a) The closed parents. (b) The open parents. (c) The F_1.

6-6. The following is a pedigree of a fairly common human hereditary trait (shaded symbols):

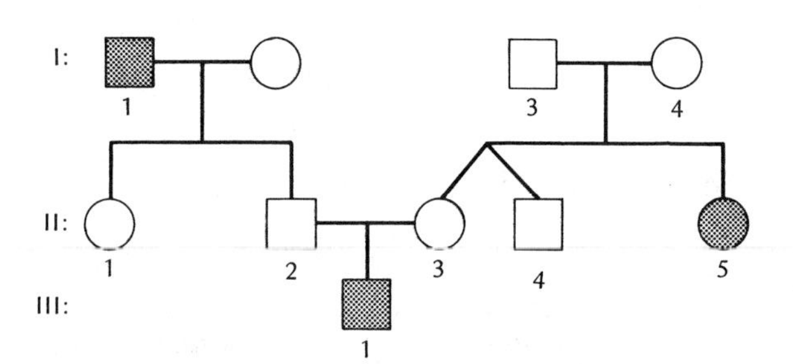

(a) Indicate whether you think the gene differences causing this effect are dominant or recessive, e.g., *A* (dominant) or *a* (recessive). (b) Designate the genotype of each individual in the pedigree, and indicate the choice of genotypes if you think more than one genotype is possible for an individual.

6–7. Indicate whether the trait in the following pedigree (affected individuals are shaded) is caused by a dominant or a recessive gene.

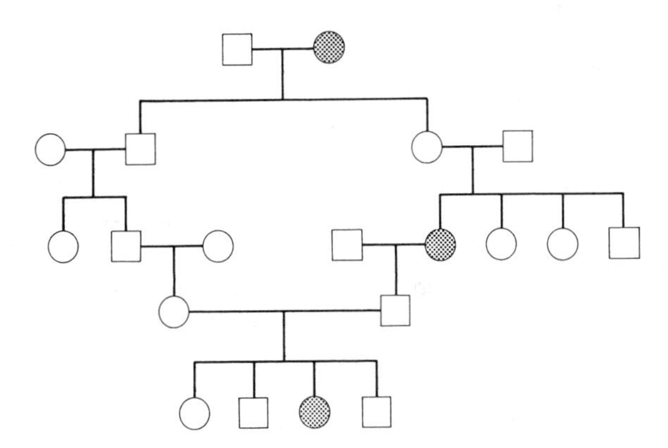

6–8. Two affected individuals suffering from a *very rare* recessive trait (shaded symbols) appeared in the pedigree shown below. If individual I–1 (circled) is considered to be a heterozygous carrier of the trait, which of the other individuals in the pedigree *must necessarily* have also been heterozygous for the gene causing the trait?

6–9. The following is a pedigree for a dominant trait, caused by gene *A*, that occurs in man. Shaded symbols show individuals affected with the trait, unshaded individuals are normal (*aa*).

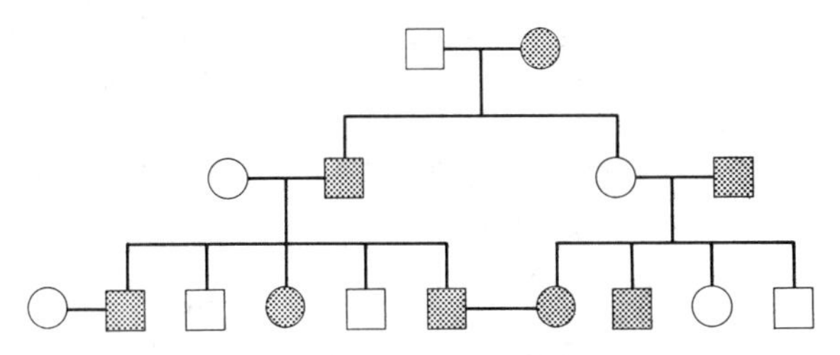

(a) For each individual write the genotype involved. (b) Among the progeny arising from the marriage of individual III–1, what proportion would be expected to show the trait? (c) Among the progeny arising from the marriage of individual III–6, what proportion would be expected to show the trait?

6–10. Brachydactyly is a rare human trait that causes a shortening of the fingers. Various investigations have shown that approximately half the progeny of brachydactyly × normal marriages are brachydactylous. What proportion of brachydactylous offspring would be expected in matings between two brachydactylous individuals?

6–11. Assume that eye color in humans is controlled by a single pair of genes of which the effect of that for *brown* (*B*) is dominant over the effect of that for *blue* (*b*). (a) What is the genotype of a brown-eyed

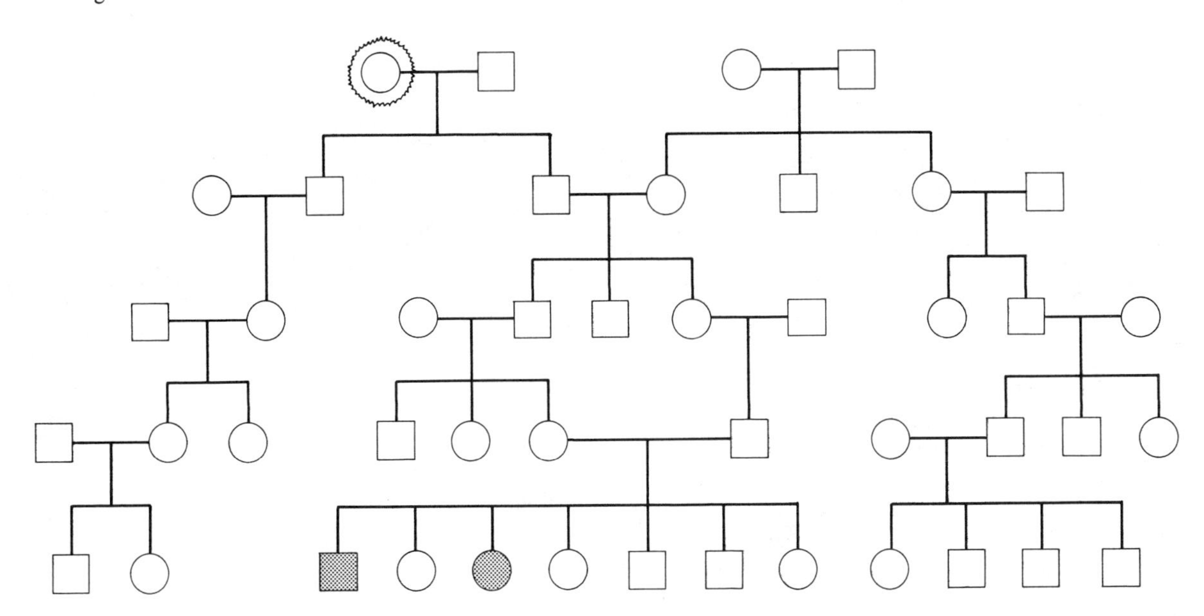

individual who marries a blue-eyed individual and produces a first offspring that is blue-eyed? (b) For the same mating as in (a), what proportions of the two eye colors are expected among further offspring? (c) What are the expected proportions of eye colors among the offspring of a mating between two brown-eyed individuals who each had one parent that was blue-eyed?

6-12. A black stallion of unknown ancestry was mated to a number of sorrel (red) mares with purebred pedigrees. These matings produced 20 sorrel offspring and 25 black offspring. (a) Which of these phenotypic characters is most likely to be caused by a recessive homozygote? (b) According to your hypothesis, what numbers of each kind would you have expected? (c) Test your hypothesis using the chi-square method (p. 146), and indicate whether you would accept or reject your hypothesis on the basis of this test.

REFERENCES

BATESON, W., 1909. *Mendel's Principles of Heredity.* Cambridge Univ. Press.

CASTLE, W. E., and J. C. PHILLIPS, 1909. A successful ovarian transplantation in the guinea pig and its bearing on problems of genetics. *Science*, **30,** 312–313.

ILTIS, H., 1932. *Life of Mendel* (Transl. by E. Paul and C. Paul). W. W. Norton, New York.

MENDEL, G., 1866. Versuch über Pflanzen-Hybriden. Mendel's classic paper, originally published in the Proceedings of the Brünn Natural History Society, has been translated into English under the title *Experiments in Plant Hybridization* and reprinted in the collections of Peters (see References, Chapter 1); Stern and Sherwood (see References, Chapter 7); and Voeller (see References, Chapter 1). It has also been published separately: 1960, Harvard Univ. Press, Cambridge; 1965, Oliver and Boyd, Edinburgh (introduction and commentary by R. A. Fisher).

7

MENDELIAN PRINCIPLES: II. INDEPENDENT ASSORTMENT

Mendel's study of single gene differences in *monohybrid* crosses led, as we have seen, to the discovery of the principle of segregation. Through use of this principle it was possible to predict the separation between two different alleles in a single gene pair and their subsequent behavior in each generation. Although this discovery, by itself, was of great value, Mendel did not stop his experiments, but went on to consider the products of individuals differing in two pairs of genes, or *dihybrid* crosses.

One cross that Mendel used for this purpose was a mating of a plant with smooth and yellow seeds to a plant with wrinkled and green seeds. As might be expected because of dominance the seeds of this mating appeared smooth and yellow. However, when the F_1 plants were grown and self-fertilized, they gave rise to 556 F_2 seeds of the following types: 315 smooth yellow, 108 smooth green, 101 wrinkled yellow, and 32 wrinkled green. In terms of proportion these numbers are very close to a 9:3:3:1 ratio, which, for 556 individuals, would ideally be 312.75:104.25:104.25:34.75. How does this ratio come about biologically?

A simple way of deciphering these proportions is to separate the results and analyze them individually with respect to each single gene pair, as follows:

P_1:	smooth × wrinkled	yellow × green
F_1:	smooth	yellow
F_2:	423 smooth: 133 wrinkled or about 3/4 smooth: 1/4 wrinkled	416 yellow: 140 green or about 3/4 yellow: 1/4 green

The F_2 ratios therefore closely fit the ratios expected from crosses involving single gene pairs in which dominance occurs. When we now combine these two sets of results in a single dihybrid cross, we find that each gene pair acts independently of the other. This means that the chances for a plant to be smooth or wrinkled do not interfere with, or are *independent of*, its chances to be yellow or green. Mathematically this relationship is expressed by multiplying the probability that a plant will have a certain characteristic, e.g., smooth, times the probability that it will have a different characteristic, e.g., yellow. Therefore, if a seed has a 3/4 chance of being smooth and a 3/4 chance of being yellow, its chances to be both smooth and yellow at the same time would be $3/4 \times 3/4 = 9/16$. We can therefore obtain the ratio of each phenotypic combination by multiplying the probabilities of the individual phenotypes:

3/4 smooth ×
3/4 yellow = 9/16 smooth yellow
3/4 smooth ×
1/4 green = 3/16 smooth green
1/4 wrinkled ×
3/4 yellow = 3/16 wrinkled yellow
1/4 wrinkled ×
1/4 green = 1/16 wrinkled green
16/16, and a ratio of 9:3:3:1

These are exactly the ratios that were observed,* testifying to the independence, or *independent assortment*, of these two gene pairs.

GENOTYPES OF DIHYBRID CROSSES

The above 9:3:3:1 ratio refers only to the phenotypes produced in the dihybrid cross. What were the genotypes? If we denote the factor or gene for smooth seed shape *S*, wrinkled seed shape *s*, yellow seed color *Y*, and green seed color *y*, the original parental strains in Mendel's cross can be symbolized $SSYY \times ssyy$. The gametes formed by each of the parental strains would therefore have either the formula *SY* or *sy*, which would combine to form an F_1 of genotype *SsYy*. Obviously the dominance relationship between *S* and *s* is not affected by that between *Y* and *y*, or vice versa, since all F_1 individuals are of smooth and yellow seed shape. What types of gametes does the F_1 produce? Clearly, according to the principle of segregation, half the gametes will contain *S* and the other half *s*. Similarly, for the *Yy* gene pair, half the gametes will contain *Y* and the other half *y*. There will then be equal proportions of *SY* and *Sy* gametes and there will also be equal proportions of *sY* and *sy* gametes. In other words, the four types of gametes produced by the F_1 are *SY*, *sY*, *Sy*, and *sy*, in a ratio 1:1:1:1 or 1/4:1/4:1/4:1/4. Since this is true for both the F_1 pollen and egg cells, the probability that each of the four types of pollen will combine with any one of the four types of egg cells is $1/4 \times 1/4 = 1/16$ (Fig. 7-1). As can easily be seen, nine of the combinations produce smooth yellow phenotypes, three are smooth green, three are wrinkled yellow, and one is wrinkled green.

TESTING DIHYBRID GENOTYPES

Since smooth is dominant over wrinkled and yellow over green, the F_2 plants that arose from wrinkled and green seeds must be homozygous for both recessives and have the genetic constitution *ssyy* (*double recessive*). Such plants, upon self-fertilization, would be expected to produce only wrinkled green seeds, and this was exactly what Mendel found. Smooth and yellow plants, on the other hand, might have varied genotypes: *SSYY*, *SsYY*, *SSYy*, *SsYy*. For example, the F_1 hybrid between the smooth yellow and wrinkled green parental strains would be expected to have the genotype *SsYy*. As a test of this assumption Men-

* For a given total number of offspring, the expected number of each phenotype is obtained by multiplying the total number by the fraction expected. Thus the expected number of wrinkled green in this cross is $1/16 \times 556 = 34.75$, wrinkled yellow $= 3/16 \times 556 = 104.25$, etc.

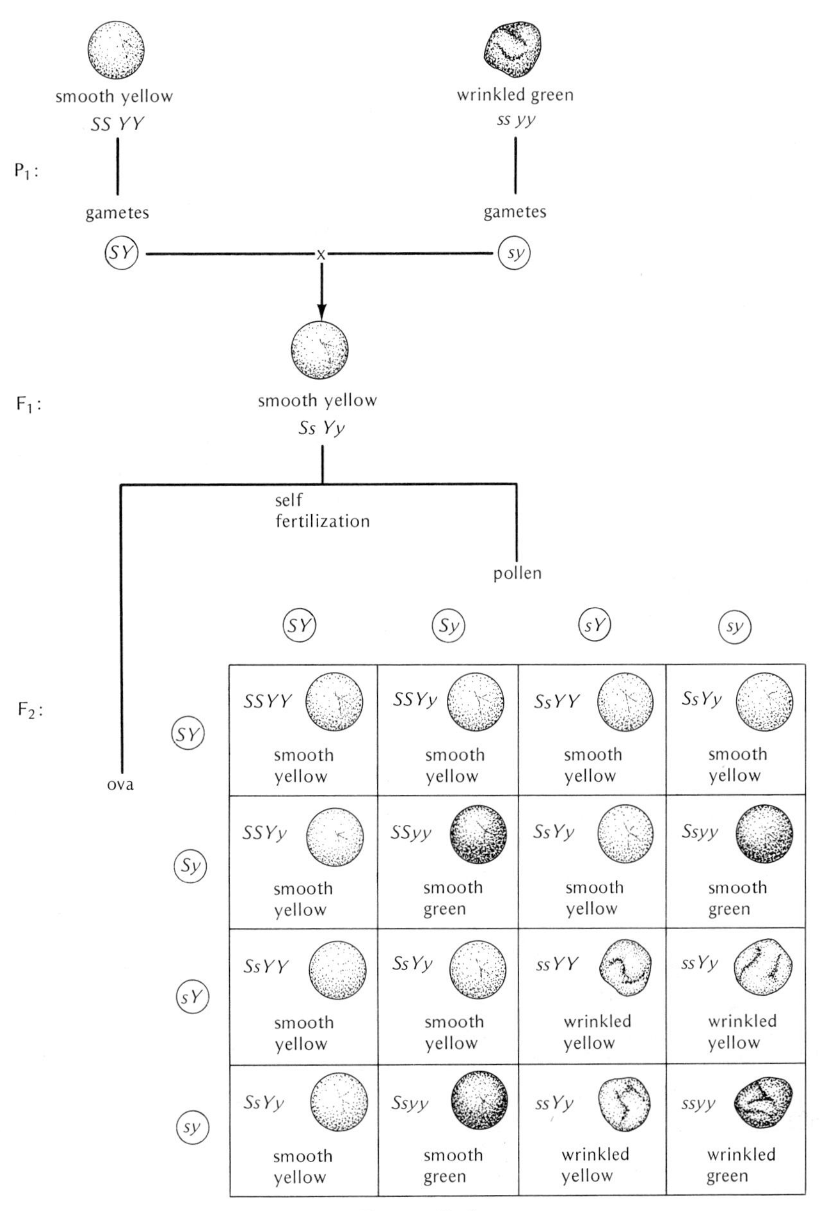

Figure 7–1

Explanation of Mendel's results for the segregation and assortment of seed shape and color.

del crossed the F_1 *SsYy* plants to each of the two parental types, as follows:

1. *SsYy* × *ssyy* (wrinkled green parent)
2. *SsYy* × *SSYY* (smooth yellow parent)

Although the F_1 is expected to produce four types of gametes in equal ratio (*SY*, *Sy*, *sY*, and *sy*), the double-recessive parental wrinkled green strain in cross 1 will produce only *sy* gametes. A combination between the two can then be diagrammed as follows:

		F_1 Gametes			
		SY	*Sy*	*sY*	*sy*
Gametes of Wrinkled Green Variety	*sy*	*SsYy*, smooth yellow	*Ssyy*, smooth green	*ssYy*, wrinkled yellow	*ssyy*, wrinkled green

Thus four phenotypic combinations of offspring would be expected, in equal proportions. Mendel scored more than 200 seeds produced from this cross, with the following results:

55 were smooth yellow and therefore of genotype *SsYy*.
51 were smooth green and therefore of genotype *Ssyy*.
49 were wrinkled yellow and therefore of genotype *ssYy*.
52 were wrinkled green and therefore of genotype *ssyy*.

As expected, the four phenotypic classes appeared in a 1:1:1:1 ratio.

The second cross, between the F_1 hybrid and the smooth yellow variety, produced offspring of only one phenotype, smooth yellow. This is also completely as expected, since the parental smooth yellow variety is homozygous, and all its gametes are *SY*. However, as might be anticipated according to the following diagram, the smooth yellow offspring should consist of four genotypes in a ratio 1:1:1:1.

		F_1 Gametes			
		SY	*Sy*	*sY*	*sy*
Gametes of Smooth Yellow Variety	*SY*	*SSYY*, smooth yellow	*SSYy*, smooth yellow	*SsYY*, smooth yellow	*SsYy*, smooth yellow

Since self-fertilization of a heterozygote causes homozygous recessives to appear in the next generation, Mendel permitted 177 of these smooth yellow F_2 plants to become self-fertilized in order to detect the presence of recessive genes. He found that, indeed, they consisted of four different types in equal proportions:

45 produced only smooth yellow seeds in the F_3 and were therefore *SSYY*.
42 produced smooth yellow and green seeds in the F_3 and were therefore *SSYy*.
47 produced smooth yellow and wrinkled yellow seeds in the F_3 and were therefore *SsYY*.
43 produced smooth yellow and green seeds, and wrinkled yellow and green seeds, and were therefore *SsYy*.

As a result of these tests the F_1 smooth yellows were therefore shown to have the genotype *SsYy*. On the other hand, the F_2 smooth yellow plants (derived from the $F_1 \times F_1$) could be expected to have four different genotypes, *SSYY*, *SSYy*, *SsYY*, and *SsYy*, in the ratio 1:2:2:4 (see Fig. 7–1). Mendel tested this assumption by selfing 301 of the original 315 smooth yellow F_2 plants (14 did not yield plants or seed) and obtained the following results:

38 produced only smooth yellow seeds and were therefore *SSYY*.
65 produced only smooth yellow and green seeds and were therefore *SSYy*.
60 produced only smooth yellow and wrinkled yellow seeds and were therefore *SsYY*.
138 produced smooth yellow and green seeds, and wrinkled yellow and green seeds, and were therefore *SsYy*.

The ratio 38:65:60:138 represented, according

to Mendel, "a very fair approximation to the (expected) ratio numbers of 33:66:66:132," or an approximate overall ratio of 1:2:2:4.

CROSSES INVOLVING THREE OR MORE GENE DIFFERENCES

The excellent correspondence that Mendel found between the results of dihybrid crosses and those expected according to independent assortment was also true for trihybrid crosses, or the assortment of three gene differences. In his trihybrid experiment he used the following three gene pairs (the first factor mentioned is dominant):

1. Smooth and wrinkled seed shape (*S* and *s*)
2. Yellow and green seed color (*Y* and *y*)
3. Violet and white flowers (*V* and *v*)

He crossed smooth yellow violet plants (*SSYYVV*) to wrinkled green and white plants (*ssyyvv*) and obtained an F_1 having the phenotype of the dominant parent (*SsYyVv*). The F_1 was then selfed and gave rise to F_2 plants that had 8 phenotypes and 27 presumed genotypes in the following approximate ratio:

27 smooth yellow violet (8 genotypes)
9 smooth yellow white (4 genotypes)
9 smooth green violet (4 genotypes)
9 wrinkled yellow violet (4 genotypes)
3 smooth green white (2 genotypes)
3 wrinkled yellow white (2 genotypes)
3 wrinkled green violet (2 genotypes)
1 wrinkled green white (1 genotype)

Both the genotypic and phenotypic F_2 ratios can be derived by considering that 8 different types of gametes are formed by the F_1 hybrid: *SYV*, *SYv*, *SyV*, *Syv*, *sYV*, *sYv*, *syV*, and *syv*. As in the dihybrid cross, each of these gametes has an equal chance of combining with any other gamete, thereby providing $8 \times 8 = 64$ expected combinations. Because of dominance only 8 different phenotypes are formed, since some phenotypic effects are produced by more than one genotype; i.e., smooth yellow violet can be genotypically produced in 8 different ways. In such crosses the number of different phenotypes is therefore always less than the number of genotypes.

When the number of gene-pair differences is more than three, the number of possible combinations between them is greatly increased. Table 7–1 gives the numbers for each of the various categories that can be expected from hybrids that are heterozygous for different numbers of gene pairs. As we have seen, the different kinds of gametes produced by an F_1 hybrid increases as the number of gene pairs for which it is heterozygous increases. This relationship is expressed as 2^n different kinds of gametes, where n is the number of gene pairs for which the hybrid is heterozygous. The number of combinations of gametes that result when F_1 hybrids are crossed is, of course, the product of multiplying the different kinds of gametes produced by each, or $2^n \times 2^n = 2^{2n} = 4^n$. This value is the "perfect" or minimum population size necessary for each genotype to be expressed in at least one individual according to its expected frequency. For example, hybrids heterozygous for 2 gene pairs (i.e., *AaBb*) will produce the homozygous double recessive (*aabb*) in 1 out of 16 combinations, or once in a perfect population size of 16. Within this perfect population size, some genotypes will be produced more than once, and the number of possible genotypes produced is 3^n.

Note that with each increase in the number of gene-pair differences, the number of different genotypes produced increases to a much greater extent, so that differences at 10 gene pairs, for example, would yield 3^{10}, or 59,059, different possible genotypes. Since considerably more than 10 gene-pair differences exist in the individuals of most species, the number of different possible genotypes is immense. Even if only 20 gene-pair differences are segregating independently in a species, the number of possible genotypes is about $3\frac{1}{2}$ billion.

It is also of interest that as the number of gene-pair differences increases, the relative proportion of homozygous genotypes decreases. For example, a cross between individuals heterozygous for a

TABLE 7–1

Effects of segregation and independent assortment in crosses between individuals heterozygous for given numbers of gene pairs

Number of Gene Pairs in which Differences Occur	Different Kinds of Gametes Produced by F_1 Heterozygotes	Number of Combinations of $F_1 \times F_1$ Gametes —"Perfect Population Size"	Different Kinds of Genotypes in F_2	Different Kinds of F_2 Genotypes that are Homozygous	Different Kinds of F_2 Genotypes that are Heterozygous	Different Kinds of Phenotypes in F_2 (Complete Dominance)
n	2^n	4^n	3^n	2^n	$3^n - 2^n$	2^n
1	2	4	3	2	1	2
2	4	16	9	4	5	4
3	8	64	27	8	19	8
4	16	256	81	16	65	16
5	32	1,024	243	32	211	32
10	1,024	1,084,576	59,049	1,024	58,025	1,024

single gene pair will produce three genotypes, two of which are homozygous (i.e., *AA* and *aa*). A cross between individuals heterozygous for two gene pairs will produce nine genotypes, four of which are homozygous (i.e., *AABB*, *AAbb*, *aaBB*, *aabb*). The number of homozygous genotypes produced for n gene-pair differences is thus 2^n, and the number of heterozygous genotypes is $3^n - 2^n$. With little doubt, most, if not all, individuals in any freely interbreeding population are therefore heterozygous for at least one gene-pair difference, and probably for many more.

Two factors contributing to the immense variability possible among sexually reproducing diploid organisms are therefore readily apparent. First, matings between a wide variety of genotypes ensure the continued production of new combinations each generation. Second, the offspring of individual heterozygotes are always segregating for a variety of gene differences. With few exceptions (i.e., identical twins) sexual-reproducing organisms can therefore easily furnish unique genotypes to every living member of the species.

HISTORY OF MENDEL'S DISCOVERY

Mendel presented his main results to the Brünn Society of Natural Science in 1865 and also published them in the proceedings of that society in 1866. Although the title of his paper was cited in a few bibliographies, there was no response to his discovery until 1900. This long delay in recognizing Mendel's contributions derived from a number of special reasons.

One reason for this neglect arose from Mendel's demonstration that the variability among the F_2 and further hybrid generations could be traced to the original variability in the first parental cross. According to Mendel, the factors that were followed did not change during the period of observation but only expressed themselves in new and different combinations among the offspring. To those biologists who were seeking for a source of variability in evolution, Mendel's findings indicated, on the contrary, an unacceptable "constancy" of hereditary factors. In addition the types of characters Mendel used in his experiment were examples of *discontinuous characters.* That is, Mendel followed differences that appeared to be "all or none," i.e., tall or short, smooth or wrinkled, yellow or green, etc. Many biologists concerned with the problems of heredity and evolution, such as Galton, Darwin, and others, were looking for gradual changes from one type to another, *continuous variation.* The concept that continuous variation in a character may actually occur through a series of small discontinuous steps, each determined by mendelian laws, was not generally adopted until much later (see Chapter 14). Furthermore, Mendel's preoccupation with probability events and mathematical ratios was an unfamiliar approach to biology that did not excite much interest. Prevailing opinions maintained that biological phenomena were much too complex to yield to mathematical analysis. If we also consider that the physical basis of heredity in terms of nuclear and chromosomal division was not clarified until later, it is easy to understand that Mendel's findings took place in a period unprepared to appreciate and develop them.

Nägeli, the famous botanist with whom Mendel corresponded, was no less subject to the many prejudices of the time than other biologists and therefore also failed to recognize Mendel's fundamental contributions. If Mendel's results had been confirmed in other organisms, it is likely that his influence would have spread much more rapidly. Unfortunately no one took up Mendel's experimental banner and Mendel himself failed to further develop his contributions in published form. His communications with Nägeli mainly concerned repeated failures to achieve a demonstration of segregation in the hawkweed, *Hieracium*. Unknown to Mendel, *Hieracium* was a poor choice as an experimental organism because many of its embryos, especially those of hybrids between two different varieties, arise directly from diploid tissue in the ovary without fertilization of gametes (apomixis). Such embryonic tissue will, of course, have exactly the same genetic constitution as its maternal parent, and will therefore not segregate for different characters among its offspring.

With the assumption of extensive duties in 1868 as abbot of his monastery, Mendel offered no further publications on heredity. However, his work on the pea plant had been distributed to many libraries and found its way into a few bibliographies concerned with hybridization.

When, toward the end of the nineteenth century, attention began to turn to investigating single characters with regular hereditary patterns, bibliographies were searched for appropriate illustrations. Finally, in 1900 Mendel's work was discovered simultaneously by three botanists, Correns, DeVries, and Tschermak. There was an immediate appreciation and response to Mendel's experiments and they rapidly became known throughout the world.

CORRESPONDENCE BETWEEN MENDELIAN FACTORS AND CHROMOSOMES

The year 1900 also marked a period when cellular studies had already shown the particulate nature of individual chromosomes and their duplication and separation in meiosis. The correspondence between chromosomes and mendelian factors was striking: Chromosomes were in pairs of maternal and paternal origin; mendelian factors were in pairs of maternal and paternal origin. Chromosomes in each pair separated, or "disjoined," during meiosis, one going to each gamete; so did mendelian factors. The spindle arrangement of the two chromosomes in a pair is independent of the arrangement of other chromosome pairs, and a gamete may contain any mixture of maternal and paternal chromosomes; the same relationship holds true for mendelian factors which assort themselves independently of each other. Fertilization restores the diploid number of chromosomes; fertilized plants show a "doubleness" for each pair of mendelian factors. In 1903 relationships between chromosomes and mendelian factors were simultaneously pointed out by Sutton and Boveri. Within a decade a number of experiments provided critical evidence that these relationships were true.

Among the first of these experiments were those relating to sexual characteristics. The chromosomal influence on sex, in certain insects, had been shown by McClung to be associated with a special sex-determining "X" chromosome. McClung proposed that a male had one X chromosome per cell (XO) and a female had two X's (XX), and that the male produced two types of gametes, X and O, in equal proportions. In the insect hemipteran *Protenor*, which was investigated by Wilson and others, the X chromosome could be clearly seen as a special dark-staining (heteropycnotic) body. As expected, counts of nuclei in *Protenor* testes showed approximately equal amounts of spermatids with and without X chromosomes. Since the female gametes all contained X chromosomes, the sex of the offspring was determined by the male gamete (see also Chapter 12). Thus segregation of the X chromosome in the male led to an equal proportion of male and female offspring. This provided an excellent correspondence between the segregation of the sex factor and the segregation of the sex chromosome.

Sometime later (1913) Carothers showed that an insect orthopteran, *Brachystola*, contained among its chromosomes a homologous pair that synapsed during meiosis, although the chromosomes were of unequal size (heteromorphic, *H* and *h*). Each member of this pair of chromosomes could be visibly recognized in male spermatids, as well as the presence or absence of the X chromosome. Thus a male segregating independently for both kinds of chromosomes should form four types of sperm in equal numbers: X*H*, X*h*, O*H*, and O*h*. Indeed, Carothers found there were 154 each of X*H*, O*h*, and 146 each of X*h*, O*H*, in a particular testis, thereby furnishing cytological evidence for the mendelian principle of independent assortment. Figure 7–2 shows the meiotic chromosomal segregations responsible for the production of the four different gametes in Mendel's F_1 smooth yellow heterozygote, *SsYy*.

SYMBOLS

In the description of Mendel's work only letter symbols have been used, such as *S* for the dominant-factor smooth seeds and *s* for the recessive-factor wrinkled seeds. There was no indication which factor represents the normal appearance of the organism or is the one most commonly found. In most experiments, however, symbols for genetic factors usually depend on whether they are normal ("wild type") or differ from normal (mutation). Symbolically the genetic factor for wild type is designated as +, and the factors for mutations by alphabetical letters derived from their descriptive names. Thus *vg* is used for the mutation-producing vestigial-winged *Drosophila* flies and *H* for hairless flies. In most instances recessive mutations are designated by small letters (*vg*) and dominant mutations by capitalizing the first letter (*H*). In diploid organisms homozygotes can be designated by repeating the symbol twice, *vg*/*vg*, or else by letting the symbol stand by itself, *vg*.

Similarly, wild-type homozygotes can be written +/+ or just +. Heterozygotes for a single gene difference such as *vestigial* and wild-type are symbolized + /*vg*. The diagonal between two symbols (*vg*/*vg*, + / +, or + /*vg*) designates a pair of homologous chromosomes. If two or more pairs of genes are being scored, each associated with a different pair of chromosomes, i.e., *vg* and *H*, a fly homozygous for both mutations is described as *vg*/*vg*, *H*/*H*, or *vg*, *H*. The wild-type/mutant heterozygote for the two gene differences, or "double heterozygote," is symbolized *vg*/+, *H*/+.

The wild-type gene can also be symbolized as a superscript for a particular mutant factor. Thus a wild-type homozygote for the vestigial factor can be either vg^+/vg^+ or +/+. If two or more pairs of genes are being traced, then designating the wild-type gene by a "+" superscript enables discrimination between different wild-type genes such as vg^+ and H^+ ($+^{vg}$ and $+^{H}$ symbols have also been used).

SEGREGATION AND ASSORTMENT IN HAPLOID ORGANISMS

The major portion of the life cycle of organisms such as algae and fungi is in the haploid stage. Such organisms become diploid during sexual conjugation but then only until meiosis forms new haploid clones. Since the presence of chromosomes in haploid condition is equivalent to the presence of genes in single unpaired condition, haploid organisms are neither homozygotes nor heterozygotes. Phenotypically, therefore, each allele in a haploid appears undisguised by an allelic mate, and segregation and assortment between genes can be expected to give different phenotypic ratios than in diploid organisms.

For example, although the single-celled alga *Chlamydomonas reinhardi* is ordinarily green, a gene influencing this character occasionally mutates, causing the appearance of a variety that is

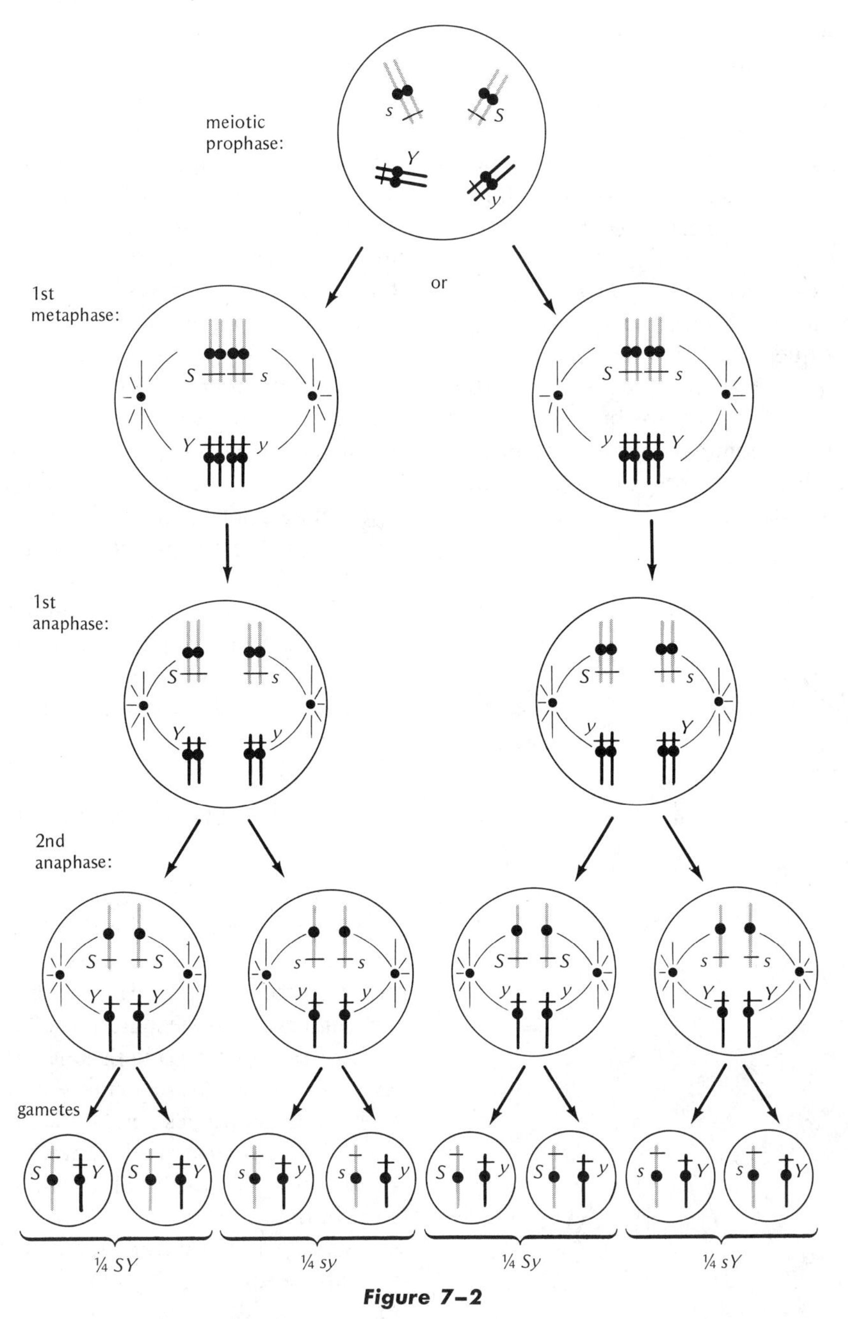

Figure 7–2

Explanation for segregation and independent assortment of seed shape and seed color in Mendel's experiments in terms of factors localized on different chromosomes. For simplicity, chiasmata and chromatid exchanges are omitted.

yellow when grown in the dark. Since *Chlamydomonas* is haploid, each cell carries only a single allele for this gene, either yellow, y, or normal, y^+. If crosses are made between two different mating types, one carrying y and the other y^+, the zygote will now be diploid $y\,y^+$ (Fig. 7–3). When meiosis occurs the haploid products of this zygote will then be of two kinds with respect to color, y and y^+, both in equal frequencies. This one-to-one segregation ratio for differences at a pair of genes can be expected in all haploids that have sexual conjugation and subsequent meiosis. Note that this ratio is identical to the 1:1 gametic segregation ratios that occur in diploid organisms, but is of course different, because of dominance, from the phenotypic ratios observed in crossing two diploid pure-breeding strains.

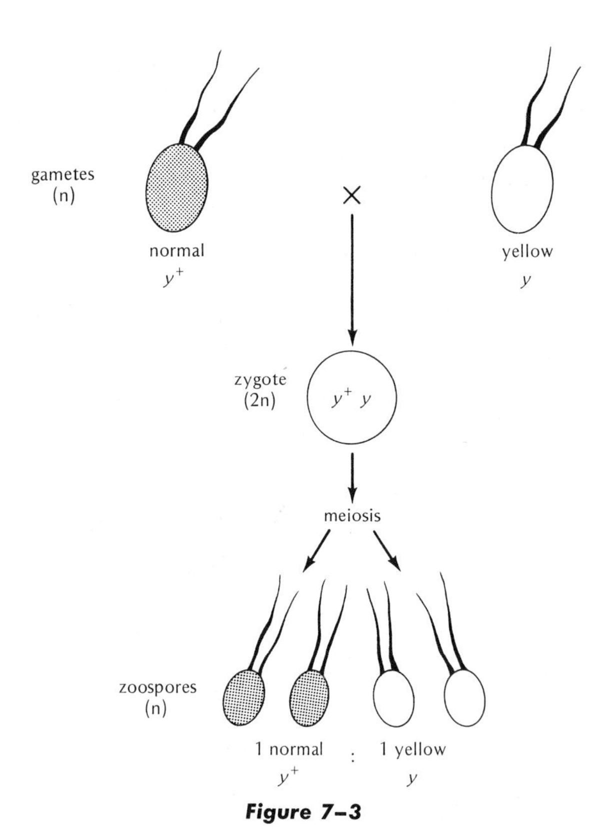

Figure 7–3

Segregation of the yellow trait in Chlamydomonas reinhardi. *n = haploid, 2n = diploid.*

When differences exist between haploid varieties at two gene pairs, independent assortment of genes can also be observed. However, the ratios again will differ from those noted in diploids. For example, the *Chlamydomonas* cross described above also differs in respect to the mating type. That is, in order for the yellow variety to mate with the green, they must each be of opposite mating types. Mating type, however, is a characteristic also inherited as a single gene difference, some strains being mt^+ and others mt^-. When y and y^+ strains are crossed, the mating type is found to segregate independently of color, and the progeny show an approximate 1 : 1 ratio of mt^+ to mt^-. However, since the presence of yellow or green does not interfere with the mt genotype of an individual, approximately half the yellow individuals are mt^+, half are mt^-, and the same mt distribution holds for green. Thus about 1/4 are y^+mt^+, 1/4 y^+mt^-, 1/4 $y\,mt^+$, and 1/4 $y\,mt^-$. Both the segregation of alleles and the independent assortment of homologous chromosomes are thereby easily demonstrated in certain haploids.

TETRAD ANALYSIS

One experimental advantage in using *Chlamydomonas* and certain other haploids such as *Neurospora* is the ability to recover all four haploid meiotic products (tetrad) of an individual zygote. Through various techniques of *tetrad analysis*, each of the four zygotic products (eight in *Neurospora*) can be separately cultured and their phenotypes and genotypes analyzed. For the *Chlamydomonas* cross described above, a tetrad analysis showed that the four haploid cells of some zygotes were in the ratio 2 y^+mt^+:2 $y\,mt^-$. Such tetrads are known as *parental ditypes* and arise because different chromosomes of the same parent have assorted to the same side of the first meiotic division spindle (left side of Fig. 7–4). However, since assortment of chromosomes is independent of which parent they come from, half the time one would expect two different chromosomes from one parent to go

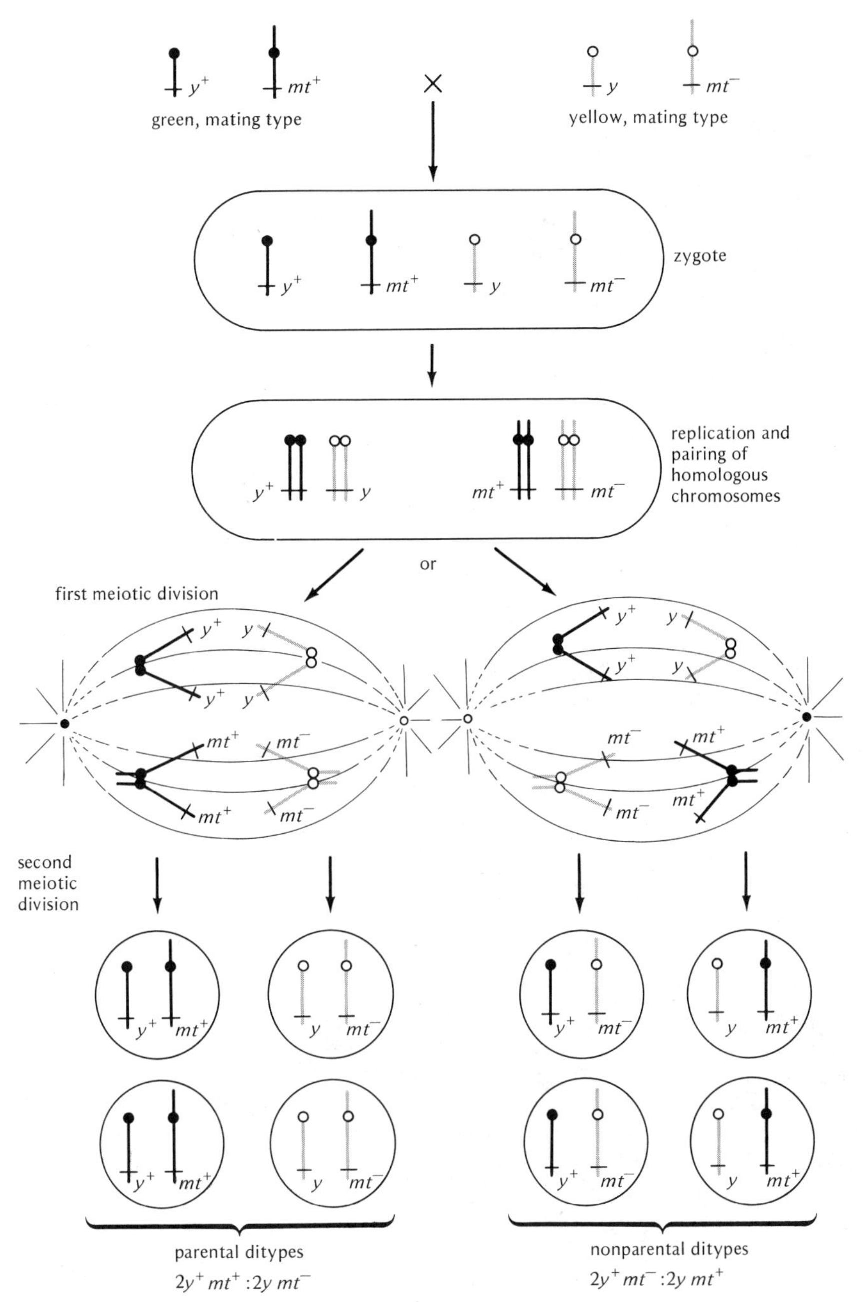

Figure 7–4

Segregation and independent assortment of the genes for color and mating type in Chlamydomonas.

to opposite poles of the spindle (right side of Fig. 7–4). This is precisely what is found, as evidenced by the observation that approximately equal numbers of tetrads contain cells in the ratio $2\,y^+mt^-:2\,y\,mt^+$, or are of the *nonparental ditype.* In Fig. 7–4 note also that for both parental and nonparental ditypes, the separation of different parental alleles, y^+ from y and mt^+ from mt^-, occurs in the first meiotic division, signifying *first-division segregation.*

In addition to these two types of tetrads, however, a third type is also found, which contains four haploid spores each genetically different: y^+mt^+, y^+mt^-, $y\,mt^+$, $y\,mt^-$. This last tetrad, called a *tetratype*, arises from crossing-over events that occur in the first meiotic division. As diagrammed in Fig. 7–5, a crossover between two chromatids of the homologues carrying y^+ and y in the region between the genes and their centromeres leads to the two different alleles remaining together in the first meiotic division products, but separating in the second division (*second-division segregation*). The results of crossing over in this case are therefore an association of both kinds of y alleles with each kind of mt allele. It can easily be noted that if a similar crossover were to occur instead between the mt^+ and mt^- chromosomes, a tetratype would also result. The proportion of tetratypes is therefore a reflection of the extent of crossing over that can occur between a gene and its centromere. In studies that will be described later (Chapter 18), such measurements have special value in calculating gene distances along a chromosome.

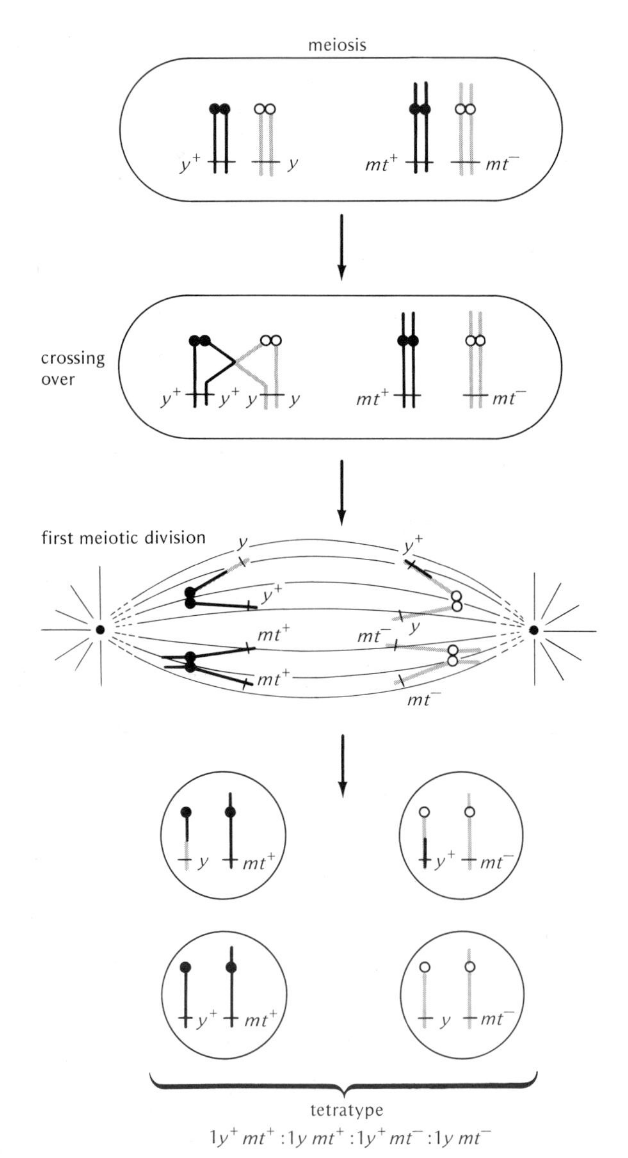

Figure 7–5

Production of a tetratype as a result of crossing over between a gene and its centromere. Note that in both this case and the example illustrated in Figure 7–4, each pair of gene differences considered individually always segregates in a 1:1 ratio. It is only the combinations resulting from the assortment of two gene pairs that mark the differences between ditypes and tetratypes.

PROBLEMS

7–1. In garden peas the effect of the tall allele (T) is dominant over that for short (t), and the effect of the smooth-seeded allele (S) is dominant over that for wrinkled (s). These two gene pairs are also known to assort independently of each other. (a) What proportions of phenotypes would you expect among the progeny of tall smooth-seeded F_1 plants crossed to each other if each such F_1 plant were derived from a cross between a pure breeding tall smooth-seeded variety (*TTSS*) and a short wrinkled-seeded variety (*ttss*)? (b) Would the proportions of phenotypes in the F_2 generation be changed if the F_1 plants were derived from a cross between a tall wrinkled-seeded variety (*TTss*) and a short smooth-seeded variety (*ttSS*)? (c) What phenotypic results would you expect if the F_1 plants in (a) were crossed to a short wrinkled-seeded plant?

7-2. In dogs dark coat color is dominant over albino, and short hair is dominant over long hair. If these effects are caused by two independently segregating gene pairs, write the most probable genotypes for the parents of each of the following crosses, using the symbols *C* and *c* for the dark and albino coat-color alleles, and *S* and *s* for the short- and long-hair alleles, respectively.

Parental Phenotypes	Phenotypes of Offspring: Dark Short	Dark Long	Albino Short	Albino Long
(a) dark short × dark short	89	31	29	11
(b) dark short × dark long	18	19	0	0
(c) dark short × albino short	20	0	21	0
(d) albino short × albino short	0	0	28	9
(e) dark long × dark long	0	32	0	10
(f) dark short × dark short	46	16	0	0
(g) dark short × dark long	29	31	9	11

7-3. In *Drosophila melanogaster* one gene pair is known to affect wing size, and the allele for normal long wings (vg^+) in this gene pair has a dominant effect over the allele for short vestigial wings (*vg*). Another independently assorting gene pair affects body color: The allele for normal grey body color (e^+) is dominant to that for ebony body color (*e*). A cross is made between a fly with normal wings and ebony body color and a fly with vestigial wings and normal body color. The normal-appearing F_1 are crossed among each other and 512 F_2 flies are raised. What phenotypes would you expect in the F_2 and in what numbers would you expect to find them?

7-4. Let us assume that you were given the parental guinea pigs in the Figure following (P_1) and crossed them to obtain F_1 individuals of the type pictured. F_1 males were then crossed with F_1 females and an F_2 was then obtained according to the numbers shown in the figure. (a) Using the simplest explanation, how many gene-pair differences would you say were involved in this cross? (b) Designate your gene pairs by alphabetical letters and indicate what phenotypic characteristics they affect and whether these genes act as dominants or

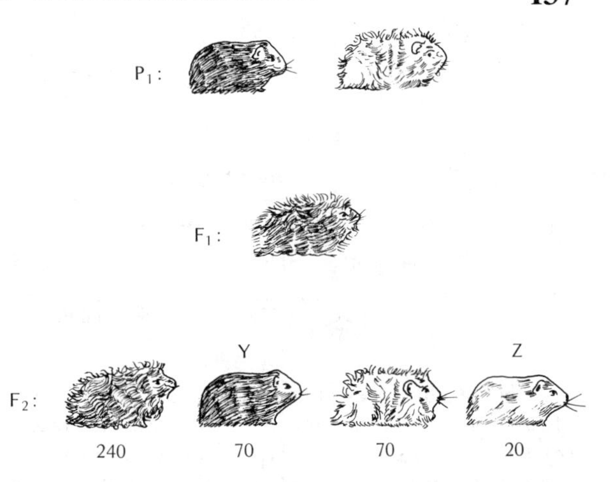

Three generations of guinea pigs described in Problem 7–4.

recessives. (c) On the basis of your gene symbols, give the possible genotypes of the F_2 phenotypes that have been marked Y and Z. (d) Using the chi-square method (p. 148), test your hypothesis and indicate whether you accept or reject it.

7-5. In corn a pair of genes determines leaf shape and another pair determines pollen shape. A ragged-leafed plant with round-pollen was crossed to a ragged-leafed plant with angular-pollen, and the resultant progeny were classified as follows:

Class 1: 186 ragged-leaf round-pollen
Class 2: 174 ragged-leaf angular-pollen
Class 3: 57 smooth-leaf round-pollen
Class 4: 63 smooth-leaf angular-pollen
Total 480

(a) Using alphabetical letters of your choice, designate the genes for the different leaf and pollen characters. (b) On the basis of the symbols given in (a), provide genotypes for the two parents. (c) According to your hypothesis what numbers would you have expected for each of the four classes of progeny? (d) Test your hypothesis statistically using the chi-square method (Chapter 8) and indicate whether you accept or reject your hypothesis.

7-6. Certain varieties of flax have been found by Flor to show different resistances to specific races of a fungus called flax rust (*Melampsora lini*). For example, flax variety "770B" is resistant to rust race 24 but susceptible to rust race 22, whereas flax variety "Bombay" is resistant to rust race 22 and susceptible to rust race 24. Flor crossed flax varieties 770B and Bombay and found the hybrid resistant to both races 22 and 24. When the F_1 hybrid was permitted to self-fertilize, it produced an F_2 that had the following phenotypic proportions:

		Rust Race 22		
		Resistant	Susceptible	
Rust Race 24	Resistant	110	43	observed F_2 numbers
	Susceptible	32	9	

(a) Propose a hypothesis to account for the genetic basis of rust resistance in flax for these particular rust races. (b) On the basis of your hypothesis, what numbers would you expect for each of the four categories in the F_2? (c) Test your hypothesis, using the chi-square method.

7-7. In 1902 Bateson reported the first postmendelian study of a cross differing in two characters. White Leghorn chickens, having large "single" combs and white feathers, were crossed to Indian Game Fowl, with small "pea" combs and dark feathers. The F_1 was white with pea combs. A cross $F_1 \times F_1$ produced the following F_2: 111 white pea, 37 white single, 34 dark pea, 8 dark single. (a) What numbers of each would you have expected? (b) Test your explanation statistically, using the chi-square method.

7-8. What phenotypes and proportions would you expect from crossing the F_1 of the previous question: (a) To the White Leghorn stock? (b) To the Indian Game fowl stock? (c) To the dark-feathered single-combed F_2?

7-9. If two gene pairs *A* and *a* and *B* and *b* are assorting independently with *A* dominant to *a* and *B* dominant to *b*, what is the probability of obtaining: (a) An *AB* gamete from an *AaBb* individual? (b) An *AB* gamete from an *AABb* individual? (c) An *AABB* zygote from a cross *AaBb* × *AaBb*? (d) An *AABB* zygote from a cross *aabb* × *AABB*? (e) An *AB* phenotype from a cross *AaBb* × *AaBb*? (f) An *AB* phenotype from a cross *aabb* × *AABB*? (g) An *aB* phenotype from a cross *AaBb* × *AaBB*?

7-10. A cross made between two plants differing in four independently assorting gene pairs, *AABBCCDD* × *aabbccdd*, produces an F_1 which is then self-fertilized. If the capital letters represent alleles with dominant phenotypic effect: (a) How many different genotypes are possible in the F_2? (b) How many of these F_2 genotypes will be phenotypically recessive for all four factors? (c) How many of these F_2 genotypes will be homozygous for all dominant genes? (d) Would your answers to (a), (b), and (c) be different if the initial cross were *AAbbCCdd* × *aaBBccDD*?

7-11. Down's syndrome (monogolian idiocy; Chapter 21) occurs in humans when a particular chromosome (no. 21) is present in triplicate instead of in the usual diploid state; i.e., such individuals have 47 chromosomes instead of the normal 46. What proportion of offspring produced by an affected mother with 47 chromosomes mated to a normal man would be similarly affected?

7-12. The transmission of two relatively rare traits is shown in the following pedigree. Trait 1 is indicated by shading of the upper half of the symbol and trait 2 is indicated by shading of the lower half. Using alphabetical symbols for the genes involved (e.g., *A* and *a* for trait 1, and *B* and *b* for trait 2), answer the following: (a) What kind of inheritance is involved in each of the traits? (b) What are the genotypes of individuals IV-4 to IV-10? (c) What phenotypes and proportions would you expect from a mating between individuals IV-3 and IV-5?

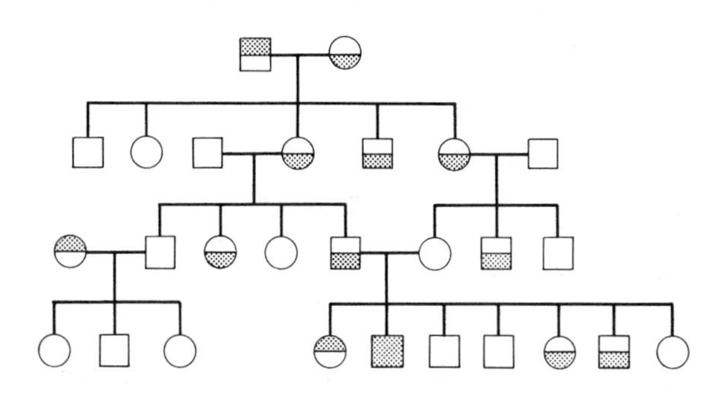

7-13. In *Neurospora* the *peach* gene (*pe*) is on one chromosome and the *colonial* gene (*col*) is on another independently assorting chromosome. What kinds of tetrads would you expect and in what proportions if two strains, $pe\ col^+ \times pe^+\ col$, were crossed and only first-division segregation occurred for both the genes?

7-14. A cross between a pink (*p*) yeast strain of mating type mt^+ and a gray strain (p^+) of mating type mt^- produced the following tetrads:

Numbers of tetrads	Kind of tetrad
18	p^+mt^+, p^+mt^+, $p\ mt^-$, $p\ mt^-$
8	p^+mt^+, $p\ mt^+$, p^+mt^-, $p\ mt^-$
20	p^+mt^-, p^+mt^-, $p\ mt^+$, $p\ mt^+$

On the basis of these results, are the *p* and *mt* genes on separate chromosomes?

7-15. Lindegren crossed a *Neurospora* strain of mating type *A* carrying the *fluffy* growth gene with a strain of mating type *a* bearing the normal growth gene. Of 872 ascospores, how many would you expect to be of the nonparental ditype if these genes assort independently of each other and no crossing over occurs between either of these genes and the centromere?

REFERENCES

BOVERI, T., 1904. *Ergebnisse über die Konstitution der chromatischen Substanz des Zellkerns.* G. Fischer, Jena.

CAROTHERS, E. E., 1913. The mendelian ratio in relation to certain orthopteran chromosomes. *Jour. Morph.* **24,** 487–511. (Reprinted in Voeller's collection; see References Chapter 1.)

DUNN, L. C., 1965. *A Short History of Genetics.* McGraw-Hill, New York.

MCCLUNG, C. E., 1902. The accessory chromosome —sex determinant? *Biol. Bull.*, **3,** 43–84. (Reprinted in Voeller's collection; see References Chapter 1.)

OLBY, R. C., 1966. *Origins of Mendelism.* Constable, London.

STERN, C. (ed.), 1950. *The Birth of Genetics.* Issued as a supplement to *Genetics*, Vol. **35.** (Contains English translations of Mendel's letters to Carl Nägeli, and of the papers by DeVries, Correns, and Tschermak, in which they first announced their discovery of the genetic laws previously discovered by Mendel.)

STERN, C., and E. R. SHERWOOD (eds.), 1966. *The Origin of Genetics, A Mendel Source Book.* W. H. Freeman, San Francisco. (Contains translations of Mendel's classic paper on garden peas, his paper on *Hieracium*, and his letters to Nägeli. Included also are relevant papers by DeVries and Correns, and a discussion of Mendel's observed ratios by R. A. Fisher and S. Wright.)

STURTEVANT, A. H., 1965. *A History of Genetics.* Harper & Row, New York.

SUTTON, W. S., 1903. The chromosomes in heredity. *Biol. Bull.*, **4,** 213–251. (Reprinted in the collections of Peters, of Gabriel and Fogel, and of Voeller; see References, Chapter 1.)

WILSON, E. B., 1905. The chromosomes in relation to the determination of sex in insects. *Science*, **22,** 500–502. (Reprinted in the collection of Gabriel and Fogel; see References, Chapter 1.)

8

PROBABILITY AND STATISTICAL TESTING

The physicist Helmholtz (1821–1894) once wrote that "all science is measurement." This statement is essentially true if we also agree that the method of measurement may differ from one science to another. In genetics much of what is measured concerns the ratios of different phenotypes and genotypes. These genetic ratios arise from probability relationships: the chance segregation and assortment of genes into gametes and their chance combination into zygotes. Since these are chance events, exact predictions cannot be made for any particular event. Just as when one tosses a coin there is no guarantee that heads must always follow tails, there is no guarantee that any particular genetic event (such as a particular genotype) will occur when a number of different types of events are possible. All we can say in advance of any chance event in genetics is that it has a certain probability of occurring.

In general terms the probability that an event will occur can be defined as the proportion of times in which that event occurs in a very large number of trials. If there are n trials and an event occurs on the average m times out of these n trials, the probability of the event may be considered to be m/n. This probability is, therefore, a fraction falling between zero and one. For example, the chance that heads will appear when an evenly balanced coin is tossed is 1/2, or a frequency of 50 percent. The closer the probability is to zero, the less probable is the event, and at zero the event presumably cannot occur. On the other hand, if the probability is one, or 100 percent, the event is considered a certainty.

Probability thus means "on the average." This "average" tends to become closer to its true expected value as the number of trials increases. A coin that is tossed 100 times may be expected to have a frequency of heads closer to 50 percent than a coin tossed only 10 times. Similarly, 100 offspring of a cross between parents that are heterozygotes for a single gene difference ($Aa \times Aa$) can be expected to show a closer approximation to a 1:2:1 AA:Aa:aa ratio than will only a few offspring from such a cross. However, even if a very large number of offspring are raised, it would be rare and indeed strange to find that such results, or the results of any experiment based on probability, fit an expected ratio perfectly. It is much more likely that some deviation from the expected ratio will occur within any single experiment, and of course different ratios will probably be obtained when the experiment is repeated. As is true for any observed numerical character of a population, such deviations are considered "statistical." The geneticist must then perform a statistical test to decide whether the observed deviations conflict with a proposed explanation, or *hypothesis*.

Essentially, therefore, the task of analyzing genetic ratios is twofold: one must determine the ratios that are expected and then determine how closely the actual results correspond to those that are expected.

PROBABILITY RULES

The determination of genetic ratios derives essentially from two basic laws of probability. The first law concerns calculation of the probability that two or more independent events will occur together. For example, Fig. 8–1 represents a population of circles half of which are shaded and the other half are not. Also half the circles are rough-edged and half are smooth-edged. If one condition has no effect on the other, a circle can be shaded and rough at the same time. Since the probability that a circle will be shaded is 1/2, and the probability that it will be rough is 1/2, then one-half the circles in Fig. 8–1 will be shaded and one-half will be rough. But note that the probability for a circle to be both shaded and rough at the same time is now $1/2 \times 1/2$, or 1/4. This stems from the law that when the probability of an event is independent of that of another event, so that the occurrence of one does not interfere with the occurrence of the other (shaded does not prevent rough or vice versa), the probability that both events will occur together is the *product* of their individual probabilities.

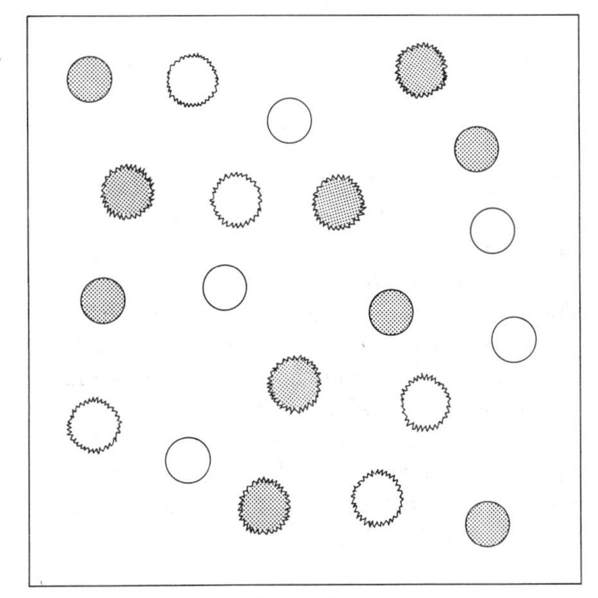

Figure 8–1

Independent events. Half the circles are shaded and half are rough edged. The chance for a circle being shaded is independent of its chance of being rough edged.

Alternatively, we may ask what is the probability that a certain event will *not* happen. For example, if the chances that a circle will be both shaded *and* rough is 1/4, what is the probability that such an event will not occur, i.e., that a circle will be *either* unshaded or smooth or both? Since the total probability for all events is 1, and 1/4 of these are shaded and rough, then $1 - 1/4$, or 3/4, represents the probability that a given circle will not be shaded *and* rough.

The second law of probability deals with occasions when the different types of events cannot occur together. If one occurs the other is excluded; that is, they are *mutually exclusive*, or *alternative*, events. In terms of the above example, this might mean that a circle cannot be both shaded or rough-bordered at the same time—if it is shaded, it is not rough, and vice versa. Therefore, if we had a group of circles of which 1/2 were rough and 1/2 were shaded, but neither of these events occurred in the same circle, they would appear as in Fig. 8–2.

Note that in this case, all circles are *either* shaded *or* rough; i.e., the probability that a circle will have one of these characteristics when these are alternative or mutually exclusive events is equal to the *sum* of their probabilities, or $1/2 + 1/2 = 1$. On the other hand, the probabilities that a circle will be *neither* shaded *nor* rough when all the circles are *either* shaded *or* rough is $1 - 1 = 0$.

CALCULATION OF GENETIC RATIOS

On the basis of mendelian principles, we already know that a diploid parent of the genetic constitution *Aa* may give rise to two types of gametes, *A* and *a*. If such an individual is self-fertilized or mates with a similar heterozygote, three different genotypes may be expected to appear among the offspring *AA, Aa,* and *aa*. If *A* is dominant over *a*, the offspring will show two different phenotypes, *A* (*AA*, *Aa*) and *a* (*aa*). On the other hand, the appearance of the same two phenotypes, *A* and *a*, may result from a mating between parents of the genetic constitution *Aa* × *aa*. Therefore, if the genetic constitution of the parents is not given, two possible explanations (hypotheses) exist which can account for the presence of these particular phenotypes *A* and *a* among the offspring. In what way can we discriminate between these two possible explanations?

Part of the answer to this question is simply to determine the ratio of genotypes and phenotypes expected according to each hypothesis. To do this in terms of probability, we make the assumption that the *Aa* parents produce two types of gametes, *A* and *a*, equally well and the *aa* parent produces only one type of gamete, *a*. Any combination between gametes to form a zygote depends on the frequency or probability of each type of gamete furnished by the parents. Thus the formation of a zygote is the result of two independent events (two gametes), each with their own probabilities, which now occur together. The probability that a particular zygote will be formed is therefore equal to the product of the probabilities of the gametes that compose it.

To calculate the values involved we can use the simple device of a checkerboard and place the gametes of each parent on one side of the checkerboard based on their probability of being formed. If two types of gamete, e.g., *A* and *a*, are formed by a parent with equal probability, then each is formed with a probability of one-half. (In the unusual case in which gametes may be formed in unequal ratios, a proposed numerical proportion can be placed next to each gamete, e.g., 4/5 *A* and 1/5 *a*). Each box of such a checkerboard then represents a zygote formed by multiplying the frequency of two gametes. For the hypotheses mentioned above, the checkerboards would look as follows:

mating: *Aa* × *Aa*

gametes		***A*** (P = 1/2)	***a*** (P = 1/2)
gametes	***A*** (P = 1/2)	*AA* Prob. 1/4	*aA* Prob. 1/4
	a (P = 1/2)	*Aa* Prob. 1/4	*aa* Prob. 1/4

mating: *Aa* × *aa*

gametes		***A*** (P = 1/2)	***a*** (P = 1/2)
gametes	***a*** (P = 1)	*Aa* Prob. 1/2	*aa* Prob. 1/2

For the *Aa* × *Aa* cross, four possible kinds of zygotes can be formed, each with a probability of 1/4. These zygotes, however, are mutually exclusive events; an *Aa* zygote cannot be *aA*, *AA*, or *aa* at the same time. Therefore, as for all mutually exclusive events, the probability that a zygote will be *either Aa* or *aA* (i.e., heterozygous) is $1/4 + 1/4$,

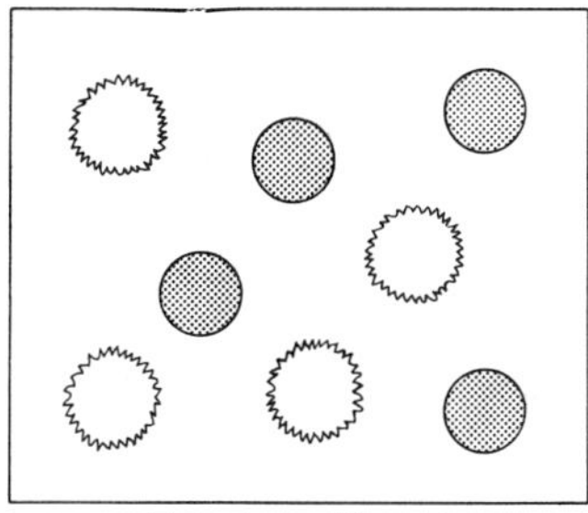

Figure 8–2

Mutually exclusive events; a circle cannot be both shaded and rough.

or 1/2; the probability that it will be either *AA*, *Aa*, or *aA* is 1/4 + 1/4 + 1/4, or 3/4; and the probability that it will be either *AA*, *Aa*, *aA*, or *aa* is, of course, equal to 1, or certainty. Note, however, that under conditions of complete dominance, the phenotypes of the homozygote *AA* and the two heterozygotes, *Aa* and *aA*, are identical. The probability that the dominant *A* phenotype will occur thus rests on three mutually exclusive events, each with a probability of 1/4, giving a total probability of 3/4.

For the *Aa* × *aa* cross, two kinds of genotypes occur with equal frequency, one producing the dominant *A* phenotype and the other producing the recessive *a* phenotype. Since each of these phenotypes has a probability of 1/2, the difference between the two hypotheses may thus be observed in the phenotypic ratios produced by each; i.e., there are 3 *A*:1 *a* phenotypes produced in one case and 1 *A*:1 *a* in the other.

RATIOS FOR TWO OR MORE SEGREGATING GENE PAIRS

Checkerboards can also be constructed for crosses which are segregating for two independent genes. For example, individuals heterozygous for the genes *A* and *B* (*AaBb*) will produce four types of gametes with equal frequency (*AB*, *Ab*, *aB*, and *ab*). A mating between two such individuals will give a checkerboard of 4 × 4 boxes, yielding a probability of 1/4 × 1/4, or 1/16, for each combination between two gametes. The use of this method was demonstrated in calculating the frequency of genotypes and phenotypes in Mendel's dihybrid crosses (Fig. 7–1).

When more than two pairs of genes are involved in a cross, the checkerboard system for determining expected ratios becomes unwieldy. A three-gene-pair difference, for example, would involve a checkerboard of 8 × 8 boxes, and additional gene differences would lead to more complicated constructions.

A shorter and simpler method than the checkerboard is to consider each pair of genes separately. For a single gene-pair difference there are six possible kinds of matings that can occur, each mating providing its own unique combination of genotypes and phenotypes, as follows:

Parents	**Offspring**	
	Genotypes	Phenotypes
1. *AA* × *AA*	all *AA*	all *A*
2. *AA* × *aa*	all *Aa*	all *A*
3. *AA* × *Aa*	1 *AA*:1 *Aa*	all *A*
4. *aa* × *aa*	all *aa*	all *a*
5. *aa* × *Aa*	1 *Aa*:1 *aa*	1 *A*:1 *a*
6. *Aa* × *Aa*	1 *AA*:2 *Aa*:1 *aa*	3 *A*:1 *a*

If the genotypes of the parents can be determined for each character, the expected genotypes and phenotypes of the offspring can be derived by a "branching" process, assuming there is no gene linkage (Chapter 16). For example, in the dihybrid mendelian experiments considered in Chapter 7, smooth (*S*) is dominant over wrinkled (*s*) and yellow (*Y*) over green (*y*). The offspring of a cross between heterozygotes for both gene pairs (*SsYy* × *SsYy*) will have genotypes and phenotypes that are combinations of both characters. Considering each character separately, each gene pair (*Ss* × *Ss*, and *Yy* × *Yy*) would produce three different genotypes in a 1:2:1 ratio or two phenotypes in a 3:1 ratio (mating No. 6 above). Since the probabilities associated with each gene pair are independent of the other gene pair, the genotypes and phenotypes that will appear (Fig. 8–3) are the result of multiplying the component probabilities.

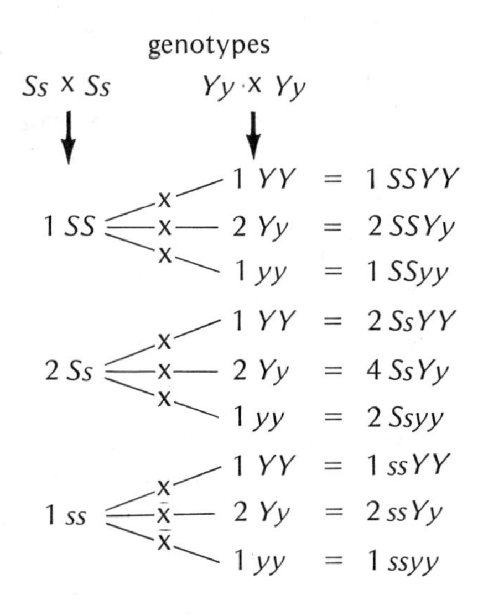

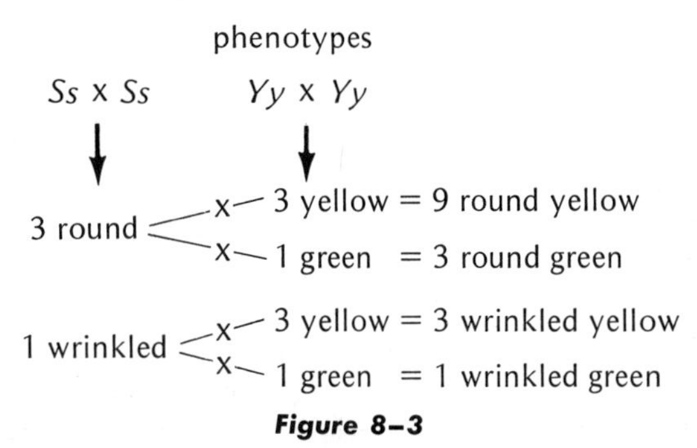

Figure 8–3

Determination of the genotypes and phenotypes produced in Mendel's dihybrid cross (seed shape and seed color) using the "branching" technique.

This same method can also be used for more than two gene differences, such as a cross between the heterozygotes *AaBbCc* × *AaBbCc*. If we use letters to designate each phenotype produced, i.e., *A* for the dominant phenotype and *a* for the recessive, the results would appear as in Fig. 8–4.

It is, of course, possible to have crosses in which both parents are not necessarily heterozygous for every gene pair but nevertheless differ genetically from each other. For example, a cross between two parents bearing the genetic constitutions

mating: *Aa Bb Cc* x *Aa Bb Cc*

Aa x *Aa* *Bb* x *Bb* *Cc* x *Cc*

3*A* — 3*B* — 3*C* = 27 *ABC*
3*A* — 3*B* — 1*c* = 9 *ABc*
3*A* — 1*b* — 3*C* = 9 *AbC*
3*A* — 1*b* — 1*c* = 3 *Abc*
1*a* — 3*B* — 3*C* = 9 *aBC*
1*a* — 3*B* — 1*c* = 3 *aBc*
1*a* — 1*b* — 3*C* = 3 *abC*
1*a* — 1*b* — 1*c* = 1 *abc*

Figure 8–4

Phenotypes produced in a cross between heterozygotes for three gene pairs where A designates the dominant phenotype for the A-a gene pair (AA or Aa) and a designates the recessive phenotype (aa). Similarly, B and C represent dominant phenotypes, and b and c represent recessive phenotypes.

AAbbCcDD × *aaBbCcDd* would, under conditions of complete dominance, produce the phenotypic results shown in Fig. 8–5.

LEVEL OF SIGNIFICANCE

Precisely determined as the above ratios may be, they are, as explained before, only expected or hypothetical. Actually observed ratios will depart to a greater or lesser extent from those expected.

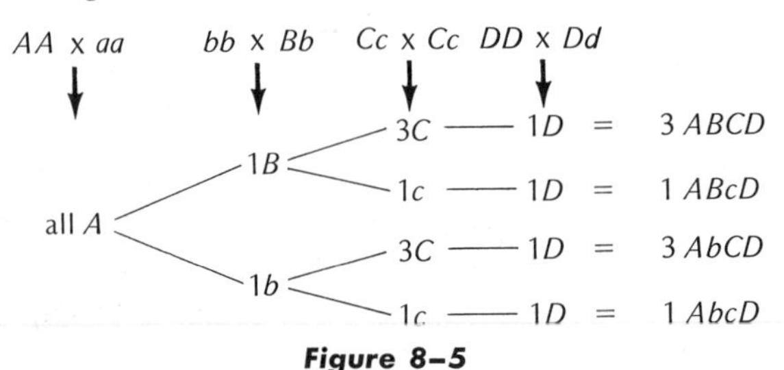

Figure 8–5

Phenotypes produced in a four-gene-pair cross.

How great must this departure be before we decide to discard our hypothesis?

For example, let us suppose that a garden pea plant, heterozygous for the gene pair *Tt*, produced 30 tall and 20 short offspring. Since the effect of the tall allele (*T*) is dominant over that for short (*t*), a 3:1 ratio would have been expected if the plant had been self-fertilized (*Tt* × *Tt*), or an exact numerical ratio of 37.5 tall to 12.5 short. Is the observed deviation from the expected ratio sufficient reason to discard the self-fertilization hypothesis and look for another explanation? For instance, must we substitute the hypothesis that the *Tt* plant was fertilized by a short plant *tt* which, ideally, would have yielded 25 tall and 25 short offspring?

To decide whether to accept or reject our hypothesis, we must therefore evaluate the size of the discrepancy between the observed and expected ratios. It is easy to see that if an experiment were repeated many times, a range of values would result, all more or less similar to the expected proportion in that particular experiment. For example, a cross between two heterozygotes should ideally produce the ratio 3 dominants:1 recessive in every experiment. Because of chance events (also known as "sampling error"), however, the observable ratios may fluctuate in size from small discrepancies from the ideal ratio to large discrepancies from this ratio. If the hypothesis is valid, and the ratio is truly 3:1, one would expect small discrepancies from this ratio much more frequently than large discrepancies. On the other hand, if the hypothesis is invalid, and the ratio is truly 1:1 or some other proportion, one would expect larger discrepancies more often than if the ratio were truly 3:1.

The size of the discrepancy therefore provides an indication of whether it can commonly or only rarely be expected; small discrepancies from a valid hypothesis are common; large discrepancies are rare. Since the large discrepancies are so rare, their occurrence in an experiment may indicate that the results fit better with other explanations that could be offered than with the proposed hypothesis. On this basis we can make the decision that small discrepancies do not cause us to reject a hypothesis (the discrepancy is "not significant"), whereas large discrepancies cause us to reject it (the discrepancy is "significant").

If we wish to assign values to these two kinds of discrepancies, we might say that large discrepancies are the largest 5 percent, and small discrepancies are the remaining 95 percent. On the basis of this arbitrary criterion we can then propose to reject the hypothesis if the discrepancy falls into the "large" class. This particular 5 percent frequency value that enables us to reject a hypothesis is called the *5 percent level of significance.* It is important to understand that the level of significance only furnishes the probability basis upon which we reject hypotheses but does not provide proof that the hypothesis is true or false. It may, for instance, still be possible to find large discrepancies from a valid hypothesis, but this is fairly rare (less than 5 percent of the time), so that we feel justified in seeking a new explanation.

We may, of course, change the level of significance so that we will less easily reject a hypothesis ("low level of significance," e.g., 1 percent) or more easily reject it ("high level of significance," e.g., 10 percent). However, the less easily we reject a hypothesis, the greater the chance that we will accept it as valid even when it is false. On the other hand, the more easily we reject it, the greater the chance that we may reject it even when it is the true explanation. As a compromise between these extremes the generally accepted level of significance for testing hypotheses is 5 percent.

DEGREES OF FREEDOM

Among the important factors that may influence the extent to which observed results depart from expected ratios is the number of independently variable classes, or *degrees of freedom*. The concept of degrees of freedom arises because, in the statistical analysis of many experimental results, such as genetic ratios, it is desirable to regard the total number of observed individuals in the experiment as a fixed or given quantity. Composing this given quantity are the contributions of component classes in the experiment, some of which are variable, or *free*, in respect to their numbers. For instance, if there are only two scored classes, such as the tall and short plants in the previous example, as soon as the number of the variable class is set, the size of the other class is automatically determined. It does not matter which particular class is considered to be variable, only that the nonvariable class furnish the needed numbers that are left over from the variable class to yield the given total number (e.g., 30 "fixed" tall + 20 "free" short = 50 total, or 30 "free" tall + 20 "fixed" short = 50 total). Thus in experiments scoring two classes of offspring, there is one degree of freedom; in experiments scoring three classes there are two degrees of freedom; and so on. The rule for experiments such as we will be concerned with is simply that the degrees of freedom equal the number of classes less one, or, notationally, df = $k - 1$ (where k equals the total number of classes).

CHI-SQUARE

Once the degrees of freedom in an experiment have been determined, and a significance level for testing a hypothesis has been decided upon, the actual measurement of the size of the discrepancy between observed and expected results remains to be done. One measure commonly used is called *chi-square* (χ^2), which, for one degree of freedom, is calculated in accord with the equation:

$$\chi^2 = \text{sum}\,\frac{[|(\text{observed} - \text{expected number})| - 1/2]^2}{\text{expected number}}$$

The reduction of 1/2 from the absolute value of the observed − expected deviation is known as the *Yates correction term*, and adds to the accuracy of chi-square determinations when the number of either of the expected classes is small. For the hypothesis that the tall plant in the above example was self-fertilized, chi-square would be calculated as in the table below.

According to statisticians, if the "expected" hypothesis is true, chi-square values have certain probabilities of occurrence, depending on the number of degrees of freedom in the experiment. On the basis of such calculations tables have been constructed that relate the number of degrees of freedom with the probability that particular groups of chi-square values will be found. For any given number of degrees of freedom the probability that large chi-square values or discrepancies will be found is, of course, much less than for small discrepancies. This relationship can be observed in Table 8–1, where, for example, a chi-square value larger than 3.84 will occur 5 percent or less of the time in experiments with one degree of freedom. Thus chi-square values of 6.64 or greater are sufficiently rare that they occur only 1 percent of the time, and values of 10.83 or greater occur only 1 out of 1000 times. On the other hand, chi-square values less than 3.84 are relatively frequent; i.e., a value of 1.07 or greater would be found in 30 out of 100 experiments if the proposed hypothesis were the true explanation for the results.

	Tall	*Short*	*Totals*
observed*	30	20	50
expected*	37.5	12.5	50
obs. − exp.	−7.5	7.5	
|(obs. − exp.)| − 1/2†	−7.0	7.0	
above no. squared	49.0	49.0	
above divided by expected	$\frac{49.0}{37.5} = 1.31$	$\frac{49.0}{12.5} = 3.92$	$\chi^2 = 1.31 + 3.92 = 5.23$

* In calculating chi-square, always use the actual numbers observed and expected. Do not use proportions or percentages.

† Note that 1/2 is deducted from the *absolute value* of the observed − expected quantity whether it is positive or negative, e.g., $|-7.5| - 1/2 = -7.0$, not -8.0.

TABLE 8–1

The probabilities of exceeding different chi-square values for degrees of freedom from 1 to 50 when the expected hypothesis is true*

	Probabilities									
df	*.95*	*.90*	*.70*	*.50*	*.30*	*.20*	*.10*	*.05*	*.01*	*.001*
1	.004	.016	.15	.46	1.07	1.64	2.71	3.84	6.64	10.83
2	.10	.21	.71	1.39	2.41	3.22	4.61	5.99	9.21	13.82
3	.35	.58	1.42	2.37	3.67	4.64	6.25	7.82	11.35	16.27
4	.71	1.06	2.20	3.36	4.88	5.99	7.78	9.49	13.28	18.47
5	1.15	1.61	3.00	4.35	6.06	7.29	9.24	11.07	15.09	20.52
6	1.64	2.20	3.83	5.35	7.23	8.56	10.65	12.59	16.81	22.46
7	2.17	2.83	4.67	6.35	8.38	9.80	12.02	14.07	18.48	24.32
8	2.73	3.49	5.53	7.34	9.52	11.03	13.36	15.51	20.09	26.13
9	3.33	4.17	6.39	8.34	10.66	12.24	14.68	16.92	21.67	27.88
10	3.94	4.87	7.27	9.34	11.78	13.44	15.99	18.31	23.21	29.59
11	4.58	5.58	8.15	10.34	12.90	14.63	17.28	19.68	24.73	31.26
12	5.23	6.30	9.03	11.34	14.01	15.81	18.55	21.03	26.22	32.91
13	5.89	7.04	9.93	12.34	15.12	16.99	19.81	22.36	27.69	34.53
14	6.57	7.79	10.82	13.34	16.22	18.15	21.06	23.69	29.14	36.12
15	7.26	8.55	11.72	14.34	17.32	19.31	22.31	25.00	30.58	37.70
20	10.85	12.44	16.27	19.34	22.78	25.04	28.41	31.41	37.57	45.32
25	14.61	16.47	20.87	24.34	28.17	30.68	34.38	37.65	44.31	52.62
30	18.49	20.60	25.51	29.34	33.53	36.25	40.26	43.77	50.89	59.70
50	34.76	37.69	44.31	49.34	54.72	58.16	63.17	67.51	76.15	86.66

← do not reject | reject →

at .05 level

* Abridged from Table IV of Fisher and Yates, *Statistical Tables for Biological, Agricultural and Medical Research*, Oliver and Boyd Ltd., Edinburgh, by permission of the authors and publishers.

Since we have agreed on a 5 percent level of significance, the chi-square value of 5.23 calculated in our example is therefore rare enough to be "significant," so that we have cause to reject our hypothesis. For convenience a line has been placed at the .05 values in Table 8–1, and chi-square values to the right of this line can be considered sufficient cause for rejection of the hypothesis at the 5 percent level of significance.

The alternative hypothesis, that the *Tt* heterozygous plant had been fertilized by a *tt* plant, also deserves consideration. According to this hypothesis, the offspring of such a mating would be expected to follow a phenotypic ratio of 1 tall:1 short. Chi-square calculations can therefore be made as follows:

	Tall	*Short*	*Totals*
observed	30	20	50
expected	25	25	50
obs. − exp.	5.0	−5.0	
|(obs. − exp.)| − 1/2	4.5	−4.5	
above no. squared	20.25	20.25	
above divided by expected	$\frac{20.25}{25} = .81$	$\frac{20.25}{25} = .81$	$\chi^2 = .81 + .81 = 1.62$

According to Table 8–1, in experiments with one degree of freedom, the probability of exceeding a chi-square value of 1.62 is between .20 and .30. This chi-square value is therefore small enough, and the probability for the occurrence of such a small value is therefore sufficiently great that we do not reject our hypothesis.

CHI-SQUARE METHOD WITH MORE THAN ONE DEGREE OF FREEDOM

When three or more classes are separately scored in an experiment, such as the four phenotypes that may occur in a dihybrid cross, then chi-square is calculated without the Yates correction factor. For example, let us suppose that two gene pairs *Aa* and *Bb*, with *A* and *B* dominant over *a* and *b*, are assumed to be segregating independently. A cross between two heterozygotes *AaBb* × *AaBb* would be expected to produce four classes of offspring in the *phenotypic* ratio 9 *AB*:3 *Ab*:3 *aB*:1 *ab*. If actually observed numbers are 1080 *AB*, 210 *Ab*, 200 *aB*, 110 *ab*, the calculations are simply those presented in the table below. Since four different classes comprise the total number, 4 – 1, or 3, of these are therefore considered variable. For three degrees of freedom the probability of a chi-square value of 97.33 or higher is less than .001. These results are therefore relatively improbable and certainly "significant" at the 5 percent level. The hypothesis that the true ratio is 9 *AB*:3 *aB*:3 *Ab*:1 *ab* can therefore be rejected.

	AB	**Ab**	**aB**
observed	1080	210	200
expected	9/16 × 1600 = 900	3/16 × 1600 = 300	3/16 × 1600 = 300
obs. – exp.	180	–90	–100
$\frac{(\text{obs.} - \text{exp.})^2}{\text{exp.}}$	36.00	27.00	33.33

	ab	***Totals***
observed	110	1600
expected	1/16 × 1600 = 100	1600
obs. = exp.	10	
$\frac{(\text{obs.} - \text{exp.})^2}{\text{exp.}}$	1.00	$\chi^2 = 97.33$

TEST FOR INDEPENDENCE (CONTINGENCY CHI-SQUARE)

It is occasionally desirable to compare one set of observations taken under particular conditions to those of a similar nature taken under different conditions. In this case there are no definite expected values; the question is whether the results are dependent (contingent upon) or independent of the conditions under which they are observed. This test is therefore called a test for independence, or *contingency test.*

In genetics, problems of this kind may be concerned with the effect of different environments or different genetic constitutions on a set of experimental observations. For example, crosses between individuals heterozygous for the same single gene difference were performed in two different experiments, A and B, and gave the following results:

		Phenotype of Offspring	
		Dominant	Recessive
Experiment	A	80	30
	B	90	25

For experiment A the dominant:recessive ratio is 2.67:1, whereas for B it is 3.60:1. Because of this difference, we may reasonably ask whether the observed results are independent of the particular experimental conditions. One statistical answer to this question depends upon the calculation of a chi-square that has only one degree of freedom.

If the calculated chi-square is less than the chi-square value at the particular level of significance desired (i.e., less than 3.84 at the 5 percent level), the hypothesis can be accepted that the observed results are statistically independent of the experimental conditions. If the calculated chi-square exceeds this value, this would indicate that the results depend upon the conditions, i.e., that there is an *interaction* between the results of a cross and its experimental condition sufficient to cause differences between the two sets of results.

To calculate a contingency chi-square, both the observed numbers and the marginal totals must be used. If we call the total number of observations N, and the individual numerical contributions to this value a, b, c, and d, respectively, then the calculations are as follows:

	Categories of Observations		
	1	*2*	Totals
A	a	b	$a + b$
B	c	d	$c + d$
	$a + c$	$b + d$	$a + b + c + d = \text{N}$

$$\chi^2 = \frac{[|ad - bc| - (1/2)\text{N}]^2\,\text{N}}{(a + b)(a + c)(c + d)(b + d)}$$

(The vertical lines on either side of $|ad - bc|$ mean the *absolute*, or *positive*, value of this difference. If this difference is zero or less than the Yates correction factor of (1/2)N in this equation, the numerator becomes zero $[(0)^2\text{N} = 0]$ and χ^2 is consequently zero.)

For the data given above, the chi-square computations are:

	Phenotypes		
	Dominant	Recessive	Totals
A	80	30	110
B	90	25	115
	170	55	225

$$\chi^2 = \frac{(|2000 - 2700| - 112.5)^2 \times 225}{(100)(170)(115)(55)}$$

$$= \frac{(700 - 112.5)^2 \times 225}{118{,}277{,}500} = .66$$

Thus, in spite of the difference in ratios between the two experiments, the chi-square is still small enough (probability between .3 and .5) to allow statistical independence of the particular experimental conditions.

HOMOGENEITY CHI-SQUARE

Because of the restricted number of entries, the example considered above is known as a 2×2 table. Of course, neither the categories of observation nor the conditions under which they are made are necessarily restricted to only two of each. In addition, a 2×2 contingency chi-square is an insufficient test in many genetic experiments where we have some prior concept of expected values for our observations and are therefore also interested in testing whether the data express a common overall ratio. Let us consider, for example, Mendel's data on five plants segregating for round and wrinkled seeds in an expected ratio of 3:1. An important question in this case is whether individual observations for each plant may be pooled to provide an overall test of the 3:1 ratio. Obviously, if the variability between the plants is high, they may not reflect common experimental conditions, and pooling their results would not be justified. A "test of homogeneity" must therefore be performed to decide whether the separate samples are sufficiently uniform to be added together.

For this purpose, four steps are necessary:

1. The chi-square of each individual sample should be calculated, based on the expected ratio. Since these chi-squares are to be added, the Yates correction factor should not be used, even though only one degree of freedom may be involved in each calculation.
2. The individual chi-squares should be summed to give a total chi-square. In this process the total chi-square accumulates a number of degrees of freedom equal to the sum of the degrees of freedom in the individual chi-squares. This total chi-square value has two components: (*a*) the chi-square contributed by the departure of the pooled data

from the expected ratio, and (*b*) the chi-square contributed by the differences between individual samples. To calculate *b*, we calculate *a* as follows and subtract this valuc from the total chi-square.

3. Calculate chi-square for the summed data of all samples. As always, the number of degrees of freedom is $k - 1$, where k is the number of phenotypic classes. However, even if there is only one degree of freedom, the Yates correction factor should not be used, since this chi-square is to be subtracted.

4. Subtract the chi-square for the summed data from the summed chi-squares (step 2) to obtain the homogeneity chi-square. Accompany this by subtracting the number of degrees of freedom between these respective values to obtain the degrees of freedom for homogeneity chi-square. If this homogeneity chi-square exceeds the value given in Table 8–1 for the prescribed level of significance, reject the hypothesis that the samples are homogeneous. In other words, a high chi-square value signifies that there are serious differences between the samples and cautions against pooling the data.

In considering the example of Mendel's plants, Table 8–2 shows the calculations necessary for the total of individual chi-squares and for the summed-data chi-square. Homogeneity chi-square is then determined as follows:

Source	Chi-square	Degrees of freedom	Probability
total	2.05	5	
summed data	.02 (.015)	1	.95–.90
homogeneity	2.03	4	.90–.70

Since the homogeneity chi-square is well within the accepted limit, there is no reason to reject homogeneity of the samples. The data can therefore be pooled and the summed data chi-square thus represents a test of a 3:1 ratio. It is obvious that the departure from this ratio is small enough to be nonsignificant.

Similar calculations can be made for crosses in which more than two phenotypes appear, except that in such instances account must be taken of thc increased number of degrees of freedom. If, for example, Mendel's five plants had been segregating for four phenotypes (i.e., dihybrid crosses giving an expected ratio of 9:3:3:1), the test for homogeneity would involve 3 degrees of freedom for each plant, or a total of 15 degrees of freedom. Deducting the 3 degrees of freedom in the summed data, there would be 12 degrees of freedom associated with the homogeneity chi-square.

THE BINOMIAL EXPANSION

Many problems in genetics concern not only the probability that a certain event will occur but also the probability that a certain combination of events will occur. For example, it might be of value to determine with what probabilities two offspring of a mating of *Aa* × *aa* will have particular genetic constitutions, i.e., both *Aa*, both *aa*, or one *Aa* and the other *aa*. Since the occurrence of any particular genotype in a single offspring is not influenced by the genotypes of other offspring, these are independent events. The probability that two *Aa* offspring will be formed from this mating is therefore equal to multiplying the probabilities of each *Aa*: $1/2 \times 1/2 = 1/4$. Thus the probabilities for each sequence of two children are as follows:

First child	Second child	Probabilities
Aa	*Aa*	$1/2 \times 1/2 = 1/4$
Aa	*aa*	$1/2 \times 1/2 = 1/4$
aa	*Aa*	$1/2 \times 1/2 = 1/4$
aa	*aa*	$1/2 \times 1/2 = 1/4$

If we summarize these values without regard to the order in which each genotype appears, we obtain the following distribution of probabilities:

probability both offspring *Aa* = 1/4
probability one *Aa* and the other *aa* = 2/4
probability both offspring *aa* = 1/4

TABLE 8–2

Calculation of chi-squares for individual and summed observations in five plants of Mendel's experiments on round and wrinkled peas

Plants		Phenotypes (3:1 expected) Round	Wrinkled	Degrees of Freedom	Chi-Squares
1	observed	45	12		
	expected	42.75	14.25		
	obs. − exp.	2.25	−2.25		
	(obs. − exp.)^2/exp.	.12	.36	1	.48
2	observed	27	8		
	expected	26.25	8.75		
	obs. − exp.	.75	−.75		
	(obs. − exp.)^2/exp.	.02	.06	1	.08
3	observed	24	7		
	expected	23.25	7.75		
	obs. − exp.	.75	−.75		
	(obs. − exp.)^2/exp.	.02	.07	1	.09
4	observed	19	10		
	expected	21.75	7.25		
	obs. − exp.	−2.75	2.75		
	(obs. − exp.)^2/exp.	.35	1.04	1	1.39
5	observed	32	11		
	expected	32.25	10.75		
	obs. − exp.	−.25	.25		
	(obs. − exp.)^2/exp.	.002	.006	1	.01
				Totals: 5	2.05
Summed data:	observed	147	48		
	expected	146.25	48.75		
	obs. − exp.	.75	−.75		
	(obs. − exp.)^2/exp.	.0038	.0115	1	.015

The pattern for this distribution is 1:2:1, which also represents the coefficients of raising two values, the "binomial" p and q, to the second power: $(p + q)^2 = 1p^2 + 2pq + 1q^2$ or, if we substitute *Aa* for p and *aa* for q, $[(Aa) + (aa)]^2 = 1\,(Aa)(Aa) + 2\,(Aa)(aa) + 1\,(aa)(aa)$. (Note again that the probability for a complementary event, such as the probability that both offspring will not be *aa*, is 1 − the probability of the particular event, or 1 − 1/4 = 3/4.) If the probabilities are calculated for the different combinations of genotypes possible among three children of the mating *Aa* × *aa*, the frequencies of each combination will be found to correspond with raising a binomial to the third power:

probability three offspring *Aa* = 1/8
probability two *Aa* and one *aa* = 3/8
probability two *aa* and one *Aa* = 3/8
probability three offspring *aa* = 1/8

or $(p + q)^3 = 1p^3 + 3p^2q + 3pq^2 + 1q^3$, or $[(Aa) + (aa)]^3 = 1\,(Aa)(Aa)(Aa) + 3\,(Aa)(Aa)(aa) + 3\,(Aa)(aa)(aa) + 1\,(aa)(aa)(aa)$.

The probability for each particular combination

TABLE 8–3

Coefficients for each term in a binomial expansion (p + q) raised to the n power when n goes from 0 to 12, and the total number of combinations for each expansion

n	*Binomial coefficients*	*Total no. combinations*
0	1	1
1	1 1	2
2	1 2 1	4
3	1 3 3 1	8
4	1 4 6 4 1	16
5	1 5 10 10 5 1	32
6	1 6 15 20 15 6 1	64
7	1 7 21 35 35 21 7 1	128
8	1 8 28 56 70 56 28 8 1	256
9	1 9 36 84 126 126 84 36 9 1	512
10	1 10 45 120 210 252 210 120 45 10 1	1024
11	1 11 55 165 330 462 462 330 165 55 11 1	2048
12	1 12 66 220 495 792 924 792 495 220 66 12 1	4096

of offspring can therefore be described by the binomial coefficient for this combination relative to the total number of possible combinations. In Table 8–3 coefficients are given for various binomial expansions [$(p + q)^n$] when n goes from 0 to 12. In sibships of four, for example, the binomial distribution shows that one would obtain an average of one sibship out of 16 with all four offspring *Aa*, or a probability for this event of 1/16 = .0625.

In general, whenever p is the probability that a particular genotype or phenotype will appear, and q or 1 − p is the probability of an alternative form, so that p + q = 1, the probability for each combination in which a succession of such events may occur is described by the binomial distribution. A formula that permits the calculation of probability for any special combination of genotypes or phenotypes, without regard to their order of appearance, is

$$\frac{n!}{w!x!}p^w q^x$$

where n refers to the total number of children in the family (sibship size), *w* to the number of children with a specific genetic or phenotypic constitution having probability p, e.g., p = probability of *Aa* offspring, and *x* to the number of children in this sibship bearing a genotype or phenotype having probability q; e.g., q = probability of *aa* offspring. The ! symbol means "factorial," or the multiplication of the number by all integers down to 1, i.e., if n = 5, n! = 5 · 4 · 3 · 2 · 1 = 120.* It should be noted that the zero factorial (0!) equals 1, and any number raised to the zero power [e.g., $(1/2)^0$] also equals 1.

If we reconsider the chances that a mating of *Aa* × *aa* will produce all *Aa* offspring in a sibship of four, then if p is the probability of *Aa*, *w* is equal to 4. Since the total number of offspring are divided among two types, *Aa* and *aa*, the numbers of each type (*w* and *x*, respectively) for any particular combination must total n; *x* is therefore equal to zero. The binomial formula is then

$$\frac{4!}{4!0!}(1/2)^4(1/2)^0 = \frac{4 \cdot 3 \cdot 2 \cdot 1}{4 \cdot 3 \cdot 2 \cdot 1(1)}(1/16)(1)$$
$$= 1/16 = .0625$$

For possible combinations of both *Aa* and *aa* children, the procedure is similar. For example, in a sibship of five children that arise from the mat-

* Values of factorials and their logarithms can be found in handbooks of mathematical tables.

ing $Aa \times aa$, the probability that three will be Aa and two aa is

$$\frac{5!}{3!2!}(1/2)^3(1/2)^2 = \frac{120}{6 \cdot 2}\,1/8 \cdot 1/4$$
$$= 5/16 = .3125$$

TESTING HYPOTHESES USING BINOMIAL DISTRIBUTIONS

The chi-square method is generally applicable when the total number of scored individuals is large. In an experiment with one degree of freedom, chi-square may be considered a satisfactory test if the total number of individuals is 30 or more when the observed proportions of each of the two classes is .5, or 50 percent. Usually, however, one of the classes will have a lower frequency than .5, and the rule is that the lower the frequency of any class, the larger the experiment should be in terms of total number. For example, an observed frequency of .2 for one of the two classes in an experiment means that the total number observed should be about 200 if chi-square is to be used satisfactorily.

In cases in which the total number of individuals scored is relatively small, a hypothesis can be tested by utilizing the binomial distribution. For this purpose the probability for all possible proportions that may occur in experiments with small numbers has been calculated by statisticians. The results of these calculations can be tabulated in a form that describes upper and lower "confidence" limits for any set of observed results (Table 8–4). For example, an experiment in which 30 was the total number of individuals scored, 5 of one class and 25 of the others, could be repeated many times over and give results other than 5 and 25, i.e., 6 and 24, etc. According to statisticians, if this same experiment were repeated often enough, 95 percent of the time the frequency of the smaller class proportion would fall between .06 and .35 and the larger class proportion would fall between .65 and .94. Each of these ranges is called a *confidence interval.* Proportions outside these limits would occur less than 5 percent of the time. According to our agreement on the level of significance (p. 145), such rare proportions are rejected as explanations for the observed results. Thus a theoretical explanation based on a probability of .01 when the observation is 5 out of 30 would be rejected, while an explanation based on a probability of .10 would not be rejected.

In the experiment in which a tall plant produced 50 offspring, 20 of which were short (p. 145), the 95 percent confidence interval for the short class falls between .27 and .55. The hypothesis of a 3:1 ratio, or 25 percent short, can therefore be rejected at the 5 percent level of significance.

MULTINOMIAL DISTRIBUTIONS

The above examples are concerned with only two possible alternatives for each event, Aa or aa offspring. It is, of course, possible to have matings in which three types of offspring appear. For example, $Aa \times Aa$ will produce AA, Aa, and aa offspring in the ratio 1:2:1, or probabilities 1/4 for AA, 1/2 for Aa, and 1/4 for aa. An additional term is therefore added to the binomial to include the third variety of offspring. The expansion term now leads to a *trinomial* distribution $(\mathrm{p} + \mathrm{q} + \mathrm{r})^{\mathrm{n}}$, where p, q, and r represent the probabilities of AA, Aa, and aa, respectively.

The formula for calculating probabilities of particular trinomial combinations does not essentially differ from the binomial, and is

$$\frac{n!}{w!x!y!}\,\mathrm{p}^{w}\mathrm{q}^{x}\mathrm{r}^{y}$$

where w, x, and y are the numbers of offspring of each of the three different types whose probabilities are p, q, and r, respectively. In a mating of $Aa \times Aa$, the probability of having exactly 1 AA homozygote, 2 Aa heterozygotes, and 1 aa homozygote in a sibship of four, would be

$$\frac{4!}{1!2!1!}(1/4)^1(1/2)^2(1/4)^1 = (24/2)(1/64) = 3/16$$

TABLE 8–4

95 percent confidence intervals, given in proportions, for the binomial distribution when sample sizes are 100 or less*

Number Observed, f	Size of Sample, n											
	10		15		20		30		50		100	
0	.00	.31	.00	.22	.00	.17	.00	.12	.00	.07	.00	.04
1	.00	.45	.00	.32	.00	.25	.00	.17	.00	.11	.00	.05
2	.03	.56	.02	.40	.01	.31	.01	.22	.00	.14	.00	.07
3	.07	.65	.04	.48	.03	.38	.02	.27	.01	.17	.01	.08
4	.12	.74	.08	.55	.06	.44	.04	.31	.02	.19	.01	.10
5	.19	.81	.12	.62	.09	.49	.06	.35	.03	.22	.02	.11
6	.26	.88	.16	.68	.12	.54	.08	.39	.05	.24	.02	.12
7	.35	.93	.21	.73	.15	.59	.10	.43	.06	.27	.03	.14
8	.44	.97	.27	.79	.19	.64	.12	.46	.07	.29	.04	.15
9	.55	1.00	.32	.84	.23	.68	.15	.50	.09	.31	.04	.16
10	.69	1.00	.38	.88	.27	.73	.17	.53	.10	.34	.05	.18
11			.45	.92	.32	.77	.20	.56	.12	.36	.05	.19
12			.52	.96	.36	.81	.23	.60	.13	.38	.06	.20
13			.60	.98	.41	.85	.25	.63	.15	.41	.07	.21
14			.68	1.00	.46	.88	.28	.66	.16	.43	.08	.22
15			.78	1.00	.51	.91	.31	.69	.18	.44	.09	.24
16					.56	.94	.34	.72	.20	.46	.09	.25
17					.62	.97	.37	.75	.21	.48	.10	.26
18					.69	.99	.40	.77	.23	.50	.11	.27
19					.75	1.00	.44	.80	.25	.53	.12	.28
20					.83	1.00	.47	.83	.27	.55	.13	.29
21							.50	.85	.28	.57	.14	.30
22							.54	.88	.30	.59	.14	.31
23							.57	.90	.32	.61	.15	.32
24							.61	.92	.34	.63	.16	.33
25							.65	.94	.36	.64	.17	.35
26							.69	.96	.37	.66	.18	.36
27							.73	.98	.39	.68	.19	.37
28							.78	.99	.41	.70	.19	.38
29							.83	1.00	.43	.72	.20	.39
30							.88	1.00	.45	.73	.21	.40
31									.47	.75	.22	.41
32									.50	.77	.23	.42
33									.52	.79	.24	.43
34									.54	.80	.25	.44
35									.56	.82	.26	.45
36									.57	.84	.27	.46
37									.59	.85	.28	.47
38									.62	.87	.28	.48
39									.64	.88	.29	.49
40									.66	.90	.30	.50
41									.69	.91	.31	.51
42									.71	.93	.32	.52
43									.73	.94	.33	.53
44									.76	.95	.34	.54
45									.78	.97	.35	.55
46									.81	.98	.36	.56
47									.83	.99	.37	.57
48									.86	1.00	.38	.58
49									.89	1.00	.39	.59
50									.93	1.00	.40	.60

* To use this table, find the total sample size along the top line (n) and then read down to the numbers of the observed class (f). For example, a sample size of 30, out of which 5 are in one class, would have a 95 percent confidence interval for this class between 6 and 35 percent. If the number of observations within a class exceeds 50, subtract this quantity from 100, find the confidence intervals for this amount, and subtract each confidence interval from 1.00. For example, a class containing 70 out of a total of 100 observations would have a confidence interval of 1.00 − lower limit of 30 to 1.00 − upper limit of 30, or from 1.00 − .21 to 1.00 − .40, or from .60 to .79.

The table can be interpolated for total sample sizes other than those given, through use of the formula

$$\text{New confidence limit} = \frac{\text{Confidence limit} \times \text{Number (using values for preceding sample size)}}{\text{Observed sample size number}}$$

For example, the lower confidence limit for an observation of 5 out of a sample size of 35 would be $.06 \times 30/35 = .05$. The upper confidence limit would be $.35 \times 30/35 = .30$.

Other multinomial distributions may also occur. For example, when a dihybrid cross is made in which four phenotypes can appear in respective frequencies p, q, r, and s, the formula would be

$$\frac{n!}{w!x!y!z!} p^w q^x r^y s^z$$

As an illustration, two gene pairs may be segregating in a pea plant in which the factor for smooth is dominant over that for wrinkled, and that for yellow is dominant over green. If a heterozygote for both gene pairs is self-fertilized, the probability of having 19 offspring of which 8 are smooth yellow (probability 9/16), 5 smooth green (probability 3/16), 4 wrinkled yellow (probability 3/16), and 2 wrinkled green (probability 1/16) would be

$$\frac{19!}{8!5!4!2!}(9/16)^8(3/16)^5(3/16)^4(1/16)^2 = .00057394$$

NORMAL DISTRIBUTION

An important feature of the binomial distribution is that as n, the number of individuals in the group, increases, and p and q remain the same, a bell-shaped curve can be drawn for the frequencies of the different combinations. For example, when p and q represent the equal probabilities (1/2) of having *Aa* or *aa* offspring, respectively, from a mating of *Aa* × *aa*, the histogram for sibships of 4, 8, and 10 would appear as in Fig. 8–6a, b, and c, respectively. When the population size becomes infinitely large, these distributions can be fitted to a continuous line known as the *normal distribution curve* (Fig. 8–6d). This curve is high in the center, representing the higher frequencies for the most common combinations and tapers off equally at both extremes for the rarer combinations.

In addition to the binomial many characteristics, such as the height of human populations, have been found to fit this curve. Some adult groups, for example, show an average height or *mean* of 68 inches, around which cluster the most commonly observed heights. Extreme heights, such as 58 inches or less, and 78 inches or more, are relatively rare.

In numerous instances the proportions of different classes in genetic experiments may also be

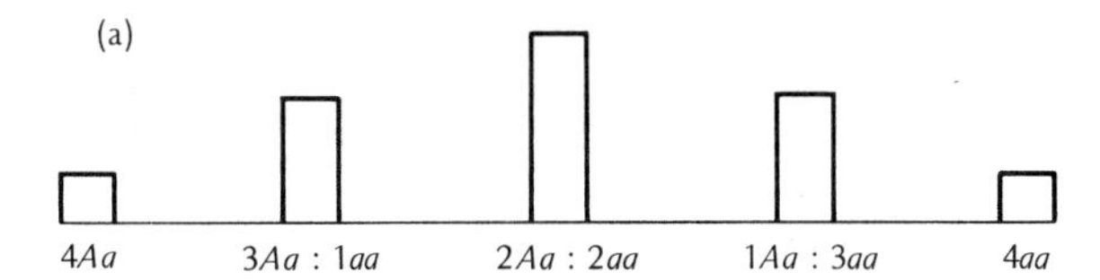

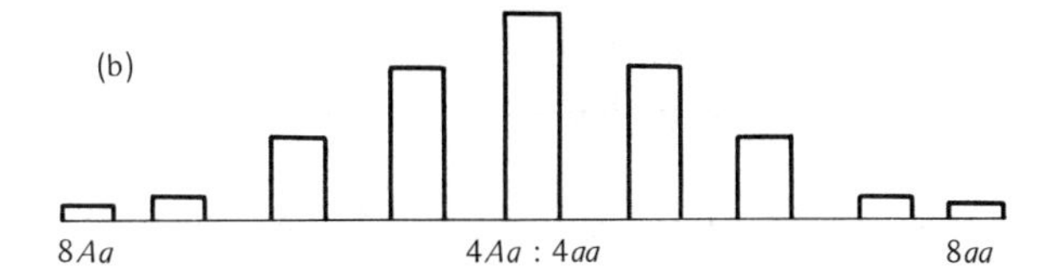

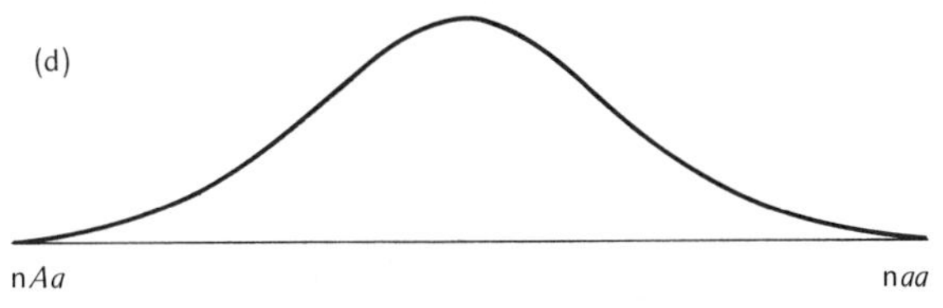

Figure 8–6

Shape of distributions for different values in the expansion $(p + q)^n$ when $p = q = 1/2$, e.g., when an Aa × aa cross produces 1/2 Aa: 1/2 aa offspring for a sibship of size n, $[(Aa) + (aa)]^n$. (a) Distribution showing the relative expected frequency (height of histogram column) for each of the five different genotypic combinations possible among sibships of four offspring from the cross Aa × aa. (b) and (c) Relative frequencies of combinations among sibships of eight and ten offspring respectively from the same cross. Only the genotypes of three of the various possible combinations are labeled. (d) Distribution when the number of offspring per sibship becomes very large, e.g., n.

considered to be normally distributed. That is, if a certain experiment were repeated many times, such as mating of *Aa* × *Aa*, the proportion of *aa* offspring would usually be close to .25. Values greater or less than .25 would be lower in frequency, and extreme values would extend into the "tails" of the normal curve.

There is, therefore, a relationship between the frequency of certain values and the area occupied by these values in the curve. The more common values (those close to the mean) occupy a larger area of the curve than the rarer values (tails). If the curve is broken into sections and the areas measured in each section, we can determine a central area that contains 95 percent of all values, while the smaller tail areas contain only 5 percent of all values. On this basis, hypotheses about a certain average expected value can be tested: Those results falling within the central 95 percent area would lead to acceptance, those results falling outside this area would lead to rejection.

The measurement of how far actual results depart from the mean expected value is done in terms of *standard deviations*, or σ, which are linear units along the horizontal axis of the curve. For experiments with proportions, $\sigma = \sqrt{pq/n}$, where p is the proportion for a particular class, q is 1 − p, and n is the total number of individuals in the experiment. When the normal curve is divided into standard deviations, practically the entire area can be included within 3.5σ to either side of the mean, and 95 percent of the area can

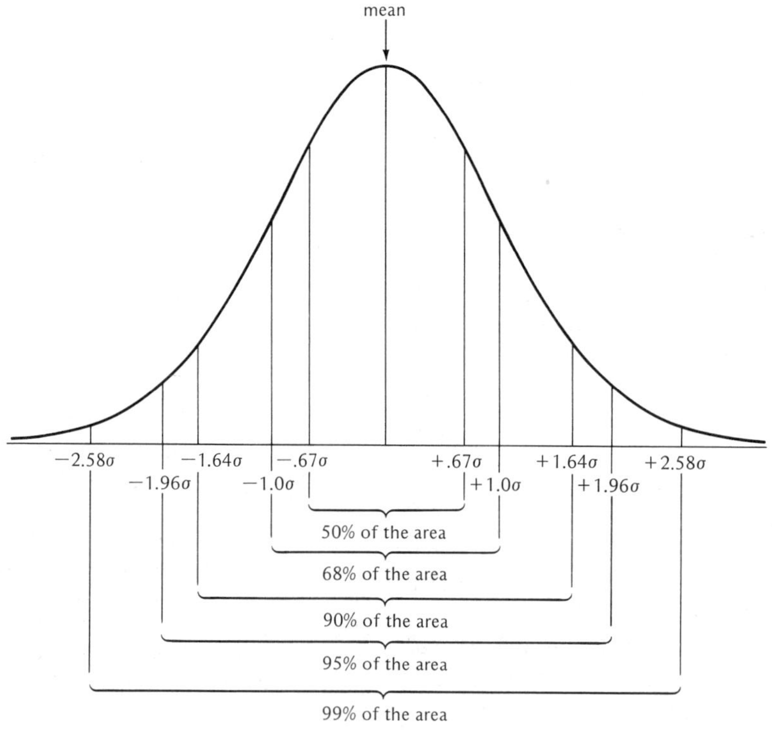

Figure 8–7

Normal distribution curve. The most frequent value (greatest height of the curve) is the mean, with smaller or greater values decreasing in frequency the more they depart from the mean. The diagram shows the distribution or area of this curve measured in terms of departures from the mean called standard deviations (σ). (After M. W. Strickberger, Experiments in Genetics with Drosophila, John Wiley, New York, 1962.)

be included within 1.96σ of the mean (Fig. 8–7). Thus observed proportions that fall outside 1.96σ of an expected mean would occur less than 5 percent of the time and the hypothesis which gave the expected value for the mean can be rejected at the 5 percent level of significance.

In the example of 20 short and 30 tall offspring produced by a heterozygous *Tt* plant, one proposed hypothesis was that the tall plant had been self-fertilized and the expected proportion of short offspring was .25. If we call the frequency of short p, and the frequency of tall q, the test of the hypothesis would be as follows:

$$\text{standard deviation } (\sigma) = \sqrt{\frac{pq}{n}} = \sqrt{\frac{(.25)(.75)}{50}}$$
$$= \sqrt{.00375} = .06$$

	Short	Tall
observed proportions	20/50 = .40	30/50 = .60
expected proportions	.25	.75
observed − expected	.15	−.15

Since the observed − expected difference for short plants is greater than the value of 1.96σ, which in this case is .12, the hypothesis may be rejected at the 5 percent level. (Note, however, that according to Fig. 8–7, it would be just barely accepted at the 1 percent level, which is 2.58σ, or .15.) For tall plants, note that the departure from the expected value is also outside 1.96σ, except that it is negative (−.15/.06 = −2.5), or to the left of the expected mean proportion.

CONFIDENCE LIMITS

When the total number of individuals (n) in an experiment is large enough, so that $n \times p \times q$ is greater than 25, the normal distribution curve itself can be used to establish a confidence interval around the observed frequency for a particular category.

The limits for this interval are calculated by considering that 1.96 standard deviations to the left and right (minus and plus) of the observed frequency will include, with 95 percent confidence, the mean of that distribution. As they were for the binomial distribution, these limits are therefore called *95 percent confidence limits.* They have a special advantage in instances where there is no exact mean value expected beforehand, but only a value that must be estimated from the observed data. The formula for the calculation of the 95 percent confidence limit is

$$\begin{matrix}\text{observed}\\ \text{mean}\end{matrix} \pm 1.96 \sqrt{\frac{(\text{observed p})(\text{observed q})}{n}}$$

As an illustration, we can consider a cross in which 70 out of 200 plants were observed to have a particular phenotype. We would like to know the confidence interval that has a high probability of including the true mean frequency of this phenotype. In this case, $n \times p \times q$ is sufficiently large (45.5) so that we can calculate the confidence interval as

$$.35 \pm 1.96 \sqrt{\frac{(.35)(.65)}{200}}$$
$$= .35 \pm 1.96(.034) = .35 \pm .07$$

With 95 percent confidence, we can therefore say that the true frequency for this phenotype lies between .28 and .42.

POISSON DISTRIBUTION

We have seen that the binomial expansion will produce a symmetrical distribution around a "central" value when the two genes or genotypes involved in the expansion are in approximately equal proportions. For example, if the probability of $A = a = 1/2$, the binomial $(A + a)^2$ will produce 1 *AA* + 2 *Aa* + 1 *aa*. As n in the expansion $(A + a)^n$ is increased, more terms are added, but the most frequent values are those occupied by genotypes in which equal numbers of *A* and *a* alleles are present. Other genotypes, such as *AA* and *aa* above, or the triploid genotypes *AAA* and *aaa* in the expansion $(A + a)^3$, are less frequent, but are nevertheless "normally" distributed, since their frequencies fall off equally on both sides of the central genotypes.

However, if the frequency of A is not equal to a, the normal bell shape of this distribution becomes distorted or "skewed," with the most common genotypes going to one side or the other of the previously "central" genotype. For example, if the proportion of A is .75 and the proportion of a is .25, the binomial expansion $(.75\,A + .25\,a)^2$ will produce three genotypes in the ratio $1(.75\,A)(.75\,A) + 2(.75\,A)(.25\,a) + 1(.25\,a)(.25\,a)$, or a distribution of .5625 *AA* + .3750 *Aa* + .0625 *aa*. As n is increased, the products of such expansions appear to be more normally distributed but will nevertheless show some degree of "skewness" or lopsidedness, depending on which of the alleles, A or a, are more frequent. The degree of lopsidedness is mainly a function of the difference in proportion between A and a.

If this difference in proportion is so large that either A or a is of the order 1/n as n is increased, the frequencies of the various combinations of A and a will appear in the form of an extremely skewed distribution known as the *Poisson distribution.* The formula of the Poisson distribution is

$$e^{-m}\left(1, m, \frac{m^2}{2!}, \frac{m^3}{3!}, \frac{m^4}{4!}, \ldots, \frac{m^i}{i!}\right)$$

TABLE 8–5

Values of e^{-m} for m ranging from 0 to 2.99*

m	*.00*	*.01*	*.02*	*.03*	*.04*	*.05*	*.06*	*.07*	*.08*	*.09*
.00	1.000	.990	.980	.970	.961	.951	.942	.932	.923	.914
.10	.905	.896	.887	.878	.869	.861	.852	.844	.835	.827
.20	.819	.811	.803	.795	.787	.779	.771	.763	.756	.748
.30	.741	.733	.726	.719	.712	.705	.698	.691	.684	.677
.40	.670	.664	.657	.651	.644	.638	.631	.625	.619	.613
.50	.607	.600	.595	.589	.583	.577	.571	.566	.560	.554
.60	.549	.543	.538	.533	.527	.522	.517	.512	.507	.502
.70	.497	.492	.487	.482	.477	.472	.468	.463	.458	.454
.80	.449	.445	.440	.436	.432	.427	.423	.419	.415	.411
.90	.407	.403	.399	.395	.391	.387	.383	.379	.375	.372
1.00	.368	.364	.361	.357	.353	.350	.346	.343	.340	.336
1.10	.333	.330	.326	.323	.320	.317	.313	.310	.307	.304
1.20	.301	.298	.295	.292	.289	.287	.284	.281	.278	.275
1.30	.273	.270	.267	.264	.262	.259	.257	.254	.252	.249
1.40	.247	.244	.242	.239	.237	.235	.232	.230	.228	.225
1.50	.223	.221	.219	.217	.214	.212	.210	.208	.206	.204
1.60	.202	.200	.198	.196	.194	.192	.190	.188	.186	.185
1.70	.183	.181	.179	.177	.176	.174	.172	.170	.169	.167
1.80	.165	.164	.162	.160	.159	.157	.156	.154	.153	.151
1.90	.150	.148	.147	.145	.144	.142	.141	.139	.138	.137
2.00	.135	.134	.133	.131	.130	.129	.127	.126	.125	.124
2.10	.122	.121	.120	.119	.118	.116	.115	.114	.113	.112
2.20	.111	.110	.109	.108	.106	.105	.104	.103	.102	.101
2.30	.100	.0992	.0983	.0973	.0963	.0953	.0944	.0935	.0926	.0916
2.40	.0907	.0898	.0889	.0880	.0872	.0863	.0854	.0846	.0837	.0829
2.50	.0821	.0813	.0805	.0797	.0789	.0781	.0773	.0765	.0758	.0750
2.60	.0743	.0735	.0728	.0721	.0714	.0707	.0699	.0693	.0686	.0679
2.70	.0672	.0665	.0659	.0652	.0646	.0639	.0633	.0627	.0620	.0614
2.80	.0608	.0602	.0596	.0590	.0584	.0578	.0573	.0567	.0561	.0556
2.90	.0550	.0545	.0539	.0534	.0529	.0523	.0518	.0513	.0508	.0503

* To obtain an exponential value look down the left-hand column to the desired first decimal place, and then across the top to the desired second decimal place; i.e., $e^{-.35}$ is at the intersection of .30 and .05 = .705.

where e is the base of natural logarithms (2.718...), m denotes the mean value of the distribution, and $i!$ is the factorial $(1 \times 2 \times 3 \times \cdots \times i)$ for the ith value in the series (where i can be any number). Each value in the series denotes the frequency of observations that are expected to fall in a particular class beginning with 0, 1, 2, 3, ... events. To eliminate computation of e^{-m}, Table 8–5 gives these values for m ranging from 0 to 2.99. For example, when m is .35, the value of e^{-m} is .705. The first term in the Poisson series (representing the class with zero events) therefore has an expected frequency of .705, and the second term in the series (the class containing a single event) has an expected frequency of $.705 \times .35$, etc.

As an illustration, let us consider rare hereditary changes in *E. coli* bacteria which confer resistance to the drug streptomycin. These changes, or mutations, can be detected by plating many bacteria on petri dishes containing antibiotic medium and then scoring the appearance of resistant colonies that arise from mutations in single bacteria. For example, 150 petri dishes of streptomycin agar were each plated with 1 million bacteria, and a total of 69 resistant colonies were scored. These 69 colonies were so distributed that 98 petri dishes had none, 40 had one colony, 8 had two colonies, 3 had 3 colonies, and 1 had 4 colonies. These data are shown in columns 1, 2, and 3 of Table 8–6. The average number of colonies per petri dish is therefore $m = 69/150 = .46$. According to Table 8–5, e^{-m} is therefore .631, which furnishes the Poisson probabilities for this distribution in column 4. Multiplication of each expected Poisson probability by the total number of petri dishes (150) gives the expected number of each kind of petri dish in column 5. Superficially the fit between observed and expected numbers is close. However, if desired, this fit can be tested by chi-square, with the proviso that the number of degrees of freedom in Poisson tests is *two* less than the number of categories, i.e., $5 - 2 = 3$ df. (An additional degree of freedom is deducted because m is an estimate of the average rather than a true expected value.) In this case $\chi^2 = 4.15$, which, for 3 df, has a probability between .20 and .30. In other words, these mutations represent a set of random events that can indeed be described by a Poisson distribution. The estimated mutation frequency is then .46 colonies per dish of one million bacteria, i.e., $.46 \times 10^{-6}$ cells.

The Poisson distribution has the property that its standard deviation is equal to the square root of its mean, or, in terms to be used later, its variance is equal to its mean. This relationship can then be used to calculate confidence intervals for the mean when the observed numbers are sufficiently high (at least 50 and preferably 100 or more). For this purpose the entire sample number can be used as the mean. That is, in the above

TABLE 8–6

Calculation of expected numbers of streptomycin-resistant colonies per petri dish according to the Poisson distribution

1. Colonies per Petri Dish	2. No. Petri Dishes	3. No. Colonies (Col. 1 × Col. 2)	4. Expected Probability (Poisson Formula)	5. Expected No. Dishes (Col. 4 × 150)
0	98	0	$.631 \times 1 = .631$	94.65
1	40	40	$.631 \times .46 = .290$	43.50
2	8	16	$.631 \times \frac{(.46)^2}{2} = .0668$	10.02
3	3	9	$.631 \times \frac{(.46)^3}{6} = .0102$	1.53
4 or over	1	4	.0020	.30
	150	69	1.0000	150.00

example, the mean mutation frequency for 150 petri dishes of bacteria is 69 colonies, and the standard deviation is $\sqrt{69} = 8.3$. Thus with 95 percent confidence we can say that the true mean lies within $\pm 1.96(8.31)$ of 69, or between 52.7 and 85.3. In terms of mutation frequency per million bacteria this value is divided by 150 to give a 95 percent confidence interval of .35 to .57 per million cells.

PROBLEMS

8-1. A woman is heterozygous for four gene pairs and homozygous for six gene pairs. (a) How many different kinds of eggs can she form? (b) If her husband had the same genetic constitution as she did, how many different kinds of genotypes are possible among their offspring? (c) If *Aa* were one of the gene pairs for which both of these parents were heterozygous, what is the probability that their first child would also be heterozygous for the same gene pair? (d) What is the probability that they would have two children, the first *Aa* and the second *aa*? (e) After the first two children were born, what is the probability that their next child would be *AA*?

8-2. The pedigree diagrammed below is for a very rare trait with affected individuals marked by shading. (a) Using the symbols *A* if the trait is caused by a dominant gene and *a* if it is caused by a recessive gene, give the genotypes for individuals I-1, I-2, II-4, III-2, IV-1, and V-1. (b) What is the probability that the brother of individual V-1 is heterozygous? (c) What is the probability that the two sisters of individual V-1 are heterozygous? (d) What is the probability that the brother of individual V-1 is *AA*? (e) If individual V-1 mated individual V-5, what is the probability that their first offspring would be affected? (f) If their first child was born and affected with this trait, what is the probability that their second child would be affected?

8-3. Infantile amaurotic idiocy is a serious mental defect occurring in individuals homozygous for a recessive gene, *a*. (a) If two normal parents had a daughter with symptoms of this disease and a normal son, what is the probability that the son is a carrier of the recessive gene? (b) If this son married a normal woman whose brother was affected by this disease, what is the probability that the first offspring of this marriage would be defective? (c) If the first child born to the mating in (b) was defective, what is the probability that the second child would also be defective? (d) If the first two children born to the mating in (b) were defective, what is the probability that the third child would be normal?

8-4. In the pedigree on the next page reported by A. Freire-Maia, a pregnant woman (III-3 in the pedigree) consulted her physician in Sao Paulo, Brazil, concerned that she might give birth to another child affected with ectrodactyly, or "lobster-claw" (shaded symbols). The physician assured her that the pattern among her living children (IV-3 to IV-7) was that one affected child is followed by one normal child, and since her last child was affected, her next child would definitely be normal. (In the pedigree, note that two sisters

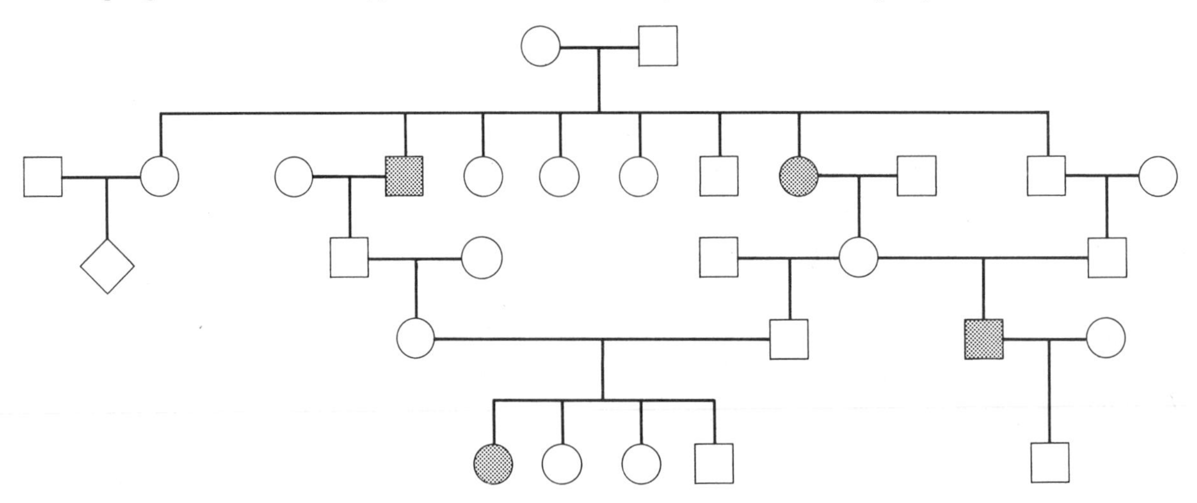

married two brothers in generation II, and again in generation III.)

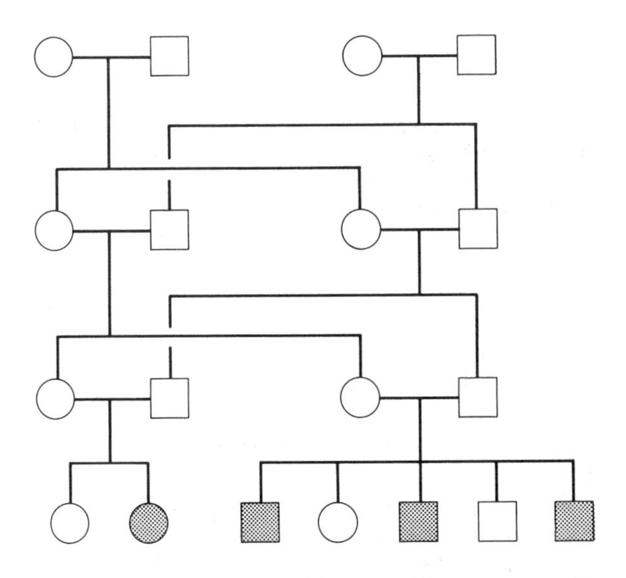

(a) Explain why you would or would not agree with the physician's prediction. (b) Assume that individual IV-1 in this pedigree would like to marry individual IV-6, and comes to you for information about the probability that this mating will produce an offspring affected with ectrodactyly. What probability would you suggest?

8-5. Each of three gene pairs *Aa*, *Bb*, and *Cc* affect a different character and assort independently of each other. If the effects caused by capital-letter genes are dominant to those caused by small-letter genes, what is the probability of obtaining:

(a) An *ABC* gamete from an *AaBbCc* individual?
(b) An *ABC* gamete from an *AABBCc* individual?
(c) An *AABBCC* zygote from a cross *AaBbCc* × *AaBbCc*?
(d) An *AABBcc* zygote from a cross *aaBBcc* × *AAbbCC*?
(e) An *ABC* phenotype from a cross *AaBbCC* × *AaBbcc*?
(f) An *ABC* phenotype from a cross *aabbCC* × *AABBcc*?
(g) An *aBC* phenotype from a cross *AaBbCC* × *AaBBcc*?
(h) An *abc* phenotype from a cross *AaBbCc* × *AaBbCc*?
(i) An *abc* phenotype from a cross *AaBbCc* × *aabbCc*?
(j) An *abc* phenotype from a cross *aaBbCc* × *AABbcc*?

8-6. A plant of the genotype *AABBCCDD* was crossed to a plant of the genotype *aabbccdd* to produce an F_1, and the F_1 was then self-fertilized. Assume that each of these four gene pairs affects a different trait, that capital letter alleles are completely dominant to lower case alleles (e.g., $A > a$, $B > b$, etc.), and that each gene pair assorts independently of the other. (a) What proportion of F_2 plants will be *phenotypically* recessive for all four traits? (b) *Phenotypically* dominant for all four traits? (c) Are there more F_2 plants that are phenotypically dominant for all four traits than plants which are phenotypically recessive for one or more of the traits? (d) What proportion of F_2 plants will be genotypically like the F_1 parent? (e) If the original parents had been *AAbbCCdd* × *aaBBccDD* would your answers to the above questions be different?

8-7. The gene producing sickle-shaped red blood cells in humans is called Hb^S and the normal allele of this gene is called Hb^A. Various investigations have tested the incidence of heavy infection of the malarial parasite *Plasmodium falciparum* in African children who are heterozygous for the sickle-cell gene (Hb^S/Hb^A) and in children who are homozygous normal (Hb^A/Hb^A). In an investigation by Allison and Clyde, the following findings were observed:

	Heavy infections	Noninfected or Lightly Infected
Hb^s/Hb^A	36	100
Hb^A/Hb^A	152	255

Test whether the heterozygotes in this sample are better protected against heavy malarial infections than the normal homozygote.

8-8. Suppose that a population of screech owls were sampled and classified for sex and plumage color as follows: red males 35, red females 70, gray males 50, gray females 45. Use the chi-square method to answer the following questions: (a) Does the sex ratio deviate significantly from 50:50? (b) Does the color ratio (for the sample as a whole) deviate significantly from 50:50? (c) Is color independent of sex in this sample?

8-9. Four of the self-fertilized F_1 plants that Mendel observed for segregation of yellow and green seed color showed the following results among their seeds:

	Plants 1	2	3	4
yellow seeds	25	32	14	70
green seeds	11	7	5	27

Test the homogeneity of the four plants for the 3:1 ratio, and determine whether the data can be summed to calculate chi-square.

8-10. Test the corn plant data (green and white

plants) given in Problem 3, Chapter 6, to see whether they are homogeneous and can be pooled.

8-11. The following results were obtained by four experimenters in a genetics laboratory class who back-crossed a stock of *Drosophila melanogaster* heterozygous for the recessive genes *black* body color and *sepia* eye color ($b/+$, $se/+$) to a stock homozygous for these recessives (b/b, se/se):

Experimenter	Phenotypes			
	Wild Type	Black	Sepia	Black-Sepia
1	42	38	38	36
2	28	30	25	25
3	102	94	100	93
4	75	80	84	72

Test these data to see whether they are homogeneous and can be pooled.

8-12. The same experimenters also crossed *black-sepia* heterozygotes among themselves ($b/+$, $se/+ \times b/+$, $se/+$) and obtained the following results:

Experimenter	Phenotypes			
	Wild Type	Black	Sepia	Black-Sepia
1	72	24	20	7
2	112	40	35	11
3	49	15	18	5
4	100	31	35	12

Test these data to see whether they are homogeneous and can be pooled.

8-13. What are the confidence limits for the frequency of the wild-type phenotype in Problem 12?

8-14. What are the confidence limits for the frequency of the black phenotype, when 42 black and 58 wild-type flies emerged from a cross between a black male and a heterozygous $b/+$ female?

8-15. The effect of gene *A* is dominant to that of its allele *a*, and *B* is dominant to *b*. (a) What is the probability that a self-fertilized plant heterozygous for these two gene pairs would produce 12 offspring, all of AB phenotype? (b) If 12 offspring were produced from a heterozygous plant and were all of the AB phenotype, what explanation would you offer?

8-16. A number of two-children families with normal marriage partners are identified by the fact that they have produced at least one albino child. (a) In what proportion of these families would you expect to find that the other child is also albino? (b) What will be the expected total frequency of albino children in these families?

8-17. If a sample of 25 five-child families had been located, each sibship containing one albino child but both parents normal, how many of the 125 children would be expected to be normal?

8-18. The ability to taste phenylthiocarbamide is an inherited trait among humans. In a certain college freshman class, all individuals were classified as follows for their taster phenotype and for sex:

taster males	60
taster females	40
nontaster males	40
nontaster females	10
	150

Is the frequency of tasters significantly different in the two sexes?

8-19. If a random sample of six males were taken from the above college freshman class, what is the probability that it will be composed of: (a) One taster, five nontasters? (b) Three tasters, three nontasters? (c) Four tasters, two nontasters? (d) All nontasters?

8-20. If the sex ratio in the college freshman class in Problem 18 represents the sex ratio at birth of an entire population, what are the probabilities that a family composed of only three children will consist of: (a) Three girls? (b) One girl and two boys? (c) At least one girl?

8-21. Answer Problem 20 if the sex ratio of male to female births is equal.

8-22. In Saxony the sex ratio at birth during the years 1876 through 1885 was approximately .515 males to .485 females. For this period, Geissler (according to Stern) recorded one million families each of which gave birth to 12 children. Among these, 655 families produced 12 sons and no daughters and 56 families produced 12 daughters and no sons. Do these values accord with those that would be expected for sibships of 12 in which one sex is absent?

8-23. A certain strain of *Diplococcus pneumoniae* mutates to penicillin resistance at a rate of 1 per 10,000,000 cells. If 200 petri dishes of media containing penicillin were plated with one million bacteria each, how many petri dishes would you expect to find under random sampling (Poisson distribution) having: (a) One resistant colony? (b) Two resistant colonies? (c) Three resistant colonies? (d) Four resistant colonies?

8-24. In an experiment with *Eberthella typhosa*,

Curcho inoculated each of 10 petri dishes of food media lacking the amino acid tryptophan with 1.3×10^8 bacteria that could not grow without tryptophan. The number of mutant colonies per plate that could not grow without tryptophan found on the 10 petri plates are shown opposite. (a) Calculate the average mutation rate. (b) Does this distribution correspond to what would be expected if mutation arose randomly?

No. mutant colonies per plate	No. plates
0	4
1	4
2	0
3	0
4	1
5	1

REFERENCES

FISHER, R. A., and F. Yates, 1963. *Statistical Tables for Biological, Agricultural and Medical Research,* 6th ed. Oliver and Boyd, Edinburgh.

GOLDSTEIN, A., 1964. *Biostatistics, An Introductory Text.* Macmillan, Inc., New York.

MATHER, K., 1951. *Statistical Analysis in Biology.* Methuen, London. (Reprinted 1965, University Paperbacks, London.)

SIMPSON, G. G., A. ROE, and R. C. LEWONTIN, 1960. *Quantitative Zoology.* Harcourt Brace Jovanovich, New York.

SNEDECOR, G. W., and W. G. COCHRAN, 1967. *Statistical Methods,* 6th ed. Iowa State Univ. Press, Ames.

STEEL, R. G. D., and J. H. TORRIE, 1960. *Principles and Procedures of Statistics.* McGraw-Hill, New York.

WOOLF, C., 1968. *Principles of Biometry.* Van Nostrand Reinhold, New York.

9

DOMINANCE RELATIONS AND MULTIPLE ALLELES IN DIPLOID ORGANISMS

The basic contribution of Mendel was his discovery of the unit nature of inheritance—that it consists of particulate factors, or genes, whose presence can be traced from one generation to another without change or dilution. The particular mode in which Mendel demonstrated the presence of genes, through the principles of segregation and assortment, have been independently demonstrated many times with many different organisms. As a rule, the F_2 mendelian ratios of 3:1 in monohybrid crosses and 9:3:3:1 in dihybrid crosses among diploids are found as expected. These particular ratios derive, as we have seen, from the segregation of two different alleles for each gene pair, one dominant and one recessive. However, in addition to these ratios, experiments have revealed the appearance of phenotypes in novel proportions that cannot be explained on the basis of simple dominance or the presence of only two kinds of alleles. As we shall see, such exceptions do not detract from Mendel's principles, but only extend and develop them.

INCOMPLETE DOMINANCE

In *simple* or *complete* dominance the heterozygote, although genetically different, has the same phenotype as one of the homozygotes (i.e., $Aa = AA$), and the presence of the recessive gene is functionally hidden. Dominance is therefore considered as a functional or physiological effect. In Mendel's examples the chosen gene differences all showed simple dominance except for one character, flowering time, for which his experiments were unfortunately incomplete.

After Mendel, the investigation of flowering time in peas was carried on by various geneticists, especially Rasmusson, who found it to be influenced by different factors, among them a pair of alleles called *A* and *a*. According to Rasmusson, there was a 5-day difference in flowering time between the homozygotes *AA* and *aa*. A cross between such parents gave an F_1 heterozygote, *Aa*, with an approximately intermediate flowering time. If we designate the *aa* flowering time as zero (early), the flowering times of the different genotypes are:

aa	0.0	early
Aa	3.7	intermediate
AA	5.2	late

Selfing the heterozygotes results in the following ratios:

$$Aa \times Aa$$
$$\downarrow$$
$$1\,AA : 2\,Aa : 1\,aa$$
late intermediate early

Note that although the genotypic ratios are exactly those we would expect in any experiment in mendelian segregation, it is the phenotypes that have changed from their usual 3:1 dominant-recessive ratio. The absence of complete dominance by one allele thus makes each genotype separately distinguishable.

Since Mendel's time many such examples of *partial* or *incomplete* dominance have been found for various traits in both plants and animals. For example, Correns found that red-flowered four-o'clocks crossed to white-flowered plants give pink heterozygotes (Fig. 9–1). If the pink F_1 is self-fertilized, the F_2 ratio is 1 red:2 pink:1 white. Certain feather color genes in birds act similarly; the blue Andalusian fowl arises from the combination of alleles for white and black feathers. A mating between two blue fowl gives offspring in the ratio 1 white:2 blue:1 black.

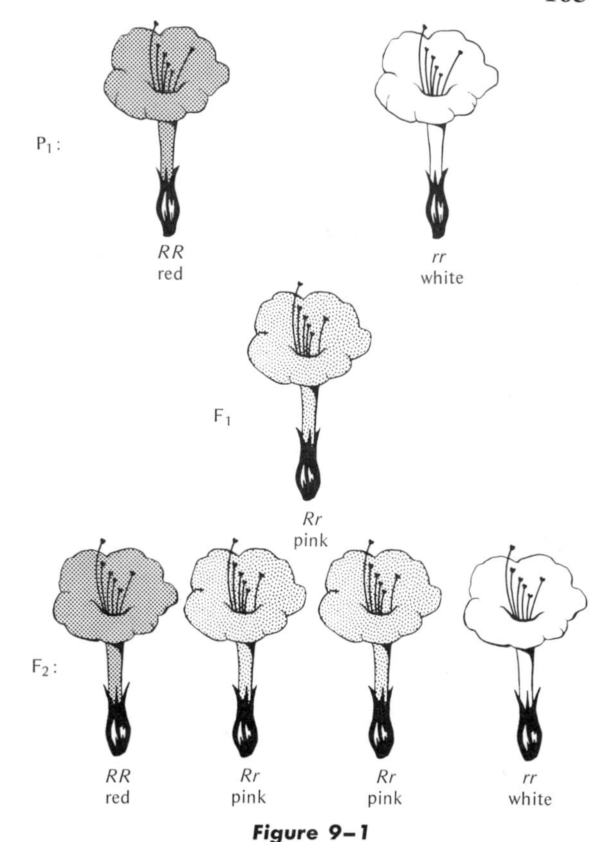

Figure 9–1

Incomplete dominance in four-o'clocks (Mirabilis jalapa). *(From Belar, after Correns.)*

Some characters show phenotypic expression of the partially dominant heterozygotes in a more quantitative manner. In the *Bar* mutation (*B*) in *Drosophila*, which reduces the number of eye facets from 800 (B^+/B^+) to about 60 (B/B), the facet number in B/B^+ heterozygotes may fall in an intermediate range between 250 and 500. Even in those instances where dominance appears to be clear-cut, it is not unusual to find some functional effect of the recessive gene in the heterozygote. This can be seen in one of Mendel's own crosses, that between smooth- and wrinkled-seeded plants. Investigations by Darbishire have shown that this phenotypic difference is closely connected with the shape and appearance of the starch grains in the seed. Because the smooth seeds have many large, round, starch grains they can retain more water and consequently appear fuller and rounder. The wrinkled seeds, on the other hand, are more sugary than starchy and lose water upon ripening. The hybrid between these, however, although smooth in appearance, has starch grains that are intermediate in type and amount.

In *Drosophila,* Ziegler-Günder and Hadorn showed that while the effects of normal eye-color genes appear dominant on superficial examination, some recessive mutations affect the amount of eye pigments in heterozygotes. Thus the *sepia* eye-color mutant, for example, acts as a recessive gene to normal red eye color, but still reduces the quantity of some of the fluorescent pteridine pigments in the heterozygote.

In general most "wild-type" alleles normally found in diploids are dominant because of their advantageous effects on the organism. On the molecular level, wild-type gene products are mostly enzymes or other proteins necessary for proper metabolic functioning of the organism, whereas mutant gene products are often nonfunctional or only partly functional. There is therefore considerable evolutionary advantage for beneficial wild-type alleles to be dominantly expressed in any genotypic combination in which they find themselves. Nevertheless, as we have seen (see also Table 9–2), mutant alleles may also be expressed phenotypically as in partial dominance, although the level at which this expression occurs may not always be obvious. Depending therefore on the level of observation, the effect of a recessive gene in heterozygotes can also be viewed quantitatively, as in Fig. 9–2. If the heterozygous phenotype (i.e., *Aa*) coincides with the phenotype of one of the homozygotes (i.e., *AA*), dominance of *A* is complete. Any lesser phenotypic effect of the heterozygote (i.e., toward *aa*) can then be termed *incomplete dominance.* According to various genetic usages discussed later (e.g., Chapters 14, 15, and 32), should the *Aa* phenotype be exactly intermediate to the two homozygotes, the effect of each allele is considered to be arithmetically equivalent

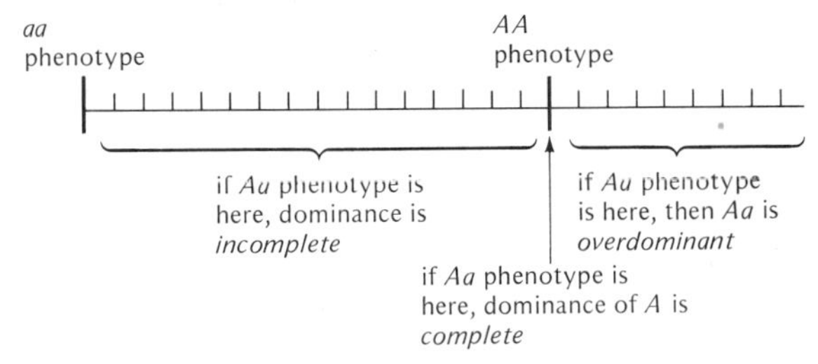

Figure 9–2

Types of dominance in relationship to the AA and aa phenotypes measured quantitatively.

or *additive*, and dominance can then be said to be *absent* or *lacking*.

OVERDOMINANCE

The phenotype of a heterozygote measured quantitatively is not always equal or intermediate to that of the homozygotes. On occasion, for some gene differences, the heterozygote may exceed the phenotypic measurement for both homozygous parents (right side, Fig. 9–2). Such heterozygotes are described as *overdominant*. For example, in *Drosophila* the white-eyed gene (w) in heterozygous condition (w^+/w) causes a marked increase in the amount of certain fluorescent pigments (sepiapteridine and himmelblaus) over both the white and wild-type homozygotes.

For the most part, however, the term overdominance has been used for characteristics concerned with biological "fitness" such as size, productivity, and viability. Crosses between homozygous individuals which were relatively poor for such fitness characteristics have, in many cases, yielded offspring that appeared overdominant to both parents. However, since many gene differences are usually involved in such crosses, it is difficult to determine the exact dominance relations of particular individual genes. This subject is of importance for understanding the genetic structure of populations and will be further discussed (Chapters 32 and 33).

CODOMINANCE AND BLOOD TYPES

Aside from its quantitative expression, a gene also has a qualitative expression in the sense that it usually affects the production of a particular substance having a special structure or function. If, let us say, the two alleles in a gene pair are each associated with different substances, *codominance* occurs when both substances appear together in the heterozygote. For example, if unique phenotypic substances X and Y are associated with the homozygotes *AA* and *aa*, respectively, the codominant heterozygote *Aa* would produce both substances, X and Y, at the same time. The blue Andalusian fowl, which exhibits incomplete dominance between black and white feather color alleles, is really a fine mosaic of black and white areas that appear to be blue. There are other examples in which visible effects are produced by each allele of a gene pair, but the detection of such qualitative differences is usually difficult.

One technique that has gained considerable favor in recognizing qualitative differences depends upon utilizing the sensitivity of an organism to material introduced from different organisms. When a foreign material, an *antigen*, enters the bloodstream of an organism, it will sometimes elicit the production of substances in the host called *antibodies* which react with this material and thereby reduce harmful effects. These antibodies are specific for a particular antigen and each different antigen causes the development of a different antibody. Thus the bloodstream (specifically, the blood serum) of an individual may contain many different kinds of antibodies for many different kinds of antigens. Human red blood cells (erythrocytes) are one type of material that can be used to elicit the production of antibodies. If such cells are washed and then injected into a rabbit, the rabbit's blood serum (*antiserum*) soon contains antibodies that can be extracted by special methods and will react with that particular type of red blood cell. The type of reaction that usually occurs is a clumping, or *agglutination*, of cells into groups. (More formal terminology describes the antigen as an *agglutinogen* and the antibody as an *agglutinin.*)

When Landsteiner and Levine tested the red blood cells of various people, they found, at first, three general types, called M, N, and MN, respectively. The M type elicited antibodies (anti-M serum) specific for M which could not agglutinate N, while the N red blood cells caused the production of antibodies specific for N (anti-N serum). Both types of antibodies however, could agglutinate the MN red blood cells.

By analyzing the relationship between people carrying the various blood types, it was found that the genes for M and N appeared to be alleles to each other. In honor of Landsteiner this gene was named *L*, and the alleles for M and N were named L^M and L^N, respectively. L^ML^M individuals had the M phenotype and produced only L^M gametes; L^NL^N individuals had the N phenotype and produced only L^N gametes; L^ML^N individuals had the MN phenotype and produced both L^M and L^N gametes. In recent years some authors omit the *L*, and these alleles are also designated simply *M* or *N*.

Since there are three different MN genotypes, there are six different possible matings, each of which, as shown in Table 9–1, produces its own particular combination of offspring. Note that among different mating combinations, some exclude the appearance of particular classes of offspring; i.e., N × N matings cannot give rise to M and MN offspring. These exclusions are occasionally important in helping to decide doubtful

TABLE 9–1

Phenotypes of offspring produced by different mating combinations of MN blood types

Parents	Offspring (Ratios)		
	M	MN	N
$L^ML^M \times L^ML^M$ or $MM \times MM$	all	—	—
$L^ML^M \times L^ML^N$ or $MM \times MN$	1	1	—
$L^ML^M \times L^NL^N$ or $MM \times NN$	—	all	—
$L^ML^N \times L^ML^N$ or $MN \times MN$	1	2	1
$L^ML^N \times L^NL^N$ or $MN \times NN$	—	1	1
$L^NL^N \times L^NL^N$ or $NN \times NN$	—	—	all

family relations in paternity cases; that is, if the blood types of offspring and suspected parents can be determined, particular individuals may, at times, be excluded as possible biological parents. For example, the father of an MN child born to a woman of N blood type must have had the *M* gene. A man of N phenotype (*NN* genotype) accused of fathering this child could therefore be excluded on a biological basis. Of course, the presence of the *M* gene would not necessarily indict an M or MN individual in this case, since many *M* genes are present in the population. Blood-grouping evidence in paternity cases is therefore useful for "excluding," but not for "accusing" (proving).

MULTIPLE ALLELES

Alleles can be defined as genes that are members of the same gene pair, each kind of allele affecting a particular character somewhat differently than the other. A particular gene pair in a diploid organism contains, by definition, only two alleles at a time, one for each member of the pair. As we have seen, this condition arises from the presence of homologous pairs of chromosomes, each of two homologous chromosomes containing one allele of the gene pair. In Mendel's experiments there were two possible kinds of alleles in a gene pair, i.e., smooth or wrinkled (*S, s*), yellow or green (*Y, y*), etc. Such a system can also be symbolized as A^1 or A^2, B^1 or B^2, and so on, with each different letter symbolizing a different gene pair and each different number symbolizing a different allele. Actually there are more than only two possible kinds of alleles in a gene pair; in some instances, hundreds of possibilities exist, i.e., A^1, A^2, A^3, . . . , A^n and B^1, B^2, B^3, . . . , B^n.* The grouping of all the different possible alleles that may be present in a gene pair is defined as a system of *multiple alleles*. Haploid organisms, of course, may also have multiple allelic systems, but instead of being present two at a time as in diploids, each of the different alleles can be present only once in any single organism.

An example of multiple alleles in *Drosophila* occurs for one of the gene pairs affecting eye color. For this gene one of the first hereditary changes recorded was a recessive mutant allele, *white*, associated with the complete absence of eye color in the normal red-eyed fly. Since that time many alleles of *white* have been found which have a quantitative effect intermediate between *white* and the wild-type red. Refined techniques for the extraction and measurement of eye pigments have furnished values for the various genotypes given in Table 9–2. Among these alleles and their combinations there are two clusters of phenotypic effects: those appearing as abnormal, with pigment ranging in amount from .0044 to .1636, and wild-

TABLE 9–2

Eye pigment in different combinations of alleles of the *w* gene in *Drosophila melanogaster*

		Relative Amount of Total Pigment
Genotypes Appearing as Nonwild		
w/w	(*white*)	.0044
w^t/w^t	(*tinged*)	.0062
w^a/w^a	(*apricot*)	.0197
w^{bl}/w^{bl}	(*blood*)	.0310
w^e/w^e	(*eosin*)	.0324
w^{ch}/w^{ch}	(*cherry*)	.0410
w^{a3}/w^{a3}	(*apricot-3*)	.0632
w^w/w^w	(*wine*)	.0650
w^{co}/w^{co}	(*coral*)	.0798
w^w/w^{col}		.1114
w^{sat}/w^{sat}	(*satsuma*)	.1404
w^{col}/w^{col}	(*colored*)	.1636
Genotypes Appearing as Wild Type		
w^{+S}/w^{+S}	(Stellenbusch wild type)	.6854
w^{a3}/w^{+C}		.8700
w^{+S}/w^{+C}		.9220
w^{+C}/w^{+C}	(Canton *S* wild type)	.9895
w^{+C}/w^{+G}		1.0546
w^{+G}/w^{+G}	(Graaf-Reinet wild type)	1.2548

From spectrophotometric data of Nolte, 1959.

* Occasionally such series are written as subscripts, e.g., A_1, A_2, A_3, etc.

type eyes with pigment quantities ranging from .6854 and higher. The data are not complete for all possible combinations but show quantitatively that the pigments of heterozygotes (i.e., w^{w}/w^{col} and w^{a3}/w^{+C}) fall between the values of their respective homozygotes. Thus dominance in this series appears to be incomplete, although it would be difficult to make such a decision without this type of refined analysis. Also of interest is the quantitative separation between the various wild-type genotypes that appear superficially as red and are not ordinarily distinguishable (i.e., w^{+S}, w^{+C}, w^{+G}). Alleles of this kind, which act within the same phenotypic range of each other, have been termed *isoalleles*. Many such isoalleles have been discovered, some within the phenotypic range of an abnormal character, *mutant isoalleles*, and some within the phenotypic range of wild type, *normal isoalleles*. A multiple allelic system may therefore be quite complex, including within it various subsidiary isoallelic systems.

Multiple allelic systems have also been found in plants, especially for self-sterility genes that prevent fertilization between individuals that are closely related to each other. As worked out by East and collaborators in the tobacco plant *Nicotiana tabacum*, a haploid pollen grain carrying a self-sterility allele, i.e., S^1, will not grow well on a diploid female style carrying the same allele, i.e., S^1S^2, but can successfully fertilize a plant carrying S^2S^3 or S^3S^4, etc. Thus successful fertilizations, in these cases, always denote the presence of an allele in the pollen that is different from the two alleles in the female plant.

Such self-sterility systems offer an advantage to new rare alleles and a disadvantage to those that are common. The reason for this is that once an allele becomes common, its frequency is reduced or limited by the many sterile mating combinations to which it is now exposed. Rare alleles, on the other hand, will have successful fertilizations with almost every female plant they meet, until they too become common. In this fashion self-sterility systems of considerable numbers of alleles can be established, reaching as high as 200 alleles or more in red clover (Bateman, 1947).

MULTIPLE-ALLELIC BLOOD-GROUP SYSTEMS

The biological uniqueness that each individual organism attains is frequently noted in the reactions that occur when it receives biological material from other organisms. In animals, tissues that are removed from one individual and grafted onto another are frequently sloughed off or rejected because of "incompatibility" between the introduced material and that of the host (see p. 174). We have taken advantage of such reactions to show how differences between the three MN blood types can be determined; that is, the ability of the rabbit to produce specific antibodies in its serum, or antiserum, enables the detection of particular antigens.

As increased numbers of different antisera are discovered, more kinds of inherited antigens are revealed, many of which are determined by multiple alleles. For example, in 1947 Walsh and Montgomery found that a certain portion of human blood serum could be used to distinguish new varieties of MN blood types. Through the analysis of family data, the inheritance of this new blood-type system, called *Ss*, was shown to be closely tied to the inheritance of the *MN* gene. According to our previous notation, there appeared to be four different codominant alleles, L^{MS}, L^{Ms}, L^{NS}, L^{Ns} (or M^S, M^s, N^S, N^s), which could give nine different phenotypic combinations (Table 9-3).

The multiple-allelic nature of the *MNSs* system, however, is not unique among blood groups. The first case of multiple alleles demonstrated in man was really that of another blood-group system, which had been discovered by Landsteiner and his students in the early 1900s. This system, called ABO, was shown by Bernstein in 1925 to consist of three alleles of a single gene, I^A, I^B, and I^O, forming four different phenotypic groups: A (I^AI^A or I^AI^O), B (I^BI^B or I^BI^O), AB (I^AI^B), and O (I^OI^O).* In this case the blood serum of man him-

* The I^O allele is also symbolized as lower case *i*, so that the O blood genotype is represented as *ii*. The three genes can also be symbolized without the *I*, i.e., *A*, *B*, and *O*.

TABLE 9–3

Antigens and reactions with different antibodies for the nine possible MNSs blood-group combinations

Genotype	Antigens Produced (Phenotype)	Reaction with Antibodies*			
		Anti-M	*Anti-N*	*Anti-S*	*Anti-s*
$L^{MS}L^{MS}$ or M^SM^S	MS	+	—	+	—
$L^{MS}L^{Ms}$ or M^SM^s	MSs	+	—	+	+
$L^{Ms}L^{Ms}$ or M^sM^s	Ms	+	—	—	+
$L^{MS}L^{NS}$ or M^SN^S	MNS	+	+	+	—
$L^{MS}L^{Ns}$ or M^SN^s; $L^{Ms}L^{NS}$ or M^sN^S	MNSs	+	+	+	+
$L^{Ms}L^{Ns}$ or M^sN^s	MNs	+	+	—	+
$L^{NS}L^{NS}$ or N^SN^S	NS	—	+	+	—
$L^{NS}L^{Ns}$ or N^SN^s	NSs	—	+	+	+
$L^{Ns}L^{Ns}$ or N^sN^s	Ns	—	+	—	+

* Symbol + indicates a positive agglutination reaction, and — indicates no agglutination.

self manufactured the antibodies that reacted with the blood-cell antigens of other individuals.

As shown in Fig. 9–3, transfusions of blood between individuals of different ABO types may result in an agglutination reaction, especially if large quantities of a different blood type are being introduced; either the antibodies of the recipient will destroy (hemolyze) the red cells of the donor or the antibodies of the donor will hemolyze the red cells of the recipient. If small quantities of blood are being introduced, however, transfusion can be considered safe as long as the serum of the recipient does not contain antibodies for the blood-cell antigens of the donor. In small transfusions, even if the donor serum transmits antibodies for the recipient's blood cells, the antibodies are usually quickly absorbed by other tissues or greatly diluted in the recipient's bloodstream. On this basis people of O blood type have been described as "universal donors" (absence of A and B antigens) and people of AB blood type as "universal recipients" (absence of A and B antibodies).

The A and B antigens are found not only on red blood cells but often in the body fluids as well. Individuals who possess such antigens in their fluids, such as the saliva, are called "secretors," and can be shown to be either homozygous (*Se Se*) or heterozygous (*Se se*) for the dominant *secretor* allele (controlled by a different gene pair than the ABO blood group genes). Whether secreted or attached to red blood cells, the ABO antigens are now known to represent only slight modifications of the sugar (saccharide) portion of either the protein-sugar compounds (mucopolysaccharides) present in red blood cells or the fat-sugar compounds (lipopolysaccharides) present in secretions. According to Kabat, Watkins, and others, it is the terminal sugars of these compounds which differ between the A and B antigens (Fig. 9–4). These differences, although small, are nevertheless significant, so that antibodies can discriminate between one antigen and the other. In fact, additional multiple alleles at the ABO locus have recently been found (A_2, A_3, A_x, A_m) which probably differ in even more minor respects. In cattle, the number of different alleles for a particular blood-type gene called "B" has reached more than 300 so far, according to Stormont.

Of the various human blood groups listed in Table 9–4, about half have turned out to be multiple-allelic systems. Unfortunately, since each newly discovered antigen receives a name that is not always immediately related to its appropriate gene, the nomenclature for these alleles has become quite complex (e.g., additional alleles for MNSs are designated *Mi*^{a}, *Vw*, *Mu*, *Hu*, *He*, etc.) and have therefore not been included in the table. Hopefully, uniform and simple methods for desig-

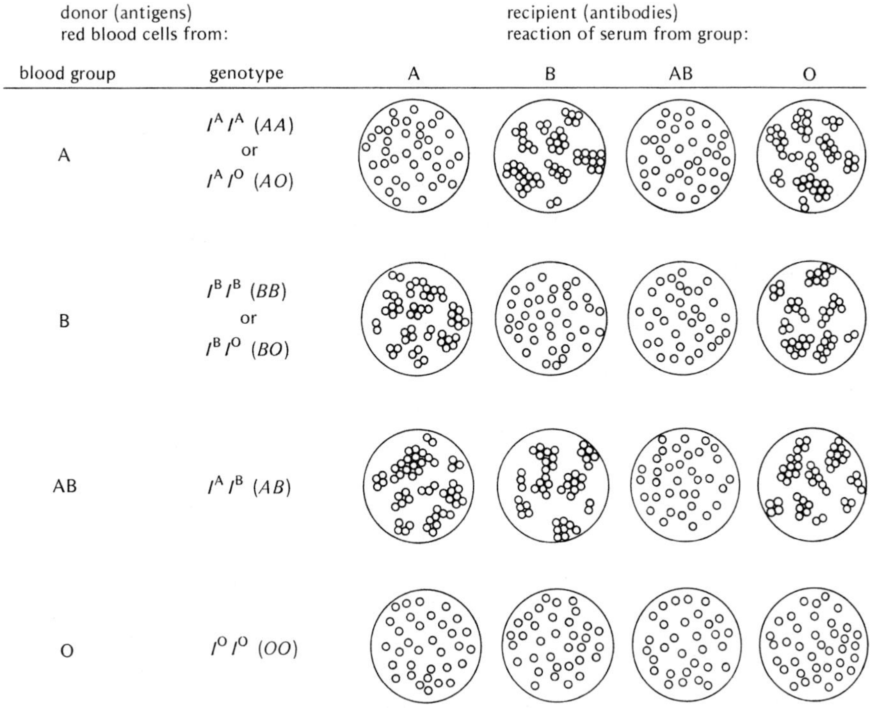

donor (antigens)
red blood cells from:

recipient (antibodies)
reaction of serum from group:

blood group	genotype	A	B	AB	O
A	$I^A I^A$ (*AA*) or $I^A I^O$ (*AO*)				
B	$I^B I^B$ (*BB*) or $I^B I^O$ (*BO*)				
AB	$I^A I^B$ (*AB*)				
O	$I^O I^O$ (*OO*)				

Figure 9–3

Action of serum from persons of various blood types (recipients) on the different kinds of red blood cells (donors). Clumping of cells indicates reaction of serum antibodies with cell antigens. The A serum carries antibodies for B and AB cells, and the B serum has antibodies for A and AB cells. AB serum has no antibodies for any of the cells, whereas the O serum carries antibodies for all types of cells except its own. (After G. Hardin, 1961. Biology: Its Principles and Implications. *W. H. Freeman, San Francisco.)*

nating these multiple alleles will be instituted. Despite this difficulty, however, the large variety of blood-group systems enables very exact descriptions of individual genotypic and phenotypic differences. Race and Sanger, for example, have found that of the various blood-group combinations found in England, none occur more frequently than once in 270 people, and some are as rare as about one in 100 million people.

It may be anticipated that this uniqueness of different genotypes is of special value in paternity problems, since, as the number of different blood groups increases, there is an increased efficiency for solving cases of disputed parentage. For example, when the test is restricted to the ABO blood types, men wrongfully accused in paternity cases have only an 18 percent chance of demonstrating their biological innocence. On the other hand, by being tested for seven of the different blood groups, approximately 60 percent of men wrongfully accused in paternity suits can be excluded from biological parentage.

RH AND ABO INCOMPATIBILITIES

The agglutination reaction that occurs when red blood cells are clumped by serum antibodies may also occur in the circulation of a mammalian embryo having a blood type antigenically different

TABLE 9–4

Fourteen of the most commonly detected blood group systems

Blood-group System	*Date Discovered*	*Probable No. Alleles*
ABO	1900	5
MNSs	1927	20+
P	1927	4
Rh	1940	30+
Lutheran	1945	3
Kell	1946	6
Lewis	1946	2
Duffy	1950	3
Kidd	1951	3
Diego	1955	2
Yt	1956	2
I	1956	2
Xg	1962	2
Dombrock	1965	2

Data of Race and Sanger, and others.

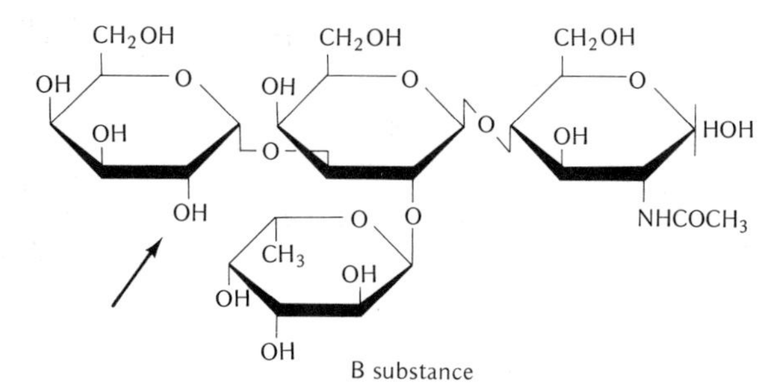

Figure 9–4

Proposed structure of the terminal sugars in human A and B antigens with arrows pointing to the only differences observed between the two structures. It has been suggested (see review of Ginsburg) that the genes responsible for these antigens produce specific enzymes which then add the sugars designated above to certain mucopolysaccharides or lipopolysaccharides that have been previously prepared by earlier-acting genes. (From Watkins.)

from its mother. In such cases embryonic antigens that cross the barrier of the mammalian placenta will cause the formation of maternal antiserum which, in turn, crosses the placenta and reacts with the red blood cells of the offspring. The first instance of such incompatibility was noted in checking the blood types of children born with serious anemia (erythroblastosis fetalis) caused by the breakdown (hemolysis) of their normal red blood cells.

As determined by Levine and others, these children had a blood type, *Rh positive*, inherited from their fathers, but antigenically different than their *Rh negative* mothers. This Rh factor, first detected in the red blood cells of Rhesus monkeys, was initially thought to be caused by a gene with only two alleles, *R* and *r*. The events leading to erythroblastosis thus arose from the Rh negative genotype of the mother (*rr*) producing antiserum against the antigens of Rh positive offspring (*Rr*). Since the *R* allele acted as a dominant to *r*, Rh positive males married to Rh negative females could have either all or half their offspring phenotypically Rh positive, depending on whether the paternal genetic constitution was respectively homozygous (*RR*) or heterozygous (*Rr*). Thus once an Rh negative mother had an Rh positive child, or once she had received transfusions of Rh positive blood, she became *immunized* against Rh positive antigens by producing anti-Rh serum. Successive pregnancies in which Rh positive children were conceived could then lead to serious hemolytic anemia among these offspring.

Further investigation soon showed that the Rh factor was genetically quite complex. Additional kinds of Rh antisera were found which could distinguish between different Rh antigens. Anti-rh′ serum, for example, reacted both with certain Rh negative and Rh positive blood types but not with all. A somewhat different pattern of differentiation between Rh blood types could be found by using a serum called anti-rh″. Correlated with genetic studies, this indicated to Wiener and others the existence of at least eight different Rh alleles.

TABLE 9–5

Agglutination reactions of different antisera to eight of the Rh alleles and to three common genotypic combinations

Notation		Agglutination with Antisera					
WIENER	FISHER*	ANTI-RH$_0$ (ANTI-D)	ANTI-rh′ (ANTI-C)	ANTI-rh″ (ANTI-E)	ANTI-hr′ (ANTI-c)	ANTI-hr″ (ANTI-e)	ANTI-hr (ANTI-f)
Alleles							
r	dce	—	—	—	+	+	+
r'	dCe	—	+	—	—	+	—
r''	dcE	—	—	+	+	—	—
r^y	dCE	—	+	+	—	—	—
R^0	Dce	+	—	—	+	+	+
R^1	DCe	+	+	—	—	+	—
R^2	DcE	+	—	+	+	—	—
R^z	DCE	+	+	+	—	—	—
Genotypes							
R^1r	DCe/dce	+	+	—	+	+	+
R^2r	DcE/dce	+	—	+	+	+	+
R^1R^2	DCe/DcE	+	+	+	+	+	—

* The "F" and "V" genes are omitted in this table. Note that an anti-d serum in this notation has not yet been discovered.

Table 9–5 shows the agglutination reactions of the products of each of these alleles with each of the three antisera. Note that most alleles produce some kind of observable antigen and are therefore codominant. Only the r allele appears fully recessive, because of the inability of its product to react with any of the three antisera. This recessiveness, however, disappears with the introduction of new antisera called anti-hr′, anti-hr″, and anti-hr (last three columns), which differentiate between r and the other Rh negative alleles.

Fisher, Race, Sanger, and other English blood-group workers, have tried to replace this Rh nomenclature by one based on three closely associated gene pairs, D, C, and E. According to their system, each Rh gene is a combination of alleles at each of these gene pairs, e.g., $r = dce$, $R^z = DCE$. As also shown in Table 9–5, the antisera can then be called anti-D, anti-C, etc. In recent years new antisera have been discovered for this blood group and more than 30 different Rh alleles are known. Proponents of the DCE system have therefore been forced to postulate new gene pairs, i.e., F and V, to account for some of these alleles, as well as to propose multiple allelic systems for some of these subsidiary genes (e.g., D, D^u, d), and this nomenclature no longer has its former simplicity. The designations offered by Wiener seem therefore more reasonable, although both systems of nomenclature still remain in use.

Because of the large number of Rh alleles and antisera, the occurrence of hemolytic disease of the newborn is not confined to incompatibility between Rh positive and negative blood types, i.e., anti-Rh$_0$ serum, but may also occur when the mother produces anti-rh′ or any other antiserum for which the child carries antigens. Thus a mother whose blood type was Rh positive (i.e., R^0r') nevertheless produced anti-rh″ serum, which caused anemia in an Rh positive child that had rh″ antigens (i.e., allele R^2).

Thorough testing of maternal and paternal blood types is therefore of considerable value in predicting the possibility of hemolytic anemia among the newborn. Certainly mothers or prospective mothers receiving transfusions should not be given blood to which they may develop a high "titer" (concentration) of antiserum. When hemolytic anemia does occur as a result of maternal-fetal incompatibility, the child's blood should be replaced as soon as possible. An Rh positive child, for example, whose blood has been seriously affected by the anti-Rh serum of its mother, should be given transfusions of blood which do not contain further Rh positive antigens and which do not contain further anti-Rh serum; i.e., the introduced blood should be from an Rh negative individual who has not been exposed to Rh positive antigens.

The Rh blood group, however, is only one of the systems which may cause mother-offspring incompatibilities. Other blood group antigens may also travel across the placenta and produce maternal antisera. As a rule, however, there are many fewer cases of newborn hemolytic anemia than would ordinarily be expected. Rh incompatibility, for example, is estimated to occur in about 10 percent of all pregnancies; nevertheless, only 1/20 to 1/50 of these incompatible offspring turn out to be affected by hemolytic anemia.

The reasons for this lowered incidence of hemolytic disease may be twofold: First, for a woman to develop antibodies against the antigens of her offspring, the fetal cells must diffuse through the placenta into the maternal circulation. This diffusion does not seem to occur too frequently and, even when it occurs, the amount of diffused antigen may be low enough so that the amount of maternal antibody production is not very high. Another cause lowering the frequency of Rh hemolytic disease may, surprisingly, result from an incompatibility between ABO blood types. For example, any embryonic A or B red blood cells which leak through the placental membranes of an O-type mother may be quickly destroyed by already existent anti-A or anti-B antibodies before anti-Rh antibodies can be formed. Thus, although Rh incompatibility may exist between mother and offspring, ABO incompatibility may prevent anti-Rh serum from developing. Evidence for this view is seen in the data of Levine (Table 9–6), showing increased incidence of Rh hemolytic disease when mother and offspring are compatible for the ABO blood groups. When they are incompatible, and the diffusing fetal cells can be destroyed by maternal ABO antibodies, the frequency of Rh hemolytic disease decreases.

TABLE 9–6

Number of cases of ABO compatible and incompatible matings in which Rh hemolytic offspring appeared

	Mother	Father	Cases of Rh Disease
ABO compatible	A	O	765
	B	O	156
	AB	O	59
	AB	A	48
	AB	B	10
			1038
ABO incompatible	O	A	319
	O	B	81
	O	AB	2
	A	AB	27
	B	AB	6
			435

From Levine, 1958.

At present the destruction of fetal cells in the maternal circulation is deliberately undertaken by the novel method of giving the Rh negative mother an injection of Rh antiserum soon after the birth of her first (and succeeding) Rh positive child. By this means the Rh positive fetal cells that cross the placenta during parturition are destroyed before the mother can produce her own Rh antiserum. The injected antiserum then disappears, so that Rh antiserum is no longer present at the next pregnancy.

HISTOCOMPATIBILITY GENES AND ANTIBODY FORMATION

Blood group incompatibility is only one possible interaction between different vertebrate individuals. Another type of interaction may occur when transplants of skin or most other organs are attempted between individuals. Unless the individuals involved are identical twins or come from an inbred stock with a high degree of genetic similarity, the transplant will usually be rejected by the host. An "immune" reaction involving the production of antibodies appears responsible for this rejection and a second transplant from the same donor is usually rejected even more rapidly than the first. The precise mechanism of antibody formation is presently of profound interest because it relates to questions of disease resistance, organ replacement, and tumor rejection.

From the genetic point of view a number of genes seem to be involved in the rejection or tolerance of tissue transplants and are collectively called *histocompatibility genes* or *loci*. The action of alleles at many of these genes seems to be codominant, and individuals will reject the tissues of donors that carry alleles which they themselves do not carry. For example, if we confine ourselves to only three pairs of genes, A, B, and C, an inbred strain of mice may bear the histocompatibility genotype A^1A^1 B^1B^1 C^2C^2. Such mice can tolerate tissue transplants from other mice of the same genotype but will reject transplants from inbred mice of genotype A^2A^2 B^2B^2 C^1C^1. The F_1 hybrid of these two strains, A^1A^2 B^1B^2 C^1C^2, will accept transplants from either parental strain, but its tissues will be rejected by either parental strain. If $F_1 \times F_1$ matings are made, the F_2 will show a wide variety of possible genotypic combinations (e.g., A^1A^1 B^1B^2 C^1C^1; A^1A^2 B^1B^2 C^2C^2; etc.) in many of which one or more F_1 alleles will be lacking. Such F_2 individuals will therefore generally reject F_1 donor tissues, while F_2 tissues will not be rejected by F_1 hosts. Greater genetic complexities occur, of course, since at least 15 genes are known to be involved in histocompatibility determination in mice, some maintaining multiple alleles, and some, such as the H–2 system, being more influential than others.

In humans histocompatibility is believed to be primarily determined by a counterpart of the mouse H–2 system, called the HL–A system, although antigens produced by the ABO blood group and other systems are also involved. So far two separate genes linked together on the same chromosome have been proposed for the HL–A system, each with about a dozen or more possible alleles, each allele specifying a particular antigen. Although originally found on human white blood cells, HL–A antigens are now also shown to be present on skin, kidney, lung, and liver tissue. Skin grafts between siblings who share the same HL–A antigens survive significantly longer than grafts between siblings whose HL–A antigens differ. Most unrelated people, however, differ markedly in their HL–A genotypes, and transplanted organs and tissues are most often rejected. To avoid rejection the transplantation recipient is usually given "immunosuppressive" drugs that inhibit the growth and cell division of antibody-forming cells. Unfortunately such treatment also inhibits antibody formation against bacterial cells as well as against the unique antigens carried by many tumor cells, thereby leading to increased susceptibility to infection and cancer.

At present antibody-producing cells in vertebrates are believed to arise in the bone marrow and are then processed by two main types of lymphatic organs to become "T" and "B" lymphocytes (Fig. 9–5). Exposure of these lymphocytes to a foreign antigen then causes their transformation into antibody-producing cells. According to the presently accepted "selective theory," the ability to produce various types of antibodies is already present in lymphocytes, and the antigen acts merely to switch on the appropriate gene or genes. However, it is still not clear whether there exists an individual gene for each type of antibody produced, or whether the products of a restricted number of genes combine in different ways to form the wide variety of antibodies observed (see also Chap. 30).

In man, diseases involving antibody-formation include those in which little or no antibody is formed (agammaglobulinemias) and those that form large excesses of only one type of antibody or part of an antibody (multiple myelomas). In mice, the ability to form specific immune responses (i.e., specific antibodies) to particular antigens has been shown to be associated with particular dominant or codominant genes. That is, a mouse bearing the appropriate gene can produce antibodies against a particular antigen, whereas a mouse lacking the gene is unable to do so. In other words, not only are the kinds of *antigens* produced by an organism influenced by genotype, but the ability to form *antibodies* against specific antigens is, in these cases, also genetically determined. In fact some of the immune response specificity genes and the histocompatibility genes appear to be closely associated in the same chromosomal area.

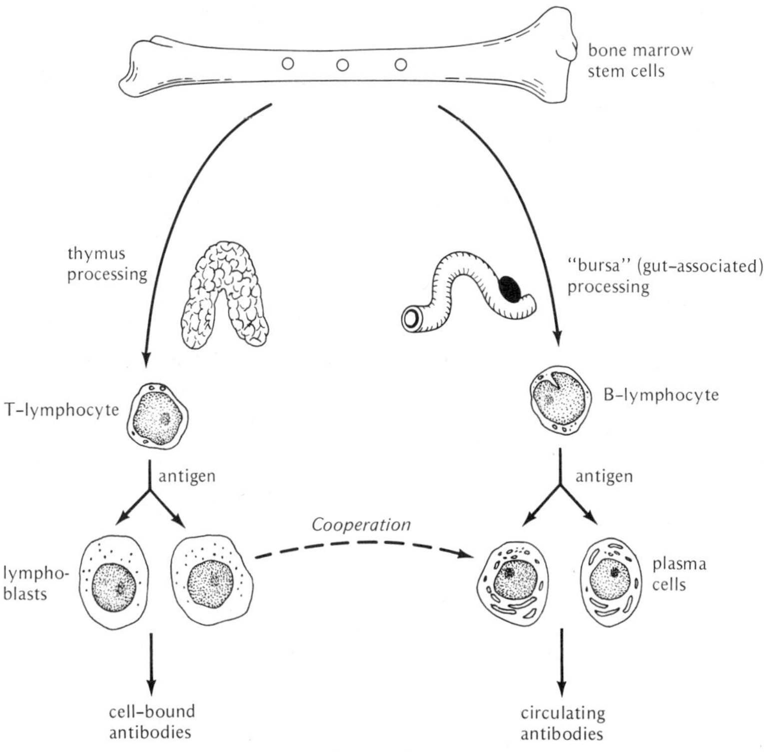

Figure 9–5

The origin and processing of antibody-producing cells. When exposed to an antigen, the T-cells do not produce circulating antibodies themselves but apparently help the B-cells to do so. In the process of development from lymphocyte to plasma cell, the amount of endoplasmic reticulum and ribosomes markedly increases and the cell synthesizes large amounts of protein. The fact that little, if any, antigen is found in the plasma cell indicates that antibodies are not formed directly on antigen "templates" but that specific genes are in some way "turned on" in tissue cells to produce specific antibodies. This is also supported by the fact that each plasma cell generally produces only one type of antibody, no matter how many antigens are present. (After Roitt.)

Complex as they may seem, antibodylike responses probably developed early in evolutionary history, and reactions of this kind can be shown to occur in all vertebrates tested as well as organisms as "primitive" as earthworms, insects, and crustaceans. In all cases the basis of such defensive mechanisms seems to lie in the need for an organism to discriminate between its own tissues and that of others that may damage it through infection or parasitization. At times this discrimination can be circumvented by the introduction of foreign tissue into an organism before it has had the opportunity of developing antibody recognition of "self" and "non-self." In mice, Medawar and his colleagues showed that the tissues of unrelated mice introduced into a recipient shortly after birth would be recognized as its own, and later transplants from the same unrelated mouse would not be rejected.

Surprisingly the other extreme also exists—the occasional rejection of an individual's own tissues. Such *autoimmune* diseases in humans range from the rejection of specific organs because of antibodies formed against thyroid glands (Hashimoto's thyroiditis) and adrenal glands (Addison's disease) to the widespread breakdown of numerous tissues (systemic lupus erythematosus, rheumatoid arthritis) including the formation of antibodies against one's own DNA! Evidence indicating that autoimmune diseases may have genetic causes includes the fact that lines of mice and chickens can be bred which develop high rates, respectively, of autoimmune hemolytic anemia and autoimmune thyroiditis.

PROBLEMS

9-1. In four-o'clock plants, the allele for red flower color has an effect that is incompletely dominant over the effect of the white color allele (p. 165). If a cross between two plants produced 18 red, 32 pink, and 15 white plants, what are the phenotypes of the parents?

9-2. What ratios of flower color in four-o'clocks would you expect among the offspring of the following crosses: (a) Red × red? (b) Red × pink? (c) White × pink? (d) Pink × pink?

9-3. In cattle, the effect of the allele producing red coat color (R) is incompletely dominant over the effect of the allele producing white coat color (r), the heterozygote being roan-colored (Rr). On the other hand, the effects of alleles for the absence of horns show complete dominance; HH and Hh are hornless or "polled," and hh is horned. On the assumption that these two gene pairs assort independently: (a) What would be the phenotype of an F_1 derived from a mating $RRHH \times rrhh$? (b) What would be the phenotypes and their proportions in an F_2 derived from crossing $F_1 \times F_1$? (c) What would be the phenotypic proportions among the progeny derived from crossing F_1 individuals to the original white horned stock?

9-4. With some modifications, the following represents the findings of Asmundsen and co-workers in mating experiments between pheasants differing in plumage color. The differences in numbers of progeny produced by the various crosses are not caused by differences in viability but by the number of sires (fathers) and dams (mothers) involved in each cross. (a) Using alphabetical letters to designate genes (e.g., A and a, etc.) what hypothesis would you propose that best explains these results in terms of kinds of numbers of progeny observed? (b) On the basis of the letters you choose, give the genotypes of the light, buff, and ring phenotypes.

Sires	Dams	Progeny		
		Light	Buff	Ring
light	light	92		
light	buff	78	70	
buff	light	14	12	
light	ring		208	
buff	buff	75	141	68
buff	ring		122	128
ring	buff		34	38
ring	ring			347

9-5. Petrakis, Molohon, and Tepper investigated the inheritance of different types of earwax (cerumen) among American Indians living in the United States and observed that an individual could have one of two types of earwax, dry or sticky, which were inherited in the fashion shown by the following data:

Parental Combination	Number of Matings	Offspring	
		Sticky	Dry
sticky × sticky	10	32	6
sticky × dry	8	20	9
dry × dry	12	0	42

On the basis of these data explain whether the inheritance of *dry* earwax is caused by dominance, recessiveness, partial dominance, codominance, or some other mode.

9-6. Skin color in humans has been explained as resulting from different gene pairs in which each gene pair has equal phenotypic effect (e.g., $A = 2$, $a = 1$; $B = 2, b = 1$), dominance between alleles is absent, and each gene pair is perhaps located on a different nonhomologous chromosome. State the number of different phenotypes possible if skin color is caused by the following numbers of gene pairs: (a) Two gene pairs. (b) Three gene pairs. (c) Four gene pairs.

9-7. In guinea pigs, one of the genes that affect coat color has a number of different alleles. In a certain strain of guinea pig, homozygous combinations of these alleles produce the following phenotypes:

CC = black c^dc^d = cream
c^kc^k = sepia c^ac^a = albino

Assuming that these alleles show complete dominance in the order, $C > c^k > c^d > c^a$, what are the phenotypes and proportions that you would expect among the offspring of the following crosses: (a) Homozygous black × homozygous sepia? (b) Homozygous black × homozygous cream? (c) Homozygous black × homozygous albino? (d) Homozygous sepia × homozygous cream? (e) The F_1 of (a) × the F_1 of (c)? (f) The F_1 of (a) × the F_1 of (d)? (g) the F_1 of (b) × the F_1 of (d)?

9-8. Using the guinea pig alleles in Problem 7, what are the most probable genotypes for the parents that produce the following offspring:

Phenotypes of Parents	Phenotypes of Progeny			
	Black	Sepia	Cream	Albino
(a) black × black	22	0	0	7
(b) black × albino	10	9	0	0
(c) cream × cream	0	0	34	11
(d) sepia × cream	0	24	11	12
(e) black × albino	13	0	12	0
(f) black × cream	19	20	0	0
(g) black × sepia	18	20	0	0
(h) black × sepia	14	8	6	0
(i) sepia × sepia	0	26	9	0
(j) cream × albino	0	0	15	17

9-9. In one cross between a black guinea pig and one that appeared to be sepia, 20 offspring were raised, 6 of which were sepia. Would you discard the hypothesis that the sepia parent is homozygous for the sepia gene? (See p. 153.)

9-10. Some investigators found that matings between horses with pale cream coat color to horses with chestnut coat color produced exclusively "palomino" individuals with intermediate coat color. A number of matings between palominos themselves produced 19 pale cream, 21 chestnut, and 44 palominos. (a) Define gene symbols and suggest genotypes for the three coat colors. (b) Test your hypothesis statistically using the chi-square method, and indicate whether you accept or reject it. (c) If a palomino horse breeder wanted to eliminate pale cream individuals both as parents and as offspring, but nevertheless wanted to obtain as many palominos as possible, what breeding method would you suggest that he follow?

9-11. Emerson found that a strain of corn with red pericarp (red tissue surrounding the seeds) crossed to other strains with different pericarp colors produced F_1 offspring all with red pericarp. In one set of crosses the $F_1 \times F_1$ cross produced an F_2 in the ratio of 3 red:1 colorless. In crosses of the same red strain with a variegated strain (red streaked with white) the $F_1 \times F_1$ cross produced an F_2 in the ratio of 3 red:1 variegated. On the basis of these results explain whether you believe that *colorless* and *variegated* are each members of separate pairs of genes or members of a multiple allelic system at one pair of genes.

9-12. The gene causing sickle-cell hemoglobin in humans (Hb^S) is codominant with the normal allele (Hb^A), so that heterozygotes (Hb^A/Hb^S) produce both normal and sickle-cell hemoglobins, described phenotypically as hemoglobins A and S. In 1950 Itano and Neel discovered two families which segregated for a new type of abnormal hemoglobin called C, distinctly different from hemoglobin S. The pedigrees, with their hemoglobin phenotypes, are as follows:

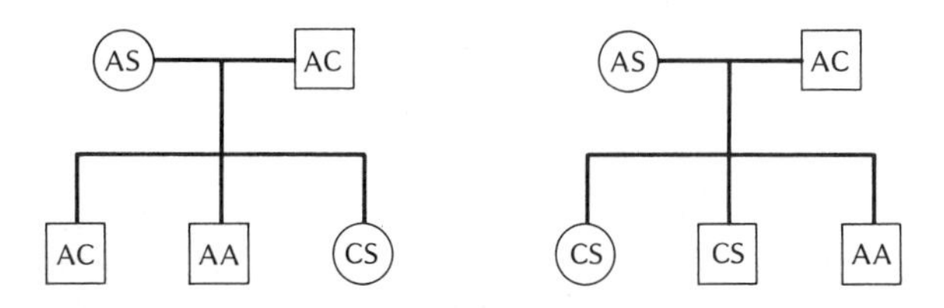

On the basis of these pedigrees would you say that hemoglobin C is caused by a gene that is allelic to that for hemoglobins A and S? Explain.

9-13. Ducks may exhibit either "restricted," "mallard," or "dusky" plumage patterns. Utilizing such differences, three types of crosses were made, with the following results: (1) restricted × mallard: F_1, all restricted; $F_1 \times F_1$, 3/4 restricted, 1/4 mallard. (2) mallard × dusky: F_1, all mallard; $F_1 \times F_1$, 3/4 mallard, 1/4 dusky. (3) restricted × dusky: F_1, all restricted; $F_1 \times F_1$, 3/4 restricted, 1/4 dusky. (a) Assume that an F_1♂ from cross (1) is crossed to an F_1♀ from cross (2). List the phenotypes with frequencies expected among the progeny of this cross. (b) Assume that an F_1♂ from cross (3) is crossed to an F_1♀ from cross (2). List the phenotypes with frequencies expected among the progeny of this cross.

9-14. Wooly, sharply curled hair is caused by a rare dominant gene in European populations. A woman with wooly hair that belongs to the blood group O marries a man with straight hair that belongs to blood group AB. (a) What are the chances that they will have a wooly-haired group B child? (b) What are the chances that they will have a normal-haired group B child? (c) If three normal-haired group A children are born to these parents, what are the chances that the next child will be wooly-haired group B?

9-15. In the following pedigree, the MN blood type is given below the symbol of each individual. In addition, Gershowitz and Fried observed a new blood type

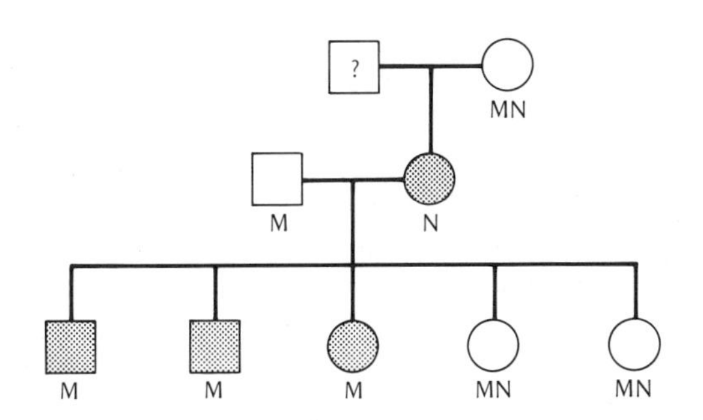

called V (shaded figures). (a) What is the mode of inheritance of trait V? (b) Explain whether V could or could not be caused by an allele of the MN gene, i.e., M^V?

9-16. The red blood cells of the following individuals are tested with the different antisera given in Table 9-5 and are agglutinated only by those antisera indicated below:

Individual	***Red blood cells agglutinated by***
(a)	anti-rh′, anti-rh″, anti-hr′
(b)	anti-rh″, anti-hr′, anti-hr″, anti-hr
(c)	anti-rh′, anti-hr′, anti-hr″, anti-hr
(d)	anti-Rh_0, anti-hr′, anti-hr″, anti-hr
(e)	anti-Rh_0, anti-rh′, anti-hr″
(f)	anti-Rh_0, anti-rh′, anti-rh″

Give the Rh genotype for each individual.

9-17. Two different antisera, α and β, react with the antigens on certain types of red blood cells causing either agglutination (+) or nonagglutination (—). In a particular population three classes of individuals were found to react with these antisera as follows:

		Antiserum	
Class	***Number of Individuals Found***	α	β
(1)	34	+	+
(2)	39	+	—
(3)	29	—	+

How many gene pairs and alleles involved in this antigenic response is it necessary to assume in order to explain these data?

9-18. If a third antiserum, γ, was found to agglutinate the blood cells of about one-half of the individuals in classes 2 and 3 of the above problem, but none of those in class 1, how would you explain this?

Note: In the following questions assume that only three alleles determine the ABO blood type. I^A produces the A phenotype, I^B produces the B phenotype, and i produces the O phenotype. The effects of I^A and I^B are dominant to i, but are codominant with each other. Similarly, assume that only two alleles, R and r, determine the Rh phenotype (R = Rh$^+$ and r = rh$^-$), with the effects of R dominant to r. Also, two alleles at another gene pair, L^M and L^N, determine the M and N blood types, respectively, but their effects are fully codominant.

9-19. What are the *genotypes* of the following parents?

	Phenotypes of Offspring (Proportions)			
Phenotypes of Parents	A	B	AB	O
(a) B × B		3/4		1/4
(b) B × AB		1/2	1/2	
(c) B × A		1/2	1/2	
(d) B × A	1/4	1/4	1/4	1/4
(e) B × AB	1/4	1/2	1/4	
(f) B × O		1		
(g) B × O		1/2		1/2

9-20. In 1901 Landsteiner published the first research paper demonstrating A, B, and O blood group differences in man. He used the blood serum of six laboratory workers (one was himself: 6 in the table below) and tested each of the six sera with red blood cells also taken from the same six individuals. In some cases agglutination of red blood cells occurred (designated by +) and in others no agglutination occurred (−). The following table represents the results he observed:

		Red Blood Cells of The Six Individuals					
		1	2	3	4	5	6
Serum of the Six Individuals	1	—	+	+	+	+	—
	2	—	—	+	+	—	—
	3	—	+	—	—	+	—
	4	—	+	—	—	+	—
	5	—	—	+	+	—	—
	6	—	+	+	+	+	—

(a) Which individuals belong to group O? (b) Which individuals are AB blood type? (c) If the blood type of individual 2 is called A, what are the blood types of 3, 4, and 5?

9-21. What phenotypes and ratios would you expect among the offspring of the following crosses: (a) $I^AI^A \times ii$? (b) $I^AI^A \times I^AI^B$? (c) $I^AI^A \times I^Bi$? (d) $I^AI^A \times I^Ai$? (e) $I^Ai \times I^Ai$? (f) $I^Ai \times I^AI^B$? (g) $I^Ai \times ii$?

9-22. What phenotypes and ratios would you expect among the progency from the following crosses: (a) $I^Ai\ Rr \times I^BI^B\ rr$? (b) $ii\ Rr \times I^Ai\ rr$? (c) $ii\ Rr \times I^Bi\ Rr$? (d) $I^AI^B\ Rr \times I^AI^B\ Rr$?

9-23. In the following types of matings, the *phenotypes* of the parents are listed together with the frequencies of phenotypes occurring among their offspring. Indicate the *genotype* of each parent.

(a) Parents: AB Rh^+ × O Rh^+
Offspring: 3/8 A Rh^+:3/8 B Rh^+:1/8 A rh^-: 1/8 B rh^-

(b) Parents: A Rh^+ × A rh^-
Offspring: 3/4 A Rh^+:1/4 O Rh^+

(c) Parents: B Rh^+ × A rh^-
Offspring: 1/4 AB Rh^+:1/4 A Rh^+:1/4 B Rh^+:1/4 O Rh^+

(d) Parents: B Rh^+ × A rh^-
Offspring: 1/8 AB Rh^+:1/8 AB rh^-:1/8 A Rh^+:1/8 A rh^-:1/8 B Rh^+:1/8 B rh^-:1/8 O Rh^+:1/8 O rh^-

(e) Parents: B Rh^+ × A Rh^+
Offspring: 3/16 AB Rh^+:3/16 A Rh^+:3/16 B Rh^+:3/16 O Rh^+:1/16 AB rh^-: 1/16 A rh^-:1/16 B rh^-:1/16 O rh^-

9-24. In a case of disputed parentage, the blood-group phenotype of the mother is A MN rh^- and her child's phenotype is B N Rh^+. List all the possible blood-group phenotypes which the father may possess.

9-25. Which of the following males would you exclude as the possible father of an offspring whose phenotype is O Rh^+ MN, when the maternal phenotype is O rh^- MN? Male phenotypes: AB Rh^+ M; A Rh^+ MN; B rh^- MN; O rh^- N.

9-26. A woman is married to one man and produces four legitimate children with the following genotypes: $ii\ RR\ L^ML^N$, $I^Ai\ Rr\ L^NL^N$, $ii\ RR\ L^NL^N$, $I^Bi\ rr\ L^ML^M$. What are the genotypes of the parents?

9-27. (a) List the *phenotypes* and their proportions among the offspring of matings between genotypes $I^Ai\ rr\ L^ML^M$♀♀ × $I^BI^B\ Rr\ L^NL^N$♂♂. (b) List the *phenotypes* and their proportions among the offspring of matings between genotypes $I^AI^B\ rr\ L^NL^N$♀♀ × $I^Bi\ Rr\ L^ML^M$♂♂. (c) If an offspring having the A blood type were produced in one of the (a) matings, could a male from one of the (b) matings possibly be the father? (d) If ABO incompatibility between mother and offspring offers an advantage against Rh hemolytic disease of the newborn, which of the above matings, (a) or (b), will produce fewer Rh hemolytic children?

9-28. The following five mothers, (a) through (e), with given phenotypes, each produced one child whose phenotype is described. For each child, select as the father one of the five males whose genotypes are given.

Mother	Maternal phenotype	Phenotype of child	Genotype of male
(a)	A M Rh^+	O M Rh^+	1. $I^Ai\ L^ML^N\ rr$
(b)	B N rh^-	O N rh^-	2. $I^Bi\ L^ML^N\ RR$
(c)	O M rh^-	A MN Rh^+	3. $ii\ L^NL^N\ rr$
(d)	A N Rh^+	AB MN Rh^+	4. $ii\ L^ML^M\ rr$
(e)	AB MN rh^-	AB M rh^-	5. $I^AI^A\ L^ML^N\ RR$

9-29. In corn, plants of type A used as pollen parents are sterile when mated to plants of their own type but produce about 50 percent type A offspring and 50 percent type C offspring when crossed as pollen parents to type B plants used as females. Plants of type B used as pollen parents are sterile with B females but produce 50 percent B offspring and 50 percent C offspring when crossed to females of type A. Interestingly, both A and B strains used as pollen parents can be crossed separately to type C females and will produce in both cases 50 percent A and 50 percent B offspring. Explain these results in terms of self-sterility alleles of a single gene pair.

9-30. In the example of histocompatibility genes in mice given on page 175, determine what proportion of F_2 individuals will accept: (a) F_1 donor tissues? (b) Donor tissues from one of the parental strains?

9-31. Answer (a) and (b) in Problem 30 if the inbred parental strains differed in four pairs of genes.

REFERENCES

BACH, F. H. (ed.), 1974. *Immunobiology of Transplantation*. Grune & Stratton, New York.

BATEMAN, A. J., 1947. Number of S-alleles in a population. *Nature*, **160,** 337.

BELAR, K., 1928. *Die Cytologischen Grundlagen der Vererbung*. Borntraeger, Berlin.

BENACERRAF, B., and H. O. MCDEVITT, 1972. Histocompatibility-linked immune response genes. *Science*, **175,** 273–279.

BERNSTEIN, F., 1925. Zusammenfassende Betrachtungen über die erblichen Blutstrukturen des Menschen. *Zeit. Induk. Abst. u. Vererbung.*, **37,** 237–270.

CORRENS, C., 1912. *Die Neuen Vererbungsgesetze*. Borntraeger, Berlin.

DARBISHIRE, A. D., 1912. *Breeding and the Mendelian Discovery*. Cassell, London.

EAST, E. M., and P. MANGLESDORF, 1925. A new interpretation of the hereditary behavior of self-sterile plants. *Proc. Nat. Acad. Sci.*, **11,** 166–171.

ERSKINE, A. G., 1973. *Principles and Practice of Blood Grouping*. Mosby, St. Louis.

FUDENBERG, H. H., J. R. L. PINK, D. P. STITES, and A. WANG, 1972. *Basic Immunogenetics*. Oxford Univ. Press, New York.

GINSBURG, V., 1972. Enzymatic basis for blood groups in man. *Adv. in Enzymol.* **36,** 131–149.

HARRIS, H., 1975. *The Principles of Human Biochemical Genetics,* 2nd ed. Elsevier, New York. Chap. 7: The blood group substances.

LANDSTEINER, K., and P. LEVINE, 1927. Further observations on individual differences of human blood. *Proc. Soc. Exp. Biol. Med.*, **24,** 941–942.

LANDSTEINER, K., and A. S. WIENER, 1940. An agglutinable factor in human blood recognized by immune sera for rhesus blood. *Proc. Soc. Exp. Biol. Med.*, **43,** 223.

LEVINE, P., 1958. The influence of the ABO system on Rh hemolytic disease. *Hum. Biol.*, **30,** 14–28.

LEVINE, P., L. BURNHAM, E. M. KATZIN, and P. VOGEL, 1941. The role of isoimmunization in pathogenesis of erythroblastosis fetalis. *Amer. Jour. Obstet. Gynec.*, **42,** 925–937.

NOLTE, D. J., 1959. The eye-pigmentary system of *Drosophila*. *Heredity*, **13,** 219–281.

RACE, R. R., and R. SANGER, 1968. *Blood Groups in Man*, 5th ed. F. A. Davis, Philadelphia.

RASMUSSON, J., 1935. Studies on the inheritance of quantitative characters in *Pisum*. I. Preliminary note on the genetics of time of flowering. *Hereditas*, **20,** 161–180.

ROITT, I. M., 1971. *Essential Immunology*. Blackwell, Oxford.

SEARLE, A. G., 1968. *Comparative Genetics of Coat Colour in Mammals*. Academic Press, New York.

STERN, C., 1973. *Principles of Human Genetics*, 3rd ed. W. H. Freeman, San Francisco. Chaps. 12 and 13.

STORMONT, C., 1962. Current status of blood groups in cattle. *Ann. N.Y. Acad. Sci.*, **97,** 251–268.

WALSH, R. J., and C. M. MONTGOMERY, 1947. A new human isoagglutinin subdividing the blood groups. *Nature*, **160,** 504.

WATKINS, W. M., 1966. Blood-group substances. *Science*, **152,** 172–181.

WIENER, A. S. (ed.), 1970. *Advances in Blood Grouping*, III. Grune & Stratton, New York.

ZIEGLER-GÜNDER, I., and E. HADORN, 1958. Manifestation rezessiver Augenfarb-Gene in Pterininventar heterozygoter Genotypen von *Drosophila melanogaster*. *Zeit. Vererbung.*, **89,** 235–245.

10
ENVIRONMENTAL EFFECTS AND GENE EXPRESSION

The relations between genes and their phenotypic effects is not always straightforward; the mere presence of a particular allele does not always ensure the presence of a particular effect. This has been shown to be true for dominance relations between alleles in which the effects of recessive genes are obscured. As may be expected, there are also other kinds of interactions.

The causes that can modify the effect of particular genes can be traced to the large gap between the genetic material, on one hand, and the phenotype of an organism, on the other. Genetic material, as we have seen, consists fundamentally of only a few thousandths of a gram of DNA transmitted between generations. The mass of a developed organism, however, is considerably larger and adult humans, for example, represent a many-trillion-fold increase over the original diploid genetic complement at fertilization. For an organism to grow, a relatively large amount of material must therefore be obtained from the surrounding environment and organized through appropriate reactions into the phenotype. Thus, although the genetic material provides the "messages" and direction for phenotypic growth and behavior, the development and continued existence of an organism must also depend upon environmental factors such as the availability of food, suitability of temperature, light, etc. The genotype thus interacts with the environment throughout an organism's lifetime to form the phenotype.

Since the developmental process has many individual steps, at each of which interaction may occur between the environment and the genotype,

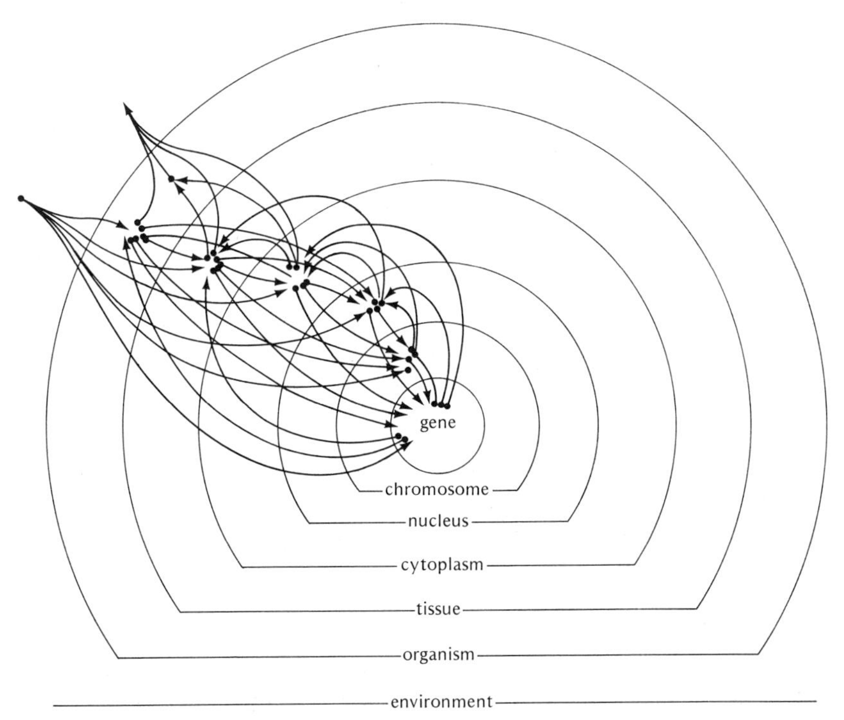

Figure 10–1

Possible interactions between some of the functional and structural levels associated with life. Further complexities exist, of course, within each level, but for diagrammatic simplicity these have not been indicated. Note that the expression of any particular level is dependent upon the effects produced by other levels. (After Weiss.)

or between various elements of the organism and the genotype (Fig. 10–1), we cannot really say that a particular gene determines a particular character. There is, for example, no gene which by itself determines blue eyes or brown hair. There are, however, genes which, given particular environments and genotypes, will influence development toward formation of blue eyes or brown hair. This distinction, although not always made, is important, because it emphasizes the fact that the phenotype appears as the result of close-knit interaction between many factors at many levels of development.

It is not surprising, then, that the appearance of an organism does not always reflect its genetic constitution. In precise terms, two kinds of effects can be noted and measured: *penetrance*, the proportion of genotypes that show an expected phenotype, and *expressivity*, the degree to which a particular effect is expressed by individuals. A dominant mutation such as *Lobe* eye in *Drosophila* may, when heterozygous, show itself in only 75 percent of certain groups of flies although all individuals are carrying the *Lobe* gene. Under these circumstances, *Lobe* has 75 percent penetrance. On the other hand, the *expression* of the *Lobe* gene among the various individuals that show its effect may vary considerably from complete absence of eyes to almost normal size (Fig. 10–2). Thus, although the *Lobe* gene itself is stable, its phenotypic consequences may be variable. The present chapter will consider some of the environmental interactions that may be responsible for differences in penetrance and expressivity primarily among diploids.

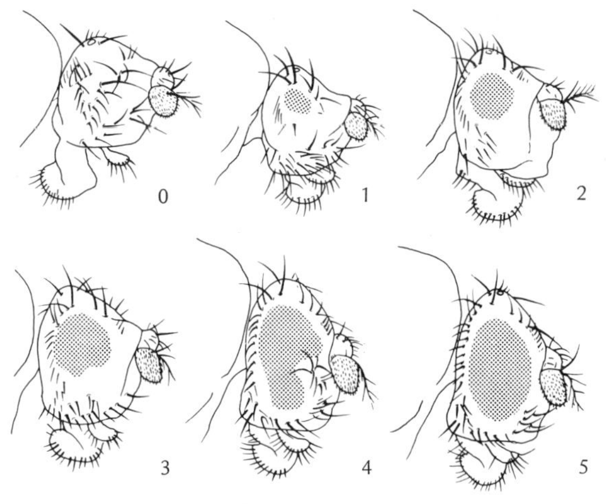

Figure 10–2

Expression of the Lobe *gene in different individuals in* Drosophila melanogaster. *Eye size ranges from grade 0 to the wild-type phenotype, grade 5. (After a drawing provided by courtesy of L. C. Dunn.)*

EFFECT OF EXTERNAL ENVIRONMENT

TEMPERATURE. The close correlation observed by chemists between the rate of chemical reactions and the temperature can surely be expected to occur in living organisms as well. Since the basic developmental effect of many genes is to control the rate of a specific reaction, a change in temperature can be expected to have widespread effects on development. In some cases these temperature effects are quite dramatic, such as the red flower color of primrose plants observed at room temperature and their white color observed at temperatures over 86°F. In barley a certain type of albinism appears only when the plants are grown outdoors. In the greenhouse the barley albinos develop the normal green color.

In some mammals, such as rabbits of the Himalayan variety and Siamese cats, certain genetic constitutions cause a darkening of the fur at the extremities (paws, nose, and ears) because of cooler temperature at these points. If the environment is changed by removing the dark fur and placing the animal or formerly dark area under warm conditions, then light fur grows back. On the other hand, light fur can be made to grow back as dark fur if it is removed and the animal placed under cold conditions. The phenotypic expression or lack of phenotypic expression of the *Himalayan* gene is therefore dependent upon temperature at the time of fur growth.

Similar to other genetic manifestations, penetrance may also be strongly influenced by environment. The mutant gene *tetraptera*, for example, causes the balancing organs of *Drosophila* to develop into wings in 35 percent of individuals raised at 25°C. This 35 percent penetrance, however, is reduced to 1 percent when the flies are raised at 17°C.

When the effect of a gene can be measured quantitatively, such as the number of facets in the *Bar*-eye mutation in *Drosophila*, temperature effects can be gauged more accurately. For example, as the temperature increases from 15° to 31°C, the number of eye facets decreases in *Bar* mutants. On the other hand, the situation reverses for an allele of *Bar*, named *Infrabar*; as the temperature increases, the number of facets also increases. We therefore see that the effect of gene-environment interactions can be reversed by simple gene substitutions, but interactions nevertheless remain.

LIGHT. Light also provides energy, and is essential for the growth and development of practically all plants. In fact seedlings grown in the dark, even if they do survive for a short period, will not develop chlorophyll, and will therefore appear as albinos, although having genes for chlorophyll production. Light can also produce unusual effects, such as the face freckling that occurs among people of certain genotypes exposed to the sun. In corn plants that are homozygous for a particular gene, an even more dramatic effect occurs when they are exposed to sunlight and turn bright red ("sunred") in appearance. According to Singleton, this effect can be prevented by screening out rays from the blue-violet end of the light spectrum so that corn ears grown with red cellophane wrapped around them (permitting only red light rays to enter) do not show the sunred trait.

NUTRITION. Food serves many functions, including providing the energy for carrying out necessary processes, and providing the material that must be incorporated into necessary structures. Genetically, however, different organisms, even of the same species, usually differ in the kind and amount of nutrients they require. This can easily be seen when organisms, because of genetic mutations, are unable to synthesize specific compounds which must then be added as extra nutrients to their diet.

The simpler the nutritional requirements, the easier such metabolic deficiencies can be traced. Among the foremost examples of this kind are the diverse nutritional mutants in *Neurospora*, first discovered by Beadle and Tatum. Normally *Neurospora* can grow on a simple medium containing inorganic salts, sugar, and the B vitamin biotin. Through a special technique of identifying nutritional mutations (p. 536), Beadle and Tatum found numerous varieties whose diet had to be supplemented by specific chemicals. At the present time the kinds of nutritional supplements required for growth in different *Neurospora* strains range from ammonia, through most of the different amino acids that compose proteins, to compounds such as adenine and uracil. As indicated previously (p. 34), bacteria also show a variety of hereditary changes that produce nutritional dependence.

Depending upon its genotype, simple dietary changes can therefore lead to important consequences for an organism. In rabbits, for example, the appearance of yellow fat depends on two factors: the presence of the recessive gene *y* in homozygous condition, and the presence of green vegetables (xanthophyll) in the diet. In eliminating greens from the rabbit diet, the appearance of yellow fat is also eliminated. Additional examples of gene-diet relationships are found in chickens for the appearance of yellow shanks, in *Drosophila*, and in numerous other organisms.

MATERNAL RELATIONS. When direct body relations between parent and offspring extend beyond fertilization, as in mammals, additional environmental interactions can occur between the fetus and the maternal environment. Blood-group incompatibilities between mother and offspring (p. 172) are one example producing special effects on the survival of particular genotypes.

Occasionally even traits that are usually considered normal may be disadvantageous under special maternal environments. In mice, for example, the gene *hair-loss* (*hl*) is associated with an absence of hair growth, and is considered deleterious. However, homozygous hair-loss females (hl/hl) mated to heterozygous normal males (hl/hl^+) do not produce the expected equal ratios of hair-loss and normal offspring but generally give birth to twice as many of the former as the latter. It appears that the heterozygous normal embryos of such mothers suffer from calcium loss and therefore usually die at birth. When the mother is normal, however, there is no calcium-loss effect and no change in the expected ratios of offspring.

EFFECTS OF INTERNAL ENVIRONMENT

Environmental influences on the effect of the genotype are not confined to external events but may also include less obvious internal changes. The distinction between the effects caused by an organism's internal and external environments, however, is not always easy to make. It is clear that for events outside the organism to effect its phenotype, some internal impression of these events must occur. Similarly internal environmental events must have, at least at some point, an external origin. In general, we can define external environmental effects as those leading to phenotypic changes that seem directly correlated with observable changes outside the organism. On the other hand, internal environmental effects are those leading to phenotypic changes that appear correlated primarily to changes within the organism. Examples of the latter are the effects of age, sex, and the presence or absence of internal substrates. Changes in each of these may affect the expression of any particular genotype.

AGE. Broadly defined, the process of aging of most organisms may be considered to begin with their origin at fertilization, so that they do not remain exactly the same during any succeeding

TABLE 10–1

Age of onset of different genetic traits in humans

Genetic Trait	Usual Age of Onset
blood-group antigens	prenatal
alkaptonuria (darkening of the urine)	birth
infantile amaurotic idiocy (Tay-Sachs disease)	4–6 months
vitamin D-resistant rickets	1 year
severe muscular dystrophy (Duchenne's)	2–5 years
juvenile amaurotic idiocy	5–10 years
periodic paralysis	10 years
pattern baldness	20–30 years
diabetes mellitus	40–60 years

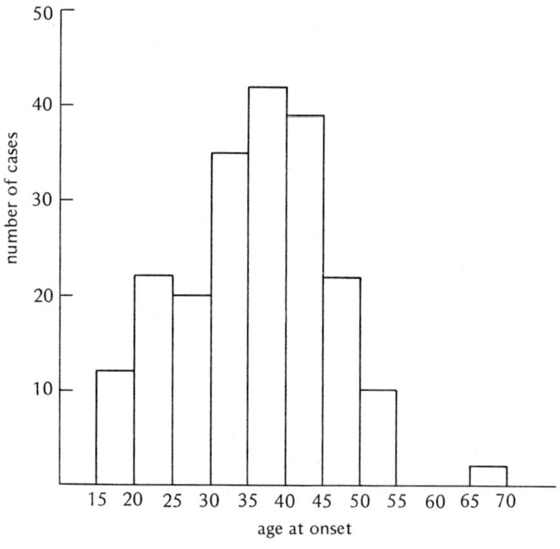

Figure 10–3

Distribution of ages at the onset of choreic (nervous, involuntary) movements in 204 cases of Huntington's chorea, a hereditary disease caused by a dominant gene and characterized by progressive mental deterioration. (Data of Reed and Chandler.)

interval. In each age interval phenotypic changes occur which permit further genotypic effects to express themselves. The effect of genes that influence feather color in chickens and coat color in mammals, for example, depends upon the appearance of these structures during development. In humans such aging effects have been documented with great care and accuracy. As shown in Table 10–1, some traits express themselves within a relatively short period of the human life-span, especially those that have their effects in early embryonic life, such as the formation of blood-group antigens. Others, such as diabetes mellitus and Huntington's chorea, may be spread out over a wide range of ages (Fig. 10–3). In all cases the actual genes responsible for the particular genetic effect are present at fertilization; it is only the appearance of the effect that depends on the age of the organism.

SEX. Sexual phenotypic differences are usually evident in the reproductive structures and specialized behavior of each sex. As will be seen later (Chapter 12), such differences are often associated with the presence of special sex chromosomes which differ in number or kind between the sexes. Some differences may therefore arise because certain genes associated with these chromosomes (*sex-linked genes*) also differ between the sexes. In other cases, however, genes affecting a particular trait are not on the sex chromosome and are present equally in both sexes, yet the trait may only appear in one sex and not in the other. Such genes, for example, are those responsible for the appearance of horns in certain species of sheep, or genes that influence the amount of milk yield in dairy cattle, or that cause the prolapse of the uterus in humans. Since gene expression is limited in these cases to one sex only, these traits are therefore called *sex-limited traits.*

Some hereditary traits, common in one sex, nevertheless still appear in the other sex, although relatively infrequently. Since characteristics of this kind are not limited to a single sex, they have been called *sex-controlled* or *sex-influenced.* Harelip (incomplete fusion of upper lip), pattern baldness, and gout (accumulation of uric acid in the tissues) are among sex-controlled human traits showing greater frequency in men than in women. Conversely spina bifida (forked spine with open spinal cord) is found more frequently in women than men.

SUBSTRATES. The kinds of reactions that take place in an organism depend to a great degree on the materials that are internally present. These materials, or substrates, are in many instances synthesized by metabolic actions of the organism, and their presence or absence may frequently be traced to genetic control. Phenylketonuria, for instance, is a rare human hereditary disease diagnosed by the presence of phenylpyruvic acid in the urine. It is caused by a recessive gene in homozygous condition and is held responsible for a variety of symptoms including early idiocy. Organically the symptoms of the disease seem to arise from an accumulation of the amino acid phenylalanine, which is normally a substrate derived from the breakdown of proteins and used in the synthesis of other products. It is presumed that phenylalanine accumulates because of the inactivation of a specific liver enzyme, phenylalanine hydroxylase, preventing the metabolism of phenylalanine to tyrosine (Fig. 26-2). Knowledge that these substrates are involved has enabled successful treatment of the disease in a number of instances, since some patients placed on low-phenylalanine diets show some improvement in their symptoms (Table 10-2).

The hereditary disease diabetes mellitus is much more common among humans than phenylketonuria and has an approximate frequency of 2 to 5 percent. Fortunately an understanding of the substrates involved enables most diabetics to lead normal lives. In diabetes, sugar levels normally controlled by the pancreatic hormone *insulin* are not properly regulated. High-blood-sugar levels and excretion of sugar in the urine are among the symptoms that may eventually lead to a change in body metabolism from the use of sugar for energy to the excessive use of fatty acids. Coma and death may ensue as a consequence. Treatment of diabetics therefore includes both a restriction in sugar and carbohydrate intake and, when necessary, the administration of insulin.

These examples demonstrate that, as we approach the level where necessary substrates can be distinguished, we have come quite close to what may be called the immediate chemical environment of genes. As we have seen, there is a double aspect to this environmental relationship. On the one hand, genes contribute to this environment and, on the other hand, they also rely on its presence for their own activity and effect. Therefore, to speak of gene activity without specifying or at least implying a particular environment has little real meaning.

PHENOCOPIES

The strength of environmental changes is sufficient to modify the effects of many genes. In some instances specific environmental changes may modify the development of an organism so that

TABLE 10-2

Some of the positive changes in intelligence scores of phenylketonurics after varying degrees of dietary control involving a reduction in phenylalanine intake

Age Diagnosed	Initial Intelligence Score	Change in Score After Treatment	Present Age in Years
14 months	62	+5	2
3 years	10	+44	4
2 years	48	+3	3
1 year	31	+19	3
5 years	26	+33	7
22 months	98	+2	3
32 months	75	+5	$5\frac{1}{2}$
5 months	80	+9	$2\frac{1}{2}$
3 weeks	?	(91)	$1\frac{1}{2}$

After Koch, Graliker, Fishler, and Ragsdale.

its phenotype simulates the effect of a particular gene, although this effect is not inherited. Such individuals are known as *phenocopies*. Diabetics dependent on insulin, for example, are phenocopies of normal individuals in the sense that the drug environment prevents the effects of the disease. Should their offspring also inherit diabetes, the phenocopy treatment with insulin may have to be administered again to achieve the normal phenotype. In no sense, therefore, is the diabetic genotype changed by the insulin treatment; there is only a phenotypic effect.

The converse to the above example also holds true. That is, specific environmental changes may cause the appearance of an abnormal trait such as diabetes in genotypically normal individuals. In humans removal, injury, or infection of the pancreas is frequently accompanied by diabetic symptoms. In experimental animals diabetes has also been induced by chemicals such as *alloxan*, by introducing antibodies against an individual's own insulin, and by pituitary and thyroid hormones. Again, such environmentally induced effects are not inherited.

One relatively recent series of tragedies resulted from the unsuspecting use of a phenocopy-producing drug by women who were still in the early stages of pregnancy. Normally, except for unusual accidents, human limb abnormalities are rare, although a number of families have been recorded in which serious limb abnormalities are inherited, including, in some instances, even the complete absence of hands and feet (*acheiropody*). In 1960 and 1961 it was noted that a considerable number of German and British Commonwealth babies were being born with shortened limbs, giving a flipperlike appearance called *phocomelia*. No hereditary basis for this sudden increase in phocomelia was apparent, and proposed causes for this epidemic originally ranged all the way from dietary changes to atomic bomb testing. However, after considerable investigation, Lenz (Germany) and McBride (Australia) showed that this trait arose from the use of a sleep-inducing drug, thalidomide, by mothers in early pregnancy. In these cases the drug seemed to produce a serious effect on skeletal formation of the fetus at the time that the early limb buds were forming. Similar thalidomide effects have since been shown to occur in rabbits, mice, and monkeys. In chickens, Landauer had previously shown that phocomelia could be induced by a phenocopy treatment with insulin or boric acid. When injected into chick eggs at specific stages in development, these chemicals produce shortened limbs similar to a dominant mutation known as "Creeper."

TWIN STUDIES

Although both environment and heredity are involved in the development of any character, a change in some aspects of the environment may change one character relatively little as compared to its effect on another character. The type of blood group, for example, seems fairly impervious to practically all environmental effects, while the diabetic may have his phenotype radically altered by a mere change in diet. Qualitatively, therefore, the relations between certain environmental changes and alterations in specific characters may be expected to give completely different values for some of the traits that can be measured; i.e., a high carbohydrate diet would have zero percent effect on blood grouping and perhaps a 50 percent increase in the usual frequency of diabetes.

Not all characters, however, are determined by simple genetic effects which have easily observable relationships with simple environmental changes. Some traits, such as intelligence in animals and crop yield for agricultural species, are probably the result of complex genetic-environmental interactions which are then measured as a single trait. Because of the complexity of such characters and many others whose genetic basis is not clearly known, it is usually difficult to define the roles of heredity and environment in exact numerical terms. Various methods have, therefore, been developed to cope with this problem, especially among experimental plants and animals (Chapter 15). In humans a technique based on the occurrence of twins was proposed for this purpose in the last century by Spath and by Galton, and has been developed and used in numerous studies since then.

The wide variety of combinations that can arise from the random union of two human gametes makes it highly unlikely that any two individuals in a single generation will have the same genotype. Identical twins, however, arising from the division of a single fertilized egg (monozygotic twins), furnish an unusual opportunity for observing two genetically homogenous human beings. Since genetic differences between these twins are absent, observed differences in phenotype can be considered as purely environmental in origin. By contrast, fraternal twins *of the same sex* coming from separate eggs (dizygotic) may differ both genetically and environmentally for any character. By measuring both kinds of twins for a particular character, we can then evaluate the roles of environment and heredity. That is, on the one hand, we have a genetically identical pair of the same age and sex raised in a single uterine environment, and, on the other hand, we have a genetically dissimilar pair of the same age and sex also raised in a common uterine environment. Presumably the difference between these two kinds of twins is only in the extent of their genetic similarity. If phenotypic similarities for a particular character are greater among identical twins than among fraternals, or, conversely, if phenotypic differences are less among identicals, we can ascribe this to the genetic similarity of the identical pairs and the genetic dissimilarity among the fraternal pairs. On the other hand, if phenotypic differences for a particular character are the same for both identical and fraternal twins, we can assume that genetic similarity or dissimilarity plays less of a role than the environmental differences which occur in both kinds of twins.

CONCORDANCE–DISCORDANCE

One of the ways phenotypic similarities and differences for a character can be measured in twins is simply to note whether the character is present or absent in one or both members. If both members of a pair act in conjunction, and if both either possess the trait or are free of it, the pair is *concordant*, or phenotypically similar. If only one member of a pair possesses the trait, the pair is *discordant*, or phenotypically dissimilar. The extent to which identicals and fraternals differ in their degree of concordance offers a measure with which to assess the relative roles of environment and heredity. If, for example, 20 pairs each of both identical and fraternal twins are evaluated for a particular trait, and 12 pairs of identicals (60 percent) show concordance as compared to only 4 pairs of fraternals (20 percent), the trait can be considered to have a strong hereditary element. Similarly, comparisons in this example between the degree of discordance for identicals (40 percent) and for fraternals (80 percent) show that the genetic dissimilarity between members of the latter group has contributed a greater share to the phenotypic dissimilarity. More equal concordance and discordance ratios between the identical and fraternal groups would, in turn, signify less emphasis on hereditary similarity in determination of the trait and more emphasis on environmental factors.

Some traits, of course, are not fully penetrant, so that not all twin pairs that bear the genotype for this character will necessarily show the affected phenotype. Although such twin pairs are concordant because both members carry a particular genotype without showing the phenotype, it is quite difficult to know exactly which pairs are of this type and which pairs do not carry the particular genotype at all. The results reported in most studies of this kind therefore include in the concordant category only instances when both twins are phenotypically affected. Also, since identical twins are always of the same sex, both members of each fraternal pair reported for these studies are in practically all cases also of the same sex.

Figure 10–4 represents a compilation of concordance ratios for a variety of conditions. Note that the first character listed, the time when infants begin to sit up, shows little difference in concordance between identical and fraternal twins. This finding can be interpreted to mean that most twin

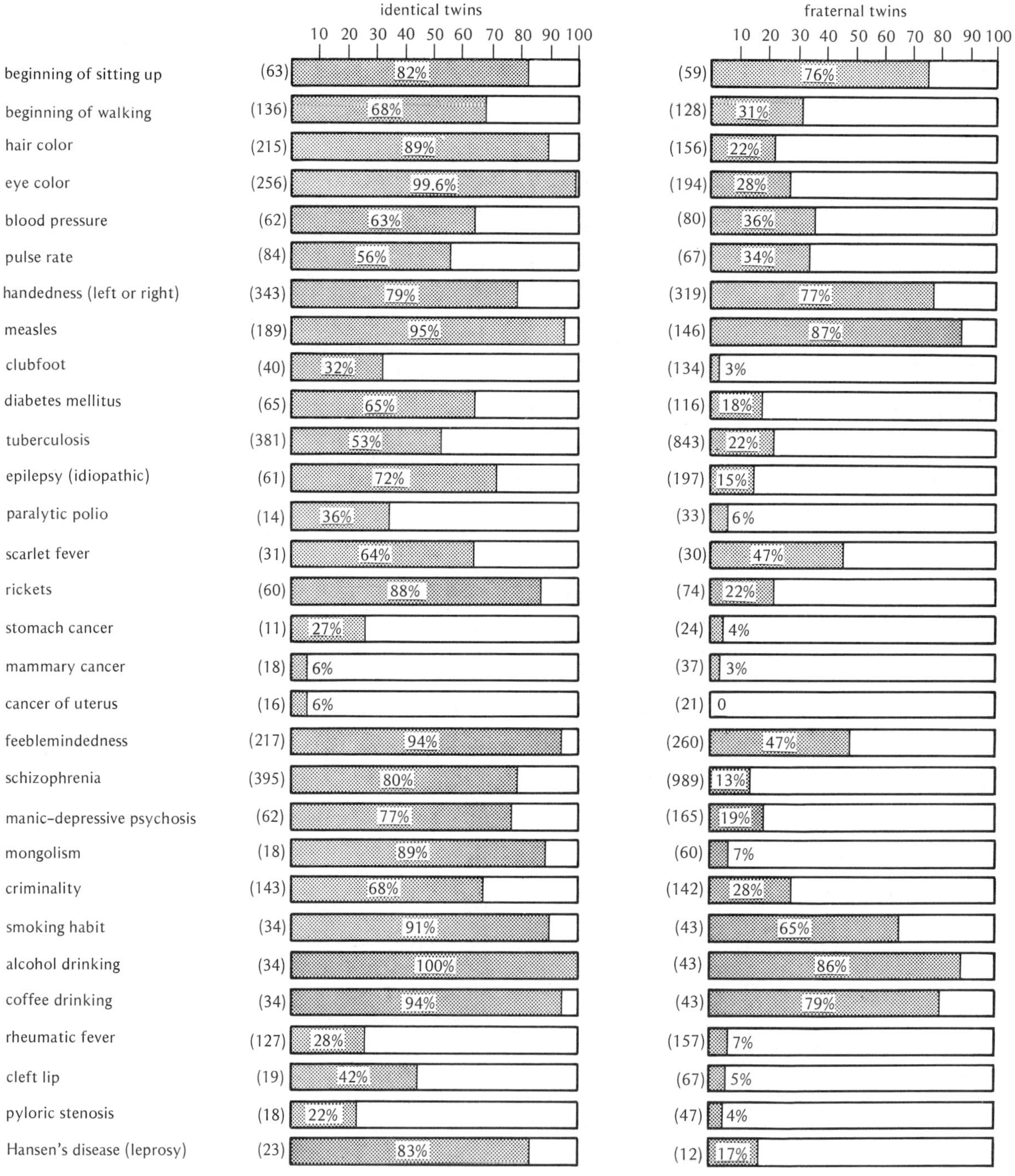

Figure 10–4

Percentage of concordance (shaded areas) among identical and fraternal twins when both are affected with a particular trait or when both are phenotypically similar within a small measurable range. The percentage of discordance (unshaded areas) is 100 — the percentage of concordance. Numbers in parentheses indicate numbers of twin pairs scored. (Derived from Stern and other sources.)

infants, no matter what their genotypic similarities or differences, will tend to begin sitting up at the same age; or, in other words, the genetic similarities between identical twins do not necessarily lead to more concordance than among fraternal twins. This trait is therefore relatively unaffected by differences in genotype and seems to be influenced by a common environmental factor, primarily age. On the other hand, the time when infants begin walking shows a much higher concordance between identical twins than among fraternals. This signifies a greater influence of genetic similarity on the concordant appearance of this trait than similarity in age or common environmental factors that occur between fraternal twins. Similarly hair color and eye color have strong genetic components, while susceptibility to measles appears more dependent on environmental similarities. That is, if one twin catches measles, the other is sure to catch it, as long as there is no special genetic immunity involved.*

Other traits with important environmental factors seem to be those of mammary cancer and cancer of the uterus, in both of which the degree of concordance between identical twins is quite low. Stomach cancer, however, shows a somewhat higher degree of concordance among identicals, emphasizing the greater influence of genetic identity on the penetrance of this trait.

On the whole most of the characters listed seem to have significant genetic elements influencing their appearance. In diabetes, rickets, feeblemindedness, and schizophrenia, concordance is noticeably large among identicals (in contrast to fraternals), signifying a high degree of penetrance for these characters; i.e., when genetically present the character tends to appear despite any environmental differences between identical twins.

However, data of this kind, although pointing to strong genetic influence on some characters, do not reveal the particular genetic mechanisms that may be responsible. For many of these traits it is not definitely known whether they are the results of single or multiple gene differences, and whether the genetic factors involved are dominant or recessive. Schizophrenia, for example, is believed by some to be influenced by a single dominant gene, and feeblemindedness to be influenced by a series of different genes (polygenes), while others suggest just the reverse (see Kidd and Cavalli-Sforza).

We must also keep in mind that the effect of environment on some of these characters may be much more subtle than on first appearance. For example, the very high degree of concordance in feeblemindedness for both identical and fraternal twins may arise from changes that have occurred in the maternal prenatal environment. Evidence for this comes from studies by Rosanoff and his colleagues, demonstrating that the nontwin siblings of fraternal twins affected with feeblemindedness show a much greater degree of discordance or phenotypic dissimilarity for this particular trait. Since there is the same degree of genetic dissimilarity between these sibs and between the two members of fraternal twin pairs in the same family (each coming from separately fertilized eggs), the higher concordance for feeblemindedness in the latter group has been explained as caused by their common uterine environment.

It may also be questioned, for some traits, whether the large differences in concordance between the identicals and fraternals need always signify the influence of genetic similarities among the identicals. It may well be that the environment of identical twins is considerably more similar than the environment of fraternals. Prenatally the birth

* To analyze whether identical twins are significantly different in their concordances and discordances than fraternal twins, a chi-square test for independence (see pp. 148–149) can be performed, such as in the following example·

		Twins		
		Identicals	Fraternals	**Totals**
Responses	Concordant	82	44	126
	Discordant	16	22	38
	Totals	98	66	164

$$\chi^2 = \frac{[(1804 - 704) - \frac{1}{2}(164)]^2(164)}{(126)(98)(38)(66)} = 5.49$$

In this case, therefore, the concordance for identicals (82/98 = 84%) is statistically significantly higher than the concordance for fraternals (44/66 = 67%), indicating that the trait has a significant genetic component.

membranes surrounding monozygotic twins as well as their placental attachments tend to be single rather than the double form common to dizygotic twins. Also, identical twins, because of their strong similarities, tend to be treated in similar fashion after being born, and to seek out similar environments on their own. Fraternal twins, because of their dissimilarities, have greater opportunity to develop within environments more different from each other than identicals.

The greater concordance of identicals in traits such as criminality may reflect this precise effect. The special social conditions in which criminality develops could be expected to find a more similar response among identicals, with their more similar behavior patterns and more similar early environmental experiences than among fraternals. If, then, despite these considerations, there is still a genetic component in criminality, it is undoubtedly not for criminality as such, but for some elements of behavior which specific environments develop into criminality. There are no "born" criminals; under different environmental conditions, the interactions that produce such phenotypes might not occur at all.

The greater concordance of identicals for mongolism, or Down's syndrome (mental defects accompanied by slanted Mongolian eye shape) is also not simply a reflection of a hereditary trait transmitted between various generations. Mongolism is now known to arise from chromosomal abnormalities (pp. 487–489) that occur more frequently in the egg cells produced by older mothers. A defective zygote of this kind that splits to form identical twins can certainly be expected to lead to concordance quite frequently, since both twins will probably contain duplicates of the chromosomal abnormality. Fraternal twins formed from two separate zygotes, on the other hand, will be concordant for mongolism only if both original egg cells had separate chromosomal abnormalities—a much rarer event. Concordance in this case, therefore, arises basically from a maternal environmental effect (age) responsible for chromosomal changes rather than from the simple transmission of this abnormality between generations.

Interpreted with caution, however, these studies of concordance between twins do present data that emphasize the important roles of both heredity and environment in many traits formerly believed to be determined by only one factor or the other.

IDENTICAL TWINS REARED APART

In addition to comparisons between identical and fraternal twins, further comparisons can be made between members of identical pairs reared in different environments. Since identical twins have the same genetic constitution, rearing each member of a pair separately may show which traits are more or less susceptible to environmental differences and thereby help reveal the extent of genetic influence on different characters. Although this method has the important advantage that environmental differences operate on identical genotypes, such cases are quite rare and only a small amount of data has been gathered.

One of the major studies on identicals reared apart was carried out by Newman, Freeman, and Holzinger. They accumulated data on 19 pairs of twins reared separately, 50 pairs each of identical and fraternal twins reared together, and 52 non-twin siblings reared together. Some of the results of this study are presented in Table 10–3 for a few physical traits and also for intelligence measured by the Stanford-Binet IQ (intelligence quotient) test. The table gives average measured differences between members of the different kinds of twin pairs and between the sibs. For example, the height differences between members of identical twin pairs reared apart ranged from 0 to 8.9 cm with an average difference of 1.8 cm. Interestingly this value is hardly different from that for members of twin pairs reared together, but is quite different from the values among the fraternal twins and their sibs. Since the same holds true for head length and width, we can confidently assume that traits such as body height and head measurements are predominantly influenced by genetic factors and relatively little by the environmental differences encountered by twin pairs reared apart. Body weight, on the other hand, shows a much larger difference for twin pairs reared apart than

TABLE 10–3
Average differences between identical and fraternal twins

Trait	*Identical Twins* Reared Together	Reared Apart	*Fraternal Twins*	*Sibs*
height (cm)	1.7	1.8	4.4	4.5
weight (lb)	4.1	9.9	10.0	10.4
head length (mm)	2.9	2.2	6.2	
head width (mm)	2.8	2.9	4.2	
intelligence (Binet IQ score points)	5.9	8.2	9.9	9.8

From Newman, Freeman, and Holzinger, 1937.

for those reared together. In fact this value is similar to body-weight differences between fraternal twins and between nontwin sibs. Evidently body weight is more environmentally flexible than body height, and the experiences of most individuals in our society undoubtedly supports this. That is not to say that there are no genetic determinants for body weight, but only that these determinants do not produce the same phenotypic effect in all environments.

NATURE, NURTURE, AND INTELLIGENCE

What was of special interest in the twin studies of Newman and co-workers were the results of the Stanford-Binet intelligence scores. Although IQ tests have numerous deficiencies, they do offer a fairly reproducible measure of some important mental attributes. For the identical twins reared apart, the results showed a significantly higher average difference in scores than for the identical twins reared together. Newman, Freeman, and Holzinger drew the conclusion that social and educational environmental differences may strongly affect intelligence despite genetic similarities. Certainly one of their twin pairs reared apart strongly supported this view. This particular twin pair differed by 24 IQ points, one member completing college and becoming a schoolteacher, and the other having no formal schooling beyond the second elementary grade. On the other hand, another twin pair reared apart differed by only 1 IQ point, although one member was raised by a well-to-do physician while his twin was raised by a truck farmer.

The finding that the identical twins reared apart do not achieve the same difference in IQ scores as fraternal twins or their sibs also points to the possibility that the genetic similarities influencing that portion of intelligence measured in this fashion are not completely disrupted by environmental differences. Support for the existence of genetic components in intelligence can also be gained from numerous studies in which the degree of relationship between tested individuals is measured along with their similarity in IQ. As shown in Fig. 10–5, the closer the relationship between individuals, the more similar their test scores. Thus the IQs of children and their natural parents are, on the average, more closely correlated than the IQs of children and their foster parents. At the same time, however, individuals with the same degree of relationship who are reared together generally show more similar test scores than when reared apart, this indicating again the presence of environmental components. Can genetic and environmental components be separated and analyzed?

Aside from the genes which we know are involved in the production of mental deficiency, such as phenylketonuria (p. 187), Tay-Sachs disease (Table 10–1), and some others, there is no present knowledge of specific genes that produce differences within the "normal" range of intelligence.

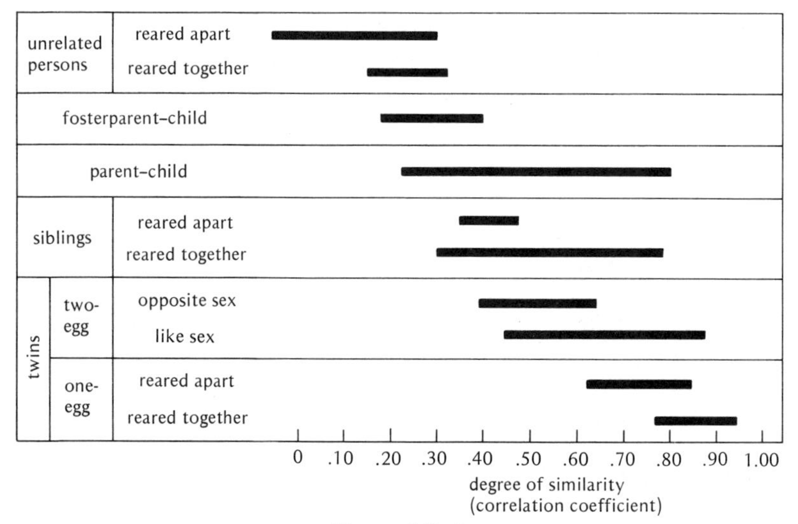

Figure 10–5

Degrees of similarity (horizontal lines) observed in various categories of individuals tested for IQ in a large number of studies. Although the ranges of some categories overlap, similarities in IQ measured as correlation coefficients (see Chapter 15) tend to increase in these studies as relationship between the tested individuals increases. (After Erlenmeyer–Kimling and Jarvik.)

The genetic components of intelligence can therefore, at best, be described only by statistical methods (see Chapter 15) and even then without precise knowledge of the kinds and numbers of genes involved. The environmental components of IQ performance, however, can more easily be investigated, and a number of them are now known:

1. THE SIZE OF THE FAMILY. A number of studies have shown that the greater the number of children in a family, the lower the IQ per child (Fig. 10–6). Such results have been interpreted to mean that IQ achievements depend to some degree upon the attention given to an individual during his development, and that the children in larger families usually have less opportunity to gain such attention. This view is supported by the fact that twins, no matter what their social class, are 5 IQ points lower on the average than their nontwin siblings. Obviously the attention given to each of two young children born at the same time is less than the attention given to each child when it is born separately. (Although other causes, such as intrauterine effects, may be offered.)

As shown in Fig. 10–6, the particular birth order position that a child occupies in a family is also strongly associated with IQ performance. For reasons that are not understood, yet that hardly can be caused by basic genetic differences, first-born children perform consistently better on IQ tests than their later-born siblings.

2. THE EFFECT OF CULTURE. IQ differences, some large and some small, have been shown to exist between different regions of the country (e.g., North and South), between different races (e.g., Orientals and Caucasians) and occasionally even between different religious groups (e.g., Jews and Gentiles). In some instances, especially in respect to racial performances, attempts have been made to ascribe IQ differences to genetic causes (for a modern example, see Jensen). However, the fact that IQ tests measure responses to questions and concepts evaluated from the

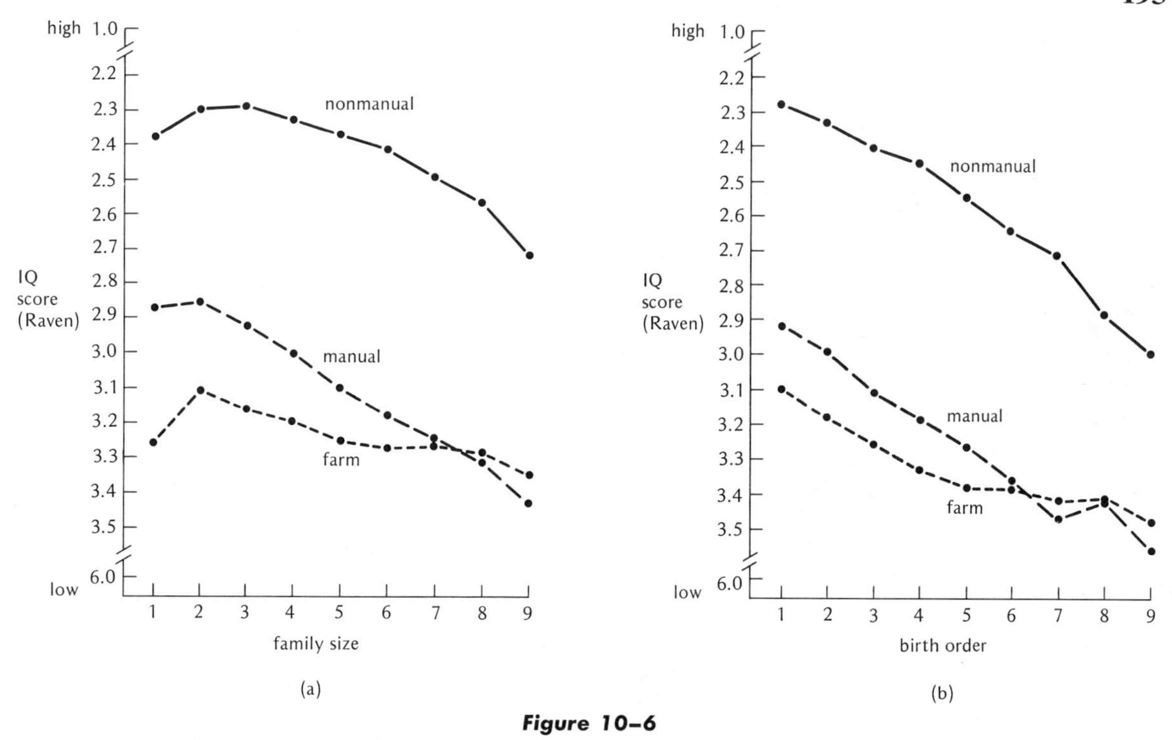

Figure 10–6

Relationship between IQ scores, family size, birth order, and social grouping. The tests were performed during 1963–1966 on about 400,000 males in the Netherlands, all 19 years old, who were examined to determine their fitness for military duty. The scores on the intelligence examination were scaled according to the "Raven Progressive Matrices," and range from a possible high of 1.0 to a possible low of 6.0. Family size (a) indicates the number of siblings in the family of the individual being tested, and birth order (b) indicates the birth position of the individual among his siblings. The social group designation is based on the father's occupation: nonmanual (professional and "white-collar" workers), manual (skilled or unskilled "blue-collar" workers), and farm (farmers or farm workers). (After Belmont and Marolla.)

viewpoint of "success" favored by a specific culture, that of the so-called modern middle class, makes cultural differences between races and groups difficult to ignore, and raises the question: To what extent does culture rather than heredity alone affect IQ scores?

A number of instances are known in which IQ scores have significantly increased when children were given social and educational opportunities unavailable to their parents, that is, *cultural modification*. For example, some Puerto Rican children of ages 7 and 8 raised in a relatively modern affluent environment have already achieved mental ages on their test scores as high as or higher than their parents raised in rural poverty (see Hunt and Kirk). The converse of this has also been demonstrated: institutionalized youngsters living under poor social and physical conditions often show a consistent gradual decline in IQ performance. Educational expectations made of children also play an important role in IQ, and there is little question that there exist differences in such expectations between different cultural groups. For example, the educational expectations made by Jewish and Oriental parents of their children, although never scientifically measured, have undoubtedly been traditionally higher than the educational expectations to which the children of black parents have been long exposed. Similarly the educational expectations of teachers undoubtedly differ in dealing with different cultural groups. In fact there has been considerable recent criticism that our present educational system, as a whole, often does little more than help to rigidify the thought patterns that children bring to school, rather than help them develop new ways of thinking.

3. EARLY ENVIRONMENT AND THE MOTHER. The development of the perceptive and problem-solving abilities that are reflected in intelligence scores often depends upon the stimuli and encouragement that infants and young children receive. It is known that experimental animals deprived of stimuli during early ages will show both intellectual and emotional retardation. In humans, as in most mammals and primates, early stimuli depend strongly upon the mother, and cultural differences that affect the maternal stimulation of the child may therefore play an important role. In a study of this type reported by Willerman and co-workers, IQ scores were taken of the children of black-white interracial couples in which the mother was black (and the father white), and cases in which the mother was white (and the father black). All of the couples had about the same levels of education. Interestingly the average IQ score of the children of black mothers was 93.7, whereas the average IQ score of the children of white mothers was 100.9. Since the children of the black and white mothers probably bear equal mixtures of genes from both races, the difference in IQ scores probably arises because of the differences in the early maternal environment.*

4. DIETARY EFFECTS. There has been little question that severe malnutrition of an embryo or infant will affect most aspects of development, especially behavior. In rats, Zamenhof and co-workers have shown that protein-deprived females will produce offspring with a reduction in the number of brain cells. This effect is accompanied also by kidney deficiencies and other debilitations sufficiently severe, so that defective females of this type who become pregnant also produce offspring with diminished brain cell numbers. Dietary deficiencies caused by social and economic impoverishment can therefore hardly be ignored as causes for IQ differences.

In summary; we can say that the separation of the effect of environment from heredity in respect to intelligence is an immensely difficult task (see also p. 308). As for other complex behavioral attributes, and perhaps even some biological ones, the trait we are looking at is the result of continuous interaction between the stimuli outside the organism and those responses that the organism is capable of making. Moreover, the behavioral responses themselves depend very often upon previous environments, some of which the individual himself influences or creates. That is, the "environment" is often psychologically incorporated into the individual, so that his responses are successively restricted by those interactions that have gone on in the past. When one is dealing with a historical process of this kind, it is certainly not clear what percentage of a psychological response can be ascribed to "nature" and what percentage to "nurture!"

For those who believe that the improvement of intelligence in modern life is socially desirable, it seems essential to beware of invidious comparisons that assert or imply that certain races are genetically more or less intelligent. Such distinctions, although unproved, often diminish the social motivations that are necessary to examine, pinpoint, and improve the environmental factors so deeply involved in the development of intelligence. It is clear that without sustained environmental improvement, the fullest development of each person's intelligence, genetically unique though he may be, will not take place.

NORM OF REACTION

The discussion until now leads to two simple and important concepts: (1) that different genotypes may react differently in the same environ-

* One could also claim that the genes for intelligence are sex-linked (see Chapter 12) and are therefore passed from mother to son, and from father to daughter. However, there was no evidence for this.

ment, and (2) that the same genotype may behave differently in different environments.

The first of these concepts, that different genotypes can differ phenotypically in a particular environment, is the mainstay of all genetic analysis. At present we separate and describe genotypes mainly on this basis, although in the future we shall hopefully also separate genotypes on the basis of their nucleotide configuration. The second concept, the effect of environmental differences, has been the main subject of the present chapter and arises because the environment of living organisms is not always the same. As we have seen, gene action does not exclusively produce a particular phenotype under all circumstances, but produces this phenotype through interaction with a particular environment. Since phenotypes may change in different environments, a full comparison between different genotypes should also include measurements in a variety of environments. According to terminology introduced early in the development of mendelian genetics and used later by Goldschmidt and others, such comparisons would measure the *norm of reaction* of individual genotypes or the extent to which they are phenotypically affected by environmental change. For some genotypes this norm of reaction is relatively constant, and for others it may be highly variable. Blood types, for example, seem to be relatively unaffected by environmental changes and, once genotypically determined, persist unchanged throughout life. Other genotypes, such as that of diabetics, may produce phenotypes quite sensitive to environmental changes (i.e., diet or insulin).

An example of a quantitative comparison between the norms of reaction of different genotypes was undertaken by Dunlop for specific characteristics associated with wool production among a number of strains of Australian sheep. For a period of years (1947 through 1955), samples of each strain were raised in three different environments (A, C, and D) with different rainfall and seasonal patterns. As shown in Fig. 10–7, the amount of wool produced depended both upon the particular environment in which the strain was raised and upon the particular strain. Environment A, for example, produced a comparatively greater amount of fleece in strain F than in any others, but this relationship was then reversed in environment D. Strain B did relatively well in environment C but poorly in environment A. Only strain A appeared to produce a relatively uniform amount of fleece in each environment. Therefore, if we were to select a genotype that

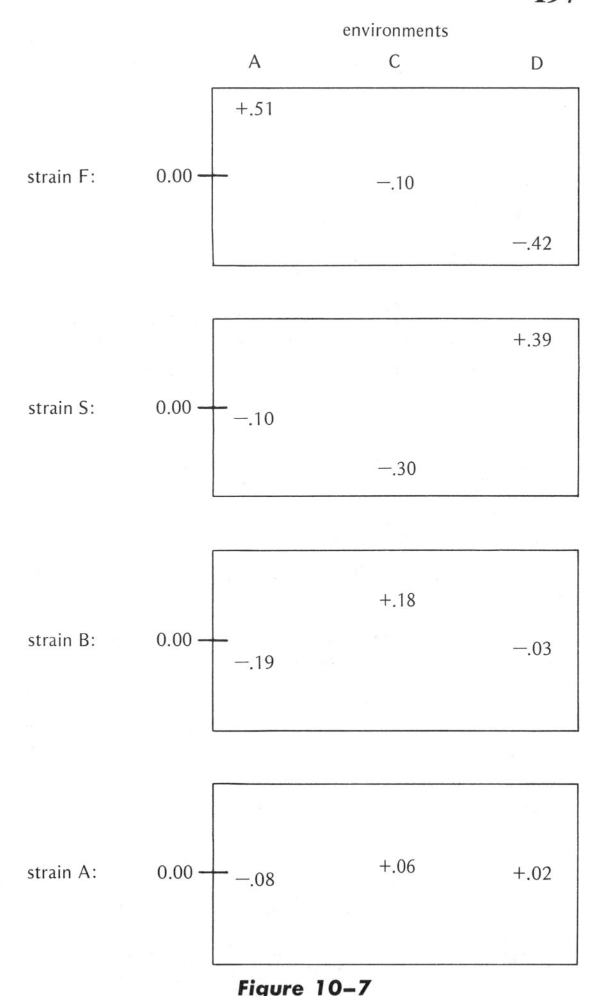

Figure 10–7

Amounts of wool produced by four different strains of Australian merino sheep in three different environments. The designation 0.00 on the left-hand scale for each strain represents the mean weight of fleece for all three environments combined. (From data of Dunlop.)

performed best in environment A, we would undoubtedly choose genotype F. However, for overall performance in all environments we might well choose genotypes B or A. It is easy to see that questions such as these are of considerable importance in dealing with population and evolutionary problems in which both environments and genotypes are not constant factors.

PROBLEMS

10-1. In humans "stiff little finger" (camptodactyly) is caused by a dominant gene with 75 percent penetrance. (a) What proportion of offspring will show the trait if heterozygotes for this gene marry normal individuals? (b) If such heterozygotes mate with each other?

10-2. Answer (a) and (b) of Problem 1 if the degree of penetrance for the trait were only 25 percent.

10-3. In the following pedigree, at least three individuals are phenotypically affected by a very rare dominant gene *A* that has 50 percent penetrance and that has been introduced by individual I-1, a heterozygote. (a) Using your own discretion, select two or more other individuals in this pedigree to be affected by this trait, and then give the genotypes of all individuals in the pedigree. (b) If, let us say, individual III-4 were phenotypically affected, what genotypes would you suggest for his parents, II-1 and II-2?

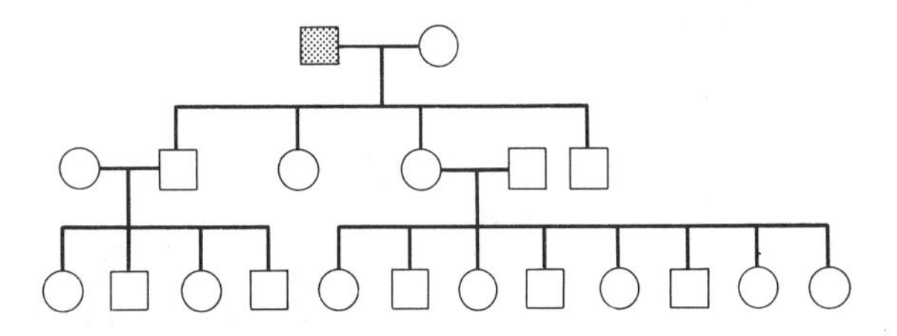

10-4. Huntington's chorea is a degenerative nerve disease caused by a rare dominant gene that usually does not show its effect until after reproductive age (see Fig. 10-3). (a) If a man with two children discovers that his own father (the children's grandfather) was suffering from this disease, what is the probability that his first-born child is carrying the gene? (b) If his first child shows symptoms of the disease, what is the probability that the second child is carrying the gene? (c) If neither child shows any symptoms, what is thc probability that one (but not both) of his two children are carrying the gene?

10-5. Phenylketonurics (homozygous *aa*) who have developed normally through maintenance treatment on a low phenylalanine diet (see p. 187) will when they marry usually produce normal heterozygous children (*Aa*), since their spouses are often normal homozygotes (*AA*). It has, however, recently been discovered that in some cases treated female phenylketonurics (*aa*) married to normal homozygous individuals (*AA*) can produce *all* mentally defective children. How would you explain this?

10-6. *Scar* is a mutant effect that appears on a localized section of the body of the flour beetle (*Tribolium castaneum*). Bell, Shideler, and Eddleman studied the effect of temperature on the appearance of this mutant phenotype in crosses between individuals homozygous for the gene *scar*. Their results were as follows:

	Phenotypes	
Temp. °C	Scar	Nonscar
33	472	145
35	405	70

Is there an effect of temperature on the penetrance of *scar*? (*Hint:* to test whether an association exists, perform a chi-square test for independence; see pp. 148–149.)

10-7. Meier, Myers, and Huebner scored the frequency of leukemia in mice which differed only by the presence of a single gene: one type was hairless, *hr/hr*, and the other was haired but heterozygous for this gene hr/hr^+. Both types showed that they carried the murine leukemia virus which is generally associated with the incidence of leukemia in mice. The results of their observations are given in the following table, in which the values in the first two columns are the total numbers of mice affected by leukemia to that age and the values in the last column are the numbers of mice tested of the given genotypes. (Not all tested mice developed leukemia by the given ages.)

		Age		
		8–10 months	18 months	Total Tested
Genotype	*hr/hr*	141	243	337
	hr/hr^+	4	56	279

(a) Would you say the incidence of leukemia is independent of age and genotype? (b) In a cross between heterozygotes, what proportion of offspring would you expect to be leukemic at 8–10 months of age according to these data? (Assume that hr^+/hr^+ homozygotes are relatively resistant to leukemia.)

10–8. In the damselfly, *Ischnura damula*, there are two alternative types of pigment pattern on the thorax "H" (heteromorphic) and "A" (andromorphic). Females of this species can appear as either "H" or "A" depending on their genotype. Males, on the other hand, can only have one color pattern, "A", no matter what their genotype.

In one experiment, Johnson crossed a particular male with three different females, of which two were "H" phenotypes and one was of the "A" phenotype, and found the following phenotypes and numbers among the female offspring of these crosses:

Cross	Phenotype of mother	Phenotypes and numbers of female offspring	
		"H"	"A"
(1)	"H"	30	11
(2)	"H"	38	None
(3)	"A"	22	17

Following the *simplest* line of reasoning, present a hypothesis that explains the number of gene pairs segregating in this cross, indicating which alleles are dominant and which recessive, and provide the genotypes of the single male parent and the three female parents.

10–9. In poultry the appearance of certain feather patterns is limited to males who can be either cock-feathered (showy) or hen-feathered (dull), depending on their genotypes. Females, on the other hand, are hen-feathered no matter what their genotype. If *F* represents the dominant gene and *f* represents the recessive gene for this character, furnish the genotype for each of the following pairs of parents:

(a) Parents: cock-feathered ♂ × hen-feathered ♀
Offspring: ♂♂: 3/4 cock-feathered: 1/4 hen-feathered
♀♀: all hen-feathered

(b) Parents: hen-feathered ♂ × hen-feathered ♀
Offspring: ♂♂: all cock-feathered
♀♀: all hen-feathered

(c) Parents: hen-feathered ♂ × hen-feathered ♀
Offspring: ♂♂: 1/2 cock-feathered: 1/2 hen-feathered
♀♀: all hen-feathered

(d) Parents: cock-feathered ♂ × hen-feathered ♀
Offspring: ♂♂: 1/2 cock-feathered: 1/2 hen-feathered
♀♀: all hen-feathered

10–10. In a continued mating between a cock-feathered male and a hen-feathered female 90 offspring were observed. Thirty-six of these were females and were hen-feathered as usual. Of the male offspring, 32 were cock-feathered and 22 were hen-feathered. (a) In respect to feather pattern, what genotypes would you propose for the two parents? (b) On the basis of a chi-square test, determine whether you will accept or reject this hypothesis at the 5 percent level of significance.

10–11. In cabbage butterflies, yellow color is sex-limited to males only, who can appear as yellow or white. Females, on the other hand, appear white no matter what their genotype. Club antennae, however, is not sex-limited, and both females and males can have either club antennae or the normal type (full). If the dominant color allele is called *C* and its recessive *c*, and the dominant antennae allele is called *A* and its recessive *a*, what are the genotypes of the parents in the following crosses:

(a) Parents: yellow club ♂ × white full ♀
Male offspring: 3/8 yellow club: 3/8 yellow full: 1/8 white club: 1/8 white full
Female offspring: 1/2 white club: 1/2 white full

(b) Parents: white full ♂ × white full ♀
Male offspring: 3/8 yellow full: 3/8 white full: 1/8 yellow club: 1/8 white club
Female offspring: 3/4 white full: 1/4 white club

(c) Parents: white full ♂ × white club ♀
Male offspring: all yellow full
Female offspring: all white full

10–12. Horns are completely absent in the Suffolk breed of sheep, but all animals are horned in the Dorset breed. When a Suffolk hornless female is mated to a Dorset horned male, the F_1 females are all hornless but the F_1 males are all horned. Identical results occur when a Dorset horned female is mated to a Suffolk hornless male. When the F_1 females from each of these crosses are mated to their brothers, an F_2 is produced of the following phenotypes:

females: 3/4 hornless: 1/4 horned
males: 3/4 horned: 1/4 hornless

Considering this as an example in which an important genetic attribute (such as dominance or recessiveness) can be modified by sex, what is the simplest explanation you can offer for these results?

10-13. Using the information in Problem 12, answer the following: (a) What phenotypic ratios will be found in each sex produced from a cross between a Suffolk hornless female and an F_1 horned male? (b) From a cross between a Dorset horned female and an F_1 horned male?

10-14. Vesell and Page scored the length of time it took for a dose of the drug antipyrine to become reduced to half of its initial level in the blood of 9 pairs of both identical and fraternal twins. If we interpret their data as representing like or unlike responses of each pair of twins (like response = half-life difference of one hour or less; unlike response = half-life difference of more than one hour), then the results can be written as:

Identical Twins		*Fraternal Twins*	
Like Response	Unlike Response	Like Response	Unlike Response
9 pairs	0 pairs	1 pair	8 pairs

Explain the relative proportions of heredity and environment you think accounts for the control of the level of this drug.

10-15. The degree of concordance for trait A is 48 percent among monozygotic twins and only 10 percent among dizygotic twins. In trait B, however, monozygotic concordance is 88 percent but dizygotic concordance is still about 10 percent. (a) What are the relative influences of environment and heredity in trait A? (b) What are the relative influences of environment and heredity in trait B? (c) How would you explain the difference in monozygotic concordance between these two traits?

10-16. Conterio and Chiarelli analyzed a number of pairs of identical and fraternal twins for various personal habits. Of the cigarette-smoking pairs among these twins, they found the following concordances and discordances for inhaling and noninhaling of tobacco smoke:

	Identical	*Fraternal*
like inhaling habits	27	17
unlike inhaling habits	3	7
	30	24

When they analyzed pairs of twins for the amount of coffee consumed, they found the following:

	Identical	*Fraternal*
like amounts of coffee consumed	24	9
unlike amounts of coffee consumed	10	34
	34	43

On the basis of these data, compare the extent of genetic and environmental influences on these two habits. (As an aid in determining whether an association exists between each of these habits and a particular type of twin, perform a chi-square test for independence; see pp. 148-149.)

10-17. The factors that determine three different human traits, (a), (b), (c), are the following: (a) is affected only by genetic factors; (b) is affected only by environmental factors; and (c) is affected exactly equally by genetic factors and environmental factors. For each of these three traits, provide values of concordances that you think would be appropriate for identical twins reared together, identical twins reared apart, fraternal twins reared together, and fraternal twins reared apart.

REFERENCES

Astauroff, B. L., 1930. Analyse der bilateralen Symmetrie im Zusamenenhang mit der selbständigen Variabiltät ähnlicher Strukteren. *Zeit. Induk. Abst. u. Vererbung.*, **55,** 183-262.

Beadle, G. W., 1945. Genetics and metabolism in *Neurospora. Physiol. Rev.* **25,** 643-663.

Belmont, L., and F. A. Marolla, 1973. Birth order, family size, and intelligence. *Science*, **182,** 1096-1101.

Collins, J. L., 1927. A low temperature type of albinism in barley. *Jour. Hered.*, **33,** 82-86.

Dunlop, A. A., 1962. Interactions between heredity and environment in the Australian merino. I.

Strain × location interactions in wool traits. *Austral. Jour. Agric. Res.*, **13,** 503–531.

ERLENMEYER-KIMLING, L., and L. F. JARVIK, 1963. Genetics and intelligence: A review. *Science*, **142,** 1477–1479.

HUNT, J. M., and G. E. KIRK, 1971. Social aspects of intelligence: Evidence and issues. In *Intelligence: Genetic and Environmental Influences*, R. Cancro (ed.). Grune & Stratton, New York, pp. 262–306.

JENSEN, A. R., 1969. How much can we boost IQ and scholastic achievement? *Harvard Educ. Rev.*, **39,** 1–123. (Reprinted with comments by 7 contributors in *Environment, Heredity, and Intelligence,* by Harvard Educ. Rev., Cambridge, 1969.)

KIDD, K. K., and L. L. CAVALLI-SFORZA, 1973. An analysis of the genetics of schizophrenia. *Soc. Biol.*, **20,** 254–265.

KOCH, R., B. GRALIKER, K. FISHLER, and N. RAGSDALE, 1963. Clinical aspects of phenylketonuria. In *First Inter-American Conference on Congenital Defects.* Lippincott, Philadelphia, pp. 127–132.

KRAFKA, J., JR., 1920. The effect of temperature upon facet number in the bar-eyed mutant of *Drosophila. Jour. Gen. Physiol.*, **2,** 409–464.

LANDAUER, W., 1948. Hereditary abnormalities and their chemically-induced phenocopies. *Growth Symposium*, **12,** 171–200.

LENZ, W., 1964. Chemicals and malformations in man. In *Proceedings of the Second International Conference on Congenital Malformations.* International Medical Congress, New York, pp. 263–276.

LUCE, W. M., 1926. The effect of temperature on *Infrabar*, an allelomorph of *Bar Eye* in *Drosophila. Jour. Exp. Zool.*, **46,** 301–316.

NEWMAN, H. H., F. N. FREEMAN, and K. J. HOLZINGER, 1937. *Twins: A Study of Heredity and Environment.* Univ. of Chicago Press, Chicago.

REED, T. E., and J. H. CHANDLER, 1958. Huntington's chorea in Michigan. I. Demography and genetics. *Amer. Jour. Hum. Genet.*, **10,** 201–225.

ROSENTHAL, D., 1971. *Genetics of Psychopathology.* McGraw-Hill, New York.

SINGLETON, W. R., 1962. *Elementary Genetics.* Van Nostrand, Reinhold, New York.

STERN, C., 1973. *Principles of Human Genetics*, 3rd ed. W. H. Freeman, San Francisco, Chaps. 16, 17, 25, 26, and 27.

WEISS, P., 1950. Perspectives in the field of morphogenesis. *Quart. Rev. Biol.*, **25,** 177–198.

WILLERMAN, L., A. F. NAYLOR, and N. C. MYRIANTHOPOULIS, 1970. Intellectual development of children from interracial matings. *Science*, **170,** 1329–1331.

ZAMENHOF, S., E. VAN MARTHENS, and L. GRAVEL, 1971. DNA (cell number) in neonatal brain: Second generation (F_2) alteration by maternal (F_0) dietary protein restriction. *Science*, **172,** 850–851.

11
GENE INTERACTION AND LETHALITY

Although Mendel did not give descriptive names to the factors involved in his crosses, but called them by alphabetical letters, later geneticists were not so reticent. Soon after the rediscovery of Mendel's paper, geneticists began naming each hereditary factor according to the name of the character it affected, i.e., *tall*, *green*, *early flowering*, etc. This served the purpose of identifying genes by a symbol other than a letter or number. Despite this advantage, however, the succession of names that soon appeared seemed to imply that there was a succession of gene effects, each individually distinct from the other. To many people, therefore, development seemed to be a "mosaic" of small, nonoverlapping individual effects, fitted into an organism like pieces of a stained-glass window.

It was, however, soon discovered by researchers, including Bateson and Punnett, that genes were not merely separate elements producing distinct individual effects, but that they could interact with each other to give completely novel phenotypes. For example, although the allele *A* could be characterized by phenotype A, and the allele *B* by phenotype B, an organism having both alleles *A* and *B* at the same time might show an entirely new phenotype, C. The occurrence of such interactions means that the phenotype we observe is not present in the genes themselves but arises from a complicated developmental process for which we have given some evidence in Chapter 10. Thus there are no genes which individually create specific organ structures, or organisms which have such structures "preformed" within the individual gene. In classical terms, the doctrine of preformationism (p. 4) is therefore replaced by more modern "epigenetic" concepts, in which development occurs through a complex network of reaction and interaction influenced by genes. Since genetic analysis is tied in with an understanding of the phenotypic ratios that result from various crosses, we may then ask: What effects do interactions have upon observed ratios?

When two gene pairs, each having two alleles, i.e., *A*, *a*, and *B*, *b*, are segregating independently in a diploid organism, the genetic consequences are, as we have seen, easily predictable. A cross between heterozygotes for both pairs, *AaBb* × *AaBb*, will give offspring carrying nine genotypes in the following ratios: 1/16 *AABB*:2/16 *AABb*:2/16 *AaBB*:4/16 *AaBb*:1/16 *AAbb*:2/16 *Aabb*:1/16 *aaBB*:2/16 *aaBb*:1/16 *aabb*.

As long as the assortment at each gene pair is independent of the other, and as long as the existence of each of these genotypes is not impaired, the above genotypic ratios will hold true for all such crosses. Genotypic ratios, however, are only inferred from phenotypic ratios, since it is only the latter that we directly observe and not the former. Because some of the different genotypes may have similar phenotypes, the phenotypic ratios need not always reflect exactly the genotypic ratios.

Table 11–1 represents some of the different relations between two gene pairs and their consequent phenotypic ratios. In this table the genotypic combinations that arise from a cross between two heterozygotes, *AaBb* × *AaBb*, are numbered from 1 to 16. Note that each of these genotypic combinations appear in each of the 15 examples listed without change; it is only the phenotypic ratios that are modified.

EACH GENE PAIR AFFECTING A DIFFERENT CHARACTER

THE CLASSICAL RATIO (EXAMPLE 1). When different characters are affected by each of two gene pairs maintaining complete dominance relations between the alleles involved, the phenotypic results of crossing heterozygotes for both gene pairs is that of the classical 9:3:3:1 ratio. As previously discussed in the example of Mendel's peas, the most common $\frac{9}{16}$ class shows the dominant phenotype for both characters. In one of the $\frac{3}{16}$ classes, the dominant phenotype for one character appears together with the recessive for the other character, but this relationship is then reversed in the other $\frac{3}{16}$ class. Only the small $\frac{1}{16}$ class shows both recessive characters, as would be expected if the frequency of each homozygote among all phenotypes were 1/4, giving a combined double recessive frequency of 1/4 × 1/4.

PARTIAL AND CODOMINANCE (EXAMPLES 2 AND 3). A change from complete to partial dominance or codominance enables the heterozygotes to be recognized as a distinct class. When this occurs at either one or both gene pairs, there is a corresponding change in the phenotypic frequencies. In cattle, for example, the roan coat color (mixed red and white hairs) occurs in the heterozygous (*Rr*) offspring of red (*RR*) and white (*rr*) homozygotes, and *r* is therefore partially dominant. Crosses between heterozygotes for this character that are also heterozygous for the hornless "polled" condition (polled being dominant over horned) will give a mixed phenotypic ratio which is 3:1 polled:horned and 1:2:1 red:roan:white. When combined, the overall phenotypic frequencies are 6:3:3:2:1:1, as shown in Example 2.

Codominance, as in human blood groups (Example 3), is effectively similar to partial dominance since the heterozygotes are also phenotypically distinct. Crosses between heterozygotes for two blood groups, i.e., ABMN × ABMN, would be expected to give nine classes of offspring in ratios which are identical to the genotypic ratios 1:2:1:2:4:2:1:2:1. The similarity between genotypic and phenotypic ratios in this case arises from the distinct phenotypes shown by each class of genotypes.

EACH GENE PAIR AFFECTING THE SAME CHARACTER

NOVEL PHENOTYPES (EXAMPLE 4). When one character is affected by two (or more)

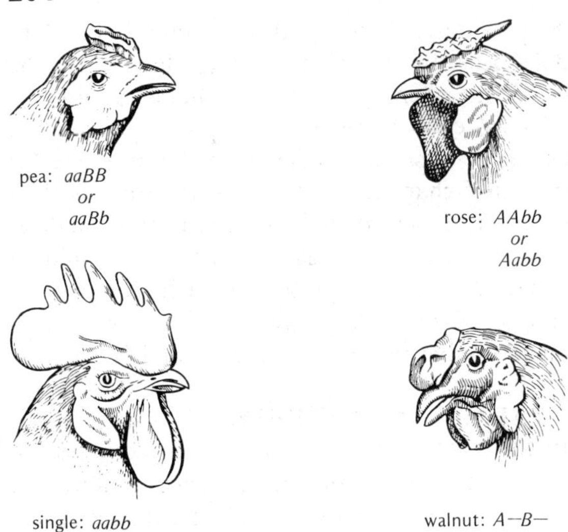

Figure 11–1

Comb shapes in poultry produced by different combinations of alleles in two pairs of genes. (After Punnett.)

gene pairs, a variety of phenotypic ratios may appear, depending upon the type and degree of interaction between the gene pairs. In some instances, such as comb shape in poultry, completely novel phenotypes occur as a result of special genotypic combinations. For example, the factors for both rose and pea combs are dominant to single combs; however, a combination of the dominant factors for rose and pea results in a different type of comb, walnut (Fig. 11–1). As worked out by Bateson and Punnett, two different gene pairs interact to give walnut combs when there are dominant alleles present in both pairs and interact to give single combs when both gene pairs are homozygous recessive. The four phenotypes derived from a cross between heterozygotes for both gene pairs (walnut appearance) are thus in a 9:3:3:1 ratio (Example 4), each phenotype having a uniquely distinct comb shape.

EPISTASIS (EXAMPLE 5). The formation of new phenotypes is only one example of interaction between gene pairs. Other examples may include instances in which a novel phenotype does not appear but one gene pair hides the effect of the other. This latter type of interaction is called *epistasis* and may be considered the counterpart of dominance relations between alleles (when one allele hides the effect of the other at the same gene pair). Epistasis may be caused by the presence of homozygous recessives at one gene pair (i.e., *aa* hides the effect of the *B* gene) or by the presence of a dominant allele at a gene pair (*A* hides the effect of the *B* gene). In addition the epistatic effect may be in only one direction, from one particular gene pair to another (effects at *A* hide effects at *B*, or vice versa), or in both directions when each gene pair is mutually epistatic to the other.

The example of mouse-coat-color determination given in Table 11–1 (Example 5) has many similarities with coat-color determination in a number of other mammals. In these animals the wild-type color, a grayish pattern called *agouti*, comes from the color combinations present in the bands along each hair. Numerous genes may modify these colors, including a gene pair that determines whether any color appears at all (see p. 212). If this latter gene pair is homozygous recessive, the animal appears as a white albino no matter what color alleles it carries at other gene pairs. In the example considered, the *AaBb* parents are agouti. Among their progeny the presence of homozygous recessives at the color gene *A* (i.e., *aa* in genotypes 11, 12, 15, and 16) is sufficient to hide whatever color occurs in the *B* gene pair (in this case agouti or black): *A* may therefore be called epistatic to *B*.

In developmental terms we may envisage the appearance of color in this example as arising from two sequential processes, the first of which is controlled by gene *A* and the other by gene *B*. When the first process is inhibited (by *aa*), the second process cannot occur.

$$\left.\begin{array}{l}\text{original}\\ \text{substance}\end{array}\right\} \xrightarrow[\text{1st process}]{A\text{ gene}} \left.\begin{array}{l}\text{intermediate}\\ \text{substance}\end{array}\right\} \xrightarrow[\text{2nd process}]{B\text{ gene}} \text{color}$$

Another clear example of this pattern of interaction can also be seen in man, where the appearance of detectable ABO blood type antigens has been shown to depend upon the presence of a gene, *H*. Individuals that are homozygous for the very rare recessive allele *h* show no such antigens and are phenotypically of blood type O ("Bom-

TABLE 11–1

Phenotypic ratios in crosses between individuals heterozygous for two gene pairs, each pair assorting independently

AaBb × *AaBb*

gametes

gametes		*AB*	*Ab*	*aB*	*ab*	
	AB	*AABB* (1)	*AABb* (2)	*AaBB* (3)	*AaBb* (4)	
	Ab	*AABb* (5)	*AAbb* (6)	*AaBb* (7)	*Aabb* (8)	genotypes
	aB	*AaBB* (9)	*AaBb* (10)	*aaBB* (11)	*aaBb* (12)	
	ab	*AaBb* (13)	*Aabb* (14)	*aaBb* (15)	*aabb* (16)	

	Offspring Expected in F_2		**Numbered Genotypes in Above Figure**
	Ratios	**Phenotypes**	
Each Gene Pair Affecting a Different Character			
1. Complete dominance at both gene pairs			
Example: Mendel's peas	9/16	yellow round	1, 2, 3, 4, 5, 7, 9, 10, 13
Gene pair *A*: (seed color) yellow dominant over green	3/16	yellow wrinkled	6, 8, 14
Gene pair *B*: (seed shape) round dominant over wrinkled	3/16	green round	11, 12, 15
	1/16	green wrinkled	16
2. Complete dominance at one gene pair; partial or codominance at the other			
Example: cattle	6/16	polled roan	2, 4, 5, 7, 10, 13
Gene pair *A*: polled (hornless) dominant over horned	3/16	polled red	1, 3, 9
	3/16	polled white	6, 8, 14
Gene pair *B*: red hair color partially dominant over white, producing "roan" heterozygotes	2/16	horned roan	12, 15
	1/16	horned red	11
	1/16	horned white	16
3. Partial or codominance at both gene pairs			
Example: human blood groups	1/16	AAMM	1
Gene pair *A*: (ABO) A and B codominant	2/16	AAMN	2, 5
Gene pair *B*: (MN) M and N codominant	1/16	AANN	6
	2/16	ABMM	3, 9
	4/16	ABMN	4, 7, 10, 13
	2/16	ABNN	8, 14
	1/16	BBMM	11
	2/16	BBMN	12, 15
	1/16	BBNN	16
Each Gene Pair Affecting the Same Character			
4. Complete dominance at both gene pairs; new phenotypes resulting from interaction between dominants, and also from interaction between both homozygous recessives	9/16	walnut	1, 2, 3, 4, 5, 7, 9, 10, 13
Example: comb shape in poultry	3/16	rose	6, 8, 14
Gene pair *A*: rose comb dominant over nonrose	3/16	pea	11, 12, 15
Gene pair *B*: pea comb dominant over nonpea	1/16	single	16

TABLE 11-1 (continued)

	Offspring Expected in F_2		Numbered Genotypes in Above Figure
	Ratios	Phenotypes	
Each Gene Pair Affecting the Same Character			
Interaction: dominants for rose and pea produce walnut comb. Homozygous recessives for rose and pea produce single comb			
5. Complete dominance at both gene pairs, but one gene, when homozygous recessive, is epistatic to the other			
Example: mouse coat color	9/16	agouti	1, 2, 3, 4, 5, 7, 9, 10, 13
Gene pair *A*: color dominant over albino	3/16	black	6, 8, 14
Gene pair *B*: agouti color dominant over black	4/16	albino	11, 12, 15, 16
Interaction: homozygous albino is epistatic to agouti and black			
6. Complete dominance at both gene pairs, but either recessive homozygote is epistatic to the effects of the other gene			
Example: sweet pea flower color	9/16	purple	1, 2, 3, 4, 5, 7, 9, 10, 13
Gene pair *A*: purple dominant over white	7/16	white	6, 8, 11, 12, 14, 15, 16
Gene pair *B*: color dominant to colorless (white)			
Interaction: homozygous recessives at either gene *A* or *B* produce white			
7. Complete dominance at both gene pairs, but one gene, when dominant, epistatic to the other			
Example: fruit color in summer squash	12/16	white	1, 2, 3, 4, 5, 6, 7, 8, 9, 10, 13, 14
Gene pair *A*: white dominant to color			
Gene pair *B*: yellow dominant to green	3/16	yellow	11, 12, 15
Interaction: dominant white hides the effect of yellow or green	1/16	green	16
8. Complete dominance at both gene pairs, but either gene, when dominant, epistatic to the other			
Example: seed capsules of shepherd's purse (*Bursa*)	15/16	triangular	1–15
	1/16	ovoid	16
Gene pair *A*: triangular shape dominant to ovoid			
Gene pair *B*: triangular shape dominant to ovoid			
Interaction: a dominant allele at either gene pair hides the effect of ovoid			
9. Complete dominance at both gene pairs, but one gene, when dominant, epistatic to the second, and the second gene, when homozygous recessive, epistatic to the first			
Example: feather color in fowl	13/16	white	1, 2, 3, 4, 5, 6, 7, 8, 9, 10, 13, 14, 16
Gene pair *A*: color inhibition is dominant to color appearance	3/16	color	11, 12, 15
Gene pair *B*: color is dominant to white			
Interaction: dominant color inhibition prevents color even when color is present; color gene, when homozygous recessive, prevents color even when dominant inhibitor is absent			
10. Complete dominance at both gene pairs; interaction between both dominants to give new phenotypes			
Example: fruit shape in summer squash			

	Offspring Expected in F_2		Numbered Genotypes in Above Figure
	Ratios	Phenotypes	
Each Gene Pair Affecting the Same Character			
Gene pair *A*: sphere shape dominant over long shape	9/16	disc	1, 2, 3, 4, 5, 7, 9, 10, 13
Gene pair *B*: sphere shape dominant over long shape	6/16	sphere	6, 8, 11, 12, 14, 15
	1/16	long	16
Interaction: dominants at *A* and *B*, when present together, form disc-shaped fruit			
11. Complete dominance at one gene pair, partial dominance at the other; first gene, when homozygous recessive, is epistatic to second gene			
Example: hair direction in guinea pigs	7/16	smooth	1, 3, 9, 11, 12, 15, 16
Gene pair *A*: rough hair dominant over smooth			
Gene pair *B*: modifies rough to smooth (*A–BB* smooth; *A–Bb* partly rough; *A–bb* full rough)	6/16	partly rough	2, 4, 5, 7, 10, 13
	3/16	full rough	6, 8, 14
Interaction: when hair is smooth (*aa*), effect of modifier does not appear			
12. Complete dominance at one gene pair, partial dominance at the other; either gene, when homozygous recessive, is epistatic to the other; when both genes are homozygous recessive, the second is epistatic to the first			
Example: body color in the flour beetle, *Tribolium castaneum*	6/16	sooty	2, 4, 5, 7, 10, 13
	3/16	red	1, 3, 9
Gene pair *A*: normal (red) is completely dominant over jet (a darker color with a tinge of red)	3/16	jet	11, 12, 15
	4/16	black	6, 8, 14, 16
Gene pair *B*: normal (red) is partially dominant over black; the black heterozygote appears sooty in color			
Interaction: when jet is homozygous recessive (*aa*), the effect of *BB* or *Bb* does not appear; the double recessive, *aabb*, is black			
13. Complete dominance at one gene pair, partial dominance at the other; the partially dominant heterozygote has the same phenotype as the recessive homozygote at the other gene pair and an additive effect when they are both combined			
Example: body color in the flour beetle, *Tribolium castaneum*	7/16	sooty	2, 4, 5, 7, 10, 11, 13
Gene pair *A*: normal (red) is completely dominant over sooty	4/16	black	6, 8, 14, 16
	3/16	red	1, 3, 9
Gene pair *B*: normal (red) is partially dominant over black; the black heterozygote is sooty in color and when combined with the sooty homozygote (*aa*) causes a dark-sooty phenotype	2/16	dark sooty	12, 15
14. Dominance at both gene pairs only if both kinds of dominant alleles are present. If the dominant allele at one gene pair is absent, the dominant allele at the other gene pair acts as a recessive			

TABLE 11–1 (continued)

	Offspring Expected in F_2		Numbered Genotypes in Above Figure
	Ratios	Phenotypes	
Each Gene Pair Affecting the Same Character			
Example: pigment glands in cotton plants (*Gossypium hirsutum*)	11/16	glandular	1, 2, 3, 4, 5, 6, 7, 9, 10, 11, 13
Gene pair *A*: glandular dominant to glandless	5/16	glandless	8, 12, 14, 15, 16
Gene pair *B*: glandular dominant to glandless			
Interaction: dominants at *A* and *B*, when present together, produce pigment glands; absence of dominant allele at one gene pair produces glandular plants only when the dominant allele at the other gene pair is homozygous (e.g., *aaBB, AAbb*)			
15. Partial dominance at both gene pairs; additive effects for each partially dominant gene			
Example: flower color in beans (Mendel)	1/16 *AABB*	purple shade 10	1
Gene pair *A*: purple flower color partially dominant to white; additive effect on color for each *A* gene (i.e., value of 3)	2/16 *AABb*	purple shade 8	2, 5
	2/16 *AaBB*	purple shade 7	3, 9
	1/16 *AAbb*	purple shade 6	6
Gene pair *B*: purple flower color partially dominant to white; additive effect on color for each *B* gene (i.e., value of 2)	4/16 *AaBb*	purple shade 5	4, 7, 10, 13
	1/16 *aaBB*	purple shade 4	11
	2/16 *Aabb*	purple shade 3	8, 14
	2/16 *aaBb*	purple shade 2	12, 15
	1/16 *aabb*	purple shade 0 (white)	16

bay" phenotype). Investigations by Watkins and others have indicated that the *H* gene is responsible for the attachment of certain subterminal sugars to those polysaccharides upon which the terminal sugars specified by the ABO genes then attach (see Fig. 9–4). Thus, *hh* individuals lack the polysaccharide organization for terminal sugar attachment, and therefore appear to lack ABO antigens.

EXAMPLE 6. The sweet pea cross given in Table 11–1 as Example 6 was perhaps the first clear illustration of two gene pairs "complementing" each other in their effect on the same character. Discovered by Bateson and Punnett while looking for the mechanism of pollen-shape inheritance, they crossed two white-flowered sweet peas having different pollen shapes and surprisingly obtained a purple-flowered F_1. The offspring of the purple-flowered plants crossed with each other appeared in the given 9 purple: 7 white ratio. These results fitted closely to expectations if flower color were affected by two independently segregating gene pairs, either of which could produce an absence of color because of homozygosity for recessive alleles. Purple thus arose as the complementary effect of dominant alleles at two different gene pairs.

Since Bateson and Punnett's experiment, numerous complementary effects between gene pairs have been observed. One of the foremost examples of this type of interaction is that of *aleurone* color in corn. Below the outer *pericarp* layer of a corn kernel (Fig. 11–2) a single layer of cells called the *aleurone* may have a variety of colors, depending on the appearance of the *anthocyanin* pigments present within it (see Fig. 26–3). As in the case of sweet peas, the appearance of any color in corn kernels depends on a complementary effect between different gene pairs. According to Emerson, however, not two, but three, gene pairs are necessary for anthocyanin production; when recessive alleles at either one or more of these are homozygous, no aleurone pigment is produced.

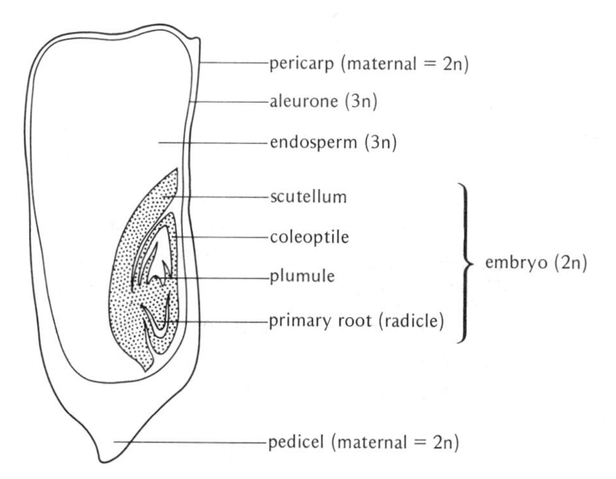

Figure 11–2

Corn kernel tissues and the number of homologues (n) in their chromosome complements. (After Briggs.)

EXAMPLES 7 AND 8. Epistasis, caused by the presence of homozygous recessives (Examples 5 and 6), may also be caused by the presence of dominant alleles. In Example 7 a white fruit color gene in summer squash has a dominant effect that hides the appearance of any color at all. If we call the white gene *A*, and a color gene *B*, then an *A* white variety (i.e., *AAbb*) crossed to a yellow variety (i.e., *aaBB*) would produce heterozygous plants that appeared all white (*AaBb*). The F_2 crosses between these heterozygotes would appear as white in every genotypic combination in which *A* was present (12/16 or 3/4). Color would appear in the remaining genotypes that were homozygous recessive *aa*, but only 1/4 of these (1/16 of the total) would be the double recessive *aabb*, in this case, green.

Of course, epistatic effects by dominant alleles cannot be confined to one gene pair. If two gene pairs affect a single character, it is certainly conceivable that dominant alleles at both gene pairs may have epistatic effects on each other. One outstanding example in which this occurs was shown by Shull in the plant *Bursa* (shepherd's-purse), in which two gene pairs affect the shape of the seed capsule (Example 8). For both genes the alleles causing triangular shape are dominant to the alleles causing an ovoid appearance, and this dominant effect occurs even when the other gene pair is homozygous for the recessive ovoid alleles. Thus a cross between triangular-appearing heterozygotes for both gene pairs, i.e., *AaBb* × *AaBb*, would produce only 1 of 16 plants which is homozygous recessive for both gene pairs and therefore ovoid.

EXAMPLE 9. Epistatic effects between two gene pairs may also be mixed in the sense that they can arise when one gene pair contains a dominant allele or when the other gene pair is homozygous recessive. For example, in fowl, color can be affected by two different gene pairs—one gene pair, e.g., *A*, inhibiting color with a dominant allele, and the other, e.g., *B*, inhibiting color with a homozygous recessive. Thus a cross between two white heterozygotes *AaBb* will only produce color in those offspring having the genotype *aaB–* (3/16). Color in all other offspring will be inhibited because of the presence of *A* (12/16) or because of the presence of the homozygous recessive *bb* (1/16).

EXAMPLE 10. In plants the origin of new phenotypes through interaction has been demonstrated by Sinnott for two gene pairs that determine fruit shape in the summer squash (*Cucurbita pepo*). For each gene pair the allele for sphere-shaped fruit is dominant to that for long shape, but when sphere-shape alleles are present at both gene pairs, a more extreme flattening of the fruit occurs, causing a disc shape. Thus a cross between two sphere-shaped varieties (i.e., *AAbb* × *aaBB*) would produce a disc-shaped F_1. Intercrossing the F_1 with itself produces F_2 phenotypes in the ratio 9/16 disc-shaped:6/16 sphere-shaped:1/16 long-shaped, as shown in Example 10 of Table 11-1.

In mammals a similar ratio can result from crosses in which coat-color genes are involved. Pigs, for example, have a wide variety of colors, of which one is light red and called *sandy*. According to work by Wentworth and Lush, there appears to be two different gene pairs that can cause sandy color when either one is homozygous recessive and the other dominant, e.g., *A–bb* or *aaB–*. When the dominant allele is present at both gene pairs, *A–B–*, the result is red coat color, and when both are homozygous recessive, *aabb,* the result is white coat color. Wentworth and Lush first crossed a breed carrying genotype *AABB* to a breed carrying *aabb.* Because of interaction with other gene pairs, such as those affecting black pigment, the results were somewhat complex, but the F_1 of this cross appeared to have a uniform *AaBb* genotype (red coat color). A cross between the F_1 heterozygotes produced a variety of phenotypes, including blacks. When the blacks were separated out because of coat-color interactions, the 37 F_2 pigs remaining had the following coat colors: 20 red, 14 sandy, and 3 white. This is quite close to the expected 9:6:1 ratio.

INCOMPLETE DOMINANCE AT ONE GENE PAIR (EXAMPLES 11, 12, AND 13). The various examples of interaction between two pairs of genes have, until now, been restricted to cases in which there are dominant alleles at both gene pairs. Complete dominance is, of course, not a rigid rule, and many examples exist in which interaction occurs between two gene pairs, only one of which has a dominant allele. Among the examples of this kind considered in Table 11–1, number 11 is perhaps the simplest, because the interaction between the gene pairs occurs in only one direction. According to Wright, the allele for rough hair in guinea pigs (e.g., *A*) is dominant to the allele causing the usual smooth coat (*a*). Another gene pair (e.g., *B*) modifies the effect of rough hair (*A*) to smooth, but cannot modify smooth hair (*aa*) to rough. The *B* gene pair has incomplete dominance, so that genotype *A–BB* is smooth, *A–bb* is full rough, and *A–Bb* is intermediate, or partly rough. Genotypes producing smooth hair coat can therefore be either *A–BB* or homozygous recessive *aa* with any *B* or *b* allele. The other genotypes produce rough or partly rough hair, depending on the particular *B* or *b* alleles present. When heterozygotes for both gene pairs (*AaBb*) are crossed, the expected ratios are therefore 7/16 smooth:6/16 partly rough:3/16 full rough, as shown in Table 11–1.

Example 12, discovered by Sokoloff in the flour beetle, *Tribolium castaneum*, is somewhat more complex, because epistatic effects are produced by both gene pairs. In this beetle normal body color is red (actually chestnut) but can be modified by various gene pairs. At one of these gene pairs the normal allele for red color (e.g., *A*) is dominant over a recessive allele, which when homozygous (e.g., *aa*) can produce a darker effect called *jet.* Another gene pair (e.g., *B*) also affects body color, but the homozygote *bb* is, in this case, much darker than jet and is called *black.* The black allele also has an intermediate effect, causing *Bb* genotypes to appear *bronze* or *sooty* in color. These genes interact with each other so that the jet *aa* genotypes hide the effect of the normal *B* alleles and the black *bb* genotypes hide the effect of the *A* alleles including the jet effect of *aa.* Each of the four phenotypes—red, jet, black, and sooty—therefore has a unique constellation of genotypes, as shown in the table. When crosses are made between two sooty heterozygotes, *AaBb* × *AaBb,* the phenotypic results appear in the ratio 6/16 sooty: 3/16 red: 3/16 jet: 4/16 black.

Another gene pair in this same flour beetle (Example 13) also produces a change in body color, but this time the effect is similar to that of the intermediate black heterozygote, previously mentioned. If we call this gene pair *A,* the dominant allele *A* causes normal red body color and the homozygous recessive *aa* causes a sooty phenotype. As stated previously, the *B* gene pair also affects body color, *BB* being normal red, *Bb* sooty, and *bb* black. When alleles at both of these *A* and *B* gene pairs are segregating in a culture, sooty phenotypes can therefore be produced in two ways, either *aaBB* or *A–Bb.* The *aaBb* genotype gets the sooty effect from both gene pairs, one adding to the other (additive effect), and appears *dark sooty,* whereas all the *bb* genotypes appear *black,* hiding the effect of the *A* gene pair. The phenotypes produced by a cross between sooty heterozygotes *AaBb* are, as seen in Table 11–1, in the ratio 7/16 sooty: 4/16 black: 3/16 red: 2/16 dark sooty.

DOMINANCE MODIFICATION (EXAMPLE 14). Whether a particular allele acts as a dominant is often a function of the effects of other genes (p. 215). Occasionally genes that appear to act as dominants may reverse their roles when changes are made at other gene pairs. In an example investigated by Fuchs, Smith, and Bird, two gene pairs were shown to affect the appearance of leaf pigment glands in the seedling stage of cotton (*Gossypium hirsutum*). Certain alleles at these two gene pairs (e.g., *A* and *B*) seemed to function equally as dominants, so that plants which contained both alleles (*A–B–*) were glandular, and the glandless plants would have been expected to be only the double homozygotes, *aabb*, i.e., an F_2 ratio of 15 glandular: 1 glandless. Surprisingly the investigators found that the F_2 ratio was 11 glandular:5 glandless, the glandless plants apparently including genotypes which had only one of the so-called "dominant" alleles (*Aabb*, *aaBb*). Thus "dominance" in this case is expressed only at one gene pair when there is another "dominant" allele present at the second gene pair. If the second gene pair does not have a dominant allele, then glandular plants are produced only when the "dominant" allele at the other gene pair is homozygous (*AAbb*, *aaBB*). One can also interpret this ratio as a "threshold" effect (see Chapter 14); that is, the plant is glandular only when two or more capital letter alleles are present, and glandless when one or no capital letter alleles are present.

EXAMPLE 15. When incomplete dominance exists at two gene pairs determining a single character, one of the simplest relations between them is that in which each allele makes a specific measurable contribution to the character (e.g., allele $a = 0$, allele $A = 3$, allele $b = 0$, allele $B = 2$). These contributions are known as *additive* effects, because the phenotype of the character is determined by adding the effects of each allele at the two gene pairs (e.g., $0 + 3 + 0 + 2 = 5$).

This additive relationship was noted by Mendel himself. He crossed two species of beans, one having white flowers and the other having purple flowers, to obtain an all-purple F_1. From these results the white flower color effect appeared recessive to purple and would be expected to appear in one-quarter of the F_2 plants if a single gene pair was involved. Mendel's actual F_2 results were far different: the heterozygous F_1 when crossed with itself produced 31 plants, only one of which was white, the others producing flowers of various purple shades ranging from light to dark. Mendel concluded that two or three pairs of factors were responsible for flower color in these species. If, as shown in Example 15 of Table 11-1, we assign specific quantitative values to each allele of two gene pairs, the variety of purple flower-color shades produced in the F_2 can be explained.

The correctness of these conclusions has been borne out by later experiments in plants and animals which show a wide variety of characters influenced by additive relations among numerous gene pairs. These characters have been called *quantitative characters* and will be discussed more fully in Chapter 14.

INTERACTION BETWEEN MORE THAN TWO GENE PAIRS

Most observable characters in organisms are end products of a long chain of biological processes. Wherever geneticists have been able to analyze these processes in detail, they have been shown to consist of many individual steps, each of which may usually be influenced by the action of specific genes. Among the numerous characteristics in which many gene pairs are involved is that of coat color in mammals. The hair composing the mammalian coat are long thin cylinders of a horny substance (keratin) often containing rows of pigmented cells in a central "medulla" surrounded by an envelope of flattened cells in the "cortex." The amount, kind, and localization of pigment in the medullary cells, the air spacing between them, the relative thickness of medulla and cortex, and the organization of various types of hair (guard, awn, down, etc.) in any section of the body determine the coat color observed.

Chemically the main pigment involved in mammalian coat color is also widespread among other living creatures (planaria, insects, molds, etc.) and is called *melanin*. In mammals it appears in either a brown-black form (eumelanin) or yellow-reddish form (phaeomelanin). The biochemical pathway to melanin has been widely investigated, and essentially begins with the amino acid tyrosine which undergoes a number of transformations leading to a long-chained polymer that contains repeating melanin subunits (Fig. 11–3). Each step is catalyzed by enzymes, some of which have been identified as the products of particular genes. Thus the *C* gene in mammals is associated with the production of tyrosinase, and the *albino* allele of this gene (*c*) causes the complete absence of melanin pigment since no detectable tyrosinase is formed. The *Himalayan* (c^h) allele of this gene does produce tyrosinase, but it is in a heat-sensitive form, so that it is less active at warm temperatures than at cold temperatures. As previously noted (p. 184), this often

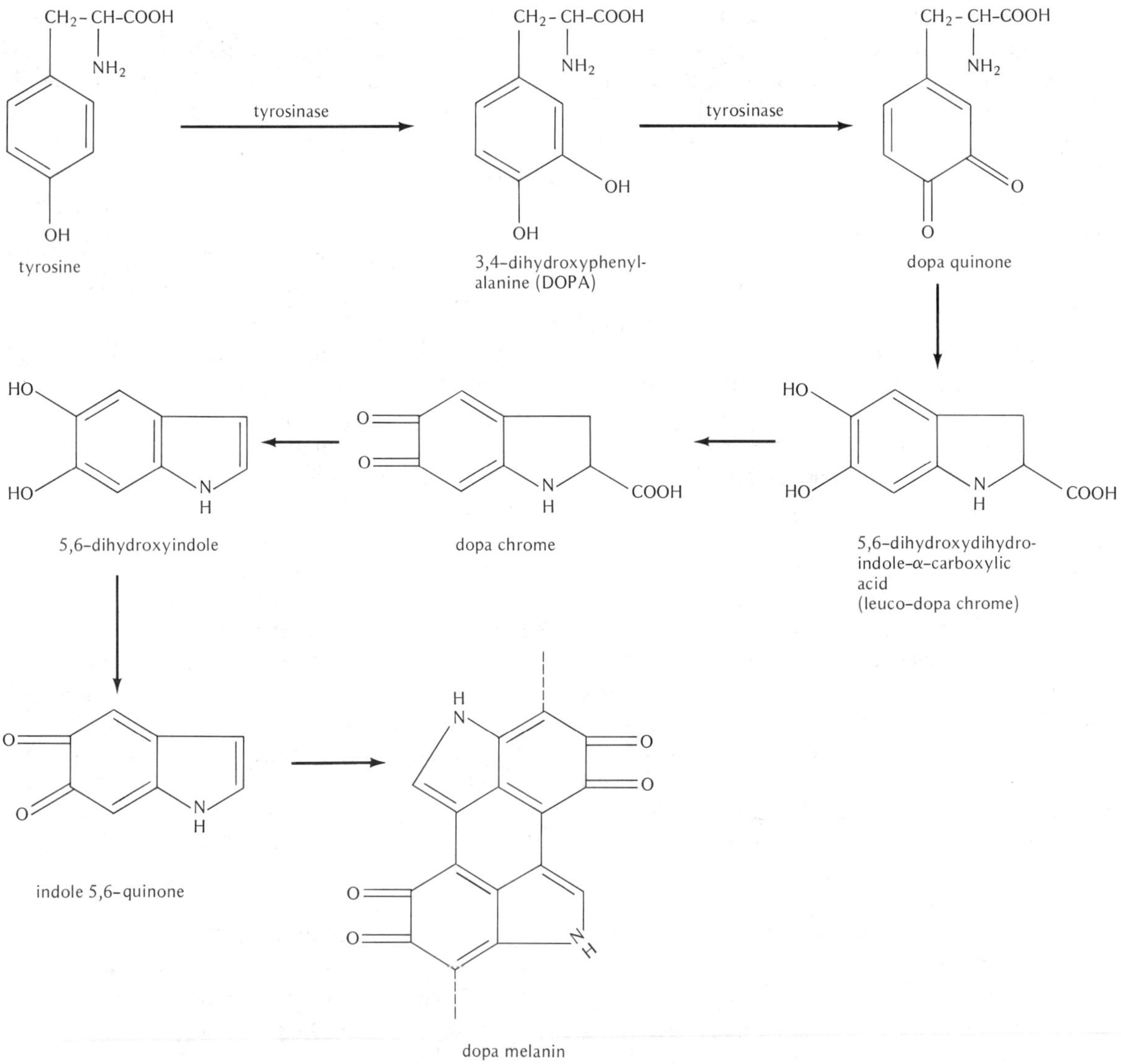

tyrosine → (tyrosinase) → 3,4-dihydroxyphenylalanine (DOPA) → (tyrosinase) → dopa quinone → 5,6-dihydroxydihydroindole-α-carboxylic acid (leuco-dopa chrome) → dopa chrome → 5,6-dihydroxyindole → indole 5,6-quinone → dopa melanin

Figure 11–3

Synthesis of melanin from tyrosine. (After Mazur and Harrow.)

leads to animals such as Siamese cats who are homozygous c^hc^h bearing darker fur at their extremities than at their warmer body parts.

The melanocyte cells in which pigment is formed arise mostly in the neural crest during vertebrate embryonic development (see Fig. 30-9). With the exception of retinal melanocytes which remain close to the optic cup, melanocytes migrate outward from the neural crest to cover the various pigmented sections of the body. Should conditions be appropriate, melanin is then synthesized and deposited within small protein-fibered vesicles, called melanosomes, which become the pigment granules. In the hair follicles the melanocytes surround the hair bulb and inject their pigment granules into the cells of the growing hair shaft.

It is obvious that a modification at almost any stage of this process can cause changes in color and pattern. For example, in rodents the gene *Splotch* and some of the *spotting*, *silver*, and *hooded* genes may cause the death of pigment cells or prevent or delay their migration from the neural crest into the skin thereby producing unpigmented white areas. The organization of the melanosome fibers may also be affected, as has been shown for the *pink-eyed dilution* and the *brown* genes. A delay in the synthesis of pigment granules, as caused by the *ruby-eyed* gene, may reduce their number and size, thereby lightening the overall color. Entry of the pigment granules into the hair shaft may be reduced by genes that cause hair abnormalities such as *brindled* and *blotchy* or may be affected by genes such as *tabby* which cause striped patterns. The kind of pigment produced, whether black, yellow, or mixed, depends strongly on the presence of a particular set of alleles (*E* genes). The products of these genes, in turn, are affected by the *Agouti* gene, which in dominant form produces the normal black-tipped hair with a yellow band, or by *somber* and *dark* modifiers that may blacken the yellow pigment. A diagrammatic representation of the stages at which some genes are believed to act in the mouse is shown in Fig. 11-4. A number of similar genes are found also in other mammals, often with similar effects and similar alleles. The following represents a small sampling of the main gene pairs known and some of their alleles:

A, agouti (wild type, black and yellow banded hairs); A^w, light-bellied agouti; A^Y, yellow; a^t, black and tan; *a*, non-agouti (black)

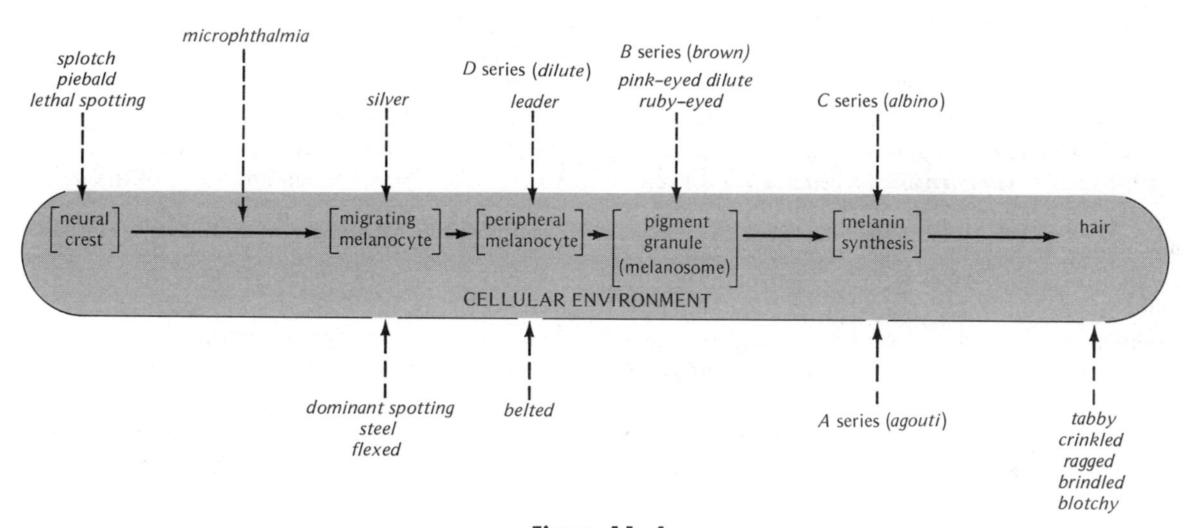

Figure 11-4

Developmental stages at which coat color genes in the mouse are believed to act. The upper part of the figure shows genes that probably act directly on the pigment cells, and the lower part shows genes that may act on the environment of these cells. (After Searle.)

B, black; B^{lt}, light; b^c, cordovan; *b*, brown
C, full color; c^{ch}, chinchilla; c^e, extreme dilution; c^h, Himalayan; c^p, platinum; *c*, albino
D, full color; d^s, slight dilution; *d*, maltese ("blue") dilution
E, black, no spotting; e^j and e^p, yellow and black; *e*, yellow

Given such interrelationships, the various kinds of interactions described in Table 11-1 are almost a matter of course. Further additional genes and gene interactions are known to affect this process as well as environmental factors such as temperature, age, and hormones. In conclusion, single characters are not solely determined by any single gene, although a single gene may have important effects on a character.

MODIFIERS

In the early literature of genetics, the phenomena of gene action and interaction were little known. To many geneticists, the activity of any particular gene appeared to produce phenotypic effects independently of all or most other genes. The discovery of instances in which the effect of one gene changed or modified that of another gene was therefore always of special interest. As more examples of gene interaction were found, a series of terms arose that were used to describe some of the different kinds of interaction. Some of these terms, such as *complementary genes*, *epistasis*, and others we have mentioned, are still in use today because they help us visualize the phenotypic behavior of particular genes. Actually, of course, the phenotypes we observe are the results of more complex processes than were first imagined. Epistasis, for example, is not to be taken literally as one gene "hiding" another, but may be a consequence of increased or decreased enzyme activity, changes in acidity that affect the appearance of color (p. 600), or various other complex events. Whatever term we use for a gene at the gross phenotypic level, the particular developmental basis for its activity must be studied separately and unraveled.

With this reservation in mind we can also mention a large group of genes that come under the general heading of *modifiers* but whose particular forms of action are not always known. Broadly defined, modifiers are genes that change the phenotypic effects of other genes in a quantitative fashion. Many of the genes considered above are of this kind. For example, the dilution and sooty factors involved in mammalian coat color modify, through *dilution* or *enhancement*, the effect of the color genes. Similarly the *inhibitor* gene that prevents the appearance of feather color in chickens (p. 209) can be considered as a modifier of the normal fowl color genes.

On the other hand, modifiers of mutant genes also exist, and in some cases they completely reduce the phenotypic expression of the mutant gene, although the latter is still present. Such genes have been called *suppressors*, and examples of their activity have been studied in numerous organisms. In *Drosophila*, for example, a particular gene, *su–Hw*, suppresses the phenotype caused by the mutant gene for *Hairy-wing*. Although *su–Hw* also reduces the mutant phenotypes caused by genes for *forked* bristles, *interrupted* wing veins, and others, there are numerous additional suppressor genes that act only on one particular mutant phenotype, such as the suppressor of *Star* eye shape (*su–S*). In some cases the suppressor of one mutant effect may enhance another mutant phenotype, such as the enhancement of the hairy-wing phenotype by the suppressor of purple eye color in *Drosophila*. Suppression and some of the mechanisms that may cause it will be discussed further in Chapter 28.

Modifiers may, of course, be dominant or recessive, and may have large or small quantitative phenotypic effects. White spotting in mice, for example, occurs in individuals homozygous for a recessive gene *s* but may be greatly enhanced by selection of many genetic factors with small effect, so that an almost completely white animal is produced.

The extent to which a gene has a dominant or recessive effect may also be influenced by modifying genes. This was dramatically illustrated by Ford in experiments with the currant moth,

Abraxas grossulariata. In this moth a single gene, *lutea*, in homozygous condition, produces yellow color instead of the normal white ground color but has an intermediate effect as a heterozygote. After four generations of selecting moths for greater and lesser expression of the *lutea* phenotype in the heterozygote, Ford was able to show that two distinct strains could be obtained; in one case *lutea* acted as a complete dominant and in the other case as a complete recessive. In each strain special modifiers had been chosen, some enhancing and some detracting from the dominance of this particular gene.

In the *lutea* example there is no evidence that the modifiers have a particular phenotypic effect of their own aside from changing dominance relations. Numerous other examples exist, however, in which gene pairs with distinct phenotypic effects of their own may also act as modifiers for an entirely different character. In *Drosophila*, Hersh has shown that the number of facets in *Bar*-eyed individuals may be significantly altered by the presence of other genes that also affect bristles, eyes, and wings. Similarly Dobzhansky (1927) has demonstrated that a gene such as that for *white* eyes in *Drosophila* may affect the shape of sperm storage organs in females as well as other structures. This multiple phenotypic effect of single genes is called *pleiotropism* and is especially valuable in helping to understand relations between different developmental processes (Chapter 30). Many genes, when investigated in thorough fashion, appear to have pleiotropic effects. Indeed the mutual interdependence of all the numerous developmental stages in an organism's growth makes it likely that most genes, if not all, affect more than one phenotypic characteristic (see, for example, Fig. 26–20).

LETHALITY

Perhaps the most serious effect a gene can have on an organism is to cause its death. Originally, at the time of discovery of mendelism, it was difficult to conceive that such lethal genes could ordinarily be present in a population. In 1905, however, a French geneticist, Cuénot, reported on the inheritance of a mouse-body-color gene, *Yellow*, that did not appear to fit the usually expected mendelian segregation pattern. Although *Yellow* was apparently a gene with dominant effect, it could not be bred as a pure strain, because crosses of *Yellow* × *Yellow* always produced the normal agouti as well as yellow offspring. By backcrossing yellow individuals to agouti, Cuénot found that all yellow mice were heterozygous for the *Yellow* gene and that no *Yellow* homozygotes could be found.

At first, Cuénot supposed that sperm carrying *Yellow* could not penetrate *Yellow* eggs. Later, however, Castle and Little offered the explanation that *Yellow* homozygotes were formed but died *in utero*. According to them, *Yellow* had a dominant phenotypic effect on coat color but had at the same time a recessive effect on lethality, so that homozygotes for yellow were inviable. A cross of *Yellow* × *Yellow* therefore always produced offspring in the ratio 2/3 yellow:1/3 agouti instead of 3/4 yellow:1/4 agouti (Fig. 11–5). The absent yellow offspring were the inviable homozygotes. Evidence by Robertson, Eaton, Green, and others has shown the correctness of this explanation; approximately 25 percent inviable embryos can be dissected out of the female parents in such crosses.

Soon after Cuénot's discovery, Baur found a similar lethal gene in plants. In the snapdragon, *Antirrhinum majus*, a yellow-leaved dominant genotype, *aurea*, when crossed *aurea* × *aurea*, always produced normal green plants as well as *aurea* heterozygotes. *Aurea* homozygotes lacked the ability to make chlorophyll and died either before germination or as seedlings.

Since that time numerous other lethal factors have been found which have a dominant phenotypic effect in the heterozygote but are lethal in homozygous condition. Genes such as *Dexter* in cattle are fully viable in heterozygous condition, producing shortened limbs similar to achondroplastic dwarfs in man. In homozygous condition the *Dexter* gene produces "bull-dog" calves with extremely reduced features that are usually aborted before birth. Heavy freckling in man, caused by the presence of the gene for *xeroderma pigmentosum*, is another dominant phenotypic

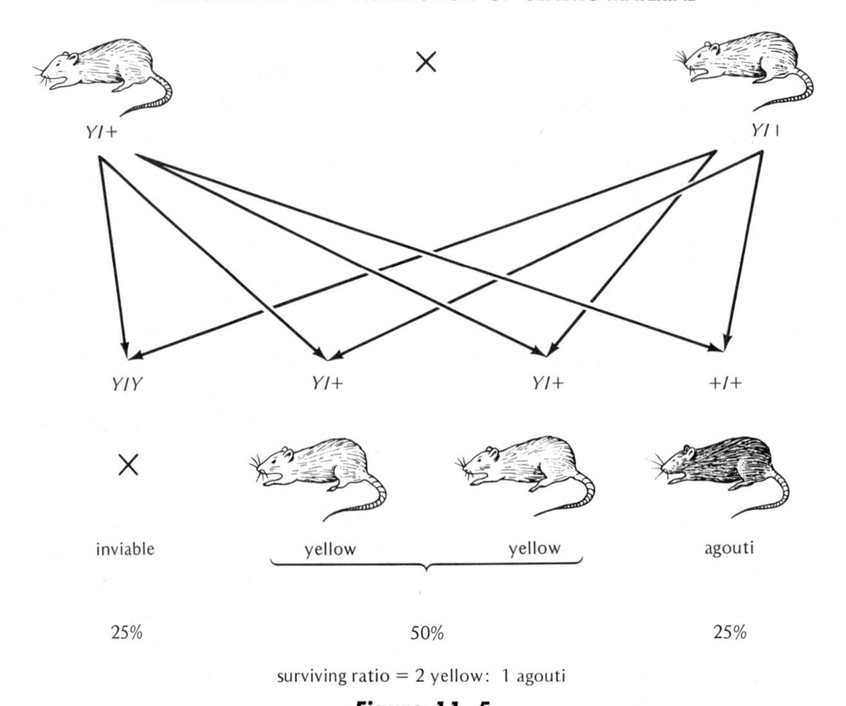

Figure 11–5

Segregation of the Yellow gene (Y) in mice in a cross Y/+ × Y/+. (After Hadorn.)

effect which becomes lethal when the gene is present in homozygous condition.

Lethality, however, is not confined to genes with a dominant phenotypic effect but can also be caused by genes whose phenotypic effects are ordinarily recessive, such as infantile amaurotic idiocy in humans. That is, in heterozygous condition, such recessive alleles have no observable phenotypic effect but produce a noticeable and eventually lethal change in homozygotes. In both instances, whether the lethal gene has a dominant or recessive phenotypic effect, it is called a *recessive lethal* as long as its lethality depends upon its presence in homozygous condition.

Dominant lethals, on the other hand, are genes whose lethal effects occur in heterozygous individuals. A single dose of the gene for *epiloia* in humans, for instance, causes abnormal skin growths, severe mental defects, and multiple tumors, so that most heterozygotes for this gene die quite young. As will be more fully demonstrated later (Chapter 24), dominant lethal effects among progeny may be caused quite easily by x-ray treatments of the reproductive organs of the parents; the greater the x-ray dosage, the greater the observed lethality. In *Drosophila*, for example, x-ray treatment of male spermatozoa may cause the death of fertilized offspring at various developmental stages.

The penetrance (p. 183) of lethals, recessive or dominant, may also vary, so that not all genotypically affected individuals are phenotypically lethal. Some lethal genes have a high degree of penetrance and expression, permitting little or no survival among affected genotypes beyond the embryonic or infant stage. Others, called *semilethals*, or *subvitals*, permit larger proportions of affected genotypes to survive. Viability is thus a character influenced by a broad spectrum of genes, ranging from complete lethality to sublethal, subvital, normal, and occasionally *supervital*, or better-than-average, genotypes. The boundaries between these categories are not always clear, and a few authors have even broadened the category of lethals to include any gene whose effects, such as that of sterility, interfere with reproduction of the next generation. As a rule, however, lethal genes are regarded as those whose effects cause death to the organism, usually in the early growth stages.

BALANCED LETHAL SYSTEMS. In 1918 Muller reported an unusual stock of *Drosophila* flies which were always heterozygous for a particular lethal gene, *Beaded*. The *Beaded* gene had a dominant effect on the shape of the wings but had been shown to be lethal in homozygous condition. Thus *Beaded* flies were always heterozygotes. A cross of *Beaded* × *Beaded*, therefore, should result in 2/3 Beaded and 1/3 normal wings, or a 2 : 1 ratio, as for the *Yellow* × *Yellow* cross among mice. In one of Muller's stocks, however, only *Beaded* flies were produced and normal-winged flies never appeared. After detailed analysis Muller found that the homologous chromosome carrying the normal allele for *Beaded* also carried a gene that was lethal in homozygous condition, i.e., *le*, but which was not an allele of *Beaded*. Thus, in addition to *Beaded* heterozygotes, the *Beaded* × *Beaded* cross produced two types of homozygous lethals, *Beaded/Beaded* and *le/le*, as shown in Fig. 11–6. Since *Beaded* and *le* are recessive lethals and are not allelic, the *Beaded* +/+ *le* individuals are viable, because each homologous chromosome carries the wild-type allele for the recessive lethal on the other homologue. The *Beaded* heterozygotes are therefore self-perpetuating as long as the homologous chromosome also contains a recessive lethal factor (see p. 504). This balancing effect between two different lethals in a self-perpetuating stock was named a *balanced lethal* system by Muller.

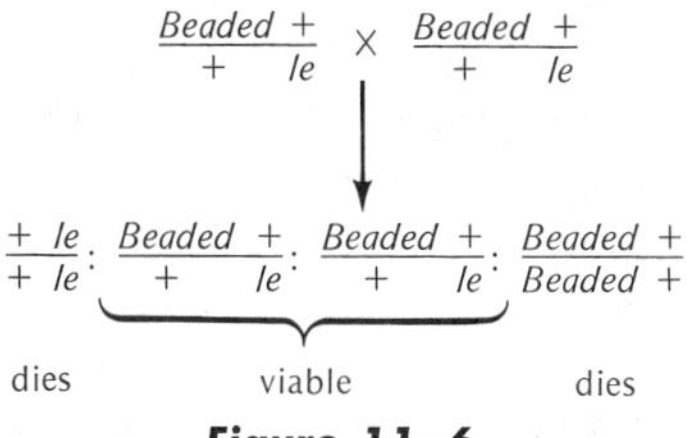

Figure 11–6

Cross between individuals in a balanced lethal stock of Drosophila melanogaster *producing viable heterozygous individuals identical to the parents.*

Among other organisms in which balanced lethal stocks occur is that of mice. The gene for *Brachyury*, when heterozygous, has a dominant effect in shortening the tail and is lethal in homozygous condition. This gene known as *T* can be balanced with another gene, recessive *tailless*, or *t*, which is also homozygous lethal, so that heterozygotes *T* +/+ *t* are always produced.

In some instances both of the recessive lethals of a balanced lethal system are associated with dominant phenotypic effects such as the *Drosophila* mutants *Curly* and *Plum*, one affecting wing shape and the other eye color. Because of the lethality of *Curly/Curly* and *Plum/Plum* flies, the *Curly/Plum* stock is as self-perpetuating as the *Beaded/le* stock. (For simplicity the + alleles are usually omitted in the notation but are understood to be present; i.e., *Curly/Plum* = *Curly* +/+ *Plum*.) In addition to *Curly/Plum*, two other independently segregating genes on another pair of homologous chromosomes, which are also balanced lethals, may be present, i.e., *Hairless* and *Stubble*. A stock of *Curly/Plum Hairless/Stubble* would therefore also be self-perpetuating and produce only heterozygotes for both sets of genes (Fig. 11–7).

Two requirements are therefore necessary for the maintenance of a balanced lethal system: (1) that each member of a pair of homologous chromosomes carry a different nonallelic recessive lethal, and (2) that each of these two different recessive lethals always remain on separate homologous chromosomes. The second requirement is fulfilled when transfer of a nonallelic lethal from one homologous chromosome to the other is prevented by a "crossover suppressor" (Chapter 22).

CHANGES IN LETHALITY. Lethality, as many other phenotypic expressions, can be influenced by the environment in many instances. A graphic demonstration of this is shown by Dobzhansky for certain genotypes of *Drosophila pseudoobscura* that are quite normal when raised at a temperature of 16.5°C but are quite lethal

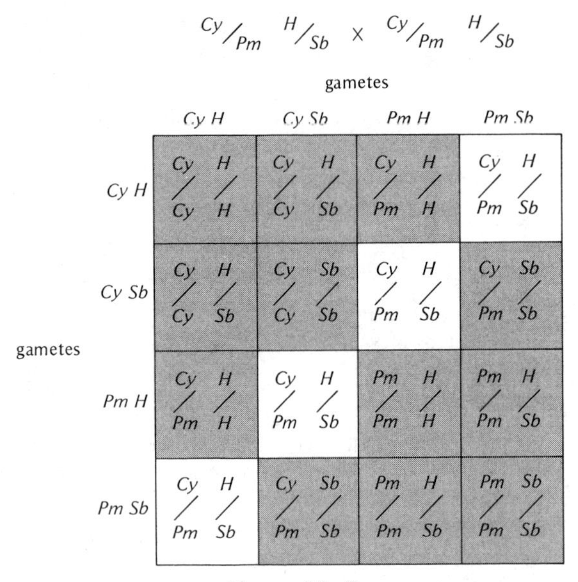

Figure 11–7

Progeny produced by a balanced lethal stock of Drosophila melanogaster *heterozygous for lethal genes on two different pairs of homologous chromosomes:* Curly (Cy) *and* Plum (Pm) *on one pair of homologues, and* Hairless (H) *and* Stubble (Sb) *on the other pair. Shaded genotypes are lethal.*

at 25.5 °C. Similarly the *kidney*-eyed mutant of the wasp *Bracon hebetor* (once called *Habrobracon juglandis*) is lethal at 30 °C, according to Whiting, but fully viable at lower temperatures. Such lethals are now called *conditional lethals* in the sense that they permit the organism to survive under "permissive" conditions and cause lethality under "restrictive" conditions.

In *Drosophila melanogaster*, Suzuki and coworkers have analyzed numerous conditional lethals and shown that the restrictive condition may often be localized to a specific ("monophasic") period of development. For example, lethality of the particular mutant shown in Fig. 11–8a occurs upon exposure to high temperature only during the latter part of the larval stage. In fact this mutant will survive even if it is raised at high temperatures during its entire life cycle as long as it is exposed to low temperature during the specific temperature-sensitive period (Fig. 11–8b). The lethal gene product that is affected by temperature, probably a protein, seems therefore to be either produced or activated only during this short interval.

Other patterns of lethality have also been found, such as (1) the occurrence of lethality by exposure of the organism to the restrictive temperature during any of two or more ("polyphasic") developmental stages; (2) possession by lethal-bearing males and females of different ("sexual dimorphic") temperature-sensitive stages; (3) the occurrence of lethality by exposure to the restrictive condition at any stage of development. As will be discussed in Chapters 26 and 27, conditional lethals have received extensive attention in bacteria and viruses because they enable the identification and analysis of many genes and proteins that would otherwise be unavailable for study.

Among other environmental conditions that may have distinct effects on lethality is that of nutrition. Phenylketonuria, discussed on page 185, has serious consequences when affected genotypes are raised on phenylalanine diets. Similarly, the inability of some strains of yeast to ferment the sugar galactose may have lethal results on a galactose medium, but no effect on a glucose medium.

The remainder of the genotype or "genetic background" in which lethal genes occur may also seriously modify lethal effects. In *Drosophila*, Gloor was able to change the penetrance of a recessive lethal factor *cryptocephal* from 50 percent to about 1 percent by selecting a special genetic background. Sturtevant has demonstrated the presence of a dominant gene in *Drosophila* called *Prune-killer* which has no observable effect of its own but causes the death of genotypes homozygous for *prune*, an eye-color gene. In humans the occurrence of certain types of cancer that may have some genetic components, such as that of the stomach, uterus, pancreas, and prostate, is believed to be of greater frequency in the blood group A genotype than in other genotypes of the ABO group.

SEGREGATION DISTORTION

As we have seen so far, it is the lethality among zygotes which results in the absence of certain

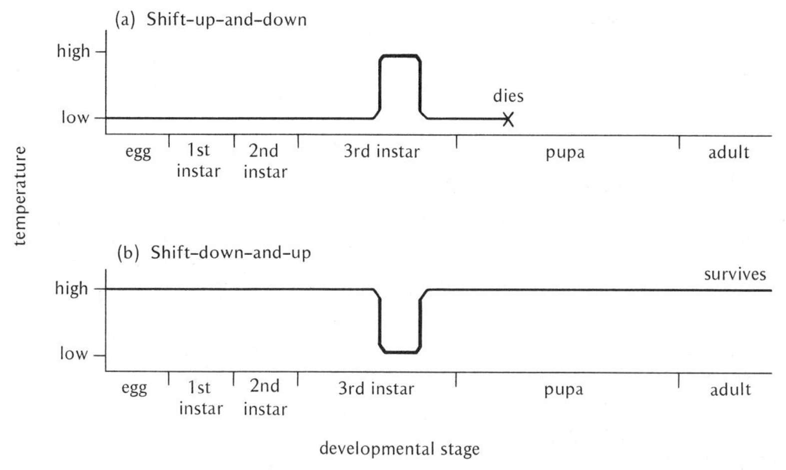

Figure 11–8

Studies of stocks of a particular temperature-sensitive mutation using temperature shifts to determine the developmental stage most sensitive to high temperature. (a) Lethality is caused by a "shift-up" in temperature during a relatively short interval during the late larval stage. Note that the lethal phase (pupal death in this case) doesn't necessarily coincide with the temperature-sensitive phase. (b) The same mutant exposed to high temperatures during the entire life cycle except for the temperature-sensitive phase when it is "shifted-down" to low temperature. Note that there is no lethality now. (After Suzuki.)

classes and the distortion of expected mendelian ratios. A much more basic distortion of such ratios arises when one or more classes of *gametes* are either lethal or unable to conjugate according to the usual meiotic segregation pattern, a phenomenon that has been termed *meiotic drive* by Sandler and Novitski. For example, certain *Drosophila* males of the genotype *cn bw*/cn^+ bw^+ (*cn* and *bw* are symbols for the recessive genes *cinnabar* and *brown*, each causing a modified eye color but interacting to produce white eyes when homozygous in the same fly), when mated to white-eyed *cn bw*/*cn bw* females, produce offspring in ratios of about 25 wild type (cn^+ bw^+/*cn bw*) to 1 white eye (*cn bw*/*cn bw*), instead of the expected 1:1. Differential mortality of zygotes does not seem to be involved in this case, and the cause for these ratios has been ascribed by Sandler and co-workers to a dominant *Segregation Distorter* (*SD*) gene carried on the cn^+bw^+ chromosome.

One explanation offered for *SD* action is that it causes the malfunction of sperm carrying the SD^+ chromosome in male SD/SD^+ heterozygotes. Thus it is primarily the *SD* chromosome bearing the genes cn^+ bw^+ in the above example which is carried in functional sperm, whereas its SD^+ homologue carrying *cn bw* is mostly lost in dysfunctional sperm. According to Hartl, this type of sperm dysfunction is actually caused by an interaction between *SD* and a specific gene on the homologous SD^+ chromosome which responds to the *SD* product. Peacock and Miklos propose that the *SD* gene may act by interfering with normal first meiotic division pairing with its SD^+ homologue, which then causes dysfunction in SD^+ sperm.

A similar segregation distortion is noted in male mice bearing certain of the *tailless t* alleles. Such heterozygotes (the *t* alleles are recessive lethals) transmit these alleles in frequencies as high as 99 percent compared to the normal allele. Experiments by Braden and by Yanagisawa, Dunn, and Bennett indicate differences in the fertilizing power of the two types of sperm produced by such heterozygotes. The cause for these differences may lie in specific antigens produced by *t* alleles on the surface of sperm (see Bennett et al.).

Although a further example of segregation distortion will be discussed in Chapter 12 (p. 244), such exceptions to Mendel's law of segregation are extremely rare. For practically all chromosomal genes the bulk of present evidence indicates that the gametes are formed and distributed according to mendelian principles.

PROBLEMS

11-1. For each example of a different F_2 ratio shown in Table 11-1, give the phenotypes and ratios expected from a backcross of the F_1 heterozygote to the parental homozygous recessive. (For example, the backcross in Example 1 of Table 11-1, using Mendel's peas, would be *AaBb* × *aabb,* and would yield a ratio of 1 yellow round : 1 yellow wrinkled : 1 green round : 1 green wrinkled.)

11-2. Numerous kinds of head combs are known in fowl, four of which are shown in Fig. 11-1. If you started out with pure-breeding stocks of pea-combed and rose-combed fowl, what matings would you perform to obtain a pure-breeding single-combed stock?

11-3. Hall and Aalders found a white-fruited form of the blueberry *Vaccinium augustifolium* which they crossed to the normal blue-fruited variety. The F_1 offspring of this cross were 30 blue-fruited plants. When the F_1 were crossed with each other, 127 blue-fruited plants and 4 white-fruited plants were produced. (a) What hypothesis would you offer to explain these results? (b) What further experiments would you undertake to test your hypothesis?

11-4. Three possible phenotypes in barley are "hooded," "long-awned," and "short-awned." An investigator crossed two pure-breeding stocks, hooded × short-awned, and obtained an F_1 which was phenotypically hooded. He then crossed $F_1 \times F_1$ and obtained the following F_2: 450 hooded, 150 long-awned, 200 short-awned. (a) How many gene pairs do you think are involved in determining these traits? (b) Using letters to designate genes, present a hypothesis to explain the observed results, and give the ratios you would expect according to your hypothesis. (c) On the basis of your answers to (a) and (b), assign alphabetically designated genotypes to the parental pure-breeding stocks and to the F_1 hooded individuals.

11-5. Nilsson-Ehle made crosses between two types of oats, one with white-hulled seeds and one black-hulled. The F_1 between them was black-hulled, and the F_2 ($F_1 \times F_1$) contained 560 plants as follows: black, 418; gray, 106; and white, 36. (a) Define gene symbols and explain the inheritance of hull color in these crosses. (b) Furnish genotypes for the white and gray F_2 individuals.

11-6. Using the chi-square method, test the fit of Wentworth and Lush's results on the pig-coat-color experiment (pp. 209–210) to a 9:6:1 ratio.

11-7. Flower color in a particular plant may be purple, red, or white, and these traits are known to assort independently of each other. Homozygous stocks of each of these colors were raised, crossed in the following fashion, and produced the following results:

cross 1: P_1: purple × red
F_1 = all purple
$F_1 \times F_1$ = 3/4 purple : 1/4 red
cross 2: P_1: purple × colorless
F_1 = all purple
$F_1 \times F_1$ = 9/16 purple : 3/16 red : 4/16 colorless
cross 3: P_1: purple × colorless
F_1 = all colorless
$F_1 \times F_1$ = 12/16 colorless : 3/16 purple : 1/16 red

List the phenotypes and their ratios expected from the following crosses: (a) If an F_1 plant from cross 1 is crossed to an F_1 plant of cross 2. (b) If an F_1 plant from cross 1 is crossed to an F_1 plant of cross 3. (c) If an F_1 plant from cross 2 is crossed to an F_1 plant of cross 3.

11-8. In man, as discussed on page 204, an example of suppression caused by homozygosity for a rare recessive gene (*h*) is known as the "Bombay" effect in which none of the ABO blood type antigens appear, and the individual is phenotypically of O blood type (no antigens). (a) In the following pedigree (blood type below each symbol), which blood types are most probably caused by the Bombay effect? (b) Explain whether the

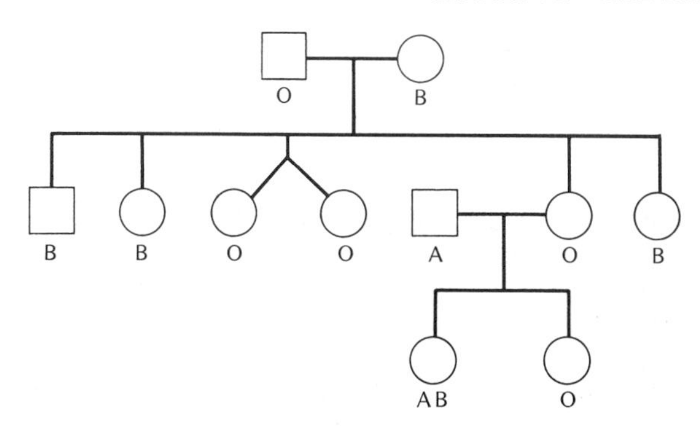

Bombay gene is or is not an allele of the other ABO genes (pedigree from Race and Sanger).

11-9. In the Japanese morning glory (*Pharbitis nil*), Hagiwara reported that purple flower color may be caused by dominant alleles at either one of two separate gene pairs, e.g., *A–bb* or *aaB–*. When dominant alleles are present at both gene pairs (*A–B–*), the flower color is blue, and when both are homozygous recessive (*aabb*), the color is scarlet. A blue F_1 was therefore produced by crossing two different purple types *AAbb* × *aaBB*. (a) What phenotypes and proportions would you expect from crossing this F_1 to either of the parental strains? (b) What phenotypes and proportions would you expect from an $F_1 \times F_1$ cross?

11-10. For each of the following crosses between different Japanese morning glories (Problem 9), give the genotypes of the parents.

Parental phenotypes	***Offspring***
(a) blue × scarlet	1/4 blue:1/2 purple:1/4 scarlet
(b) purple × purple	1/4 blue:1/2 purple:1/4 scarlet
(c) blue × blue	3/4 blue:1/4 purple
(d) blue × purple	3/8 blue:4/8 purple:1/8 scarlet
(e) purple × scarlet	1/2 purple:1/2 scarlet

11-11. Brewbaker found that inbred F_1 plants that had been derived from crosses between two strains of white-flowered "white clover" plants (*Trifolium repens*) gave rise to an F_2 generation of 5 red:82 white. No lethality was indicated. (a) Using the simplest explanation, how many gene pair differences are involved in these crosses? (b) Using alphabetical letters, define the alleles involved and give the genotypes of the red F_2 plants.

11-12. The *Gossypium hirsutum* experiment performed by Fuchs, Smith, and Bird (see Example 14, p. 211) was to cross a glandular strain (*AABB*) with a glandless strain (*aabb*), and then intercross the glandular F_1 (*AaBb*) with itself to obtain an F_2. In one of the experiments they performed, the F_2 data showed 89 glandular plants and 36 glandless plants. (a) Using the chi-square test, how well do these data fit the 11:5 proposed ratio? (b) Test whether these data exclude explanations using other known ratios, such as 15:1, 13:3, 9:7.

11-13. Two homozygous strains of yellow plants, W and Y, were each crossed to the same homozygous green strain, Z, and to each other, with the following results:

cross 1: P_1: yellow W × green Z
F_1 = all yellow
$F_1 \times F_1$ = 3/4 yellow:1/4 green
cross 2: P_1: yellow Y and green Z
F_1 = all green
$F_1 \times F_1$ = 3/4 green:1/4 yellow
cross 3: P_1: yellow W × yellow Y
F_1 = all yellow
$F_1 \times F_1$ = 13/16 yellow:3/16 green

(a) Do the alleles involved in all these crosses express dominance relationships that are complete or incomplete? (b) Using an alphabetical letter (beginning with A, B, C, . . .) for each gene pair, write the alleles that are *segregating* in cross 1 and the alleles that are *segregating* in cross 2.

On the basis of the symbols used in answering (b), write the complete genotypes for the following phenotypes: (c) Parental yellow W of cross 1. (d) Yellow F_1 of cross 1. (e) Parental yellow Y of cross 2. (f) Green F_2 of cross 1. (g) Yellow F_2 of cross 2. (h) Green F_2 of cross 3.

11-14. In the snapdragon, *Antirrhinum majus*, Baur crossed a white-flowered plant of normal flower shape (*personate*) to a red-flowered plant with peloric-shaped flowers. The F_1 of this cross was pink and normal-shaped. An $F_1 \times F_1$ cross, however, produced the following:

	Number of plants
pink normal	94
red normal	39
white normal	45
pink peloric	28
red peloric	15
white peloric	13

(a) Explain these results. (b) Test your hypothesis statistically.

11-15. In some plants, such as corn, there is a recessive gene on one chromosome which, when homozygous, produces white plant color, and also a dominant gene on another chromosome which produces white whether heterozygous or homozygous. What proportion of white plants will occur among the offspring of a self-fertilized plant that is heterozygous for both of these genes?

11-16. At the foot of this page is a pedigree for the appearance of a particular form of deafness in humans (shaded figures). Give the most likely genotypes for the numbered individuals, using capital letters for genes with dominant effect and small letters for recessives.

11-17. The leaves of pineapples can be classified into three types: spiny, spiny-tip, and piping (nonspiny). In crosses made by Collins and Kerns, the following results appeared:

Parent phenotypes	F_1 phenotype	F_2 phenotype
(a) spiny-tip × spiny	spiny-tip	3 spiny-tip : 1 spiny
(b) piping × spiny-tip	piping	3 piping : 1 spiny-tip
(c) piping × spiny	piping	12 piping : 3 spiny-tip : 1 spiny

Using alphabetical letters, define gene symbols and explain these results in terms of the genotypes produced and their ratios.

11-18. Using your analysis of the previous problem, give the phenotypic ratios you would expect if you crossed: (a) The F_1 progeny from piping × spiny to the spiny parental stock. (b) The F_1 progeny of piping × spiny to the F_1 progeny of spiny × spiny-tip.

11-19. In *Drosophila melanogaster* a gene *ebony* produces a dark body color when homozygous, and an independently assorting gene *black* has a similar effect. (a) What would be the appearance of the F_1 progeny from a cross between a homozygous black parent and a homozygous ebony parent? (b) What phenotypes and proportions would occur in the progeny of an $F_1 \times F_1$ cross? (c) What phenotypic ratios would you expect to find in a backcross between the F_1 and the ebony parental stock? (d) What phenotypic ratios would you expect from a backcross between the F_1 and the black parental stock?

11-20. When inbreeding various strains of summer squash, Sinnott and Durham found that lines of white-fruited plants may occasionally give rise to yellow or green plants, and that lines of yellow plants occasionally produce green plants, but that yellow or green lines never produce white-fruited plants nor do green lines produce yellow plants. After inbreeding these different lines to the point where they were all homozygous and bred true, crosses were made between the different lines, and the F_1 products were then crossed $F_1 \times F_1$ with the following results:

Cross	Parents	F_1	F_2
(a)	green × yellow	yellow	81 yellow, 29 green
(b)	white × yellow	white	155 white, 40 yellow, 10 green

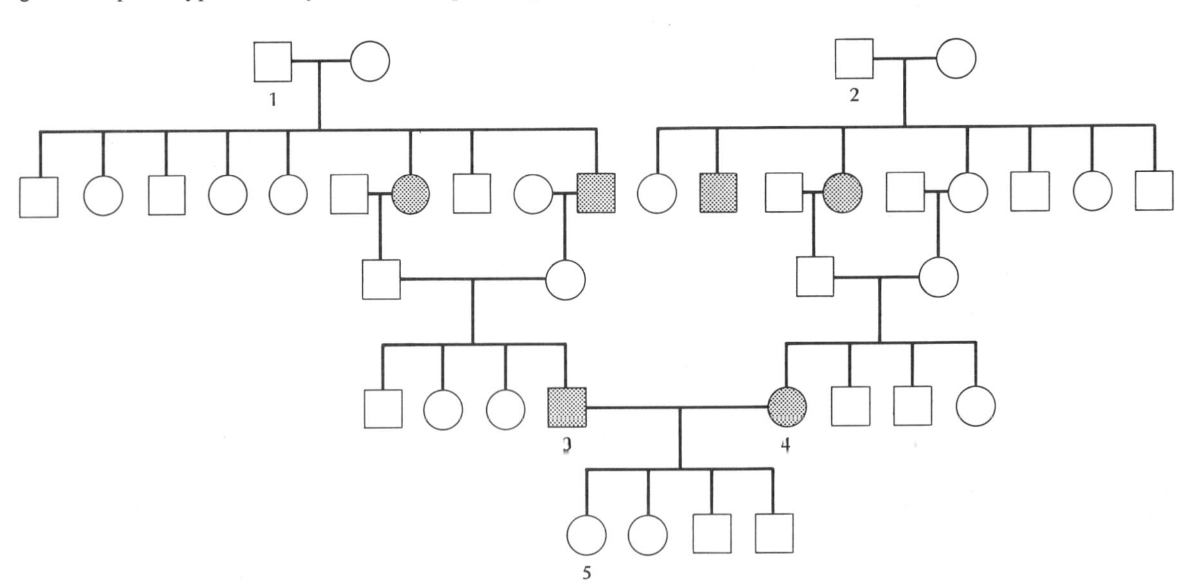

Define the symbols for the genes involved and explain these results.

11-21. Two dwarf corn plants, dwarf 1 and dwarf 2, each had a different origin but were phenotypically identical. When crossed with a variety of tall plants that were known to be homozygous for all genes determining size, both dwarf strains produced an F_1 of only tall plants. When any of these F_1 plants was self-fertilized, the F_2 appeared in the ratio 3 tall:1 dwarf. In crosses between the parental dwarf strains, dwarf 1 × dwarf 2, only tall plants appeared in the F_1. However, when the F_1 of this latter cross was self-fertilized, the proportion of dwarf plants that appeared in the F_2 was 7/16. (a) Define the gene symbols and explain these results in terms of the genotypes involved. (b) What phenotypic ratios would you expect if you backcrossed the F_1 from the latter set of crosses (dwarf 1 × dwarf 2) to the dwarf 1 parent? (c) To the dwarf 2 parent? (d) To the F_1 plants from the dwarf 1 × tall cross?

11-22. According to experiments by Murray with the snail *Cepaea nemoralis*, the gene producing pink shell color has an effect dominant to that for yellow shell color, and the gene for dark bands has an effect dominant to that for transparent bands. Murray found a snail of this species which had a cream shell color and transparent bands. When this animal was mated to a yellow snail with dark bands, the offspring were all pink with dark bands. In a cross between two of the F_1, only a single offspring was found, and this showed the original cream shell color and transparent bands. Define the gene symbols and suggest genotypes for each of the kinds of snails mentioned.

11-23. Three green-seeded plant strains, superficially identical in phenotype and designated X, Y, and Z, were individually crossed to homozygous yellow-seeded plants. The F_1 in each case were yellow and were self-fertilized to produce an F_2. Each green strain was then found to produce a different F_2 segregation as follows:

Cross 1: X produced an F_2 of 27/64 yellow:37/64 green
Cross 2: Y produced an F_2 of 3/4 yellow:1/4 green
Cross 3: Z produced an F_2 of 9/16 yellow:7/16 green

(a) How many gene pairs are segregating in cross 1? (b) In cross 2? (c) In cross 3?

11-24. If the genes determining green seed color in the various plant strains of the previous question assort independently of each other, answer the following: (a) What is the phenotype of the F_1 produced by crossing strain X to strain Y? (b) If this F_1 is self-fertilized, what proportion of the F_2 is expected to be green? (c) What is the phenotype of the F_1 produced by crossing strain X to strain Z? (d) If this F_1 is self-fertilized, what proportion of the F_2 is expected to be yellow?

11-25. In corn the dominant alleles of three independently assorting gene pairs, *Aa*, *Cc*, and *Rr*, must be present to produce seed (kernel) color. Another gene pair, called "purple" (*Pp*), can produce red kernel color when homozygous recessive *p/p*, or purple color when the dominant allele *P* is present.

If the F_1 derived from a cross between a red plant *AACCRRpp* and a colorless plant *AACCrrPP* were then inbred ($F_1 \times F_1$), what would be the ratios of the F_2 phenotypes and genotypes?

11-26. The normal eye color in *Drosophila melanogaster* is red, but it may be changed to purple by the presence of an autosomal gene, *pr*, in homozygous condition. This purple effect, however, may be suppressed by an independently segregating autosomal gene, *su-pr*, when the latter is in homozygous condition. In what proportions would you expect red and purple phenotypes to appear among the offspring of a cross between red-eyed flies heterozygous for both genes?

11-27. In foxes a *platinum* mutation arose in Norway in 1933 that appeared to be a dominant allele of the *silver* gene. The platinum mutant was then mated to silver animals to obtain an F_1. About 50 matings were then made between F_1 platinum and F_1 platinum which produced an F_2 of about 100 platinum foxes and 50 silver. How can this ratio be explained?

11-28. Asmundsen crossed two varieties of fowl and then intercrossed the F_1 among each other. A total of 763 F_2 eggs were raised, of which 716 produced normal chicks. In the other 47 eggs, however, all embryos were malformed and none reached the hatching stage. How would you explain these results?

11-29. Among the various coat colors in mice, there may be yellow, agouti, or black. Crosses made between certain mice provided the following results:

	Parents	F_1 offspring
cross 1	yellow × yellow	2/3 yellow:1/3 agouti (see p. 215)
cross 2	yellow × agouti	1/2 yellow:1/2 agouti
cross 3	yellow × black	1/2 yellow:1/2 black
cross 4	black × agouti	all agouti

What phenotypic ratios would you expect from the following crosses: (a) F_1 yellow (cross 1) × F_1 yellow (cross 2)? (b) F_1 agouti (cross 4) × F_1 yellow (cross 2)? (c) F_1 agouti (cross 4) × F_1 yellow (cross 3)?

11-30. Each member of each generation of a stock of *Drosophila* flies shows the two different phenotypic effects A and B, produced, respectively, by two different dominant genes, *A* and *B*. When this stock is crossed to wild-type *Drosophila*, the F_1 offspring are in the ratio 1/2 A:1/2 B, but A and B do not occur together in the same F_1 individual. Explain what you think accounts for this.

11-31. Pollen from a pure-breeding virescent tomato plant (yellowish because of a deficiency of chlorophyll) was used to fertilize a normal green plant. All the hybrids were normal green. When one of these hybrids was crossed with a virescent plant the progeny had 112 green and 72 virescent plants. (a) What do you think accounts for these results? (b) What further tests would you perform to check your hypothesis?

11-32. The gene for yellow coat color in the house mouse A^Y is dominant to the normal wild-type gene. The independently assorting gene for short tail *T* (*Brachyury*) is also dominant to the normal wild-type gene. Embryos that are homozygous for either or both of these dominant genes die before birth. (a) What phenotypic ratios will be expected to result from a cross between two yellow short-tail individuals? (b) If the normal litter number in mice is eight offspring, what average number would you expect to find in such crosses?

11-33. In *Drosophila melanogaster*, a dominant gene, *Hairless* (*H*), has a marked effect in reducing bristle number and is lethal in homozygous condition. An independently assorting gene, *Suppressor of Hairless* (*Su–H*), ordinarily has no effect on bristle number, except in the presence of *Hairless*. When they are present together, a single dose of *Su–H* will suppress the *Hairless* effect. In homozygous condition, *Su–H* is lethal. (a) What will be the phenotype of the F_1 produced by a mating between a stock carrying *Hairless* and a stock carrying *Su–H* (without *Hairless*)? (b) What phenotypic ratios would you expect from a backcross of this F_1 to the *Hairless* parental stock? (c) To the *Su–H* parental stock? (d) To a wild-type non-*Su–H* stock? (e) What phenotypic ratios would you expect from an $F_1 \times F_1$ cross?

11-34. If the chromosome carrying the *Suppressor of Hairless* gene in Problem 33 also carried a *Segregation Distorter* gene (see p. 219), what changes in phenotypic ratios would you expect among the offspring of the given crosses?

REFERENCES

BARTLETT, A. C., A. E. BELL, and D. SHIDELER, 1962. Two loci controlling body color in flour beetles. *Jour. Hered.*, **53,** 291–295.

BATESON, W., and R. C. PUNNETT, 1905, 1906, and 1908. Experimental studies in the physiology of heredity. *Reports to the Evolution Committee of the Royal Society, II, III, and IV.* Harrison and Sons, London. (Excerpts reprinted in the collection of Peters; see References, Chapter 1.)

BAUR, E., 1907. Untersuchungen über die Erblichkeitsverhältnisse einer nur in Bastardform lebenshäfigen Sippe von *Antirrhinum majus. Ber. Dtsch. Bot. Ges.*, **25,** 442–454.

BENNETT, D., E. GOLDBERG, L. C. DUNN, and E. A. BOYSE, 1972. Serological detection of a cell-surface antigen specified by the *T* (*Brachyury*) mutant gene in the house mouse. *Proc. Nat. Acad. Sci.*, **69,** 2076–2080.

BRADEN, A. W. H., 1958. Influence of time of mating on the segregation ratio of alleles at the T locus in the house mouse. *Nature*, **181,** 786–787.

BRIGGS, R. W., 1966. Recognition and classification of some genetic traits in maize. *Jour. Hered.*, **57,** 35–42.

CASTLE, W. E., and C. C. LITTLE, 1910. On a modified mendelian ratio among yellow mice. *Science*, **32,** 868–870.

CUÉNOT, L., 1905. Les races pures et leur combinaisons chex les souris. *Arch. Zool. Exp. et Genet.*, **3,** 123–132.

——— 1908. Sur quelques anomalies apparentes des proportions mendeliennes. *Arch. Zool. Exp. et Genet.*, **9,** 7–15.

DOBZHANSKY, TH., 1927. Studies on manifold effect of certain genes in *Drosophila melanogaster. Zeit. Induk. Abst. u. Vererbung.*, **43,** 330–388.

——— 1946. Genetics of natural populations. XIII. Recombination and variability in populations of *Drosophila pseudoobscura. Genetics*, **31,** 269–290.

DUNN, L. C., and D. R. CHARLES, 1937. Studies on spotting patterns. I. Analysis of quantitative variations in the pied spotting of the house mouse. *Genetics*, **22,** 14–42.

EATON, G. J., and M. M. GREEN, 1962. Implantation and lethality of the yellow mouse. *Genetica*, **33,** 106–112.

FORD, E. B., 1940. Genetic research in the *Lepidoptera. Ann. Eugenics*, **10,** 227–252.

FOSTER, M., 1965. Mammalian pigment genetics. *Adv. in Genet.*, **13**, 311–339.

FUCHS, J. A., J. D. SMITH, and L. S. BIRD, 1972. Genetic basis for an 11:5 dihybrid ratio observed in *Gossypium hirsutum*. *Jour. Hered.*, **63**, 300–303.

GLOOR, H., 1945. Zur Entwicklungsphysiologie and Genetik des Letalfaktors *crc* bei *Drosophila melanogaster*. *Arch. Jul. Klaus-Stiftg. Vererb-Forschung*, **20**, 209–256.

HADORN, E., 1961. *Developmental Genetics and Lethal Factors*. John Wiley, New York.

HARTL, D. L., 1973. Complementation analysis of male fertility among the segregation distorter chromosomes of *Drosophila melanogaster*. *Genetics*, **73**, 613–629.

HERSH, A. H., 1929. The effect of different sections of the X-chromosome upon bar eye in *Drosophila melanogaster*. *Anat. Rec.*, **68**, 378–382.

MAZUR, A., and B. HARROW, 1971. *Textbook of Biochemistry*, 10th ed. Saunders, Philadelphia.

MULLER, H. J., 1918. Genetic variability, twin hybrids and constant hybrids in a case of balanced lethal factors. *Genetics*, **3**, 422–499.

PEACOCK, W. J., and G. L. G. MIKLOS, 1973. Meiotic drive in *Drosophila:* New interpretations of the segregation distorter and sex chromosome systems. *Adv. in Genet.*, **17**, 361–409.

PUNNETT, R. C. 1911. *Mendelism*. Macmillan, New York.

ROBERTSON, G. G., 1942. An analysis of the development of homozygous yellow mouse embryos. *Jour. Exp. Zool.*, **89**, 197–231.

SANDLER, L., and E. NOVITSKI, 1957. Meiotic drive as an evolutionary force. *Amer. Naturalist*, **91**, 105–110.

SANDLER, L., Y. HIRAIZUMI, and I. SANDLER, 1959. Meiotic drive in natural populations of *Drosophila melanogaster*. I. The cytogenetic basis of segregation distortion. *Genetics*, **44**, 233–250.

SEARLE, A. G., 1968. *Comparative Genetics of Coat Colour in Mammals*. Academic Press, New York.

SHULL, G. H., 1914. Duplicate genes for capsule form in *Bursa bursa-pastoris*. *Zeit. Induk. Abst. u. Vererbung.*, **12**, 97–149.

SINNOTT, E. W., 1927. A factorial analysis of certain shape characters in squash fruits. *Amer. Naturalist*, **61**, 333–344.

SOKOLOFF, A., 1962. Linkage studies in *Tribolium castaneum* Herbst. V. The genetics of bar eye, microcephalic and microphthalmic and their relationships to black, jet, pearl, and sooty. *Canad. Jour. Genet. Cytol.*, **4**, 409–425.

STURTEVANT, A. H., 1956. A highly specific complementary lethal system in *Drosophila melanogaster*. *Genetics*, **41**, 118–123.

SUZUKI, D. T., 1970. Temperature-sensitive mutations in *Drosophila melanogaster*. *Science*, **170**, 695–706.

WENTWORTH, E. N., and J. L. LUSH, 1923. Inheritance in swine. *Jour. Agr. Res.*, **23**, 557–581.

WHITING, P. W., 1934. Mutants in Habrobracon, II. *Genetics*, **19**, 268–291.

WRIGHT, S., 1963. Genic interaction. In *Methodology in Mammalian Genetics,* W. J. Burdette (ed.). Holden-Day, San Francisco, pp. 159–192.

YANAGISAWA, K., L. C. DUNN, and D. BENNETT, 1961. On the mechanism of abnormal transmission ratios at T locus in the house mouse. *Genetics*, **46**, 1635–1644.

12
SEX DETERMINATION AND SEX LINKAGE IN DIPLOIDS

As a rule the two parental gametes that unite during fertilization are physically distinct from each other (i.e., sperm and eggs) and are formed by different organs (i.e., testes and ovary). In most plants and many lower animals these different organs are combined within single *hermaphroditic* or *monoecious* individuals, each capable of producing both types of sexual gametes. On the other hand, some plant species and most higher animals consist of separate male and female (*bisexual* or *dioecious*) individuals, that produce sperm or eggs but not both. If we presently restrict ourselves to instances in which sexual differences reside in separate diploid individuals, we can ask, how is sex itself inherited?

THE SEX CHROMOSOME

The inheritance of sex, as that of other characteristics, was long thought to stem from factors unrelated to material units of heredity. Legends attributed sex determination to phases of the moon, time of day during fertilization, wind direction, whether the right or left testis was involved, and other such causes. With the discovery of mendelism in 1900 the search for mechanisms of sexual inheritance shifted to more material and cytological events. This had been foreshadowed by the suggestion of Mendel in a letter to Nägeli that sex determination might follow the same segregational pathways as other inherited characteristics.

Soon after 1900 the possible association between the sex characteristic and the presence or absence of a particular chromosome was put forth in definite form by McClung (see p. 131). Observations made of certain diploid male insects had shown the presence of an "accessory" chromosome which stained darkly (heterochromatic) and occurred in half of the spermatozoa but was absent in the other half. McClung then pointed out that since the sex ratio of males to females in every generation is approximately 1:1, the chromosomal mechanism for determining sex would have to fulfill the following requirements:

1. The sex element should be located on a particular chromosome that behaves normally in the special cell divisions that form the gametes.
2. Until gametes are formed, all cells of the animal that determine sex (i.e., the male) should contain this element.
3. In the formation of gametes the number of spermatozoa that contain this element should be equal to the number of spermatozoa that do not. If the element is associated with maleness or femaleness, this distribution among the offspring would ensure a 1:1 sex ratio.

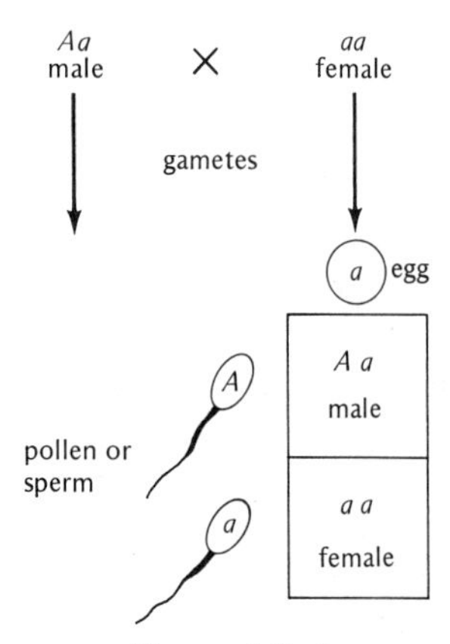

Figure 12–1

Backcross of a male heterozygous for a sex-determining factor (Aa) to a female homozygote (aa).

McClung, however, was not able to follow the presence of this accessory, or sex, chromosome in females and believed that only males contained the sex chromosome. Finally, in 1905, Wilson and Stevens showed that this chromosome occurred in both males and females. In the bug *Protenor*, for example, the male had one sex or "X" chromosome and the female had 2 X's. In meiosis each gamete formed by females contained one of the X chromosomes. The males, on the other hand, had an X chromosome in only half their gametes. Thus fertilization between male and female gametes always produced zygotes with one X chromosome from the female, but only 50 percent of the zygotes had an additional X chromosome from the male. In this way 1X and 2X individuals would be formed in equal proportions, the former being males and the latter females. The nonsex chromosomes, called *autosomes*, show no sexual differences and are therefore alike in number in both males and females.

Parallel to the cytological discovery of sex chromosomes by Wilson and by Stevens was the discovery by Correns in 1907 that the pollen grains of certain plants were sex-determining. According to Correns the eggs were sexually all of one type, whereas the pollen grains were of two kinds, equal in number. When pollen of one kind fertilized an egg, a male plant was produced; the other kind of pollen produced females. Graphically (Fig. 12–1) the scheme was similar to that of the mendelian "backcross," in which a heterozygote (i.e., male in this case) producing two kinds of gametes is crossed to a homozygous recessive (i.e., female) producing one kind of gamete, thereby forming 50 percent homozygotes (females) and 50 percent heterozygotes (males).

THE Y CHROMOSOME

Once different types of sex chromosomes could be distinguished, further investigations by Wilson, Stevens, Montgomery, and others showed a wide variety of chromosomal sex-determining mechanisms. The example of *Protenor* was essentially the simplest, because only the presence or absence of a single chromosome was the determining factor, and this was consequently called the XO–XX type. The male *Protenor*, forming two types of gametes, with and without the X chromosome, was designated the *heterogametic* sex. The female, producing gametes all of the same type, was termed *homogametic*.

It was soon found that the single X chromosome of the male in many species pairs in meiosis with another chromosome, called the Y. Although similar in appearance to the X in some cases, the Y is usually morphologically distinct. In *Drosophila melanogaster*, for example, the Y has a J shape, as compared to the rod-shaped X (Fig. 12–2). The XY–XX system thus results in the same number of chromosomes in both males and females. According to this terminology, if we designate a haploid set of autosomes collectively by the letter A, the male produces two types of sperm, X + A and Y + A, and the female one type, X + A. Fertilization may, therefore, be of two kinds:

$$\underset{\text{sperm}}{(X + A)} + \underset{\text{egg}}{(X + A)} = XX + 2A = \text{female}$$

$$\underset{\text{sperm}}{(Y + A)} + \underset{\text{egg}}{(X + A)} = XY + 2A = \text{male}$$

In *Drosophila melanogaster* there are three autosomes in each haploid set, yielding a total diploid number of eight chromosomes in each sex.

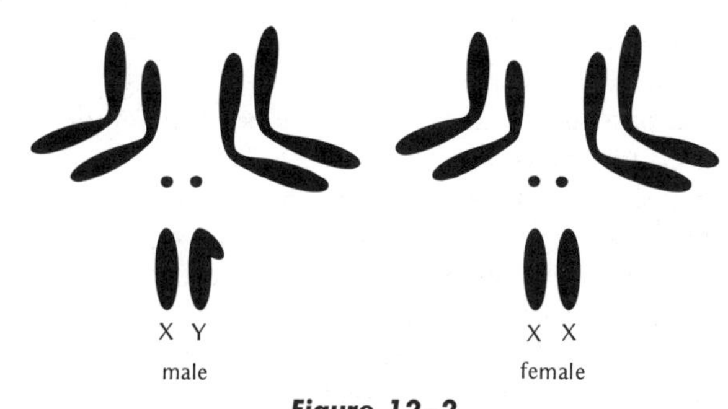

Figure 12–2

Chromosome constitutions (karyotypes) of males and females in Drosophila melanogaster.

COMPOUND SEX CHROMOSOMES

Although the X is most commonly found as a single chromosome or single homologous pair of chromosomes, some species maintain *compounds* of more than one kind of X chromosome that act together as a sex-determining group. In such species there may be large differences in the number of chromosomes between males and females. For example, in the nematode *Ascaris incurva* the number of autosome pairs is 13 but the number of compound X chromosomes is 8. A male therefore forms two types of gametes, 8X + A (21 chromosomes) and Y + A (14 chromosomes), while the female forms only the former type. On fertilization, the diploid number of chromosomes is therefore 35 in males and 42 in females:

$$\underset{\text{sperm}}{(8X + A)} + \underset{\text{egg}}{(8X + A)} = 16X + 2A = \text{female}$$

$$\underset{\text{sperm}}{(Y + A)} + \underset{\text{egg}}{(8X + A)} = 8X + Y + 2A = \text{male}$$

The Y chromosome may also exist as a compound group, and, in some instances, both compound X's and compound Y's may be found together in the same species. An extreme example of compound sex chromosomes occurs in the beetle *Blaps polychresta*, where the male has 12 X's and 6 Y's in addition to 18 autosomes. As a rule, however, the X and Y chromosomes of most species are in a single form and the male is the heterogametic sex.

SEX DETERMINATION

The almost universal presence of sex chromosomes among sexual species does not necessarily mean that these are the only chromosomes that affect sexual development. Since sex is a complex developmental character, it is usually affected by numerous autosomal genes as well. Indeed, in many cases, the potentiality for both male and female development exists at the time of fertilization no matter what the chromosomal complement. The function of the sex chromosome is to act as part of the "switch" mechanism that directs development into one of the possible paths that the organism is capable of following.

The influence of autosomal genes on sex is most dramatically illustrated in *Drosophila melanogaster* by the effect of the *transformer* gene discovered by Sturtevant. *Transformer*, or *tra*, is a recessive with no observable effect in heterozygous condition on either males or females. However, in homozygous condition, *tra/tra* individuals that would otherwise be female (XX) are completely transformed into phenotypic, but sterile, males. A gene of this type, but with opposite effect, occurs in humans and may also be autosomal. It causes the condition known as *testicular feminization* in which chromosomal XY individuals, who would otherwise be males, often have feminine characteristics such as breasts and vagina, yet maintain internal degenerate testes and are sterile.*

Other switch mechanisms are, of course, also possible. In bees, for example, it was known from the discoveries of Dzierzon at the time of Mendel (see p. 8) that fertilization of an egg affected its sex. Later studies then showed that sex appeared to be determined by the number of sets of chromosomes a bee receives. Fertilized eggs produce diploid females, while unfertilized eggs divide asexually (*parthenogenesis*) to produce haploid but fertile males. The male bees produce sperm by mitosis rather than meiosis, yielding one functional sperm per spermatocyte. The cycle then continues when the queen bee uses this sperm to produce females, and also lays some unfertilized eggs to produce males.

In another hymenopteran, the wasp *Bracon hebetor*, the males are also haploid and the females diploid. Since the males are believed to arise exclusively from unfertilized eggs, paternal characteristics, as in bees, would not be expected in male offspring. Whiting, however, found that homozygous orange-eyed females (*o/o*) mated to wild-type black-eyed males (+) produced not only orange-eyed sons (*o*) and black-eyed daughters (*o*/+) but also some black-eyed sons. These exceptional males were mostly sterile and turned out to be diploid (*o*/+). Their origin was puzzling until sex determination in this species, as in many other Hymenoptera, was found to be based on a multiple-allelic system rather than exclusively on diploidy or haploidy.

According to Whiting, there are nine alleles of a particular *X* gene in this wasp, called *Xa*, *Xb*, *Xc*, etc. A combination between any two different alleles produces a female (i.e., *Xa/Xb*, *Xa/Xc*, *Xb/Xc*, etc.). On the other hand, the presence of either a single allele in haploid condition (*Xa*, *Xb*, etc.) or of a homozygote in diploid condition (*Xa/Xa*, *Xb/Xb*, etc.) produces a male. Since most matings are between unrelated individuals, e.g., *Xa* × *Xb/Xc*, the diploids produced are all female. Some matings, however, such as those observed above for the orange-eyed cross, may occur between related individuals, e.g., *Xa* × *Xa/Xb*, and thereby produce a significant number of diploid but mostly sterile males. In honey bees diploid males homozygous for sex-determining genes may also be formed from matings between a queen bee and a related male, but such larvae are usually cannibalized by adult female workers (see review by Rothenbuhler et al.).

In contrast to genetic mechanisms, environmentally dependent sex-determining mechanisms are known in the sea worms *Bonnelia* and *Dinophilus* and in the horsetail plant, *Equisetum*. In *Bonnelia*, larvae that are free-swimming and settle on the sea bottom develop into females, each with a long proboscis. Larvae that land on the female proboscis develop into tiny males that lack digestive organs and exist in parasitic fashion in the genital ducts of the female. In *Dinophilus* the sex-determining mechanism seems to be related to the size of the eggs; large eggs produce females, small ones males. *Equisetum* will show female characteristics when raised under good growth conditions and male characteristics when raised under poor

* D. D. Federman (1973, *Progr. Med. Genet.*, **9**:215–235) suggests that this defect may arise through the inability of the male hormone, testosterone, to attach to nuclear chromatin and exert its normal regulatory functions in turning off some genes and turning on others. [Perhaps because of the malfunction of a hormone receptor (p. 717).] Thus, development of the male Wolffian duct system (seminal vesicles, vasa deferentia, and epididymides) does not occur in testicular feminization, and the common estrogen secretions that produce female secondary characteristics are not counteracted by male gene products normally induced by testosterone.

conditions. In some plants considerable changes in sex expression can occur through changes in day length and temperature which affect various hormonal and chemical processes. For example, the appearance of female flowers in the cucumber (*Cucumis sativus*) and the muskmelon (*C. melo*) has been shown to be highly correlated with production of the chemical ethylene (Byers et al.).

Other factors that affect sex determination will be discussed later in this chapter as well as in Chapter 21. In general, however, the most common forms of sex-determining mechanisms in diploids seem to be chromosomal, associated with differences between two types of sex chromosomes, X and Y.

MEIOTIC BEHAVIOR OF SEX CHROMOSOMES

When the X chromosome of the heterogametic sex is accompanied by a Y, they usually act as pairing mates in cell division. Even when the Y is morphologically distinct, the occurrence of pairing between X and Y is considered evidence that each chromosome has a mutual *pairing segment*. This pairing segment may vary in size from a very small area, in which very little contact occurs between the X and Y, to almost the entire length of both chromosomes. The unpaired area is known as the *differential segment*. Presumably the pairing segment represents that section of the X and Y chromosomes which is homologous and in which meiotic crossing over can occur if this segment is long enough.

In the first "reductional" division the X and Y chromosomes separate, each going to an individual pole. Thus, when the second meiotic division occurs, the haploid gametes that are formed consist of equal numbers of X- and Y-containing cells. In species with compound X chromosomes, the first meiotic division also separates the Y from the multiple X group, and gamete formation is essentially the same as for single X chromosomes (Fig. 12–3). Compound Y chromosomes may, of course, act similarly.

In XO–XX species, the X chromosome does not have a pairing mate during meiosis in the heterogametic sex and, in many cases, precedes the autosomes to one of the poles during the first meiotic division. It then divides at the second meiotic division, thereby forming two X-bearing gametes out of four, or a 1:1 ratio of male- to female-determining gametes. Occasionally, as in *Protenor*, the

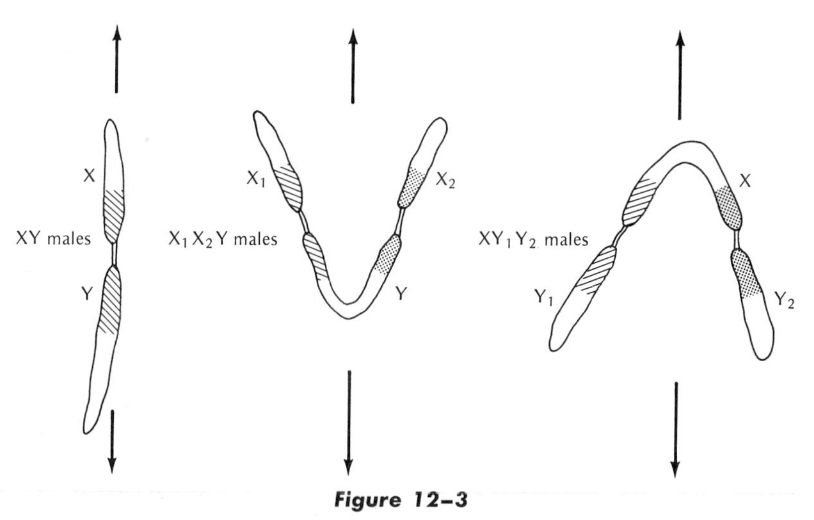

Figure 12–3

Reductional division of sex chromosomes in males with different sex chromosome constitutions. Hatched and shaded areas represent pairing segments.

first meiotic division is equational for the X chromosome, but this, too, leads to a 1:1 ratio of X:O gametes.

NONDISJUNCTION

Occasionally homologous chromosomes do not separate or "disjoin" during meiosis into individual gametes. This phenomenon is called *nondisjunction* and appears to arise from the lack of proper pairing between homologous chromosomes during the metaphase period. When this occurs the centromeres of the two homologues are not on opposite sides of the metaphase plate and the two unpaired homologues then separate randomly, going either to opposite poles or to the same pole. The migration of both sex chromosomes to the same pole during a meiotic division may lead to the presence of both chromosomes in one gamete and none in another.

Males

1st meiotic division

2nd meiotic division

XX YY

primary spermatocyte

XX YY

secondary spermatocytes

X + X + Y + Y

sperm

Females

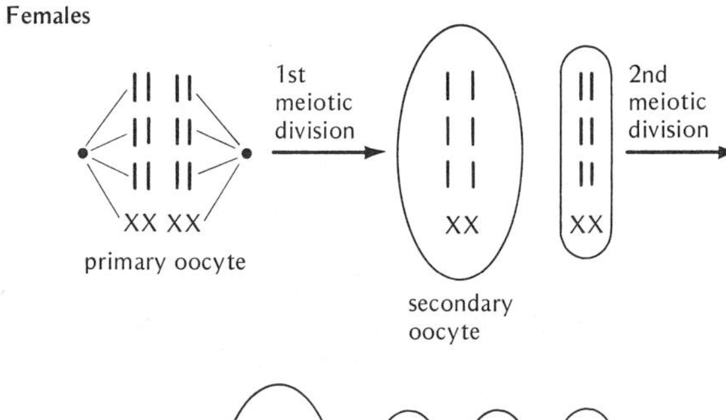

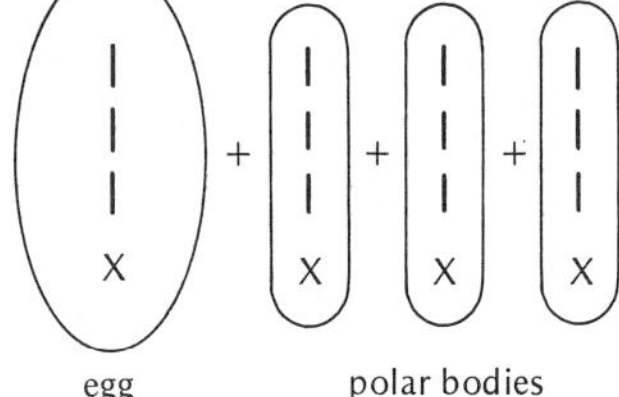

Figure 12–4

Normal disjunction for the sex chromosomes during meiosis in males (XY) and females (XX).

For example, in a species in which the males are XY and the females XX, normal disjunction will produce the kinds of gametes that are diagrammatically presented in Fig. 12–4. If nondisjunction occurs in males in the first meiotic division, the X and Y chromosomes will then go to the same pole and form two XY and two O ("nullo-X") gametes during the second meiotic division (Fig. 12–5a). Nondisjunction in the second meiotic division has various alternatives and may produce XX, YY, and O sperm, depending on which sex chromosome "nondisjoins" (Fig. 12–5b). In females nondisjunction in either the first or second meiotic divisions produces XX or O eggs (Figs. 12–5c, d).

SEX LINKAGE

The relationship between sex determination and the presence of particular chromosomes, as discussed above, was discovered in the early twentieth century, shortly after the rediscovery of mendelism. An equally important and fundamental problem was the relationship between sex and various nonsexual characteristics of an organism.

Kölreuter, who had pioneered in plant-hybridization experiments in the seventeenth century,* pointed out that it made little difference in the appearance of an F_1 hybrid whether the male or female parent was of one variety or another. This finding was taken to mean that although varieties might differ considerably in many characters, these characters entered the F_1 equally from both sides regardless of parental sex. Numerous exceptions

*See Voeller's collection, References Chapter 1.

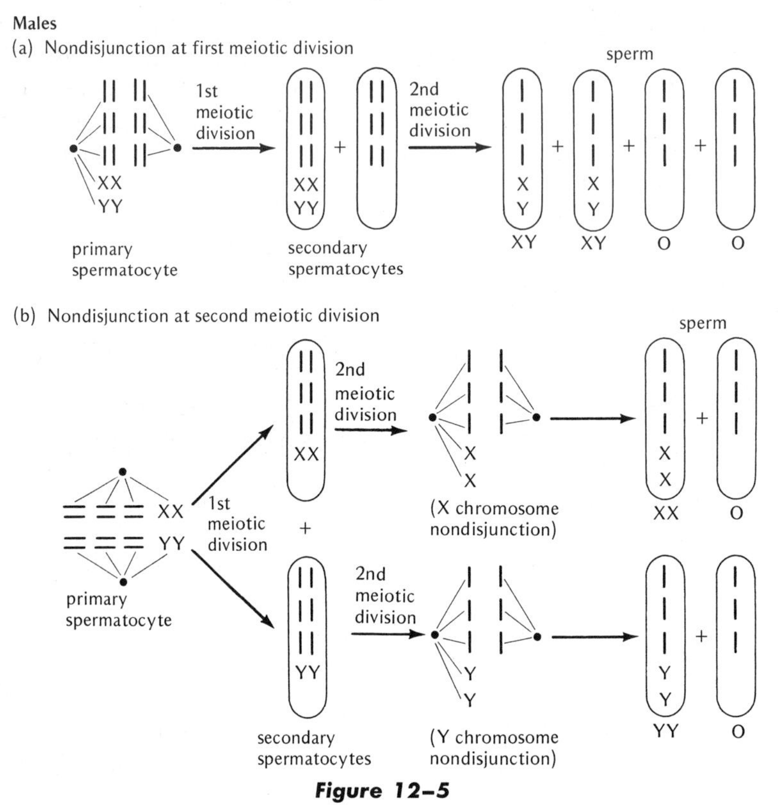

Figure 12–5

Effects of nondisjunction on the distribution of sex chromosomes in the first and second meiotic divisions of males (a, b) and females (c, d).

to this rule existed, however, or were soon found. One of the earliest known instances of the "linkage" of a specific character to sex was that of "bleeding," or hemophilia, a disease in which the blood fails to clot normally. The Jews, in their practice of circumcision of male infants, had come across instances where bleeding could not be stopped. According to the Jewish Talmud (the book of Yebamoth, written before A.D. 600), this was regarded as a hereditary defect if it occurred in two children of the same mother. The law then excused further offspring of this female from circumcision. In an instance where three sisters produced male children that bled upon circumcision, male offspring of the fourth sister were excused from this ritual. No restriction of circumcision was made, however, for the children of any of the male relatives involved.* Implicit in this law, therefore, was the recognition that the bleeding defect was carried by females, although only the males appeared to be affected.

Although other traits of this kind were suspected before 1900, the actual experimental evidence for sex linkage had to await observations that followed the rediscovery of Mendel's work. In 1910 T. H. Morgan presented clear-cut evidence that a specific character in *Drosophila melanogaster*, white eyes, was linked to the inheritance of sex and most likely associated with a particular chromosome, X. According to Morgan's data, a white-eyed male had appeared in a culture of normally red-eyed flies and was then bred to its red-eyed sisters. The

*I am indebted to Rabbi J. Gniwesch for this information.

Females

(c) Nondisjunction at first meiotic division

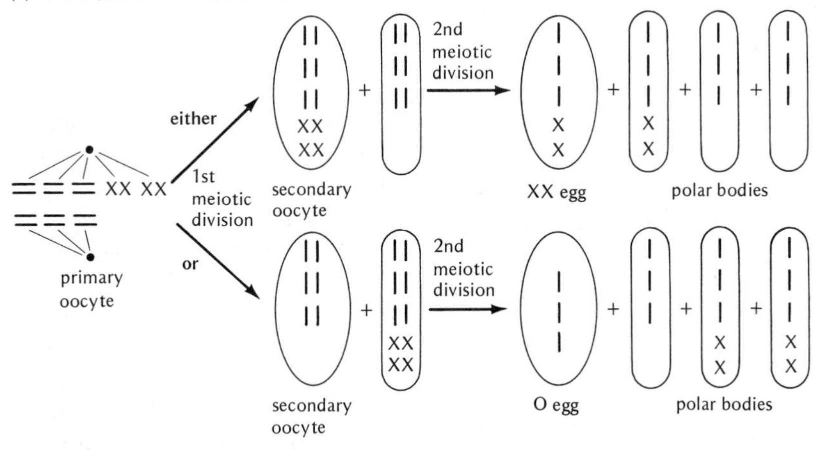

(d) Nondisjunction at second meiotic division

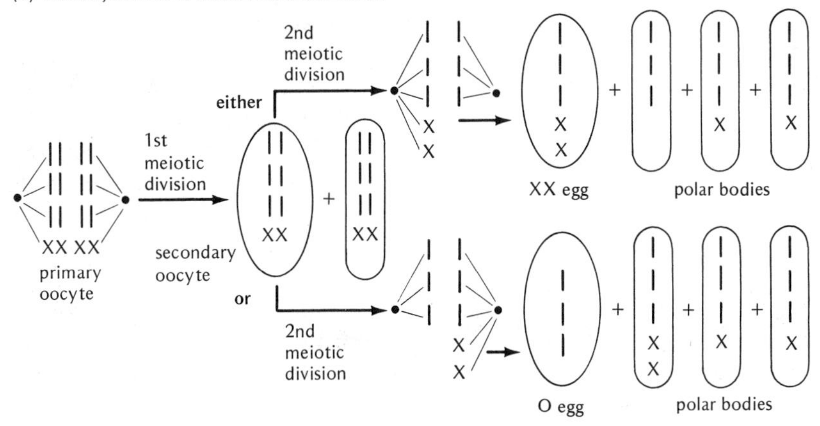

Figure 12–5 *(continued.)*

F_1 of this cross were red-eyed and when crossed among themselves produced red-eyed and white-eyed males, as well as red-eyed females, *but no white-eyed females*. White-eyed females (along with red-eyed females and white-eyed and red-eyed males) did, however, appear when F_1 red-eyed females were backcrossed to their white-eyed male parent.

To explain these results Morgan proposed that the F_1 red-eyed females were heterozygous for the white-eyed character, which was recessive. Thus white-eyed females could only occur when they were homozygous for the *white* gene. A male offspring, on the other hand, had an equal chance of being red-eyed or white-eyed if his mother was heterozygous for the *white* gene, no matter what the genotype of his father. These results all fall into neat order by considering them in the light of the well-substantiated findings of Wilson, Stevens, and others that the male had only one X chromosome; the female, two. Accordingly, the original white-eyed male had the *white* gene located on its single X chromosome. Although *white* is recessive, this male appeared as white-eyed because of the absence of an additional normal X chromosome. In females, however, eye color depends upon the genes carried in both its X chromosomes, and white eyes can only appear when

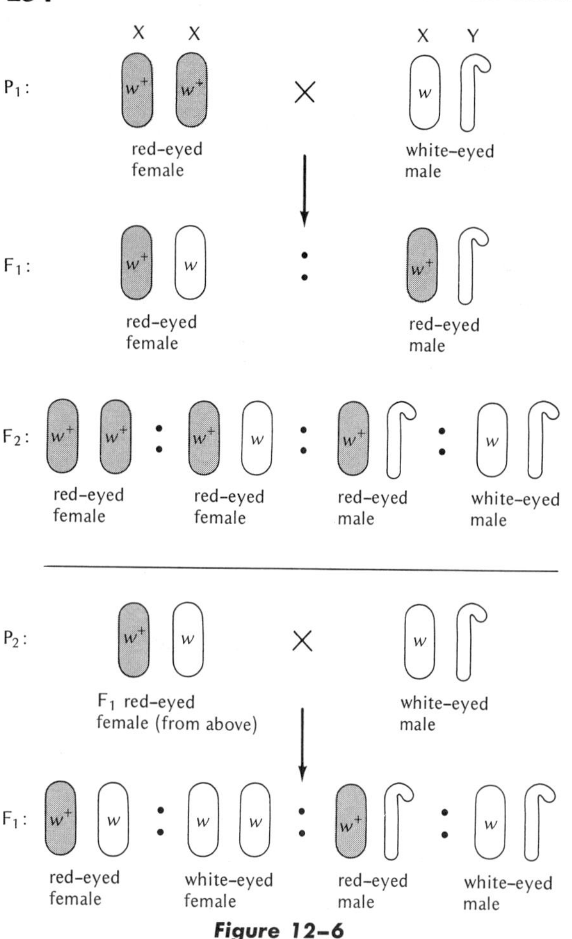

Figure 12–6

Diagram of sex linkage of the white gene demonstrated in Morgan's experiments.

both the X chromosomes carry the recessive *white* gene.

Diagrammatically Morgan's crosses are presented in Fig. 12–6, with the white-eye color gene, *w*, located on the X chromosome. Since the Y chromosome does not contain any eye color genes, a single *w* allele is expressed in the male (by the phenomenon of pseudodominance; see p. 496). Therefore practically all sex-linked genes, whether dominant or recessive, make their phenotypic appearance in males which are considered *hemizygous* for sex-linked genes. In the females, of course, the two X chromosomes provide for normal dominant-recessive relationships, and two of the recessive *white* alleles are necessary for white eyes to appear. Note that males inherit their X chromosomes only from their mothers, whereas females obtain an X chromosome from each parent. Thus the X chromosome follows "crisscross" transmission, beginning, for example, with a P_1 male, passing to the F_1 daughter, and thence to the F_2 sons.

If Morgan's hypothesis were correct, a series of further crosses should substantiate it, as follows:

1. A cross between white-eyed males and white-eyed females should produce only white-eyed male and female offspring.
2. The red-eyed F_1 females of Morgan's cross are heterozygous for the *white* allele and should always produce about 50 percent white-eyed male offspring in matings to either red-eyed or white-eyed males as well as 50 percent white-eyed females when mated to white-eyed males.
3. When mated to a white-eyed female, the red-eyed F_1 males, although descended from a white-eyed father, should produce only red-eyed daughters and white-eyed sons.
4. The F_2 red-eyed females in his experiments should be of two kinds: homozygotes for red eye, and heterozygotes. When crossed individually to white-eyed males, the difference between these two types of females should be observed in the production of only red-eyed offspring in the former case and of both white-eyed and red-eyed males and females in the latter case.

Each of these crosses was experimentally performed by Morgan, and the results clearly supported the sex-linkage hypothesis for the *white* gene. The *white* gene was thus a *marker* for the X chromosome.*

*A piece of genetic material that bears or produces a distinctive feature is generally considered to be a *marker*. The marker is usually a mutant gene, and can be either dominant or recessive. Some authors use the terms *mutant gene* and *marker* interchangeably. In referring to chromosomes, a *marker chromosome* can also be considered as a distinctive gene arrangement such as an inversion that carries a particular identifying mutation (for example, the *Curly* and *Plum* chromosomes in Fig. 11–7, and the Muller-5 chromosome in Fig. 23–1).

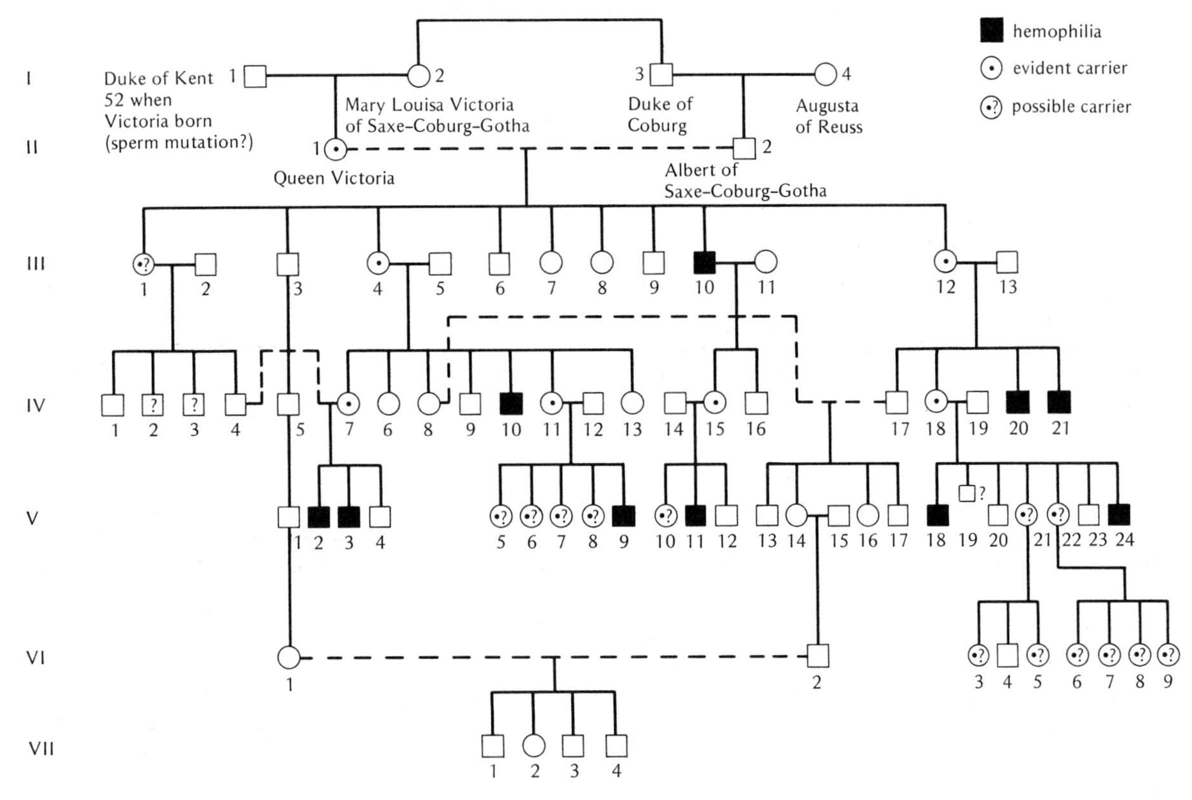

Figure 12–7

Pedigree of Queen Victoria and some of her descendants. Although inbreeding has occurred at various stages of the pedigree (dashed-line connections between symbols) the transmission of hemophilia is clearly sex-linked. Note that individuals VII–1 to VII–4 (present English royal family) are free of the hemophilia gene in spite of inbreeding. (From Textbook of Human Genetics *by Max Levitan and Ashley Montagu. Copyright © 1971 by Oxford University Press, Inc. Reprinted by permission.*)

III–1, Princess Victoria, wife of Emperor Frederick III of Germany (III–2);
III–3, King Edward VII of England;
III–4, Princess Alice, wife of Grand Duke Ludwig IV of Hesse-Darmstadt (III–5);
III–10, Prince Leopold, Duke of Albany, died, age 31, of hemorrhage after a fall;
III–12, Princess Beatrice, wife of Prince Henry Maurice of Battenberg (III–13);
IV–1, Kaiser Wilhelm II of Germany;
IV–2 and IV–3, Prince Sigismund (d. age 2) and Prince Waldemar (d. age 11) of Prussia;
IV–5, King George V of England;
IV–7, Princess Irene of Hesse, wife of Prince Henry of Prussia (IV–4);
IV–8, Princess Victoria of Hesse, wife of Prince Louis Alexander of Battenberg (IV–17), founder of English Mountbatten family;
IV–10, Prince Friedrich of Hesse, died as a child of hemorrhage after a fall;
IV–11, Princess Alix, later Queen Alexandra, wife of Tsar Nicholas II of Russia (IV–12);
IV–15, Princess Alice, wife of Alexander, Prince of Teck (IV–14);
IV–18, Princess Victoria Eugenié of Battenberg, wife of King Alfonso XIII of Spain (IV–19);
IV–20, Prince Leopold of Battenberg, died, age 33, presumably of hemorrhage, after surgery;
IV–21, Prince Maurice of Battenberg, died, age 23, in Battle of Ypres;
V–1, King George VI of England;
V–2, Prince Waldemar of Prussia, lived to be 56 but had no children;
V–3, Prince Henry of Prussia, died age 4;
V–9, Tsarevitch Alexis of Russia, executed 1918 at age 13;
V–11, Rupert, Lord Trematon, died of hemorrhage following auto accident;
V–14, Alice Mountbatten, married to Prince Andrew of Greece (V–15);
V–18, Alfonso Pio, Prince of Austurias, died age 31, of hemorrhage after auto accident;
V–24, Prince Gonzalo of Spain, died age 20, of hemorrhage after auto accident;
VI–1, Queen Elizabeth II of England, married to Prince Philip Mountbatten, Duke of Edinburgh (VI–2), slightly more closely related than third cousins;
VII–1, Prince Charles of England.

The sex-linked mode of inheritance for some human traits was therefore readily explained. Hemophilia, for example, had been carefully pedigreed in European royalty and generally showed transmission through females who were unaffected to some of their male offspring who were hemophilic. A dramatic lineage of this kind, as shown in Fig. 12–7, began with Queen Victoria of England, who passed the hemophilic gene on to at least three of her nine children. Since about half her sons were hemophilic, she was undoubtedly heterozygous for this gene, carrying it on one of her X chromosomes. Note that, as in most such sex-linked pedigrees, the females are the unaffected carriers of the disease; the disease appears in the sons of such carriers although their mating partners are normal; and the males who are free of the disease do not pass it on.

The finding that it was always the males who showed hemophilia but not the females led to the general belief that females bearing two hemophilic genes on their X chromosomes (and therefore producing only hemophilic sons) were probably inviable. In 1950, however, Brinkhous and Graham showed that hemophilia among dogs is also sex-linked and, through appropriate crosses, hemophilic but viable homozygous females could be produced. Since then hemophilic human females have been located, although their frequency is quite rare, as might be expected from the low frequency of the hemophilia gene (see Chapter 31).

At present, as a result of the discoveries of Aggeler, Biggs, Rosenthal, and others, three general types of hemophilia are known, each affecting the production of thromboplastin, a substrate necessary for blood clotting (Fig. 12–8):

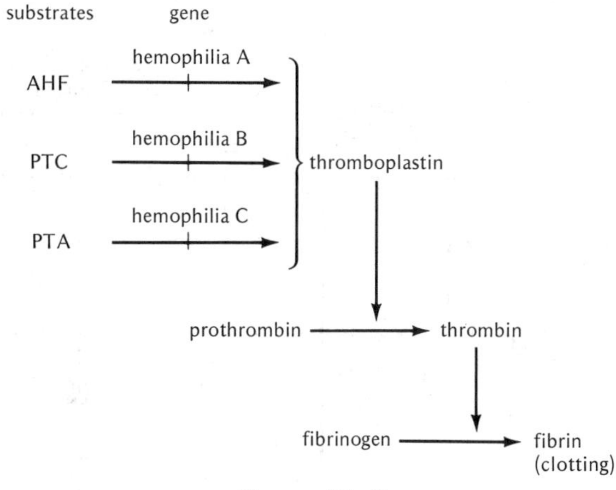

Figure 12–8

Diagram of blood-coagulation mechanism as affected by hemophilias A, B, and C. (Modified from Aggeler, Hoag, Wallerstein, and Whissell.)

1. The "classic" sex-linked type, or hemophilia A, which can be detected by a reduction in the amount of a substance known as *antihemophilic factor* (AHF or factor VIII), accounts for approximately 80 percent of known hemophilia.

2. Another sex-linked type, hemophilia B, or "Christmas disease," is caused by a reduction in the amount of *plasma thromboplastin component* (PTC) and accounts for about 20 percent of hemophiliacs. The gene for hemophilia B is not allelic to that of hemophilia A.

3. A rare autosomal gene, causing hemophilia C, interferes with the production of *plasma thromboplastin antecedent* (PTA), and is responsible for less than 1 percent of hemophiliacs. Mild forms of these three hemophilias are also known in which the concentrations of these necessary substrates are reduced but not completely absent. These may be caused either by alleles with milder effects or by modifiers at other genes or by both.

Sex linkage has now been investigated in a wide variety of animals and involves many traits. In humans sex-linked traits include colorblindness, ocular albinism and Xg blood type. Interestingly some of the same traits are sex-linked in other mammals. For example, production of certain forms of glucose-6-phosphate dehydrogenase (an enzyme that removes a hydrogen atom from phosphorylated glucose sugars) is sex-linked in man, horse, donkey, European hare, and mouse. Not only hemophilia A, but also hemophilia B, is sex-linked in both man and dog. Some coat color traits, such as "orange," are sex-linked in both the cat and the hamster. Findings of this kind indicate that many mammals have shared at least part of the same X chromosome during their evolution.

BRIDGES' DEMONSTRATION OF NONDISJUNCTION

Morgan's hypothesis that *white* was a sex-linked character located on the *Drosophila* X chromosome was cytologically substantiated by Bridges in 1916. Bridges noted that white-eyed females crossed to red-eyed males occasionally produced exceptional offspring (about 1 per 2000 flies) that were phenotypically exactly like their parents rather than the red-eyed females and white-eyed males normally expected from such a cross. Although the details of this phenomenon were not known at the time, Bridges showed that nondisjunction (p. 231) of the two X chromosomes must have been responsible for the observed results. As diagrammed in Fig. 12–9, white-eyed female parents occasionally produced exceptional eggs containing both X chromosomes and nullo-X eggs.

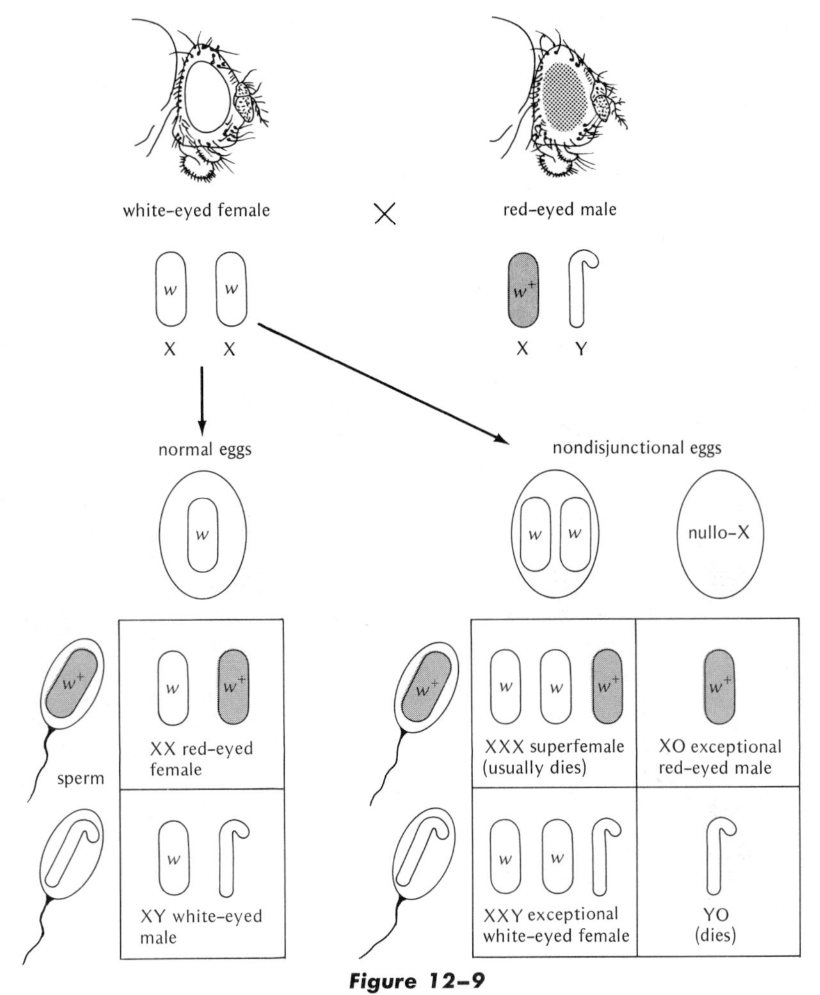

Figure 12–9

Origin of exceptional types in Drosophila melanogaster *through nondisjunction of X chromosomes in white-eyed females. (After E. W. Sinnott, L. C. Dunn, T. Dobzhansky, 1958. Principles of Genetics, 5th ed. McGraw-Hill, New York.)*

Fertilization of the eggs with 2X chromosomes by the male X and Y gametes could lead to two types of zygotes, XXX or XXY females. The XXX female, or "superfemale," is quite weak and usually does not emerge from the pupal case. The XXY female is fertile and quite normal in appearance, except that it has white eyes because both X's are of maternal origin. Two types of male offspring are also produced when the nullo-X eggs are fertilized by the male sperm: the first is an XO male with a paternal X chromosome and therefore red-eyed, and the second type inherits only a paternal Y chromosome and no X chromosomes. The XO male is fully viable but sterile, whereas the YO male is usually inviable and dies in the egg stage.

If this interpretation were correct, the exceptional XXY females can, in turn, produce further nondisjunctional gametes containing two X

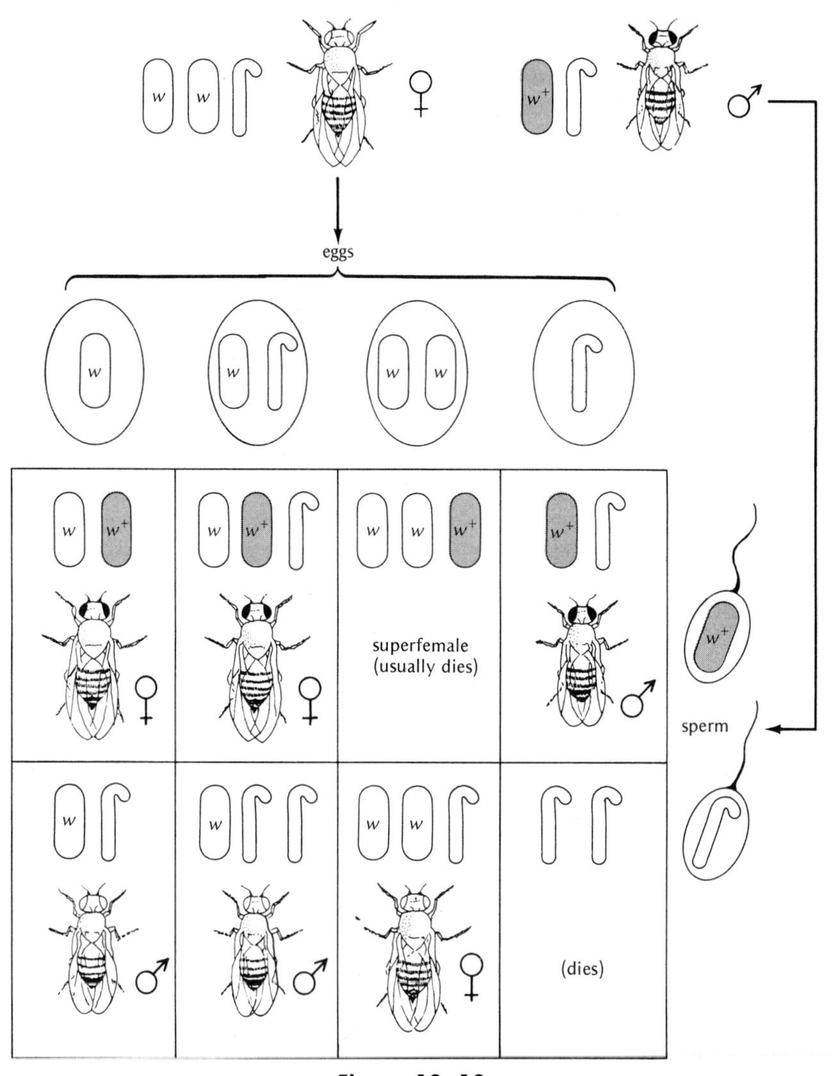

Figure 12–10

Results of secondary nondisjunction in a cross between an XXY white-eyed female and an XY red-eyed male. (After Sinnott, Dunn, and Dobzhansky.)

chromosomes in some eggs and a single Y chromosome in others, as well as the expected XY and X eggs. In contrast to the original nondisjunction in normal females, called *primary nondisjunction*, this derived type was called *secondary nondisjunction*, and proved to be much more frequent than the primary type. XXY white-eyed females mated to normal red-eyed males produced about 4 percent offspring that were exceptional, that is, derived from XX eggs or Y eggs. As shown in Fig. 12–10, eight classes of offspring were actually produced, including white-eyed XYY males who proved to be fertile. The unique chromosome constitution of the exceptional XO, XXX, XXY, and XYY individuals were all identified under the microscope by Bridges, thereby demonstrating the first direct correlation between chromosomal and genetic inheritance. Bridges later demonstrated that nondisjunction can occur for other sex-linked characters as well as for *white*.

ATTACHED-X

A few years after Bridges' paper, L. V. Morgan (T. H. Morgan's wife), discovered a strain of flies in which 100 percent nondisjunction occurred. This strain consisted of females that, when mated to males carrying sex-linked mutant genes, always produced female offspring exactly like their mother and males exactly like their father. For example, a white-eyed male crossed to these females always produced white-eyed sons but normal daughters. This inheritance pattern suggested that the female offspring of these crosses received both X's from their mother, whereas the males received a single X directly from their fathers. Eventually stocks were made up in which all such nondisjunctional females were white-eyed and the males were red-eyed. That is to say, each female transmitted its two *white*-bearing X chromosomes only to its daughters and the male transmitted its normal X chromosome only to its sons.

The production of only nondisjunctional offspring by females of this strain was explained as resulting from a physical attachment between both maternal X chromosomes. Similar to Bridges' XXY female, L. V. Morgan's females were also XXY, but the X's were so connected that they were inherited as a unit ("attached X," or $\widehat{XX}$). Such females therefore produced only two types of gametes, $\widehat{XX}$- and Y-bearing eggs. Fertilization of the $\widehat{XX}$ eggs by Y sperm produced fertile attached-X females, whereas fertilization of such eggs by X-bearing sperm produced XXX females that usually died before emergence. The Y-bearing eggs produced inviable YY males when fertilized by Y sperm but produced normal males when fertilized by X-bearing sperm. In this fashion, all female offspring received the maternal X chromosomes and the paternal Y, while the reverse situation occurred in male offspring. Thus sex-linked characters such as white eyes could be directly transmitted from mother to daughter or father to son, as shown in Fig. 12–11. The cytological evidence for this phenomenon soon followed in the observation that the two X chromosomes in these exceptional females were physically connected at the centromere region. This striking correlation between the inheritance of particular genes and particular chromosomes offered further evidence for the correctness of the chromosome theory of heredity.

SEX LINKAGE IN MOTHS AND BIRDS

Preceding Morgan's white-eye *Drosophila* experiment by a few years, Doncaster and Raynor in 1906 had already noted the sex linkage of a body-color character in the currant moth *Abraxas grossulariata*. They were not, however, able to associate this character with a specific sex chromosome as Morgan did, since the sex-chromosome mechanism in moths was not cytologically determined until 1913. In this moth a light-colored variety called *lacticolor* occurs which can be crossed to the darker *grossulariata*. In a cross between a *grossulariata* male and a *lacticolor* female, the offspring were always of the darker variety, while the reciprocal cross between *grossulariata* females and *lacticolor* males produced *laticolor* females and *grossulariata* males.

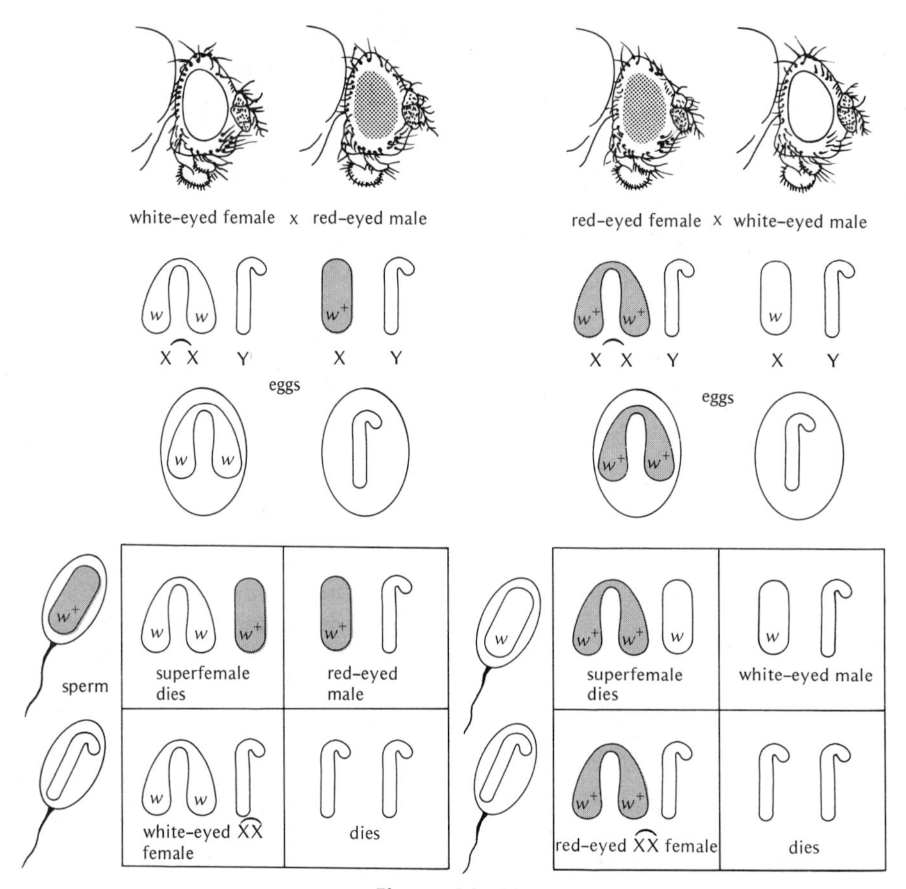

Figure 12–11

Consequences of mating an attached-X Drosophila *female to a male that differs from it in a sex-linked trait. A chromosome consisting of two homologous chromosome arms connected together by a common centromere is called an* isochromosome *(see p. 476).*

In chromosomal terms we now understand that in moths and butterflies (*Lepidoptera*) the female is heterogametic XO or XY and the male is homogametic XX. Since the observed body-color character in *Abraxas* is sex-linked, the phenotypes of female offspring in these crosses are always similar to the paternal parent from whom they obtain their single X chromosome. On the other hand, each male in the F_1 obtains X chromosomes from both parents, and the dominant *grossulariata* therefore hides the *lacticolor* effect. Spillman, in 1909, found a similar pattern for the sex-linked inheritance of barred feather color in poultry, males being XX and females XO (Fig. 12–12).

However, in addition to simple sex-chromosome ratios, other sex-determining factors may, of course, be involved in female heterogametic species. One such example, studied extensively by Goldschmidt, is that of the gypsy moth, *Lymantria dispar*. Goldschmidt found that, ordinarily, crosses between gypsy moths of the same geographic locality yielded expected results. That is, male (XX) × female (XY) produced normal sex ratios of male and female offspring. However, crosses between geographically separate strains yielded, in some cases, unusual progeny. For instance, crosses between females from Korea to males from Tokyo produced normal males but also produced females

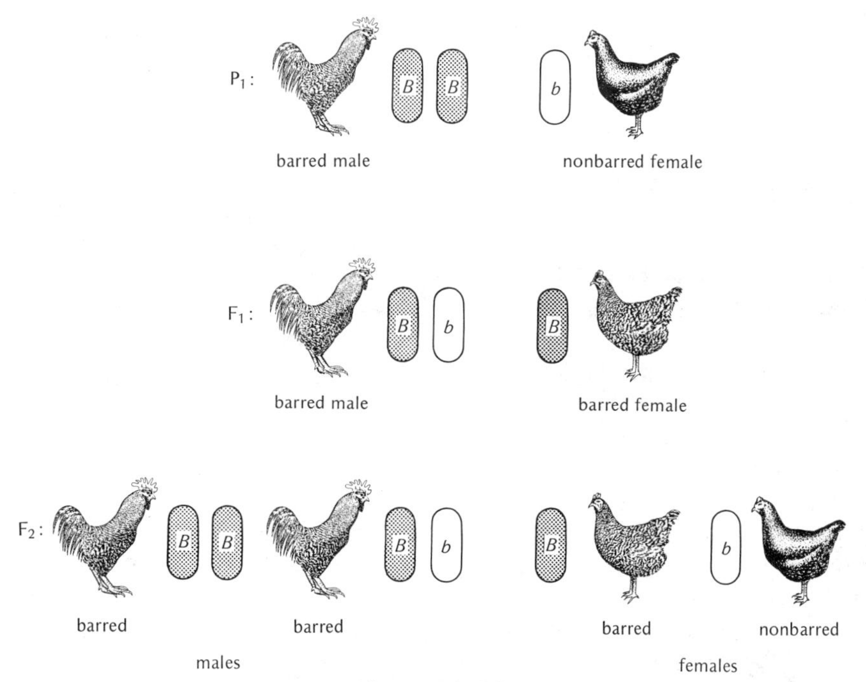

Figure 12–12

Inheritance of the barred and nonbarred effect in a cross between a barred Plymouth Rock male and a black Langshan female. In the reciprocal cross, Langshan male × Plymouth Rock female, the F_1 females are black and the F_1 males are barred. Nonbarred is therefore caused by a sex-linked recessive gene, and the females in poultry may be considered to have only a single X chromosome.

that showed a considerable amount of male characteristics, called *intersexes*. On the other hand, the reciprocal cross produced normal females and normal males. These results were explained by Goldschmidt as caused by differences in the "strengths" of female (F) and male (M) factors between each locality.

According to Goldschmidt, the F factors are carried on the Y chromosome and the M factors on the X chromosome. The strength of these factors is balanced in local matings, so that one dose of F factors on the Y chromosome is ordinarily sufficient to produce a female in the presence of one dose of M factors on the X. However, there are many kinds of M and F factors with different sexual potencies, some "strong" (M_s, F_s) and some "weak" (M_w, F_w). Accordingly, the above cross that produces female intersexes can be symbolized

$$P_1:\ F_wM_w\ \text{♀} \times M_sM_s\ \text{♂}$$

(Korean) (Japanese)

↓

$$F_1:\ M_wM_s \quad : \quad F_wM_s$$

(male) (intersex)

The female intersex is therefore caused, in this case, by the inability of the weak female factor (F_w) to fully overcome the effects of the strong male factor (M_s). However, the transformation of males into females is also noted, indicating that female-producing factors are present in such males, either in the autosomes or in the cytoplasm.

To explain intersexes, Goldschmidt proposed the "time law" of development: that intersexes begin their development as one or the other sex and then, at a critical "turning point," switch their development into that of the opposite sex. However, the findings of Seiler with the moth *Solenobia triquetrella* indicate that intersexes maintain both male and female tissue from the beginning (see also gynandromorphs, p. 482) and there is no crucial "turning point." As yet the exact mechanisms involved in intersex development are not clear.

DETECTION OF SEX-LINKED DISEASES IN HUMANS

At least 80 human traits are now known which are caused by genes carried on the X chromosome. In some instances, such as colorblindness and the Xg blood type, sex-linked genes have little if any observable viability effect. In other instances, however, sex-linked genes result in highly debilitating diseases (e.g., hemophilia; see p. 236), and the lives of affected individuals are preserved only with difficulty. One of the most dreaded sex-linked recessive diseases is the convulsive Lesch-Nyhan disorder, in which affected children (hemizygous males) often mutilate themselves to the point where they have literally chewed their lips and fingers to the bone.

To prevent the occurrence of these diseases it would therefore be of great value if the heterozygous female carriers of the gene could be identified before producing affected offspring. Fortunately biochemical techniques now enable identification of various carriers to be made with a fair degree of certainty. About 85 percent of the female carriers of hemophilia, for example, can be identified by a significant reduction in one of the clotting factors (factor VIII). A majority of the carriers of sex-linked muscular dystrophy (Duchenne's) show an increased amount of the enzyme, creatine kinase, in blood compared to normal females. Females heterozygous for the gene causing the Lesch-Nyhan syndrome often give rise to two different populations of cells in tissue culture: in one population the cells are normal and able to incorporate and metabolize the purine hypoxanthine, whereas the other population is unable to incorporate hypoxanthine. However, although carrier detection is now possible for

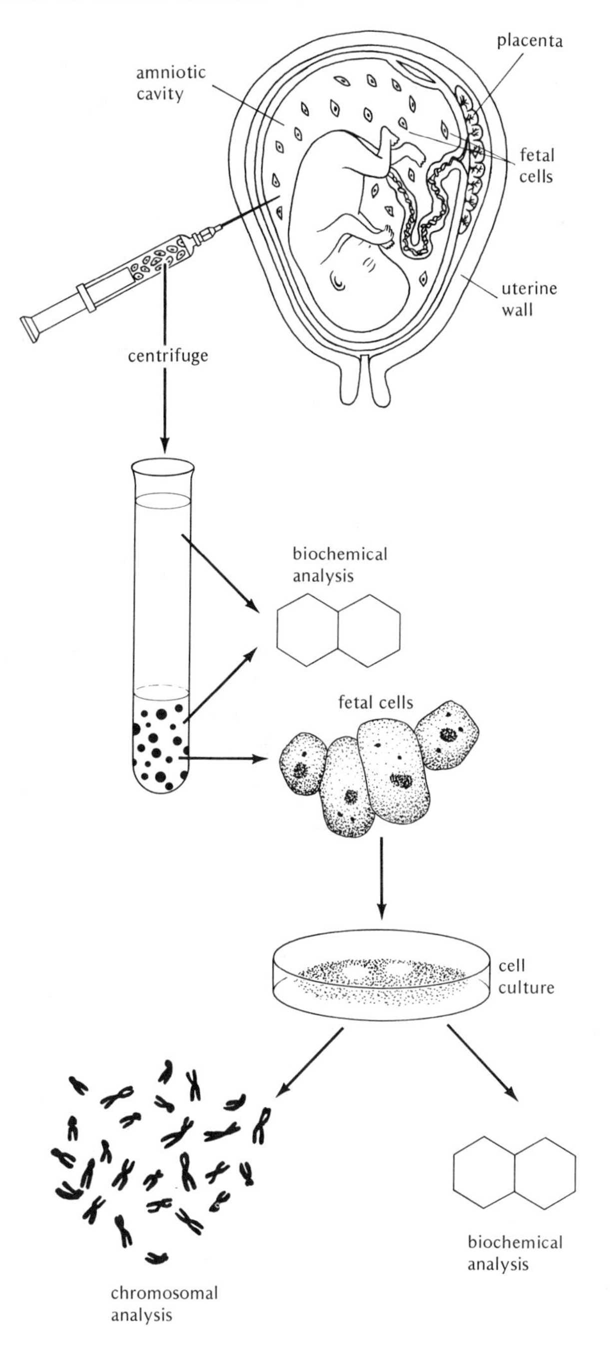

Figure 12–13

Amniocentesis procedure. The withdrawal of fluid is usually performed when the amniotic cavity is relatively large, about the sixteenth week of pregnancy.

more than 70 sex-linked and autosomal diseases, the techniques are at present difficult and expensive. So far only relatively few potential carriers have been tested—usually the relatives of people who have already produced offspring affected by genetic disorders.

In those cases where a woman who has previously produced an affected child becomes pregnant, a method known as *amniocentesis* ("amnion puncture") has been developed to determine the state of the unborn fetus with respect to the disease. As shown in Fig. 12–13, a small amount of amniotic fluid is withdrawn between the fifteenth and twentieth weeks of pregnancy from the membranes that surround the fetus. Within this fluid are free-floating embryonic cells that can be tissue-cultured for biochemical or cytological analysis, or the amniotic fluid itself examined directly. This information then enables further decisions to be made regarding survival or abortion of the fetus. For example, a pregnant woman who had already produced a son with the Lesch-Nyhan syndrome chose to have an abortion when the male fetus she was carrying showed the biochemical symptoms of the disease. In another instance a male fetus was shown to be unaffected, and the pregnancy was continued. Similar amniotic analyses can be made for many autosomal diseases such as cystic fibrosis (accumulation of unusual staining fatlike "metachromatic" granules) and Tay-Sachs disease (p. 186; the absence of β-D-N-acetylhexosanimidase), as well as chromosomal defects such as Down's syndrome (see pp. 192 and 487–489).

SEX RATIO

Normally, the equal proportions of the male- and female-producing gametes formed by the heterogametic sex in a species result in approximate equal proportions of the two sexes at birth. In some instances, however, the proportions of male and female births are far from equal. Examples of this kind have been found among many plants and animals in which either males or females may be the more numerous sex. One such instance occurs in certain stocks of *Drosophila melanogaster* (e.g., Muller's *ClB* stock), in which one of the X chromosomes carries a lethal gene *l*. Males inheriting this X chromosome die before emergence as adults, while the heterozygous females are fully viable, since they also carry one X chromosome bearing the normal allele for *l*, i.e., l^+. Such *ClB* females, therefore, produce sex ratios of two females to one male (Fig. 12–14).

On the other hand, sex-linked lethals may also affect the viability of females, so that relatively more males are produced. One interesting example of this is the recessive gene *bobbed* (symbolized *bb*, producing shortened bristles and abnormalities of the abdomen) carried on the X chromosome of *D. melanogaster*. According to the findings of Stern, this X chromosome gene is the only one known

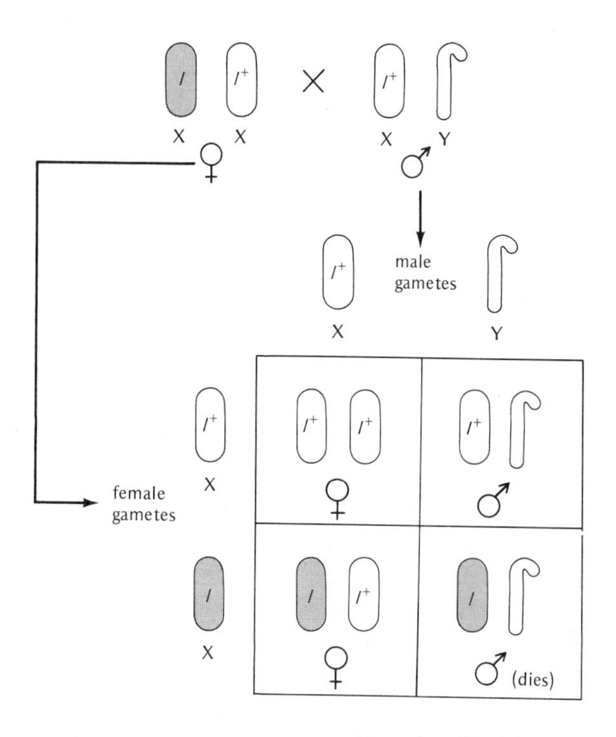

Figure 12–14

Modified sex ratio produced by a female carrying a sex-linked lethal gene, l.

that is also carried on the male Y chromosome. Thus a male carrying a *bb* X chromosome will appear as wild type if its Y chromosome is bb^+, whereas females can be homozygous *bb*/*bb* or phenotypically bobbed. If a lethal allele of the *bb* gene is used, i.e., bb^l, a cross between heterozygotes carrying this gene on one of their X chromosomes will produce a ratio of two males to one female.

Abnormal sex ratios can also be produced when part of the *gametes* are inviable or ineffective in fertilization because of segregation distortion (p. 219). According to Policansky and Ellison, males in certain "sex ratio" stocks of *Drosophila pseudoobscura* produce only half the number of sperm of the normal male. Presumably such gametes bear primarily X chromosomes rather than Y chromosomes, and thereby produce a majority of female offspring. Other factors that may affect the sex ratio will be discussed in Chapters 13 and 24.

PROBLEMS

12-1. Baur and Shull crossed broad-leaved females of the plant species *Lychnis alba* to narrow-leaved males and obtained an F_1 consisting of males and females that were all broad-leaved. When the F_1 plants were crossed among themselves the F_2 females were all broad-leaved, but the F_2 males were of two kinds—broad-leaved and narrow-leaved. Explain how you can determine which of these sexes is heterogametic (XY) and which homogametic (XX).

12-2. In experiments by Correns on two plant species of the genus *Bryonia*, one species, *Bryonia dioica*, was dioecious and when crossed with other members of the same species produced offspring in the ratio 1 female to 1 male. The other species, *Bryonia alba*, was monoecious and when crossed with other *B. alba* individuals always produced monoecious hermaphroditic offspring. In crosses between the two species, if *B. dioica* was used as the female parent and *B. alba* as the male, the offspring were always female. If *B. alba* was the female parent and *B. dioica* the male, their offspring were in an approximate ratio 1♂:1♀. How do you explain these results?

12-3. In the normally dioecious hemp plant (*Cannibus sativa*), the male is the heterogametic sex but occasionally bears a few flowers that can function as females. On the basis of this observation, McPhee self-fertilized some of these male plants and found that they produced a sex ratio 3♂:1♀. How do you explain these results?

12-4. In the plant *Fragaria* the female is the heterogametic sex, and no matter how much pollen is used the sex ratio produced is always 1♀:1♂. In *Lychnis* Correns found that pollination with an excess of pollen produced more females than males, but the sparse use of pollen produced a 1:1 sex ratio. How do you explain these results?

12-5. In the salamander genus *Ambystoma*, R. R. Humphrey removed a testis from a male embryo and inserted in its place the ovary of a female. During further development of this male embryo, the inserted female ovary became transformed into a functioning male testis but maintained its nuclear origin (its female chromosome constitution did not change). When this ovary was fully transformed into a functioning testis, Humphrey removed the other testis. This phenotypic male was therefore producing sperm, all of which were now derived from the cells of the former female ovary. He then mated this male to a normal female and obtained both male and female progeny. How would you explain these results in terms of sex determination in this species?

12-6. In the beetle *Blaps polychresta* there are 18 autosomes in addition to 24 X chromosomes in the female and 12 X and 6 Y chromosomes in the male. What are the chromosome constitutions of the two types of gametes produced by the male?

12-7. The sex-determining mechanism in *Drosophila* is XX♀ and XY♂, but this mechanism is reversed in the sexes of birds. In some *Drosophila* species (*D. parthenogenetica*, *D. mangabeira*, and others) individuals that arise by parthenogenesis are almost always females, but birds (e.g., turkeys) that arise by parthenogenesis are almost always male. (a) How would you account for this difference? (b) How would you explain the rare parthenogenetically produced males in *Drosophila*?

12-8. Hollander found a dominant sex-linked trait in pigeons called *faded* that produced a light gray color in the heterozygous *faded*/+ male and in the hemizygous *faded* female. The cross *faded*/+ × *faded* produced the expected types of females, but some of the males showed a new phenotype, white, caused by the *faded*/*faded* genotype. How would you use this gene to "sex" newborn pigeons automatically so that the males are always recognizably different than the females?

12-9. If the son of a female wasp in *Bracon hebetor* were crossed with its mother, what would be the sex ratio among the offspring if one third of the eggs are unfertilized?

12-10. In the fly *Sciara coprophila*, male somatic tissue is XO, and female somatic tissue is XX. Chromosomes in the gametes of this species, however, show a reversed composition: sperm have two X chromosomes and eggs only one. As diagrammed in Fig. 12-15, reasons for these changes include elimination of a single chromosome from the germ lines and somatic tissues of each sex; elimination of an additional X chromosome from the somatic tissues of males; and X-chromosome nondisjunction in male meiosis. According to Metz, Crouse, and others, it seems likely that the sex-determining mechanism resides completely in the female: one type of female (X′X) produces only females, and the other type (XX) produced only males. (a) What sex ratio of males to females would you expect to find in these flies? (b) Would it make a difference in the sex ratio as to which of the parental X chromosomes were eliminated, maternal or paternal?

12-11. What phenotypic sex ratios would you expect from: (a) matings between *Drosophila* females heterozygous for the transformer (*tra*) gene (see p. 229) to males homozygous for this gene (*tra*/*tra*)? (b) matings between flies heterozygous for the *transformer gene* (*tra*/+ × *tra*/+)? (Indicate the proportions of sterile males you would expect to find in each cross.)

12-12. Jones and Emerson have shown that corn, a hermaphroditic plant, can be converted into a dioecious plant in which the male and female organs are carried by separate individuals. Genes that can be used for this purpose include *silkless* (*sk*), a recessive gene which, when homozygous, suppresses female ovary development in the ears (silks), and *Tassel seed*$_3$ (*Ts*$_3$), an independently assorting dominant gene that converts the male tassel into a fertile female flower. Using these two genes, show how a stable sex-determining system can be developed in which the female is the heterogametic sex.

12-13. Another *tassel seed* allele, *ts*$_2$, also converts the

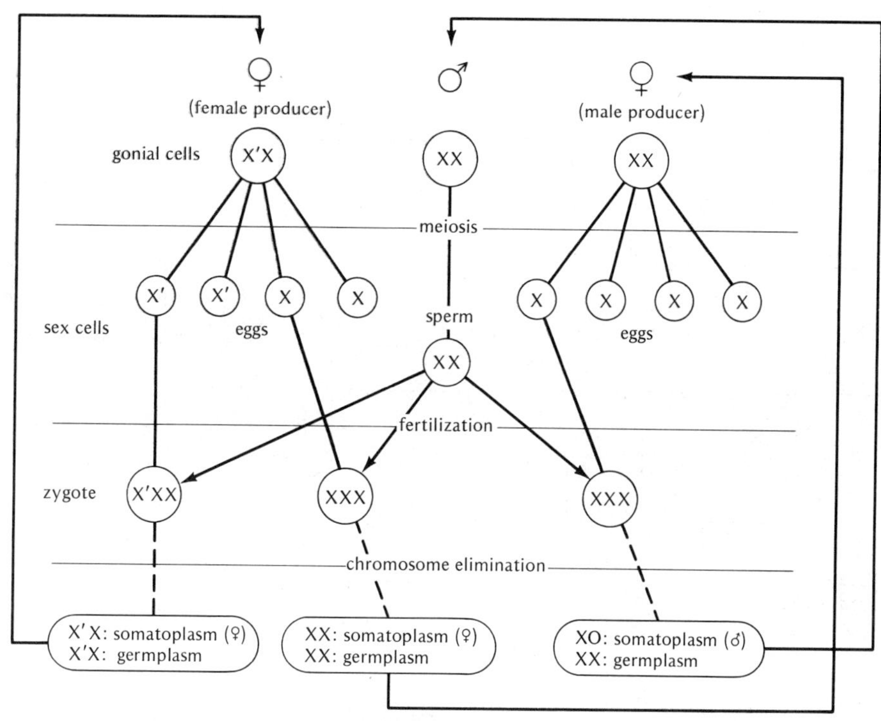

Figure 12-15

Sex determination mechanism in Sciara coprophila*; see Problem 12-10. (After J. F. Anderson, "Sex Determination in Insects," in* The Biology of Sex*, ed. by A. Allison, Penguin Books, 1967.)*

male tassel of corn into a fertile female flower, but does this only when homozygous. The ts_2/ts_2 homozygote also enables the *silkless* (*sk/sk*) homozygote to be female fertile and produce functioning ovaries in both the corn ear and tassel. Using these two genes, show how a stable sex-determining system can be developed in which the male is the heterogametic sex.

12–14. Abnormal eye shape in *Drosophila* may be caused by a variety of mutant genes, dominant or recessive, sex-linked or autosomal. One normal-eyed *Drosophila melanogaster* male from a true-breeding normal stock was crossed to two different abnormal-eyed females with the following results:

	Progeny of female 1		Progeny of female 2	
	♀	♂	♀	♂
normal-eyed	108	0	51	49
abnormal-eyed	0	102	53	50

Explain the differences between these two sets of results and furnish genotypes for the male and the two females.

12–15. The following results were obtained by Shaw and Barto in tests for different variants of an enzyme which metabolizes 6-phosphate sugars in deer mice. They scored each animal for one of three phenotypes of this enzyme, *a*, *b*, or *ab*.

Phenotypes of Parents		Phenotypes of Offspring					
		a		*b*		*ab*	
Female	Male	♀	♂	♀	♂	♀	♂
b	*b*			4	2		
a	*ab*	1	3			1	1
ab	*b*			2	1		4
ab	*ab*	1	2		2	2	5

Explain whether you believe this gene is sex-linked or autosomal.

12–16. The following diagram represents a partial pedigree for a disease known as microphthalmia, causing blindness (shaded symbols). What is the probable mode of inheritance of the gene that may be causing this disease?

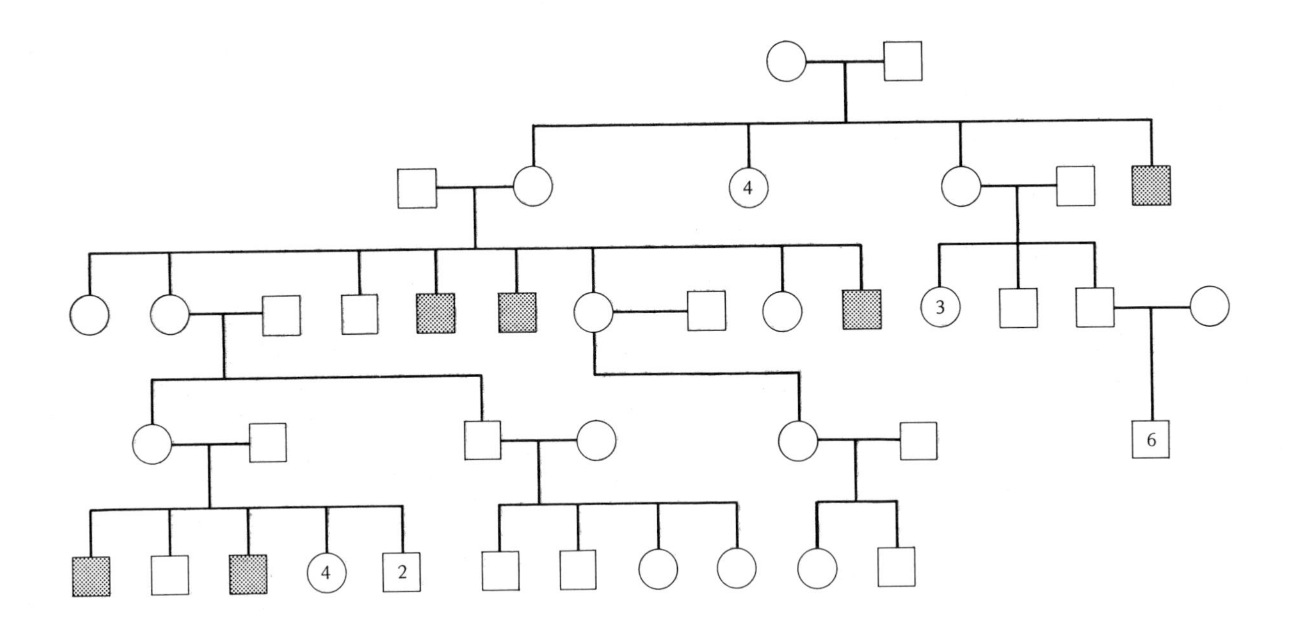

12–17. The following pedigree of individuals affected with muscular dystrophy (shaded symbols) was produced from matings between a particular woman and two different men in two separate marriages. Assuming that the gene is completely penetrant, describe the kind of gene you think is responsible for this defect (dominant, recessive, sex-linked, autosomal, etc.) and give reasons for your opinion.

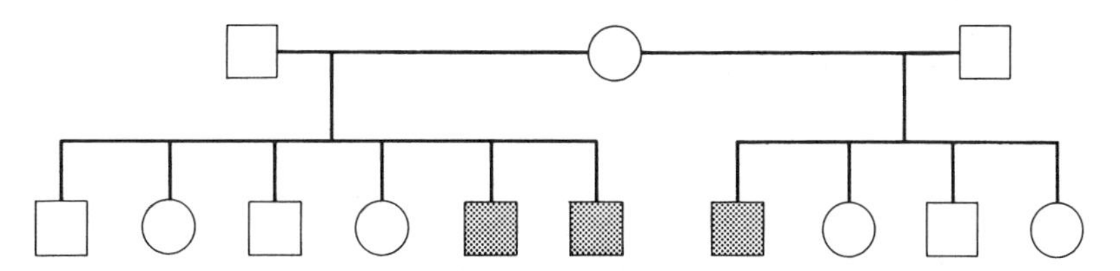

12-18. The following pedigree is for a family with a rare trait called hypophosphatemia, which is usually accompanied by skeletal rickets resistant to vitamin D. What mode of inheritance accounts for the transmission of this trait? (Assume that each individual in the pedigree who produces offspring is mated to a person with normal genotype whose symbol is not shown.) (After Winters and co-workers.)

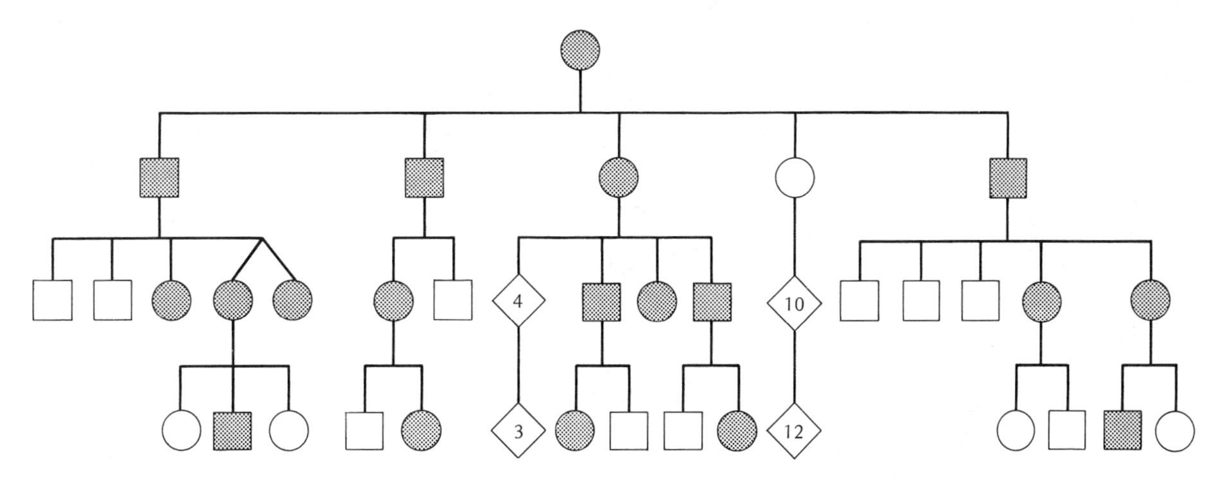

12-19. In the following pedigree of a particular rare trait (shaded symbols), state whether you believe it is caused by a sex-linked recessive gene or by an autosomal dominant gene that is expressed only in males.

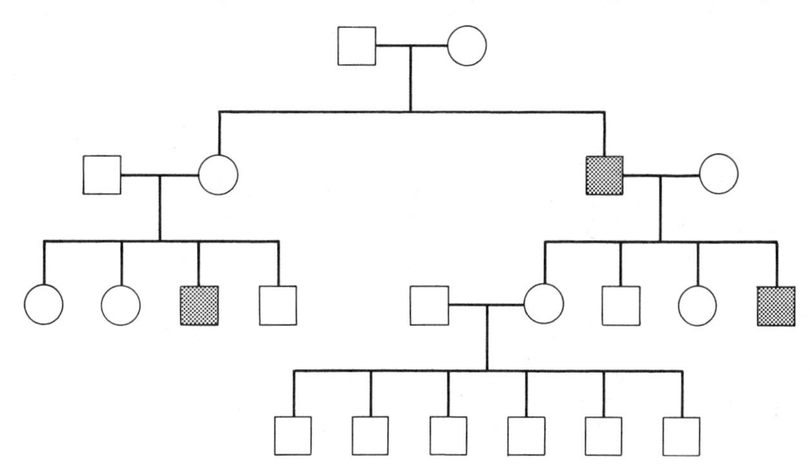

Each of the following three pedigrees (Problems 12-20, 12-21, and 12-22) represents a different human trait (affected individuals are shaded). Each trait can be considered to be very rare and can be assumed to have been introduced into each pedigree only in generation I. For each of the three pedigrees choose one of the following five genetic mechanisms to explain the inheritance of the trait and provide for each of the other mechanisms an explanation of why you have rejected it: (a) X-linked recessive; (b) X-linked dominant; (c) Y-linked dominant; (d) Autosomal dominant; (e) Autosomal recessive.

12–20.

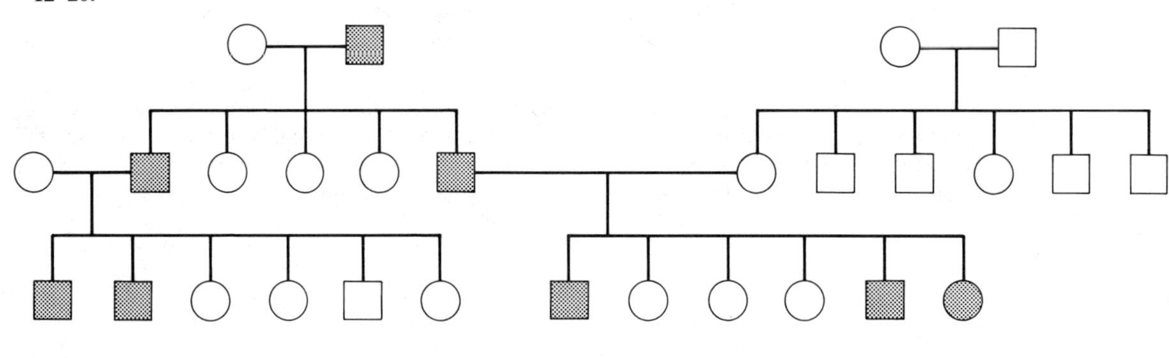

12–21.

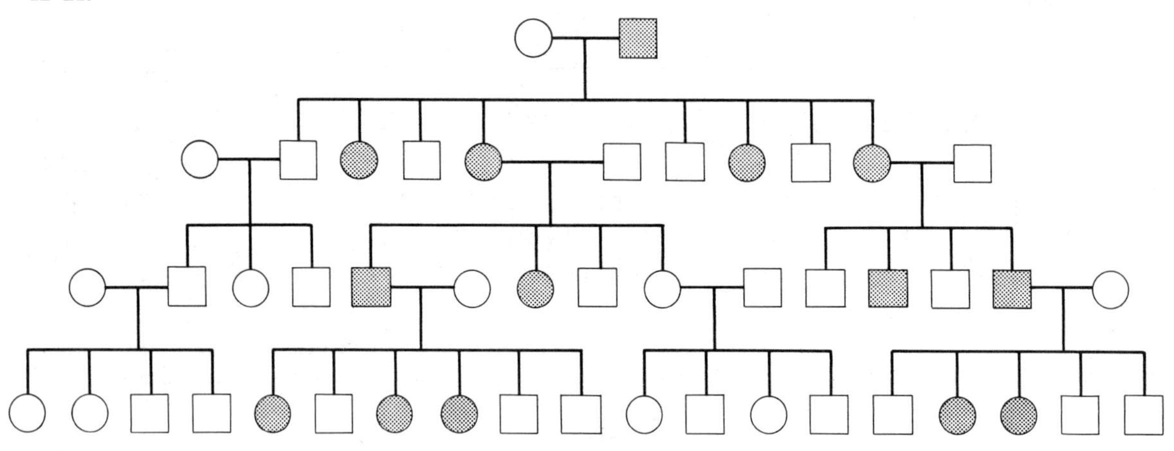

12–22.

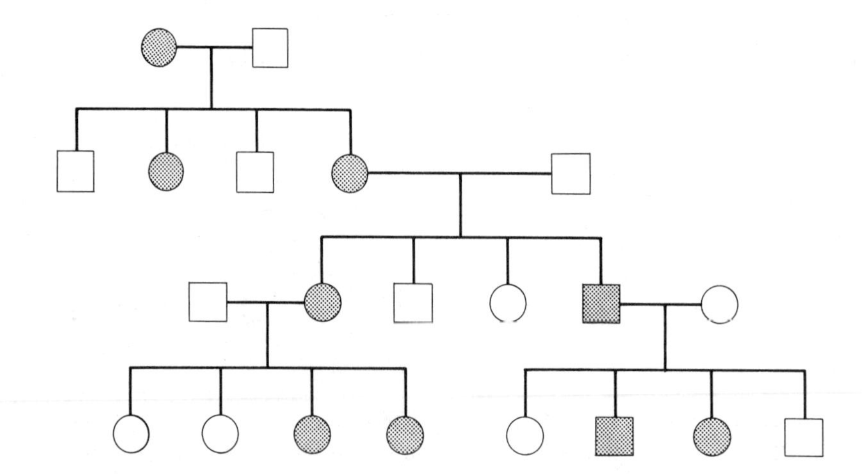

12-23. The following three pedigrees represent a single Sardinian family (i.e., the same persons are represented in all three pedigrees) investigated by Siniscalco and Filippi, segregating for three sex-linked traits: colorblindness of the deutan type, deficiency for the enzyme glucose-6-phosphate dehydrogenase (g-6-pd), and Xg blood type. Using the gene symbols c for the allele causing colorblindness and c^+ for its normal allele, g for the allele causing g-6-pd deficiency and g^+ for its normal allele, a^- for the allele causing the *non-appearance* of Xg blood type and a^+ for the allele causing Xg blood type, give the most probable genotypes of individuals I-1, I-2, II-1, II-2, III-1, III-5 for all three genes. (First determine the mode of inheritance of each trait and then assign a genotype for each trait to the individual involved. For example, a hypothetical female V-1 might have the constitution cc from the top pedigree, g^+g^+ from the center pedigree, and a^-a^- from the bottom pedigree.)

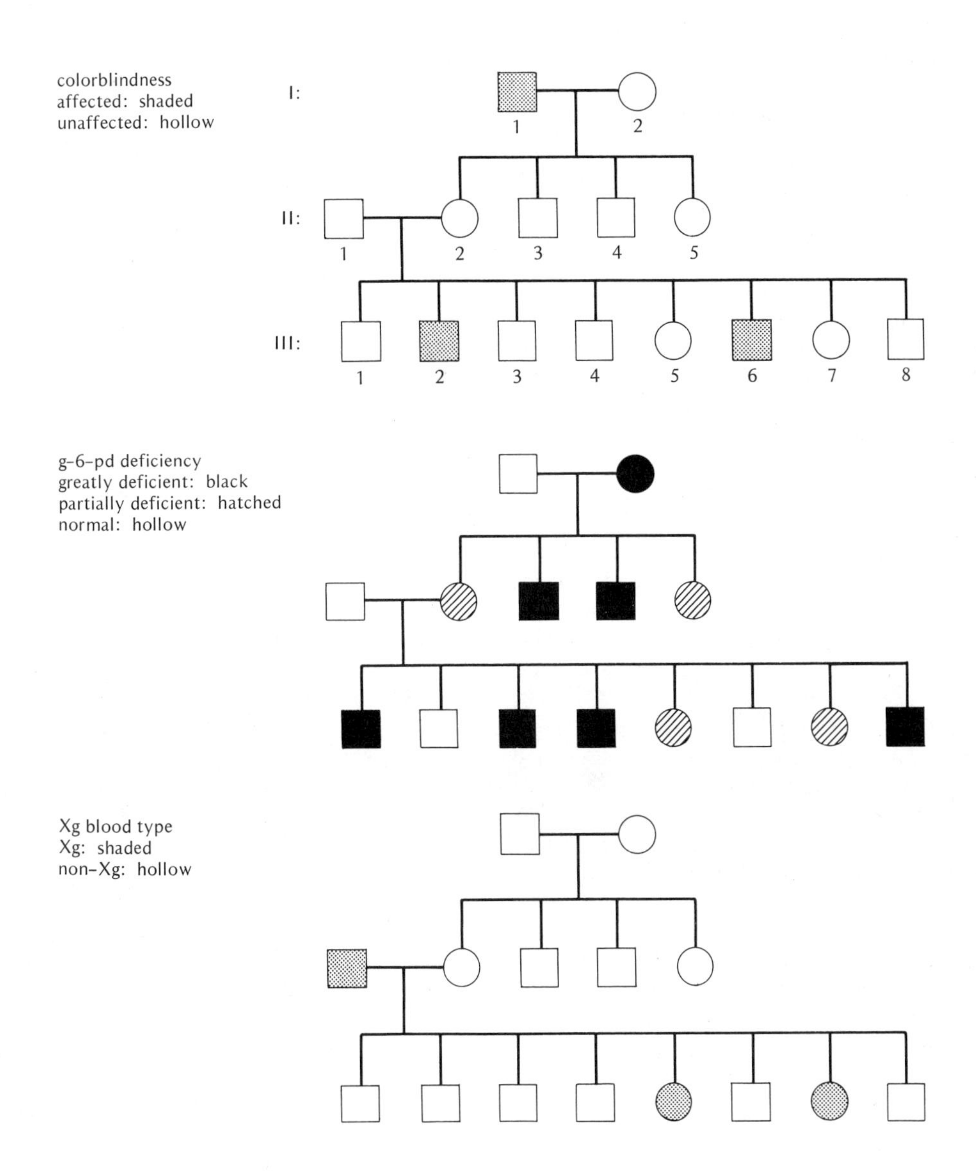

12–24. In a species of moth in which the female sex chromosomes are XO and the males are XX: (a) How would you explain the occasional transmission of a sex-linked recessive gene from mother to daughter? (b) How would you explain the occurrence of such mother-to-daughter transmission in all matings of a particular stock?

12–25. In man, mouse, and *Drosophila*, males are the heterogametic sex; in the amphibian *Ambystoma* and the moth *Bombyx*, females are heterogametic. Assuming that gametes may occasionally result from nondisjunction of the sex chromosomes in the females of each species, what would be the sex-chromosome complements of individuals that arise from the fertilization of these exceptional gametes?

12–26. Because of nondisjunction in some stocks of *Drosophila melanogaster*, about 1 egg in 2500 receives both X chromosomes from the mother, and about 1 egg out of 600 receives no X chromosomes at all. Assuming that meiosis in males is always normal, what is the expected frequency of XXY females and XO males arising from any counted number of eggs?

12–27. Hemophilia A is caused by a sex-linked recessive gene in humans and in dogs. (a) What proportions (and sexes) among their offspring will be hemophiliacs if a hemophilic male is mated to a homozygous nonhemophilic female? (b) If a daughter produced by the mating in (a) is mated to a normal male, what proportions (and sexes) will be hemophilic among their offspring?

12–28. A common kind of red-green color blindness in humans is caused by the presence of a sex-linked recessive gene *c*, whose normal allele is c^+. Using these genes, what are all the possible colorblind genotypes and their corresponding phenotypes in males and in females?

12–29. (a) Can two colorblind parents produce a normal son? (b) A normal daughter? (c) Can two normal parents produce a colorblind son? (d) A colorblind daughter?

12–30. (a) Can a normal daughter have a colorblind father? (b) A colorblind mother? (c) Can a colorblind daughter have a normal father? (d) A normal mother?

12–31. (a) Can a normal son have a colorblind mother? (b) A colorblind father? (c) Can a colorblind son have a normal mother? (d) A normal father?

12–32. (a) Can a colorblind brother and sister have another brother who is normal? (b) Can a colorblind brother and sister have another sister who is normal? (c) Can a colorblind brother and sister have parents who differ in their color-vision phenotype (i.e., one normal, the other colorblind)?

12–33. A colorblind woman marries a normal-visioned man. They have two children, a boy and a girl. (a) What will be the genotype and phenotype of the boy? (b) What will be the genotype and phenotype of the girl?

12–34. A normal woman whose father was colorblind marries a colorblind man. They produce a son and a daughter. (a) What is the probability that the son is colorblind? (b) What is the probability that the daughter is colorblind?

12–35. A normal woman whose mother was colorblind has a son. Nothing is known of the color-vision phenotype of the father. What is the probability that the son will be colorblind?

12–36. A woman of A blood type and normal color vision produced five children as follows: (a) male, A blood type, colorblind; (b) male, O blood type, colorblind; (c) female, A blood type, colorblind; (d) female, B blood type, normal color vision; (e) female, A blood type, normal color vision. Of the two men that may have mated with this woman at different times, no. 1 had AB blood type and was colorblind, and no. 2 had A blood type with normal color vision. Which of these men is the most probable father in each case?

12–37. *Drosophila melanogaster* females from a homozygous *white* stock that is otherwise wild type are crossed to males from a homozygous *ebony* stock that is also otherwise wild type. *White* is a sex-linked recessive gene that affects eye color; *ebony* is an autosomal recessive gene that affects body color. (a) What results would you expect in the F_1 of this cross? (b) What results would you expect among the progeny of $F_1 \times F_1$?

12–38. In an experiment using the crosses in Problem 37, a genetics laboratory student found the following phenotypic results in the F_2:

Males	Females
197 white	206 white
64 white ebony	82 white ebony
215 wild type	213 wild type
75 ebony	68 ebony

(a) Considering each sex separately, what numbers would you have expected for each phenotype? (b) Is the deviation between the observed and expected numbers significant in each case?

12–39. In *Drosophila melanogaster* the normal wild-type eye color is red, and white eye color is produced by a sex-linked gene, *white* (*w*). A hypothetical eye color, grape, is produced by an autosomal gene (*Grape*, *G*) whose effect is dominant to wild type. The effect of the *white* gene in homozygous (♀) or hemizygous (♂) condition is epistatic to all other eye colors, hiding their effect. From a cross homozygous *Grape* ♀ × *white* ♂ (both homozygous wild type for all other genes), the F_1 females are crossed to the F_1 males. (a) What are the phenotypes of the F_1? (b) What are the F_2 phenotypic ratios in each of the sexes?

12-40. In *Drosophila*, the recessive genes *vermilion* (*v*) and *cinnabar* (*cn*) have the same phenotypic effect, a lighter and brighter eye color than the usual brick-red wild type. *Vermilion*, however, is located on the X chromosome and *cinnabar* on one of the autosomes. (a) If a *vermilion* male is crossed to a homozygous *cinnabar* female, what are the genotypes and phenotypes of their offspring? (b) If these F_1 flies are permitted to breed with each other, what proportions of each different phenotype will be found in the F_2?

12-41. A *Drosophila* cross of white-eyed males × pink-eyed females produced an F_1 all with normal red eyes. The F_1♀ × F_1♂ cross produced the following results:

Phenotype	**No.**
red-eyed ♀♀	890
red-eyed ♂♂	455
white-eyed ♂♂	605
pink-eyed ♂♂	140
pink-eyed ♀♀	310

Define the gene symbols and provide a genetic interpretation that accounts for the F_1 and F_2 results. (Hint: break down the data into phenotypic frequencies for males and females separately.)

12-42. When curly-winged *Drosophila* (males or females) are crossed with normal straight-winged flies, the offspring are in the ratio 1 curly:1 straight. When curly are crossed with each other, the offspring are in the ratio 2 curly:1 straight. When females bearing a certain *Minute* bristle effect are crossed to normal long-bristled males, the offspring are in the ratio 2 females:1 male; half the females are minute and the other half long-bristled, while all the males are long-bristled. If a curly minute female is crossed to a curly male, what adult phenotypic ratios would you expect: (a) Among the female progeny? (b) Among the male progeny?

12-43. Two phenotypically wild-type *Drosophila* females, (a) and (b), were mated to the same wild-type male and produced the following offspring:

	Females		**Males**	
	White Eye	Wild Type	White Eye	Wild Type
female (a)	0	73	32	37
female (b)	0	68	33	0

Explain these results.

12-44. A particular allele of the *Drosophila* sex-linked gene *singed* produces sterility in homozygous females, but males bearing this *singed* allele are fertile. Explain how such an allele could be used to establish a "balanced" stock that would perpetuate itself when combined with a sex-linked lethal *l* such as that in Muller's *ClB* chromosome (see p. 243).

12-45. In 1922 Haldane proposed the rule that when a particular sex is either absent, rare, or sterile among the progeny from a cross between two strains, that sex is the heterogametic sex. (a) What evidence supports this rule? (b) What kinds of mutations would circumvent such a rule?

REFERENCES

AGGELER, P. M., M. S. HOAG, R. O. WALLERSTEIN, and D. WHISSELL, 1961. The mild hemophilias. *Amer. Jour. Med.*, **30**, 84–94.

ALLISON, A. (ed.), 1967. *The Biology of Sex*. (*Penguin Science Survey 1967*). Penguin Books, Middlesex.

BIGGS, R., A. S. DOUGLAS, R. G. MACFARLANE, J. V. DACEY, W. R. PITNEY, C. MERSKEY, and J. R. O'BRIEN, 1952. Christmas disease: A condition previously mistaken for haemophilia. *Brit. Med. Jour.*, **2**, 1378–1382.

BRIDGES, C. B., 1916. Nondisjunction as proof of the chromosome theory of heredity. *Genetics*, **1**, 1–52, 107–163. (An extract is reprinted in the collection of Voeller; see References, Chapter 1.)

BRINKHOUS, K. M., and J. B. GRAHAM, 1950. Hemophilia in the female dog. *Science*, **111**, 723–724.

BYERS, R. E., L. R. BAKER, H. M. SELL, R. C. HERNER, and D. R. DILLEY, 1972. Ethylene, a natural regulator of sex expression of *Cucumis melo. Proc. Nat. Acad. Sci.*, **69,** 717–720.

CORRENS, C., 1907. *Die Bestimmung und Vererbung des Geschlechtes.* Gebruder Borntraeger, Berlin.

DONCASTER, L., and G. H. RAYNOR, 1906. On breeding experiments with *Lepidoptera. Proc. Zool. Soc. Lond.*, **1,** 125–133.

GOLDSCHMIDT, R. B., 1955. *Theoretical Genetics.* Univ. of California Press, Berkeley, Part IV.

LEVITAN, M., and A. MONTAGU, 1971. *Textbook of Human Genetics.* Oxford Univ. Press, New York.

LEWIS, K. R., and B. JOHN, 1968. The chromosomal basis of sex determination. *Int. Rev. Cytol.*, **23,** 277–379.

MITTWOCH, U., 1973. *Genetics of Sex Differentiation.* Academic Press, New York.

MONTGOMERY, T. H., 1906. Chromosomes in the spermatogenesis of the Hemiptera Heteroptera. *Trans. Amer. Phil. Soc.*, **27,** 97–173.

MORGAN, L. V., 1922. Non-criss-cross inheritance in *Drosophila melanogaster. Biol. Bull.*, **42,** 267–274.

MORGAN, T. H., 1910. Sex limited inheritance in *Drosophila. Science*, **32,** 120–122. (Reprinted in the collections of Levine, of Peters, and of Voeller; see References, Chapter 1.)

MORGAN, T. H., A. H. STURTEVANT, H. J. MULLER, and C. B. BRIDGES, 1922. *The Mechanism of Mendelian Heredity,* 2nd ed. Holt, New York.

OHNO, S., 1967. *Sex Chromosomes and Sex-linked Genes.* Springer-Verlag, Berlin.

POLICANSKY, D., and J. ELLISON, 1970. "Sex ratio" in *Drosophila pseudoobscura*: Spermiogenic failure. *Science*, **169,** 888–889.

ROSENTHAL, R. L., O. H. DRESKIN, and N. ROSENTHAL, 1953. New hemophilia-like disease caused by deficiency of a third plasma thromboplastin factor. *Proc. Soc. Exp. Biol. Med.*, **82,** 171–174.

ROTHENBUHLER, W. C., J. M. KULINČEVIĆ and W. E. KERR, 1968. Bee genetics. *Ann. Rev. Genet.*, **2,** 413–438.

SEILER, J., 1965. Sexuality as a developmental process. In *Genetics Today,* Vol. II, Proc. Intern. Congr. Genet., S. J. Geerts (ed.). Pergamon, Oxford, pp. 199–207.

SPILLMAN, W. J., 1909. Barring in barred Plymouth Rocks. *Poultry*, **5,** 708.

STERN, C., 1926. Vererbung im Y-chromosom von *Drosophila melanogaster. Biol. Zentralbl.*, **46,** 344–348.

STEVENS, N. M., 1905. Studies in spermatogenesis with especial reference to the "accessory chromosome." *Carneg. Inst. Wash. Publ. No. 36*, Washington, D.C. (An extract is reprinted in the collection of Voeller; see References, Chapter 1.)

STURTEVANT, A. H., 1945. A gene in *Drosophila melanogaster* that transforms females into males. *Genetics*, **30,** 297–299.

WESTERGAARD, M., 1958. The mechanism of sex determination in dioecious flowering plants. *Adv. in Genet.*, **9,** 217–281.

WHITE, M. J. D., 1973. *Animal Cytology and Evolution*, 3rd ed. Cambridge Univ. Press, Cambridge.

WHITING, P. W., 1939. Multiple alleles in sex determination of *Habrobracon. Jour. Morph.*, **66,** 323–355.

WILSON, E. B. 1905. Studies on chromosomes. II. The paired microchromosomes, idiochromosomes and heterotrophic chromosomes in *Hemiptera. Jour. Exp. Zool.* **2,** 507–545.

———, 1906. Studies on chromosomes. III. The sexual difference of the chromosome groups in *Hemiptera,* with some considerations on the determination and inheritance of sex. *Jour. Exp. Zool.*, **3,** 1–40.

13
MATERNAL EFFECTS AND CYTOPLASMIC HEREDITY

Previous chapters have confined themselves to cases in which the mode of inheritance of a particular character is tied to the behavior of the nucleus or, more specifically, to the chromosomes. That is, transmission of a character and its appearance in individuals can be predicted from knowledge of chromosome segregation and assortment. This assumption is difficult to challenge so far, since we have considered DNA as the basic genetic material and since practically all cellular DNA is localized in the chromosomes.

On the other hand, DNA does not initiate development of biological characteristics in the absence of all other cellular components. Certainly "raw" DNA does not produce an organism alone but obviously depends upon an already existing medium in order to function. Thus, as we have seen, genotypic effects may be considerably modified by environment. In the cell itself one important source of environmental effect is the cytoplasm immediately surrounding the nucleus. The components of the cytoplasm may vary between individuals, so that it would hardly be surprising to find that a genotype in one cytoplasm would function somewhat differently than the same genotype placed directly into another cytoplasm. Although such experiments are obviously difficult to perform mechanically, differences may be observed, however, when the cytoplasmic contribution of one parent in the form of an egg is considerably larger than the cytoplasmic contribution of the other parent in the form of a sperm. For example, cross fertilization between two strains *A* and *B*, capable of producing both sperm and eggs, leads to a contribution of *A* genotype and *a* egg cytoplasm in one case and of *B* genotype and *b* egg cytoplasm in the other. Thus the cross-fertilized diploid zygotes have two possible genotypic-cytoplasmic combinations, *AB–a* and *AB–b*, depending upon which strain provides the egg. If each such cytoplasm produces a unique phenotype, genotypic similarities may be disguised by the effects of different cytoplasms (*AB–a* appears different from *AB–b*), or, on the other hand, genotypic differences may be disguised by the effects of similar cytoplasms; i.e., *AB–b* would have the same *b* phenotype as *BB–b* individuals. Since such effects are produced by the egg cytoplasm, they are known as *maternal effects*.

MATERNAL EFFECT

A clear illustration of maternal effect was found by Caspari in the flour moth, *Ephestia kuhniella*. This moth has dark-brown eyes and various pigmented larval parts which owe part of their color to the presence of a pigment precursor, *kynurenine*, produced by the dominant gene *A*. However, when the organism is homozygous for a recessive allele of this gene, *a*, kynurenine is absent, causing red eye color and a lack of larval pigmentation. As can normally be expected, when a male heterozygous for *Aa* fertilizes an *aa* recessive female, their off-

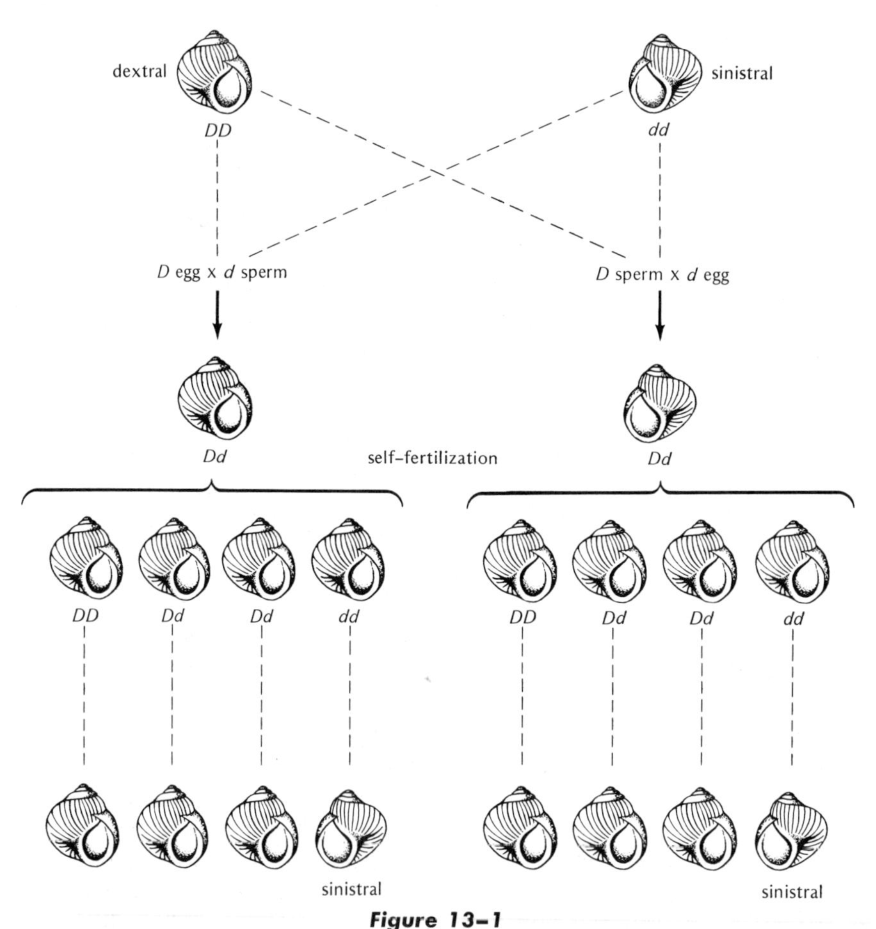

Figure 13–1

Inheritance of coiling in Limnaea peregra.

spring will appear in the ratio 1 *Aa*:1 *aa*. Phenotypically the larvae will consequently be in the ratio 1 pigmented:1 nonpigmented, and will then mature into dark-brown and red-eyed individuals, respectively.

The results are, however, quite different when the reciprocal cross is made, *Aa* female × *aa* male. In this case all early larvae appear pigmented, as though kynurenine were present in all individuals. However, when the larvae mature, only half of them are dark-brown-eyed, the other half are red-eyed. The explanation for these unexpected pigmented larvae is simple, and is based on the egg cytoplasm contributed by the heterozygous mother. That is, all eggs of the dark-brown-eyed female contain kynurenine no matter what their genotype and therefore begin development as pigmented larvae. Half the larvae, however, are *aa* and unable to synthesize further kynurenine. Such larvae consequently develop into red-eyed adults as the kynurenine is used up.

In some cases the maternal effect does not diminish during development but lasts throughout adult life. Such instances occur when development is started in a particular direction that cannot be reversed. One striking example of this occurs in a species of the snail *Limnaea*, in which the direction of coiling is determined by a single pair of genes. Snails can coil to the left (sinistral) or right (dextral), and the allele for dextral coiling (*D*) is dominant to that for sinistral coiling (*d*). Most interestingly, in crosses between these snails the direction of coiling of the offspring is always determined by the *genotype* of the maternal snail, even when this is completely different from the offspring's genotype. Thus a self-fertilized dextrally coiled heterozygote (*Dd*) will always produce eggs that develop into dextrally coiled offspring even though some fertilized eggs are homozygous recessive (*dd*) and should be sinistrally coiled (Fig. 13-1). Similarly the products of a self-fertilized sinistrally coiled heterozygote (heterozygous offspring of a *dd* mother) will all be dextrally coiled even though some eggs are again homozygous *dd*. These maternal effects, however, last only one generation, for in the following generation sinistrally coiled offspring are produced by the homozygous *dd* maternal parents even though they themselves are dextrally coiled.

According to Conklin and others, the embryological reason for shell coiling appears to lie in the direction of the cleavage pattern initiated by the first few cell divisions of the fertilized egg. In mollusks and some other invertebrates this cleavage pattern is spiral, meaning that the cell-division spindles are somewhat displaced in respect to the axis of the egg (Fig. 13-2). Apparently the direction of displacement, to right or left, is determined by the maternal genotype. Such cases therefore represent a maternal effect transmitted through the egg cytoplasm for only one generation, since in the following generation new egg cytoplasm will be formed according to the pattern of the new maternal genotype.

Occasionally fertility and survival also come under the influence of maternal effect, such as effects caused by *grandchildless* genes in *Drosophila*. In *D. subobscura*, for example, Spurway found that although a female homozygous for *grandchildless* is fertile, her offspring are all sterile. The reason for this arises from cytoplasmic-dependent pathways for the development of many *Drosophila* organs: the egg cytoplasm formed by a *Drosophila* female is not uniform, and various parts of the egg appear to be specifically assigned for the formation of different tissues. Thus the fate of *Drosophila* germ cells to produce either ovaries or testes is determined early in development when a number of embryonic nuclei migrate to the posterior pole of the egg. These "polar cap cells" later differentiate into an embryonic (or imaginal) disc which will eventually become the mature gonad during metamorphosis of the larvae into the adult form. Eggs produced by *grandchildless* homozygotes apparently have the cytoplasm of their posterior poles modified, so that nuclei that enter this area cannot produce functional gonads.*

*The egg cytoplasm may also interact with a specific chromosomal constitution of the zygote. For example, in *D. melanogaster,* the autosomal *daughterless* (*da*) gene causes the lethality of daughters (XX eggs) born to females that are homozygous *da/da*. Since Bell and others have shown that this lethal effect occurs even when the progeny are XX males homozygous for the *transformer* gene (see p. 229) or XX inter-

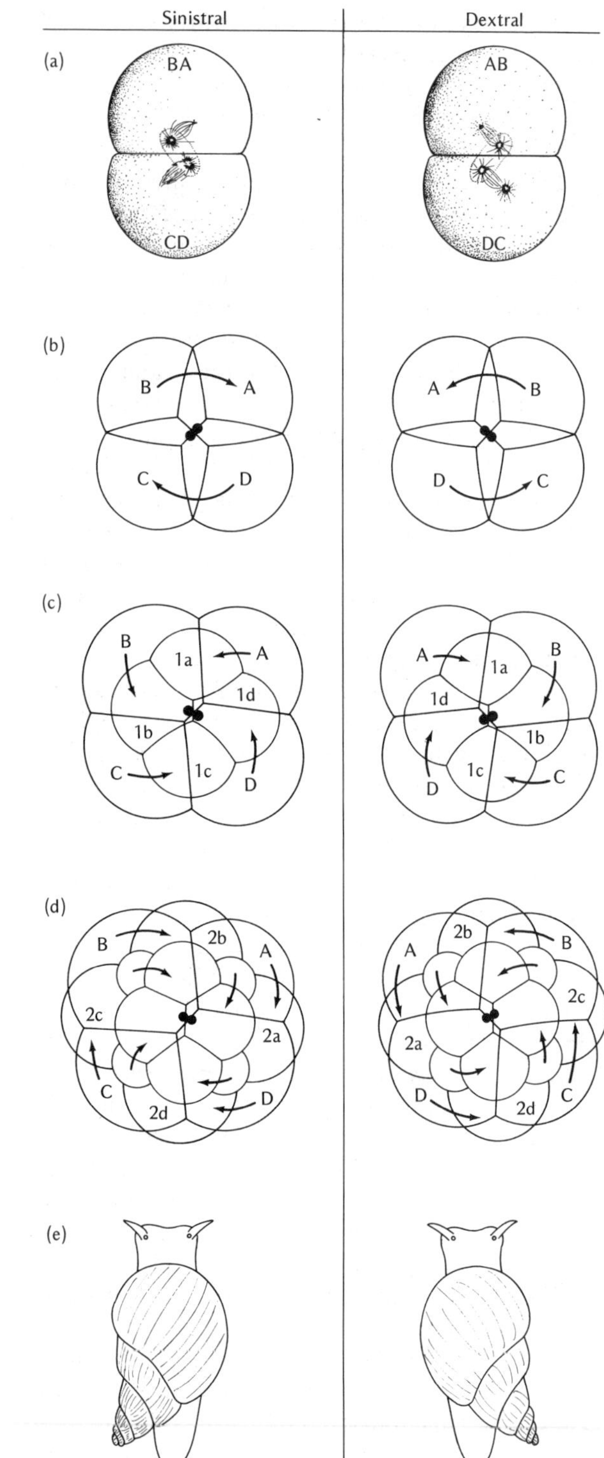

Figure 13–2

Development of the coiling pattern in Limnaea. The sequence leading to sinistral coiling is on the left, and to dextral coiling on the right. (a)–(d) Early egg cleavages showing the spiral patterns. Arrows indicate the direction toward which the mitotic spindles incline during the cleavage divisions. Some of the cells marked by letters and numbers may later give rise to specific tissues. (e) Adult snails. (After Conklin 1922.)

Maternal effects thus arise from egg cytoplasm which has been modified by chromosomally transmitted genes.† Its distinguishing characteristic is the difference in the results of reciprocal crosses, so that cytoplasm produced by a particular genotype acts differently on a developing zygote than cytoplasm produced by a different genotype; i.e., there is a difference in the phenotypes of offspring A♀ × a♂ and a♀ × A♂. Of course, sex linkage (Chapter 12) may also produce differences in the results of reciprocal crosses, but in those cases the phenotype can be predicted from the sex of the parents and offspring. In the maternal effect, phenotypic changes appear because of differences in egg cytoplasm rather than differences in sex chromosomes and often (although not always) affect both male and female offspring equally.

sexes homozygous for *doublesex* (see p. 479), it seems clear that the *daughterless*-produced cytoplasm must directly affect the two X chromosomes or their products rather than affecting the sexual phenotype. Sandler has suggested that the *daughterless*-affected genes are probably in the darkly stained heterochromatic regions of the X chromosome, since the lethality can be modified by using sex chromosomes with different amounts of heterochromatin (see Mange and Sandler). Male offspring are certainly not excluded from specific cytoplasmic interactions, and their absence (*sonless,* see Colaianne and Bell) or reduction in frequency (*abnormal oocyte,* see Sandler) is known to occur in the progeny of *Drosophila* females homozygous for these genes.

†Among mammals, maternal effects can be more broadly defined, since mammalian mothers may affect the development of their offspring not only through the egg cytoplasm but also through the uterine environment. We have seen that the maternal blood group genotype may be the cause of serious defects in fetal development because of Rh incompatibility (Chapter 9). Other maternal effects are known or suspected among humans, such as the embryonic defects caused by maternal diabetes, the maternal influence on left- or right-handedness, and the maternal effect on birth weight. These effects seem to arise at least partially because of interactions caused by the maternal genotype.

EXTRANUCLEAR (CYTOPLASMIC) INHERITANCE

The examples of maternal effect considered above indicate the presence of cytoplasmic factors that exist for a time independent of the nucleus and produce phenotypic effects of their own, such as kynurenine in *Ephestia*. As we have seen, such a factor is not self-perpetuating and disappears unless replaced by the effect of an appropriate nuclear gene (e.g., *A*). Some cytoplasmic factors, however, show evidence for both self-perpetuation and independent transmission, and may therefore be considered as genetic units fully equal to those in the chromosomes. Because of their location outside the nuclear chromosomes, these "extranuclear" or "extrachromosomal" genetic factors have been called "plasmagenes," "plasmons," "plasmids," "cytogenes," etc.

In general, when crosses can be made between organisms carrying traits caused by extrachromosomal factors, the traits are not transmitted according to expected mendelian ratios. Most commonly this departure from chromosomal patterns is indicated by differences in the results of reciprocal crosses. In contrast to the maternal effect, however, the differences caused by extrachromosomal factors do not usually disappear after one generation but may persist as long as the extrachromosomal factor can perpetuate itself.

CHLOROPLASTS

The first example of cytoplasmic genes inherited solely through the maternal parent was reported by Correns in 1909. He found that a certain variety of *Mirabilis jalapa,* the four-o'clock plant, had branches that produce either green, white, or mixed green-white (*variegated*) leaves. In crosses between flowers of these branches, the offspring are all green if the maternal parent is a flower from a green branch. Furthermore such offspring remain green throughout subsequent generations as long as the maternal plant is green. Similarly, as long as the maternal parent is from a white branch, the offspring are all white. (Pure white plants, however, die for lack of photosynthesis.) If the maternal parent is mixed green and white, or variegated, then the offspring are similarly variegated. The phenotype thus depends upon a factor within the maternal cytoplasm which seems to be self-perpetuating.

In *Mirabilis* the extranuclear factor is not hard to find and is most probably the chloroplast containing the green chlorophyll. As in most other green plants such chloroplasts occasionally occur in a white defective form in which chlorophyll is absent. From the results of Correns' experiments and those of numerous others, it is extremely likely that there is genetic continuity between chloroplasts just as there is genetic continuity between nuclei. Evidence for this includes direct observations of dividing chloroplasts in *Micromonas*, an alga that has only one chloroplast and one mitochondrion. In *Spirogyra*, an abnormal form containing one mutant chloroplast lacking the pyrenoid starch body and one normal chloroplast was observed to divide so that each daughter cell received both of the two kinds of chloroplast. The fact that chloroplasts have their own DNA (Fig. 13–7) strongly supports the idea that at least some changes in chloroplast structure can be perpetuated similar to the changes governed by nuclear DNA. However, since chloroplasts are found only in the cytoplasm, their transmission to new generations occurs, as we have seen, mainly through the maternal gamete.

These results do not mean that nuclear genetic material has no effect on chloroplast production. The interrelationship that must occur between all sections of a phenotype precludes any notion that any single part of an organism develops in complete isolation from all others, especially from the effect of nuclear genes. Beginning with the investigations of Baur, numerous instances are known in which green- and white-leaved factors segregate according to a mendelian pattern; i.e., self-fertilized F_1 heterozygotes produce F_2 plants in a ratio 3 green : 1 white. As support, von Wettstein has described five stages in chloroplast development that may be affected by nuclear genes, beginning with the immature "proplastid" and ending with the mature chloroplast. In *Chlamydomonas,* Levine, Goodenough, and others have identified a wide array of nuclear genes which block specific steps in photosynthesis.

However, even when cytoplasmic factors appear to be completely independent of the nucleus, some relationship between them may nevertheless exist or have existed in the past. For example, Rhoades has demonstrated that the presence of the nuclear gene *iojap* in homozygous recessive condition in corn will cause the appearance of white-leaved sections ("striped" plants) in which the chloroplasts are in some way affected. Once formed by this nuclear process, such striping characteristics are now unaffected by the nuclear genotype and transmitted as though they were completely determined by a cytoplasmic particle inherited only through the maternal cytoplasm (Fig. 13-3). According to present concepts, it seems likely that it is the proplastid that functions as the physical link between chloroplast generations.

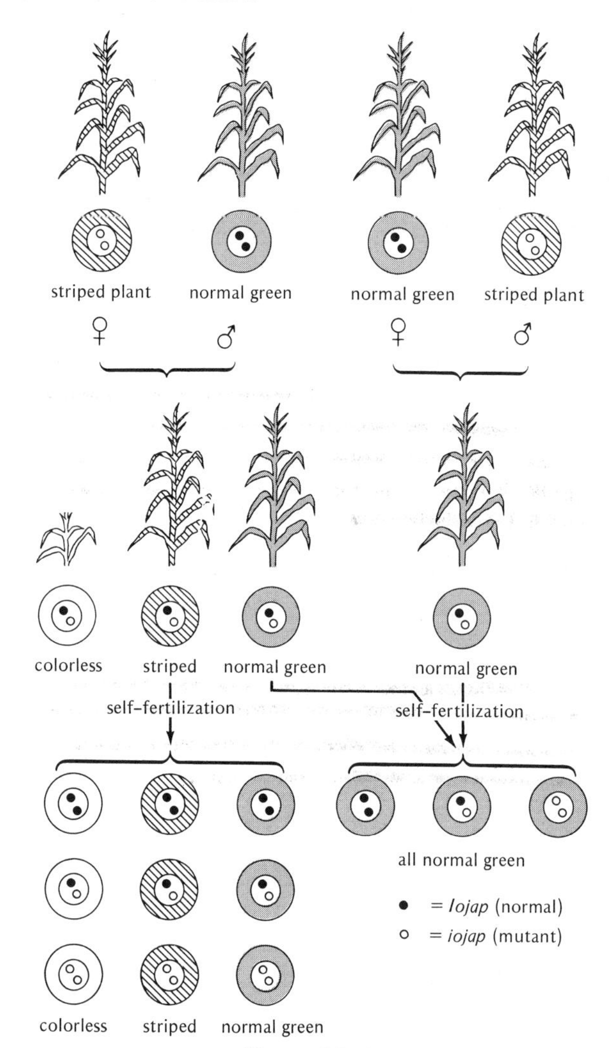

Figure 13-3

Inheritance of striping in corn. Crosses on the left show that striped female plants (ij/ij) produce three types of offspring: colorless, striped, and normal green. F_1 female plants (Ij/ij) also produce the three types of offspring, two of which are colorless or striped although their genotypes are normal Ij/Ij. On the other hand, pollen from a striped plant do not pass on the striped effect when mated to a normal green female (crosses on the right). This chloroplast effect is therefore carried by the female cytoplasm, although the initial striped effect occurs in plants homozygous for the ij genes. (After Sager.)

STREPTOMYCIN RESISTANCE IN *CHLAMYDOMONAS*

Whatever their origin, the existence of extrachromosomal factors has been demonstrated through differences between reciprocal crosses in a variety of organisms. Strangely, such reciprocal differences may occur even when the two uniting gametes are not obviously different in size. In *Chlamydomonas,* for example, the uniting gametes of each zygote are superficially identical (p. 37), and it might appear as though their contributions are equal. Sager, however, has found that an extrachromosomal factor conferring resistance to the drug streptomycin is inherited through only one particular mating type and not through the other.

Sager's findings arose from the observation that streptomycin resistance varied considerably between strains, some strains being resistant, *sr*, and some strains sensitive, *ss*. In many of the *sr* strains, resistance to streptomycin appears to be inherited in regular mendelian fashion, so that crosses of resistant strains with *ss* varieties produce 1/2 *sr* and 1/2 *ss* offspring. One resistant strain, however, *sr*-500, acts quite differently: If the *sr* parent is the "plus" mating type ($sr\ mt^+ = sr+$) and the *ss* parent is the "minus" mating type ($ss\ mt^- = ss-$), all offspring are *sr*. However, on the reciprocal cross, $sr- \times ss+$, all offspring are *ss* (Fig. 13–4). This effect does not merely last for one generation but continues as long as the plus parent is of the given streptomycin type. Even after four generations of backcrossing *sr*+ cells to an *ss*− stock, no *ss* offspring appear. Since such results cannot be explained on the basis of chromosomal segregation, in which a 1 *sr*:1 *ss* ratio would be expected each generation, they are ascribed to an extrachromosomal factor transmitted only through the plus mating type.

The mechanism that causes such uniparental inheritance is not known. It would certainly be of advantage to an organism if the number of extrachromosomal particles in the cytoplasm were re-

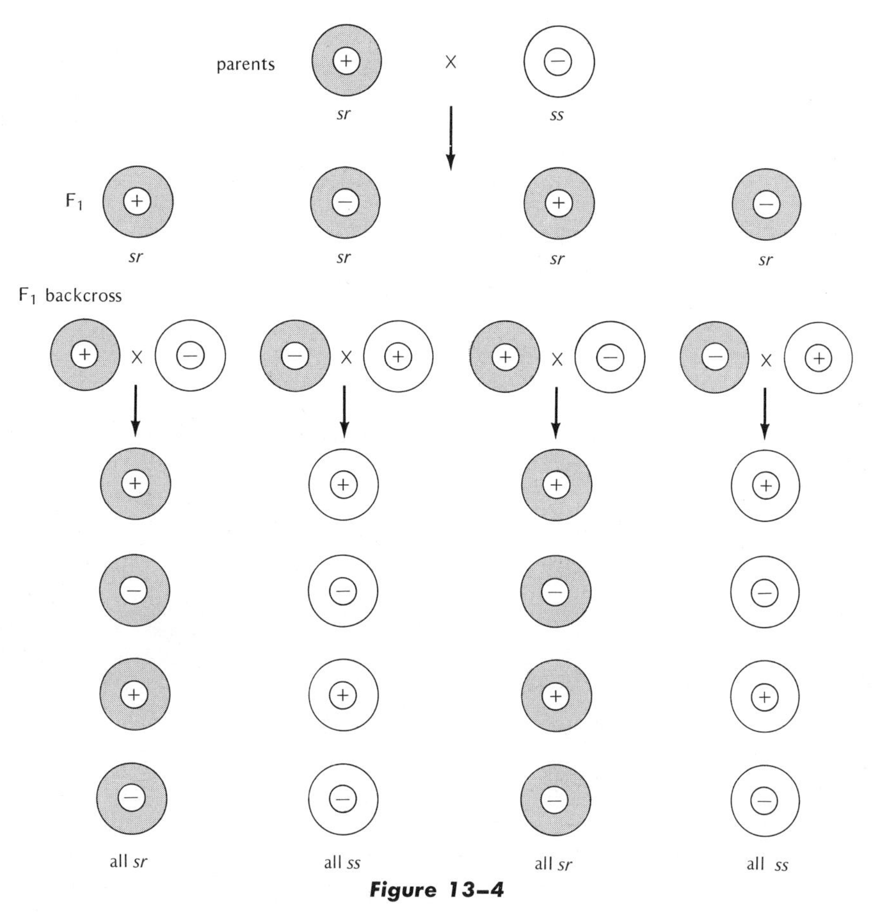

Figure 13–4

Inheritance of streptomycin resistance in Chlamydomonas reinhardi *strain* sr-500. *The two mating types, designated by + and −, are inherited chromosomally, as can be seen from the normal 2:2 mendelian ratios that segregate from a cross between the two mating types (see also Figs. 7–4 and 7–5). Streptomycin resistance, however, is usually only inherited through the + individual in the mating. (From Sager and Ryan.)*

duced before or during fertilization, so that doubling of such particles did not occur each generation. As we know, reduction division serves this purpose for chromosomes. Jinks (1964) has presented the interesting hypothesis that size differences between male and female gametes in most species serve this purpose for extrachromosomal particles. That is, the number of such particles are reduced or absent in the male gamete, and the zygote therefore receives only a single dose of such particles from the female gamete.

Since Sager's discovery of *sr*-500, a sizable number of cytoplasmic genes have been found in *Chlamydomonas*, among which are mutants that require acetate, mutants resistant to (or dependent upon) antibiotics, mutants that can only grow very slowly, and temperature-sensitive mutations. All of these mutants appear to be associated with maternal (mt^+) transmission of the single chloroplast in the species, although occasional exceptions arise with a frequency of less than 1 percent in which both parents appear to transmit their cytoplasmic genes to the zygote. Sager and Ramanis have found that ultraviolet irradiation of the mt^+ parent will convert the usual maternal inheritance to biparental inheritance in about 50 percent of the zygotes.

Biparental inheritance of cytoplasmic genes in *Chlamydomonas* as well as in yeast (see Bolotin et al.) has offered an unusual opportunity to detect segregation and assortment between such factors. Since these genes lie outside the usual chromosomal division mechanism, the segregation of different cytoplasmic genes from each other in heterozygotes is, of course, not restricted to meiosis. Instead the segregation of cytoplasmic factors seems to take place primarily after meiosis, during the mitotic divisions that form each clone of zoospores. So far a number of experiments have shown that cytoplasmic factors will segregate from each other in more or less the ratios that would be expected if two kinds were present in about equal numbers in "heterozygotes." Thus *Chlamydomonas* cells bearing both of the *acetate* cytoplasmic factors *ac1* and *ac2* will form clones maintaining each of these genes separately in about 1:1 ratio. A somewhat similar ratio is found for the segregation of streptomycin resistance (*sr*) from streptomycin sensitivity (*ss*).

RESPIRATORY DEFICIENCIES

In baker's yeast, *Saccharomyces cerevisiae*, Ephrussi and collaborators have discovered varieties that are defective in the ability to utilize oxygen in the metabolism of carbohydrates. When the sugar glucose is present, they grow to only very small-sized colonies, and are therefore called "petites." Enzyme analyses have shown that they lack components necessary for respiratory activity (cytochromes b, c_1, and cytochrome oxidase a, a_3) that are normally associated with the inner membranes of mitochondria. Such deficiencies, in addition to preventing growth on carbon sources that must be respired, also prevents petites from producing spores, a process that seems to be dependent upon oxygen respiration. Thus petite diploid cells cannot sporulate, nor can a mating between two haploid petites produce a zygote that will sporulate. Petites, however, can be maintained indefinitely in the vegetative state, either as diploid or haploid, and can be mated with normal yeast cells. When such matings are carried out, three petite varieties can be classified:

1. *Segregational* (*nuclear*) petites. Such petites, when crossed to wild type, produce ascospores which segregate in the ratio 1 petite:1 normal (Fig. 13–5a). In other words, the petite characteristic follows ordinary nuclear mendelian inheritance.
2. *Neutral* petites. Only wild-type ascospores and colonies arise from matings between such petites and normal yeast strains (Fig. 13–5b). In further generations, the petite characteristic never reappears and seems to have been physically lost. This nonmendelian behavior is certainly difficult to explain on the basis of nuclear genes and indicates that such petites are caused by an extrachromosomal particle.
3. *Suppressive* petites. Such petites seem to suppress normal respiratory behavior in

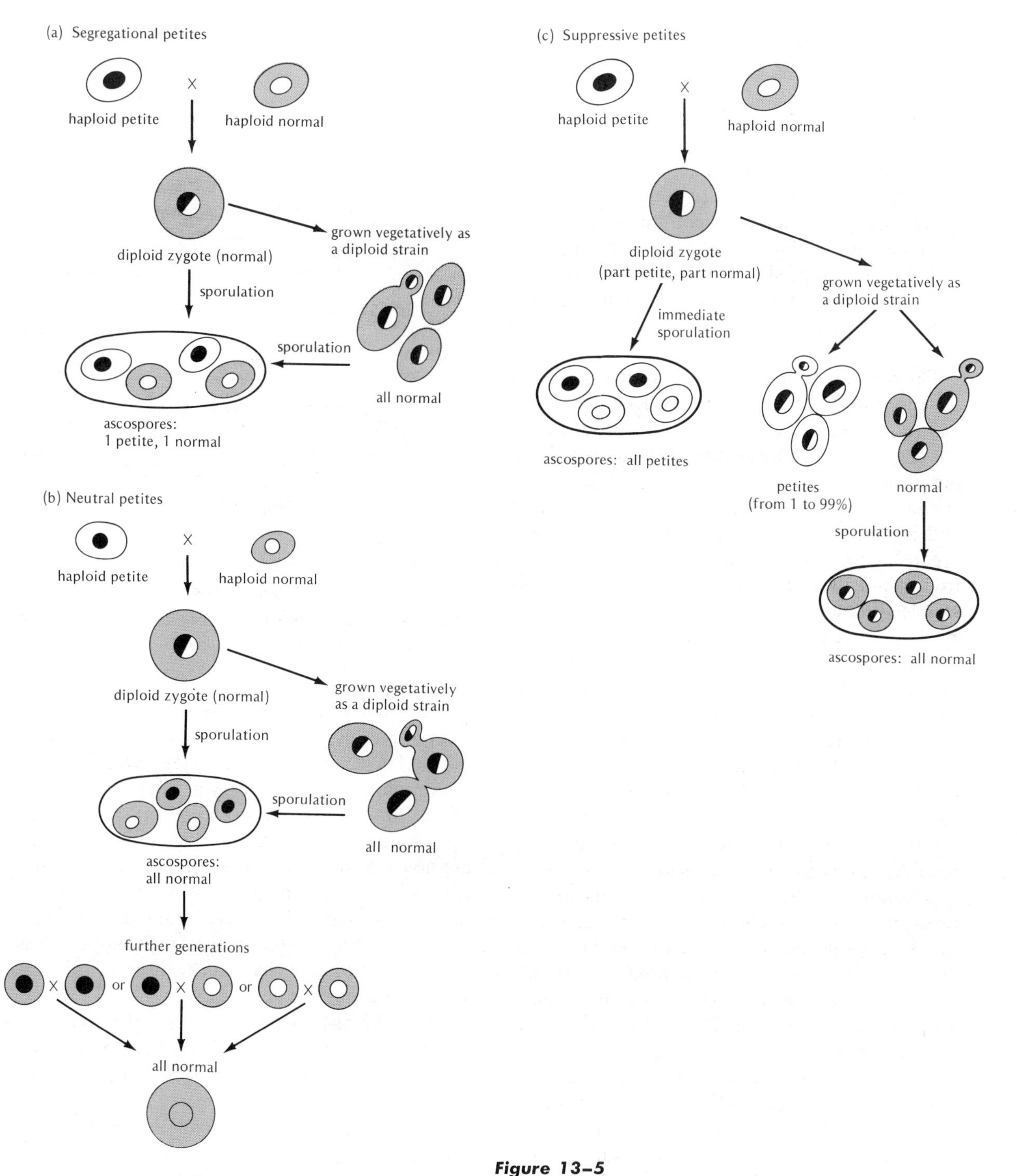

Figure 13–5

Results of crosses between different kinds of "petites" and normal strains of yeast. (Adapted from Wilkie.)

crosses with normal strains so that, in some cases, as much as 99 percent of the diploid cells derived from a zygote consists of petites (Fig. 13–5c). When suppressive × normal zygotes are induced to sporulate in a special environment, most of them give rise to asci in which all four spores are petites. The suppressive petite factor therefore acts as a "dominant," although the ratios among the diploid strains produced from such crosses are hardly mendelian and may vary from 99 percent to 1 percent petites.

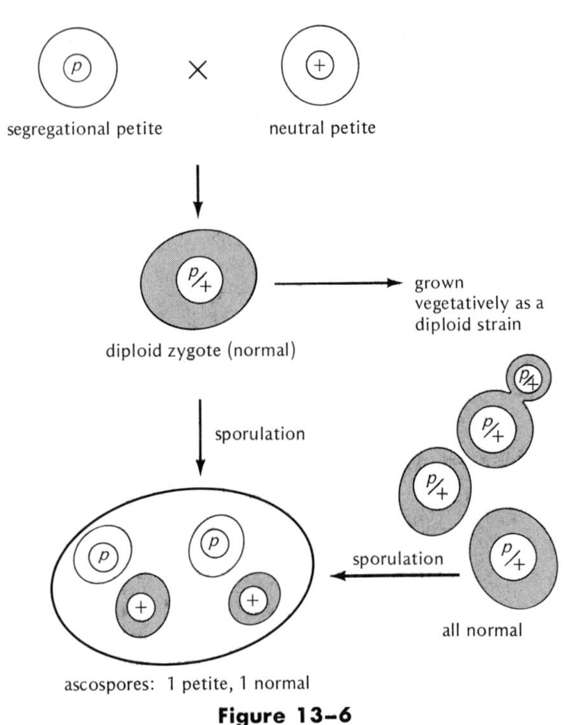

Figure 13–6

Results of a cross between a segregational petite and a neutral petite. (Adapted from Wilkie.)

Thus there seem to be two distinct genetic causes for respiratory deficiency in yeast: nuclear and extrachromosomal. On this basis a neutral petite, although cytoplasmically affected, may well have the nuclear genes for normally functioning mitochondria. A cross between a segregational petite and a neutral petite should therefore produce a zygote that can utilize the normal nuclear genes from the neutral petite and respire normally. This is indeed found to be the case (Fig. 13–6): diploid zygotes as well as diploid colonies derived from such zygotes function normally and are of normal size. Only when such diploid cells are induced to sporulate does the petite character reappear, and then in the 1:1 ratios expected from mendelian segregation. Since the neutral petite strains obviously carry the normal nuclear genes for these respiratory enzymes, the cytoplasmic factor causing the neutral petite character to appear seems relatively independent of nuclear control.

Additional evidence that strongly indicates the involvement of extrachromosomal inheritance in the petite character includes the observation that neutral petite strains are easily produced by subjecting normal strains to low doses of dyes called *acriflavines*. Since nuclear changes are very rarely noted at such doses, the high frequency of petites that are found indicates the occurrence and transmission of extranuclear changes. Also important is the finding that, like other fungi, yeast can form "heterocaryons" in which haploid nuclei from different "homocaryons" share a common cytoplasm without combining into diploid nuclei. As shown by Wright and Lederberg, when such heterocaryons are formed between neutral petites and normal strains, the heterocaryons function normally. However, when the heterocaryotic nuclei are reisolated into homocaryons identical to those of the original parent strains, some of the cells carrying the nuclei from the original petite strains are now also normal and remain normal in future generations, and some of the cells carrying normal nuclei now show the petite character. In other words, there has been a phenotypic change in loss or gain of the petite character with very little possibility of a nuclear change.

As will be discussed below, it is now also known that mitochondria have their own DNA, and the cellular division of mitochondria has been observed under the light microscope in *Micromonas* (see p. 257) and under the electron microscope in mice (Tandler et al.). The cellular continuity of mitochondria and mitochondrial DNA therefore explains the cytoplasmic continuity of the neutral and suppressive petites.

According to M. B. and H. K. Mitchell, the respiratory deficiencies observed in yeast seem to be paralleled in *Neurospora* by certain strains called "poky." These strains also exhibit slow growth, and show a reduction in some of the same respiratory cytochrome components that are missing in the *petites*. As in yeast this trait seems to be inherited in certain strains in nonmendelian fashion, indicating its extrachromosomal nature. In contrast to yeast, however, the *Neurospora* gametes are of unequal size, and the extrachromosomal nature of *poky* can be observed in reciprocal crosses. When the "female" gamete contributing the cytoplasm (protoperithecial parent) is normal, and the "male" gamete (microconidial parent) is *poky*, the resultant ascospores all produce normal colonies. When the protoperithecial parent is *poky* and the microconidial parent is normal, the resultant colonies are all *poky*. Once a colony is *poky*, it no longer segregates for normal, and vice versa. Also, as in yeast, heterocaryotic tests show that the *poky* character is associated with the cytoplasm rather than the nucleus. Nevertheless the exact defect caused by *poky* remains unknown, although recent studies have invariably localized the *poky* trait to changes involving either mitochondrial ribosomes, cytochromes, or structural proteins (see, for example, Rifkin and Luck).

CYTOPLASMIC DNA

The discovery of DNA in the cytoplasm of eucaryotes in the early 1960s offered for the first time a clear genetic basis for explaining at least some examples of extranuclear inheritance. To date, cytoplasmic DNA has been identified by Feulgen staining (p. 59), autoradiography (p. 62), electron microscopy, and density gradient centrifugation (p. 73). In the latter technique small amounts of DNA are noted which, because of their different guanine and cytosine contents, have different densities and therefore band at different points on the centrifuge tube than does nuclear DNA (see Fig. 13–7). Cytoplasmic DNA also differs from nuclear DNA in nearest neighbor frequencies (p. 78) and, when observed under the

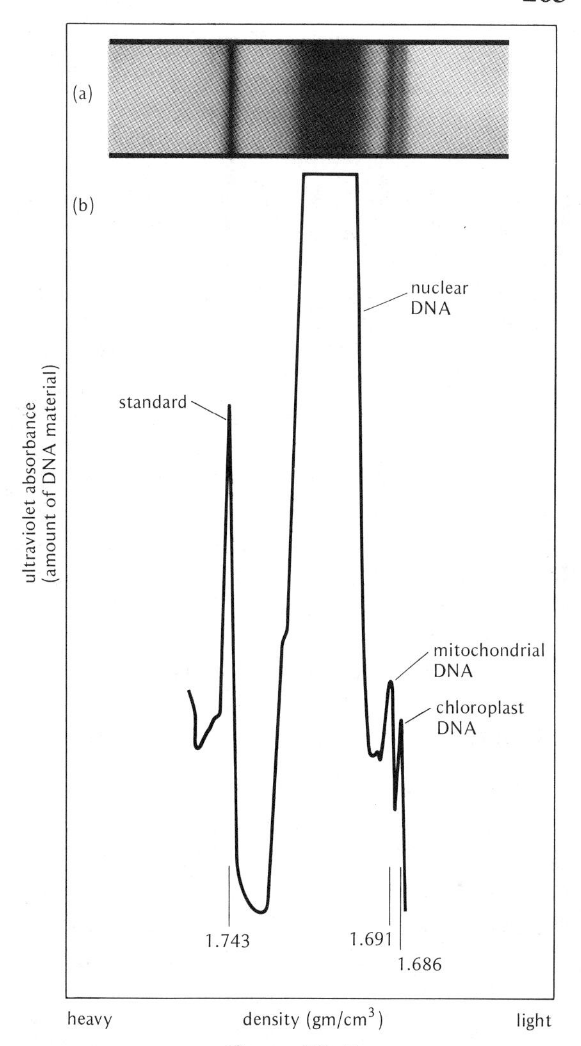

Figure 13–7

(a) The banding observed in a centrifuge tube containing DNA from Euglena gracilis (center and right), and DNA from bacteriophage SP8 used for comparison as a standard (left). (b) Ultraviolet analysis of the banding as recorded on a spectrophotometer. The 1.743 density band is the SP8 standard, and the remainder of the DNA is that of Euglena. Most of the Euglena DNA is from the nucleus and is in the central main band. The mitochondrial DNA bands at 1.691, and the chloroplast DNA at 1.686. Present estimates of the size of a molecule of Euglena chloroplast DNA are in the neighborhood of 92 × 10^6 molecular weight. (From Edelman et al.)

electron microscope, it is often considerably thinner than the protein-covered DNA molecules of the eucaryotic nuclear chromosomes. Because of its thin fibrous appearance, cytoplasmic DNA is believed to be "naked" nonhistone DNA similar to that found in procaryotes (e.g., bacteria and viruses).

The sources of cytoplasmic DNA include, so far, the chloroplasts of plants and algae, the mitochondria of both plant and animal cells, and the mitochondrialike kinetoplasts (basal flagellar organs) of some protozoan parasites.* With some exceptions, cytoplasmic DNA is often only a small percentage of cellular DNA, and is composed of molecules that range in size from $1\text{–}7 \times 10^7$ molecular weight in mitochondria to molecules about 100 times as large in chloroplasts. Both chloroplast and mitochondrial DNAs are double-stranded and have been shown to replicate semiconservatively. In fact the circular mitochondrial DNA often found in animal and fungal cells have been observed in various species to show "growing forks" similar to those seen in bacteria and viruses (pp. 87–90).

Unfortunately the exact relationship between the presence of cytoplasmic DNA and its effects is still not clear. For example, in some petite mutants very little mitochondrial DNA is discerned, whereas others seem to have a full complement. Also, some petite mutants may show a significant alteration in their guanine plus cytosine content, even to the point where the mitochondrial DNA is almost entirely composed of adenine-thymine, whereas other petite mutants show the same nucleotide composition as do normal mitochondria. Nevertheless there is little question at present that cytoplasmic DNA is active, and its transcription into RNA can be seen in electron micrographs that show RNA "bushes" being formed along the circular DNA molecules of motochondria. Some of the products of this transcription include both the ribosomal and transfer RNA molecules of mitochondria, since some of these are unique in base composition and "hybridize" (see p. 101) with mitochondrial DNA. Similar hybridizations have been observed between the RNA of ribosomes in *Euglena* chloroplasts and the chloroplast DNA.

Some messenger RNA used in the translation of proteins is probably also made on cytoplasmic DNA templates, although only a few proteins are so far known to be derived exclusively through this process. From all indications the protein-synthesizing systems controlled by the ribosomes of mitochondria and chloroplasts are mostly concerned with the inner membranous structure of these organelles. However, the extent of this synthesis is relatively small; for example, less than 10 percent of mitochondrial proteins are probably synthesized on mitochondrial ribosomes. Interestingly the transcriptional and translational processes in these organelles are inhibited by chemicals (rifampicin, chloramphenicol) similar to those that inhibit the same processes in bacteria but, again like bacteria, they are resistant to chemicals that inhibit these processes in eucaryotes (α-amanitin, cycloheximide).

*The presence of DNA in centrioles is still seriously disputed, with different experiments giving different results.

CRITERIA FOR EXTRACHROMOSOMAL INHERITANCE

On the basis of our discussion so far, several criteria emerge as useful in determining the extrachromosomal nature of an inherited trait:

1. Is there a difference between reciprocal crosses and does the inheritance of the trait continually follow the maternal line no matter what its genotype?
2. If there is no observable difference in cytoplasmic size of the uniting gametes, as in *Chlamydomonas* and yeast, do nonmendelian segregations nevertheless occur?
3. When heterocaryons can be made and homocaryons reisolated, does the trait change according to its cytoplasmic association rather than according to the nuclear genotype?
4. Can the trait be permanently changed by agents which do not affect the nucleus?
5. Can genetic material, in the form of extrachromosomal nucleic acid, be shown to be associated physically with the trait?

Although these tests cannot be done for all traits, a preponderance of positive answers to these questions in any one case would indicate extrachromosomal inheritance. The most convincing evidence would, of course, be a positive answer to the last question, since the DNA (or viral RNA) mechanism would be sufficient to account for both the appearance and self-replication of the trait. However, even if the answer to this question is negative, a trait might still be extrachromosomally determined for the following reason.

As we have seen so far, the genetic code seems to be linear—in other words, a one-dimensional code (Chapters 5 and 26). That is, information from a linear sequence of nucleotides in DNA is translated into a linear sequence of amino acids in proteins (see also Chapter 28). In addition to this type of coding, however, there might also be information transferred from one surface to another, i.e., a two-dimensional code. For example, if the configuration of the surface of a newly formed organelle is an exact counterpart of a previously existing organelle, we may conceive that the surface configurations were transferred from one to the other by contact rather than through DNA coding. Just as the growth of crystal layers may follow a pattern laid down by previous layers, similar events may occur in cells.

Recently Sonneborn and collaborators (see Sonneborn, 1970) have shown that the arrangement of granules on the cortex of *Paramecium aurelia* can be modified. These changes will maintain themselves in cells of *different* genotypes through numerous sexual and asexual divisions. It therefore seems reasonable to conclude that the parental pattern of cortical granules serves as a template for that of the offspring. Here, then, is a rare example where the usual separation between a genotype and its phenotypic expression is removed, and both "genotype" and "phenotype" appear to be identical structures.

INFECTIVE HEREDITY IN *PARAMECIA*

In 1943 Sonneborn demonstrated the inheritance of two alternative traits in strains of *Paramecium aurelia*. A strain, no. 51, carrying the trait called *killer*, had the property of destroying other strains, such as no. 47, bearing the trait called *sensitive*. The destruction of sensitives occurred through secretion of a toxic substance (paramecin) into the culture medium. Upon ingestion this substance is believed to break down the food vacuole membranes of sensitive cells. However, in spite of this toxic relationship, conjugation could nevertheless be accomplished between killers and sensitives, since *Paramecia* are resistant to killing during mating.

As previously shown (Chapter 3), such conjugation proceeds through the establishment of a narrow cytoplasmic bridge between the two conjugants and a mutual exchange of haploid nuclei across this bridge. The two resultant exconjugants are therefore diploid, each bearing identical nuclear complements. Cytoplasmically, however, very little exchange has normally occurred, and the exconjugants may be regarded as having maintained their original cytoplasm unchanged. When Sonneborn performed the killer-sensitive mating experiments described above, the exconjugants formed separate killer and sensitive clones, depending upon the parent from whom they had been derived (Fig. 13–8a). Furthermore, when autogamy or self-fertilization (p. 35) was induced in these exconjugants, segregation for nuclear differences had no effect on appearance of the killer trait. Also, when conjugation was prolonged (Fig. 13–8b), massive transfer of cytoplasm occurred between conjugants, and, in such cases, the killer trait could be transferred to the no. 47 sensitive strain. It therefore seemed evident that the killer trait was cytoplasmically inherited, and Sonneborn gave the name *kappa* to the extrachromosomal particle responsible for the trait.

Despite its cytoplasmic nature, however, the persistence of *kappa* was soon found to depend upon the presence of at least one nuclear gene. This could be seen in crosses between the killer strain, no. 51, and a different sensitive strain, no. 32. When there was no cytoplasmic exchange (Fig. 13–9a), only one of the exconjugant clones possessed *kappa*, as in previous crosses with no. 47. In the case of no. 32, however, induced autogamy

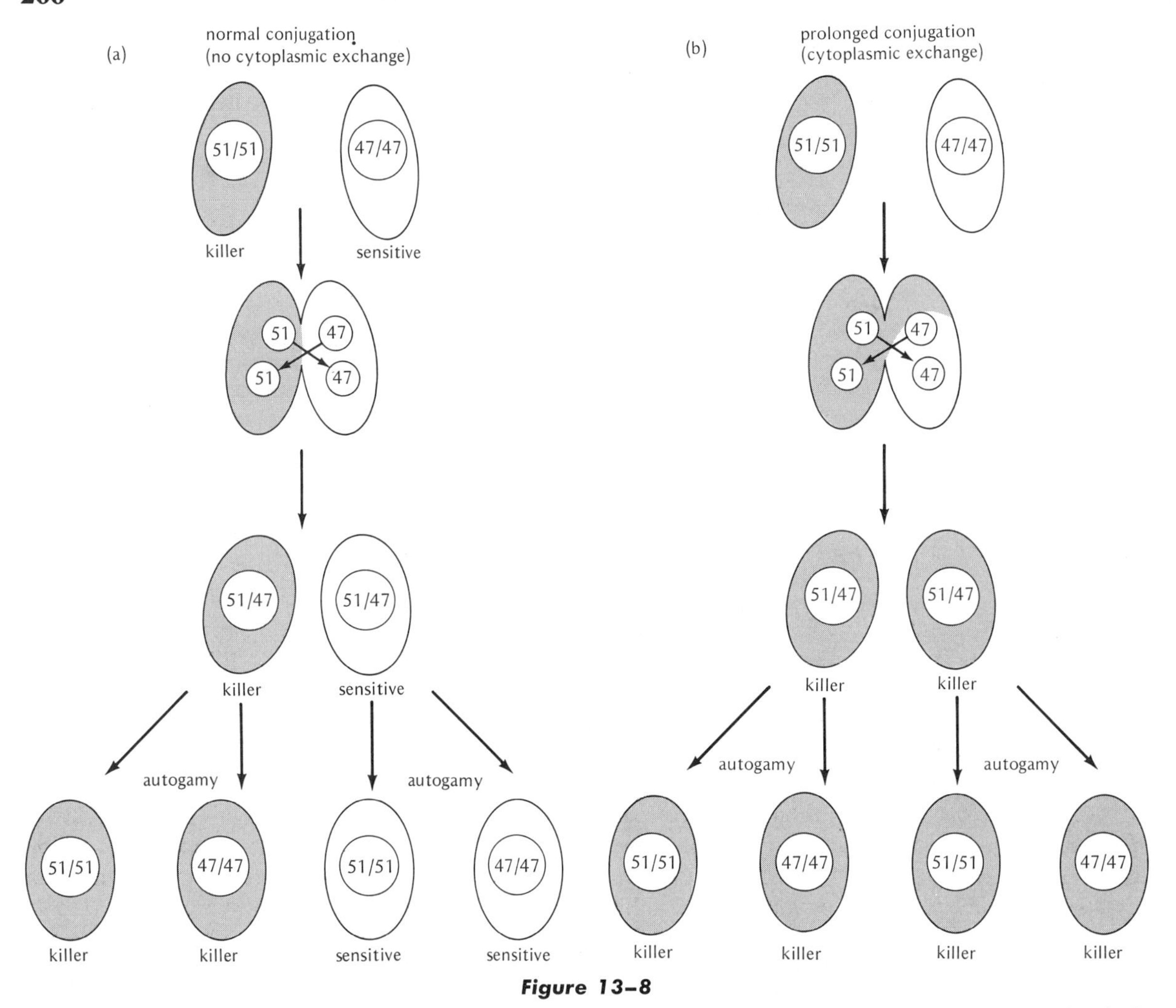

Figure 13–8

Results of crosses between killer (no. 51) and sensitive (no. 47) strains of Paramecium aurelia *without (a) and with (b) cytoplasmic exchange.*

gave rise to a sensitive clone among the offspring of this heterozygous killer exconjugant. Similarly, when a large amount of cytoplasm was exchanged and both exconjugants showed the killer trait (Fig. 13–9b), autogamy again gave rise to sensitive clones. The killer trait, or *kappa*, was thereby shown to be maintaining itself in the presence of a no. 51 or no. 47 nuclear gene, called *K*, but not in the presence of a homozygous no. 32 nuclear gene. Since both the no. 51 × no. 32 exconjugants can maintain *kappa*, *K* was apparently dominant. Thus strains 51 and 47 could be described as *KK* and strain 32 as *kk*.

Although present, however, *K* could not by itself initiate the appearance of *kappa*; i.e., the *KK* strain 47 remained sensitive until no. 51 cytoplasm was introduced. The presence of *kappa*, in addition to the gene *K*, was therefore necessary for a strain to act as a killer. When *kappa* was removed from killer strains by high temperatures, x-rays, and other techniques, these strains became sensitive. Interestingly they could revert to killer strains by

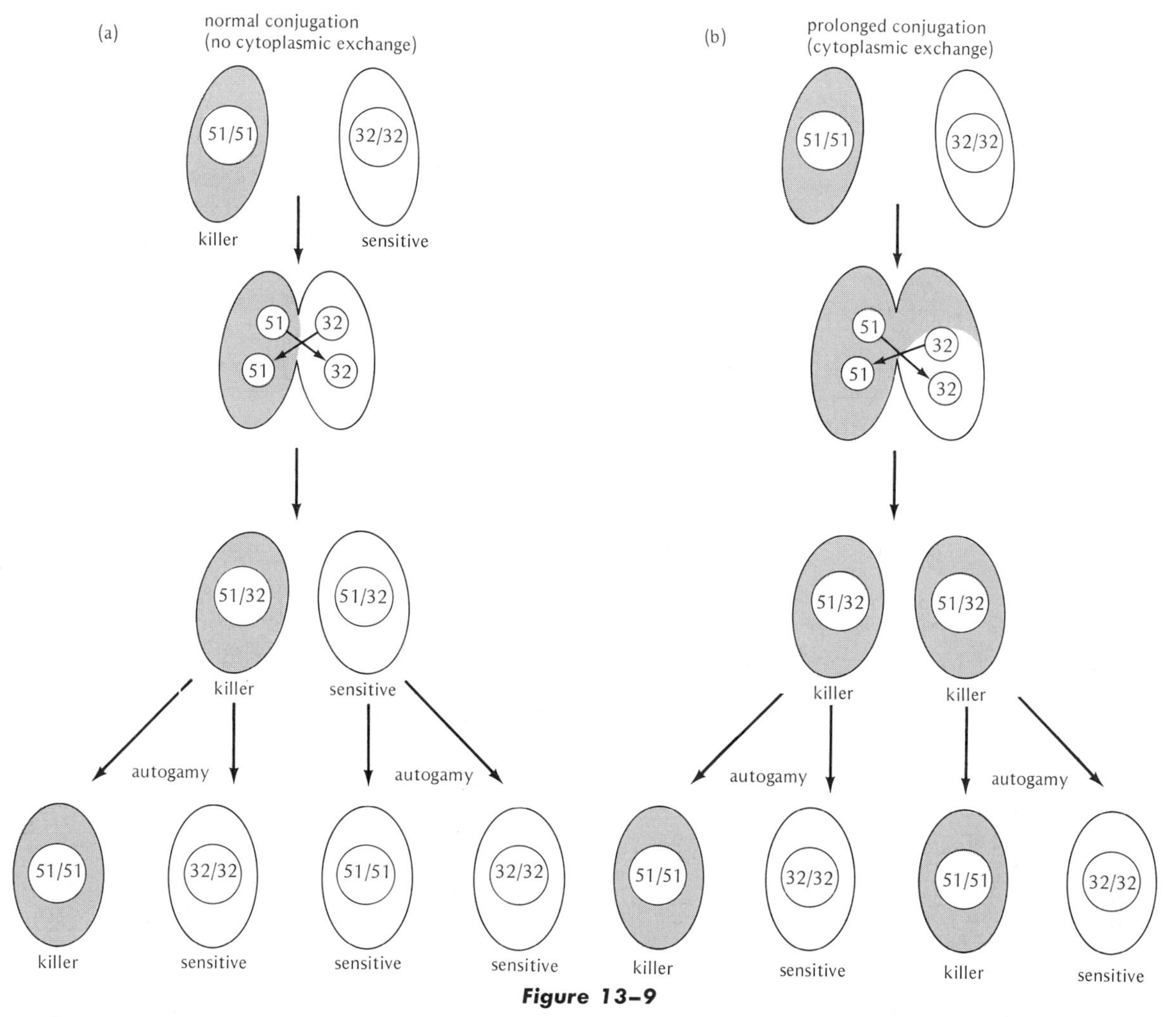

Figure 13–9

Results of crosses between killer and sensitive strains of Paramecium aurelia *when the sensitive strain (no. 32) is homozygous for a gene (k) unable to maintain killer particles. (a) Without cytoplasmic exchange. (b) With cytoplasmic exchange.*

ingesting chopped-up killer animals. Thus, although relying upon nuclear genes for maintenance, *kappa* seemed to be an infective particle with a hereditary continuity of its own.

A partial answer to the strange behavior of *kappa* was found by Preer and others who identified these particles physically. They discovered that killer animals could maintain as many as 1600 of these particles, each on the order of .2 micron in diameter or the approximate size of a small bacterium. When stained with Feulgen (p. 59), each *kappa* particle was found to contain a small amount of DNA, indicating some hereditary independence. Also, the cytochrome pigments that *kappa* utilizes in oxygen respiration were significantly different from those of its host, but similar to those of certain groups of bacteria. Thus, since *Paramecium* exists quite well without *kappa*, its presence seems to be that of an accessory organism or *symbiont*. It has therefore been hypothesized that *kappa* may have had its initial origin independent of *Paramecium*, but that at some time in the past it established this unique relationship with certain strains of these ciliates.

Surprisingly the symbiotic relationship between *kappa* and its protozoan host may be echoed by a somewhat similar relationship between a bacteriophage and the *kappa* particle itself. According to Preer and co-workers, viruses have been isolated from *kappa*, and they suggest that these are "temperate" bacteriophages (see p. 32, and Chapter 19) which occasionally multiply vegetatively and cause lysis of the bacterium. It is the bacterial host products formed by this vegetative growth of the virus which may then produce the toxic agents responsible for the killer activity (see also "phage conversion," p. 425). Thus, in a sense, cytoplasmic agents may exist within cytoplasmic agents!

Since the discovery of *kappa*, other particles have been found in *Paramecium* and given names such as *gamma*, *lambda*, and *mu*. Cytoplasmically determined killer strains have also been reported in yeast (see, for example, Somers and Bevan). Although all these particles contain nucleic acids and seem to be extrachromosomally inherited, they too, like *kappa*, depend upon the presence of particular nuclear genes.

INFECTIVE HEREDITY IN *DROSOPHILA*

One of the first examples of cytoplasmic inheritance in *Drosophila* was found by L'Héritier and Teissier for carbon dioxide (CO_2) sensitivity. In *Drosophila* most strains can be anaesthetized by exposure to prescribed levels of CO_2 without any serious aftereffects. Such strains are considered physically *resistant* to ill effects. Other strains, however, will not recover after CO_2 exposure and are called *sensitives*. This sensitivity to CO_2 is inherited on a maternal basis, since no matter to whom mated, sensitive mothers will mainly give rise to sensitive offspring. When rare resistant daughters appear in such lines, sensitivity is lost, and future offspring of such daughters are all resistant. Some sensitivity may also be transmitted through matings between sensitive males and resistant females. In such cases, however, sensitivity is quickly lost after the first generation. Transmission through sperm cytoplasm is thus relatively limited and ineffective.

This nonmendelian inheritance points strongly to an extrachromosomal basis for CO_2 sensitivity. Further evidence in this respect is the observation that elevated temperatures can, under certain circumstances, "cure" a strain of CO_2 sensitivity without affecting nuclear genes.

CO_2 sensitivity is also infective, and extracts of sensitive flies will induce sensitivity when introduced into resistant flies. This induced sensitivity shows an incubation period, increasing in strength with time, similar to the growth cycle of many viruses. L'Héritier has therefore postulated that CO_2 sensitivity is caused by a small viruslike particle named *sigma*. Although DNA determinations have not yet been made, some physical characteristics of *sigma* are known. It seems to be a particle approximately .07 micron in diameter, much smaller than *kappa*. *Sigma* has been transferred to different *Drosophila melanogaster* strains and to different species of the genus *Drosophila*. In these cases, however, the induced CO_2 sensitivity is not always of the same kind and may vary between strains and between species. Also, attempts to transfer *sigma* to other insect genera have been unsuccessful. Because of this, it seems likely that the maintenance of *sigma*, just as the maintenance of *kappa*, depends upon the presence of appropriate genes.

A second example of extrachromosomal inheritance in *Drosophila* was found by Magni among flies of *D. bifasciata* collected in Italy. A small percentage of wild females of this species produced offspring consisting of 95 percent or more daughters when raised at temperatures 21°C or lower. These predominantly female producers, called *sex-ratio* females, passed this trait on to their daughters but not to their sons. In future generations the sex-ratio condition persisted in transmission through the female line but not through the males. The deficiency of male offspring among the progeny of sex-ratio females seemed to result from the death of male embryos and pupae. However, breeding of eggs and pupae at higher temperatures (26°C) increased the frequency of surviving males. Thus sex-ratio seemed to be caused by an extrachromosomal particle similar, in some respects, to *sigma*.

The discovery of an agent that could be responsible for the sex-ratio trait arose mainly from studies of Malogolowkin, Poulson, and Sakaguchi in *D. willistoni*. They found that this trait was infective and could be transferred by injection into previously normal *willistoni* females. As in *sigma*, an incubation period is necessary for the trait to manifest itself, indicating a growth cycle of an independent particle. This particle was physically isolated and found to be a protozoan spirochete, probably of the genus *Treponema*. Although flourishing and reproducing itself in both sexes, its predominantly lethal effect in sex-ratio lines occurred in male zygotes. In spite of its self-reproducing ability, however, the spirochete does show dependence upon nuclear genes for its persistence. It will not, for example, persist in the presence of particular genes nor infect certain strains and certain species. Even when it does infect a particular strain, its ability to cause a disturbance in sex ratio may depend upon the genotype of the strain.

CONCLUSIONS

We can perhaps divide extrachromosomal inheritance into two main types: one based on the transmission of surface configurations such as those discussed by Sonneborn (p. 265), and the other based on the transmission of nucleic acids in extrachromosomal particles. The first form of transmission, that of cortical patterns, seems unaffected by both chromosomal and extrachromosomal genes, and appears therefore as an extreme departure from our standard genetic system. Nevertheless it exists in *Paramecium*, *Tetrahymena* (see Nanney), and some other unicellular organisms. As Sonneborn (1970) points out, cortical inheritance may serve a general developmental function, to enable "existing molecular assemblies to determine the placement and orientation of new ones." Because such a transmission mechanism may (except for its component substructures) be largely independent of other cellular constituents, it can provide an unchanging pattern, and thereby enable an organism to preserve complex features that must be transmitted without error. However, the extent to which non-DNA patterns are transmitted has been hardly explored, and very little is yet known of their importance in multicellular organisms.

Extrachromosomal inheritance based on nucleic acid transmission has provided, so far, the main area of investigation, and two general forms can be distinguished: (1) instances where the inherited cellular particles are necessary for the normal functioning of the cells, such as chloroplasts and mitochondria; and (2) instances where the inherited extrachromosomal particles are not normally necessary for cell function, such as *kappa, sigma,* and *sex-ratio* spirochetes. Some of the differences between the two forms reside in their mode of inheritance: not only are particles such as *kappa, sigma,* and *sex ratio* self-reproducing, but they are also *infective* to some degree since it is through this means that they can be introduced into new hosts. [This infectivity has also been found to be a property of certain viruslike particles, called *episomes* (discussed in Chapter 19), which are known to occur in bacteria and are capable of attaching themselves to chromosomes.] It has therefore been seriously debated whether to consider these particles as parasites and symbionts or as extrachromosomal organelles. It is certainly possible, according to explanations by Darlington and others, that some of these particles may begin as parasites and symbionts and eventually be incorporated into the normal functioning of the cell as organelles.

Recently Margulis, Stanier, and others have furthered the claim that the many strong similarities that mitochondria and chloroplasts bear to procaryotes such as bacteria and algae probably arise because these cytoplasmic organelles are descendants of ancient procaryotes captured by early eucaryotic cells. As in procaryotes, mitochondrial and chloroplast DNA do not contain histones, are often in circular form, and their transcriptional and translational processes are affected by similar agents. In addition the DNA of these organelles is unbounded by a nuclear membrane and their ribosomes are mostly smaller in size than those of eucaryotes, resembling, in some instances, the ribosomes of procaryotes. Schlanger and co-workers have also recently shown that the relationship between some antibiotic-resistance genes on the chloroplast DNA of *Chlamydomonas* may be similar to the relationship between genes with the same function in the procaryote *Bacillus subtilis.* John and Whatley point out that the aerobic bacterium *Paracoccus denitrificans* possesses many enzymatic, membranous, and respiratory mitochondrion-like features.

These resemblances and others have suggested that at least some of the oxidative functions of mitochondria and photosynthetic functions of chloroplasts are being maintained through the preservation of procaryotic systems in present eucaryotic cells. According to this conjectured lineage, early eucaryotic cells were predatory (heterotrophic) and unable to utilize oxygen (anaerobic). Then, through ingestion and subsequent symbiosis with an aerobic procaryote, some eucaryotic cells developed the ability to utilize oxygen via these new mitochondrionlike organelles. One branch of these aerobic eucaryotes then set up a similar symbiotic relationship with a chloroplastlike algal cell and thus became ancestors of present-day autotrophic plants. Symbiotic relationships between blue-green algae and various protozoa are even now occasionally discovered; numerous such opportunities must have existed in the past. However, acceptance of this phylogeny depends strongly on the anaerobic origin of eucaryotes, and contrary views have been presented by Raff and Mahler and by Uzzell and Spolsky.

In summary, although documented instances of extrachromosomal inheritance exist, and we may even envisage their origin, such modes of transmission are probably rare compared to the usual chromosomal heredity. One of the reasons for this is simply that normal cell division mechanisms ensure equal partition of hereditary chromosomal determinants (nuclear genes) and their transmission to all daughter cells. Ensured transmission of independent cytoplasmic factors to daughter cells, on the other hand, depends upon the possesion of a large number of such factors, so that cell division will not leave either daughter cell without at least one cytoplasmic representative. Cytoplasmic factors depending upon such events may more easily be lost than chromosomal factors, especially since male gametes in higher organisms have very little cytoplasm, and such transmission relies almost entirely on the maternal cytoplasm. Cytoplasmic factors may, of course, utilize the cell-division spindle apparatus to ensure their transmission as a kind of independent chromosome. This, however, demands development of an elaborate mechanism, as has been observed in the division of mitochondria in scorpion spermatogenesis (Wilson, 1931). As yet it is doubtful that such exact division mechanisms operate for the transmission of cytoplasmic factors other than mitochondria and chloroplasts. Thus, although extrachromosomal inheritance is of extreme interest, it presents serious difficulties in ascertainment and analysis.

PROBLEMS

13-1. A particular strain of chickens shows an unusual trait, *T.* Show what procedures you would follow to determine whether this trait is caused by genes on nuclear chromosomes, or by cytoplasmic genes, or involves a "maternal effect," or is caused solely by the environment.

13-2. How would you differentiate between a sex-linked trait and a cytoplasmically inherited trait?

13-3. Bateson, Saunders and Punnett found two kinds of pollen in sweet peas (*Lathyrus odoratus*): long type and round type. When a pure-breeding plant of the long type was crossed to a pure-breeding plant of the round type, the F_1 all showed long pollen. A cross, $F_1 \times F_1$, produced an F_2 in the proportion of three-quarters long pollen to one-quarter round pollen plants. Since this trait is apparently caused by a single pair of genes, why does the F_1 produce only long pollen if half of the pollen is carrying the allele for long, and the other half is carrying the allele for round?

13-4. In crosses between the two closely related

Drosophila species, *D. neorepleta* and *D. repleta*, Sturtevant found that *D. neorepleta* carried a dominant autosomal gene that could transform XX individuals into malelike intersexes if both X chromosomes came from the *repleta* stock. Interestingly, the intersexes do not have to carry the dominant gene themselves but become intersexes when their mothers carry this gene. (a) If a female is heterozygous for this gene and carries *repleta* X chromosomes, what are the expected sexual phenotypes among her offspring if she is mated to a *repleta* male? (b) What are the expected sexual phenotypes if her male offspring are mated to *D. repleta* females?

13-5. Reciprocal crosses between two types of evening primrose, *Oenothera hookeri* and *O. muricata*, were shown by Renner to produce different effects on plastids.

hookeri ♀ × *muricata* ♂ → yellow plastids
muricata ♀ × *hookeri* ♂ → green plastids

Since the chromosome constitution is the same in both cases, explain the difference. (As will be discussed in Chapter 22, chromosome complications exist in *Oenothera*, but these are omitted here for the sake of simplicity.)

13-6. In plants from the above *muricata* ♀ × *hookeri* ♂ cross, Renner found occasional sections of yellow tissue which were not caused by mutation. These yellow sections produced flowers, and could then be bred so their cytoplasm was combined with a *hookeri* nucleus. When this nucleocytoplasmic combination occurred, the plastids were green. How would you explain the origin of the yellow sections in the *muricata* ♀ × *hookeri* ♂ cross?

13-7. If yellow color in a plant (in contrast to the normal green) can be caused by either a recessive autosomal gene or by a cytoplasmic factor, what results would you expect for each type of inheritance from the following: (a) Reciprocal crosses between pure-breeding yellow and green strains? (b) Backcrosses between the products of each of these reciprocal crosses to the parental strains? (c) Self-fertilization of the products of each reciprocal cross? (d) Cross-fertilization between the products of the two reciprocal crosses?

13-8. In crosses using *Lychnis* species, Winge found that the male plants did not generally show variegation (see p. 257), apparently because of the presence of the Y chromosome. Other chromosomes however also affected variegation, and a number of autosomal dominant inhibitors of variegation were found. In one case, Winge crossed a variegated female plant with a green male and obtained offspring of which approximately one-quarter of the females were variegated, three-quarters of the females were not, nor were any of the males. How would you explain these results?

13-9. In another case using *Lychnis*, Winge crossed a green female plant with a green male and obtained 2458 offspring which consisted of 1216 females and 1242 males. Of the females, 127 were variegated, whereas none of the males were variegated. Confining yourself to independently assorting chromosomal genes, how would you explain these results?

13-10. A streptomycin-resistant strain of *Chlamydomonas* carrying resistant factors located both in the nucleus and in the cytoplasm was crossed to a streptomycin-sensitive strain. (a) What results would you expect if the resistant strain was the plus parent and the sensitive strain the minus parent? (b) If the reciprocal cross was made?

13-11. To show that cytoplasmic genes could "segregate" from each other in those instances in which mt^- as well as mt^+ parents contributed their cytoplasm to zygotes (see p. 260), Sager and Ramanis made use of two kinds of cytoplasmic mutants in *Chlamydomonas: streptomycin-sensitive* (*ss*) in which growth cannot occur in the presence of streptomycin, and *streptomycin-dependent* (*sd*) in which growth cannot occur in the absence of streptomycin. Their crosses were of the following kinds: (1) *sd* mt^+ × *ss* mt^-; (2) *ss* mt^+ × *sd* mt^-. Each cross was performed on streptomycin-containing agar and on streptomycin-free agar. Explain in which crosses and on which media you would expect to isolate zoospores that are heterozygous for the two cytoplasmic genes.

13-12. As discussed in this chapter, Sonneborn showed that a cross between the killer strain of *Paramecium* (no. 51) and the sensitive strain (no. 32) produced an F_1 that were killers (*Kk* + *kappa*) and sensitives (*Kk*). What results would you expect from: (a) A cross between two F_1 killers? (b) A cross between an F_1 killer and an F_1 sensitive? (c) Autogamous reproduction of the F_1 sensitive? (d) Crosses of a *KK* sensitive [produced in (c)] to an F_1 killer?

13-13. Humphrey has discovered a gene, *o*, in the axolotl that is indistinguishable from wild type in the heterozygous condition but produces marked effects in homozygotes. Males that are *o/o* are sterile, and females that are *o/o* produce abnormal embryos no matter to whom they are mated. Homozygous *o/o* females, however, will produce normal offspring from a +/+ normal ovary that is transplanted into their abdomen. Explain what effect is responsible for these results.

13-14. Corn is a hermaphrodite producing both pollen and ovules on the same plant. One strain of corn obtained by Rhoades, however, was almost entirely male-sterile, yielding little or no pollen. Backcrosses of this male-sterile line to male-fertile lines did not restore fertility, although all chromosomes were soon of male-fertile origin. When the rare pollen produced by the male-sterile line was used to fertilize male-fertile plants, the offspring were all male-fertile. How would you explain these results?

13–15. A chromosomally located *Restorer* gene found in corn acts as a dominant gene in suppressing the effect of the male-sterility factor described in Problem 14 without modifying the factor itself. (a) If a plant homozygous for this *Restorer* gene was crossed to a male-sterile plant, what would be the genetic and phenotypic constitution of the F_1 with respect to sterility? (b) If the F_1 plants were fertilized by the pollen of plants that did not bear the *Restorer* gene, what results would you expect? (c) If the F_1 plants were self-fertilized, what results would you expect?

13–16. Reciprocal crosses between the two *Drosophila* species *D. melanogaster* and *D. simulans* were found by Sturtevant to produce the following results:

melanogaster ♀ × *simulans* ♂ = only females
melanogaster ♂ × *simulans* ♀ = males + few or no females

How would you explain these findings?

13–17. Jollos and others have demonstrated that persistent abnormalities ("dauermodifikationen") can be induced in plants and animals by chemical and environmental treatment and carried for several generations by the female line. In the bean *Phaseolus vulgaris*, for example, Hoffman found that morphological abnormalities in leaf shape induced by chloral hydrate persisted in the female line for almost six generations before disappearing. How would you attempt to explain the origin of these modifications and the fact that they do not persist beyond the early generations?

13–18. One of the consequences of maternal influence on egg cytoplasm (see pp. 254–256) may be lethality, as seen in the offspring produced by *Drosophila melanogaster* females homozygous for the autosomal gene, *bicaudal*. The embryos produced by such females are grossly misshapen, with their head and thoracic structures replaced by a second abdomen. (The replacement of one body part by a different one is called *homoeosis*, and mutations causing such effects are called *homoeotic*.) In the case of *bicaudal*, the embryonic lethal phenotype depends only on the maternal genotype, and not on that of the offspring. How would you establish a stock that would continually perpetuate the *bicaudal* gene without the need for selecting special mating pairs each generation?

REFERENCES

Barath, Z., and H. Kuntzel, 1972. Cooperation of mitochondrial and nuclear genes specifying the mitochondrial genetic apparatus in *Neurospora crassa*. *Proc. Nat. Acad. Sci.*, **69**, 1371–1374.

Beale, G. H., 1965. Genes and cytoplasmic particles in *Paramecium*. In *Cellular Control Mechanisms and Cancer*, P. Emmelot and O. Muhlbock (eds.). Elsevier, Amsterdam, pp. 8–18.

Bell, A. E., 1954. A gene in *Drosophila melanogaster* that produces all male progeny. *Genetics*, **39**, 958.

Beisson, J., and T. M. Sonneborn, 1965. Cytoplasmic inheritance of the organization of cell cortex in *Paramecium aurelia*. *Proc. Nat. Acad. Sci.*, **53**, 275–282.

Bolotin, M., D. Coen, J. Deutsch, B. Dujon, P. Netter, E. Petrochilo, and P. P. Slonimski, 1971. La recombinaison des mitochondries chez *Saccharomyces cerevisiae*. *Bull. Inst. Pasteur*, **69**, 215–239.

Boycott, A. E., and C. Diver, 1923. On the inheritance of sinistrality in *Limnaea peregra*. *Proc. Roy. Soc. Lond.* (B), **95**, 207–213.

Caspari, E., 1936. Zur analyse der Matroklinie der vererbung in der α-serie der augenfarbenmutationen bei der Mehlmotte (*Ephestia kühniella*). *Zeit. Induk. Abst. u. Vererbung*, **71**, 546–555.

Chun, E. H. L., M. H. Vaughan, Jr., and A. Rich, 1963. The isolation and characterization of DNA associated with chloroplast preparations. *Jour. Mol. Biol.*, **7**, 130–141.

Cohen, S., 1973. Mitochondria and chloroplasts revisited. *Amer. Sci.*, **61**, 437–445.

Colaianne, J. J., and A. E. Bell, 1972. The relative influence of sex of progeny on the lethal expression of the *sonless* gene in *Drosophila melanogaster*. *Genetics*, **72**, 293–296.

Conklin, E. G., 1903. The cause of inverse symmetry. *Anat. Anz.*, **23**, 577–588.

———, 1922. *Heredity and Environment,* 4th ed. Princeton Univ. Press, Princeton, N.J.

CORRENS, C., 1909. Vererbungsversuche mit blass (gelb) grünen und bunt blättrigen Sippen bei Mirabilis, Urtica, und Lunaria *Zeit. Induk. Abst. u. Vererbung.,* **1,** 291–329.

EDELMAN, M., J. A. SCHIFF, and H. T. EPSTEIN, 1965. Studies of chloroplast development in *Euglena.* XII. Two types of satellite DNA. *Jour. Mol. Biol.,* **11,** 769–774.

EPHRUSSI, B., 1953. *Nucleo-Cytoplasmic Relations in Microorganisms.* Oxford Univ. Press, London.

JINKS, J. L., 1964. *Extrachromosomal Inheritance.* Prentice-Hall, Englewood Cliffs, N.J.

JOHN, P., and F. R. WHATLEY, 1975. *Paracoccus denitrificans* and the evolutionary origin of the mitochondrion. *Nature,* **254,** 495–498.

KIRK, J. T. O., and R. A. E. TILNEY-BASSET, 1967. *The Plastids: Their Chemistry, Structure, Growth, and Inheritance.* W. H. Freeman, San Francisco.

LEVINE, R. P., and U. W. GOODENOUGH, 1970. The genetics of photosynthesis and of the chloroplast in *Chlamydomonas reinhardi. Ann. Rev. Genet.,* **4,** 397–408.

L'HÉRITIER, Ph., 1970. *Drosophila* viruses and their role as evolutionary factors. *Evol. Biol.,* **4,** 185–209.

MAGNI, G. E., 1953. Sex-ratio, a non-mendelian character in *Drosophila bifasciata. Nature,* **172,** 81.

MALOGOLOWKIN, C., and D. F. POULSON, 1957. Infective transfer of maternally inherited abnormal sex-ratio in *Drosophila willistoni. Science,* **126,** 32. (Reprinted in the collection of Levine; see References, Chapter 1.)

MANGE, A. P., and L. SANDLER, 1973. A note on the maternal effect mutants *daughterless* and *abnormal oocyte* in *Drosophila melanogaster. Genetics,* **73,** 73–86.

MARGULIS, L., 1970. *Origin of Eukaryotic Cells.* Yale Univ. Press, New Haven.

MITCHELL, M. B., and H. K. MITCHELL, 1952. A case of "maternal" inheritance in *Neurospora crassa. Proc. Nat. Acad. Sci.,* **38,** 442–449.

NANNEY, D. L., 1968. Cortical patterns in cellular morphogenesis. *Science,* **160,** 496–502.

NASS, M. M. K., S. NASS, and B. A. AFZELIUS, 1965. The general occurrence of mitochondrial DNA. *Exp. Cell. Res.,* **37,** 516–539.

OEHLKERS, F., 1964. Cytoplasmic inheritance in the genus *Streptocarpus. Adv. in Genet.,* **12,** 329–370.

POULSON, D. F., 1963. Cytoplasmic inheritance and hereditary infections in *Drosophila.* In *Methodology in Basic Genetics.,* W. J. Burdette (ed.). Holden-Day, San Francisco, pp. 404–424.

POULSON, D. F., and B. SAKAGUCHI, 1961. Nature of sex-ratio agent in *Drosophila. Science,* **133,** 1489–1490. (Reprinted in the collection of Levine; see References, Chapter 1.)

PREER, J. R., Jr., 1950. Microscopically visible bodies in the cytoplasm of the "killer" strains of *Paramecium aurelia. Genetics,* **35,** 344–362.

———, 1971. Extrachromosomal inheritance: Hereditary symbionts, mitochondria, chloroplasts. *Ann. Rev. Genet.* **5,** 361–406.

PREER, J. R., Jr., L. B. PREER, and A. JURAND, 1974. *Kappa* and other endosymbionts in *Paramecium aurelia. Bact. Rev.,* **38,** 113–163.

PREER, J. R., Jr., L. B. PREER, B. RUDMAN, and A. JURAND, 1971. Isolation and composition of bacteriophage-like particles from *kappa* of killer *Paramecia. Molec. Gen. Genet.,* **111,** 202–208.

RAFF, R. A., and H. R. MAHLER, 1972. The nonsymbiotic origin of mitochondria. *Science,* **177,** 575–582.

RAVEN, P. H., 1970. A multiple origin for plastids and mitochondria. *Science,* **169,** 641–646.

RHOADES, M. M., 1946. Plastid mutations. *Cold Sp. Harb. Symp.,* **11,** 202–207.

RIFKIN, M. R., and D. J. L. LUCK, 1971. Defective production of mitochondrial ribosomes in the *poky* mutant of *Neurospora crassa. Proc. Nat. Acad. Sci.,* **68,** 287–290.

RIS, H., and W. PLAUT, 1962. Ultrastructure of DNA-containing areas in the chloroplast of *Chlamydomonas. Jour. Cell Biol.,* **13,** 383–391.

SAGER, R., 1972. *Cytoplasmic Genes and Organelles.* Academic Press, New York.

SAGER, R., and Z. RAMANIS, 1967. Biparental inheritance of nonchromosomal genes induced by ultraviolet in radiation. *Proc. Nat. Acad. Sci.,* **58,** 931–937.

SAGER, R., and F. J. RYAN, 1961. *Cell Heredity.* John Wiley, New York.

SANDLER, L., 1972. On the genetic control of genes located in the sex-chromosome heterochromatin of *Drosophila melanogaster. Genetics,* **70,** 261–274.

SCHLANGER, G., R. SAGER, and Z. RAMANIS, 1972. Mutation of a cytoplasmic gene in *Chlamydomonas* alters chloroplast ribosome function. *Proc. Nat. Acad. Sci.,* **69,** 3551–3555.

SEECOF, R., 1968. The sigma virus infection of *Drosophila melanogaster. Curr. Topics Microbiol. and Immun.* **48,** 59–93.

SOMERS, J. M., and E. A. BEVAN, 1969. The inheritance of the killer character in yeast. *Genet. Res.,* **13,** 71–83.

SONNEBORN, T. M., 1959. Kappa and related particles in *Paramecium. Adv. in Virus Res.*, **6,** 229–356.

———, 1970. Gene action in development. *Proc. Roy. Soc. Lond.* (*B*), **176,** 347–366.

SPURWAY, H., 1948. Genetics and cytology of *Drosophila subobscura.* IV. An extreme example of delay in gene action, causing sterility. *Jour. Genet.*, **49,** 126–140.

STANIER, R. Y., 1970. Some aspects of the biology of cells and their possible evolutionary significance. *Symp. Soc. Gen. Microbiol.*, **20,** 1–38.

STURTEVANT, A. H., 1923. Inheritance of direction of coiling in *Limnaea. Science*, **58,** 269–270.

TANDLER, B., R. A. ERLANDSON, A. L. SMITH, and E. L. WYNDER, 1969. Riboflavin and mouse hepatic cell structure and function. *J. Cell Biol.*, **41,** 477–493.

UZZELL, T., and C. SPOLSKY, 1974. Mitochondria and plastids as endosymbionts: A revival of special creation? *Amer. Sci.*, **62,** 334–343.

WETTSTEIN, D. VON, 1957. Genetics and the submicroscopic cytology of plastids. *Hereditas*, **43,** 303–317.

WILKIE, D., 1964. *The Cytoplasm in Heredity.* Methuen, London.

WILSON, E. B., 1931. The distribution of sperm-forming materials in scorpions. *Jour. Morph.*, **52,** 429–483.

WOLFE, S. L., 1972. *Biology of the Cell.* Wadsworth Publ. Co., Belmont, Calif., Chap. 16.

WRIGHT, R. E., and J. LEDERBERG, 1957. Extranuclear transmission in yeast heterokaryons. *Proc. Nat. Acad. Sci.*, **43,** 919–923.

14
QUANTITATIVE INHERITANCE

Most of the observations upon which early genetic laws were based arose from obvious differences in the general appearance of phenotypes, such as those found between short and tall plants, green and yellow seeds, etc. These qualitative differences in color, size, and other attributes at first appeared to be the main function of gene activity. Of course, smaller and less striking *quantitative* differences were also noted, especially with regard to measurable characters such as size. Mendel's own short pea plants were not uniformly small, nor were his tall plants uniformly large. Instead gradations could be observed from short-"short" to tall-"short" and from short-"tall" to tall-"tall." These quantitative variations were obviously difficult to analyze genetically, since their phenotypic range appeared to be *continuous*, without distinct steps that could be accounted for by distinct genes. In fact most measurable characters follow a normal distribution curve ranging from low to high values but without sharp separation in values between the many classes within this range (Fig. 8–7). It was easy to conceive that these small differences arose because of slight environmental differences between individuals who may really have been genetically alike. Early geneticists such as Bateson and DeVries therefore proposed that continuous characters were not genetically produced and were therefore not inherited; only large *discontinuous* variations or differences were caused by genes.

At the same time this view was being promulgated, a different approach had also established its roots among a number of English geneticists and statisticians. Galton, a cousin of Charles Darwin, had, in the last half of the nineteenth century, founded a group which sought to explain the origin of observable differences between humans. In respect to both physical and mental characteristics, Galton noted that many of these differences were of small, barely perceptible steps. Furthermore, such measurable or, "metrical," characteristics appeared to be inherited, so that taller individuals, for example, produced taller children on the average. To measure the degree to which such characters were inherited, new biometrical techniques such as correlation and regression (discussed in Chapter 15) were invented. Thus, although the segregation and assortment of individual hereditary factors could not be determined, the "biometricians" were able to demonstrate statistically that there were nevertheless likenesses between relatives with respect to numerous continuously distributed quantitative traits. Their analyses indicated that individuals seemed to be a mixture of parental characteristics, who were in turn a mixture derived from their grandparents, etc. In the absence of detailed knowledge about genes, it therefore appeared that many of these continuous traits were determined by a mixing or "blending" process.

Soon after the rediscovery of Mendel's experiments, these different points of view formed the basis for two main groups among geneticists: (1) "mendelians," who proposed that all evolutionarily important heritable differences are qualitative and discontinuous; (2) "biometricians," who proposed that heritable variation is basically quantitative and continuous and who denied that genes existed as separate units. It seemed difficult to reconcile these views, at first, since there was no evidence that genes, which are presumably particulate and discontinuous, could account for the inheritance of phenotypes that are intermediate and continuous. Similarly how could the inheritance of discontinuous traits be explained by the same blending mechanisms that presumably produced continuous traits?

JOHANNSEN'S PURE LINES

One of the first steps taken to resolve the conflict between mendelians and biometricians arose from a detailed breeding analysis of a quantitative character in beans (*Phaseolus vulgaris*). Johannsen, a Danish geneticist, measured the weight of princess beans in a commercial seed lot and found numerous sizes ranging from "light" (15 centigrams) to "heavy" (90 centigrams). On first appearance there were ostensibly no genetic differences between light and heavy seeds, since some of the light seeds produced as wide a range of weights among their offspring as some of the heavy seeds. However, by performing a unique experiment, Johannsen demonstrated that, although superficially hidden, many genetic differences actually existed between these seeds. Johannsen's technique, based on inbreeding, was to separate the beans into various distinctive lines.

Since beans are highly self-pollinating, it was easy to establish different highly inbred stocks by self-fertilizing successive generations of plants that were derived initially from single seeds. As pointed out by Mendel years before, selfing should lead to a rapid increase in homozygosity for all genes and the consequent establishment of a stock or "pure line" with a high degree of genetic uniformity. In Johannsen's experiment 19 pure lines were established, derived from 19 single seeds. In contrast to previous findings from mixed seeds, each pure line produced average seed weights that were individually distinct (Table 14–1). Line 1, for example, was the heaviest, with an average weight of 64 centigrams. Line 19 was the lightest, with a mean weight of 35 centigrams. The genetic difference between lines could also be seen when comparing the offspring of different pure lines with the same parental seed weight. For instance, parent beans of line 2, with a seed weight of 40 centigrams, produced offspring with an average weight of 57.2, while line-19 parents of 40 centigrams produced offspring with an average weight of 34.8. In general different parental sizes within a particular line produced offspring with uniform mean values. That is not to say that all phenotypes of individual plants in each pure line were exactly the same. Some degree of variability existed within each pure line that was apparently caused by environment, as can be seen in the variety of seed weights produced by a single sample line, that of number 13 in Table 14–2.

On the basis of Johannsen's experiment a popu-

TABLE 14–1

Mean weights of offspring produced by princess beans of selected sizes ("parent beans") taken from 19 pure lines

Pure Line Number	*Weight of Parent Beans Taken From Pure Lines, Centigrams* 20 × 20 ↓	30 × 30 ↓	40 × 40 ↓ Weight of Offspring, cg	50 × 50 ↓	60 × 60 ↓	70 × 70 ↓	Average Weight of Offspring
1					63.1	64.9	64.2
2			57.2	54.9	56.5	55.5	55.8
3				56.4	56.6	54.4	55.4
4				54.2	53.6	56.6	54.8
5			52.8	49.2		50.2	51.2
6		53.5	50.8		42.5		50.6
7	45.9		49.5		48.2		49.2
8		49.0	49.1	47.5			48.9
9		48.5		47.9			48.2
10		42.1	46.7	46.9			46.5
11		45.2	45.4	46.2			45.4
12	49.6			45.1	44.0		45.5
13		47.5	45.0	45.1	45.8		45.4
14		45.4	46.9		42.9		45.3
15	46.9			44.6	45.0		45.0
16		45.9	44.1	41.0			44.6
17	44.0		42.4				42.8
18	41.0	40.7	40.8				40.8
19		35.8	34.8				35.1
							47.9 = mean of all lines

After Babcock and Clausen, from data of Johannsen.

lation of individuals that varied in respect to a quantitative characteristic was conceived to consist of a number of genetically different groups. Within each group the quantitative characteristic would have a range of measurements because of environmental differences between the component individuals. Some degree of overlapping would therefore occur between the ranges of different groups,

TABLE 14–2

Numbers of offspring of different seed weights produced in Johannsen's pure line, number 13

Weight of Parents	*Weight Classes of Offspring, Centigrams* 17.5	22.5	27.5	32.5	37.5	42.5	47.5	52.5	57.5	62.5	Total
27.5			1	5	6	11	4	8	5		40
32.5				1	3	7	16	13	12	1	53
37.5		1	2	6	27	43	45	27	11	2	164
42.5	1		1	7	25	45	46	22	8		155
47.5			5	9	18	28	19	21	3		103
52.5		1	4	3	8	22	23	32	6	3	102
57.5			1	7	17	16	26	17	8	3	95
	1	2	14	38	104	172	179	140	53	9	712

After Babcock and Clausen, from data of Johannsen

so that the separation between them was not distinct (Fig. 14–1). Johannsen's experiment thus helped to explain that the continuous variation noted for quantitative characteristics was the result of both genotype and environment. However, although basically genetic, the distinction between Johannsen's pure lines could not be ascribed to particular genes, since none were individually identified. The genetic basis for quantitative variation thus remained disputed and unresolved.

MULTIPLE FACTORS

In 1906 Yule suggested that continuous quantitative variation might be produced by a multitude of individual genes, each with a small effect on the measured character. This suggestion had also been tentatively put forth by Mendel to account for the wide variety of flower colors in beans that arose from a *white* × *purple* cross (p. 211). Soon after Yule's proposal experiments by Nilsson-Ehle demonstrated an actual segregation and assortment of genes with quantitative effect. According to Nilsson-Ehle, there were three individual gene pairs involved in the determination of grain color in wheat, i.e., *Aa*, *Bb*, *Cc*, with genes for red (*ABC*) dominant over genes for white (*abc*). Each of these three gene pairs segregated in predictable mendelian fashion, so that the products of heterozygotes for any one pair, i.e., *Aa* × *Aa*, produced offspring in the ratio 3 red (*A*–):1 white (*aa*). When two gene-pair differences were segregating at the same time in Nilsson-Ehle's experiments, i.e., *AaBb* × *AaBb*, the results also followed mendelian principles, producing a ratio 15 red (*A*–*B*–, *A*–*bb*, *aaB*–):1 white (*aabb*). Similarly a cross between heterozygotes for three gene pairs produced a close fit to the predicted ratio 63 red:1 white. It made little difference which of the red genes was segregating, *A*, *B*, or *C*, since all three appeared to act similarly.

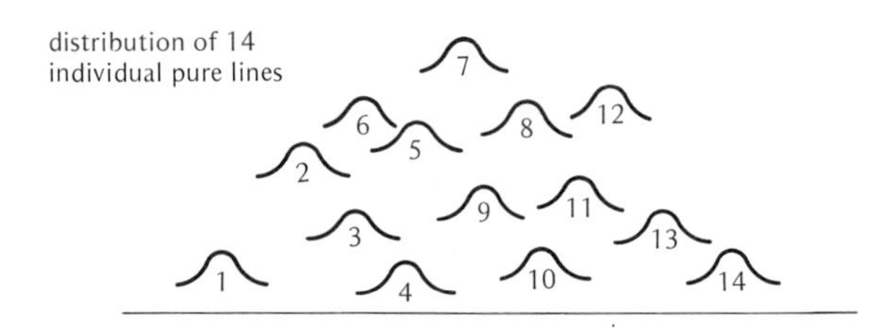

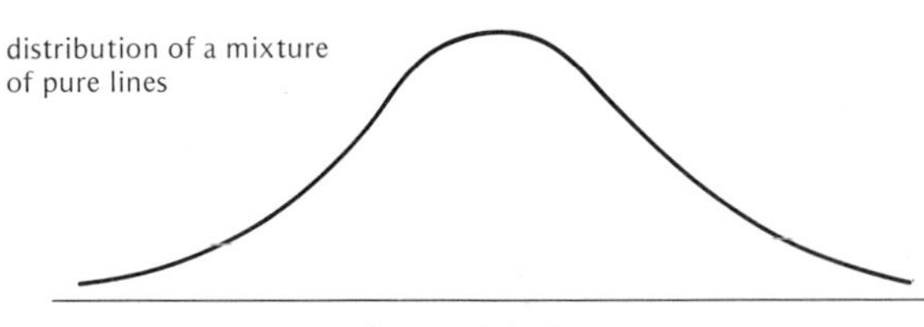

Figure 14–1

Individual distributions of a number of pure lines (above) and combined distribution of a mixture between them (below).

On this basis we can conclude that a variety of red genotypes occur, some containing more red genes than others. If we imagine further that the action of each red gene is to add a small degree of redness to the plant, then the range of red phenotypes observed by Nilsson-Ehle in his various crosses, corresponds to the range of red genotypes. As shown in the bottom half of Fig. 14–2, extreme phenotypes among the F_2 are expected to be quite rare, while the intermediate phenotypes, especially those near the average or mean, should be more frequent.

We can thus see that the type of phenotypic distribution that results from the segregation of three such independent genes, each with definite effect, begins to approximate the normal distribution expected for a quantitative character. The reason for this "normalcy" arises from the random fashion in which genes segregate and assort. As we have seen previously (Chapter 8), the frequencies of offspring segregating for two kinds of gene differences can be calculated on the basis of the binomial distribution. In an individual heterozygous for n genes, 2^n types of gametes will be formed that will be distributed according to the coefficients of the binomial $(1/2 + 1/2)^n$. In the present example the expansion of the binomial $(1/2 \text{ red} + 1/2 \text{ white})^3$ will produce 8 types of gametes from one individual in the ratio 1/8 (3 red genes):3/8 (2 red genes, 1 white gene):3/8 (1 red gene, 2 white genes):1/8 (3 white genes). For the products of two such individuals the distribu-

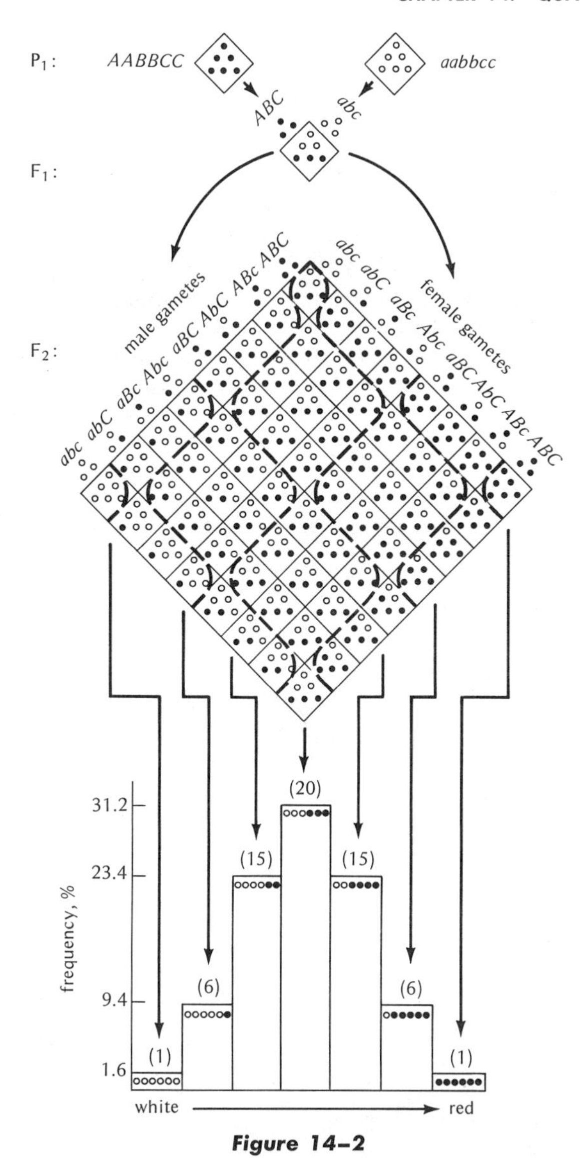

Figure 14–2

Results of crosses between two strains differing in three gene pairs that determine grain color in wheat. The F_2 distribution of color frequencies is shown in the histogram (each black dot represents a red color gene).

tion of color differences among the zygotes follows the coefficients of the product:

$$(1/2 \text{ red} + 1/2 \text{ white})^3 \times (1/2 \text{ red} + 1/2 \text{ white})^3 = (1/2 \text{ red} + 1/2 \text{ white})^6$$

If we restrict our attention to the number of red genes in the zygotes, the coefficients of this last expression produce the ratio 1 (6 red):6 (5 red):15 (4 red):20 (3 red):15 (2 red):6 (1 red):1 (0 red), exactly as shown in the bottom half of Fig. 14–2. Although there are still obvious steps in this distribution, these steps are gradually reduced in size as the number of independently segregating gene pairs increase until finally the distribution approaches a normal curve (Fig. 14–3). It is this behavior of genes with quantitative effect which was held sufficient to account for the normal bell-shaped distribution of many quantitative characteristics. Since there are usually many genes of this type for any one quantitative character, they were appropriately named *multiple factors*.

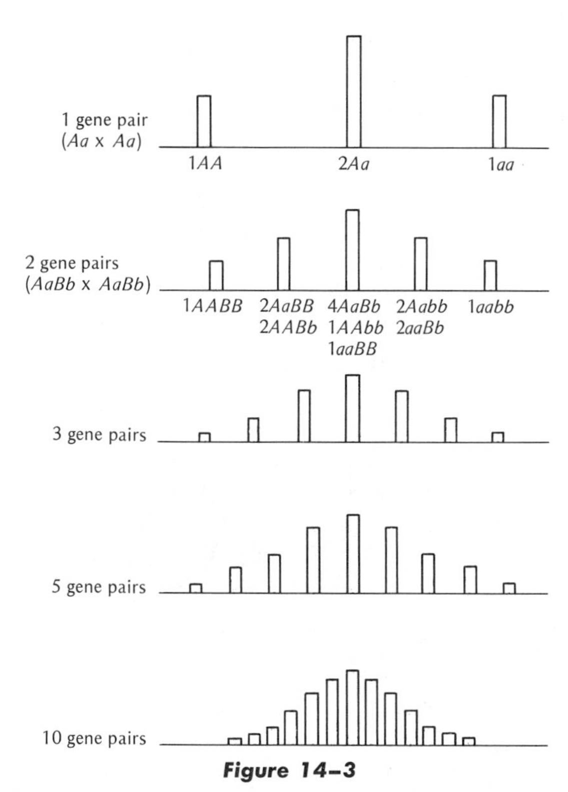

Figure 14–3

Relative frequencies (length of columns) of genotypes produced from crosses between individuals heterozygous for various numbers of independently segregating gene pairs. Illustrations using genes Aa and Bb are given for the top two conditions.

Many multiple factors, however, cannot be individually isolated because their effects are usually too small to be traced. Also, their observed quantitative effects may arise as "side effects" of genes that are concerned with other processes. Because of these and other limitations, it is generally agreed that many of these genes are not identifiable as factors in the original sense in which the term was used. "Multiple factor" has therefore been replaced in recent years by the term *polygene* proposed by Mather. Polygenes are defined as genes with a small effect on a particular character that can supplement each other to produce observable quantitative changes. Some of these quantitative effects can be considered *additive* if they can be added together to produce phenotypes which are the sum total of the negative and positive effects of individual polygenes.

Genes are, of course, only one source of phenotypic influence. Environmental effects, although not inherited, may also produce modifications in the expected phenotypes of any generation. For example, segregation of a single gene difference with observable quantitative effects would be ex-

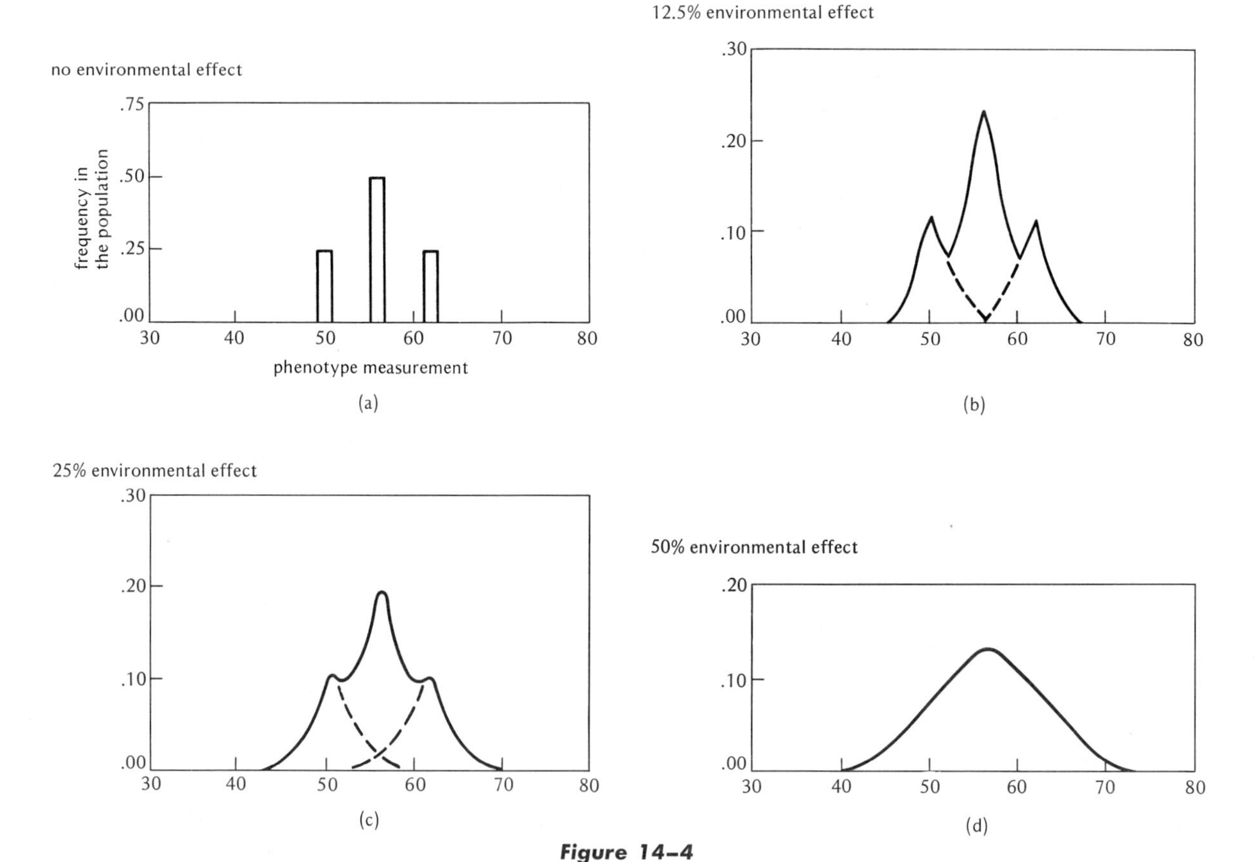

Figure 14–4

Effect of different degrees of environmental influence on the phenotypic expression of F_2 genotypes produced by segregation of a single gene difference. The two homozygous genotypes in (a) have respective measurements of 50 and 62, and the heterozygotes are exactly intermediate at 56. Environmental differences (or "environmental variance," see Chapter 15) will modify the measurements of a genotype so that the greater this effect, the more varied will be the measurements. For example, homozygous genotypes whose measurements were all at 50 in the absence of an environmental effect (a) will range in measurements from about 43 to 57 when the environmental effect or variance is 25 percent (c). (After Allard.)

pected to produce three distinct phenotypes in the F_2 in the absence of dominance (Fig. 14–4a). If the environment has a small effect on changing the appearance of these genotypes, a number of additional phenotypes will result (Fig. 14–4b). As the environmental effects increase, a greater variety of phenotypes are formed, until the single gene character appears to be normally distributed.

Two factors are therefore operating to produce continuity of measurement for quantitative characters: increased numbers of segregating gene pairs, and environmental variations between individuals. As indicated in the experiments of Johannsen, the only effective way of discriminating between these possibilities is to perform breeding experiments. Environmental variation that produces a continuous distribution of phenotypes should produce similar distributions for similar genotypes. The occurrence of different distribution curves for groups raised in the same environment (e.g., Johannsen's "pure" lines) is therefore indicative of genetic differences between them. However, although such genetic differences can be shown to exist, the precise distribution and number of polygenes involved are usually difficult to evaluate without controlled breeding techniques. One of the first such investigations was performed by East in experiments on flower length in *Nicotiana longiflora*.

East started with two self-pollinating varieties of this tobacco plant that differed in respect to flower length, one with an average of 40.5 millimeters and the other with an average of 93.3 millimeters (Fig. 14–5). Each group had been long inbred and was presumably homozygous at the start of the experiment. Although minor variations within each group were apparently caused by environmental differences, the large difference between the two groups was undoubtedly genetic. East then crossed these two lines and found that the F_1 flowers were intermediate in size to both parents, as would be expected of genes that have quantitative additive effects. Also, since the variability ("spread") of the F_1 was similar to each parent, the F_1 genotypes were, as their parents, probably all genetically uniform. That is, although heterozygous, the F_1 were all uniformly heterozygous for the same genes. Thus, as in the parent lines, the slight range of differences between F_1 individuals could be ascribed to environmental causes.

When East inbred the F_1, however, entirely different results occurred; the F_2 had a wider distribution than the F_1, indicating that differences were no longer merely environmental but also genetic. Evidence of genetic differentiation could also be seen when different F_2 individuals were used to produce an F_3. In almost every case the mean value of each F_3 group depended upon the F_2 individual that was chosen as a parent. The F_2 was therefore not a "pure" line, but a "mixed" line consisting of many genotypes. What were these genotypes and in how many gene pairs did they differ?

A possible estimate of the number of gene differences can be made if we consider that as the number of segregating gene pairs increase, the relative proportion of individuals exactly like their original parents decreases. For example, from an initial parental cross between homozygotes for one gene difference (i.e., $AA \times aa$), one-fourth of the F_2 progeny would be like each original parent ($1/4\,AA:2/4\,Aa:1/4\,aa$). For two gene-pair differences ($AABB \times aabb$), the ratio of F_2 progeny bearing a particular parental genotype falls to 1/16. For three gene differences (i.e., Nilsson-Ehle's wheat-color genes), this ratio is now 1/64. A cross between F_1 individuals heterozygous for n gene-pair differences will therefore produce $(1/4)^n$ offspring that have the same genotypes as one of the initial homozygous parents. Thus, for four gene-pair differences, 1/256 of the F_2 would be expected to have a parental genotype. Since East raised 444 F_2 plants without obtaining recovery of the parental genotypes, it can be concluded that probably more than four gene-pair differences are involved in flower length.

Other means for estimating the number of gene pairs involved in quantitative characters have been proposed with varying success. Thoday and coworkers, for example, have developed a technique using marker chromosomes in *Drosophila* that has enabled them to identify some genes that have fairly large effects on bristle number. The identifi-

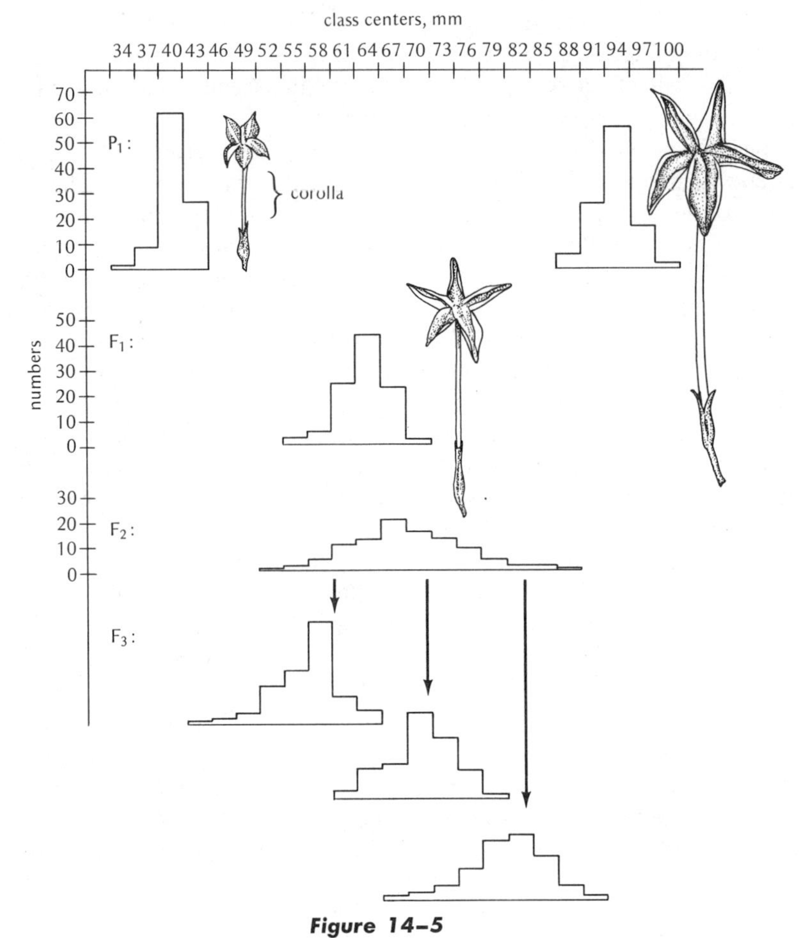

Figure 14–5

Measurements of corolla length in different varieties of Nicotiana longiflora *and in different generations of crosses between them. (Data of East.)*

cation of polygenes with relatively small effect, however, is still quite difficult, and genetic analysis can only be accomplished by statistical methods given in the following chapter and in some of the references cited there (e.g., see Mather and Jinks).

SKIN-COLOR DIFFERENCES IN HUMANS

Skin pigmentation in humans appears to be a quantitative characteristic varying from light to dark. If we exclude albinos, the phenotypic extremes in skin color occur among "white" Caucasians, on the one hand, and "blacks" of African ancestry, on the other. Actually, a wide array of color differences are present in both whites and blacks, although the distinction between the two groups is usually easily observable.

One of the first investigations of the genetic basis of skin-color differences was undertaken by G. C. and C. B. Davenport soon after the multiple-factor hypothesis was demonstrated by Nilsson-Ehle. The Davenports measured the grades of skin color by matching them to a rotating "color wheel" in which various proportions of different colors could be combined. Depending upon the size of the black area in their color wheel, they assigned a value of five grades, from 0 to 4, to different individuals. A number of marriages between whites and blacks were then investigated and found to produce an F_1 exactly intermediate in skin color to both parents (Fig. 14–6). Skin-color differences thus appeared to arise from quantitative and additive genetic effects. How many gene differences were involved? The Davenports felt this question was answered by their data for the F_2 generation. Crosses between F_1 individuals produced five grades of color among the offspring, distributed according to the ratios shown in Fig. 14–6. If we accept the Davenport classification of skin color into five grades, each F_2 grade is believed to represent genotypes formed from the segregation and assortment of two pairs of genes. For example, the first 0 grade is *aabb*, the 1 grade is either *Aabb* or *aaBb*, and so on. On the basis of two pairs of genes, the grades in the expected F_2 distribution should be in the ratio of 1:4:6:4:1. However, although the observed proportions agree to some extent with those expected on a two-gene basis, it appears that the Davenports' classification of skin-color grades was a highly arbitrary one. Measurements taken since with modern photometric methods show a continuous distribution in skin color from light to dark without the Davenports' stepwise distinctions. The distribution of the F_2 data may therefore be arranged in a variety of categories besides the five grades.

A wider sample of measurements taken from the American black population has shown that probably more than two pairs of genes are involved in skin color. The solid line in Fig. 14–7 represents the observed frequencies of blacks with various degrees of pigmentation. Since it is estimated that in some areas of the United States as much as 30 percent of American black genes are derived from

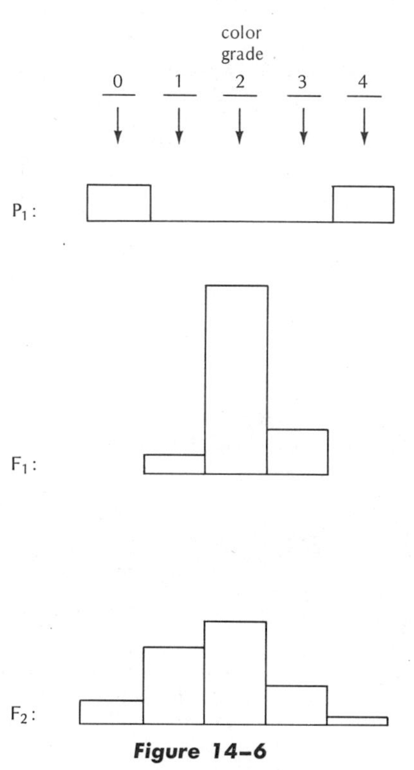

Figure 14–6

Distributions of skin-color measurements using the Davenports' "color wheel" among the progeny of white-black matings. In the F_1, 29 offspring were measured. In the F_2 (progeny of $F_1 \times F_1$), 32 offspring were measured. (Data of G. C. and C. B. Davenport.)

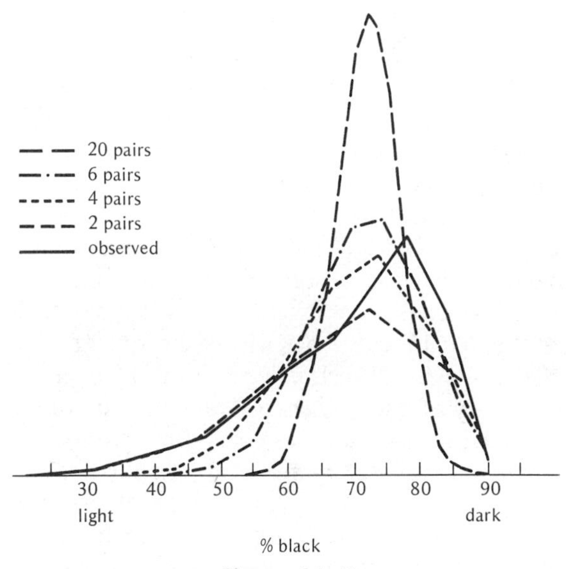

Figure 14–7

Frequencies (height of line) of individuals in the American black population with different degrees of skin pigmentation (horizontal scale) observed and expected according to different numbers of independently segregating additive genes. (After Stern, 1953.)

the white population (see Chapter 32), the expected distribution of skin color genotypes among such blacks can be represented by the binomial $(.30 + .70)^n$, where n is the number of gene pairs involved in color. If we accept the simplifying assumptions that have been made to arrive at this formula (complete mixing of all genes and random mating between blacks of different color), the observed distribution seems to be best fitted by a curve based on four or five pairs of genes for color. In addition some matings show the segregation of modifying genes that appear to affect the expression of the color genes.

In any case skin color is a quantitative character undoubtedly influenced by polygenes with at least partially additive effect. Thus two parents, each carrying a few dark polygenes, may produce a child as dark as or somewhat lighter or darker than themselves. However, legends that matings between light individuals can produce a "black" or dark child would have to be explained on the basis of fully recessive color genes, and this view is not supported by reliable evidence.

EFFECT OF DOMINANCE

Most distributions of quantitative characters appear to follow the normal distribution curve, with increased frequencies of individuals that have values intermediate to both extremes. This type of distribution and the continuity between its values indicates, as we have seen, that the appearance of such characters is not determined by "all or none" effects, but by polygenes with small quantitative effects. Ideally all such polygenes should act independently of each other, so that the addition of one such gene should have the same effect in all individuals, whatever their genotype. Because of this simple relationship, a normal distribution means more than the dependence of each phenotypic class upon the number of polygenes carried within it. Such a distribution also means that the *frequencies* of particular classes will be related to the number of polygenes carried. That is, in a mating between heterozygotes, the class with the intermediate phenotype carrying the intermediate number of polygenes is the most frequent and the other classes are "symmetrically" distributed in frequency around this mean value according to the number of polygenes carried (e.g., Fig. 14–2). Of course, not all phenotypic characters are caused by genes with small additive effect, nor are all gene effects independent of each other. Nonadditivity and interactions of various kinds will cause the distribution curve to appear asymmetric or "skewed," with a much greater frequency of phenotypes at one end of the curve than at the other.

Dominance between alleles is one type of effect that may cause a skewed distribution. For example, a cross between two inbred lines, one line homozygous mostly for dominants affecting a measurable phenotype, and the other line homozygous mostly for the recessive alleles, will produce an F_1 with an average phenotype closer to that of the dominant parent. If, for purposes of illustration, we confine ourselves to the four gene pairs in Fig. 14–8, most F_1 phenotypes are in the high-value range. The presence of additive and environmental effects would, of course, make the distribution "smoother" and more continuous but it would nevertheless still be skewed toward the dominant parent.

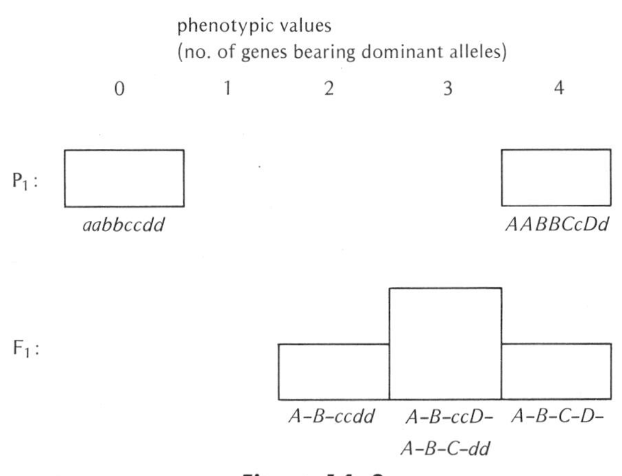

Figure 14–8

Distribution of genotypes bearing different numbers of dominant alleles among the progeny of a cross between two strains, one homozygous recessive for four gene pairs and the other bearing six dominant alleles and two recessives at these gene pairs.

In the F_2 distribution, however, the appearance of dominance is not always easily noted, but depends upon the number of genes involved. As shown in Fig. 14–9, segregation of a single pair of genes, i.e., *Aa* × *Aa*, will produce an F_2 distribution .75 *A*:.25 *a*. As the number of segregating gene pairs increases, the distribution becomes less skewed, until at 10 gene pairs the distribution is difficult to distinguish from that of additive genes.

GENES WITH MULTIPLYING EFFECTS AND THE SCALE OF MEASUREMENTS

Some genes act by multiplying certain underlying effects, so that when many such genes are present, the distribution of phenotypes departs considerably from those that would be expected if the gene effects were simply additive. For example, each gene increasing the number of facets in a certain stock of *Bar*-eyed *Drosophila* may act by doubling the number of facets already present, so that adding two such genes would quadruple the facet number. If an inbred stock of this kind had a large number of such multiplicative alleles (and therefore a high mean value), the variability of the stock would be much higher than the same stock carrying few such alleles (with a low mean value). This follows from the fact that the presence or absence of one multiplicative allele when their numbers are high has a much greater phenotypic effect than when their numbers are low. For example, if the phenotypic height of a stock of plants was 50 inches, the multiplicative allele of a gene that doubles the height of a plant would produce a value of 100 inches when present and 50 inches when absent—a difference of 50 inches. On the other hand, the presence or absence of this allele in a 5-inch stock would only cause differences in 5 inches, or one tenth the previous differences when measured on the same scale as the larger plants. Thus, when variability differs between strains that have high and low mean values, genes with multiplying effect can be held responsible. On the other hand, additive genes only increase the measure of a character by a constant arithmetic amount and show no greater variability when their numbers are high or low (Fig. 14–10).

Nevertheless, whether additive or multiplicative, both kinds of genes still produce quantitative changes. For this reason a scale of measurements should be chosen so that the relative contributions of each gene are not obscured by the particular process through which these effects occur. When the gene effect is merely additive, the scale used is simply the observed phenotypic scale in arith-

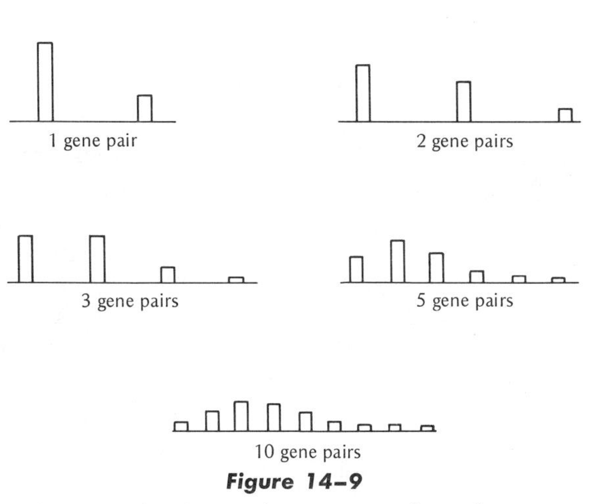

Figure 14–9

F_2 frequency distributions for varying numbers of gene-pair differences at which dominance is present.

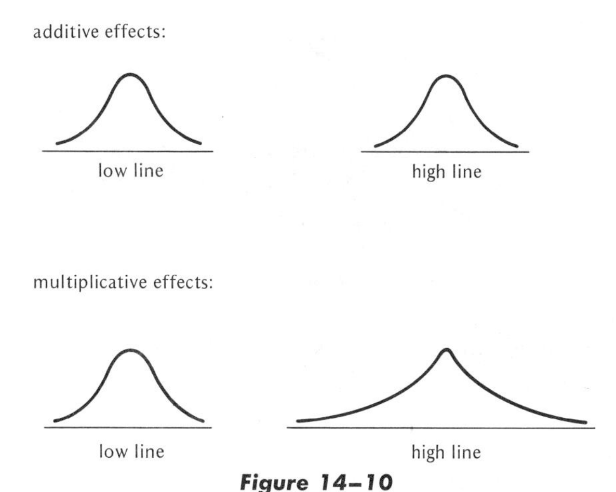

Figure 14–10

Distribution of phenotypes in low and high inbred lines when the gene effects are additive and multiplicative.

metical units. When the gene effect is multiplicative (as determined by the wider range in the high-line inbred group) a scale which converts observed measurements into units which are multiples of arithmetic numbers should be used. For example, a quantitative relationship in which the values are doubled by each additional gene should have similar distances between 1-2-4-8-16. In other words, the effect of each multiplicative gene would appear to be additive on such a scale and provide a distribution similar to that of genes whose additive effects are measured on an arithmetical scale.

Logarithms, which are the exponents of a particular base number, are valuable for this purpose, because the addition of logarithms of two numbers is the same as multiplying the numbers themselves. That is, the increasing distance between successive numbers in the series 1-2-4-8-16 $\cdots$, is represented by the constant distance between logs in the series $\log 1 + \log 2 + \log 2 + \log 2 \cdots$. When such logarithmic transformations are carried out for multiplicative effects, gene action in some traits produces a normal distribution (Fig. 14-11). Often, when the variation is proportional to the mean, we can reasonably assume that some multiplicative effect is involved which may then be eliminated by an appropriate logarithmic transformation of the scale of measurement.

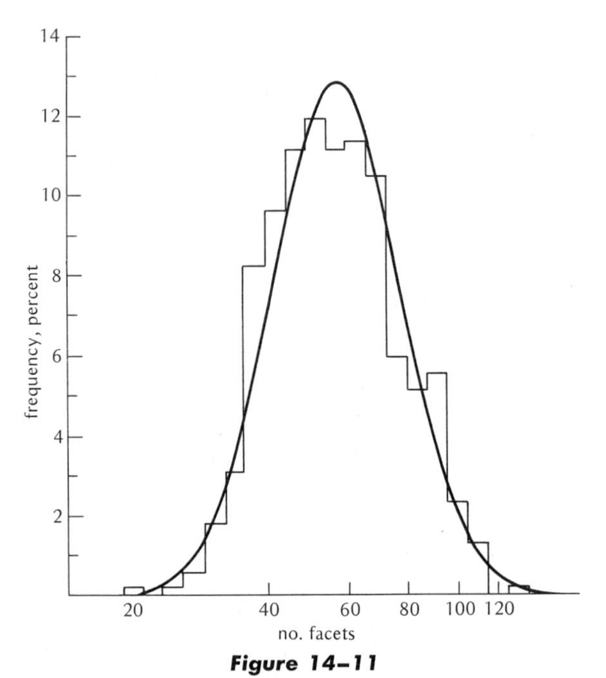

Figure 14-11

Number of facets in mutant Bar females. Frequency of individuals (in percent) is on an arithmetic scale, and facet number is on a logarithmic scale. (From Falconer, after data of Zeleny.)

POLYGENES IN DISCONTINUOUS TRAITS

In many cases discontinuous or stepwise distinctions between phenotypes inevitably accompany measurements for a particular characteristic. For example, the number of vertebrae in some species of chordates may differ between individuals, but the difference is generally classified on the basis of whole numbers of vertebrae, or, as in resistance to disease, the character is expressed in an "all or none" fashion. Although lacking a continuous distribution of intermediate values, such characters may nevertheless be influenced by numerous polygenes.

The relationship between polygenes and the expression of discontinuous characters comes about through the establishment of "thresholds." That is, those polygenically determined genotypes that have values below the threshold show no expression of the character. Expression only occurs when the genotypes have values above this threshold. Since the genotypic values are polygenically determined (e.g., by the number of capital-letter alleles in the gene pairs *Aa*, *Bb*, *Cc*, *Dd*), the frequency of phenotypes above and below a particular threshold depends upon the distribution of polygenic genotypes above and below the same threshold. Thus there are two separate scales involved: the underlying polygenic distribution, which is continuous, and the superficial phenotypic distribution, which is discontinuous.

The relationship between these two scales was first demonstrated by Wright in crosses between strains of guinea pigs that differed from each other in the number of toes on the hind leg. One strain (no. 2) had the normal amount of three toes, and the other (D) was polydactylous, with four toes. Without exception their F_1 offspring were three-toed. Four-toed individuals, however, appeared in the F_2 in a frequency of 188 three-toed to 45 four-toed. Since this ratio did not differ too greatly from a 3:1 ratio, it seemed as though these crosses were segregating for only a single gene difference in the F_2, with the three-toed allele dominant to the four-toed allele. This view seemed supported by the observation that backcrosses of the F_1 to the presumed recessive four-toed stock produced approximate equal proportions of three-toed to four-toed individuals. However, if this one-gene hypothesis were true, the three-toed products of this backcross should be heterozygous. Wright, on the contrary, found that backcrossing these three-toed "heterozygotes" to the four-toed stock resulted in 77 percent four-toed to 23 percent three-toed offspring—a ratio that hardly fits the single-gene hypothesis.

As a result of these and other tests Wright therefore proposed that approximately four pairs of polygenes, or eight alleles, were involved in polydactyly. Individuals which had about five or more polydactylous alleles out of eight had exceeded the "threshold" and appeared as four-toed. The initial four-toed stock was therefore entirely of this type. The parental three-toed stock, however, had its distribution centered far below the threshold (Fig. 14–12). Thus the F_1 hybrids were mostly three-toed, since only about one-fifth carried more than the four polydactylous alleles inherited from the gametes of their four-toed parents. In the F_2 the distribution of genotypes widened considerably and produced a noticeable number of individuals that carried five, six, seven, and eight polydactylous alleles, thereby exceeding the four-toed threshold.

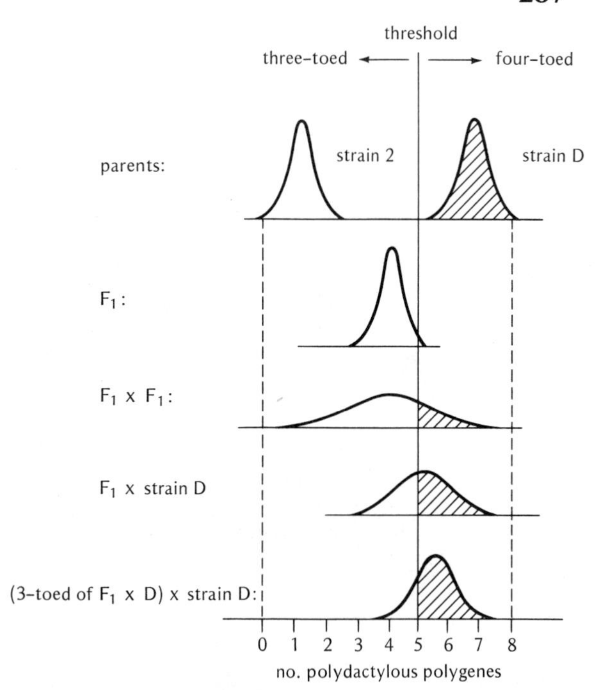

Figure 14–12

Distributions of different generations in crosses between guinea pig strains differing in the number of toes. The parental strains are assumed to be on either side of a threshold which determines the three-toed or four-toed state. The F_1 of this cross is entirely three-toed, but later generations produce various frequencies of four-toed progeny, depending upon the number of genotypes that cross the threshold. (After Stern, 1973, modified.)

PROBLEMS

14-1. Two full-grown plants of a particular species are given to you which have extreme phenotypes for a quantitative character such as height, e.g., 1 foot tall and 5 feet tall. (a) If you had only a single set of environmental conditions in which to conduct your experiment (e.g., one uniformly lighted greenhouse), how would you determine whether plant height is environmentally or genetically caused? (b) If genetically caused, how would you attempt to determine the number of gene pairs that may be involved in this trait?

14-2. A cross between two inbred plants that had seeds weighing 20 and 40 centigrams, respectively, produced an F_1 with seeds that weighed uniformly 30 centigrams. An $F_1 \times F_1$ cross produced 1000 plants; four had seeds weighing 20 centigrams, four had seeds weighing 40 centigrams, and the other plants produced seeds with weights varying between these extremes. How many gene pairs would you say are involved in the determination of seed weight in these crosses?

14-3. Two homozygous inbred strains differing with respect to five gene pairs with additive effects on a quantitative trait were crossed to produce an F_1. What proportion of the F_2 ($F_1 \times F_1$) would be expected to resemble one of the homozygous inbreds?

14-4. Clausen and Hiesey made crosses between two California races of *Potentilla glandulosa*. One was a coastal population active in growth all year round, and the other was an alpine population dormant during the winter months. Measurements were then taken of the degree of winter growth activity for the F_1 and F_2 of these crosses, as shown in the Table below.
(a) What type of gene effects are involved in the determination of winter activity? (b) How many gene pairs would you suggest are segregating for this trait?

14-5. Hayes observed the leaf numbers in the Table below in two strains of tobacco plants (Cuban and Havana), and in the F_1 and F_2 of crosses between them. (a) Why does the F_1 have less variability than the F_2? (b) Why is the range of the F_2 greater than that of both of the parental strains? (c) Explain your views about the effectiveness of establishing a strain of plants maintaining an extreme leaf number (e.g., 30 leaves). (d) Explain whether you would expect an F_3 raised from crosses between high-leaf-numbered plants in the F_2 (e.g., 30 and 31) to give still higher leaf-numbered plants in the F_3 (e.g., 35 to 40)?

14-6. If skin color is caused by additive genes, explain your answers to the following: (a) Can matings between individuals with intermediate colored skins give birth to lighter-skinned offspring? (b) Can such matings produce dark-skinned offspring? (c) Can matings between individuals with light skins produce dark-skinned offspring?

14-7. The data shown in the Table below were collected by Punnett and Bailey and show the results of crosses between Hamburg and Sebright strains of fowl and their F_2 when the individuals are measured in grams. Provide an explanation for these data in terms of

Table for Problem 14-4

	Degree of Winter Growth				
	Fully Active	Fairly Active	Intermediate	Fairly Dormant	Completely Dormant
P_1 (coastal)	+				
P_2 (alpine)					+
F_1 ($P_1 \times P_2$)			+		
F_2 ($F_1 \times F_1$) no. of plants	43	139	601	178	14

Table for Problem 14-5

	Leaf No.																			
	14	*15*	*16*	*17*	*18*	*19*	*20*	*21*	*22*	*23*	*24*	*25*	*26*	*27*	*28*	*29*	*30*	*31*	*32*	*33*
Cuban			1	7	16	37	36	35	12	4	1	1								
Havana				3	22	44	42	22	10	6	1									
F_1		1	1	3	8	39	60	30	7			1								
F_2 ($F_1 \times F_1$)	3	4	8	8	20	18	30	24	25	17	16	5	4	3	1	1	1	1	2	1

Table for Problem 14-7

	Grams									
	500	*600*	*700*	*800*	*900*	*1000*	*1100*	*1200*	*1300*	*1400*
Sebright	1	1								
Hamburg									1	
F_1 ♂(S × H)							5	2	1	
F_2 ♂($F_1 \times F_1$)		1	4	7	15	26	19	29	9	2

number of gene pairs. (Do all the genes involved affect weight additively? Why or why not?)

14-8. Assume that two pairs of genes with two alleles each, *Aa* and *Bb*, determine plant height additively in a population. The homozygote *AABB* is 50 centimeters high, the homozygote *aabb* is 30 centimeters high. (a) What is the F_1 height in a cross between these two homozygous stocks? (b) After an $F_1 \times F_1$ cross, what genotypes in the F_2 will show a height of 40 centimeters? (c) What will be the F_2 frequency of these 40-centimeter plants?

14-9. If three independently segregating genes, each with two alleles, determine height in a particular plant, e.g., *Aa*, *Bb*, *Cc*, so that the presence of each capital-letter allele adds 2 centimeters to a base height of 2 centimeters: (a) Give the heights expected in the F_1 progeny of a cross between homozygous stocks *AABBCC* (14 centimeters) × *aabbcc* (2 centimeters). (b) Give the distribution of heights (phenotypes and frequencies) expected in an $F_1 \times F_1$ cross. (c) What proportion of this F_2 progeny would have heights equal to the parental stocks? (d) What proportion of the F_2 would breed true for the height shown by the F_1?

14-10. If each capital-letter allele in Problem 9 acted as a dominant, e.g., *A–B–C–* = 8 centimeters what would be the answers to (a), (b), (c), and (d) in Problem 9?

14-11. If each capital-letter allele in Problem 9 acted to multiply the existing height, e.g., *Aabbcc* = 4, *AAbbcc* = 8, *AABbcc* = 16 centimeters, etc., what would be the answers to (a), (b), (c), and (d) in Problem 9?

14-12. If gene *A* in Problem 9 acted as a dominant for height but genes *B* and *C* acted additively, answer (a), (b), (c), and (d) in Problem 9.

14-13. If gene *A* in Problem 9 acted multiplicatively and genes *B* and *C* acted additively, answer (a), (b), (c), and (d) in Problem 9.

14-14. If the capital-letter alleles in Problem 9 acted as complete dominants, e.g., *A–B–C–* = 8 centimeters, give one set of possible genotypes for the parents in the following crosses (proportions of phenotypes of offspring in parentheses):

Parents	Offspring
(a) 8 cm × 8 cm	(3) 8 cm:(1) 6 cm
(b) 8 cm × 8 cm	(9) 8 cm:(6) 6 cm:(1) 4 cm
(c) 8 cm × 6 cm	(1) 8 cm:(1) 6 cm
(d) 8 cm × 6 cm	(3) 8 cm:(4) 6 cm:(1) 4 cm
(e) 8 cm × 6 cm	(3) 8 cm:(1) 6 cm
(f) 8 cm × 6 cm	(9) 8 cm:(15) 6 cm:(7) 4 cm: (1) 2 cm
(g) 6 cm × 6 cm	all 8 cm
(h) 6 cm × 6 cm	(1) 8 cm:(2) 6 cm:(1) 4 cm
(i) 6 cm × 6 cm	(3) 8 cm:(7) 6 cm:(5) 4 cm: (1) 2 cm
(j) 4 cm × 4 cm	(1) 6 cm:(1) 4 cm

REFERENCES

ALLARD, R. W., 1960. *Principles of Plant Breeding*. John Wiley, New York.

BABCOCK, E. B., and R. E. CLAUSEN, 1927. *Genetics in Relation to Agriculture*. McGraw-Hill, New York.

DAVENPORT, C. B., 1913. Heredity of skin color in Negro-white crosses. *Carneg. Inst. Wash. Publ. No. 554*, Washington, D.C.

EAST, E. M., 1916. Studies on size inheritance in *Nicotiana*. *Genetics*, **1**, 164–176.

FALCONER, D. S., 1960. *Introduction to Quantitative Genetics*. Oliver and Boyd, Edinburgh, Chaps. 6, 17, and 18.

GALTON, F., 1889. *Natural Inheritance*. Macmillan, Inc., New York.

JOHANNSEN, W., 1903. *Über Erblichkeit in Populationen und in reinen Linien*. G. Fischer, Jena. (Excerpts have been reprinted in the collection of Peters; see References, Chapter 1.)

MATHER, K., 1943. Polygenic inheritance and natural selection. *Biol. Rev.*, **18**, 32–64.

NILSSON-EHLE, H., 1909. Kreuzungsuntersuchungen an Hafer und Weizen. *Lunds Univ. Aarskr. N.F. Afd., Ser. 2, Vol. 5*, No. 2, pp. 1–122.

STERN, C., 1953. Model estimates of the frequency of white and near-white segregants in the American Negro. *Acta Genet. et Stat. Medica*, **4**, 281–298.

———. 1973. *Principles of Human Genetics*, 3rd ed. W. H. Freeman, San Francisco, Chap. 18.

THODAY, J. M., 1967. New insights into continuous variation. *Proc. Third Intern, Congr. Hum. Genet.*, J. F. Crow, and J. V. Neel (eds.), Johns Hopkins Press, Baltimore, pp. 339–350.

WRIGHT, S., 1934. The results of crosses between inbred strains of guinea pigs, differing in number of digits. *Genetics*, **19**, 537–551.

YULE, G. U., 1906. On the theory of inheritance of quantitative compound characters on the basis of Mendel's laws—A preliminary note. *Rept. 3rd Intern. Confr. Genet.*, pp. 140–142.

15

ANALYSIS OF QUANTITATIVE CHARACTERS

Quantitative differences between genotypes may, as we have seen, be strongly affected by environment. For example, poor growth conditions for a genotype with many genes for increased height may produce an individual no taller, or even smaller, than a genotype that has fewer such genes but is grown in a more nutritive environment. Even when environmental differences among individuals are small enough to enable discrimination among genotypes that are widely different, they may not be small enough to discriminate easily among genotypes with only slight differences. Thus, as a result of environmental action, the different genotypic classes for quantitative characteristics become "blended" into a distribution of phenotypes which may or may not correspond with the underlying distribution of genotypes. The degree of this correspondence will obviously depend upon the relative roles of genotype and environment; the greater the environmental effect on the character, the less reliable is the relationship between phenotypic distribution and genotypic distribution. Thus, for example, small phenotypes may not necessarily represent small genotypes if height is easily subject to environmental change.

Because of these complexities the usual analysis in which the genotype is evaluated in terms of separately traced genes is ineffective, since we can hardly be certain of the exact genotype of a particular phenotype or make predictions as to the exact ratio in which a particular phenotypic class is to be expected. Since it is so difficult to handle quantitative characteristics with ordinary mendelian methods, how then can we analyze their genetic basis?

To enable a genetic analysis to be made, we must first consider means by which a quantitative characteristic may be described. In general the data in quantitative experiments consist of a number of measurements that constitute a *sample* set of observations. This sample is theoretically derived from a much larger group of potential observations, called a *population*. Ideally a sample should consist of all the measurements that could possibly be made in the entire population; i.e., the sample and the population should be the same. If this occurred, the conclusions drawn from the sample would be identical to those that could be drawn from the population, and there would be no "sampling error." It is obvious, however, that for many experiments the size of the population may far exceed the number of measurements it is practical to make. For instance, the infinite number of offspring that can potentially be produced by matings between individuals differing in a particular height can never actually exist.

The sample that is actually scored in a particular experiment, therefore, is usually only a representative of this larger potential population. The extent to which this sample is a *true* representative of this large population depends upon various factors, but most importantly it depends upon whether the sample measurements are taken *randomly* from the many potential measurements that are possible. For example, if a quantitative characteristic is potentially normally distributed (Chapter 8) under a particular set of circumstances, a random set of sample measurements would also be expected to be normally distributed. If a quantitative characteristic has a certain potential average value under these conditions, the random sample would follow closely.

Randomness implies that the selection of one sample observation from the population in no way influences the selection of the next sample observation. It has important advantages, since living organisms possess many variables which, if not randomly distributed during an experiment, would cause a bias in one direction or another. For example, an experiment in which the effect of a difference in diet is being measured on weight production in cattle has little meaning if one diet is given to a group of heavy cattle and the other diet is given to a group of light cattle. A final difference in weight between the two types of cattle may then signify nothing more than the initial difference in weight between the two groups.

It is therefore essential to ensure that only the experimental factor or factors that are being tested, e.g., diet, are responsible for the observed differences, and that all other variable factors, such as initial weight, are randomly or uniformly distributed between the experimental groups. In other words, any individual being subjected to experimental treatment should have the same chance to be assigned to a particular treatment group as any other individual in the entire experiment. There will, of course, be some differences between the sample and the population, since the sample is only part of the population, but these differences will not be too important if the sample is large enough and if it is selected randomly. In statistical terms, all values computed from sample data are known as *statistics*, while values of the larger population from which the sample has been taken are known as *parameters*.

THE MEAN

The most common and useful measurement describing the central value of a set of sample observations is the *mean*. This value is simply the sum of all observations divided by the number of observations, N.* If each observation is denoted by X, i.e., $X_1, X_2, X_3, X_4, \ldots X_N$, the mean ($\bar{X}$) is formulated as

$$\bar{X} = \frac{X_1 + X_2 + X_3 + X_4 + \cdots + X_N}{N}$$

Statisticians make use of notations which sim-

* Other measures of central tendency are the *median* (the value for which 50 percent of the observations are on either side) and the *mode* (the value which occurs most frequently). In most observed distributions, the mean, the median, and the mode are not identical values. It is the mean, however, which is considered a more efficient and precise measure of central tendency, since it does not vary as much from sample to sample of a population as do the median and the mode.

plify such formulas by using the Greek letter Σ (sigma) for "sum of," and by using X_i to symbolize any value of a particular measurement, X, within the array of observed values (i.e., X_1, X_2, X_3, X_4, $\cdots$, X_N). Thus, statistically, the mean is written

$$\bar{X} = \frac{\sum_{i=1}^{N} X_i}{N}$$

which signifies that the mean is equal to the sum of all X_i's divided by N, where i goes from 1 to N. If 20 observations are included in the sample, the mean is

$$\bar{X} = \frac{\sum_{i=1}^{20} X_i}{20} \quad \text{or, in shorthand,} \quad \frac{\sum X_i}{20}$$

A set of 100 sample observations whose sum equals 1700 is given in Table 15-1, and the mean is therefore calculated as

$$\frac{\sum X_i}{N} = \frac{1700}{100} = 17.0$$

TABLE 15-1

A set of 100 sample observations whose sum equals 1700.0

10.8	15.7	16.9	18.3
11.9	15.8	17.0	18.4
12.8	15.9	17.0	18.4
13.0	15.9	17.0	18.7
13.7	15.9	17.1	18.8
13.8	16.0	17.1	18.8
14.0	16.0	17.2	18.9
14.2	16.0	17.2	19.0
14.3	16.1	17.2	19.1
14.7	16.2	17.3	19.2
14.7	16.2	17.4	19.2
14.8	16.3	17.4	19.2
14.8	16.4	17.6	19.3
14.8	16.4	17.7	19.4
15.0	16.6	17.7	19.4
15.0	16.6	17.8	19.4
15.0	16.6	17.8	19.8
15.1	16.7	17.9	20.0
15.1	16.7	18.0	20.0
15.3	16.7	18.0	20.4
15.4	16.8	18.1	20.8
15.6	16.8	18.1	21.0
15.6	16.8	18.2	21.3
15.6	16.9	18.2	22.1
15.7	16.9	18.3	22.3

However, although the mean tells us the central value for this set of numbers, it tells us little else. We would, for example, like to know whether all values are close to each other at the mean or whether some are further apart, i.e., how uniform these values are. In a rough way, this can be observed by inspection of Table 15-1, but a more exact method is to collect these observations into groups, or *class intervals*, of specified length. In this fashion we obtain frequencies for each group, i.e., so many measurements in groups A, B, C, . . . , etc. When these frequencies are collected together we have a *frequency distribution*.

Class intervals in such a distribution can, of course, be large or small, depending upon the needs of the problem at hand. If too small, calculations become somewhat tedious to make, and the overall picture of the distribution is lost in its numerous subdivisions. On the other hand, if the class intervals are too large, considerable information may be hidden by such nondiscriminate classification.

Table 15-2 provides a convenient method whereby observations can be grouped into classes whose intervals are determined as a certain proportion of the distance between the two extreme values ("range"). For the distribution we have been using, the range is 10.8 to 22.3, or 11.5. Considering that there are 100 observations in our sample, this range is divided by the 12.5 value in Table 15-2 to yield a value approximately equal to 1. The midpoint of each class interval is then one unit from the midpoint of the next class interval, i.e., 11, 12, 13, . . . , etc. We can then designate the limits for each class interval as extending half a unit on each side, so that the limit for the 11 class, for example, is from 10.5 to 11.5. If an even

TABLE 15-2

Method of obtaining the class intervals for different sample sizes*

Sample Size	Range to be Divided by
20	9.2
30	10.2
50	10.7
70	12.0
100	12.5
150	13.2
200	13.7
300	14.5
400	14.7
500	15.2
700	15.7
1000	16.2

* For the sample number observed (column 1) find the difference between the highest and lowest values, the *range*. Then divide the range by the number in column 2 to obtain the class interval.

TABLE 15-3

Frequency table and histogram for data given in Table 15-1

X (Class Midpoint)	f	fX
11	1	11
12	1	12
13	2	26
14	5	70
15	12	180
16	18	288
17	23	391
18	16	288
19	13	247
20	4	80
21	3	63
22	2	44
	$\Sigma f = 100$	$\Sigma fX = 1700$

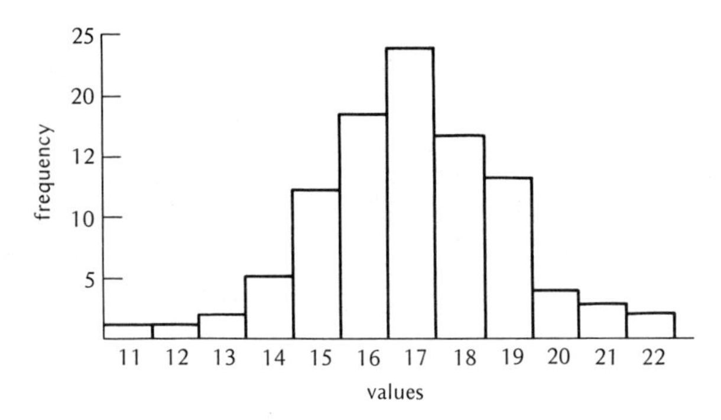

number of values falls exactly on one limit, half are assigned to one class and half to the other. If the number of such values is odd, the odd value is randomly assigned to one of the two classes.

On the basis of this method the previous observations (Table 15-1) have been tabulated according to frequency in Table 15-3. The first column, X, represents the midpoint of the class interval; the next column, f, represents the frequency or number of observations in that interval; and the last column, fX, represents the product of these two values for each interval. Accompanying this frequency distribution is a histogram showing its graphic appearance. Note that the values appear to be normally distributed in the familiar bell-shaped curve. Note also that Σ fX in Table 15-3 is equal to the ΣX_i previously calculated when the observations were added individually. This result is not surprising, since the frequency table represents the same number of observations as those taken individually ($\Sigma f = 100$). Occasionally, however, there may be a slight difference between Σ fX and ΣX_i if the frequency table is constructed of observations that depart from the class midpoints. The values in the present example have been so selected that this difference does not occur. However, even when it does occur, the difference between Σ fX and ΣX_i is usually small and can easily be ignored. Because of this identity, the mean of a frequency table can therefore be calculated at $\Sigma fX/\Sigma f$, or $1700/100 = 17.0$, in the present case.

VARIANCE AND DEVIATION

The frequency distribution of a set of values thus enables these values to be described more graphically than can be done by the mean itself. In Fig. 15-1 all the distributions have the same mean, yet depart considerably from each other with respect to the shape of their distributions. Figures 15-1a, b, c, and d are obviously non-normal distributions, while Figs. 15-1e, f, and g

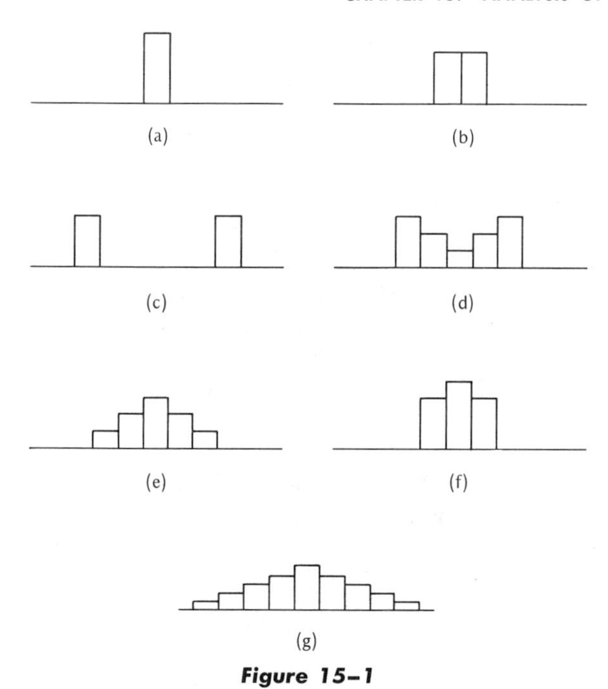

Figure 15–1

Different kinds of distributions for a quantitative measurement, all having the same mean. The horizontal axis signifies measurement, while the vertical axis signifies the relative frequency of each class of observations.

appear to have values that can be considered normally distributed, although to different degrees.

To help describe the shape of a distribution and permit its comparison with other distributions, a measure has been devised which indicates the extent to which values within this distribution depart from a mean. Such a measure, called the *variance*, is relatively large when values are spread out (e.g., Fig. 15–1g) and small when values are close together (e.g., Fig. 15–1e). Its symbol is s^2 for a sample (σ^2 is the symbol for this value in the population from which this sample was taken), and it is defined as the sum of squared differences between each individual value and the mean, divided by one less than the total number of observations (N).*

$$s^2 = \frac{(X_1 - \bar{X})^2 + (X_2 - \bar{X})^2 + (X_3 - \bar{X})^2 + \cdots + (X_N - \bar{X})^2}{N - 1}$$

$$= \frac{\sum_{i=1}^{N}(X_i - \bar{X})^2}{N - 1}$$

To eliminate subtraction of each value from the mean and provide greater ease in computation, the variance can also be calculated according to the formula

$$s^2 = \frac{\sum_{i=1}^{N} X_i^2 - \left[\left(\sum_{i=1}^{N} X_i\right)^2 / N\right]}{N - 1}$$

$$= \frac{N \sum_{i=1}^{N} X_i^2 - (\Sigma X_i)^2}{N(N - 1)}$$

This method of calculation considers each observation as an individual value (X_i) and would be suitable for calculating the variance for the values in Table 15–1. For the frequency distribution in Table 15–3, however, numerous observations (f) can be lumped within each representative value or class interval (X) and the variance is then calculated as

$$s^2 = \frac{\Sigma f(X^2) - [(\Sigma fX)^2/N]}{N - 1}$$

$$= \frac{N \Sigma f(X^2) - (\Sigma fX)^2}{N(N - 1)}$$

As shown in Table 15–4, this value is 4.10. Because of the manner in which it is obtained, the variance is a measure described in square units. It has the advantage that it is additive, in the sense that the variance of two independent distributions is the sum of the variances of both when each element (or number) in one distribution is summed in all possible ways with each element in the other distribution. Thus, if each distribution had a variance of 2, the combined distribution would have a variance of 4. Also, the variance of

*For the derivation of this and other statistical formulas used in this chapter, the reader is referred to Mather and other sources cited in the references of Chapter 8.

TABLE 15–4

Calculation of variance for the data in Table 15–1 arranged according to class frequencies

X	f	fX	f(X²)
11	1	11	121
12	1	12	144
13	2	26	338
14	5	70	980
15	12	180	2,700
16	18	288	4,608
17	23	391	6,647
18	16	288	5,184
19	13	247	4,693
20	4	80	1,600
21	3	63	1,323
22	2	44	968
	100	1700	29,306

$$N = \Sigma f = 100 \qquad \bar{X} = \frac{\Sigma fX}{N} = \frac{1700}{100} = 17.0$$

$$s^2 = \frac{N\Sigma f(X^2) - (\Sigma fX)^2}{N(N-1)}$$

$$= \frac{2{,}930{,}600 - 2{,}890{,}000}{9{,}900} = \frac{40{,}600}{9{,}900} = 4.10$$

a distribution can be broken into individual components enabling the causes responsible for variance to be analyzed (p. 302). However, since the mean of the distribution is not in square units, descriptive association between the mean and the variance is usually computed in terms of the square root of the variance, or the *standard deviation* s. For the previous data, $s = \sqrt{4.10}$, or approximately 2.0. The distribution can then be described as 17.0 ± 2.0 units. In contrast to variance, standard deviations are not additive, e.g., the two distributions with variances of 2 each mentioned above would not have a combined standard deviation of $\sqrt{2} + \sqrt{2} = 2.8$ but would have a standard deviation that is the square root of the sum of the variances $\sqrt{2+2} = \sqrt{4} = 2$.

When it is desired to compare the variability of different samples or experiments, a useful measure is known as the *coefficient of variation*, or C.V. This measure is defined as percent C.V. $= (s/\bar{X}) \times 100$, where $\bar{X}$ and s are, respectively, the mean and standard deviation in a single sample. Since both s and $\bar{X}$ have the same units, division of one by the other cancels these units and produces a numerical value that is independent of the scale used for measurement (i.e., centimeters, feet, etc.). Also, division of s by $\bar{X}$ produces an estimate of variability that is independent of the size of the sample measurements. Thus large means with large standard deviations may now be compared with small means and small standard deviations. Using the mean and standard deviation calculated above, the coefficient of variability for this sample is (2.00 units/17.0 units) $\times$ 100 = 11.8 percent.

VARIANCE AND STANDARD ERROR OF THE MEAN

The variance calculated above (Table 15–4) refers specifically to the individual observations in the sample, and it is only this group of observations, representing a larger potential population, that has a variance of 4.10. In addition to the individual observations, however, the mean may also be considered to be a sample of a population of many means with their own mean and variance. It is, however, obvious that the variance of a distribution of means would be smaller than the variance of all the many dispersed individual observations that make up a particular sample. Statistically, the variance of the mean is calculated as 1/N of the variance of a sample set of observations, or $s^2_{\bar{X}} = s^2/N$. The square root of this value is called the *standard error* of the mean. For the data in Table 15–4, the mean can be described as having a standard error of $\sqrt{4.10/100} = \sqrt{.041} = .202$.

As in other cases the variance of the mean is additive, and the variance of the means of two distributions which are independent of each other is equal to the sum of the variances of the means of each. This additivity is not affected by either the addition of the means of the two distributions or the subtraction of one from the other. The reason for this is that both positive and negative deviations from the mean produce the variance, and these positive and negative deviations are only interchanged but not altered when the means are subtracted. Thus, if two different samples with means $\bar{X}$ and $\bar{Y}$ have respective variances of the mean $s^2_{\bar{X}}$ and $s^2_{\bar{Y}}$, the variance of the sum ($s^2_{\bar{X}+\bar{Y}}$) or of the difference ($s^2_{\bar{X}-\bar{Y}}$) of the two means are calculated in the same fashion:

$$s^2_{\bar{X}+\bar{Y}} = s^2_{\bar{X}-\bar{Y}} = s^2_{\bar{X}} + s^2_{\bar{Y}} = \frac{s^2_{\bar{X}}}{N_X} + \frac{s^2_{\bar{Y}}}{N_Y}$$

where N_X and N_Y refer to the number of observations in the X and Y samples, respectively.

Again, the standard error is the square root of the variance and

$$s_{\bar{X}+\bar{Y}} = s_{\bar{X}-\bar{Y}} = \sqrt{\frac{s^2_{\bar{X}}}{N_X} + \frac{s^2_{\bar{Y}}}{N_Y}}$$

If the variances of the X and Y samples are assumed to be identical, $s^2_{\bar{X}} = s^2_{\bar{Y}} = s^2$,

$$s^2_{\bar{X}+\bar{Y}} = s^2_{\bar{X}-\bar{Y}} = s^2\left(\frac{1}{N_X} + \frac{1}{N_Y}\right),$$

$$\text{and } s_{\bar{X}+\bar{Y}} = s_{\bar{X}-\bar{Y}} = s\sqrt{\frac{1}{N_X} + \frac{1}{N_Y}}$$

TESTING THE DIFFERENCE BETWEEN MEANS

Knowledge of the variance and standard error of the mean enables us to test whether a particular mean is significantly different from another. As for many distributions, it is easy to see that a mean taken from a population of means will usually be close to the central value of that population of means. Only rarely will a randomly chosen mean be far from the central value and be located in the "tails" of the distribution. We have seen previously (p. 156) that a normal bell-shaped distribution will include 95 percent of the entire distribution within (plus or minus) approximately two standard deviations of the mean. Thus values outside this two-standard deviation limit can be considered extremely rare and can be rejected at the 5 percent level of significance.

In the present case a difference between two means can be tested by a similar procedure. That is, the difference between two means $\bar{X}$ and $\bar{Y}$, taken from the same population of means, would be expected to be zero on the average, $\bar{X} - \bar{Y} = 0$ (i.e., have a distribution with a mean of zero). The extent to which the value of this difference departs from a zero mean can therefore be used to indicate how significantly different the two means really are. Means which differ from each other by more than a certain number of standard errors are probably not drawn from the same population of means. In other words, we calculate a standard error for the difference between two means and compare this standard error to the observed difference according to the ratio $(\bar{X} - \bar{Y})/s_{\bar{X}-\bar{Y}} = t$. If t is large, the difference between the means includes many standard errors, and is therefore significant. If this ratio is small, the difference between means is nonsignificant and can be considered to be caused by chance variations in sampling.

To determine the significance of t we can use the area given by the normal distribution and consider that differences between means greater than $\pm$ 1.96 standard errors are significant. However, if $\bar{X}$ and $\bar{Y}$ have been determined on the basis of small sample numbers, the difference between their means is not normally distributed. In such cases the significance of t depends upon a distribution discovered by W. S. Gosset, who published under the pseudonym "Student." "Student's" distribution for t is given in Table 15–5 for different degrees of freedom. To find the probability of t in this table, the number of degrees of freedom are calculated according to the formula $df = (N_X - 1) + (N_Y - 1)$, where N_X and N_Y designate, as before, the sample-size number for $\bar{X}$ and $\bar{Y}$, respectively. For instance, if $N_X = 10$ and $N_Y = 10$, $df = 18$. Under this condition, a t that is equal to or greater than 2.10 signifies that, at the 5 percent level of significance, we can reject the hypothesis that both $\bar{X}$ and $\bar{Y}$ are derived from the same population of means.

An example of this method is given in Table 15–6 for the heights of samples of two plant populations grown in the same field. In this case note that t is equal to 1.74 and therefore has a proba-

TABLE 15–5

Distribution of *t* values ("Student's" distribution)*

	Probabilities							
df	*.90*	*.70*	*.50*	*.30*	*.10*	*.05*	*.01*	*.001*
1	.16	.51	1.00	2.00	6.31	12.70	63.66	636.62
2	.14	.44	.82	1.39	2.92	4.30	9.92	31.60
3	.14	.42	.76	1.25	2.35	3.18	5.84	12.92
4	.13	.41	.74	1.19	2.13	2.78	4.60	8.61
5	.13	.41	.73	1.16	2.02	2.57	4.03	6.87
6	.13	.40	.72	1.13	1.94	2.45	3.71	5.96
7	.13	.40	.71	1.12	1.90	2.36	3.50	5.41
8	.13	.40	.71	1.11	1.86	2.30	3.36	5.04
9	.13	.40	.70	1.10	1.83	2.26	3.25	4.78
10	.13	.40	.70	1.09	1.81	2.23	3.17	4.59
11	.13	.40	.70	1.09	1.80	2.20	3.10	4.44
12	.13	.40	.70	1.08	1.78	2.18	3.05	4.32
13	.13	.39	.69	1.08	1.77	2.16	3.01	4.22
14	.13	.39	.69	1.08	1.76	2.14	2.98	4.14
15	.13	.39	.69	1.07	1.75	2.13	2.95	4.07
16	.13	.39	.69	1.07	1.75	2.12	2.92	4.02
17	.13	.39	.69	1.07	1.74	2.11	2.90	3.96
18	.13	.39	.69	1.07	1.73	2.10	2.88	3.92
19	.13	.39	.69	1.07	1.73	2.09	2.86	3.88
20	.13	.39	.69	1.06	1.72	2.09	2.84	3.85
21	.13	.39	.69	1.06	1.72	2.08	2.83	3.82
22	.13	.39	.69	1.06	1.72	2.07	2.82	3.79
23	.13	.39	.68	1.06	1.71	2.07	2.81	3.77
24	.13	.39	.68	1.06	1.71	2.06	2.80	3.74
25	.13	.39	.68	1.06	1.71	2.06	2.79	3.72
26	.13	.39	.68	1.06	1.71	2.06	2.78	3.71
27	.13	.39	.68	1.06	1.70	2.05	2.77	3.69
28	.13	.39	.68	1.06	1.70	2.05	2.76	3.67
29	.13	.39	.68	1.06	1.70	2.04	2.76	3.66
30	.13	.39	.68	1.06	1.70	2.04	2.75	3.65
40	.13	.39	.68	1.05	1.68	2.02	2.70	3.55
60	.13	.39	.68	1.05	1.67	2.00	2.66	3.46
120	.13	.39	.68	1.04	1.66	1.98	2.62	3.37
∞	.13	.38	.67	1.04	1.64	1.96	2.58	3.29

* Abridged from Table III of Fisher and Yates, *Statistical Tables for Biological, Agricultural and Medical Research*, published by Oliver and Boyd Ltd., Edinburgh, and by permission of the authors and publishers.

bility of approximately 10 percent, according to Table 15–5, at 18 degrees of freedom. Thus, since this probability is greater than the 5 percent level of significance, the difference between $\bar{X}$ and $\bar{Y}$ can be considered as sufficiently frequent to be caused by chance rather than by a real difference between the two populations from which the samples were taken.

CONFIDENCE LIMITS OF THE MEAN

Knowledge of the standard error of the mean also enables us to estimate, with a definite degree of confidence, the range within which the true mean may be found. In Table 15–6 $\bar{X}$ is 13 and $s_{\bar{X}} = \sqrt{.82} = .905$. As in the calculation of other variances, the number of degrees of freedom that

TABLE 15–6

Calculation of t for the heights of 10 individual plants taken from each of two populations X and Y

X	Y	$(X - \bar{X})^2$	$(Y - \bar{Y})^2$
15	12	4	1
12	9	1	4
8	13	25	4
14	10	1	1
16	8	9	9
16	12	9	9
9	13	16	4
15	14	4	9
11	9	4	4
14	10	1	1
130	110	74	46

$\bar{X} = 13.0 \quad \bar{Y} = 11.0 \quad s_X^2 = \frac{74}{9} = 8.22 \quad s_Y^2 = \frac{46}{9} = 5.11$

$s_{\bar{X}}^2 = \frac{8.22}{10} = .82 \quad s_{\bar{Y}}^2 = \frac{5.11}{10} = .51$

$$s_{\bar{X}-\bar{Y}}^2 = .82 + .51 = 1.33$$

$$s_{\bar{X}-\bar{Y}} = \sqrt{1.33} = 1.15$$

$$t = \frac{\bar{X} - \bar{Y}}{s_{\bar{X}-\bar{Y}}} = \frac{13 - 11}{1.15} = \frac{2}{1.15} = 1.74$$

remain is N − 1, or 9. Since the observed $\bar{X}$, is only a sample mean in the distribution of the true $\bar{X}$, it has a 95 percent chance of being included in a distribution which is 2.26 (the value of t at 9 df, according to Table 15–5) standard errors to the right or left of the true mean. Conversely, 2.26 $s_{\bar{X}}$ to the right or left of the observed mean has a 95 percent chance of including the true mean. Thus we can say with 95 percent confidence that the true mean lies within the range $\bar{X} \pm 2.26 s_{\bar{X}} = 13 \pm 2.26\ (.905) = 13 \pm 2.05$, or from 10.95 to 15.05.

CORRELATION

If measurements are taken of two characteristics (X and Y) of each of a series of individuals, i.e., $X_1X_2\cdots$ and $Y_1Y_2\cdots$, it may be of importance to know whether these two characteristics are, in some way, correlated with each other. That is, does a change in one characteristic correspond with a particular directional change (i.e., increase or decrease) in the other, or do these two characteristics vary independently of each other? To measure any relationship present, a *correlation coefficient*, r, can be calculated based on the variance relationships of these two sets of measurements.

To do this a new variance component is calculated, known as the *covariance* (cov) of X and Y:

$$\text{cov}(X, Y) = \frac{\sum_{i=1}^{N}(X_i - \bar{X})(Y_i - \bar{Y})}{N - 1}$$

Note that the difference between the covariance and the variance is that in the former the summation sign includes the products of both X *and* Y, whereas in the latter only products of X *or* Y are included. The correlation coefficient is then calculated as

$$r = \frac{\text{cov}(X, Y)}{s_X s_Y} = \frac{\dfrac{\Sigma(X - \bar{X})(Y - \bar{Y})}{N - 1}}{\sqrt{\dfrac{\Sigma(X - \bar{X})^2}{N - 1}\dfrac{\Sigma(Y - \bar{Y})^2}{N - 1}}}$$

$$= \frac{\dfrac{\Sigma(X - \bar{X})(Y - \bar{Y})}{N - 1}}{\dfrac{1}{N - 1}\sqrt{\Sigma(X - \bar{X})^2\,\Sigma(Y - \bar{Y})^2}}$$

$$= \frac{\Sigma(X - \bar{X})(Y - \bar{Y})}{\sqrt{\Sigma(X - \bar{X})^2\,\Sigma(Y - \bar{Y})^2}}$$

For arithmetical simplicity, the three basic sections of this formula can be computed in the following fashion:

$$\Sigma(X - \bar{X})(Y - \bar{Y}) = \Sigma XY - \frac{\Sigma X \Sigma Y}{N}$$

$$\Sigma (X - \bar{X})^2 = \Sigma X^2 - \frac{(\Sigma X)^2}{N}$$

$$\Sigma (Y - \bar{Y})^2 = \Sigma Y^2 - \frac{(\Sigma Y)^2}{N}$$

$$r = \frac{\Sigma XY - (\Sigma X \Sigma Y/N)}{\sqrt{(\Sigma X^2 - [(\Sigma X)^2/N])(\Sigma Y^2 - [(\Sigma Y)^2/N])}}$$

TABLE 15–7

Calculation of correlation between weight and height of 10 males

Individual	Height in Inches, X	Weight in Pounds, Y	X^2	Y^2	XY
1	66	128	4,356	16,384	8,448
2	68	141	4,624	19,881	9,588
3	64	118	4,096	13,924	7,552
4	70	153	4,900	23,409	10,710
5	69	138	4,761	19,044	9,522
6	73	170	5,329	28,900	12,410
7	68	135	4,624	18,225	9,180
8	67	130	4,489	16,900	8,710
9	65	125	4,225	15,625	8,125
10	72	167	5,184	27,889	12,024
	682	1405	46,588	200,181	96,269

$$r = \frac{\Sigma XY - (\Sigma X \Sigma Y/N)}{\sqrt{\left[\Sigma X^2 - \frac{(\Sigma X)^2}{N}\right]\left[\Sigma Y^2 - \frac{(\Sigma Y)^2}{N}\right]}} = \frac{96{,}269 - \frac{(682)(1{,}405)}{10}}{\sqrt{\left[46{,}588 - \frac{(682)^2}{10}\right]\left[200{,}181 - \frac{(1405)^2}{19}\right]}}$$

$$= \frac{448}{\sqrt{(76)(2779)}} = \frac{448}{\sqrt{211{,}204}} = \frac{448}{459.57} = .97$$

Calculated in this manner r can range from $+1$ to -1. At values of r greater than 0, Y is said to be *positively correlated* with X, meaning that an increase in X is associated with an increase in Y. At values of r less than 0, Y is *negatively correlated* with X and an increase in X is associated with a decrease in Y. At a zero value of r, no evidence of either type of correlation is found to exist, and Y is considered to vary independently of X.

An illustration of the calculation of r is shown in Table 15–7 for heights and weights of 10 human males. In this case r is close to $+1$ and it is evident that there is a high positive correlation or an increase in weight accompanies an increase in height. The same data are also shown in Fig. 15–2a as a series of points plotted on a graph in which height is the X axis and weight is the Y axis. In this graph, each individual is represented as a single point; e.g., the lowest point is that of individual 3 (height 64 inches, weight 118 pounds). Note that an imaginary straight line can be drawn in this figure (dashed line) which is quite close to many of the points and has a shape that extends from low values of X and Y to high values of X and Y. Such a shape is called a *positive slope* and corresponds to the positive value of r. When the correlation coefficient between X and Y is negative, the slope is correspondingly negative, as shown for a distribution of points in Fig. 15–2b. In the absence of correlation between X and Y, i.e., $r = 0$, plotting of the data will not show an obvious slope (Fig. 15–2c).

However, even when values of r differ from zero, a detectable correlation may or may not exist, depending upon how large r is and upon the number of observations on which it is based. To test the significance of r, a quantity can be calculated which has the same distribution as Student's t. In this case,

$$t = \frac{r}{\sqrt{(1 - r^2)/(N - 2)}}$$

where r is the absolute value of the correlation coefficient (ignoring sign) and N is the number of paired observations that were made. If t is larger

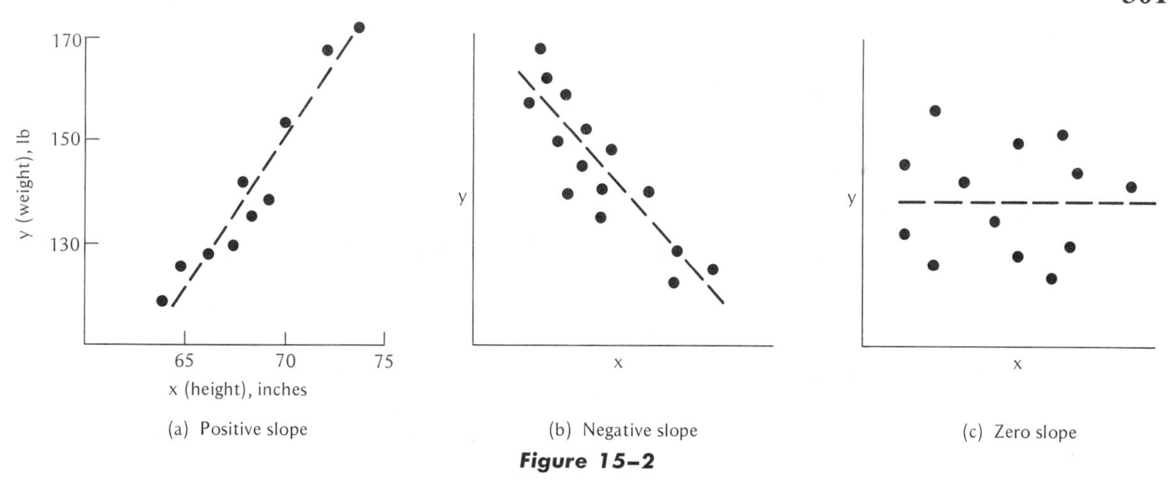

Figure 15–2

Regression slopes (dashed lines) plotted for various data.

than the value given in Table 15–5 at the 5 percent level of significance for N − 2 degrees of freedom, r can be considered as significantly different from zero and we can then assume that there is a correlation between X and Y.

For the previously calculated data,

$$t = \frac{.97}{\sqrt{(1 - .94)/8}} = \frac{.97}{\sqrt{.0075}}$$

$$= \frac{.97}{.0866} = 11.2$$

At N − 2 = 8 degrees of freedom, this value of t is far greater than the .05 probability level in Table 15–5, or even the .01 level. The value of r can therefore be considered significantly different from zero, and X and Y are thus positively correlated.

Despite the significance of r, however, correlations must be interpreted with care. Not in all cases is there a causal relationship between X and Y, but rather X and Y may be correlated with a third factor which produces an apparently direct correlation between them. Increases in population size, for example, have resulted in yearly changes in numerous factors such as the number of professors and the output of whiskey. If we calculated r for these two latter quantities, the increase in number of professors is positively correlated with an increase in the production of whiskey. However, in spite of the professorial consumption of whiskey, it may be considered doubtful that a simple cause-and-effect relationship between these two variables is responsible for the high value of r.

REGRESSION

The existence of a significant correlation between two variables such as X and Y has further importance if we want to predict the value of one variable from a given value of the other. For the data on heights and weights, the straight line running close to all points in Fig. 15–2a can be used for this purpose. Given a specific height X, for example, a corresponding weight, Y, can be expected to appear close to the point where X intercepts the dashed line. This line is known as a *regression* line if it is so constructed that the squared distance between itself and all points on the graph is minimized. For these data the regression line for predicting Y from X has the formula

$$Y = \bar{Y} + b(X - \bar{X})$$

where Y and X are, respectively, the two values along the Y and X axes for any one point on the line; $\bar{Y}$ and $\bar{X}$ are, respectively, means of all Y and X values; and b is a quantity known as the *slope* or *coefficient of regression* of Y on X and is defined as

$$b = \frac{\text{cov}(X, Y)^*}{s_X^2} = \frac{\Sigma (X - \bar{X})(Y - \bar{Y})}{\Sigma (X - \bar{X})^2}$$
$$= \frac{\Sigma XY - (\Sigma X \Sigma Y/N)}{\Sigma X^2 - (\Sigma X)^2/N}$$

For the data on height and weight (Table 15–7), these factors are already known, and

$$b = 448/76 = 5.9$$

A slope of 5.9 means that for each unit increase in X (height), there is an approximate six-unit increase in Y (weight). This relationship can be seen by marking off a one-unit increase in X from any point on the regression line in Fig. 15–2a and observing that the new value of Y that corresponds to this point is approximately six units greater than the previous value of Y.

The full equation of the regression line is then

$$Y = \bar{Y} + b(X - \bar{X})$$
$$= 140.5 + 5.9(X - 68.2) = -261.9 + 5.9X$$

Once such a regression equation is formulated, expected values of Y can be estimated by substituting given values of X. For example, at a height of 5 feet (X = 60), the expected weight of an individual would be approximately

$$Y = -261.9 + 5.9(60) = 92.1 \text{ pounds}$$

If it is desired to make this prediction within certain confidence intervals, the standard deviation for the predicted Y value can be obtained according to methods described by Steel and Torrie and other statistical textbooks mentioned in the references for Chapter 8.

COMPONENTS OF PHENOTYPIC VARIANCE

The variations observed between individuals when they are measured for most quantitative traits may be caused by genetic differences between individuals as well as by environmental differences, and occasionally by interaction between genotype and environment. Interaction occurs when the genotypes act differently in different environments, so that accurate quantitative predictions cannot be made by considering genotype and environment alone (see, for example, Fig. 10–5). The breakdown of observed variation into its components is commonly done by methods known as *analysis of variance*, described in many statistical textbooks. For our purposes, we may consider that the observed phenotypic variance, called V_P, is composed of three main sources of variation: genotypic, environmental, and interaction.

$$V_P = V_G + V_E + V_{GE}$$

The environmental variance (V_E) may include such obviously variable factors as growth in different localities or in different years, or even less obvious ones, such as different maternal environments, different order in birth sequences, and differences between conditions in one side of the body and the other. It includes all variation not caused by genetic differences. Genotype-environmental interactions (V_{GE}) are those in which the relative performances of different genotypes vary with the environments in which the genotypes are placed.

Genotypic variance (V_G), as a cause for phenotypic differences may arise from a variety of sources. Some genes, for example, produce *additive* effects, so that the substitution of a single allele in a gene pair corresponds with an increase

---

*The relationship between regression and correlation can be seen by comparing the equations

$$r = \frac{\text{cov}(X, Y)}{s_X s_Y} \quad \text{and} \quad b = \frac{\text{cov}(X, Y)}{s_X^2}$$

In other words,

$$r = \frac{\text{cov}(X, Y) s_X}{s_X^2 s_Y} = b\frac{s_X}{s_Y}$$

or decrease in phenotypic value, e.g., $A_1A_1 = 4$, $A_1A_2 = 5$, $A_2A_2 = 6$. In this case the numbers signify measurements in which the addition of an A_2 allele causes an increment of 1. The contribution of alleles at more than one gene pair may also be additive, as we have seen in the case of Nilsson-Ehle's red wheat color in Chapter 14. On the other hand, alleles that interact with each other in the same gene pair may produce *dominant* effects. In this case the presence of only a single allele can cause a particular phenotype to appear, e.g., $A_1A_1 = 4$, $A_1A_2 = 6$, $A_2A_2 = 6$. Essential to the notions of both additivity and dominance is the understanding that the changes caused by allelic substitutions at these genes are not interfered with by alleles at other gene pairs. However, *epistatic* or *interaction* effects are also known (see Chapter 11) in which the phenotypic expression caused by one pair of genes depends upon alleles at another gene pair; e.g., gene pair *Aa* shows additive effects in the presence of *BB* but shows dominance in the presence of *bb*.

As a result genotypic variance can be divided as follows:

$$V_G = V_A + V_D + V_I$$

where V_A, V_D, and V_I represent, respectively, the additive, dominance, and epistatic (interaction) variances. Phenotypic variance can thus be written

$$V_P = V_A + V_D + V_I + V_E + V_{GE}$$

The values of some of these variance components may be estimated in appropriate experimental designs. If, for example, the genotypes of all individuals in an experiment are identical, the only variability remaining is that caused by environmental differences. Thus Clayton, Morris, and Robertson measured abdominal bristle number in genetically mixed and genetically uniform *Drosophila* populations and found that the phenotypic variance was reduced by 61 percent in the uniform population. Since this 61 percent of the variance is the component eliminated by genetic uniformity, it therefore represents the genetic variance. The remaining portion, 39 percent, may be considered as containing the variance caused by environment and genotype-environmental interaction.

When "pure" lines can be isolated that are presumably homozygous for quantitative genes, additional components of variance can be estimated by measuring the variance in the parental and filial generations as well as in the backcross hybrids. The justification for such estimates lies in analysis of the variance components contributed by each of these generations. Since the F_1 of a cross between two inbred lines, for example, is genetically uniform, the variance of both kinds of parents and the F_1 can be used as estimates of the environmental variance (which is presumed to be the only variance remaining). Also, since the F_1 mean value would be expected to be exactly intermediate to both parents if the genetic effects are additive, departures from this midparental value would indicate the effect of dominance.

These relationships are graphically presented in Fig. 15–3 for three possible genotypes at a single locus. In these notations *d* represents the departure from the midparental value caused by dominance. The dominance effect may be toward either a phenotypic increase (A_2A_2) or decrease (A_1A_1). Beginning from the midparental value and going toward each of the homozygotes is an additive increment, *a*, so that the phenotypic difference between parental values is 2*a*. When dominance is complete, $d = a$, and the phenotypic value of the heterozygote is identical to either A_1A_1 or A_2A_2. The *degree of dominance* calculated as d/a

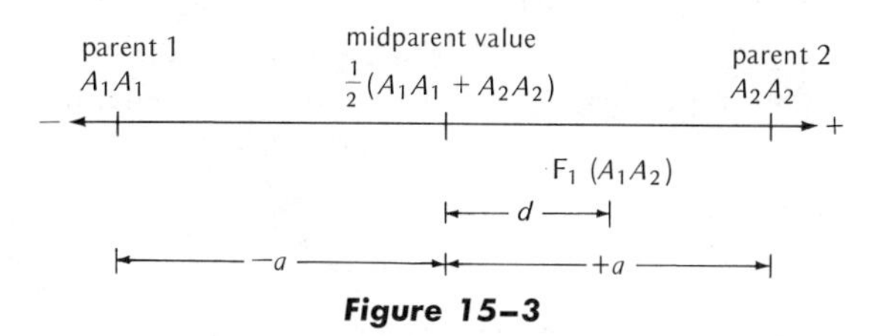

Figure 15–3

Parental quantitative values and departures from these values caused by additivity (a) and dominance (d). When dominance is absent the F_1 is found at the midparental value. When dominance has a value between zero and a the F_1 is between the midparental value and one of the parental values. When d = a dominance is complete and the F_1 is situated at one of the parental phenotypes.

is then equal to 1. If dominance is lacking, the relationship between these alleles is wholly additive, $d = 0$, and the degree of dominance is zero. In such instances the A_1A_2 mean value is exactly at the midparental line and a can then be considered as the additive effect caused by substitution of a single allele.

The determination of the extent of additivity (a) might therefore seem to be most easily obtained by measuring the departure from the A_1A_2 mean to either of the parents, or halving the parental difference. In actual crosses, however, numerous genes are involved in a single quantitative trait, and observation of the F_1 at the midparental line may arise from dominance effects at different gene pairs canceling each other out (i.e., A_1A_2 is dominant in a plus direction and B_1B_2 is dominant in a minus direction). For this reason dominance and additive components are estimated from the variances of F_2 and further generations according to methods developed by Mather. For example, since the F_2 is segregating for genetic differences among its members, the F_2 variance components will receive contributions from each genotype and will also retain the environmental variance. When a single gene pair (A_1 and A_2) is segregating, there are three distinct genotypes in the F_2 in the following proportions:

$$1/4\, A_1A_1 : 1/2\, A_1A_2 : 1/4\, A_2A_2$$

If we consider the phenotypic measurement of each one of these as a departure from the midparental value in accordance with Fig. 15-3, then $A_1A_1 = -a$, $A_1A_2 = d$, and $A_2A_2 = +a$. Taking into account the proportions of each genotype, the average measurement, or F_2 mean, will be

$$1/4\,(-a) + 1/2\,(d) + 1/4\,(+a) = 1/2\, d$$

Since the variance contributed by each genotype is its squared departure from the mean multiplied by its frequency $[f(X - \bar{X})^2]$, the F_2 genetic variance is

$$\begin{aligned} &1/4\,(-a - 1/2\, d)^2 + 1/2\,(d - 1/2\, d)^2 \\ &\quad + 1/4\,(a - 1/2\, d)^2 = 1/4\,(a^2 + ad + 1/4\, d^2) \\ &\quad + 1/2\,(1/4\, d^2) + 1/4\,(a^2 - ad + 1/4\, d^2) \\ &= 1/2\, a^2 + 1/4\, d^2 \end{aligned}$$

If we designate a^2 as A, d^2 as D, and the environmental component as E, the total F_2 variance is $1/2\, A + 1/4\, D + E$. Note that these components are, in reality, the additive (V_A), dominance (V_D), and environmental (V_E) variances; i.e., for the F_2, $1/2\, A = V_A$, $1/4\, D = V_D$, $E = V_E$. Variance components for other generations are shown in Table 15-8.

An illustration of how these variances can be computed can be gathered from data of Smith for flower length in crosses between *Nicotiana langsdorfii* and *N. sanderae*. To eliminate multiplicative effects (see Chapter 14), the values given in Table 15-9 are based on the logarithms of individual measurements. Since the environmental variance is the only component present in the variances of the parental and F_1 generations, E can be considered equal to the average value of these three variances, or $(48 + 32 + 46)/3 = 42$. Subtraction of E from V_{F_2} and $2E$ from $V_{B_1} + V_{B_2}$ then leaves only additive and dominance variances as follows:

$$\begin{aligned} V_{F_2} &= 1/2\, A + 1/4\, D + E - E \\ &= 1/2\, A + 1/4\, D = 130.5 - 42 = 88.5 \\ V_{B_1} + V_{B_2} &= 1/2\, A + 1/2\, D + 2\, E - 2\, E \\ &= 1/2\, A + 1/2\, D = 184 - 84 = 100 \end{aligned}$$

These two equations can then be solved as simul-

TABLE 15-8

Variance components for measurements in different generations*

$V_{P_1} = E$
$V_{P_2} = E$
$V_{F_1} = E$
$V_{F_2} = 1/2\, A + 1/4\, D + E$
$V_{B_1} = 1/4\, A + 1/4\, D + E$
$V_{B_2} = 1/4\, A + 1/4\, D + E$
$V_{B_1} + V_{B_2} = 1/2\, A + 1/2\, D + 2E$

* P_1 and P_2 refer to the two initial parental stocks and F_1 is the resultant progeny. F_2 is the progeny of a cross $F_1 \times F_1$. B_1 and B_2 refers to the respective backcrosses $F_1 \times P_1$ and $F_1 \times P_2$.

TABLE 15–9

Results obtained in crosses between tobacco species (*Nicotiana*) differing in flower length

	No. Individuals	*Mean Measurement*	*Variance*	
P_1 [*N. sanderae* (S)]	47	1292	48	
P_2 [*N. langsdorfii* (L)]	62	37	32	
F_1 (S × L)	38	742	46	
B_1 (F_1 × S)	24	1045	99	85.5
	91	1085	72	
B_2 (F_1 × L)	159	317	90	98.5
	120	312	107	
F_2 (F_1 × F_1)	139	568	125	130.5
	238	729	136	

From Smith, 1937.

taneous equations by doubling both sides of the top equation and subtracting one from the other:

$$\begin{aligned} A + 1/2\,D &= 177.0 \\ 1/2\,A + 1/2\,D &= 100.0 \\ \hline 1/2\,A &= 77.0 \\ A &= 154.0 \end{aligned}$$

Substitution of this value of A into either equation then gives $D = 46$. Since $d^2 = D$, and $a^2 = A$ (above), the degree of dominance shown for flower length is

$$\frac{d}{a} = \sqrt{\frac{D}{A}} = \sqrt{\frac{46}{154}} = .55$$

Thus, although not complete, dominance is obviously present, as may be gathered from the observation that the F_1 mean was somewhat higher (742) than the midparental value (664.5). Knowledge of the values of D and A then enables calculation of the F_2 variance components as

$$\begin{aligned} V_A &= 1/2\,A = 77.0 \\ V_D &= 1/4\,D = 11.5 \\ V_E &= E = 42.0 \\ \hline V_P &= 130.5 \end{aligned}$$

This illustration, perhaps one of the simplest, does not consider the variance caused by interaction. For calculation of variance components under other circumstances and with other methods, the works of Mather, Hayman, Kempthorne, and others should be consulted.

HERITABILITY

The genetic variance ($V_A + V_D$) estimated above is of considerable importance in evaluating and predicting the degree of resemblance between one generation and another. Such predictions, of course, depend greatly upon the amount of environmental variance. The accuracy of prediction would be low, for example, for a trait whose measurements show little genetic variance but a relatively high degree of environmental variance. In such instances, the parental environment may differ sufficiently from that of the offspring so that the observed phenotype of the offspring depends little, if at all, upon the measurements of the parents. For parents to transmit characteristics to their offspring in some predictable degree, it is obvious that environmental variance should be low, and genetic variance high; i.e., the trait should be only slightly affected by environmental differences but greatly affected by genetic differences.

In the extreme, a trait such as ABO blood group shows practically no environmental effect, and its predicted appearance among offspring is highly correlated with its appearance among parents. However, even for the ABO blood group, predictability is not perfect because of dominance between genes; parents that are of the A and B phenotype ($AO \times AO$, $BO \times BO$, $AO \times BO$) may easily give rise to O phenotypes.

On the other hand, predictability is greatly enhanced for genetically determined traits that lack dominance, such as red and white coat color in cattle. If environmental variance is low for such additively determined traits, and interaction is absent, the phenotype of a parent is a fairly accurate indication of its genotype. Mating between two particular phenotypes then produces offspring similar to the parents. Red × red matings among cattle, for example, can reliably be expected to produce only red offspring. Thus the additive portion (V_A) of phenotypic variance is of greater importance in the resemblance between relatives than the dominant portion (V_D). Because of this, the relationship between the additive variance and the total phenotypic variance is generally used as a measure of the inheritance or *heritability* of traits:

$$\frac{V_A}{V_P} = \text{heritability} = h^2 *$$

For the previous flower-length experiment, $h^2 = 77.0/130.5 = .590 = 59.0$ percent. In other words, 59 percent of the observed variability of this trait in this experiment is caused by additive genetic differences.

To the plant and animal breeder such heritability values are of obvious value because they indicate that the selection of parents bearing particular measurements will produce offspring of a similar phenotype. Predicting the selective improvement of many agriculturally valuable traits therefore depends upon a fairly accurate evaluation of the heritability of the trait. As shown in Table 15–10, heritabilities may vary for different traits in the same organism and for the same traits in different organisms (and even for the same traits in different populations of the same organism). Generally, traits that are most essential to the survival of an organism (e.g., conception rate or litter size) show little heritability, since only few heritable differences can be tolerated and more than that decrease survival. On the other hand, traits that do not affect survival to any great degree (e.g., spotting in cattle) may tolerate considerable genetic differences without causing death or extinction.

Since heritability is a measure of the resemblance between relatives, it can also be estimated by evaluating the degree of such resemblance; i.e.,

TABLE 15–10

Estimates of heritability (in percent) for various characters in different species and varieties of farm and laboratory animals

	Percent Heritability
CATTLE	
birth weight (Angus)	49
gestation length (Angus)	35
calving interval (Angus)	4
milk yield (Ayrshire)	43
conception rate (Holstein)	3
white spotting (Friesians)	95
SHEEP	
birth weight (Shropshire)	33
weight of clean fleece (Merino, 22 months)	47
length of wool fiber (Rambouillet, 14 months)	36
multiple birth (Shropshire)	4
CHICKENS	
body weight (Plymouth Rock, 8 weeks)	31
shank length (New Hampshire)	50
egg production (White Leghorn)	21
egg weight (White Leghorn)	60
hatchability (composite)	16
MICE (FROM FALCONER, 1960)	
tail length (6 weeks)	60
body weight (6 weeks)	35
litter size	15
DROSOPHILA MELANOGASTER	
abdominal bristle number	52
thorax length	47
wing length	45
egg production	18

*This is the common statistical symbol for heritability. It *does not* signify the square of heritability.

the correlation between the phenotypes of relatives can be used to obtain a correlation coefficient or regression slope (b), which indicates additive variance. If the regression slope is large, additivity and heritability are considered high. If the regression slope is close to zero, heritability is considered low or absent.

In these determinations, however, it is important to distinguish among correlations produced by similar genes and similar environments. Where possible, therefore, phenotypic relationship between offspring and father (sire) is used to calculate heritability, since there is usually less environmental relationship between these relatives than between others. For example, offspring of the same two parents (full sibs) usually share a common maternal and family environment and may therefore show a greater correlation between each other in phenotype. Environmental factors peculiar to a mother (dam) and her offspring may also produce phenotypic correlations in the sense that, among mammals, a large mother may provide a better environment for the growth of her offspring and thereby produce larger offspring. Where maternal environment is not important and need not be taken into account, mother-offspring correlation and regression can also be used to yield fairly reliable estimates of heritability.

For the regression of offspring on one parent, heritability is estimated as twice the regression coefficient, $2b = h^2$ or $b = 1/2\, h^2$. When the midparent value is used (one-half of each combined set of paternal and maternal measurements), heritability is equal to the regression coefficient, $b = h^2$. In both instances the regression analysis follows the method outlined on page 301, in which we consider Y as the measurement of individual offspring and X as the measurement of the male or female parent or as the midparental value. If more than one offspring is measured for each parent or set of parents, the average measurement of this group of offspring can be used.

An illustration of this method is shown in Table 15-11 for a hypothetical set of bristle counts on a particular segment of *Drosophila* thorax. Y in this case represents the average number of bristles in four female offspring (daughters) and X represents the number of bristles in the mother (dam) of each set of daughters. The slope, b, is therefore known as the "daughter-dam regression," and yields a value of .145. Since $b = 1/2\, h^2$ for this regression, heritability is therefore equal to 29 percent.

Other considerations may enter into the estimate of heritability, and numerous analytical methods have been devised. The most important of these are reviewed by Falconer (1960), who also presents methods for evaluating the precision of heritability estimates. In this respect it is essential to keep in mind that such estimates apply only to a particular population at a particular time. It is obvious that if environmental differences among individuals increase, the additive genetic variance upon which heritability is based will occupy a smaller share of the total phenotypic variance than previously; i.e., heritability will decrease. Similarly genetic changes may occur in a population (Chap-

TABLE 15-11

Calculation of the regression coefficient, *b*, for bristle numbers in *Drosophila* when measurements have been taken for a number of mothers (dams) and their respective groups of four daughters

Group	X (Dam)	Y (Average of 4 daughters)	XY	X^2
1	8	7	56	64
2	7	7	49	49
3	7	9	63	49
4	8	9	72	64
5	9	8	72	81
6	7	8	56	49
7	6	8	48	36
8	9	7	63	81
9	6	6	36	36
10	9	8	72	81
	$\Sigma X = 76$	$\Sigma Y = 77$	$\Sigma XY = 587$	$\Sigma X^2 = 590$

$$b = \frac{\Sigma XY - (\Sigma X \Sigma Y/N)}{\Sigma X^2 - [(\Sigma X)^2/N]} = \frac{587 - [(76)(77)/10]}{590 - (5776/10)}$$

$$= \frac{587 - 585.2}{590 - 577.6} = \frac{1.8}{12.4} = .145$$

ters 33 and 34) which will affect the proportion of additive genetic variance and consequently change the heritability of a trait.

Among humans twin studies have offered one of the more common techniques to arrive at heritability estimates. As discussed in Chapter 10, greater phenotypic similarity for a particular trait among identical twins than among fraternal twins indicates that differences between the two kinds of twins are primarily caused by genetic factors. Heritability-like estimates based on this approach have been presented for various traits (see Cavalli-Sforza and Bodmer), and range from .05 for body weight to .90 for cephalic index (head width/head length). Among the underlying assumptions necessary to derive these estimates are (1) Only the genetic variance differs between the two types of twins but the environmental variance is the same. (2) There are no interactions between particular genotypes and particular environments which would exaggerate the genetic variance. Although such assumptions may be partially valid for simple traits in some populations, the validity of heritability estimates for some complex traits has been seriously challenged.

An important area of dispute is intelligence. Claims have been made by Jensen and others that heritability for intelligence among humans is high ($\cong$.80) and differences in IQ scores between black and white racial groups reflect genetic differences. To many geneticists, however, the Jensen view is not supported because of the many serious but unknown effects caused by environmental and cultural differences. As pointed out previously (pp. 194–196), there may be considerable differences in the kinds and amounts of environmental variances between races, and there is an obvious lack of clarity about many factors responsible for IQ differences within a race. In view of all this ignorance it certainly seems presumptious to make racial genetic comparisons about intelligence. (Discussion on this topic, *pro* and *con*, can be found in Eaves and Jinks, Lewontin, Jensen, Morton, Thoday, Jencks, Herrnstein, Layzer, Kamin, Feldman and Lewontin, and in the review by Loehlin, Lindzey, and Spuhler.)

PROBLEMS

15-1. The following are measurements of a quantitative character made on 10 individuals: 8, 12, 10, 14, 11, 13, 12, 9, 9, 7. Calculate: (a) The mean. (b) The standard deviation. (c) The standard error of the mean. (d) The 95 percent confidence interval within which the true mean for this character can be found.

15-2. If the distribution of a particular quantitative trait is normal with a numerical mean of 50 and a standard deviation of 4, state the percentage of the population that has the following numerical measurements: (a) More than 58. (b) Between 42 and 58. (c) Between 50 and 58. (d) Between 42 and 46. (e) Less than 38.

15-3. Three selected inbred stocks (A, B, and C) of a single species of plant were measured for size (in centimeters) and showed the following means and variances:

Stock	*Mean*	*Variance*
A	70	49
B	193	841
C	315	3969

(a) Calculate the coefficients of variation. (b) Indicate the types of genes you think are causing the size differences in these stocks.

15-4. Ten individuals each of two strains (A and B) of a species were measured for a particular trait in the same environment with the following results:

strain A: 10, 12, 8, 14, 12, 9, 7, 7, 13, 6
strain B: 13, 9, 14, 12, 10, 13, 15, 11, 14, 12

Is there a significant difference in this trait between the two strains?

15-5. Two strains of mice (A and B) were grown under the same environmental conditions and fed the same rations. Twenty adult mice in each strain of the same age and sex were then measured for body weight (in grams) and the results were as follows:

Strain	*Mean body weight*	*Variance*
A	16	16
B	28	25

Is there a significant difference in body weight between the two strains?

15-6. The two members of each of the 19 pairs of identical twins reared apart that were studied by Newman, Freeman, and Holzinger were scored for differences in IQ and for educational differences, as follows:

Case no. of twin pair	IQ difference (Binet test)	Educational difference
1	12	15
2	12	32
3	−2	12
4	17	22
5	4	11
6	8	7
7	−1	9
8	15	14
9	6	7
10	5	10
11	24	37
12	7	19
13	1	11
14	−1	12
15	1	9
16	2	8
17	10	15
18	19	28
19	9	9

(a) Is there a significant correlation between education and intelligence tested in this fashion? (b) If the two members of a pair of identical twins showed no difference in educational opportunities, what difference between them in IQ would you expect according to these data?

15-7. As an example of regression, Galton reported the following average seed sizes for a group of 100 progeny derived from seven different parental classes of sweet peas. (Values are seed diameter given in hundredths of an inch.)

	Parental size						
	15	*16*	*17*	*18*	*19*	*20*	*21*
average progeny size	15.3	16.0	15.6	16.3	16.0	17.3	17.5

(a) Using progeny size as the dependent variable Y, calculate the regression coefficient of Y on X. (b) What progeny size would you expect for a parental size of 23? (c) What progeny size would you expect for a parental size of 12?

15-8. A cross was made between two strains of plants differing in size, that had the respective quantitative values 20 and 14 centimeters. The F_1 of this cross had the value 18 centimeters. Calculate the degree of dominance.

15-9. A randomly breeding population of *Drosophila melanogaster* shows a variance of .366 for thorax length measured in units of .01 millimeter. On the other hand, the F_1 derived from crosses between highly inbred strains of *Drosophila* show a variance of only .186 for the same character. (a) What components of variance can be estimated from these results? (b) What proportion of the variance in the randomly breeding population is attributable to genetic differences between individuals?

15-10. The means and variances of heading date (flowering time) in two parental varieties of wheat and the progeny of their subsequent crosses were studied by Allard and found to be as follows.

	Mean	Variance
P_1 (Ramona, early heading)	12.99	11.036
P_2 (Baart, late heading)	27.61	10.320
F_1 ($P_1 \times P_2$)	18.45	5.237
F_2 ($F_1 \times F_1$)	21.20	40.350
B_1 ($F_1 \times P_1$)	15.63	17.352
B_2 ($F_1 \times P_2$)	23.88	34.288

Using the variance components, calculate: (a) The degree of dominance. (b) Heritability.

15-11. Shank length and neck length are measured in a particular species of mammal and the following variances observed:

shank length	
phenotypic variance	310.2
environmental variance	103.4
additive genetic variance	103.4
dominance variance	103.4
neck length	
phenotypic variance	730.4
environmental variance	365.2
additive genetic variance	182.6
dominance variance	182.6

Explain which of these two characters would be more rapidly changed by selection.

15-12. Heritability for various traits in cattle estimated from data on the degree of resemblance between identical twins is usually much higher than heritability estimates using parent-offspring resemblances. What factors do you think account for these differences?

REFERENCES

BREWBAKER, J. L., 1964. *Agricultural Genetics*. Prentice-Hall, Englewood Cliffs, N.J.

CAVALLI-SFORZA, L. L., and W. F. BODMER, 1971. *The Genetics of Human Populations*. W. H. Freeman, San Francisco, Chap. 9.

CLAYTON, G. A., J. A. MORRIS, and A. ROBERTSON, 1957. An experimental check on quantitative genetical theory. I. Short-term responses to selection. *Jour. Genet.*, **55,** 131–151.

EAVES, L. J., and J. L. JINKS, 1972. Insignificance of evidence for differences in heritability of IQ between races and social classes. *Nature*, **240,** 84–88.

FALCONER, D. S., 1960. *Introduction to Quantitative Genetics*. Oliver and Boyd, Edinburgh, Chaps. 7–10.

FELDMAN, M. W., and R. C. LEWONTIN, 1975. The heritability hang-up. *Science,* **190,** 1163–1168.

FISHER, R. A., 1918. The correlation between relatives on the supposition of mendelian inheritance. *Trans. Roy. Soc. Edinburgh*, **52,** 399–433.

HAYMAN, B. I., 1960. The theory and analysis of diallel crosses. III. *Genetics*, **45,** 155–172.

HERRNSTEIN, R. J., 1973. *IQ and the Meritocracy*. Little, Brown, Boston.

JENCKS, C., M. SMITH, H. ACLAND, M. J. BANE, D. COHEN, H. GINTIS, B. HEYNS, and S. MICHELSON, 1973. *Inequality: A Reassessment of the Effect of Family and Schooling in America*. Basic Books, New York.

JENSEN, A. R., 1973. *Educability and Group Differences*. Methuen, London.

KAMIN, L. J., 1974. *The Science and Politics of I.Q.* Halsted Press, John Wiley, New York.

KEMPTHORNE, O., 1957. *An Introduction to Genetic Statistics*. John Wiley, New York.

LAYZER, D., 1974. Heritability analyses of IQ scores: science or numerology. *Science*, **183,** 1259–1266.

LEWONTIN, R. C., 1970. Race and intelligence. *Science and Pub. Aff.*, **26,** 2–8.

———, 1974. The analysis of variance and the analysis of causes. *Amer. Jour. Hum. Genet.*, **26,** 400–411.

LOEHLIN, J. C., G. LINDZEY, and J. N. SPUHLER, 1975. *Race Differences in Intelligence*. W. H. Freeman, San Francisco.

LUSH, J. L., 1945. *Animal Breeding Plans*. Iowa State Coll. Press, Ames.

MATHER, K., and J. L. JINKS, 1971. *Biometrical Genetics*, 2nd ed. Chapman and Hall, London.

MORTON, N. E., 1974. Analysis of family resemblance. I. Introduction. *Amer. Jour. Hum. Genet.*, **26,** 318–330.

SMITH, H. H., 1937. The relations between genes affecting size and color in certain species of *Nicotiana*. *Genetics*, **22,** 361–375.

SOKAL, R. R., and F. J. ROHLF, 1969. *Biometry*. W. H. Freeman, San Francisco.

THODAY, J. M., 1973. Educability and group differences. *Nature*, **245,** 418–420.

PART

ARRANGEMENT OF GENETIC MATERIAL

Verily not by design do the first beginnings of things station themselves each in his right place, occupied by keen-sighted intelligence, . . . but because after trying motions and unions of every kind, at length, they fall into arrangements such as these out of which this our sum of things has been formed.

LUCRETIUS
On the Nature of Things

16
LINKAGE AND RECOMBINATION

When Sutton first proposed the relationship between mendelian factors and chromosomes in 1903, he also pointed out that the number of chromosomes in any organism is probably considerably less than the number of individual mendelian factors. This conclusion is most obvious in organisms such as *Drosophila*, which carry only a few pairs of chromosomes but, at the same time, show numerous separately inherited traits. Sutton and others therefore suggested that some genes, specifically those on the same chromosome, would be expected to adhere or "link" together in a group and be transmitted as a single unit.

Many early genetic studies, however, showed independent assortment between genes but did not show the expected group transmission or linkage. Mendel's experiments with peas, for example, showed the inheritance of seven different characters each transmitted independently of the other. The absence of linkage between genes was therefore believed by some biologists to arise from the complete breakdown of chromosomes during meiosis into smaller sections consisting only of individual genes. After random assortment of these particles, they were reaggregated into chromosomes. Chromosomes, according to this explanation, were therefore of secondary importance and served only to tie the genes together for short periods of time.

Determination of whether linkage between genes actually existed had to await examples demonstrating the failure of gene pairs to assort independently and the association between a number of different gene pairs and a specific chromosome. As long as different gene pairs assorted independently of each other, their behavior could be explained either by their presence on separate chromosomes or by the breakdown of chromosomes during meiosis.

As we have previously seen, independent assortment between two gene pairs, each with complete dominance and affecting different characters, should result in a 9:3:3:1 ratio among the offspring of heterozygotes for both pairs; i.e., *AaBb* × *AaBb* should produce four phenotypes in a ratio 9 *A–B–* :3 *A–bb* :3 *aaB–* :1 *aabb*. The first example of a departure from this ratio was found by Bateson and Punnett in crosses between different varieties of sweet pea. They found that plants of different flower color and pollen shape, when crossed together, gave rise to F_1 and F_2 offspring in which the genes for flower color (*A* and *a*) and pollen shape (*B* and *b*) did not assort independently but were "tied together" so that F_2 offspring appeared in ratios which contained too many of the original parental genotypes (e.g., *A–B–* and *aabb*) and too few of the newly combined genotypes (e.g., *A–bb* and *aaB–*). It was as though the original parental varieties in this case (*AABB* and *aabb*) produced gametes in which the *AB* and *ab* genes remained preferentially associated through each generation.

To account for this phenomenon and for other characters that also showed linkage, Bateson and Punnett proposed that in such cases certain gametes multiplied preferentially after meiosis. For example, an F_1 hybrid derived from a cross between *AABB* × *aabb* would produce *AB*, *Ab*, *aB*, and *ab* gametes, but the parental-type gametes (*AB* and *ab*) would then further multiply mitotically. There were serious objections to this view, among them the lack of evidence that such mitotic replications occur once gametes are formed. Within a short time, however, the true meaning of linkage was clarified by the experimental data that began to appear from genetic studies of *Drosophila*.

LINKAGE GROUPS

In *Drosophila melanogaster*, Morgan and coworkers found numerous hereditary characteristics, many of which were associated together in groups. Besides white eyes, for example, there were other sex-linked characters such as miniature wings, yellow body, singed bristles, and so on, that constituted a sex-linked group and tended to be inherited together. For example, females heterozygous for *yellow* (*y*) and *white* (*w*), in which these genes came from one parent and the wild-type alleles from another, produced male offspring that were mostly either yellow-bodied and white-eyed (*yw*) or wild type (y^+w^+). The new combinations yw^+ and y^+w were quite rare. This indicated a strong connection or linkage between *y* and *w* on one of the female X chromosomes and y^+ and w^+ on the other. Similarly, numerous non-sex-linked characters were found that could, in turn, be divided into three additional groups of characters which remained associated during transmission. For example, Bridges found that the autosomal genes *dachsous* (shortened wings) and *black* (dark body color) tended to be transmitted together. Also associated with *dachsous* were *Star* (small rough eyes), *aristaless* (certain bristles reduced), and other genes.

In all, *D. melanogaster* was found to have four groups of linked genes. It was then only one further step to point out that each linkage group in this organism corresponded to one of its four pairs of chromosomes. The association of sex-linked characters with the X chromosome was inferred from their special mode of transmission (Chapter 12), and they were labeled the "first" linkage group. The three non-sex-linked autosomal groups of genes were associated with the three autosomal pairs of chromosomes and were labeled the "second," "third," and "fourth" linkage groups. Departures from independent assortment could therefore be explained as resulting from the nonrandom assortment of genes that were on the same chromosome during meiosis.

When listings of the various genes on the four linkage groups were compiled, one important fact emerged—the finding that relatively few genes were found in the fourth linkage group as compared to the others (see Fig. 17–3). This disproportion, however, can be explained by identifying the fourth linkage group with the very tiny dot chromosome. If, as expected, the number of genes is somehow related to the physical size of the chromosome, the "dot" chromosome should be carrying the smallest number of all. To date, this relationship holds true, and only about a dozen genes are known for this chromosome compared with 150 or more for each of the other three chromosomes in *Drosophila.*

Numerous other organisms have shown this same relationship between the number of linkage groups and the haploid number of chromosomes. In corn 10 linkage groups are known, corresponding to the 10 pairs of chromosomes (see Neuffer and Coe). In peas (*Pisum sativum*) seven linkage groups correspond to seven pairs of chromosomes (see Blixt). In no instance among genetically well-studied organisms are there more linkage groups than haploid number of chromosomes, although in some organisms where genetic data are more difficult to obtain, such as humans, the number of presently known linkage groups is less than the haploid number of chromosomes (see p. 342ff. and Table 17–5).

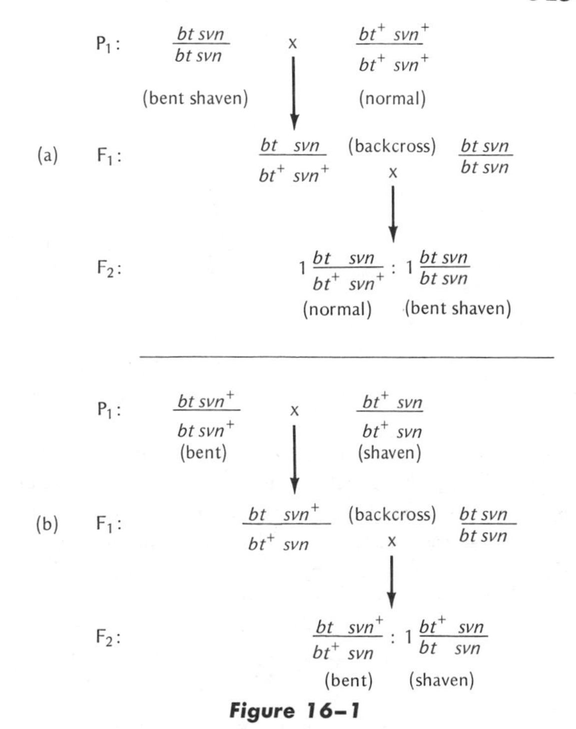

Figure 16–1

Complete linkage between alleles of the bent and shaven genes in two types of Drosophila crosses. Notationally, linked genes are usually symbolized by placing them on the same side of a horizontal line; symbols above the line represent alleles on one chromosome; symbols below the line represent alleles for the same gene pairs on the homologous chromosome.

COMPLETE LINKAGE

When genes are so closely associated that they are always transmitted together upon coming from the same parent, linkage between them is considered *complete.* For example, practically all the fourth chromosome mutations of *Drosophila melanogaster* show little or no assortment from each other during transmission. A fourth chromosome mutant carrying genes producing bent-wings and shaven-bristles (*bt svn/bt svn*) crossed to a normal fly ($bt^+\ svn^+/bt^+svn^+$) would produce a normal-appearing F_1 that is heterozygous for the mutant genes (Fig. 16–1a). When such F_1 flies are crossed to the homozygous *bent shaven* stock, the offspring are almost all phenotypically either bent and shaven or normal for both characters; practically no bent-winged normal-bristled or normal-winged shaven-bristled flies appear. The almost complete absence of independent assortment between these two gene pairs is evidence of very strong linkage between them.

This linkage exists when the *bent* and *shaven* alleles are associated together on the same homologue and also when they are each associated with normal alleles on the same homologue. For example, an F_1 heterozygote derived from a cross *bent* × *shaven* could also be backcrossed to flies from *bent shaven* stock. Here again the F_2 offspring would almost exclusively have the same phenotypes as the original parents, but in this case the phenotypes are bent-winged normal-bristled and normal-winged shaven-bristled (Fig. 16–1b).

Complete linkage is noted for all gene pairs in each of the other linkage groups when *Drosophila* males are used as the heterozygous parent in a testcross. For example, the genes for both purple eyes and vestigial wings are on the second linkage group and are recessive to those for normal red eyes and long wings. When a homozygous *purple vestigial* fly is crossed to a normal red-eyed long-

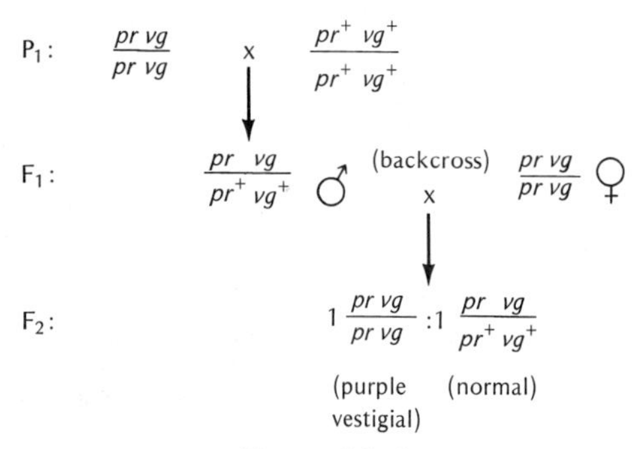

Figure 16–2

Complete linkage between alleles of purple and vestigial genes when only males are used as the heterozygous parents in the backcross to the purple vestigial stock. Note that the pr and vg alleles remain together on one homologue in the heterozygous males as do the pr^+ and vg^+ alleles on the other homologue, and their progeny inherit only one of these combinations (pr-vg) or the other (pr^+-vg^+). Mixed combinations of alleles in a gamete—in this case, pr-vg^+ and pr^+-vg—are not formed in the male as long as the genes involved are on the same linkage group.

winged fly, the heterozygous F_1 offspring are all normal. A backcross of an F_1 male to a homozygous *purple vestigial* female produces only two phenotypes: the homozygote *purple vestigial* and the heterozygote normal, as shown in Fig. 16–2.

This absence of any assortment between all genes in a linkage group appears to be a peculiarity of male *Drosophila* and of the heterogametic sex in a few other species (e.g., female silkworms). Independent assortment, nevertheless, does occur in these cases, but only between genes of *different* linkage groups. For example, *Drosophila* males heterozygous for both *purple* on the second linkage group and *bent* on the fourth linkage group (pr/pr^+; bt/bt^+) will produce all four possible combinations of gametes in accord with the law of independent assortment. Such males can be backcrossed to homozygous *purple bent* females and equal frequencies are observed for each of the four phenotypic classes of offspring that result (Fig. 16–3).

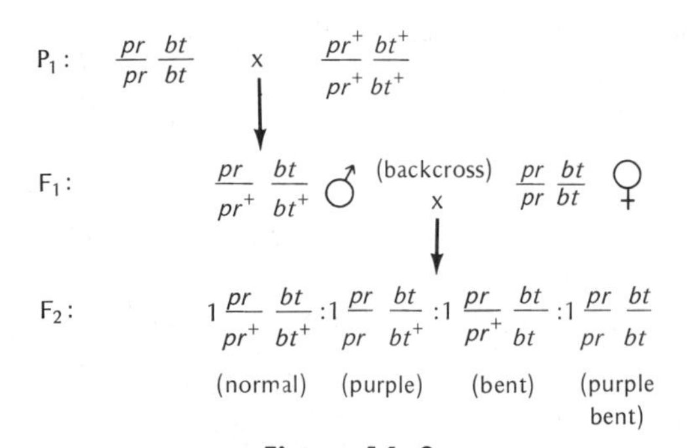

Figure 16–3

Evidence for independent assortment of genes on different linkage groups in male Drosophila. Such males show complete linkage of two different genes only when they are located on the same linkage group.

INCOMPLETE LINKAGE AND RECOMBINATION

Complete linkage between genes on the same chromosome, however, is a rarity in most sexually reproducing species. As a rule linkage is not complete, and the gene pairs in most linkage groups assort at least partially independently of each other. This was first clearly noted by Morgan in crosses between white-eyed and miniature-winged *Drosophila* stocks. Both gene pairs affecting these traits are sex-linked and have recessive effects, so that hemizygous males with only a single X chromosome show the phenotype of either or both recessive genes although each is only present in single dose. When Morgan mated a *white miniature* male to a homozygous normal wild-type female, the F_1 females were, of course, wild type, since one of their two X chromosomes carried normal alleles ($w\ m/w^+\ m^+$). Their F_2 male offspring, however, could be either wild type ($w^+\ m^+$), *white miniature* ($w\ m$), *white* ($w\ m^+$), or *miniature* ($w^+\ m$), depending on the X chromosome they received from their F_1 maternal parent. If assortment between *white* and *miniature* on the X chromosome were completely independent, the F_2 males should show these four phenotypes in equal frequency. However, if linkage between the *white* and *miniature* genes were complete, only the original parental phenotypes should appear among the males. The F_2 results that Morgan actually observed fell between complete linkage and random assortment for these two gene pairs. Of 1190 F_2 males, 37.6 percent represented the two new recombinant types, *white* and *miniature*, while 62.4 percent represented the parental types, either wild type or *white miniature* (Fig. 16–4).

Two main conclusions can be derived from Morgan's experiments. First, there was no apparent breakdown of chromosomes during meiosis followed by a random reassortment of genes into new chromosomes. Rather, sex-linked genes remained linked to X chromosomes through all generations tested, although not necessarily to the same parental X chromosome. This same constancy of chromosome association was also found to be true for each of the other linkage groups. Second, the occurrence of recombinant types (*recombination*) signified that a *crossing over* or genetic exchange had taken place in the F_1 females between one parental chromosome and its homologue from the *other* parent. Morgan hypothesized that genes occurred in linear order along the length of the chromosome. Linkage was therefore a physical relationship between genes that could be modified by a physical crossover during meiosis between gene pairs on homologous chromosomes (Fig. 16–5). This crossover theory fitted well with a theory proposed by Janssens in 1909 that a cytologically-observed chiasma (Chapter 2) represented an exchange point between homologous chromosomes.

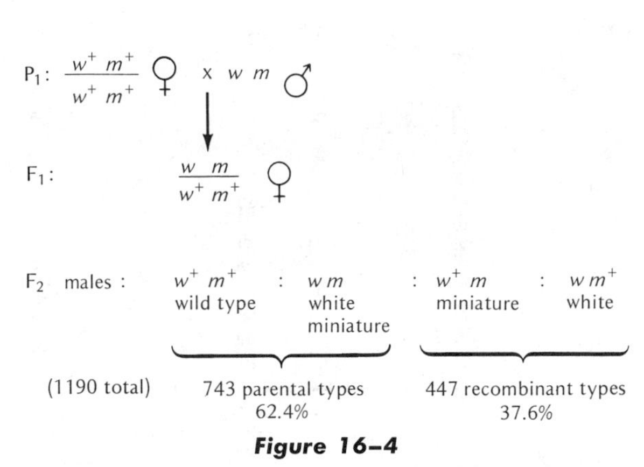

Figure 16–4

Recombination between white and miniature as evidenced by recombinant classes among F_2 males. (Data of Morgan, 1911b.)

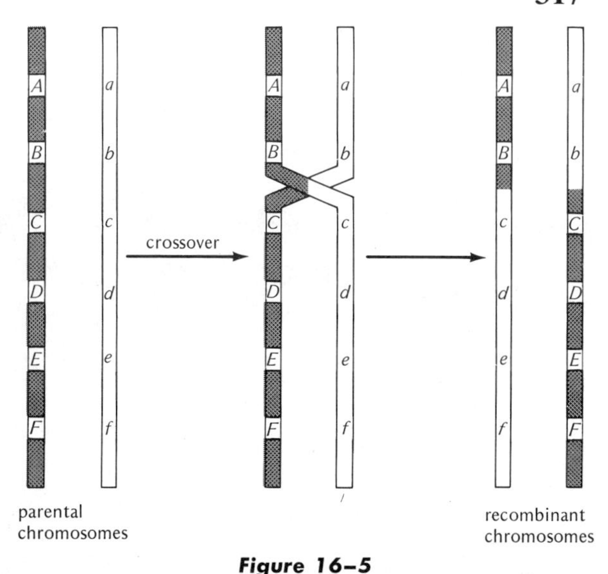

Figure 16–5

Crossing over between gene pairs arranged linearly on homologous chromosomes to yield recombinant arrangements of alleles of the same gene pairs. For simplicity, only one chromatid from each homologous chromosome is shown.

FOUR-STRAND CROSSING OVER

If Morgan's theory on physical crossing over were correct, the question arises whether recombination occurred between entire homologous chromosomes or between the individual chromatids of these homologues (Fig. 16–6). In the former case each crossover at the two-strand stage should yield four new recombinants among the gametes. In the latter case crossing over between two chromatids at the four-strand stage should yield two recombinant products and two parental combinations. To decide between these alternatives, all four gametic products from a crossover event should be recovered. Fortunately, in *Drosophila*, the occurrence of attached-X isochromosomes (p. 239) permits observation of two homologous chromosomes that are transmitted as a unit. Any two-stranded crossover between one attached

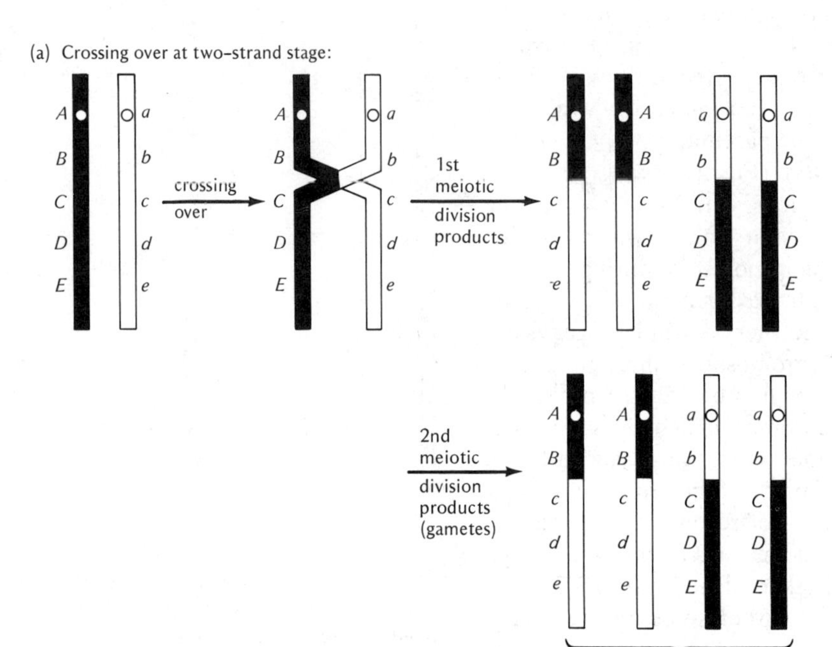

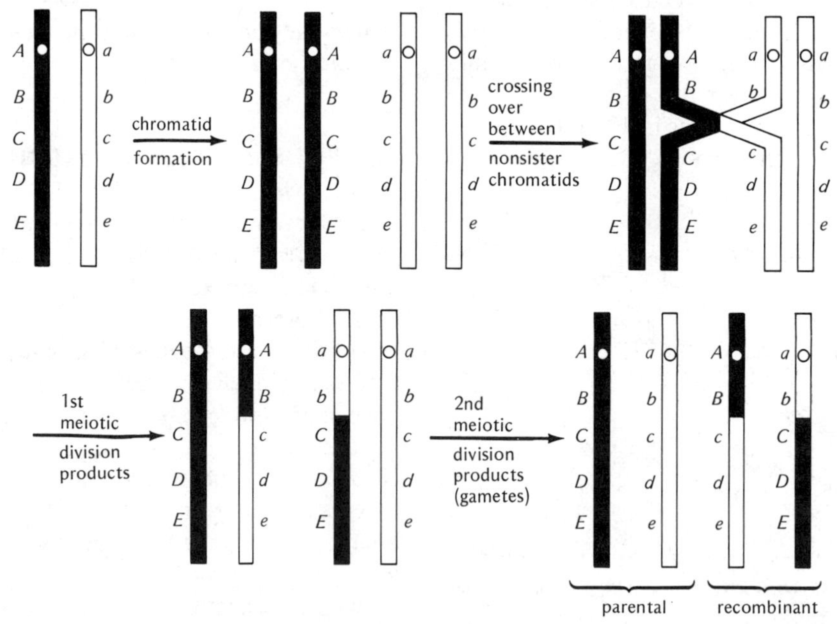

Figure 16–6

Difference in consequences between crossing over at the (a) two-strand and (b) four-strand stages.

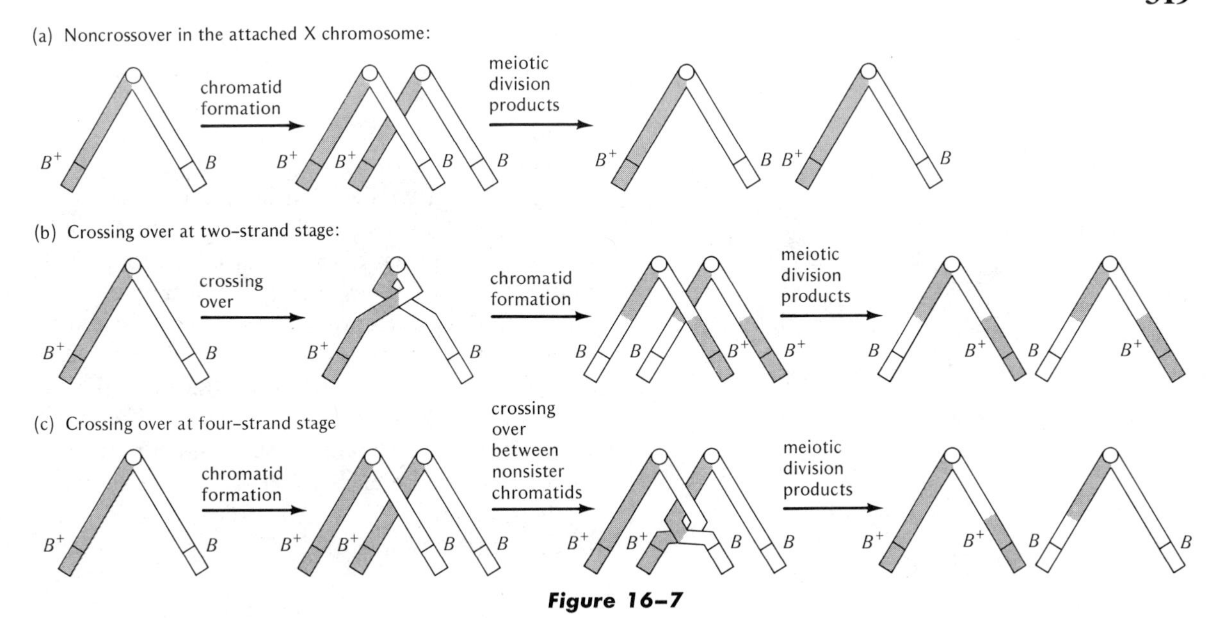

Figure 16–7

Consequences of noncrossing over and crossing over at two- and four-strand stages in a female Drosophila *carrying attached-X chromosomes that are heterozygous* B^+/B.

homologue and the other in a heterozygote for a gene such as *Bar*, would lead to the same phenotypes as if exchange were absent (Fig. 16–7a, b). On the other hand, exchange at the four-strand stage would yield completely new combinations, i.e., attached-X gametes that are homozygous wild type or homozygous *Bar* (Fig. 16–7c). Since *Bar* is a semidominant gene, either homozygous type can be phenotypically distinguished from the heterozygotes when fertilized by sperm carrying a Y chromosome to yield $\widehat{XX}Y$ daughters. Experiments with heterozygous attached-X females were initially performed by Anderson, and the appearance of new combinations demonstrated that crossing over occurred during the four-strand stage.

In some haploid organisms all tetrad products of a single meiosis can be recovered, enabling a detailed analysis of each recombinational event. When tetrad analysis is performed in the green alga *Chlamydomonas*, we have seen that tetratype tetrads result from crossing over between a gene and its centromere on two of the four meiotic strands (p. 136). In the fungus *Neurospora*, since each set of meiotic products (ascospores) is lined up in linear fashion, the particular meiotic division at which genetic exchange occurs can be determined by dissection of the ascus and individual growth of each ascospore. For example, a cross between an *albino* (*al*) and normal (al^+) strain would, in the absence of crossing over, result in a linear arrangement of four albino spores and four normal spores (Fig. 16–8). If crossing over were to occur between the *al* gene and the centromere at the two-strand stage, the linear arrangement of four albino and four normal spores would not be changed. However, if crossing over occurs between two nonsister chromatids at the first meiotic division (Fig. 16–9), the usual spore arrangement is altered if the crossover region is between the *albino* gene and the centromere. Depending on the orientation of these first meiotic division products (dyads) at metaphase of the second meiotic division, the arrangement of spores in the ascus will be either in pairs of alternating genotypes or two pairs of similar genotypes

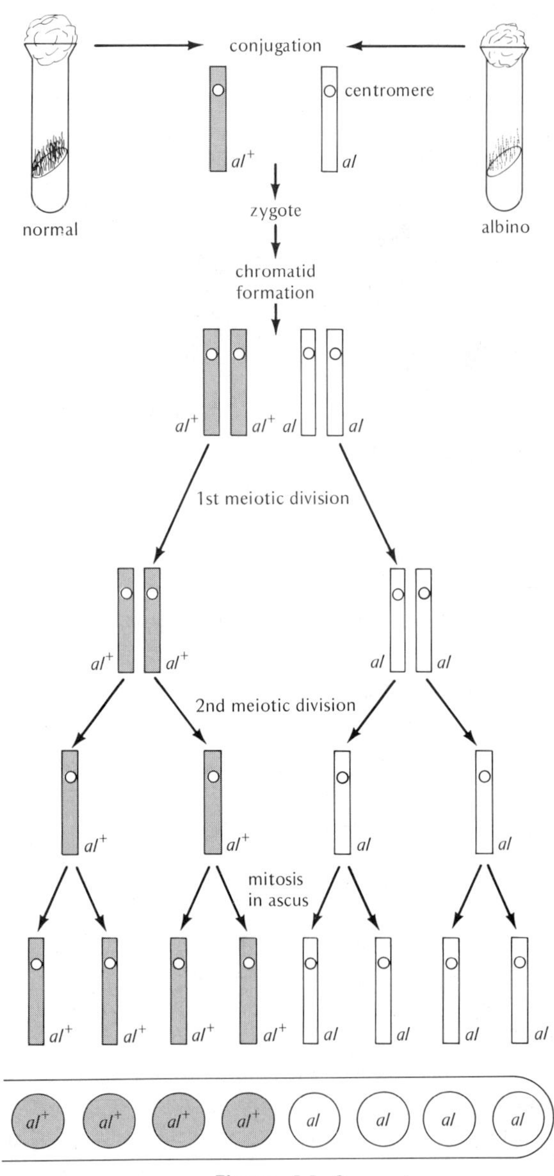

Figure 16–8

Results of a cross between a normal (al^+) *and albino* (al) *strain of* Neurospora *in which crossing over does not take place between the* al *gene and its centromere. In the absence of such a crossover, the chromatids containing* al *are separated from those containing* al^+ *at the first meiotic division; that is, first division segregation occurs (see also p. 136).*

bordered by single pairs of the alternative genotype. Thus the occurrence of these new linear arrangements was presented by Lindegren as evidence that crossing over had occurred between two nonsister chromatids at the four-strand stage.

Additional evidence from other organisms (Chapter 17) has helped support the theory that recombination involves an actual physical transfer of blocks of genes between homologous chromosomes, and that this transfer occurs between only two homologous chromatids out of the four chromatids in a bivalent. Crossing over at any particular point along the chromosome is thus limited to only 50 percent of the chromatids. Since each crossover event will therefore produce one pair of recombinant chromatids and one pair of parental chromatids, recombination events for a particular interval cannot be expected more than 50 percent of the time, even if a crossover occurs within that interval in every meiosis.*

DETECTION OF LINKAGE

To detect whether two gene pairs are linked or assorting randomly, the simplest method is to compare the number of individuals observed in each phenotypic class with those expected on the basis of independent assortment, and then to test the deviation between these values by chi-square. For example, a mating between a heterozygote *AaBb* and a double-recessive homozygote *aabb* should produce four classes of offspring: *AaBb*, *Aabb*, *aaBb*, and *aabb*. If assortment between *Aa* and *Bb* is independent, each of the four classes should be in equal ratios of 1/4, and departures from these ratios can be detected by the chi-square test. Linkage between the *Aa* and *Bb* gene pairs, however, is not the only cause that can be expected to affect the frequencies of certain classes. Other effects, such as viability differences, may change

*As shown in Chapter 18 (Fig. 18–3), four-strand double crossovers that occur in the interval between two linked pairs of genes will produce four recombinant chromatids from a single meiosis. Although theoretically this might indicate that the number of recombinants can exceed 50 percent, such four-strand double crossovers are usually balanced by a similar or even greater number of two-strand double crossovers (see p. 375) which lead to the parental combination in all four chromatids. Thus the recombinational limit between any two points remains at 50 percent.

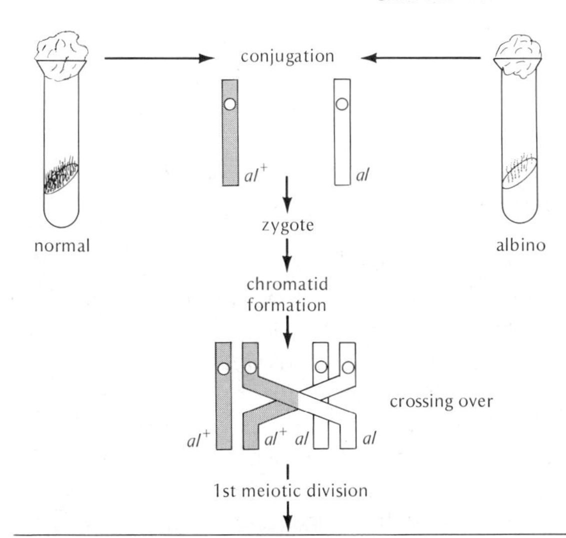

Figure 16–9

Results of crossing over between a normal and albino strain of Neurospora in which crossing over takes place between the gene and its centromere. Different orientations of the first meiotic division products provide different arrangements of genotype pairs in the ascus: either + − + − as in (a), or + − − + as in (b), or also − + + − (not shown). In these cases full separation between chromatids containing al and those containing its al⁺ allele is not achieved until the second meiotic division, i.e., second division segregation occurs for this gene (see also p. 136).

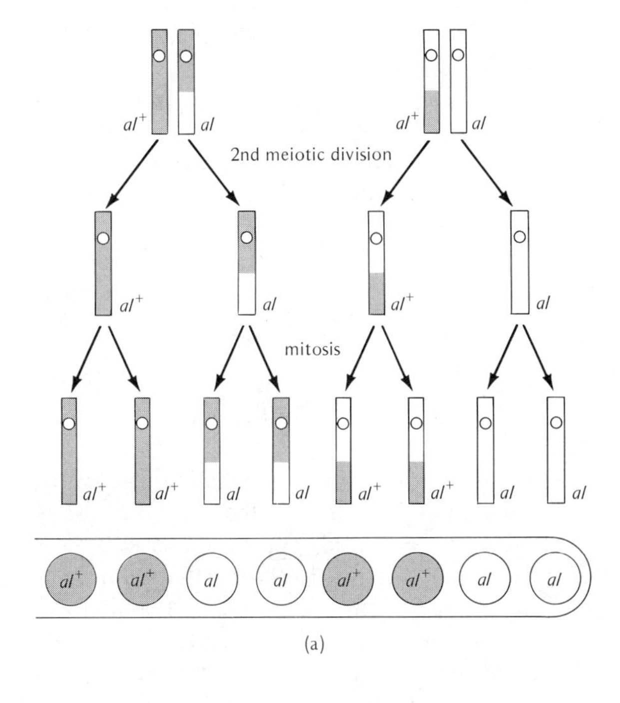

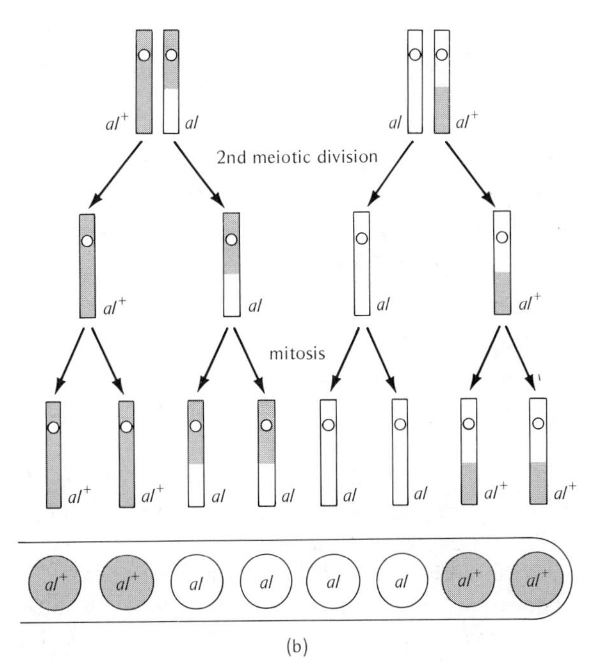

the ratios for individual gene pairs so that, for example, a cross $Aa \times aa$ does not produce the expected 1/2 Aa:1/2 aa because some or all of the aa genotypes are lethal. To distinguish between these effects, the chi-square procedure must first test whether each individual gene pair segregates as expected, and then whether both gene pairs segregate independently of each other. Effectively, therefore, three chi-square values must be calculated: one each for the segregation of the two individual gene pairs and a third for their joint segregation.

As an illustration, a cross of $AABB \times aabb$ produces an F_1 of genotype $AaBb$ that is backcrossed to the homozygous double-recessive parent. Let us assume that the four classes of offspring from this backcross appear in the following numbers, whereas their expected numbers are in the ratio of 1:1:1:1:

	AaBb	*Aabb*	*aaBb*	*aabb*	Total
observed	140	38	32	150	360
expected	90	90	90	90	360

The over-all chi-square for three degrees of freedom calculated according to the method given in Chapter 8 is 135.19. Evidently the departure from the expected ratios is highly significant. What is responsible for this large discrepancy? When chi-square is calculated for the expected *Aa* and *aa* phenotypes (1/2 *Aa*:1/2 *aa*) the results are as follows:

	Aa	*aa*	Total
observed	140 + 38 = 178	32 + 150 = 182	360
expected	180	180	360

The chi-square value for the segregation of the *Aa* gene is .025, which at one degree of freedom, shows excellent agreement with the expected ratios. Similarly, the segregation of the *Bb* gene can be calculated:

	Bb	*bb*	Total
observed	140 + 32 = 172	38 + 150 = 188	360
expected	180	180	360

Again, a chi-square value of .62 for one degree of freedom is certainly within the bounds of a normally expected chance deviation.

On the other hand, calculation of chi-square for random assortment gives entirely different results. Since the original cross was *AABB* × *aabb*, the parental chromosomal combinations in the F_1 are *AB* and *ab*. The double-recessive parent to whom this F_1 heterozygote is backcrossed produces only *ab* gametes. Thus each of the four backcross progeny classes can be characterized on the basis of the gene combinations received from their F_1 parent, whether parental (*AB*, *ab*) or new (*Ab*, *aB*): *AaBb* and *aabb* are parental types, *Aabb* and *aaBb* are new. If assortment is independent among these gene pairs, the number of new combinations should equal the number of parental types as follows:

	Parental	Recombinant	
	AB + *ab**	*aB* + *Ab**	Total
observed	290	70	360
expected (1:1)	180	180	360

It is now obvious, even without a chi-square calculation, that the discrepancy from random assortment is large. The actual chi-square value of 133.22 for one degree of freedom is very highly significant. Thus the cause for departure from a 1:1:1:1 ratio among the backcross progeny in this case can be confidently ascribed to linkage.

In matings that would produce expected 9:3:3:1 ratios if the genes were assorting independently, tests to detect linkage are similar to those described above. For example, a cross between heterozygotes for two gene pairs, *AaBb* × *AaBb*, produced 180 *AB* phenotypes, 30 *Ab*, 60 *aB*, and 10 *ab*. When tested by chi-square, these data show an obvious departure from the expected 9:3:3:1 ratio, as follows:

	AB	*Ab*	*aB*	*ab*	Total
observed	180	30	60	10	280
expected (9:3:3:1)	157.5	52.5	52.5	17.5	280

In this example chi-square is 17.14, signifying a probability less than .001 for three degrees of freedom.

Here, again, there are three possible causes for this large chi-square: (1) a departure for the *A* gene from the expected segregation ratio 3 *A*–:1 *aa* phenotype, perhaps caused by the relative inviability of one of these classes; (2) a similar departure for the *B* gene from the expected ratio 3 *B*–:1 *bb* phenotype; and (3) a departure from independent assortment caused by linkage. As shown in Table 16–1, the first two of these possibilities can be tested by chi-square for a 3:1 ratio. The test for linkage involves a chi-square test for "independence" (p. 148), since the question is es-

*Many geneticists have used the term "coupling" for the double-dominant and double-recessive *AB* and *ab* combinations, and "repulsion" for the mixed dominant-recessive *Ab* and *aB* combinations.

TABLE 16–1

Chi-square tests for segregation ratios and linkage when the observed offspring of a cross *AaBb* × *AaBb* are in the phenotypic ratio 180 *AB*:30 *Ab*:60 *aB*:10 *ab*

a. Segregation at the *A* Gene: $Aa \times Aa \rightarrow 3\,A\!-:1\,aa$

	A–	*aa*	Total
observed	180 + 30 = 210	60 + 10 = 70	280
expected	210	70	280

$\chi^2_{1\,df} = .00$

b. Segregation at the *B* Gene: $Bb \times Bb \rightarrow 3\,B\!-:1\,bb$

	B–	*bb*	Total
observed	180 + 60 = 240	30 + 10 = 40	280
expected	210	70	280

$\chi^2_{1\,df} = 17.14$

c. Independence Test for Linkage

	B	*b*	Totals
A	180 (*AB*)	30 (*Ab*)	210
a	60 (*aB*)	10 (*ab*)	70
	240	40	280

$$\chi^2 = \frac{[|(180)(10) - (30)(60)| - 1/2(280)]^2 \times 280}{(240)(40)(210)(70)}$$

$$= \frac{[|0| - 1/2(280)]^2 \times 280^*}{(240)(40)(210)(70)}$$

$$= .00$$

* Note that if the absolute value of the difference in cross products is zero, as it is in this example, then the numerator of this equation becomes zero, and chi-square is therefore equal to zero (see p. 149).

sentially whether the association between the phenotypes of one gene pair and those of the other gene pair are completely random. Even distorted segregation ratios will not affect this test for linkage. In the present example (Table 16–1c) we can see that these relationships are in the ratios 6 *AB*:1 *Ab* and 6 *aB*:1 *ab* for the associations of the *A* gene phenotypes and of 3 *BA*:1 *Ba* and 3 *bA*:1 *ba* for associations of the *B* gene phenotypes. It is obvious that these associations are completely random and chi-square is zero. Chi-square is also zero for the segregation ratio at the *A* gene (Table 16–1a), but is 17.14 for the segregation ratio at the *B* gene (Table 16–1b). Thus the departure from a 9:3:3:1 ratio in this case is not caused by linkage but by the relative scarcity of *bb* phenotypes.

In *Drosophila* an additional way to test for linkage is to take advantage of the absence of crossing over in males. As previously mentioned, genes associated on the same chromosome will always demonstrate complete linkage when male *Drosophilia* are used as the heterozygous parents. The only independent assortment that occurs in males is between nonhomologous chromosomes (different linkage groups). In the example cited on page 322, partial linkage between the *A* and *B* genes will appear as complete linkage when F_1 males are backcrossed to homozygous recessive females. In other words, *AB*/*ab* ♂ × *ab*/*ab* ♀ would produce only *AaBb* and *aabb* genotypes in the F_2. The ascertainment of linkage among different gene pairs on the same chromosome is therefore considerably simplified in *Drosophila*, and this method can be used to determine to which particular linkage group a new mutation belongs.

When a mutation is present in a stock in which previously known genes are absent, the detection of the mutant linkage group demands different methods. One common technique for this purpose makes use of a "balanced stock" carrying dominant markers such as *Cy*/*Pm H*/*Sb* (see p. 217). *Cy* and *Pm* are located on the two homologous 2nd chromosomes of *Drosophila melanogaster* and *H* and *Sb* are located on the 3rd chromosomes. A female homozygous for the mutation in question (e.g., *m*/*m*) is mated to a male of this marker strain and the F_1 progeny examined. If the mutation is a sex-linked recessive, all F_1 males will bear the mutant phenotype, while F_1 females will appear wild type for the mutation. If the mutation is recessive but not sex-linked, none of the F_1 will show the mutant phenotype. In this case F_1 males are selected bearing two of the dominant markers, e.g., *Cy* and *Sb*, and backcrossed to homozygous mutant females *m*/*m*. Because of the absence of crossing over in males, the *m* gene carried by the F_1 males cannot cross over to the *Cy* and *Sb* chromosomes. Thus, if the *m* gene is located on the 2nd chromosome homologue of *Cy*, its wild-type (dominant) allele is located on the *Cy* chromosome, and the F_1 male can form only *Cy* and

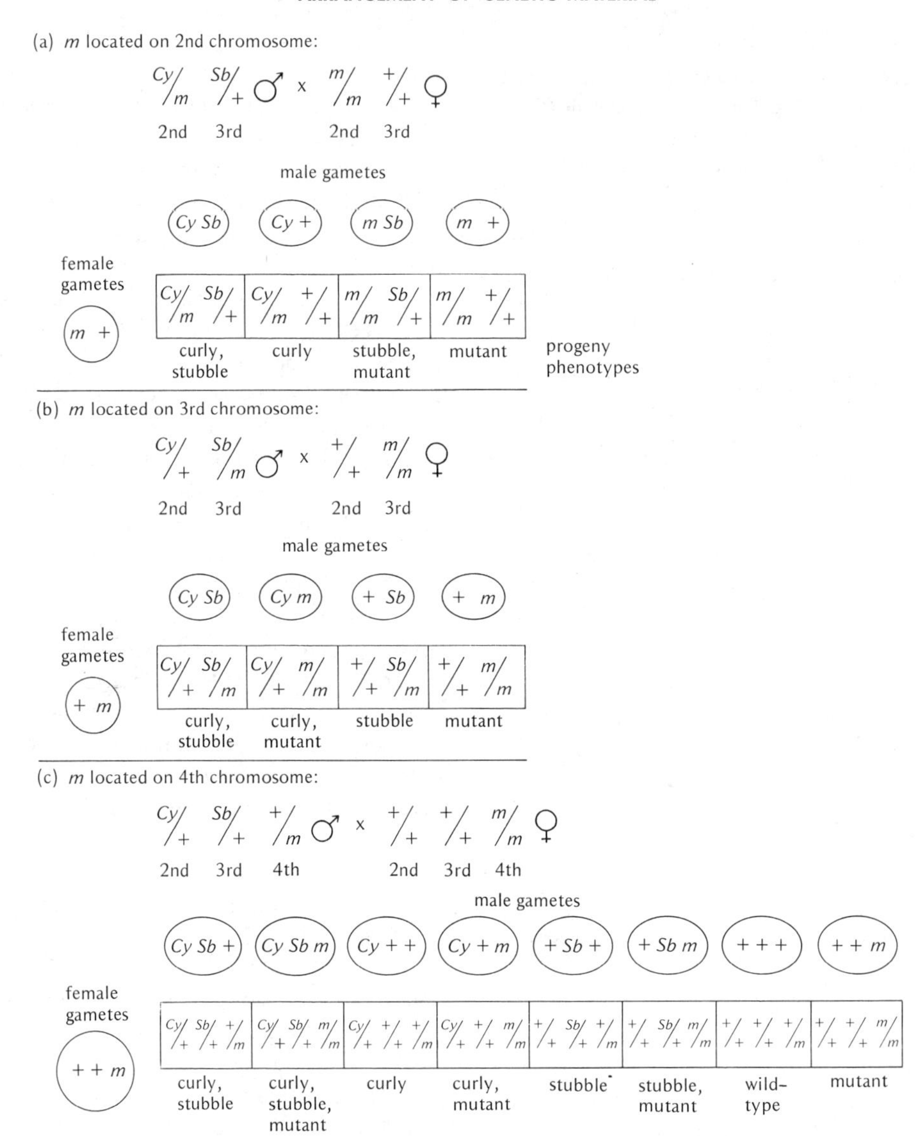

Figure 16–10

Method for detecting the autosomal linkage group of a mutant gene m in Drosophila melanogaster. The homozygous mutant stock is crossed to the marker stock (Cy/Pm H/Sb) and the Curly Stubble F_1 male offspring are backcrossed to the mutant stock. Depending on which chromosome m is located, the progeny of this backcross will show specific associations between the mutant phenotype and the phenotypes of the dominant genes, Curly and Stubble, as shown below the various genotypic combinations in (a), (b), and (c).

m gametes for this chromosome. The fertilization of female *m* eggs will, therefore, never produce *Cy* flies that are also *m* in phenotype (Fig. 16–10a). On the other hand, *Sb* is not so restricted, since the 3rd chromosome assorts independently of the 2nd; half the time *Sb* will be associated with the *m* phenotype and half the time it will not. If the *m* gene is located on the 3rd chromosome the *m* phenotype will never appear in the same fly that carries *Sb* (Fig. 16–10b). If *m* is located on the small dotlike 4th chromosome, its phenotype will appear in equal association with both *Cy* and *Sb* (Fig. 16–10c).

On the other hand, if the unknown mutation is dominant (e.g., *M*), all the F_1 flies of both sexes will show the mutant phenotype. One then chooses F_1 males bearing two of the dominant markers, e.g., *Cy* and *Sb*; these are now mated with wild-type nonmutant females. If *M* is sex-linked, only the F_2 females will show the mutant phenotype, since only females inherit the paternal X chromosome. If *M* is on the 2nd chromosome, the *M* phenotype will never appear in the same F_2 fly as its *Cy* homologue, since it segregates from *Cy* during meiosis. Similarly the location of *M* on the 3rd chromosome will produce F_2 flies that do not bear the *Sb* and *M* phenotypes together. Location of *M* on the fourth chromosome will yield equal association between the *M* phenotype and *Cy* and *Sb*. Once the linkage group of the mutant gene is determined, stocks are available carrying a variety of markers to localize the gene still further.

In humans test matings of this kind cannot be made, and statistical methods to detect linkage have been devised by Bernstein, Penrose, Morton, and others. At present, however, more success has been achieved through the sex-linked pedigree methods and somatic cell hybridizations described in Chapter 17.

CALCULATION OF AUTOSOMAL RECOMBINATION FREQUENCIES BY BACKCROSSING TO HOMOZYGOUS RECESSIVES

The proportion of recombinant types between two gene pairs as compared to the sum of all combinations is known as the *recombination frequency*. Thus, in the example considered on page 322, 70 of 360 were new combinations, representing a recombination frequency of .19. As a rule, recombination experiments using the same two gene pairs give fairly repeatable results. Between different sets of gene pairs, however, recombination values in a particular cross may vary from 0 to 50 percent. The lower limit, of course, signifies absence of recombination between the two gene pairs, while the upper limit signifies a degree of recombination equal to independent assortment.

In diploids recombination frequencies between two genes can be most simply calculated by backcrossing or testcrossing an F_1 heterozygote to a homozygous double recessive. This procedure produces new combinations whose individual frequencies are immediately apparent. For example, an F_1 heterozygote with parental combinations $A\,B/a\,b$ ("coupling") crossed to homozygous $a\,b/a\,b$ will, upon crossing over, produce two recombined F_2 types $A\,b/a\,b$ and $a\,B/a\,b$. Similarly an F_1 heterozygote with the parental combination $A\,b/a\,B$ ("repulsion") will produce the recombinant types $a\,b/a\,b$ and $A\,B/a\,b$ in a testcross.

In many instances the *a* and *b* genes are recessive mutations while the dominants *A* and *B* are wild type. For example, a *Drosophila* female may be heterozygous for the linked 2nd-chromosome recessive mutations, *vestigial* (*vg*), producing reduced wings, and *cinnabar* (*cn*), producing light eye color, in the form $vg\ cn/vg^+\ cn^+$. Upon backcross to homozygous *vestigial cinnabar* males, this female will produce the expected four phenotypes if large numbers of offspring are raised. Two of these four, having the vestigial ($vg\ cn^+/vg\ cn$) and cinnabar ($vg^+\ cn/vg\ cn$) phenotypes, represent recombinant classes, and the sum of their frequencies is therefore the recombination frequency between these two genes.

An example where *A* and *B* represent dominant mutations and *a* and *b* their recessive wild-type alleles is the 2nd-chromosome dominant mutations in *Drosophila*, *Jammed* (*J*) and *Lobe* (*L*), which produce narrow wings and reduced eyes, respectively. A female heterozygous for these in the form $J\,L/J^+\,L^+$, mated to a recessive normal male, $J^+\,L^+/J^+\,L^+$, will produce recombinant classes that are phenotypically jammed ($J\,L^+/J^+\,L^+$) and lobe-eyed ($J^+\,L/J^+\,L^+$).

Similarly, *A* may represent a dominant mutation and *b* a recessive mutant gene, while *a* and *B* are wild-type alleles for these two gene pairs. For example, recombinant classes produced by a female carrying the *Lobe* and *vestigial* mutations in the form $L\,vg/L^+\,vg^+$ can be detected by mating her to a male that is normal-eyed but vestigial-winged, $L^+\,vg/L^+\,vg$. Such a mating will produce the recombinant classes having the lobe ($L\,vg^+/L^+\,vg$) and vestigial ($L^+\,vg/L^+\,vg$) phenotypes.

RECOMBINATION FREQUENCIES FOR SEX-LINKED GENES

In male heterogametic species such as *Drosophila* the male offspring obtain their X chromosomes only from their mothers. Any X chromosome recombination in the female parent is therefore immediately noted in her sons, who, because of their single X chromosome, show the effects of both recessive and dominant sex-linked genes. If we confine our attention to the male offspring, we can consider Morgan's experiment on the sex-linked *white* and *miniature* genes (p. 317) which gave 37.6 percent recombination as an example of this kind. Other sex-linked genes and *white* will most likely differ in recombination frequency. As an illustration, when yellow-bodied and white-eyed females ($y\,w/y\,w$) are bred to normal gray-bodied and red-eyed males (y^+w^+), the F_1 females are gray-bodied and red-eyed ($y\,w/y^+\,w^+$), as expected. No matter to which males the F_1 females are mated, the F_2 male offspring appear in the following approximate ratios:

Yellow white ($y\ w$)	**Gray red ($y^+\ w^+$)**
49.25%	49.25%
98.5% parental types (noncrossovers)	
Yellow red ($y\ w^+$)	**Gray white ($y^+\ w$)**
0.75%	0.75%
1.5% new combinations (crossovers)	

(The F_2 females will also appear in this same ratio if their mothers have been crossed to the recessive *yellow white* males.) Thus, as compared to *white* and *miniature*, the crossover frequency between *white* and *yellow* is significantly lower.

This same low crossover frequency can be noted when the recessive *white* and *yellow* enter the F_1 female from opposite parental chromosomes ("repulsion"). When a yellow red-eyed female ($y\,w^+/y\,w^+$) is crossed to a gray white-eyed male (y^+w), the F_1 females ($y\,w^+/y^+w$) will produce male offspring in the ratios

Yellow red ($y\ w^+$)	**Gray white ($y^+\ w$)**
49.25%	49.25%
98.5% parental types (noncrossovers)	
Yellow white ($y\ w$)	**Gray red ($y^+\ w^+$)**
0.75%	0.75%
1.5% new combinations (crossovers)	

RECOMBINATION FREQUENCIES IN $F_1 \times F_1$ CROSSES

When autosomal linkage is being investigated and test crosses to homozygous recessive stocks cannot be easily achieved, recombination frequencies can be calculated by scoring F_2 ratios in a cross between F_1 heterozygotes. For a cross involving two gene pairs at which complete dominance exists, $A > a$, $B > b$, the F_1 heterozygotes ($AaBb$) may either derive AB from one parent and ab from the other (coupling) or derive Ab from one parent and aB from the other (repulsion). In both cases the F_2 phenotypes expected from an $F_1 \times F_1$ cross according to independent assortment are AB:Ab:aB:ab in the ratio 9:3:3:1. Should a significant departure from this ratio occur caused by linkage (pp. 322–323), the percentage recombination can be estimated according to the following method devised by Fisher and others:

1. Assign values of a_1, a_2, a_3, and a_4 to the numbers in each category of F_2 phenotypes as follows:

$$F_1 \times F_1$$
$$\downarrow$$
$$F_2{:}AB(a_1){:}Ab(a_2){:}aB(a_3){:}ab(a_4)$$

2. Calculate the ratio z, which is the (product of recombinant types)/(product of parental types). That is, if the F_1 heterozygotes were in coupling, $AB/ab \times AB/ab$,

$$z = \frac{a_2 \times a_3}{a_1 \times a_4}$$

If the F_1 heterozygotes were in repulsion, $Ab/aB \times Ab/aB$,

$$z = \frac{a_1 \times a_4}{a_2 \times a_3}$$

3. Look up the z value obtained in steps 2 or 3 in Table 16–2 and determine the recombination percentage for this particular z value for the coupling or repulsion phase used.

For example, a cross between F_1 heterozygotes in repulsion produced 1031 *AB*:473 *Ab*:465 *aB*: 31 *ab* F_2 phenotypes. The value of z is then $(1031 \times 31)/(465 \times 473) = .15$. According to Table 16–2, this signifies an approximate recombination frequency of 25 percent.

As a rule, if r is the crossover or recombination frequency between two linked gene pairs, *A,a* and *B,b*, the gametes formed by parents heterozygous for both gene pairs will have frequencies that are dependent upon the original gene combinations. As shown in Table 16–3, when the genes are in "coupling," *AB/ab*, a total of r recombinant gametes are formed of which $1/2\, r$ are *Ab* and $1/2\, r$ are *aB*. The total frequency of parental gametes, formerly 1, will thus be reduced by r; or each of the two parental types (*AB*, *ab*) will be formed in a frequency $1/2 - 1/2\, r$. Similarly heterozygous genes in "repulsion," *Ab/aB*, will form r recombinant gametes (*AB*, *ab*) and lose a frequency of r parental gametes (*Ab*, *aB*). Knowledge of recombination frequencies together with the coupling or repulsion phase of the genes involved thus enables the prediction of genotypic frequencies among the progeny of heterozygotes for linked genes. For example, if recombination is 10 percent between two linked genes, *Aa* and *Bb*, and the heterozygote carries these genes in coupling, the frequency of *AB*, *Ab*, *aB*, and *ab* gametes are, respectively, .45, .05, .05, and .45. A 4 × 4 checkerboard will then give the frequencies of each of the 16 possible combinations; e.g., the frequency of the homozygous double-recessive *aabb* will be .45 × .45 = .2025.

TABLE 16–2

The percent recombination in crosses between F_1 individuals heterozygous for two gene pairs *Aa* and *Bb* in coupling, *AB/ab* × *AB/ab*, or repulsion, *Ab/aB* × *Ab/aB*

	Percent Recombination			Percent Recombination	
z	*Coupling*	*Repulsion*	*z*	*Coupling*	*Repulsion*
.0010	2.7	2.2	.095	21.8	20.6
.0015	3.3	2.7	.100	22.3	21.1
.0020	3.7	3.2	.11	23.2	22.0
.0025	4.2	3.5	.12	24.0	22.9
.0030	4.6	3.9	.13	24.8	23.7
.0035	4.9	4.2	.14	25.5	24.5
.0040	5.2	4.5	.15	26.2	25.2
.0045	5.5	4.7	.16	26.9	25.9
.005	5.8	4.9	.17	27.5	26.6
.006	6.4	5.5	.18	28.1	27.2
.007	6.8	5.9	.19	28.7	27.9
.008	7.3	6.3	.20	29.3	28.5
.009	7.7	6.7	.22	30.4	29.6
.010	8.1	7.0	.24	31.3	30.7
.012	8.8	7.7	.26	32.3	31.7
.014	9.4	8.3	.28	33.2	32.6
.016	10.0	8.9	.30	34.0	33.5
.018	10.6	9.4	.32	34.8	34.3
.020	11.1	9.9	.34	35.5	35.1
.022	11.6	10.3	.36	36.2	35.8
.024	12.1	10.8	.38	36.9	36.6
.026	12.5	11.2	.40	37.6	37.2
.028	12.9	11.6	.42	38.2	37.9
.030	13.3	12.0	.44	38.8	38.5
.035	14.2	13.0	.46	39.4	39.1
.040	15.1	13.8	.48	39.9	39.7
.045	15.9	14.6	.50	40.5	40.3
.050	16.7	15.3	.55	41.8	41.6
.055	17.4	16.0	.60	42.9	42.8
.060	18.0	16.7	.65	44.0	43.9
.065	18.7	17.4	.70	45.0	45.0
.070	19.2	18.0	.75	46.0	45.9
.075	19.8	18.5	.80	46.9	46.9
.080	20.3	19.1	.85	47.7	47.7
.085	20.9	19.6	.90	48.5	48.5
.090	21.4	20.1	.95	49.3	49.3
			1.00	50.0	50.0

Adapted from Stevens, 1939.

TABLE 16–3

Proportions of gametic frequencies expected when the recombination frequency between two gene pairs (*Aa* and *Bb*) is equal to *r*

	Gametic Frequencies			
	AB	**Ab**	**aB**	**ab**
AB/ab ("coupling")	$1/2 - 1/2\,r$	$1/2\,r$	$1/2\,r$	$1/2 - 1/2\,r$
Ab/aB ("repulsion")	$1/2\,r$	$1/2 - 1/2\,r$	$1/2 - 1/2\,r$	$1/2\,r$

PROBLEMS

16–1. Two 3rd-chromosome mutations in *Drosophila melanogaster* are *Stubble* (*Sb*), a gene with dominant effect producing shortened bristles, and *curled* (*cu*), a gene with recessive effect producing curved wings. If a female heterozygous for these genes in the form $Sb\ cu/Sb^+\ cu^+$ is to be mated to detect recombinants among her offspring, what male genotype would you choose as a mate?

16–2. Males from a *Drosophila* stock homozygous for the recessive genes *bumpy* and *tinted* are crossed to the balanced marker stock carrying the dominant genes *Curly/Plum* on the 2nd chromosome and *Hairless/Stubble* on the 3rd chromosome. F_1 males are selected that are curly and stubble in phenotype and backcrossed to homozygous *bumpy tinted* females. The progeny produced by this last cross are as follows:

Phenotype	Males	Females
bumpy	0	104
bumpy tinted	96	0
bumpy curly	0	96
bumpy curly tinted	104	0
stubble curly	0	109
stubble curly tinted	105	0
stubble	0	91
stubble tinted	95	0

(a) On which chromosome is *bumpy* located? (b) On which chromosome is *tinted* located? (HINT: consider *bumpy* and *tinted* separately.)

16–3. Through use of electrophoresis (mobility of compounds in an electric field), slight enzyme variations can be discovered that prove to be of an inherited nature. One such example in *Drosophila melanogaster* includes variants of the enzyme, alcohol dehydrogenase (*Adh*), in which three forms have been distinguished: Adh^{Fast}, Adh^{Slow}, and a "hybrid" enzyme Adh^{Fast}/Adh^{Slow}. The genes producing these enzymes are allelic, the first two phenotypes being produced by the respective homozygotes Adh^F/Adh^F and Adh^S/Adh^S, and the hybrid enzyme is produced by the heterozygote Adh^F/Adh^S. To discover the location of the gene determining this enzyme behavior, Grell, Jacobson, and Murphy crossed a stock of wild-type males carrying Adh^S to a balanced dominant marker stock *Cy/Bl Ubx/Vno* that was Adh^F. (*Cy* and *Bl* are 2nd-chromosome markers; *Ubx* and *Vno* are 3rd-chromosome markers.) The F_1 male offspring of this cross that had the dominant marker constitution $Cy/+\ Ubx/+$ were then backcrossed to wild-type Adh^S females. The F_2 progeny were then scored for the presence of dominant markers and analyzed for *Adh* enzymes. In all cases flies bearing *Cy* showed the hybrid enzyme Adh^F/Adh^S, while the absence of *Cy* was accompanied by the presence of only Adh^S. On the other hand, half of the flies bearing *Ubx* showed the Adh^F/Adh^S enzyme, while the other half showed the Adh^S enzyme. On which chromosome is the *Adh* factor located, and why?

16–4. In *Drosophila melanogaster* the *black* recessive gene for dark body color is located on the 2nd chromosome, and the *pink* recessive gene for light eye color is located on the 3rd chromosome. (The 1st chromosome is the X, determining sex.) In this species an animal with short legs was discovered and a stock homozygous for this new recessive mutation, *short*, was then established. A homozygous *black pink short* stock was then bred, and a female of this multimutant stock was then mated to a wild-type male. The F_1 males of this cross were then backcrossed to homozygous *black pink short* females, and the phenotypes of the progeny appeared in the following numbers:

Phenotype	Males	Females
wild type for all characters	25	26
black only	26	24
pink only	25	28
short only	27	25
black and pink	27	26
pink and short	23	25
black and short	26	25
black, pink, and short	25	24

On which chromosome is *short* located? (After an experiment by Muller.)

16-5. Two gene pairs, *Aa* and *Bb*, are located on the same chromosome arm in *Neurospora*. On the assumption that the *A* locus is closest to the centromere, diagram the crossover points at the four-strand stage that produced the following asci in a cross $AB \times ab$. (The ascopores of each ascus are given according to their linear position, going from left to right.)

(a) *AB AB Ab Ab aB aB ab ab*
(b) *AB AB ab ab AB AB ab ab*
(c) *AB AB aB aB Ab Ab ab ab*
(d) *Ab Ab aB aB AB AB ab ab*
(e) *aB aB Ab Ab AB AB ab ab*
(f) *AB AB ab ab aB aB Ab Ab*
(g) *Ab Ab ab ab aB aB AB AB*
(h) *aB aB aB aB Ab Ab Ab Ab*

16-6. What sort of matings would you look for in humans to demonstrate that the gene determining the ABO blood type is inherited independently of the gene determining the MN blood type?

16-7. In corn the seeds can be colored or white, nonshrunken or shrunken. Each of these characteristics is determined by a separate pair of genes, *C* and *c* and *Sh* and *sh*. Hutchison crossed a homozygous colored shrunken strain (*CC shsh*) to a homozygous white nonshrunken strain (*cc ShSh*) and obtained a heterozygous colored nonshrunken F_1. The F_1 was backcrossed to a homozygous recessive white shrunken stock and the progeny were as follows:

	No. plants
colored shrunken	21,379
white nonshrunken	21,096
colored nonshrunken	638
white shrunken	672

What is the recombination frequency between these two genes?

16-8. How would you use an XXY *Drosophila* female (nonattached-X chromosomes) heterozygous for two sex-linked gene pairs to demonstrate that crossing over occurs at the four-strand stage?

16-9. Using the frequency of recombinants as one proportion and the frequency of nonrecombinants as another, calculate the 95 percent confidence limits for the recombination frequency between *C* and *Sh* in Problem 16-7 (see Chapter 8).

16-10. In addition to colored or white, corn kernels can be starchy or waxy, each of these characteristics being caused by the separate gene pairs *C* and *c* and *Wx* and *wx*. Bregger crossed a homozygous colored starchy strain (*CC Wx Wx*) to a homozygous white waxy strain (*cc wx wx*) and then crossed $F_1 \times F_1$ to obtain an F_2. His F_2 phenotypic data were as follows:

	No. plants
colored starchy	1774
white waxy	420
colored waxy	263
white starchy	279

What is the percentage of recombination between these two gene pairs?

16-11. In rabbits an allele for spotted pattern is dominant to one for self-colored, and an allele for short hair is dominant to one for long hair (Angora). An English spotted short-haired rabbit is mated to a long-haired Angora self-colored rabbit and the F_1 animals are then backcrossed to the Angora self-colored stock. If the backcross produces 26 spotted Angora, 144 self-colored Angora, 157 spotted short-haired, and 23 self-colored short-haired, what is the percentage of recombination between these two genes?

16-12. In corn colored aleurone (*R*) is dominant to colorless (*r*) and yellow plant color (*y*) recessive to green (*Y*). The phenotypes and frequencies of progeny following test crosses of two heterozygous plants to homozygous recessives are listed below:

Phenotype	Progeny of plant 1	Progeny of plant 2
colored green	88	23
colored yellow	12	170
colorless green	8	190
colorless yellow	92	17
	200	400

(a) What are the genotypes of heterozygous plants 1 and 2? (b) What is the observed recombination percentage? (c) If the two heterozygous plants had been crossed, what is the expected frequency of colorless, yellow progeny?

16–13. In the mouse a strain of homozygous recessive *frizzy*-haired individuals are mated to a strain of homozygous recessive *shakers*. Each of these strains is homozygous for the dominant allele of the recessive gene in the other strain (i.e., $ffSS \times FFss$). The F_1 of this cross is then mated with homozygous *frizzy shakers* to produce an F_2. (a) If both these autosomal recessive genes are located on the same chromosome and the recombinational frequency between *frizzy* and *shaker* is 15 percent, what are the expected *phenotypes* and their *ratios* in the progeny of this cross? (b) What phenotypic ratios in the F_2 would you expect if the same F_1 above were mated $F_1 \times F_1$?

16–14. In a particular organism such as *Drosophila*, the effect of autosomal gene *A* is dominant over *a*; and at another autosomal gene pair, the effect of *B* is dominant over *b*. A cross is made between stocks homozygous $AAbb \times aaBB$, and the resultant F_1 females are then crossed to homozygous *aabb* males. The following shows two hypothetical sets of results, for each of which you are to provide the cause responsible for the departure of observed numbers from those expected according to independent assortment.

	Numbers of F_2 Phenotypes Observed				
	AB	*Ab*	*aB*	*ab*	**Total**
Case (a)	140	80	120	60	400
Case (b)	45	155	145	55	400

16–15. If the dominance relationships between the alleles at two different gene pairs, *Aa* and *Bb*, are such that the effect of *A* is dominant to that of *a*, and *B* is dominant to *b*, explain why you could or could not detect linkage between the two genes in the offspring of cross: (a) $Aabb \times aaBb$; (b) $AaBb \times Aabb$.

16–16. Individual plants homozygous for various combinations of the linked genes *A* and *B* (i.e., *AABB*, *aaBB*, *AAbb*, and *aabb*) were crossed to produce F_1 plants which were then self-fertilized to produce an F_2. If the effect of *A* is dominant to that of *a*, and *B* dominant to *b*, state the chromosomal constitution (e.g., *AB*/*aB*) for each F_1 parent of the following F_2 phenotypic distributions.

Self-fertilized F_1 Plant	F_2 Phenotypes			
	AB	*Ab*	*aB*	*ab*
(a)		98		32
(b)	64	10	8	18
(c)	106	34		
(d)			104	36
(e)	50	25	23	2
(f)	99		31	

16–17. Calculate the recombination frequency between genes *A* and *B* in Problem 16.

16–18. Fisher and Snell crossed mice homozygous for the *ruby* gene to mice homozygous for the *jerker* gene. The wild-type F_1 were intercrossed among each other and produced 157 offspring in the following phenotypic classes: 86 wild type, 31 jerker, 35 ruby, and 5 jerker ruby. Is there linkage between the two genes?

16–19. Hertwig crossed mice homozygous for *oligodactyly* to *albino* mice and then intercrossed the wild-type F_1 among each other. The F_2 were 334 wild type, 107 oligodactylous, 188 albino, and 13 oligodactylous albino. What factors account for the departure from the expected values?

16–20. Genes *A* and *B* are linked on one chromosome, while gene *C* is on an independently assorting nonhomologous chromosome. At each of these genes, there are two alleles such that *A* is dominant to *a*, *B* to *b*, and *C* to *c*. A cross is then made $ABC/ABC \times abc/abc$ to produce an F_1, and the F_1 is then backcrossed to the homozygous recessive parent abc/abc. What are the F_2 phenotypic ratios you would expect: (a) If the linkage between *A* and *B* is complete and no recombination occurs between them? (b) If there is 20 percent recombination between *A* and *B*?

16–21. A stock homozygous for three recessive autosomal genes, *a*, *b*, and *c*, not necessarily linked on the same chromosome, is mated to a stock homozygous for their dominant wild-type alleles, *A*, *B*, and *C*. F_1 females are then backcrossed to males of the homozygous recessive parental stock and the following results are observed:

Phenotype	**Number**
abc	211
ABC	209
aBc	212
AbC	208

(a) Which of these genes are linked? (b) What is the recombination frequency between the linked genes?

16-22. An experiment performed in an identical fashion as in the above problem, but using three different genes (with capital letter alleles also dominant over small letter alleles) gave the following results among the progeny of the F_1 backcross:

Phenotype	Number	Phenotype	Number
def	42	*deF*	6
DEF	48	*DEf*	4
dEF	46	*dEf*	5
Def	44	*DeF*	5

Answer questions (a) and (b) in the preceding problem.

16-23. In *Drosophila, autosomal* gene *A* has an effect dominant to that of *a*, and *autosomal* gene *B* has an effect dominant to that of *b*. Keeping in mind that crossing over does not occur in male *Drosophila*: (a) What phenotypic ratios would you expect from a cross between two F_1 heterozygotes bearing two linked genes "in repulsion": *Ab/aB* ♀ × *Ab/aB* ♂? (b) Would the extent of recombination between *A* and *B* change this phenotypic ratio?

16-24. If you began with two stocks of *Drosophila*, one homozygous *aaBB* and the other homozygous *AAbb*, what crosses would you make to obtain a stock homozygous for the double recessive *aabb*: (a) If the two gene pairs are linked autosomally? (b) If the two gene pairs are sex-linked?

16-25. If the recombination frequency between the linked autosomal genes *A* and *B* in *Drosophila* is 15 percent, what phenotypic frequencies would you expect from a cross *AB/ab* ♀ × *AB/ab* ♂?

16-26. If genes *A* and *B* in *Drosophila* are sex-linked with a recombination frequency between them of 15 percent, what phenotypic frequencies would you expect among the sexes resulting from the following crosses:

(a) *AB/ab* ♀× *ab* ♂?
(b) *Ab/aB* ♀× *ab* ♂?
(c) *AB/ab* ♀× *AB* ♂?
(d) *AB/ab* ♀× *Ab* ♂?
(e) *Ab/aB* ♀× *Ab* ♂?

16-27. If you had two different temperature-sensitive mutant stocks, both of which are inviable at high temperature (29°C) but able to grow at low temperature (22°C, see p. 218), how would you detect crossing over between the gene pairs involved if both gene pairs are located close together on the same chromosome?

REFERENCES

ANDERSON, E. G., 1925. Crossing over in a case of attached X chromosomes in *Drosophila melanogaster*. *Genetics*, **10**, 403–417.

BATESON, W., E. R. SAUNDERS, and R. C. PUNNETT, 1905. Experimental studies in the physiology of heredity. *Reports to the Evolution Committee Royal Society*. II. Harrison and Sons, London. (Excerpts reprinted in the collection of Peters; see References, Chapter 1.)

BLIXT, S., 1974. The pea. In *Handbook of Genetics*, Vol. 2, R. C. King (ed.). Plenum Press, New York, pp. 181–221.

BRIDGES, C. B., and K. S. BREHME, 1944. The mutants of *Drosophila melanogaster*. *Carneg. Inst. Wash. Publ. No. 552.* Washington, D.C.

LINDEGREN, C. C., 1933. The genetics of *Neurospora*. III. Pure bred stocks and crossing-over in *N. crassa*. *Bull. Torrey Bot. Club*, **60**, 133–154.

MATHER, K., 1951. *The Measurement of Linkage in Heredity*. Methuen, London.

MORGAN, T. H., 1911a. Random segregation versus coupling in mendelian inheritance. *Science*, **34**, 384. (Reprinted in the collections of Gabriel and Fogel, and of Voeller; see References, Chapter 1.)

———, 1911b. An attempt to analyse the constitution of the chromosomes on the basis of sex-limited inheritance in *Drosophila. Jour. Exp. Zool.*, **11**, 365–414.

MORTON, N. E. 1962. Segregation and linkage. In *Methodology in Human Genetics*, W. J. Burdette (ed.). Holden-Day, San Francisco, pp. 17–47.

NEUFFER, M. G., and E. H. COE, JR., 1974. Corn (maize). In *Handbook of Genetics*, Vol. 2, R. C. King (ed.). Plenum Press, New York, pp. 3–30.

STEVENS, W. L., 1939. Tables of the recombination fraction estimated from the product ratio. *Jour. Genet.*, **39**, 171–180.

17
GENE MAPPING IN DIPLOIDS

In 1911, when Morgan discovered linkage in *Drosophila*, he expressed the opinion that there was probably a relationship between the recombination frequency of linked genes and their linear distance along the chromosome. In Morgan's terms, "the proportions that result are not so much the expression of a numerical system as of the relative location of the factors in the chromosomes."

According to this view, the degree of separation between genes *A* and *B* and between genes *B* and *C* should enable the prediction of the degree of separation between genes *A* and *C*, if all three genes were in a linear order on the same chromosome. Thus, if the linear arrangement or "linkage map" is *A-B-C*, and recombination frequency is used as a measure of separation or distance, the sum of recombination frequencies between *A* and *B* and *B* and *C* should equal the recombination frequency between *A* and *C* (Fig. 17–1). This relationship becomes apparent if we consider the opportunity for recombination as a function of distance. If a particular distance along a chromosome shows a particular recombination frequency, twice that distance should show twice the frequency. In the above example the order could also be *A-C-B*, in which case the sum of recombination frequencies *A-C* and *C-B* should equal that of *A-B*, and the expected *A-C* recombination frequency (as well as that of *C-B*) is now therefore smaller than that of *A-B*. Numerous experiments by Sturtevant, Muller, Bridges, and others have shown that linkage relationships are indeed of this additive kind.

One common method used in mapping genes has been to measure recombination frequencies between three gene pairs segregating in a single experiment, the "three-point cross." For example, Bridges and Olbrycht

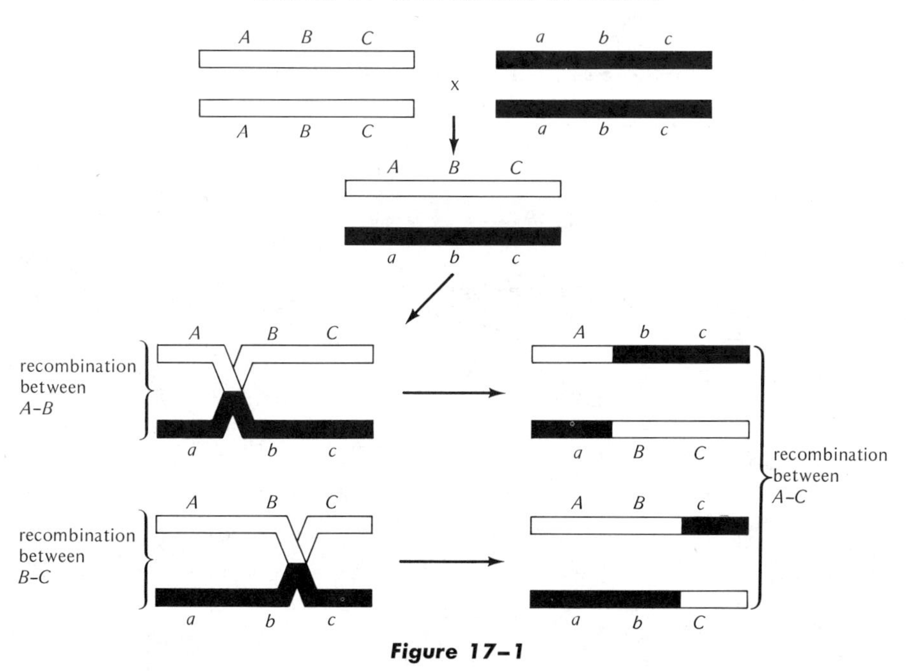

Figure 17–1

If three genes are in linear order A-B-C, single crossovers between A and B and between B and C will appear also as recombinations between A and C. (For simplicity, only the recombinant chromatid of each chromosome is shown and the parental noncrossover chromatid is omitted (see Fig. 16–6b).)

crossed a mutant stock of *Drosophila melanogaster* homozygous for three sex-linked genes, *echinus* (*ec*, rough eye), *scute* (*sc*, certain bristles missing), and *crossveinless* (*cv*, certain wing veins missing) to wild type. The heterozygous F_1 females were then crossed to mutant males of the parental stock. Crossovers in the F_1 females could then be detected phenotypically among the single X chromosomes of the F_2 male offspring. In addition the F_2 female phenotypes also demonstrated crossovers, since they carried a paternal mutant X chromosome, *ec sc cv*, which enables any mutant genes transmitted by the maternal chromosome to be observed. The results of a particular set of cultures are shown in Table 17–1.

Of a total of 2635 F_2 flies scored, note that 2108, or 80.0 percent, show no recombination at all, and linkage between the three genes in this case is apparently complete both for the mutant chromosome (*ec sc cv*) and for the wild-type chromosome (+ + +). In 20.0 percent, however, crossovers occurred, but these were of different types. One type resulted from crossing over between *echinus* and its original companion gene *scute*, so that the two mutants, or their wild-type alleles, are no longer on the same chromosome. This occurred in 239 flies, or 9.1 percent of the total, as shown by the phenotypes *sc* and *ec cv*. A second type of crossover is between *crossveinless* and *echinus* or between the wild-type alleles for each, so that *cv* and *ec* are no longer on the same chromosome. In all, 288 flies of this type (*cv* and *ec sc* phenotypes) were found, giving a recombination frequency of 10.9 percent. A third type of crossover, however, also occurred, that between *sc* and *cv*, so that these two genes are now present in new combinations, *sc* with cv^+, and sc^+ with *cv*. This latter type of crossover has occurred in all four recombinant phenotypes and therefore has a recombination frequency of 527/2635, or 20.0 percent.

TABLE 17–1

Recombination data for three sex-linked genes in *Drosophila melanogaster*: *echinus*, *scute*, and *crossveinless*

P_1*: $\frac{ec\ sc\ cv}{ec\ sc\ cv}$ ♀ × + + + ♂

F_1: $\frac{+\ +\ +}{ec\ sc\ cv}$ ♀ × *ec sc cv* ♂

F_2 Phenotypes	F_2 Genotypes Male	F_2 Genotypes Female	Observed No.	
echinus scute crossveinless	*ec sc cv*	$\frac{ec\ sc\ cv}{ec\ sc\ cv}$	934	$\frac{2108}{2635} \times 100 = 80.0\%$ nonrecombinants
wild type	+ + +	$\frac{+\ +\ +}{ec\ sc\ cv}$	1174	
scute	+ *sc* +	$\frac{+\ sc\ +}{ec\ sc\ cv}$	140	$\frac{239}{2635} \times 100 = 9.1\%$ recombinants: *ec*· · ·*sc*
echinus crossveinless	*ec* + *cv*	$\frac{ec\ +\ cv}{ec\ sc\ cv}$	99	
crossveinless	+ + *cv*	$\frac{+\ +\ cv}{ec\ sc\ cv}$	124	$\frac{288}{2635} \times 100 = 10.9\%$ recombinants: *ec*· · ·*cv*
echinus scute	*ec sc* +	$\frac{ec\ sc\ +}{ec\ sc\ cv}$	164	
			2635	

From cultures 4 to 14 of Table 3 of Bridges and Olbrycht, 1926.

* Note that the three genes in the parental chromosomes of this table have been written in the sequence ec-sc-cv. This written sequence, however, as any other written sequence of these genes, has no validity until the linkage relationships between them are determined. For example, the three genes could have initially been written as sc-cv-ec without in any way affecting the subsequently determined gene order, sc-ec-cv. As explained in the text, it is only the recombination frequencies in the three intervals, ec-sc, sc-cv, ec-cv, that determines the true gene order.

We can see that there are three crossover regions involved in this cross, one between *echinus* and *scute*, 9.1 percent; a second between *echinus* and *crossveinless*, 10.9 percent; and a third between *scute* and *crossveinless*, 20.0 percent. The largest recombination frequency, indicative of the longest map distance, is between *scute* and *crossveinless*. Therefore, if we wish to arrange the genes in linear order, *echinus* must be situated in the middle, with *scute* and *crossveinless* at opposite ends:

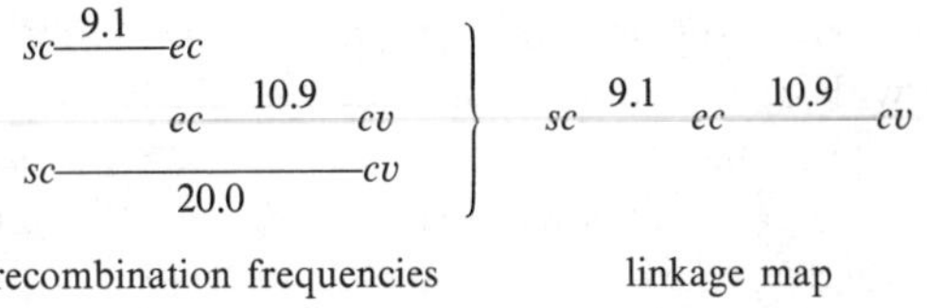

sc —9.1— *ec*
ec —10.9— *cv*
sc —20.0— *cv*

recombination frequencies

sc —9.1— *ec* —10.9— *cv*

linkage map

Because the genes have now been located on the chromosome in the form of a *linkage map*, each is called a *locus* (plural: *loci*), and all alleles of a gene, whether wild type or mutant, occur at that locus. Thus *white-eosin*, *white-coral*, and other *white* alleles (p. 168) map at the *white* locus.

In the above experiment all three mutations, *sc*, *ec*, and *cv*, entered the F_1 female on the same chromosome and all the wild-type alleles for each entered on another chromosome. This produced two nonrecombinant or parental classes, one of which was wild type. Since the viability of wild type is usually much greater than that of mutant phenotypes, the relationship between the two parental classes is often numerically distorted (note that there were 1174 wild type and only 934 of the triple mutant *sc ec cv* class). To reduce this distortion, one can obtain two nonrecombinant classes that are *both* mutant by deriving F_1 females, for example, from a cross between a homozygous *scute crossveinless* stock and a homozygous *echinus* stock (+ *ec* + /*sc* + *cv*). Such a female, when mated to an *sc ec cv* male, will produce more equal frequencies of parental classes, as shown in Table 17-2.

Note that although there is a smaller fraction of recombinants than in the previous experiment, the map relationships of the three loci are similar. In this case, the two original X chromosomes of the F_1 female contained *sc* on one chromosome together with the wild-type allele for *ec*, and the wild-type allele for *sc* together with *ec* on the other chromosome. Thus crossovers between the *sc* and *ec* loci produce individuals in which both *sc* and *ec* are present together or in which their wild-type alleles are present together. A total of 287 flies, or a frequency of 6.6 percent, are of this recombinant type. Similarly, crossovers between the *ec* and *cv* loci result in individuals having both *ec* and *cv*, or ec^+ cv^+, present together. This crossover type is seen in 444 flies, or 10.2 percent of the total. To calculate the crossover distance between *sc* and *cv* we must now take into account that, of the two X chromosomes that entered the F_1 female, one contained both mutant alleles, *sc* and *cv*, and the other contained the wild-type alleles for each. Crossing over between these two loci therefore produces phenotypes in which *sc* is present without *cv* and vice versa. This category includes all four crossover phenotypes in the experiment, or a frequency of 731/4338 = 16.8 percent. Again, the linear relationship between the three loci is obviously *sc-ec-cv*, since the *sc-cv* distance is the sum of the individual *sc-ec* and *ec-cv* distances. It is possible to map linkage relationships by testing two genes at a time, *A-B*, *B-C*, and *A-C*, and

TABLE 17-2

Recombination data for three sex-linked genes in *Drosophila melanogaster*: scute, echinus, and crossveinless

$$P_1: \frac{sc + cv}{sc + cv} ♀ \times + ec + ♂$$

$$F_1: \frac{+ ec +}{sc + cv} ♀ \times sc\ ec\ cv ♂$$

F_2 Phenotypes	F_2 Genotypes Male	F_2 Genotypes Female	Observed No.	
echinus	+ *ec* +	$\frac{+ ec +}{sc\ ec\ cv}$	1779	$\frac{3607}{4338} \times 100 = 83.2\%$ nonrecombinants
scute crossveinless	*sc* + *cv*	$\frac{sc + cv}{sc\ ec\ cv}$	1828	
scute echinus	*sc ec* +	$\frac{sc\ ec +}{sc\ ec\ cv}$	148	$\frac{287}{4338} \times 100 = 6.6\%$ recombinants: *sc*···*ec*
crossveinless	+ + *cv*	$\frac{+ + cv}{sc\ ec\ cv}$	139	
echinus crossveinless	+ *ec cv*	$\frac{+ ec\ cv}{sc\ ec\ cv}$	227	$\frac{444}{4338} \times 100 = 10.2\%$ recombinants: *ec*···*cv*
scute	*sc* + +	$\frac{sc + +}{sc\ ec\ cv}$	217	
			4338	

From cultures 1 to 3, 6 to 16 of Table 5 of Bridges and Olbrycht, 1926.

observing which is the longest distance, but this method consumes more time than the three-point cross because it involves three separate experiments. In some cases it is also not as reliable because some variability exists in recombination frequencies and an *A-C* distance, for example, may occasionally be smaller in a two-point cross than an *A-B* distance, although the true gene order established by other tests is *A-B-C*.

DOUBLE CROSSING OVER

Recombination among three loci, such as *A B C*, should theoretically result in 2^3, or 8, combinations, i.e., *A B C*, *A b C*, etc. Note, however, that in the above crosses a particular pair of combinations is missing. That is, if single crossovers occur between *A* and *B* and between *B* and *C*, some chromosomes should be formed which contain both crossovers at the same time, i.e., *double crossovers* as shown in Fig. 17–2. In the last example, double-crossover chromosomes should produce individuals that are either completely wild type (+ + +) or completely mutant (*sc ec cv*). However, because of the small distances involved between these three loci, double crossovers are found only rarely, and large numbers of F_2 individuals had to be examined to obtain a reasonable estimate of their frequency. As shown in Table 17–3, Bridges and Olbrycht combined many recombination experiments and observed five such double crossovers, or 0.02 percent, among the more than 20,000 flies scored.

The recombination frequency between each pair of genes is now increased by the frequency of these two double-crossover classes. That is, recombination between sc^+ and *ec* (or *sc* and ec^+) in Table

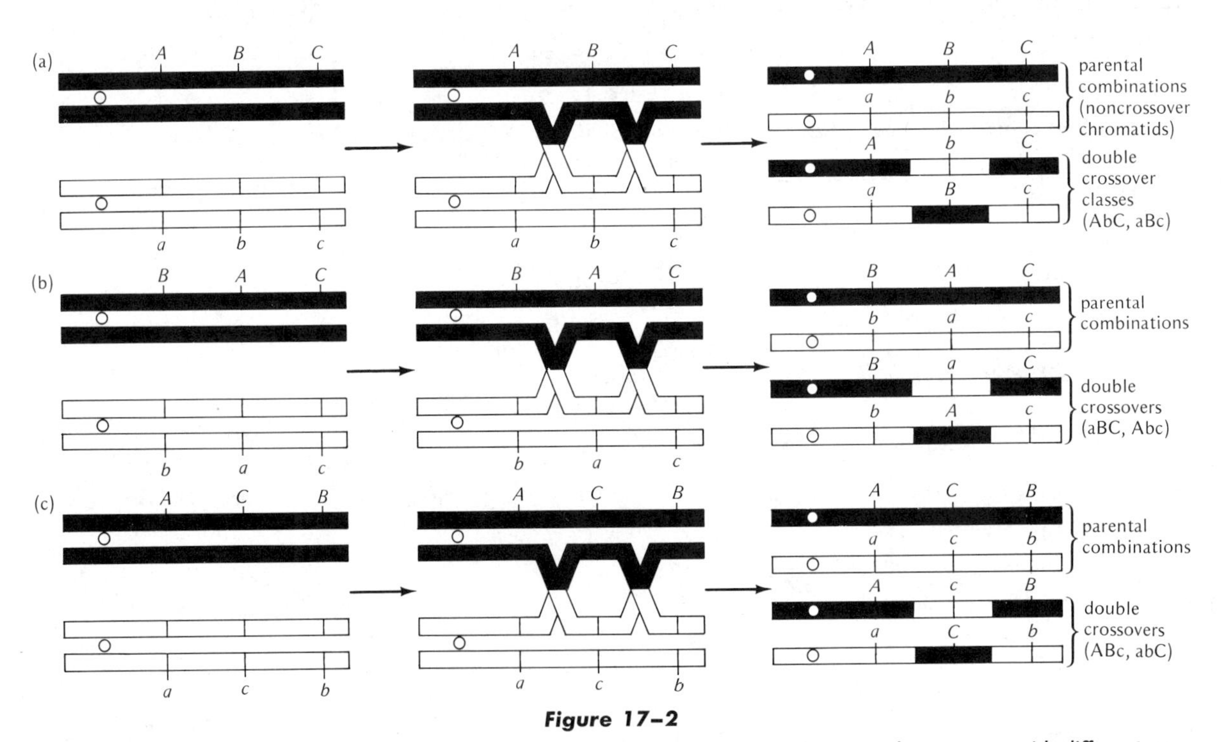

Figure 17–2

Diagram of crossover events that can lead to the formation of double crossover classes in chromosomes with different gene orders. Note that each of the three possible gene orders, A-B-C, B-A-C, and A-C-B, produces a unique pair of double crossover chromosomes. Because only two nonsister chromatids are involved in each case, these are called "two-strand" double crossovers. Three- and four-strand double crossovers also occur, and are discussed in the following chapter (see, for example, Table 18–1 or Fig. 18–3).

TABLE 17–3

Recombination data including double-crossover classes

F_1: $\frac{+\ ec\ +}{sc\ +\ cv}$ ♀ × *sc ec cv* ♂

F_2 Phenotypes	F_2 Genotypes Male	F_2 Genotypes Female	Observed No.	
echinus	+ *ec* +	$\frac{+\ ec\ +}{sc\ ec\ cv}$	8,576	$\frac{17{,}384}{20{,}785} = 83.64\%$ nonrecombinants
scute crossveinless	*sc* + *cv*	$\frac{sc\ +\ cv}{sc\ ec\ cv}$	8,808	
scute echinus	*sc ec* +	$\frac{sc\ ec\ +}{sc\ ec\ cv}$	681	$\frac{1397}{20{,}785} = 6.72\%$ recombinants: *sc* · · · *ec*
crossveinless	+ + *cv*	$\frac{+\ +\ cv}{sc\ ec\ cv}$	716	
echinus crossveinless	+ *ec cv*	$\frac{+\ ec\ cv}{sc\ ec\ cv}$	1,002	$\frac{1999}{20{,}785} = 9.62\%$ recombinants: *ec* · · · *cv*
scute	*sc* + +	$\frac{sc\ +\ +}{sc\ ec\ cv}$	997	
scute echinus crossveinless	*sc ec cv*	$\frac{sc\ ec\ cv}{sc\ ec\ cv}$	4	$\frac{5}{20{,}785} = .02\%$ double crossovers: *sc* · · · *ec* · · · *cv*
wild type	+ + +	$\frac{+\ +\ +}{sc\ ec\ cv}$	1	
			20,785	

recombination frequency *sc-ec* = 6.72 + .02 = 6.74
recombination frequency *ec-cv* = 9.62 + .02 = 9.64
recombination frequency *sc-cv* = 6.72 + 9.62 + 2(.02) = 16.38

From Table 5 of Bridges and Olbrycht, 1926.

17–3 now encompasses four classes, including *sc ec cv* and + + +, thereby producing a *sc-ec* distance of 6.72 + .02, or 6.74 percent. Similarly, the recombination distance between *ec* and *cv* includes the four classes in which these genes have been separated, or a distance of 9.62 + .02 = 9.64 percent. As before, the *sc-cv* recombination distance includes, at first approximation, four phenotypic classes (*sc ec*, *cv*, *ec cv*, *sc*) with a frequency of 6.72 + 9.62, or 16.34 percent. Once again the gene sequence is obviously *sc-ec-cv*. Note, however, that for this gene sequence the *sc ec cv* and + + + phenotypes are double-crossover classes, in which *sc* has been separated from *cv* by two crossovers in each case. The *sc-cv* recombination frequency in these last two classes is therefore twice their phenotypic frequency, or 2 × .02 = .04 percent. When added to 16.34, this gives a total of 16.38 percent for *sc-cv*, or the exact sum of the *sc-ec* (6.74) and *ec-cv* (9.64) distances.

Thus the center gene in a three-point sequence is the one separated by the double-crossover classes from its former parental combination with the two outside genes. In general, if the real order between three loci is *a c b*, and the chromosome constitution of the F_1 heterozygotes is *a c b*/+ + +, the F_2 combinations, + *c* + and *a* + *b*, can only arise from simultaneous crossovers between *a* and *c* and between *c* and *b*. On the other hand, separations of either *a* or *b* from their parental combination will arise, in addition, from crossovers in just single regions. Since double crossovers are generally much rarer than single crossovers, the least frequent combinations in this case will be $+ c +$ and $a + b$. As a rule, therefore, the gene showing the lowest frequency of separation from its parental combination is the center gene in a three-point cross.

One method of conducting a three-point linkage analysis is to proceed as follows:

1. Using the data of parental and recombinant classes in the F_2, calculate the recombination frequency between each pair of genes to obtain their map distance. For example, an F_1 cross $A\,b\,C/a\,B\,c \times a\,b\,c/a\,b\,c$ would produce F_2 progeny that could be classified as recombinants for the *A-B* distance (all *A B* – and *a b* – phenotypes), *B-C* distance (all – *B C* and – *b c* phenotypes), and *A-C* distance (all *a* – *C* and *A* – *c* phenotypes). The recombination frequency for each gene pair is the proportion of observed recombinants for those two genes over the total number of F_2 individuals scored.

2. Determine the map order between the three genes based on the three distances. The largest recombination frequency represents the longest distance, i.e., the two "outside" genes.

3. This map order now enables you to recognize which among the progeny are the double-crossover classes. For example, if the F_1 heterozygote had the chromosome constitution $A\,b\,C/a\,B\,c$ as given above, and the map order determined in steps 1 and 2 is *B-A-C*, then the F_1 can be written as $B\,a\,c/b\,A\,C$. The two double-crossover classes in the F_2 would therefore be *B A c* and *b a C* (or *A B c* and *a b C*).

4. Note that in this cross the two double-crossover classes would have been included in calculating the frequencies of crossing over between *A* and *B*, and between *A* and *C*, but not in calculating the *B-C* distance. Since each double-crossover individual represents *two* crossovers in the *B-C* distance, twice the frequency of the double-crossover classes must be added to the *B-C* recombination frequency calculated in step 1.

5. When step 4 is completed, the *B-C* distance should be the exact sum of the *A-B* and *A-C* distances in this particular example.

COINCIDENCE AND INTERFERENCE

As we have seen, the low frequency of double crossovers observed points to the low probability of separating the center gene in a three-point cross from its neighbors on either side. However, double crossovers generally occur even more rarely than expected. If crossing over between one pair of loci did not affect crossing over between a neighboring pair, we would expect double crossovers to appear in a frequency which would be the product of both. (The probability of two independent events occurring together equals the probability of one event multiplied by the probability of the other.) In diploid organisms observed double-crossover frequencies, however, are usually much less than the multiple of individual crossover frequencies. In the cross considered in Table 17–3, Bridges and Olbrycht examined a total of 20,785 flies, of which a proportion of .0674 were crossovers between *sc* and *ec*, and .0964 were crossovers between *ec* and *cv*. The probability of simultaneous crossing over in both regions should therefore be $.0674 \times .0964$, or .00650. In numbers this would therefore be $.00650 \times 20{,}785 = 135$ double-crossover individuals. However, they observed only five double crossovers, or a frequency of .00024.

According to a definition suggested by Muller, the ratio of the observed frequency of double crossovers to the expected frequency of double crossovers is known as the *coincidence coefficient*. For Bridges and Olbrycht's data, the coincidence coefficient is .00024/.00650 or 5/135, meaning that only 3.7 percent of the expected number of double crossovers have occurred. Since crossing over in one region apparently inhibits or interferes with crossing over in a neighboring region Muller termed this phenomenon *interference*. In the above case interference is so large that it prevents 96.3 percent of the expected double crossovers from occurring. The degree of interference is therefore 1 – the coefficient of coincidence.

TABLE 17–4

Recombination data for crossing over among the sex-linked genes *vermilion, cut,* and *garnet*

F_1: $\dfrac{v\ ct\ g}{+\ +\ +}$ ♀ × $v\ ct\ g$ ♂

F_2(*Males and Females*)		Observed No.	
vermilion cut garnet	($v\ ct\ g$)	1015	$\frac{2385}{3249} = 73.4\%$ nonrecombinants
wild type	(+ + +)	1370	
cut	(+ ct +)	249	$\frac{503}{3249} = 15.5\%$ recombinants: $v \cdots ct$
vermilion garnet	(v + g)	254	
garnet	(+ + g)	185	$\frac{344}{3249} = 10.6\%$ recombinants: $v \cdots g$
vermilion cut	($v\ ct$ +)	159	
cut garnet	(+ $ct\ g$)	8	$\frac{17}{3249} = .5\%$ double crossovers: $ct \cdots v \cdots g$
vermilion	(v + +)	9	
		3249	

recombination frequency v-ct = 15.5 + .5 = 16.0
recombination frequency v-g = 10.6 + .5 = 11.1
recombination frequency ct-g = 15.5 + 10.6 + 2(.5) = 27.1

From Table 3, Bridges and Olbrycht, 1926.

As can be expected, the extent of interference is not the same for every three-point cross. For example, a cross between an F_1 female heterozygous for the sex-linked genes *vermilion* (v, bright scarlet eyes), *cut* (ct, cut wings), *garnet* (g, purplish eye color), and their wild-type alleles produced the F_2 shown in Table 17–4 when backcrossed to a $v\ ct\ g$ male. For these loci, the sequence is ct-v-g, since the ct-g recombination rate is obviously the sum of the v-ct and v-g distances. The expected frequency of double crossovers is therefore $.160 \times .111 = .0178$. Since the observed double-crossover frequencies are only .005, coincidence is .005/.0178, or .28. Interference is therefore 1 − .28 = .72, or 72 percent, a much lower value than the 96.3 percent value observed for *sc ec cv*.

As a rule the value of coincidence falls and the value of interference rises when the distance between loci decreases. Thus three loci with very short crossover distances between them would be expected to show little or no double crossing over. This effect led to speculations that the crossover chromatids may have a certain amount of physical rigidity that prevents "bending" or crossing over twice within a short distance.

LINKAGE MAPS

On the basis of these demonstrated linear relationships between genes, extensive linkage or recombination maps have been established by testing groups of genes that have at least one locus in common. For example, a three-point linkage test of the genes *echinus crossveinless cut* by Bridges and Olbrycht established the order *ec-cv-ct* with the *cv-ct* recombination frequency equal to 8.4 percent. When these results are combined with those obtained for *cut vermilion*, and *garnet*, the linkage map for the six genes in these experiments appears as shown at the foot of the page if we consider *scute* as locus 0.0 and the other loci as having linkage positions relative to *scute*.

The map distances are given as recombination frequencies expressed as percentages and represent relationships based entirely on genetic crossover data. Some geneticists use the term *morgan* for these units, 1 morgan being equal to 1 percent recombination, and a centimorgan being .01 morgan, whereas others refer to them as map units. As we shall see later (Chapter 22), gene order can also be derived from studies of the genes' physical positions on chromosomes as determined by changes in chromosome structure. Both methods

recombination intervals:		(6.7)		(9.6)		(8.4)		(16.0)		(11.1)	
gene:	*sc*	—	*ec*	—	*cv*	—	*ct*	—	*v*	—	*g*
linkage position:	0.0		6.7		16.3		24.7		40.7		51.8

of gene location, genetic and cytological, provide the exact same gene orders, although the map distances established by the two methods do not exactly correspond.

After many tests involving numerous sex-linked genes, the entire X chromosome of *Drosophila melanogaster* was mapped and found to have a length of 66 map units (Fig. 17–3). This, however, does not mean that there is a recombination frequency of 66 percent in a linkage experiment between *yellow* and *bobbed*, which are at the extreme ends of the X chromosome. As explained previously, no more than 50 percent recombination is expected between any two loci, since only two of four chromatids in a meiotic tetrad are involved in any particular crossover point. In fact the actual recombination value observed between *yellow* and *bobbed* may be even less than 50 percent in a linkage experiment for the simple reason that not every X chromosome bivalent may have a crossover in that particular interval.*

In general, as the average number of physical crossover events increases and the map distance gets longer, the observed number of recombinants will also increase but will still never exceed 50 percent, and recombination frequency will remain smaller than the actual map distance. The reason for this lies in the fact that long distances between two loci may have more than one crossover event between them, yet such multiple crossover events may be quite difficult to detect. For example, if we restrict our attention to multiple crossovers between two nonsister chromatids (Fig. 17–4), double and even-numbered crossovers are not discernible between loci if only those two loci are scored in the experiment. Odd-numbered multiple crossovers will produce crossover products, but again, the number of crossover events cannot be known unless exchanges between the "intermediate" loci can be scored. The presence of multiple crossovers thus makes it difficult to estimate large map distances accurately, whereas small map distances with fewer multiple crossovers are estimated more reliably. The most accurate linkage maps are therefore those obtained from recombination frequency data for a sequence of many small gene intervals.

To approximate actual map distances, observed recombination frequencies may be corrected, with the correction becoming greater as recombination frequency increases. Based on numerous experiments with *Drosophila*, corn, and *Neurospora*, the relationship between recombination frequency and actual map distance assumes the form of a curve (Fig. 17–5), showing anticipated increased map distances for larger recombination frequencies. According to this relationship, the map distance between two loci is the point along the horizontal axis of Fig. 17–5 at which the observed recombination frequency (vertical axis) intersects the curve.

As expected, the coincidence coefficient, which is a measure of double crossing over, also increases as map distance lengthens. Figure 17–6 shows this relationship between coincidence and map distance in the form of a curve based on a large amount of *Drosophila* data. Taken as a rough approximation, the coincidence coefficient for double crossing over among three loci within a given map distance can be determined as the point on this curve that intersects the given map distance (horizontal axis).

In *Drosophila melanogaster*, as is shown in Fig. 17–3, linkage map distances have been obtained for all chromosomes except the Y. Although it was known that the Y chromosome was necessary for male fertility (XO males are sterile), no Y-linked genes were found until 1926, when Stern discovered that the Y carried the normal allele for *bobbed* (see p. 244). However, aside from fertility factors and the *bobbed* gene, no other genes are

*For example, even if there is an average of one physical crossover event for a particular interval, some bivalents may have no crossovers for that interval, whereas others have one, two, three, etc. According to the Poisson distribution for random events (p. 158), the chances for a bivalent to have no crossovers within such an interval is equal to $e^{-1}(1) = 1/2.718$, or .3679. Thus, theoretically 37 percent of such bivalents will produce only parental combinations, and the remaining 63 percent will produce both recombinant *and* parental combinations in about equal frequency. This would mean that a linkage interval that has an average of one physical crossover event in each meiosis will show only $1/2 \times 63$ or 31.5 percent recombinant products whereas 50 percent recombination would have been expected (see p. 320). If we designate the distance between two loci in terms of map units based on the *actual number* of crossover events in that interval it is obvious that the percent of recombinant products observed in an experiment will generally be less than the physical map distance.

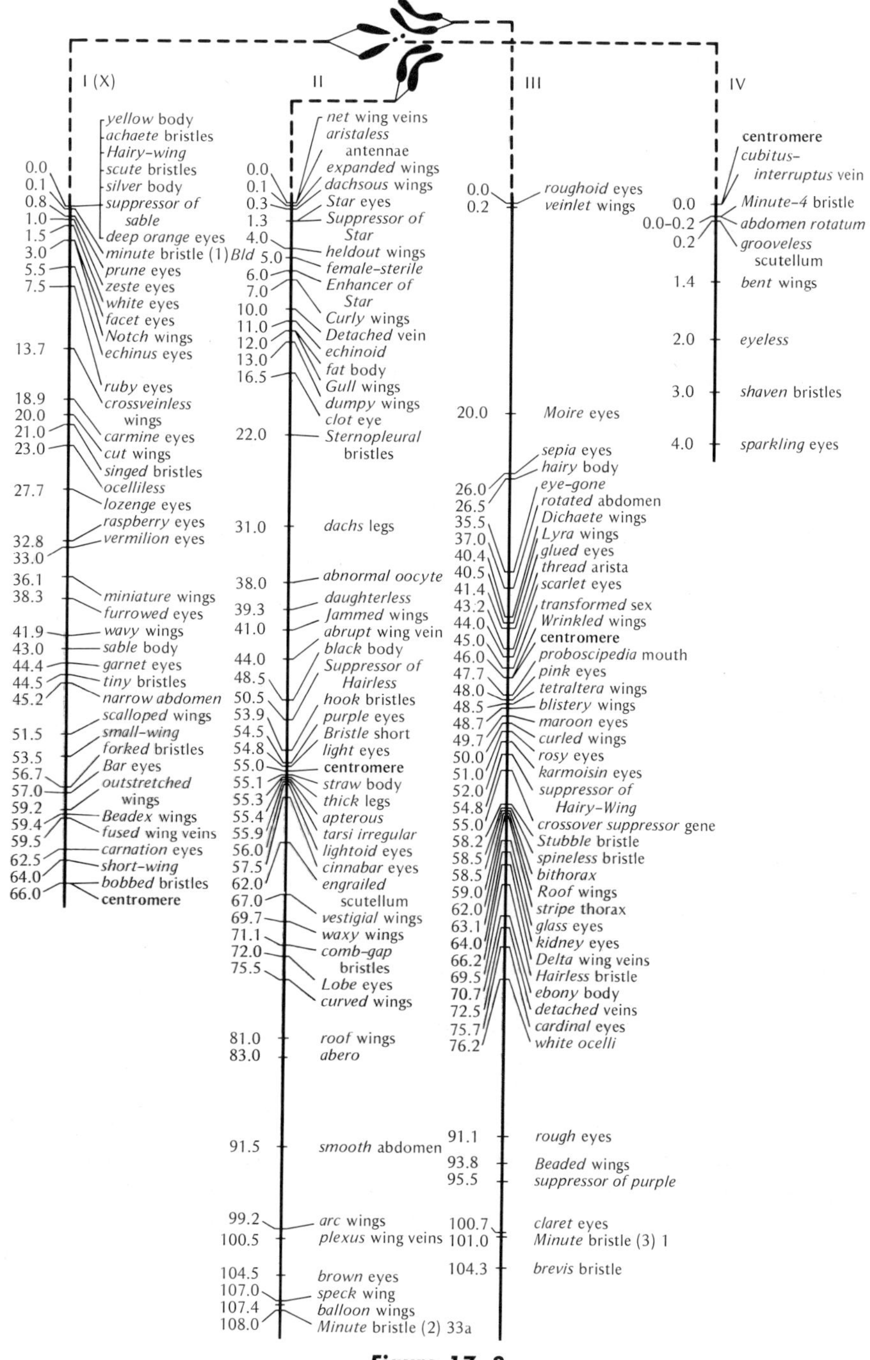

Figure 17–3

Linkage map of some of the important genes in the four chromosomes of Drosophila melanogaster. *Note that a variety of different genes may affect a specific character such as eye color, wing shape, bristles, etc. As discussed in Chapter 11 (see also Chapter 26), this indicates that many steps exist in the development of a particular function, each step governed or capable of being modified by separate and different genes. (Extensive descriptions of* Drosophila melanogaster *mutations and their linkage map positions can be found in Lindsley and Grell.)*

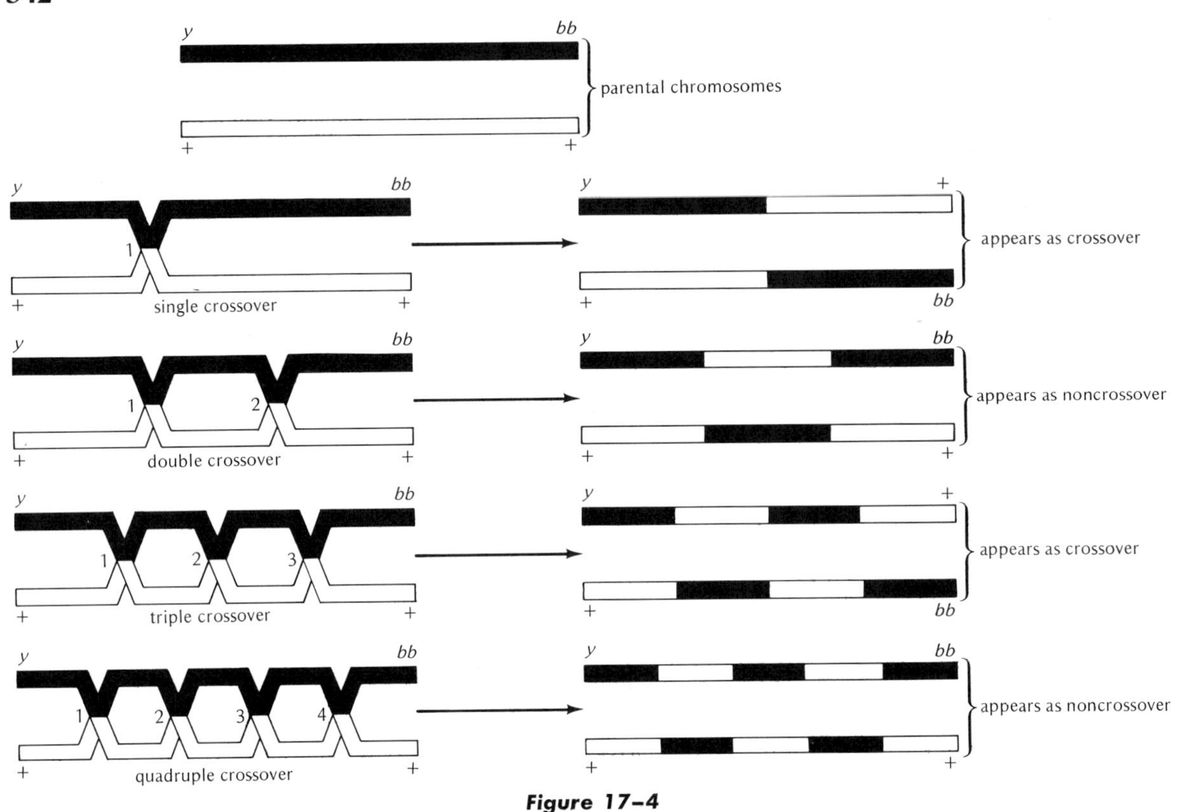

Figure 17–4

Diagram of crossovers between two nonsister strands in which the meiotic products are scored only in respect to two loci, yellow (y) and bobbed (bb), and intermediate loci are not scored. The diagram shows that only single, or odd-numbered, multiple crossovers between y and bb produce new combinations of the two genes. No matter how many odd-numbered multiple crossovers occur, only one pair of recombinant products appear. Double, or even-numbered, multiple crossovers produce only parental combinations. (If three or four chromatids were involved in multiple crossovers, the results would be somewhat different, as explained in Chapter 18. But in any case, the average number of observed recombinants between two scored loci would not be expected to exceed 50 percent.)

known on the Y chromosome, although it may produce effects when transposed to other chromosomes or when more Y chromosomes are added (pp. 546–547).

LINKAGE IN HUMANS

In addition to *Drosophila*, fairly extensive linkage maps are now available for a number of diploid species such as corn and mice in which recombination experiments can be performed. In humans, however, the estimation of linkage between genes can only be accomplished very roughly from family data, since matings cannot be controlled and sample sizes of a given mating are therefore very small. Also, the presence of 23 pairs of chromosomes makes it likely that the loci which can be investigated by such means will be on different chromosomes. Nevertheless family studies in man have enabled linkage determinations in a number of instances where different pairs of genes are being transmitted together in particular pedigrees. To date, autosomal linkages are known between the *nail-patella* syndrome (abnormal nails and kneecaps) and the ABO blood-group locus;

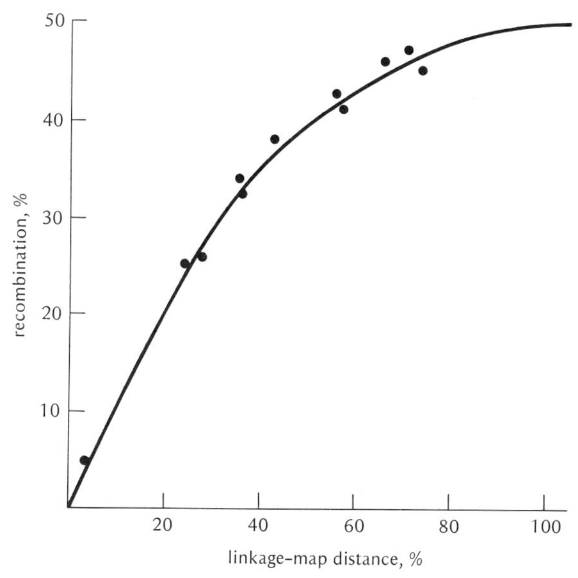

Figure 17–5

Relationship between observed recombination frequency (vertical scale) for linked loci and their linkage map distance (horizontal scale). Based on Drosophila, corn, *and* Neurospora *data. (After Perkins.)*

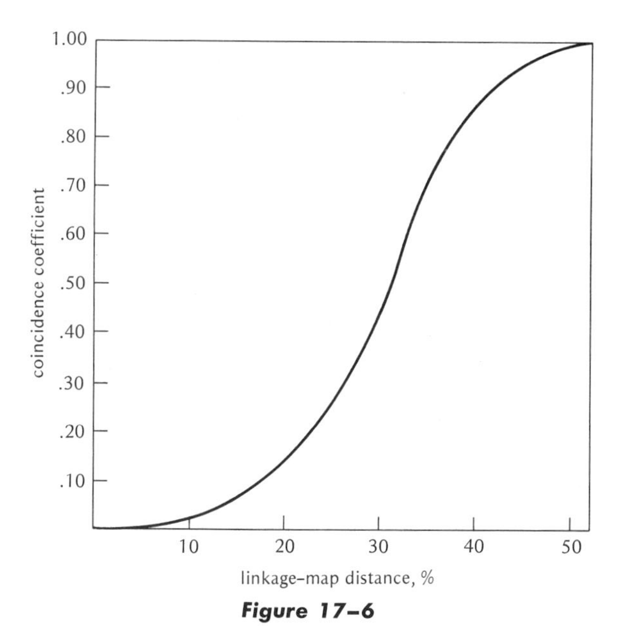

Figure 17–6

Relationship between coincidence coefficients and linkage map distance in Drosophila. *(Calculations of Weinstein from data of Morgan, Bridges, and Schultz.)*

between *elliptocytosis* (oval red blood cells caused by a dominant gene) and the Rh blood-group locus; between the Lutheran blood-group locus and the ABO *secretor* gene controlling the appearance of ABO antigens in water-soluble form (see p. 170); between the beta and delta hemoglobin chains (see Chapter 26); and between two kinds of protein found in blood serum, serum bisalbumin and the group specific component (GC). A number of linkages between other autosomal genes are suspected, with linkage distances ranging from a few percent to as much as 20 percent.

In the human X chromosome, at least eight loci have been mapped through family studies with somewhat greater accuracy than can be achieved for autosomes. The reason for this is that X-chromosome recombination occurs only in the female (2 X's), and the recombinant products are immediately obvious in her male sons, who carry only a single X chromosome which is of maternal origin. Thus, if the genetic constitutions of the mothers are known and the number of their male offspring is sufficiently numerous, recombination frequencies can be estimated from relatively few pedigrees.

This can be illustrated by considering only the first four sons of a single female segregating for two sex-linked recessive traits, hemophilia B (Christmas disease, p. 236) and colorblindness. As shown in Fig. 17–7, this female was descended from a father normal for these traits, married a normal male, and produced sons that were either affected by hemophilia B, colorblind, or both.

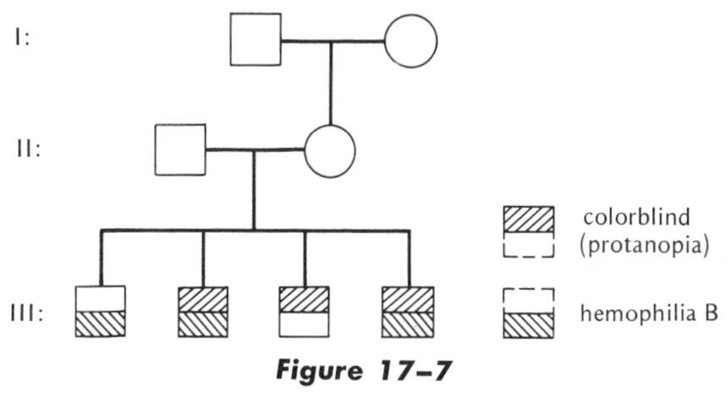

Figure 17–7

Pedigree of a female (II–2) that has produced sons segregating for two sex-linked genes, colorblindness and hemophilia B.

Since the X chromosome obtained from her normal father did not bear either hemophilia B or colorblindness, she must have carried those two genes on the single X chromosome from her mother. If we call the hemophilia B and colorblind alleles h^B and c, respectively, her X chromosome constitution can be described as $h^{B+}\,c^+/h^Bc$. Thus individuals III–2 and III–4 inherited their two sex-linked genes in the original combination (h^Bc), and are therefore nonrecombinant. Individuals III–1 and III–3, on the other hand, have arisen through crossovers between the two maternal X chromosomes (h^Bc^+, $h^{B+}c$). The recombination frequency of 50 percent indicates that these genes are sufficiently far apart to enable free recombination between them. Through this same method other loci have been shown to have lesser crossover distances, and a tentative map of the human X chromosome is presented in Fig. 17–8.

The limitations in finding appropriate pedigrees enabling linkage analysis in humans have been partly overcome in recent years by a technique that permits "hybridization" between different lines of somatic cells raised in tissue culture. Grown together on appropriate medium, cells of different strains of the same mammalian species (Barski and co-workers) or even cells of different mammalian species (Harris and Watkins; Ephrussi and Weiss) can be shown to fuse followed by the subsequent loss of some of the parental chromosomes.*

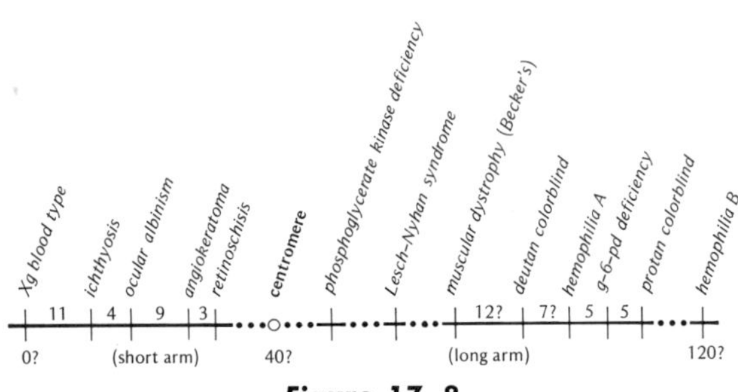

Figure 17–8

A tentative map of the human X chromosome showing approximate distances (above the line) in the form of recombination frequencies observed between some known sex-linked genes. Although other arrangements of these genes are possible from pedigree data (e.g., angiokeratoma on the far left end of the chromosome followed by ocular albinism, ichthyosis, etc.), it is now fairly certain that the dotted lines separate at least three "clusters" of sex-linked genes: Xg, hemophilia A, and hemophilia B. The loci at the left of this map are presumed to be located somewhere on the short arm of the X chromosome (see Fig. 21–15), and the loci on the right may reside on the long arm.

In those cases involving human cell lines crossed with the somatic cells of mice or hamsters, the chromosomes most rapidly lost in the dividing hybrid cells are the human ones. Localization of a particular gene to a particular chromosome is then accomplished by noting whether the retention or loss of a particular cellular function or enzyme is associated with the presence or absence of a recognizable human chromosome.

In 1967 Weiss and Green first used this technique to localize the gene involved in the production of the enzyme thymidine kinase, which transforms the nucleoside thymidine into the nucleotide thymine. First, they mixed together in a common culture medium a strain of thymidine kinase-deficient mouse cells and a strain of normal enzyme-producing human cells. This mixture was then placed in a culture containing the drug aminopterin as well as the purine hypoxanthine, and the pyrimidine thymidine ("HAT" medium). Aminopterin prevents the cellular synthesis of purines and pyrimidines, so that such compounds must be provided in the culture medium in order for DNA replication to proceed. Because of this exposure, the mouse thymidine kinase-deficient cells soon die, since they are unable to metabolize the thymidine in the medium, whereas the human nondeficient parental cells as well as the hybrid cells bearing complements of mouse and human chromosomes survive.

Within a short time, the vigorously growing hybrid cells form distinct colonies (Fig. 17–9) which rapidly overgrow the cultures and replace

* Recently Carlson and co-workers have shown that even the cells of different plant species can be made to hybridize by removing the cell walls from leaf tissue, and growing them under conditions that permit only fused hybrid cells to survive. Furthermore, as discussed later (see Fig. 21–3), these hybridized cells (*Nicotiana glauca* × *N. langsdorfii*) can be raised to full-grown plants and shown to bear characteristics different from either parent.

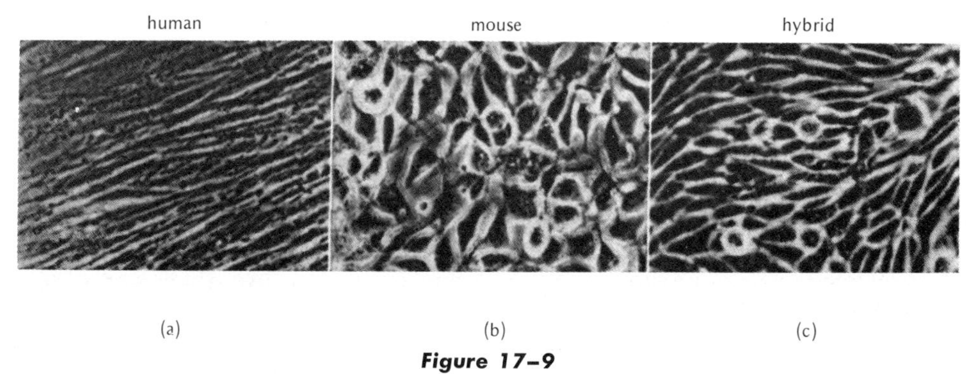

Figure 17–9

Photomicrographs of human, mouse, and human-mouse hybrid cells grown in tissue culture. Each cell type is recognizable by the shape of its cells and their refractive index. (From Weiss and Green.)

all of the human cells. Many of the human chromosomes are lost from these hybrid cells in subsequent divisions, but those hybrid cells that continue to grow in HAT medium must obviously retain the human chromosome bearing the gene for thymidine kinase. Furthermore, this chromosome can also be identified by the fact that it must be absent from hybrid cells which grow in a medium containing 5-bromodeoxyuridine (BUDR), since thymidine kinase permits incorporation of the lethal BUDR nucleotide into DNA. Using both methods of identification, work by Migeon, Miller, and others has shown conclusively that the thymidine kinase gene is located in a human chromosome of the E group, specifically No. 17 (see Fig. 21–15).

The technique used by Weiss and Green has since been elaborated so that the degree of cell fusion between different cell lines is increased one hundred- to one thousandfold through use of irradiated Sendai (parainfluenza) virus or the chemical lysolecithin. A large number of genes, mostly recognized by their association with specific enzymatic products, are now assigned to specific human chromosomes (Table 17–5) which are, in turn, identified by peculiarities in size, shape, banding, and other characteristics discussed in Chapter 21. And in some instances linkages between two or more different genes on the same chromosome have been established based on their concordant association with a particular human chromosome. Also, techniques involving the use of chromosome aberrations (Chapter 22) now ena-

TABLE 17–5

Linkage assignments of genes to human chromosomes by somatic hybridization studies

Chromosome	*Genes Assigned*
1	phosphoglucomutase-1; peptidase C; 6-phosphogluconate dehydrogenase
2	cytoplasmic isocitrate dehydrogenase; cytoplasmic malate dehydrogenase; red cell acid phosphatase
4–5	adenine-B production (enzyme unknown)
5	hexosaminidase B
6	indolephenoloxidase
7	mitochondrial malate dehydrogenase
10	glutamate oxaloacetate transaminase
11	lactate dehydrogenase A; esterase A_4; lysosomal acid phosphatase
12	lactate dehydrogenase B; peptidase B; glutamic pyruvic transaminase B; serine hydroxymethylase
13	esterase D
14	nucleoside phosphorylase
15	pyruvate kinase; mannose phosphate isomerase; hexosaminidase A
16	adenine phosphoribosyl transferase
17	thymidine kinase
18	peptidase A
19	glucosephosphate isomerase
20	adenosine deaminase
21	indolephenoloxidase A; antiviral response factor
23(X)	glucose-6-phosphate dehydrogenase; hypoxanthine-guanine phosphoribosyl transferase; phosphoglycerate kinase; α-galactosidase

Assignment of gene loci to human chromosomes can be found in Ruddle 1973, and in "Reports to the Rotterdam Conference: Second International Workshop on Human Gene Mapping," *Cytogenetics,* **14:**161–480 (1975).

ble detection of the particular section of a chromosome carrying a specific gene. For example, strains of somatic cells bearing a "deletion" of a particular section of a chromosome may be associated with absence of a gene, whereas its presence in other cell lines may be associated with the "translocation" of that particular section to another chromosome. Information of this kind can then be combined with pedigree studies to show specific linkage relationships. Thus the P blood group in humans can now be assigned to chromosome No. 20, since families segregating for P blood group alleles can be shown to have these differences linked to enzymatic differences in adenosine deaminase which have been localized to that chromosome.

EXPECTED RECOMBINATION FREQUENCIES

The existence of linkage maps and knowledge of coincidence coefficients enable one to predict the proportions of recombinant phenotypes to be expected in various crosses. For example, if three gene pairs, *A*, *B*, and *C*, have map distances of 15 units between *A* and *B* and 20 units between *B* and *C*, the expected genotypes among 200 offspring of a cross *A B C*/*a b c* × *a b c*/*a b c* would be calculated as follows.

1. If crossing-over interference between neighboring regions were lacking, i.e., if the coincidence coefficient were 1.00, the number of double-crossover individuals would be .15 × .20 × 200, or six. Interference, however, can be expected to reduce this value. In the present case the expected coincidence coefficient for a total map distance of 35 is approximately .70, as shown in Fig. 17–6. This means that only about .7 of the expected double crossover individuals will occur, or .7 × 6 = 4. These individuals will have the genotypes *A b C*/*a b c* and *a B c*/*a b c*, and we would expect two of each.

2. *A-B* recombinant genotypes will be of four kinds, the two kinds of double crossovers mentioned above (each double indicating one crossover between *A-B*) and the single-crossover genotypes, *A b c*/*a b c* and *a B C*/*a b c*. In all, 15 percent *A-B* recombinants are expected, or .15 × 200 = 30. Since 4 of these are double crossovers, the remaining 26 individuals can be divided among the two single-crossover genotypes, or 13 *A b c*/*a b c* and 13 *a B C*/*a b c*.

3. *B-C* recombinant genotypes will also be of four kinds: the same double crossovers mentioned previously and two single-crossover genotypes, *A B c*/*a b c* and *a b C*/*a b c*. Since 20 percent *B-C* recombination is expected, this will amount to .20 × 200 = 40 individuals. As before, 4 of these are double-crossover genotypes, so we expect 36 single crossovers. These are divided equally among the two single-crossover genotypes, giving 18 *A B c*/*a b c* and 18 *a b C*/*a b c*.

4. The number of nonrecombinant genotypes, *A B C*/*a b c* and *a b c*/*a b c*, will be the difference between the number of recombinants and the total number scored, or 200 − 66 = 134 individuals. Divided among the two nonrecombinant genotypes, this yields 67 *A B C*/*a b c* and 67 *a b c*/*a b c*. The ideal expectation for the 200 progeny of cross *A B C*/*a b c* × *a b c*/*a b c* is therefore:

Genotype	Number	
A B C/*a b c*	67	nonrecombinants
a b c/*a b c*	67	
A b c/*a b c*	13	single crossovers
a B C/*a b c*	13	
A B c/*a b c*	18	
a b C/*a b c*	18	
A b C/*a b c*	2	double crossovers
a B c/*a b c*	2	
	200	

FACTORS THAT AFFECT RECOMBINATION FREQUENCIES

Recombination frequencies between specified loci may vary between experiments. One factor that can influence the amount of crossing over to a greater or lesser degree is sex. An extreme example of this can be found in those *Drosophila* species (e.g., *melanogaster*), in which crossing over is completely suppressed in males. In general the heterogametic sex of a species has been found to have lower crossover frequencies.

In addition to sex, numerous factors that affect recombination frequencies have been found in *Drosophila*, including the following:

MATERNAL-AGE EFFECT. Bridges showed that as maternal age increases, crossing over tends to decrease.

TEMPERATURE EFFECT. Starting from a median temperature of about 22°C, both cooler and warmer temperatures have been found by Plough to increase the rate of crossing over.

CYTOPLASMIC EFFECT. Experiments by Thoday and Boam and by Lawrence have shown the existence of crossing-over factors carried in the cytoplasm; females selected for reduced recombination frequencies will pass this trait on through their daughters.

NUTRITIONAL, CHEMICAL, AND RADIATION EFFECTS. Crossing-over frequencies also seem to be affected by the concentration of metallic ions, such as calcium and magnesium. When R. P. Levine fed young *Drosophila* females on a high-calcium diet, crossing over between X chromosome genes was decreased. When fed a medium containing chelating agents to remove metallic ions, crossing over was increased. Neel has found that larval starvation at certain ages generally increases crossing over. Suzuki has shown an increase in crossing over in flies injected with antibiotics such as mitomycin C and actinomycin D. X-ray irradiation has been shown by Whittinghill and others to cause crossover increases not only in *Drosophila* females but also in *Drosophila* males!

GENOTYPIC EFFECTS. Crossing-over frequencies between the same two loci in different strains of the same species show variation presumably because of numerous gene differences between the strains. This has been shown in *Drosophila pseudoobscura* by R. P. and E. E. Levine, and is undoubtedly a partial cause for some of the variable results that are found when linkage experiments are repeated. A 3rd-chromosome gene (*c3G*) in *Drosophila melanogaster*, which completely suppresses crossing over when homozygous in females, was first noted by J. W. and M. S. Gowen, and Hinton has since reported that this same gene enhances recombination when heterozygous. According to Meyer (see Moses reference, Chapter 2), the oocytes of *c3G* homozygotes do not possess the synaptinemal complexes usually found in female meioses. Further investigations by Sandler and co-workers have shown that a large number of mutant genes can be individually identified that affect recombination frequencies in *D. melanogaster* females. Some of these genes seem to affect the preconditions for exchange, that is, chromosome pairing; while others seem to act after pairing has been completed, during the actual exchange process.

CHROMOSOME-STRUCTURE EFFECTS. The order of genes on a chromosome is subject to change by various structural chromosome rearrangements. (See p. 19 and further discussion in Chapter 22.) When the genetic effects of these changes were first noted about 1915, their structural causes were not known, but their negative effect on crossing over caused many of them to be designated as *crossover suppressors,* or *C* factors. In heterozygous condition, a *C* factor reduces crossing over in the particular pair of homologues in which it is present. However, Sturtevant, Schultz, Redfield, and others have shown that such aberrations in one pair of chromosomes will increase the frequency of crossing over in other normal nonhomologous chromosomes.

CENTROMERE EFFECT. Through the use of structural chromosome aberrations, the position of genes relative to the centromere of the chromosome can be changed. When a section of chromosome containing two or more marker genes is placed close to the centromere, the genes tend to show reduced crossing over, although the physical distance between them is not changed. This centromere effect can also be observed in Fig. 17-3, in which numerous genes appear clumped together near the centromere regions of the 2nd and 3rd chromosomes, although the relative physical distances between them are actually longer (see also Fig. 22-29).

RELATIONSHIP BETWEEN GENETIC AND CYTOLOGICAL CROSSING OVER

Among the most important experiments establishing the chromosomal basis of genes and gene order were those that demonstrated that genetic crossing over is paralleled by an actual physical crossing over between homologous chromosomes. Experiments of this kind were first reported in 1931, both by C. Stern for *Drosophila melanogaster* and by Creighton and McClintock for corn. Both the *Drosophila* and corn experiments employed a cytologically visible "marker" on one member of a pair of homologous chromosomes which also differed with respect to the alleles of certain gene pairs. Genetic crossing over therefore resulted in unique phenotypes, and was found to be accompanied in each case by a transfer of the cytological marker and a recognizable change in chromosomal structure.

In Stern's experiment, stocks carrying structural chromosomal changes were crossed to produce a female which had a section of Y chromosome attached to one of its X chromosomes. The other X chromosome was composed of two approximately equal fragments, each with its own centromere. Each of the two X chromosomes was therefore uniquely recognizable under the microscope (Fig. 17-10). In addition, one of the X chromosome fragments carried the mutant alleles for *carnation* (*car*, recessive light eye color) and *Bar* (*B*, semidominant narrow eye shape), while the long X-Y chromosome carried the normal alleles for these genes. In the absence of crossing over, such females would be expected to produce only *car B* or + + gametes. When fertilized by a carnation-eyed male (*car* +), the female noncrossover offspring of the mating should therefore appear as carnation and bar-eyed (*car B*/*car* +), or wild type (+ +/*car* +). Genetic crossing over, however, will produce different female phenotypes, carnation (*car* +/*car* +) and bar (+ *B*/*car* +), but, more interestingly, will also produce uniquely recognizable chromosomes if chromosomal exchange accompanies genetic crossing over. As shown in Fig. 17-10, the two crossover products produced by the female are one long X chromosome and a short X fragment with attached Y section. When combined with the normal long X chromosome of the male, the crossover female offspring are thus cytologically distinct from the noncrossovers. Stern found the predicted genotype-X chromosome correlation in almost every crossover female examined.

RELATIONSHIP BETWEEN CROSSING OVER AND CHIASMA FORMATION

Stern's experiments, as well as those of Creighton and McClintock, indisputably demonstrated that genetic crossing over between homologous chromosomes is accompanied by a physical exchange between these same chromosomes. This physical crossing over, however, may or may not be identical with the chiasmata observed during meiotic prophase. To explain the relationship between crossing over and cytological chiasmata, two main theories have been proposed, and a discussion of these theories may be helpful in evaluating some of the mechanisms that may be involved.

The *classical* interpretation has been that there is no necessary one-to-one relationship between chiasmata and crossing over. When chiasmata are formed, according to this view, they signify an accidental physical crossing of homologous chromatids which produces an *equational* separation of chromosome segments caused by the close pairing of a chromatid from one homologue with that of another. The gap between two chiasmata should therefore be considered to separate equivalent chromatid sections representative of both chromosomes rather than separate paired chromatid sections of one homologue from paired chromatid sections of the other, as in *reductional* separation. Such chiasmata may or may not lead to breakage and subsequent genetic crossing over (left side of Fig. 17-11). In any case, when genetic

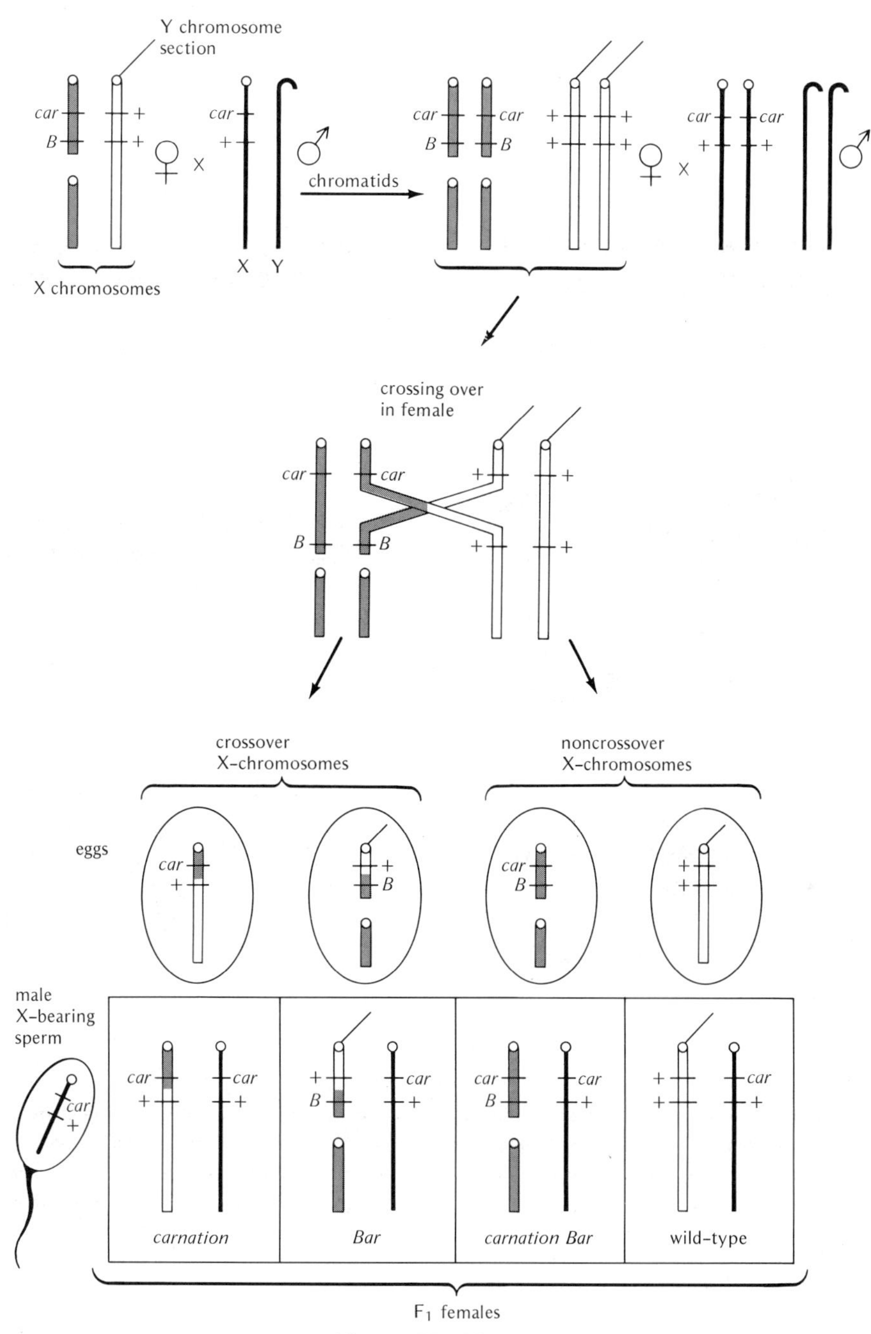

Figure 17–10

Diagram of Stern's experiment in Drosophila melanogaster *showing the correlation between genetic and cytological crossing over between two sex-linked loci,* carnation *and* Bar. *In this experiment recombinant female offspring carried X chromosome combinations uniquely different from their noncrossover sisters.*

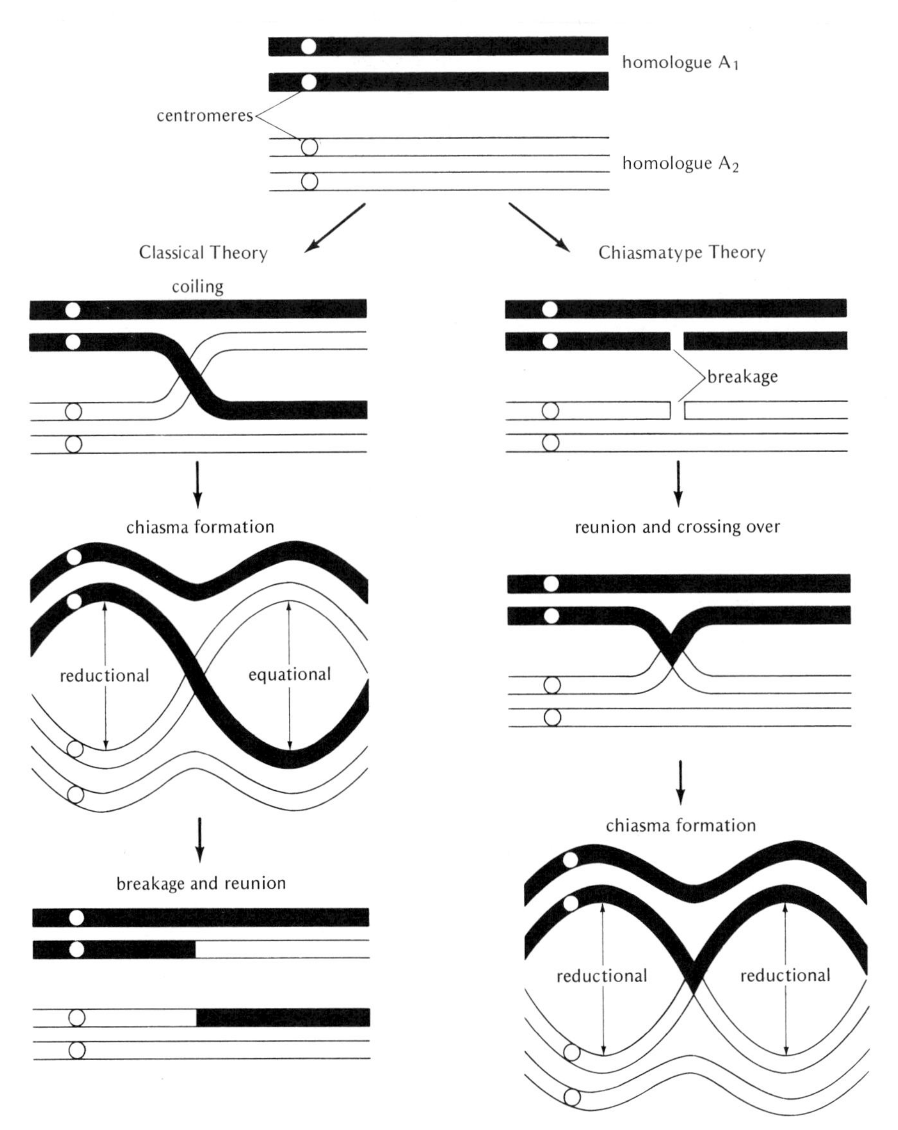

Figure 17–11

Differences between the classical theory of chiasma formation (left side) and the chiasmatype theory (right side) in explaining the origin of observed chiasmata in pairing between two homologous chromosomes. In the classical theory breakage and reunion (genetic and cytological crossing over) follow chiasma formation, and one side of a chiasma is presumed to represent an equational division of chromatids. In the chiasmatype theory breakage and reunion precede chiasma formation, and both sides of a chiasma represent a reductional division of chromatids.

crossing over occurs according to the classical view, it is as a consequence of the physical strains imposed by chiasma formation. Thus the observation of chiasmata at the diplotene stage of meiosis presumably precedes the occurrence of genetic crossing over, which occurs between diplotene and the first anaphase.

An alternative interpretation, called the *chiasmatype* theory, was presented by Janssens and later developed more fully by Belling and by

Darlington. This theory proposes that crossing over occurs sometime during the early meiotic stages, perhaps pachytene, when homologous chromatids are closely paired. As the meiotic cell moves toward metaphase and reductional division, a chiasma is formed at the point where crossing over has occurred (Fig. 17–11, right side). Thus, according to the chiasmatype theory, each chiasma represents one genetic crossover and there is always *reductional* separation on both sides of a chiasma.

From a genetic point of view the chiasmatype theory had the advantage of suggesting that crossing over at any one point occurs only between two chromatids, one from each homologue, rather than between entire chromosomes. This could therefore account for the considerable data showing that each crossover event involves only two out of four strands (p. 317). It also helped to explain the observation that recombinational events are "reciprocal"; that is, a crossover between two gene loci produces *both* kinds of recombinant products. As shown consistently in all previous illustrations and examples, the two crossover chromatids that result from matings such as $A\ B \times a\ b$ bear the two reciprocal genotypes $A\ b$ and $a\ B$ (e.g., Fig. 17–1).

Evidence presented by Darlington and others has indicated that the chiasmatype theory can explain additional observations that are inexplicable according to the classical theory. Of these, one of the most important is the mode of pairing that occurs in organisms carrying three sets of homologous chromosomes (triploids). With only rare exceptions (see Comings and Okada) it has been cytologically observed that no matter how many homologues are carried in an organism, a chromosome will only associate during meiosis with one other homologue at a time, either along its entire length or over a limited region. Thus, if there are three homologues of one chromosome, A, B, and C, A can pair fully lengthwise with either B or C but not with both at the same time, and similarly B can pair with either A or C.

In triploids, occasional trivalent structures are noted in which chiasmata link together three homologous chromosomes. The only way these trivalents can be explained according to the classical theory is to assume that complete synapsis has occurred between the three chromosomes in the same regions, i.e., A_1-B_1-C_1 and A_2-B_2-C_2 (Fig. 17–12, left side). Only this type of pairing would permit equational division on both sides of each chiasma. Since simultaneous pairing among three homologues is unsupported by all other evidence, the repeated observation of such trivalents makes the classical theory extremely doubtful. The chiasmatype theory, on the other hand, offers the more acceptable explanation that pairing among such trivalents has occurred in a given region between only two homologous chromosomes at a time, and the separation at each chiasma is reductional (Fig. 17–12, right side).

The chiasmatype theory should receive further support if the numbers of chiasmata and crossovers can both be scored. For example, if the chiasmatype theory were correct, the presence of a single chiasma in each meiosis of a bivalent would represent a map length of 50 crossover units, since it means that recombination between the two pairs of end genes will be observed about 50 percent of the time. Similarly, chiasma frequencies of two and three per chromosome would represent genetic map lengths of 100 and 150 units respectively. The large meiotic chromosomes of corn permit chiasma frequencies to be more accurately observed than in *Drosophila*, and in this organism there does appear to be a correlation between the genetic map length of some chromosomes and their length calculated according to observed chiasma frequencies (Table 17–6).

Experiments by Beadle and by Brown and Zohary also offer strong support for the one-to-one correspondence between chiasmata and genetic crossing over. In Beadle's study a region of the 9th chromosome in corn which produced 12 percent genetic recombination showed a chiasma frequency of 20 percent. Since each chiasma involves only two of the four chromatids, it would take a 24 percent chiasma frequency to produce the 12 percent observed recombinants. The actual 20 percent chiasma frequency is therefore fairly close to that expected. Chiasma frequencies have therefore been used to estimate recombinational distances for organisms in which only chiasmata can

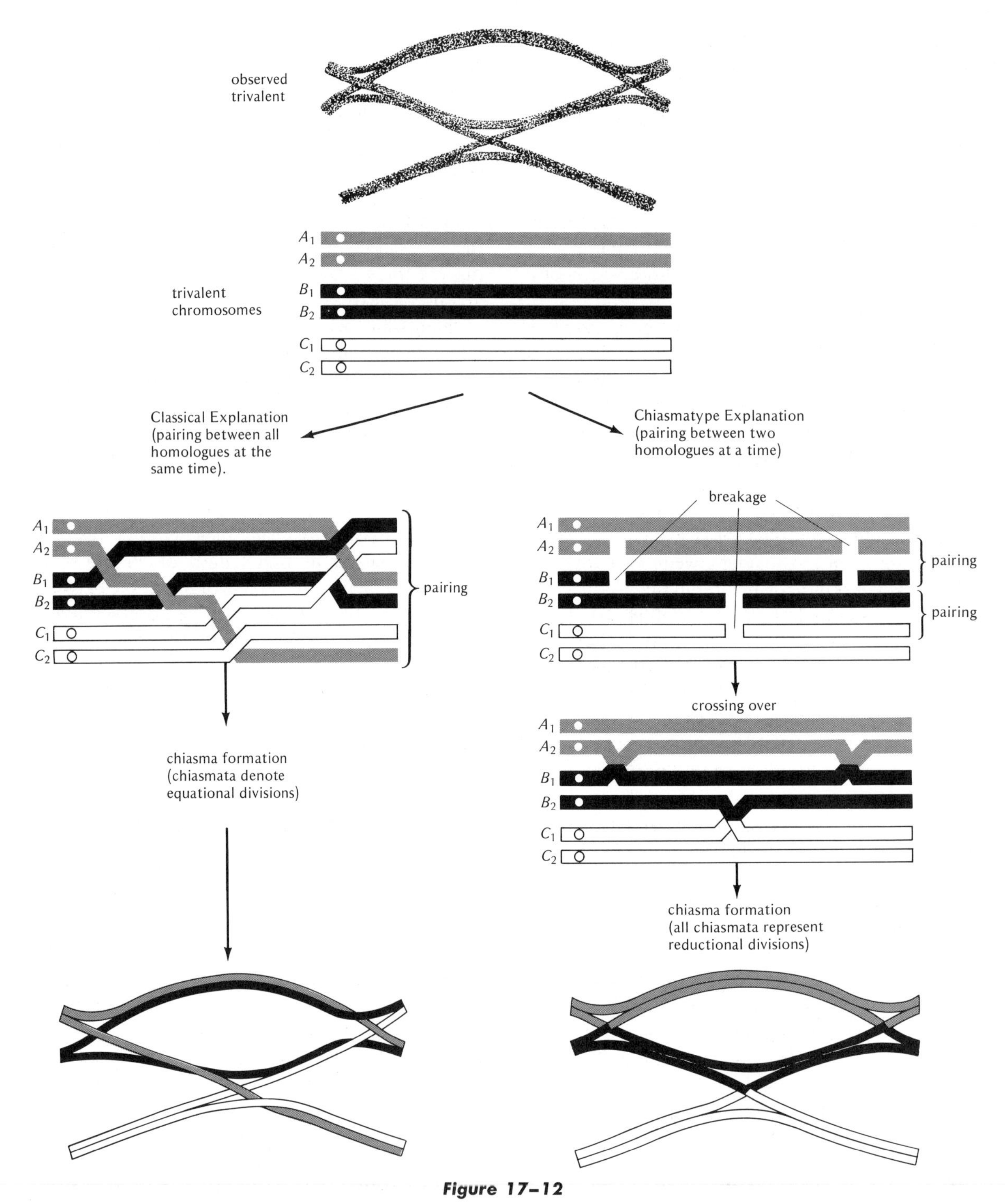

Figure 17–12

Classical (left) and chiasmatype (right) explanations for observed chiasmata in a trivalent (meiotic combination of three homologous chromosomes). (Trivalent figure from White.)

TABLE 17–6

Correlation among the physical length of corn chromosomes, observed chiasma frequencies, and observed map distances

Chromosome No.	Chiasma Frequency	Relative Physical Length	Calculated Map Distance (50 × Chiasma Frequency)	Observed Map Distance (1950)
1	3.65	100	183	156
2	3.25	86	162	128
3	3.00	78	150	121
4	2.95	76	148	111
5	2.95	76	148	72
6	2.20	53	110	64
7	2.45	61	123	96
8	2.45	61	123	28
9	2.20	53	110	71
10	1.95	44	98	57

From Swanson, 1957, after Darlington and after Rhoades.

be observed. In humans, for example, Ford and Hamerton have scored a mean chiasma frequency of 56 per cell, or approximately two chiasmata per chromosome. This corresponds to about 100 recombinational units per chromosome, or a possible total map length of 2300 units.

The proposed correspondence between chiasmata and genetic crossover has also helped to explain certain other data. For example, the limitation in observed number of chiasmata per chromosome might help to account for genetic interference, the reduced double-crossover frequencies that are noted when distances between genes are small. Also, a lack of genetic interference between loci on opposite sides of the centromere that has been observed in corn and in *Drosophila* might be explained by the observation that chiasma interference across the centromere is not found in some organisms.

The chiasmatype theory has nevertheless had objections raised against it. One of the most serious of these are observations by Kaufmann and by Cooper demonstrating that chiasmata are formed in some tissues of male *Drosophila*, including spermatocytes. Since genetic crossing over is absent in male *Drosophila*, these chiasmata cannot be explained by the chiasmatype theory. However, Slizynski has indicated that the chiasmata observed in *Drosophila* males may really be surface associations rather than cytological exchanges. The chiasmatype theory therefore remains at present the most accepted explanation for the relationship between genetic crossing over and cytologically observed chiasmata.

THEORIES OF CHROMOSOMAL CROSSING OVER

It has proved quite difficult to learn how genetic material is actually exchanged between chromosomes. Darlington, the main proponent of the chiasmatype theory, proposed that exchanges arise because of attractive and repulsive forces that act between chromosomes during cell division. This theory, known as the "precocity" theory, suggests that the difference between mitosis and meiosis lies in the degree of chromosome replication that has taken place. When cell division occurs in mitosis, chromosome replication takes place before prophase, and the "pairing need" of each chromosome caused by cell division is satisfied by the newly replicated strands. In meiosis, however, prophase is "precocious," arising before chromosome replication has taken place. Thus the pairing need for each chromosome can only be satisfied by pairing or synapsis with its homologue.

As meiosis proceeds, the mechanical strains that

result from the separation of paired homologous chromosomes during pachytene are presumably accompanied by additional strains from the separation of the sister chromatids in each chromosome (Fig. 17–13). These strains and torsions then lead to breaks in identical positions of homologous chromatids, and to subsequent crossovers when the broken ends of nonsister chromatids unite to produce chiasmata. By the end of pachytene only repulsive forces are acting between homologous chromosomes and, according to Darlington, it is only the chiasmata which then hold the homologues together until the first meiotic anaphase. Darlington's theory therefore correlates close pairing between homologous chromosomes (generally observed during meiotic prophase) and crossing over. Thus, where pairing between homologues is interfered with, such as in chromosome structural changes (Chapter 22), crossing over is restricted.

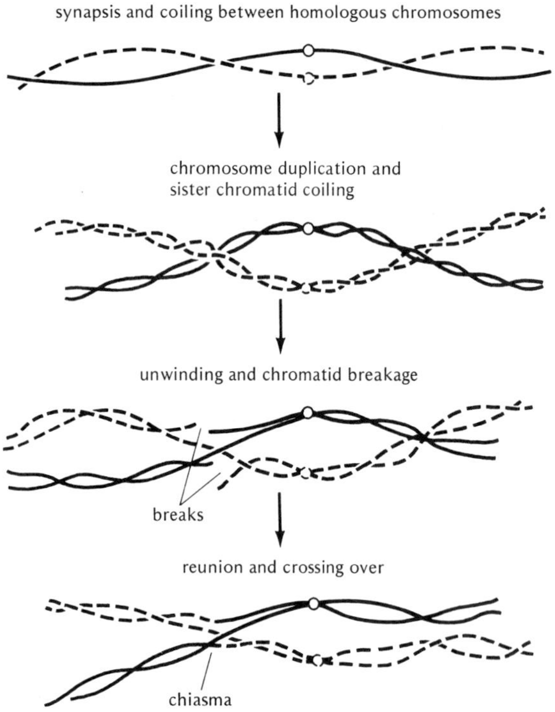

Figure 17–13

Diagram of Darlington's theory of chromosomal crossing over as a result of breaks caused by unwinding of chromosomes and chromatids. (After Darlington.)

We can see that in Darlington's theory, pairing between homologous chromosomes is necessary for normal disjunction between these homologues to take place at the first meiotic anaphase. If there is no pairing, the two homologues will not align themselves opposite each other on the metaphase plate and may consequently go to the same pole at anaphase, causing nondisjunction (see Chapter 12). Recently, Grell has pointed out that this aspect of Darlington's theory should be expanded (at least in *Drosophila*) to include two types of chromosome pairing during meiosis. One type, called *exchange pairing*, involves homologous pairing and consequent crossing over between homologues. The other type, called *distributive pairing*, is unaccompanied by crossing over. Chromosomes that undergo exchange pairing disjoin normally during meiosis, while those in which exchange pairing is absent are forced to rely upon distributive pairing for normal meiotic segregation. Distributive pairing, according to her view, may easily lead to nondisjunction if it occurs between nonhomologous chromosomes. For example, the two homologous 4th chromosomes of *Drosophila melanogaster* are very small and rarely undergo exchange pairing. They do, however, normally undergo distributive pairing and consequent normal disjunction. If a small extra fragment of the X chromosome containing its own centromere is now introduced, distributive pairing may occur between this fragment and one of the 4th chromosomes rather than between the two 4th chromosomes. When this takes place, the two 4th chromosomes may go to the same pole rather than disjoining normally. The mechanism of distributive pairing does not, therefore, appear to be caused by the usual mechanism of homologous attraction used in exchange pairing. Grell has hypothesized that chromosomes that are unable to undergo exchange pairing, such as the tiny 4th chromosomes and those bearing many structural changes ("nonexchange chromosomes"), undergo distributive pairing on the basis of their common size.

So far, at least one *Drosophila* mutation has been isolated which seems to interfere with the disjunction of chromosomes whose relationships

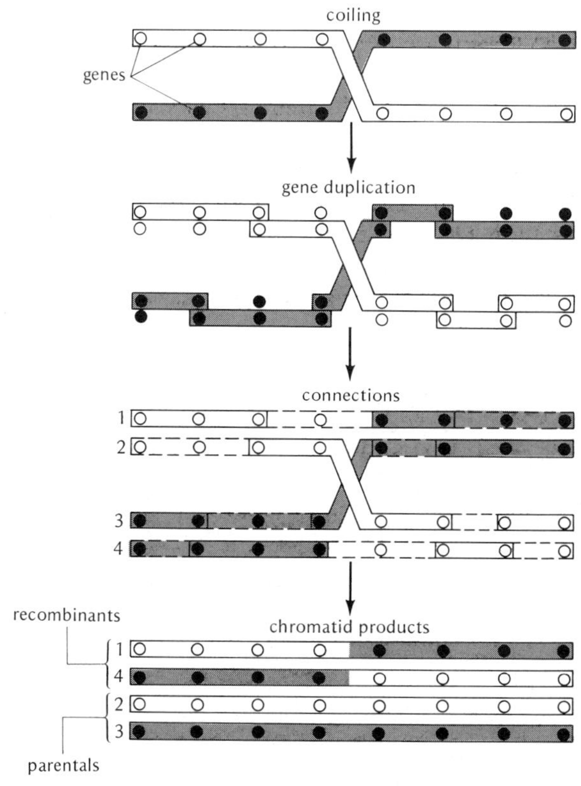

Figure 17–14

Diagram of Belling's theory of crossing over as a result of new connections made between duplicated genes on two separate chromosomes that have temporarily intercrossed. (After White.)

normally follow the distributive pattern. This mutant called *nod* (*no distributive disjunction*) has been shown by Carpenter to increase by more than 800-fold the frequency of gametes that are nondisjunctional for the 4th chromosome, and to significantly increase the nondisjunction of other nonexchange chromosomes. Since *nod* has no effect on recombination and on the subsequent process of exchange pairing and disjunction, its activity seems to support Grell's hypothesis for the existence of a separate distributive process. However, it remains to be demonstrated whether similar mechanisms operate in other higher organisms.

A somewhat different theory of chromosomal crossing over than Darlington's was proposed by Belling. According to Belling, breakage and reunion between chromosomes is not the cause for crossing over, but, rather, crossovers occur because of novel attachments that are formed between newly duplicated genes. According to this theory, chromosome duplication consists of two stages: (1) the replication of genes, and (2) the formation of new connections between these genes (Fig. 17–14). The first stage, gene duplication, occurs in pachytene and results in one daughter gene of each pair lacking connection to either one or both of its neighboring genes. Later, when gene interconnections are formed, the previously unconnected daughter genes are now connected to their closest neighbors. If homologous chromosomes intercross when these connections are formed, the new chains of genes will show crossovers. (A modified version of Belling's theory, called "copy choice," was espoused by various microbial geneticists and will be discussed in Chapter 20.)

Although both the Darlington and Belling theories are advantageous in explaining some of the observed facts, there are indications that neither of these theories is completely satisfactory. Objections raised against the Darlington theory include demonstrations by Hughes-Schrader that pairing among all four chromatids exists in the tetrads of certain insects without any observable chiasmata. Objectors to Belling's theory point out that there is no evidence for a two-stage formation of chromosomes and considerable evidence that the genetic material of chromosomes has a continuous linear organization.

Chromosome-labeling experiments by Taylor bear directly on both of these theories, since autoradiography (p. 62) permits the incorporation of labeled DNA to be correlated with its later distribution in chromosomes during cell division. From studies mentioned previously (p. 92), Taylor and co-workers had demonstrated that chromosomes and chromatids divide as though they were "double" structures composed of two subunits. Furthermore, in the grasshopper *Romalea*, physical exchanges marked by tritium-labeled DNA were shown to occur during meiosis between the chromatid of one homologue and the chromatid of another. Later analysis indicated that the number of such DNA exchanges was not significantly different from the number of chiasmata observed. Interestingly, in experiments with root cells of the plant *Bellevalia*, Taylor has shown that exchanges between one chromatid and another involve the exchange of both subunits in each, and each of the subunits seem to be oriented differently from each other (i.e., ⇄).

These results therefore are in good accord with semiconservative replication of the chromosome at the DNA level and with the possibility that a chromosome consists of a single double helix of DNA with strands of opposite polarity. Although results that were not in agreement with those of Taylor were obtained by La Cour and Pelc, most of the recent data support Taylor's conclusions (see Peacock, 1973), and also indicate a numerical correspondence between the number of chiasmata observed and the number of exchanges of radioactive label.

CROSSING OVER ON THE MOLECULAR LEVEL

In recent years, the events associated with recombination in eucaryotes have been investigated with considerable success in the lily (*Lilium*) by H. Stern and his colleagues. Microsporocytes in the anthers of these plants are synchronized to go through each prophase stage for an extended period of two to five days. This provides large numbers of cells all at the same meiotic stage, which have the additional advantage that they can be grown *in vitro* (explanted) on a nutrient medium and analyzed biochemically. These experiments and other indicate, so far, the following pattern of meiotic events:

1. The decision on whether the cells of plants such as *Lilium* are going to enter meiosis and undergo subsequent pairing and recombination appears to be made at some point in the early leptotene prophase stage. Microsporocytes that are explanted into nutrient medium before this stage go into mitosis rather than meiosis, and show neither chiasmata nor other accompaniments of close pairing. Only cells explanted in the early leptotene stage seem to be committed to undergo meiosis. At this stage in *Lilium*, DNA synthesis is not detectable.

2. Upon entering into the zygotene stage, a small amount of DNA synthesis occurs in *Lilium* to the extent of 0.3 percent of the nuclear DNA. Interestingly, if this DNA synthesis is inhibited, chromosome pairing is prevented, and the chromosomes may be fragmented. This finding indicates the important role that the zygotene-synthesized DNA plays in aligning homologous chromosomes and the probably scattered distribution of this DNA along the entire length of the chromosome.

3. Associated with the zygotene stage is also the appearance of the synaptinemal complex (see p. 20). This complex may originate at various points for any particular pair of homologues, but rapidly extends to form its characteristic pattern of lateral and central elements along the length of the bivalent. The importance of this complex to recombinational events is evident from the fact that *Drosophila melanogaster* females homozygous for the *c3G* gene which suppresses genetic crossing over (see p. 347) lack the synaptinemal complex, nor can it be found in *Drosophila* males in which crossing over is absent.

4. Protein synthesis at the end of the zygotene stage also appears to be necessary for the formation of chiasmata, since chiasmata are eliminated from cells subjected to inhibitors of protein synthesis or to elevated temperatures that change the molecular configuration of proteins. It has been suggested that a protein involved in this process is one known to bind preferentially to denatured single-stranded DNA molecules and to catalyze their renaturation into double-stranded helixes. This "binding protein" isolated from *Lilium* is remarkably similar to the "gene 32 protein" necessary for genetic recombination in T4 bacteriophage (Chapter 20), and to a nuclear protein isolated from the early prophase stages of various mammalian cells, including those of humans. Therefore at least part of the recombination mechanism appears to be alike in both procaryotes and eucaryotes.

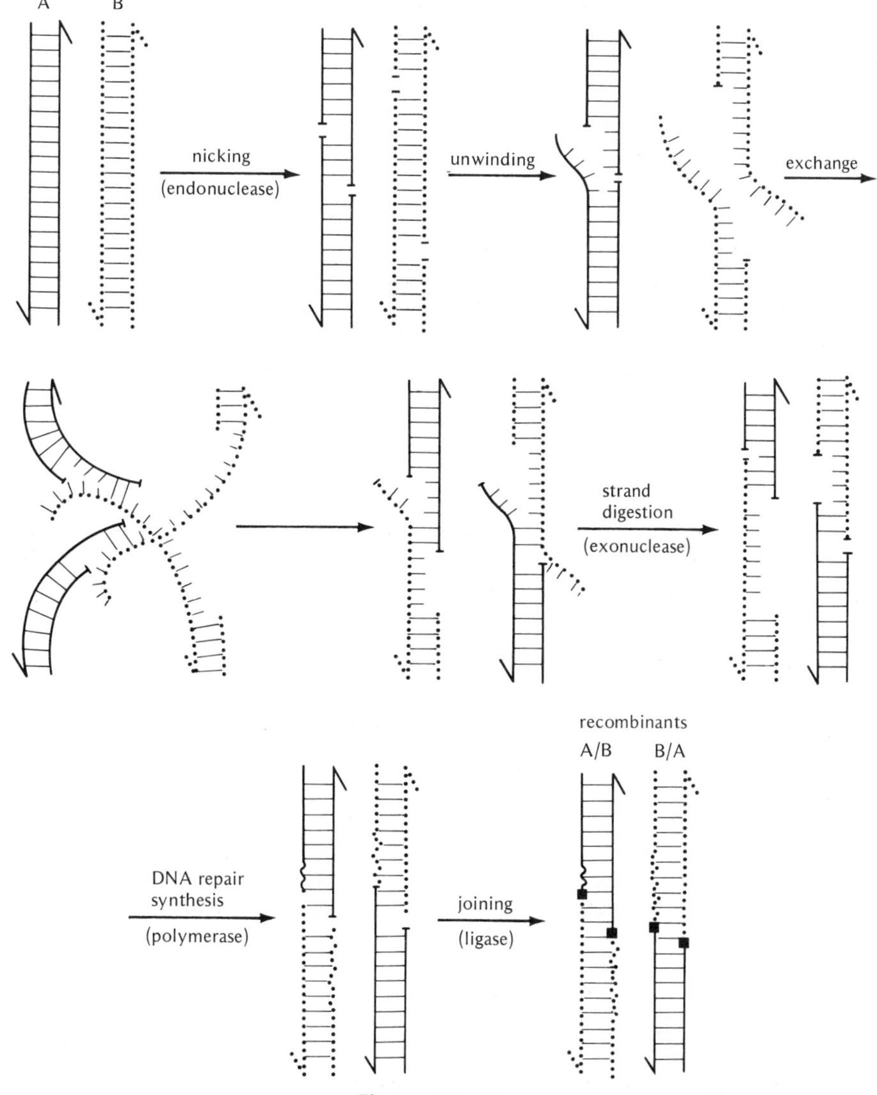

Figure 17–15

This model represents in diagrammatic fashion one possible sequence of events leading to recombination during the late zygotene-early pachytene stages of meiosis. A and B represent individual DNA double helixes, each from one homologous pair of chromosomes. The cross bars between strands represent many complementary nucleotide base pairs, and the half arrow at the end of each DNA strand indicates the polarity of the sugar-phosphate backbone. In this sequence, pairing of homologous chromosomes is followed by "nicking" of each DNA strand in the two paired double helixes. This allows unwinding and rotation of each helix leading to a substitutional exchange of strands of the same polarity. The new recombinational complex formed by this exchange contains a short section of "hybrid" double-stranded DNA, perhaps 100 or more nucleotides long, composed of one complementary DNA strand from each parent. Unpaired DNA ends are digested by an exonuclease enzyme, and unpaired gaps are filled in with complementary base-paired sections synthesized by a polymerase enzyme. A ligase enzyme then joins the broken strands into continuous covalently bonded nucleotide chains. In addition to the nicks pictured above, there are probably numerous others along the chromosome which are unaccompanied by exchanges yet where single DNA strands are partially digested and new DNA is synthesized. Combined together, the DNA synthesized in exchanged and unexchanged areas is believed to account for the DNA synthesis observed at this stage of meiosis. (Modified after models of Whitehouse, Holliday, Howell and Stern, and others.)

5. Genetic crossing over is associated with an actual breakage of "parental" DNA strands from homologous chromosomes and their subsequent reunion into recombinant products. Although the best evidence for this derives from experiments with virus (Chapter 20), the works of Taylor, Peacock, Jones, and others have provided good reason to believe that DNA breakage and reunion is also responsible for genetic recombination in eucaryotes. In fact, Howell and Stern have recently shown that an enzyme system appears during zygotene and pachytene of meiotic prophase which is fully capable of performing the tasks necessary for breakage and reunion of DNA. Included in this system is an endonuclease that can produce nicks in single strands of DNA (see p. 87) and a ligase that can rejoin broken DNA strands (see p. 90). It is possible that the alignment of single strands of DNA from different homologous chromatids into new recombinant double-stranded molecules is accomplished with the aid of the binding protein mentioned in step 4 above.

6. During the pachytene stage a small burst of DNA synthesis occurs which seems to be of the type involved in breakage and reunion. That is, there is no net increase of DNA but only the "repair" of DNA sections that may have been degraded during the breakage events that occurred upon the appearance of the endonuclease enzyme at zygotene. Some of these repair activities may be accompanied by or lead to exchanges of DNA molecules, and result in genetic crossing over.

A conjectured scheme that takes into account some of these biochemical findings is diagrammed in Fig. 17–15. It is based on the breakage of parental DNA strands and their subsequent pairing over a short distance to form "hybrid" or "heteroduplex" DNA molecules containing one strand from each parent. Note that the recombinant strands produced by such breakage contain equal amounts of the two parental sequences; that is, recombination has been "reciprocal" (p. 351). This type of recombination can be achieved in various ways only one of which is diagrammed: (1) enzymatic "nicking" of each of the four recombining DNA strands; (2) enzymatic digestion of each DNA strand followed by DNA synthesis in the resulting gaps. Although it is not clear at present which methods are used in eucaryotic cells, eucaryotes do show mostly reciprocal recombinant products. In those instances where nonreciprocal genetic products are also found (see "conversion," Chapter 18), they can be explained as arising from occasional unequal exchanges on the molecular level within sections that normally produce reciprocal exchanges (see Fig. 18–8).

PROBLEMS

17-1. Can you distinguish between two gene loci located on the same chromosome that have 50 percent crossing over between them and two gene loci each located on different nonhomologous chromosomes? Explain.

17-2. Stadler crossed a stock of corn homozygous for the linked recessive genes *colorless* (*c*), *shrunken* (*sh*), and *waxy* (*wx*) to a stock homozygous for the dominant alleles of these genes (*C Sh Wx*) and then backcrossed F_1 plants to the homozygous recessive stock. The progeny were as follows:

Phenotype	Number	Phenotype	Number
C Sh Wx	17,959	*C Sh wx*	4,455
c sh wx	17,699	*c sh Wx*	4,654
C sh wx	509	*C sh Wx*	20
c Sh Wx	524	*c Sh wx*	12

(a) Draw a linkage map for the three genes. (b) Calculate the degrees of coincidence and interference with respect to double crossovers.

17-3. In corn a strain homozygous for the recessive genes *a* (*green*), *d* (*dwarf*), and *rg* (*normal leaves*) was crossed to a strain homozygous for their dominant alleles *A* (*red*), *D* (*tall*), and *Rg* (*ragged leaves*). Offspring of this cross were then backcrossed to homozygous recessive plants, *green dwarf normal leaves*. Listed below are the phenotypes produced from the backcross:

Phenotype	Number	Phenotype	Number
A D Rg	265	*A d rg*	90
a d rg	275	*a D Rg*	70
A D rg	24	*A d Rg*	120
a d Rg	16	*a D rg*	140

(a) Which of the above classes represent crossovers between *a* and *d*? *d* and *rg*? *a* and *rg*? (b) Propose a linkage map with distances between the three genes. (c) What is the coefficient of coincidence here? (d) What degree of interference occurred in this experiment? (e) If interference were increased in this experiment, what change would you expect in the results? (f) If an *a Rg*/*A rg* plant were crossed with an *a rg*/*a rg*, what *phenotypes* and *ratios* would you expect among their offspring?

17-4. In the tomato the mutant genes *o* (*oblate* = flattened fruit), *p* (*peach* = hairy fruit), and *s* (*compound inflorescence* = many flowers in a cluster) were found to be in chromosome 2. From the following data (testcross mating of an F_1 heterozygote for all three genes × homozygous recessive for all three genes) determine: (a) The sequence of these three genes in chromosome 2. (b) The genotypes of the homozygous parents used in making the F_1 heterozygote. (c) The recombination distances between the genes. (d) The coefficient of coincidence.

Phenotypes of testcross progeny	Number
+ + +	73
+ + *s*	348
+ *p* +	2
+ *p s*	96
o + +	110
o + *s*	2
o p +	306
o p s	63

17-5. Gregory found the following linkage distances between four genes (*B*, *G*, *S*, *L*) on one of the chromosomes of the Chinese primrose, *Primula sinensis*: *S-B*, 7.6; *S-G*, 33.5; *S-L*, 37.0; *B-G*, 31.0; *B-L*, 35.7; *G-L*, 3.3. Using these data construct a linkage map.

17-6. Two genes, *A* and *B*, known to be linked together on the same chromosome, are tested for recombination in a cross *A B*/*a b* × *a b*/*a b*. The phenotypes of the offspring of this cross appear in the ratio 28 *A B*:32 *a b*:23 *A b*:17 *a B*. (a) On the basis of this ratio, what is the recombination distance between the two genes? (b) If the actual map distance between the two genes is 80, what percentage of double crossovers probably occurred in the above cross? (c) Which phenotypes of the above offspring are carrying these double crossovers? (d) If triple crossovers also occurred, which phenotypes of the above offspring would be carrying them?

17-7. In corn three gene pairs linked on the same chromosome are those for plant color, in which *yellow* (*y*) is recessive to green (*Y*); endosperm shape, in which *shrunken* (*sh*) is recessive to full (*Sh*); and seed color, in which *colorless* (*c*) is recessive to colored (*C*). Three different plants, I, II, and III, each heterozygous for these three gene loci, although not necessarily heterozygous in the same fashion as the others (i.e., *y sh c*/*Y Sh C*, *Y Sh c*/*y sh C*, etc.), were backcrossed to homozygous triple recessives, *y sh c*/*y sh c*, and produced progeny which had the following phenotypic frequencies:

Phenotype of offspring	Numbers Produced by Heterozygous Plants		
	I	*II*	*III*
y sh c	95	368	22
Y Sh C	100	387	23
y Sh C	3	21	390
Y sh c	2	24	365
y sh C	20	4	96
Y Sh c	25	1	99
y Sh c	375	98	0
Y sh C	380	97	5
	1,000	1,000	1,000

(a) Using the progeny from any one of the heterozygous plants, propose the gene order and the recombination distances between the three genes. (b) On the basis of (a), compute the coefficient of coincidence. (c) For the plant analyzed in (a) and (b), describe the frequency changes you would expect among its progeny if interference were increased. (d) Give the exact genotypes for each of the two homologous chromosomes bearing the three genes in heterozygous plants, I, II, and III.

17-8. Three linked genes, *a*, *b*, *c*, occupy the following linkage positions on a particular chromosome:

gene:	*a*	*b*	*c*
position:	0	10	30

Assuming that double crossovers do not occur within the *a-b* region or within the *b-c* region, but that double crossovers may occur without interference between *a* and *c*, what proportions of germ cells going through meiosis are expected: (a) To show two chiasmata in the *a-b-c* region, one between *a* and *b*, the other between *b* and *c*? (b) To show a single chiasma in the *a-b-c* region, between *a* and *b*? (c) To show no chiasmata at all in the *a-b-c* region? (To answer these questions, first determine the relationship between chiasmata and recombination frequencies.)

17-9. Solve Problem 8 assuming a coincidence coefficient of .5.

17-10. If the double-crossover frequency for the genes in Problem 8 shows a coincidence coefficient of .5, what phenotypic ratios would you expect to find in the F_2 progeny produced by the following crosses: (a) $P_1: a\,c/a\,c \times + +/+ + \rightarrow F_1 \times a\,c/a\,c \rightarrow F_2$? (b) $P_1: a + /a + \times + c/ + c \rightarrow F_1 \times a\,c/a\,c \rightarrow F_2$?

17-11. In *Drosophila, Dichaete* is a 3rd-chromosome mutation with dominant effect on wing shape, and *pink* and *ebony* are 3rd-chromosome recessive mutations that affect, respectively, eye color and body color. Flies from a *Dichaete* stock were crossed to homozygous *pink ebony* flies, and the F_1 flies with dichaete phenotype were backcrossed to the *pink ebony* homozygotes. The results were as follows:

Phenotypes	Number
dichaete	824
pink ebony	878
dichaete ebony	214
pink	187
dichaete pink	11
ebony	18
dichaete pink ebony	28
wild type	31

(a) What are the recombination distances and linkage order between each of these genes? (b) What is the coefficient of coincidence? (c) On the basis of this coincidence coefficient and these linkage distances, how many double crossovers would you expect if these same crosses produced 1000 offspring?

17-12. Stern and Bridges crossed a stock carrying the dominant eye mutation *Star* on the 2nd chromosome to a stock homozygous for the 2nd-chromosome recessive mutations *aristaless* and *dumpy*. The F_1 *Star* females were then backcrossed to homozygous *aristaless dumpy* males and the following phenotypes observed:

Phenotype	Number
aristaless dumpy	918
star	956
aristaless star	7
dumpy	5
aristaless	132
star dumpy	100

(a) What are the recombination distances and the order of loci for these three genes? (b) What classes of phenotypes are missing, and why?

17-13. If three recessive mutations, 10 map units apart, are linked in the order *a-b-c*, and their wild-type alleles are dominant (designated as +), which *phenotypes* would you expect to be least frequent among the progeny of crosses: (a) $a\,b\,c/+ + + \times$ homozygous triple recessive? (b) $a + c/+ b + \times$ homozygous triple recessive? (c) $a\,b +/+ + c \times$ homozygous triple recessive? (d) $+ b\,c/a + + \times$ homozygous triple recessive?

17-14. Answer questions (a) to (d) in Problem 13 for each of the following conditions: (I) All three mutations are dominant, i.e., *A B C*, and their wild-type alleles are recessive. (II) Only the first mutation is dominant, i.e., *A*, and its wild-type allele is recessive. (III) Only the center mutation is dominant, i.e., *B*, and its wild-type allele is recessive. (IV) Only the first two mutations are dominant, i.e., *A B*, and their wild-type alleles are recessive. (V) Only the two outside mutations are dominant, i.e., *A C*, and their wild-type alleles are recessive.

17-15. In the same experiment described in Problem 3, Chapter 16, Grell, Jacobson, and Murphy crossed wild-type flies bearing the Adh^S enzyme to Adh^F flies homozygous for a number of linked recessive genes (*b*, *el*, *rd*, and *pr*). The F_1 heterozygous females were then backcrossed to males homozygous for Adh^F, *b*, *el*, *rd*, and *pr*. The three regions, *b-el*, *el-rd*, and *rd-pr*, were analyzed for single crossovers among the backcross progeny, and the following associations were found between recombinants and the presence of *Adh* alleles:

Adh allele	Recombinant	Number
Adh^S	*b + + +*	10
Adh^F	*b + + +*	0
Adh^S	*+ el rd pr*	0
Adh^F	*+ el rd pr*	6
Adh^S	*b el + +*	3
Adh^F	*b el + +*	25
Adh^S	*+ + rd pr*	17
Adh^F	*+ + rd pr*	2
Adh^S	*b el rd +*	0
Adh^F	*b el rd +*	5
Adh^S	*+ + + pr*	5
Adh^F	*+ + + pr*	0

(a) In which of the three regions is *Adh* located? (b) If the map locations of the four recessive mutations are $b = 48.5$, $el = 50.0$, $rd = 51.0$, $pr = 54.5$, where is the exact location of *Adh*?

17–16. A student in a genetics laboratory class discovered a sex-linked recessive lethal gene, *l*, which was kept in heterozygous condition with the Muller-5 chromosome carrying the semidominant gene *Bar* and a "crossover suppressor" that prevented *l* from transferring to the Muller-5 chromosome. Aside from its lethal effect in males and in homozygous females, the chromosome carrying *l* appeared to be of the wild type in all other respects. To determine the location of *l*, the student crossed a Muller-5/*l* female to a male carrying the sex-linked recessive mutations *scute* (*sc*), *crossveinless* (*cv*), *vermilion* (*v*), and *forked* (*f*), and then crossed the F_1 *l*/*sc cv v f* females having the wild-type phenotype to their Muller-5 brothers. Of the sixteen possible classes of F_2 males, the numbers scored were as follows:

	Number		Number
$sc\ cv\ v\ f$	191	$sc\ cv\ v^+ f^+$	0
$sc^+ cv^+ v^+ f^+$	0	$sc^+ cv^+ v\ f$	26
$sc\ cv\ v\ f^+$	43	$sc\ cv^+ v\ f^+$	1
$sc^+ cv^+ v^+ f$	20	$sc^+ cv\ v^+ f$	0
$sc\ cv\ v^+ f$	0	$sc\ cv^+ v^+ f$	6
$sc^+ cv^+ v\ f^+$	7	$sc^+ cv\ v\ f^+$	11
$sc\ cv^+ v\ f$	2	$sc\ cv^+ v^+ f^+$	0
$sc^+ cv\ v^+ f^+$	0	$sc^+ cv\ v\ f$	37

(a) Determine in which gene interval *l* is located. (b) Using the linkage map positions for *sc*, *cv*, *v*, and *f* in Fig. 17–3, give the exact linkage position of *l*.

17–17. The location of another sex-linked recessive lethal gene was tested by mating females heterozygous for this gene to males carrying the sex-linked recessives *ruby* (*rb*) and *cut* (*ct*). F_1 females of the constitution *l*/*rb ct* were then crossed to *ruby cut* males and the following percentages of phenotypes observed:

Phenotypes	Male	Female
wild type	3.90	43.75
ruby	12.40	6.25
cut	.10	6.25
ruby cut	83.60	43.75
	100.00	100.00

(a) What is the linkage order of the three genes? (b) What is the numerical locus of the lethal gene? (c) What is the coefficient of coincidence and degree of interference in this cross?

17–18. The following is a linkage map for three recessive genes located on the same chromosome in a certain species, measured in percent recombination frequencies:

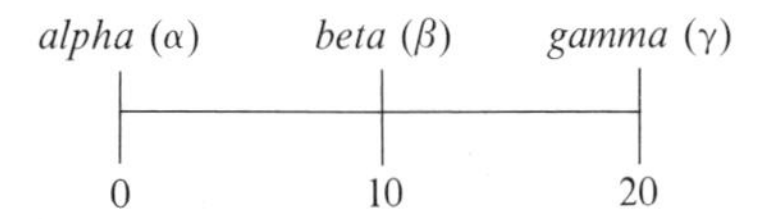

If the coefficient of coincidence is .60 in this case, determine the frequencies of phenotypes expected among *1000* offspring of a cross $\alpha\,\beta\,\gamma/+\,+\,+ \times \alpha\,\beta\,\gamma/\alpha\,\beta\,\gamma$.

17–19. Solve Problem 18 if the cross made is

$+\,\beta\,\gamma/\alpha\,+\,+ \times \alpha\,\beta\,\gamma/\alpha\,\beta\,\gamma$

17–20. A stock homozygous for four different recessive mutations *a*, *b*, *c*, and *d*, linked together on one chromosome (but not necessarily in that order), was mated to a stock homozygous for their dominant wild-type alleles. F_1 females were then backcrossed to parental homozygous recessive males, and 2000 of the offspring were scored phenotypically as follows:

Phenotype	Number	Phenotype	Number
a b c d	668	+ *b* + +	1
+ + + +	689	*a* + *c* +	69
a b c +	97	+ *b* + *d*	76
+ + + *d*	98	*a b* + +	1
a b + *d*	5	+ + *c d*	1
+ + *c* +	5	*a* + + +	145
a + *c d*	2	+ *b c d*	143

(a) determine the recombination distances between the four genes and draw their positions on a linkage map. (b) What is the degree of interference between the genes in the first and third positions? (c) Between the genes in the second and fourth positions? (d) Are the number of triple crossovers equal to those expected?

17–21. Considering that man has 22 pairs of autosomes, what are the chances that any two different autosomal traits chosen randomly will be linked?

17–22. The two human pedigrees at top of p. 362 show the transmission of two sex-linked recessive traits. (a) Indicate each individual whose genotype has unquestionably been produced by recombination between the *a* and *b* loci. (b) Based on the number of recombinants you have scored, what is the percentage

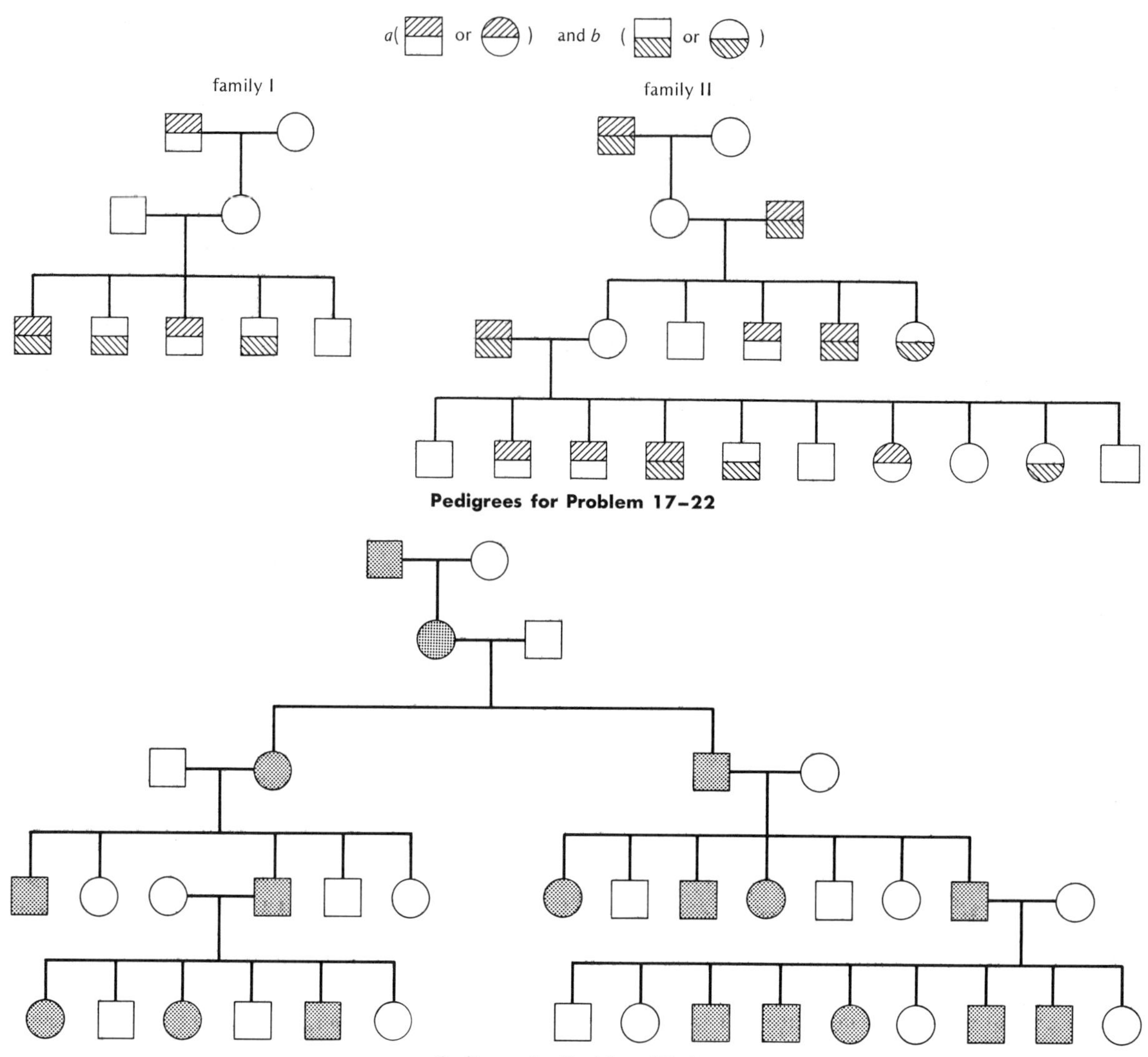

Pedigrees for Problem 17-22

Pedigree for Problem 17-23

of recombination between genes *a* and *b* in each pedigree? (c) In both pedigrees combined?

17-23. The common pairing segment of the X and Y chromosomes in certain species such as man suggests, at least theoretically, the possibility of genetic crossing over between these two chromosomes. In the above pedigree of a sex-linked dominant trait (shaded symbols) which individuals would you suggest might have arisen as a result of X-Y crossovers?

17-24. Two types of sex-linked colorblindness in man are distinguished in the pedigree diagram by hatched

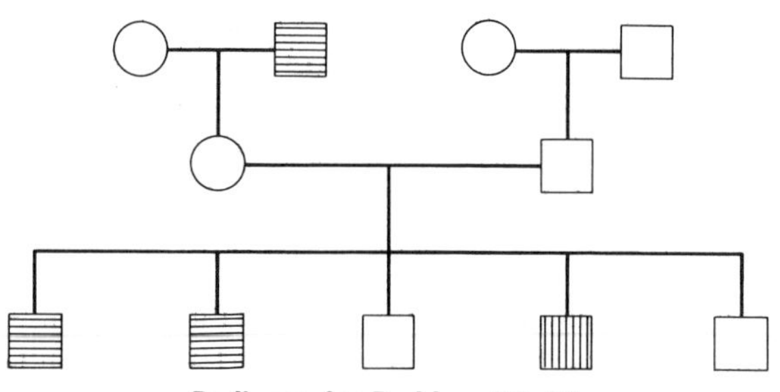

Pedigree for Problem 17-24

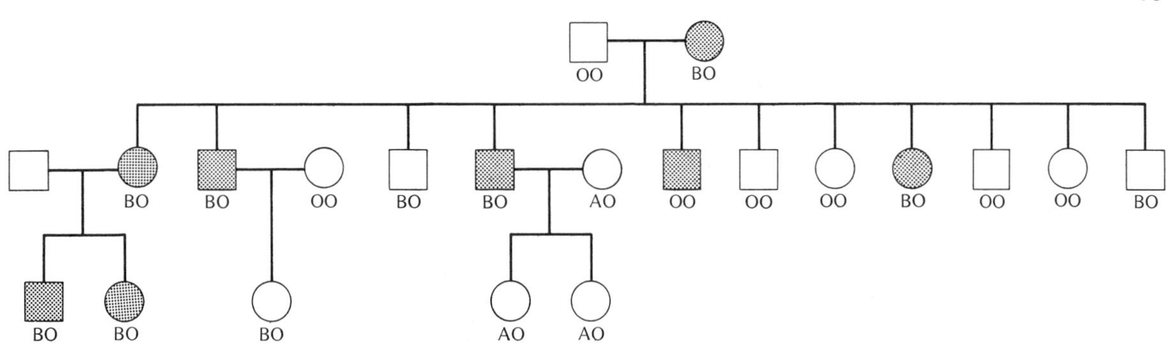

Pedigree for Problem 17–25

symbols: horizontal hatching ▤ = type A, vertical hatching ▥ = type B. (Females in this pedigree showed neither type.)

If the mother of the color blind sons (II–1) is considered to be heterozygous for both types, explain whether this pedigree indicates that the two genes are allelic to each other, or whether they are at separate loci. (Modified from Vanderdonck and Verriest.)

17–25. The above human pedigree shows the inheritance of the rare nail-patella syndrome, which causes misshapen nails and kneecaps (affected individuals are shaded symbols). In this pedigree, the genotype for the ABO blood group is given beneath the symbol of each person. (After L. S. Penrose, 1963. *Outline of Human Genetics,* 2nd ed. John Wiley, New York; from Renwick and Lawler.)

(a) If the grandfather of this family (I–1) is not a carrier of the nail-patella syndrome, indicate whether you believe this trait is dominant or recessive, and whether it is sex-linked or autosomal. (b) On the basis of this pedigree, would you infer that there is linkage between the gene for ABO blood group and the gene for nail-patella syndrome? (c) If you believe that linkage exists, draw the linked genes on each of the two homologous chromosomes of both grandparents of the family (I–1, I–2). (d) According to your proposed linkage scheme, which offspring represent crossovers? (e) What is the recombination frequency between the two genes? (f) Explain whether this pedigree indicates that crossing over can occur in human males.

17–26. If individual III–1 in the pedigree of Problem 25 mated a normal female of O blood type, what is the probability that their first offspring will be: (a) Of B blood type with nail-patella syndrome? (b) Of O blood type with nail-patella syndrome?

17–27. A normal woman whose father was normal, married to a colorblind man who is otherwise normal, produces one son who is colorblind and has hemophilia A, one daughter who is colorblind, and one daughter who is normal. If all of the colorblind individuals in this family are of the deutan type, use the linkage map for these genes (Fig. 17–8) to determine: (a) The probability that the colorblind daughter may have hemophilic sons. (b) The probability that the normal daughter may have hemophilic sons.

17–28. For the pedigree shown in Problem 23, Chapter 12, which, if any, individuals have genotypes that have unquestionably been produced by recombination between the genes for colorblindness, g-6-pd deficiency, and Xg blood type?

17–29. Graham and co-workers discovered the following pedigree, in which a woman bearing the sex-linked genes below gave birth to five male offspring. Since the woman's parents could not be examined, the coupling relationships between the genes on her X chromosomes were not known. Nevertheless the investigators hypothesized that a double crossover must have occurred. Why?

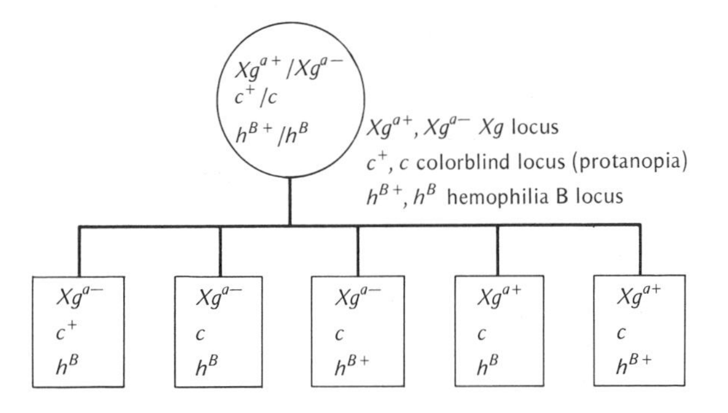

REFERENCES

Barski, G., S. Sorieul, and F. Cornefert, 1960. Production dans des cultures *in vitro* de deux souches cellulaires en association, de cellules de caractere "hybride." *C. R. Acad. Sci., Paris,* **251,** 1825–1827.

Beadle, G. W., 1932. The relation of crossing-over to chromosome association in Zea-Euchlaena hybrids. *Genetics,* **17,** 481–501.

Belling, J., 1933. Crossing-over and gene rearrangement in flowering plants. *Genetics,* **18,** 388–413.

Bridges, C. B., and T. M. Olbrycht, 1926. The multiple stock "Xple" and its use. *Genetics.* **11,** 41–56.

Brown, S. W., and D. Zohary, 1955. The relationship of chiasmata and crossing over in *Lilium formosanum. Genetics,* **40,** 850–873.

Carlson, P. S., H. H. Smith, and R. D. Dearing, 1972. Parasexual interspecific plant hybridization. *Proc. Nat. Acad. Sci.,* **69,** 2292–2294.

Carpenter, A. T. C., 1973. A meiotic mutant defective in distributive disjunction in *Drosophila melanogaster. Genetics,* **73,** 393–428.

Comings, D. E., and T. A. Okada, 1972. Architecture of meiotic cells and mechanisms of chromosome pairing. *Adv. in Cell and Mol. Biol.,* **2,** 309–384.

Cooper, K. W., 1949. The cytogenetics of meiosis in *Drosophila.* Mitotic and meiotic autosomal chiasmata without crossing-over in the male. *Jour. Morph.,* **84,** 81–122.

Creighton, H. B. and B. McClintock, 1931. A correlation of cytological and genetical crossing over in *Zea mays. Proc. Nat. Acad. Sci.,* **17,** 492–497. (Reprinted in the collections of Gabriel and Fogel, of Levine, of Peters, and of Taylor; see References, Chapter 1.)

Darlington, C. D., 1932. *Recent Advances in Cytology.* Churchill, London.

———, 1958. *The Evolution of Genetic Systems.* Basic Books, New York.

Ephrussi, B., 1972. *Hybridization of Somatic Cells.* Princeton Univ. Press, Princeton.

Ephrussi, B., and M. C. Weiss, 1965. Interspecific hybridization of somatic cells. *Proc. Nat. Acad. Sci.,* **53,** 1040–1042.

Ford, C. F., and J. L. Hamerton, 1956. The chromosomes of man. *Nature,* **178,** 1020–1023.

Gowen, M. S., and J. W. Gowen, 1922. Complete linkage in *Drosophila melanogaster. Amer. Naturalist,* **56,** 286–288.

Green, M. C., 1963. Methods for testing linkage. In *Methodology in Mammalian Genetics,* W. J. Burdette (ed.). Holden-Day, San Francisco, pp. 56–82.

———, 1966. Mutant genes and linkages. In *Biology of the Laboratory Mouse,* 2nd ed., E. L. Green (ed.). McGraw-Hill, New York, pp. 87–150.

Grell, R. F., 1964. Distributive pairing; the size-dependent mechanism responsible for the regular segregation of the fourth chromosomes in *Drosophila melanogaster. Proc. Nat. Acad. Sci.,* **52,** 227–232.

———, 1969. Meiotic and somatic pairing. In *Genetic Organization,* Vol. I, E. W. Caspari and A. W. Ravin (eds.). Academic Press, New York, pp. 361–492.

Harris, H., and J. Watkins, 1965. Hybrid cells derived from mouse and man: Artificial heterocaryons of mammalian cells from different species. *Nature,* **205,** 640–646.

Hinton, C. W., 1966. Enhancement of recombination associated with the *c3G* mutant of *Drosophila melanogaster. Genetics,* **53,** 157–164.

Hotta, Y., M. Ito, and H. Stern, 1966. Synthesis of DNA during meiosis. *Proc. Nat. Acad. Sci.,* **56,** 1184–1191.

Howell, S. H., and H. Stern, 1971. The appearance of DNA breakage and repair activities in the synchronous meiotic cycle of *Lilium. Jour. Mol. Biol.,* **55,** 357–378.

Hughes-Schrader, S., 1943. Meiosis without chiasmata in diploid and tetraploid spermatocytes of the mantid *Callimantis antillarum* Saussare. *Jour. Morph.,* **73,** 111–141.

Janssens, F. A., 1909. Spermatogénèse dans les Batraciens. V. La théorie de la chiasmatypie. Nouvelles interprétation des cinèses de maturation. *Cellule,* **25,** 387–411.

Jones, G. H., 1971. The analysis of exchanges in tritium-labeled meiotic chromosomes. II. *Stethophyme grossum. Chromosoma,* **34,** 367–382.

Kaufmann, B. P., 1934. Somatic mitoses in *Drosophila melanogaster. Jour. Morph.,* **56,** 125–155.

Kushev, V. V., 1974. *Mechanisms of Genetic Recombination.* Consultants Bureau, New York.

LaCour, L. F., and S. R. Pelc, 1958. Effect of colchicine on the utilization of labelled thymidine during chromosomal reproduction. *Nature,* **182,** 506–508.

Lawrence, M. J., 1958. Genotypic control of crossing over on the first chromosome of *Drosophila melanogaster. Nature,* **182,** 889–890.

LEVINE, R. P., 1955. Chromosome structure and the mechanism of crossing over. *Proc. Nat. Acad. Sci.*, **41,** 727–730.

LEVINE, R. P., and E. E. LEVINE, 1955. Variable crossing over arising in different strains of *Drosophila pseudoobscura. Genetics*, **40,** 399–405.

LINDSLEY, D. L., and E. H. GRELL, 1968. *Genetic Variations of Drosophila melanogaster. Carneg. Inst. Wash. Publ. No. 627.* Washington, D.C.

MIGEON, B. R., and C. S. MILLER, 1968. Human-mouse somatic cell hybrids with single human chromosome (Group E): link with thymidine kinase activity. *Science*, **162,** 1005–1006.

MULLER, H. J., 1916. The mechanism of crossing over. II. *Amer. Naturalist*, **50,** 284–305.

NEEL, J. V., 1941. A relation between larval nutrition and the frequency of crossing over in the third chromosome of *Drosophila melanogaster. Genetics*, **26,** 506–516.

PEACOCK, W. J., 1971. Cytogenetic aspects of the mechanism of recombination in higher organisms. *Stadler Genet. Symp.*, **2,** 123–152.

———, 1973. Chromosome structure and units of function in higher organisms. In *The Biochemistry of Gene Expression in Higher Organisms*, J. K. Pollak and J. Wilson (eds.). D. Reidel, Dordrecht, pp. 3–20.

PERKINS, D. D., 1962. Crossing over and interference in a multiply marked chromosome arm of *Neurospora. Genetics*, **47,** 1253–1274.

PLOUGH, H. H., 1921. Further studies on the effect of temperature on crossing-over. *Jour. Exp. Zool.*, **32,** 187–202.

RENWICK, J. H., 1971. The mapping of human chromosomes. *Ann. Rev. Genet.*, **5,** 81–120.

RUDDLE, F. H., 1973. Linkage analysis in man by somatic cell genetics. *Nature*, **242,** 165–169.

SANDLER, L., D. L. LINDSLEY, B. NICOLETTI, and G. TRIPPA, 1968. Mutants affecting meiosis in natural populations of *Drosophila melanogaster. Genetics*, **60,** 525–558.

SCHULTZ, J., and H. REDFIELD, 1951. Interchromosomal effect on crossing over in Drosophila. *Cold. Sp. Harb. Symp.*, **16,** 175–197.

SLIZYNSKI, B. M., 1964. Chiasmata in spermatocytes of *Drosophila melanogaster. Genet. Res.*, **5,** 80–84.

STERN, C., 1931. Zytologisch-genetische Untersuchungen als Beweise Für die Morgansche Theorie des Faktorenaustauchs. *Biol. Zentralbl.*, **51,** 547–587.

STERN, H., and Y. HOTTA, 1973. Biochemical control of meiosis. *Ann. Rev. Genet.* **7,** 37–66.

STURTEVANT, A. H., 1913. The linear arrangement of six sex-linked factors in *Drosophila*, as shown by their mode of association. *Jour. Exp. Zool.*, **14,** 43–59. The first paper on gene mapping. (Reprinted in the collections of Peters, of Taylor, and of Voeller; see References, Chapter 1.)

SUZUKI, D. T., 1965a. Effects of actinomycin D on crossing over in *Drosophila melanogaster.* Genetics, **51,** 11–21.

———, 1965b. Effects of mitomycin C on crossing over in *Drosophila melanogaster. Genetics*, **51,** 635–640.

SWANSON, C. P., 1957. *Cytology and Cytogenetics.* Prentice-Hall, Englewood Cliffs, N.J., Chap. 8.

TAYLOR, J. H., 1958. The organization and duplication of genetic material. *Proc. 10th Intern. Congr. Genet.*, **1,** 63–78.

———, 1969. The structure and duplication of chromosomes. In *Genetic Organization*, Vol. I, E. W. Caspari and A. W. Ravin (eds.), Academic Press, New York, pp. 163–221.

THODAY, J. M., and T. B. BOAM, 1956. A possible effect of the cytoplasm on recombination in *Drosophila melanogaster. Jour. Genet.*, **54,** 456–461.

WEINSTEIN, A., 1958. The geometry and mechanics of crossing-over. *Cold Sp. Harb. Symp.*, **23,** 177–196.

WEISS, M. C., and H. GREEN, 1967. Human-mouse hybrid cell lines containing partial complements of human chromosomes and functioning human genes. *Proc. Nat. Acad. Sci.*, **58,** 1104–1111.

WHITE, M. J. D., 1973. *Animal Cytology and Evolution*, 3rd ed. Cambridge University Press, Cambridge, Chap. 6.

WHITTINGHILL, M., 1937. Induced crossing over in *Drosophila* males and its probable nature. *Genetics*, **22,** 114–129.

18
RECOMBINATION IN FUNGI

The special biology of haploids offers a number of advantages that have made them extremely useful for understanding recombination mechanisms. One factor is the presence of only a single allele per gene in any one haploid individual which removes the complication of dominance in analyzing meiotic products. Second, certain haploid organisms, especially fungi, have a life cycle which enables each of the four products of a *single* meiosis to be analyzed. Furthermore, in fungi such as *Neurospora*, the meiotic products are linearly arranged in the form of "ordered tetrads" so that distinction can be made between first- and second-division segregation. Thus, as shown previously in Fig. 16–8, the *Neurospora* allele *A* that has entered the zygote from one parent can be observed to segregate from its contrasting allele *a* at the first division if no crossing over has occurred between this locus and the centromere. However, if crossing over has occurred in the region between *A* and the centromere. the ultimate separation of *A* and *a* is postponed until the second meiotic division (Fig. 16–9). This fact enables distances between centromere and gene to be mapped by merely observing whether first- or second-division segregation has occurred.

For example, *Neurospora* crosses between *albino* (*al*) and wild type (al^+) by Houlahan, Beadle, and Calhoun showed that first-division segregation had occurred in 129 tetrads ($al\ al\ al^+\ al^+$) and that second-division segregation had occurred in 141 tetrads ($al\ al^+\ al\ al^+$ or $al\ al^+\ al^+\ al$). Since each second-division segregation ascus is the result of crossing over between a gene and its centromere in only two out of four strands, this means that the remaining two strands are nonrecombinant. Since only half the strands in second-division asci have recombined, the recombination percentage is only half the percentage of second-division asci. Thus the data in this case indicate that the distance between *albino* and its centromere is $1/2 \times [141/(141 + 129)] = .26$, or 26 percent recombination.

When a locus is sufficiently far from its centromere so that single crossovers occur between these two points in every tetrad, second-division segregation tetrads would be expected to have a frequency of 100 percent. This would, of course, yield a recombination frequency of 50 percent when multiplied by one-half. However, when a locus is that far from its centromere, double crossovers within this region are also common, and some of them (two-strand and four-strand doubles, see Table 18–1) will produce first-division segregation. As a rule second-division segregation does not exceed 67 percent of all tetrads, no matter how far a locus is from its centromere. Thus recombination frequencies of 33 percent ($1/2 \times 67$), calculated in this fashion, do not necessarily reflect linkage distances of 33 units but may reflect distances of 50 units or more. As in diploids (see p. 340), reliable linkage distances must therefore be calculated on the basis of smaller recombination frequencies, i.e., shorter distances.

LINKAGE DETECTION

The method for detecting linkage between genes in haploids is principally the same as for diploids. That is, one compares the frequency of parental types to recombinant types, and looks for a significant reduction in the latter from the frequencies expected on the basis of independent assortment. In terms of tetrads, the parental ditypes (PD, see p. 134) indicate preservation of the parental combinations, while the nonparental ditypes (NPD) and tetratypes (T) indicate recombination. Equal frequencies of both PD and NPD indicate independent assortment among the genes involved. For example, a cross between an adenine-requiring strain of *Neurospora* (*ad*-4) of mating type *A* and a wild type strain (+) of mating type *a* (i.e., *ad A* × + *a*), produced the following tetrads:

Parental ditypes	**Nonparental ditypes**	**Tetratypes**
ad A	*ad a*	*ad a*
ad A	*ad a*	*ad A*
+ *a*	+ *A*	+ *a*
+ *a*	+ *A*	+ *A*
10	9	1

Note that the T tetrad contains parental and nonparental progeny in equal ratio. The distinction between linkage and random assortment in the present example therefore depends upon the numerical difference between 10 PD and 9 NPD. By observation, this difference is small. If desired, it can be demonstrated to be nonsignificant by calculating chi-square on the basis that each of the two classes PD and NPD would be expected to have a number equal to one-half the sum of both ditypes or 9.5 (if Yates' correction is used, this would yield a chi-square of zero).

For this method of detecting linkage, it is important to consider each tetrad as a unit, either PD or NPD, rather than separately scoring the meiotic products of each tetrad. The reason for this is that each tetrad is formed as a result of an independent meiotic event, crossing over or its absence, and comparisons may appropriately be made between one kind of tetrad and another. On the other hand, the four meiotic products of a tetrad are not formed independently of each other; i.e., if the genotypes of two products of the tetrad are given, the other two genotypes are thereby determined. Thus a tetrad ratio of 10 PD:6 NPD would show a nonsignificant difference ($\chi^2 = 1.00$), indicating independent assortment, while the ratio of their meiotic products, 40 PD:24 NPD, would erroneously imply linkage ($\chi^2 = 4.00$).

The tetratype frequencies do not enter into the calculations, since they can occur for both linked and independently assorting genes. In independent assortment, tetratypes usually arise through second-division segregation between a gene and its centromere. In the present example, both genes, *ad* and *a*, are apparently close to their respective centromeres, thereby reducing the frequency of tetratype tetrads. If the centromere-gene distance were longer, the frequency of tetratypes would be greater.

When there is linkage between two loci, the frequency of NPD tetrads falls below the frequency of the PD tetrads. For example, crosses between an albino strain (*al* 15300) of mating type *A* and a wild-type strain of mating type *a* produced the following asci:

PD	NPD	T		
al A	*al a*	*al a*	*al a*	*al a*
al A	*al a*	+ *a*	*al A*	+ *A*
+ *a*	+ *A*	*al A*	+ *a*	*al A*
+ *a*	+ *A*	+ *A*	+ *A*	+ *a*
24	3	27		

Therefore, for *albino* and *mating type*, the difference between parental (24) and nonparental (3) ditypes is statistically significant ($\chi^2 = 14.82$) and indicates linkage between these two loci. On this basis the PD frequency represents primarily nonrecombinant tetrads, the NPD frequency represents tetrads in which four-strand *double* exchanges have occurred, and the T frequency represents tetrads in which *single* exchanges have mostly occurred (see Fig. 18–1) or, less often, three-strand *double*, *triple*, and higher degrees of crossing over. Generally, if the NPD frequency in recombination between two linked genes is low (meaning relatively few four-strand double crossovers), the number of tetratypes will be much larger and signifies mostly singles, since we would not expect many doubles or higher numbers of crossovers between closely linked genes. The NPD:T ratio can therefore also be used to discriminate between linkage and independent assortment by the following argument.

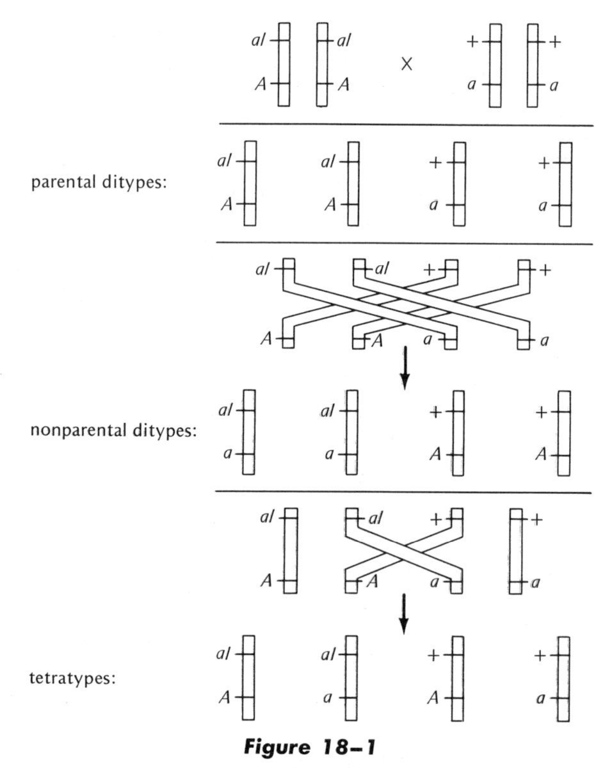

Figure 18–1

Explanation for the origin of the different kinds of tetrads in the cross al A × + a. (Tetratypes also originate from three-strand doubles and some other multiple crossover events, but these are much rarer than single crossovers.)

In independent assortment tetratypes arise from crossing over between at least one of the genes and its centromere (p. 136), while approximately half of all ditypes are of the NPD variety. If the two nonlinked genes are close to their respective centromeres, the frequency of T will be low, while the frequency of NPD will be relatively high (approximately equal to PD), as discussed on the previous page. If the genes are far from their centromeres, the frequency of T increases and may reach any value relative to NPD. On the other hand, crossing over between two linked genes will produce NPD tetrads mainly through double exchanges and produce tetratypes through single exchanges as well as multiple exchanges. In this case, even if the distance between the two genes is great and the frequency of double exchanges is very high, we can still be fairly sure that the frequency of NPD will not ordinarily rise above one-fourth the frequency of T, no matter how long the distance between them. In other words, independent assortment may produce a high NPD:T ratio if the genes are close to their centromeres, while linkage will probably not produce an NPD:T ratio greater than 1:4. Thus good evidence for independent assortment is found in the approximate equal frequencies of PD:NPD, and in an NPD:T ratio that is significantly greater than 1:4.

An illustration of how this last method is used

can be seen in the results of a cross between the *peach* (*pe* +) and *leucine* (+ *leu*) mutants of *Neurospora crassa*, in which 19 PD:15 NPD:18 T appeared. Although the ratio of PD:NPD is not exactly 1:1, the NPD:T ratio is significantly higher than 1:4 indicating that these two genes are on separate chromosomes.

LINKAGE DISTANCE IN TWO-POINT CROSSES

The determination of the actual linkage distance between two genes can be done on the basis of centromere relationships if the order of meiotic products is known for each tetrad. That is, for two genes on the same chromosome, the distance between them is either the sum or the difference of their respective distances from the centromere, depending on whether the genes are on opposite sides or on the same side of the centromere.

The differently ordered tetrads that may arise as a result of these possibilities, including that of independent assortment, are shown in Table 18–1 for the hypothetical dihybrid cross *AB* × *ab*. As we have seen previously, equality between the first two segregation classes in this table (PD and NPD) indicates 50 percent recombination or independent assortment, while a relative decrease in class ② (NPD) indicates linkage. In the case of linkage we can differentiate whether the two genes are on opposite chromosome arms or on the same arm by scoring the frequencies of classes ⑤ and ⑥. Note that when the genes are on opposite arms, classes ⑤ and ⑥ are both formed by double exchanges, which differ only in whether two or four strands are involved. Since there is no strong indication that one type of double exchange is formed in preference to the other, the equality between both types of double exchange (class ⑤ = class ⑥) would therefore indicate that the two genes are on opposite sides of the centromere. On the other hand, for two genes on the same side of the centromere, class ⑤ is formed by a single exchange and class ⑥ by a triple exchange. As compared to single exchanges in any given region, triples are quite rare and we would then expect class ⑤ to have a much greater frequency than class ⑥.

Thus the position of the centromere is fairly easy to determine if a sufficient number of ordered tetrads have been scored to enable a comparison between classes ⑤ and ⑥. Once the centromere position is determined, the linkage distance between two genes on the same chromosome arm is simply the difference between the two centromere distances, while for two genes on opposite arms it is simply the sum of the two centromere distances.

An example of this kind can be found in another cross by Houlahan, Beadle, and Calhoun (37803 *a pdx* × 5531 *A pan*) between a *Neurospora* strain auxotrophic for pyridoxine (*pdx*) but wild type for pantothenic acid synthesis (pan^+), and a strain auxotrophic for pantothenic acid (*pan*) but wild type for pyridoxine synthesis (pdx^+). The tetrad analysis for this cross, given in Table 18–2, can be interpreted in terms of the tetrad classes of Table 18–1 if we consider the *pdx* pan^+ parental strain as *AB*, and the pdx^+ *pan* strain as *ab*. Note that the ratio of PD:NPD tetrads in Table 18–2, or the ratio of tetrad classes ① to ②, is of the order of 15:1, which is far above the equality expected if assortment were random between the two genes. The type of linkage between these genes can then be determined by noting the relatively large number of tetrads in class ⑤ (13) and the complete absence of class ⑥. On this basis the two genes must be located on the same chromosome arm, and the linkage distance between them is the difference between the two centromere distances.

Which gene is then closest to the centromere? From the data given, only 16 tetrads show second division segregation for *pdx* (classes ④, ⑤, ⑦), whereas there are 32 tetrads of this type for *pan* (classes ③, ⑤, ⑦). Obviously *pan* is further from the centromere than *pdx*, since more crossovers occurred in the *pan*-centromere interval, and *pan* can be considered as locus *B* on the right side of Table 18–1. In terms of exact values, the *pan*-centromere distance will therefore now include half the proportion of single crossovers within this interval (tetrad classes ③ and ⑤) added to the entire proportion of tetrads showing double cross-

TABLE 18–1

Crossovers accounting for tetrad order when crosses are made in "coupling," *AB* × *ab*. For crosses made in "repulsion," *Ab* × *aB*, the crossovers that account for segregation classes 1 and 2 are reversed (i.e., when markers are on different arms of the same chromosome, a four-strand double exchange in I causes the appearance of segregation class 1), and the crosses that account for segregation classes 5 and 6 are also reversed

SEGREGATION CLASS	DIVISION IN WHICH SEGREGATION OCCURS: GENE *Aa*	GENE *Bb*	ORDER OF PRODUCTS IN TETRAD*	CROSSOVERS ACCOUNTING FOR TETRAD ORDER: GENES ON DIFFERENT CHROMOSOMES (*A* × *a*, *B* × *b*)	GENES ON THE SAME CHROMOSOME: DIFFERENT ARMS (*A B* × *a b*)	SAME ARM (*A B* × *a b*)
①	1st	1st	*A B*, *A B*, *a b*, *a b*	I, II: *A*, *a*; *B*, *b* — none	I, II: *A B*, *a b* — none	I, II: *A B*, *a b* — none
②	1st	1st	*A b*, *A b*, *a B*, *a B*	none	four-strand double exchange: both in I or both in II	four-strand double exchange: both in II
③	1st	2nd	*A B*, *A b*, *a B*, *a b*	two-strand single exchange in II	two-strand single exchange in II	two-strand single exchange in II

overs (classes ②, ④, and ⑦)*. In other words, the *pan*-centromere distance is 1/2 (30/49) + 4/49 = .387. No double crossovers are detectable in the *pdx*-centromere distance (locus *A* on the right side of Table 18-1) and recombination within this interval will therefore only be marked by tetrad classes ④, ⑤, ⑥, and ⑦. These tetrads for single crossovers between *pdx* and the centromere yield a distance of 1/2 (16/49) = .163. The *pan*-*pdx* interval is therefore .387 − .163 or 22.4 percentage units of recombination. (The *pan-pdx* distance can also be calculated directly by scoring single- and double-crossover tetrads within this interval according to Table 18–1: (1/2) (17 + 1 + 2)/49 + 1/49 = .224.) A linkage map can therefore be drawn in the form:

centromere —— 16.3 —— *pdx* —— 22.4 —— *pan*

Ordered tetrads are not available in many orga-

*Single crossover tetrads represent, as explained previously, one crossover between only two out of four strands, while a double crossover represents two crossovers in four strands, or the equivalent of the entire tetrad taking part in a crossover event. Had there been any tetrads in class six these would have represented triple crossovers between *pan* and the centromere and would have affected this distance by one and a half times their proportion. Thus the recombination value between two points in terms of the proportions of tetrads is: 1/2 (singles) + 1 (doubles) + 1½ (triples) + 2 (quadruples), etc.

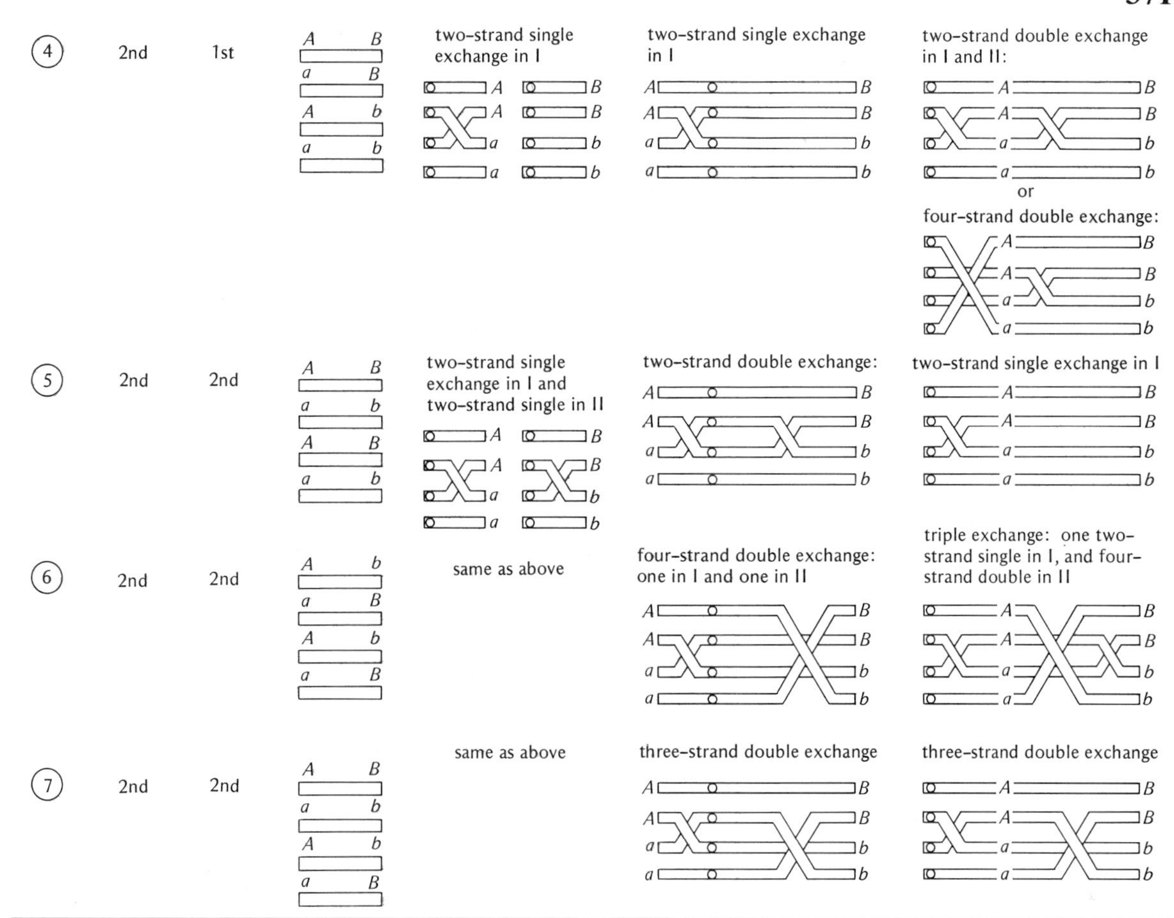

*Note that the given order of products in the tetrad can be reversed but would still signify the same meiotic events, i.e., *AB Ab aB ab* = *ab aB Ab AB*.

nisms that produce tetrads, such as yeast, *Chlamydomonas*, and others. In such cases, linkage distances between the two genes may have to be determined through their recombination values rather than through their centromere distances. For unordered tetrads, there are only three discernible types, PD, NPD, and T, of which the two recombinant classes, NPD and T, represent exchanges between all four strands and exchanges mostly between two of four strands, respectively (Fig. 18–1). The recombination frequency among all strands produced in the cross is obtained by adding the number of NPD to half the number of T and dividing by the total number of strands, or (NPD + 1/2 T)/(PD + NPD + T). As an illustration, the data of the previous example may be considered as a number of unordered tetrads in which classes ① and ⑤ represent PD, class ② represents NPD, and classes ③, ④, and ⑦ represent T. The recombination frequency between *pan* and *pdx* can therefore be calculated as [1 + 1/2(20)]/49, or approximately 22.4 percent.

Although second division segregations cannot be identified in unordered tetrads, centromere locations can nevertheless be determined from such data if one of the genes on a cross is known to be close to its centromere. Such genes, called "centromere markers," can be found by looking

TABLE 18–2

Tetrads obtained from a cross between a *pdx pan*$^+$ strain of *Neurospora crassa* and a *pdx*$^+$ *pan* strain

①*		②		③	
pdx	+	*pdx*	*pan*	*pdx*	+
pdx	+	*pdx*	*pan*	*pdx*	*pan*
+	*pan*	+	+	+	+
+	*pan*	+	+	+	*pan*
15		1		17	
④		⑤		⑦	
pdx	+	*pdx*	+	*pdx*	+
+	+	+	*pan*	+	*pan*
pdx	*pan*	*pdx*	+	*pdx*	*pan*
+	*pan*	+	*pan*	+	+
1		13		2	

Data from Houlahan, Beadle and Calhoun, 1949.

* The classes of tetrads (circled numbers) are designated according to the classes in Table 18–1 considering the first strain (*pdx* +) as A B and the second strain (+ *pan*) as *a b*.

for tetratype crosses in which two loci are segregating independently and in which tetratypes are rare or absent. For example, if *A* and *B* were on separate chromosomes, a cross between *AB* × *ab* would produce parental (*A B A B a b a b*) and nonparental (*A b A b a B a B*) ditypes with equal frequency. Tetratypes (*A B A b a B a b*) would occur only when an exchange took place between at least one locus and its centromere, causing second-division segregation. The absence of tetratypes would therefore indicate that both loci occupy centromere positions in their respective linkage groups, and this information could then be used to map other loci.

When tetrad data of any kind are unobtainable, the individual meiotic products ("random spores") can be used for mapping on the same statistical basis as in diploids. That is, for haploids, samples of spores are plated and scored for whether they each represent parental or crossover combinations. As in diploids, the frequency of nonparental combinations in percent among all strands is the recombination frequency. For example, the data of Table 18–2 can be considered an experiment in which 392 individual spores (49 asci × 8 spores each) show four possible combinations, *pan* +, + *pdx*, *pan pdx*, and + +. The first two combinations are parental and the last two recombinational. Linkage between *pan* and *pdx* is then noted by the excess of parental (304) over recombinational (88) spores, or a linkage distance of $88/392 = 22.4$ percent. Some information is lost by this method because neither double crossovers nor centromere position can be detected, and its statistical accuracy is restricted to the number of independent events based on the number of asci (49), and not on the number of spores (392). On the other hand, the saving in the labor of isolating spores from individual asci may warrant its use.

When nutritional mutants are used, such as in the above example, crosses can be made so that the recombinant spores are automatically selected and can then be easily scored. This method, known as *selective plating*, involves crossing two mutant auxotrophs, i.e., *pan* + × + *pdx*, and then plating a known number of spores on a ("minimal") medium permitting only the prototroph, + +, to survive (see p. 536). If assortment between *pan* and *pdx* is independent, about 25 percent of the spores should be prototrophic. If the two genes are linked, this proportion will be less than 25 percent. The actual recombination frequency will then be twice the frequency of prototrophs, since we can assume that one reciprocal, but undetectable, double mutant (*pan pdx*) has been formed for each + + prototroph. In the present example, the frequency of prototrophs is 44/392, or 11.2 percent, and the distance between *pan* and *pdx* is twice this amount, or 22.4 percent.

THREE-POINT LINKAGE DISTANCES

When three gene differences are involved in a cross between two strains, ordered tetrad analysis of even relatively few tetrads may furnish considerable linkage information. Table 18–3 shows the genetic composition of 61 tetrads produced by a cross between a methionine auxotroph of mating type *a* (*me* + *a*) and an adenine auxotroph of mating type *A* (+ *ad A*). Linkage among these three loci is evident from the high frequency of parental spores (*me* + *a*, + *ad A*) compared to all

TABLE 18–3

Tetrads obtained from a cross *me* + *a* × + *ad A*

Class	Strand 1	Strand 2	Strand 3	Strand 4	Number
①	*me* + *a*	*me* + *a*	+ *ad A*	+ *ad A*	46
②	*me* + *a*	+ + *a*	*me ad A*	+ *ad A*	6
③	*me* + *a*	*me* + *A*	+ *ad a*	+ *ad A*	4
④	*me* + *a*	+ *ad a*	*me* + *A*	+ *ad A*	2
⑤	*me* + *a*	+ + *A*	*me ad a*	+ *ad A*	2
⑥	*me* + *a*	+ + *A*	*me ad A*	+ *ad a*	1
⑦	*me* + *a*	+ + *A*	*me ad a*	+ *ad A*	0
⑧	*me* + *a*	+ *ad A*	*me* + *a*	+ *ad A*	0
⑨	*me* + *a*	+ *ad a*	*me ad A*	+ + *A*	0
⑩	+ *ad a*	+ + *a*	*me ad A*	*me* + *A*	0

From data of Barratt et al., 1954.

others. Class ① represents the parental combination and can therefore be explained as arising from the absence of crossovers in any of the linked areas.

The next step, learning the gene order, relies upon the principle of arranging the three genes so that each class of tetrads arises through the fewest possible number of crossovers. On this basis the most frequent recombinant class, ②, can be explained most simply as arising from a single crossover between *me* and *ad*, and the next most frequent class, ③, arises from a single crossover in the region *ad-A*. Note that class ④ also arises from a single exchange between *ad* and *A*, but has, in addition, second-division segregation for *me* and *ad*. In other words, class ④ represents a crossover between *me* and *ad* and their centromere, which has also produced recombination between *ad* and *A*. In toto, there are six single exchanges between *ad* and *A* (classes ③ and ⑤). Note, however, that each *me-ad* exchange is accompanied by an exchange between *me* and *A* but not between *ad* and *A*. On the other hand, each *ad-A* exchange is accompanied by an exchange between *me* and *A* but not between *me* and *ad*. The simplest gene order is therefore *me-ad-A*, and the centromere position, on the basis of class ④, is probably between *ad* and *A*. Table 18–4 shows how each of the different classes can be explained in terms of the postulated gene order, with the three intermediate crossover regions labeled I, II, and III. The zero classes (⑦ to ⑩) have been added as a sample of additional tetrads that may occasionally occur.

As a rule gene order can be determined in three-point linkage tests by observing the frequency of the double-crossover classes. If these classes are more frequent than any of the presumed single crossover classes, the gene order is suspect and should be changed. Table 18–5 interprets four types of tetrads for the three possible gene orders in a three-point cross, *a b c* × + + +. For convenience, second-division segregation has been omitted, since it merely identifies the position of the centromere but does not affect the gene order. Note that each of the tetrads ②, ③, and ④, represents the products of a single crossover event for two of the three gene orders and a two-strand double crossover event for the third. The decision as to gene order can therefore usually be made by comparing the frequency of these tetrads and choosing the gene order for which the tetrad with the lowest frequency is the double-crossover class. Other tetrads are, of course, possible but none of these represents any of the commonly frequent single-crossover classes for the three gene arrangements, and are therefore not often as efficient in distinguishing gene order.

Once the gene order is determined, the calculation of recombination frequencies then proceeds as follows: In the previous example (Table 18–4), there are three crossover regions, I, II, and III, for which information is available among a total of 61 tetrads. The number of tetrads showing exchange in region I is 9 (classes ②, ⑤, and ⑥), or an exchange frequency per tetrad of $9/61 = 14.7$ percent. However, since each such exchange tetrad represents crossing over in only two of four strands, the actual recombination frequency per strand is half this value, or 7.3. Similarly, the recombination frequency in region II can be calculated as 1.5, in region III as 5.7. Using methods such as these, all seven linkage groups of *Neurospora crassa* can be mapped (Fig. 18–2).

TABLE 18–4

Explanation of the data in Table 18–3 in terms of crossing over in regions I, II, and III among the four meiotic chromatids

	I	II		III
Parental chromatids	*me*	+		*a*
1st				
2nd				
	me	+	↑ centromere ↓	*a*
	+	*ad*		*A*
3rd				
4th				
	+	*ad*		*A*

Exchanges in Tetrad Classes

① none	② single in I	③ single in III	④ single in II	⑤ two-strand double in I and II
46	6	4	2	2
⑥ three-strand double in I (2nd and 3rd) and III (2nd and 4th)	⑦ two-strand double in I and II	⑧ two-strand double in II and III	⑨ three-strand double in I (2nd and 3rd) and II (2nd and 4th)	⑩ four-strand double in I and II
1	0	0	0	0

TABLE 18–5

Determination of gene order in a three-point cross by observation of the relative frequencies of different tetrad classes, three of which (②, ③ and ④) represent single-crossover tetrads for two of the three gene orders and a two-strand double-crossover tetrad for the third*

Parental strains $a\ b\ c \times +\ +\ +$

	Tetrad Classes			
	①	②	③	④
	a b c	*a b c*	*a b c*	*a b c*
	a b c	*a b* +	*a* + *c*	*a* + +
	+ + +	+ + *c*	+ *b* +	+ *b c*
	+ + +	+ + +	+ + +	+ + +
Gene Order				
a-b-c	no exchange	single crossover (*b-c*)	double crossover (*a-b, b-c*)	single crossover (*a-b*)
a-c-b	no exchange	double crossover (*a-c, c-b*)	single crossover (*c-b*)	single crossover (*a-c*)
b-a-c	no exchange	single crossover (*a-c*)	single crossover (*b-a*)	double crossover (*b-a, a-c*)

* For example, the gene order is most likely *a-b-c* if class ③ (double crossover) is less frequent than classes ② and ④ (single crossovers).

CHROMATID AND CHIASMA INTERFERENCE

In the three-point linkage data given above, only two types of double-exchange tetrads were noted (classes ⑤ and ⑥), and no triple or quadruple exchanges could be observed. This is not surprising if we consider the relatively small number of multiple exchanges usually noted in recombination experiments and the small number of tetrads analyzed in the present experiment. It is of special interest, however, that for the double-exchange tetrads we have observed, two kinds of information can be obtained. One is information concerning which strands are involved, and the other is information concerning which regions are involved.

The first kind of information on strands gives us the opportunity to note whether the relationship between strands, or chromatids, is affected by crossing over. That is, if a crossover occurs between two strands in a particular region, we may ask what nonsister strands (chromatids from the different homologues) are involved in crossing over in an adjacent region. On a random basis, if there were no effect of the chromatids in one exchange on the chromatids of another exchange, four possible combinations of nonsister chromatids should be equally likely to occur in double exchanges. As shown in Fig. 18–3, one of these is a two-strand double exchange, two are three-strand double exchanges, and one is a four-strand double exchange. A departure from this 1:2:1 ratio would indicate that there is some form of *chromatid interference.*

For example, if crossing over between two chromatids in one locality prevents crossing over between these same chromatids in an adjacent locality, we would have relatively few two-strand doubles but an excess of four-strand doubles, or "positive" interference. Note that an excess of four-strand doubles between two loci would increase the frequency of nonparental ditypes, since each such tetrad would produce recombination in all four meiotic products. On the other hand, if crossing over between two chromatids increases the chances of crossing over between these same chromatids in an adjacent region, we would expect an excess of two-strand doubles, or "negative" chromatid interference.

The double-exchange data in the previous experiment (Table 18–3) is unfortunately small, and does not allow definite conclusions to be drawn. However, when summarized with other data, Emerson (1963) has found that of approximately 2000 double-exchange tetrads, 29 percent are two-strand, 48 percent are three-strand, and 23 percent are four-strand. Thus there seems to be a slight excess of two-strand double exchanges, implying some degree of negative chromatid interference. Attached-X data in *Drosophila* also permit some distinction to be made between different double exchanges, and these too have shown a slight excess of two-strand doubles. Although not a general rule, indications exist that the excess of two-strand doubles may be greater when the map distances between two loci are smaller (see Emerson, 1969).

A second form of interference also occurs in double exchanges between adjacent regions. However, instead of arising through chromatid or strand interference, it is caused by the same type of *chiasma interference* observed in *Drosophila* (p. 338), in which a crossover in one region interferes in some way with crossing over in an adjacent region. However, in contrast to *Drosophila*, tetrad analysis in fungi enables chromatid interference to be easily calculated, and, as shown above, it is almost always absent. The main cause for interference between adjacent regions in fungi can therefore be ascribed to chiasmata. The test then is simple: If chiasma interference does not exist, the frequency of exchange in two adjacent regions should be the product of exchange values in each individual region.

An example of such calculations is obtained from the data of Bole-Gowda, Perkins, and Strickland for a cross in which six loci were mapped on the first linkage group of *Neurospora crassa.* As shown in Table 18–6, products between exchange frequencies in two regions are either the same as, or greater than, the observed double exchanges. Coincidence values, calculated as the ratio of ob-

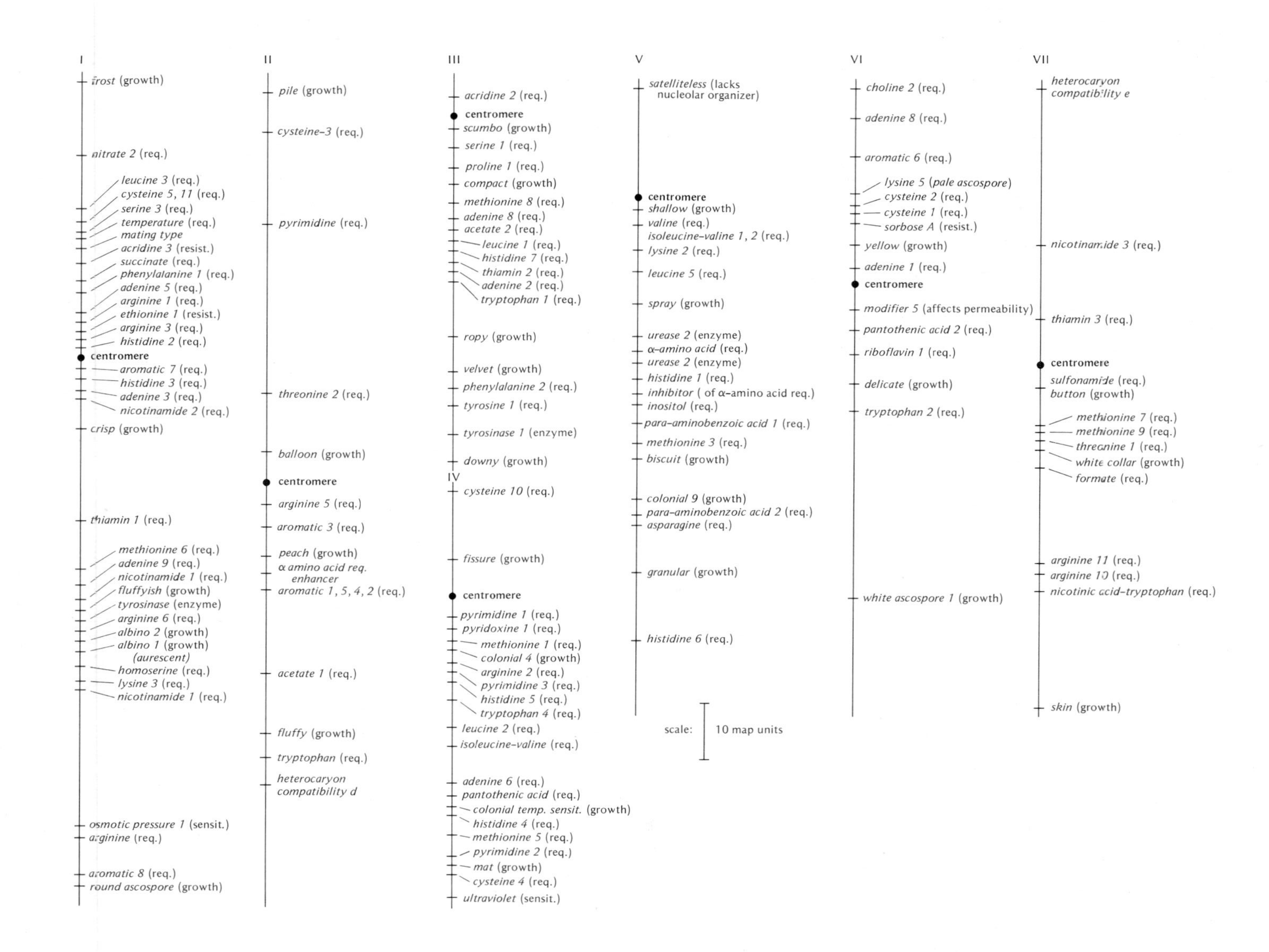
I
frost (growth)
nitrate 2 (req.)
leucine 3 (req.)
cysteine 5, 11 (req.)
serine 3 (req.)
temperature (req.)
mating type
acridine 3 (resist.)
succinate (req.)
phenylalanine 1 (req.)
adenine 5 (req.)
arginine 1 (req.)
ethionine 1 (resist.)
arginine 3 (req.)
histidine 2 (req.)
centromere
aromatic 7 (req.)
histidine 3 (req.)
adenine 3 (req.)
nicotinamide 2 (req.)
crisp (growth)
thiamin 1 (req.)
methionine 6 (req.)
adenine 9 (req.)
nicotinamide 1 (req.)
fluffyish (growth)
tyrosinase (enzyme)
arginine 6 (req.)
albino 2 (growth)
albino 1 (growth)
(aurescent)
homoserine (req.)
lysine 3 (req.)
nicotinamide 1 (req.)
osmotic pressure 1 (sensit.)
arginine (req.)
aromatic 8 (req.)
round ascospore (growth)
II
pile (growth)
cysteine-3 (req.)
pyrimidine (req.)
threonine 2 (req.)
balloon (growth)
centromere
arginine 5 (req.)
aromatic 3 (req.)
peach (growth)
α amino acid req.
enhancer
aromatic 1, 5, 4, 2 (req.)
acetate 1 (req.)
fluffy (growth)
tryptophan (req.)
heterocaryon
compatibility d
III
acridine 2 (req.)
centromere
scumbo (growth)
serine 1 (req.)
proline 1 (req.)
compact (growth)
methionine 8 (req.)
adenine 8 (req.)
acetate 2 (req.)
leucine 1 (req.)
histidine 7 (req.)
thiamin 2 (req.)
adenine 2 (req.)
tryptophan 1 (req.)
ropy (growth)
velvet (growth)
phenylalanine 2 (req.)
tyrosine 1 (req.)
tyrosinase 1 (enzyme)
downy (growth)
IV
cysteine 10 (req.)
fissure (growth)
centromere
pyrimidine 1 (req.)
pyridoxine 1 (req.)
methionine 1 (req.)
colonial 4 (growth)
arginine 2 (req.)
pyrimidine 3 (req.)
histidine 5 (req.)
tryptophan 4 (req.)
leucine 2 (req.)
isoleucine-valine (req.)
adenine 6 (req.)
pantothenic acid (req.)
colonial temp. sensit. (growth)
histidine 4 (req.)
methionine 5 (req.)
pyrimidine 2 (req.)
mat (growth)
cysteine 4 (req.)
ultraviolet (sensit.)
V
satelliteless (lacks
nucleolar organizer)
centromere
shallow (growth)
valine (req.)
isoleucine-valine 1, 2 (req.)
lysine 2 (req.)
leucine 5 (req.)
spray (growth)
urease 2 (enzyme)
α-amino acid (req.)
urease 2 (enzyme)
histidine 1 (req.)
inhibitor (of α-amino acid req.)
inositol (req.)
para-aminobenzoic acid 1 (req.)
methionine 3 (req.)
biscuit (growth)
colonial 9 (growth)
para-aminobenzoic acid 2 (req.)
asparagine (req.)
granular (growth)
histidine 6 (req.)
scale:
10 map units
VI
choline 2 (req.)
adenine 8 (req.)
aromatic 6 (req.)
lysine 5 (pale ascospore)
cysteine 2 (req.)
cysteine 1 (req.)
sorbose A (resist.)
yellow (growth)
adenine 1 (req.)
centromere
modifier 5 (affects permeability)
pantothenic acid 2 (req.)
riboflavin 1 (req.)
delicate (growth)
tryptophan 2 (req.)
white ascospore 1 (growth)
VII
heterocaryon
compatibility e
nicotinamide 3 (req.)
thiamin 3 (req.)
centromere
sulfonamide (req.)
button (growth)
methionine 7 (req.)
methionine 9 (req.)
threonine 1 (req.)
white collar (growth)
formate (req.)
arginine 11 (req.)
arginine 10 (req.)
nicotinic acid–tryptophan (req.)
skin (growth)

Figure 18–2

Linkage map for genes on the seven chromosomes of Neurospora crassa. *The genes listed are restricted to those whose relative positions are fairly well established. Many of the mutations indicate requirements for particular substances* (e.g., leucine, serine, aromatic amino acids, etc.), *some refer to the shape, form, or color of growth* (e.g., frost, balloon, albino, colonial, etc.), *others refer to the absence or modification of known enzymes* (e.g., tyrosinase, urease, etc.), *and others indicate resistance or sensitivity to particular chemicals or conditions* (e.g., acridine, ethionine, osmotic pressure, etc.). *Note that, as in* Drosophila *and other organisms (see Fig. 17–3), there are many genes in the* Neurospora *genome that may affect a particular function* (e.g., leucine 1, leucine 2, etc.). *Further phenotypic descriptions of the effects of various mutations can be found in a listing by R. W. Barratt and A. Radford, 1970, in* Handbook of Biochemistry, *2nd edition, H. A. Sober (ed.),* Chemical Rubber Co., *Cleveland, (Adapted from Fincham and Day, after data of Barratt, Perkins, and others.)*

served to expected double exchanges, range from .2 to approximately 1.0. (Note that exchange values per tetrad, as explained previously, are approximately twice the recombination values per strand.) The main distinction between regions having low and high coincidence values in these data seems to reside in whether these regions are on the same side or on opposite sides of the centromere. From this and other experiments it appears evident that there is considerable chiasma interference between loci on the same side of the centromere but almost no interference between loci on opposite sides of the centromere. The centromere thus seems to act as a physical block to chiasma interference.

In the mold *Aspergillus nidulans* studies by Pritchard and others showed either the absence of interference in some cases or the significant presence of *negative interference* in others. That is, chiasma interference in most fungi and higher organisms seemed to be positive for the linkage distances examined (coincidence coefficients less than 1, see p. 338), but in *Aspergillus* the occurrence of an exchange in one chromosomal section did not restrict and may even have enhanced the occurrence of other exchanges in adjacent sections (coincidence coefficients greater than 1).

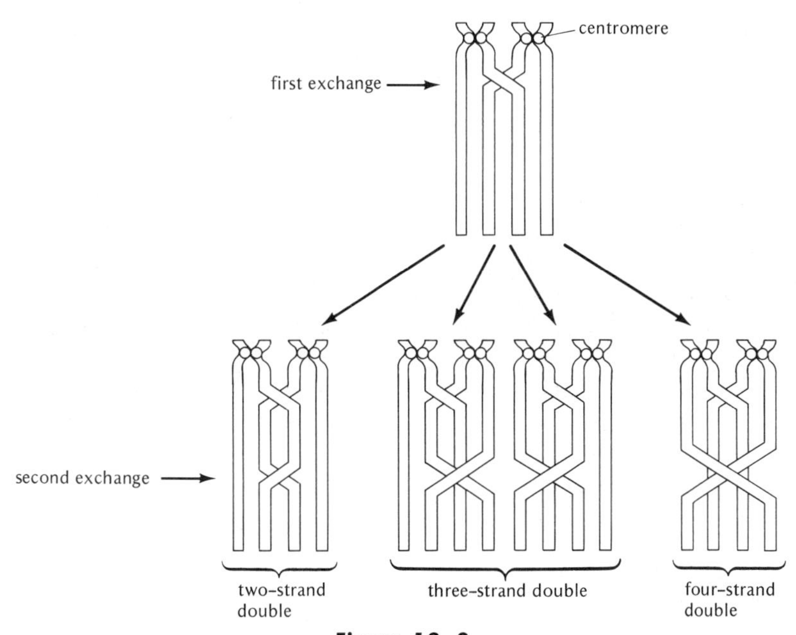

Figure 18–3

Possible alternative types of double exchanges between nonsister chromatids.

TABLE 18–6

Exchange frequencies in 2945 tetrads in a cross between mutants of the first linkage group in *Neurospora*

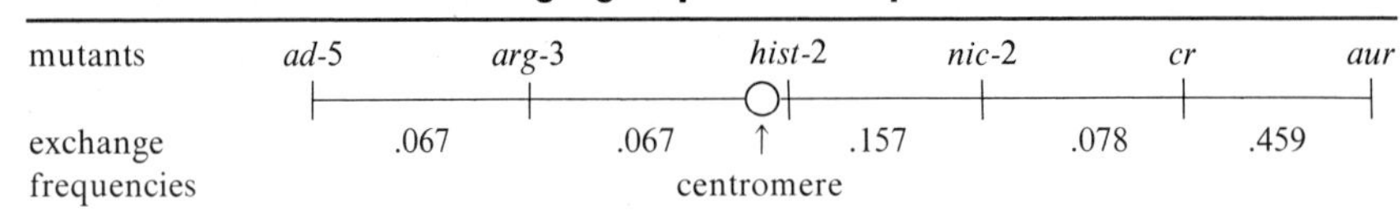

Interval	Double Exchanges Expected	Double Exchanges Observed	Coincidence (Observed/ Expected)
on the same sides of the centromere			
ad-arg-hist	13.1	2	.2
hist-nic-cr	35.7	6	.2
nic-cr-aur	100.6	72	.7
	149.4	80	.54
on opposite sides of the centromere			
ad-arg, hist-nic	30.9	27	.9
ad-arg, nic-cr	15.1	17	1.1
ad-arg, cr-aur	86.9	78	.9
arg-hist, hist-nic	31.0	21	.7
arg-hist, nic-cr	15.3	13	.8
arg-hist, cr-aur	87.2	92	1.1
	266.4	248	.93

From Bole-Gowda, Perkins, and Strickland, 1962.

According to the hybrid DNA model of recombination previously mentioned (p. 358) and further discussed below (p. 385), it is believed likely that at least some cases of negative interference arise because of mismatching events between parental DNA strands over very short linkage distances. Perhaps enzymatic systems that function during crossing over can produce a cluster of apparent recombinational events within the immediate area of nucleotide mismatching (see also p. 459). In support of this view are findings in bacteriophage, *Neurospora*, yeast, and other organisms, showing that when extremely short crossover intervals can be examined by means of closely placed gene markers, the likelihood of detecting nonreciprocal exchanges and negative interference is increased. The specific causes for these events, however, are still unknown.

MITOTIC RECOMBINATION

In 1936 Stern demonstrated that, in addition to meiotic crossing over, genetic exchange may also occur between homologous chromatids during ordinary somatic mitosis. The organism in which he demonstrated this phenomenon was *Drosophila*, but it has since been found to occur in numerous fungi including yeast. Stern's classical experiments utilized morphological mutants such as *yellow* and *singed* which had visible effects on *Drosophila* body color and bristles. Crossing over in such somatic tissues (enhanced by the presence of genes such as *Minute*) could be detected by observing differences between adjacent sections. For example, the sex-linked genes *yellow* and *singed* produce, respectively, yellow body color and bent blunted bristles. Since these genes are recessive, a heterozygous female is, of course, wild type, and the transmission of these genes between somatic cells is not expected to produce any visible differences (Fig. 18–4a). However, if crossing over occurs during mitosis between nonsister strands in the *y–sn* region, a yellow skin spot appears composed of cells that trace their ancestry to one of the crossover products (Fig. 18–4b).

One may argue, of course, that such spots are caused by mutation of the wild-type gene to *y* ("somatic mutation") rather than by recombination. Stern demonstrated that recombination was a more likely explanation, however, by finding sections of yellow tissue lying adjacent to equal-sized sections of singed tissue. According to the mutation theory, such "twin spots" would demand simultaneous mutation at the two loci in adjacent cells, while according to recombination they could be explained by the occurrence of crossing over in the *sn*-centromere region (Fig. 18–4c).

The relative high frequency of twin spotting, even greater than that for single yellow spots, therefore indicated a recombinational explanation. This was also supported by the very low frequency of single singed spots, apparently caused by the rarity of double crossovers (Fig. 18–4d). In other words, the frequency of spotting explained by mitotic recombination reflected previously determined meiotic recombinational data, with the *y-sn* distance shorter than the *sn*-centromere distance, and very few double crossovers (see also Garcia-Bellido). It was also found that, as in meiotic recombination, crossing over occurs in the four-strand stage.

Drosophila data, unfortunately, were limited by the few observable morphological differences between twin spots, by the difficulty of finding such small spots, and by the inability to isolate recombinational somatic tissues and reproduce them further. In numerous fungi, especially *Aspergillus*, such limitations do not exist, and mitotic recombination has been used either as an aid or as a primary source for the detection and measurement of linkage.

Since the normal vegetative state of most fungi is haploid, mitotic recombination involves, first of all, the establishment of a state in which homologous chromosomes may exist in stable diploid condition. Such strains were discovered by Roper (1952) in *Aspergillus nidulans* through occasional fusion between nuclei of different strains. As shown in Fig. 18–5, two different strains plated together will develop some *heterocaryons* which contain the nuclei of both strains. Although these haploid nuclei remain separate in the heterocaryon they may occasionally fuse to form diploid nuclei. Since the haploid and diploid nuclei of the heterocaryon form haploid and diploid conidia, respectively, and each conidium contains only one nucleus, these two types can be separated in conidial form if they are visibly distinct, (e.g., white, yellow, green). Also, diploid colonies can be established by plating the conidia on a medium in which only diploids can survive. For example, the conidia of heterocaryons carrying the nuclei of two nutritional mutants lacking the ability to synthesize a or b may be plated on a medium in which only the diploid heterozygote $a\,b^+/a^+\,b$ can survive. Once formed, diploids can be recognized by the larger volume of their conidia. Also, as expected, their DNA content per nucleus is twice that of haploids.

The course followed by diploids may then proceed along two paths: either revert to the haploid stage or remain as a diploid. The first path, *haploidization*, can be used to distinguish between markers on the same and on different linkage groups. This is, if a diploid organism is heterozygous for markers a and b, and they are both on the same linkage group, reduction to haploidy will cause them to segregate together as a pair since somatic haploidization, unlike gamete formation in meiosis, is not usually associated with crossing over. For example, the haploids from a "coupling" heterozygote $a\,b/a^+\,b^+$ will all be either $a\,b$ or $a^+\,b^+$, except for the very rare crossovers. Similarly, haploidization of a "repulsion" heterozygote, $a^+\,b/a\,b^+$, will produce individuals which are almost all a^+b and $a\,b^+$. On the other hand, if the two markers are on different nonhomologous chromosomes, the diploid heterozygote (a/a^+, b/b^+) will produce haploids in which all four combinations are equally likely ($a^+\,b^+$, $a^+\,b$, $a\,b^+$, $a\,b$). In this fashion the linkage group of any marker can be determined by placing it in heterozygous condition into a diploid that is carrying markers whose linkage groups are already known. Nonrandom assortment with the known marker indicates the same linkage group for both of them, and random assortment indicates that the unknown marker is located on a different chromosome. In *Aspergillus nidulans* eight linkage groups have been established so far by this method.

Occasionally the process of haploidization is

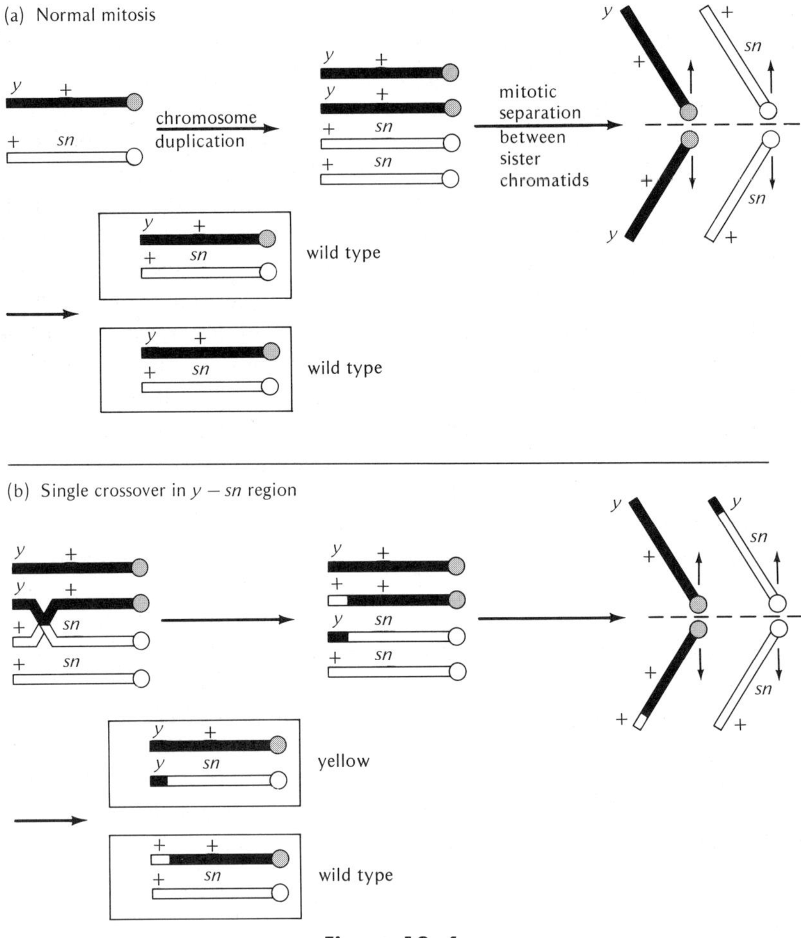

Figure 18–4

Diagram of normal mitosis and mitotic crossover events in females of Drosophila melanogaster *carrying two X chromosomes that differ with respect to the yellow (y) and singed (sn) genes.*

accompanied by mitotic crossing over between markers in the same linkage group and leads to the production of recombinant haploids which are analogous to the recombinant gametes produced by meiosis. This process, which begins with the diploid and ends with a recombinant haploid, has therefore been termed by Pontecorvo a *parasexual* cycle, since it produces genetic variability in the same fashion as sex.

As a rule, however, mitotic recombination between linked markers occurs only rarely in haploidization, and the mapping of such loci is undertaken through analysis of the diploid recombinant products. Among the selective methods that have had to be devised to enable the recognition and selection of diploid mitotic recombinants is a visual selective method in which a diploid strain heterozygous for *yellow* appears green, but forms yellow diploid sectors if mitotic recombination has occurred. This is illustrated in Fig. 18–6, showing that of the two recombinant products, one is distinctly visible. Similarly, a diploid that is heterozy-

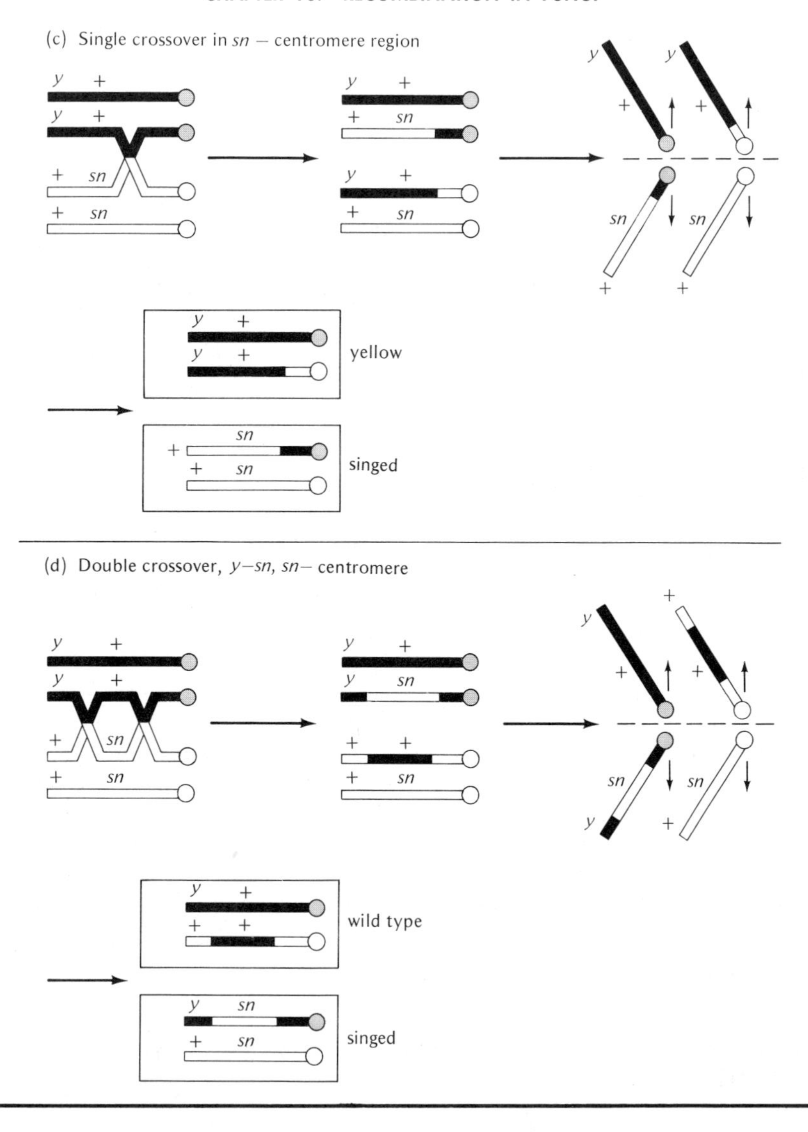

gous for *white* will produce white diploid sectors when mitotic crossing over occurs in that chromosome.

Once recognized, such mitotic recombinants are isolated and tested for recombination with other markers known to be located on the same linkage group. The determination of linkage relationships then depends upon the relationship between homozygosis for one marker and homozygosis for another after further somatic recombination. For example, a diploid heterozygous for the linked markers *a*, *b*, and *c* may produce recombinant homozygotes for various combinations of markers, i.e., *aa*, *bb*, *cc*, *aabb*, *aacc*, etc., depending upon the gene order. If the gene order is centromere-*a*-*b*-*c*, then *c*, being furthest from the centromere, will have the greatest likelihood of becoming recombinant, and some *cc* homozygotes will be formed without being accompanied by homozygosis for *a* and *b*. On the other hand, for this gene order a recombinant for *b* will also include recombination and homozygosis for *c* unless a very rare

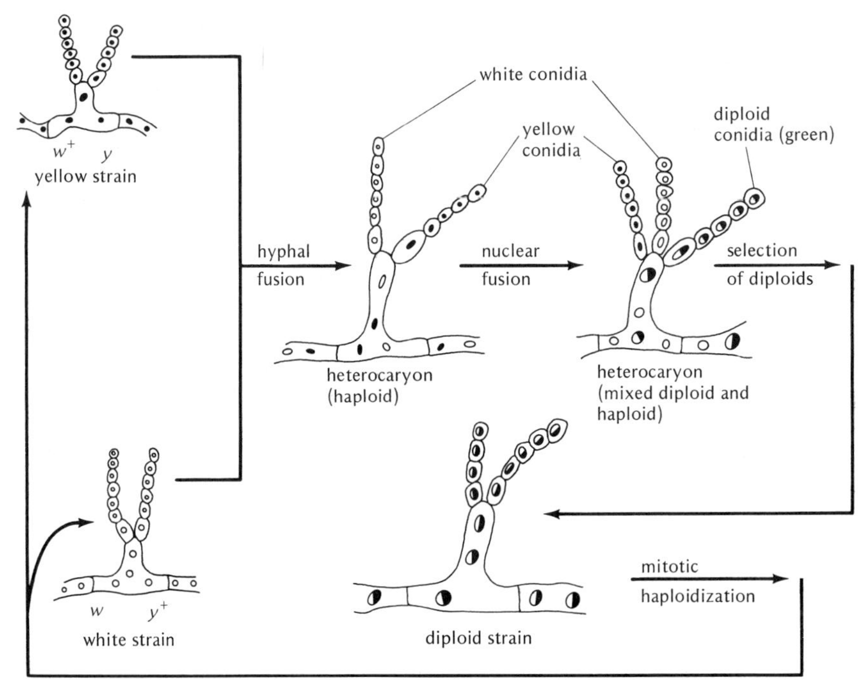

Figure 18–5

The process of diploid formation in heterocaryons of Aspergillus *using visible markers. The white and yellow conidia are haploid and distinctly recognizable. The diploid conidia are green and larger than haploid conidia. Haploid conidia produced by such diploid strains generally occur through a failure in mitosis rather than through the usual meiotic divisions.*

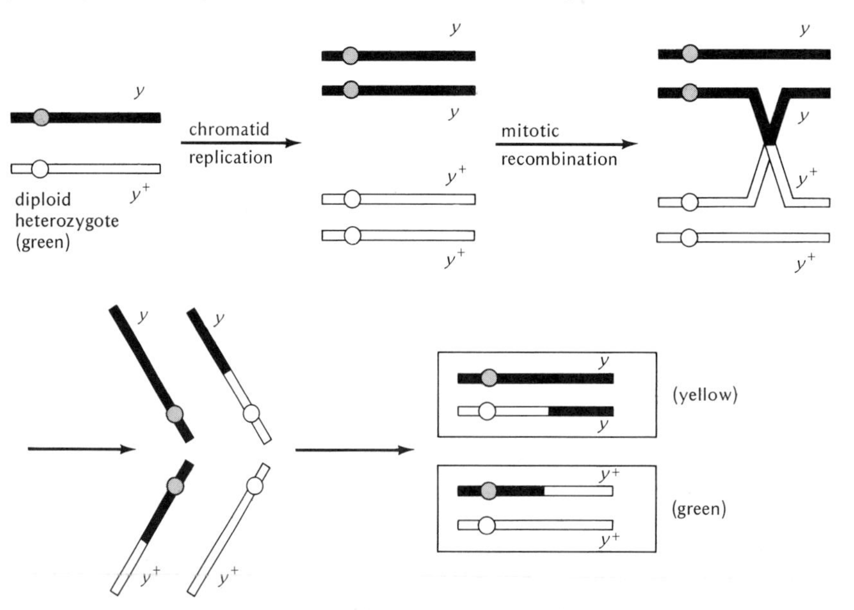

Figure 18–6

Production of yellow sectors in Aspergillus *as a result of mitotic crossing over.*

double crossover takes place. Similarly, recombination for *a* will also include recombination and homozygosis for *b* and *c*.

An actual example of such gene-order determination is given by Pontecorvo and Kafer for a diploid strain which had the heterozygous genotypes *ad*-14 + + +/+ *pro*-1 *pab*-1 *y*. Homozygotes for the adenine-requiring (*ad*-14), proline-requiring (*pro*-1), and *p*-aminobenzoate-requiring (*pab*-1) mutations could all be recognized by their inability to grow on media deficient for the various compounds. The *yellow* (*y*) mutation had no nutritional effect but was visible in homozygous condition, enabling Pontecorvo and Kafer to select recombinants for this particular linkage group. In all, they selected 371 yellow homozygotes, of which 96 showed no homozygosis for the other mutations; i.e., they were prototrophic and could grow on minimal medium. However, 245 of the yellow recombinants required *p*-aminobenzoate, and 30 more required proline as well as *p*-aminobenzoate. Thus recombination for *yellow* was not necessarily accompanied by recombination for the other mutations in 96 cases, and *yellow* could therefore be assigned a position furthest from the centromere (Fig. 18–7a).

The *p*-aminobenzoate homozygotes, although accompanied by yellow homozygotes, were unaccompanied by proline homozygotes in a large majority of cases (245/275) and therefore had a position between *pro* and *y* (Fig. 18–7b). The recombination that produced the proline homozy-

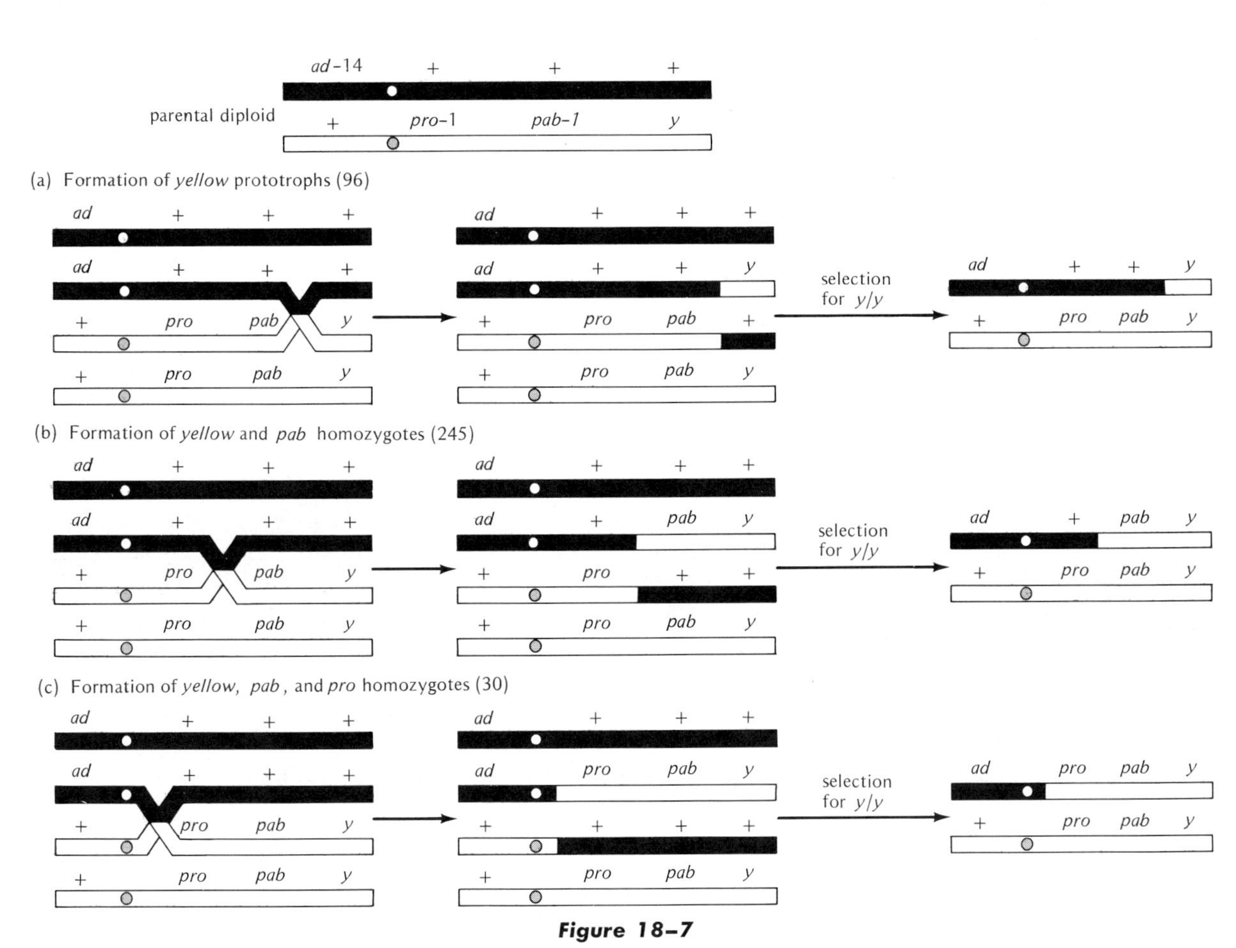

Figure 18–7

Explanation for gene-order determination using mitotic crossing over in diploids of Aspergillus nidulans.

gote also produced homozygosis for *pab* and *y*, and *pro* was therefore closest to the centromere (Fig. 18–7c). On the other hand, no *ad*-14 homozygotes appeared, meaning that this mutation is either to the "right" of the centromere, but so close that crossing over cannot occur, or that it is to the "left" of the centromere, and does not recombine with loci on the other arm because simultaneous mitotic crossing over in both arms of a chromosome is a rare event. This matter was decided by other experiments in which a diploid strain having *ad*-14 and *y* in "coupling" (*ad*-14 *y*/+ +) underwent mitotic recombination and produced *ad*-14/*ad*-14 homozygotes which were nevertheless heterozygous *y*/+. If the centromere were to the left of *ad*-14, a crossover producing an *ad*/*ad* homozygote should also have produced a *y*/*y* homozygote. Since this did not occur, *ad*-14 must be to the left of the centromere.

Map distances for mitotic recombinant loci can be estimated on the basis of the relative incidence of recombination in one area as compared to recombination in all areas. For the above example, a total of 371 recombinants were detected, 96 in the *pab*-*y* interval, 245 in the *pro*-*pab* interval, and 30 in the centromere-*pro* region. The relative frequencies are therefore 26 percent, 66 percent, and 8 percent, respectively. These values do not correspond to meiotic recombination values, which are 36, 18, and 46, respectively. This disparity, however, is not surprising, since neither process need occur with equal constancy over the entire chromosome length.

The most notable difference between mitotic and meiotic recombination is the frequency with which they occur. As a rule mitotic recombination is a rare event, occurring in *Aspergillus* at a rate of about 1 in 100 mitotic cell divisions. Also, once a mitotic crossover occurs in one chromosome, the chance that another one will occur in the same chromosome or even in different chromosomes is quite small. Meiotic recombination, on the other hand, occurs in almost every chromosome, and double exchanges are quite common. However, although its recombination frequency is lower, the mechanism of mitotic recombination appears to follow in principle the mechanism of meiotic recombination. That is, a physical "mating" between parts of homologous chromosomes must occur in which homologous sections lie close enough to enable crossing over to take place.

GENE CONVERSION

Thus far, recombination has been considered a reciprocal event involving some form of mutual exchange between one homologous chromosome and another. For example, the transfer of gene *A* from homologue 1 to homologue 1′ is accompanied by the reciprocal transfer of the allele of *A*, i.e., *a*, from homologue 1′ to homologue 1. In fungi the normal segregation of two alleles, whether transferred between chromosomes or not, should therefore yield tetrads in which each allele is present twice, or a 2:2 ratio. However, departures from this ratio have been found, initially by Lindegren in yeast and later by Mitchell in *Neurospora*.

Mitchell's crosses used genes that affect the synthesis of the nutritional substance, pyridoxine. At one locus there is a mutant allele causing a requirement for pyridoxine that is remedied by a change in acidity (pH) and is called *pdxp*. At another locus, very closely linked to the first, there is a non-pH-sensitive pyridoxine-requiring mutation called *pdx*.* In one experiment, Mitchell crossed the two mutant strains + *pdxp* × *pdx* + and dissected out almost a thousand asci. As shown in Table 18–7, four of these asci showed single pairs of spores that were completely wild type (+ +) for pyridoxine, indicating a recombinational event between the two loci. Contrary to expectations, however, the double-mutant reciprocal products of these recombinational events (*pdx pdxp*) were not recovered, although analyses were made which would have identified them had they existed. Also, these events could not be explained as arising from mutation because their frequency was much higher than the normal mutation rate for these genes.

* Indications exist that these two loci are probably sufficiently close together to be considered a "complex locus"—a topic discussed in Chapter 25.

TABLE 18–7

The four asci of Mitchell's experiment with *Neurospora crassa* (+ *pdxp* × *pdx* +) in which wild-type pyridoxine spores (+ +) appeared*

Spore Pairs	Asci 1	2	3	4
1st	+ *pdxp*	*pdx* +	+ +	*pdx* +
2nd	+ +	*pdx* +	+ *pdxp*	+ *pdxp*
3rd	+ *pdxp*	+ +	*pdx* +	+ +
4th	*pdx* +	+ *pdxp*	*pdx* +	*pdx* +

From data of M. B. Mitchell, 1955.

* Note that of the four spore pairs (eight spores) in each ascus only one pair is completely wild type. In the first ascus this has diminished the expected frequency of *pdx* and in the other asci it has diminished the expected frequency of *pdxp*.

These unusual events were termed *gene conversions*, a term first suggested by Winkler to indicate the apparent conversion of one allele into another (e.g., *pdx* into pdx^+ in ascus 1, Table 18–7). In general many cases of gene conversion seem to be accompanied by recombination for markers outside the conversion area. The outside marker recombination, however, is of the usual reciprocal kind.

Unexpected tetrad ratios have also been found in other fungi, such as *Sordaria* and *Ascobolus*, and conversion phenomena have also even been noted in attached-X chromosome data for *Drosophila melanogaster* gathered by Smith, Finnerty, and Chovnick. In some of the fungal asci, these unusual ratios are not merely of the 3:1 spore-pair type observed by Mitchell, i.e., 6:2 spores, but include also 5:3 and 7:1 ratios. Such odd numbers mean that only one member of a pair of mitotic products immediately following meiosis has become converted. However, because of their frequency, such conversions could not have taken place during mitosis, i.e., through somatic mutation, but must have occurred during meiosis. This means that each of the four chromatids in a tetrad (bivalent) during meiosis probably consists of one duplex (two-stranded) DNA molecule, as suggested by Taylor for the root cells of *Bellevalia* (p. 356).

Gene conversion has nevertheless been found to occur in mitotic events, but there too it is often associated with recombination of outside markers. In general, gene conversion is restricted to relatively short sections of the chromosome although simultaneous conversion of two closely linked loci is also known (see, for example, Fogel and Mortimer). In a number of cases ranging from *Ascobolus* to *Neurospora*, *Aspergillus*, and yeast, gene conversion also seems to occur in a polarized direction favoring the conversion of one genetic site more than an adjacent one (e.g., if more *pdxp* were converted into $pdxp^+$ than *pdx* into pdx^+).

Among the most popular theories that account for gene conversion are those which interpret conversion events through the formation of hybrid DNA ("heteroduplex" structures; see also p. 446); that is, formation of a DNA double helix that contains a single-stranded section from one parental chromatid paired to a single-stranded section from the other parental chromatid. Models of this kind were first proposed by Whitehouse and Hastings and by Holliday, and one such scheme is presented in Fig. 18–8 to illustrate how both gene conversion and crossing over can occur simultaneously (see also reviews of Radding and of Stadler). The model shows that the crossover event caused by DNA "nicking" and unwinding of two homologous chromatids (see also Fig. 17–15) is also accompanied by "mismatching" of one section of hybrid DNA because of the lack of complementary base pairing. Excision of one strand from the hybrid mismatched region then allows the

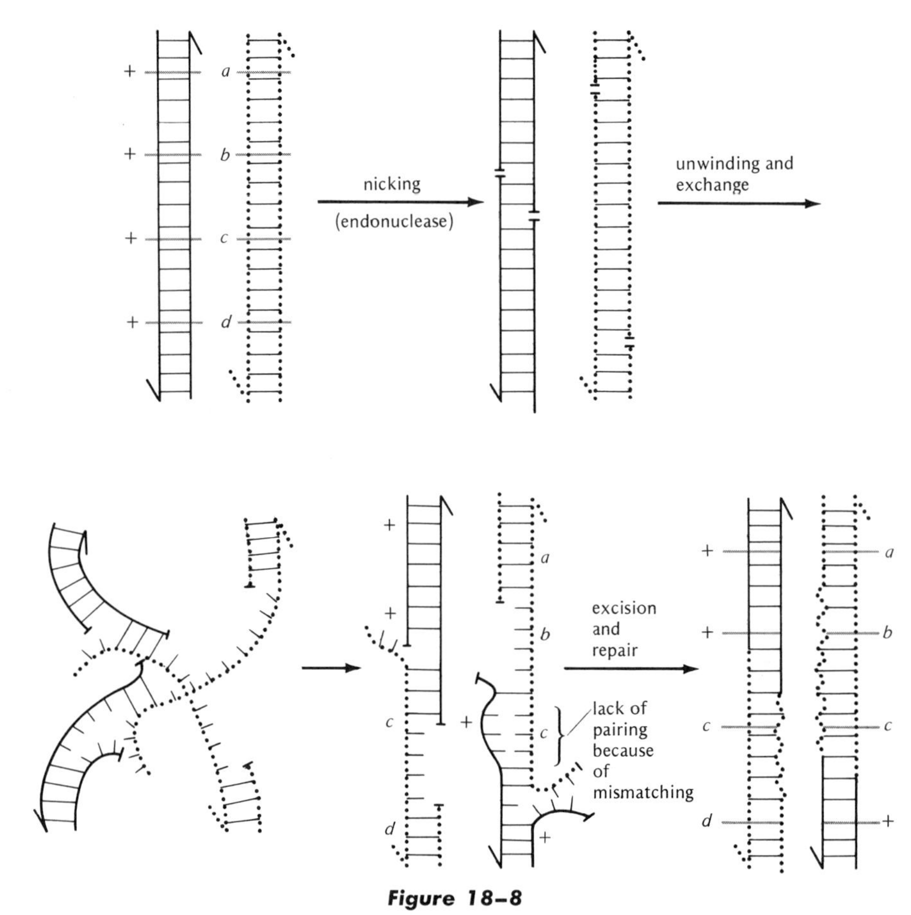

Figure 18–8

Hybrid DNA model showing the occurrence of a conversion event ($c^+ \rightarrow c$) because of the excision of a mismatched section of single-stranded DNA from one of the parental chromatids. As in Fig. 17–15, the solid lines represent a DNA double helix from one chromatid, and the dotted lines represent a DNA molecule from its homologue. The half arrows at the end of each strand indicate the polarity of the sugar-phosphate backbone. There may be hundreds of complementary base pairs between the loci a, b, c, and d, but to simplify the diagram relatively few crossbars between strands have been drawn.

synthesis of a section of DNA double helix which is identical to the homologous section on the other crossover chromatid (locus *c* in Fig. 18–8). In conjunction with the two noncrossover chromatids (+ + + +, *a b c d*), the meiotic products of this tetrad will therefore be

$$\begin{array}{cccc} + & + & + & +, \\ + & + & c & d, \\ a & b & c & +, \\ a & b & c & d. \end{array}$$

The mitotic replication of this tetrad will then give the familiar conversion ratio of 6:2 for the *c* locus.

The possibility also exists that excision of the mismatched section of the chromosome carrying the c^+ gene does not occur, but instead this section remains in the form of a hybrid DNA molecule with + on one strand of DNA and *c* on the other. Mitotic replication of the two crossover chromatids will then produce three DNA molecules (chromatids) bearing *c*, and only one bearing the + allele. When these four chromatids are added to the replication products of the two noncrossover chromatids (+ + + +, *a b c d*), the *c*: + ratio will be 5:3, and would therefore help account for some of the unusual conversion ratios mentioned above.

To account for the polarity effect, the hybrid DNA model assumes that the nicks leading to unwinding tend to occur at specific localities. Since it is these regions which lead to the presence of hybrid DNA and subsequently to conversion, polarity would be explained as a conversion gradient that goes from greater to lesser frequency as loci recede from the initial nick. Satisfactory as the hybrid DNA model is, however, not all events are yet explained by it, although there is no question that crossing over and conversion are frequently associated.

CYTOPLASMIC GENE MAPPING

Linkage and recombination is, of course, not restricted to nuclear chromosomal genes. The existence of cytoplasmic genes associated with the DNA of organelles such as mitochondria and chloroplasts offers the possibility of additional linkage groups. The first indication of recombination between cytoplasmic genes was noted in *Chlamydomonas* by Sager and Ramanis and by Gillham, and similar studies have now been made for a number of mitochondrial gene markers in yeast (see, for example, Avner et al.). It's not yet clear how such genetic exchange occurs in yeast except that the mitochondria seem to be of two kinds, ω^+ and ω^-, and the mitochondrial gene markers of one kind are preferentially incorporated into mitochondrial progeny when the cross is $\omega^+ \times \omega^-$.

In *Chlamydomonas*, a primary difficulty in measuring linkage between cytoplasmic genes was their usually strict maternal inheritance which produced very few biparental progeny. Sager's discovery (see p. 260) that treating mt^+ gametes with ultraviolet light prior to mating would markedly enhance the frequency of biparental zygotes led to an increase in cytoplasmic recombinational opportunities. For example, one set of crosses between *Chlamydomonas* strains sensitive to the antibiotics neamine and streptomycin (*nea-s* sm_2-*s*) and resistant strains (*nea-r* sm_2-*r*) produced 1200 progeny of which about 20 percent were recombinant (*nea-s* sm_2-*r*; *nea-r* sm_2-*s*). By these and other means a tentative linkage map for the *Chlamydomonas* cytoplasmic genes shown in Fig. 18–9 has been proposed by Sager. Interestingly the linkage map she offers is circular indicating that all of these genes are linked in sequence to each other as though the chromosome on which they are located were circular. Note, for example, that every gene on the map is no greater than about 30 recombinational units from any other gene although the entire map length is about 60 units. Although some aspects of Sager's analysis have been questioned (Gillham 1974), the presence of circular linkage maps is not unusual for small DNA chromosomes, and their presence in various bacteria and viruses will be discussed in Chapters 19 and 20.

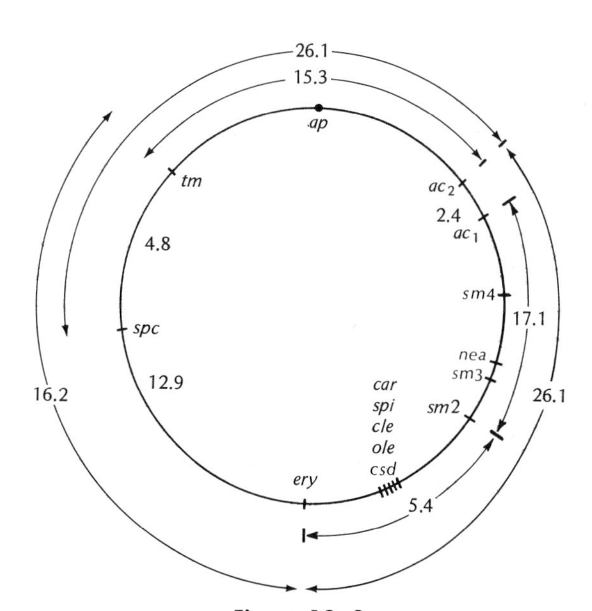

Figure 18–9

Tentative linkage map for cytoplasmic genes (chloroplast) for Chlamydomonas reinhardi. Numbers represent recombination frequencies Symbols; ap, attachment point; ac2 and ac1, acetate-requirement; sm4, streptomycin dependence; nea, neamine resistance; sm3, low level streptomycin resistance; sm2, high level streptomycin resistance; ery, erythromycin resistance; csd, conditional streptomycin dependence; car, carbomycin resistance; spi, spiramycin resistance; cle, cleosine resistance; ole, oleandomycin resistance; spc, spectinomycin resistance; and tm, temperate sensitivity. (From Sager, 1972).

PROBLEMS

18-1. For each of the following traits in *Neurospora* in which first- and second-division segregation was observed, state the gene-centromere distance (data of Lindegren).

Trait	No. Asci: First-Division Segregation	Second-Division Segregation
(a) Mating type (*A* vs. *a*)	331	51
(b) *Pale* conidial color vs. orange	73	36
(c) *Fluffy* growth form vs. normal	42	67

18-2. The following are the results of ordered tetrad analysis from a cross between a *Neurospora* strain carrying *albino* (*al*) that was also unable to synthesize *inositol* (*inos*) and a wild-type strain (+ +) (adapted from data of Houlahan, Beadle, and Calhoun, 1949):

①	②	③	④	⑤	⑥	⑦
al inos	*al* +	*al* +	*al* +	*al inos*	*al* +	*al* +
al inos	*al* +	*al inos*	+ +	+ +	+ *inos*	+ *inos*
+ +	+ *inos*	+ +	*al inos*	*al inos*	*al* +	+ +
+ +	+ *inos*	+ *inos*	+ *inos*	+ +	+ *inos*	*al inos*
4	3	23	36	15	16	22

(a) Determine whether these two genes are linked and, if so, the linkage distance between them. (b) Which gene has the longest gene-centromere distance?

18-3. A cross between a *Neurospora* stock bearing the mutant gene *snowflake* (*sn*), which was otherwise wild type, to a *colonial temperature-sensitive* stock (*cot*), which was otherwise wild type, produced a number of regular tetrads showing first- and second-division segregation (see top of next column; from data of M. B. Mitchell). Determine whether these genes are linked, and their gene-centromere distances.

sn cot	*sn* +	*sn cot*	*sn cot*	+ +	+ *cot*	*sn* +
sn cot	*sn* +	*sn* +	*sn* +	+ *cot*	+ +	*sn cot*
+ +	+ *cot*	+ *cot*	+ +	*sn cot*	*sn* +	+ +
+ +	+ *cot*	+ +	+ *cot*	*sn* +	*sn cot*	+ *cot*
25	16	11	6	8	6	8

18-4. A *Neurospora* stock which was adenine-requiring (*ad*) and tryptophan-requiring (*tryp*) was mated to a wild type stock (+ +) and produced the following tetrads:

①	②	③	④	⑤	⑥	⑦
ad tryp	*ad* +	*ad tryp*	*ad tryp*	*ad tryp*	*ad* +	*ad tryp*
ad tryp	*ad* +	*ad* +	+ *tryp*	+ +	+ *tryp*	+ +
+ +	+ *tryp*	+ *tryp*	*ad* +	*ad tryp*	*ad* +	+ *tryp*
+ +	+ *tryp*	+ +	+ +	+ +	+ *tryp*	*ad* +
49	7	31	2	8	1	2

Determine whether these two genes are linked and, if so, draw a linkage map including the centromere.

18-5. In addition to the *ad* and *tryp* loci in the above problem, there are other such loci in the *Neurospora* genome which also affect the requirements for adenine and tryptophan. In a cross similar to the one in Problem 18-4 but using different *ad* and *tryp* genes, the same classes of tetrads were observed but the numbers of tetrads in each class were different, as follows: ① = 36, ② = 1, ③ = 39, ④ = 21, ⑤ = 1, ⑥ = 1, ⑦ = 1. Determine whether these *ad* and *tryp* genes are linked and, if so, draw a linkage map that includes the centromere.

18-6. In a cross between a *Neurospora* strain requiring histidine (*hist*-2) and a strain requiring alanine (*al*-2), Giles and co-workers analyzed 646 unordered tetrads with the following compositions and numbers.

Number	Tetrad Composition
115	2 *hist* al^+:2 $hist^+$ *al*
45	2 *hist al* :2 $hist^+$ al^+
486	1 *hist al* :1 *hist* al^+:1 $hist^+$ *al*:1 $hist^+$ al^+

Determine whether the two genes are linked and, if so, the linkage distance between them.

18-7. In an experiment with yeast (unordered tetrads) a cross was made between two strains each differing in respect to three different genes *a*, *b*, *c*, not necessarily linked on the same chromosome. The cross was $+ b c \times a + +$, and produced the following tetrads:

Number	Tetrad Composition
407	2 + + *c* :2 *a b* +
395	2 + *b c* :2 *a* + +
104	1 + *b c* :1 + *b* + :1 *a* + *c*: 1 *a* + +
92	1 + + *c* :1 + + + :1 *a b c*: 1 *a b* +
1	2 + *b* + :2 *a* + *c*
1	2 + + + :2 *a b c*

From these data determine which genes are linked, if any, and the linkage distances.

18-8. The test performed by Stern to detect mitotic recombination in *Drosophila* utilized two X chromosomes, one *y* +, and the other + *sn* (Fig. 18-4). Draw the necessary diagrams to determine whether mitotic recombination could be detected in females carrying these mutations in "coupling", i.e., *y sn*/+ +.

18-9. Mitotic recombination analysis in *Aspergillus* revealed the linked order *riboflavine* (*ribo*-1)–*adenine* (*ad*-14)–centromere. Other tests revealed a linked order for two other mutations: *biotin* (*bi*-1)–*proline* (*pro*-1)–centromere. Mitotic recombination alone revealed no linkage between these two groups, but haploidization showed that they segregated together; i.e., they were all on the same chromosome. What is the centromere position in respect to these genes?

18-10. Two diploid strains of *Aspergillus nidulans,* Y and Z, were each constructed of two linked mutations, *white* (*w*) and *Acriflavine* (*Acr*), so that Y was $Acr\,w/Acr^+\,w^+$ (coupling) and Z was $Acr^+\,w/Acr\,w^+$ (repulsion). Pontecorvo and Kafer found that the white diploids (*w*/*w*) that arose through mitotic recombination in strain Y were all homozygous *Acr*/*Acr*, and the white diploids that arose through mitotic recombination in strain Z were all homozygous Acr^+/Acr^+. On the other hand, the homozygous *Acr*/*Acr* individuals produced by mitotic recombination in strain Y were 87 percent homozygous *w*/*w* and 13 percent heterozygous w/w^+. Similarly the homozygous *Acr*/*Acr* recombinants in strain Z were 82 percent homozygous w^+/w^+ and 18 percent heterozygous w/w^+. What is the linkage relationship between these genes and the centromere?

18-11. A *Neurospora* cross was made between an *albino* strain of *A* mating type that was wild type for leucine requirements (*al* + *A*) to a leucine-requiring strain of *a* mating type (+ *leu a*). Derive the linkage order and linkage distances for the three genes from the following 1000 ordered tetrads that were obtained from this cross:

Tetrad	Spore pair 1	Spore pair 2	Spore pair 3	Spore pair 4	Number
①	*al* + *A*	*al* + *A*	+ *leu a*	+ *leu a*	591
②	*al* + *A*	+ + *A*	*al leu a*	+ *leu a*	142
③	*al* + *A*	*al leu A*	+ + *a*	+ *leu a*	134
④	*al* + *A*	*al leu a*	+ + *A*	+ *leu a*	123
⑤	*al* + *A*	+ *leu a*	*al* + *A*	+ *leu a*	1
⑥	*al leu a*	+ + *A*	*al* + *A*	+ *leu a*	1
⑦	*al* + *A*	+ *leu a*	+ + *A*	*al leu a*	1
⑧	+ + *A*	*al leu a*	+ + *A*	*al leu a*	1
⑨	*al* + *A*	+ *leu A*	*al* + *a*	+ *leu a*	1
⑩	*al leu A*	+ + *A*	*al* + *a*	+ *leu a*	1
⑪	*al* + *A*	+ *leu A*	*al leu a*	+ + *a*	1
⑫	*al* + *A*	*al leu a*	+ *leu A*	+ + *a*	1
⑬	*al* + *A*	*al* + *a*	+ *leu A*	+ *leu a*	1
⑭	+ *leu A*	*al* + *a*	+ + *A*	*al leu a*	1

18-12. A diploid strain of *Aspergillus* that was heterozygous for the mutations for *yellow* (*y*), *white* 3 (*w*-3), *putrescine* (*pu*), *adenine* 1 (*ad*), *small* (*sm*), and *phenylalanine* 2 (*phen*-2) produced haploids of which 41 yellow phenotypes were selected and tested. These yellow haploids had the following genotypes and numbers:

7 *y w*-3 *pu ad*-1 *sm phen*-2
11 *y w*-3 *pu ad*-1 + +
16 *y* + + + *sm phen*-2
7 *y* + + + + +

What are the linkage relationships among these genes?

REFERENCES

AVNER, P. R., D. COEN, B. DUJON, and P. P. SLONIMSKI, 1973. Mitochondrial genetics IV. Allelism and mapping studies of oligomycin resistant mutants in *S. cerevisiae. Molec. Gen. Genet.*, **125**, 9–52.

BARRATT, R. W., D. NEWMEYER, D. D. PERKINS, and L. GARNJOBST, 1954. Map construction in *Neurospora crassa. Adv. in Genet.*, **6**, 1–93.

BOLE-GOWDA, B. N., D. D. PERKINS, and W. W. STRICKLAND, 1962. Crossing over and interference in the centromere region of linkage group I of *Neurospora. Genetics*, **47**, 1243–1252.

EMERSON, S., 1963. Meiotic recombination in fungi with special reference to tetrad analysis. In *Methodology in Basic Genetics*, W. J. Burdette (ed.). Holden-Day, San Francisco, pp. 167–206.

———, 1969. Linkage and recombination at the chromosome level. In *Genetic Organization*, Vol. I, E. W. Caspari and A. W. Ravin (eds.). Academic Press, New York, pp. 267–360.

FINCHAM, J. R. S., and P. R. DAY, 1971. *Fungal Genetics*, 3rd ed. Blackwell, Oxford.

FOGEL, S., and R. K. MORTIMER, 1969. Informational transfer by meiotic gene conversion. *Proc. Nat. Acad. Sci.*, **62**, 96–103.

GARCIA-BELLIDO, A., 1972. Some parameters of mitotic recombination in *Drosophila melanogaster. Molec. Gen. Genet.*, **115**, 54–72.

GILLHAM, N. W., 1965. Linkage and recombination between non-chromosomal mutations in *Chlamydomonas reinhardi. Proc. Nat. Acad. Sci.*, **54**, 1560–1567.

———, 1974. Genetic analysis of the chloroplast and mitochondrial genomes. *Ann. Rev. Genet.*, **8**, 347–391.

HOLLIDAY, R., 1964. A mechanism for gene conversion in fungi. *Genet. Res.*, **5**, 282–304.

HOULAHAN, M. B., G. W. BEADLE, and H. G. CALHOUN, 1949. Linkage studies with biochemical mutants of *Neurospora crassa. Genetics*, **34**, 493–507.

KAFER, E., 1958. An 8-chromosome map of *Aspergillus nidulans. Adv. in Genet.*, **9**, 105–146.

LINDEGREN, C. C., 1953. Gene conversion in *Saccharomyces. Jour. Genet.*, **51**, 625–637.

LISSOUBA, P., and G. RIZET, 1960. Sur l'existence d'une génétique polarisée ne subissant que des échanges non réciproques. *C. R. Acad. Sci., Paris*, **250**, 3408–3410.

MITCHELL, M. B., 1955. Aberrant recombination of pyridoxine mutants of *Neurospora. Proc. Nat. Acad. Sci.*, **41**, 215–220.

OLIVE, L. S., 1959. Aberrant tetrads in *Sordaria fimicola. Proc. Nat. Acad. Sci.*, **45**, 727–732. (Reprinted in the collection of Taylor; see References, Chapter 1.)

PONTECORVO, G., and E. KAFER, 1958. Genetic analysis based on mitotic recombination. *Adv. in Genet.*, **9**, 71–104.

PRITCHARD, R. H., 1955. The linear arrangement of a series of alleles of *Aspergillus nidulans. Heredity*, **9**, 343–371.

RADDING, C. M., 1973. Molecular mechanisms in genetic recombination. *Ann. Rev. Genet.*, **7**, 87–111.

ROPER, J. A., 1952. Production of heterozygous diploids in filamentous fungi. *Experientia*, **8**, 14–15.

———, 1966. Mechanisms of inheritance. 3. The parasexual cycle. In *The Fungi*, Vol. II. G. C. Ainsworth and A. S. Sussman (eds.). Academic Press. New York, pp. 589–617.

SAGER, R., 1972. *Cytoplasmic Genes and Organelles.* Academic Press, New York.

SAGER, R., and Z. RAMANIS, 1965. Recombination of nonchromosomal genes in *Chlamydomonas. Proc. Nat. Acad. Sci.*, **53**, 1053–1061.

SMITH, P. D., V. G. FINNERTY, and A. CHOVNICK, 1970. Gene conversion in *Drosophila*: Non-reciprocal events at the maroon-like cistron. *Nature*, **228**, 442–444.

STADLER, D. R., 1973. The mechanism of intragenic recombination. *Ann. Rev. Genet.*, **7**, 113–127.

STERN, C., 1936. Somatic crossing over and segregation in *Drosophila melanogaster. Genetics*, **21**, 625–730.

WHITEHOUSE, H. L. K., 1949. Multiple allelomorph heterothallism in the fungi. *New Phytologist*, **48**, 212–244.

———, 1972. Chromosomes and recombination. *Brookhaven Symp. Biol.*, **23**, 293–324.

WHITEHOUSE, H. L. K., and P. J. HASTINGS, 1965. The analysis of genetic recombination on the polaron hybrid DNA model. *Genet. Res.*, **6**, 27–92.

WINKLER, H., 1930. *Die Konversion der Gene.* G. Fischer, Jena.

19
RECOMBINATION IN BACTERIA

The understanding of linkage relationships between genes in bacteria and viruses was made possible by the discovery of genetic recombination in these organisms. Gene recombination, coupled with the short generation time of these organisms, the numerous conveniently recognized mutants that they possessed, and the ease with which recombinant colonies could be plated and scored gave bacteria and viruses a decided advantage for recombinational studies over those possessed by "higher organisms." At the present time research utilizing bacterial and viral recombination is one of the most active areas in all genetics.

TRANSFORMATION

The first evidence of genetic recombination or exchange of hereditary material in bacteria was noted by Griffith in the transformation of harmless pneumococci into virulent ones (Chapter 4). As fully demonstrated by Avery, MacLeod, and McCarty in 1944, the transforming agent responsible for the observed genetic change turned out to be pure DNA that could be extracted from donor strains. Recent studies have shown that during the process of extraction, the donor DNA is broken into smaller transforming molecules which are about 1/200 of the total donor DNA. These transforming molecules are usually large, containing on the average about 20,000 nucleotide pairs. Smaller DNA pieces can also be absorbed by the recipient cell, but a minimum length of about 450 base pairs seems to be essential for transformation to occur.

When recipient bacteria are exposed to these molecules, some bacteria may become transformed, although the frequency is usually low. In bacterial cultures that are "competent" (transformable), the frequency of transformation usually stays at about 1 percent, but under optimal conditions in *D. pneumoniae* it may reach 10 percent. The reason for this low frequency of transformation seems to reside in the relatively small number of transforming molecules (10 or so) that a recipient bacteria can absorb. It has been proposed that a restricted number of entry "sites" exist on the bacterial cell wall through which transforming molecules may enter. Since transformation may be prevented by agents that affect enzymes, such as those that block protein synthesis (chloramphenicol) or block energy production (dinitrophenol), it is suggested that these entry sites are probably areas involving active enzymatic processes rather than merely "holes" in the cell wall. In *D. pneumoniae* there is evidence indicating that the growing zone of the cell envelope is the area actively involved in absorbing donor DNA (see Tomasz, Zanati, and Ziegler).

Whatever the entry mechanism, transforming or donor DNA appears to enter in double-stranded form and is then broken down within the cell into single strands. In *D. pneumoniae* it is generally believed at present that only one of the two strands remains intact and is incorporated into the recipient DNA, while the other degenerates into smaller fragments. In *Hemophilus influenzae*, Stuy has suggested that incorporation of transforming DNA is in double-stranded form, whereas Goodgal and co-workers have shown that single-stranded DNA is also capable of transformation. However, whether single- or double-stranded, Lacks and others have shown that this incorporation is physically direct, since labeled transforming DNA produces labeled recipient DNA. Also, as demonstrated by Fox and others, incorporation of transforming DNA proceeds in the absence of DNA synthesis, thereby testifying to the physical insertion of transforming DNA into the recipient DNA molecule.

This incorporation of transforming DNA, however, does not take place through its insertion as a duplicate piece of material which is in addition to the normal DNA complement. Instead, the transforming DNA is substituted or exchanged for a homologous section of recipient DNA. Evidence for this comes from a demonstration by Harriet Taylor that transformation is reversible. She transformed chain-producing pneumococci, fil^+, into the more usual unchained form, fil^-, and then transformed these once again into fil^+. If the initial transformation ($fil^+ \rightarrow fil^-$) were caused by the "addition" of new fil^- DNA (Fig. 19–1a), the transformation of fil^- to fil^+ would have to be explained as arising from the "loss" of fil^--DNA rather than from its transformation to fil^+ DNA (Fig. 19–1b). Since there is no evidence that fil^+ transforming DNA causes the removal of fil^--DNA, the changes accompanying the many reversible transformations that have been observed are more easily explained as the *substitution* of a section of DNA of one alternative form for another (Fig. 19–1c).

Transformation thus seems to arise from some form of recombination mechanism which produces gene exchange similar to that produced by sexual recombination. As shown in Fig. 19–2, at least some complementary base pairing between single-stranded DNA sections appears necessary in transformation. Thus, if "foreign" donor DNA is used which bears little or no homology with the recipient, successful transformation will not take place. Only in cases of transfection by viruses (p. 66) can extracted nonhomologous nucleic acids successfully replicate themselves in a recipient cell.

LINKAGE

The chain of 20,000 nucleotide pairs in the average transforming DNA molecule certainly seems sufficiently long to contain within it more than one gene. For example, a transforming molecule carrying gene *A* may also carry gene *B*. If these genes are closely linked, there is a good likelihood that transformation at the *A* locus produced by a single DNA molecule would also produce transformation at the *B* locus (*double transformation*). If *A* and *B* are not linked within one transforming DNA

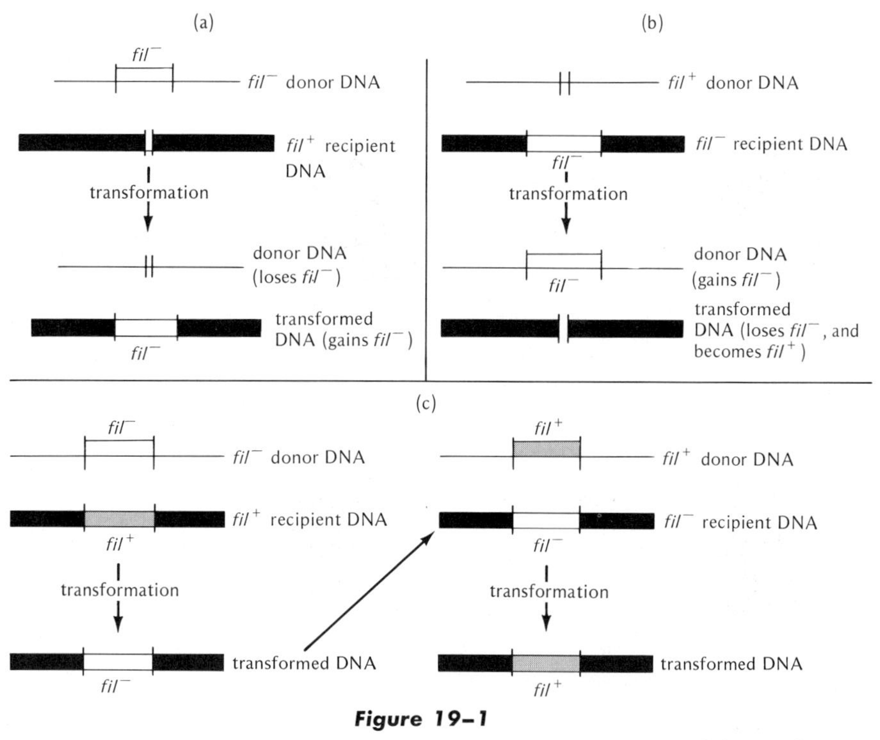

Figure 19–1

Two alternative explanations for transformation in D. pneumoniae. In (a) and (b) transformation is explained as the loss of a portion of chromosome by the donor and its consequent gain by the recipient. Note that in (b) this means the loss of a formerly acquired section of chromosome when fil^- is the recipient, and a gain of this section by the donor chromosome. Such an explanation is obviously contradictory. A more reasonable explanation is presented in (c), in which transformation means an exchange of sections between donor and recipient chromosomes.

molecule, the frequency of double transformation will depend upon the presence of *two* different transforming DNA molecules. In the latter case, double transformation is caused by two independent events; i.e., its frequency is the product of both. Thus closely linked genes will produce a much higher frequency of double transformants than those that are not linked or only distantly linked. The first example of linkage between two genes in transformation was shown by Hotchkiss and Marmur for streptomycin-resistant and mannitol-utilizing strains in *D. pneumoniae*. Since then linkage relationships have been analyzed in *Hemophilus* and in *Bacillus subtilis*.

An illustration of how linkage is now determined can be taken from the work of Nester, Schafer, and Lederberg. In *Bacillus subtilis*, two known genes affect the synthesis of tryptophan (trp_2^-) and tyrosine (tyr_1^-), respectively, so that double-mutant stocks are unable to grow on a medium unsupplemented with both of these amino acids. (For the detection of nutritional mutations, see p. 522.) This double mutant, however, may be transformed by the DNA of other strains so that both single ($trp^+ \, tyr^-$ and $trp^- \, tyr^+$) and double transformants ($trp^+ \, tyr^+$) can be selected and scored. In one experiment the transforming donor DNA was given as a mixture of two separate types, $trp_2^+ \, tyr_1^-$ and $trp_2^- \, tyr_1^+$ to recipient $trp_2^- \, tyr_1^-$ cells. As shown in Table 19–1, three

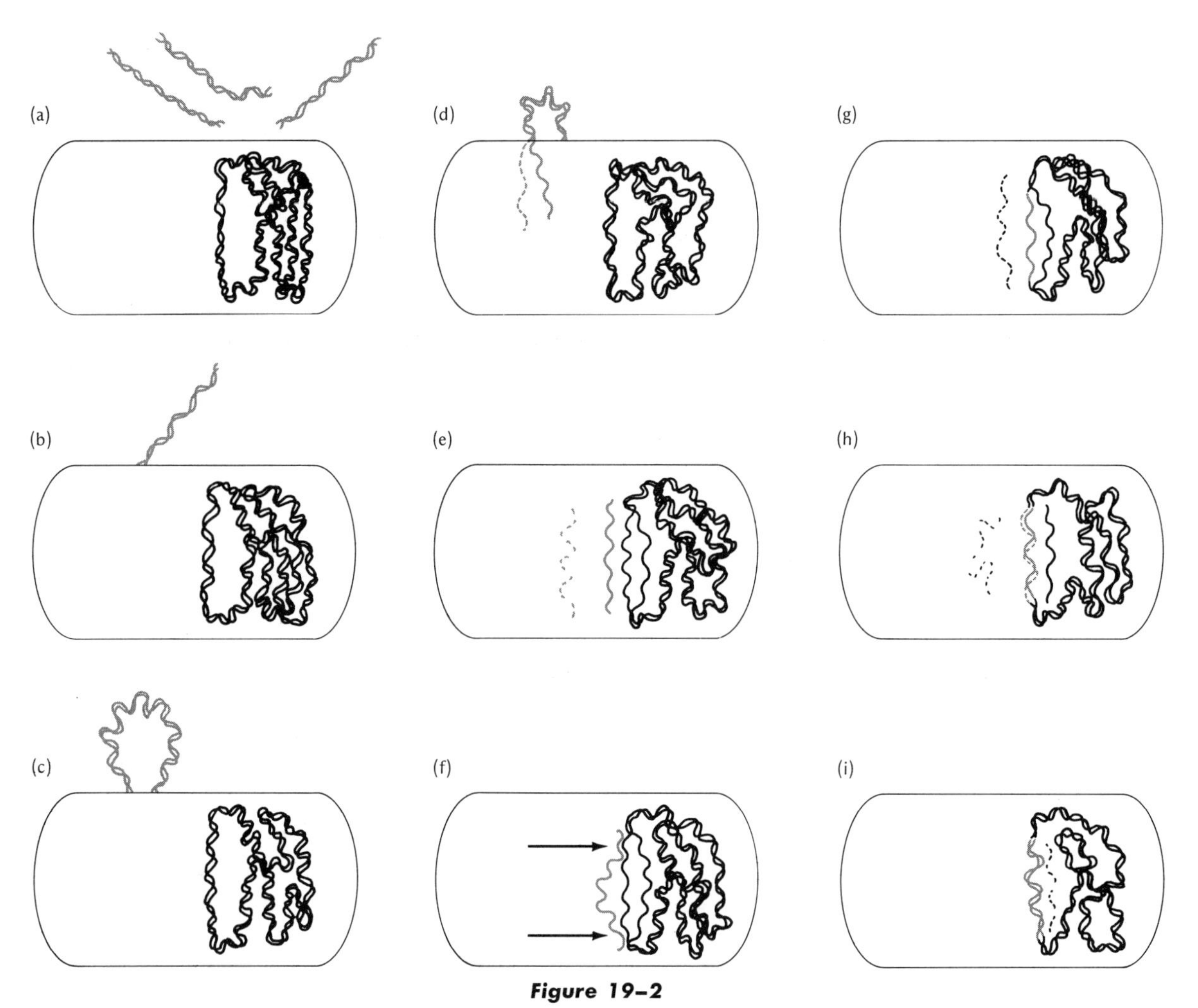

Figure 19–2

The sequence of transformation events according to some present molecular concepts. (a) Donor DNA molecules moving randomly near a bacterial cell. (b) The binding of one end of a donor DNA double helix to an entry site on a recipient cell. (c) Possible subsequent binding of the opposite end of the donor molecule. (d) Entry of one end of the donor molecule through the cell wall. As this occurs, one of the two DNA strands is usually degraded and only a single DNA strand remains. This may be caused by an exonuclease enzyme which degrades a DNA strand in a particular direction (e.g., 5′ to 3′, see p. 85) while simultaneously "drawing in" the complementary nondegraded strand. (e) "Activation" of the recipient chromosome to form one or more single-stranded sections. In many cases reported so far the activation process does not seem to be associated with DNA replication. In fact, Hotchkiss and Gabor point out: "Possibly the whole art of transformation consists of persuading cells to prepare for an easy-going regime in which there will be minimal need for DNA synthesis or repair systems, then suddenly confronting them with an information-carrying segment of DNA so nearly homologous that they are not prepared to resist it." (f) Synapsis between homologous single-stranded sections of donor and recipient molecules by complementary base-to-base pairing (arrows). This is followed by recombination events at each of the two paired sections. (g) Recombination has led to integration of the donor strand and replacement of a recipient strand. (h) The donor strand "replicates" forming a double helix while the remaining unpaired recipient strand is excised. (i) Completed transformation process. As in other molecular recombinational sequences (Figs. 17–15 and 18–8), the excision and degradation of molecules, patching of "gaps," and end-to-end joining of strands is accomplished through the various nuclease, polymerase, and ligase enzymes. (Derived from various sources including Shoemaker and Guild, and Williams and Green.)

TABLE 19–1

Number of single- and double-transformant classes arising from the transformation of trp_2^- tyr_1^- cells in *Bacillus subtilis*

Donor DNA		Recipient Cells	Transformant Classes	No. Colonies Scored
			trp^+ tyr^-	190
trp_2^+ tyr_1^- / trp_2^- tyr_1^+	×	trp_2^- tyr_1^-	trp^- tyr^+	256
			trp^+ tyr^+	2
			trp^+ tyr^-	196
trp_2^+ tyr_1^+	×	trp_2^- tyr_1^-	trp^- tyr^+	328
			trp^+ tyr^+	367

Corrected data of Nester, Schafer, and Lederberg, 1963.

transformant classes arose as a result of this treatment, of which two were single transformants and one was the double type. Note that the double-transformant class is present in very low frequency, since recombination with two separate transforming DNA molecules was necessary to obtain it. On the other hand, when transforming DNA is used from a trp_2^+ tyr_1^+ donor, the frequency of double transformants rises significantly. Both wild-type alleles for these markers are apparently carried on the same transforming molecule, and it is therefore obvious that these two genes are linked.

The extent of linkage can be calculated on the basis of recombination between the two markers of the donor DNA as a proportion of all the bacteria that were transformed. Note that when trp_2^+ tyr_1^+ DNA is used as a donor, the recombinant classes are the single transformants, trp^+ tyr^- and trp^- tyr^+. In this sense recombination means an exchange within the interval between the two genes so that only part of the transforming DNA molecule has been incorporated (Fig. 19–3). As might be expected from other recombination mechanisms, the longer the distance between the two genes, the easier it is for single transformants to arise; the shorter the distance, the more likelihood that both genes will be transformed together. In this experiment, the only transformed class in which the two gene markers have not recombined is the double transformant trp^+ tyr^+. The frequency of recombination between these two genes is then the proportion of recombinants among all transformants, or $(196 + 328)/(196 + 328 + 367) = 58.8$ percent. This value, called q, thus represents the approximate transformation linkage distance between trp_2 and tyr_1. Unfortunately transformation linkage distances are not as repeatable as sexual recombination linkage distances, and considerable variation may occur between experiments. On the whole, however, they furnish a general estimate permitting rough comparisons with other transformation linkage distances.

When differences at three or more linked loci are simultaneously involved in a transformation experiment, not only recombination distances, but also linkage order between the genes, can be calculated. One such experiment performed by Nester, Schafer, and Lederberg consisted of a cross between transforming DNA from a prototrophic strain able to synthesize tryptophan, histidine, and tyrosine (trp_2^+ his_2^+ tyr_1^+) to an auxotrophic strain unable to synthesize any of these amino acids (trp_2^- his_2^- tyr_1^-). The numbers of each of the seven transformant classes produced from this cross are given in Table 19–2. On the basis of these results three individual linkage distances can be calculated, trp_2-his_2, trp_2-tyr_1, and his_2-tyr_1. As

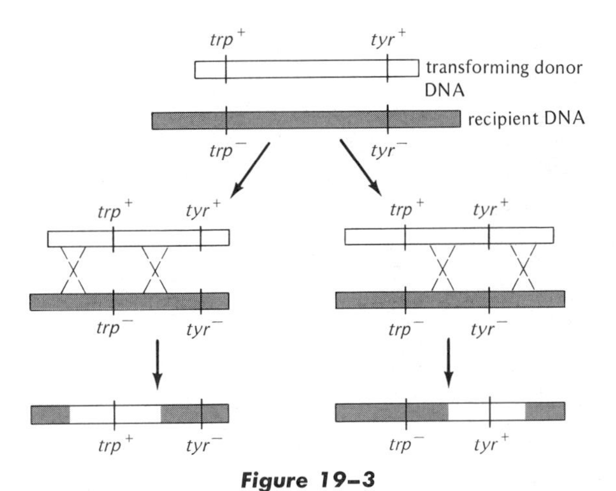

Figure 19–3

Mode of production of single transformants (trp^+ tyr^- or trp^- tyr^+) when the donor DNA is trp^+ tyr^+ and the recipient is trp^- tyr^-.

TABLE 19–2

Number of colonies in different transformant classes from a cross of trp_2^+ his_2^+ tyr_1^+ as the donor with trp_2^- his_2^- tyr_1^- as the recipient

trp^-	trp^-	trp^-	trp^+	trp^+	trp^+	trp^+
his^-	his^+	his^+	his^-	his^-	his^+	his^+
tyr^+	tyr^-	tyr^+	tyr^-	tyr^+	tyr^-	tyr^+
685	418	3660	2600	107	1180	11,940

From Nester, Schafer, and Lederberg, 1963.

TABLE 19–3

Calculation of linkage distances and a linkage map between the three loci trp_2, his_2, and tyr_1, on the basis of transformant classes given in Table 19–2

Distance	Transformants		No. Colonies	q
trp_2---his_2	trp^+ his^-	recomb.	2600 + 107	
	trp^- his^+	recomb.	418 + 3660	6785/19,905 = .34
	trp^+ his^+	parental	1180 + 11,940	
trp_2---tyr_1	trp^+ tyr^-	recomb.	2600 + 1180	
	trp^- tyr^+	recomb.	685 + 3660	8125/20,172 = .40
	trp^+ tyr^+	parental	107 + 11,940	
his_2---tyr_1	his^+ tyr^-	recomb.	418 + 1180	
	his^- tyr^+	recomb.	685 + 107	2390/17,990 = .13
	his^+ tyr^+	parental	3,660 + 11,940	

trp —— *his* —— *tyr*
← .34 → ← .13 →
← .40 →

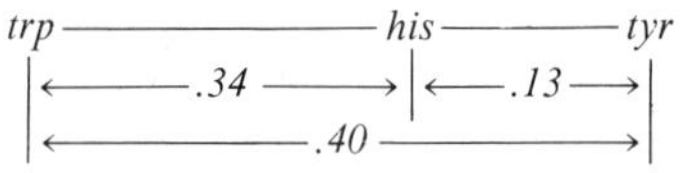

shown in Table 19–3, these distances yield a linkage relationship in which the distance trp_2-tyr_1 is larger than either of the intermediate distances. In other words, the linkage order is most likely trp_2-his_2-tyr_1. Further evidence for this can be gained by noting that the phenotype of the rarest transformant class is trp_2^+ his_2^- tyr_1^+. If the gene order is as given, more simultaneous crossovers (four) are necessary to produce this particular class than any other, as shown in the diagram below, and its low frequency is therefore fully expected. As a rule, therefore, the rarest phenotype in a triple-transformation experiment represents the quadruple recombinant class, and the order of the three genes can be determined accordingly.

SEXUAL CONJUGATION

Transformation, as discovered by Griffith in 1928, was only the first recombination mechanism found in bacteria. The second, discovered by Lederberg and Tatum approximately 20 years later, was that of sexual conjugation in *Escherichia coli*. As explained previously (p. 34), this discovery hinged upon the appearance of prototrophs, i.e.,

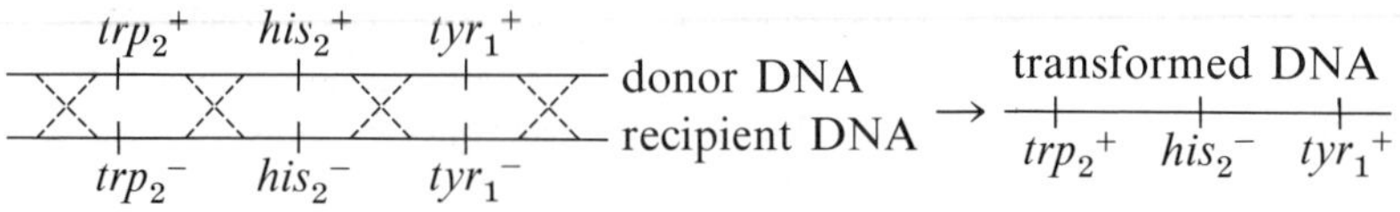

trp_2^+ his_2^+ tyr_1^+ donor DNA
trp_2^- his_2^- tyr_1^- recipient DNA
→ transformed DNA
trp_2^+ his_2^- tyr_1^+

$T^+L^+B_1^+B^+Ph^+C^+$, from a mixed culture of two kinds of auxotrophs, i.e., $T^-L^-B_1^-B^+Ph^+C^+ \times T^+L^+B_1^+B^-Ph^-C^-$.* Since these prototrophs only arose through mutual cell contact between the strains, and it was unlikely that mutation of all three markers could occur simultaneously at the observed frequencies, transformation and mutation as possible explanations for the conjugation experiment could reasonably be excluded.

To explain these events it was first proposed that both auxotrophic cells fused physically to form a zygote. This assumption, however, was soon shown to be erroneous by Hayes, since the continued viability of only one of the auxotrophic cells was necessary for the fertility of the mating. Only strain A, for example, appeared to be incorporated entirely into zygotes, and its death led to the loss of recombinant classes. The complementary mating strain B, on the other hand, could enter into sexual conjugation and then die without in any way diminishing the frequency of recombinants. Thus strain A acted as the "recipient" of zygotic material, while the function of Strain B seemed to be confined to the role of "donor." Once classified as either donor or recipient, a strain appeared to remain fairly consistent in its sexual behavior; recipients never became donors spontaneously, although donors occasionally became recipients.

THE F FACTOR

The difference in sexual behavior between donors and recipients was found to be caused by a small transmissible factor called F, the sex or fertility factor. Donor cells, containing F (called F^+), could transmit this factor in high frequency to recipient cells (F^-), which became, in turn, donor cells. Furthermore, the transmission of F seemed to be independent of the transmission of chromosomal genes. That is, incubation of a recipient strain with a donor strain for about 1 hour caused the conversion of most of the recipients into donors but produced only a low frequency of genetic recombinants (about one per million cells). F thus acted as a cytoplasmic inclusion that was itself transmissible by contact but could also occasionally cause the transmission of chromosomal material.

It was therefore of considerable interest when Cavalli and also Hayes discovered that some F^+ populations gave rise to donor strains which could transmit chromosomal genes in high frequency. These strains, called Hfr (high-frequency recombination), increased the frequency of recombinants to about one per thousand cells or better. An additional difference between Hfr and F^+ were the frequencies in which particular segments of the donor chromosome entered into recombinations. F^+ strains, for example, could produce recombinants for all genes, e.g., from A to Z, each in approximately equally low frequency. Hfr strains, on the other hand, showed a gradation in the frequencies in which various donor loci appeared among the recombinant products. The A^+ gene of a particular Hfr strain, for example, appeared most frequently among the recombinants of a cross Hfr $(A^+B^+C^+D^+\ldots) \times F^-$ $(A^-B^-C^-D^-\ldots)$, the D^+ gene less frequently, the J^+ gene still less frequently, until one locus, e.g., Z, showed itself least frequent of all among the recombinants. If we can consider that genes A to Z are on one linkage group, it appeared as though the Hfr donor gene A consistently entered into the recipient first and therefore had the greatest opportunity for recombination. Genes B to Z entered sequentially in that order, and their recombination frequencies were determined by the order of entry. Since F^+ strains showed no special sequence of recombination, their donor chromosome could be considered to enter the recipient at any point.

Furthermore, in contrast to F^+ strains, most of the recombinants produced by Hfr remained F^- recipients and only a few inherited the Hfr state. However, those few recombinants that became Hfr were also recombinant for the genes that had the lowest recombination frequency, i.e., Z. In other words, if the bacterial chromosome in an Hfr strain has the sequence A to Z, with A entering the recipient before the others and Z at the end, the Hfr

*Different symbols for these markers are also used (see Fig. 19-8 and Table 19-4).

factor appears to be directly connected to the "hind," or terminal, end of the chromosome. Since the entire chromosome must be transmitted for genes at this end to appear in the recipient, the low frequency of transmission of such genes indicates that the chromosome is probably broken off before the terminal genes enter.

Not all Hfr strains, however, initiated the transfer of the bacterial chromosome at the same point. As shown in Fig. 19–4, linkage maps for different strains could be established based on recombination frequencies, showing that the sequence of markers were transmitted in different orders in each strain. Thus strain H in this figure transmits the gene *thr* (threonine synthesis) in the highest frequency and the gene *thi* (thiamine synthesis) in the lowest. This order is reversed in strain J4. Note, however, that if both ends of each chromosome are connected to form a circle, the sequence is identical for each strain (lower half of Fig. 19–4). From this and other evidence (see Fig. 5–4) it has therefore been postulated that the bacterial chromosome is circular and the Hfr factor can break this chromosome and orient its transfer at different points, yielding different Hfr strains. Since a particular Hfr strain, once established, maintains the same sequence of marker transmission, the Hfr factor seems to have established a stable attachment at a particular point on the chromosome. In strain H, for example, this point is near *thr* and in strain J4 it is near *thi*. With an F^+ strain, by contrast, chromosome transfer is random both in respect to point of origin and polarity.

As a result of findings by Jacob, Wollman, Hayes, Adelberg, and others, the following main features of sexual conjugation in *E. coli* (variety K12) seem to be fairly well established at the present time.

1. The F factors are composed of double-stranded DNA of a molecular length of about 30 microns or about 100,000 nucleotide pairs. (The bacterial chromosome is about 40 times longer.) One segment, called the origin (O), enters the recipient cell first and everything else follows sequentially in a linear order.

2. In F^+ strains, the F factor is circular, and only one such F factor is usually present in each bacterial cell. This factor replicates midway through each bacterial chromosome replication cycle, thereby maintaining its frequency, and also appears to inhibit the replication of additional F factors that may be present. The exclusion of other F factors may arise from incompatibility products produced by the residing F factor, as well as from its attachment to a unique replicating site on the bacterial membrane, which can be occupied by only a single F particle.*

3. F^+ and Hfr cells produce a new surface component that enables them to form a union with cells that do not contain these factors. This union, called *effective contact*, is then followed by formation of a cytoplasmic connection which has the appearance of a conjugation tube in some electron micrographs (see Figure 3–5). For material to be transferred through this connection, it must be prepared or *mobilized*, and energy provided. There are numerous indications that all these functions are controlled by genes in the sex factors which are also responsible for the appearance of one or two distinctive rodlike structures called "F-pili." A pilus of this kind is quite thin and flexible, and often extends from the bacterial cell wall for a distance of ten or more times the length of the cell. It has been suggested that these pili function as conveyors along which the donor chromosome is transmitted, or as tentacles to attract or bind together mating pairs of bacteria. F-pili are also unique in serving as the specific attachment points for the icosahedral RNA bacteriophages such as f2, MS2, and $Q\beta$, or for the filamentous DNA phages such as fd, f1, and M13. This phage-susceptibility is specific for donor cells, although some phages are known (e.g., T7) to which F^- cells are more susceptible.

4. When conjugation occurs between an F^+ donor and an F^- recipient, two possible states may

*Because of the incompatibility that prevents two different F particles from occupying the same cell, the gene exchange necessary for recombinational mapping cannot take place. Some F particle gene mapping has nevertheless been accomplished through the complementation and deletion methods explained in Chapter 25, as well as through electron microscope studies which enable physical comparisons to be made between F factors with and without deletions (see Achtman).

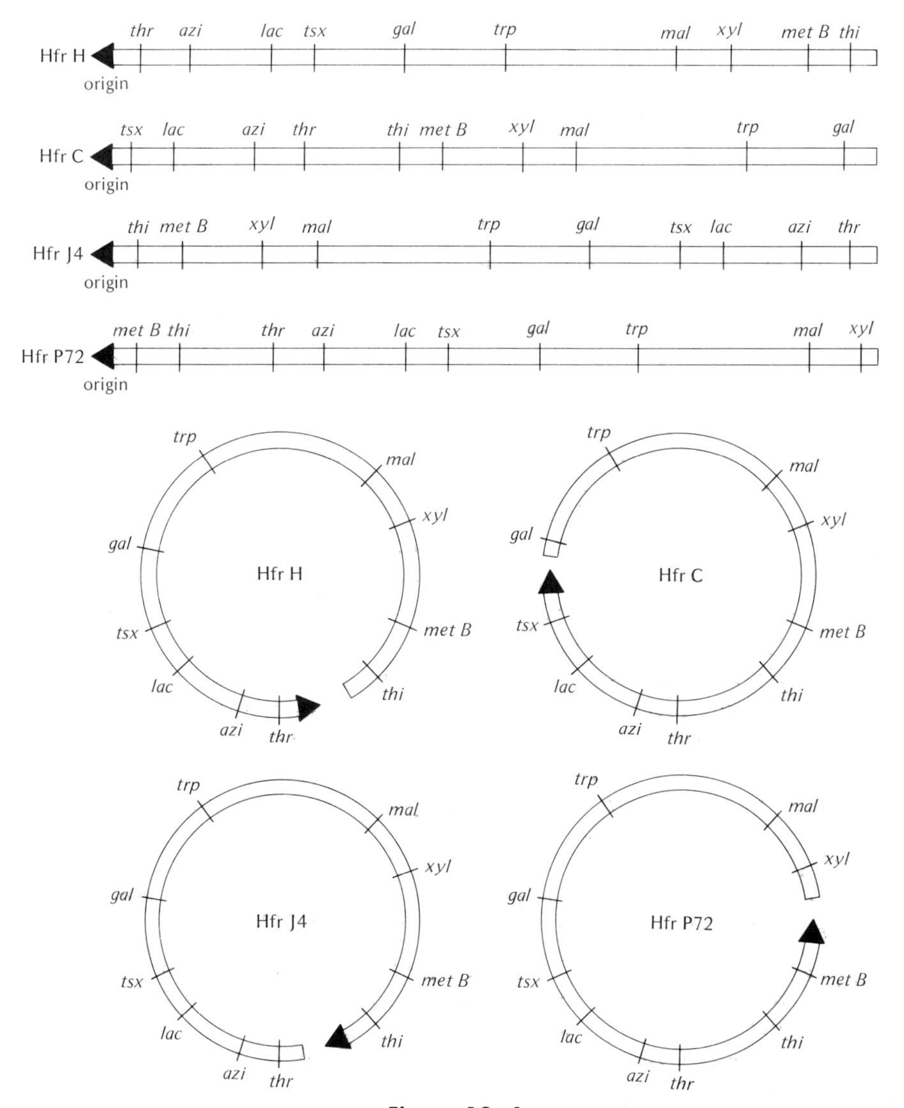

Figure 19–4

Sequence of transmission of genes in four Hfr strains of E. coli *and the explanation of these differences in terms of breakage at different points of a circular chromosome. (Arrowheads indicate genes entering first.) Since each linear array is an arrangement, or permutation, of the same circular gene order, such arrays are also called "circular permutations." All of the genes in this figure are also listed with their effects in Table 19–4. (Linear arrays after Hayes, Jacob, and Wollman.)*

exist for the F factor: (a) It remains outside the bacterial chromosome as an independent cytoplasmic inclusion, or (b) it recombines with a section of the bacterial chromosome, so that the chromosome now has a directional orientation (Fig. 19–5). In the first case only the F factor will enter the recipient cell, thereby producing a new F^+ cell, which may now, in turn, act as a donor. In the case of attachment to the bacterial chromosome, the chromosome itself now acts as a directionally oriented factor entering the recipient cell.

5. The difference between F^+ and Hfr cells resides in the chromosomal integration of the sex factor that each carries. In F^+ strains the sex factor

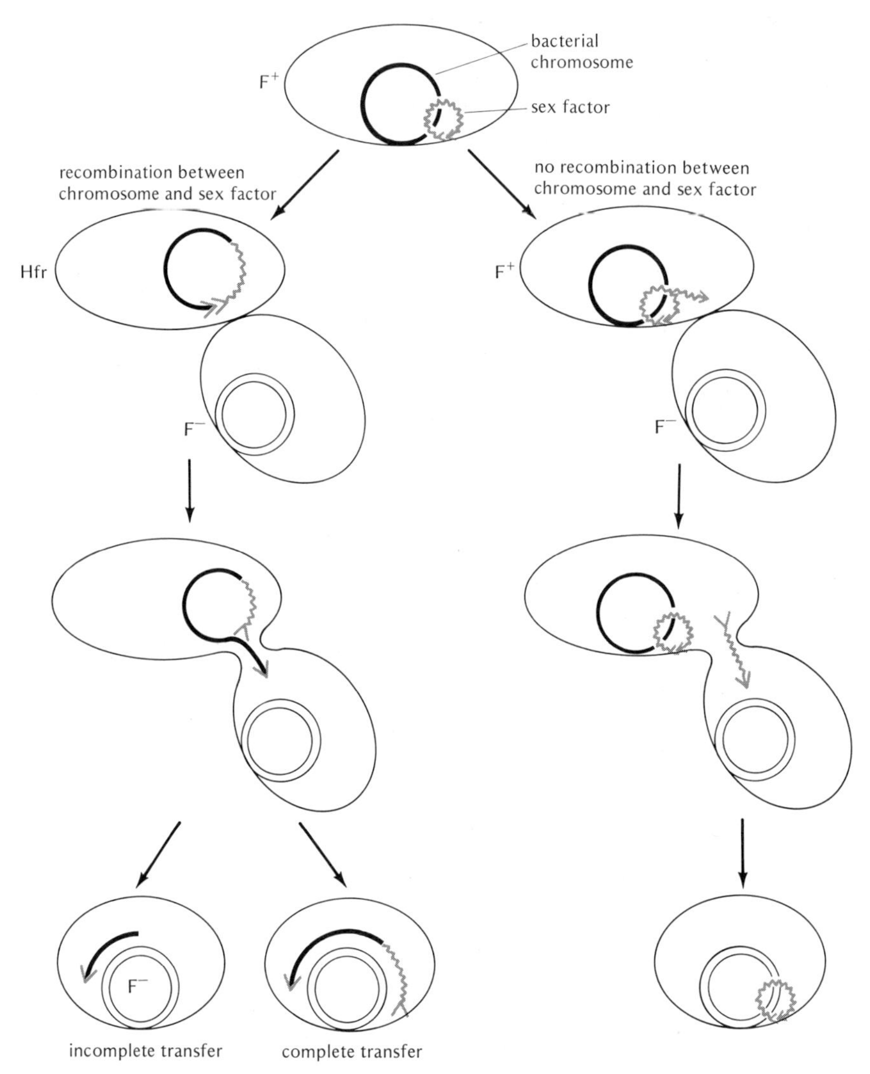

Figure 19–5

Two possible consequences of conjugation in an F^+ cell. Left side: integration into the chromosome, thereby forming an Hfr cell and consequent complete or incomplete transfer of the chromosome into the recipient. Right side: transfer of the F factor only.

lies in the cytoplasm and only very rarely combines with the bacterial chromosome to form an *integrated* F factor that is capable of orienting and transferring the bacterial chromosome to an F^- recipient. For reasons not yet clear, some bacterial strains allow the integration of an F factor into their chromosomes more easily than others. When such integration occurs the cell becomes converted to an Hfr cell. It is such Hfr cells, isolated in individual clones, that are the origin of Hfr strains. Occasionally the F factor will leave its chromosomal position in an Hfr cell and reconvert the cell to an F^+ cell, different Hfr strains showing differences in the frequency in which this occurs.

6. According to proposals by Jacob, Brenner, Cuzin, and others, the F factor acts as a replicating

unit that begins replication of DNA upon sexual contact and transfers only one of the replicating donor DNA strands to the recipient (Fig. 19–6). Evidence for this includes the fact that when the recipient carries a temperature-sensitive mutation that blocks DNA synthesis, only single strands of the donor DNA can be shown to accumulate at the restrictive temperature. Other evidence (see, for example, Siddiqui and Fox) indicates that the donor-transmitted DNA strand is specifically the one which enters the recipient at its 5′ end. This strand then acquires a complementary strand through DNA synthesis in the recipient before undergoing recombination with the recipient chromosome. As yet, details of the recombination process are not fully known although it may follow a pattern similar to that previously observed for transformation (Fig. 19–2) and other recombinational processes (Fig. 17–15, Fig. 18–8)—that is, homologous pairing → single-strand exchange → integration of donor DNA → complementary strand formation. In any case, when the F factor is in the cytoplasm outside the chromosome (i.e., F^+) it transfers only one of its own DNA strands, but when F is incorporated into the bacterial chromosome it transfers part or all of one of the bacterial chromosome strands.

7. If chromosome transfer by the male cell is complete, the integrated F factor is transmitted along with the bacterial chromosome, and the recipient chromosome may then recombine with the F portion of the donor chromosome and become converted into an Hfr cell. If transfer is incomplete, the recipient cell remains in F^- condition unless infected by the F factor of another F^+ cell. In both cases transmission of donor chromosomal material enables recombination to take place with recipient chromosomal material and new gene combinations to be formed.

8. Whether through F^+ or Hfr conjugation, a donor chromosome that is in the process of transfer to a recipient cell may break anywhere along its length so that only partial transfer occurs. The chances for breakage are not confined to any particular points, but occur randomly throughout the length of the chromosome. Thus the longer the length of chromosome that is being transferred, the more points along its length that are breakable and the greater the chances for incomplete transfer. Long-length chromosome transfers are therefore much rarer than short-length transfers. This means that genetic recombination between donor and recipient chromosomes is limited to the length of chromosome that was transferred to the zygote; i.e., it is limited *prezygotically*. Plating conjugating strains on a solid medium decreases the probability of chromosomal breakage and permits longer chromosome lengths to be transferred.

9. However, even when a donor chromosome is present in the recipient cytoplasm, this does not necessarily mean that it is going to be incorporated or recombine with the recipient chromosome. For example, donor chromosomes carrying the prophage *lambda* (λ) enter the recipient cytoplasm in certain crosses at a frequency that affects approxi-

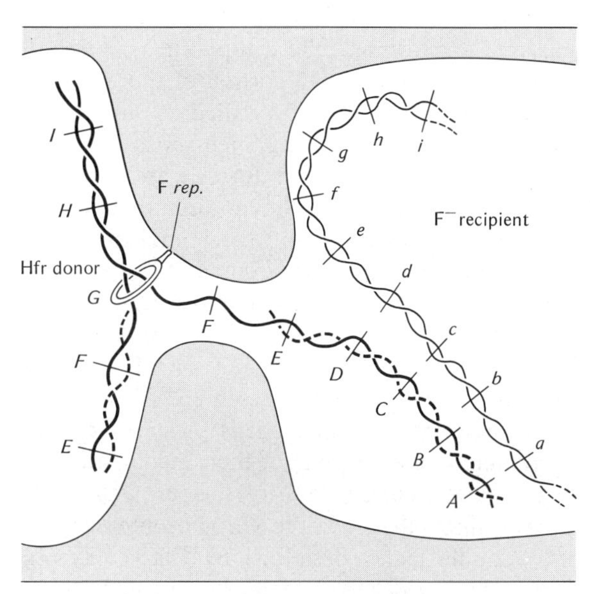

Figure 19–6

Transfer of a strand from the replicating DNA molecule of an Hfr cell to an F^- cell. The replication system of F (F rep.) is presumed to initiate both replication and transfer of the Hfr chromosome at the same point. Upon entering the recipient, a complementary strand is synthesized, and pairing then occurs between homologous sections of the donor and recipient chromosomes (e.g., A-a, B-b, C-c, . . .). (After Jacob, Brenner, and Cuzin, modified.)

mately 70 percent of all recipients. [Evidence for entry is gained by testing whether λ viral reproduction can be induced in these cells (p. 32).] However, of these recipients, only about half show recombination with a gene marker (*galactose*) that is close to the prophage on the donor chromosome. In other words, once a donor chromosome marker has entered the recipient cell, its chances of appearing on the recipient chromosome are about 50 percent. This *postzygotic coefficient of integration* has been found to be approximately the same for most chromosomal markers. At the extreme, some recipient strains show little or no integration of any donor markers into their chromosomes. Such recombination-deficient strains, called *rec*$^-$, have been found to bear mutations at one of three separate loci, *rec A*, *B*, *C*, presumably involved in the production of recombination enzymes and proteins.

10. Occasionally when the sex factor of an Hfr cell loses its integrated state on the bacterial chromosome and reenters the cytoplasm as a free F particle, it incorporates a portion of the bacterial chromosome. This F is now called a "substituted sex factor," or F′ (F-prime). Such substituted factors may enter a recipient F$^-$ cell and produce recombination between the portion of chromosome they are carrying and the homologous section on the recipient chromosome. As does F, an F′ particle replicates autonomously and, as explained later (p. 422), has a circular structure. However, whether F or F′, only one F factor is normally found per cell. When a cell is shared by more than one such factor it is probably caused by the elimination of one of their incompatibility genes. In those cases both of the sex factors are usually integrated into the chromosome such as the "double male" described by Clark, or both factors are fused into a single element such as the fused F′ factors described by Willetts and Bastarrachea. Electron micrographs of two fused F′ factors show, as expected, a single circular element instead of two. In some instances, an F′ particle was found to replicate in an Hfr strain, but this was apparently caused by a mutation in the incompatibility gene of the integrated Hfr sex factor.

11. In Hfr strains the initiation of effective contact and conjugation with a recipient cell causes the donor chromosome to open at the F-factor location and enter the recipient in an oriented direction. As in F$^+$ strains, if chromosome transfer is complete, recombinants may be produced which include the integrated F factor and are therefore Hfr cells. If chromosome transfer is incomplete, Hfr cells are not produced, but recombination may nevertheless occur with other genes that have been transferred. Usually only part of the donor chromosome is transferred, and the recipient, for at least a short period of time, is diploid for some donor and recipient genes. In this stage it is called a *partial zygote* or *merozygote* (also *merodiploid* or *merogenote*).

CONJUGATION MAPPING

The observation that there was a particular point at which the donor Hfr chromosome broke and entered the recipient cell offered the opportunity for mapping the gene sequence in the donor chromosome by timing the entry of different genes. An important method developed for this purpose was the "interrupted mating technique" of Jacob and Wollman. They mixed Hfr cultures of a particular strain, e.g., HfrH, and recipient F$^-$ cells, permitting them to conjugate for a short period of time. Then, at periodic intervals afterward, they removed samples from this mixture and subjected them to violent agitation in a Waring Blendor. This had the effect of separating the conjugating bacteria and breaking the donor chromosome between them. Thus the length of donor chromosome that entered the recipient cell could be controlled by timing the interval between the onset of conjugation and the agitation treatment.

As a rule an 8-minute interval after conjugation seemed to be necessary for chromosome transfer to begin. Once it had begun, short time intervals thereafter permitted only short chromosome lengths to enter, and longer time intervals enabled more genes to be transferred. The relationships between genes and their position on the chromosome could therefore be mapped in terms of *time units*, in which one time unit is equal to 1 minute.

An early experiment illustrating conjugation mapping was performed by Wollman and Jacob, who crossed an HfrH prototrophic strain to an F^- recipient that was auxotrophic for threonine (*thr*$^-$) and leucine (*leu*$^-$), sensitive to sodium azide (*azi*$^-$), and to bacteriophage T1 (*ton A*), and unable to ferment lactose (*lac*$^-$) and galactose (*gal*$^-$). In all, six gene differences were involved between the two strains, which were then mapped according to the time at which each appeared to enter the recipient. To ensure that only recombinant cells would be scored from this cross, the HfrH donor cells and any nonrecombinant F^- recipient cells were eliminated after mating by the use of genes called *selective markers*. In this experiment the donor cells carried the gene for streptomycin sensitivity, and the recipient cells were streptomycin-resistant. After conjugation between the two strains was interrupted, selection for recombinant cells was made on a streptomycin medium that lacked threonine and leucine. This medium therefore selectively killed the HfrH donors and permitted the growth of only those F^- recipients that had incorporated *thr*$^+$ and *leu*$^+$ genes. *Thr* and *leu* were actually the first of the HfrH genes that entered the recipient, and they produced *thr*$^+$ and *leu*$^+$ recombinants even when mating was interrupted within $8\frac{1}{2}$ minutes.

As shown in Fig. 19–7, samples of the *thr*$^+$ *leu*$^+$ recombinants were selected and tested for the presence of other donor genes after the parental strains had been permitted to conjugate for various given periods of time. Thus, after about 9 minutes of mating between HfrH and F^-, the *azi* gene entered the recipient, since some samples of *thr*$^+$ *leu*$^+$ recombinant cells were now *azi*$^+$. If mating between the parental strains was permitted to continue beyond 9 minutes, the *azi*$^+$ gene increased in frequency among the *thr*$^+$ *leu*$^+$ recombinants, until samples of the recombinants taken after 20 minutes of mating were 90 percent *azi*$^+$. The *ton* gene entered at about 10 minutes, since any mating period of less than 10 minutes would prevent the appearance of recombinants that were resistant to bacteriophage T1. By similar reasoning we can see from Fig. 19–7 that *lac* and *gal* were next in order of appearance, and also increased

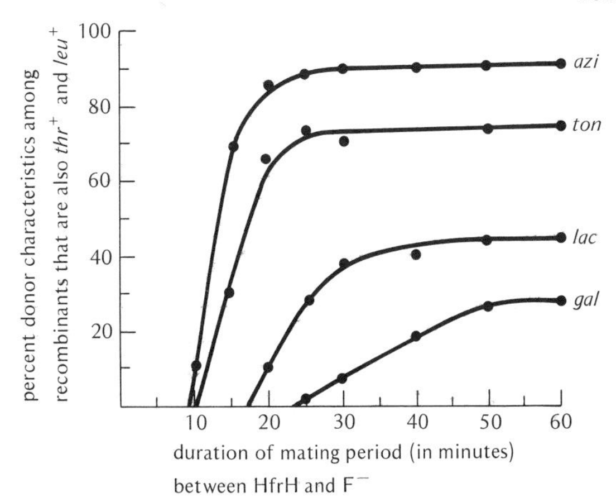

Figure 19–7

Order of appearance of recombinants for azi, ton A, lac, and gal when mating is interrupted at different times in a cross between HfrH and F$^-$. All these recombinants are also thr$^+$ and leu$^+$, since prototrophs for these two genes are selected to demonstrate that recombination between the two chromosomes has occurred. (After Wollman, Jacob, and Hayes.)

in frequency as the conjugation period was lengthened. The recombination levels reached after long periods of mating indicate that a gene located relatively far from the origin (i.e., *gal*) has less chance of being incorporated into the recipient chromosome even if there is sufficient time for it to have entered the cell (see p. 401).

Since these early experiments, Taylor and Thoman have shown that the time of initial chromosome entry after conjugation may be even more rapid than first believed. In some experiments they have found genes that enter the recipient cell only 5 minutes after mating has begun. On the basis of these and other findings, the entire bacterial chromosome can be mapped as a complete length of 90 time units (Fig. 19–8 and Table 19–4). Such mapping is advantageous, since the relationship between genes is more easily determined as a function of time rather than as a function of recombination frequencies; the earliest appearance of a gene among recombinants is sufficient to indicate its location on the chromosome, and frequencies do not have to be scored. For genes that are three or more time units apart, time-unit mapping

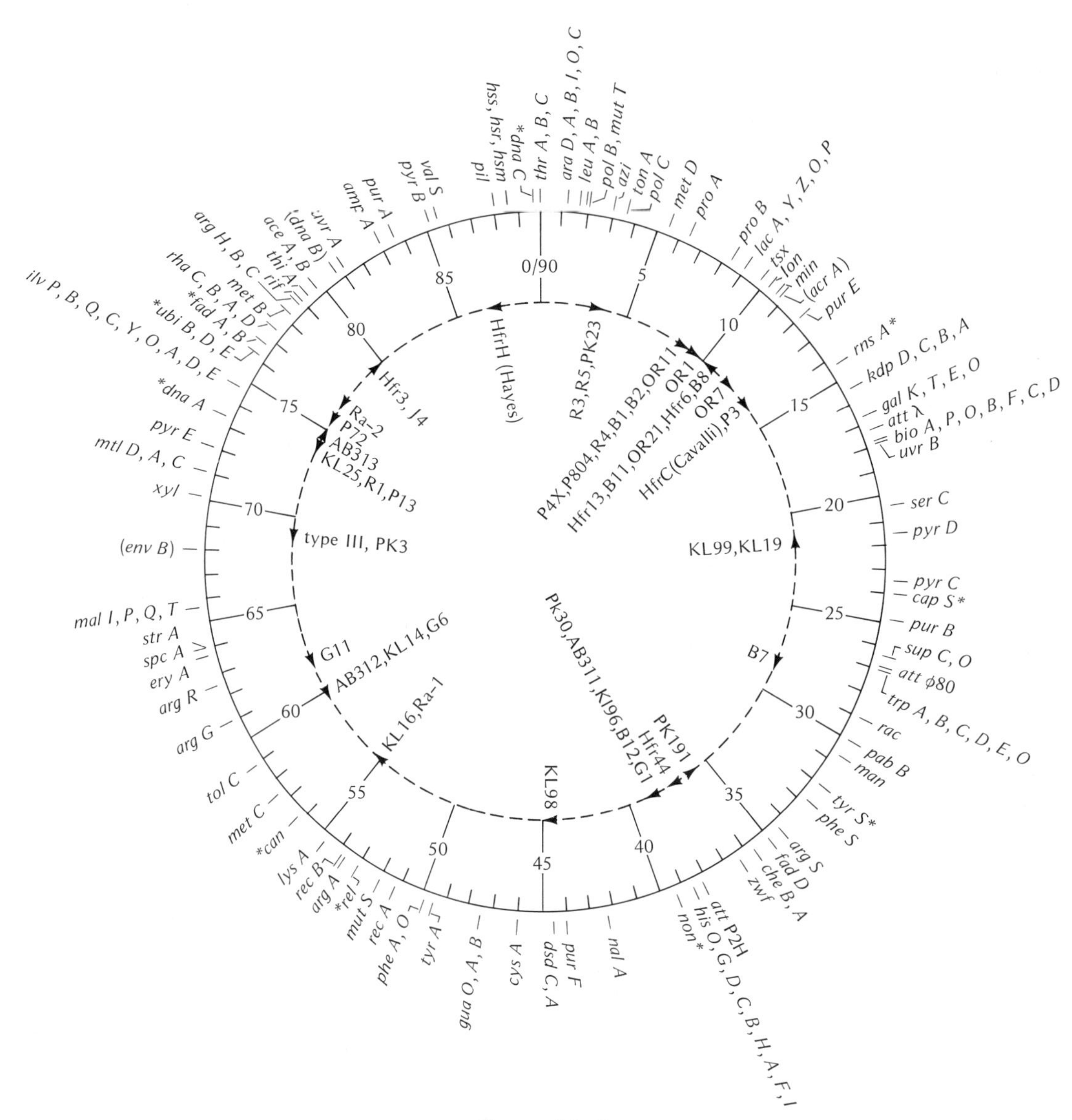

Figure 19–8

Circular map of a sample of E. coli *K12 mutations mapped in time units (outer circle), showing also the origin and direction of transfer of the chromosomes of a variety of Hfr strains (inner circle). The origin of the HfrH chromosome (at approximately the 87-minute position), for example, is located somewhat after the* methionine B *locus (78 minutes) and causes the chromosome to enter the recipient in a counterclockwise direction from the 87-minute position onward. In other words, the* met B *locus is transferred among the last when the HfrH strain is used as donor. On the other hand, it is transferred among the first genes when the clockwise-oriented Hfr strain J4 (79-minute position) is used as donor. The gene loci in this figure are identified further in Table 19–4, and represent some of the 460 loci presently known. Asterisks designate genes whose positions are only approximate, yet which are more precisely located than genes enclosed in parentheses. (Outer circle abridged from Taylor and Trotter; inner circle after Low.)*

TABLE 19–4

A representative list of *E. coli* genes and their linkage positions including those shown in Figure 19–8.

Symbol	Map Position (Minutes)	Mutation (Substance or Function)
ace A, B, E, F	80, 80, 2, 2	acetate
acr A	10	acridine sensitivity
ade	see *pur*	adenine
amp A	83	penicillin resistance
ara A, B, C, D, E, I, O	1, 1, 1, 1, 55, 1, 1,	arabinose
arg A, B, C, D, E, F, G, H, I, P, R, S,	54, 79, 79, 64, 79, 7, 61, 79, 85, 56, 63, 35	arginine
att λ	17	integration site for λ prophage
att P2H, II	38, 77	integration sites for P2
att φ*80*	27	integration site for φ80
azi	2	azide resistance
bio A, B, C, D, F, H,	18, 18, 18, 18, 18, 66	biotin
can	56	canavine resistance
cap S	24	capsule synthesis
che A, B, C	36, 36, 37	chemotactic motility
cys A, B, C, E, G	47, 27, 52, 73, 65	cysteine
dna A, B, C, E, F, G	73, 81, 89, 4, 42, 63	DNA synthesis
dsd A, C	45, 44	d-serine
env A, B	2, 68	cell envelope formation
ery A, B	84, 10	erythromycin resistance
fad A, B, D	77, 77, 35	fatty acid degradation
gal E, K, O, T, R, U	17, 17, 17, 17, 55, 27	galactose
gua A, B, C, D	47, 47, 89, 47	guanine
his A, B, C, D, E, F, G, H, I, O	all at 39	histidine
hsm, hsr, hss	89	host specificity
ilv A, B, C, D, E, F	75, 75, 75, 75, 75, 48	isoleucine-valine
kdp A, B, C, D	16	potassium dependence
lac A, I, O, P, Y, Z	9	lactose
leu A, B, S	2, 2, 14	leucine
lon	10	filamentous growth
lys A, C	55, 80	lysine
mal B, I, P, Q, T	81, 66, 66, 66, 66	maltose
man	31	mannose
min	10	minicell (no DNA)
met A, B, C, D, E, F	80, 78, 58, 6, 76, 78	methionine
mtl A, C, D	71, 71, 71	mannitol
mut L, S, T	83, 52, 2	mutator genes
nal A, B	42, 51	nalidixic acid resistance
non	39	blocks mucoid capsule
pab A, B	65, 30	para-aminobenzoate
phe A, O, S	50, 50, 33	phenylalanine
pil	88	absence of pili
pol A, B, C	76, 2, 4	DNA polymerases I, II, III
pro A, B, C	7, 8, 10	proline
pur A, B, C, D, E, F, G, H, I	84, 25, 47, 79, 12, 44, 47, 79, 48	purines
pyr A, B, C, D, E, F	1, 85, 24, 21, 72, 27	pyrimidines
rac	29	recombination activation
rec A, B, C, F, G, H	51, 54, 54, 73, 74, 52	ultraviolet sensitivity and genetic recombination
rel	53	RNA synthesis
rha A, B, C, D	77, 77, 77, 77	rhamnose
rif	79	rifampycin resistance
rns A, B, C	14, 57, 74	ribonucleases
ser A, B, C, O, S	56, 90, 20, 20, 20	serine
spc A	64	spectinomycin resistance
str A, B	64, 7	streptomycin resistance
sup A, B, C, D, E, F	79, 15, 26, 38, 15, 26	suppressors of various mutations
thi A, B, O	79, 78, 78	thiamine (vitamin B_1)
thr A, B, C	0	threonine
tol A, B, C, D	17, 17, 59, 20	tolerance to various colicins
ton A, B	3, 27	resistance to phages
trp A, B, C, D, E, O, P, R, S	(*A–P*) 27, 90, 65	tryptophan
tsx	10	T6 resistance
tyr A, R, S	50, 28, 32	tyrosine
ubi A, B, D, E, F, G	83, 76, 76, 76, 15, 42	ubiquinone
uvr A, B, C, D, F	81, 18, 36, 75, 73	repair of ultraviolet damage
val S	85	valine
xyl	70	xylose
zwf	36	glucose-6-phosphate dehydrogenase

offers a fairly precise estimate of the linkage order. However, for genes that are only one or two time units apart, different experiments may occasionally show differences in linkage, especially when the genes are far from the point of origin on the Hfr donor chromosome.

RECOMBINATION MAPPING

To detect linkage order between genes separated by smaller distances than three time units, the more traditional method of scoring recombination frequencies can be used. As in transformation, the

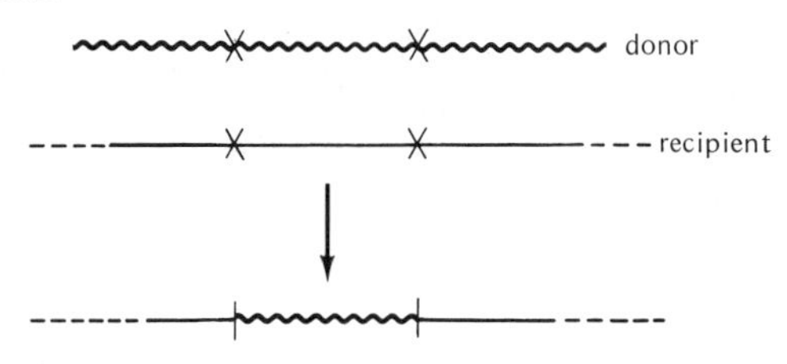

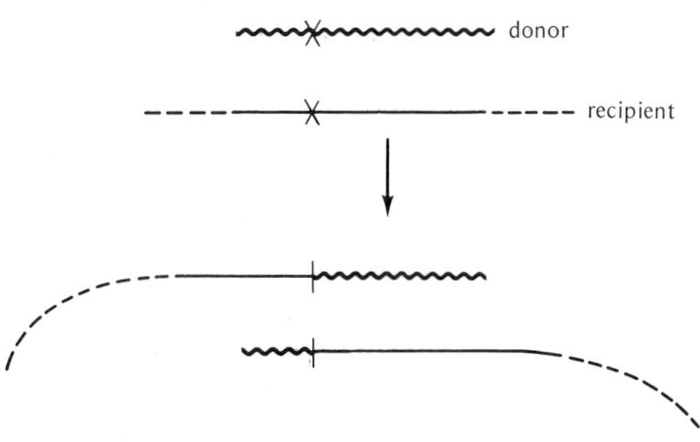

Figure 19–9

Consequences of a two-break exchange (above) or a one-break exchange (below) between donor and recipient chromosomes.

appearance of recombinants depends upon a physical exchange between donor and recipient chromosomes; an even number of breaks must take place between the two chromosomes to enable the insertion of a donor section into a recipient chromosome. If only one or an odd number of breaks occur, the complete chromosome structure is severed, and chromosome fragments and duplications are formed (Fig. 19–9).

In addition to this prerequisite, recombination among sexual conjugants is also obviously limited to the length of chromosome transferred. Donor genes that lie outside a transferred piece of chromosome have no chance to appear among the recombinants. Thus linkage measurements in bacterial recombinations must ensure that all genes being scored lie within the transferred length of chromosome. The method by which this is done is to choose a gene marker which enters the recipient after the entry of the gene, or genes, whose linkage relationships are to be observed. The appearance of recombinants for this distal gene marker then indicates that the opportunity has existed for recombination among the proximal genes as well. For example, in certain interrupted mating experiments, the adenine marker (*ade*) appears to be more distal from the origin than the marker for lactose (*lac*). Recombinants for *ade* in a cross in which the donor is *lac*$^+$ *ade*$^+$ and the recipient *lac*$^-$ *ade*$^-$ will appear as *lac*$^+$ *ade*$^+$ if no crossing over between *lac* and *ade* has occurred (Fig. 19–10). However, if crossing over between the two loci has taken place and recipients are

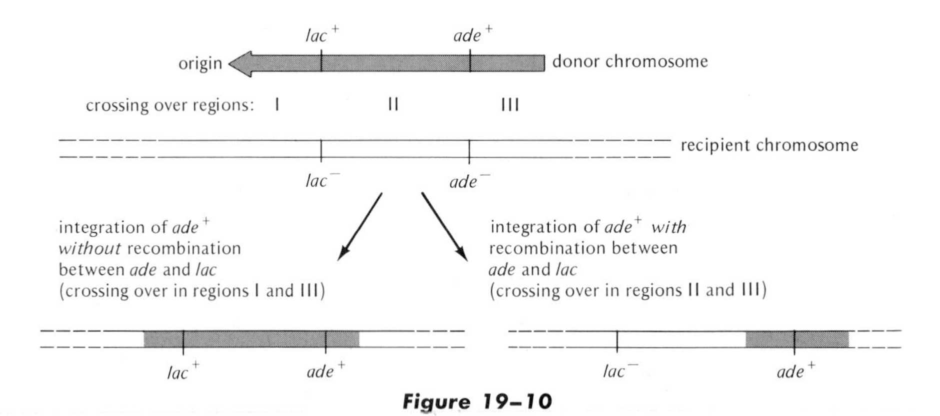

Figure 19–10

Recombination resulting in the incorporation of an adenine marker into a recipient chromosome. On the left side the recombinant section includes the lac$^+$ *donor marker. On the right side the recombinant section includes only the ade*$^+$ *marker.*

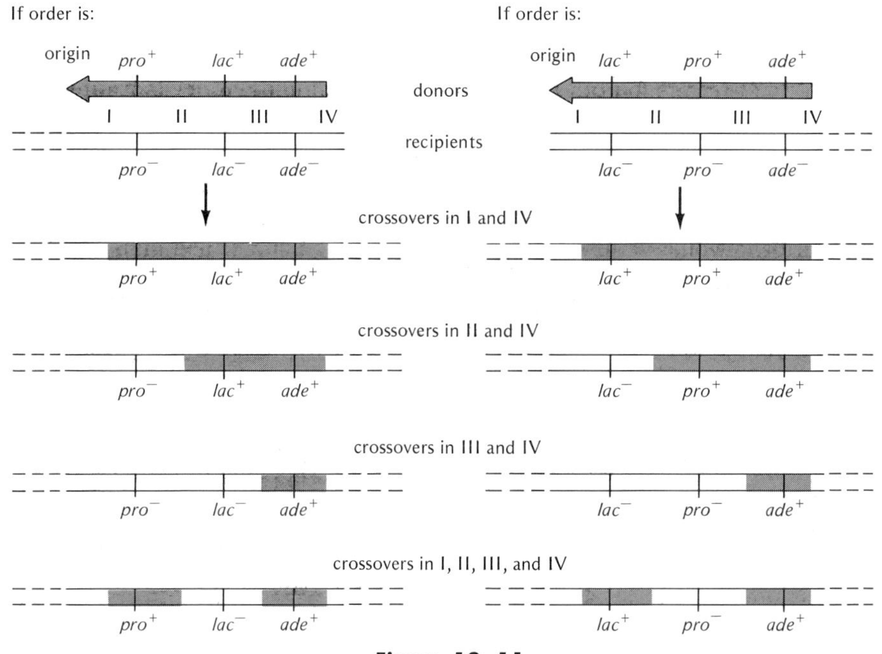

Figure 19–11

Consequences expected for different linkage orders in recombination between prototrophic and auxotrophic bacteria. The least frequent class, arising from a quadruple crossover, differs for each linkage order.

selected bearing the ade^+ phenotype (by growing on a nonadenine containing medium), then such *lac-ade* recombinants will appear as lac^- ade^+. The proportion lac^- ade^+/ade^+ then furnishes an estimate of recombination frequency between the two genes. For *lac-ade* this proportion is about .22, or a recombination frequency of 22 percent.

To detect the order among three genes, the relative frequency of particular recombination classes is used to indicate the linkage relationships. As shown in Fig. 19–11, a linkage order *pro-lac-ade* will demand different crossovers for the appearance of the lac^+ pro^- ade^+ recombinant class than if the linkage order were *lac-pro-ade* (*ade* enters the recipient after *lac* and *pro*). In one case (*pro-lac-ade*), the appearance of this class is explained by only two crossovers. In the other case (*lac-pro-ade*), this recombinant class appears only through a quadruple crossover.

A method that takes advantage of this difference in frequency between crossover classes is the use of *reciprocal* matings in which one gene order gives similar results in both matings, while the second gene order gives different results in the two matings. For example, reciprocal crosses between pro^- lac^+ $ade^+ \times pro^+$ lac^- ade^- strains will produce differences in the frequencies of prototrophs (pro^+ lac^+ ade^+) between them if the gene order is *lac-pro-ade* (Fig. 19–12). According to Jacob and Wollman, the gene order has been established as *pro-lac-ade*, and recombination frequencies between *pro-lac* and *lac-ade* are calculated as 20 and 22 percent on the basis of crossing over in regions II and III, respectively.

When related to time units, recombination map units seem to be of the order of approximately 20 recombination percentage units to 1 time unit. Genes separated by a length of more than 3 minutes in conjugation time appear unlinked in recombinational analysis. Since there are approximately 90 time units in the DNA chromosome, the entire length of the chromosome is 1800 recom-

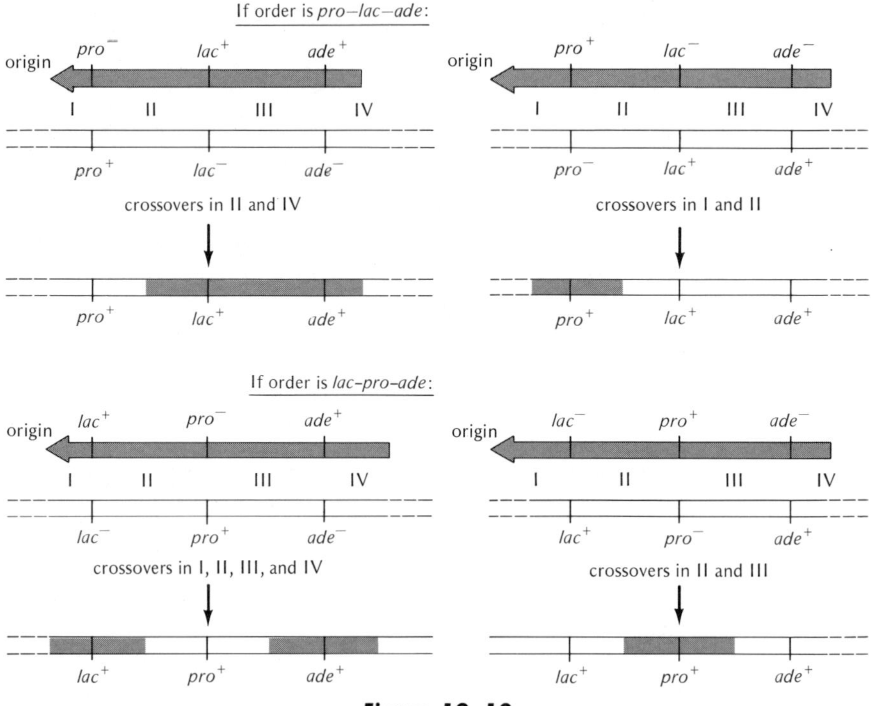

Figure 19–12

Crossovers necessary to produce prototrophs (pro^+ lac^+ ade^+) from reciprocal crosses for two alternative gene orders. Note that if the gene order is lac-pro-ade, a quadruple crossover is necessary to produce prototrophs from one of the reciprocal crosses.

binational units. In physical terms, both time units and recombination map units can be translated into nucleotide base pairs because of the known length of the *E. coli* chromosome (see Fig. 5–4) and the 3.4 angstrom distance between base pairs. Thus, *E. coli* has 1100 microns × 2941 (base pairs per micron) = 3.235×10^6 base pairs, and the numbers of base pairs transmitted by conjugation per minute is $(3.235 \times 10^6)/90 = 36{,}000$.

TRANSDUCTION

The relationship between bacteria and the bacteriophages that parasitize them was assumed to be a harsh but simple host-parasite relationship in which the phage can only replicate through bacterial destruction (lysis). However, beginning in the 1920s and continuing since then, work by Bordet, the Wollmans, Lwoff, and others has shown that, for some bacteria and for some phages, this relationship is partly a symbiotic one; i.e., a phage may be carried within a bacterium without causing immediate lysis. Such bacterial strains are called *lysogenic*, since they are nevertheless capable of being lysed through occasional proliferation of the viruses they carry. This is in contrast to *nonlysogenic* bacterial strains, which do not carry viruses within them and will therefore not lyse unless newly infected by phage. The phages involved in lysogenic relationships are called *temperate*, while phages that infect and destroy bacteria but cannot lysogenize them are called *virulent*. The "inducing" trigger that produces lysis in lysogenic strains, although not yet exactly known, involves a change in the activity of the phage from a quiescent *prophage* state to a proliferative *vegetative* state. Experimentally, induction of the vegetative state and consequent lysis can be produced by small doses of ultraviolet (UV) light. If large doses of UV are used, the few surviving bacterial cells may be "cured" of their prophages and become nonlysogenic (Fig. 19–13).

Lysogeny is an advantage to bacterial cells, since the presence of the prophage enables them to withstand infection and prevent vegetative growth by virus particles of the same variety. An *E. coli* cell lysogenized with λ prophage, for example, can be placed into a medium containing free λ particles and will not lyse. On the other hand, nonlysogenic *E. coli* placed in the same medium will soon be destroyed by the virus. Some nonlysogenic cells may nevertheless escape destruction under such conditions, but these are invariably found to have become lysogenic; i.e., they now bear the λ prophage and produce lysogenic colonies that are immune to further lysis by new infections of λ. (As in other lysogenic strains, however, such colonies may now produce λ phage by induction.)

In 1952 Zinder and Lederberg discovered that temperate phages may also act as carriers for genes between one bacterial cell and another. These findings furnished evidence for the third important recombination mechanism in bacteria, known as *transduction*. Initially their experiments began as a test to discover whether the genetic exchange previously demonstrated in *E. coli* by Lederberg and Tatum also existed in the mouse typhus bacterium *Salmonella typhimurium*. For this purpose they combined various strains of auxotrophs on a minimal medium and then looked for new prototrophic combinations; e.g., strains unable to synthesize phenylalanine (phe^-) and tryptophan (trp^-) might combine with strains unable to synthesize methionine (met^-) and histidine (his^-) to

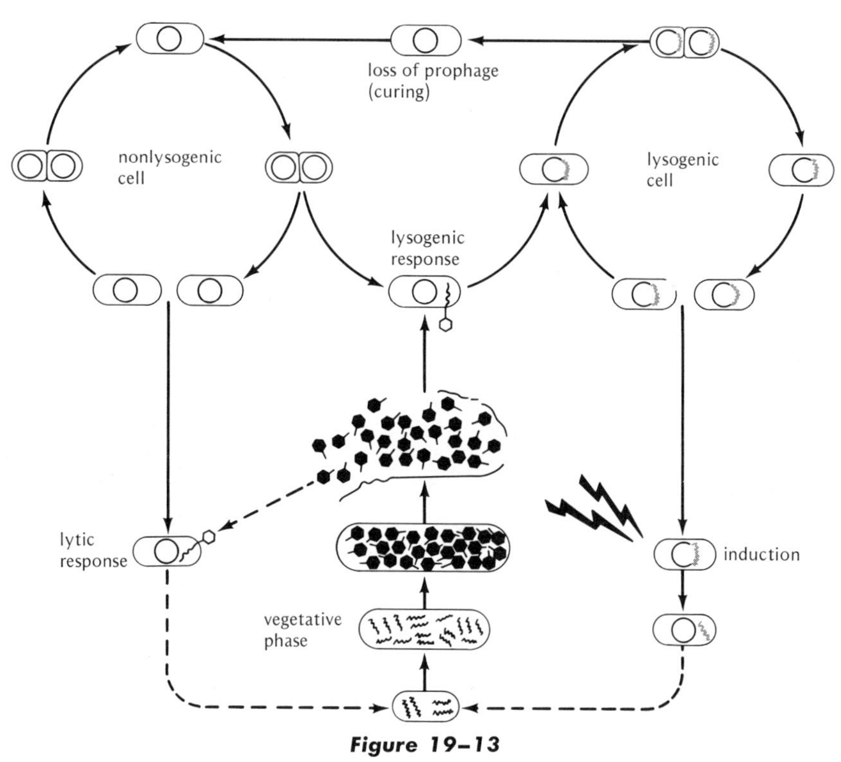

Figure 19–13

Lysogenic cycle showing the lysogenic response, the induction of phage growth in a lysogenic cell, the lytic response of nonlysogenic cells, and the "curing" of lysogenic bacteria. (From Stent, after Lwoff.)

form wild-type prototrophs (*phe*$^+$ *trp*$^+$ *met*$^+$ *his*$^+$). They found that such prototrophs were indeed formed. Moreover, since these new combinations occurred only among mixtures between strains and did not occur within the auxotrophic strains themselves, it seemed evident that genetic exchange was responsible for their formation. This exchange, however, did not occur with the same frequency for each strain examined, and some gave very few "recombinants." A combination of strains LA22 and LA2 appeared to be the most "fertile," and the ability for genetic exchange between these two strains extended to numerous additional characteristics such as the utilization of galactose, xylose, mannitol, maltose, and streptomycin resistance.

To test the mode of exchange between these two strains and to discover whether cell contact was necessary, Zinder and Lederberg placed each strain in one of the two arms of a Davis U-tube (Fig. 19–14), separated by a glass filter. This filter prevents the transfer of bacterial cells but permits the free passage of growth medium between the two arms through the application of suction and pressure. Thus the two auxotrophic strains, although separate, were raised in the same medium. The results of these experiments showed that prototrophs appeared in the arm containing LA22 but none appeared in the arm containing LA2. In other words, a genetically active "filterable agent" (FA) arose in connection with LA2 that could produce prototrophs in LA22. This agent, however, was produced only by LA2 when LA2 had shared the same growth medium with LA22. When grown separately, away from contact with LA22 medium, LA2 did not produce active FA. Additional peculiarities of FA were its ability to withstand treatment with deoxyribonuclease and its notably larger size than the usual DNA molecules. FA was therefore not an ordinary transforming agent or DNA molecule.

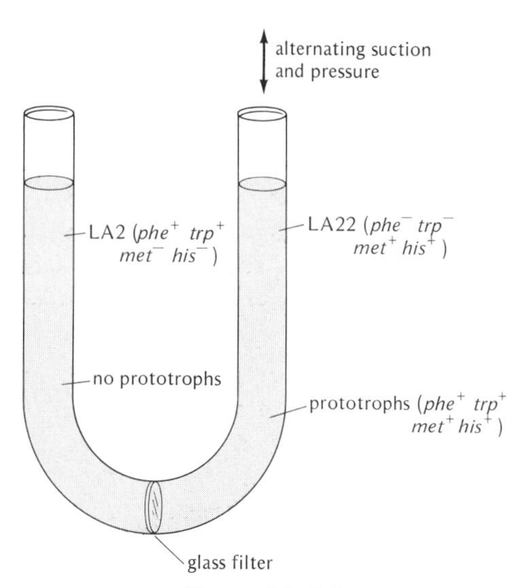

Figure 19–14

Transducing experiment performed by Zinder and Lederberg in 1952. Two auxotrophic strains of Salmonella typhimurium, *LA22 and LA2, were each placed into a separate arm of a Davis U-tube. These arms are separated by a glass filter that prevents cellular exchange but permits the liquid medium to move from one arm to the other. Prototrophs appeared only in the arm containing LA22 but not in that containing LA2.*

In constitution FA was soon shown to correspond with a temperate phage that was able to move through the culture medium from one bacterial strain to the other. This phage, called P22, was ordinarily associated with, or lysogenized, the LA22 strain of *Salmonella*. Occasionally the prophages of some of these cells passed into a vegetative state, proliferated, and lysed their bacterial hosts. Because of their small size, these phage particles then passed through the glass filter and infected the nonlysogenic LA2 strain. Lysis of LA2 then occurred with a consequent production of new phage particles that passed into the medium and entered LA22 cells. This time, however, the entering phage particles were associated with genetic material from LA2, some of which was wild type for the mutant genes carried by the LA22 strain and could therefore produce prototrophs by recombination between the LA2 material and the LA22 chromosome. According to the terms used, the phage particles had *transducing activity* and the LA22 strain was *transduced.* Since LA2 was nonlysogenic and did not carry a temperate phage, this helped explain why it could only transfer its genetic material in the presence of lysogenic LA22. Further identification between FA and P22 was demonstrated by the fact that they both had the

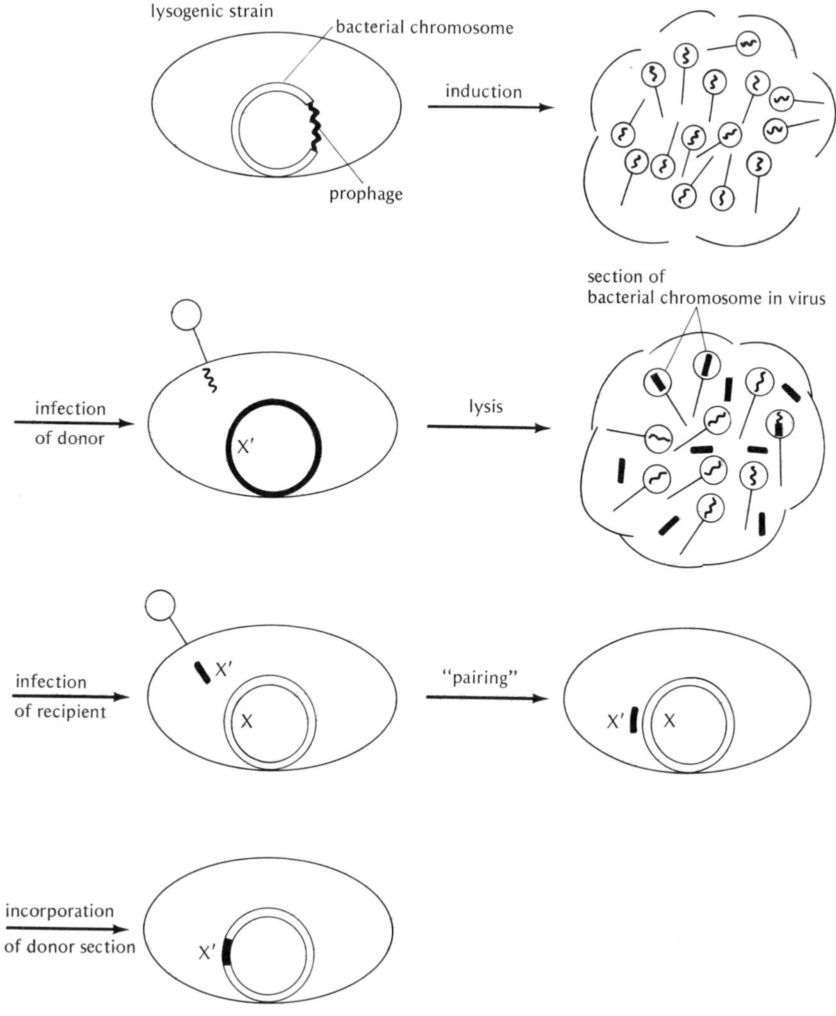

Figure 19–15
Transduction events.

same physical dimensions, were both destroyed by exposure to anti-P22 serum, and were both active at the same time. Furthermore, transducing activity by FA was confined to *Salmonella* strains to whose cell walls P22 phage could become attached.

Transduction technique in *Salmonella* therefore resolved itself into the growth of P22 phage on sensitive nonlysogenic bacteria with consequent harvest of the resultant phage products. These phage products were then incubated with bacteria that were to be transduced. If the introduced P22 phage carried a section of the donor chromosome and did not produce lysis in the recipient, the opportunity existed for this donor chromosome section to become incorporated (probably through recombination) into the recipient chromosome (Fig. 19–15). The frequency of transduction, however, was quite low; only one of 10^5 to 10^7 bacterial cells were transduced. Also, usually only one genetic marker was transduced at a time; e.g., a cross between phage cultured on trp^+ gal^+ xyl^+ bacteria and lysogenic trp^- gal^- xyl^- bacteria produced *either* trp^+ gal^- xyl^-, *or* trp^- gal^+ xyl^-, *or* trp^- gal^- xyl^+. Zinder and Lederberg therefore proposed that the rare transducing phage particles can carry only a small piece of bacterial chromosome material.

Within a few years of Zinder and Lederberg's discovery, transduction was also found to take place in *E. coli*. Investigations by Lennox and by Jacob uncovered two temperate phages, P1 and 363, that could be used to transfer markers from one *E. coli* strain to another. As in *Salmonella*, however, such transductions were limited to an exceedingly short section of the bacterial chromosome. For example, transductional studies showed that the selection of threonine prototrophs in a cross in which the donor is wild type for threonine, leucine, and sodium azide (thr^+ leu^+ azi^+) and the recipient is auxotrophic (thr^- leu^- azi^-) produced only a few percent that were also leu^+ and none that were also azi^+. In other words, the donor chromosome fragment beginning with the *thr* locus occasionally extended to the *leu* locus but never extended from *thr* to *azi*. This length of chromosome, when mapped in conjugation studies, is only about 1/100 of the entire length of the bacterial chromosome, or about equal in length to the phage chromosome.

Surprisingly, not merely nutritional markers, but lysogeny itself, was found to be transducible. The λ prophage, for example, was known through conjugation studies to occupy a specific chromosomal section in *E. coli* near a galactose (*gal*) locus. When a prototrophic strain lysogenic for λ (thr^+ lac^+ gal^+ λ^+) was used as a donor in the transduction of a nonlysogenic auxotrophic strain (thr^- lac^- gal^- λ^-) by means of tranducing phage 363, none of the new *thr* and *lac* prototrophs were lysogenic for λ. However, about 6 percent of the gal^+ transductants turned out to be carrying λ, as might be expected from the proximity of the two loci (Fig. 19–8, 17-minute position). Since practically any locus can be transduced by P22, P1, 363, and similarly tested viruses, this phenomenon has been given the name *generalized transduction.*

In almost all cases of generalized transduction, very little or none of the phage DNA is carried in the transducing particle, which consists primarily of host chromosome material surrounded by a phage envelope. One consequence of this is that the transducing particle is often unable to establish a lysogenic relationship with the host and to replicate along with it. Thus, transducing material that is not incorporated into the host chromosome through recombination may remain in the cytoplasm as a free but nonreplicating particle. Phenotypically this causes a phenomenon called "abortive transduction". That is, many "minute" colonies of transductants are formed in which single bacterial cells carrying a transducing portion of the donor chromosome divide so that the transducing fragment is transmitted to only one of the two daughter cells. Thus only one daughter cell in each generation of bacterial cells carries the transducing fragment and all other cells are of the nontransduced genotype. However, since the functional products (e.g., enzymes) produced by the transducing fragment may remain active in the cytoplasm for a few generations before they become diluted or used up, a few nontransductant daughter cells may also show the donor phenotype. The general appearance of the colony is therefore largely of the nontransduced phenotype, with tiny "abortive" exceptions that appear as "minute" or "micro" colonies (Fig. 19–16).

There is now considerable data that show that most generalized transductional events are of the abortive type, occurring with a frequency of 10 to 20 times that of complete chromosomal integration. Ebel-Tsipis and co-workers suggest that the relative rarity of chromosomal integration during transduction may lie in the fact that integrated DNA in transduction is double stranded in comparison to the single-stranded integration that occurs in transformation. The integration of double-stranded DNA into the bacterial chromosome may be more difficult than for single strands.

In spite of its rarity, however, transduction is perhaps the most commonly used method for mapping bacterial gene relationships over short distances since the length of chromosome incorporated into a viral particle is necessarily small. Two or more loci that are sufficiently linked to be transduced together (*cotransduced*) can therefore be considered to lie within a short distance of each

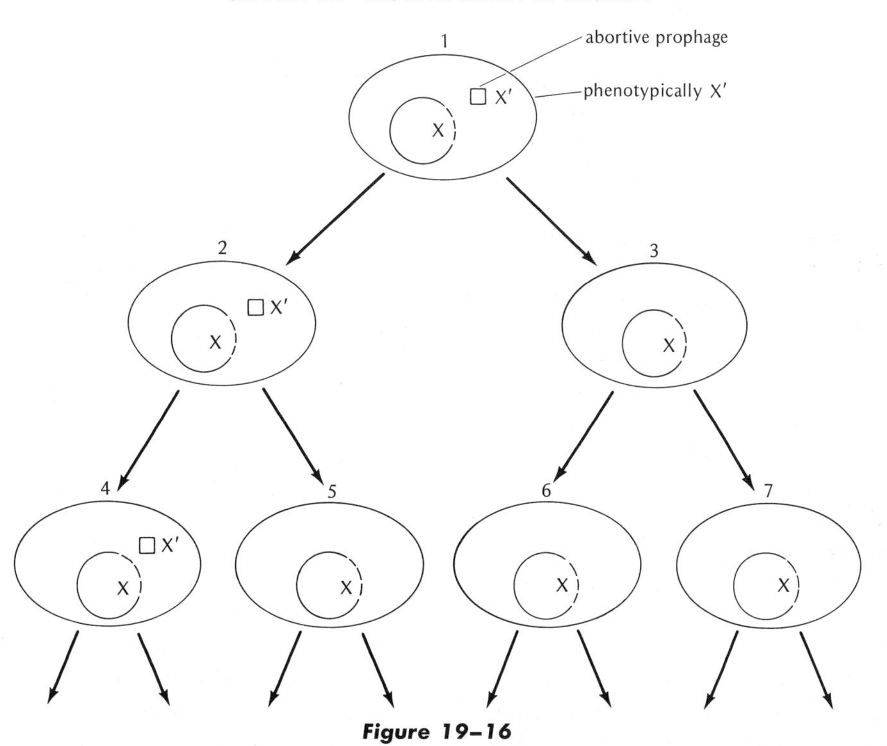

Figure 19–16

Diagram of events in abortive transduction for a hypothetical trait X that is being transduced to X'. Because the prophage bearing X' is unable to replicate properly, division of transduced cells produces one X'/X and one X genetic product. The X-containing product (e.g., 3) may be phenotypically X or phenotypically X', depending upon the residual effect of the X' gene on the cytoplasm. (See maternal effect, Chapter 13.) Sooner or later, however, the effect of the X' gene on the cytoplasm will wear off and the X-containing cells will appear phenotypically X (e.g., 6 and 7). Thus the general phenotype of the colony is only weakly X', since the transduction effect resides in relatively few cells (e.g., 4).

other, and their relationship to other nearby loci can be fairly easily determined. For example, Lennox found that *thr* and *leu* may or may not be transduced together with a third locus *ara* (governing ability to ferment the sugar arabinose). In crosses using thr^+ leu^+ ara^+ as a donor strain and thr^- leu^- ara^- as recipient, he found that 75 per cent of the ara^+ transductants were also leu^+ but none of them were thr^+. Obviously the *ara* locus was more closely linked to *leu* than to *thr*. The gene order could therefore be either *thr-ara-leu* or *thr-leu-ara*. If it was the former, transductants for both thr^+ and leu^+ should generally include ara^+, while if *ara* was on the outside, thr^+ leu^+ transductants might include ara^+ only rarely, if at all. He found that thr^+ leu^+ transductants were also ara^+ in 85 per cent of the cases, indicating a gene order *thr-ara-leu*. More precise tests for obtaining the linkage order of three genes can be performed by analyzing the results of reciprocal transduction according to the methods described on pages 405 ff. In general, maps constructed from transduction experiments agree with those constructed from conjugation experiments but offer considerably more detail.

SPECIALIZED TRANSDUCTION

In 1956 Morse and E. and J. Lederberg, looking for transducing viruses in *E. coli*, found that λ phage could also serve this function but that its transducing activity was restricted to the *galactose* locus. For example, wild-type prototrophs used as donors in λ-mediated transduction could only affect recipient auxotrophs by changing gal^- to gal^+ but never transduced other loci, such as *thr*, *trp, lac,* etc. This phenomenon, known as *specialized transduction*, has been mostly investigated in *E. coli*, although it has also been found in other bacteria, such as *Salmonella*. The technique using λ is not dissimilar from other methods of specialized transduction except that a special medium is used for the detection of gal^+ transductants. The λ phages are harvested from the donor and then used to infect recipient bacteria which are plated on an "indicator" agar [eosin–methylene blue (EMB)] to which galactose has been added. Although both gal^- and gal^+ bacteria can grow on this medium, the gal^+-transduced colonies form darkly pigmented colonies, while the gal^- colonies are lighter colored.

The isolation of transduced gal^+ colonies by these methods showed important novelties. First, in contrast to generalized transduction, the gal^+ colonies used as donors had to be *lysogenic* for the prophage, which was then *induced* by irradiation. If they were not carrying the λ prophage but were instead directly infected and lysed with λ particles, the λ phages harvested from these bacteria had no transducing power. Second, in contrast to normal gal^+ colonies, the transduced gal^+ colonies were unstable, producing about one gal^- cell for each thousand bacterial divisions. Third, those of the transductants that remained gal^+ were all lysogenic for λ, either able to produce λ upon induction, or immune to lysis by new infective λ particles. On the other hand, about 90 percent of the gal^- segregants formed by unstable transductants were no longer lysogenic for λ but appeared to have lost the prophage. Fourth, when the transductant gal^+ colonies were induced and the λ phage harvested, about *one-half* of the phage particles were now able to transduce gal^- bacteria, in contrast to the initial transduction efficiency of one out of a million.* This phage harvest, with its increased transducing efficiency, is called an *Hft* (high-frequency transduction) *lysate*.

To explain these observations the following mechanisms have been proposed: In ordinary gal^+ cells lysogenic for λ, about one in a million exchange their gal^+ region for a portion of the prophage chromosome. When such cells are induced they release phage particles carrying the gal^+ gene. Upon infection of gal^- recipient bacteria, the gal^+/gal^- partial diploid combination, now called a *heterogenote*, is considered to be composed of an *endogenote* (gal^-) on the bacterial chromosome, and an *exogenote* (gal^+) that is now part of the viral prophage chromosome. The heterogenote, phenotypically gal^+, may then remain in this state until induced and will then produce numerous λ progeny that are genetic copies of the gal^+ exogenote. Since these λ chromosomes are now exclusively gal^+, it is this that accounts for the high transducing activity of the Hft lysate. On the other hand, the gal^+ exogenote, along with its λ chromosome carrier, may be lost, thereby explaining the reversion of gal^+ transductants to the gal^- nonlysogenic state. A third possibility is recombination in the *gal* region between the exogenote and endogenote leading to the insertion of the gal^+ gene directly into the bacterial chromosome. In such cases gal^+ endogenotes are formed which are as stable as the original gal^+ bacterial strain and no longer give off gal^- segregants. Approximately one-third of the gal^+ transductants fall into this category. A still further possibility reflects the occasional finding that recombination may produce a gal^- exogenote paired with a gal^- endogenote, thereby forming a gal^-/gal^- *homogenote*. Such diploid homogenotes are similar to heterogenotes in producing Hft lysates, but their transducing behavior is now gal^- rather than gal^+. Underlying all of these recombinational events are the unique and intricate relationships between the viral chromosome and its lysogenic host. To understand the genetics of these organisms, it is therefore essential to understand their specific interactions.

PROPHAGE INTEGRATION, EXCISION, AND TRANSDUCTION

The specialized transduction observed in bacteria lysogenized by λ and other such phages indi-

*As will be discussed later (p. 419), the infectivity and lysogenicity of transducing λ particles carrying the *gal* locus depend upon the presence of "helper" phages.

cates that these prophages are restricted to specific areas on the bacterial chromosome at which, or near which, genetic exchange may occur. This localized relationship may result from an initial pairing between homologous sections in the bacteriophage and bacterial chromosomes. However, until fairly recently it was not clear whether the prophage was merely paired to the bacterial chromosome as an extraneous body or physically inserted into it. Stimulated by a hypothesis first developed by Campbell, investigators have gathered considerable evidence indicating that various types of prophages act as linear extensions of the bacterial chromosome, and are inserted between two bacterial loci. In the case of λ, some of the known events in the lysogenization sequence are as follows:

The λ chromosome is a double-stranded DNA molecule 46,500 nucleotides long with a single-stranded projection of 12 nucleotides at each 5′ end. These projections are complementary to each other, so that when injected into an *E. coli* cell as a linear molecule the chromosome rapidly circularizes because of its "cohesive" ends (Fig. 19–17). Circularity protects the λ chromosome against degradation by host exonuclease enzymes (see Pilarski and Egan), and may also offer advantages in replication. In this form two possible alternatives present themselves that will be discussed in greater detail in Chapter 29: (1) Genes enabling the vegetative reproduction of λ can begin to function, leading to the proliferation of λ particles and eventual lysis of the host cell. (2) The genes involved in vegetative growth may be *repressed* by a particular protein produced by the λ *cI* gene. If repression is successful, then the opportunity for lysogenization is now made manifest: that is, the integration and insertion of λ into the host chromosome.

Whether λ follows the lytic or lysogenic pathways is influenced by various factors including the physiological state of the host. Assuming appropriate conditions prevail, insertion into the host chromosome then depends on whether *attachment* regions are present in both host and viral chromosomes. If such regions are present, as shown in Fig. 19–18a, b ($BB' \leftrightarrow PP'$), then pairing may occur and recombination is possible. A specific gene on the phage chromosome (*int*) has been shown responsible for the recombinational event which then leads to integration of the viral (now prophage) chromosome (Fig. 19–18c, d). Once integrated, the prophage continues to produce its *cI* repressor, which prevents its other genes from functioning, and also confers immunity upon the host cell against the replication of other infecting λ particles. Should the continued function of the λ repressor be impaired (induction), then a number of λ gene products are rapidly manufactured, one of which (*xis* gene product) permits the "excision" of the prophage from the host chromosome, which aids it in entering the lytic pathway.

Occasionally mutants of λ arise, such as λ*b2*, which are defective in their attachment region and unable to pair with the bacterial chromosome. Although such mutants are repressed from multiplying vegetatively, they are still unable to integrate into the host chromosome, and therefore remain as independent bodies in the cytoplasm. Evidence for this is the fact that Hfr cells lysogenic for λ*b2* do not transmit the λ*b2* DNA to F^- recipients, whereas bacterial lysogens of normal λ easily transmit integrated prophage (see p. 450). Moreover, λ*b2* is not replicated along with the bacterial chromosome. Thus only one daughter cell at each bacterial cell division carries λ*b2*, and lysogenization by this mutant in a growing culture remains only with the initially infected cells. This restriction of lysogeny to a "micro" colony is very much like the abortive transduction previously discussed (p. 412) and is therefore called "abortive lysogenization."

Evidence that successful lysogenization of λ depends upon the chromosomal integration of prophage has now been obtained from a variety of sources. Among the earliest evidence was the finding that the genetic map of the integrated λ prophage is different from the genetic map of the vegetative phage. As shown in Fig. 19–18, the linkage order of three phage genes is *J-cIII-R* in vegetative phage crosses (recombination in sensitive nonlysogenic bacteria; see Chapter 20), and *cIII-R-J* in prophage crosses (bacterial matings followed by induction). According to Fig. 19–18,

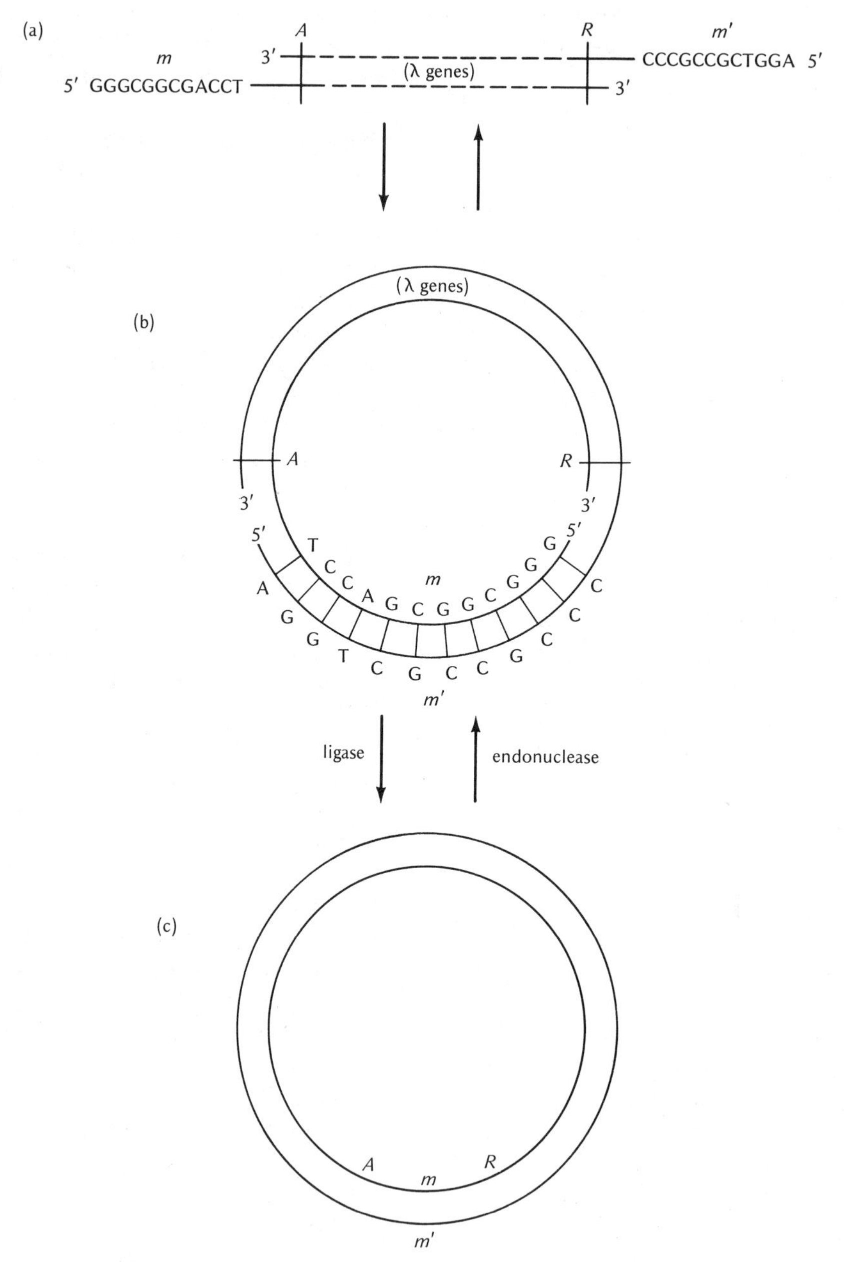

Figure 19–17

Diagram of the process by which the λ chromosome may become circular or linear through pairing or nonpairing of the bases at its cohesive ends. (a) Linear structure of λ showing its single-stranded ends and their nucleotide base structure. The "left" end near gene A is called m, and the "right" end near R is called m′. (b) Pairing of the ends leading to a double helical DNA structure and circularization of the λ chromosome. (c) The phosphoester gaps remaining between nucleotides because of circularization are chemically connected (covalent bond) by the ligase joining enzyme. The m — m′ base sequence is unique on the λ chromosome and can apparently be recognized by a specific endonuclease enzyme which can insert a nick on each DNA strand 12 nucleotides apart. This acts to reverse the circularization process, and leads back to a linear chromosome.

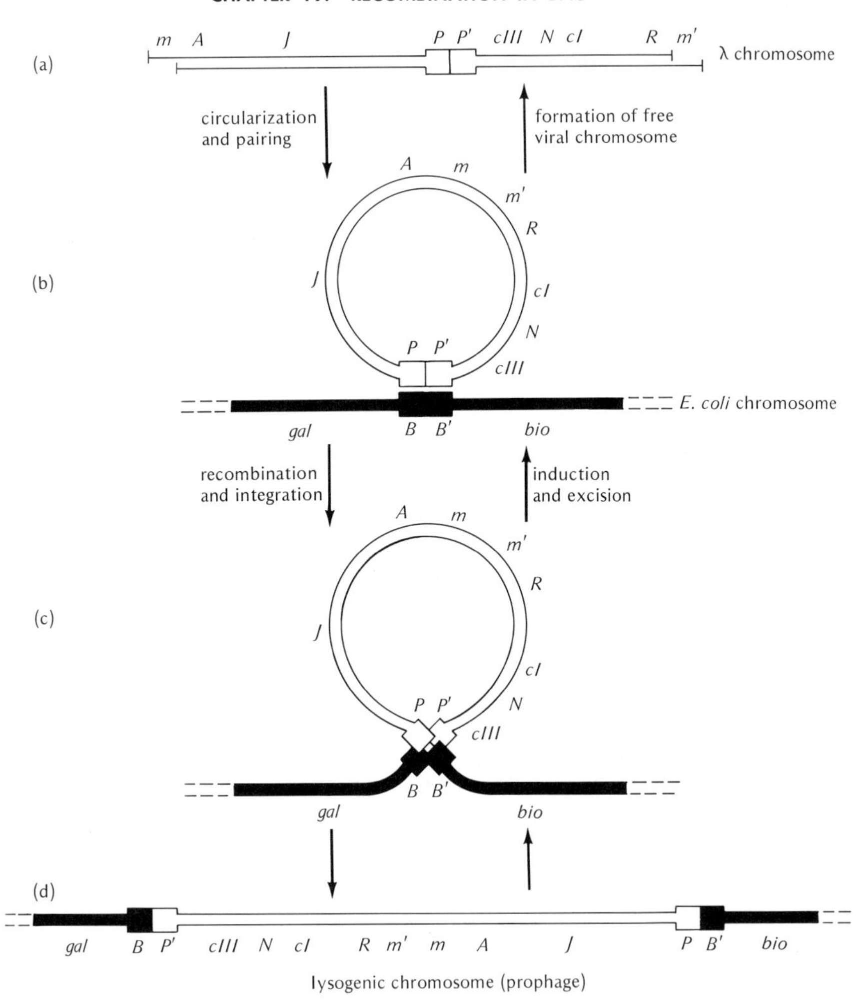

Figure 19–18

Lysogenization pathway for λ, leading from (a) the infective linear λ chromosome, to (b) circularization and pairing, to (c) recombination, to (d) integration. Note that the attachment site for λ on the bacterial chromosome (BB′) lies between bacterial genes governing galactose (gal) and biotin (bio) metabolism. (Other genes are also found in this area but are not indicated.) Through induction, the lysogenization process may be reversed, (d) → (c) → (b) → (a), dependent upon coiling of the prophage to enable resumed pairing between viral (PP′) and bacterial (BB′) attachment sites, and consequent excision of the viral chromosome.

it is clear that this permutation of gene order depends upon the attachment region of the λ virus being located between two of these genes (*J* and *cIII*). Also as expected, the flanking bacterial markers (*gal* and *bio*) are moved farther apart after lysogenization as demonstrated by their reduced frequency in joint transduction by generalized transducers such as phage P1 (see Rothman). In the case of another temperate phage, φ80, Franklin and co-workers showed that certain bacterial chromosome deletions extend into the phage chromosome, as though one were a linear exten-

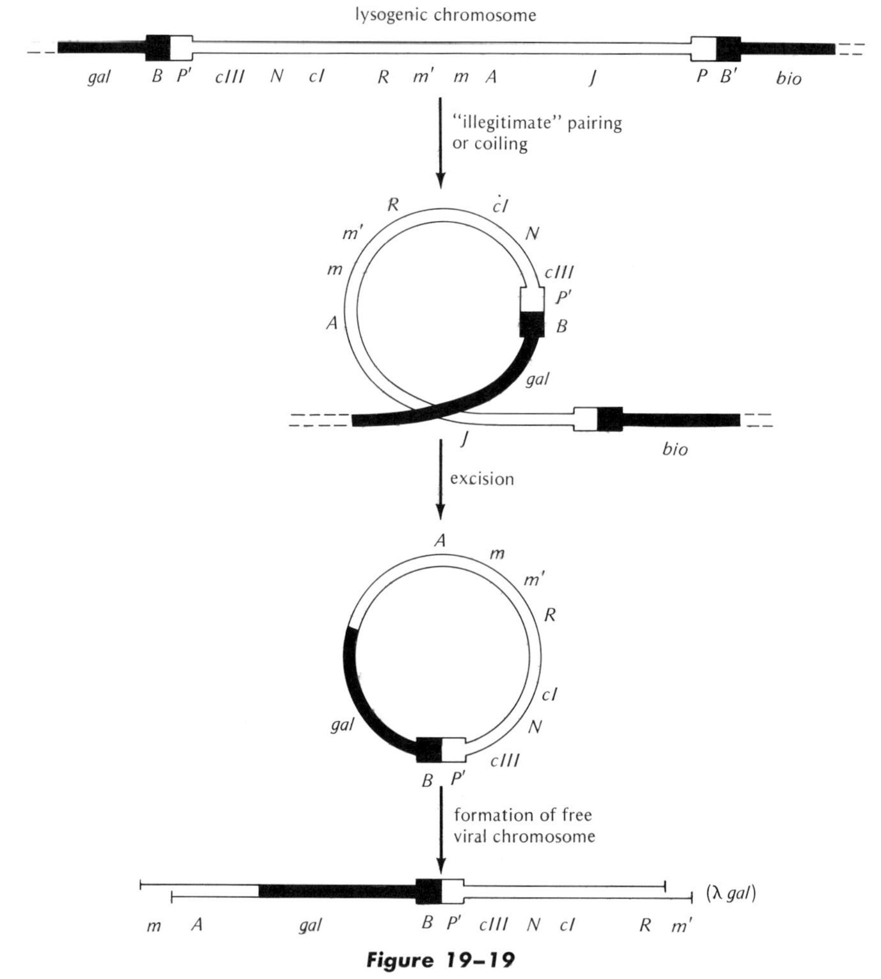

Figure 19–19

Events leading to the excision of a λ transducing particle which has incorporated the bacterial gene, gal.

sion of the other. More recently Sharp, Hsu, and Davidson have obtained electron micrographs showing the physical insertion of the λ prophage into an F particle bearing the bacterial attachment site for λ.*

For λ and other such viruses, the process of lysogenization is apparently an intermediate but necessary step in the formation of specialized transductants. The reason for this is simply that only through lysogenization can such transducing particles pick up a particular section of the bacterial chromosome. In the case of λ, its proximity to the bacterial *gal* and *bio* loci allows rare errors to occur in which induction leads to the excision of viral particles which include one or the other of these loci. For example, as shown in Fig. 19–19, a λ prophage may occasionally "coil" in such a way that the recombinational event that leads to the excision of a circular viral chromosome in-

* Possession of a localized attachment point on the bacterial chromosome, as occurs for λ and φ80, is not universal for all temperate phages. Phage P2, and some others, may integrate into more than one site on the bacterial chromosome, and phage P1 can even lysogenize bacteria without insertion into the host chromosome.

cludes the *gal* locus. When this aberrant chromosome is excised, it is of approximately the same length as a normal λ chromosome but is missing those genes that were replaced by the substituted bacterial regions. It is therefore defective and unable to perform some or many λ wild-type functions. If the *gal* gene is incorporated, then the transducing particle is called λ*gal* (or λ*dg*). If the incorporated bacterial gene is *bio*, then the particle is called λ*bio* (or λ*db*).

These defective λ particles can be induced (by UV) to reproduce themselves in a bacterial cell, causing lysis of its host, but the phage products of this lysis cannot establish lysogenic relationships and, with few exceptions, are unable to further infect bacterial cells. Therefore, when a suspension of λ*dg* particles are tested on nonlysogenic bacteria, the presence of λ*dg* cannot be observed either through the formation of plaques or by lysogenization of the bacterial host. On the other hand, once inside a bacterial cell, the presence of λ*dg* confers immunity upon its host so that it cannot be lysed by other λ particles. Thus, although λ*dg* is the carrier for the *E. coli gal* locus and apparently the main transducing agent in the Hft lysates (p. 414), it is, by itself, unable to transduce. What then accounts for the high transducing activity of the Hft lysates?

The transducing activity of λ, according to Campbell and others, occurs mainly in association with a nontransducing active λ phage, or "helper."* According to present concepts, diagrammed in Fig. 19–20, full pairing at the attachment site of the bacterial host is accomplished by the normal (nontransducing) λ helper phage ($PP' \leftrightarrow BB'$). It is only when this has taken place that a special attachment site is formed (BP') which now enables pairing and recombination between λ*gal* and the host chromosome. In other words, the bacterial host is often "doubly lysogenized" during specialized transduction, and can now be considered a partial diploid for the *gal* region (gal^+/gal^-), or a merozygote (p. 402). The diploid nature of the transduced heterogenote is therefore readily explained, as well as the fact that occasional excision and loss of the λgal^+ exogenote leads to the gal^- phenotype remaining in the endogenote.

It is also interesting to note that excision of the λ chromosome upon induction may now remove the gal^- region rather than gal^+ if the coiling event leading to excision is to the left of gal^+ in Fig. 19–20; that is, if the extruded λ chromosome includes the interval between the λ *A* gene and a point left of the bacterial gal^- gene. Such an event will permit gal^+ to remain on the host chromosome, leading to a gal^+ endogenote. Furthermore, if this process occurs during replication of the bacterial chromosome, the gal^- exogenote formed by this new "illegimate" incident may then pair with that daughter chromosome that still carries a gal^- gene, and insert itself to form a gal^-/gal^- homogenote. The many events that may occur during the insertion and excision of prophage chromosomes can therefore explain all of the specialized transductional phenomena previously mentioned.

SEXDUCTION

In 1959 Adelberg and Burns discovered novel F factors in *E. coli* which seemed to have been formed as a result of recombination with the bacterial chromosome. Whereas a normal Hfr strain transfers its sex factor only at the end of sexual conjugation, one of their Hfr strains transferred its sex factor with the same high efficiency as if it were an F^+ strain. On the other hand, this apparent F^+ factor continued to transfer the bacterial chromosome with the same directed orientation as in its initial Hfr state, but with lower efficiency than the usual Hfr. When transferred to F^- cells, it converted them to F^+ cells, but again it transferred their chromosomes according to the previously determined Hfr orientation. The sex factor, there-

* Helper phage may also aid in the transfection of cells by other phage DNA (see p. 66) by attaching the transfecting DNA to the cohesive ends of the invading helper DNA if the terminal nucleotide sequences in the two DNAs are complementary. So far, a number of different phages (φ80, 82, 21, 424, 434) have cohesive end sequences that can pair with those of λ, whereas a group of "non-lambdoid" phages (P2, P4, 186, 299, φD) have end sequences different from the lambdoid group but similar to each other (see, for example, Murray and Murray).

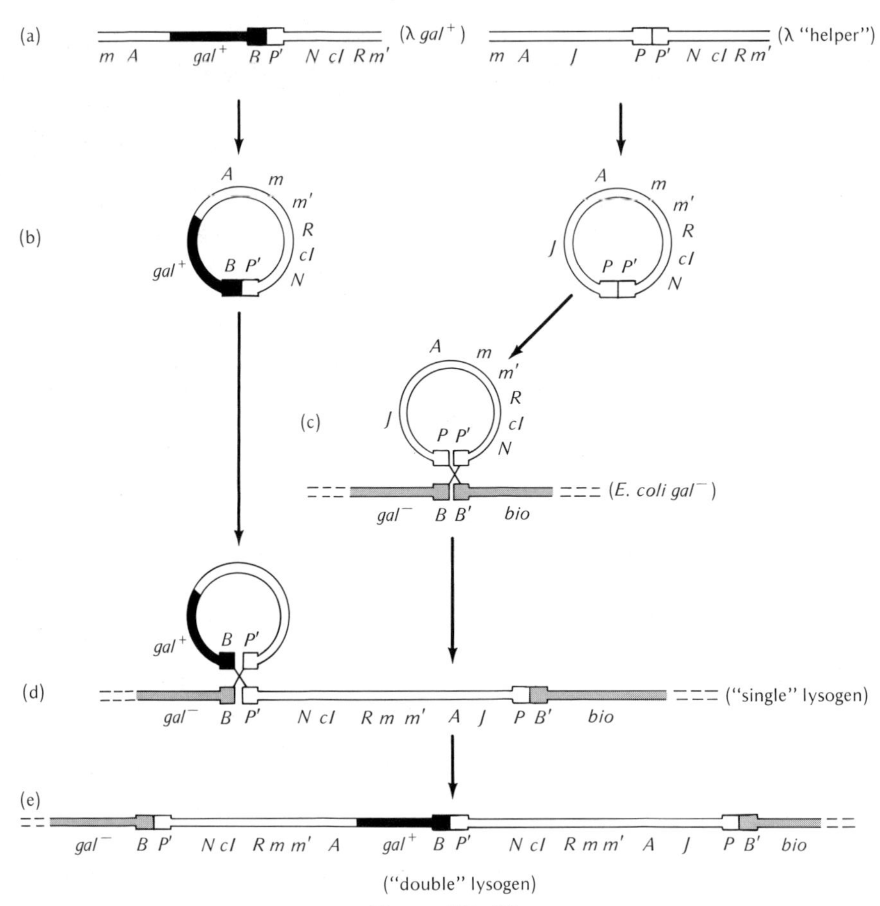

Figure 19–20

Events leading to transduction of an E. coli gal⁻ recipient by a λ phage carrying the gal⁺ gene. (a) Chromosomes of λgal⁺ and normal λ "helper." (b) Circularization of phage chromosomes infecting E. coli. (c) Pairing and recombination between the helper phage and bacterial chromosome at the λ attachment site. (d) Pairing and recombination between the λgal⁺ chromosome and the BP′ site formed by the previous integration. (e) Gene sequence in the transduced doubly lysogenic heterogenote.

fore, seemed to have changed from the stable Hfr state to the cytoplasmic F⁺ state, but nevertheless maintained a "memory" for breaking and orienting the donor chromosome at the exact same point and in the same direction as in its initial Hfr state. Furthermore, when this novel strain was treated with acridine orange, a chemical that "cures" the cells of the sex factor, the introduction of a new wild-type F⁺ factor caused these cells to become genetic donors. Instead of the low-frequency genetic donors formed when ordinary F⁻ cells are infected with F⁺, these new F⁺ cells were high-frequency donors of genetic material.

On the basis of these findings, Adelberg and Burns hypothesized that recombination had taken place in the original Hfr strain between the sex factor and the *E. coli* chromosome at the specific point to which the sex factor is normally attached in this strain. This produced a bacterial chromosome carrying part of the sex factor, and a sex factor carrying part of the bacterial chromosome. Because of this exchange, the sex factor is no longer as stable as in the Hfr state but behaves in a fashion similar to a cytoplasmic F^+ factor. When transferred to an F^- cell this recombined sex factor (called F') will "recognize" the same position on the host chromosome that is homologous to the position of the *E. coli* chromosome section it is carrying. During subsequent conjugation the F' factor will produce breakage and orientation of the chromosome at this precise point. Removal of the F' factor from the initial strain by acridine orange does not remove that portion of the sex factor which has been inserted into the bacterial chromosome. Thus the introduction of wild-type F^+ factors provides an opportunity for pairing between the F^+ factor and the bacterial chromosome at the specific point at which they both share homologous sections. During conjugation, breakage and oriented transfer of the chromosome can take place readily at this point, although the F' factor has now been substituted by F^+.

Of special importance in this sequence of events is the retention of a portion of bacterial chromosome within the F' factor itself. Thus the F' factor may act as an infective particle, similar to λ, which can transduce F^- cells. This process is known as *sexduction*, or *F-duction*. According to Jacob and Adelberg, F' particles capable of sexduction can be obtained by interrupted mating of conjugants between Hfr and F^- cells before transfer of the Hfr sex factor is ordinarily completed. Selection is then made for those recipients which show the phenotype for a gene close to the terminal portion of the donor chromosome. For example, the B8-Hfr strain regularly transfers the bacterial chromosome in the sequence *ade*· · ·*lac*-F. Thus in a mating between lac^+ Hfr and lac^- F^- cells, lac^+ does not ordinarily enter the F^- cell until about 2 hours after conjugation has begun. If mating is interrupted within 1 hour, however, rare recombinant cells occur that are nevertheless lac^+. These lac^+ cells have a number of unusual attributes, among which are the following:

1. Their only heritage from the Hfr parent is the F' factor and the lac^+ characteristic. Since this F' lac^+ combination can be serially transmitted to a number of other F^- strains, the F' factor and lac^+ thus seem to be combined in the same particle.

2. Transfer of the F' lac^+ factor from donor cells to recipient cells can occur independently of the transfer of other gene markers on the donor chromosomes. Occasionally, however, transfer of the donor chromosome to the recipient will also occur as in normal conjugation, but the chromosome transfer will usually be oriented as in the ancestral Hfr strain, *ade*· · ·*lac*. The F' factor is thus an infective particle that can either be independently transferred to recipient cells in the same fashion as an F^+ particle, or can orient chromosome transfer in the same fashion as an Hfr factor.

3. The lac^+ colonies formed by these F' lac^+ particles are unstable, segregating lac^- cells (like the F^- parent) at a rate of one per thousand cell divisions. The lac^+ colonies are therefore partial diploids or "heterogenotes" (lac^-/F' lac^+), similar to the diploid genotypes produced by λ transduction (p. 414) and sexual conjugation (p. 402).

4. Occasional lac^- homogenotes are produced that bear F' lac^- particles ($lac^-/$ F' lac^-). Conjugation between such cells and F^- lac^+ strains will then produce heterogenotes that are lac^+/F' lac^-. Recombination between the F particles and the bacterial chromosome may therefore produce a variety of combinations.

5. The presence of F' factors can increase the amount of a particular phenotypic product. For the *lac* gene, Jacob, Perrin, Sanchez, and Monod showed that the amount of β galactosidase, an enzyme involved in lactose utilization, increases two to three times because of F' lac^+ factors. Similarly Garen and Echols have shown that an F' factor carrying the gene for alkaline phosphatase doubles the production of this enzyme as compared to haploid cells. Thus probably one or two F' factors are present in each cell.

6. Pairing between the F' factor and the homologous section on the chromosome can

be surmised from the observation that an F′ *lac* factor will orient transfer of a bacterial chromosome only when a *lac* region is present on this chromosome. If the bacterial chromosome is missing this region (deletion, see Chapters 22 and 25), the F′ *lac* factor will no longer orient chromosome transfer at this point but will act as an ordinary F^+ factor.

Extending the insertion hypothesis of Campbell (p. 415) to F factors, it has been proposed that F′ arises from a recombinational event in which an F factor that has been inserted into the bacterial chromosome recombines to form an extruded loop that now contains a portion of the bacterial genome. According to this hypothesis, the DNA of the F factor, as of λ, is circular and bears regions of homology with the bacterial chromosome. When in the free F^+ state, the F factor may become inserted into any portion of the bacterial chromosome by a single recombinational event (Fig. 19–21a). Once F is inserted, the bacterial chromosome acts as an Hfr cell and will transfer its chromosome in the direction determined by the F factor (solid arrows, Fig. 19–21). Occasionally recombination between the pairing sites which led to the initial insertion of the F factor will occur in the Hfr cell. When this takes place, the F factor is extruded into the cytoplasm and the cell is converted into an F^+ cell (Fig. 19–21b). More rarely, homologous pairing occurs between portions of the bacterial chromosome on either side of the F factor that leads to a recombinational event in which a section of bacterial chromosome is included in the extruded F factor (Fig. 19–21c, d). It is this event which is presumed to be the origin of F′ particles.

This hypothesis helps explain the fact that recipients infected with F′ particles are now easily converted to Hfr cells with the same orientation as the original Hfr. This is because the F′ particle can now pair easily with the homologous portion of bacterial chromosome and become inserted into the chromosome by a single recombinational event. The necessity for homologous pairing as a prelude to recombination also explains the observation that a deletion of the bacterial chromosome for its homologous region in F′ will cause the F′ particle to act as an ordinary F^+ factor and prevent the same kind of oriented chromosome transfer as in the original Hfr strain.

EPISOMES AND PLASMIDS

We have seen that the F′ factor shares with λ phage the ability to transfer a portion of the donor chromosome to a recipient cell. In addition, together with all F factors, F′ also shares with λ the ability to exist independently in the cytoplasm or to attach itself to the bacterial chromosome. F and λ nevertheless differ in a number of respects: F^+ can usually attach to the bacterial chromosome at many different loci, whereas λ is limited to a specific site; and λ may cause cellular lysis by vegetative multiplication outside the bacterial chromosome, whereas F does not. They also use different modes of infection to enter new cells; the F factor is dependent upon cell contact, whereas λ retains its own infective apparatus, the viral protein envelope. In both cases, however, the presence of the particle, either as temperate λ prophage or as F, prevents the vegetative replication of similar particles in the bacterial cytoplasm. Also, in both cases the particle is replicated along with the replication of the bacterial chromosome.

The basic identity between these two particles prompted Jacob and Wollman to propose a category in which they would both be included, that of *episomes.* They defined episomes as particles which are *in addition* to the normal bacterial chromosome and which cannot arise by bacterial mutation but must be introduced from without. This introduction is *infectious*, coming either through cell contact in conjugation or through extracellular viral particles. Moreover, these two particles may exist in *alternative* states, either cytoplasmically or chromosomally attached.

The infective nature of these particles has been used to transfer genetic material not merely from one strain to another but from one species to another. *E. coli* sexual donors, for example, can infect *Shigella* species (includes dysentery-causing bacteria) with the F factor, which may be accompanied by sections of *E. coli* chromosome. Although the

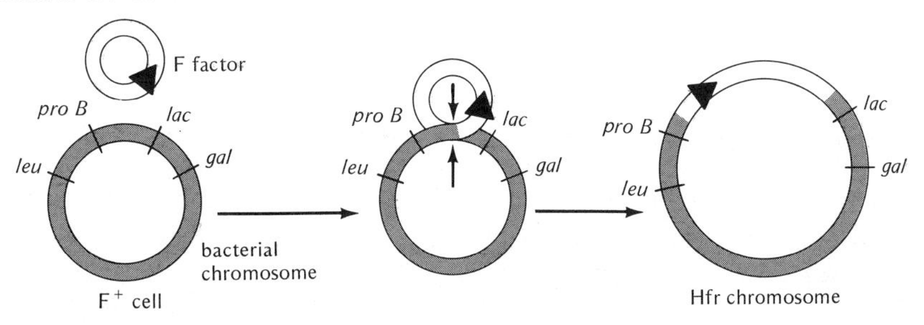

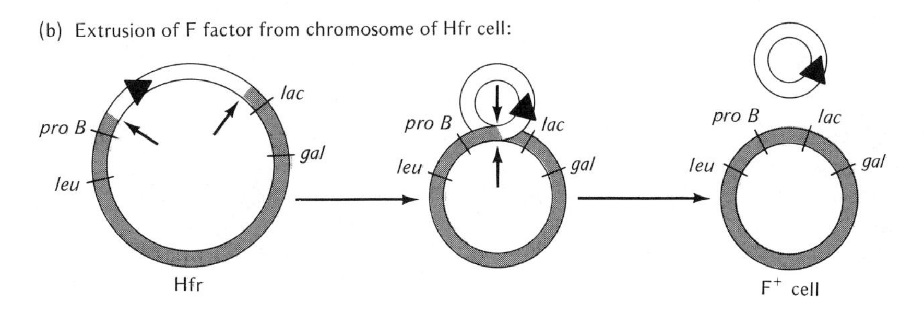

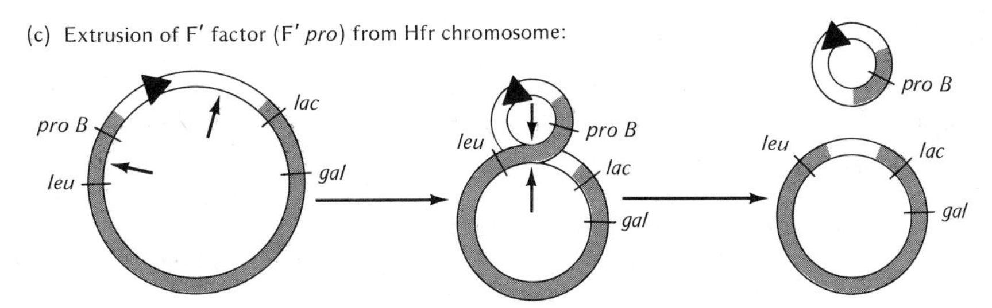

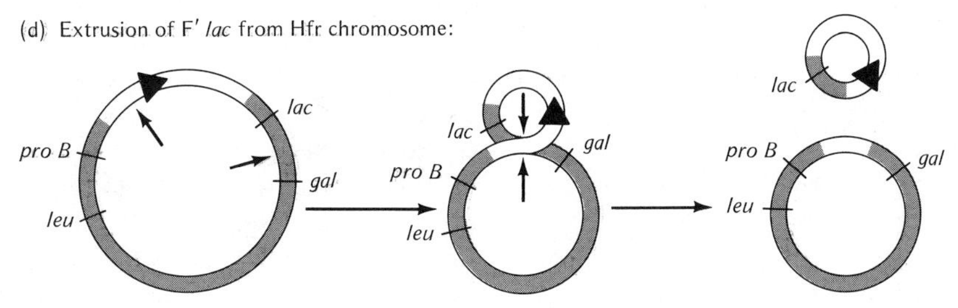

Figure 19–21

Proposed Hfr, F+, and F′ formation by means of single recombinational events involving the circular F particle and the circular bacterial chromosome. In (a) homologous pairing occurs between the F particle and the bacterial chromosome (homologous sections indicated by short arrows) leading to insertion of the F particle. In (b) the F particle is extruded by recombination between the indicated homologous sections. In (c) homologous pairing between sections that include a portion of the bacterial chromosome causes formation of an F′ particle that carries the bacterial gene pro B. In (d) pairing leads to extrusion of an F′lac factor. The solid arrowhead of the sex particle represents the origin and direction of chromosome transfer in sexual conjugation. [Not shown are instances in which bacterial loci at both ends of the F factor (e.g., pro and lac) are incorporated into the F′ factor simultaneously (e.g., F′ pro lac). In such cases the F′ factor may induce transfer of a bacterial chromosome at either of these loci.]

E. coli chromosome is unable to recombine with the *Shigella* chromosome, the F factor itself may nevertheless remain in the *Shigella* cytoplasm. Such *Shigella* cells may then act as male conjugants which cannot transfer their chromosome but can transfer the F factor, thereby transforming *E. coli* F^- cells into F^+. The same relationship has been found to exist between the *E. coli* F factor and *Serratia marcescens* (red-pigmented bacteria) and *Pasteurella pestis* (bubonic plague).

Conjugation between *E. coli* F^+ cells and *Salmonella* species is more successful, since the *E. coli* factor placed in the *Salmonella* cell may now mobilize the *Salmonella* chromosome for transfer. In fact such *Salmonella* F^+ cells give rise to Hfr strains which transfer the *Salmonella* chromosome in an oriented direction to either *Salmonella, Shigella,* or *E. coli* recipients. Thus F factors, normally absent in species groups such as *Salmonella, Shigella, Serratia,* and *Pasteurella,* are able, with varying degrees of success, to survive, replicate, and infect. Sexual conjugation in other bacterial species, mediated by fertility factors that seem to act in a fashion similar to F, have been found in *Pseudomonas aeruginosa* (pus bacteria) and in *Vibrio comma* (cholera).

In addition to λ and F factors, a third form of DNA-containing particle is the *colicinogenic factor* (*col*). Factors of this kind are carried by particular strains of bacteria which occasionally cause the death of some cells with subsequent release of proteins (*colicins*) capable of killing susceptible strains. Different colicins are found (*E1, E2, E3, I, K,* etc.), each distinguished by specificity for particular susceptible bacterial strains through recognition of special receptor sites on the cell surface. Once attached to a sensitive cell membrane, their bacteriocidal action seems to derive from the release of enzymes that cause breakdowns in various processes. For example, *col E1* appears to affect membrane-bound oxidative phosphorylation and interfere with ATP production; *col E2* either degrades DNA or prevents DNA synthesis; and *col E3* causes ribonuclease activity leading to ribosomal alterations and inhibition of protein synthesis.

Colicinogenic cells (col^+) are cells carrying the colicinogenic factor in inhibited or repressed form and, as in lysogenic cells, are themselves immune to death by colicins of their own type. Such colicinogenic cells may, however, transfer colicinogenic factors to col^- bacterial recipients either through cell conjugation or through transduction, thereby converting them to col^+. The achievement of conjugational transmission may be dependent upon the presence of an F factor, so that some *col* factors such as *col E2-P9* cannot be transmitted between F^- strains but can be transferred from an F^+ to an F^- strain. In other instances, *col* factors can achieve transmission independently of F, and have become associated with sex factors which differ from the F factor. For example, *col I* and many other *col* factors produce a particular kind of sex pilus (*I*) that is attacked by bacteriophages (If1, If2) that differ from those attacking pili produced in F^+ cells. Other differences between sex factors lie in their cross-incompatibilities. That is, sex factors often prevent sex factors of their own type from existing in the cell but do not affect sex factors of other types. On this basis six different groups of sex factors have, so far, been described.

At present a number of reports exist showing that *col* factors and their associated sex factors may help to bring about bacterial chromosome transfer from their hosts to recipients. In most cases, however, there is little or no evidence that *col* factors ever become stably integrated into the bacterial chromosome. They thus exist independently in the cytoplasm, and this extrachromosomal stability gives them the name *plasmids* (see p. 257) rather than episomes. Nevertheless, when their genetic material is isolated, it too, like F and λ, is composed of a circular DNA double helix whose length is characteristic of that particular *col* factor. It has therefore been suggested that these plasmids, like some episomes, may have their origin in temperate bacteriophages that became defective. As evidence, some of the bacteriocidal proteins produced by *col* factors and similar plasmids in other bacteria (*Pseudomonas aeruginosa, Bacillus subtilis*) are of high molecular weight and closely resemble the tail structures of bacteriophages (see review of Meynell).

Similar sexual activity and chromosome transfer has been noted for a group of plasmids called *resistance factors* (R). These were first detected in Japan in 1957 when strains of *Shigella* were found to be resistant to many of the antibiotics used during a dysentery epidemic. It was noted that simultaneous resistance to as many as four or more antibiotics (chloramphenicol, streptomycin, sulfonamide, tetracycline, and others) could be transferred extrachromosomally by cell-to-cell contact both within and between species such as *Shigella* and *E. coli*. The fact that such transfer was independent of whether the cells were F^+ or F^- indicated that a particle other than the F factor was involved. Since then, studies by Watanabe and others have shown that, like F, such R particles are circular cytoplasmic self-replicating DNA structures which produce sex pili in their host, but differ from F in lacking the ability to integrate themselves stably into the bacterial chromosome. They appear to owe their origin to events that occurred long before antibiotics were used, and can be found in rare cells among bacterial strains that were never exposed to any of the modern bacteriocidal agents. Their present success is probably the result of the wide use of medicinal and agricultural antibiotics causing a selective increase in R frequency by destroying nonresistant cells.

The production of unique phenotypic effects by episomes and plasmids is not unusual (see also Chapter 13). In lysogenized cells, the changes associated with the presence of phage, called *phage conversion*, are dramatically illustrated in the relationship between the diphtheria bacillus, *Corynebacterium diphtheriae*, and a specific phage. According to investigations by Freeman, Groman, and others, the diphtheria bacillus produces its poisonous toxin only when it is lysogenized by this particular phage. More specifically, the toxin is produced by a particular phage gene shown also to possess mutations which produce nontoxic proteins. Somewhat similar findings have been made in *Clostridium botulinum* type C: production by these bacteria of a highly virulent toxin causing "botulism" is dependent upon the presence of a particular bacteriophage.

In multicellular organisms, an especially serious effect is the conversion of normal cells into uncontrolled proliferating tumor or cancer cells as a result of infection by polyoma, SV40, Rous sarcoma, Shope papilloma, and other such "oncogenic" DNA and RNA viruses (see p. 106). Although it is suspected that aspects of the cell-virus relationship in some tumors may be similar to lysogeny, the precise mode of chromosomal integration and replication for these viruses is not yet known.

HOST RESTRICTION AND MODIFICATION

In spite of their unicellular structure and small size, bacteria are not always at the mercy of invading bacteriophages, episomes, and plasmids, that have managed to penetrate the cell wall. Early studies showed that the success with which invading bacteriophages could grow on a particular host often depended on the previous host from which they had arisen. For example, most λ phages grown on their usual host, *E. coli* strain K12, are "restricted," since they are unable to grow on *E. coli* K12 strains which are lysogenic for phage P1, called K12 (P1). Such restricted λ DNA can enter the host cell but is subject to breakage and rarely survives. Only the progeny of those few λ phages which somehow escape this restriction can now successfully infect other K12 (P1) bacteria and are called "unrestricted." On the other hand, practically all of these unrestricted λ phages will produce restricted progeny if they are later grown on K12 bacteria nonlysogenic for P1. Obviously the presence of the P1 temperate phage in the host is responsible for restricting the entering DNA of λ phage previously grown on K12, but can nevertheless modify λ DNA grown in K12 (P1) enabling it to become unrestricted. This modification, however, just as the restriction, lasts for only one generation.

As a result of studies by Arber and others (see reviews of Arber and Boyer), the mechanisms responsible for host restriction and modification of DNA have been tied to various enzymes, among which are endonucleases that can cleave a DNA

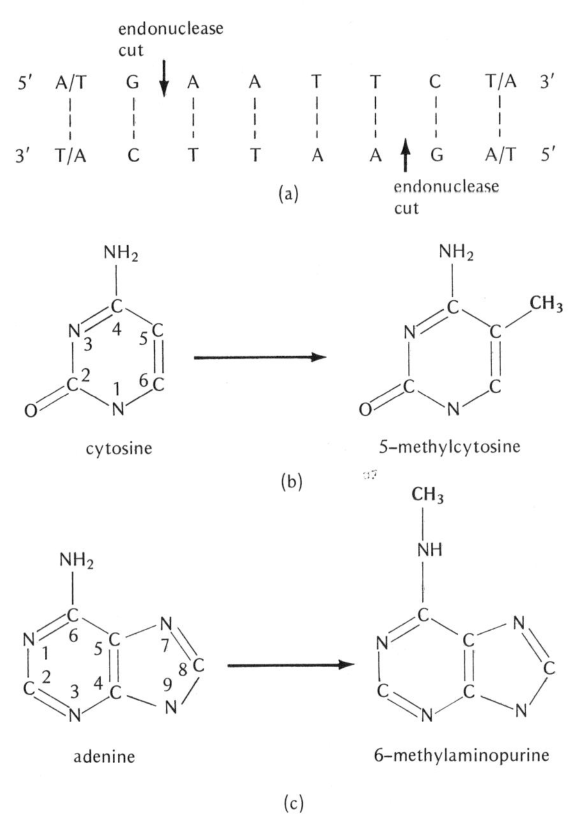

Figure 19–22

Effects of host restriction and modification activities on DNA. (a) Action of an E. coli restriction endonuclease on λ DNA that has been grown in cells not carrying this enzyme. Note that the sequence attacked by the endonuclease is "palindromic"; that is, the upper strand going in one direction (e.g., 5′ to 3′) has the same nucleotide sequence as the lower strand going in the opposite direction (5′ to 3′). There are apparently five such sites on λ DNA. Also interesting is the fact that the endonuclease cleavages produce single strands which have overlapping "cohesive" ends (see p. 415). (After Hedgpeth, Goodman, and Boyer.) The action of methylase enzyme is shown on (b) cytosine, and (c) adenine. For any host restriction system DNA methylation has been found only at those positions attacked by the endonuclease. Thus, the palindromic sequences shown in (a) are methylated in E. coli cells carrying the R factor called fi$^+$, and thereby protected against degradation.

molecule at specific nucleotide positions, and methylases that can add methyl groups to these nucleotides and thereby prevent cleavage. As shown in Fig. 19–22a, a restriction endonuclease of *E. coli*, *Eco* RI, can recognize a specific nucleotide sequence on a λ DNA strand (AATTCA and AATTCT) and cleave these strands at staggered positions, so that the continuity of the λ double helix is broken at five such sites. Thus, λ DNA grown on K12 would be degraded upon entering *E. coli* cells that carry this restriction enzyme. However, that DNA which escapes restriction is modified or protected against the endonuclease by the methylation of cytosine and adenine nucleotides at these sites to 5-methylcytosine and 6-methylaminopurine (Fig. 19–22b, c). The fact that both nuclease action and methylation can occur at the same site or neighboring sites is probably because both activities are part of the same molecule (see Haberman et al.) or because the enzymes involved can recognize similar sites.

Since there are many specific kinds of restriction and modification genes, some carried on the bacterial chromosome and some on temperate viruses and plasmids, "foreign" DNA entering a bacterium may have to endure a gantlet of enzymatic attacks. Host modification via methylation is therefore an important device enabling a bacterial cell to "recognize" its own DNA.

From a genetic point of view, the discovery of restriction endonucleases has opened a number of important areas of research:

1. CLEAVAGE MAPPING. Since restriction enzymes permit the cleavage of DNA molecules at specific sites, they enable such molecules to be partitioned into segments which can then be individually identified and placed in sequential order. For example, the DNA of the first virus analyzed by this method, SV40, is split by *Hemophilus influenzae* restriction endonuclease into 11 fragments which can then be identified by their molecular weights. If enzymatic digestion by the endonuclease is halted before completion, larger fragments are also obtained, and the entire DNA molecule can be ordered into a "cleavage map" by observing which fragments are contained within larger sections.

Thus if a chromosome contains fragments A, B, C, D, . . . , and these are ordered in sequence

A-D-C-B---, a section of length A---B (#1) would, after further enzymatic digestion, be found to contain within it segments ACD (#2), BCD (#3), AD (#4), BC (#5), and CD (#6). The only linear order that can consistently explain the contents of these sections is then derived by simply arranging them so that similar fragments overlap:

Segment No.	Order within segment
4	A-D
2	A-D-C
6	D-C
3	D-C-B
5	C-B
1	A-D-C-B

This "overlapping" method of analysis, used also in the sequencing of proteins (Chapter 26) and RNA (Chapter 28), recently enabled Danna, Sack, and Nathans to obtain a cleavage map of the SV40 virus without the need for a prior linkage map of individual genes (Fig. 19–23). It is important to note that their procedure reverses the usual order of analysis in which gene relationships are derived first, and the genes are then later localized to specific physical sections of the chromosome (Chapter 22). In the case of SV40, by contrast, the cleavage map is being used to help establish a linkage map by observing which cleavage fragments of DNA hybridize with messenger RNA (see p. 101) from different parts of the SV40 life cycle. Thus, Khoury and co-workers have found that genes acting early during lysis of the cellular host appear to be restricted to fragments A, H, I, and B, whereas later-acting genes are in the remaining cleavage map interval from A to J.

2. DNA PROBES. Hybridization techniques add considerably to the utility of restriction enzymes since specific DNA fragments formed by such enzymes can be radioactively labeled and used as probes for the presence of viral genomes in host cells. Thus, the DNA of some adenoviruses that cause the transformation of cells into tumors have been "dissected" by restriction endonucleases, and only particular fragments of viral DNA have been found to hybridize with the DNA of tumor cells. In the case of adenovirus type 2, rat cells transformed into tumors by this virus seem to contain only one end of the viral genome, about one seventh of its length (Gallimore, Sharp, and Sambrook). Along with other data, this information strongly suggests that only a small section of adenovirus DNA is necessary for carcinogenic function and the "oncogene" hypothesis (p. 106) may apply to these DNA viruses.

3. ARTIFICIAL RECOMBINATION. Perhaps the most far-reaching effect of restriction endonucleases is their use in cutting different DNA molecules at the *same* nucleotide positions leaving identical "cohesive" ends (see p. 415). This allows

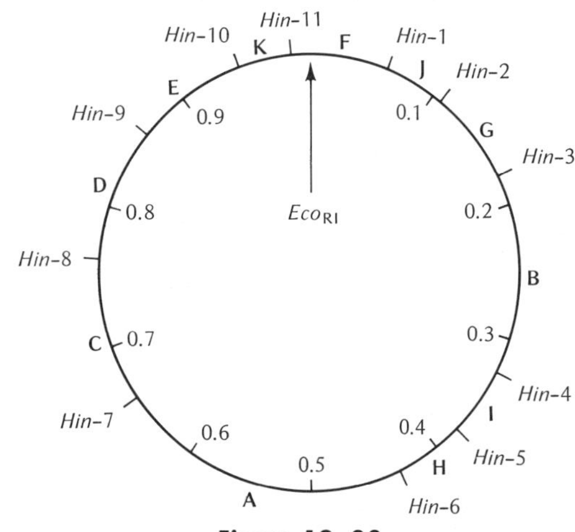

Figure 19–23

Cleavage map of the SV40 genome showing the sites at which cleavage occurs upon exposure to Hemophilus influenzae *restriction endonuclease (labeled* Hin-1, Hin-2, *etc.), and the fragments produced by this cleavage (A, B, C, . . .).* Eco_{RI} *is the single cleavage site for* E. coli *restriction endonuclease RI, and marks the zero point for the map. The physical map distances are given as decimals representing the fractional length of the SV40 DNA molecule in clockwise direction from* Eco_{RI}*. A third restriction enzyme, that of* Hemophilus parainfluenzae*, divides the genome into three major segments. (After Danna, Sack, and Nathans.)*

different DNA molecules to pair together end-to-end, or enables a "donor" DNA molecule to be inserted into a "recipient". As shown in Fig. 19–24, a circular viral or plasmid DNA molecule broken at a single site by restriction endonuclease, *Eco* RI, can open up to form cohesive bonds with each end of a linear segment of donor DNA that has been cleaved by the same enzyme. After such pairing, the receptor molecule can circularize once more, and ligase enzyme can then "join" the two paired molecules into a single stable and functioning unit.

By these means, Cohen and co-workers have inserted the DNA of genes that confer resistance to various antibiotics into *E. coli* plasmids, and shown that they then functioned to confer resistance upon their hosts. The "splicing" of eucaryotic DNA into procaryotes soon followed, and genes from both *Xenopus laevis* (Morrow et al.) and *Drosophila* (Wensink et al.) have now also been incorporated into *E. coli* plasmids. In the case of *Drosophila melanogaster*, DNA segments from the eucaryotic genome were spliced into a circular DNA of the plasmid pSC101 which carried a gene for tetracycline resistance. The recombinant plasmids were then used to transform *E. coli* bacteria sensitive to the antibiotic, thus enabling the isolation of resistant bacteria carrying the plasmid DNA. When portions of DNA isolated from these clones were transcribed into RNA copies labeled with tritium, they could be shown to hybridize with specific sections of *Drosophila* chromosomes by the technique of Pardue and Gall (pp. 98–99).

These discoveries therefore indicate that "genetic engineering" techniques are being perfected that will permit preferential incorporation into cells of whatever genes can be identified, isolated (Chapter 29), and properly cleaved. Bacterial cells, for example, may eventually be used as genetic hosts to provide an inexpensive, highly efficient source for production of many necessary eucaryotic hormones and enzymes, including insulin for diabetics, blood factors missing in hemophilics (Fig. 12–8) and other proteins. Restriction endonuclease techniques might also offer the opportunity of therapeutically "infecting" the cells of organisms and humans deficient for particular genes with plasmids or viruses carrying these genes.

Figure 19–24

Events that take place during the splicing of a length of donor DNA into a circular DNA molecule by means of restriction endonuclease, Eco RI. (a) The enzyme cleaves the two DNA molecules at the positions indicated by short arrows. The circular plasmid used as a recipient DNA molecule by Cohen et al. had only one such cleavage site, and this is also now known to be true for some mutant strains of λ. (b) After cleavage, the molecules possess exposed "cohesive" single-stranded ends. Depending on the distances between the enzyme recognition sites, cleaved fragments from a donor molecule may be several thousand nucleotides long. (c) Hydrogen bond formation can then occur between cohesive ends of donor and recipient, leading to a circular structure once again. The final step in the process is the formation of covalent bonds between adjacent donor and recipient nucleotides by the use of DNA ligase enzyme.

Nevertheless, experiments with these enzymes are accompanied by various potential hazards, some of which have recently been enumerated by a group of geneticists (Berg et al.) as follows:

> Antibiotic-resistance genes might be introduced into pathogenic bacteria that were not formerly resistant, or genes for toxin production might be introduced into bacteria that were not formerly pathogenic.
>
> The introduction of DNA from tumor-producing viruses into other viruses or plasmids, such as those carried by *E. coli* bacteria in the intestines of animals and humans, might increase the incidence of cancer.
>
> Eucaryotic DNA spliced into viral or plasmid chromosomes that are to be "grown" in bacteria such as *E. coli* might also carry DNA sequences that represent the chromosomal "provirus" form of tumor-producing RNA viruses (see p. 106).

It is presently difficult to evaluate the importance of each of these dangers but there is no question that they exist, and it is possible that some experiments may lead to tragedy or that

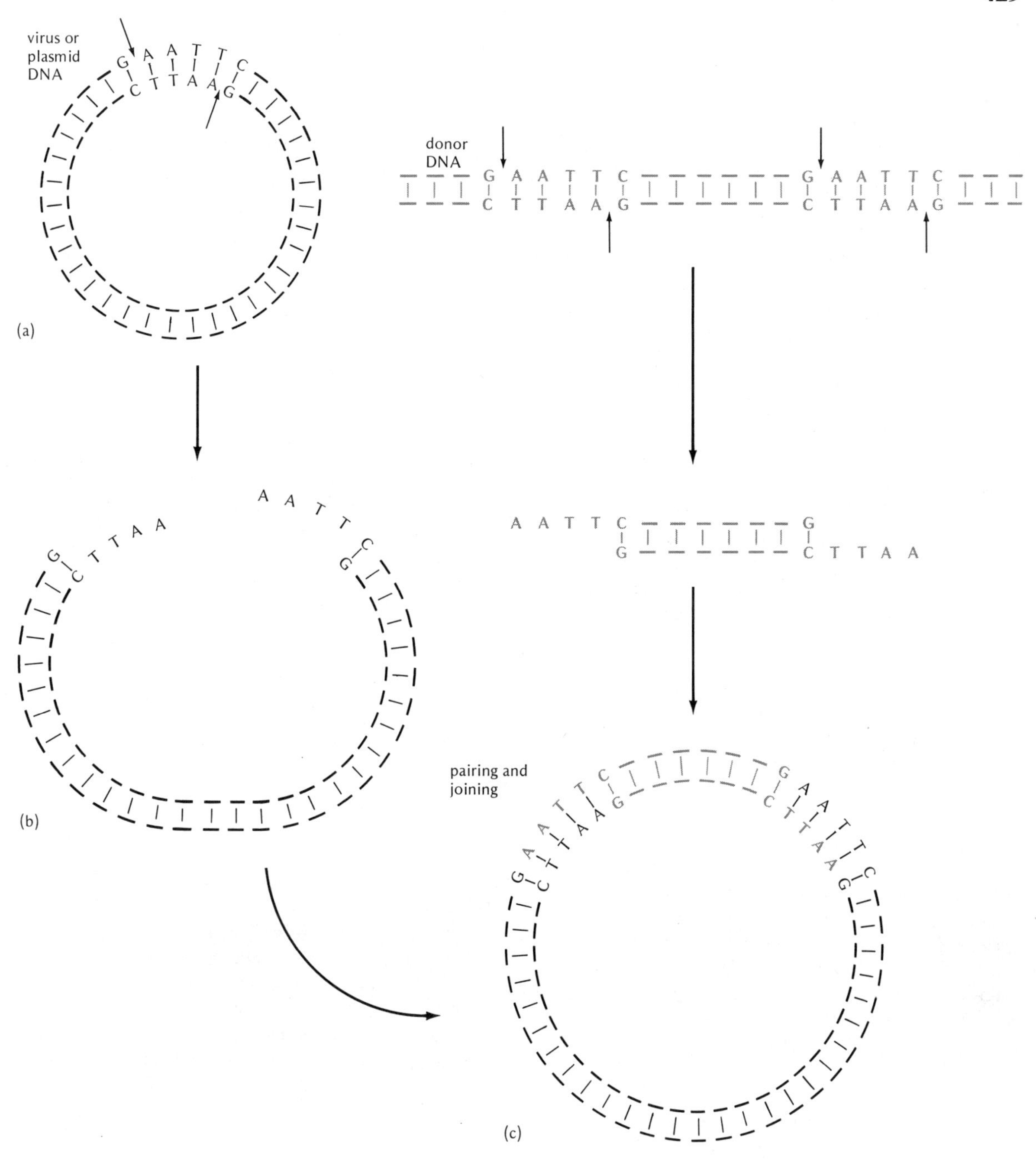

some governments might purposely use these techniques to increase the lethality of biological warfare. The need to ensure the use of science for the benefit of mankind is obviously a serious social problem, and such concerns have been expressed by various geneticists (see p. 685 and Chapter 36).

PROBLEMS

19-1. If a bacterial strain bearing a mutation (X) that prevents it from growing on medium unsupplemented by amino acid X is to be changed into wild type, what experiments would you perform to ensure that this change occurs through: (a) Transformation? (b) Conjugation? (c) Transduction?

19-2. If, as indicated on page 392, each *Hemophilus* bacterium can absorb only 10 molecules of transforming DNA, and a donor chromosome is broken into approximately 200 molecules of transforming DNA, what maximum frequency of transformants would you expect to find in a *Hemophilus* population that is being transformed for one particular gene?

19-3. Using the transformation technique, Anagnostopoulos and Crawford made a large number of crosses in *B. subtilis* between strains differing in two factors at a time. In the following data, for example, the recipient strain was unable to synthesize histidine (*his*) but was wild type for the single mutant gene which each of the three donor strains carried (*ant*, *trp*, *ind*). Transformants wild type for both donor and recipient genes were then scored:

Donor	Recipient	Frequency of Wild Type Transformants (+ +)
ant +	+ *his*	.450
trp +	+ *his*	.190
ind +	+ *his*	.263

Propose a linkage map for these four genes.

19-4. DNA was extracted from a wild-type strain of bacteria and used to transform a mutant strain unable to synthesize the amino acids alanine (*ala*), proline (*pro*), and arginine (*arg*). The number of colonies produced in the different transformant classes were as follows: 8400 ala^+ pro^+ arg^+, 840 ala^+ pro^- arg^-, 2100 ala^+ pro^- arg^+, 1400 ala^+ pro^+ arg^-, 420 ala^- pro^+ arg^+, 840 ala^- pro^+ arg^-, 840 ala^- pro^- arg^+. (a) What are the linkage distances between these genes? (b) What is the linkage order?

19-5. If a donor-transforming strain of bacteria is wild type for three genes a^+, b^+, and c^+, and the recipient strain carries the mutations a^-, b^-, and c^-, which transformant classes would you expect to be least frequent: (a) If a and b are linked together, but c is relatively unlinked? (b) If a and c are linked, but b is unlinked? (c) If all three genes are linked in the order *a-c-b*? (d) If they are linked in the order *c-a-b*?

19-6. Lederberg made crosses between various *E. coli* auxotrophs that were resistant (V^r) or sensitive (V^s) to phage T1. Wild-type prototrophs were then selected and scored for their resistance or sensitivity to T1. One set of results is as follows:

Cross	Number of prototrophs ($B^+ M^+ T^+ P^+$) V^r	V^s
$B^- M^- T^+ P^+ V^r \times B^+ M^+ T^- P^- V^s$	49	8
$B^- M^- T^+ P^+ V^s \times B^+ M^+ T^- P^- V^r$	5	19

Test whether the gene for viral resistance assorts independently of the others.

19-7. Using the technique of interrupted mating, five Hfr strains (1, 2, 3, 4, 5) were tested for the sequence in which they transmitted a number of different genes (F, G, O, P, Q, R, S, W, X, Y) to an F$^-$ strain. Each Hfr strain was found to transmit its genes in a unique sequence, as follows (only the first six genes transmitted were scored for each strain):

Order of transmission		Hfr Strain 1	2	3	4	5
	(first)	*Q*	*Y*	*R*	*O*	*Q*
	(second)	*S*	*G*	*S*	*P*	*W*
	(third)	*R*	*F*	*Q*	*R*	*X*
	⋮	*P*	*O*	*W*	*S*	*Y*
		O	*P*	*X*	*Q*	*G*
		F	*R*	*Y*	*W*	*F*

(a) What is the gene sequence in the original strain from which these Hfr strains derived? (b) For each of these Hfr strains, state which donor-gene marker should be selected in the recipients after conjugation to obtain the highest proportion of recombinants that will be Hfr?

19-8. A streptomycin-sensitive Hfr strain of *E. coli* of genotype a^+ b^+ c^+ d^+ e^+ was mated with an F$^-$ strain of genotype a^- b^- c^- d^- e^- for a period of 30 minutes and then subjected to streptomycin treatment. Prototrophs of the e^+ type were then selected out from the surviving recipients, and the following frequencies of other + genes were found:

70% were a^+
No b^+ prototrophs were found
85% were c^+
10% were d^+

What are the relative positions of the four genes a, b,

c, d, in respect to the origin of the donor chromosome (enters F^- recipient first)?

19-9. To enable recombination between conjugants to be detected, a donor marker must be selected that identifies the recombinants. On the other hand, the continued presence of the donor strain must be prevented to enable the recombinants to be selected without selecting the donor cells themselves at the same time. In other words, the donor strain should also carry a special marker which enables it to be "contraselected." For example, the donor strain can be *streptomycin-sensitive*, so that it can be eliminated by growing the conjugants on a streptomycin medium. If an Hfr strain were streptomycin-sensitive, at which end of the chromosome would you suggest the gene be located—at the origin or at the distal end?

19-10. Linkage analysis of mutations that affect the synthesis of the amino acid tryptophan in *Salmonella typhimurium* reveals that most such mutations are clustered in one general area of the bacterial chromosome near the *cysteine B* locus. These tryptophan mutations can be classified into four groups, *A*, *B*, *C*, and *D*, each group affecting a different step in tryptophan metabolism (Chapter 26). Wild-type (ability to synthesize tryptophan) depends upon the presence of normal (+) alleles in each of these groups. In one set of experiments M. and Z. Demerec made reciprocal crosses between $cys\ B^+\ trp\ B\ trp\ D^+$ and $cys\ B\ trp\ B^+\ trp\ D$ using P22 transducing phage. They found that when $cys\ B\ trp\ B^+\ trp\ D$ was the donor, more than half the resultant wild-type tryptophan colonies ($trp\ B^+\ trp\ D^+$) were mutant for *cysteine B*. On the other hand, when $cys\ B^+\ trp\ B\ trp\ D^+$ was the donor, a much lower frequency of transduced bacteria that were wild type for tryptophan were mutant for *cysteine B*. On this basis, decide whether the linkage order is *cys B - trp B - trp D* or *cys B - trp D - trp B*.

19-11. *E. coli,* as many bacteria, is usually multinucleate, carrying two to four chromatin bodies. Growth in a medium lacking phosphate, however, will cause *E. coli* to become uninucleate. Driskell has shown that when F^+ bacteria have been made uninucleate and mixed with F^- recipients, they will lose their sex factor and become F^- at the same rate as the recipients become F^+. What does this finding signify?

19-12. Figures 19-21c and 19-21d show the formation, respectively, of F′ *pro* and F′ *lac* factors from an Hfr strain, e.g., strain A. Assuming that each of these factors inserts into normal bacterial chromosomes to form two separate Hfr strains, B and C, respectively, will there be a difference in the directional orientation of the chromosome markers in B and C as they enter the recipient?

19-13. Mutations at particular loci of *E. coli*, called *rec*, are known to reduce recombination throughout the *E. coli* genome (see p. 402). However, some differences exist between these loci: *rec A* mutants degrade much of their DNA into small pieces during bacterial growth, and the amount of degraded DNA increases after ultraviolet irradiation. On the other hand, *rec B* mutants (and also *rec C*) do not degrade their DNA during growth and show relatively little DNA degradation after ultraviolet irradiation. Summarizing these data, combinations of these genes show the following results in respect to DNA degradation:

$rec\ A^+\ B^+$ = normal, no degradation
$rec\ A^-\ B^+$ = considerable degradation
$rec\ A^+\ B^-$ = very little degradation
$rec\ A^-\ B^-$ = very little degradation

Which of the gene products, *rec A* or *rec B*, is responsible for most of the DNA degradation observed?

REFERENCES

ACHTMAN, M., 1973. Genetics of the F sex factor in Enterobacteriaceae. *Curr. Topics in Microbiol. and Immunol.*, **60**, 79-123.

ADELBERG, E. A., and S. N. BURNS, 1960. Genetic variation in the sex factor of *Escherichia coli. Jour. Bact.*, **79**, 321-330.

ARBER, W., 1974. DNA modification and restriction. *Progr. Nuc. Acid Res. and Mol. Biol.*, **14**, 1-37.

BERG, P., D. BALTIMORE, H. W. BOYER, S. N. COHEN, R. W. DAVIS, D. S. HOGNESS, D. NATHANS, R. ROBLIN, J. D. WATSON, S. WEISSMAN, and N. D. ZINDER, 1974. Potential biohazards of recombinant DNA molecules. *Proc. Nat. Acad. Sci.*, **71**, 2593-2594.

BOYER, H. W., 1971. DNA restriction and modification mechanisms in bacteria. *Ann. Rev. Microbiol.*, **25**, 153-176.

CAMPBELL, A. M., 1969. *Episomes*. Harper & Row, New York.

CLARK, A. J., 1963. Genetic analysis of a "double male" strain of *Escherichia coli* K-12. *Genetics,* **48,** 105–120.

COHEN, S. N., A. C. Y. CHANG, H. W. BOYER, and R. B. HELLING, 1973. Construction of biologically functional bacterial plasmids *in vitro*. *Proc. Nat. Acad. Sci.*, **70,** 3240–3244.

DANNA, K. J., G. H. SACK, Jr., and D. NATHANS, 1973. Studies of simian virus 40 DNA. VII. A cleavage map of the SV40 genome. *Jour. Mol. Biol.*, **78,** 363–376.

EBEL-TSIPIS, J., M. S. FOX, and D. BOTSTEIN, 1972. Generalized transduction by bacteriophage P22 in *Salmonella typhimurium*, II. Mechanism of integration of transducing DNA. *Jour. Mol. Biol.*, **71,** 449–469.

ECHOLS, H., 1972. Developmental pathways for the temperate phage: Lysis vs. lysogeny. *Ann. Rev. Genet.*, **6,** 157–190.

ECHOLS, H., and D. COURT, 1971. The role of helper phage in *gal* transduction. In *The Bacteriophage Lambda,* A. D. Hershey (ed.). Cold Spring Harbor Laboratory, Cold Spring Harbor, pp. 701–710.

FOX, M. S., 1960. Fate of transforming deoxyribonucleate following fixation by transformable bacteria, II. *Nature*, **187,** 1004–1006.

FRANKLIN, N. C., W. F. DOVE, and C. YANOFSKY, 1965. The linear insertion of a prophage into the chromosome of *E. coli* shown by deletion mapping. *Biochem. Biophys. Res. Comm.*, **18,** 910–923.

FREEMAN, V. J., 1951. Studies on the virulence of bacteriophage-infected strains of *Corynebacterium diphtheriae. Jour. Bact.,* **61,** 675–688.

GALLIMORE, P. H., P. A. SHARP, and J. SAMBROOK, 1974. Viral DNA in transformed cells: II. A study of the sequences of adenovirus 2 DNA in nine lines of transformed rat cells using specific fragments of the viral genome. *Jour. Mol. Biol.*, **89,** 49–72.

GAREN, A., and H. ECHOLS, 1962. Genetic control of induction of alkaline phosphatase synthesis in *E. coli. Proc. Nat. Acad. Sci.*, **48,** 1398–1402. (Reprinted in the collection of Adelberg; see References, Chapter 1.)

GOODGAL, S. H., and E. H. POSTEL, 1967. On the mechanism of integration following transformation with single-stranded DNA of *Hemophilus influenzae. Jour. Mol. Biol.*, **28,** 261–273.

GROMAN, N. B., 1953. Evidence for the induced nature of the change from nontoxigenicity to toxigenicity in *Corynebacterium diptheriae* as a result of exposure to specific bacteriophage. *Jour. Bact.*, **66,** 184–191. (Reprinted in Adelberg's collection; see References, Chapter 1.)

GROSS, J. D., 1972. DNA replication in bacteria. *Cur. Topics in Microbiol. and Immunol.*, **57,** 37–74.

GROSS, J. D., and E. ENGLESBERG, 1959. Determination of the order of mutational sites covering L-arabinose utilization in *Escherichia coli* B/4 by transduction with phage Plbt. *Virology*, **9,** 314–331. (Reprinted in the collection of Adelberg; see References, Chapter 1.)

HABERMAN, A., J. HEYWOOD, and M. MESELSON, 1972. DNA modification methylase activity of *Escherichia coli* restriction endonucleases K and P. *Proc. Nat. Acad. Sci.*, **69,** 3138–3141.

HAYES, W., 1968. *The Genetics of Bacteria and Their Viruses*, 2nd ed. John Wiley, New York.

HAYES, W., F. JACOB, and E. L. WOLLMAN, 1963. Conjugation in bacteria. In *Methodology in Basic Genetics*, W. J. Burdette (ed.). Holden-Day, San Francisco, pp. 129–156.

HEDGPETH, J., H. M. GOODMAN, and H. W. BOYER, 1972. DNA nucleotide sequence restricted by the RI endonuclease. *Proc. Nat. Acad. Sci.*, **69,** 3448–3452.

HOTCHKISS, R. D., and M. GABOR, 1970. Bacterial transformation, with special reference to recombination process. *Ann. Rev. Genet.*, **4,** 193–224.

HOTCHKISS, R. D., and J. MARMUR, 1954. Double marker transformations as evidence of linked factors in desoxyribonucleate transforming agents. *Proc. Nat. Acad. Sci.*, **40,** 55–60.

JACOB, F., 1955. Transduction of lysogeny in *Escherichia coli. Virology*, **1,** 207–220. (Reprinted in the collection of Stent; see References, Chapter 1.)

JACOB, F., and E. A. ADELBERG, 1959. Transfer of genetic characters by incorporation in the sex factor of *Escherichia coli. C. R. Acad. Sci. Paris*, **249,** 189–191. (Reprinted in the collection of Adelberg; see References, Chapter 1.)

JACOB, F., S. BRENNER, and F. CUZIN, 1963. On the regulation of DNA replication in bacteria. *Cold Sp. Harb. Symp.*, **28,** 329–348. (Reprinted in the collections of Abou-Sabé and of Adelberg; see References, Chapter 1.)

JACOB, F., D. PERRIN, C. SANCHEZ, and J. MONOD, 1960. L'opéron: groupe de gènes à expression coordonée pare un opérateur. *C. R. Acad. Sci. Paris*, **250,** 1727–1729. (Translated into English and reprinted in Adelberg's collection; see References, Chapter 1.)

Jacob, F., and E. L. Wollman, 1958. Les épisomes, élémentes génétiques ajoutés. *C. R. Acad. Sci. Paris,* **247,** 154–156.

———, 1959. The relationship between the prophage and the bacterial chromosome in lysogenic bacteria. In *Recent Progress in Microbiology*. Almquist and Wiksells, Uppsala, Sweden, pp. 15–30.

———, 1961. *Sexuality and the Genetics of Bacteria.* Academic Press, New York.

Khoury, G., M. A. Martin, T. N. H. Lee, K. J. Danna, and D. Nathans, 1973. A map of simian virus 40 transcription sites expressed in productively infected cells. *Jour. Mol. Biol.,* **78,** 377–389.

Lacks, S., 1962. Molecular fate of DNA in genetic transformation of *Pneumococcus. Jour. Mol. Biol.,* **5,** 119–131.

Lederberg, J., 1947. Gene recombination and linked segregations in *Escherichia coli. Genetics,* **32,** 505–525. (Reprinted in the collections of Adelberg and of Taylor; see References, Chapter 1.)

Lennox, E. S., 1955. Transduction of linked genetic characters of the host by bacteriophage P1. *Virology,* **1,** 190–206.

Low, K. B., 1972. *Escherichia coli* K-12 F-prime factors, old and new. *Bact. Rev.,* **36,** 587–607.

Meynell, G. G., 1972. *Bacterial Plasmids.* Macmillan, London.

Morrow, J. F., S. N. Cohen, A. C. Y. Chang, H. W. Boyer, H. M. Goodman, and R. B. Helling, 1974. Replication and transcription of eucaryotic DNA in *Escherichia coli. Proc. Nat. Acad. Sci.,* **71,** 1743–1747.

Morse, M. L., E. M. Lederberg, and J. Lederberg, 1956. Transduction in *Escherichia coli* K12. *Genetics,* **41,** 142–156. (Reprinted in the collection of Adelberg; see References, Chapter 1.)

Murray, K., and N. E. Murray, 1973. Terminal nucleotide sequences of DNA from temperate coliphages. *Nature,* **243,** 134–139.

Nester, E. W., M. Schafer, and J. Lederberg, 1963. Gene linkage in DNA transfer: A cluster of genes concerned with aromatic biosynthesis in *Bacillus subtilis. Genetics,* **48,** 529–551.

Notani, N. K., and J. K. Setlow, 1974. Mechanism of bacterial transformation and transfection. *Progr. Nuc. Acid Reg. and Mol. Biol.,* **14,** 39–100.

Ozeki, H., 1956. Abortive transduction in purine-requiring mutants of *Salmonella typhimurium. Cold Sp. Harb. Symp.,* **612,** 97–106. (Reprinted in the collection of Adelberg; see References, Chapter 1.)

Pilarski, L. M., and J. B. Egan, 1973. Role of DNA topology in transcription of coliphage λ *in vivo.* II. DNA topology protects the template from exonuclease attack. *Jour. Mol. Biol.,* **76,** 257–266.

Rothman, J. L., 1965. Transduction studies on the relation between prophage and host chromosome. *Jour. Mol. Biol.,* **12,** 892–912.

Sharp, P. A., M. Hsu, and N. Davidson, 1972. Note on the structure of prophage λ. *Jour. Mol. Biol.,* **71,** 499–501.

Shoemaker, N. B., and W. R. Guild, 1972. Kinetics of integration of transferring DNA in *Pneumococcus. Proc. Nat. Acad. Sci.,* **69,** 3331–3335.

Siddiqui, O., and M. S. Fox, 1973. Integration of donor DNA in bacterial conjugation. *Jour. Mol. Biol.,* **77,** 101–123.

Stent, G. S., 1971. *Molecular Genetics, An Introductory Narrative.* W. H. Freeman, San Francisco.

Stuy, J. H., 1965. Fate of transforming DNA in the *Haemophilus influenzae* transformation system. *Jour. Mol. Biol.,* **13,** 554–570.

Taylor, A. L., and M. S. Thoman, 1964. The genetic map of *Escherichia coli* K-12. *Genetics,* **50,** 659–677. (Reprinted in Adelberg's collection; see References, Chapter 1.)

Taylor, A. L., and C. D. Trotter, 1972. Linkage map of *Escherichia coli* strain K-12. *Bact. Rev.,* **36,** 504–524.

Taylor, H. E., 1949. Transformations réciproques des formes R et ER chez le pneumocoque. *C. R. Acad. Sci., Paris,* **228,** 1258–1259.

Tomasz, A., E. Zanati, and R. Ziegler, 1971. DNA uptake during genetic transformation and the growing zone of the cell envelope. *Proc. Nat. Acad. Sci.,* **68,** 1848–1852.

Watanabe, T., 1971. Infectious drug resistance in bacteria. *Curr. Topics in Microbiol. and Immunol.,* **56,** 43–98.

Wensink, P. C., D. J. Finnegan, J. E. Donelson, and D. S. Hogness, 1974. A system for mapping DNA sequences in the chromosomes of *Drosophila melanogaster. Cell,* **3,** 315–325.

Willetts, N., and F. Bastarrachea, 1972. Genetic and physicochemical characterization of *Escherichia coli* strains carrying fused F′ elements derived from KLF1 and F57. *Proc. Nat. Acad. Sci.,* **69,** 1481–1485.

Williams, G. L., and D. M. Green, 1972. Early extracellular events in infection of competent *Bacillus subtilis* by DNA of bacteriophage SP82G. *Proc. Nat. Acad. Sci.,* **69,** 1545–1549.

Wollman, E. L., F. Jacob, and W. Hayes, 1956. Conjugation and genetic recombination in *Escherichia coli* K-12. *Cold Sp. Harb. Symp.*, **21**, 141–162. (Reprinted in the collection of Adelberg and in *Selected Papers in Biochemistry,* Vol. 1; see References, Chapter 1.)

ZINDER, N. D., and J. LEDERBERG, 1952. Genetic exchange in *Salmonella, Jour. Bact.*, **64,** 679–699. (Reprinted in the collection of Peters; see References, Chapter 1.)

20
RECOMBINATION IN VIRUSES

In 1946, contemporary with the discovery of genetic exchange in bacteria by Lederberg and Tatum, genetic recombination was found to occur between different strains of bacteriophages by Delbruck and Bailey and by Hershey. Their investigations have since given rise to some of the most detailed studies of genetic exchange yet undertaken in a single organism. However, before dealing with these advances, it is important to consider three special topics:

1. Phenotypic differences which enable the diagnosis of genetic differences among the relatively invisible virus particles.
2. An explanation that accounts for those phenotypic exchanges between different viruses that are not caused by genetic recombination.
3. The general mode of replication of the viral chromosome.

PHAGE PHENOTYPES

In bacteriophage, phenotypic differences are most easily recognized by different effects on the bacterial host cells rather than by differences in the appearance of the virus itself. As explained previously (Chapters 3 and 4), bacteriophage infection proceeds through the attachment of part of the phage protein envelope (the tail) to the bacterial cell wall and consequent insertion of phage DNA. Once inside the host cell, this DNA enters a vegetative growth phase in which phage proteins are made as well as new phage DNA strands. As this vegetative DNA "pool" grows, strands are continually being removed and condensed in order to enter the newly formed phage protein envelopes. In this process, enzymes, called lysozymes, are produced which are capable of destroying the bacterial cell wall. After a certain interval, a sufficient amount of lysozymes has accumulated, and lysis of the bacterial host takes place with the release of many mature and infective viral particles.

Ordinarily, in the absence of phage infection, a culture of bacteria appears turbid (translucent but not transparent) because of the light scattering caused by the many individual cells. When the number of virulent bacteriophages added are sufficient to infect each bacterium with at least one phage, the entire culture later becomes clear as a result of lysis of all cells. When relatively few phages are "seeded" on a bacterial culture that is spread on a solid layer of food medium (nutrient agar), each single phage functions as a separate center of infection and produces a "plaque" or circle of lysed bacteria. The size and appearance of the plaque depend both upon the type of phage used and the type of bacterial host. Usually the plaque continues to grow as long as physiological conditions are appropriate. When T2 bacteriophages are seeded on *E. coli*, for example, small plaques are formed whose outer perimeters appear as diffuse turbid halos. This halo represents areas in which bacteria have become "superinfected" by successive later generations of T2 particles that have been derived from lysis of bacteria in the center of infection. Because of this superinfection, there is interference with the progress of the initial phage infection of the bacterium, thereby causing inhibition and delay of the usual lytic pattern. Thus, in this case, although many bacteria in the plaque have been lysed, a sufficient number of superinfected but unlysed bacteria are present to produce some turbidity in the plaque.

However, not all plaques are of this particular wild-type pattern. Occasionally plaques can be formed which are larger than wild type and which do not show the turbid halo but instead have sharp edges. Such plaques are caused by phages that can superinfect bacteria without inhibition or delay of lysis. Hershey therefore named these phages *rapid lysis*, or *r* mutants, in contrast to the r^+ wild type.

A second phenotypic character often used in phage recombination studies is *host range*. Wild-type T2 phage, for example, will infect and lyse *E. coli* strain B, but will not infect *E. coli* strain B/2 ("B-bar-2"). The reason for this difference lies in a change in the cell surface of *E. coli* B/2, preventing T2 phage from adsorbing onto it. Thus, when wild-type T2 phage is plated on a mixture of B and B/2, the latter cells are not lysed, and the entire plaque appears turbid because of their presence. Occasional phages arise, however, whose host range has been extended and which can adsorb to and lyse B/2. Such mutant phages, called *h*, will produce clear plaques with turbid haloes when plated on a mixture of B and B/2, since both types of bacterial cells have been lysed in the center of the plaque, and inhibition of lysis (because of superinfection) only takes place at the perimeter.

PHENOTYPIC MIXING

For each of the many kinds of bacteriophages (e.g., T1, T2, T3, T4, etc.), resistant bacteria are usually found that are capable of preventing adsorption of the phage to the bacterial cell wall (e.g., B/1, B/2, B/3, B/4, etc.). When some of these phages are grown together, a "phenotypic mixing" occurs in which the adsorption characteristics of one phage (e.g., T2) becomes attached to the genotype of another (e.g., T4). However, this mixing phenomenon, first noted by Delbruck and Bailey and later studied by Novick and Szilard and by Streisinger, is not synonymous with genetic recombination. It arises because of the replicating mechanism of bacteriophages and other viruses in which the protein envelope and cell-wall-attachment organs are made separately from the viral chromosome. Only after they are separately synthesized are the viral chromosomes enclosed within the protein envelopes.

Therefore, as expected, the protein envelope and the viral chromosome are of the same type when only one type of virus is replicating in a cell. However, if two kinds of viruses infect a cell, e.g., T2 and T4, two types of protein envelopes and two types of viral chromosomes are formed. Thus some

T2 viral chromosomes may find their way into T4 envelopes, and vice versa. Such "mixed" particles may reproduce well for a single generation on bacterial cultures that are resistant to the phage chromosome genotype, but will no longer grow on the same culture in the next generation. Conversely such particles are unable to reproduce on bacterial cultures resistant to the viral protein, but after one generation of growth on nonresistant bacteria they can infect cells that were formerly resistant.

For example, Novick and Szilard isolated particles of this type containing the T2 chromosome covered with the T4 protein. Because of their T4 phenotype they could not be grown on B/4 bacteria alone. However, by infecting the B bacteria in a B and B/4 mixture, these particles produced a new generation of T2 phages which infected the B/4 bacteria, thereby forming clear plaques. Since only T2 phages are produced, phenotypic mixing thus lasts for one generation only (Fig. 20–1), and is essentially similar to the mixed "reconstituted" tobacco mosaic virus discussed in Chapter 4. Therefore, to ensure correct determination of the genotype, traits subject to phenotypic mixing should be diagnosed after bacteriophages have *singly* infected bacteria for at least one generation.

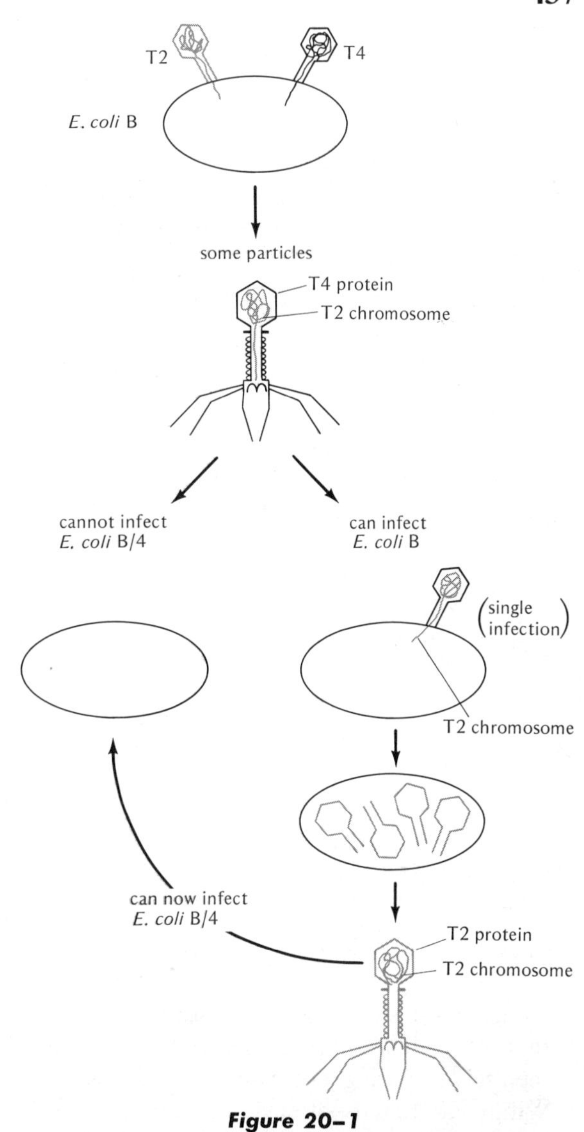

Figure 20–1
Explanation for "phenotypic mixing" observed among some of the phage progeny of mixed infections with T2 and T4.

VIRAL CHROMOSOME REPLICATION

In contrast to the sexual process in other organisms, viruses must exchange genetic material in a host cytoplasm. Since many copies of the viral chromosome are present in any one cell, genetic exchange may occur between members of a numerous population of chromosomes rather than being restricted to the usual two- or four-chromosome strands. Moreover, not all viral chromosomes formed in the host cell become embodied into viral progeny, only samples taken from a large chromosomal "DNA pool." The usual zygote techniques were therefore inadequate to analyze recombination in viruses, and different methods had to be devised. The first important question, however, concerned the origin of a viral population of chromosomes—by what general mechanism were they formed? This question was first answered by Luria in 1951 in an experiment distinguishing between the following three modes of chromosome replication (Fig. 20–2):

1. *Geometric* or *clonal* replication is that in which an infecting viral chromosome repli-

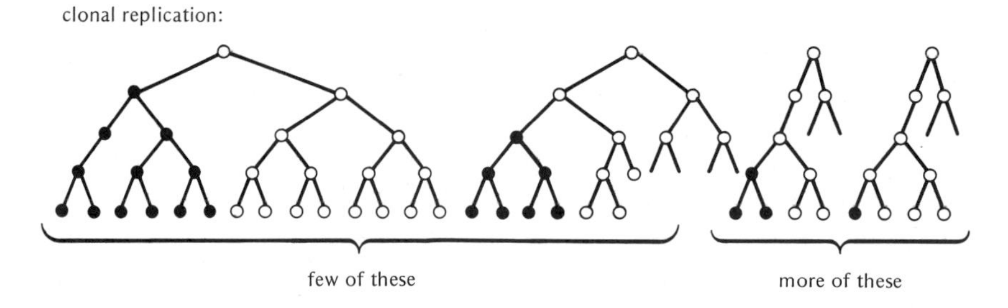

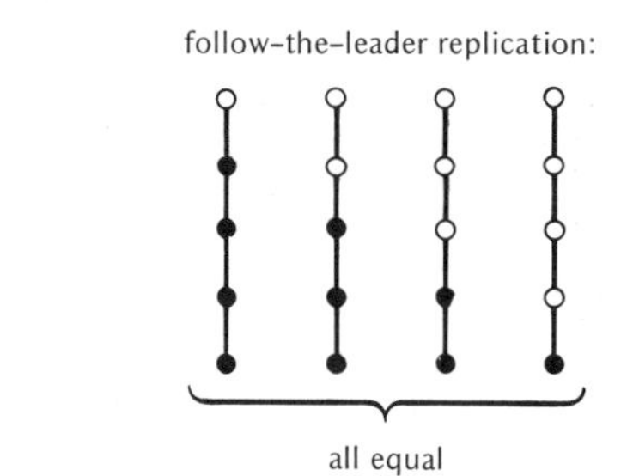

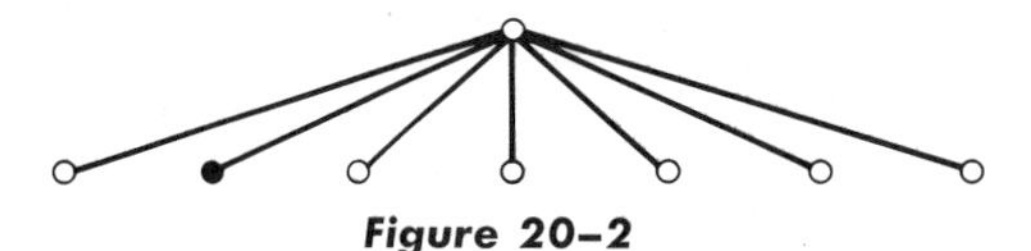

Figure 20–2

Three possible modes of replication of phage particles in cells. Black circles represent bacteriophages carrying a mutation. (From Stent, after Luria.)

cates to produce two progeny, each of which replicates to produce two more, etc. In this model, viral chromosome replication is like that of cellular chromosomes that double themselves during successive cell divisions. Within 10 rounds of such replication, over 1000 viral chromosomes would have been produced, sufficient to account for the size of practically any viral DNA pool.

2. *Follow-the-leader* replication is that in which one phage replicates to produce only one other, which then replicates to produce one other, etc. For 1000 viral chromosomes to be formed by this conservative method, 1000 consecutive replications are necessary.

3. *Printing-press* replication is that in which the initial infecting virus produces a template which then "prints" or stamps out all the necessary copies. Here, too, 1000 separate replications are necessary to produce 1000 chromosomes, but they are all formed from a single chromosome.

The method by which Luria distinguished among these three mechanisms was to consider the effects of a rare random mutation, such as $r^+ \rightarrow r$ (*rapid lysis*), for each possibility. If we regard each bacterium as containing a population of phages derived from one single parent phage, we may further consider that the mutation frequency of $r^+ \rightarrow r$ is low enough, so that only a single phage mutation at most occurs during replication in each bacterium.

For geometric or clonal replication, a mutation occurring early will give rise to a large clone of mutant phages, while a later mutation will give rise to a smaller clone. Since mutations can occur at any time, early or late, the frequency of different-sized clones should be about equal. However, one additional point must be kept in mind, and that is that the chances for a mutation to occur are much higher among many replicating phages than among a few. Thus phages that have replicated in a bacterium a sufficient number of times to produce many progeny will have an increased chance of producing at least one mutant phage in further replications. On this basis a large number of bacteria carrying phage mutations will be carrying only one or a few phage mutants, while among the others there will be different-sized clones of phage mutants but each size will be in almost equal frequency. When such a relationship is plotted graphically, it assumes the proportions of an "exponential curve," the frequency of bacteria carrying phage mutations rising rapidly at low numbers of phage mutants. For convenience, such curves are usually plotted on logarithmic scales (multiple units rather than arithmetic units, see Chapter 14), in order to appear as a straight line. In Fig. 20–3, line *a* shows this expected exponential relationship.

For follow-the-leader replication, only one phage is replicating at a time. Since the chances of mutation are the same at any one of these replications, the frequencies of bacteria carrying many phage mutations should equal the frequencies of bacteria carrying only few. Line *b* in Fig. 20–3 shows this expected relationship as an almost horizontal line.

For printing-press replication only one phage is replicating, but it is always the same phage. Since the mutation frequency of $r^+ \rightarrow r$ is low, it is very unlikely that this replicating phage will produce more than one mutant phage, as shown on line *c* of Fig. 20–3. Thus practically none of the bacteria carrying mutant phages would be expected to carry more than one such mutant.

Luria's experiment was, first, to infect each of 23,000 bacteria with individual r^+ phages. He then distributed the infected cells among about 3,000 separate tubes and permitted the phages to replicate until their progeny burst each bacterium. Then he plated the contents of each tube on appropriate bacteria and observed the plaques formed by each phage to determine whether it was r^+ or r. Because of his experimental method, the presence of phage mutants in practically any particular bacterium could be detected. In all, about 90 of the infected bacteria were found to give rise to 700 mutant phages, and the results are plotted as individual points in Fig. 20–3. Note that they follow closely the values predicted according to clonal replication.

However, the mode of replication of phages with single-stranded genomes, such as ϕX174, may be different from phages with double-stranded nucleic acids, such as T2. As discussed previously (p.

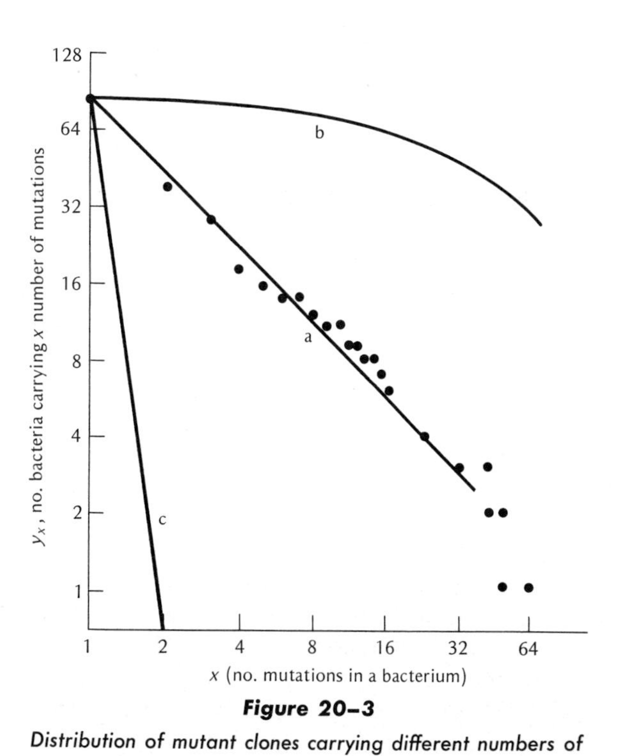

Figure 20–3

Distribution of mutant clones carrying different numbers of mutations, each line representing a different expected distribution. (a) Expected distributions for clonal replication; the most common bacteria are those carrying few mutations, and there is an exponential fall in frequency of bacteria carrying many mutations. (b) Expected distributions for follow-the-leader replication; there are almost equal frequencies of bacteria carrying few or many mutations. (c) Expected distributions for printing-press replication; almost no bacteria are expected to show more than one mutation. Black dots represent the observed distributions of mutant clones. (From Stent, after Luria.)

90), during a portion of its life cycle ϕX174 replicates so that new copies of the initial single strand, called "plus," are produced on a complementary single-stranded template called "minus." This probably arises through the "rolling circle" mechanism in which single strands are continuously generated complementary to a circular single-stranded "printing press" template (see p. 88). Single-stranded RNA phages such as Qβ, MS2 and others, also show evidence for the synthesis of minus strands which are then used as templates for the printing press production of new plus strands. For reasons that are presently not clear, RNA phage replication is much more efficient than the replication of DNA phages, and a host cell may have its entire weight converted into RNA phage progeny.

RECOMBINATION AND MAPPING

One of the first examples demonstrating recombination in phage made use of the *host range* and *rapid lysis* mutations previously described. In 1949 Hershey and Rotman, using these mutations, reported the results of infecting *E. coli* bacteria with a series of different pairs of T2 phages. In one set of experiments one member of the infecting pair always carried the same *host range* mutation (could infect *E. coli* B/2) but was wild type for the *rapid lysis* gene and therefore symbolized as $h\,r^+$. The other member of this pair was wild type for *host range*, h^+, but had the mutant phenotype *r*. (Since the *r* mutation arose not only once but many times, a variety of different $h^+\,r$ stocks was available for this purpose, i.e., $h^+\,r1$, $h^+\,r2$, etc.) Recombination in this set of experiments would therefore be expected to produce $h^+\,r^+$ and $h\,r$ genotypes in addition to the $h\,r^+$ and $h^+\,r$ parental classes.

In another set of experiments they tested recombination between the same genes but used the cross $h^+\,r^+ \times h\,r$. Recombinant classes would thus be $h^+\,r$ and $h\,r^+$. In both sets of experiments the number of phages introduced was sufficient to infect each bacterium with about five of each type. After 1 hour, most or all the bacteria were lysed

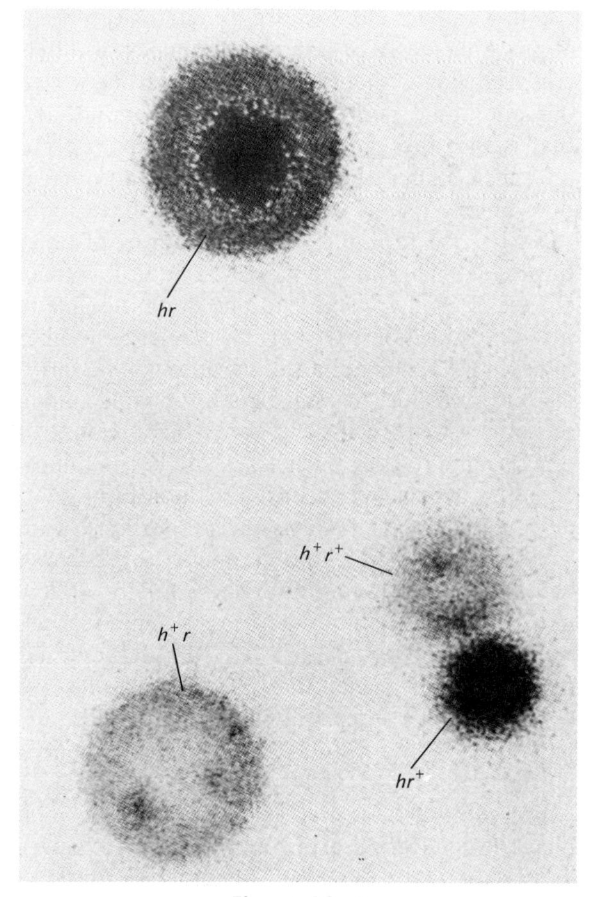

Figure 20–4

Phenotypes of different phage genotypes on a mixture of E. coli *B and B/2. On this mixed indicator, the h mutants infect B/2 as well as B, producing clear plaques as compared to the turbid plaques produced by h⁺. The r mutants enable the size of the plaque to be larger than r⁺. (From Hershey and Chase.)*

and samples of the phage progeny from about 40,000 bacteria of each cross were plated on special indicator plates consisting of a mixture of *E. coli* B and B/2. Since each of the four genotypes produces a distinct phenotype on this mixed indicator (Fig. 20–4), their relative frequencies are easily scored.

The Hershey and Rotman results are shown in Table 20–1 for crosses utilizing *h* and the three *r* mutants *r1*, *r7*, and *r13*. Note that in all cases the two recombinant classes of each cross appear in

TABLE 20–1

The data of Hershey and Rotman (1949) for six crosses between the *h* gene and three different *r* genes of T2 bacteriophage infecting *E. coli*

Cross (Approximately 50 Percent of Each)	Progeny, Percent			
	$h^+ r^+$	$h\,r^+$	$h^+ r$	$h\,r$
$h\,r1^+ \times h^+ r1$	12	42	34	12
$h\,r1 \times h^+ r1^+$	44	14	13	29
$h\,r7^+ \times h^+ r7$	5.9	56	32	6.4
$h\,r7 \times h^+ r7^+$	42	7.8	7.1	43
$h\,r13^+ \times h^+ r13$	0.74	59	39	.94
$h\,r13 \times h^+ r13^+$	50	.83	.76	48

approximately equal frequency. This can be interpreted to mean that genetic exchange, as commonly noted in higher organisms, is a *reciprocal* event involving a physical exchange between two chromosomes, both of which are then recovered. Second, note that the recombination frequency varies from about 1 to 2 percent (*h-r13*) to about 25 to 30 percent (*h-r1*). In classical terms, this indicates some form of linkage relationship between these genes. (*Rapid lysis* mutations located in different regions of the viral chromosome have been designated by roman numerals, e.g., *rI*, *rII*, etc.)

Of course, the above crosses are not by themselves sufficient to produce a linkage map. Hershey and Rotman therefore made additional linkage determinations between the *r* mutants themselves and between these and a gene called *minute* (*m*), which produced very small plaques. They discovered two interesting facts: First, that although there was a variety of linkage distances, no distance was ever greater than 30 percent. Second, there were numerous genes linked together by small recombination frequencies which, as groups, always showed the same 30 percent linkage distance with certain other genes (i.e., *m*). On this basis Hershey hypothesized that there were three linkage "groups" in T2, and that "independent assortment" between these linkage groups was characterized by 30 percent recombination rather than by the 50 percent recombination expected in higher organisms (Fig. 20–5).

The smaller recombination frequency produced by independent assortment in phages arises because different kinds of phage genomes introduced into a cell can "mate" homozygously as well as heterozygously. For example, a bacterial cell infected with $a\,b$ phage as well as $a^+\,b^+$ phage will form many chromosomes of each kind. Matings between chromosomes may then be of the $a\,b/a\,b$, $a\,b/a^+\,b^+$, and $a^+\,b^+/a^+\,b^+$ varieties. If such matings are random, the heterozygous variety, $a\,b/a^+\,b^+$, will have approximately a 50 percent chance of occurring. However, if a and b are widely separated or are not linked at all, these two loci will assort independently and the recombinational products ($a^+\,b$ and $a\,b^+$) of such heterozygous matings will be as frequent as the nonrecombinational products ($a\,b$ and $a^+\,b^+$). Thus only 1/2 of 1/2, or 1/4, of the mating products will be recombinants after the first round of mating.

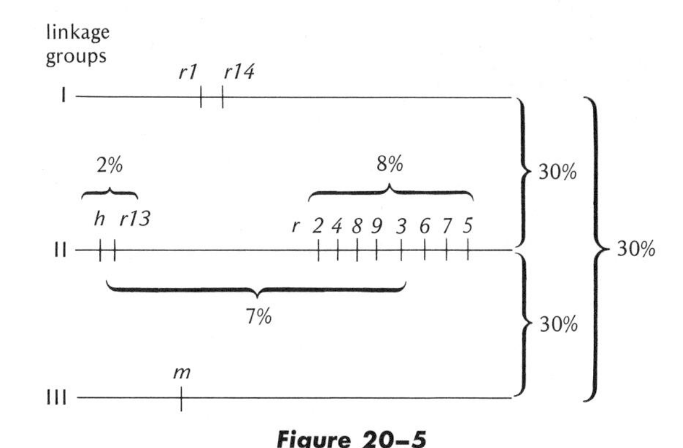

Figure 20–5

Linkage relations found by Hershey and Rotman in bacteriophage T2. (After Hershey and Rotman.)

As more rounds of mating take place within a single host cell, mating between the different parent phage chromosomes is again possible, and more recombinants may be formed. On the other hand, recombinants that are already formed may recombine with similar products in the next or future rounds of mating (i.e., $a^+ b/a b^+$) to furnish the original parental combinations once again. If the rounds of mating could proceed indefinitely, a 50 percent equilibrium of recombinants would eventually be reached. However, mating possibilities are undoubtedly limited by the restricted number of rounds of replication, as well as by the continuous removal of chromosomes from the DNA pool during maturation and their incorporation into viral particles. Equilibrium frequency for widely separated gene markers is therefore never reached. In T2 the largest recombination values fall, as we have seen, near 30 percent; in T4 they reach approximately 40 percent; and in λ phage they are no higher than about 15 percent (probably indicating restricted opportunities for mating).

To some extent recombination values may therefore be altered by changing the number of rounds of mating. This has been done experimentally by either inducing early lysis or delaying it through superinfection. According to experiments by Levinthal and Visconti, such modifications change the *h-r13* recombination frequency from 2 percent after 20 minutes of infection to as high as 9 percent after 80 minutes of infection. A comparison between recombination frequencies in phage crosses must therefore be performed under uniform conditions of phage maturation.

The potentiality for repeated rounds of mating to occur between viral chromosomes in a host cell was further demonstrated by Hershey and Chase for simultaneous crosses between *three* different strains of phage T2. As shown in Table 20–2, crosses in which each of the three strains bears a different independently assorting mutant gene (*h*, *m*, *r*) give rise to phages that have all three mutants together (*h m r*). Similarly, when each of the three strains have different wild-type genes (h^+, m^+, r^+), complete wild-type phages are formed ($h^+ m^+ r^+$). Such events can only take place by genetic exchange between all three strains, and can occur in either of two ways: (1) through two successive recombinations in the same cell, the first between chromosomes of two strains and the second between the resultant recombinant and a chromosome of the third strain; and (2) through "group mating" between all three chromosomes at a time. The exact mechanism responsible for these results is not yet known.

The events described above, as well as other findings, demonstrate at least one important difference between recombination in viruses and recombination in other organisms. Specifically viral recombination is not confined to simple pairing between two homologous chromosomes and to only one set of recombinational events in the "zygotes." On the contrary, the same viral chromosomes may pair and exchange genetic material more than once and may possibly even mate with more than one chromosome at a time. It thus seemed clear that viral recombination was a "population" phenomenon whose events would have to be described statistically rather than through the

TABLE 20–2

Progeny of crosses among three strains of T2 differing in respect to three independently assorting genes

		Progeny, percent							
Cross		$h^+m^+r^+$	h^+m^+r	$h\ m^+r^+$	$h^+m\ r^+$	$h\ m^+r$	$h\ m\ r^+$	$h^+m\ r$	$h\ m\ r$
$h\ m^+\ rl^+ \times h^+\ m\ rl^+ \times h^+m^+rl$	(1)	25	22	17	12	9	5	7	2
	(2)	25	15	18	20	4	10	5	3
$h\ m\ rl^+ \times h^+\ m\ rl \times h\ m^+\ rl$	(1)	3	5	6	10	17	19	14	26
	(2)	2	4	9	9	14	26	15	20

From Hershey and Chase, 1951.

usual simple descriptions of meiosis and recombination in higher organisms. That is, it would be necessary to explain the relationship between the number of rounds of mating that occur in a host cell and the number of recombinants that are formed. Since many phages could infect a single bacterium, it was also necessary to understand the different recombinational results caused by equal and unequal proportions of the different kinds of infecting parental phages (*equal* and *unequal* inputs).

For these purposes a statistical theory was proposed by Visconti and Delbruck and developed further by Bresch, Steinberg and Stahl, and others. Some of their formulas have been used to estimate the number of rounds of mating in a host cell. For T2 and T4 it appears as though the rounds of mating generally keep pace with the rounds of replication. Visconti and Delbruck estimated approximately five rounds of mating for T2. Phage λ, more restricted in its ability to recombine, seems to have only one round of mating. On the whole, however, these theories do not yet enable exact predictions of many recombinational events. It seems likely that, in addition to the variable number of generations that affect phage recombination in a single cell, there are additional complicating factors, such as the variable distance between the DNA pools of the different infecting phages, the variable proportions of pair mating and group mating in cells, the variable abilities of different phages to recombine, and the variable effect of the host cell itself on recombination. Phage maps can nevertheless be constructed which give consistent linear relationships between genes, although recombination frequencies may vary in different experiments.

THE CIRCULAR MAP

The presence of three linkage groups in T2 was found to be paralleled by a similar number of linkage groups in T4 by Doermann and Hill. These conclusions were based, in both instances, on the apparent absence of linkage between each of the three individual groups in the usual crosses. In 1960, however, Streisinger and Bruce demonstrated a sensitive technique that enabled the three presumed linkage groups within each species to be combined into single linkage maps. Their experiments utilized crosses differing in three genes, e.g., $a\, b\, c \times a^+\, b^+\, c^+$. From previous experiments they knew that linkage existed between two of these genes, e.g., *a-b*, but not apparently between these two and the third, *c*.

If this were true, and there was no linkage with *c*, crossing over between the *a* and *b* loci should not affect the presence of *c*. As shown in Fig. 20–6a, selection for the $a\, b^+$ recombinant class should yield approximately equal frequencies of c and c^+. On the other hand, if linkage exists with *c*, there are two possible ways the three genes can be associated: either *a b c* or *c a b*. As shown in Fig. 20–6b and c, these two possibilities result in different consequences. If *c* is to the right of *b*, selection of $a\, b^+$ recombinants should yield more c^+ genotypes than of the double recombinants

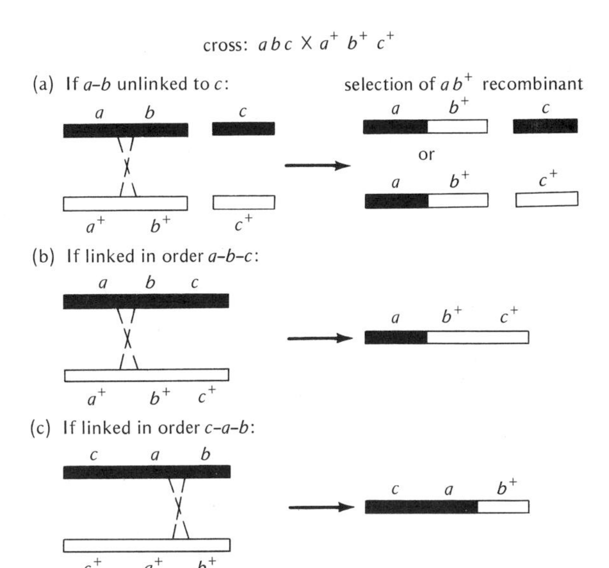

Figure 20–6

Consequences of linkage and the absence of linkage between a gene c and two known linked genes (a and b) when the cross is a b c × a⁺ b⁺ c⁺ and the a b⁺ recombinants are selected. The frequency with which c and c⁺ associates with these recombinants will determine its relative position.

$a\, b^+ c$. If c is to the left of a, selection of $a\, b^+$ recombinants should yield more c genotypes than double-recombinant genotypes $a\, b^+ c^+$.

By utilizing different proportions of each parent and forcing premature lysis of the bacterial cells (to limit the frequency of recombination), they demonstrated in each case that linkage existed between the previously separated linkage groups. Moreover, the maps of T2 and T4 were not merely single straight lines but could be shown by this same method to be connected end to end in the form of a circle.

This view of a circular map has been reinforced by studies of Edgar, Epstein, Baylor, and others, who have discovered numerous new mutations that map in areas formerly thought to be "empty." In both T2 and T4, these new genes are spread over a definite continuum which connects the previously separated linkage groups (Fig. 20–7).

Mapping of gene loci within this circular map can be done in various ways but is usually most reliable in crosses between stocks differing in at least three gene loci. According to Streisinger, gene order between the hypothetical genes a, b, and c is most easily obtained from crosses between the three possible double mutants and the remaining single mutant, i.e., $a\, b\, c^+ \times a^+ b^+ c$, $a\, b^+ c \times a^+ b\, c^+$, and $a^+ b\, c \times a\, b^+ c^+$. The detection of gene order

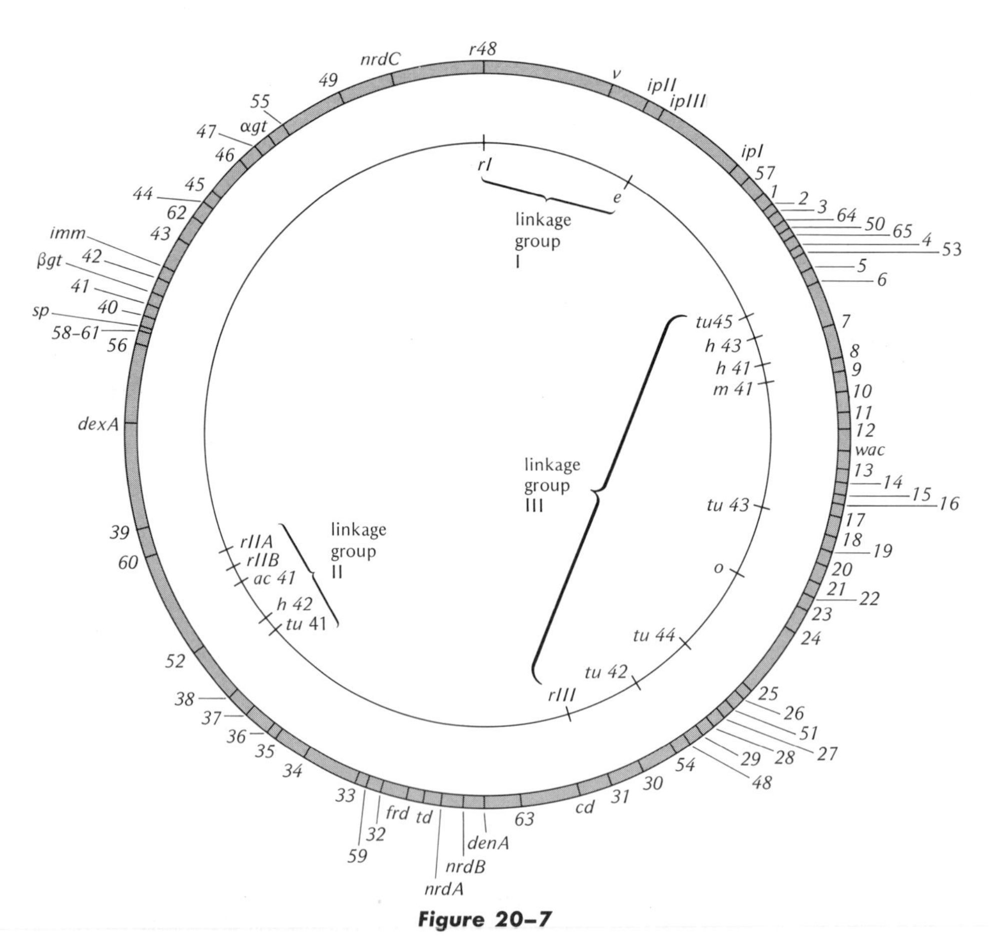

Figure 20–7

Composite linkage map of bacteriophage T4 including the former linkage groups of Hershey and Rotman (inside circle) and the later conditional mutations discovered by Edgar, Epstein, and others (outer circle). (Outer circle after Wood.)

then relies upon the fact that the frequency of double recombinants in most crosses is usually much less than the frequency of single recombinants. If c is located between a and b, the first cross, $a\,b \times c$, will yield the fewest wild-type recombinants, since two crossovers, one at each side of c, are necessary for wild type to appear. On the other hand, the other crosses can furnish wild-type recombinants with only single crossovers for this same sequence. Similarly, if b is in the center, fewest wild-type recombinants appear in the cross $a\,c \times b$, and if a is in the center, this reduction in wild-type products occurs in the cross $b\,c \times a$. Thus the true gene order can be distinguished by simply ascertaining which cross produces the fewest wild-type recombinants.

HETEROZYGOTES

In 1951, while performing their experiments demonstrating triparental genetic exchange (p. 442), Hershey and Chase noted that a certain proportion of plaques, formed by infecting bacteria with both r and r^+ phages, had a mixed or "mottled" appearance. This mottling was apparently caused by "clumps" of r and r^+ phages sticking together and forming single centers of infection. The r^+ phage produced sectors of the plaque which were turbid, while the r mutants produced sectors which were clear. Most unusual, however, was the finding that a certain proportion of the plaques formed by the *progeny* of r and r^+ infections were also mottled. In this case there were various indications suggesting that mottling was not caused by clumps of r and r^+ phages, but by single phage particles called *heterozygotes*, which seemed to carry both the r and r^+ genes.

One such indication of heterozygosity arose from infecting bacteria with $h\,r$ and $h^+\,r^+$ phages. Hershey and Chase found that only about 3 percent of the plaques that showed mottling for the r and r^+ characteristics in such experiments were also mottled for the h and h^+ characteristic. If clumps of the two different phages were involved in the mottling, then all or most r/r^+ plaques should also be h/h^+. The fact that these two kinds of mottling were independent of each other, each arising at a frequency of about 2 or 3 percent, helped demonstrate that heterozygosity at one locus was not necessarily accompanied by heterozygosity at another locus. Only if two gene loci were close together could heterozygotes for both genes also be found (double heterozygotes).

Of additional interest was the finding that the proportion of heterozygotes formed seemed to be constant. For r and r^+ infections the progeny were always approximately 2 percent heterozygous. Also, this proportion remained the same no matter when bacterial lysis occurred, early or late. The fact that heterozygotes do not increase in frequency during the phage growth cycle indicates they probably do not reproduce themselves as such (otherwise their frequency would continually increase). Much more likely is the probability that they are formed at a certain rate in the phage DNA pool of the bacterial host and are then withdrawn into phage particles. How then do heterozygotes originate in the DNA pool?

One clue to their source was observed by Hershey and Chase in crosses of the type $h\,r \times h^+\,r^+$. When the h locus was sufficiently far from the r locus that recombinations could occur freely, a very large proportion of phages in the mottled plaques formed by heterozygotes were recombinant for the two loci. A cross of $h\,rI \times h^+\,rI^+$, for example, gave heterozygous mottled plaques of which *94 percent* showed the presence of recombinants, $h\,r^+$ and $h^+\,r$, while only *6 percent* showed no recombinant phages! It was therefore thought that the presence of a heterozygote was closely followed by the segregation of recombinants; i.e., the heterozygote might well be an intermediate stage in phage recombination.

A further elaboration of heterozygote structure by Levinthal arose from experiments designed to answer the question whether a heterozygote at a particular locus represents merely the addition of a new physical fragment at that locus or whether it is accompanied by a recombinational structure involving other genes alongside it. Instead of restricting himself to the two loci scored by Hershey and Chase, Levinthal added one more gene difference and then scored the genotypes found in

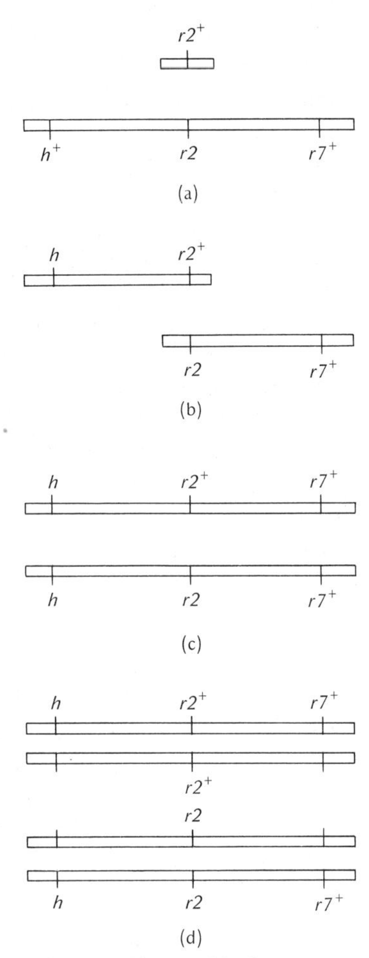

Figure 20–8

Parts (a) and (b) show two possible structures of a heterozygote for the r2 gene. In (a) the heterozygote is nonrecombinant for any associated loci. In (b) the heterozygote is recombinant for loci on one side. Part (c) shows the "heteroduplex" structure believed to be responsible for the observed heterozygote. In (d) the heteroduplex has replicated, producing two normal molecules.

plaques that were heterozygous (mottled) for the middle locus. In a cross $h\,r2^+\,r7 \times h^+\,r2\,r7^+$, for example, heterozygotes for *r2* might be of the two following kinds:

1. An *r2* fragment is added to another phage chromosome, i.e., $r2^+/h^+\,r2\,r7^+$, yielding heterozygosis for *r2* but no change in the h^+ -$r7^+$ combination, since these outside genes are provided by only one parent (Fig. 20–8a).
2. The two parental chromosomes overlap at the *r2* region in the form $h\,r2^+/r2\,r7^+$, so that the heterozygotes are also recombinant for the *h* and *r7* loci, each of which is provided by a different parent (Fig. 20–8b).

Levinthal found that the heterozygotes were generally recombinant for the outer loci, indicating that the second alternative was correct and that both parents contributed other loci to the heterozygote in addition to the one that was heterozygous. He proposed that the heterozygous structure was two-stranded, both strands bearing the *h* and $r7^+$ loci, for example, but one of the strands being *r2* and the other $r2^+$ (Fig. 20–8c). This structure, called a *heteroduplex*, can be considered to exist on the level of the DNA molecule itself, so that the *r2* and $r2^+$ sections do not have the usual complementary base pairing. Replication, however, proceeds normally, each strand producing a two-stranded nonheterozygous structure (Fig. 20–8d). Since then, considerable evidence has accumulated indicating that the formation of a significant proportion of T4 heterozygotes is tied to the circularity of the T4 genetic map. What do T4 circular maps and T4 heterozygotes have in common?

MAKING ENDS MEET

If we consider each phenomenon individually, the existence of heterozygosity and map circularity in phage T4 entails two seeming paradoxes: (1) What unique mechanisms could *consistently* produce DNA double helixes bearing two different alleles of the same gene? (2) What accounts for the fact that phage T4 chromosomes are *linear* both in the bacterial host cell and in their viral envelopes, yet show a *circular* genetic map? In the 1960s Streisinger, Stahl, and others, offered an explanation for these difficulties based on two important features of the T4 chromosome: its circular permuted nature, and its terminal redundancy.

In defining these terms we need only keep in

mind that permutations of gene order on a linkage map are merely different arrangements of the same genes. "Circular" permutations are those arrangements which may be linear in appearance, yet which are permutations of gene order derived from the same circle (see also Fig. 19–4 for Hfr genes). For example, if a chromosome had 9 genes in order 1 through 9, circular permutations of this order would be 1-2-3-4-5-6-7-8-9, 2-3-4-5-6-7-8-9-1, 3-4-5-6-7-8-9-1-2, etc.

The existence of such permutations can be demonstrated by a technique devised by Thomas and

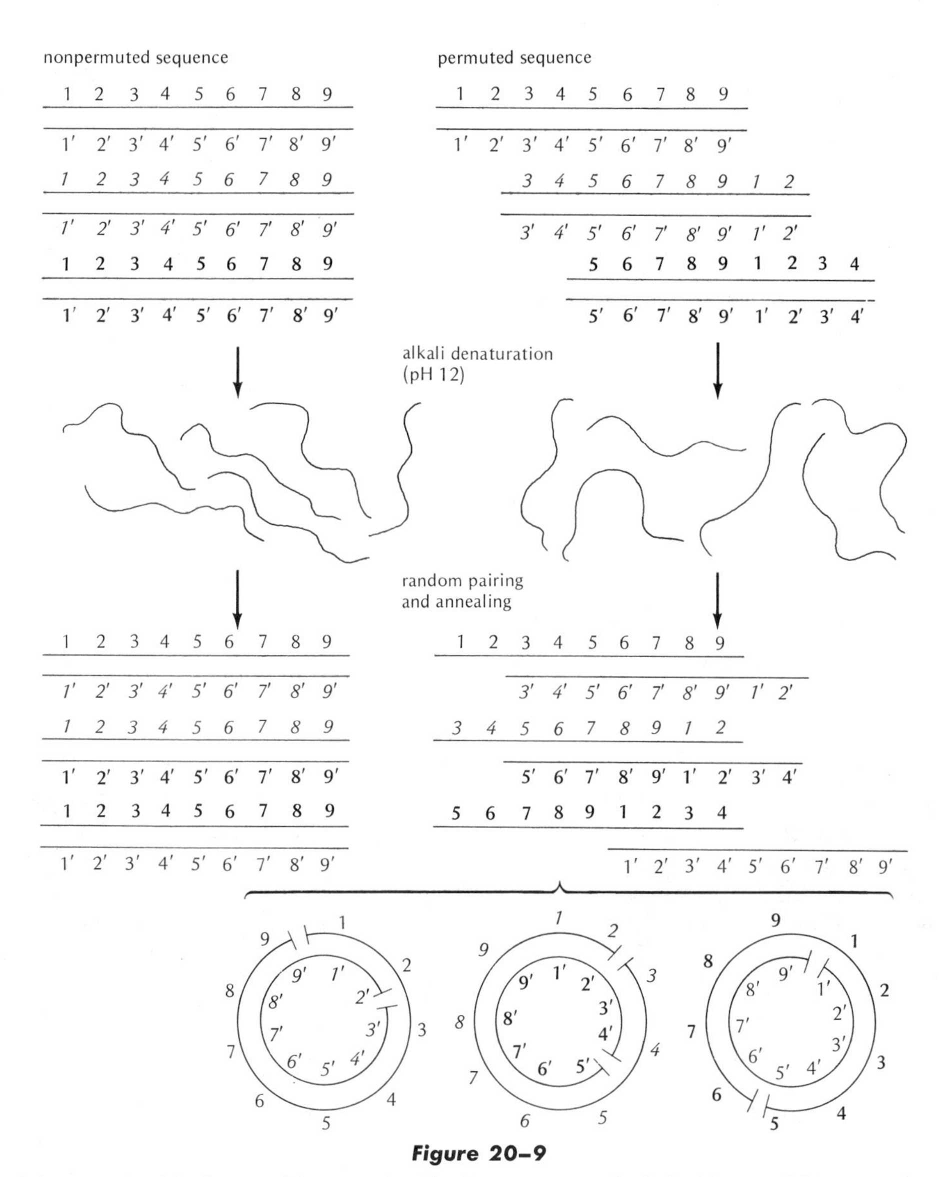

Figure 20–9

The effects of denaturation (single-strand formation) and subsequent annealing (double-strand formation) on nonpermuted and circularly permuted sequences. This procedure leads only to linear double helixes if the sequences are not permuted (left side), whereas at least some circles are formed if the sequences are circularly permuted (right side). (After Thomas.)

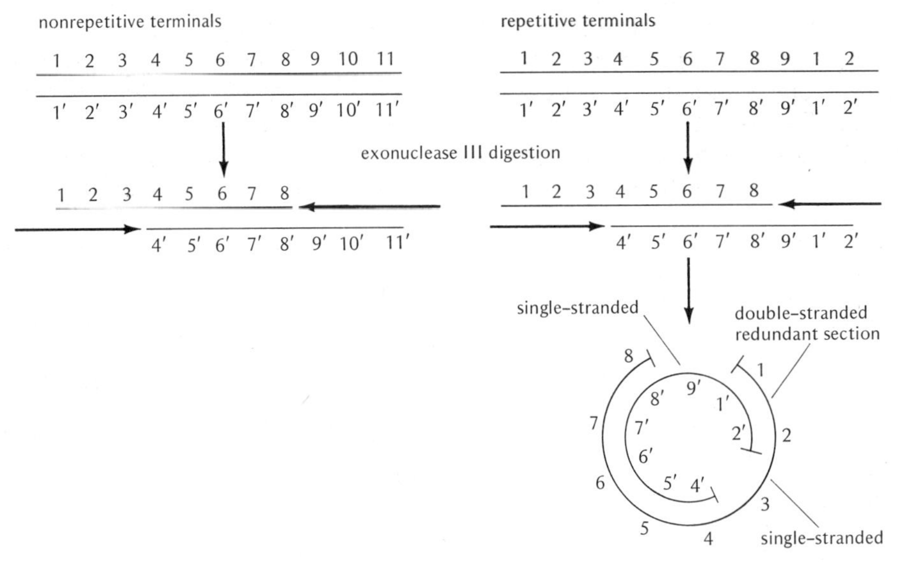

Figure 20–10

The effects of exonuclease III digestion on DNA double helixes with nonrepetitive (left side) or repetitive terminals (right side). This exonuclease "nibbles" double helical DNA from the 3′ to 5′ direction thus exposing 5′ single strands at each end. If the molecule has a terminal redundancy, these ends can now anneal to form a complementary double helix which thus transforms the molecule into a circle. Furthermore, if exonuclease digestion is permitted to proceed long enough, the complementary double-stranded section of the terminal redundancy is now bounded on each side by a single-stranded unpaired section. These differences can be observed under the electron microscope and allow estimates to be made of the length of the terminal redundancy. (After Thomas.)

others which begins by denaturing DNA double helixes into single strands. These are then randomly "annealed" into double strands and observed under the electron microscope. As shown in Fig. 20–9, random annealing of single strands from nonpermuted sequences will produce only linear double helixes. On the other hand, sequences that are circularly permuted will form circles if their strands anneal with others derived from different permutations.* Experiments of this kind have shown that a number of phages including T2 and T4 bear circular permutations, whereas phages such as T5 and T7 are nonpermuted. The paradox of a circular genetic map arising from linear chromosomes can therefore readily be explained by the permuted gene order. For example, the gene order *A-B-C-D*, will also exist in permuted chromosomes as *B-C-D-A*, *C-D-A-B*, *D-A-B-C*; that is, there will be no "end" genes, and every gene will at times be linked to genes that come both "before" and "after" it.

The second quality shown by phage T4 was the repetition of genes at the ends of its linear chromosome, called "terminal redundancy." As diagrammed in Fig. 20–10, this terminal repetition can be discerned from the fact that partial enzymatic digestion of the two 3′ ends of a double helix can lead to the formation of a circle if the ends bear repetitive sequences. Fig. 20–10 also shows that the length of this redundancy can be measured in electron microscope studies by noting that single-stranded sections are formed on each side of the redundant double-stranded section when enzyme digestion has been allowed to pro-

*As expected according to this theory, when chemically bonded crosslinks are artificially made between the two strands of the same DNA molecule (nitrous acid), annealing of strands between different molecules cannot occur, and no circles appear.

ceed slightly past the limits of the redundancy. These findings indicate that terminal redundancies are extremely common among viruses, and range in length from the 12 exposed nucleotides at each end of the λ chromosome (see p. 415) to the 2000–6000 terminally redundant nucleotide pairs observed in phages T2 and T4.

The existence of terminal redundancies thus offered a simple explanation for the consistent production of heterozygotes, since a T4 viral chromosome is essentially diploid for its terminal genes. It can therefore easily be heterozygous for these genes if chromosomes from two different parental strains have crossed over, e.g., *A-B-C-D-A* × *a-b-c-d-a* to yield *A/a* heterozygotes such as *A-B-c-d-a* and *a-b-C-D-A*. Since T4 is circularly permuted, a small proportion of its chromosomes are always diploid for any locus, and heterozygosity of this type would not be an unusual event. Note, however, that terminal heterozygosity is a phenomenon different from the "internal heterozygosity" or "heteroduplex structure" previously described. That is, heterozygotes may also arise when recombination has produced a single DNA strand bearing a section with one nucleotide sequence paired to a section carrying the sequence of a different allele. Such internal heterozygotes are not truly diploid, and obviously derive from the overlapping of single strands resulting from recombination between different parental chromosomes (see Fig. 18–8).*

The presence of circular permutations of genes in phage T4 and their terminal redundancy suggested to Streisinger a simple and common origin for both. He proposed that phage chromosomes infecting a cell produce long DNA molecules that are composed of at least a few end-to-end repetitions of the entire original gene sequence. These linear aggregations of repeated sequences, called "concatemers" (also concatenates), are then cut into prescribed chromosome lengths by a specific enzyme. For example, a concatemer composed of a repeating sequence of nine "genes," 1-2-3-4-5-6-7-8-9-1-2-3-4-5-6-7-8-9-1-2-3-4. . . , might be cut into "chromosomes" of 11 units each to produce the circularly permuted terminally redundant sequences: 1-2-3-4-5-6-7-8-9-1-2, 3-4-5-6-7-8-9-1-2-3-4, 5-6-7-8-9-1-2-3-4-5-6, ... The length of DNA cut by this enzyme is presumably governed by the "head" size of the viral particle carrying the chromosome.

Evidence for this "headful" hypothesis has been obtained from observations showing that long chromosome deletions in T4 do not affect the length of chromosome transmitted: as expected, they only increase the extent of the terminal redundancy. Moreover, electron micrographs have shown (Fig. 20–11) that each infected bacterial cell has only one large complex of phage T4 DNA with fibers of sufficient length to account for the entire phage DNA pool. The molecules in this complex often reflect the presence of concatemers; that is, they are 5 or 6 times the length of a single phage DNA.

Whence do concatemers arise? In phage T4 one likely origin is from recombination between unit

---

* Genetic demonstrations for the existence of the two types of heterozygosity, internal and terminal, have been provided by Séchaud et al. and others. They derive from an earlier observation by Nomura and Benzer that crosses between deletion mutants for the *r* gene in phage T4 (further discussed in Chapter 25) are responsible for a reduced number of heterozygotes. Since deletions of nucleotide sections do not permit pairing to occur in those areas, this finding suggested that the deletions reduce the number of internal heterozygotes (heteroduplexes) which can only arise through some type of pairing and recombination. "Point" mutations, on the other hand, affect only one or a few nucleotides and would certainly not be expected to seriously inhibit the formation of internal heterozygotes. In either case, whether a mutation is a point or a deletion, the frequency of *terminal* heterozygotes should not change, since these are diploid and do not depend upon heteroduplex pairing.

Given this distinction, experiments were performed exposing bacterial cells to an inhibitor of DNA synthesis, fluorodeoxyuridine (FUDR), before the replication and recombination cycles were completed. This had the effect of preventing heteroduplexes that were already formed from replicating further and "losing" their internal heterozygosity. Thus the frequency of internal heterozygotes would be expected to increase as a result of FUDR treatment. The frequency of terminal heterozygotes, however, would not be expected to change, since their formation is based on terminal redundancies whose occurrence is not inhibited by FUDR. Various experiments showed that the presence or absence of FUDR had no effect at all on terminal heterozygotes arising from crosses between deletion mutants, but had a marked effect on heterozygotes arising from crosses between point mutants (include both internal and terminal). One set of data for crosses between point mutants showed 63 percent heterozygotes among wild-type recombinants in the presence of FUDR, and only 16 percent in its absence. As expected, FUDR therefore significantly increased the frequency of internal heterozygotes and showed that these could be separately, although indirectly, identified.

chromosomes which then recombine with other recombination products to form long continuous strands:

1-2-3-4-5-6-7-8-9-1-2
×
1-2-3-4-5-6-7-8-9-1-2
↓
1-2-3-4-5-6-7-8-9-1-2-3-4-5-6-7-8-9-1-2
×
1-2-3-4-5-6-7-8-9-1-2
↓
1-2-3-4-5-6-7-8-9-1-2-3-4-5-6-7-8-9-1-2-3-4-5-6-7-8-9-1-2

Skalka has suggested that λ concatemers may arise as a result of "rolling circle" replication (see p. 88), since they can be formed in cells unable to support recombination. Various other viruses such as T5, T7, P22, φX174 also show the presence of concatemers.

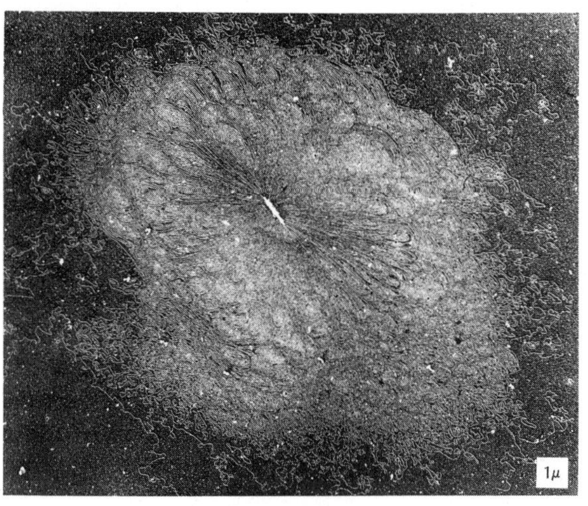

Figure 20–11

Electron micrograph showing a large number of T4 DNA fibers radiating from a central point in an E. coli cell which had been infected for a 30-minute period by T4 bacteriophage. The total amount of DNA in these fibers is equivalent to about 6000 microns, or more than 100 lengths of phage DNA. (Courtesy of J. Huberman.)

TEMPERATE PHAGES

Lysogeny, as discussed previously (Chapter 19), involves a symbiotic relationship between a temperate phage and a bacterial host. In this form, the prophage is replicated as part of the bacterial chromosome conferring immunity upon the bacterium to the vegetative growth of other similar phages. The K12 strain of *E. coli*, for example, was found by E. M. Lederberg to bear the prophage which protected it against lysis by λ. Such strains, however, could occasionally yield λ-phage particles that produced plaques on nonlysogenic strains of K12 that had been "cured" of their protective prophages.

Since these findings took place in *E. coli*, they offered the advantage of permitting bacterial conjugation between lysogenic and nonlysogenic strains. In such crosses lysogeny was shown to segregate among the progeny as though the prophage itself were directly attached to the bacterial chromosome. Specifically the point of attachment seemed to be closely connected to the *E. coli gal* locus that determines the ability of bacteria to metabolize the sugar galactose to glucose. This was demonstrated most clearly in experiments utilizing Hfr strains of *E. coli*: An Hfr nonlysogenic strain (λ^-) mated to an F^- lysogenic strain (λ^+) produced nonlysogenic recombinants in the recipient cells only when the *gal* locus of the donor chromosome had also entered. In this fashion Jacob and Wollman mapped prophage locations for 14 different temperate phages. In each case they found specific attachment sites on the bacterial chromosome to which a particular prophage became attached.

The immunity to bacterial lysis rendered by the prophage could be further demonstrated by making crosses between lysogenic donors and nonlysogenic recipients. In this case almost no lysogenic recombinants were recovered, since the prophage seemed to enter the vegetative stage immediately upon entry into the nonlysogenic host. This type of "induction" thus occurred as a result of chromosome transmission and has been appropriately named *zygotic induction.* Since zygotic induction and subsequent lysis only occur when the lysogenic donor chromosome enters a nonlysogenic recipient but does not occur when such a chromosome enters a lysogenic recipient, it is apparent that something within the cytoplasm of the lysogenic F^- recipient is capable of repressing viral multiplication. This cytoplasmic material, named the *repressor substance* by Jacob and Monod, is manufactured by the prophage, and inhibits the vegetative multiplication of λ, thus conferring immunity against lysis upon the host cell (see also p. 415). When lysogenic strains are subjected to external inducing agents such as ultraviolet light, x-rays, or chemicals such as nitrogen mustard, the repressor activity is destroyed causing induction of the prophage and consequent lysis of the host.

Because of the complexity of this entire lysogenic process, various opportunities exist for a genetic change in the relationship between a temperate phage and its host. Under normal circumstances a culture of λ particles infecting a nonlysogenic strain of *E. coli* will form a plaque similar to that formed by other bacteriophages except that the plaque will be turbid throughout, because of numerous bacterial cells that have become lysogenic and immune to lysis. Occasionally, however, *clear* plaques arise that are caused by phages unable to establish lysogenic relationships. These phages cause lysis of all infected bacteria and, since they transmit this attribute to future generations, are considered to bear mutations at one of the three *c* genes, *cI*, *cII*, and *cIII*. As will be discussed further in Chapter 29, we now know that it is the *cI* gene, in its wild-type form, which produces the repressor substance. This substance, a protein, prevents function of the *N*, *O*, and *P* genes which are necessary for vegetative reproduction. Genes *cII* and *cIII*, in turn, affect the function of the *cI* gene. Some of the early designations given for these genes and others that cause distinguishable changes in plaque morphology were *cocarde* (*co*), *medium* (*m*), *small* (*s*), and *minute* (*mi*).

In the 1950s Kaiser and others found that these mutations could be mapped by analyzing either two-factor or three-factor crosses according to standard recombination methods. For this purpose bacterial cells are infected with two different kinds of parental phage and the progeny emerging through lysis are then plated on sensitive bacterial cells. The results of one such set of experiments are given in Tables 20–3 and 20–4. For the two factor crosses (Table 20–3), note that the sum of the two recombination frequencies co_1-*mi* and *s*-co_1 is equal to the frequency of *s*-*mi* recombinants. For the three-factor crosses (Table 20–4), the *s*-*mi* recombination frequency is obviously larger than either the *s*-*co* or *co*-*mi* frequencies, thereby confirming the linkage order *s*-*co*-*mi*. As in diploids (p. 336), the *s*-*mi* recombination frequency includes twice the frequency of the double-crossover classes, yielding values which in this case are almost identical to those of the two-factor crosses. The gene order can also be determined by observing that the least frequent classes in a three-factor cross are the double-crossover classes, and then choosing a gene order that satisfies this requirement. A linkage map of λ using more recent gene designations is presented in Fig. 20–12.

NEGATIVE INTERFERENCE

The unique events that give rise to recombination in phages also produce unusual effects when recombination is scored in two adjacent phage regions. For example, three loci in sequence *a b c*, with the interval 1 between *a* and *b*, and interval 2 between *b* and *c*, might show recombination

TABLE 20–3

Results of two factor crosses between mutants of λ, utilizing experiments in which more than 10,000 plaques were scored

Parents	Total No. Scored	Progeny, Percent: Parentals	Progeny, Percent: Recombinants
co_1 + × + *mi*	12,324	41.89 *co* +	2.52 + +
		52.82 + *mi*	2.77 *co mi*
		94.71	5.29
s + × + co_1	13,266	53.53 *s* +	1.09 + +
		44.11 + *co*	1.27 *s co*
		97.64	2.36
s mi × + +	28,515	46.48 *s mi*	3.59 *s* +
		45.88 + +	4.05 + *mi*
		92.36	7.64

From data of Kaiser, 1955.

TABLE 20–4

Results of a three-factor cross between λ mutants *s* + *mi* × + *co*$_1$+

Total No. Scored	Progeny, Percent							
	+ + +	*s co mi*	*s* + +	+ *co mi*	*s co* +	+ + *mi*	*s* + *mi*	+ *co* +
12,324	.31	.19	2.21	2.58	.91	.98	51.84	40.98

recombination distances

s-co: .31 + .19 + .91 + .98 = 2.39
co-mi: .31 + 19 + 2.21 + 2.58 = 5.29
s-mi: 2.21 + 2.58 + .91 + .98 = 6.68 + 2 × doubles
= 6.68 + 2(.50) = 7.68

From data of Kaiser, 1955.

frequencies of f_1 and f_2 for these respective intervals. Under standard conditions of recombination, a mating $a\ b\ c \times a^+\ b^+\ c^+$ would therefore be expected to produce double recombinants ($a\ b^+\ c$, $a^+\ b\ c^+$) in the approximate frequency $f_1 \times f_2$. In most organisms considered so far, we have seen that there is some degree of "positive" interference (pp. 338 and 377), which prevents crossing over in regions adjacent to one in which crossing over has occurred.

In many phage crosses, however, the reverse is true, and interference is "negative," as evidenced by an increased frequency of double recombinants relative to those expected. That is, genetic exchange in one region is accompanied by an increase in genetic exchange in adjacent regions, i.e., coincidence coefficients occur that are greater than 1. Kaiser, for example (Table 20–4), found crossover frequencies of .0239 in the *s-co* region and .0529 in the *co-mi* region. On this basis double crossovers for both regions would be expected at a frequency of .0239 × .0529, or .00126. The actual observed frequency of double crossovers, however, was .005, or about four times the expected value.

Reasons for this negative interference in phage arise because of two peculiarities of phage chromosome reproduction:

1. Two different phages infecting a single bacterium produce two clones of progeny that mate within clones as well as between clones. As we have seen previously (p. 441), only half the time can matings be expected to occur between phage chromosomes of different origin. Thus if f_1 and f_2 are both .2, the observed frequency of double recombinants (.2 × .2 = .04) would be halved, or approximately equal to .02. On the other hand, the frequency of single recombinants is also

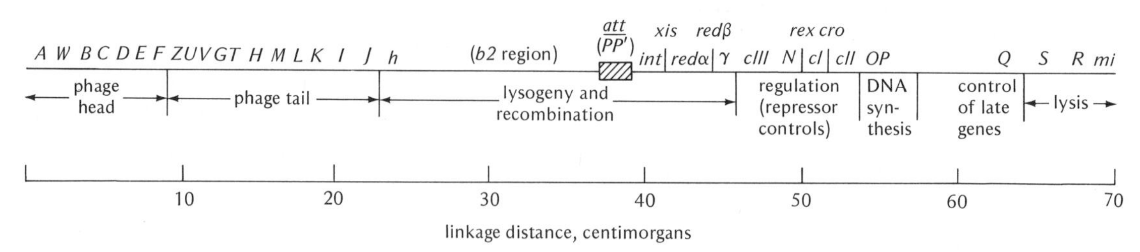

Figure 20–12

Linkage map of λ, showing genes (above), their general function (center), and approximate linkage distances (below). A map of λ providing information on the regulation and control of some of these genes is given in Chapter 29 (Fig. 29–7). (After Campbell.)

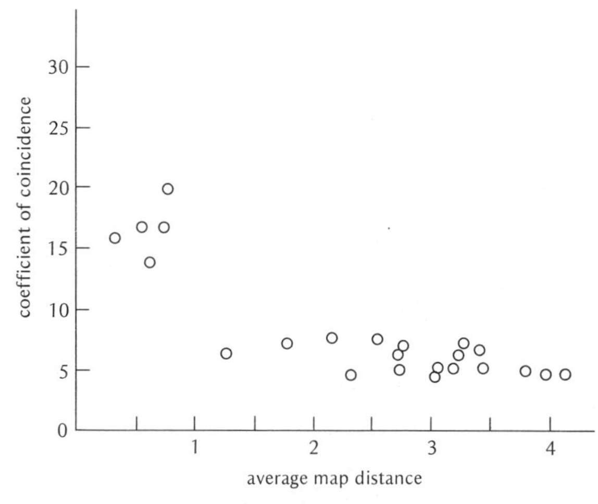

Figure 20–13

Relationship between coefficient of coincidence and map distance in three-point crosses between different r mutants of bacteriophage T4. Note that as map distances became less than 1, the coefficient of coincidence increased, i.e., interference decreased and became more "negative." Compare these data to the curve in Fig. 17–6 which shows positive interference. (After Hershey, 1958, from data of Chase and Doermann.)

halved, so their apparent frequencies are both .1. On the basis of such apparent single recombinant frequencies, the expected double-recombinant frequency, $.1 \times .1 = .01$, is therefore one half of that observed, leading to an apparent excess of observed double recombinants.

2. Because more than one round of mating is possible among phage chromosomes, a chromosome in which a single recombinational event has occurred in one interval may mate again and take part in another recombinational event in an adjacent interval. For example, an already recombinant $a\ b^+\ c^+$ chromosome may mate with an $a\ b\ c$ or $a^+\ b\ c$ chromosome, to produce the apparent double recombinant $a\ b^+\ c$. Thus two single recombinant events are now scored as a double recombinant.

Note, however, that in both the above cases the relative increase in double recombinants is only apparent, and does not arise from an actual increase in simultaneous genetic exchange in the two adjacent intervals. Such apparent, but not actual, negative interference, first noted by Visconti and Delbruck and commonly observed in phage crosses, has been termed *low negative interference* since its effect is relatively slight.

Of considerably greater magnitude and much more difficult to explain is the phenomenon known as *high negative interference*. In this case the excess of double recombinants may reach as high as 30 times or more than expected. For example, three-point crosses made by Chase and Doermann between different *r* mutations in T4 gave excesses of double recombinants ranging from 5 to 32 times that of the expected frequencies. As shown in Fig. 20–13, there was a distinct increase in negative interference as recombination frequencies became smaller in the two adjacent intervals. In other words, pairing and exchange over a small chromosome section seem to increase the probability for additional genetic exchanges within that section (see also p. 378).

BREAKAGE AND REUNION, OR COPY CHOICE?

Theories explaining the mechanism of recombination in phage (as well as in other organisms) may be divided into three general groups:

1. *Breakage and reunion.* The two pieces of parental chromosome information that are combined in the recombinant chromosome arise from physical breaks in the parent chromosomes with subsequent physical exchange. (Such exchanges have already been noted on page 356 for *Bellevalia* chromosomes bearing radioactive labels, although they have not been correlated with genetic exchange.) In this case recombination does not depend upon reproduction or synthesis of new DNA molecules, since the information entering the recombinant is merely transferred from the parentals (Fig. 20–14a).

2. *Breakage and copying.* The recombinant chromosome is formed by utilizing the physi-

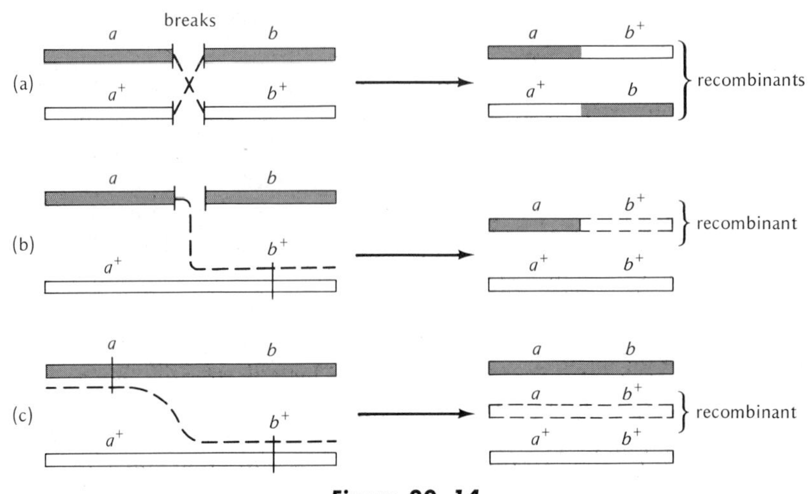

Figure 20–14

Three possible explanations for recombination events. (a) Breakage and reunion. (b) Breakage of one chromosome followed by copying (dashed lines) of another. (c) Copy choice.

cal section of only one parental chromosome and by copying the other. For example, two parent chromosomes $a\,b$ and $a^+\,b^+$ can produce a recombinant by permitting the a section of one parent to become attached to a newly synthesized section copied from the b section of the other parent (Fig. 20–14b).

3. *Complete copy choice.* The entire recombinant arises from newly synthesized sections which have copied part of their sequence from the a section of one parent and part of their sequence from the b section of the other (Fig. 20–14c, and see also Belling's hypothesis, p. 355). In this case, as well as in breakage and copying, DNA synthesis must necessarily accompany recombination.

Evidence with respect to these theories is presently as follows. The concept of breakage and reunion traditionally used to explain genetic exchange was cast into considerable doubt by three basic findings in viral recombination. The first such finding was the observation of "nonreciprocal recombination" by Hershey and Rotman in crosses between phages differing at the h and r loci (p. 440). Although they observed equal proportions of reciprocal recombinants in the experiment as a whole, they also noted numerous instances in which lysates from single bacteria ("single bursts") did not contain such equal proportions.

For example, the cross $h\,rl^+ \times h^+\,rl$ gave equal proportions of $h^+\,r^+$ and $h\,r$ recombinants for the entire experiment (Table 20–1), but single bursts produced cases in which the yield of one recombinant was considerably higher than the yield of the other. As shown in Fig. 20–15, many single bursts produced as much as 20 percent or more $h\,r$ but less than one half that proportion of $h^+\,r^+$. Since breakage and reunion imply a reciprocal exchange between the two parental chromosomes and equal yield of both types of recombinant offspring, the decreased proportions of one recombinant in these experiments seemed to indicate that a copy-choice mechanism was responsible for the results. That is, since recombination by copy choice produces only one recombinant product, nonreciprocal events might be expected. This view was supported by the nonreciprocal gene-conversion events shown to occur in fungi by Mitchell, Lindegren, and others (pp. 384–387).

The second apparent departure from breakage and reunion was the occurrence of viral heterozygotes as intermediates in recombination (p. 445). Although the exact mechanism for the formation of the heterozygote was not known, it seemed difficult to explain on the basis of the usually pictured breakage and reunion. Levinthal, in fact, suggested a possible copy-choice method diagrammed in Fig. 20–16.

The third argument against breakage and reunion arose from the Chase and Doermann experiments, showing high negative interference (p. 453). If recombination were exclusively by breakage and reunion, the rodlike bendings involved in breakage would generally be expected to cause positive interference or, at best, no interference. The high negative interference values observed seemed, therefore, difficult to explain. On the other hand, if recombination were by copy choice, the "switch" in copying from one parental strand to the other need not inhibit, or might even enhance, further "switches" nearby.

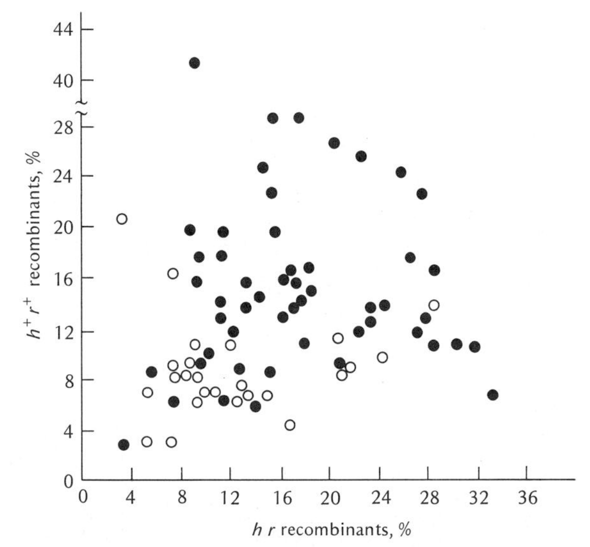

Figure 20–15

Percent of recombinants in the bursts from single bacteria infected by T2 phages of the type $h\ r1^+ \times h^+\ r1$. Each point represents the viral progeny of a single bacterium. (The open circles indicate bursts which produced proportionately more of one parental combination than the other.) (From Hershey and Rotman.)

Despite this evidence, a recombination mechanism based exclusively on copy choice has been difficult to accept. One important reason for this is the observation that recombination may occur in the absence of DNA synthesis. This was shown most clearly in experiments of Tomizawa and Anraku demonstrating that potassium cyanide treatment of phage-infected cells interfered with DNA synthesis but did not interfere with phage recombination. Furthermore, when such treated cells are infected jointly with radioactive T4 of normal light density and nonradioactive T4 of heavy density, they produce DNA molecules that are both radioactive and of heavy density. In other words, in the absence of DNA synthesis, structures occur which contain the labeled DNA of both parents. Similarly, Fox and others have shown (p. 392) that labeled donor DNA used as a transforming agent enters physically into the recipient chromosome. Only breakage and reunion mechanisms could account for these exchanges and incorporations (see also Oppenheim and Riley).

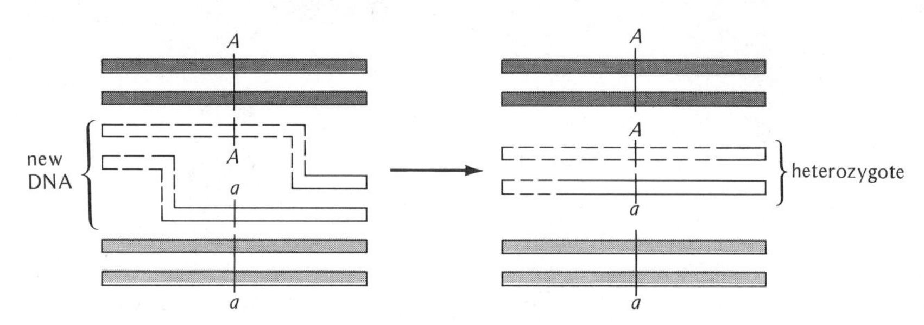

Figure 20–16

Explanation of heterozygote formation (A/a) as a result of copy choice. (After Levinthal, 1959.)

At present the most compelling evidence for breakage and reunion derives from experiments by Meselson and Weigle. Using λ phage differing at the *c* and *mi* loci, they labeled one of the parents with both heavy carbon (^{13}C) and nitrogen (^{15}N) and then observed labeling among the offspring as they were produced in an unlabeled medium. Because of method and sensitivity, the experiments enabled distinction between three main classes of offspring: (1) those in which both DNA strands were labeled, (2) those in which one DNA strand was labeled, and (3) those that bore no label.

As shown in Fig. 20–17, a cross between labeled *c mi* and unlabeled c^+ mi^+ produced *c mi* nonrecombinants which fell into all three classes. Some were like their parents, with both DNA strands labeled, and had apparently not replicated during infection. Others, however, were descendants of parental phages which had replicated once or more and therefore had one or no strands labeled. As expected, the c^+ mi^+ offspring were for the most part nonrecombinant and therefore unlabeled like their parents, since they had replicated in unlabeled medium. Of great interest was the finding that the *c* mi^+ recombinants fell into three main classes, with peaks I, II, and III. The first two of these peaks were somewhat lower in density than that of the *c mi* parental types, indicating that heavy labeling did not completely cover both strands at peak I or one strand at peak II. If we consider that the *c* locus has a position approximately 85 percent distant from one end of the chromosome and that the *mi* locus is even further along the chromosome (Fig. 20–12), the simplest interpretation of the three peaks is given in the bottom portion of Fig. 20–17. That is, crossing over in the *c-mi* interval had produced the insertion of a small unlabeled section (bearing mi^+) into the nonreplicated structure at peak I and into the replicated structure at peak II. Further replication of these recombinant chromosomes in unlabeled media then produced the completely unlabeled recombinants of peak III. Based on this same linkage relationship, the absence of heavy labeling in the c^+ *mi* recombinants (peak IV) is easily explained: Since only the *mi* locus comes from the labeled parent, only a small section, at most, of the recombinant chromosome is labeled.

Although breakage must have caused insertion of the parental heavily labeled *c* section to account for the appearance of peaks I and II, one could still argue that the nonlabeled mi^+ section in these recombinants arose from copying rather than through insertion of parental nonlabeled DNA. These experiments could therefore be explained on the basis of breakage and copying as well as by breakage and reunion. To distinguish between these two alternatives, Meselson extended these experiments to recombination between the *h* and *c* loci of λ. Since these loci are toward the center of the λ chromosome, breakage and reunion among the recombinants could be expected to give results different from breakage and copying. This can be seen if we simultaneously infect cells with labeled λ bearing *h* and *c* and labeled wild-type λ h^+ c^+, and then note the type of *h* c^+ recombinants that are formed in an unlabeled medium. Theoretically, such recombinants may bear one or more of the five following proportions of labeled chromosome under breakage and reunion (shaded bars labeled, unshaded bars unlabeled):

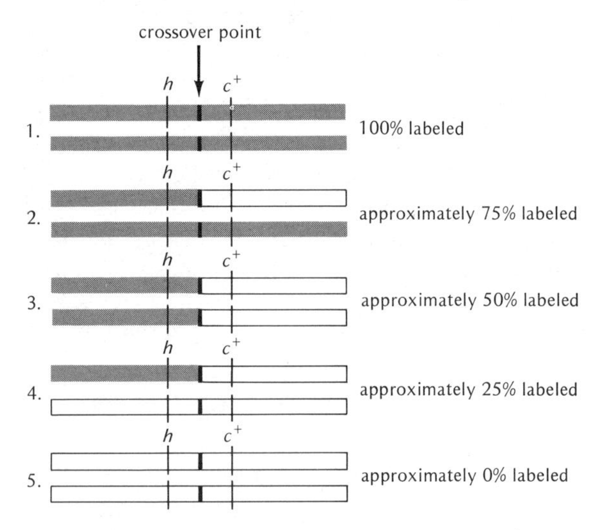

In some cases such as 3, 4, and 5, one cannot distinguish between breakage and reunion and breakage and copying. However, for full labeling and three-quarter labeling of recombinants (1, 2), the only possible explanation is that of insertion

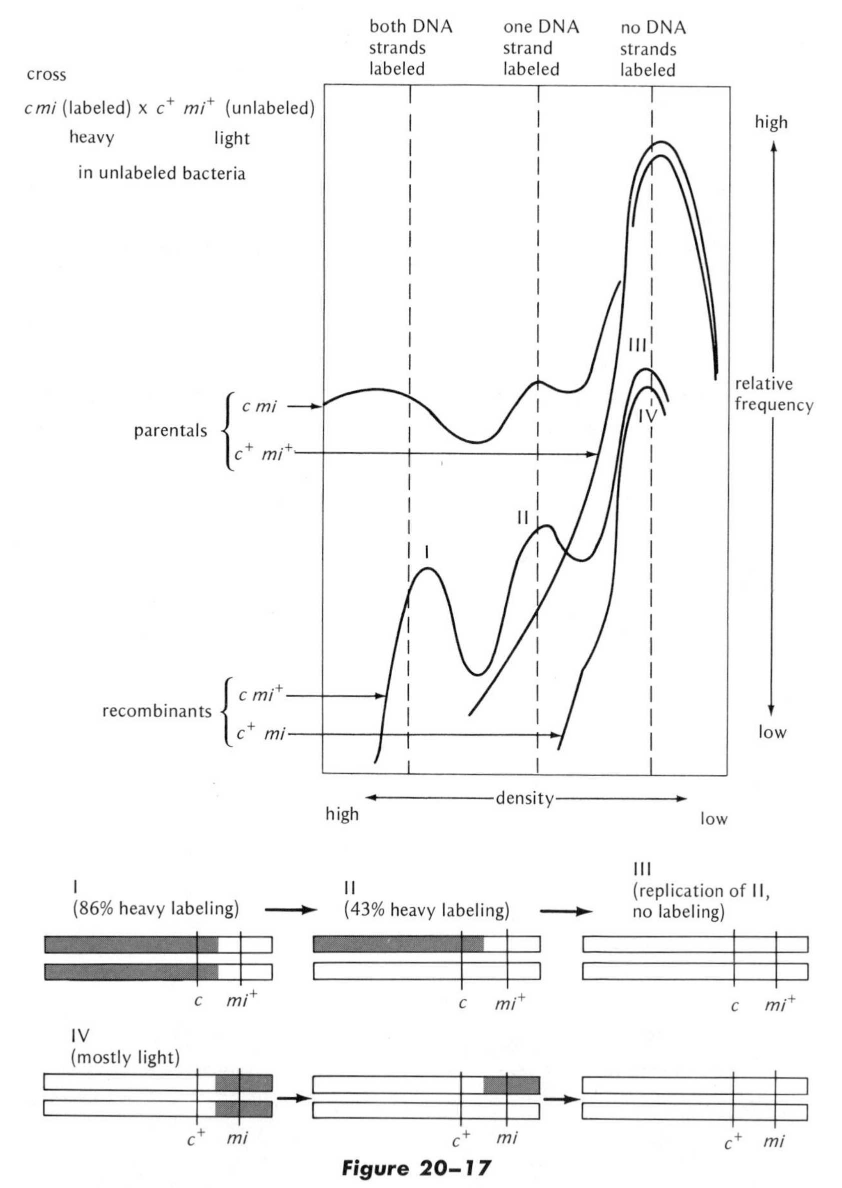

Figure 20–17

Results observed (above) and explanation of different peaks (below) in the Meselson and Weigle experiment. (After Meselson and Weigle.)

of a labeled c^+ region into a labeled strand. Since the labeled strand cannot arise through copying in an unlabeled medium, the labeled c^+ section is most easily explained by a physical insertion of a portion of a wild-type chromosome by breakage and reunion.

Since Meselson found the existence of types 1 and 2, there is now little question that recombination occurs through breakage and reunion. It is, however, the particular form of this mechanism that remains to be discovered. So far, any seriously proposed recombination model must take into ac-

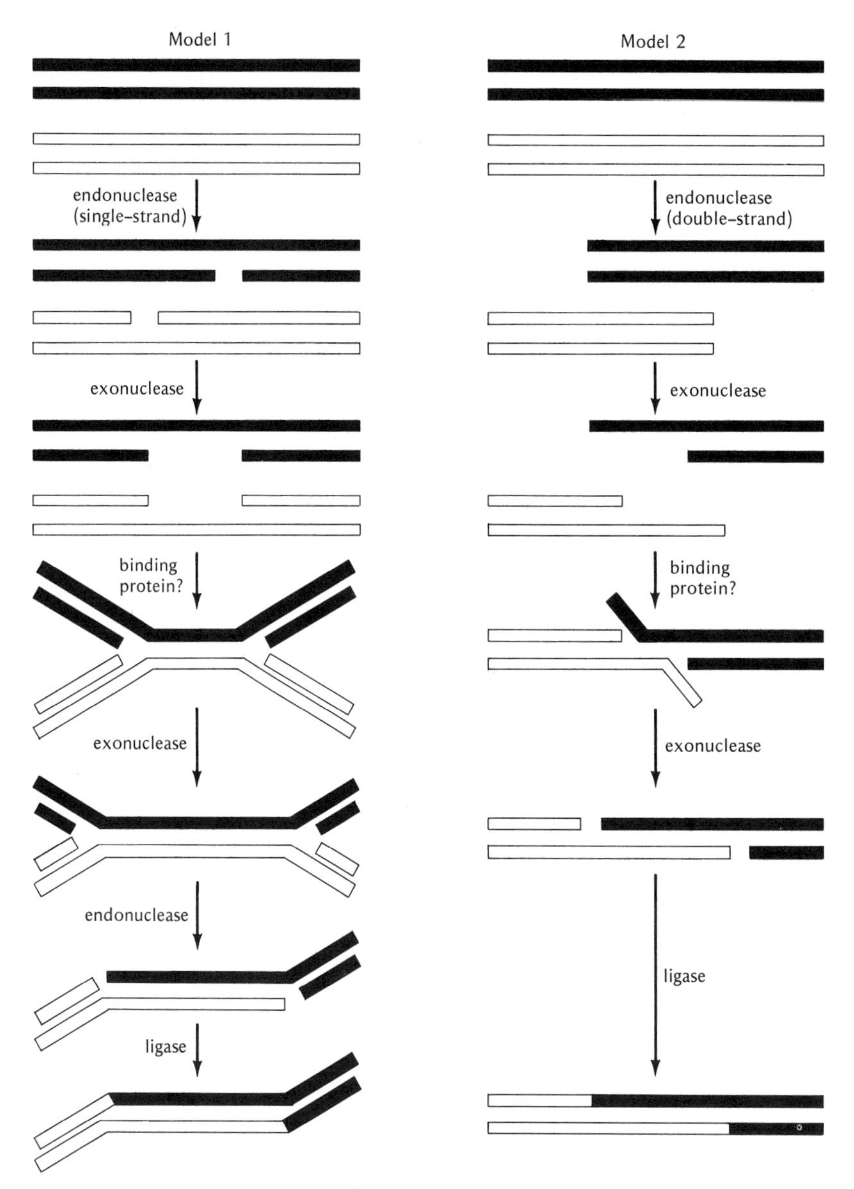

Figure 20–18

Two models that have been offered to explain recombinational events between bacteriophages. In model 1, single-strand cuts in the parental double-stranded DNA molecules lead to pairing between the parental molecules followed by an extension of the paired hybrid region. Endonucleases then attack the points at which the branches are joined to the hybrid region leaving only a DNA double-stranded recombinant heteroduplex. In model 2, double-stranded cuts in the parental molecules are followed by exonuclease "nibbles" which produce single-strand terminals. These terminals then pair to form a hybrid region which is lengthened by the removal of adjacent nucleotides through the action of exonuclease. As in model 1, the final step involves the formation of a chemically bonded continuous DNA molecule by ligase action. (Modified after Clark based on models of Broker and Lehman, and Cassuto and Radding.)

count phenomena that seem quite difficult to explain by a single mechanism, i.e., nonreciprocal recombination, phage heterozygosity, and high negative interference. To be explained through breakage and reunion, these events demand considerable flexibility in the recombining structure (perhaps involvement of only one of the complementary DNA strands) as well as the potential loss of sections of parental strands during recombination.

We have seen previously (Fig. 18–8) that recombination models exist which can explain orthodox recombinational events as well as nonreciprocal conversions and heteroduplex structures. The two examples presented in Fig. 20–18 show how extensively these models depend upon the widely prevalent exo- and endonucleases, polymerases, and ligases: enzymes that "cut and splice" DNA molecules in almost any order (see also review of Signer). A phenomenon such as high negative interference may therefore arise from a localized concentration of one or more of these enzymes, or from mechanical considerations not yet understood.

In the testing of some of these models, bacteriophages have offered considerable advantages because of the ease with which recombinants can be detected (e.g., phage heterozygotes). In phage T4 a variety of genes so far are known to be involved in controlling recombination, some of which also affect the appearance of the various DNA enzymes and other proteins. For instance, one gene, #32, produces a "binding protein" which binds preferentially to single DNA strands and appears to be necessary to allow recombinational pairing between them. A remarkably similar protein with similar function has been found in eucaryotes as well (see p. 356). In λ at least two recombinational pathways exist, one of which is centered around the *int* gene and is specific for recombination of the integration locus necessary for lysogeny (see p. 415). Another pathway, called *red,* involves two genes (*red* α, *red* β), which produce an exonuclease enzyme and another protein essential for recombination. In summary, various recombinational pathways exist in organisms, each probably sensitive to the effects of many genes. Since recombination lies at the base of so many sexual processes, and is responsible for so much of the variation among organisms, the unraveling of these pathways has become one of the most interesting and essential tasks of modern genetics.

PROBLEMS

20–1. One environmental condition enabling T-even phages to adsorb onto host cells is the amino acid tryptophan, and such phages are called tryptophan-requiring (c). Some phages, however, have mutated to the tryptophan-independent condition (c^+). Interestingly, when bacteria are infected with c and c^+ phage, approximately half of the tryptophan-independent progeny are shown upon further testing to be of genotype c. How would you explain this finding?

20–2. Streisinger found that certain viral products produced by mixed infections with T2 and T4 phage were capable of infecting *both* B/2 and B/4 bacteria (see p. 436). After another generation, however, the viruses produced were either pure T2 (could only infect B/4) or pure T4 (could only infect B/2). (a) Propose a hypothesis to explain these events by considering the possibility that more than one site on viral protein may be involved in determining the host range. (b) What proportion of the progeny of a mixed infection produced by equal amounts of T2 and T4 would you expect to be of this type (capable of infecting *both* B/2 and B/4)?

20–3. In an experiment by Thomas, radioactively labeled single strand fragments of the ends of T2 and T5 chromosomes were "hybridized" with single strands derived from the entire chromosomes of these species. That is, T2 "ends" × T2 "complete" strands, and T5 "ends" × T5 "complete" strands. In which of these hybridizations would you expect to find the hybridized "complete" DNA strands more heavily labeled?

20–4. If the DNA of bacteriophage T2 (permuted) were first fragmented into sections each one-fourth or less of the total chromosomal length, and then subjected to exonuclease treatment (see Fig. 20–10), explain whether you would or would not expect to find these fragments in the form of circles.

20-5. Explain whether you would expect an increase in the frequency of T2 heterozygotes for a particular gene (e.g., h/h^+) when the chromosome is also carrying a large deletion at another section.

20-6. The following two-factor crosses represent a sample of those made by Kaiser to detect linkage relationships between four genes in phage λ: co_1, *mi, c,* and *s.*

Parents	Progeny
(a) co_1 + × + *mi*	5162 *co* +, 6510 + *mi*, 311 + +, 341 *co mi*
(b) *mi* + × + *s*	502 *mi* +, 647 + *s*, 65 + +, 56 *mi s*
(c) *c* + × + *s*	566 *c* +, 808 + *s*, 19 + +, 20 *c s*
(d) *c* + × + *mi*	1213 *c* +, 1205 + *mi*, 84 + +, 75 *c mi*

Determine the recombination frequencies in each cross and draw a linkage map of these genes.

20-7. Doermann infected *E. coli* cells with two strains of T4 virus: one that was mutant for *minute* (*m*), *rapid lysis* (*r*), and *turbid* (*tu*), and one that was wild type for all three markers. The lytic products of this infection were plated and classified as follows:

Genotype	Number of Plaques
m r tu	3,467
+ + +	3,729
m r +	853
m + *tu*	162
m + +	520
+ *r tu*	474
+ *r* +	172
+ + *tu*	965
	10,342

(a) Determine the linkage distances between *m-r*, *r-tu*, and *m-tu*. (b) What linkage order would you suggest for the three genes? (c) What is the coefficient of coincidence in this cross, and what does it signify?

20-8. According to electron micrographs, the phage DNA pool produced by T2 infections is generally localized within specific areas in the bacterial cell. In phage-T5 infections, the DNA pool is distributed more widely throughout the cytoplasm. In phage-λ vegetative growth, the DNA pool cannot be seen. How do these observations reflect on recombination frequencies in each of these phages?

20-9. In phage T4 we have seen that a linear chromosome can generate a circular genetic map (p. 443). The λ chromosome, on the other hand, replicates vegetatively as a circle (see p. 415, Chapter 19), and probably recombines with other λ phages in this form. How would you explain the fact that λ generates a linear rather than a circular map?

20-10. A cross is made between two strains of λ phage differing in respect to genes *I* and *J*, whose positions on the λ chromosome are both approximately one-third from the "left" end (see Fig. 20-12). One strain, $I^+ J^+$, has its DNA labeled with both high density isotopes ^{13}C and ^{15}N, whereas the other strain, *I J*, is unlabeled with these isotopes and therefore of light density. The *E. coli* cells infected with these two kinds of phages are carrying normal unlabeled carbon and nitrogen. Let us assume that there was only one round of mating between these two kinds of phages; that, at most, only single crossovers could occur between them; and that each phage chromosome was replicated no more or less than once. On this basis, what percentage of high density labeled DNA would you expect to find in the chromosomes of the $I^+ J$ recombinants: (a) According to the "breakage and reunion" hypothesis? (b) According to the "copy-choice" hypothesis? (c) Would the answers to (a) and (b) be different if you were to look at the high-density labeling of the $I J^+$ recombinants?

REFERENCES

ARBER, W., G. KELLENBERGER, and J. WEIGLE, 1957. La défectuosité du phage lambda transducteur. *Schweiz. Zeit. f. Path. Bakt.*, **20,** 659–665. (Translated into English and reprinted in the collection of Adelberg; see References, Chapter 1.)

BAYLOR, M. B., A. Y. HESSLER, and J. P. BAIRD, 1965. The circular linkage map of bacteriophage T2H. *Genetics*, **51,** 351–361.

BRESCH, C., 1959. Recombination in bacteriophage. *Ann. Rev. Microbiol.*, **13,** 313–334.

CAIRNS, J., G. S. STENT, and J. D. WATSON (eds.), 1966. *Phage and the Origins of Molecular Biology*. Cold Spring Harbor Laboratory Quantitative Biology, Cold Spring Harbor, New York. (A volume dedicated to Max Delbruck containing a collection of personal accounts by more than 30 scientists of their participation in important discoveries concerned with phage and molecular genetics.)

CAMPBELL, A., 1971. Genetic structure. In *Bacteriophage Lambda,* A. D. Hershey (ed.). Cold Spring Harbor Laboratory, Cold Spring Harbor, pp. 13–44.

CHASE, M., and A. H. DOERMANN, 1958. High negative interference over short segments of the genetic structure of bacteriophage T4. *Genetics,* **43,** 332–353.

CLARK, A. J., 1971. Toward a metabolic interpretation of genetic recombination of *E. coli* and its phages. *Ann. Rev. Microbiol.,* **25,** 437–464.

DELBRUCK, M., and W. T. BAILEY, JR., 1946. Induced mutations in bacterial viruses. *Cold Sp. Harb. Symp.,* **11,** 33–37. (Reprinted in the collection of Levine; see References, Chapter 1.)

DOERMANN, A. H., 1965. Recombination in bacteriophage T4 and the problem of high negative interference. In *Genetics Today,* Vol. II, Proc. XI Intern. Congr. Genet., S. J. Geerts (ed.). Pergamon, Oxford, pp. 69–79.

DOERMANN, A. H., and M. B. HILL, 1953. Genetic structure of bacteriophage T4 as described by recombination studies of factors influencing plaque morphology. *Genetics,* **38,** 79–90.

EDGAR, R. S., G. H. DENHARDT, and R. H. EPSTEIN, 1964. A comparative genetic study of conditional lethal mutations of bacteriophage T4D. *Genetics,* **49,** 635–648.

HERSHEY, A. D., 1946. Spontaneous mutations in bacterial viruses. *Cold Sp. Harb. Symp.,* **11,** 67–76.

———, 1958. The production of recombinants in phage crosses. *Cold Sp. Harb. Symp.,* **23,** 19–46.

——— (ed.), 1971. *The Bacteriophage Lambda.* Cold Spring Harbor Laboratory, Cold Spring Harbor.

HERSHEY, A. D., and M. CHASE, 1951. Genetic recombination and heterozygosis in bacteriophage. *Cold Sp. Harb. Symp.,* **16,** 471–479. (Reprinted in the collections of Stent and of Taylor and in *Selected Papers in Biochemistry,* Vol. 2; see References, Chapter 1.)

HERSHEY, A. D., and R. ROTMAN, 1949. Genetic recombination between host-range and plaque-type mutants of bacteriophage in single bacterial cells. *Genetics,* **34,** 44–71. (Reprinted in the collection of Stent, and in *Selected Papers in Biochemistry,* Vol. 2; see References, Chapter 1.)

HUBERMAN, J. A., 1968. Visualization of replicating mammalian and T4 bacteriophage DNA. *Cold Sp. Harb. Symp.,* **33,** 509–524.

JACOB, F., and E. L. WOLLMAN, 1957. Genetic aspects of lysogeny. In *The Chemical Basis of Heredity,* W. D. McElroy and B. Glass (eds.). John Hopkins Press, Baltimore, pp. 468–500.

KAISER, A. D., 1955. A genetic study of the temperate coliphage λ. *Virology,* **1,** 424–443.

LEDERBERG, E. M., and J. LEDERBERG, 1953. Genetic studies of lysogenicity in *Escherichia coli. Genetics,* **38,** 51–64. (Reprinted in the collection of Levine; see References, Chapter 1.)

LEVINTHAL, C., 1954. Recombination in phage T2: Its relationship to heterozygosis and growth. *Genetics,* **39,** 169–184. (Reprinted in *Selected Papers in Biochemistry,* Vol. 2; see References Chapter 1.)

———, 1959. Bacteriophage genetics. In *The Viruses,* Vol. II, F. M. Burnet and W. M. Stanley (eds.). Academic Press, New York, pp. 281–318.

LEVINTHAL, C., and N. VISCONTI, 1953. Growth and recombination in bacterial viruses. *Genetics,* **38,** 500–511.

LURIA, S. E., 1951. The frequency distribution of spontaneous bacteriophage mutants as evidence for the exponential rate of phage reproduction. *Cold Sp. Harb. Symp.,* **16,** 463–470. (Reprinted in the collection of Stent, and in *Selected Papers in Biochemistry,* Vol. 2; see References, Chapter 1.)

MESELSON, M., 1964. On the mechanism of genetic recombination between DNA molecules. *Jour. Mol. Biol.,* **9,** 734–745.

MESELSON, M., and J. J. WEIGLE, 1961. Chromosome breakage accompanying genetic recombination in bacteriophage. *Proc. Nat. Acad. Sci.,* **47,** 857–868. (Reprinted in the collections of Abou-Sabé, of Taylor, and of Stent, and in *Selected Papers in Biochemistry,* Vol. 2; see References, Chapter 1.)

NOMURA, M., and S. BENZER, 1961. The nature of the "deletion" mutants in the *r* II region of phage T4. *Jour. Mol. Biol.,* **3,** 684–692.

NOVICK, A., and L. SZILARD, 1951. Virus strains of identical phenotype but different genotype. *Science,* **113,** 34–35.

OPPENHEIM, A. B., and M. RILEY, 1966. Molecular recombination following conjugation in *Escherichia coli. Jour. Mol. Biol.,* **20,** 331–357.

SÉCHAUD, J., G. STREISINGER, J. EMRICH, J. NEWTON, H. LANFORD, H. REINHOLD, and M. M. STAHL, 1965. Chromosome structure in phage T4. II. Terminal redundancy and heterozygosis. *Proc. Nat. Acad. Sci.,* **54,** 1333–1339. (Reprinted in *Selected Papers in Biochemistry,* Vol. 2; see References, Chapter 1.)

SIGNER, E., 1971. General recombination. In *The Bacteriophage Lambda,* A. D. Hershey (ed.). Cold Spring Harbor Laboratory, Cold Sp. Harb., pp. 139–174.

Skalka, A., 1971. Origin of concatemers during growth. In *The Bacteriophage Lambda,* A. D. Hershey (ed.). Cold Spring Harbor Laboratory, Cold Spring Harbor, pp. 535–547.

Steinberg, C., and F. Stahl, 1958. The theory of formal phage genetics. *Cold Sp. Harb. Symp.*, **23,** 42–45. (Published as an appendix to the paper by Hershey, 1958. Reprinted in *Selected Papers in Biochemistry*, Vol. 2; see References, Chapter 1.)

Stent, G. S., 1963. *Molecular Biology of Bacterial Viruses.* W. H. Freeman, San Francisco.

Streisinger, G., 1956. Phenotypic mixing of host range and serological specificities in bacteriophages T2 and T4. *Virology*, **2,** 388–398.

———, 1966. Terminal redundancy, or all's well that ends well. In *Phage and the Origins of Molecular Biology,* J. Cairns, G. S. Stent, and J. D. Watson (eds.). Cold Spring Harbor Laboratory, Cold Spring Harbor, pp. 335–340.

Streisinger, G., and V. Bruce, 1960. Linkage of genetic markers in phages T2 and T4. *Genetics*, **45,** 1289–1296.

Streisinger, G., R. S. Edgar, and G. H. Denhardt, 1964. Chromosomes structure in phage T4. I. Circularity of the linkage map. *Proc. Nat. Acad. Sci.*, **51,** 775–779. (Reprinted in Stent's collection; see References, Chapter 1.)

Thomas, C. A., Jr., 1967. The rule of the ring. *Jour. Cell. Physiol.*, **70** (Suppl. 1, Part 2): 13–33. (Reprinted in *Selected Papers in Biochemistry*, Vol. 2; see References, Chapter 1.)

Tomizawa, J., and N. Anraku, 1964a. Molecular mechanisms of genetic recombination in bacteriophage. I. Effect of KCN on genetic recombination of phage T4. *Jour. Mol. Biol.*, **8,** 508–515.

———, 1964b. Molecular mechanisms of genetic recombination in bacteriophage, II. Joining of parental DNA molecules of phage T4, *Jour. Mol. Biol.*, **8,** 516–540.

Visconti, N., and M. Delbruck, 1953. The mechanism of genetic recombination in phage. *Genetics*, **38,** 5–33.

Weissmann, C., M. A. Billeter, H. M. Goodman, J. Hindley, and H. Weber, 1973. Structure and function of phage RNA. *Ann. Rev. Biochem.,* **42,** 303–328.

Wood, W. B., 1974. Bacteriophage T4. In *Handbook of Genetics,* Vol. 1, R. C. King (ed.). Plenum Press, New York, pp. 327–331.

Zinder, N. D. (ed.), 1975. *RNA Phages.* Cold Spring Harbor Laboratory, Cold Sp. Harb.

PART IV

CHANGE AND STRUCTURE OF GENETIC MATERIAL

If nature be once seized in her variations, and the cause be manifest, it will be easy to lead her by art to such deviation as she was first led to by chance; and not only to that but others, since deviations on the one side lead and open the way to others in every direction.

FRANCIS BACON
Novum Organum

21
CHROMOSOME VARIATION IN NUMBER

In our development of many of the principles considered up to now, we have assumed the constancy of genetic material during the period of observation. This assumption has made it easier to derive the various genetic laws without having to be concerned about unexplained changes that may occur in the midst of an experiment. At the same time, however, we have seen that genetic changes do occur, and it is these alterations that provide the variation between individuals that furnishes the basis for deriving genetic laws. What kinds of genetic alterations are there and what are their origins?

In general, genetic changes, when they arise, have been called *mutations.* For ease of classification they may be divided into mutations that are cytologically visible in the nucleus as chromosome changes, and into cytologically invisible "gene," or "point," mutations, which nevertheless have an observable developmental effect on the phenotype of an organism. By convention the term mutation is commonly used for gene changes, and the more obvious chromosomal changes are known as chromosomal variations, or aberrations. Chapters 21 and 22 will concern themselves with chromosomal variations and Chapters 23 and 24 will deal mainly with gene mutation.

Among chromosomal variations the easiest to observe are usually those that involve changes in number. These may be of two types: *euploidy,* variations that involve entire sets of chromosomes; and *aneuploidy,* variations that involve only single chromosomes within a set.

EUPLOIDY

Some of the euploid variations are briefly described in Table 21-1 with each homologous set of chromosomes designated by n and a numerical prefix for the number of chromosome sets or "ploidy." Among eucaryotes, most sexually reproducing organisms are diploids (2n), although various algae and fungi exist as haploids throughout the majority of their life cycles (see Chapter 3). Haploids also occur routinely among some diploid species of insects, mites, rotifers, and others that, through parthenogenesis, produce haploid males in addition to the sexually produced diploid females (see p. 229).

In most diploid animal species, however, haploid individuals usually develop abnormally. Amphibian haploids, for example, have been induced by various means, but the embryos rarely reach the adult stage. In plant species, such as *Datura*, fully viable monoploids can be formed, but meiosis among such haploids usually produces gametes with less than the necessary complement of chromosomes. The reason for this is that meiosis will produce an equal distribution of chromosomes in daughter cells only when they begin meiotic division homologously paired. In the absence of homologous pairing, such as in haploids, the meiotic process will produce a random distribution of chromosomes in the daughter cells. A cell with the haploid number of 8, for example, may produce at the first meiotic division two daughter cells with chromosome numbers anywhere from 0 to 8 (see also p. 17).

TABLE 21-1
Variations involving entire sets of chromosomes

Euploid Type	No. Homologues Present (n)	Example
monoploid	one (n)	*A B C*
diploid	two (2n)	*AA BB CC*
polyploid	more than 2	
triploid	three (3n)	*AAA BBB CCC*
tetraploid	four (4n)	*AAAA BBBB CCCC*
pentaploid	five (5n)	*AAAAA BBBBB CCCCC*
hexaploid,	six (6n),	
septaploid	seven (7n),	
octoploid, etc.	eight (8n), etc.	

Polyploid refers to any organism in which the number of complete chromosome sets exceeds that of the diploid. When such variations arise within a species, the extra chromosome sets may pair homologously with those already present. This situation is called *autopolyploidy*, since it is a "self" (auto) increase in ploidy. Among the ways in which polyploidy can occur in sexual organisms are the fertilization of an egg by more than one sperm leading to a zygote nucleus with three or more sets of chromosomes; a failure of mitosis that multiplies the number of somatic chromosomes in a sex organ, thereby increasing the number of gametic chromosomes; and a failure in meiosis that produces a diploid gamete instead of a haploid. The failure of normal mitotic division in asexual organisms may also produce nuclei with doubled sets of chromosomes.

Triploids, containing three sets of chromosomes (3n), may arise by various means. If normal meiosis fails, for example, a diploid gamete (2n) will be formed that may then be fertilized by a haploid gamete (n) of the same species to produce an autotriploid (3n) or triploid. Triploids, however, are quite rare, since even if the triploid is viable, as it is in most plants, it is usually sterile because of defective gamete formation. Since meiotic pairing in any region is limited to only two homologues at a time, the third homologue of a triploid, especially if small, may fail to pair with either of the other two, thereby becoming a univalent (Fig. 21-1a). Even if it does pair over short regions to form a trivalent (Fig. 21-1b), it may still be randomly distributed to the gametes, so that some gametes have two homologues and others only one. Also, the presence of a third homologue may interfere with pairing resulting in three univalents, which are then randomly distributed to the gametes (Fig. 21-1c).

Thus Darlington and Mather found that a triploid strain of hyacinths having the constitution $3n = 8 + 8 + 8 = 24$ produces a variety of gametic chromosome numbers ranging from 8 to 16.

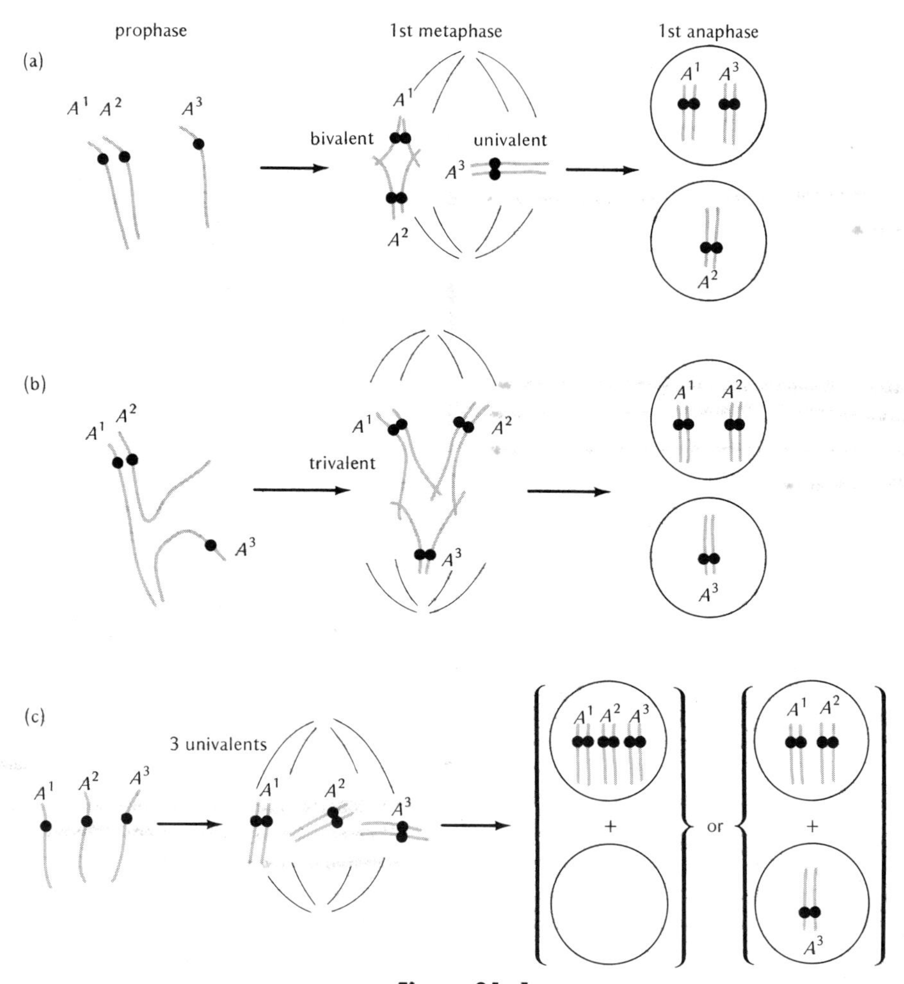

Figure 21–1
Three possible types of pairing combinations and their consequences in the meiosis of triploids.

Many of these gametes undoubtedly contain "unbalanced" numbers of homologous chromosomes, i.e., three of one and zero of another, so that union between them usually leads to the presence of some chromosomes in multiple dosage and the relative absence of others. Since normal development of embryos depends upon their possession of the proper kinds and amounts of genetic materials, such unevenly balanced zygotes easily bring about impaired development.

The sterility found in triploids holds for practically all sexually reproducing individuals that have odd-numbered sets of chromosomes, i.e., pentaploids (5n), septaploids (7n), etc. On the other hand, organisms with even-numbered sets of chromosomes, have a better likelihood of achieving an equal separation of chromosomes during meiosis, since univalents and trivalents may be absent. For example, in autotetraploids of *Datura*, an equal separation of chromosomes during meiosis occurs in the majority of cases. Also, among certain autotetraploid species of tulip and lotus, gamete formation appears to be normal, because only one chiasma is formed for any one pair of homologues, so that each tetraploid group of four homologues divides as two bivalents. On the whole, however, autopolyploids usually produce some proportion of sterile gametes and most are usually found to propagate asexually.

Despite these difficulties, autopolyploidy is not necessarily rare, especially among plants. The "giant" mutation in *Oenothera lamarckiana*, first noted by DeVries, turned out to be an autotetraploid. Autopolyploids have also been found among mosses, plums, tomatoes, corn, *Datura*, rye, and others. In the common eating banana, which is triploid, sterility is not considered commercially disadvantageous, because normal banana seeds are hard and inedible. Other polyploids, such as triploid winesap apples and European pears, and tetraploids such as alfalfa, potato, coffee, and peanuts, are usually larger or more vigorous than the diploid varieties. Because of these and other desirable features, techniques have been developed to produce autopolyploids by inhibiting normal mitosis (Fig. 21–2).

The larger nucleus and cell size of polyploids may, in some cases, be responsible for usually observed "giant" effects. In other cases, however, no such increase in cell size occurs, or, if it does, some form of compensation occurs, so that the large size of polyploid cells is accompanied by a reduction in the number of cells. For example,

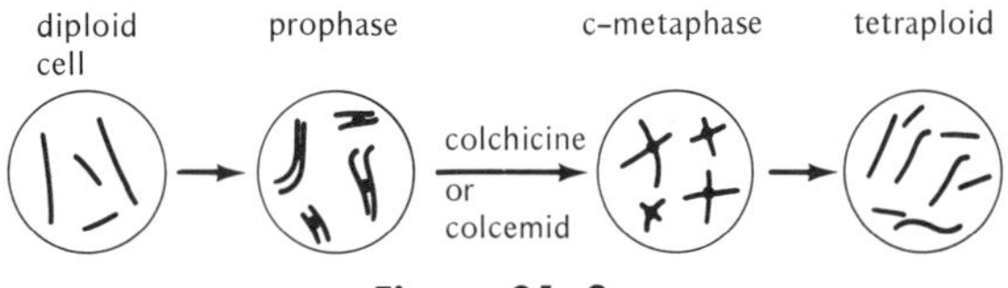

Figure 21–2

Use of the chemical colchicine (an alkaloid derived from the autumn crocus Colchicum) or colcemid (a synthetic equivalent) to prevent the separation of chromosomes during mitosis and thereby produce tetraploid cells. Note that since colchicine acts by preventing the formation of spindle fibers, the metaphase chromosomes (colchicine metaphase or c-metaphase) remain scattered in the cytoplasm and do not move to a metaphase plate. If colchicine applications are continued, reduplication of the chromosome complement may occur. In fast-growing onion cells that have been bathed in colchicine for four days, as many as 1000 chromosomes have been found in a single nucleus.

polyploid salamanders investigated by Fankhauser maintain the same body and organ size, although the cell sizes are noticeably increased.

In addition to autopolyploidy, polyploids may also be produced by the multiplication of chromosome sets that are initially derived from two different species. For example, a gamete from an ancestral species *A* may fertilize a gamete from ancestral species *B*, giving rise to a diploid with chromosome complement *AB*. This diploid, of course, will usually be partly or completely sterile, since, as in haploids, the absence of homologues between the A and B chromosome sets will prevent pairing and can easily cause the uneven meiotic distribution of these chromosomes. That is, if the A_1 and B_1 chromosomes do not pair, some gametes will contain one or both of these chromosomes and others none. However, should such a diploid undergo a failure of cell division producing a tetraploid, meiosis may now proceed on a regular basis. That is, a tetraploid with chromosomes $A_1A_1B_1B_1$ will produce pairs of two meiotic bivalents (A_1A_1 and B_1B_1), and each gamete will then contain a sample of each chromosome (A_1 and B_1). This mode of chromosome multiplication is known as *allopolyploidy*, since the chromosome sets are of different ("allo") origin.*

The proposal by Winge in 1917 that polyploidy of this kind was a means by which new species with greater chromosome numbers may originate was soon followed by experiments of Clausen and Goodspeed, who employed this mechanism to synthesize a new species. Using tobacco plants, they crossed gametes of *Nicotiana tabacum* (2n = 48, gametes n = 24) with gametes of *N. glutinosa* (2n = 24, gametes n = 12) and obtained a few viable but mostly sterile F_1 plants. One of the F_1 plants, however, produced some fertile F_2's which, upon cytological analysis, proved to contain a tetraploid number of 72 chromosomes, or the sum total of the diploid numbers of both original parents. They pointed out that this species, named *Nicotiana digluta,* begins its existence homozygous

* An allotetraploid of the kind described, *AABB*, with *A* chromosomes from one species and *B* chromosomes from another is also called an *amphidiploid.*

for all homologous chromosomes, since each chromosome of the two parental sets has been exactly duplicated in the tetraploid. Since then, numerous allopolyploids have been produced including some which are now formed by parasexual techniques (see p. 344). That is, somatic cells of different species are treated so that the cell walls are removed, and the resultant "protoplasts" may then fuse into hybrid allopolyploids (Fig. 21–3).

The combining species that form such allopolyploids may, of course, be so related that pairing occurs between the chromosomes of the two sets, thereby forming quadrivalents rather than merely bivalents. The distinction between auto- and allopolyploidy may therefore be blurred. However, if pairing is restricted to the newly risen homologues and does not occur between chromosomes of the two species, an allopolyploid may also be considered to function as an ordinary diploid. Thus, at one stroke, a new species can be created through interspecific cross-fertilization if propagation can occur through asexual means until fertile tetraploids are formed. These situations are almost exclusively restricted to the plant kingdom, in which numerous allopolyploids have been synthesized or discovered. In fact, more than one-third of all flowering plants (angiosperms) are polyploid, and in some groups, such as grasses, this proportion is as high as 70 percent.

As shown in Fig. 21–4, polyploids may originate by various means, and some show evidence of

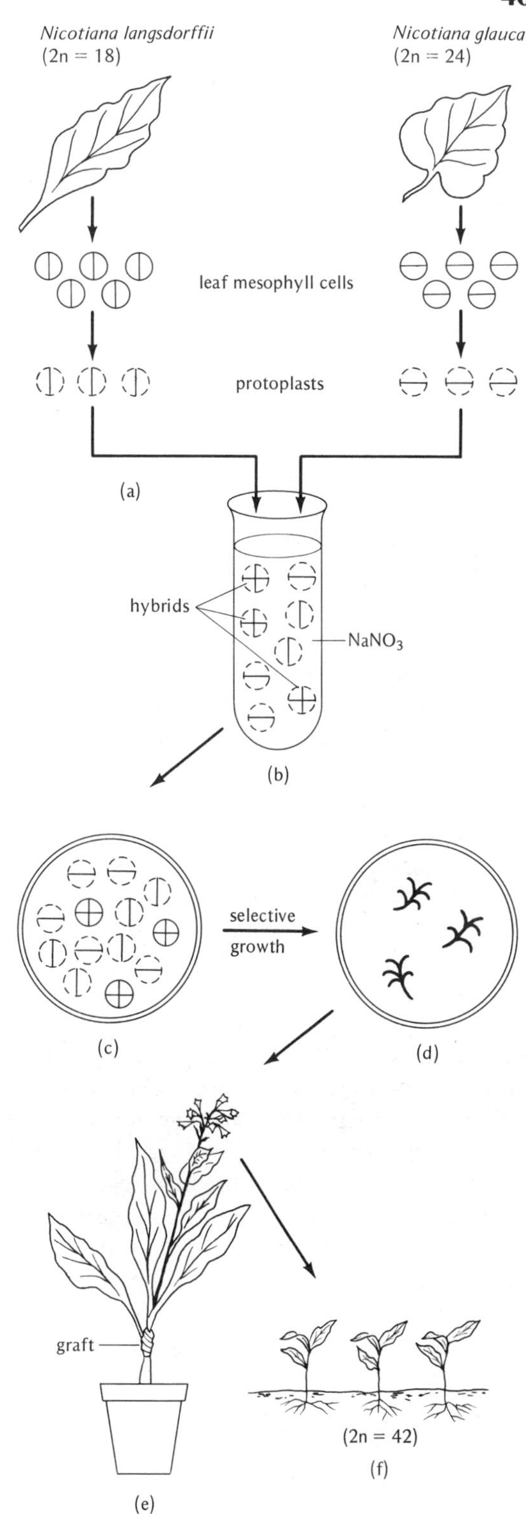

Figure 21–3

Diagram of somatic cell hybridization technique for the parasexual formation of Nicotiana *allopolyploids. (a) Leaf cells of the two parental* Nicotiana *species are treated with enzymes to digest the cell walls. (b) The protoplasts that result are then placed in a sodium nitrate solution which encourages cellular fusion to form allopolyploid hybrids. (c) After being centrifuged, the cells are plated on a medium that permits only hybrid cells to grow. (d) The hybrids differentiate into stems and leaves that are then grafted (e) onto one of the parent plants. (f) Fertile flowers produced by the grafted hybrid plant yield seeds which then germinate to produce seedlings and fully mature allopolyploid plants. (After Smith based on experiments by Carlson and coworkers; see p. 344.)*

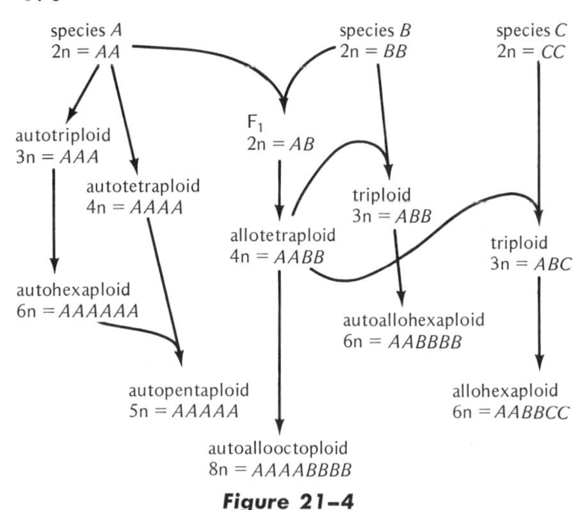

Figure 21–4

Origin and relationships of different kinds of polyploids. (After Stebbins, and after Lewis and John.)

having been derived from as many as three different species. One such example is the modern wheat species *Triticum aestivum*. This crop plant is a hexaploid that maintains 21 pairs of chromosomes and is thought to be derived from three ancestral species designated as A, B, and D, each of which originally contributed 7 chromosome pairs. Surprisingly, although meiotic pairing is normally absent between wheat chromosomes of different origin, it has been found that A, B, and D chromosomes can be made to pair together (e.g., chromosome 1 of $A \times$ chromosome 1 of B, $1A \times 1D$, $1B \times 1D$, $2A \times 2B$, etc.) as long as chromosome $5B$ is absent. Since chromosomes from the A, B, and D groups that pair together are called *homoeologous*, Riley and co-workers have called the gene on chromosome $5B$ which regulates this pairing relationship, *pairing homoeologous* or *Ph*. It has been suggested that this gene determines the positions of the chromosomes on the nuclear membrane at the beginning of meiosis, and thereby affects their pairing behavior.

In animals polyploidy is greatly restricted, since the XY mechanism of sex determination can be easily upset when chromosome sets are duplicated. For example, a male tetraploid of sex-chromosome constitution XXYY would produce XY gametes, while a female XXXX tetraploid would produce XX gametes. The union between these gametes gives rise to a zygote XXXY which, in some species, is neither completely male nor completely female, and is consequently sterile. Therefore, when polyploidy is found in some animal species among snout beetles (*Curculionidae*), psychid moths (*Solenobia*), crustacean shrimp (*Artemia*), and sow bugs (*Trichoniscus*), it is usually accompanied by some form of asexual reproduction, i.e., parthenogenesis. Only rarely are sexually reproducing polyploid species found among animals (e.g., some anuran amphibians), and these usually seem to be of the autopolyploid type.

Allopolyploidy among animals is probably extremely limited by the rarity of interspecific cross-fertilization (see isolation mechanisms, Chapter 35), and by developmental defects in such hybrids should fertilization nevertheless occur. However, even when fully viable animal hybrids are formed, they are unable to reproduce vegetatively and therefore cannot last long enough for the rare chromosome-doubling event to occur that forms allotetraploids. Also, even if tetraploid tissue is formed in an interspecific hybrid, the genes in two different chromosome sets may interact with each other to cause sterility. This can be seen in those male hybrids between *Drosophila pseudoobscura* and *D. persimilis* that carry tetraploid testicular tissue but nevertheless produce defective gametes. Thus, when animal species that appear to be allopolyploid are found among some fishes (*Poeciliopsis*), salamanders (*Ambystoma*), and lizards (*Cnemidophorus*), they are generally parthenogenetic.

In general, while polyploidy is more common among plants than animals, there is little question that it has occurred among animals as well. In both cases a polyploid species may confer an evolutionary advantage upon its descendants by giving them the opportunity to evolve a number of different functions for gene pairs that were originally present only once. Thus, if a particular gene codes for a particular protein, one gene pair of the polyploid can now produce that protein, leaving one or more duplicate gene pairs to evolve a related or entirely new function by mutation. As will be discussed in Chapter 35, the evolution of duplicate gene pairs may explain a number of similarities between seemingly different proteins.

SEGREGATION AND LINKAGE IN POLYPLOIDS

The presence of extra chromosomes in polyploids has important effects on the genotypic and phenotypic ratios produced among their offspring. In triploids, for example, an organism recessive for a gene at one locus near the centromere, *AAa*, would in theory, produce haploid and diploid gametes in the ratio 2 *A* : 1 *AA* : 2 *Aa* : 1 *a* (Fig. 21–5). Selfing such an organism might then be expected to yield diploid and triploid genotypes in the ratio 26 *A* : 1 *a*. (If the tetraploid combinations are included, this ratio is 35 *A* : 1 *a*.) The segregation in this case is based only on the separation of chromosomes, and is termed "random chromosome segregation." If, however, the locus in question is sufficiently far from the centromere so that crossing over can occur, the gametic ratios are changed, and one of the gametes produced is a double recessive, formed by the segregation of crossover chromatids to the same pole. This is one form of "random chromatid segregation," which, upon

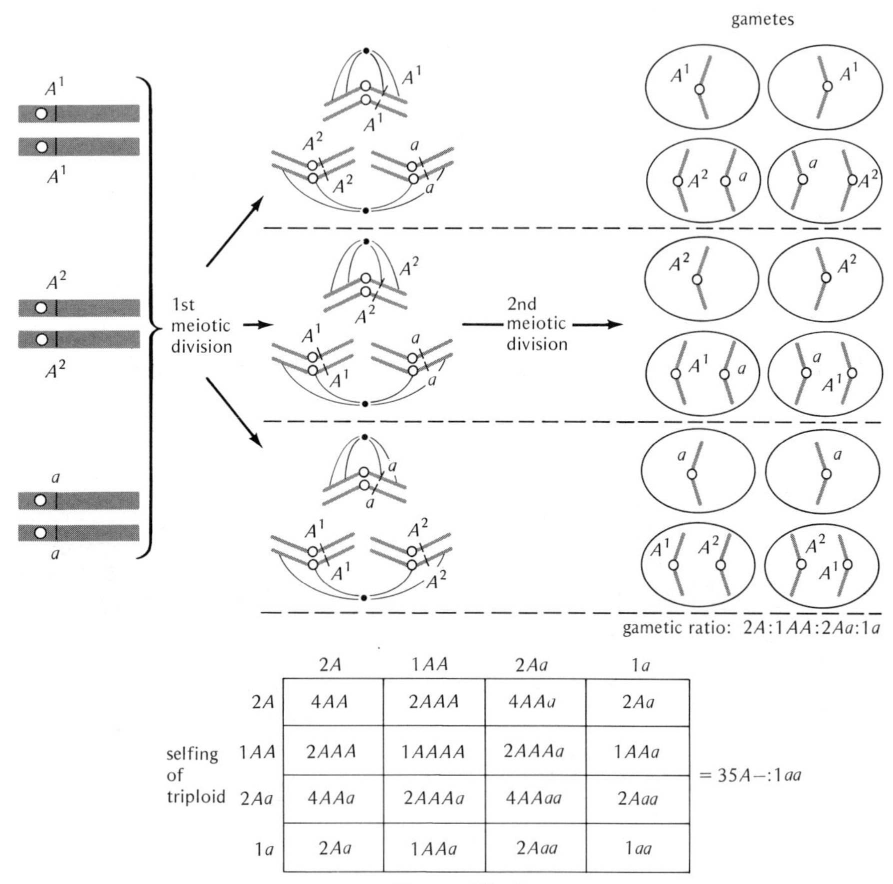

	2A	1AA	2Aa	1a
2A	4AA	2AAA	4AAa	2Aa
1AA	2AAA	1AAAA	2AAAa	1AAa
2Aa	4AAa	2AAAa	4AAaa	2Aaa
1a	2Aa	1AAa	2Aaa	1aa

Figure 21–5

Production of gametes in a triploid that is heterozygous AAa, and consequent production of 35 A : 1 a phenotypes upon selfing. As shown, the three alternative orientations of the chromosomes at meiotic metaphase I are assumed to be equally probable.

selfing, will produce diploid and triploid phenotypes in the ratio 17 A:1 a. (The inclusion of tetraploid combinations modifies this ratio to 22 A:1 a.)

Linkage relations between mutant genes in triploids, first investigated in *Drosophila*, are not always identical to that of diploids. For example, Bridges and Anderson found that crossing over between the *yellow* and *ruby* region on the *Drosophila melanogaster* X chromosome is normally about 7.5 units in diploids but is almost double that in triploids. In general, small linkage distances in diploids are increased in triploids, while large linkage distances in diploids are decreased in triploids. Since the diploid linkage map is somewhat distorted in respect to the actual physical distances (see Fig. 22–29), the triploid linkage distance would yield perhaps a truer picture of the distance between genes. However, because of the many unbalanced gametes produced, expected phenotypic ratios among triploids are rarely observed, and triploid linkage information is consequently difficult to obtain.

In autotetraploids, meiotic pairing between homologues is more regular than in triploids and leads to more predictable gametic ratios. For example, random chromosome segregation in an autotetraploid that is heterozygous *AAaa* will produce diploid gametes in the ratio 1 *AA*:4 *Aa*:1 *aa* (Fig. 21–6). Self-fertilization of such a tetraploid yields a phenotypic ratio 35 A:1 a, provided the presence of one or more A genes is sufficient to produce the A phenotype. As with triploids, how-

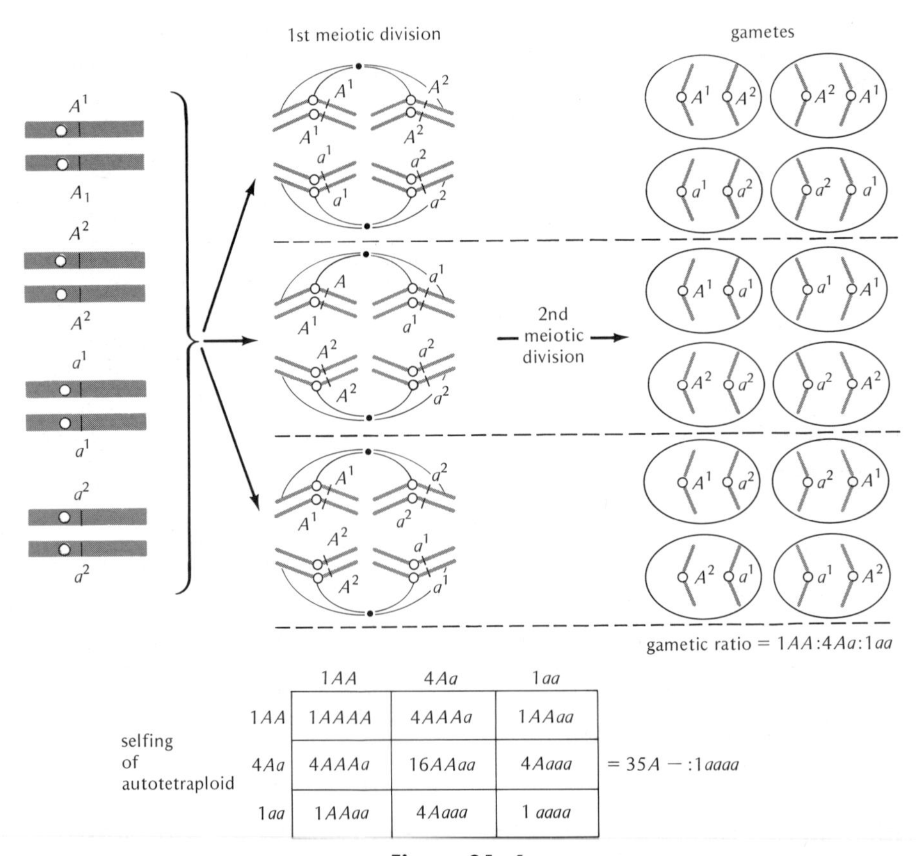

	1*AA*	4*Aa*	1*aa*
1*AA*	1*AAAA*	4*AAAa*	1*AAaa*
4*Aa*	4*AAAa*	16*AAaa*	4*Aaaa*
1*aa*	1*AAaa*	4*Aaaa*	1 *aaaa*

Figure 21–6

Production of gametes in a heterozygous autotetraploid, AAaa, and consequent production of 35 A:1 a phenotypes upon selfing.

ever, if the gene is located sufficiently far from the centromere to permit crossing over, the gametic ratios will be changed. The total gametic ratio for all crossover possibilities in a tetraploid of this type is 2 *AA*:5 *Aa*:2 *aa*.

Table 21-2 presents some of the other expected ratios for triploids and tetraploids heterozygous at one locus when this locus is closest to the centromere ("chromosome segregation") and when a maximum of crossover combinations occur ("complete chromatid segregation"). In reality, however, genes do not segregate according to either of these extremes. In *Datura stramonium*, for example, Blakeslee, Belling, and Farnham found that observed phenotypic ratios for tetraploids segregating for the gene producing armed capsules (*A*) usually fell between chromosome and chromatid segregation values. Genes that control pairing between homoeologous chromosomes (p. 470) may obviously affect segregation ratios.

In these ratios and in those of previous tables, the frequency of homozygous recessive phenotypes is considerably less than for ordinary diploid segregations and these phenotypes only become appreciably noticeable when many recessives are already present. Therefore, if such homozygous recessives are desirable, they may be quite difficult to obtain through polyploid segregation. For example, if it is advantageous for a plant to be homozygous for two pairs of recessive genes, *aa bb*, a diploid heterozygote (*Aa Bb*) will form such homozygotes in the ratio 1:16 when selfed, but a selfed autotetraploid (*AAaa BBbb*) will produce recessive homozygotes at these two loci in a ratio of only 1:1296. In selfing an autotetraploid trihybrid *AAaa BBbb CCcc*, the frequency of the multiple recessive *a–b–c–* drops to 1:46,656.

TABLE 21-2

Gametic frequencies and phenotypes produced by autotriploids and autotetraploids heterozygous at a single locus*

	Chromosome Segregation (No Crossing Over Between Centromere and A Locus)	
Genotype	Gametic Frequencies	Phenotypes Produced by Selfing
AAa	1 *AA*:2 *Aa*:2 *A*:1 *a*	26 *A*:1 *a*
Aaa	2 *Aa*:1 *aa*:1 *A*:2 *a*	2 *A*:1 *a*
AAAa	1 *AA*:1 *Aa*	all *A*
AAaa	1 *AA*:4 *Aa*:1 *aa*	35 *A*:1 *a*
Aaaa	1 *Aa*:1 *aa*	3 *A*:1 *a*
	Complete Chromatid Segregation (Formation of Quadrivalents and Single Crossovers Between the A Gene and Centromere)	
Genotype	Gametic Frequencies	Phenotypes Produced By Selfing
AAa	5 *AA*:6 *Aa*:1 *aa*:8 *A*:4 *a*	17 *A*:1 *a*
Aaa	1 *AA*:6 *Aa*:5 *aa*:4 *A*:8 *a*	1.77 *A*:1 *a*
AAAa	13 *AA*:10 *Aa*:1 *aa*	575 *A*:1 *a*
AAaa	2 *AA*:5 *Aa*:2 *aa*	19.2 *A*:1 *a*
Aaaa	1 *AA*:10 *Aa*:13 *aa*	2.4 *A*:1 *a*

* Only the phenotypic ratios of diploid and triploid offspring are given for the autotriploids.

Crossing over in the autotetraploid is also more complicated as compared to the diploid. In the diploid there are two homologous chromosomes which may segregate during meiosis as bivalents. In the tetraploid, however, there are four homologous chromosomes which may segregate during meiosis as univalents, bivalents, trivalents, or quadrivalents. Since crossing over is associated with homologous pairing, the frequency of recombinant classes in tetraploids depends not only upon linkage distance but also upon the distribution of the different genes among the four chromosomes and their mode of segregation. If, for example, only bivalents are formed, a tetraploid carrying gene differences at two linked loci, *A* and *B*, that is, *A b*, *a B*, *a b*, *a b*, will have crossover products that show no recombination (e.g., $a\,B/a\,b \rightarrow a\,b;\ a\,B$) as well as the more obvious crossover products (e.g., $A\,b/a\,B \rightarrow A\,B;\ a\,b$). If univalents, trivalents, and quadrivalents are also formed, the frequency of recombination is even further obscured, since some combinations may or may not produce obvious crossover products.

The simplest case of recombination between two tetraploid loci occurs when two dominants are on one chromosome and only recessives are on all others, e.g., *A B*, *a b*, *a b*, *a b*. If we assume that pairing between tetraploid chromosomes in such individuals occurs only in the form of bivalents and that all crossover products pass to opposite poles, pairing of the *A B* chromosome with any of the three others will produce recognizable crossover products ($A\,B/a\,b \rightarrow A\,b$; $a\,B$). If we call r the recombination frequency between the *A* and *B* loci, then, according to De Winton and Haldane, the different gametes produced by such individuals will be in the proportions

$$\frac{(1-r)}{2} A\,B/a\,b : \frac{r}{2} a\,B/a\,b : \frac{r}{2} A\,b/a\,b : \frac{(1-r)}{2} a\,b/a\,b$$

r is then the sum of the frequencies of the two recombinant classes $A\,b/a\,b$ and $a\,B/a\,b$.

An example of this kind occurs in the tetraploid *Primula sinensis*, in which the effect of the gene for short style (*S*) is fully dominant over that for long style (*s*) at one locus, and green style (*G*) is fully dominant over red style (*g*) at another locus. A plant bearing the chromosomes *S G*, *s g*, *s g*, *s g* fertilized a completely homozygous recessive plant and produced 636 offspring with the following phenotypes: 204 *S G*:126 *S g*:113 *s G*:193 *s g*. In accord with our previous formula, the percent recombination between *S* and *G* can then be calculated as (126 + 113)/636 = 38 percent.

The calculation of linkage values in polyploids becomes considerably more difficult when genotypes other than those above are mated and when crossing over can occur between a locus and the centromere to produce additional new gametic combinations. It is even further complicated in higher autopolyploids such as autohexaploids. Some of these complexities have been discussed by Fisher and others.

ANEUPLOIDY

The 1916 discovery by Bridges of *Drosophila* lacking an X chromosome (XO males) or bearing an additional sex chromosome (XXY females) gave rise to a long series of studies of *aneuploid* organisms that can be defined as individuals whose chromosome numbers differ from normal by an amount that is *not* an exact multiple of some basic monoploid number (e.g., n + 1, 2n + 1, 2n − 1, 2n + 2, etc.). Bridges' example, as discussed previously (p. 237), was explained on the basis of nondisjunction; that is, both members of a pair of homologous chromosomes fail to disjoin properly during meiosis, so that some gametes are formed containing both homologues while other gametes contain none. Fertilization of such irregular gametes then produces zygotes that either have an additional chromosome (2n + 1) or lack a chromosome (2n − 1). In modern terminology the former individuals are known as *trisomics*, since they possess three homologues of one chromosome, while the latter individuals are called *monosomics*, since they have only one homologue of a particular chromosome. Table 21–3 presents other possible variations of the ordinary diploid *disomic* condition.

TABLE 21–3

Variations involving individual chromosomes within a diploid set, *aneuploidy*

Type	No. Chromosomes Present	Example
disomic (normal)	2n	*AA BB CC*
monosomic	2n − 1	*AA BB C*
nullisomic	2n − 2	*AA BB*
polysomic	extra chromosomes	
(a) trisomic	2n + 1	*AA BB CCC*
(b) double trisomic	2n + 1 + 1	*AA BBB CCC*
(c) tetrasomic	2n + 2	*AA BB CCCC*
(d) pentasomic	2n + 3	*AA BB CCCCC*
hexasomic, septasomic, etc.	2n + 4, 2n + 5, etc.	

As expected aneuploid chromosomes behave differently from disomics during meiosis and produce distributions that are similar to those of polyploids. For example, a trisomic can usually be expected to produce two types of gametes, one with only one homologue and the other with two. Segregation for individual genes present in triplicate then follows the same pattern of segregation as in triploids. In *Drosophila*, for example, a fly trisomic for the 4th chromosome, carrying one recessive gene *eyeless*, would form gametes in the ratio 1(+ +):2(+ *ey*):2(+):1(*ey*), as shown in Fig. 21–7. When mated to a fly homozygous for *eyeless*, the zygotic combinations are:

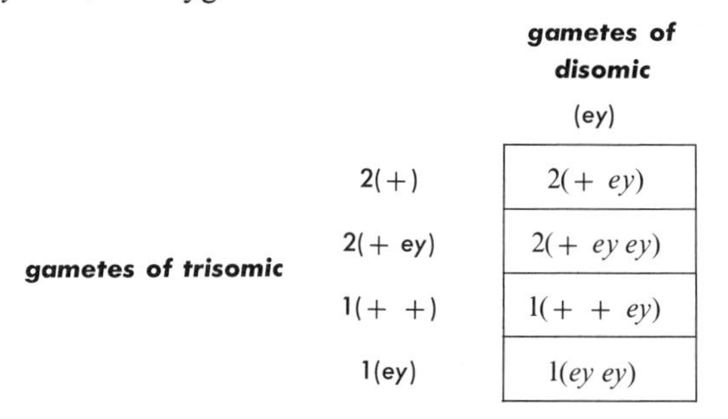

		gametes of disomic (ey)
gametes of trisomic	2(+)	2(+ *ey*)
	2(+ ey)	2(+ *ey ey*)
	1(+ +)	1(+ + *ey*)
	1(ey)	1(*ey ey*)

If the wild-type gene, *eyeless*$^+$, is fully dominant in any dosage, the phenotypes of the offspring are in the proportion 5 wild type:1 eyeless.

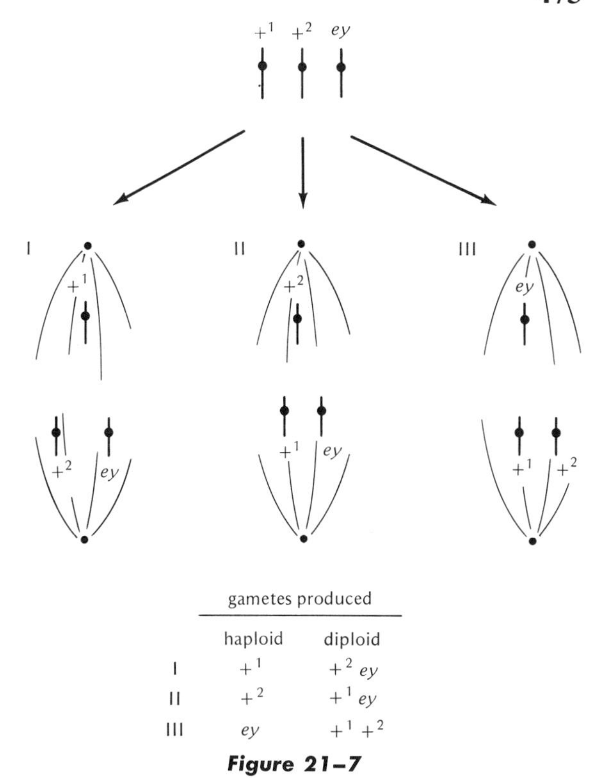

Figure 21–7

Segregation in triplo-IV Drosophila melanogaster *of genotype + + ey, when two 4th chromosomes go to one pole and one goes to the other pole. To distinguish between the two wild-type 4th chromosomes, they have been numbered 1 and 2.*

In plants aneuploids have been studied most extensively in the Jimson (Jamestown) weed *Datura stramonium*. This weed, also known as the "thorn apple" because of its spiked seed capsule, has long been noted for its toxic effect on the nervous system. Blakeslee, after initiating a breeding program with this plant in 1913, soon observed unusual transmission ratios for some of its mutants. The "globe" mutant, for example, appeared to be dominant but was inherited mainly through the female, which gave only about 25 percent globe progeny when selfed or crossed to normal plants. In all, 12 mutants were found to produce these peculiar transmission characteristics. In 1920 Belling discovered that each of these 12 mutant types was trisomic for 1 of the 12 *Datura* chromosomes. In other words, instead of the normal diploid complement of 24 chromosomes, each mutant had 1 extra chromosome, or 25, but a *different* extra chromosome for each phenotypically different mutant. If we name each of the 12 different chromosomes by letter from *A* to *L*, each mutant effect can be represented as caused by a different letter chromosome. Thus an additional dose of the *A* chromosome produces *rolled* leaves, and an additional dose of the *B* chromosome produces *glossy* leaves, etc. Figure 21–8 presents the various effects of the different trisomics on the seed capsule.

Trisomy for these chromosomes helped explain the unusual ratios that Blakeslee had observed. First, many of the extra chromosomes were lost during meiosis, so that the proportion of n + 1 gametes were less than the 50 percent expected. Second, trisomic zygotes usually did not fare as well in germination as disomics. Third, male n + 1

Figure 21–8

Seed capsule of the normal diploid Datura stramonium, *at the top, and those of the twelve kinds of primary trisomics, below. (After Blakeslee.)*

pollen were usually much slower in pollen-tube growth than the n pollen, and fertilized very few ovules. All these factors reduced the frequency of trisomic offspring below the expected ratios.

Since *Datura* normally has 12 chromosomes, the number of such *primary trisomics* was also expected to be limited to 12. However, soon after Belling's discovery, many *secondary trisomics* were found, each producing phenotypes that further accentuated characteristics noted in the initial set of trisomics. Upon cellular examination, the secondaries were cytologically similar to the ordinary 2n + 1 types, but acted quite differently during meiosis. In meiosis a primary trisomic usually forms a trivalent that is in the shape of an open V or in the shape of a bivalent with an attached rod-shaped univalent. The secondary trisomics, however, formed trivalents which were a complete ring of three chromosomes or a bivalent with a doughnut-shaped univalent. To explain these events Belling proposed that the extra chromosome in a secondary trisomic was composed of two identical half-chromosomes attached to a common centromere, a combination later termed an *isochromosome* (see also Fig. 12–11). If we consider that the normal *A* chromosome has chromosome arm 1 on one side of the centromere and chromosome arm 2 on the other side and the normal *B* chromo-

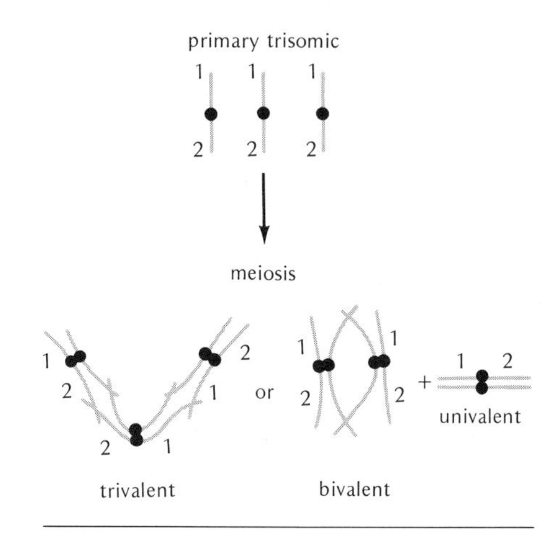

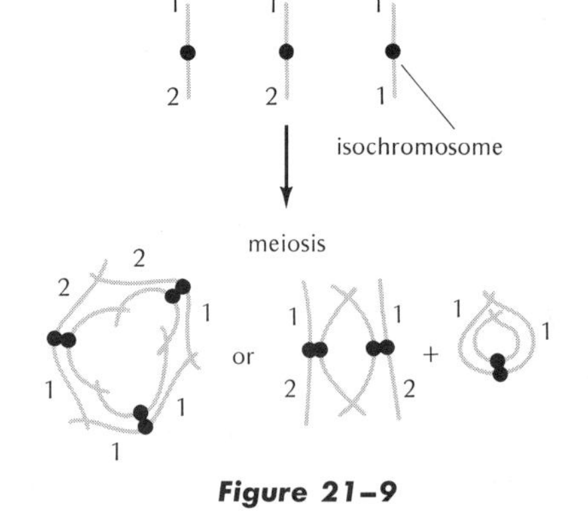

Figure 21–9

Meiotic appearance of homologous chromosomes during meiosis in a primary trisomic (above) and a secondary trisomic (below). Note that the presence of the isochromosome 1 · 1 causes a ring-shaped trivalent or univalent in the secondary trisomic.

TABLE 21–4

The twelve primary trisomics of *Datura stramonium* (names in capital letters) and the known secondary trisomics

Chromosome Name	Chromosome Arms of Primary Trisomic (2n + . . .)	Chromosome Arms of Secondary Trisomic (2n + . . .)	Trisomic Name	Phenotype
		1·1	polycarpic	small plants; narrow leaves; tiny capsules
A	1·2		ROLLED	narrow, in-rolled leaves
		2·2	sugar loaf	large, pointed capsules
		3·3	smooth	small, nearly smooth capsules
B	3·4		GLOSSY	glossy, dark green leaves
		5·5	strawberry	erect habit; compressed capsules
C	5·6		BUCKLING	large leaves with surfaces distorted
		6·6	areolate	leaves with irregular pale areas
		7·7	undulate	spreading habit; narrow, wavy leaves
D	7·8		ELONGATE	erect habit; elongated leaves
		9·9	mutilated	broad, dark leaves; capsules malformed
E	9·10		ECHINUS	large capsules; very long spines
		10·10	thistle	capsules narrow; spines wavy, slender
		11·11	wedge	spreading habit; capsules tapered
F	11·12		COCKLEBUR	spreading habit; small, narrow capsules
		13·13	marbled	leaves mottled
G	13·14		MICROCARPIC	small, depressed capsules
		14·14	mealy	capsules and young leaves glaucous
		15·15	scalloped	spreading habit; narrow, irregular leaves
H	15·16		REDUCED	spines, short, blunt
		17·17	dwarf	small plants
I	17·18		POINSETTIA	leaves long, clustered
		19·19	divergent	very spreading habit
J	19·20		SPINACH	leaves dark, puckered
K	21·22		GLOBE	capsules globose
L	23·24		ILEX	small capsules, leaves sharply toothed

From Avery et al., 1959.

some has arms 3 and 4, etc., an *A* isochromosome may have either chromosome arms 1 and 1 (1·1) or 2 and 2 (2·2), and a *B* isochromosome may have the chromosome formula 3·3 or 4·4. As shown in Fig. 21–9, the activity of an isochromosome during meiosis leads to ring-shaped pairing between the three trisomic homologues or to the formation of an ordinary bivalent and a V-shaped or O-shaped univalent. Table 21–4 gives the chromosome-arm constitutions and phenotypes for the various known primary and secondary trisomics.

(Belling also found trisomics which accentuated characteristics of two different primary trisomics. In this case it appeared as though there had been an exchange of arms between two different nonhomologous chromosomes, e.g., 1·2 and 3·4, to produce a trisomic which had, for example, the chromosome constitution 2·3. Thus chains of five chromosomes could be observed, e.g., 2·1–1·2–2·3–3·4–4·3. This type of trisomic was called *tertiary* and arose through a *translocation*, a mechanism for structural change that will be more fully discussed in Chapter 22.)

GENIC BALANCE

A detailed examination of the plants in Table 21–4 generally shows that the primary trisomic for any chromosome is phenotypically intermediate to the two secondaries. For example, leaf shape as well as capsule size for the primary trisomic for the *A* chromosome is intermediate to that of the secondary trisomics. Extra doses of the 1 arm leads to smaller capsules, narrower leaves, and more erect plants, and extra doses of the 2 arm leads to larger capsules, broader leaves, and shorter plants. The phenotypic effect caused by trisomics thus arises from a change in the number of a particular group of genes localized in a particular chromosome arm. It is as though the normal development of an organism is "balanced" by the presence of gene products in appropriate amounts. An extra dose of one group of genes without added doses of others shifts the balance and produces a new phenotype. Therefore polyploids, although morphologically different from diploids, will not usually show as wide a shift in development as trisomics, since all genes are duplicated equally in the polyploid. For example, capsule size in the *Datura* tetraploid is more globular than that of the diploid but not as large and round as in the 2n + *K* trisomic. Similarly the addition of extra chromosomes to a diploid often has a more drastic "unbalancing" effect than the addition of extra chromosomes to the tetraploid, which has a larger number of homologues (Fig. 21–10).

This relationship of balance between the numbers of different genes or chromosomes was first expressed by Bridges in 1922. In crossing triploid *Drosophila* females to normal males, he obtained a variety of combinations: flies that had three sets of autosomes and one X chromosome ("supermales"); flies that had two sets of autosomes and three X chromosomes ("superfemales"); and an unusual group of flies, neither completely male nor female, with three sets of autosomes and two X chromosomes ("intersexes"). It was evident that two X chromosomes by themselves were insufficient to produce a female, and that the sex of an individual depended on the number of X chromosomes relative to the number of sets of autosomes. The Y chromosome appeared to have very little effect on sex. In Table 21–5, the sexual phenotypes in *Drosophila* are expressed in terms of the ratios of X chromosomes (X) to autosome sets (A). According to Bridges' explanation, the X chromosomes were female-determining and the autosomes were male-determining: an X:A ratio of 1.00 or higher produces females, and an X:A ratio of .50 or lower produces males. The intersex phenotypes result from ratios between these two values.

As may be expected, however, intersex phenotypes can be modified both by environment and by the presence of particular modifying genes. Intersexes raised at high temperatures are more "female-like," while those at low temperatures are more "male-like." Also, different stocks of triploid

TABLE 21–5

Sexual phenotypes of *Drosophila melanogaster* bearing different ratios of X chromosomes and autosomes

X Chromosomes	Autosome Sets (A)	X:A Ratio	Sex
3	2	1.50	superfemale
4	3	1.33	metafemale
4	4	1.00	normal female
3	3	1.00	normal female
2	2	1.00	normal female
2	3	.67	intersex
1	2	.50	normal male
1	3	.33	supermale

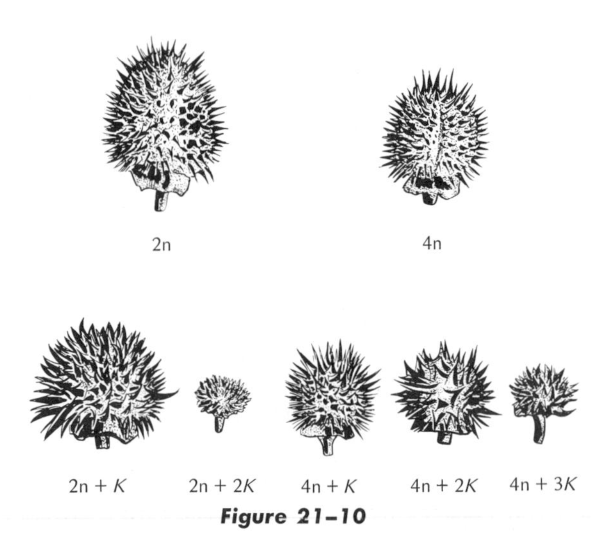

Figure 21–10

Effect on seed-capsule shape of the addition of extra chromosomes to diploids and tetraploids of Datura stramonium. *(From Blakeslee.)*

females can be selected that consistently produce either more feminine or more masculine intersexes among their offspring. Thus, in addition to the 2X:3A cause for intersexes, specific chromosomal genes can doubtlessly modify these phenotypes. In fact, Hildreth discovered a third chromosome gene in *Drosophila melanogaster*, called *doublesex*, which when homozygous transforms individuals that would otherwise be normal females and males (2X:2A, XY:2A) into intersexes. Considerable flexibility in sex-determining mechanism therefore exists, and determination through the sex-chromosome ratio is not necessarily true for all species (see pp. 217–218, 241, and 486).

DETECTION OF LINKAGE GROUPS WITH ANEUPLOIDS

The change in genetic transmission ratios caused by aneuploids has had special value in discovering which genes are associated with which chromosomes. Through the use of nullisomics (completely lacking both homologues of a chromosome), the association between a gene and a chromosome may be quite easy to detect. For example, if the mutant gene is recessive, a homozygous stock of the mutant crossed to all the different kinds of nullisomics of a species will produce a series of F_1 aneuploids, each monosomic for a different chromosome, e.g., $AABB \times AAOO \rightarrow AABO$, $AABB \times OOBB \rightarrow AOBB$, etc., where O designates the absence of a chromosome. If the mutant gene is located on chromosome B (e.g., B^m), the F_1 of the cross with nullisomic B ($AAOO$) will all bear the mutant phenotype (AAB^mO) and the F_1 of crosses with other nullisomics will be wild type (e.g., AOB^mB). Similarly a mutation located on A will produce mutant phenotypes when crossed with the A nullisomic ($OOBB$). In general, a recessive phenotype will appear in the F_1 of any cross made between a mutant stock and a stock nullisomic for the chromosome on which the mutant is located.

If, however, the recessive phenotype is not expressed in hemizygous monosomic condition, F_2 crosses must be made and cytologically analyzed. For example, if the previously mentioned mutation located on chromosome B (B^m) is not phenotypically expressed in the F_1, a cross between F_1 monosomics $AAB^mO \times AAB^mO$ will produce some disomic individuals, AAB^mB^m, which will show the mutant phenotype. Although mutant phenotypes are also produced in crosses derived from stocks of other nullisomic chromosomes, mutants in the B chromosome crosses will always be disomic (Fig. 21-11). As an illustration Sears found that the recessive gene for *sphaerococcum* in wheat did not appear in any of the F_1's derived from crosses with the 21 different nullisomic wheat strains, but did appear in the F_2's in a ratio 3 wild:1 *sphaerococcum*. He then scored the chromosome numbers in the F_2 mutant phenotypes from the different lines and found that these were all disomic in the chromosome 16 line, while in the others only about one-fourth of the recessive phenotypes were disomic. With a high degree of confidence, the gene for *sphaerococcum* could therefore be assigned to chromosome 16.

To detect the linkage group of a dominant mutant gene using nullisomics, crosses are made between the mutant stock and each different nullisomic stock. As diagrammed in Fig. 21-12, the F_1 will show the dominant phenotype in all cases, but depending upon which linkage group it belongs to, this dominant gene will be located on only one of the different monosomic F_1 chromosomes. If the F_1's are then inbred, the family monosomic for the dominant mutant gene will produce mainly mutant phenotypes, while the others will segregate for both mutants and wild type in a ratio 3:1. The low frequency of wild-type phenotypes among the offspring of the monosomic mutant arises because these phenotypes are all nullisomic, relatively less viable, and are therefore formed in considerably less frequency than the monosomic and disomic wild type produced by the other crosses. Unrau, for example, reported that the F_1 from a cross between nullisomic I white-glumed wheat and mutant red-glumed wheat produced 528 red-glumed and 38 white-glumed, or a 93.3 percent frequency of the dominant. The summarized data for the other 16 chromosomes tested showed, on the other hand, 10,975 red to 3,462 white, or a 76

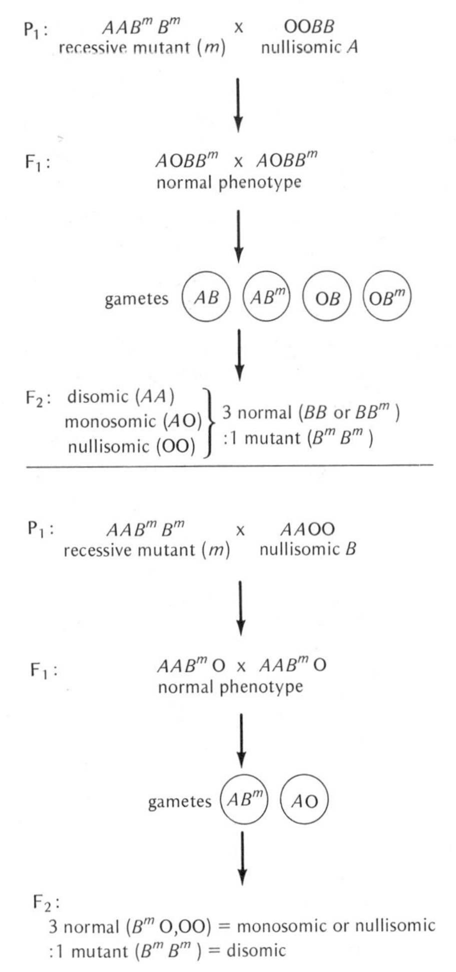

Figure 21–11

Test performed to detect the mutant linkage group when the hemizygous mutant (e.g., B^mO) appears wild type. If, for example, the mutation is located on chromosome B, a cross with a nullisomic for any other chromosome will produce an F_2 in which the mutant phenotype may be either disomic, monosomic, or nullisomic (top). However, a cross with the nullisomic for the linkage group containing the mutant will produce an F_2 in which the mutant phenotype is always disomic (bottom).

percent dominant frequency. The exceptional white-glumed phenotypes that appeared in the F_2 of Unrau's nullisomic I family were undoubtedly nullisomics, as has been cytologically demonstrated by Sears for other such crosses.

In most animals and many plant species, however, nullisomics are extremely difficult to use in these crosses because of their reduced fertility and viability.* In tobacco, for example, nullisomics are completely inviable, and techniques for the detection of linkage with monosomics have been used instead by Clausen, Cameron, and others. Analysis with monosomics differs little from nullisomic analysis and depends upon the availability of strains monosomic for most or all of the chromosomes of a species. When such a variety of monosomic strains is obtained, a mutant disomic strain is crossed to each of these monosomics, and the linkage group of the mutant gene can be detected either in the F_1 or F_2. If the mutant gene is recessive, it will show its effect in those F_1 individuals that are monosomic for the particular chromosome carrying the mutant. Such monosomics have mutant phenotypes since they derive the mutant gene from their disomic parent but do not receive a wild-type dominant allele from their monosomic parent. If the mutation is dominant, then, as with nullisomics, all the F_1 show the mutant phenotype, and further crosses must be made. The procedure for dominants is to inbreed only the monosomic plants of the F_1 and to score the F_2 ratios. The F_2 family derived from the monosomic strain on which the dominant is located (the "critical family") will not show the expected 3 dominant:1 recessive ratio but will, instead, show mostly dominant phenotypes and only a few recessive nullisomics. In tobacco, since nullisomics are inviable, only dominant phenotypes appear in the F_2 critical family.

Because of the unusual genetic ratios they produce, trisomics have also been used to identify genes with a particular linkage group. This identification is made by comparing the observed ratios to those that would be expected if the mutation

*The loss of both members of a pair of chromosomes in a diploid usually means the loss of essential genes located on these chromosomes. When nullisomics are viable, therefore, one can usually assume that their remaining chromosomes contain replicates of the missing genes. Wheat, for example, is an allohexapolyploid (see p. 470) in which the loss of one pair of homologous chromosomes in the nullisomic is apparently compensated by the presence of two similarly acting (homoeologous) chromosome pairs that remain.

were on a disomic chromosome or on a trisomic chromosome. For example, a disomic heterozygote *Rr* in corn, when selfed, should produce offspring in the ratio 3 colored (*R*–):1 colorless (*rr*). Also, when backcrossed to a homozygous recessive, such a heterozygote should produce a ratio 1 colored (*Rr*):1 colorless (*rr*). On the other hand, if the *R* mutation is located on a trisomic combination, e.g., *RRr*, both selfing and backcrossing of this plant to the homozygous recessive will produce completely different ratios. As shown in Table 21–6, McClintock and Hill found a 10:1 ratio when a strain, trisomic for chromosome 10 and also heterozygous *RRr*, was selfed. Furthermore, backcrosses of this trisomic strain to homozygous *rr* yielded 4:1 and 2:1 ratios rather than the expected 1:1 ratio. Control crosses using a disomic strain, *Rr*, produced normal results, and showed that the distorted ratios were not caused by the *R* gene but by the trisomic combination. In this fashion the *R* gene could be definitely located on chromosome 10.

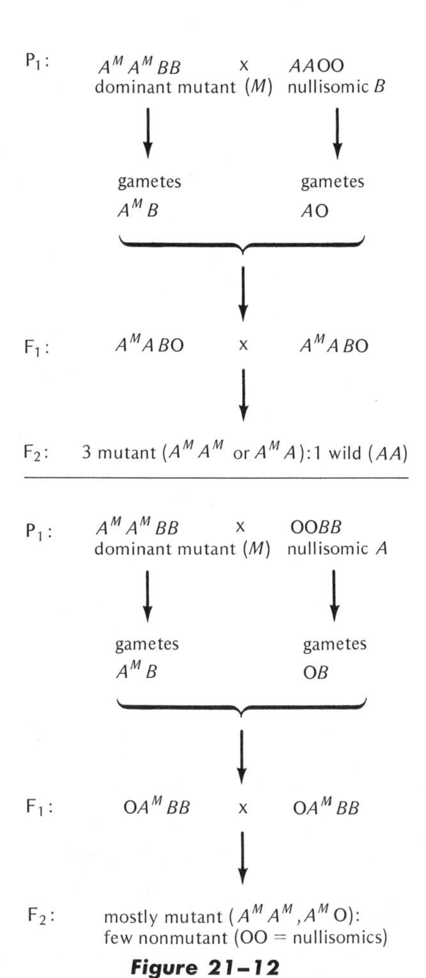

Figure 21–12

Test performed to detect the mutant linkage group of a dominant mutation. If, for example, the mutation is located on chromosome A, a cross with a nullisomic for any other chromosome will produce an F_2 in which the mutant phenotype appears in the ratio 3/4 mutant: 1/4 wild type (top part). If the cross is made with a nullisomic for the mutant linkage group, the F_2 will be mostly mutant, and the only nonmutants will be the nullisomics, which are relatively less viable than the monosomics and disomics (bottom).

TABLE 21–6

Results of crosses for corn plants trisomic for chromosome 10 and heterozygous for R and r, together with crosses of disomic controls

	Progeny		Colored: Colorless Ratio
	Colored	Colorless	
RRr selfed	396	41	10:1
RRr ♀ × *rr* ♂	819	213	4:1
rr ♀ × *RRr* ♂	949	486	2:1
controls			
Rr selfed	608	204	3:1
Rr ♀ × *rr* ♂	1,161	1,196	1:1
rr ♀ + *Rr* ♂	132	135	1:1

After Burnham from data of McClintock and Hill, 1931.

MOSAICS AND CHIMERAS

A change in chromosome number is not necessarily restricted to zygotic changes which affect all cells of an individual. Occasionally a chromosome change occurs after the zygote has been formed which may lead to sections of tissue growing side by side bearing different chromosome constitutions. Such individuals are known as *mosaics* or *chimeras* (after a mythical monster with the head of a lion, the body of a goat, and a serpent's tail), and arise either spontaneously or through chemi-

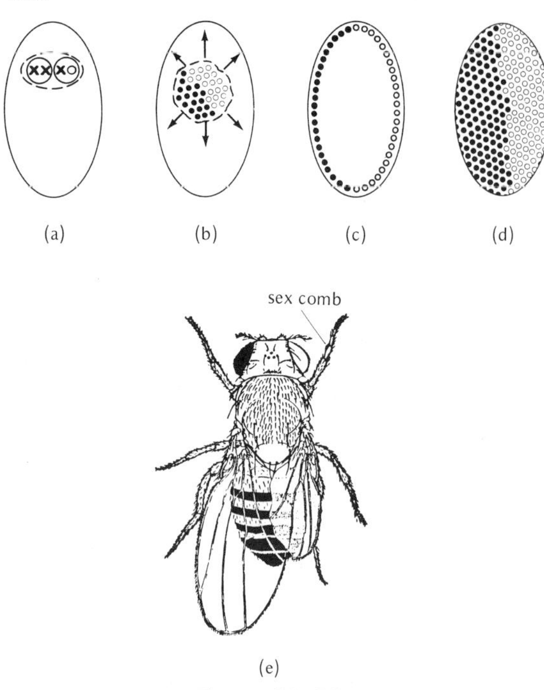

Figure 21–13

Formation of a gynandromorph in Drosophila melanogaster *mosaic for cells carrying two X chromosomes (left side) and one X chromosome (right side). (a) The loss of one X chromosome from one of the two diploid nuclei formed soon after the fertilization of an X-bearing egg by an X-bearing sperm. In (b–d) the XX "female" nuclei are shown as small black circles, and are composed of one X chromosome that carries recessive mutant alleles affecting eye color and wing shape (*white-eosin *and* miniature*), and another X chromosome that carries the dominant wild-type alleles for these mutations. The open circles represent "male" XO nuclei in which the wild-type X-chromosome has been lost and which therefore bear only the mutant X chromosome. (b) The fertilization nuclei undergo nine mitotic divisions while clustered in the center of the egg. (c) These nuclei then migrate to the egg surface and undergo three further divisions. This is the "blastoderm" stage of* Drosophila *embryogenesis, and cell membranes soon form around each nucleus (see also p. 724). (d) View of the blastoderm surface seen from the outside and one of the possible mosaic patterns that occurs. (e) Adult fly showing the XX heterozygous wild-type phenotype on the left, and the XO hemizygous mutant phenotype (mutant eye, miniature wing) on the right. Note that the male sex comb is only present on the right foreleg. [(b–d) after Hotta and Benzer, (e) after drawing by E. M. Wallace in Morgan and Bridges.]*

cal and physical treatment.* In some instances tissue differences in chromosome number follow a uniform developmental pathway such as the "chromosomal diminution" that regularly occurs in *Ascaris* (see p. 26) and *Sciara* (see Fig. 12–15), or the chromosome increase often found in the liver cells of mammals (see Table 4–2). Also, as shown by Lewis and co-workers, supernumerary chromosomes may accumulate more easily in some plant tissues than in others. Most mosaics, however, occur only as a result of relatively rare accidents such as the mosaic blood types caused by occasional embryonic transfer of blood-forming tissue in mammals between fraternal dizygotic twins.

Mosaics with striking appearance can be found in *Drosophila* when one of the sex chromosomes is lost or eliminated (XX → XO) during cell division in the female embryo. If this nondisjunctional event happens early enough in development, as much as one-half of the body may be XO and therefore of the male sexual phenotype (see pp. 238 and 478). Such sexual mosaics are known as *gynandromorphs,* or *gynanders.* If the two initial X chromosomes differ genetically, the male tissue may also show sex-linked traits that differ from female tissue (Fig. 21–13). The sharp distinction between the two types of tissue is especially notable in insects because of the usual lack of hormones which would modify tissues in different localities. Insect tissues of unique phenotypes therefore usually develop "autonomously," unaffected by the presence of surrounding tissues. (An exception to autonomous development found by Sturtevant for the gene *vermilion* laid the basis for the important work of Beadle and Ephrussi on gene action discussed in Chapter 26.)

The frequency with which gynandromorphs

* Some authors restrict the term "chimera" to plants and use "mosaic" only for animals. Others use these terms according to the varieties of tissue mixture that have taken place. Since there is no commonly accepted distinction between these terms, they can both be used to describe the presence of genetically distinct tissues in a single individual. Recently, *DNA chimeras* have been obtained through the insertion of DNA from various species into plasmid DNA (see p. 428)

occur in *Drosophila* can be considerably increased by use of females bearing one of their X chromosomes in the shape of a *ring*. Because of this abnormality, the ring X is frequently lost during embryonic cell divisions leading to the side-by-side presence of male and female tissue in various parts of the body.

Recently Mintz has shown that mosaic mice can be produced by a direct mixing of cells from two genotypically different fertilized eggs. As illustrated in Fig. 21–14, these mixtures are then introduced into the uterus of a "foster mother," and development is completed. Depending on the genotypes of the fused eggs, such mosaic mice may show sharp differences in pigmentation as well as sexual gynandromorphism.

Other causes for gynandromorphs include the "double fertilization" by two sperm of a binucleate egg containing one egg nucleus and one polar body. Such a multinucleate egg may originate through meiotic abnormalities caused by recessive genes, as has been shown for some strains of *Drosophila* and *Bombyx mori*. By whatever mode produced, gynandromorphs differ from intersexes in the sense that gynandromorphs are obviously mosaic, while intersexes presumably have the same chromosome constitution in all cells (see discussion pp. 241–242).

In plants chimeras are more commonly observed than in animals and can easily be produced through grafting or treatment with colchicine. They offer an advantage in determining the embryological origin of different tissues, since the different genotypes of a chimera may each be associated with a distinctly shaped nucleus and cell. Through the colchicine treatment of *Datura*, for example, Satina has "tagged" each of three different germ layers with cells of a distinct size. The growth of the different layers into different tissues was then easily followed.

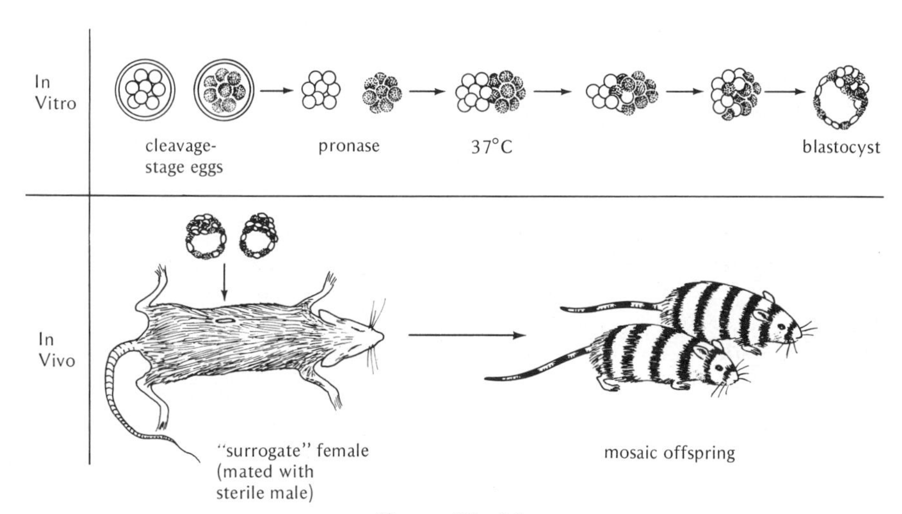

Figure 21–14

Procedure for the production of mosaic ("allophenic") mice from developing eggs that are in early cleavage stages. The membranes of two different eggs are digested away by a pronase solution and the cells are then mixed together into aggregates. These are cultured until the blastocyst stage is reached and then surgically implanted into a pseudopregnant female who had been mated to a sterile (vasectomized) male. If the initial aggregates derive from differently pigmented strains, the progeny produced may show dramatic patterns consisting of transverse bands of alternating pigmentation. (From Mintz.)

CHROMOSOMES OF MAN

The search for the exact number of chromosomes in human cells began in the nineteenth century with estimates ranging from 16 to more than 40. The question was not fully resolved until 1956 when Tjio and Levan demonstrated that there are 46 chromosomes in man, or 23 pairs, of which 22 pairs are autosomes and one pair are sex chromosomes (Fig. 21–15). In males the sex chromosomes consist of an X and Y, and in females they consist of two X's. Through improved tissue culture and cytological techniques, it was found that the shape and size of these chromosomes could be fairly well determined. This is now commonly done by removing and culturing tissue (usually the circulating lymphocytes of venous blood) so that chromosome division figures appear. Since somatic tissue is usually used, these are mitotic cells with a full complement of 46 chromosomes. In mitosis, however, each chromosome has duplicated to form two chromatids that remain attached to the common centromere until anaphase. Observation of such cells at metaphase therefore reveals X-shaped chromosomes if the centromere is in a median position (metacentric) and V-shaped if the centromere is near one end (acrocentric). Depending upon size and centromere position, the 46 chromosomes have been divided into seven groups, as shown in Table 21–7.

Within groups, distinctions between chromosomes have been made on the basis of morphological features such as the secondary constrictions (nucleolar organizers) in some of the long chromosomal arms (e.g., chromosome 9), and the dotlike satellites on the short arms of some acrocentrics. More reliable distinctions, however, can now be achieved by observing the unique banding patterns that appear when chromosomes are stained with Giemsa dye or with various quinacrine dyes

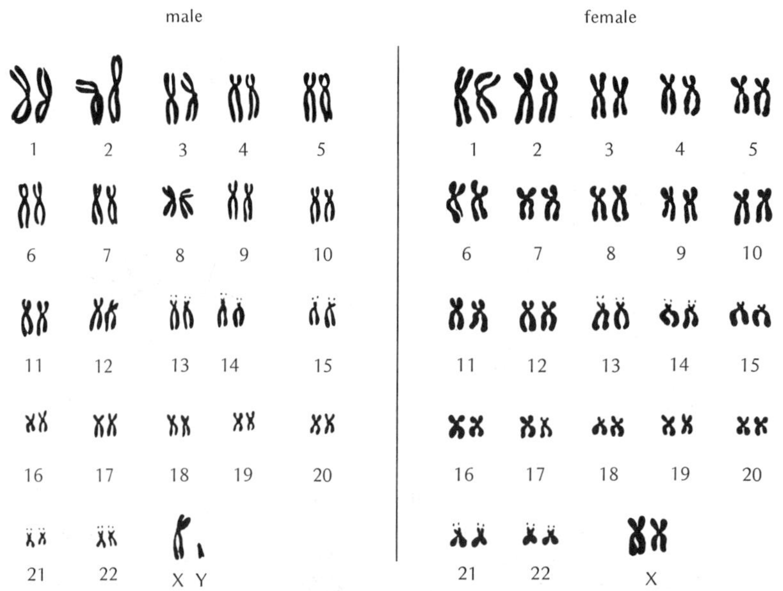

Figure 21–15

Metaphase chromosomes of mitosis in a human male and female arranged according to homologous pairs. (Victor A. McKusick, © 1964. Human Genetics. *By permission of Prentice-Hall, Englewood Cliffs, N.J.)*

TABLE 21–7

Human chromosomes grouped according to size and centromere location

Chromosome Group	Group Designation	Size	Centromere Location
1–3	A	large	approximately median
4–5	B	large	submedian
6–12 and X	C	medium	submedian
13–15	D	medium	acrocentric
16–18	E	short-medium	median or submedian
19–20	F	short	approximately median
21–22 and Y	G	very short	acrocentric

that fluoresce under ultraviolet light. These staining techniques can also be supplemented by differences between chromosomes in their uptake of radioactive labels: when tritiated thymidine is added to the culture medium toward the end of cell division, only the "late replicating" chromosomes or chromosome sections are labeled. By these means it is now possible to make distinctions between every human chromosome, and between many of the chromosomal arms (see Miller et al.).

SEX-CHROMOSOME NUMBERS

In addition to these chromosomal distinctions, there is also an important sex difference in the staining of many somatic nuclei. According to Barr and Bertram, who discovered this in mammals in the 1940s, there was one dark-staining spot in many female cell nuclei, but none in those of males (Fig. 21–16). These spots, named *Barr bodies*, or *chromatin-positive bodies* (also *sex chromatin*), had one unusual feature, which was their unexpected appearance or absence in many cases of sex abnormalities. People affected with *Turner's syndrome,* for example, are externally females but are nevertheless chromatin-negative. On the other hand, those with *Klinefelter's syndrome* are externally males, yet are chromatin-positive. The fact that both these syndromes prevent proper sexual functioning indicated that such sex abnormalities were probably caused by nuclear or chromosomal changes. Evidence for this view was finally provided when chromosome numbers could be scored in affected individuals. Those with Turner's syndrome were demonstrated to have one X chromosome missing (XO) and were therefore monosomic with only 45 chromosomes (Fig. 21–17a). Those with Klinefelter's syndrome had the full male XY complement but were polysomic for additional X chromosomes (Fig. 21–17b). For instance, persons with trisomic Klinefelter's have a complement of 47 chromosomes including XXY; those with tetrasomic Klinefelter's have 48 chromosomes including XXXY, etc.

Thus, contrary to *Drosophila*, the Y chromosome in humans and other mammals is necessary to produce the male phenotype; XO individuals lacking a Y chromosome are female (Turner's syndrome), although sexually nonfunctioning, and XXY individuals are males because of the presence of a Y chromosome, although sterile. No matter how many additional X chromosomes are present, and as many as four have been observed, the presence of a single Y chromosome is sufficient to shift development toward masculinity. However, the presence of two Y chromosomes in males, XYY, does not seem to have any obvious effect on sexual phenotype, although some of these

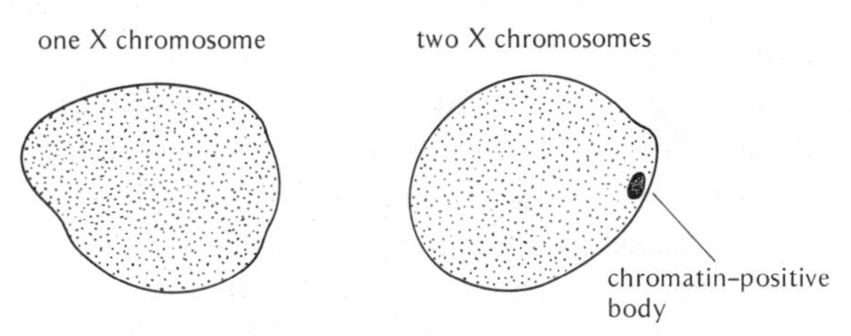

Figure 21–16

Interphase nuclei from buccal smears showing chromatin-negative appearance in normal males (left) and chromatin-positive appearance in normal females (right). Surprisingly, females with Turner's syndrome were found to be chromatin-negative as in normal males, whereas males with Klinefelter's syndrome were chromatin-positive as in normal females.

males have been found to be relatively tall, aggressive, and of less than average intelligence. According to some studies, the percentage of such XYY individuals in prisons and prison hospitals is higher than in the nonprison population (see review of Hook). In general the explanation for these various sex chromosome abnormalities in humans is similar to nondisjunctional explanations offered for aneuploidy found in other organisms: about 2 or more per thousand children arise from gametes in which some form of sex chromosome nondisjunction has occurred (see Jacobs et al., 1974).

BARR BODIES AND HETEROCHROMATINIZATION

The observation that, for each sexual karyotype, the maximum number of Barr bodies was one less than the number of X chromosomes (Table 21–8) demanded explanation. According to the hypothesis presented by Lyon and by Russell (known generally as the "Lyon hypothesis"). Barr bodies result from the inactivation, or "heterochromatinization," of all but one of the X chromosomes. That is, a mammalian female with two X chromosomes will have one of these chromosomes somatically inactive (one Barr body), so that the dosage relationship between her sex chromosomes and autosomes will be the same as in the somatic tissues of the male. Individuals with more than two X chromosomes have all but one of the X chromosomes inactivated in many of the tissues.

In general the inactivation of the additional X chromosome or chromosomes seems to be random, so that some tissues are mosaic for different active X chromosomes. This phenomenon is striking in "calico" cats, in which a female bearing an X chromosome with a black gene (*Y*) and one with a yellow gene (*y*) produces random patches of black and yellow fur. Support for this view is also gained cytologically from a study by Mukherjee and Sinha. They noted that when the leukocytes of a female mule were labeled with tritiated thymidine, about half had an active heavily labeled X chromosome shaped like that of the donkey parent and the other half had an active chromosome shaped like that of the horse parent. In other words, only one X chromosome in each leukocyte appeared to be active, and it was a matter of chance whether this chromosome was from the parental donkey or horse. Since then, other experiments have shown that individual cells of the mule produce clones, each of which is unique for either the maternal or paternal form of the enzyme glucose-6-phosphate dehydrogenase (G-6-PD). Since the gene producing this enzyme is sex-linked, and no single hybrid cell produces both horse and donkey forms of the enzyme, it is clear that each cell has only one X chromosome functional with respect to this gene product.

Quantitatively the effect of X-chromosome inactivation can be detected by measuring the amount of a particular protein produced by a sex-linked gene in individuals carrying different numbers of X chromosomes. In humans such determinations show that the enzymatic activity of glucose-6-phosphate dehydrogenase is the same in individuals of karyotypes XY, XX, XXY, and XXX, although a gene for the production of this enzyme occurs on each X chromosome. Similar phenomena were noted long ago in *Drosophila* by Muller, who used the term *dosage compensation* to describe the fact that the phenotypic effects of sex-linked genes are similar whether the fly carries one (male) or two (female) gene doses. In *Drosophila*, however, the mechanism which compensates for sex-linked gene dosage differences is not the same as in mammals. Rather than only one X chromosome functioning, evidence exists that both *Drosophila* X chromosomes function in female somatic cells, since females heterozygous for sex-linked recessive genes do not show mosaicism (see, for example, Seecof et al.). As yet, the dosage compensation mechanism in *Drosophila* is not known, but a number of regulatory models that might explain it have been suggested (see Lucchesi).

In humans, although many individuals with additional X chromosomes are seriously affected with traits ranging from some degree of physical

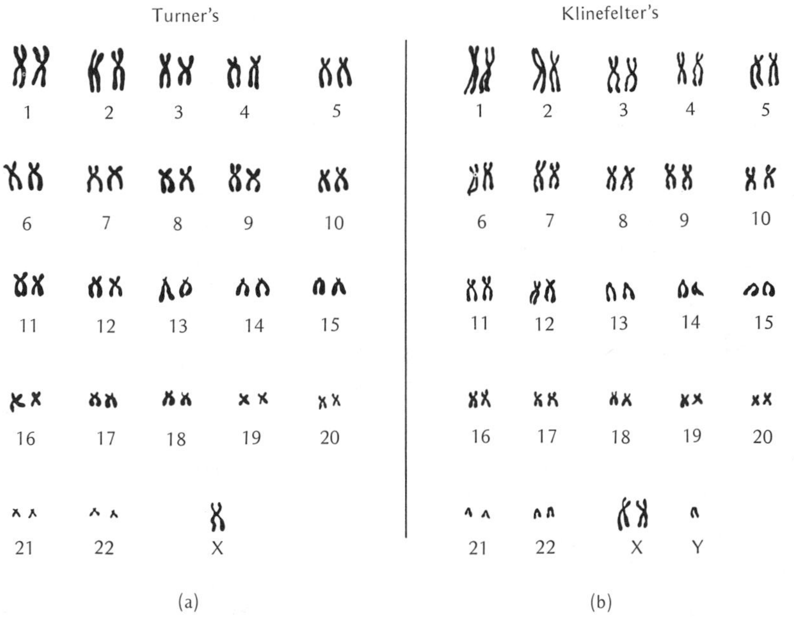

Turner's

Klinefelter's

(a)

(b)

Figure 21–17

(a) Chromosome constitution (karyotype) of an individual with Turner's syndrome showing only one X chromosome without a Y. Such individuals are phenotypic females, but their ovaries do not develop properly. In addition, they are usually characterized by a webbed neck, short stature, and impaired intelligence. (b) Karyotype of an individual with Klinefelter's syndrome bearing two X chromosomes with a Y. Such persons are phenotypic males but, as in Turner's syndrome, have underdeveloped gonads. People affected with Klinefelter's syndrome usually have long limbs, sparse body hair, and mental deficiency. (From photographs provided courtesy of P. Monteleone.)

malformation to mental retardation, they are nevertheless viable because of the dosage compensation mechanism, which seems to reduce but not entirely eliminate the somatic effect of additional X chromosomes. Individuals completely without X chromosomes are never found, probably because of inviability.

TRISOMY FOR CHROMOSOME 21

Aneuploidy for autosomal chromosomes in humans can also occur but is probably often lethal, since a change in autosomal dosage lacks the inactivation mechanism that seems to operate for additional X chromosomes. A few autosomal trisomics are nevertheless found, the most common being for the small chromosome 21. The clinical syndrome associated with this trisomy was first described by Seguin in 1844 but has since been known as mongolian idiocy, or Down's syndrome. It is probably the most common congenital abnormality, with a frequency of one out of 600 births. It is characterized by mental retardation, a short body with stubby fingers, swollen tongue, monkeylike skin ridges on the extremities, and eyelid folds resembling those of Mongolian races. One of its unusual features, known since the end of the nineteenth century, is the higher incidence of the syndrome among children born of older

TABLE 21-8

Sexual aneuploids in man, their sexual phenotypes, and the maximum number of observable Barr bodies*

Sex Chromosomes	Sexual Phenotypes	No. Barr Bodies
FEMALE		
XO (monosomic)	Turner's syndrome	0
XX (disomic)	normal	1
XXX (trisomic)	metafemale (mental abnormalities)	2
XXXX (tetrasomic)	metafemale (mental abnormalities)	3
XXXXX (pentasomic)	metafemale (mental abnormalities)	4
MALE		
XY (disomic)	normal	0
XYY (trisomic)	normal	0
XXY (trisomic)	Klinefelter's syndrome	1
XXYY (double trisomic)	Klinefelter's syndrome	1
XXXY (tetrasomic)	extreme Klinefelter's	2
XXXXY (pentasomic)	extreme Klinefelter's	3

* In this table the disomic condition is considered to be XX in the female and XY in the male.

mothers (Fig. 21-18) but not of older fathers. Because of this characteristic, it was hypothesized by some that the disease was purely environmental in origin, arising from the depletion of nutrient substances in the eggs or uteri of older mothers. In 1959, however, Lejeune, Gautier, and Turpin, as well as others, showed that trisomy for chromosome 21 ($2n + 1 = 47$) was an almost constant corollary of Down's syndrome (Fig. 21-19), and like some of the sex chromosome abnormalities, probably arose through nondisjunction.* In about 5 percent of people with Down's syndrome, the normal disomic complement of 46 chromosomes was found, but in those instances the additional trisomic 21 chromosome had become attached to one of the larger chromosomes. This attachment, or translocation, was one of the first such transfers of chromosomal material found in humans and will be further discussed in Chapter 22.

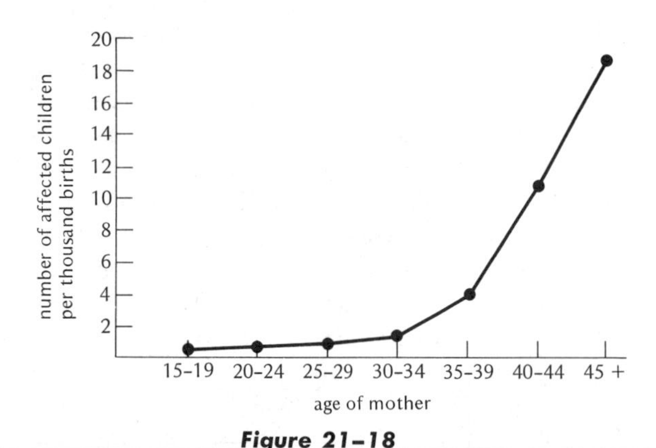

Figure 21-18

Incidence of Down's syndrome in the children of mothers of different ages. (After Lejeune, from data of Carter and Evans.)

Since the chromosomal explanation for Down's syndrome seemed to exclude an environmental origin, the question remained of what would account for the prevalence of this chromosomal defect among the progeny of older mothers. Among the various hypotheses put forth are some based on the fact that each egg of a human female is already partly formed while she is still an unborn fetus. This embryonic egg immediately enters into

*Recently a young chimpanzee showing a large number of traits similar to those shown by human children with Down's syndrome was also reported to be trisomic for one of the small acrocentric chromosomes in the G group (21?).

the prophase of the first meiotic division, but does not divide further until it is released during ovulation. Since ovulation first begins with menstruation during the adolescent period, an egg released by a human female has remained in this prophase stage for no less than 11 or 12 years and as much as 40 or more years in older mothers. A period of 40 years may certainly be long enough to permit weakening of the spindle fibers or other structures necessary for the proper disjunction of chromosomes. Also, older females often undergo dramatic hormonal changes with possible side effects on cellular division.

FURTHER HUMAN ANEUPLOID VARIATIONS

In addition to Down's syndrome, trisomics for other autosomes have also been found, but their effects are usually more serious than for the 21 trisomy. Patau's syndrome, for example, involves a large number of malformations including harelip and cleft palate, along with serious cerebral, ocular, and cardiovascular defects; all associated with trisomy for chromosome 13 in the D group. Compared to the 16 year average survival time for children with Down's syndrome, children with Patau's syndrome live about 3 months. Trisomy for the somewhat small chromosome 18 (Edwards' syndrome) has less drastic effect, but nevertheless reduces an infant's average survival time to 6 months.

Autosomal monosomics, on the other hand, are not known, although deficiencies have been recorded for parts of a chromosome. One well-known example of this kind is the "Philadelphia" chromosome, a G group chromosome (no. 22) which has lost a large portion of its long arm. This chromosome abnormality is found in the bone marrow of 90 percent of patients suffering from chronic myeloid leukemia. Such leukemic individuals are therefore mosaics with respect to the long arm of chromosome 22, and bear both disomic and monosomic tissues. Similarly other mosaics have been observed for both the sex chromosome and autosomal abnormalities (i.e., XO:XX, XO:XY, XXY:XX, 21-21:21-21-21, etc.). Many of these mosaics have symptoms intermediate to that of the pure chromosomal types and appear to have arisen by nondisjunction in the early cell divisions of the zygote.

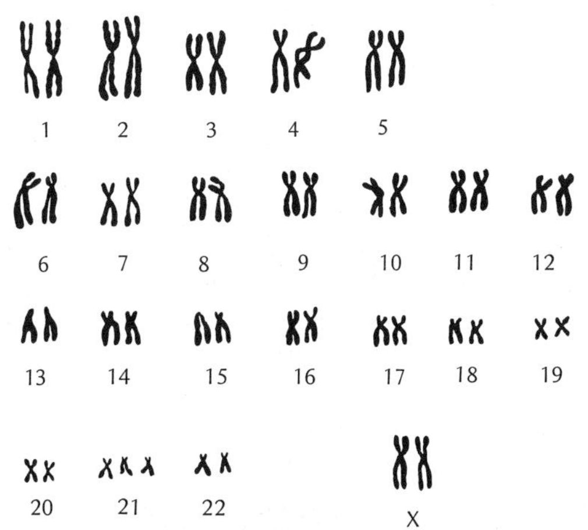

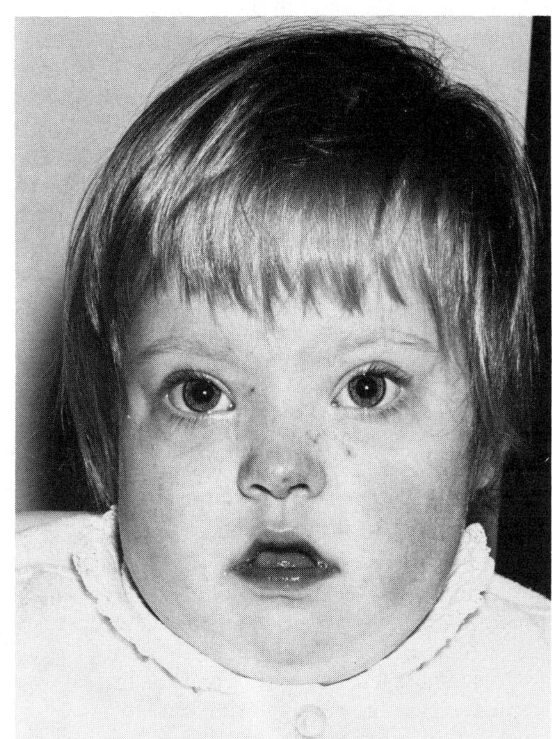

Figure 21–19

Karyotype and photograph of a female with Down's syndrome showing three 21 chromosomes. (Courtesy P. Monteleone.)

It is obvious that almost any human chromosome can be subject to the various nondisjunctional events that lead to aneuploidy, although only a small proportion of such aberrations are viable. Present estimates suggest that perhaps 4 percent of all human conceptions are chromosomally abnormal at fertilization, and only 10 percent of these survive to be born. Of those embryos that do not reach birth, many die because of inability to implant themselves in the uterus (e.g., autosomal monosomics), and others are aborted even if implantation is successful (e.g., most autosomal trisomics). Euploid variations also occur, but are known almost entirely from aborted fetuses, although one triploid has been reported which survived birth for 4 hours. Recent data indicate that about 20 percent of all spontaneous abortions bear chromosomal defects.

PROBLEMS

21-1. A triploid is formed through fertilization of a diploid gamete containing 10 chromosomes with a haploid gamete containing 5 chromosomes. If the average number of chromosomes in the gametes of this triploid is 6, how many of its chromosomes are lost on the average during meiosis?

21-2. Many years ago in Kew Gardens, London, two ornamental primrose plants were crossed, *Primula verticillata* and *P. floribunda*. This cross produced an exceedingly vigorous hybrid that was unfortunately sterile but could be propagated vegetatively by cuttings. After many propagations the hybrid yielded a branch that bore fertile seeds and was then widely distributed under the name *Primula kewensis*. (a) How would you explain the origin of fertile *P. kewensis*? (b) If the two original parents had haploid gametes containing 9 chromosomes in each, how many bivalents would you expect to find in *P. kewensis* during meiosis?

21-3. A fertile allopolyploid, originally produced from a semisterile hybrid by Karpechenko and named *Raphanobrassica*, consists of chromosome sets derived from a radish (*Raphanus sativa*) and a cabbage (*Brassica oleracea*), and can be coloquially called the "rabbage." (Although of scientific interest, the rabbage is an agronomic failure: it has the roots of the cabbage and the leaves of the radish.) If the rabbage has a total number of 36 chromosomes and its radish parent has a gametic number of 9 chromosomes, what is the gametic number in the cabbage parent?

21-4. Two individuals of a tetraploid plant species have the following genotypic constitutions at a particular locus: (a) *AAAa*, (b) *Aaaa*. What are the proportions of the different diploid gametes that each will produce assuming the *A* gene is close to the centromere and that the sister chromatids formed by each chromosome move to opposite poles?

21-5. Among the plants considered in Problem 4, the presence of one gene *A* is sufficient to produce the dominant *A* phenotype. Under these circumstances, what will be the phenotypic ratios among the progeny of each of the given plants if they are self-fertilized?

21-6. An autotetraploid was heterozygous for two gene loci, *AAaa* and *BBbb*, each locus affecting a different character and located on a different set of homologous chromosomes. If dominance is complete, i.e., *Aaaa* = *A* phenotype, *Bbbb* = *B* phenotype, what phenotypic ratios would you expect among the progeny of this autotetraploid when it is self-fertilized?

21-7. Because of the small size of the *Drosophila melanogaster* 4th chromosome, flies can be monosomic or trisomic for this chromosome and still be viable. (a) If a fly disomic for the 4th-chromosome recessive gene *eyeless* were mated to a fly monosomic for this chromosome but normal, what would be the appearance of the F_1? (b) What would be the phenotypic ratios produced by crossing the different F_1 phenotypes?

21-8. If a fly trisomic for the 4th chromosome, and carrying the gene *bent* in all three of its 4th chromosomes, is crossed to a normal fly that is monosomic for the 4th chromosome: (a) What phenotypic and genotypic ratios would you expect in the F_1? (b) What phenotypic ratios would you expect from each different possible type of cross between the F_1 genotypes?

21-9. The n + 1 female gametophytes (embryo sacs) produced by trisomic plants are usually more viable than the n + 1 male gametophytes (pollen grains). If 50 percent of the functional embryo sacs of a selfed trisomic plant are n + 1, but only 10 percent of the functional pollen grains are n + 1, what percentage of the offspring will be: (a) Tetrasomic? (b) Trisomic? (c) Diploid?

21-10. The following pairing relationships between certain chromosomes are observed in two different organisms during meiosis:

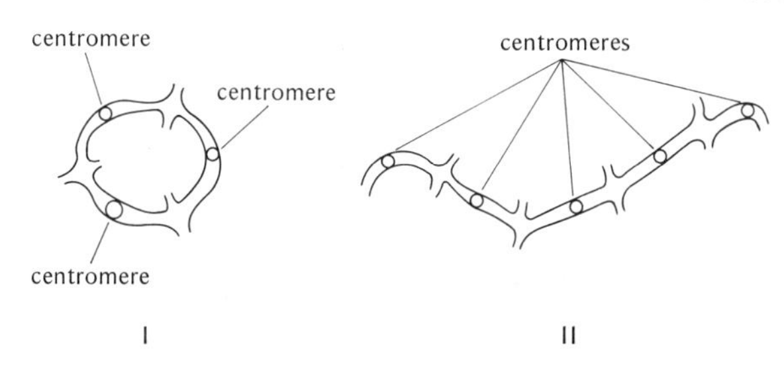

(a) Assign alphabetical letters to different chromosome arms or sections, and give the constitution of each chromosome. (b) For each organism, state the type of abnormality diagrammed.

21-11. According to the sex-determination mechanism proposed by Bridges, what are the sexual phenotypes of the following tetraploid *Drosophila*, all with four sets of autosomes but carrying different numbers of X and Y chromosomes: (a) XXYY? (b) XXXY? (c) YYYY? (d) XXY? (e) XXXXY?

21-12. In the tomato a cross is made between a normal female plant trisomic for chromosome 6 and a disomic potato leaf male (*c/c*). (a) Assuming the gene *c* is located on chromosome 6, give the ratio of normal-leaved diploid plants to potato-leaved plants when a trisomic F_1 is backcrossed to a potato-leaved male. (b) Give the progeny of the same cross assuming that the *c* gene is not located on chromosome 6.

21-13. A tobacco plant homozygous for the genes producing red flowers, and monosomic for a certain chromosome "X," is crossed with a normal plant homozygous for a recessive factor for white flowers, *w/w*. Give the expected results of this cross if: (a) The locus of *w* is on chromosome "X." (b) The locus of *w* is *not* on chromosome "X."

21-14. In the same set of experiments by McClintock and Hill mentioned on page 481, a strain trisomic for chromosome 10 but homozygous for the recessive sugary gene *su* was crossed to a disomic plant homozygous for the dominant starchy allele *Su*. The trisomic plants produced by this cross were then selfed and gave rise to 1758 starchy progeny and 586 sugary progeny. Is the *Su* locus on chromosome 10?

21-15. *Nicotiana tabacum* is presumed to be an allotetraploid originating from a cross between *N. sylvestris* (the *S* genome) and *N. tomentosa* (the *T* genome). Each of the two parental species (2n = 24) contributed 12 chromosomes (*S* and *T*), which then doubled to form the 24 pairs of chromosomes (2n = 48) observed in *N. tabacum* (formula: *SSTT*). A monosomic *N. tabacum* plant is crossed with *N. sylvestris* and produces offspring, some with 36 chromosomes and some with 35 chromosomes. Under cytological examination at meiosis, the 35-chromosome types show 11 bivalents and 13 univalents. (a) Is the monosomic in the *S* or *T* genome? (b) How many bivalents and univalents would you have expected if the monosomic had been in the other genome?

21-16. In *Nicotiana tabacum* there are two gene loci that influence the appearance of the *yellow burley* character, i.e., *A* and *B*. When both of these loci are homozygous for their recessive alleles, i.e., *aabb*, the plant is yellow. When a dominant allele, i.e., *A* or *B*, is present at either locus, the plant is green. (a) If a cross were made between a homozygous green strain monosomic for one of these chromosomes (*ABB*) and a disomic yellow strain (*aabb*), what phenotypes would you expect among the monosomics and disomics of the F_1? (b) If the F_1 monosomics and disomics were inbred (mono × mono, di × di), what ratio of F_2 phenotypes would you expect from each cross?

21-17. Clausen and Cameron crossed stocks of green plants of the *Nicotiana* species above that were monosomic for various chromosomes with a disomic *yellow burley* variety, and then backcrossed each kind of monosomic F_1 plant with the *yellow burley* stock; their results for nine monosomic strains (M–U) were as follows:

Monosomic Strain Lacking Chromosome:	F_2 Phenotypes	
	Green	Yellow-Burley
M	36	9
N	28	8
O	19	17
P	33	9
Q	32	12
R	27	12
S	27	4
T	28	8
U	37	8

Which chromosome contains a *yellow burley* gene? Why?

21-18. A diploid strain of corn homozygous for a dominant gene giving resistance to rust disease (*SS*) was crossed by Rhoades to eight different strains of corn, each trisomic for a different chromosome but all susceptible to rust disease. Trisomic plants were selected from each of these eight progenies and backcrossed separately to disomic susceptible plants (*ss*). The F_2 were then tested for resistance and susceptibility to rust disease and the results were as follows:

Chromosome for Which the Initial Susceptible Strain was Trisomic	Resistant	Susceptible
2	239	217
3	246	271
5	279	286
6	375	319
7	191	183
8	198	225
9	279	284
10	418	763

Explain on which chromosome the gene for rust resistance is probably located.

21-19. How would you explain the origin of a gynandromorph bee whose female parts are a hybrid of both parental genotypes, but whose male parts are: (a) Of paternal origin? (b) Of maternal origin?

21-20. What proportion of offspring affected by mongolism (Down's syndrome) will be produced by the mating of two individuals that are trisomic for chromosome 21?

21-21. A woman claimed to produce a child by parthenogenesis. (a) What should its sex be? (b) How would you further test the truth of her claim?

21-22. If an individual with Klinefelter's syndrome (XXY) was found to be fertile in a mating with a normal female, what proportion of his sons would be expected to have Klinefelter's syndrome?

21-23. Although a recently identified XYY individual (human) appears as a normal male, he produced some abnormal offspring. Enumerate the different possible sexual *genotypes* and *phenotypes* and their expected frequencies among the progeny resulting from matings between XYY individuals to normal females.

21-24. If a red-green colorblind child with Klinefelter's syndrome was born to phenotypically normal parents, in which parent and at which meiotic division did the nondisjunctional event occur?

21-25. In mice the sex-determination mechanism is similar to that of man, but XO aneuploid individuals are functional females. (a) What sexual genotypic ratios would you expect to be produced in a mating between an XO female and a normal male? (b) If an XXY mouse were fertile, what genotypic ratios would result from a mating between the XO female and XXY male?

21-26. An individual with Turner's syndrome was found in a family that was also segregating for red-green colorblindness as follows (colorblind individuals are shaded):

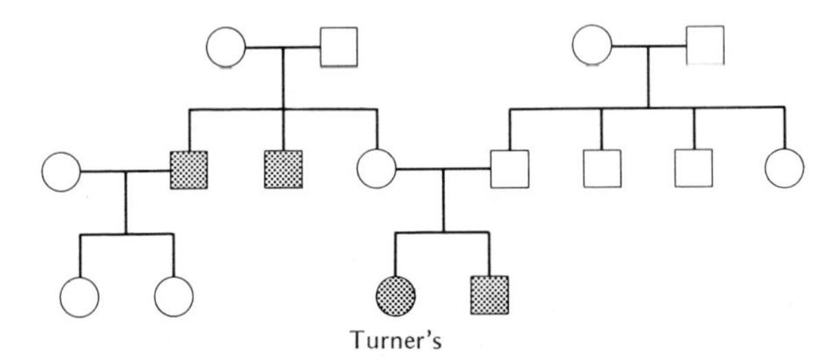

Which X chromosome, maternal or paternal, is missing in the individual with Turner's syndrome?

21-27. Control over black and yellow coat colors in cats is known to be exercised at a sex-linked locus by a pair of alleles *Y* and *y*. Females heterozygous for both alleles are mottled black and yellow, or "calico," and this mosaicism may be explained according to the Lyon hypothesis (p. 486). Male "calico" cats nevertheless occur but are very rare and are always sterile. Can you explain their origin by a maternal nondisjunctional event?

21-28. The following represents the pedigree of a *male* calico cat (shaded symbol):

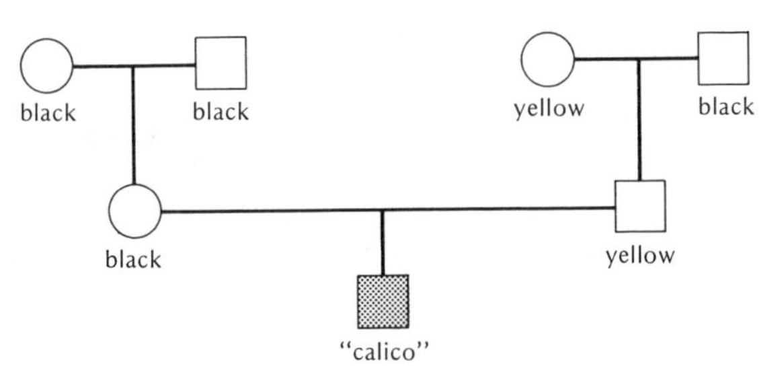

Explain in which parent the nondisjunctional event occurred.

21-29. Philip and co-workers have reported an unusual human pedigree in which the father of two monozygotic twin sisters was colorblind, but these otherwise normal twins are discordant for colorblindness. How would you explain this discordance in the simplest fashion?

REFERENCES

AVERY, A. G., S. SATINA, and J. RIETSEMA, 1959. *Blakeslee: The Genus Datura.* Ronald Press, New York.

BARR, M. L., 1966. The significance of the sex chromatin. *Intern. Rev. Cytol.*, **19**, 35–95.

Barr, M. L., and E. G. Bertram, 1949. A morphological distinction between neurones of the male and female, and the behavior of the nucleolar satellite during accelerated nucleoprotein synthesis. *Nature*, **163,** 676–677.

Blakeslee, A. F., 1934. New Jimson weeds from old chromosomes. *Jour. Hered.*, **25,** 80–108.

Blakeslee, A. F., J. Belling, and M. E. Farnham, 1923. Inheritance in tetraploid *Daturas*. *Bot. Gaz.*, **76,** 329–373.

Bridges, C. B., 1925. Sex in relation to chromosomes and genes. *Amer. Naturalist*, **59,** 127–137. (Reprinted in Peters' collection; see References, Chapter 1.)

Bridges, C. B., and E. G. Anderson, 1925. Crossing over in the X chromosomes of triploid females of *Drosophila melanogaster*. *Genetics*, **10,** 418–441.

Brown, W. V., 1972. *Textbook of Cytogenetics*. C. V. Mosby, St. Louis.

Burnham, C. R., 1962. *Discussions in Cytogenetics*. Burgess, Minneapolis, Chaps. 6–9.

Caspersson, T., and L. Zech, (eds.), 1973. *Chromosome Identification*. Academic Press, New York.

Clausen, R. E., and D. R. Cameron, 1944. Inheritance in *Nicotiana tabacum*. XVIII. Monosomic analysis. *Genetics*, **29,** 447–477.

Clausen, R. E., and T. H. Goodspeed, 1925. Interspecific hybridization in *Nicotiana*. II. A tetraploid *glutinosa-tabacum* hybrid. An experimental verification of Winge's hypothesis. *Genetics*, **10,** 278–284. (Reprinted in Levine's collection; see References, Chapter 1.)

Cohen, M. M., and M. C. Rattuzi, 1971. Cytological and biochemical correlation of late X-chromosome replication and gene inactivation in the mule. *Proc. Nat. Acad. Sci.* **68,** 544–548.

Darlington, C. D., and K. Mather, 1949. *The Elements of Genetics*. Allen and Unwin, London.

Dawson, G. W. P., 1962. *An Introduction to the Cytogenetics of Polyploids*. F. A. Davis, Philadelphia.

De Winton, D., and J. B. S. Haldane, 1931. Linkage in the tetraploid *Primula sinensis*. *Jour. Genet.*, **24,** 121–144.

Fankhauser, G., 1955. The role of nucleus and cytoplasm. In *Analysis of Development,* B. H. Willier, P. A. Weiss, and V. Hamburger (eds.). Saunders, Philadelphia, pp. 126–150.

Fisher, R. A., 1947. The theory of linkage in polysomic inheritance. *Phil. Trans. Roy. Soc. Lond.* (B), **233,** 55–87.

Garber, E. D., 1972. *Cytogenetics: An Introduction*. McGraw-Hill, New York.

Hamerton, J. L., 1971. *Human Cytogenetics*, 2 vols. Academic Press, New York.

Hildreth, P. E., 1965. Doublesex, a recessive gene that transforms both males and females of *Drosophila* into intersexes. *Genetics*, **51,** 659–678.

Hook, E. B., 1973. Behavioral implications of the human XYY genotype. *Science*, **179,** 139–150.

Hotta, Y., and S. Benzer, 1972. Mapping of behavior in *Drosophila* mosaics. *Nature*, **240,** 527–535.

Jacobs, P. A., M. Melville, S. Ratcliffe, A. J. Keay, and J. Syme, 1974. A cytogenetic survey of 11,680 newborn infants. *Ann. Hum. Genet.*, **37,** 359–376.

Jacobs, P. A., W. H. Price, and P. Law (eds.), 1970. *Human Population Cytogenetics*. Edinburgh Univ. Press, Edinburgh.

Khush, G. S., 1973. *Cytogenetics of Aneuploids*. Academic Press, New York.

Lejeune, J., 1964. The 21 trisomy—current stage of chromosomal research. *Progr. Med. Genet.*, **3,** 144–177.

Lejeune, J., M. Gautier, and R. Turpin, 1959. Étude des chromosomes somatiques de neuf infants mongoliens. *C. R. Acad. Sci., Paris*, **248,** 1721–1722. (Translated into English and reprinted in Boyer's collection; see References, Chapter 1.)

Lewis, K. R., and B. John, 1963. *Chromosome Marker*. Churchill, London.

Lewis, W. H., R. L. Oliver, and T. J. Luikart, 1971. Multiple genotypes in individuals of *Claytonia virginica*. *Science*, **172,** 564–565.

Lucchesi, J. C., 1973. Dosage compensation in *Drosophila*. *Ann. Rev. Genet.*, **7,** 225–237.

Lyon, M. F., 1961. Gene action in the X-chromosome of the mouse (*Mus musculus* L.), *Nature*, **190,** 372–373.

———, 1962. Sex chromatin and gene action in the mammalian X-chromosome. *Amer. Jour. Hum. Genet.*, **14,** 135–148. (Reprinted in Boyer's collection; see References, Chapter 1.)

———, 1972. X-chromosome inactivation and developmental patterns in mammals. *Biol. Rev.* **47,** 1–35.

McClintock, B., and H. E. Hill, 1931. The cytological identification of the chromosome associated with the R-G linkage group in *Zea mays*. *Genetics*, **16,** 175–190.

McKusick, V. A., and R. Claiborne, (eds.), 1973. *Medical Genetics*. HP Publ. Co., New York. See Section I: "Chromosomes and their disorders," containing papers on human chromosome identification, chromosome abnormalities, and chromosomes and cancer.

MILLER, O. J., D. A. MILLER, and D. WARBURTON, 1973. Application of new staining techniques to the study of human chromosomes. *Progr. Med. Genet.*, **9,** 1–47.

MINTZ, B., 1967. Gene control of mammalian pigmentary differentiation. I. Clonal origin of melanocytes. *Proc. Nat. Acad. Sci.*, **58,** 344–351.

MORGAN, T. H., and C. B. BRIDGES, 1919. Contributions to the genetics of *Drosophila melanogaster*. I. The origin of gynandromorphs. *Carnegie Inst. Wash. Publ. No. 278*, Washington, D.C., pp. 1–122.

MUKHERJEE, B. B., and A. K. SINHA, 1964. Single-active-X hypothesis: Cytological evidence for random inactivation of X-chromosomes in a female mule complement. *Proc. Nat. Acad. Sci.,* **51,** 252–259.

RILEY, R., 1974. Cytogenetics of chromosome pairing in wheat. *Genetics*, **78,** 193–203. (Proc. XIII Intern. Congr. Genet.)

RUSSELL, L. B., 1961. Genetics of mammalian sex chromosomes. *Science*, **133,** 1795–1803.

SATINA, S., 1959. Chimeras. In *Blakeslee*: *The Genus Datura,* A. G. Avery, S. Satina, and J. Rietsema, Ronald Press, New York, pp. 132–151.

SEARS, E. R., 1953. Nullisomic analysis in wheat. *Amer. Naturalist*, **87,** 245–252.

SEECOF, R. L., W. D. KAPLAN, and D. G. FUTCH, 1969. Dosage compensation for enzyme activities in *Drosophila melanogaster. Proc. Nat. Acad. Sci.*, **62,** 528–535.

SMITH, H. H., 1974. Model systems for somatic cell plant genetics. *Bioscience*, **24,** 269–276.

STEBBINS, G. L., 1950. *Variation and Evolution in Plants.* Columbia Univ. Press, New York.

STERN, C., 1968. *Genetic Mosaics and Other Essays.* Harvard Univ. Press, Cambridge.

TJIO, J. H., and A. LEVAN, 1956. The chromosome number of man. *Hereditas*, **42,** 1–6. (Reprinted in the collections of Boyer and of Levine; see References, Chapter 1.)

UNRAU, J., 1950. The use of monosomes and nullisomes in cytogenetic studies of common wheat. *Scient. Agric.*, **30,** 66–89.

22
CHANGES IN CHROMOSOME STRUCTURE

The presence and arrangement of many genes on a single chromosome permit a change in genetic information to occur not only through a change in chromosome number but also through a change in chromosome structure. In these cases the number of chromosomes usually remains the same, but their genetic material becomes altered through the loss, gain, or rearrangement of particular sections.

In origin, such structural changes are now generally believed to be caused by breaks in the chromosome, or in its subunit, the chromatid. Each break produces two ends which may then follow three alternative paths:

1. They may remain *ununited*, thereby leading to the eventual loss of that chromosomal segment which does not contain the centromere.
2. Immediate reunion or *restitution* of the same broken ends may occur, leading to reconstitution of the original chromosome structure.
3. One or both ends of one particular break may join those produced by a different break, causing an *exchange*, or *nonrestitutional union*.

Depending upon the number of breaks, their locations, and the pattern in which broken ends join together, a wide variety of structural changes are possible. The present chapter will consider some of these changes with emphasis on their genetic consequences.

DEFICIENCIES

Deficiencies or *deletions* represent a loss in chromosomal material and were the first chromosome aberrations indicated by genetic evidence. This evidence, presented by Bridges in 1917 in *Drosophila melanogaster*, showed a deletion of the X chromosome that included the *Bar* locus. Bridges crossed an XXY female bearing *white-eosin* (w^e) on one of its X chromosomes to a male carrying both *white* and *Bar* ($w\ B$). Since w is an allele of w^e and *Bar* is semidominant, half the resultant females arising through normal disjunction of their maternal X chromosomes should appear as eosin-eyed and intermediate bar-eyed ($w^e +/w\ B$), while the other half should be simply intermediate bar-eyed ($+ +/w\ B$). Females that arise through secondary nondisjunction (p. 239) should be wild type. Among 160 females that Bridges examined from this cross, all conformed to the expected phenotypes, except one that was eosin-eyed but of normal eye shape ($w^e +/w\ ?$). This female had obviously inherited its w gene from its father but had not inherited his *Bar* gene. Moreover, when outcrossed to a wild-type male, none of the offspring of this exceptional female showed the *Bar* effect.

That this was a deletion rather than a change to B^+ was indicated by the following points:

1. The sex ratio among the offspring of the exceptional female was 2 ♀ : 1 ♂, as though the males inheriting the chromosome with the missing *Bar* gene were lethal. This is not surprising, since the male is hemizygous for the X, and lethality would be expected if an appreciable portion of its X chromosome was missing. The white locus carried by the parental $w^e +/w -$ female was, of course, mutant but viable and could cross over to the normal X chromosome to produce viable white-eyed sons. However, since such white-eyed sons could only arise through crossing over, their frequency would be expected to be rarer than that of the eosin-eyed sons produced by the normal noncrossover chromosome. This was indeed what Bridges found; less than half the viable sons produced were white-eyed and apparently restricted to crossover products. Thus "the deficiency was then not simply for *Bar* but was also a deficiency for one or more genes whose action is essential to the life of the fly."

2. Bridges then reasoned that if genes were missing in the deleted area, females bearing a normal X chromosome with the recessive alleles for these missing genes, in addition to a deleted chromosome, should now show the recessive phenotype. In other words, such heterozygous females should actually be hemizygous for the deficient area. He therefore crossed females bearing the *Bar* deletion to males carrying recessive genes near the *Bar* locus, such as *forked*, *rudimentary*, and *fused*. The crosses using *rudimentary* and *fused* resulted in the appearance of only wild-type females, signifying that the deficient chromosome carried wild-type alleles for these genes. In the cross using *forked*, however, half the females produced were phenotypically forked, and these females were also those carrying the deficient chromosome! Thus the *forked* locus was included within the deficiency, and the phenotypic appearance of forked bristles in these females was caused by a missing wild-type allele. Since the recessive nature of *forked* had not changed, this phenomenon was called *pseudodominance*.

3. Crossing over was completely absent in the deficient region, as evidenced by the observation that females heterozygous for the deficiency never produced crossover products between *forked* and *Bar*. One could, of course, argue that male crossover products would be lethal because they would still carry a small section of the deficiency (Fig. 22–1a). In such cases, however, female crossover products should be viable, since they carry a normal X chromosome (Fig. 22–1b), but these too were never found. Outside the deficient area, linkage relationships were hardly affected.

Since Bridges' discovery, further deficiencies have been located in *Drosophila*, corn, and other

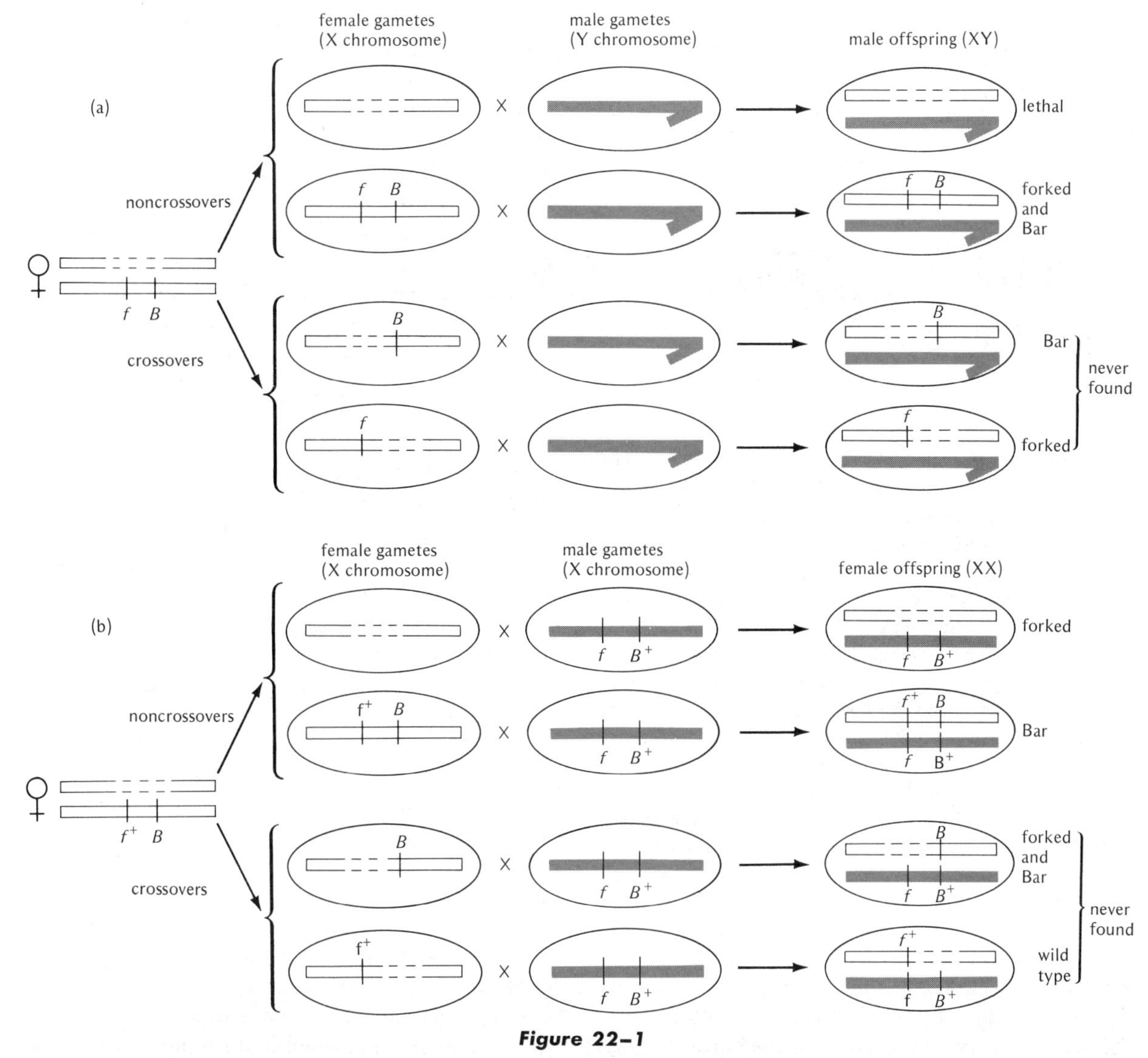

Figure 22–1

Diagram showing attempts to discover crossovers in the deficient forked-Bar region (dashed lines) of the X chromosome. In (a) male offspring of a female carrying one deficient chromosome and a forked-Bar X chromosome are scored, but recombinants are not observed. In (b) a female bearing the deficient chromosome and an f^+ B chromosome is mated to a forked male (f B^+), and recombinant female offspring are not observed.

organisms. *Drosophila* has the special advantage of permitting detailed analysis of the banding patterns in the giant salivary chromosomes, thereby physically localizing particular deficiencies. As shown in Fig. 22–2a, b, larval heterozygotes for deficiencies show a "buckling" of the normal homologue in the deficient area because of the close pairing between the two homologues in salivary nuclei. These bucklings denote *intercalary* (*interstitial*) deficiencies in which a nonterminal portion of the chromosome has been lost. Occasionally, as shown in Fig. 22–2c, d, *terminal*

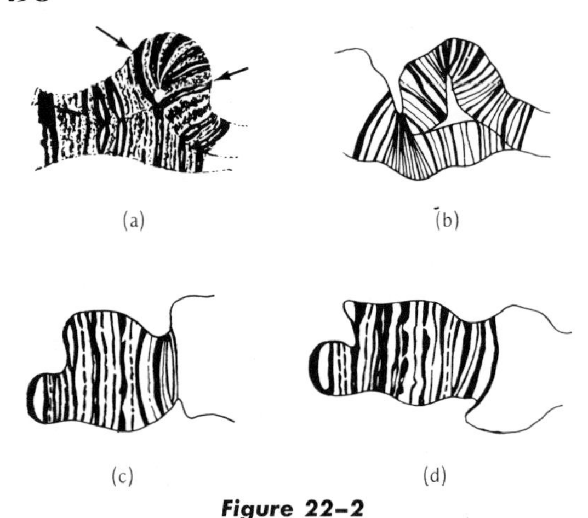

Figure 22–2

Appearance of various deficiencies, both intercalary and terminal, in the salivary chromosomes of Drosophila *larvae heterozygous for the deficiency and a normal chromosome. (a) The* Notopleural *deficiency involving a loss of about 50 bands in the right arm of the 2nd chromosome. The normal chromosome shows "buckling" with arrows pointing to the region missing in the deficient chromosome (after Bridges, Skoog, and Li). (b) A* Notch *deficiency involving a loss of about 45 bands on the X chromosome (after Slizynska). (c) A terminal deficiency on the X chromosome involving a loss of about 8 bands, including the loci* yellow *and* achaete. *This deficiency is lethal in males and homozygous deficient females. (d) A fully viable sex-linked deficiency involving a loss of only 4 bands. (c and d after Demerec and Hoover.)* P. A. Roberts *(1975,* Genetics, **80:** *135–142) suggests that these and other "terminal" deficiencies are capped by minute telomeres, and are really intercalary.*

deficiencies are noted which can be defined as a loss of the tip of the chromosome. Muller has hypothesized that the very tip of the chromosome, called the *telomere*, has a unique function in preventing adhesion between chromosome ends which would otherwise stick together. According to Muller, all deficiencies that are capable of being maintained in stocks must therefore be intercalary and cannot include loss of the telomere. The relative rarity of terminal deficiencies seems to support Muller's argument, although the apparent absence of telomeres in some terminal deficiencies may indicate that a "sticky" end may occasionally "heal."

In general, whether deficiencies are intercalary or terminal, they can produce unique phenotypic effects of their own. This has been observed for the *Drosophila* deficiencies known as *Beaded*, *Delta*, *Gull*, *Minute*, *Notch*, *Notopleural*, *Plexate*, and others which act as dominant genes in heterozygous condition. *Notch*, for example, produces a distinct indentation of the wing margin in heterozygous females and is lethal in males, thereby acting as a sex-linked recessive lethal with dominant phenotypic effect. Its precise X chromosome location can be gathered from the observation that all the many *Notch* deficiencies produce pseudodominance of the *facet* gene. Since these *Notch* deficiencies either overlap or are confined to a particular salivary region, the $3C_7$ band (Fig. 22–3), the *facet* gene must therefore be localized within this band. Studies such as this have led to the conclusion that each band, which is really a chromomere multiplied many times over, or the "interband" between such bands, may represent a distinct gene. Since there are approximately 5000 observable chromosome bands in *Drosophila melanogaster* (see p. 26), this species has been considered to possess on the order of 5000 genes.

A particular phenotypic effect associated with a deficiency is not, however, restricted to a particular locus or even to a particular chromosome. The *Minute* effect, for example, is associated with a large number of deficiencies found throughout the entire *Drosophila* genome. These deficiencies usually produce short, fine bristles and small body size as heterozygotes, but are lethal in homozygous or hemizygous condition.

In humans the chromosomal deficiencies known so far usually have identifiable and deleterious effects. This includes the "Philadelphia" 22 chromosome associated with chronic myelogenous leukemia (p. 489), and a deletion in the short arm of chromosome 5 first noted by Lejeune in 1963 and since reported many times. This latter deficiency is recognized by an entire set of symptoms called the "cri-du-chat" syndrome. As can be discerned from its name, it includes a mewing catlike cry during infancy as well as widely spaced eyes with epicanthic folds, unique facial features, and various physical and mental retardations.

In corn and other seed plants, deficiencies have

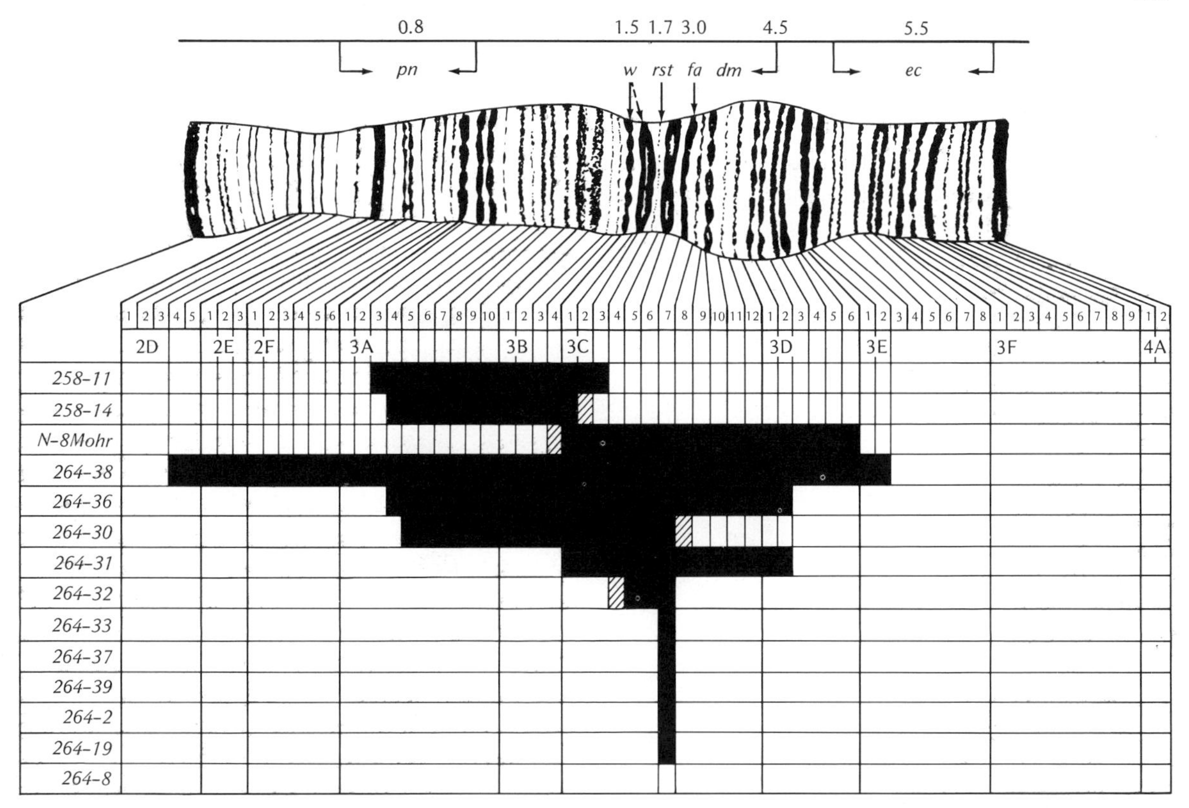

Figure 22–3

Genetic and salivary chromosome maps of the white-Notch region of the X chromosome of Drosophila melanogaster. *At the bottom of the figure are areas missing (black) in 14 deficiencies in that region. Pseudodominance occurs when a female heterozygous for a deficiency shows a mutant phenotype if she is carrying a recessive gene on her normal X chromosome that is covered by that deficiency. Thus deficiency* N-8, 264–38, *and others show pseudodominance for* facet, *indicating that the* $facet^+$ *gene was eliminated by the deficiency. In deficiencies* 264–33, 37, 39, 2, *and* 19, *only a deficiency for the* $3C_7$ *salivary band is observed. Since such individuals can show pseudodominance for* facet, *the* facet *gene is most likely located in band* $3C_7$. *(After Slizynska.)*

been observed (Fig. 22–4) but are usually restricted in their transmission because of pollen sterility. That is, the male haploid gametophyte generation is easily rendered ineffective when chromosome sections are missing. The female egg, on the other hand, has numerous nuclei (Fig. 3–12) which can supplement the deficiency. Nevertheless both Creighton and McClintock have noted occasional small deficiencies in corn that are homozygous viable and act as recessive mutations.

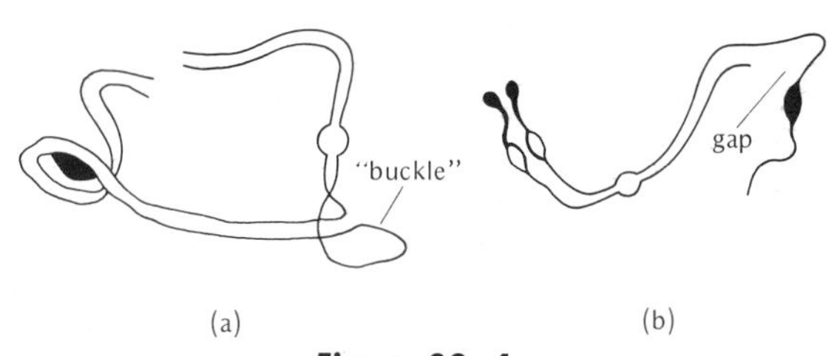

Figure 22–4

In corn, paired homologues at the pachytene stage of meiosis may show a buckling in the deficient area (a) or an actual gap that is either intercalary or terminal (b). (After Burnham.)

Not too surprisingly procaryotes also produce deficiencies, and these can be mapped by techniques that show the simultaneous loss of gene activity for a number of closely linked genes. In *E. coli*, deletions of about 1 percent or more of the bacterial chromosome are recorded, whereas deletions of as much as 20 percent have been observed in viruses such as λ. Some of these deletions can be dramatically visualized on the molecular level through electron micrographs of "heteroduplexes" (p. 446) consisting of a strand of normal DNA "annealed" to a DNA strand from the deleted strain. The molecule formed by this process appears to be double-stranded except for the deleted section where the normal DNA strand cannot find complementary bases for pairing. At those points the DNA is single-stranded and looks like a collapsed loop or "bush," analogous to the "bucklings" observed in heterozygotes for *Drosophila* and corn deletions (Fig. 22–5).

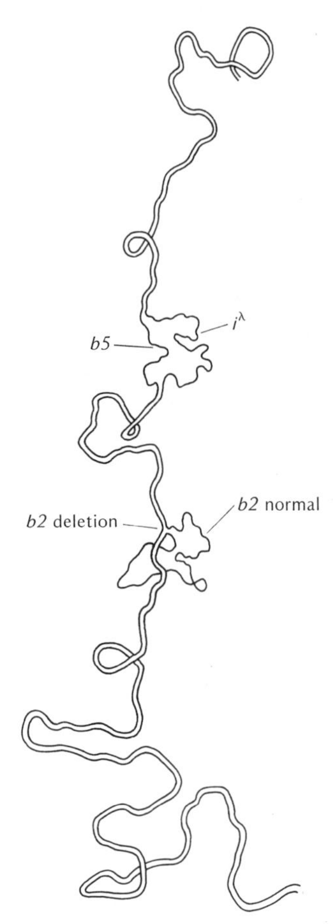

Figure 22–5

Diagram derived from an electron micrograph showing a heteroduplex λ DNA molecule in which one strand carries a deletion for the b2 section. A single-strand loop is formed in this area because the wild-type DNA strand is unable to pair with the deleted strand. Lack of pairing also occurs in the b5–iλ section because the nucleotide sequences of the two strands are not homologous at this point. (After Westmoreland, Szybalski, and Ris.)

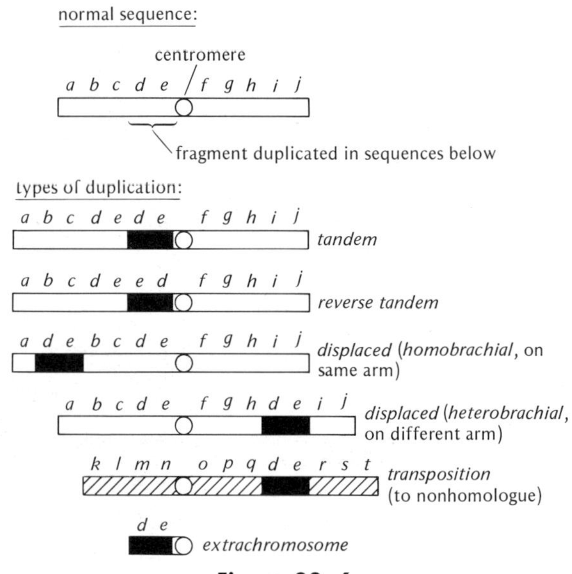

Figure 22–6

Different types of duplication possible for a particular section of chromosome material. (Nonduplicated sequence shown above.)

DUPLICATIONS

The presence of a section of a chromosome in excess of the normal amount is known as a *duplication.* The repeated section of chromosomal material may be present in one pair of homologous chromosomes or may have been transposed to a nonhomologue or, on occasion, may even exist independently with its own centromere (Fig. 22–6).

The first duplication discovered was for a section

of *Drosophila* X chromosome carrying the wild-type allele for *vermilion* (v^+) that had been transposed to an X chromosome carrying the mutant *vermilion* allele (v). Bridges found that this stock, instead of being phenotypically vermilion, was wild type. In other words, equal proportions of v and v^+ alleles produced the wild-type effect even when two doses of each allele were present. However, when such duplication females were crossed to nonduplication vermilion males, the daughters were all vermilion and the sons were all wild type. This *vermilion duplication* ($v^{+\,dup}$) in a single dose was therefore not sufficient to overcome the phenotypic effect produced by two mutant v genes in the female (Fig. 22–7).

In numerous cases discovered since then, however, the presence of only a single wild-type allele in a duplication is usually sufficient to produce a wild-type effect in an individual that is otherwise homozygous recessive. For example, an attached-X female in *Drosophila*, homozygous for various recessive sex-linked genes, will generally show the dominant phenotypes for those wild-type genes that it carries in an extra X-chromosome fragment contributed by its wild-type father. As used by Dobzhansky and others, this method has enabled the detection of duplications for the entire range of genes on the X chromosome, some of these duplications possessing their own centromeres and others having been transposed to other chromosomes (Fig. 22–8).

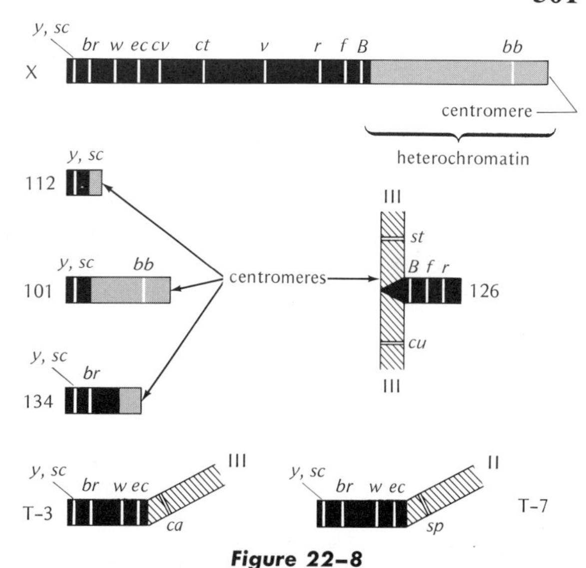

Figure 22–8

Diagrammatic representation of the normal X chromosome of Drosophila melanogaster (above) and a few of the duplications found as a result of x-ray treatment (below). Duplications 112, 101, and 134 have their own centromere. Duplications T–3 and T–7 are attached to the 3rd and 2nd chromosomes, respectively, and duplication 126 is inserted into the heterochromatic section of the 3rd chromosome between curled and scarlet. (After Dobzhansky, 1934.)

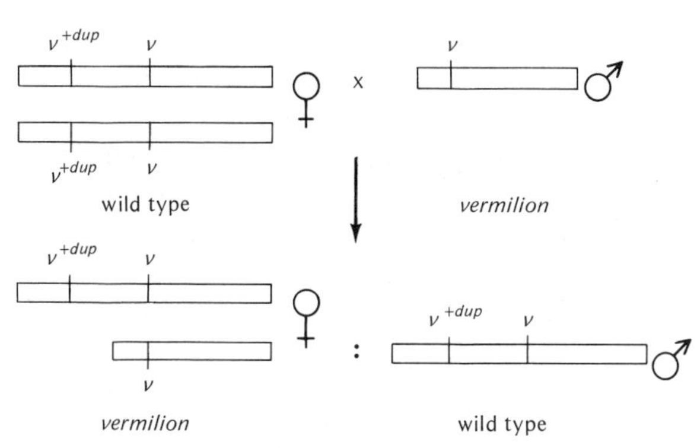

Figure 22–7

Results of a cross of females homozygous for vermilion and for vermilion-duplication × vermilion males. (From data of Bridges, 1919.)

In salivary chromosome analysis, duplications usually became observable either as bucklings in the duplication heterozygote or as cross-pairing between sections of different chromosomes. The notable frequency of duplicate banding patterns, or "repeats," in salivary chromosomes (see, for example, the *Bar* duplication, Fig. 25–3) indicates that duplications have not been uncommon in the past and have occasionally been incorporated in homozygous condition as part of the normal chromosome. It has, in fact, been suggested that all cases of "complementary" genes (Chapter 11) or "multiple factors" (Chapter 14) in which different gene pairs affect the same character in similar fashion arose initially as duplications of single genes (see also pp. 470–471).

BRIDGE–BREAKAGE–FUSION

Of special importance in helping to explain some of the behavior of broken chromosomes has been the discovery of systems in which such chromosomes are regularly produced. One such system involves a reverse tandem duplication of chromosome 9 in corn which, because of crossing over, regularly produces dicentric chromosomes. As shown in Fig. 22–9, such crossovers may occur either between nonsister chromatids (two homologues, left side of figure) or sister chromatids (one homologue, right side of figure). In both cases a dicentric chromosome is produced that "bridges" the two cell-division poles. Breakage may then take place at any point along this bridge, causing each of the separating chromosomes to retain one broken end. These broken ends, however, do not "heal" in the gametophyte or endosperm tissue but instead remain "sticky." Thus replication of the chromosome during prophase produces two paired sister chromatids that are sticky at the same end. These ends then fuse, thereby causing the reappearance of the bridge-breakage-fusion cycle.*

Note that if the bridge breaks at exact midcenter of the dicentric section (right side, Fig. 22–9), the genetic constitution of the resultant products at interphase and prophase is similar to that of the normal chromosome (*a b c d*). However, breakage may also occur offcenter (left side, Fig. 22–9) and produce products that are either deficiencies (*c d*) or duplications (*d c b a a b*) for particular sections. McClintock therefore found that, in addition to cytological tests, some of these aberrations could be detected by observing the transmission of genetic markers located on these sections. She crossed, for example, pollen from a plant bearing the 9th-chromosome dominant gene *C* for colored aleurone, but heterozygous for a 9th-chromosome duplication, to female tissue homozygous for colorless aleurone, *cc*. Since the normal triploid endosperm tissue should have the constitution *Ccc*, and appear phenotypically colored, loss of the paternal *C* gene could be easily detected as a spot of colorless tissue. Such spotted or "variegated" kernels were indeed found, each spot indicating that the bridge-breakage-fusion cycle had produced a chromosome deficient for *C*.

*For physiological reasons as yet unknown, "sticky" ends heal in the sporophytic tissues of the plant, and bridge-breakage-fusion cycles are not found there.

INVERSIONS

The structural changes noted so far encompass only additions or deletions of chromosome sections. However, even when the amount of chromosomal material remains the same, rearrangements of this material can cause new effects. One such rearrangement involves a change in gene order within a chromosome, and was first noted in comparisons between *Drosophila melanogaster* and *D. simulans* gene sequences.

These two species, although rarely interbreeding in nature, can be made to hybridize in the laboratory, making it possible to test whether a mutation in one is allelic to a mutation in the other. For example, the 3rd-chromosome mutation *scarlet* (*st*) in *D. melanogaster*, when combined with a *D. simulans* 3rd-chromosome carrying the *D. simulans scarlet*, will produce a scarlet-eyed hybrid (*st/st*) that is just as phenotypically distinctive as the scarlet-eyed homozygote in either species. On the other hand, a hybrid bearing *scarlet* from one species and *claret*, a different eye-color mutation, from the other (*st* +/+ *ca*) is wild type in eye color. Based on tests of this kind for various 3rd chromosome genes and on linkage studies performed with each species individually, Sturtevant and Plunkett were able to present the following 3rd-chromosome maps for each species:

D. melanogaster	*se*	*st*	*p*	*Dl*	*H*	*ca*
D. simulans	*se*	*st*	*H*	*Dl*	*p*	*ca*

They pointed out that these maps could most reasonably be explained as an *inversion*, i.e., reversal in order, of the gene sequence *p Dl H* between the two species. However, which gene order was primary and which secondary was not known.

The genetic evidence for inversions was further

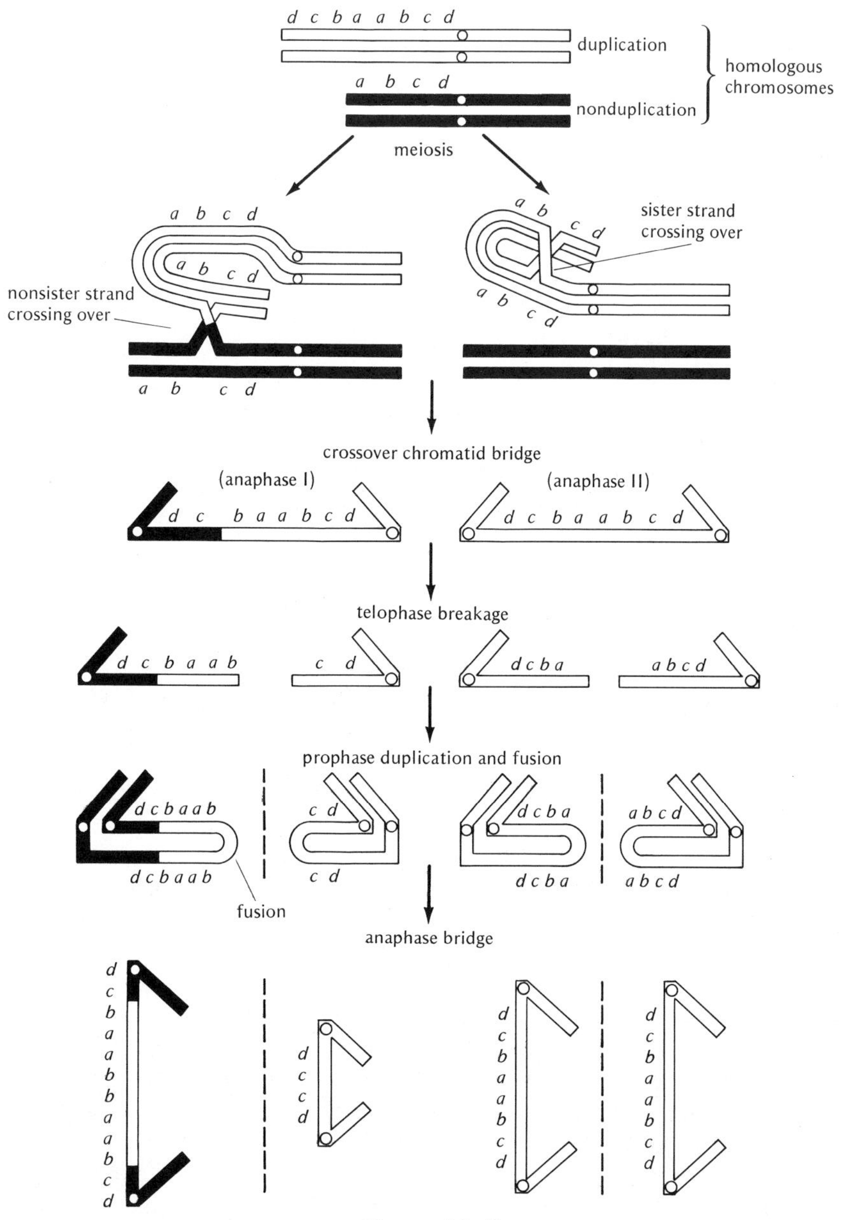

Figure 22–9

Sequence of events resulting from crossing over between duplicated sections carried by a corn chromosome. Dicentric chromosomes are produced which may break at various positions as they are being pulled apart. Fusion of the broken ends then occurs, leading to a further cycle of bridge–breakage–fusion.

clarified by Sturtevant in 1926 through analysis of "crossover suppressors" long known to occur in *D. melanogaster* (p. 347). These crossover suppressors (called *C*) had generally been localized to specific chromosome arms and to specific localities within these arms. C_{IIIB}, for example, reduced crossing over in the right arm of the 3rd chromosome and seemed to be located on the side of the *stripe* locus away from the centromere (see Fig. 17-3). Surprisingly, however, these *C* factors only reduced crossing over in heterozygous females (C/C^+) but had no effect on these same loci in homozygous females (C/C). Furthermore, when linkage studies were made in homozygous C/C females, the gene sequences in the crossover suppression areas were different than in normal C^+/C^+ stocks. For C_{IIIB}, the 3rd chromosome sequence was *scarlet* (*st*)-*stripe* (*sr*)-*claret* (*ca*)-*rough* (*ro*)-*ebony* (*e*) instead of the usual *st-sr-e-ro-ca*. In other words, the C_{IIIB} chromosome represented an inversion of the *e-ro-ca* section. Since this inversion did not include the centromere within its break points the name *paracentric* inversion ("beside" the centromere) was given to this type of aberration.

According to Sturtevant and others, it was the changed gene order of an inversion that caused its effect on crossing over. Certainly, if the inversion was small enough, crossing over would be inhibited because of the difficulty of close pairing between opposing inverted sites in the heterozygotes. On the other hand, if the paracentric inversion was long enough to permit pairing and crossing over with chromosomes bearing the normal gene sequence, a single crossover would yield dicentric and acentric chromosomes with consequent unbalanced gametes and inviable zygotes (Fig. 22-10).

In *Drosophila* crossing over was suppressed in males, so that such unbalanced genetic complements in sperm were not expected. In *Drosophila* females, however, single crossovers in some inversion heterozygotes would thus be expected to produce a noticeable proportion of inviable eggs. The relative absence of mortality among the progeny of such females therefore demanded explanation. To answer this question, Beadle and Sturtevant proposed that the meiotic products of *Drosophila* eggs were linearly oriented. That is, the first meiotic division separates the two resultant nuclei on a single plane (O ↔ O) and is then followed by a second meiotic division along the same plane (O ↔ O — O ↔ O). Of the four meiotic products, only one of the two end nuclei becomes the functional egg nucleus and the two inner nuclei are always relegated to the polar bodies. Since the dicentric products of a single crossover act as bridges between two poles, their nuclei obviously cannot travel as far as those without bridges. Thus, as shown in Fig. 22-11, the abnormal products are invariably localized in the two central nuclei of the linear quartet and consigned to polar bodies.

In a series of specially designed supporting experiments, Sturtevant and Beadle demonstrated that crossing over does indeed occur in female inversion heterozygotes with close to expected frequencies, but such crossovers remain "hidden" under normal conditions because of the linear meiotic division mechanism. The cytological appearance of this mechanism was demonstrated by Carson in the fly *Sciara* and conformed with that proposed to take place in *Drosophila*. In other words, inversions do not "suppress" crossing over in *Drosophila*, but their crossover products are usually eliminated before incorporation into functional egg nuclei.

Drosophila inversions can therefore be used to maintain a sequence of genes on a particular chromosome and "prevent" them from crossing over with their alleles on a homologous chromosome carrying a different gene order. Thus, for example, an inversion heterozygote carrying a two-gene combination, *A b*, on one chromosome and *a B* on its inverted homologue will rarely produce the recombinant gametes *A B* and *a b*. Furthermore, if each of the small letter alleles in this example designates a recessive lethal gene, mating between *A b/a B* inversion heterozygotes will only produce viable heterozygotes like themselves, since *A b/A b* and *a B/a B* genotypes are lethal. The viable homozygotes *A B/A B* are also ordinarily not produced, since the inversion prevents the formation of *A B* gametes. It is through such means that the "balanced lethal" systems described in Chapter 11 are maintained.

However, crossover products between inverted sequences are not always prevented or eliminated. In organisms such as corn, in which crossing over occurs in male tissue, all four meiotic products are

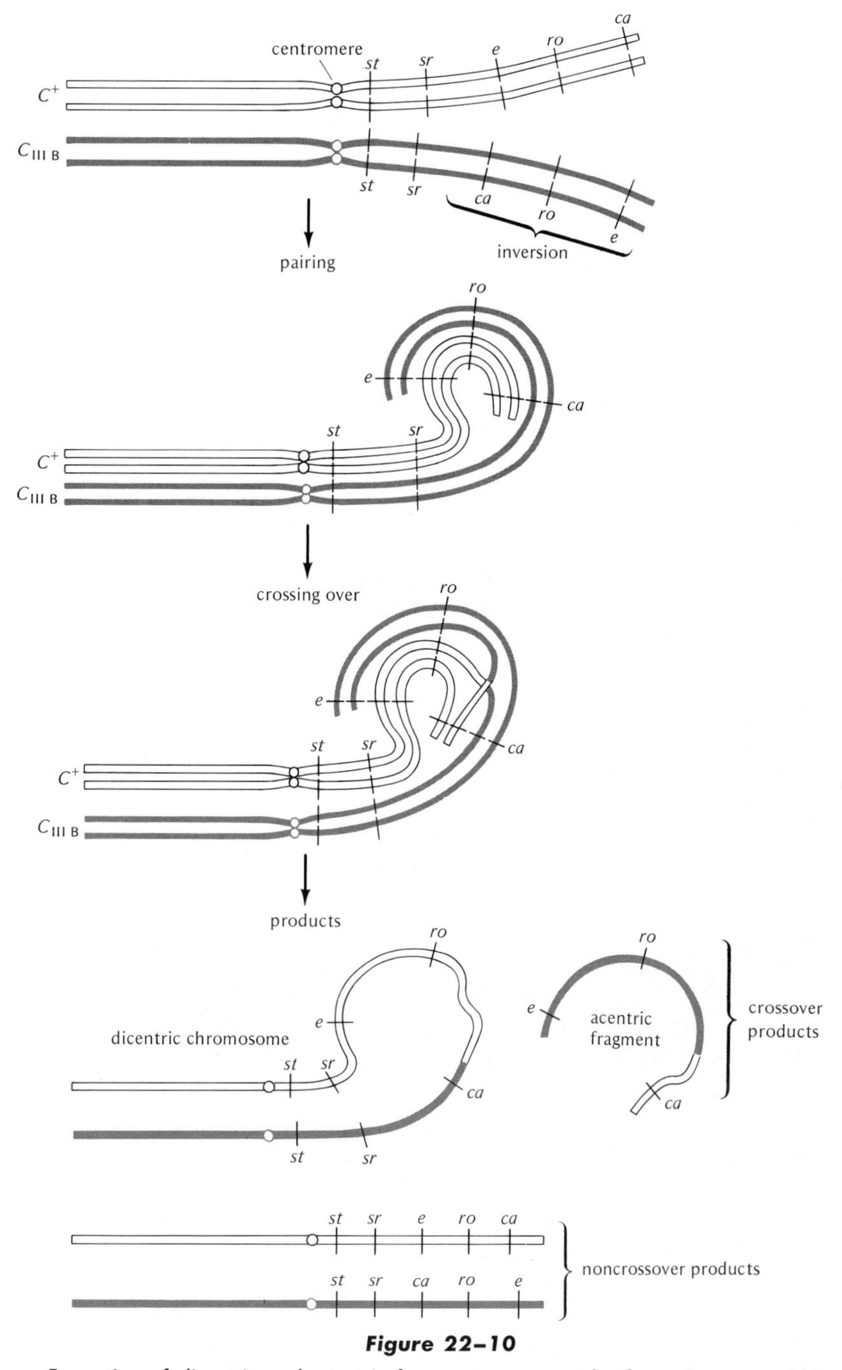

Figure 22–10

Formation of dicentric and acentric fragments as a result of crossing over within a paracentric inversion.

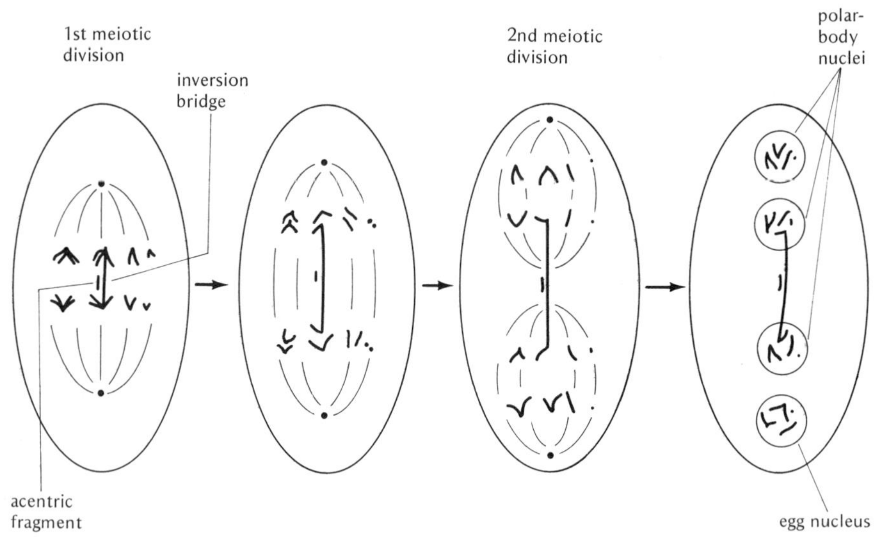

Figure 22–11

Explanation of why single exchanges in inversion heterozygotes in Drosophila melanogaster *are not included in the functional egg nucleus. As shown, the second meiotic division in* Drosophila *has two spindles whose axes are in an approximate straight line. A bridge between the two centromeres keeps the exchanged chromosomes in the two inner nuclei. Since only an outer nucleus becomes the functioning egg nucleus, bridges and fragments are excluded from affecting the zygote.*

incorporated into gametes. Bridges and acentric fragments formed by inversion heterozygotes may then produce considerable sterility in the male gametophytes, i.e., pollen. In female corn tissue, the dicentric bridges formed by such crossovers may lead to the same type of bridge-breakage-fusion cycle noted previously.

If the paracentric inversion is sufficiently long to enable double crossovers to appear regularly, the apparent "suppression" of crossing over in heterozygotes may be diminished relative to the suppression expected from only single crossovers. As shown in Fig. 22–12, the reason for this lies in the normal products formed by double crossing over between the same two strands within the inversion (crossovers ① and ②). In three-strand doubles (①, ③), the crossover products are similar to those of single crossovers and have similar difficulties of transmittance to the zygote. In four-strand doubles (①, ④) all chromatids of both homologues produce bridges and acentric fragments. Thus *Drosophila* females heterozygous for long X chromosome inversions may also give rise to some nullo-X eggs, since all the four-strand double-crossover products may remain in the polar bodies. Fertilization of such nullo-X eggs will then be expected to produce XO "patroclinous" males (see Chapter 12), and such males have indeed been found. Novitski, however, has noted that XO males are formed only when the X chromosomes of the inversion heterozygote are telocentric or acrocentric. Addition of long heterochromatic sections (long arm of the Y chromosome) to the X chromosomes, making them metacentric, eliminates patroclinous males. It seems as though such heterochromatin acts to strengthen the anaphase movement of the centromere, so that the chromatid bridges are broken and their fragments incorporated into the zygotic egg nucleus. Such zygotes, bearing portions of the maternal X chromosome, will no longer be patroclinous.

DETECTING THE BREAK POINTS OF INVERSIONS

Since we can expect to find no single crossovers within inversions in *Drosophila* and very few doubles, the genetic limits (or break points) of an inversion can be detected by noting the frequencies of different crossover classes produced by an inversion heterozygote segregating for many mutants on this chromosome. The rarer crossover classes would be expected from recombination within the inversion area, and the more frequent crossover classes from recombination outside this area. For example, Sturtevant crossed the previously mentioned C_{IIIB} inversion stock, which was wild type in appearance, to a normal 3rd chromosome stock that was homozygous for the *st sr e ro ca* mutations. Heterozygous female products of this cross were then backcrossed to the homozygous mutant stock and their offspring classified. As shown in Fig. 22–13, there were no single crossovers in the *e-ro* and *ro-ca* regions, indicating that the inversion covered this area. Nearby, between *sr-e*, the great reduction in single crossovers (from an expected 8 percent to .15 percent) signified that the inversion probably extends partly into this region. Somewhat further from the inversion, between *st* and *sr*, crossing over is reduced (from an expected 18 percent to 9.6 percent) but is nevertheless high enough to indicate that this area is definitely outside the inversion. Only two unusual genotypes are recorded, and both of these can be explained as resulting from double crossovers within the inversion limits.

Inversions have been detected cytologically by observing the inversion loops formed because of close heterozygous pairing in pachytene stages in corn and other organisms and in salivary chromosomes in *Drosophila*. The opportunity for banding analysis in *Drosophila* has, in addition, enabled detailed cytological studies of the inversion breakage points. In the 3rd chromosome of *D. pseudoobscura*, for example, inversions of the

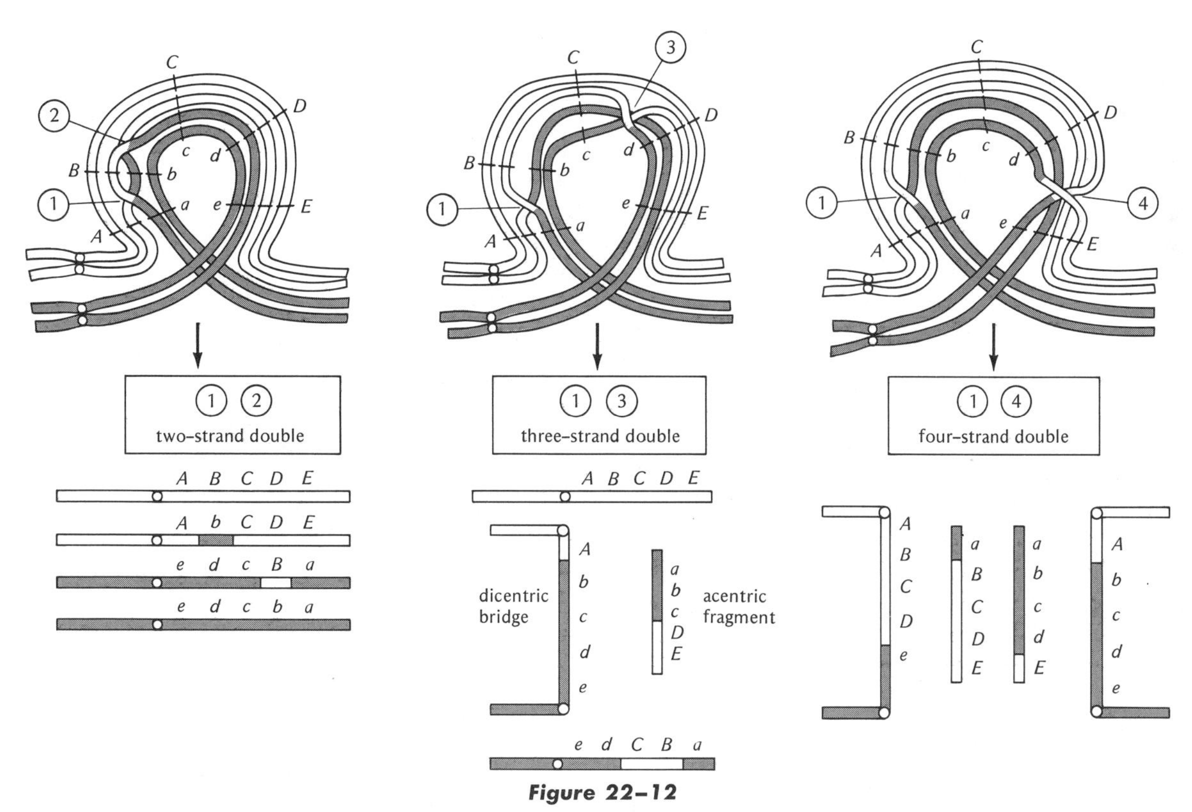

Figure 22–12

Products of double crossing over within an inverted region involving different combinations of strands.

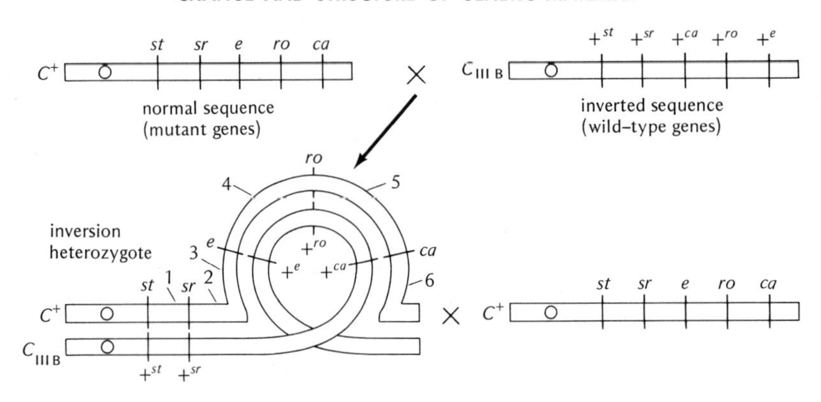

recovered chromosomes in backcross progeny	crossover regions	number
$+^{st}$ $+^{sr}$ $+^{ca}$ $+^{ro}$ $+^{e}$	no crossovers	2214
st sr e ro ca	no crossovers	2058
st $+^{sr}$ $+^{ca}$ $+^{ro}$ $+^{e}$	1 (single)	238
$+^{st}$ *sr e ro ca*	1 (single)	219
st sr $+^{ca}$ $+^{ro}$ $+^{e}$	2 (single)	4
$+^{st}$ $+^{sr}$ *e ro ca*	2 (single)	3
$+^{st}$ $+^{sr}$ $+^{ca}$ *ro e*	5,3 (double)	1
st sr e $+^{ro}$ $+^{ca}$	4,6 (double)	1
		4738

Figure 22–13

Results obtained by Sturtevant in the progeny of a Drosophila melanogaster *cross between an* st sr e ro ca *stock and females heterozygous for an inversion with the gene order* st sr ca ro e. *Note that almost all crossover products are outside the inversion area. For diagrammatic simplicity, only the paired chromosomes without chromatids are shown in the inversion heterozygote.*

Standard (ST) sequence (63-64-65-66-· · · -81) may occur at any number of points along the chromosome. Two such inversions, Arrowhead (AR) and Pikes Peak (PP), are shown in Fig. 22–14 in heterozygous condition with the Standard arrangement. Note that both AR and PP differ from the ST sequence by single loops. For AR the 70–76 region has been inverted, and for PP the inverted region covers 65–75. Thus when AR and PP are combined, two inversion differences are present rather than one, and consequently two loops are formed, both included within the 65–76 region (*overlapping inversions*, Fig. 22–15).

This type of cytological analysis is of value, since it enables determination of the evolutionary sequence of chromosome arrangements in natural populations. It is obvious, for example, that the relationship between AR and PP demands four simultaneous break points (76, 75, 70, 65) if one arrangement is to be derived directly from the other. On the other hand, the relationship between AR and ST and between PP and ST demands only two break points for each. Thus the relationship between the three arrangements is most likely AR-ST-PP (rather than AR-PP-ST or PP-AR-ST). In this relationship ST might have originated first

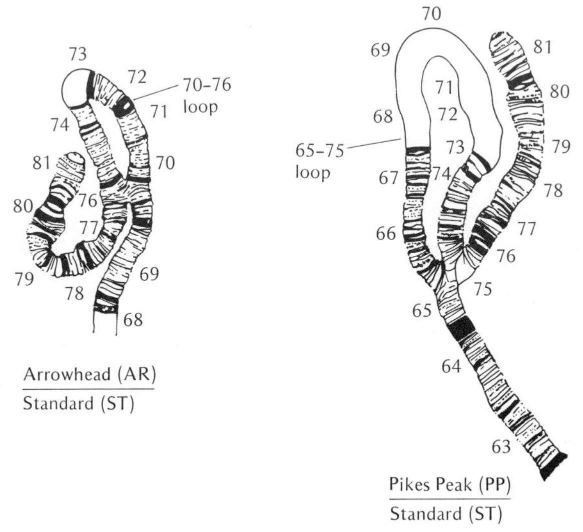

banding sequences:

ST: 63-64-65-66-67-68-69-70-71-72-73-74-75-76-77-78-79-80-81
AR: 63-64-65-66-67-68-69-76-75-74-73-72-71-70-77-78-79-80-81
PP: 63-64-75-74-73-72-71-70-69-68-67-66-65-76-77-78-79-80-81

Figure 22–14

Salivary chromosomes of D. pseudoobscura larvae heterozygous for the 3rd-chromosome arrangements Arrowhead/Standard and Pikes Peak/Standard, with numbered banding sequences. Inverted regions are underlined. (After Dobzhansky, 1944.)

and then produced AR and PP, or the initial arrangement might have been either AR or PP and then evolved the other arrangement through ST as intermediary.

The decision on the actual sequence followed in evolution depends upon other data gathered from the analysis of numerous additional arrangements. When such analyses are undertaken, chromosome phylogenies can be established indicating relationships between all known arrangements. As shown in Fig. 22–16, both *D. pseudoobscura* and *D. persimilis* share the Standard sequence, which, because of its phylogenetic position and wide distribution, is considered among the ancestral arrangements in this pair of species. Interestingly the "Hypothetical" arrangement has never been found, but must certainly have existed at one time to account for arrangements such as Santa Cruz and others which have utilized the "Hypothetical" inversion as part of their own banding sequences.

PERICENTRIC INVERSIONS

Inversions that include the centromere within the inverted region are called *pericentric* inversions

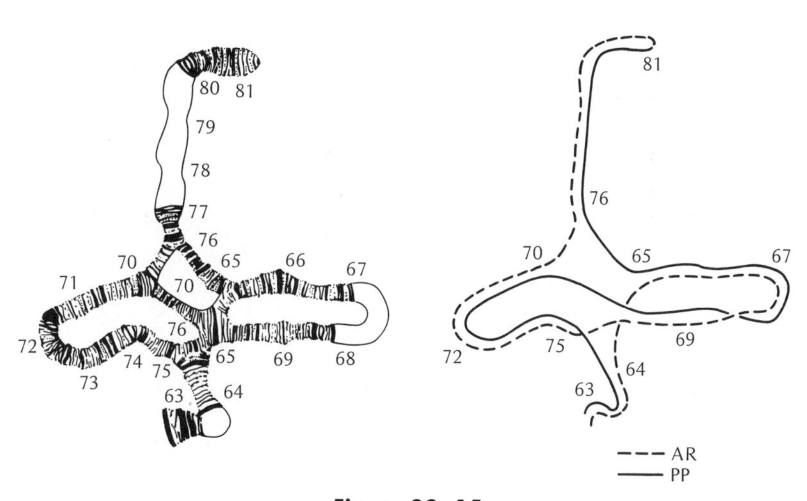

Figure 22–15

Salivary 3rd chromosome of an AR/PP heterozygote; banding configuration on the left and interpretation on the right. (Banding after Dobzhansky, 1944.)

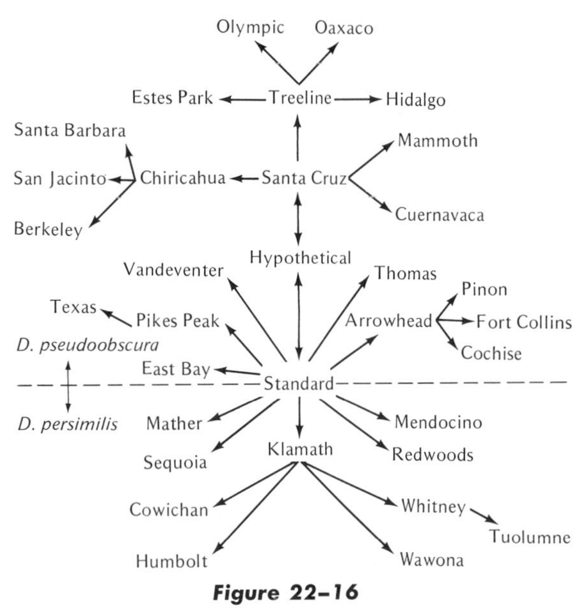

Figure 22–16

Relationships between the various 3rd-chromosome gene arrangements found in natural populations of D. pseudoobscura *and* D. persimilis. *It is generally assumed that either Standard, Hypothetical, or Santa Cruz was the ancestral arrangement and that evolution of the other 3rd-chromosome arrangements proceeded from one of these. (After a drawing provided courtesy of Wyatt Anderson.)*

("around" the centromere). Although single crossovers within a pericentric inversion do not produce the dicentric and acentric chromosomes of paracentric crossovers, they nevertheless produce duplications and deficiencies (Fig. 22–17). Only two-strand double crossovers in pericentric inversions will produce nondeficient and nonduplicated chromosomes. In general, therefore, heterozygotes for pericentric inversions can be expected to produce a significant frequency of unbalanced gametes and inviable zygotes. Also, since these duplications and deficiencies are unaccompanied by chromatid bridges, many of the egg nuclei of *Drosophila* females heterozygous for these inversions should be abnormal. According to Alexander, however, there is some indication that there is less mortality than expected among the zygotes of such *Drosophila* females, probably because of reduced pairing and consequent reduced crossing over between homologues in the inverted region. In general, pericentric inversions are much less frequently found than the paracentric type, and when found as a common feature of a species (certain grasshoppers) they seem to be "protected" from crossing over because of lack of pairing.

Because they include the centromere, pericentrics may also produce morphological changes in the appearance of a chromosome if the breaks extend unequally along the chromosome arms. For example, a metacentric chromosome with unequal pericentric breaks can be transformed into an acrocentric chromosome (Fig. 22–18). Should such a pericentric inversion be viable as a homozygote but semisterile as a heterozygote, it may lead to the separate evolution of a group of mutually fertile and cytologically identifiable pericentric homozygotes that do not breed well with the remainder of the species. Cytological evidence for such evolutionary mechanisms have been found in *Drosophila* and other organisms including mammals. One striking example is the deer mouse genus *Peromyscus*, in which the chromosome number is constant in all species ($2n = 48$), but some species (e.g., *P. collatus*) are entirely composed of metacentric chromosomes, and others (e.g., *P. boylei*) are almost entirely composed of acrocentrics.

TRANSLOCATIONS

The transfer of a section of one chromosome to a nonhomologous chromosome is known as a *translocation*. In *Drosophila* such events were first recognized genetically by the unusual behavior of a particular 2nd-chromosome gene known as *Pale*, which had the phenotypic effect of diluting certain eye colors. Although *Pale* was lethal in homozygous condition, Bridges found that its lethality as well as its phenotypic effect could be suppressed by the presence of another gene discovered at the same time on the 3rd chromosome, which was also lethal in homozygous condition. The lethality of the latter, in turn, was suppressed by the presence of the former. Linkage analysis soon showed that the *Pale* effect was caused by a deficiency for a

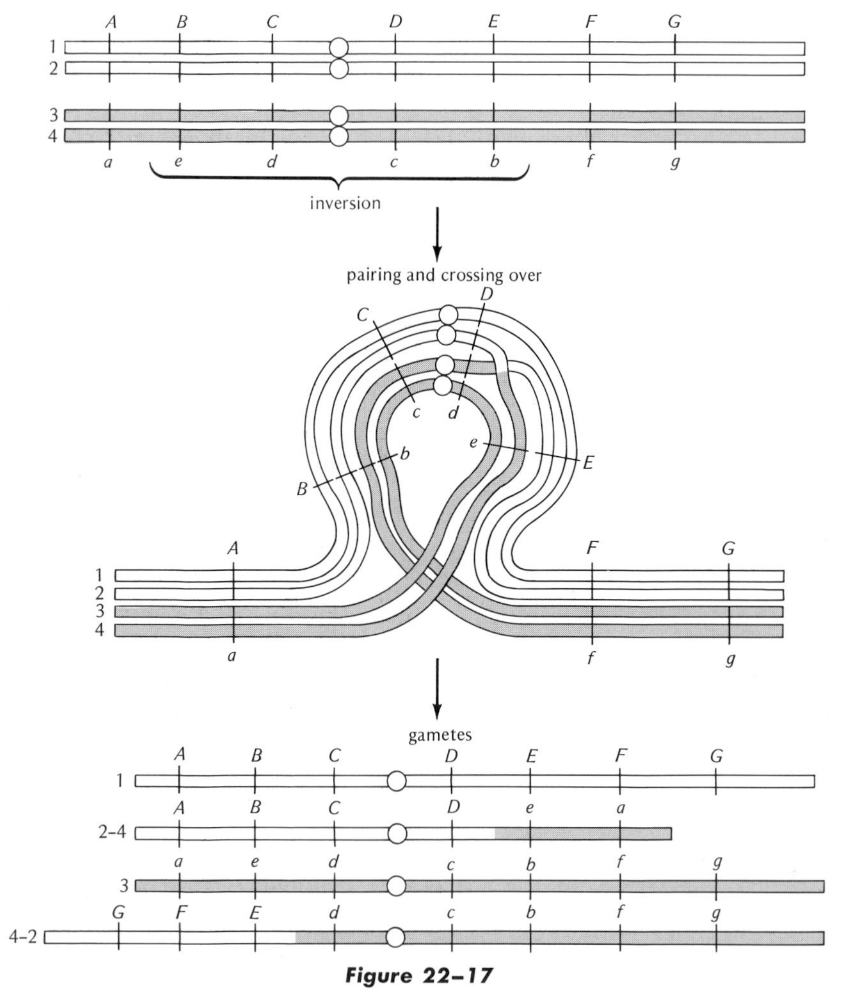

Figure 22–17

Pairing of a heterozygote for a pericentric inversion, and the meiotic products of a single crossover within the inversion limits.

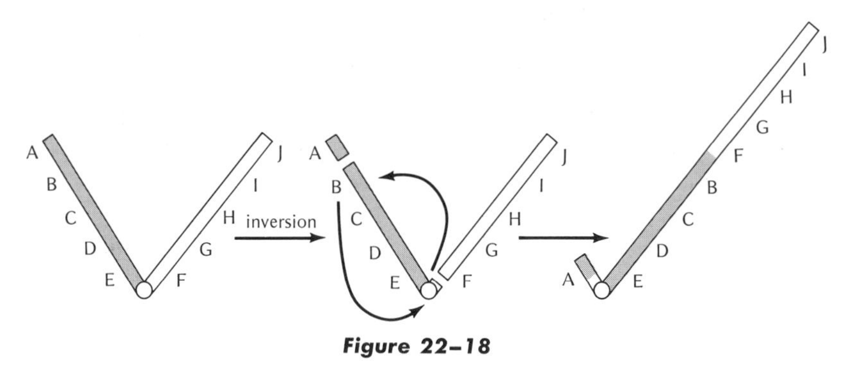

Figure 22–18

Transformation of a V-shaped metacentric chromosome into a rod-shaped acrocentric chromosome by a pericentric inversion.

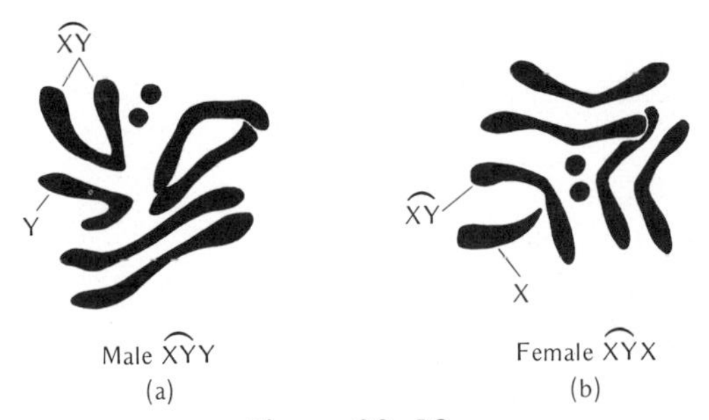

Figure 22–19

Metaphase chromosomes of Drosophila melanogaster bearing a translocation between the X and Y chromosomes. For the normal karyotype of this species see Fig. 12–2.

small section of genes on the tip of the 2nd chromosome (including *plexus* and *balloon*) which was now linked to the 3rd-chromosome genes between *ebony* and *rough*. In other words, genes from a deficiency in one chromosome had been translocated to another chromosome. In 1926 Stern gave the first cytological demonstration of such events when he showed that the genetic translocation of the *bobbed*$^+$ allele on the Y chromosome to the X chromosome was accompanied by an observable lengthening of the X chromosome (Fig. 22–19).

At present a variety of translocations are recognized, among which are the following:

1. *Simple translocations*, which involve a single break in the chromosome and the transfer of a broken piece of this chromosome directly onto the end of another (Fig. 22–20a). However, the presumed presence of "nonsticky" telomeres on the unbroken chromosome may account for the fact that such terminal chromosome attachments are rarely, if ever, found.
2. *Shifts* are more common and are translocations involving three breaks, so that a two-break section of one chromosome such as the *Pale* translocation is inserted within the break produced in a nonhomologous chromosome (Fig. 22–20b).*
3. *Reciprocal translocations*, or *interchanges*, are probably the most frequent and best-studied translocations. They occur when single breaks in two nonhomologous chromosomes produce an exchange of chromosome sections between them (Fig. 22–20c).

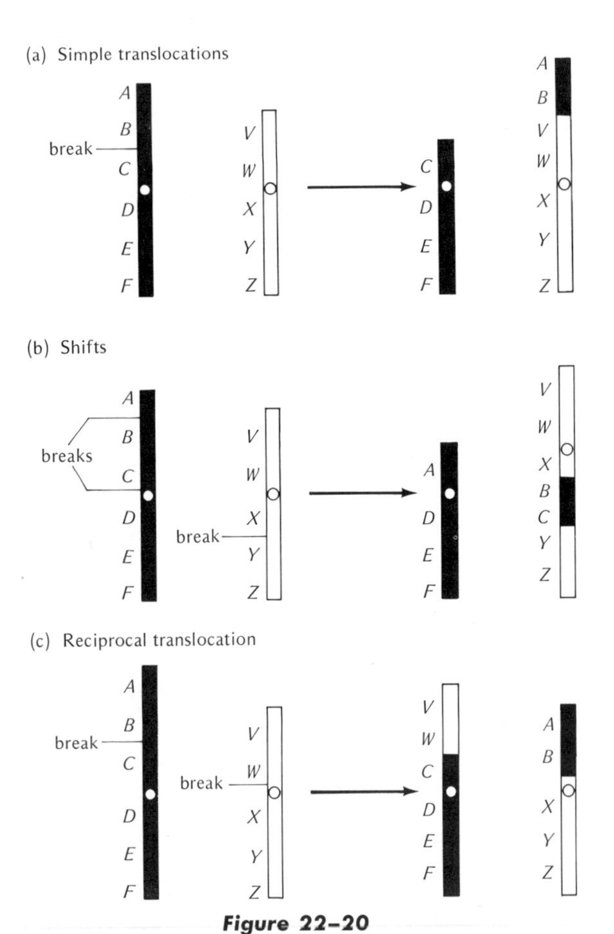

Figure 22–20

Three different types of translocations that may occur between two nonhomologous chromosomes, A B C D E F and V W X Y Z.

By whatever mechanism produced, the translocated regions may nevertheless act normally with respect to homologous pairing. Thus a transloca-

*Some authors (Kaufmann, 1954) use the term *transposition* when the three breaks take place within the same chromosome, so that a section is changed from one position to another in respect to the same centromere. Transpositions of this kind may also occur as a result of two successive inversions in the same chromosome arm which overlap, e.g., *ABCDEFGH* → *ABCGFEDH* → *ABEFGCDH*. In this way, a section of chromosome such as *CD* may become transposed to the opposite chromosome end.

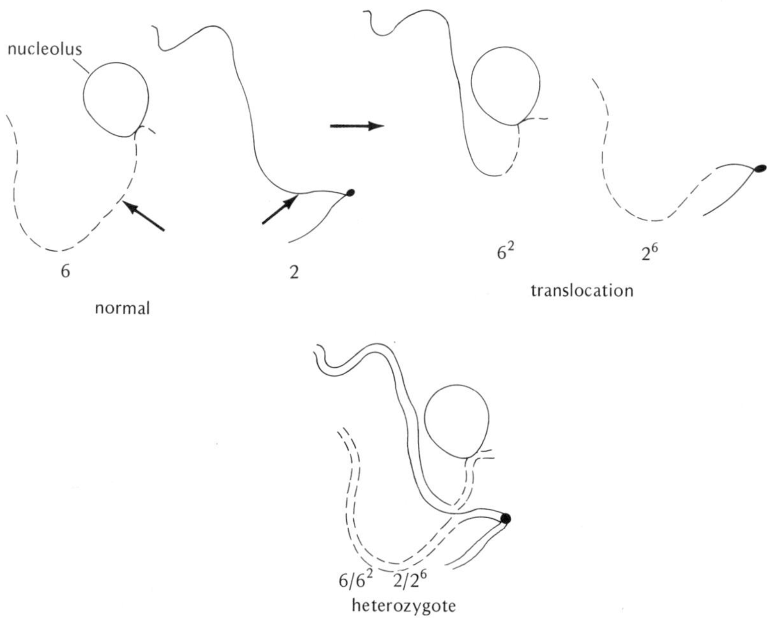

Figure 22–21

Pachytene configuration of a heterozygote derived from a translocation between the 6th and 2nd chromosomes in corn with a "cross" at the breakage points. Arrows indicate the breakage points, and the large circle represents the nucleolus carried at a specific locus on the 6th chromosome. (After Burnham.)

tion homozygote forms the same number of homologous pairs as the normal homozygote as long as centromeres have not been lost. Meiotic disjunction between homozygous translocated chromosomes is normal, and each gamete receives a full complement of genes.

The results of pairing and meiosis, however, are quite different in translocation heterozygotes bearing the two translocated chromosomes and their two normal counterparts. Such individuals can form a connected group of nonhomologous chromosomes tied together by the translocated sections. In Fig. 22–21, for example, a 6–2 translocation in corn forms a four-chromosome complex at the pachytene stage. Note that if only homologous sections pair, the position of the "cross" indicates the breakage points of the translocation. As might be expected, *Drosophila* salivary chromosomes can show such translocation crosses in great detail because of close somatic pairing between homologous sections.

The consequences of translocations are profound; they lead not only to a change in gene linkage in the translocated sections but also to easily incurred meiotic abnormalities. Let us, for example, consider a reciprocal translocation in which homologous pairing occurs between the four involved chromosomes in the heterozygote, but the centromeres can be distinguished as to whether they are homologous, e.g., 1 1′, and 2 2′, or nonhomologous, e.g., 1 2, 1′ 2, 1 2′, 1′ 2′. Let us also make a distinction between the chromosomes bearing the original nontranslocated genes, e.g., 1 and 2 and the translocation chromosomes, e.g., 1′ and 2′. As diagrammed in Fig. 22–22, chiasmata between these chromosomes may form a quadrivalent which can then disjoin in three different segregation patterns in the first meiotic division:

1. *Alternate* segregation. Opposite or alternate nonhomologous centromeres go to the same pole in a zigzag fashion, so that the

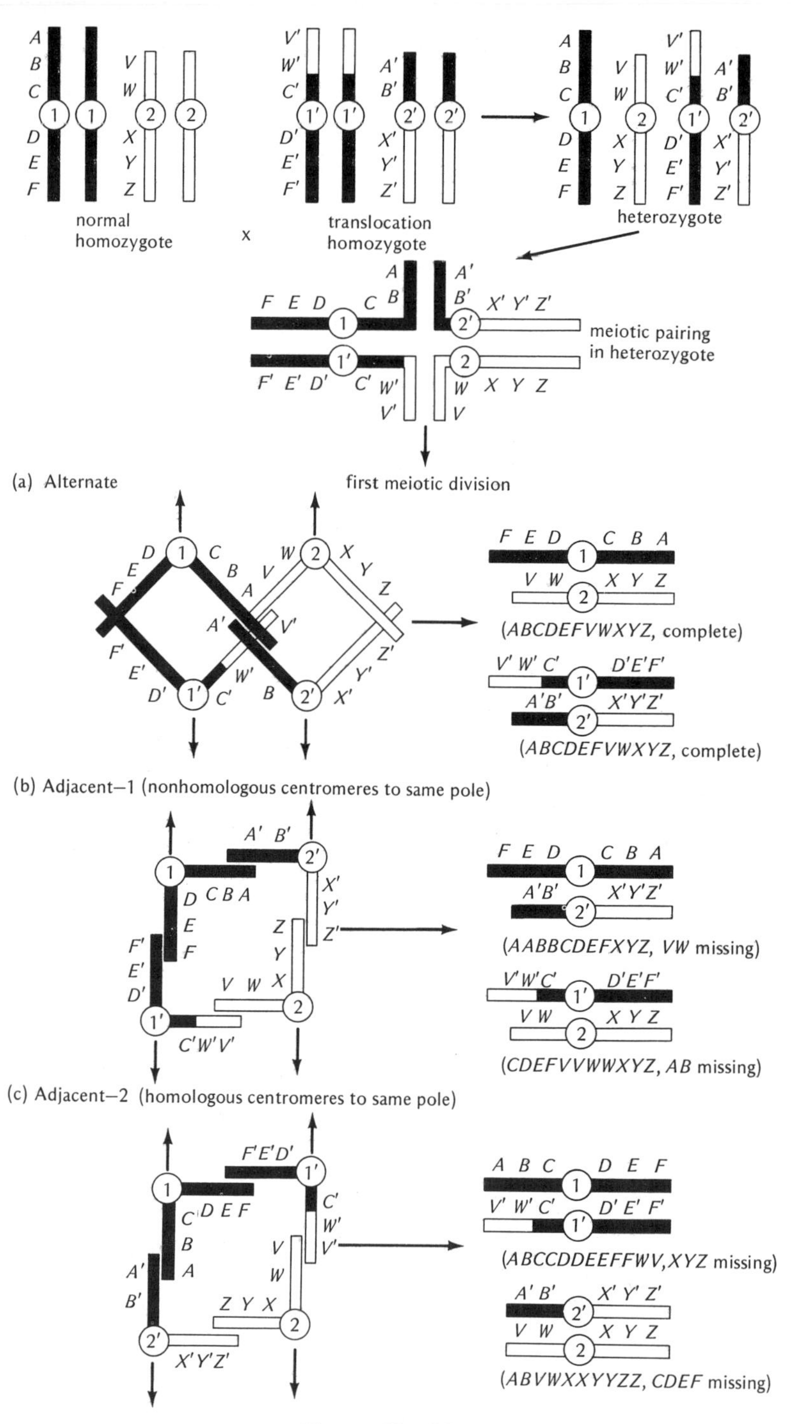

Figure 22–22

Meiosis in a translocation heterozygote showing the three types of segregation at the first meiotic division. For convenience the chromosomes are not shown in duplicated condition (chromatids). Only the gametes produced by alternate segregation are balanced.

nontranslocated (1 2) and translocated (1′ 2′) chromosomes are in separate gametes. Each gamete thus has a complete "balanced" complement of genes without duplications and without deficiencies.

2. *Adjacent-1* segregation. Nonhomologous adjacent chromosomes go to the same pole, but each gamete contains both a translocated and nontranslocated chromosome (1 2′, 1′ 2). Note that, in this case, there is both a duplication and deficiency in each gamete, or an unbalanced gametic complement.

3. *Adjacent-2* segregation. Adjacent centromeres again go to the same pole, but these are now homologous as well as containing both translocated and nontranslocated chromosomes (1 1′, 2 2′). Again, duplications and deficiencies produce an unbalanced complement of genes.

Since both adjacent-1 and adjacent-2 segregations produce unbalanced gametes that usually have lethal effects, we can expect that the fertile gametes of translocation heterozygotes will be mostly restricted to those produced by alternate segregation. Note, however, that in alternate segregation the translocated and nontranslocated chromosomes are carried in separate gametes. One of the consequences of this segregation is that independent assortment between genes on nonhomologous chromosomes will be seriously inhibited. Thus, as shown in Fig. 22–23a, *Drosophila* males heterozygous for a translocation between the 2nd and 3rd chromosomes (2-3), and also carrying the genes *brown* and *ebony* on the nontranslocated homologues, will produce only wild-type and brown-eyed ebony-bodied flies when backcrossed to the homozygous recessive stock. Because of duplications and deficiencies neither of the single mutant phenotypes brown eyes or ebony body will appear in the offspring. In similar crosses, a translocation between the X and 2nd chromosome (X-2) will produce only brown-eyed males and wild-type females, since both wild-type males and brown-eyed females will be lethal (Fig. 22–23b).

Another consequence of segregation in translocation heterozygotes is lower fertility. In plants the unbalanced gametes themselves are usually inviable, since they must function for a period of time in haploid condition as a separate gametophyte generation. Thus gametophytes act as a "screen," eliminating deleterious translocations, so that those passing through to the sporophyte generation are usually viable in homozygous condition. In animals, on the other hand, the gametic nuclei are rarely functional until after fertilization, so that unbalanced gametes usually produce inviable zygotes rather than inviable gametes. Thus many translocations in animals (about half of those found in *Drosophila*) have not been "screened" out in the gametes and must be maintained in heterozygous condition.

Translocation inviability effects pertain, of course, to duplications and deficiencies that involve essential loci. However, if the extent of duplication and deficiency is small, unbalanced gametes and even unbalanced zygotes may not necessarily be lethal, although they may be seriously affected. One example of a duplication effect concerns human translocation heterozygotes carrying part of chromosome 21 attached to chromosome 15 in addition to the normal 15 and 21 chromosomes. Such individuals can produce gametes bearing a normal 21 chromosome and a 15^{21} translocation. When such gametes unite with those produced by a normal individual, mongoloid offspring are produced that are trisomic for a major part of the 21 chromosome (p. 488). A deficiency effect in humans can also be noted when one of the parents is a translocation heterozygote in whom the short arm of chromosome 5 is translocated to a chromosome of another group (e.g., 15 in the D group). Although this parent is normal, he or she will nevertheless produce some unbalanced gametes which carry only the deficient member of the translocation pair. Children inheriting this deletion-5 gamete show the cri-du-chat syndrome previously described (p. 498).

However, even if the translocated sections are long enough to cause lethality because of major duplications and deficiencies, matings between two heterozygotes for the same translocation may occasionally produce gametic combinations that complement each other to yield viable zygotes. As

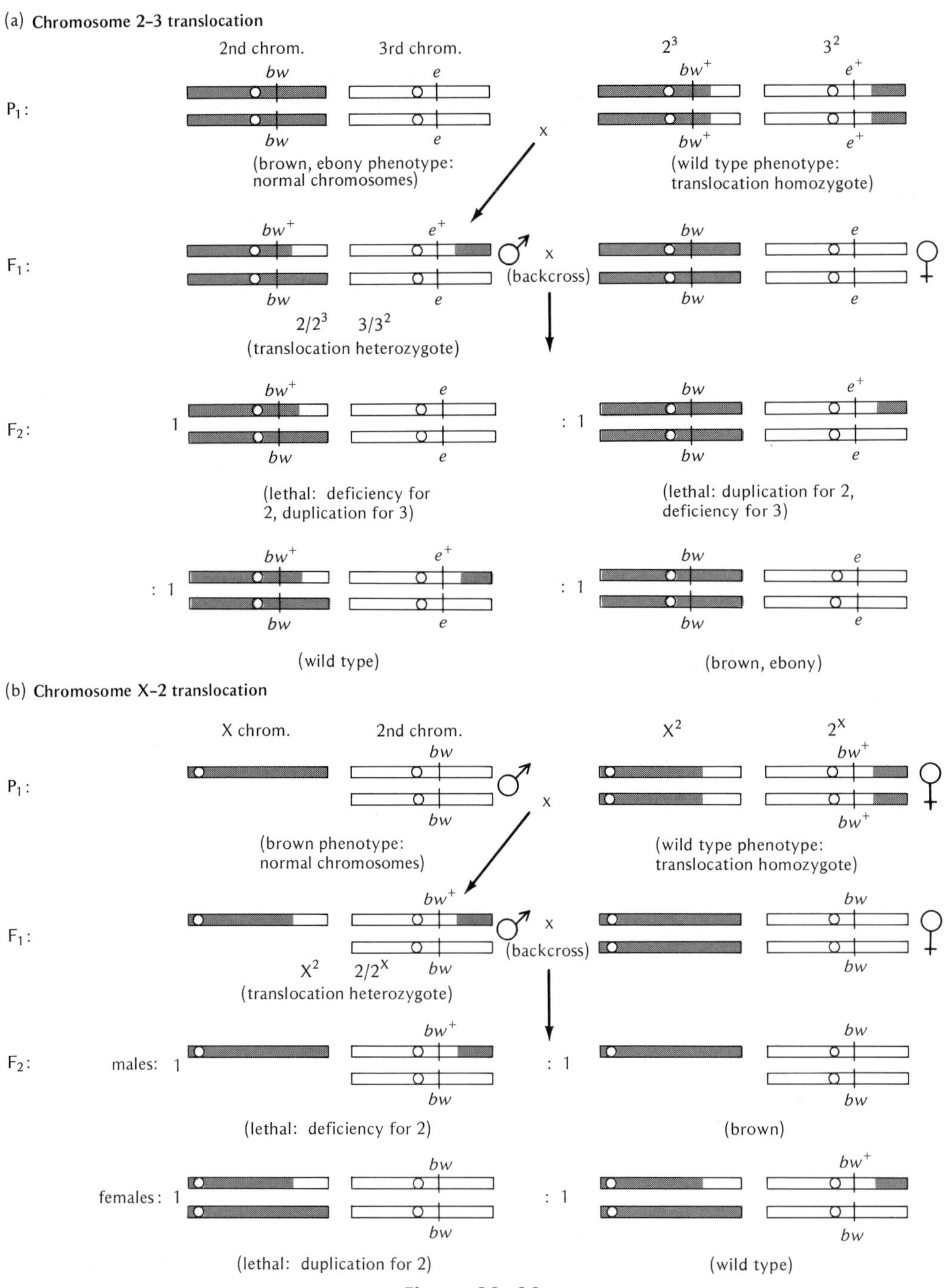

Figure 22–23

Results of mating Drosophila melanogaster *males heterozygous for translocations with homozygous recessive females. In (a) a chromosome 2–3 translocation is involved, and only the parental phenotypes appear in the backcross progeny. In (b) a chromosome X–2 translocation has occurred, and the viable F_2 males and females have different phenotypes.*

		Sperm					
		alternate		adjacent-1		adjacent-2	
Eggs		1 2	1^2 2^1	1 2^1	1^2 2	1 1^2	2^1 2
alternate	1 2	normal homo.	transl. het.				
	1^2 2^1	transl. het.	transl. homo.				
adjacent-1	1 2^1				transl. het.		
	1^2 2			transl. het.			
adjacent-2	1 1^2						transl. het.
	2^1 2					transl. het.	

Figure 22–24

Results expected from matings between heterozygotes carrying a reciprocal translocation between chromosomes 1 and 2 (designated 1^2 and 2^1). The inviable combinations are left blank, and the only chromosome constitutions indicated are those of the viable zygotes. Note that one of the combinations ($1^2 1^2 2^1 2^1$) is homozygous for the translocation and that one combination (1 1 2 2) reconstitutes the normal configuration. The translocation homozygote, if viable, can be used to derive a translocation stock. Note also that some unbalanced gametes may produce balanced zygotes in appropriate complementary combinations, e.g., 1 $2^1 \times 1^2$ 2.

shown in Fig. 22–24, of the 36 possible combinations produced in a mating between translocation heterozygotes, eight are viable, and four of these arise from a union between unbalanced but complementary gametes.

CROSSING OVER IN TRANSLOCATION HETEROZYGOTES

Close meiotic pairing between homologous sections of translocated chromosomes enables crossing over to occur in each of the four arms of such quadrivalents. A linkage map of a translocation heterozygote should therefore appear four-armed, with one common center, rather than linear. Because of this arrangement, the consequences of crossing over will depend greatly upon which segments are involved. In general a distinction can be made between *interstitial* segments that include the area between the centromeres and the break points of the translocation and the *pairing* segments that are outside this area. In Fig. 22–25 the interstitial segments are confined to the area of the *C* locus, while all other chromosome sections are shown as pairing segments. From this illustration it can be seen that crossing over in the pairing segments has no effect on the segregation pattern; there is merely an exchange of one section for a homologous section (e.g., F ↔ F′). In fact, the cross or ring-shaped configuration of the quadrivalent is probably caused by chiasmata or crossovers in these pairing segments.

However, if crossing over occurs in an interstitial segment, nonhomologous sections are also transferred (e.g., *C B A* ↔ *C′ W′ V′*). Both alternate and adjacent segregation now lead to the formation of some unbalanced gametes, and may be a

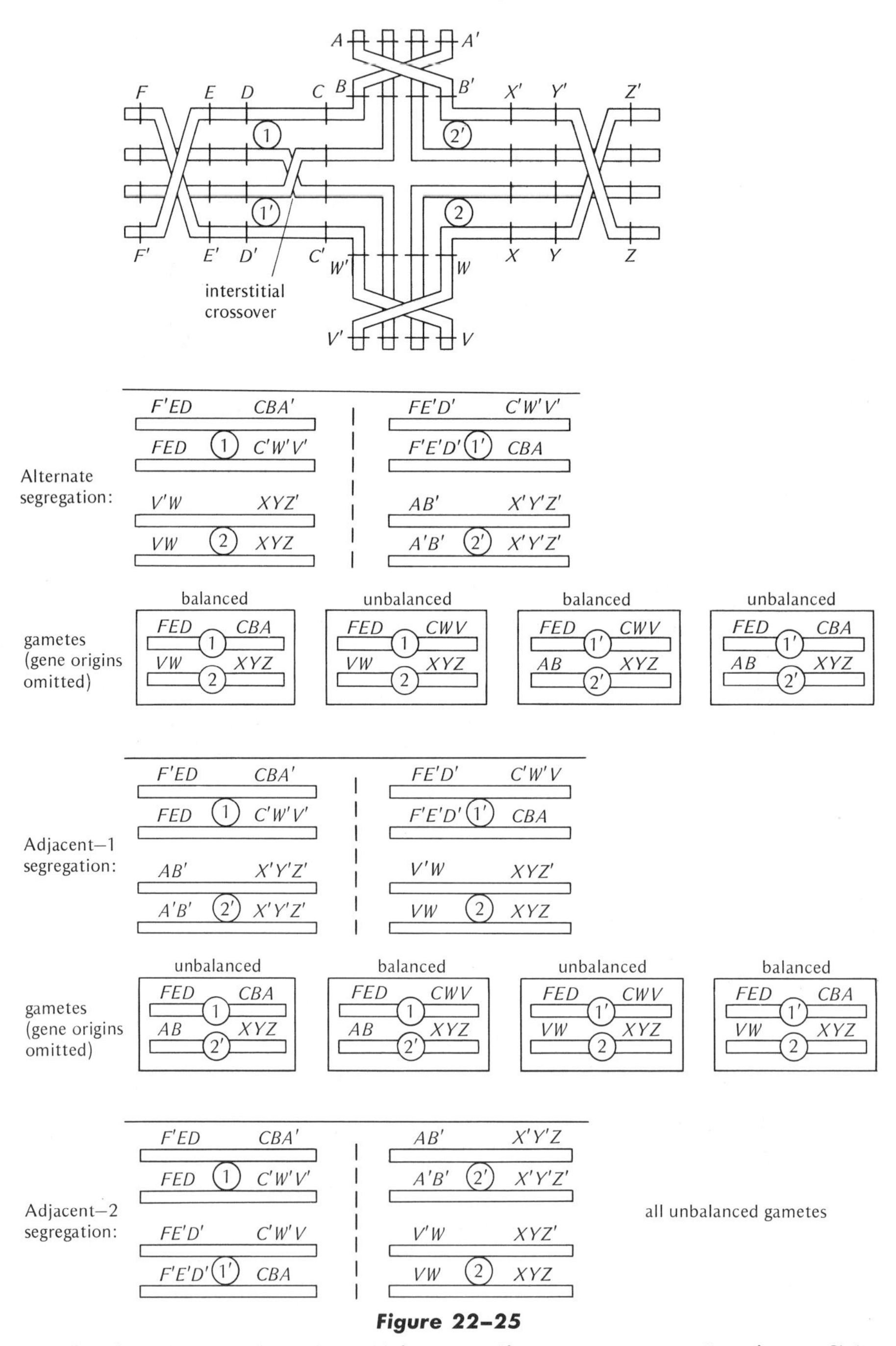

Figure 22–25

Results of crossing over in an interstitial segment (between centromere 1 and gene C) in a translocation heterozygote. Shown are the three types of chromosomal segregation that can occur during the first meiotic division, along with the gametes produced by each type. All gametes containing crossover products derived from alternate segregation of centromeres are unbalanced, while the crossover-containing gametes of adjacent-1 segregation are balanced.

serious cause of sterility in translocation heterozygotes that rely mainly upon alternate segregation patterns. Interestingly, in cases of interstitial crossovers (Burnham, Glass, and others), adjacent-2 segregation seems absent, because crossing over causes disjunction almost exclusively between nonhomologous centromeres.

The usual finding nevertheless is a restriction in crossing over in interstitial segments because of three reasons: (1) Translocation heterozygotes in which alternate segregation is most common would have many of their crossover products in unbalanced gametes, as shown in Fig. 22–25; (2) homologous pairing in interstitial segments may be less efficient than in the pairing segments; and (3) the interstitial segments are small. The low sterility of translocation heterozygotes in plant species such as *Datura*, *Oenothera*, wheat, and barley is probably a result of restricted crossing over in interstitial segments coupled with alternate segregation.

MULTIPLE TRANSLOCATION SYSTEMS

When translocations involve more than two nonhomologous pairs of chromosomes, meiotic rings containing six, eight, or more chromosomes can be found (Fig. 22–26). Surprisingly, such events are not always rare, and certain plant species and genera almost exclusively maintain multiple translocation systems. Foremost among these are many of the breeding groups or races in the genus *Oenothera*, the evening primrose. It was during his investigations of the peculiarities of *O. lamarckiana* that DeVries first proposed the term "mutation" to account for the appearance of many new hereditary types in this species. Since then these plants have been studied by Cleland, Emerson, Renner, and many others (reviewed in Cleland, 1972).

The puzzling behavior of *Oenothera* centers, in brief, around three unusual features:

1. Some of its races produce new hereditary types at a frequency that is much higher than that commonly expected for mutations. Furthermore, some of these new types involve a change in many plant characteristics simultaneously, rather than the simple changes usually caused by single gene mutations.
2. Many *Oenothera* races, such as *O. lamarckiana*, produce seeds that are about 50 percent lethal when ordinarily self-pollinated but fully viable when outbred to other races, such as *O. biennis* and *O. muricata*.
3. Although practically all *Oenothera* races have seven pairs of chromosomes, the first meiotic metaphase configurations range from seven individual bivalents (*O. hookeri*) through various combinations of rings and bivalents to a single ring of 14 chromosomes (*O. biennis*, *muricata*, and many others).

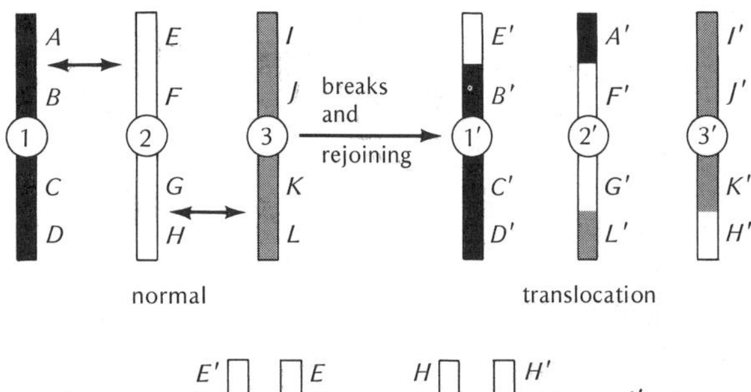

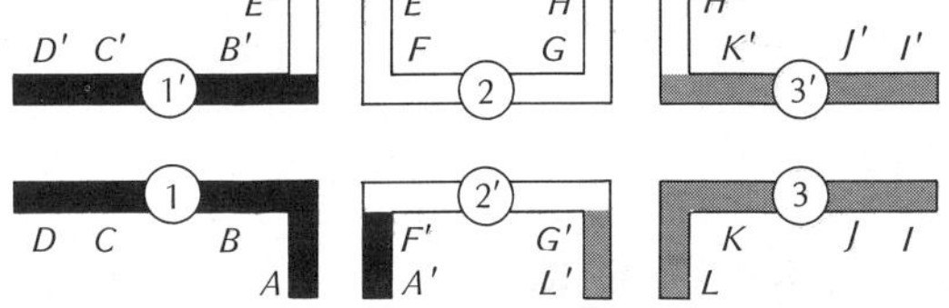

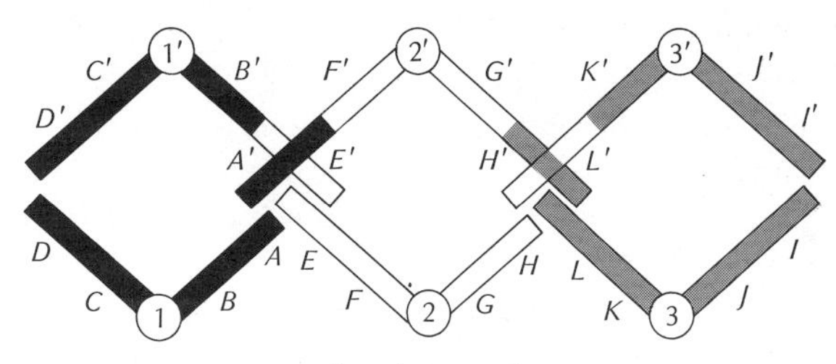

Figure 22–26

Pairing and alternate segregation in a multiple translocation heterozygote involving three pairs of nonhomologous chromosomes.

To explain these events, Belling and others proposed that translocations between numerous chromosomes were involved. If we symbolize the arms of each of the seven pairs of metacentric chromosomes of the *O. hookeri* race as 1·2, 1·2, 3·4, 3·4, 5·6, 5·6, . . . , 13·14, 13·14, each of the other races can be described as bearing different combinations of translocations between these chromosomes. *O. lamarckiana*, for example, has the combination 1·2, 1·2, 3·4, 4·12, 12·11, 11·7, 7·6, 6·5, 5·8, 8·14, 14·13, 13·10, 10·9, 9·3. Note that each chromosome arm appears in diploid condition, but, with the exception of the first chromosome pair (1·2, 1·2), a numbered arm (e.g., 4) is borne on two chromosomes whose accessory arms are different (e.g., 3·4, 4·12). Thus, as shown in Fig. 22–27, meiotic pairing between these chromosomes produces one bivalent and a ring of 12. Since alternate segregation was almost exclusively observed for these rings, duplications and deficiencies were generally absent and entire translocation *complexes* segregated as a unit in each gamete. In *O. lamarckiana* alternate segregation in the ring of 12 gave two complexes: 3·4, 12·11, 7·6, 5·8, 14·13, 10·9, and 4·12, 11·7, 6·5, 8·14, 13·10, 9·3, each also bearing the 1·2 chromosome of the segregating bivalent. Any other type of segregation in the heterozygote would of course produce unbalanced gametes. Thus each complex of six chromosomes in this case would be considered as a single linkage group. According to terminology invented by Renner, the two complexes above were named *velans* and *gaudens*, respectively.

What was both interesting and surprising, however, was that *O. lamarckiana* bred true to these complexes and did not produce either *velans*/*velans* or *gaudens*/*gaudens*, although both homozygotes are chromosomally balanced. (Such homozygotes should be cytologically unique by possessing seven bivalents during meiosis.) The answer to this puzzle came from Muller's observation of balanced lethal genes in *Drosophila* (p. 217). Apparently recessive lethals were maintained in both the *velans* and *gaudens* complexes, so that homozygous combinations were lethal. In *lamarckiana* this lethality affected the zygotes, so that half the seeds did not germinate. In other *Oenothera* complexes, gametophytes were affected so that only the male gametophytes of one complex were viable in the female gametophytes of the other; i.e., only translocation heterozygous

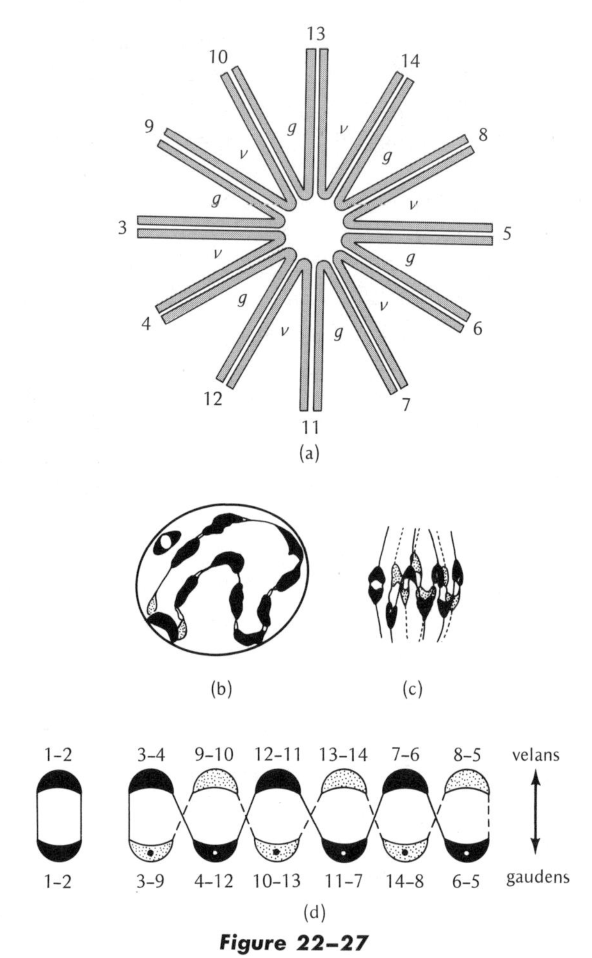

Figure 22–27

Chromosome configuration during meiosis in O. lamarckiana. (a) Diagrammatic representation of pairing between the six pairs of translocated chromosomes to produce a ring of twelve. (b) Appearance of the twelve chromosomes and the one bivalent during diakinesis. (c) Alternate segregation of the chromosomes in the ring during anaphase. (d) Interpretation of chromosome segregation in terms of the chromosome arms of the two complexes velans and gaudens, which are represented as v and g in a. (a, b, and c after Emerson, 1935; d after Emerson, 1936.)

combinations could be produced. Mechanisms such as this helped explain why outcrossing a race such as *lamarckiana* to other races produced fully viable seeds. Evidently different recessive lethals were involved in different complexes. Moreover, outcrossing *velans/gaudens* to different *Oenothera* races such as *hookeri* produced two different F_1 phenotypes, since both *velans* and *gaudens* could now cause their own distinctive effects. Since there was no alternate segregation mechanism for the seven bivalents of the *hookeri* race, it could be considered homozygous and free of recessive lethals, i.e., *hookeri/hookeri*.

Interbreeding different *Oenothera* races also provided the opportunity to discover which chromosome ends were homologous. Since the chromosomes were all metacentric and of about equal size, they could not be superficially distinguished from each other and their identification had to be made through the following ingenious technique developed by Emerson and Sturtevant.

It was obvious that many alternative chromosome translocations were possible between the 14 different chromosome arms (e.g., 1·2, 1·3, 1·4, etc.) By accepting the *hookeri* sequence as "standard," or nontranslocated, other sequences could then be determined by mating different complexes, both with *hookeri* and each other. *Flavens* × *hookeri*, for example, produced a ring of four, and five bivalents. To name the chromosome arms it could be assumed that the *flavens/hookeri* ring was caused by a translocation in the first two chromosomes (1·2, 3·4). Since *hookeri* bore no translocations, the *flavens* complex could therefore be described as 1·4, 2·3, in addition to the five nontranslocated chromosomes (i.e., 5·6, . . . , 13·14). However, the *velans/hookeri* combination also formed a ring of four and five bivalents, but *velans/flavens* formed two rings of four and three bivalents. *Velans* thus has two nontranslocated chromosomes which *flavens* does not have (i.e., 1·2 and 3·4), and also has two translocated chromosomes which *flavens* does not have (thereby accounting for one ring with *hookeri* and two with *flavens*). Again, in order to symbolize them, the latter two chromosomes could be designated as a translocation from 5·6, 7·8 to 5·8, 6·7. Thus the entire *velans* complex could be described as 1·2, 3·4, 5·8, 6·7, 9·10, 11·12, 13·14.

It is easy to see that once a translocation ring of four has been established and is being maintained by balanced lethals in each of the two complexes, further translocations between the nontranslocated chromosomes and those on the ring will produce larger rings which are also maintained by the same balanced lethals. Each complex therefore tends to grow in the number of chromosomes involved, so that many *Oenothera* races now consist of complexes including all seven chromosomes, i.e., rings of 14. These seven-chromosome translocation complexes are, of course, not identical, e.g., *curvans* is 1·14, 2·3, 4·6, 5·13, 7·12, 8·9, 10·11, and *flectens* is 1·4, 2·3, 5·7, 11·13, 6·10, 8·9, 12·14. In all, almost 400 complexes, bearing 91 variations of the 14 chromosome ends, have been analyzed. The complex whose chromosome ends are represented most frequently throughout all the *Oenothera* races is one called *johansen*, bearing the formula 1·2, 3·4, 5·6, 7·10, 9·8, 11·12, 13·14. It has therefore been proposed that the *johansen* complex was ancestral to all others, and translocations of this sequence gave rise to the entire *Oenothera* group.

Although maintained by alternate segregation and balanced lethals, the chromosomes of the translocation complexes do not always disjoin in perfect fashion. Occasionally new translocations are formed, polyploids or trisomics occur, or crossing over between different complexes takes place. The fact that each such event can be accompanied by a radical change among the resultant offspring led to DeVries' conclusion that "mutation" in *Oenothera* was a frequent event with widespread effects. To demonstrate that such "mutations" really involve a reshuffling of already existing genes, let us consider the consequences of just two such events: crossing over near the centromeres of translocated chromosomes, and crossing over in one of their pairing segments.

As shown in Fig. 22–28, close pairing between homologous sections near the centromeres of chromosomes 13·14 and 11·7 in *O. lamarckiana* can produce an exchange which would yield segregation of an entirely new complex, 1·2, 3·4,

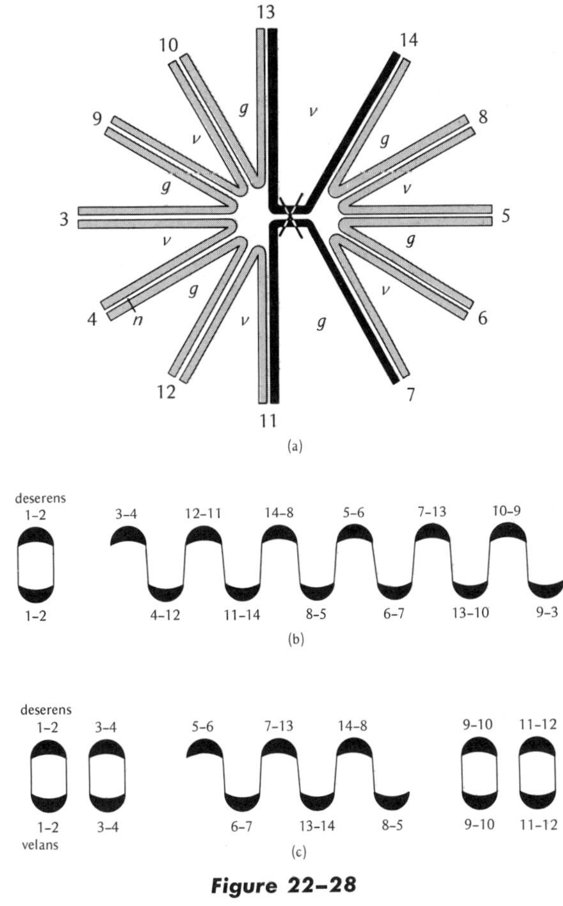

Figure 22–28

Illustration of how an interstitial crossover in a translocation heterozygote can lead to an entirely new translocation complex. (a) The velans-gaudens complexes of Oenothera, showing an interstitial crossover (X). (b) The resulting anaphase configuration, showing a new complex at one pole, deserens (1-2, 3-4, 12-11, 14-8, 5-6, 7-13, 10-9). (c) Pairing of the deserens complex with velans, named as the mutation rubrinervis. (After Emerson, 1935.)

12·11, 14·8, 5·6, 7·13, 10·9. This complex, named *deserens*, has chromosomes from both *velans* and *gaudens*, in addition to a new chromosome 7·13. It forms viable combinations with *velans*, producing a meiotic configuration of one ring of six chromosomes and four bivalents. The latter combination in turn segregates to produce 1 *deserens/deserens* (seven bivalents):2 *deserens/velans*:1 *velans/velans* (lethal). The origin of this *deserens/velans* combination at a frequency of about 6 per 10,000 plants is therefore misleading as a "mutation" but completely understandable as a rare crossover. Similarly crossing over in the pairing segment of chromosome arm 4 leads to a transfer of the *nanella* gene (*n* in Fig. 22–28) from the *gaudens* complex to the *velans* complex. Such *velans* gametes that fertilize *gaudens* will now produce phenotypically distinct homozygotes for *nanella*. The *nanella* "mutant" phenotype thus arises through crossover events at a frequency (about 3 per 1000 plants) much higher than expected for ordinary mutations.

On the whole the *Oenotheras* are a successful, widespread group in spite of their many reciprocal translocations, lethals, and self-pollination. In fact we may consider that the combination of these three deleterious attributes in a single plant compensates for each of them. Certainly the maintainance of translocation complexes that include many chromosomes enables considerable heterozygosity to be maintained. The plant thus takes advantage of "hybrid vigor" from a number of chromosomes simultaneously (see Chapter 33). In addition the presence of lethals enforces heterozygosity, and self-pollination ensures a plentiful supply of pollen and prevents outcrossing to other races, which might break up the translocation complex.

In the plant genus *Rhoeo* (represented by only one species, *R. discolor*), the advantages of translocation complexes have led to a complete loss of strains in which bivalents can be found. Thus, all 12 of its chromosomes always form a single ring in meiosis. Other plants (*Campanula percicifolia*, *Clarkia*, *Paeonia*) maintain rings, but these have not been stabilized by lethals, and homozygous bivalents can be formed.

COMPARISON BETWEEN GENETIC AND CYTOLOGICAL MAPS

The opportunity to locate genes cytologically in deficiencies and duplications as well as at the breakage points of translocations and inversions permits cytological maps of gene loci to be constructed. Thus *Drosophila* maps can be drawn based on mitotic metaphase chromosomes, on salivary chromosomes, and on genetic linkage distances. In Fig. 22–29 a comparison between all three such maps in *Drosophila melanogaster* shows that the linear order of genes is the same in each, but the relative distances between the genes are markedly different. This is especially true for genes lying near the centromeres. The second chromosome, for example, has a linkage distance of 108 map units, with the centromere at locus 55. The *black* gene (*b*) is recombinationally 6.5 units to the left of this centromere, or only about 1/10 of the linkage distance from the centromere to the end of the left arm. It is, however, more than 1/4 of this distance in the salivary chromosome, and about midway along this arm in the metaphase chromosome. *Cinnabar* (*cn*), which lies recombinationally only 2.5 crossover units to the right of this same centromere, is cytologically located about halfway along the right arm of the metaphase chromosome. Thus two genes that are separated by 9 recombinational units out of a total of 108 for the chromosome are cytologically separated by half or more of the chromosome length. The remaining 99 recombinational units in the second chromosome are thus confined to genes in the remaining halves of each arm that are distal to the centromere. This therefore indicates that crossing over between genes near the centromeres is relatively restricted compared to genes that are cytologically further away. According to Mather this restriction is probably caused by the centromere itself.

Note that in Fig. 22–29 the areas near the centromeres are mainly heterochromatic and contain many fewer genes than equivalent euchromatic segments. The Y chromosome, which has only a single known gene and perhaps a few fertility factors, shows no recombination and is entirely heterochromatic. Thus, in *Drosophila*, as in many other organisms, heterochromatinization seems to have arisen as a mechanism that permits genes to become "inert" in areas where crossing over is restricted. However, as we shall see in Chapter 23, heterochromatin can nevertheless exert an influence on surrounding genes through "position effects."

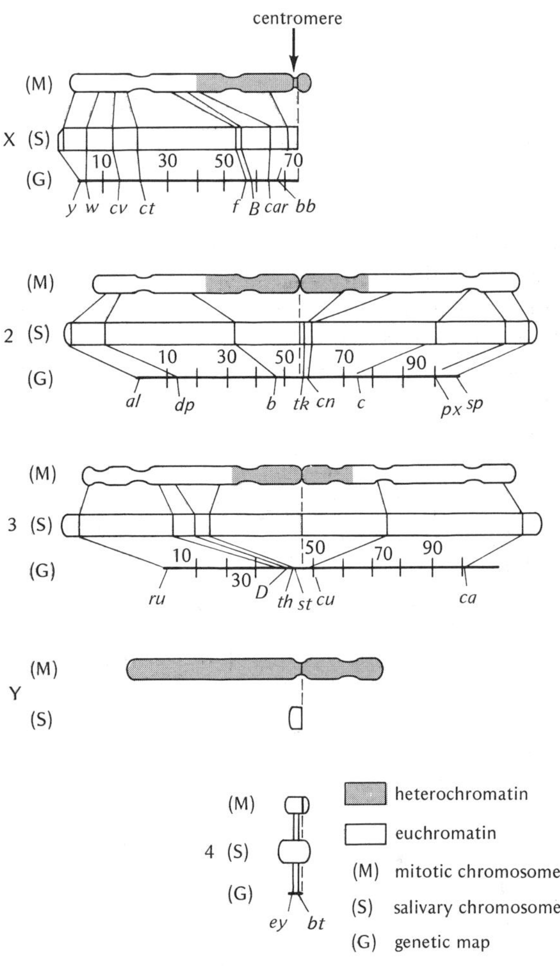

Figure 22–29

Comparison of gene locations of Drosophila melanogaster *mitotic chromosomes, salivary chromosomes, and linkage maps. The size of the salivary chromosomes compared to the mitotic chromosomes is considerably reduced, but the relationships between different salivary chromosomes are the same as are found in the salivary cell. The amount of heterochromatin is much less in salivary cells than in other cells and is almost entirely restricted to the chromocenter in which the centromeres of all salivary chromosomes are united. The Y chromosome, being almost entirely heterochromatic, is almost invisible in the salivary cell. Since there are no crossover data for the Y chromosome, its genetic map is omitted. (After Hannah.)*

STRUCTURAL CHANGES THAT LEAD TO CHANGES IN NUMBER

In addition to the changes in the configuration of individual chromosomes, a further visible consequence of chromosome rearrangements can be found in changes in chromosome number. Such numerical differences may arise from aberrations that lead to the fusion of two formerly separate nonhomologous chromosomes into a single chromosome, or to the separation of formerly joined sections of a single chromosome into two chromosomes.

As shown in Fig. 22–30, fusion is possible when a reciprocal translocation occurs between two nonhomologous acrocentrics leading to formation of a new metacentric chromosome and a small "dot" chromosome. Since the new dot chromosome will most likely be composed of nonfunctional heterochromatin associated with the previous centromeres, it probably can be lost without ill effect. Fusions of this type, called *centric fusions*, thus lead easily to a reduction in chromosome number.

The reverse process, splitting of a metacentric into two acrocentrics, may occur through perhaps two mechanisms. One mechanism, called *dissociation* by White, is also shown in Fig. 22–30, and involves the translocation of a metacentric chromosome arm to a recipient centromere located on an available dot or supernumerary chromosome. Another proposed mechanism involves the direct *fission* of the centromere of a metacentric chromosome by transverse rather than longitudinal splitting (called *misdivision* by Darlington) leading to

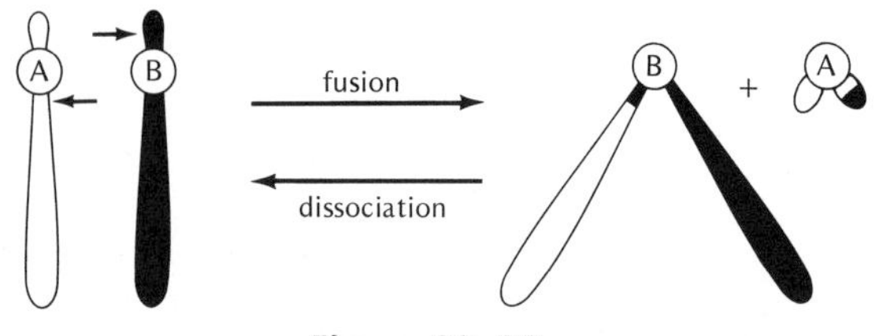

Figure 22–30

Diagram of translocations (small arrows) that lead to fusion between the arms of two acrocentric nonhomologous chromosomes (A and B) to form one metacentric chromosome and a dot chromosome carrying only a small amount of chromosomal material. If the dot chromosome is lost, a reduction of 2 in the chromosome number will take place if the frequency of individuals with the metacentric increases sufficiently in subsequent generations to form metacentric homozygotes. The reverse process of dissociation can also occur, and involves reciprocal translocations between the B metacentric and the A dot chromosome leading to A and B acrocentrics.

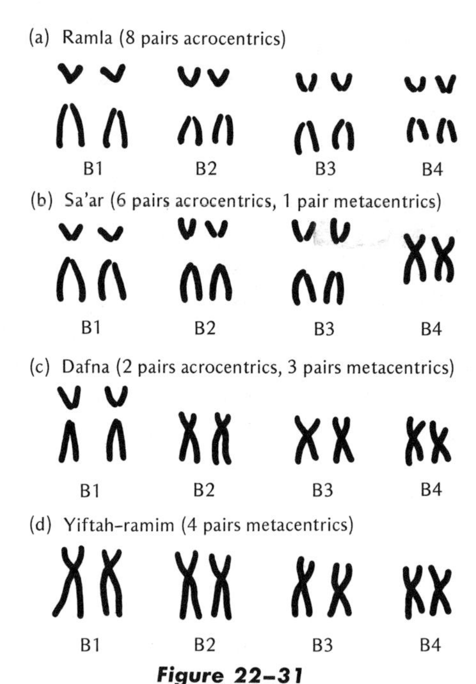

Figure 22–31

Some of the metaphase chromosomes found in each of four different Israeli populations (a–d) of the rodent Spalax ehrenbergi. Actually these rodents each carry more than 25 pairs of chromosomes, but only the "B" group of chromosomes are shown. Within this group, significant changes in number have occurred as acrocentrics are transformed into metacentrics, or vice versa. Thus, the B1 pairs of acrocentrics found in Ramla, Sa'ar, and Dafna populations are believed to correspond to the chromosome arms found in the B1 metacentrics of the Yiftah-ramim population, and the B2 acrocentrics in Ramla and Sa'ar correspond to the B2 metacentrics in Dafna and Yiftah-ramim, etc. This correspondence between metacentrics in one population to acrocentrics in another was first pointed out by Robertson in other species, and the mechanism responsible has consequently been called "Robertsonian changes" or "Robertsonian translocations." (After Wahrman, Goitein, and Nevo.)

two independent telocentric chromosomes each associated with a section of the former centromere. However, since the existence of purely telocentric chromosomes has been questioned by many cytologists (see pp. 14, 498), it is possible that some additional chromosomal material is added by further rearrangements to "cap" bare telocentric centromeres should they arise this way.

We can thus see that the cause for aneuploid changes in chromosome number need not be restricted to nondisjunctional events (Chapters 12 and 21). In fact the relatively high frequency of chromosome translocations and rearrangements found in most populations (one or more per thousand gametes) may be sufficient to account for at least some, if not a majority, of the aneuploid differences observed among populations and species. For example, Israeli populations of the mole rat, *Spalax ehrenbergi*, have differences in chromosome number ranging from 52 to 60. However, as shown by Wahrman and co-workers (Fig. 22-31), acrocentrics in some populations appear to correspond with the arms of metacentrics in other populations. Thus, although the number of chromosomes is different between these rodent populations because of fusion or splitting, the number of chromosome arms apparently remains the same.

Perhaps the most drastic change in chromosome number, which is at least partially caused by chromosome rearrangements, is that among deer species. Most deer have 25 or more pairs of chromosomes, whereas the Indian species, *Muntiacus muntjak*, has one population consisting of only three pairs of chromosomes. Obviously considerable chromosome fusion or splitting must have occurred, since it is unlikely that either large differences in the amount of essential genetic information exist between these species, or that tenfold polyploidy has taken place.

PROBLEMS

22-1. A plant that is homozygous dominant for a particular gene, i.e., *AA*, is treated with x-rays and fertilized by pollen from a homozygous recessive plant, i.e., *aa*. Out of 500 offspring, two showed the recessive phenotype. What explanation would you offer to account for this and how would you test your hypothesis?

22-2. A corn plant heterozygous for a deficiency in chromosome 9 is also heterozygous for the 9th-chromosome gene for colored aleurone; the deficient chromosome carries the dominant gene *C* producing coloring, while the normal 9th chromosome carries the recessive colorless allele *c*. Since the deficient chromosome cannot be passed through the pollen (see p. 499), what mechanism would explain the occurrence of 10 percent colored kernels from a cross between such a heterozygous plant used as pollen parent and a *cc* plant used as female parent?

22-3. *Drosophila melanogaster* females bearing the *Notch* deficiency (see p. 498), but otherwise wild type, produce some prune-eyed female offspring when mated to males carrying the sex-linked recessive gene *prune*. What sexual and phenotypic ratios would you expect from the mating of such prune-eyed females to normal males?

22-4. Draw a picture of pairing between two homologous chromosomes in the salivary nuclei of *Drosophila*, one of which has the sequence 1·234567, and the other has the sequence 1·265437. (The dot between 1 and 2 represents the centromere.)

22-5. If a crossover occurred during meiosis between loci 3 and 4 in such a heterozygote, what would the crossover products look like?

22-6. Draw a picture of pairing between two *Drosophila* salivary chromosomes, one of which has the sequence 1·23456789 and the other the sequence 1·27843659.

22-7. Two chromosomes pair in meiosis to produce the following configuration:

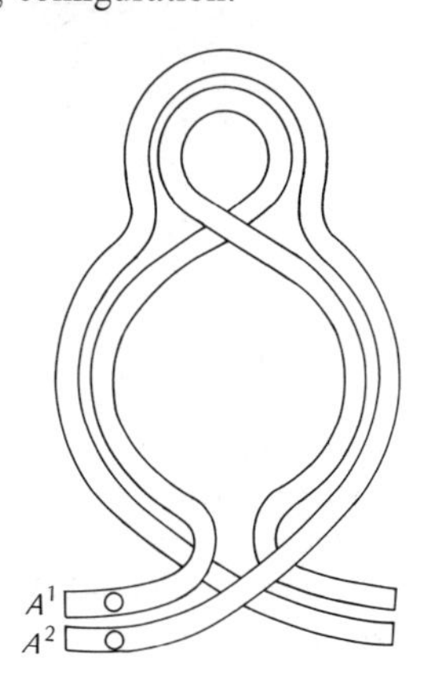

(a) If chromosome A^1 has the sequence 1 · 23456789 . . ., what is the sequence of chromosome A^2? (b) If chromosome A^1 possesses the original sequence and chromosome A^2 possesses the final derived sequence, what inversions occurred, and in what probable order?

22–8. The normal linkage-map positions of five 2nd-chromosome loci in *Drosophila melanogaster* is *purple*, 54.5; *vestigial*, 67.0; *Lobe*, 72.0; *plexus*, 100.5; *speck*, 107. However, a cross + + *L* + + /*pr vg* + *px sp* ♀ × *pr vg* + *px sp*/*pr vg* + *px sp* ♂ produced the following recombination percentages: *pr-vg*, 9.0; *vg-L*, 0.1; *L-px*, 1.3; *px-sp*, 4.0. If the heterozygous female was an inversion heterozygote, what areas are involved?

22–9. Investigations of gene sequences on a particular chromosome show the presence of the following three arrangements: (1) *ABCDEFGHIJ*, (2) *ABCHGFIDEJ*, and (3) *ABCHGFEDIJ*. Explain the evolutionary relationships between these arrangements.

22–10. A particular chromosome in a stock of *Drosophila* is known to have the gene sequence *ABCDEFGHIJ*. After many generations of treatment designed to induce chromosomal aberrations, three chromosomes are examined and found to have the following gene sequences: (a) *ABHGFEDCIJ*, (b) *ADCBEFGIHJ*, and (c) *AFEIHGBCDJ*. For each of these three chromosomes, indicate the specific changes that most probably occurred to produce the observed sequence.

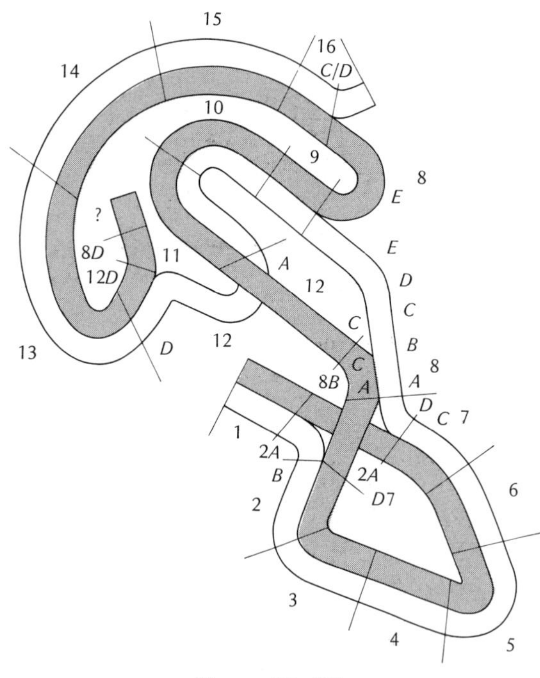

Figure 22–32

Two paired homologous salivary chromosomes in Drosophila subobscura, *differing from each other by inversions. Numbers indicate main sections of the chromosome; letters indicate subunits within such sections. (After W. Götz, 1965. Zeit. Vererbung.,* ***96,*** *285–296.)*

22–11. Figure 22–32 contains a diagrammatic representation of salivary chromosome pairing between two homologous chromosomes in *Drosophila subobscura*. Considering the white chromosome as the original sequence (1-2-3-4-. . . .-16), indicate the succession of inversions that must have occurred in this chromosome to obtain the gray chromosome.

22–12. Humans have been found who are normal in appearance and whose parents appear normal but whose no. 21 chromosomes are both attached by a common centromere (isochromosome). (a) Explain whether this karyotype most likely derives from a mitotic accident in the embryonic stages of the observed isochromosome individuals, or from a meiotic accident in the parental gametes. (b) If a normal individual carrying a 21/21 isochromosome were to mate with an individual of normal karyotype, what proportion of the offspring would you expect to show Down's syndrome?

22–13. The following is part of a pedigree in which Down's syndrome (shaded symbols) appears in two related individuals:

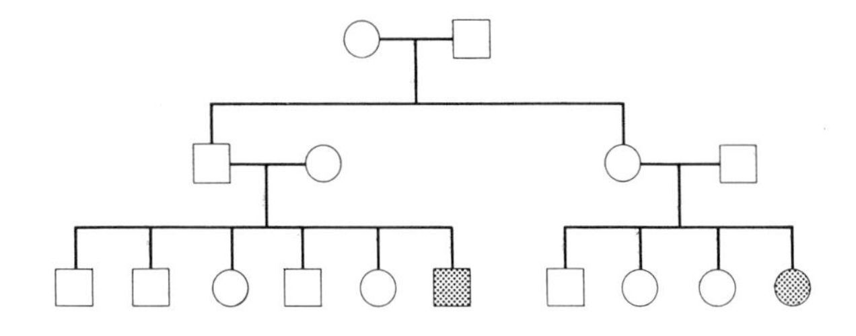

(a) Assuming that persons coming from outside the pedigree are fully normal (i.e., II–2 and II–4), what common chromosomal cause would you suggest to account for the existence of this defect in both related individuals? (b) What are the probable chromosome numbers in individuals II–1 and II–3?

22–14. Diagram an inversion loop at pachytene showing all four strands and indicate the crossovers needed to give the following structures at the first anaphase: (a) One loop and one fragment. (b) Two bridges and two fragments.

22-15. A corn plant heterozygous for a long paracentric inversion has an equal number of single and double crossovers within the inversion limits, but no noncrossovers. Assuming that chromatid interference is absent (see Chapter 18, p. 375), what proportions of cells with various abnormal chromosome configurations (dicentrics, acentrics) would you expect to find among the meiotic products of this plant at anaphase I?

22-16. "Ring" chromosomes have been found in man and other organisms, and consist of a single circular chromosome with a single centromere (see p. 483). In *Drosophila*, ring-X chromosomes are viable both in hemizygous (male) and homozygous (female) condition. If a *Drosophila* female were heterozygous for a rod-X and a ring-X, what type of eggs would she produce after the following exchanges: (a) A single crossover between the two X chromosomes? (b) A two-strand double exchange? (c) A four-strand double exchange?

22-17. The ring-X chromosome in *Drosophila* is often irregularly distributed during mitosis, causing mosaicism. It can also produce mosaicism by undergoing somatic crossing over (see Chapter 18). Design an experiment, using the *yellow* and *singed* sex-linked markers, to determine what percentage of mosaicism is caused by each process.

22-18. Normally in corn the genes *waxy* and *virescent* are linked. In a certain stock, however, Burnham found that these two genes assort independently. (a) Which of the chromosomal aberrations discussed in this chapter would you suggest to explain these results? (b) Which chromosomal aberration would produce the reverse effect, that is, interfere with expected independent assortment between genes?

22-19. Semisterility in plants may be the result of either a heterozygous translocation or a heterozygous lethal gene. Discuss the methods you would use to distinguish between these two situations and the findings you would look for.

22-20. *Drosophila melanogaster* males heterozygous for the following three mutations: the 2nd-chromosome dominant *Bristle* (*Bl*), the 3rd-chromosome dominant *Dichaete* (*D*), and the 4th-chromosome recessive *eyeless* (*ey*) were crossed individually to females homozygous for *eyeless*. Most of these crosses produced the expected eight phenotypic classes of offspring, *Bl D* +, *Bl D ey*, *Bl* + +, *Bl* + *ey*, + *D* +, + *D ey*, + + +, + + *ey*. Six males, however, produced a limited number of phenotypes, as follows:

(a) *Bl D* +, *Bl D ey*, + + +, + + *ey*
(b) *Bl D ey*, *Bl* + +, + *D ey*, + + +
(c) *Bl D ey*, *Bl* + *ey*, + *D* +, + + +
(d) *Bl* + +, *Bl* + *ey*, + *D* +, + *D ey*
(e) *Bl D* +, *Bl* + *ey*, + *D* +, + + *ey*
(f) *Bl D* +, *Bl* + +, + *D ey*, + + *ey*

For each of these males give the simplest explanation to account for the restricted number of phenotypes among their progeny.

22-21. A *Drosophila* male, heterozygous for the genes *Bristle* and *Dichaete*, was crossed to an attached-X female ($\widehat{XX}Y$) and found to produce male offspring that were either *Bl D* or *Bl* +, and female offspring that were either + + or + *D*. When these *Bl D* male offspring were mated to attached-X females, the offspring produced were the same as in the parental cross. What accounts for these results?

22-22. A *Drosophila* male heterozygous for *Bristle* and *Dichaete* mated to an attached-X female produced males that were either *Bl* + or + + and females that were either *Bl D* or + *D*. When mated to normal wild-type males, however, the + *D* females always produced + *D* sons and wild-type daughters. What accounts for these results?

22-23. A group of workers at the University of Texas mated wild-type *Drosophila* males to attached-X *yellow* females ($\widehat{yy}$) that were also homozygous for the recessive autosomal genes *brown* (2nd chromosome), *ebony* (3rd chromosome), and *eyeless* (4th chromosome). The F_1 males produced from this cross were then backcrossed individually to homozygous mutant females from the maternal stock. Phenotypes of the F_2 progeny were then scored to discern the kinds of translocations that may have occurred in the parental wild-type males. What F_2 phenotypes would you expect to find as a result of the following kinds of translocation (numbers signify chromosomes involved): (a) 2;3? (b) 3;4? (c) 2;3;4? (d) X;3? (e) X;2;4? (f) X;2;3;4?

22-24. Brink and Cooper obtained a stock of corn plants heterozygous for a translocation between two chromosomes and therefore semisterile. They crossed these plants to a chromosomally normal stock homozygous for the recessive gene *brachytic* on chromosome 1 and then backcrossed the semisterile F_1 progeny to the *brachytic* parental stock. The F_2 showed the following phenotypes:

Wild Type		Brachytic	
Semisterile	Fertile	Semisterile	Fertile
334	27	42	279

(a) What ratios would you have expected if the chromosome carrying *brachytic* was not involved in the translocation? (b) What is the linkage distance between *brachytic* and the translocation point?

22-25. The same translocation heterozygote stock as above was also crossed to a normal stock homozygous for two different recessive genes on chromosome 2, *liguleless* (*lg*) and *virescent*-4 (*v*-4). F_1 semisterile translocation heterozygotes were then backcrossed to the 2nd-chromosome mutant stock, and the F_2 progeny obtained were as follows:

	Semisterile	Fertile
wild type	163	25
liguleless	126	18
virescent	14	124
liguleless and *virescent*	13	105

What are the linkage positions of the translocation point and the two genes on chromosome 2?

22-26. Two individual plants, with a total of four chromosomes each, show all chromosomes connected in a multiple translocation ring during meiosis. When these plants are crossed with each other, three-fourths of their progeny show translocation rings of four chromosomes, and one-fourth show the four chromosomes paired as two bivalents. Using the 1·2, 3·4 system of naming chromosome arms, give the chromosome constitutions of each plant.

22-27. Two plants each have a ring of six chromosomes at meiosis. Using the 1 · 2, 3 · 4 system of naming chromosome arms, indicate the members of each ring if these two plants on crossing yield progeny:

1/4 with three bivalents
1/4 with a ring of four and one bivalent
1/2 with a ring of six

22-28. In the telocentric X chromosome of *Drosophila melanogaster*, a particular gene order is centromere-*a*-*b*-*c*-*d*-*e*. If each of these loci is genetically observed to be five recombinational units from one another, and crossing over rarely takes place close to the centromere, make a drawing comparing the genetic distances along this chromosome to the physical distances.

REFERENCES

ALEXANDER, M. L., 1952. The effect of two pericentric inversions upon crossing over in *Drosophila melanogaster*. *Univ. Texas Publ.*, **5204**, 219–226.

BEADLE, G. W., and A. H. STURTEVANT, 1935. X-chromosome inversions and meiosis in *Drosophila melanogaster*. *Proc. Nat. Acad. Sci.*, **21**, 384–390.

BELLING, J., 1927. The attachment of chromosomes at the reduction division in flowering plants. *Jour. Genet.*, **18**, 177–205.

BRIDGES, C. B., 1917. Deficiency. *Genetics*, **2**, 445–465.

———, 1919. Vermilion-deficiency. *Jour. Gen. Physiol.*, **1**, 645–656.

———, 1923. The translocation of a section of chromosome II upon chromosome III in *Drosophila*. *Anat. Rec.*, **24**, 426–427.

BRIDGES, C. B., E. N. SKOOG, and J. C. LI, 1936. Genetical and cytological studies of a deficiency (*Notopleural*) in the second chromosome of *Drosophila melanogaster*. *Genetics*, **21**, 788–795.

BURNHAM, C. R., 1962. *Discussions in Cytogenetics*. Burgess, Minneapolis, Chaps. 2–5.

CARSON, H. L., 1946. The selective elimination of inversion dicentric chromatids during meiosis in the eggs of *Sciara impatiens*. *Genetics*, **31**, 95–113.

CATCHESIDE, D. G., 1940. Structural analysis of *Oenothera* complexes. *Proc. Roy. Soc. Lond.* (B), **128**, 509–535.

CLELAND, R. E., 1972. *Oenothera: Cytogenetics and Evolution*. Academic Press, London.

CREIGHTON, H. B., 1937. White seedlings due to homozygosity of a deficiency in chromosome IX of *Zea mays*. *Genetics*, **22**, 189–190.

DEMEREC, M., and M. E. HOOVER, 1936. Three related X-chromosome deficiencies in *Drosophila*. *Jour. Hered.*, **27**, 207–212.

DEVRIES, H., 1901 and 1903. *Die Mutationtheorie*. Veit, Leipzig. (Translated into English and reprinted 1909, Open Court Publ. Co., Chicago.)

DOBZHANSKY, TH., 1934. Studies on chromosome conjugation. III. Behaviour of duplicating fragments. *Zeit. Induk. Abst. u. Vererbung.*, **68**, 134–162.

———, 1944. Chromosomal races in *Drosophila pseudoobscura* and *D. persimilis*. *Carneg. Inst. Wash. Publ. No. 554*, Washington, D.C., pp. 47–144.

EMERSON, S. H., 1935. The genetic nature of DeVries' mutations in *Oenothera lamarckiana. Amer. Naturalist*, **69,** 545–559.

———, 1936. The trisomic derivatives of *Oenothera lamarckiana. Genetics*, **21,** 200–224.

HANNAH, A., 1951. Localization and function of heterochromatin in *Drosophila melanogaster. Adv. in Genet.*, **4,** 87–125.

KAUFMANN, B. P., 1954. Chromosome aberrations induced in animal cells by ionizing radiations. In *Radiation Biology,* Vol. 2, A. Hollaender (ed.). McGraw-Hill, New York, pp. 627–711.

MCCLINTOCK, B., 1941. The stability of broken ends of chromosomes in *Zea mays. Genetics*, **26,** 234–282.

MATHER, K., 1939. Crossing over and heterochromatin in the X chromosome of *Drosophila melanogaster. Genetics*, **24,** 413–435.

MULLER, H. J., 1938. The remaking of chromosomes. *Collect. Net*, **8,** 182–195.

NOVITSKI, E., 1952. The genetic consequences of anaphase bridge formation in *Drosophila. Genetics*, **37,** 270–287.

SLIZYNSKA, H., 1938. Salivary gland analysis of the *white-facet* region of *Drosophila melanogaster. Genetics*, **23,** 291–299.

STERN, C., 1926. Eine neue Chromosomenaberration von *Drosophila melanogaster* und ihre Bedeutung für die Theorie der linearen Anordnung der Gene. *Biol. Zentralbl.*, **46,** 505–508.

STURTEVANT, A. H., 1921. A case of rearrangement of genes in *Drosophila. Proc. Nat. Acad. Sci.*, **7,** 235–237.

———, 1926. A crossover reducer in *Drosophila melanogaster* due to inversion of a section of the third chromosome. *Biol. Zentralbl.*, **46,** 697–702.

STURTEVANT, A. H., and G. W. BEADLE, 1936. The relation of inversions in the X chromosome of *Drosophila melanogaster* to crossing over and disjunction. *Genetics*, **21,** 554–604.

STURTEVANT, A. H., and C. R. PLUNKETT, 1926. Sequence of corresponding third-chromosome genes in *Drosophila melanogaster* and *D. simulans. Biol. Bull.*, **50,** 56–60.

WAHRMAN, J., R. GOITEIN, and E. NEVO, 1969. Geographic variation of chromosome forms in *Spalax*, a subterranean mammal of restricted mobility. In *Comparative Mammalian Cytogenetics,* K. Benirschke (ed.). Springer Verlag, New York and Heidelberg, pp. 30–48.

WESTMORELAND, B. C., W. SZYBALSKI, and H. RIS, 1969. Mapping of deletions and substitutions in heteroduplex DNA molecules of bacteriophage lambda by electron microscopy. *Science*, **163,** 1343–1348. (Reprinted in the collection of Zubay and Marmur; see References, Chapter 1.)

WURSTER, D. H., and K. BENIRSCHKE, 1970. Indian muntjac, *Muntiacus muntjak:* A deer with a low diploid chromosome number. *Science*, **168,** 1364–1366.

23
GENE MUTATION

Darwin's theory of evolution presented in 1859 was based on the hypothesis that natural selection acted upon the hereditary variability present in a species to preserve only those types that were better adapted to their environment. In support of his theory Darwin gathered many examples showing that such hereditary variability occurred. For example, in domesticated animals, where breeding records had long been kept, numerous instances had been recorded of novel types called "sports" suddenly appearing. Short-legged varieties of sheep and dogs, tailless cats, and similar inherited anomalies could occasionally be traced to their origin in single animals. However, Darwin himself, and most of his followers, believed that evolution proceeded in a more gradual quantitative fashion than in large "sudden steps." Evidence for the "gradual" view of evolution could be seen in the many small differences between varieties of the same species and sometimes even between related species. The search of most evolutionists was, therefore, for small effects, or "continuous variation."

In the late nineteenth century, prior to the discovery of Mendel's work, DeVries and others had put forth a theory that hereditary changes in nature, rather than having been small and continuous, may well have been large and discontinuous. DeVries called these large effects *mutations* and gathered evidence for the high frequency of such mutations from that most extraordinary plant, the evening primrose. As we have seen previously, *Oenothera lamarckiana* maintains numerous chromosomes that are translocation heterozygotes. When crossing over occurs between translocation heterozygotes, recombination may produce individuals that are strikingly different from their immediate parents. In essence, therefore, such individuals result from recombination between old genetic material, although the novelty of their appearance seems to indicate a completely new or mutant effect. However, neither DeVries nor his contemporaries were aware that the "mutations" they observed had this recombinational basis. The controversy that ensued between the "mutationists" and those who believed in small steps of continuous variation lasted until considerable evidence demonstrated that many inherited characters were affected by numerous factors, each with small quantitative effects (Chapter 14). Finally, the concept of mutation was broadened to include the origin of hereditary effects, large and small, that could be traced to unique changes within the genetic material. Although this was a notable step forward, it included many broad categories of effects, such as euploidy, aneuploidy, translocations, and so on (Chapters 21 and 22). In recent years the term mutation has generally been restricted to processes that result in a direct alteration of gene contents unrelated to observable chromosome changes. Thus *point mutation* or *gene mutation* has come to mean the process by which new alleles of a gene are produced. It is the detection and measurement of gene mutation which will be mainly discussed in the present chapter.

DETECTION OF MUTATION IN *DROSOPHILA*

Beginning with the experiments of Mendel, detection of the presence of a gene has generally depended upon the phenotypic changes produced when one allele of a gene is different from the other. Specifically it is the phenotypic changes caused by different alleles which enables us to identify the gene by observing its pattern of inheritance. Certainly, if all alleles of a particular gene were alike in their effect, such a gene would lack any unique phenotypic identification but would always appear as part of the "normal" phenotype, which includes the effects of numerous yet unknown genes. In other words, genes that have no allelic differences are not detectable by mendelian genetic techniques; it is only the presence of such allelic differences that enabled us to infer the existence of a particular gene. The search for gene mutation has therefore been oriented toward detecting alleles that produce new phenotypic effects.

In *Drosophila*, although hundreds of phenotypic differences were found by Morgan and co-workers, the detection of new mutations was not systematized in any quantitative fashion until various ingenious techniques were developed by Muller.

One of the most important of Muller's inventions involved a method by which certain *Drosophila* females could produce two types of males in equal proportions: one son carrying a marker X chromosome with *Bar* eye, and the other son carrying a wild-type X chromosome inherited through his mother from his male grandparent. Muller found that if the wild-type X chromosome derived from a sperm of the male grandparent contained recessive lethal genes, the wild-type grandsons would not appear in these crosses because of their lethality (see Fig. 12–14). Although lethal mutations are phenotypically more drastic than visible mutations, the changes in genetic material that produce lethals can be used as a partial indicator of the changes that produce visible mutations. Furthermore the detection of lethal genes rather than of viable visible mutations offers at least two advantages.

First, by restricting observations to whether wild-type males appear in the F_2, the presence or absence of lethal mutations can easily be scored. Such scores are therefore "objective," since they do not depend on the skill of the experimenter in noting the appearance of new visible mutant phenotypes such as subtle differences in eye color or body shape. Second, lethal mutations are much more frequent than viable visible mutations, as would be expected in species that have already undergone a high degree of evolution under present circumstances and are finely "tuned" to their environment. Random mutation will therefore little improve a species, since most mutations would have been tried throughout evolutionary history and the advantageous mutations already incorporated. Thus most new random mutations can be expected to be either deleterious or lethal.

A modern version of Muller's technique makes use of a stock called *Basc* or *Muller-5* which carries X chromosomes that contain *Bar* and *white-apricot* (Fig. 23–1). Included with these genes on the X chromosome are inversions that suppress recombination between the Muller-5 chromosome and any wild-type X chromosome. Thus Muller-5 females mated to wild-type males will produce F_1 females in which crossing over in the X chromosome does not occur. Such F_1 females consequently produce two types of males, Muller-5 and wild type, the latter inheriting an X chromosome derived from a sperm produced by the male grandparent. If these wild-type X chromosomes carry lethal mutations, the F_2 wild-type males will not appear. In this fashion numerous wild-type X chromosomes can be tested, each from an individual F_1 female in an individual culture, and the frequency of lethal mutations accurately scored in large samples.

A method that made use of attached-X females was devised for detecting the less frequently visible mutations in *Drosophila*. Such females, as described previously (p. 239), carry two X chromosomes connected by a single centromere, and an additional Y chromosome. When fertilized by normal males, four types of offspring are produced, of which two, XXX females and YY males, are

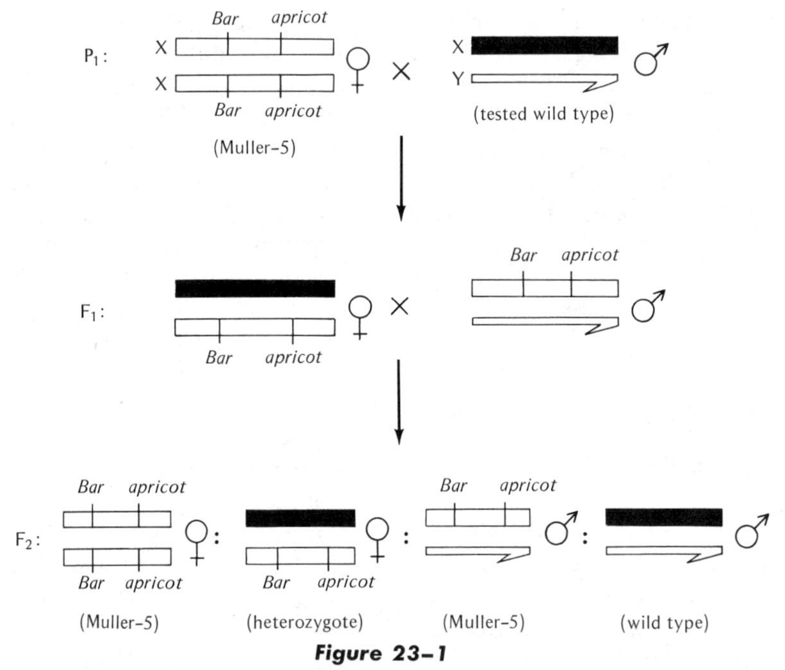

Figure 23–1

Mating system used to detect recessive lethals on the X chromosome. The wild-type parental male may, as a result of mutation, produce many kinds of X chromosomes in his sperm, some of which bear recessive lethal genes. An F_1 female is protected against sex-linked recessive lethals by the presence of the lethal-free Muller-5 (Bar-apricot) chromosome and will produce Muller-5 and wild-type male offspring in equal proportions if the wild-type X chromosome is fully viable. However, if the wild-type chromosome carried by an F_1 female bears a recessive lethal, no F_2 wild-type males will appear. To ensure a test of the complete wild-type X chromosome, the Muller-5 chromosome bears a number of inversions that prevent crossing over between the two X chromosomes in the F_1 female.

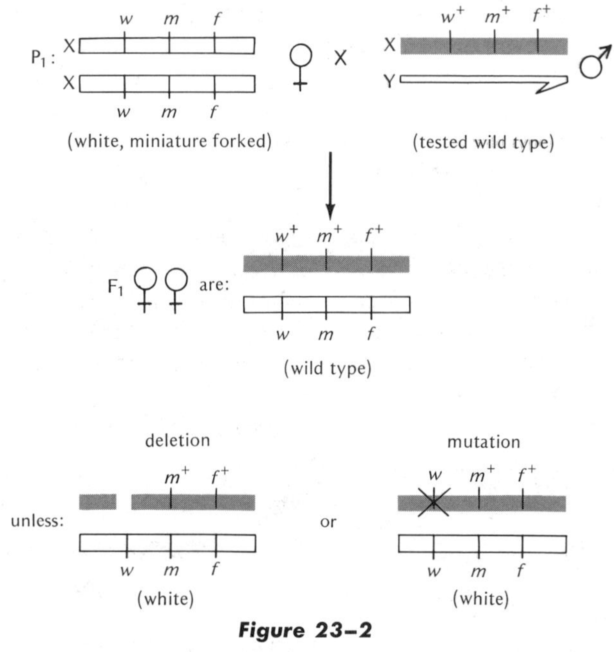

Figure 23–2

Technique used to detect mutations or deletions at three specific loci on the X chromosome, white, miniature, and forked. The parental wild-type male will generally produce wild-type X chromosomes in his sperm except for those in which mutation and deletions occur. If these mutations and deletions affect any of the three loci, w, m, and f, one or more F_1 females will show a mutant phenotype.

abnormal or inviable. The remaining two types are attached-X females and XY males, the males inheriting their Y chromosome from their mother and the X from their father. (The "detachment" of maternal attached-X chromosomes that have been transmitted to male offspring can be detected by using attached-X stocks homozygous for various sex-linked markers.) Thus F_1 males produced from a cross of this type will reveal all sex-linked recessive traits that may have been formed in the testes of their fathers. In contrast to the Muller-5 technique, which tests only a single X chromosome in each culture, the attached-X method tests numerous X chromosomes at one time in each culture, since each individual male offspring now represents a separate X chromosome contributed by its wild-type father. Considerable experience, however, is required to detect visible mutations, especially since many of them are not always strikingly different from wild type and are not always as viable.

When it is desired to test for mutations at specific loci, such as recessive mutations at the sex-linked loci, *white* (*w*), *miniature* (*m*), and *forked* (*f*), crosses can be made to a mutant stock of this type and the phenotype of the F_1 females observed. As is shown in Fig. 23–2, each F_1 female is formed from an X chromosome of the mutant stock, *w m f*, and one wild-type X chromosome. If mutations at any of these three loci have caused deletions or produced the recessive alleles *w*, *m*, or *f*, these can easily be noted by the appearance of the recessive phenotype among the F_1 females. Phenotypic mutations at other sex-linked loci can of course be tested, and Muller and his co-workers have devised many intricate stocks and techniques for such purposes.

In addition to sex-linked mutations tests have also been devised to detect autosomal mutations. For example, 2nd-chromosome recessive lethals can be detected by making such chromosomes completely homozygous. A balanced lethal stock carrying the dominant mutant markers *Curly* and *Plum* in each of the two second chromosomes, respectively, can be crossed to the wild-type stock that is to be tested (Fig. 23–3). The F_1 will consist of individual curly-winged and plum-eyed flies each constituted of a marker 2nd chromosome and a wild-type 2nd chromosome. Through the presence of inversions, the *Curly* marker chromosome is so constructed that recombination with the wild-type chromosome is suppressed. When individual curly-winged flies are backcrossed to the *Curly/Plum* stock, the offspring of each cross will consist of numerous flies heterozygous for the same wild-type 2nd chromosome. Crosses between two such heterozygotes (e.g., *Curly*/+ × *Curly*/+) should then furnish 2 *Curly*/+ : 1 *Curly/Curly*: 1 +/+. Since the *Curly* homozygotes are lethal, the phenotypic ratios of curly-winged flies to wild type should be 2:1 (instead of 3:1). Evidently any recessive deleterious gene in the wild-type chromo-

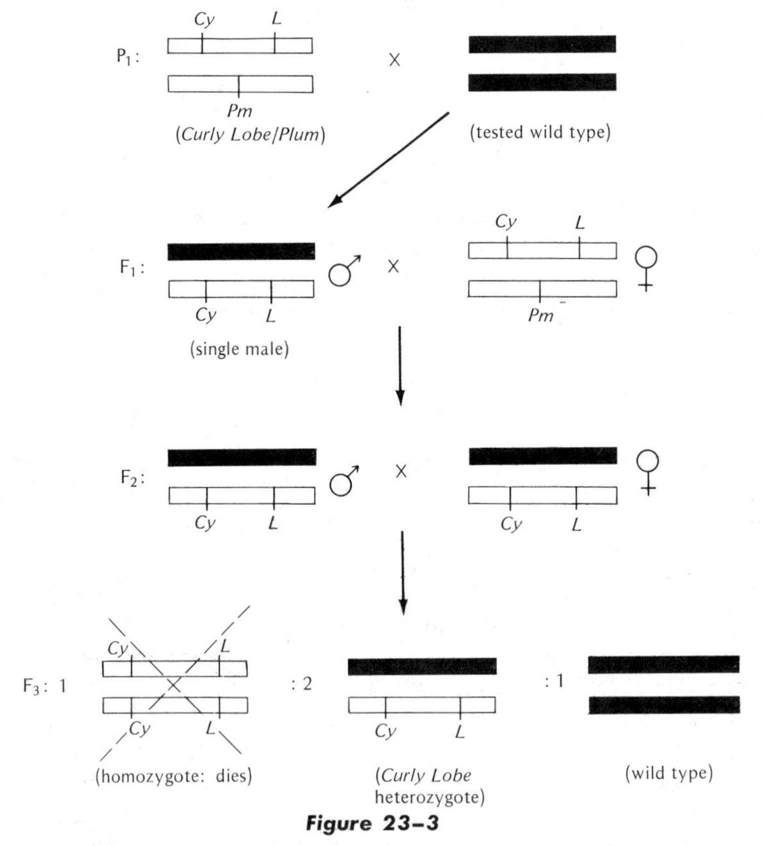

Figure 23–3

Test to detect recessive lethals on the 2nd chromosome of Drosophila melanogaster. The F_1 cross produces F_2 progeny, all bearing the same wild-type 2nd chromosome. An $F_2 \times F_2$ cross should then produce wild-type homozygotes in the ratio 1 wild type : 2 Curly Lobe heterozygotes if the tested chromosome is free of recessive lethals.

some will be made homozygous and result in lowering the 33.3 percent expected frequency of wild-type homozygotes. For example, recessive semilethal effects may produce only about 15 or 16 percent wild type. Completely lethal recessive genes on the second chromosome will result in the complete absence of the F_3 wild-type homozygotes.

Recessive autosomal mutations at specific loci can be detected by mating the wild-type stock to be tested to a mutant stock homozygous for these recessives and then observing the F_1. The appearance of mutant phenotypes in the F_1 indicates the presence of such recessive mutations in the wild-type parent. In mice, for example, a test stock is maintained homozygous for recessive mutations at seven different loci (*a*, *nonagouti*; *b*, *brown*; c^{ch}, *chinchilla*; *p*, *pink eye*; *d*, *dilute*; *se*, *short ear*; *s*, *piebald*). Wild-type males of a pure-bred stock are then mated to females of this mutant stock and the production of mutant offspring for any of these seven genes indicates the occurrence of a mutation in the parental male.

DETECTION IN HUMANS

In humans, such methods cannot, of course, be used, and the detection of mutations therefore depends upon pedigree analysis and birth statistics. In general the precise origin of autosomal recessive mutations cannot be detected, since it is quite difficult to ensure that the appearance of a recessive trait may not have arisen because of matings between two heterozygotes rather than through a mutation in only one. The origin of dominant mutations, on the other hand, is much easier to detect as long as they are fully penetrant. In such cases the appearance of a dominant mutation in a pedigree in which the parents do not possess this trait is indicative of a new mutation. If, however, the dominant mutation is incompletely penetrant, either of the parents may have been carrying the gene in an unexpressed form, and the precise origin of the trait is again difficult to determine.

Because of the presence of a single X chromosome in human males, the origin of sex-linked recessive mutations can occasionally be detected by pedigree analysis, since a female heterozygous for such a newly arisen gene will produce half of her male offspring with the affected trait. However, even for sex-linked recessive traits considerable sources of error exist because of the inability to determine whether the female heterozygote is carrying a newly arisen gene or one that may have been passed on from previous ancestors. Most reliable studies of mutations in man are therefore restricted to the novel appearance of dominant genes. Calculation of their mutation rates then follows according to methods discussed on page 539.

Another recently suggested method for the detection of mutation in man and other organisms is to screen various proteins or enzymes such as glucose-6-phosphate dehydrogenase (see p. 486) for slight biochemical variations. The techniques include subjecting a protein to an electric charge, and then observing the relative positions occupied by its components after they have completed their migration in the electric field (see Figs. 26–14 and 33–13). Through this method of *electrophoresis*, even slight variations between individuals for a particular protein (e.g., G-6-PD *fast* and G-6-PD *slow*) can be shown to be heritable, and their transmission can sometimes be traced between various pedigree generations. Although such variations may produce no obvious visible morphological or behavioral effects, their presence usually indicates that the amino acid sequence of the protein (see Chapters 4 and 26) has been altered by mutation. Unfortunately one serious limitation is that not all amino acid changes can be detected by electrophoresis, and mutation rates for a particular protein would probably be underestimated. Nevertheless, if such methods can be widely applied, they could offer considerably more information on the prevalence, kinds, and rates of mutations than is presently available (see Neel et al.).

DETECTION IN *NEUROSPORA*

A remarkable technique for the screening and detection of nutritional mutations in the bread mold *Neurospora crassa* was published by Beadle and Tatum in 1945 and is one of the landmark experiments in biochemical genetics (see Chapter 26). Their method (Fig. 23–4) made use of the fact that practically all *Neurospora*, including most defective types, could be grown on a defined "complete" food medium which consisted of special salts and numerous added vitamins and amino acids. The normal wild-type *Neurospora*, however, produced a sufficient array of its own chemical compounds so that the complete medium was not necessary. Wild type could therefore grow on a "minimal" culture medium consisting of sugar, some inorganic acids and salts, a nitrogen source (ammonium nitrate and tartrate), and the vitamin biotin.

Nutritional mutations could therefore be detected by attempting to grow a strain of *Neurospora* on a minimal medium. If the strain could not grow on such media it was evidently defective in its ability to synthesize all necessary compounds. The particular compound missing could then usually be detected by further attempts to grow the strain on different cultures of minimal media to which single vitamins or amino acids were added. Its growth in one of these cultures signified that the strain lacked the particular added vitamin or amino acid. Once discovered, such mutant strains could then be crossed to wild-type *Neurospora* of opposite mating type to observe the inheritance of the defect. Normal segregation in such a cross should yield spores that produce wild-type: mutant phenotypes in a ratio of 1:1. The observation of such ratios signified that the nutritional mutation was of nuclear origin, and its relationship to other mutations could then be mapped through linkage tests (see Fig. 18–2). This detection technique has also been used for yeast and other fungi.

DETECTION IN BACTERIA AND VIRUS

In principle the detection of mutations in bacteria is similar to the *Neurospora* method. Samples taken from bacterial clones grown on a complete medium can be placed on a minimal medium. If growth occurs, this means that the bacteria is wild type (prototroph) and can synthesize the necessary vitamins and compounds. If growth fails to occur, the strain is considered a mutant. If growth can occur by supplementing the medium with specific vitamins or compounds, the mutant strain is called an auxotroph (p. 34).

An important variation of this technique developed by Davis and by Lederberg and Zinder makes use of the fact that bacterial cells sensitive to penicillin are only killed by the drug during their active growth period. Penicillin inhibits cell-wall formation in growing bacteria, causing them to burst open. Thus, of the bacteria treated with penicillin on a minimal medium, *only* the wild-type cells will be killed, and the mutant cells will remain alive although unable to grow. If specific types of nutritional mutations are desired, e.g., histidine-requiring mutants, a solution of all nutritional substances, with the exception of histidine, can be added. By this means only nongrowing mutants will survive. After all but the desired mutant cells have been killed, the penicillin can be washed off or removed by penicillinase, and the mutant cells cultured or tested further for the presence of histidine-requiring strains.

Viruses cannot be grown on chemically defined media but must, of course, be grown on cultures of their hosts. As explained previously (Chapter 20), when bacteriophages are used the detection of mutation usually takes the form of observing phenotypic changes in the kinds of plaques produced by the viral destruction of bacteria. Such plaque differences can also be noted for animal viruses that can be cultured on animal-cell tissues (tissue cultures). In plant viruses, such as tobacco mosaic virus (p. 66), mutants can express their phenotype through the size, shape, and color of the lesions produced. Host range mutations, or the changed ability to infect formerly immune hosts, are another product of viral mutation.

All of these microorganisms not only offer considerable ease in deciding whether a mutation is present but they also enable large numbers of organisms to be tested at one time. Thus a million or more bacteria can be plated on a single petri

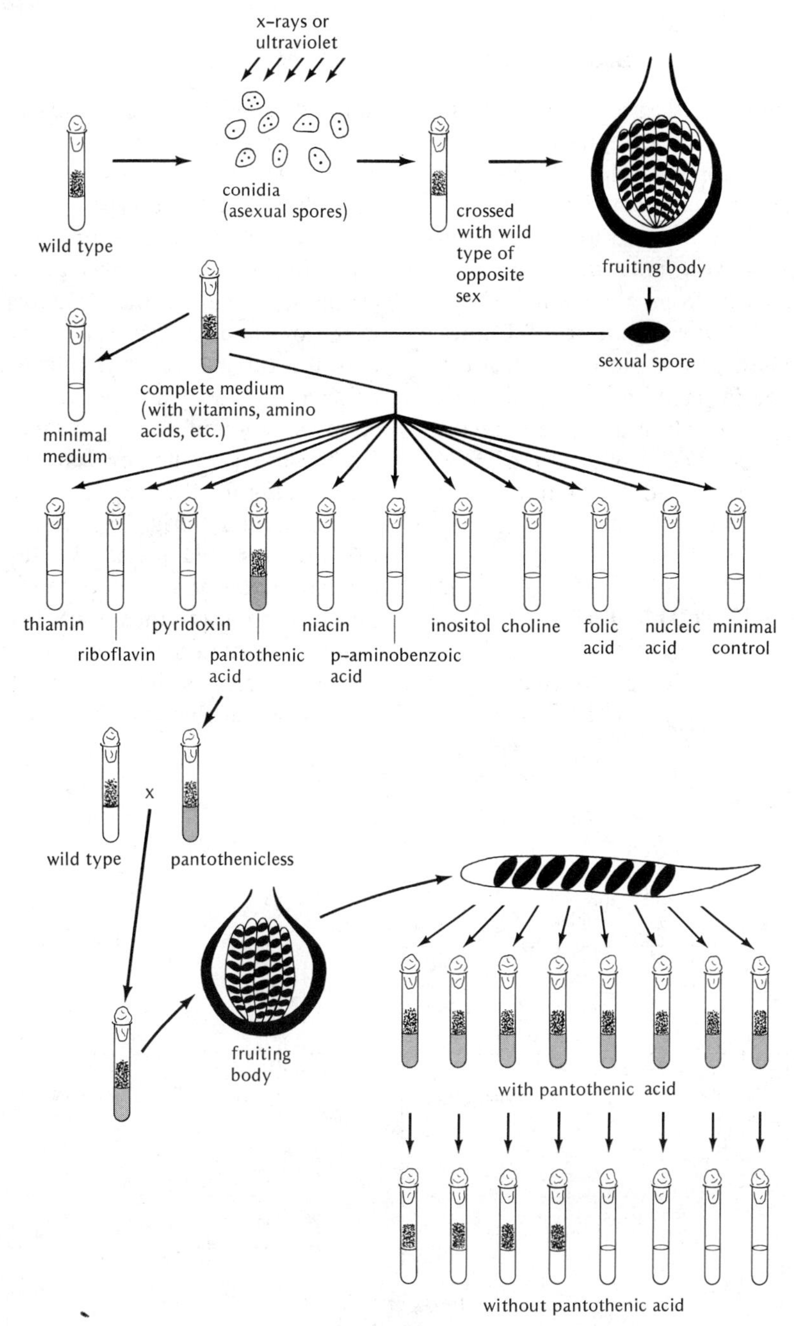

Figure 23–4

Technique for testing the appearance of nutritional mutations in Neurospora. *Conidia of a particular strain are exposed to agents causing mutation (x-rays or ultraviolet, see Chapter 24) and crossed to wild type of the opposite sex. Haploid spores of this cross are then isolated and grown on a complete medium. Inability of such an isolated strain to grow on a minimal medium indicates a growth defect. Attempts are then made to discover the source of this defect by growing the aberrant strain on a minimal medium supplemented with various additives. In the illustrated example, pantothenic acid added to a minimal medium enables the strain to grow, indicating that the mutant strain is* pantothenicless. *Observation of the expected 4 wild type : 4* pantothenicless *segregation ratio in a cross with wild type indicates that the mutation is of nuclear origin. (After G. W. Beadle, 1946.* American Scientist, **34,** *31–53.)*

dish containing a medium that selects only mutations, enabling the detection of mutation rates of 10^{-8} or even less (see, for example, pp. 159–160).

Interestingly selective plating methods used with microorganisms can also be used to detect mutations in somatic cells grown in tissue cultures that were originally derived from multicellular organisms. For example, a tissue culture of Chinese hamster cells exposed to the purine analogue 8-azaguanine in a special medium will lead to the survival of only those newly mutant cells unable to metabolize purines. The proportion of these cells can then be scored, and the mutation rate determined (see Sharp et al.).

REVERSE MUTATION

Most mutant events considered above consist of a change from wild type or normal to a new mutant genotype. Such events have been termed *forward mutations*, in contrast to rarer *back* or *reverse* mutations, in which the mutant genotype now changes to wild type.

In identifying reverse mutations, care must be taken to distinguish between a "true" reverse mutation that occurs at the same locus as the forward mutation, and a *suppressor mutation* that occurs at some other locus but hides the effect of the original mutant genotype. In microorganisms the suppressor mutation can usually be distinguished by backcrossing the revertant stock to wild type and looking for the reappearance of mutant types in the progeny (Fig. 23–5). If mutant types are present, this means that the suppressor and the original mutation have separated from each other, permitting the reappearance of mutants. The absence of mutants in the progeny of this cross strongly indicates the absence of a suppressor.

Reverse biochemical or nutritional mutations

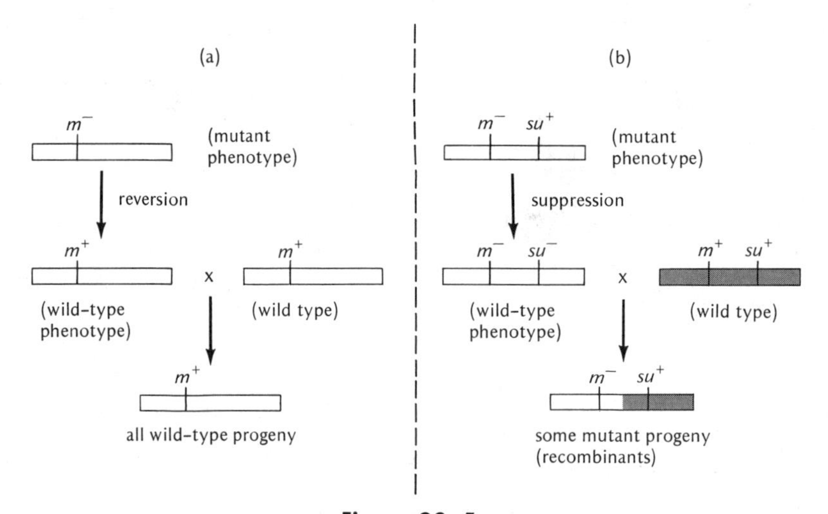

Figure 23–5

Two mechanisms accounting for the appearance of reverse mutations. In (a) the mutant locus itself, m^-, undergoes a reversion (m^+) which results in the appearance of a wild-type phenotype. After being crossed to a wild-type m^+ strain, the progeny are all m^+ and therefore wild type (except for very rare new m mutations). In (b) a change in the mutant phenotype is caused by a mutation at a suppressor locus (su). When the suppressed m^- su^- strain is crossed to wild-type m^+ su^+ some of the progeny will be m^- su^+ (mutant phenotype) because of recombination, and they will usually appear at a frequency much higher than the mutation rate of m.

have special value in microorganisms since they are easily selected and scored. For example, *E. coli* mutants that lack the ability to synthesize the amino acid tryptophan (trp^-) may be seeded on a medium in which tryptophan is absent. Under these conditions only newly appearing (trp^+) revertants will grow. When the presence of suppressors is ruled out, a large number of true revertants are found.

The occurrence of reverse mutations usually helps distinguish point mutations from large mutational effects such as deletions. That is, since a deletion involves a loss of genetic material, a reverse mutation restoring this exact portion of genetic material would be most unlikely. On the other hand, a point mutation caused by a small chemical change (nucleotide base substitution) without a significant gain or loss of genetic material would be more easily reversible. Reverse mutations therefore usually rule out deletions or insertions.

MUTATION RATES

The chances for a gene mutation to occur depends upon various factors, among which are a variety of external agents which will be discussed in Chapter 24. For the present we shall consider only those changes that take place under ordinary conditions of growth and environment, *spontaneous mutations*. How frequent are these mutations?

In some stocks of *Drosophila melanogaster*, sex-linked recessive lethals occur at a spontaneous mutation rate of approximately .1 percent, or about one chromosome per thousand tested gametes. For the *Drosophila* second chromosome, the spontaneous lethal mutation rate is approximately five times higher, or .5 percent. A similarly high mutation rate is present in the 3rd chromosome, which, in length, is about equal to the 2nd chromosome. As may be suspected, the mutation rate in the 4th "dot" chromosome is relatively small because of its size. Thus the total lethal mutation rate in *Drosophila* is about 1 percent, or one gamete per hundred carries a newly arisen lethal gene. If we also include mutation rates that produce the more numerous semilethal and subvital effects, we reach a total mutation rate for deleterious effects in *Drosophila* of approximately 5 percent, or 1 per 20 gametes.*

This 5 percent value includes, of course, mutations at many chromosomal loci, so that the mutation rate for any one locus may be expected to be considerably less. In *Drosophila*, tests for mutation rates at single loci vary (Table 23–1) but, according to Muller, can be computed to be on the order of about 1 mutation per 100,000 replicates of a particular gene in any generation. Other organisms represented in Table 23–1 also show varied mutation rates for different genes, ranging, in the case of corn, from an undetectable rate for the *waxy* gene to as much as 49 mutations per 100,000 gametes for the R^r gene.

In man mutation-rate estimates depend upon the detection of dominant and sex-linked effects in pedigrees which show no history of these genes. A dominant mutation rate, for example, can be estimated as follows. Since humans are diploid, each newly detected dominant mutation represents a change in one of two gametes that form the individual. One dominant mutation appearing in 1000 individuals therefore represents 1 new dominant gene in 2000 gametes. Thus the proportion of dominant mutations in a sample must be multiplied by 1/2 to obtain the mutation rate, i.e., $u = 1/1000 \times 1/2 = 1/2000$. For the dominant mutation causing retinoblastoma, a lethal disease, Neel and Falls have reported that, in a Michigan population, 49 children were affected out of a total number of 1,054,984 births, or an apparent mutation rate of $49/1{,}054{,}984 \times 1/2 = .000023$. According to data collected by Neel from various sources shown in Table 23–1, fairly reliable mutation-rate estimates have been given for nine different human genes. When averaged, the over-all mutation rate per gene is about 4 per 100,000

* Some estimates run considerably higher. Mukai has proposed that the spontaneous mutation rate of polygenes (see Chapter 14) with small viability effects may be as high as 35 percent per gamete in *Drosophila*.

TABLE 23–1

Spontaneous mutation rates at specific loci for various organisms*

Organism	Trait	Mutation per 100,000 Gametes†
bacteria (from many sources)		
E. coli (K12)	to streptomycin resistance	.00004
	to phage T1 resistance	.003
	to leucine independence ($leu^- \rightarrow leu^+$)	.00007
	to arginine independence	.0004
	to tryptophan independence	.006
	to arabinose dependence ($ara^+ \rightarrow ara^-$)	.2
Salmonella typhimurium	to threonine resistance	.41
	to histidine dependence	.2
	to tryptophan independence	.005
Diplococcus pneumoniae	to penicillin resistance	.01
Neurospora crassa (from Kolmark and Westergaard, and Giles)	to adenine independence	.0008–.029
	to inositol independence	.001–.010
	(one *inos* allele, JH5202	1.5)
Drosophila melanogaster males (from Glass and Ritterhoff)	y^+ to *yellow*	12
	bw^+ to *brown*	3
	e^+ to *ebony*	2
	ey^+ to *eyeless*	6
corn (from Stadler)	*Wx* to *waxy*	.00
	Sh to *shrunken*	.12
	C to *colorless*	.23
	Su to *sugary*	.24
	Pr to *purple*	1.10
	I to *i*	10.60
	R^r to r^r	49.20
mouse (from Schlager and Dickie)	a^+ to *nonagouti*	2.97
	b^+ to *brown*	.39
	c^+ to *albino*	1.02
	d^+ to *dilute*	1.25
	ln^+ to *leaden*	.80
	reverse mutations for above genes	.27
Chinese hamster somatic cell tissue culture (from Chu et al.)	to azaguanine resistance	.0015
	to glutamine independence	.014
man (from Neel, 1962)	epiloia	
	England (Gunther and Penrose; Penrose)	.4–.8
	retinoblastoma	
	England (Philip and Sorsby)	1.2
	USA, Michigan (Neel and Falls)	2.3
	USA, Ohio (Macklin)	1.8
	Germany (Vogel)	1.7
	Switzerland (Böhringer)	2.1
	Japan (Matsunaga and Ogyu)	2.1

TABLE 23–1 (*continued*)

Organism	Trait	Mutation per 100,000 Gametes†
	aniridia	
	Denmark (Mollenbach)	.5
	USA, Michigan (Shaw, Falls, and Neel)	.5
	achondroplasia (chondrodystrophy)	
	Denmark (Mørch)	4.2
	North Ireland (Stevenson)	14.3
	Sweden (Böök)	7.0
	Japan (Neel, Schull, and Takeshima)	12.2
	partial albinism with deafness	
	Holland (Waardenburg)	.4
	Pelger's anomaly	
	Germany (Nachtsheim)	2.7
	Japan (Handa)	1.7
	neurofibromatosis	
	USA, Michigan (Crowe, Schull, and Neel)	13.0–25.0
	microphthalmos–anophthalmos	
	Sweden (Sjögren and Larsson)	.5
	Huntington's chorea	
	USA, Michigan (Reed and Neel)	.5

* Mutations to independence for nutritional substances are from the auxotrophic condition (e.g., *leu*$^-$) to the prototrophic condition (e.g., *leu*$^+$).

† Mutation rate estimates of bacteria, *Neurospora*, and Chinese hamster somatic cells are based on cell counts rather than gametes.

gametes, or about four times higher than the gene mutation rate in *Drosophila*.

It should be realized that the mutation rates given in Table 23–1 refer to specific phenotypic effects that may involve genes of different sizes or even genes at more than one locus (e.g., chondrodystrophy). Attempts have therefore also been made to estimate mutation per nucleotide base pair, but so far, these data are only best obtained for procaryotes. Drake, for example, has estimated a considerably greater base-pair mutation rate for particular genes in viruses (2×10^{-8}) than in bacteria (3×10^{-10}). These differences, however, are not so significant if account is taken of the relatively larger numbers of base pairs in bacteria. In fact in terms of their entire DNA complements all procaryotes seem to have about the same frequency of mutation per replication (2×10^{-3}), whether large or small. Some aspects of the relationship between mutation and DNA replication will be further discussed in the following chapter.

Mice and corn appear to have mutation rates of the same order as *Drosophila* and humans, while among microorganisms the mutation rates are appreciably lower. This decrease in rate is perhaps associated with a shortened life cycle. The mutation rate in organisms such as bacteria is scored on the basis of changes that occur in a single cell division, while in more complex organisms such as man the generation time includes many successive cell divisions, at each of which mutation may occur.

However, when reduced to cell-division cycles, it is likely that mutation rates are probably of the same order of magnitude for most organisms, as might be expected if an optimum degree of mutation were necessary for species survival. That is, since most mutations are deleterious, too-frequent mutation would result in considerable disadvantage to their carriers. On the other hand, too low a mutation rate might fail to provide the adaptive novelties necessary for evolutionary advance under new conditions. From this point of view present mutation rates are probably optimal, and any imposed change in mutation rate would probably be harmful (see also discussion of mutation healing, p. 559).

NUMBER OF GENES FROM MUTATION RATE ESTIMATES

Using the *Drosophila* data, we have two sets of values; a mutation rate of 5 percent for the genome as a whole (U), and a mutation rate of 1 per 100,000 for single genes (u). Obviously the mutation rate for the entire genome consists of the sum of mutation rates for all genes. Thus the relationship is $U = Nu$ or $N = U/u$, where N is the total number of genes, u their individual mutation rate, and U the total mutation rate for all genes together. A broad estimate of the number of genes in *Drosophila* therefore amounts to .05/.00001, or approximately 5000 genes. Interestingly this number corresponds to the number of bands in the salivary chromosomes, and has been used as evidence for the notion that each band or interband represents a single gene (see p. 498). On the other hand, some workers consider this value as conservative and have offered estimates as high as 15,000 genes, although most suggest 10,000 gene pairs per *Drosophila*.

Since spontaneous mutation rates cannot be obtained for the entire human genome but only for particular genes, the estimation of gene number in humans is considerably more speculative than in *Drosophila*. Most geneticists nevertheless believe that the general order of magnitude for gene number is probably the same in both organisms. This is not to say that man and *Drosophila* are equally simple or complex, but only that their development may be determined by a similar *quantity* of informational units, although of considerably different *quality*.

If we accept an estimate of 10,000 genes for man, we can use this to calculate an approximate total mutation rate (U) by multiplying this value (N) by the mutation rate per individual gene (u) obtained previously. This method gives $10{,}000 \times .00004$, or a total mutation rate of .4, meaning that approximately 40 percent of all gametes carry a newly arisen mutation. If we consider that a zygote carries two gametes, each of which has a .4 probability of bearing a new mutation, the chances that either one or both gametes will bear such a mutation is 1 − the probability that *neither* gamete will be mutant, i.e., $1 - (.6 \times .6) = .64$. In terms of numbers of individuals this value indicates that about two billion people in the present world population of three billion carry some sort of newly arisen and probably deleterious mutation! Even though such values are considered too high by some geneticists, who believe the average mutation rate is below 4×10^{-5}, and too low by others, who believe that the number of human genes is far in excess of 10,000, since man has about 25 times more DNA than *Drosophila*, it is apparent that the number of new spontaneous mutations easily extends into millions per generation on a worldwide scale. Thus the creation of new genetic material through the mutation process leads to a sizable "genetic load" of deleterious effects carried by human populations (Chapter 36).

MUTATOR GENES

The observation that mutation rates may change for different loci of an organism extends also to differences between the sexes and different culture conditions. Mutation rate is therefore not of constant value, but appears to be dependent on numerous physiological factors, among which are those caused by genetic differences. In *Drosophila*, for example, mutation-rate frequencies of recessive lethals have been shown to vary for different laboratory stocks, ranging from less than .1 percent to more than 1 percent, a more-than-tenfold difference (Table 23–2). Analysis of the Florida strain of *Drosophila melanogaster* by Demerec showed that its high mutability could be laid to the presence of a recessive mutator gene on the 2nd chromosome. When this gene was removed, the frequency of sex-linked recessives in this stock fell to .07 percent.

Since then, further mutator genes have been found in *Drosophila* (see Green), and in a variety of other organisms including procaryotes. Many of these are probably associated with DNA polymerase or similar enzymes involved in DNA replication and repair. For example, the Treffers mutator gene of *E. coli*, first described in 1954 (Treffers et al.) is found to be linked very closely to the gene for DNA polymerase II (see locus 2, Fig. 19–8) and is now known to cause an occasional adenine-thymine base pair to become changed during replication into a cytosine-guanine base pair (see Chapter 24).

In corn mutator systems involving the number and positioning of various genes causing mutations have been extensively analyzed by Rhoades, McClintock, Brink, and others. The system found by Rhoades, the first of this kind to be noted, centered around a 3rd-chromosome gene, *a*, that belongs to the A_1 locus. When homozygous, *a* causes the absence of anthocyanin pigment production and consequent lack of purple color in the corn-kernel endosperm and other plant parts. However, if the 9th-chromosome dominant gene *Dt* is also present, *a* can mutate to other A_1 alleles that permit anthocyanin production in this newly formed heterozygote (*Aa*). This mutation of *a* to other A_1 alleles occurs also in somatic tissues as well as in germinal tissues, so that visible colored spots are produced in the corn-kernel aleurone layer or colored stripes appear in the plant stalk and leaves. The kernel spots are usually similar in size, each supposedly arising from a mutation of *a* to *A* in a single cell at a definite period near the end of aleurone development. Since the aleurone layer (formed from the fertilized endosperm nucleus) is triploid, the effect of as many as three doses of the *Dt* gene can be observed on the mutability of *a*, as follows:

aaa dt dt dt = 0 mutations per kernel
aaa dt dt Dt = 7.2 mutations per kernel
aaa dt Dt Dt = 22.2 mutations per kernel
aaa Dt Dt Dt = 121.9 mutations per kernel

Although the number of mutations is not proportionate to the number of *Dt* alleles, it nevertheless increases with each *Dt* addition. A more

TABLE 23–2

Spontaneous mutation rates for recessive sex-linked lethals in different stocks of *Drosophila melanogaster* raised at 22–25°C

Stock	No. Chromosomes Tested	No. Lethals	Mutation Rate, Percent
Florida inbred	2,108	23	1.09
Wooster	1,266	8	.63
Oregon R	3,049	2	.07
Florida No. 10	916	10	1.09
Lausanne	955	2	.21
Leningrad	8,614	14	.16
Sukhami	2,309	24	1.04

From Wagner and Mitchell, 1964, after other sources.

proportionate relationship occurs when the number of *Dt* alleles are held constant, but the number of *a* alleles are increased. In this case there is a proportional increase in mutation rate for each *a* allele added to the triploid endosperm.

More complex and dramatic than the single-unit *Dt* effect shown above is a two-unit system in corn, analyzed by McClintock. She found, on the one hand, a *Dissociation* (*Ds*) gene in the 9th chromosome that caused chromosome breakage at the point where it was located (Fig. 23-6). On the other hand, she found that *Ds* was only active in the presence of an *Activator* (*Ac*) gene that could be located on any of the other chromosomes. When both *Ds* and *Ac* were present, the loss of part of the 9th chromosome often led to observable effects. For example, the dominant allele *I* at the *C* locus on the 9th chromosome inhibits color formation, so that *I* kernels are colorless. An *ii* ovum fertilized by *I*-bearing pollen would therefore form an *iiI* colorless endosperm. If, however, the pollen carried *Ds* at or near the *C* locus, in addition to an *Ac* gene, many kernels would be formed in which colored spots appeared. These spots represented the loss of the *I* gene at a specific stage in development, thereby permitting color formation. In contrast to the dosage effect of *Dt*, an increase in the dosage of *Ac* produced no corresponding increase in pigmentation. In fact three doses of *Ac* led to the formation of kernels which showed only very few tiny spots. This was interpreted to mean that as the dosage of *Ac* increased, the chromosome breakage initiated by *Ds* was postponed to later stages of development (Fig. 23-7).

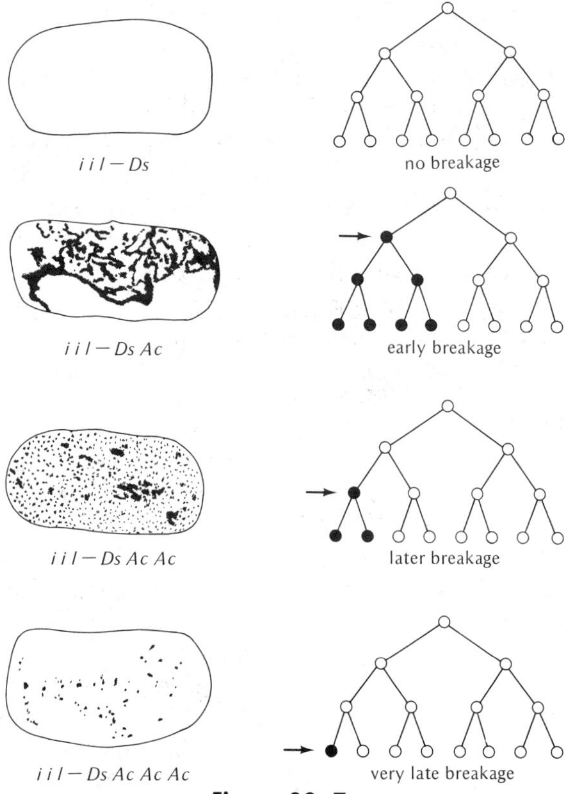

Figure 23-7

Corn kernels of different genotypes showing different degrees of spotting. On the right is an interpretation of this variegation in terms of the period in development at which chromosome breakage (arrows) is presumed to have occurred. As more Ac alleles are added, breakage is delayed to later development times, and the spots are correspondingly smaller. (After McClintock.)

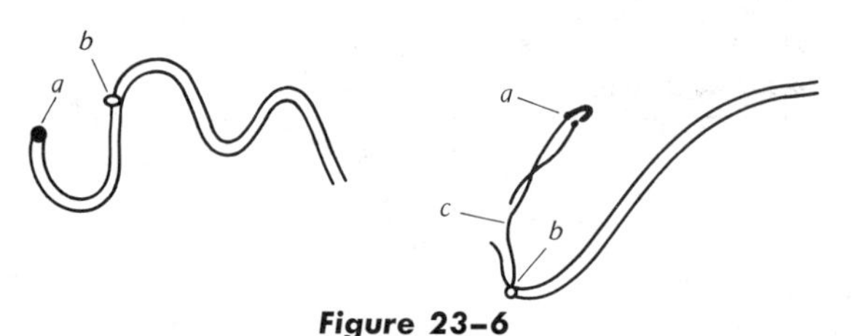

Figure 23-6

Left side: Diagram of the normal chromosome 9 of corn in bivalent condition during pachytene, (a) points to knob terminating this chromosome and (b) is the centromere. Right side: Diagram showing break (c) at the Ds locus in the heterozygote. (After McClintock.)

An especially interesting attribute of *Ds* and *Ac* was their "transposability." Both these units could physically shift position within the same chromosome or between chromosomes. In practically all such cases the new position of *Ds* led to mutation effects in its newly transposed region. Although many of these mutation effects involved chromosome breakage, some, however, appeared to cause gene mutation without observable chromosome changes.

Because of their effect on other genes, McClintock called these units "controlling elements." In one case (*Ds*) the controlling element acts on its immediate genetic neighbors, and in the other case (*Ac*) the controlling element can act at a distance. As shown by the dosage effect of *Ac*, a controlling element can also determine the time at which gene activity occurs. In their broader aspects McClintock pointed out that controlling elements are genes that regulate the activity of other genes at very basic levels in addition to mutability. (A similar hypothesis of gene control and regulation was later presented by Jacob and Monod and will be discussed in Chapter 29.)

Brink's controlling element in corn, found later, was called "modulator," *Mp*. When present at the *P* (pericarp) locus on the first chromosome, *Mp* causes the mutation of the dominant red allele P^r to colorless P^w. Mottled or *variegated* kernels are thus formed with mixed red and white tissue covering the endosperm. As with McClintock's *Ac* element, *Mp* can also be transposed to other sites. However, when *Mp* leaves the P^r gene, the mutation rate of P^r is no longer increased and the pericarp remains red. In its new site *Mp* may also act as a mutator causing, for example, the mutation of the *starchy* 9th-chromosome allele to *waxy* and vice versa. Through the process of transposability, additional *Mp* elements can be accumulated in a genome, although the more *Mp* elements that are present the more stable and less transposable they become. Thus transposition of these elements and consequent mutation seem to be regulated by the elements themselves.

HETEROCHROMATIN AND POSITION EFFECTS

Various studies indicate that the mutator genes described in corn are associated with sections of dark-staining heterochromatic material usually found interspersed between lighter euchromatic sections along the length of most chromosomes. As shown by Heitz and others, the staining differences between heterochromatin and euchromatin derive at least partially from the degree of coiling of chromosome material during the cell cycle: heterochromatin usually remains compactly coiled, whereas euchromatin undergoes a cycle of condensation and unraveling (e.g., during prophase) which is recognizably different from heterochromatin. The two substances also commonly differ in the period in which they replicate their DNA: heterochromatin often replicates its DNA later in the "S" period (see p. 92) than does euchromatin. As pointed out previously (p. 523), it is often the centromere regions that show most heterochromatin, although many species possess an entire chromosome such as the Y which remains completely heterochromatic. In both cases, centromere regions and Y chromosome, considerable evidence exists indicating an association between heterochromatin and the absence of gene activity (pp. 340–342, 486, 523).

However, not all heterochromatin acts in a constant fashion. Although some heterochromatic regions remain condensed through the entire cell cycle (*constitutive heterochromatin*), other heterochromatic regions change their staining behavior either in different cell stages or in different cells (*facultative heterochromatin*). Perhaps the most obvious example of facultative heterochromatin is the Barr body or sex chromatin effect described in Chapter 21 which leads to the unique staining behavior of only one of the X chromosomes in the cells of mammalian females: in some cells it is the paternal X chromosome that appears heterochromatic, and in other cells the maternal X chromosome is heterochromatinized. Correlated to this heterochromatinization is the gene inactivation originally suggested by Lyon (p. 486).

At the extreme, this facultative heterochromatic effect has been shown to extend to an entire haploid set of chromosomes. In some of the coccid scale insects *all* of the paternally inherited chromosomes are heterochromatinized in the somatic cells of males but not in females (Fig. 23–8). Again, evidence exists that these heterochromatinized chromosomes are inactive, since damage to the paternal chromosomes mainly affects the viability of the daughters in whom these chromosomes are euchromatic rather than that of the sons in whom they are heterochromatic. Taken

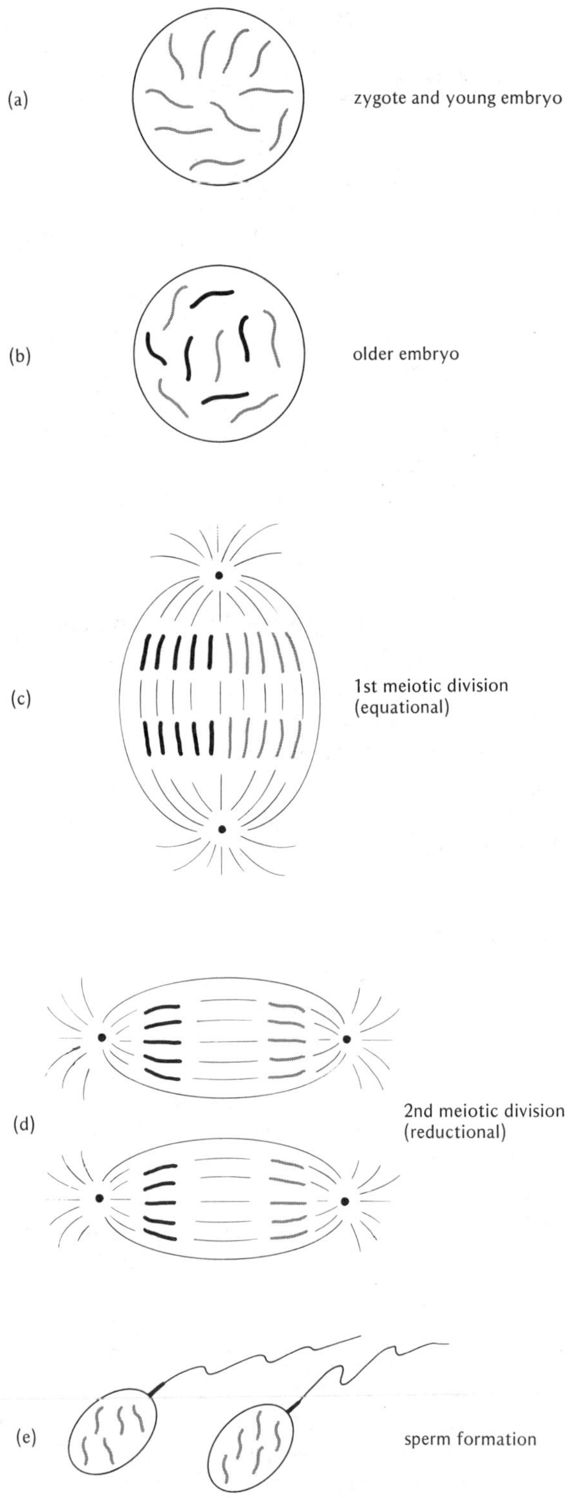

(a) zygote and young embryo

(b) older embryo

(c) 1st meiotic division (equational)

(d) 2nd meiotic division (reductional)

(e) sperm formation

Figure 23–8

Schematic diagram of chromosome cycle in a male mealy bug such as Planococcus citri. *(a) Until about the blastula stage of development, all chromosomes of the male embryo appear similar. (b) One set of chromosomes, that of the male parent, becomes relatively heterochromatinized from the blastula stage onward. At times, this paternal set appears as a dense solid mass. (c) Gametogenesis in males begins with an equational division, which is then followed by a reductional division. (d) As a result of meiosis, the paternal and maternal chromosome sets are separated from each other. (e) Only the euchromatic maternal sets of chromosomes are incorporated into sperm, and the paternal heterochromatic sets degenerate. This cycle continues and, in each generation, the chromosome set contributed paternally becomes heterochromatinized in male offspring.*

together, these findings indicate that at least two or more types of chromosome material may appear as heterochromatin. On one hand is the constitutive type which often, but not invariably, includes the repetitive DNA sequences found in satellite DNA (see p. 99). On the other hand is the facultative type that involves the heterochromatinization of what was previously normally functioning chromosome material. As yet, the mechanism responsible for the production of heterochromatin is unknown, although there is little question that heterochromatin can have profound effects on gene activity.

In *Drosophila* mutator effects associated with genes located within or near heterochromatic regions can be followed closely, since salivary banding analysis allows fairly exact physical positioning. A *white-mottled* effect, for example, is caused by either placing a heterochromatic region near the $white^+$ locus on the X chromosome or placing the $white^+$ locus near a heterochromatic region (Fig. 23–9). Females that bear heterochromatic regions next to this locus will, when heterozygous for the *white* allele *w* (w/w^+), show variegated eye color of red and white patches. As with *Ac* and *Mp* dosages in corn, an increase of heterochromatin in *Drosophila* through the addition of Y chromosomes seems to inhibit variegation in these cases. On the other hand, instances are also known in which genes normally located in the hetero-

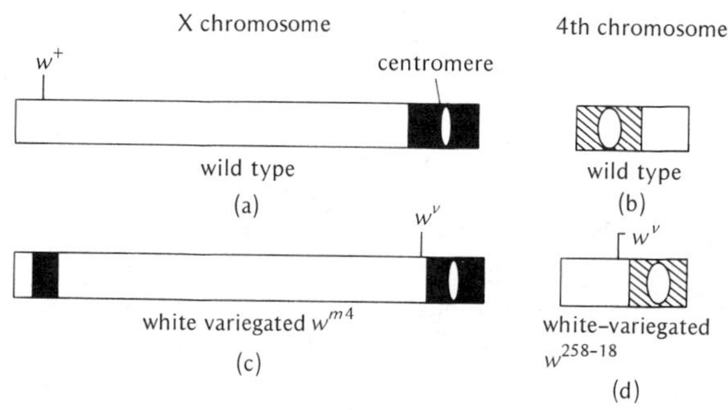

Figure 23-9

Diagrammatic representation of abnormal chromosome arrangements compared to normal chromosomes in two of the Drosophila melanogaster *stocks producing variegation for white. Parts (a) and (b) represent the normal X and 4th chromosomes, respectively. Black shading signifies heterochromatin of the X, and hatched shading heterochromatin of the 4th chromosome. In (c) an inversion has displaced the normal white allele (w^+) close to the X centromere, and it now produces a variegated effect. Similarly, a variegated effect occurs when a translocation has shifted the heterochromatin of the 4th chromosome to the left end of the X chromosome (d). Variegation is symbolized by superscript v; a female showing variegation for white eyes would therefore be w/w^v. (After Lewis.)*

chromatic areas will, when transferred to euchromatic areas, produce variegated effects. In such cases variegation is increased by the addition of Y chromosomes. In mice, Russell has found that heterochromatinization of one of the two female X chromosomes normally occurring in mammals (see p. 486) will cause variegation in autosomal genes when an X chromosome is translocated to one of the autosomes.

All these phenomena have been grouped under the title *variegated position effects.* When heterochromatin is responsible, the effect seems to be of a spreading nature, decreasing linearly along the chromosome. For example, Sutton found that the insertion of two ends of a section of X chromosome into the heterochromatic region of the 4th chromosome produced variegated effects only at the ends of the inserted section near the heterochromatin, not at the center (Fig. 23-10). This spreading effect implies that perhaps a heterochromatic product interferes with normal gene activity, and that the amount of this product is greatest near the heterochromatin. Although the cause for this is still unknown, it is likely that the quantities or kinds of proteins in heterochromatin (e.g., histones, see p. 716) can affect gene transcription and subsequent function.

Accordingly, a variegated position effect may not be a mutation in the sense of an actual physical change of gene substance, but rather a regulatory change that affects the ability of a gene to make or distribute its product. Evidence for this view is found in *Drosophila* (Hannah, 1951) and in some plants (Catcheside, 1947), where transfer of an apparently mutated gene to another position or removal of heterochromatin from the gene's proximity results in the reestablishment of the original nonmutated allele. Excluding known position effects, it may therefore be difficult to distinguish whether an observed gene mutation arises from an actual physical change of its own substance or from a position effect caused by a change in a neighboring gene. However, since our present criterion of gene mutation is based on phenotypic effect, either case may be considered to represent a gene mutation.

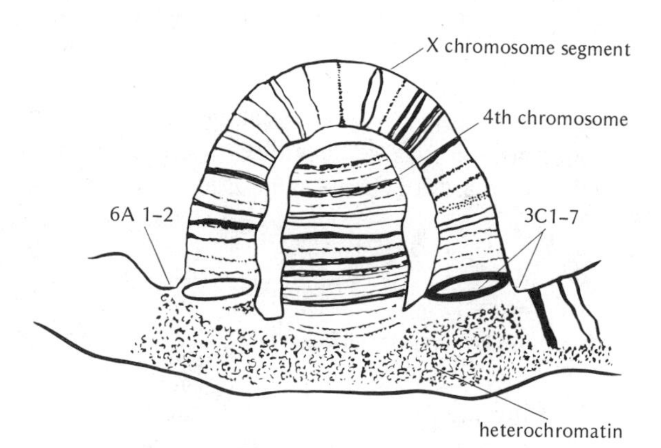

Figure 23-10

Appearance of the salivary chromosome in Drosophila melanogaster *in which a section (6A-3C) of the X chromosome has been inserted into the heterochromatic region of the 4th chromosome. Variegation occurs for the genes at the inserted ends but not for those in the center of the inserted section. (After Sutton.)*

PARAMUTATION

One of the basic assumptions of genetics has been that alleles are unaffected by the presence of other alleles in heterozygotes. This assumption, supported by considerable evidence for many genes in a multitude of organisms (see, for example, p. 121) was first seriously questioned by Brink for a particular locus in corn. This locus, *R*, bears a number of alleles concerned with anthocyanin pigment production, of which R^r is that for self-color, R^{st} is for the stippled effect, and R^{mb} is for the marbled effect. Ordinarily one dose of R^r produces dark mottling and two or three doses produce full color. Brink, however, found that when R^r alleles from localities outside the Andes Mountains are reextracted from heterozygous condition with either R^{st} or R^{mb} alleles from other localities, R^r by itself now produces considerably lighter pigment than before. Since this change in R^r is passed on to future generations, Brink has named this affect *paramutation*, and the R^{st} and R^{mb} alleles are said to be *paramutagenic*.

Reports of similar irregular behavior in a variety of other plants (*Oenothera*, tomatoes, ferns) indicate that paramutation, although rare, may provide an explanation for some exceptional events. Certain aspects of the segregation-distortion phenomenon in *Drosophila* (p. 219) may also fall into this category, since the presence of an *SD* gene is found to induce heritable changes in the male X chromosome so that it inhibits segregation distortion in males of future generations.

RANDOM AND ADAPTIVE MUTATION

The findings so far have indicated that although mutations may be affected by various factors, they do not arise as a phenotypic response to the "needs" of the external environment; e.g., green light does not produce green mutations, etc. In this sense the exact type of mutation that occurs in any environment is a factor of chance and not predictable. However, despite plentiful demonstration of random mutation events in *Drosophila*, corn, and other organisms, many bacteriologists long refused to accept this chance mutation concept for bacteria. Their argument was based on the long-standing observation that bacterial cultures will adapt rapidly to practically any new selective condition, such as streptomycin resistance, phage resistance, etc. It appeared as though the selective condition would itself induce changes in the "plastic heredity" of bacteria, so that adaptive bacterial types always arose as a response to the environment. This concept was a Lamarckian one (p. 5) and was termed *postadaptive mutation*, since these new "mutations" presumably arose after being placed in the selective environment. The contrasting theory, termed *preadaptive mutation*, indicated that random mutation had occurred prior to exposure to the new environment and that selection only *chose* the new successful types but did not produce them.

It was, at first, quite difficult to prove either theory, since bacteria are much too small to enable direct observation of the precise moment at which a mutant bacterial cell arises. However, in 1943 Luria and Delbruck devised a statistical method to help distinguish between pre- and postadaptive mutation in *E. coli*. This method made use of the fact that the number of new mutations in any culture of bacteria may vary from zero to many, depending upon how early the mutation occurred in the history of the culture. Evidently mutations that occur early will leave a clone of many descendants as compared to mutations that occur later. Thus, if many samples of *different* cultures are tested for the presence of mutations in a new environment, a large degree of variability should exist between the samples if these mutations had been present previously (preadaptive). On the other hand, if mutations arise as a response to a new environment, each different sample exposed to the new environment should have a small but equal chance of developing a mutation. In this case the variability, measured as variance, is about equal to the average number of mutations according to the Poisson distribution (p. 159).

In the experiment performed by Luria and Delbruck, they tested cultures for the appearance of bacterial mutations that were resistant to bacteriophage T1. As a control they took 10 samples of approximately the same size from a *single* culture to demonstrate that any particular culture is fairly uniform in the number of phage-resistant mutations that it contains. As shown on the left side of Table 23-3, these control experiments were repeated three times and gave variances approximately equal to the means. These are the types of results that would be expected according to the postadaptive theory for *different* cultures. However, when the experiment was performed with samples from different cultures (right side Table 23-3), the variances were many times higher than the means. This denotes a considerable variability between different cultures in the number of phage-resistant mutations they contained—a result supporting the preadaptive mutation theory.

Within 10 years of this experiment a more graphic proof for preadaptive mutation was presented by J. and E. M. Lederberg. They plated a colony of bacteria so that it uniformly covered the surface of a petri dish ("master plate") and then *replica-plated* samples of this master plate on petri dishes impregnated with bacteriophage in agar. The replica plating was done by the ingenious device of pressing a piece of velvet against the master plate and then pressing it against the bacteriophage agar. In this way transfers of bacteria are made on each hair of the velvet pile from a specific location in the original colony to a specific new location. If resistance to bacteriophage is the result of a preadaptive mutation, specific spots on the bacteriophage agar which show such resistance should mark the location of bacterial mutants on the original petri dish. In other words, new replica platings on bacteriophage agar should show at least some spots in the same location as on the previous bacteriophage agar. This indeed was found, both in experiments with bacteriophage (Fig. 23-11) and with streptomycin. Further experiments by Cavalli-Sforza, Lederberg, and others have firmly established the principle of preadaptive mutation.

TABLE 23-3

Numbers of bacteriophage-resistant colonies of *E. coli* produced from samples from single cultures and samples from different cultures

***Control* (*10 Samples Each From the Same Culture*)**				***Experimental* (*Samples From Different Cultures*)**			
	RESISTANT COLONIES				RESISTANT COLONIES		
Sample	Expt. 3	Expt. 10a	Expt. 11a	Culture	Expt. 1	Expt. 11	Expt. 15
1	4	14	46	*1*	10	30	6
2	2	15	56	*2*	18	10	5
3	2	13	52	*3*	125	40	10
4	1	21	48	*4*	10	45	8
5	5	15	65	*5*	14	183	24
6	2	14	44	*6*	27	12	13
7	4	26	49	*7*	3	173	165
8	2	16	51	*8*	17	23	15
9	4	20	56	*9*	17	57	6
10	7	13	47	*10*		51	10
mean	3.3	16.7	51.4		26.8	62	26.2
variance	3.8	15	27		1,217	3,498	2,178

From Luria and Delbruck, 1943.

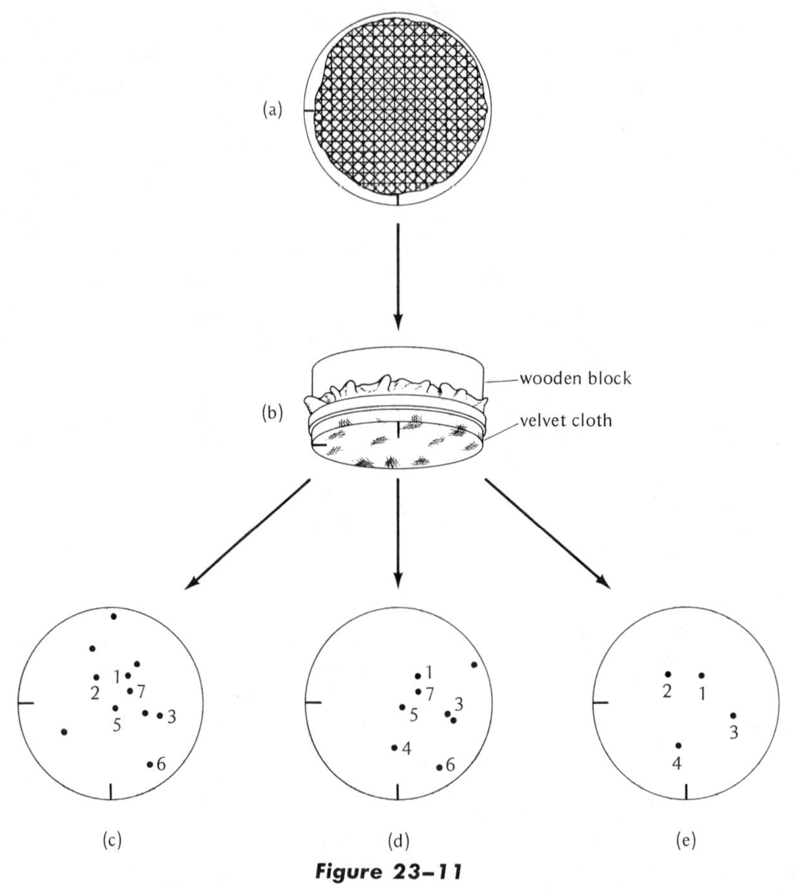

Figure 23–11

Replica plating technique to test the location of clones of E. coli *resistant to phage T1. (a) Master plate showing diffuse bacterial growth on a nonphage medium. Replicas are made by pressing a velvet-covered wooden block (b) against the master plate, and then pressing this to the surface of petri dishes containing culture medium mixed with phage T1. (c-e) Replica plates showing occurrence of resistant colonies (e.g., 1, 3, 4) in identical locations indicating that resistance to phage T1 must have been present at each of these positions in the master plate. (Adapted from Lederberg and Lederberg.)*

PROBLEMS

23–1. If a short-tailed mouse appeared in a normal colony of mice, how would you determine whether this trait is caused by a single dominant or recessive gene, by interaction between various genes, or was environmentally induced?

23–2. If you were working with *Oenothera lamarckiana* and a new mutant phenotype appeared in your stock, how would you determine whether it is caused by mutation or recombination?

23–3. A wild-type *Drosophila* male is crossed to a female homozygous for *white*, *miniature*, and *forked*, and one of his daughters is found to possess the *miniature* phenotype. How would you determine whether this effect is caused by a point mutation or deletion?

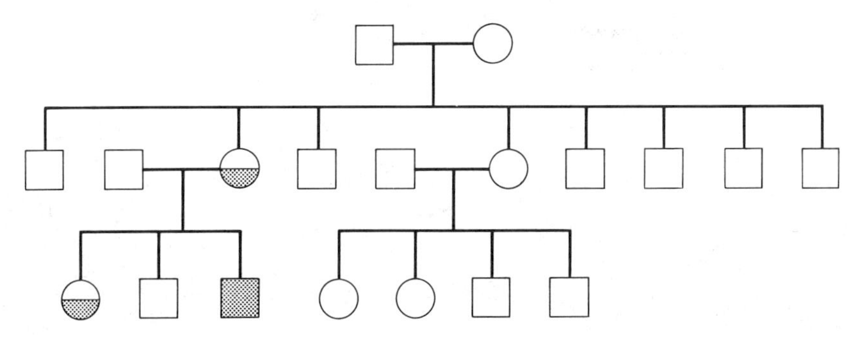

23–4. The above is a partial pedigree of a particular male (shaded symbol) showing a sex-linked recessive disease, the neurologically damaging Lesch-Nyhan syndrome (see p. 242), discovered by Itiaba and co-workers in a Canadian family. Examination of females II–3 and III–1 showed that they were carriers of the gene (half-shaded symbols) but the remaining females in the pedigree were homozygous normal. Explain whether the origin of the disease in this pedigree was caused by a newly arisen mutation or by transmission from previous generations.

23–5. Would you suggest the use of attached-X females in *Drosophila* to detect the occasional presence of sex-linked recessive lethals arising in male sperm? Explain.

23–6. How would you utilize the penicillin selection technique (p. 536) to isolate strains of bacterial double mutants? For example, use this technique to isolate bacteria unable to synthesize *both* arginine (arg^-) and leucine (leu^-).

23–7. Normal bacterial cells that are sensitive to streptomycin can be plated on streptomycin medium and the only growing colonies will be mutants that have become streptomycin-resistant. How would you distinguish between streptomycin-resistant bacteria that can grow without the antibiotic (streptomycin-independent) and streptomycin-resistant bacteria that can only grow in the presence of the antibiotic (streptomycin-dependent)?

23–8. A *Neurospora* strain dependent for growth upon the addition of arginine to the medium (*arginineless*) produces a new mutant colony which is arginine-independent. A cross is then made between a wild-type strain and the revertant strain. (a) What results would you expect if the new mutation is a true reversion? (b) What results would you expect if the new mutation is a suppressor of *arginineless* located on a different chromosome? (c) What results would you expect if the new mutation is a suppressor of *arginineless* 20 recombination units away?

23–9. A child affected by chondrodystrophic dwarfism is born to normal parents. (a) What probably accounts for this appearance? (b) If such a child marries and produces one chondrodystrophic offspring who in turn marries and produces three normal children, what accounts for the disappearance of the trait? (c) If one of the normal children marries and produces an albino child, what is the probable source for the appearance of albinism?

23–10. A form of acrocephalosyndactyly, called Apert's syndrome, affects the head, hands, and feet of humans in a noticeable manner, and is inherited as a dominant gene. In Northern Ireland, Blank reported 2 infants born with such a syndrome out of 322,182 births. What is the mutation rate per gamete for this gene?

23–11. If man had 50,000 haploid genes instead of the estimated 10,000 (p. 542) and a mutation rate per gene of 4/100,000, what would be the total mutation rate per gamete?

23–12. W. K. Baker found that when *Drosophila melanogaster* females heterozygous for the genes *white* (w/w^+) and *split* (spl/spl^+) were so constructed that the wild type genes were translocated to a heterochromatic region in the order $w^+ - spl^+$ – heterochromatin, variegation occurred in which the white-eye-color facets always had the split phenotype but the split facets were both wild type (red) and white. On the other hand, when the $spl^+ - w^+$ section was inserted into the middle of a heterochromatic region so that each of the two genes were equidistant from the heterochromatin, white-eye facets had both split and nonsplit phenotypes. How would you explain the difference?

23–13. Demerec found the following numbers of streptomycin-resistant *E. coli* bacteria in single samples taken from 20 independent cultures and in 15 samples taken from a single culture:

Samples From Independent Cultures

Culture	No. Resistant Colonies	Culture	No. Resistant Colonies
1	67	11	56
2	159	12	91
3	135	13	123
4	291	14	97
5	75	15	48
6	117	16	52
7	73	17	54
8	129	18	89
9	86	19	111
10	101	20	164

Samples From Single Culture

Sample	No. Resistant Colonies	Sample	No. Resistant Colonies
1	142	11	110
2	155	12	125
3	132	13	135
4	123	14	121
5	140	15	112
6	146		
7	141		
8	137		
9	128		
10	121		

(a) What is the mean and variance in each of these two sets of experiments? (b) Which of these two variances would you consider to be more indicative of chance variation and which to be more indicative of real differences between the samples? (c) On the basis of your answer in (b), what opinion would you offer as to the validity of pre- or postadaptation?

23-14. In their test distinguishing between pre- and postadaptation in bacteria (p. 549), Luria and Delbruck used bacterial mutation toward resistance to a *virulent* phage (T1). What results would you have expected if they had used a *temperate* phage instead (e.g., λ)?

23-15. How would you use the replica-plating method of the Lederbergs to distinguish whether Hfr strains (Chapter 19) arise spontaneously from F^+ cells or arise because of exposure of F^+ cells to F^- cells?

23-16. Many experiments have shown that resistance of bacteria to penicillin treatment appears to occur through a "training" process; bacteria selected to resist low penicillin concentrations can then be further selected to resist higher penicillin concentrations until eventually a bacterial population can be selected which will survive any penicillin concentration to which it is exposed. At first bacteriologists believed that this training process was caused by gradual stepwise changes in the "plastic" heredity of bacteria induced by the penicillin itself (postadaptive mutation). In 1948, however, Demerec showed that each step in the penicillin-resistance training process was caused by a specific different gene mutation, each mutation occurring at a frequency of about one per 10 million cells. Thus the occurrence of one penicillin-resistant mutation allowed survival at a relatively low penicillin concentration; bacteria carrying two penicillin-resistant mutations could survive higher penicillin concentrations, and so on. These mutations were therefore preadaptive, and penicillin acted only as the selective agent and not as the inducing agent.

Based on these findings, explain whether you would suggest that a patient suffering from a bacterial infection should receive gradually increasing doses of penicillin during his illness, or a massive dose of penicillin at the start.

REFERENCES

AUERBACH, C., 1962. *Mutation. Part I. Methods.* Oliver and Boyd, Edinburgh.

BAKER, W. K. 1968. Position-effect variegation. *Adv. in Genet.*, **14,** 133–169.

BEADLE, G. W., 1947. Genes and the chemistry of the organism. *Science in Progress,* **7,** 166–196. (Reprinted in Taylor's collection of papers; see References, Chapter 1.)

BEADLE, G. W., and E. L. TATUM, 1945. *Neurospora.* II. Methods of producing and detecting mutations concerned with nutritional requirements. *Amer. Jour. Bot.*, **32,** 678–686.

BRINK, R. A., 1954. Very light variegated pericarp in maize. *Genetics*, **39,** 724–740.

———, 1973. Paramutation. *Ann. Rev. Genet.*, **7,** 129–152.

BROWN, S. W., 1966. Heterochromatin. *Science*, **151,** 417–425.

CATCHESIDE, D. G., 1947. The P-locus position effect in *Oenothera. Jour. Genet.*, **48,** 31–42.

CAVALLI-SFORZA, L. L., and J. LEDERBERG, 1956. Isolation of pre-adaptive mutants in bacteria by sib selection. *Genetics*, **41,** 367–381. (Reprinted in Adelberg's collection; see References, Chapter 1.)

CHU, E. H. Y., P. BRIMER, K. B. JACOBSON, and E. V. MERRIAM, 1969. Mammalian cell genetics. I. Selection and characterization of mutations auxotrophic for L-glutamine or resistant to 8-azaguanine in Chinese hamster cells *in vitro*. *Genetics*, **62,** 359–377.

DAVIS, B. D., 1948. Isolation of biochemically deficient mutants of bacteria by penicillin. *Jour. Amer. Chem. Soc.*, **70,** 4267.

DEMEREC, M., 1937. Frequency of spontaneous mutations in certain stocks of *Drosophila melanogaster*. *Genetics*, **22,** 469–478.

DRAKE, J. W., 1969. Comparative rates of spontaneous mutation. *Nature*, **221,** 1132.

FINCHAM, J. R. S., and G. R. K. SASTRY, 1974. Controlling elements in maize. *Ann. Rev. Genet.*, **8,** 15–50.

GILES, N. H., 1951. Studies on the mechanism of reversion in biochemical mutants of *Neurospora crassa*. *Cold Sp. Harb. Symp.*, **16,** 283–313.

GLASS, B., and R. K. RITTERHOFF, 1956. Spontaneous-mutation rates at specific loci in *Drosophila* males and females. *Science*, **124,** 314–315.

GREEN, M. M., 1973. Some observations and comments on mutable and mutator genes in *Drosophila. Genetics*, **73** (Supplement): 187–194.

HANNAH, A., 1951. Localization and function of heterochromatin in *Drosophila melanogaster*. *Adv. in Genet.*, **4,** 87–125.

KOLMARK, G., and M. WESTERGAARD, 1949. Induced back-mutations in a specific gene of *Neurospora crassa*. *Hereditas*, **35,** 490–506.

LEDERBERG, J., and E. M. LEDERBERG, 1952. Replica plating and indirect selection of bacterial mutants. *Jour. Bact.*, **63,** 399–406. (Reprinted in Adelberg's collection; see References, Chapter 1.)

LEDERBERG, J., and N. ZINDER, 1948. Concentration of biochemical mutants of bacteria with penicillin. *Jour. Amer. Chem. Soc.*, **70,** 4267.

LEWIS, E. B., 1950. The phenomenon of position effect. *Adv. in Genet.*, **3,** 73–116.

LURIA, S. E., and M. DELBRUCK, 1943. Mutations of bacteria from virus sensitivity to virus resistance. *Genetics*, **28,** 491–511. (Reprinted in Adelberg's collection; see References, Chapter 1.)

MCCLINTOCK, B., 1951. Chromosome organization and genic expression. *Cold Spring Harb. Symp.*, **16,** 13–47.

MUKAI, T., 1964. The genetic structure of natural populations of *Drosophila melanogaster*. I. Spontaneous mutation rate of polygenes controlling viability. *Genetics*, **50,** 1–19.

MULLER, H. J., and I. I. OSTER, 1963. Some mutational techniques in *Drosophila*. In *Methodology in Basic Genetics,* W. J. Burdette (ed.). Holden-Day, San Francisco, pp. 249–274.

NEEL, J. V., 1962. Mutations in the human population. In *Methodology in Human Genetics,* W. J. Burdette (ed.). Holden-Day, San Francisco, pp. 203–219.

NEEL, J. V., T. O. TIFFANY, and N. G. ANDERSON, 1973. Approaches to monitoring human populations for mutation rates and genetic disease. In *Chemical Mutagens,* Vol. 3, A. Hollaender (ed.). Plenum Press, New York, pp. 105–150.

RHOADES, M. M., 1941. The genetic control of mutability in maize. *Cold Sp. Harb. Symp.*, **9,** 138–144.

RUSSELL, L. B., 1963. Mammalian X-chromosome action: Inactivation limited in spread and region of origin. *Science*, **140,** 976–978.

SCHLAGER, G., and M. M. DICKIE, 1967. Spontaneous mutations and mutation rates in the house mouse. *Genetics*, **57,** 319–330.

SHARP, J. D., N. E. CAPECCHI, and M. R. CAPECCHI, 1973. Altered enzymes in drug-resistant variants of mammalian tissue culture cells. *Proc. Nat. Acad. Sci.*, **70,** 3145–3149.

STADLER, L. J., 1942. Some observations on gene variability and spontaneous mutation. The Spragg Memorial Lectures (Third Series), Mich. State College.

SUTTON, E., 1940. The structure of salivary gland chromosomes of *Drosophila melanogaster* in exchanges between euchromatin and heterochromatin. *Genetics*, **25,** 534–540.

TREFFERS, H. P., V. SPINELLI, and N. O. BELSER, 1954. A factor (or mutator gene) influencing mutation rates in *Escherichia coli*. *Proc. Nat. Acad. Sci.*, **40,** 1064–1071.

WAGNER, R. P., and H. K. MITCHELL, 1964. *Genetics and Metabolism*, 2nd ed. John Wiley, New York, Chap. 4.

24
INDUCED GENETIC CHANGES

Until 1927 an understanding of the causes of mutation as well as the accumulation of new mutations was greatly handicapped by the generally low spontaneous mutation rates observed at practically all loci. Despite many trials, attempts to increase the mutation rate by artificial means could not be successfully demonstrated until two factors were brought together: (1) sensitive detection methods, such as those discussed in Chapter 23, and (2) discovery and effective employment of mutation-producing agents, or *mutagens*. In 1927 Muller demonstrated the first artificial induction of mutations through measurement of the effect of large doses of x-rays on the mutation rate of sex-linked lethals and visibles in *Drosophila*. Contemporary with Muller's discovery were similar observations of Stadler on barley, which were soon followed by demonstrations in other organisms. Since then, high-energy radiation has proved to be one of the most effective mutagens and it is used in numerous present-day studies. Let us consider what radiation is.

The visible light we observe is only a small part of the electromagnetic spectrum, which consists of energy in the form of a variety of wavelengths (Fig. 24–1). As the wavelengths become shorter, the energy they contain becomes stronger and more "penetrating." X-rays, which range in wavelength from about 10 angstroms to about .1 angstrom, were discovered in the nineteenth century by Röntgen and were rapidly adapted for use in diagnostic medical procedures because of their penetrating abilities. In the process of penetration, however, high-energy irradiation also produces *ions* by colliding with atoms and releasing electrons, which in turn collide with other atoms releasing further electrons, etc. The change in electron number transforms a stable atom or molecule into the reactive *ionic* state. Thus, along the track of each high-energy ray, a train of ions is formed which can initiate a variety of chemical reactions. Such irradiation is therefore appropriately called *ionizing radiation* and can be produced by machine-made x-rays, protons, neutrons, and also by alpha, beta, and gamma rays from radioactive sources, e.g., radium, cobalt-60, etc.

Measurement of most ionizing radiation is done in terms of *roentgen units* (*r*), each of which produces one electrostatic unit of charge in 1 cubic centimeter of air.* In water or in tissue, approximately two ionizations are produced by 1 *roentgen* in a volume of 1 cubic micron. *Nonionizing radiation* such as ultraviolet rays are far less penetrating and do not produce ion tracks.

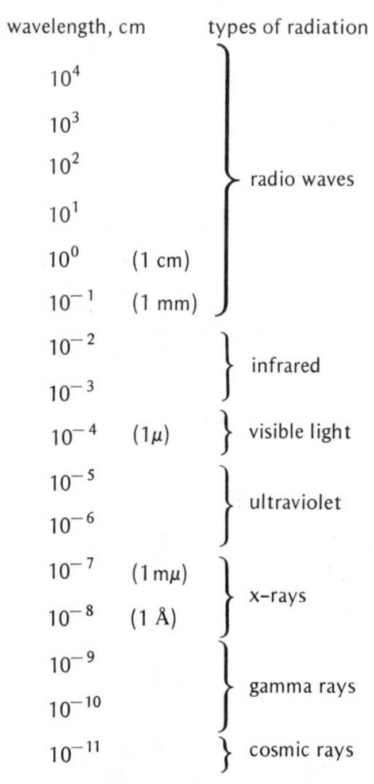

Figure 24–1

Spectrum of electromagnetic waves ranging from long radio waves to extremely short cosmic rays. (Units of length are now also designated in terms of 1 micrometer (μm) = 1 micron (μ), *and 1* nanometer (nm) = 1 millimicron (mμ)).

DOSAGE-MUTATION RELATIONSHIPS AND THE TARGET HYPOTHESIS

Experiments with high-energy radiation showed soon after 1927 that induced mutation rates depended strongly upon *radiation dosage*, or *r*; the greater the dosage, the greater the mutation rate. When this relationship is graphed, as for sex-linked lethals in *Drosophila* (Fig. 24–2), a straight line is formed with its origin at a very low mutation rate for zero dosage extending to a very high mutation rate at a dosage of 6000 *r*. In this *linear* relationship there is a proportionate increase in mutation rate for each increase in dosage, i.e., approximately 3 per cent mutation rate for each 1000 *r*.

This simple relationship between mutation and radiation has been interpreted by Timoféeff-Ressovsky, Lea, Catcheside, and others to mean that the gene is a "target" and its mutation is caused by a single "hit" of radiation. According to this hypothesis, the high-energy rays themselves or the ionizing particles produced by irradiation impinge as "hits" upon genes and chromosomes, each "hit" having a very high probability of causing a distinct mutation. Thus there is a proportionate relationship between the amount of irradiation and ionization (dosage) and the number of mutations (rate). The straight-line nature of the graph in Fig. 24–2 is therefore easily explained.

On the other hand, if more than one hit were required to produce a mutation, the relationship between dosage and mutation frequency would not be linear but would be expected to follow a curve. For example, if two hits were required, many single non-mutation-producing hits would have to occur until the dosage increased sufficiently to ensure double hits. Each succeeding increase in dosage would then produce a proportionately greater increase in mutation frequency (more double hits) than at lower dosages (more

---

*Another unit, the *rad*, measures irradiation in terms of energy absorbed by material. Exposure to 1 *rad* produces only a slightly larger value (100 ergs) than the energy absorbed by a similar volume of material exposed to 1 roentgen (93 ergs). To enable comparisons to be made between the effects of different kinds of radiation such as gamma rays, neutrons, etc., the *rem* is defined as that radiation dose absorbed by a particular tissue which causes the same biological effect as one rad of x-rays.

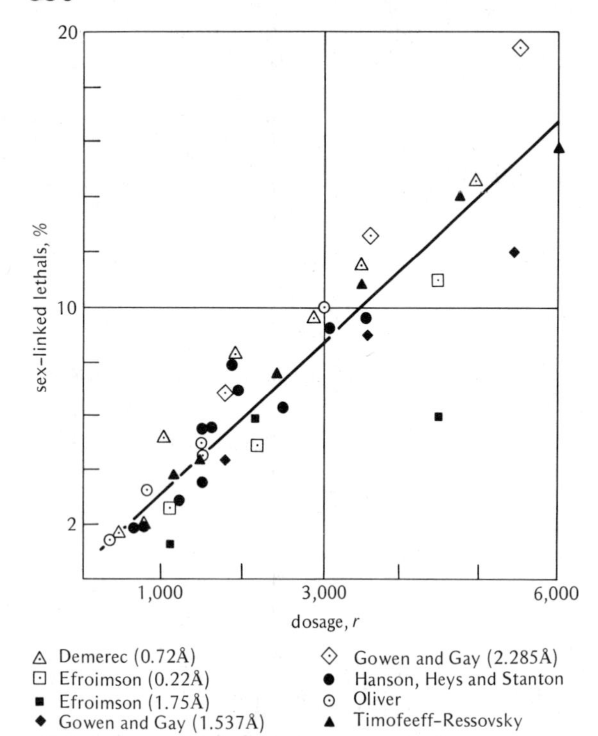

Figure 24–2

Relationship between percentage of sex-linked lethals in Drosophila and various doses of x-rays measured in roentgens. (After Schultz, 1936; from various sources.)

single hits). For two-or-more-hit mutations, the expected relationship is therefore *exponential*, or multiplicative (see Chapter 14), beginning with low mutation rates at low dosage and curving up rapidly at high dosages. To demonstrate such a relationship, however, we must turn from gene or "point" mutations in which distinctions between one- and two-hit mutations are difficult to make, to more easily distinguishable chromosome aberrations.

Depending usually upon the division stage at which cells are irradiated, a chromosome aberration may include either one or both chromatids. Irradiation at early interphase, before DNA synthesis has begun, usually causes breaks that appear later as having occurred when the chromosome was as yet unreplicated (*chromosome breaks*). Breaks induced in the interphase period after DNA synthesis has begun usually appear separately in each of the two chromatids of a chromosome (*chromatid breaks*) (Fig. 24–3). In both cases terminal deletions can be considered as caused by single hits, while subterminal (intercalary) deletions are generally two-or-more-hit mutations. Using these criteria, Chu, Giles, and Passano have found that single-hit mutations show, as expected according to the target theory, a linear relationship to dosage, while the two-hit varieties, also as expected, follow an exponential curve (Fig. 24–4).

Instead of simple physical breaks, Revell has suggested that irradiation causes chromosomal "lesions" which then stimulate exchanges between parts of the same chromosome or different chromosomes, leading, in turn, to deletions, translocations, and other chromosomal aberrations. Thus as shown in Fig. 24–5, the chromatids of an irradiated chromosome may overlap at the point where

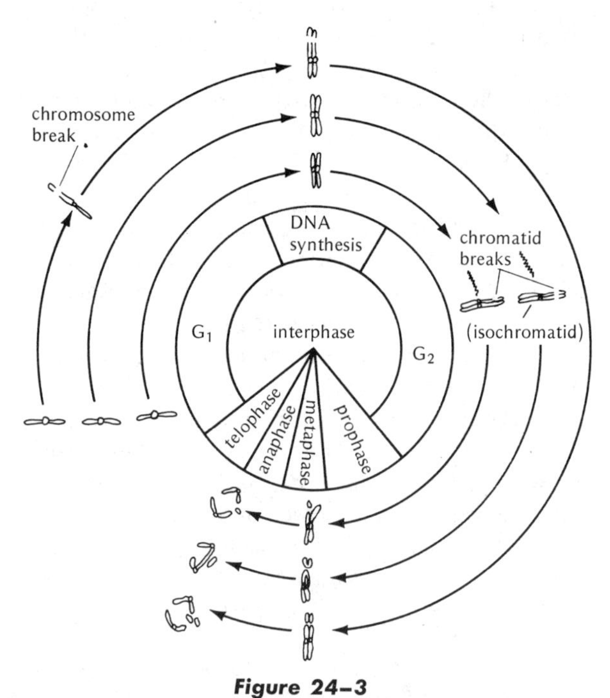

Figure 24–3

Types of terminal deletions produced by single breaks at the one-strand (G_1) or two-strand (G_2) stages during a mitotic cycle. Note that breaks of both chromatids at once (isochromatids) can also be distinguished. Stages of the cell cycle are not drawn to scale (see Fig. 5–16). (After Whittinghill.)

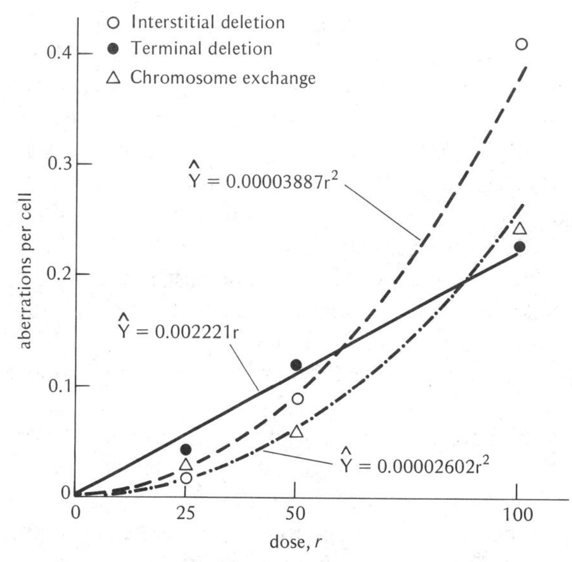

Figure 24–4

Chromosomal aberrations in human-tissue-culture cell lines detected 42 hours after irradiation. There is a straight-line relationship between single-break aberrations (terminal deletions) and dosage, and an exponential relationship between two-break aberrations (interstitial deletions and chromosome exchanges) and dosage. The equations for the curves best fitting each relationship are indicated. (From Chu, Giles, and Passano.)

two lesions coincide, leading either to complete or incomplete exchanges. If the exchange is complete, no obvious chromosomal damage is observed, whereas an incomplete exchange leads to a loss of material from one or both of the chromatids. Similarly x-ray-induced exchanges may cause inversions or, should they occur between nonhomologous chromatids, translocations. Although evidence exists both for and against the Revell hypothesis (see, for example, Heddle, Whissell, and Bodycote) the fact that recombinational mechanisms are often involved in radiation repair (see p. 569) certainly indicates an opportunity for the type of exchanges proposed by Revell.

DOMINANT LETHALS

In *Drosophila* lethals that are effective in single dose can be detected by observing the viability of immature stages. An irradiated male, for example, may produce sperm carrying dominant lethals which will prevent the fertilized zygote from developing beyond the egg stage. A comparison between the hatchability of eggs fertilized by irradiated and nonirradiated males will then indicate the extent of dominant lethals induced in male sperm. Of course, nonhatchability may have been caused by other factors besides those located on chromosomes: absence of fertilization, impaired

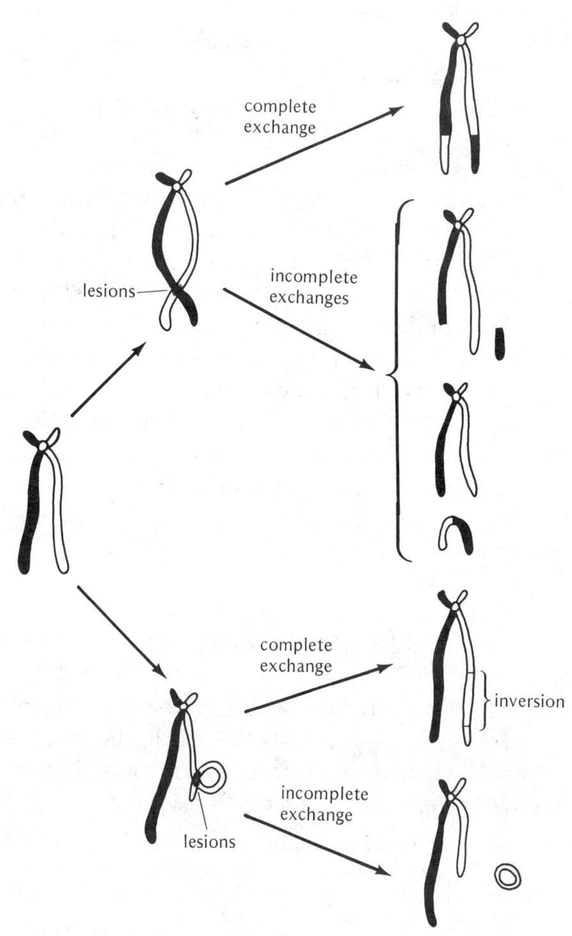

Figure 24–5

Some of the consequences of radiation-induced lesions according to the exchange hypothesis of Revell. Two chromatids, one shaded and the other unshaded, are shown united at their centromeres before division. Depending upon whether one or both chromatids are involved in the exchange, and the sections affected, deletions, inversions, and other abnormalities can be explained.

spindle formation because of an injured male centriole, etc. Because of this, other developmental stages are also used to evaluate the effect of dominant lethals.

A different technique for dominant lethal detection makes use of the haploid-diploid sexes in the wasps *Bracon hebetor* (see p. 229) and *Mormoniella.* Since unfertilized eggs in these organisms develop into males, the dominant lethal effect of irradiated sperm can be detected by observing the number of daughters produced by mating irradiated males to nonirradiated females. That is, dominant lethals carried by the paternal sperm will cause the death of female embryos but have no effect on the unfertilized male embryos. As shown in Fig. 24–6, normal matings between nonirradiated individuals in *Bracon hebetor* produce approximately two thirds daughters and one third sons. As the radiation dosage to male sperm is increased, the proportion of daughters produced decreases until none are present at 10,000 *r*. Since there is no accompanying increase in the number of sons produced, this means that the decrease in number of daughters is caused by lethality rather than by lack of fertilization. Note that the increase in dominant lethals is nonlinear, indicating, according to the target theory, that some such lethals are caused by two hits or more.*

RADIATION INTENSITY

Further support for the target theory has been garnered from observations that the mutation rate is independent of the rate at which the radiation dosage is applied, i.e., *radiation intensity*. Any increase in ionization apparently results in an increase in mutation frequency, and it matters little whether the same amount of irradiation is delivered in a single high-energy dose or spread over small doses. In other words, a number of hits will produce a number of point mutations no matter whether the hits occur all at once or one at a time. However, for those chromosome aberrations where two hits are required to produce an effect, it does make a difference whether the irradiation is continuous or spread over many doses (Fig. 24–7). Apparently chromatid and chromosome breaks can "heal" if the second break does not occur immediately after the first. Continuous radiation therefore ensures additional breaks before healing of the first break occurs. In the plant *Tradescantia* this healing process is apparently quite rapid for chromatid breaks, taking place within 4 minutes or less, but 20 minutes are needed for chromosome breaks. In *Drosophila* spermatozoa the rate of induced chromosome aberrations is independent of radiation intensity, indicating that healing does not take place while the chromosomes are in the sperm head. In fact there is evidence that breaks induced in *Drosophila* spermatozoa may remain open long enough to unite with spontaneous breaks that have occurred among the chromosomes of the egg nucleus.

Mice, as shown by Russell, provide an important

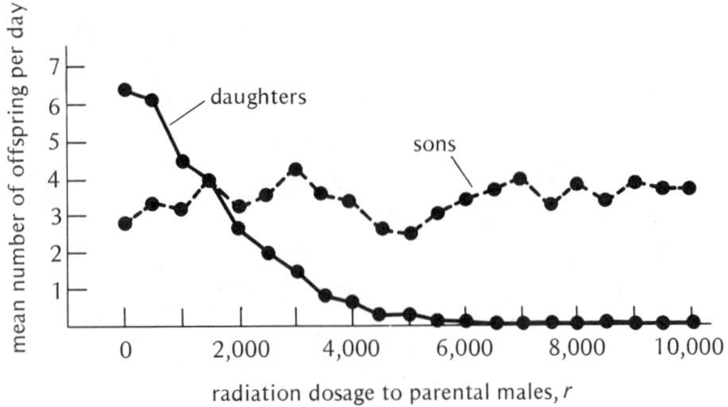

Figure 24–6

Effect of different radiation dosages on dominant lethals induced in male sperm of Bracon hebetor. *Measurements are taken in terms of male (unfertilized) and female (fertilized) offspring produced by nonirradiated females to whom the irradiated males are mated. (Data of Heidenthal.)*

* In recent years the induction of dominant lethals has been used in attempts to control insect pest populations. The most dramatic success of this kind has been the extensive eradication of the screwworm fly in various parts of the United States. This fly, *Cochliomyia hominovorax*, is a pest that formerly accounted for a considerable loss of livestock because its larvae grow in the skin wounds of cattle. Males of this fly are now bred in large numbers, heavily irradiated, and then released over infested areas where they then mate with the normal females that are already present. The progeny of these matings are lethal, and the population soon declines. Other genetic methods of insect control have since been suggested including the induction of sterility and lethality by introducing translocations or conditional lethal mutations (see Smith and von Borstel, and the review of Davidson).

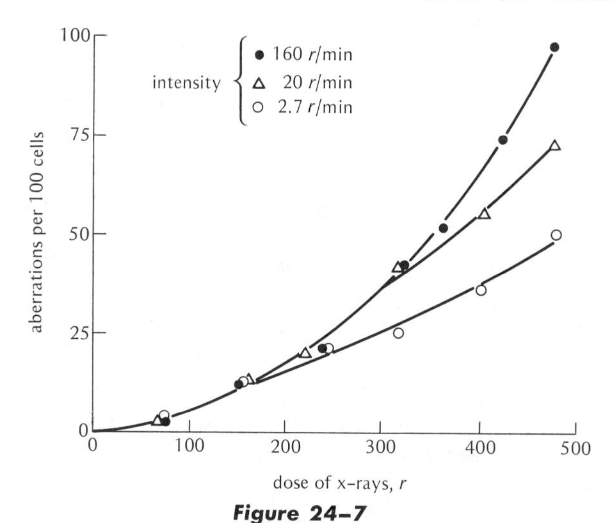

Figure 24–7

Relationship between percentage chromosome changes observed in Tradescantia and intensities at which x-ray dosages are given. Note that the highest intensity (160 r/minute) produces the greatest yield of chromosome aberrations. (From Giles, after data of Sax.)

exception to the rule that point mutations are independent of radiation intensity. Using a technique which measured mutations induced at seven different gene loci (see p. 535), Russell also varied the time interval between irradiation of the male and the time of mating to the test female. He found that when the interval between the time of irradiation and the time of mating is short (irradiation of already formed spermatozoa) there was no effect of changes in radiation intensity. However, if the interval was long enough to have permitted the irradiated spermatogonial stages to produce sperm, a healing process occurred when radiation intensity was reduced (Table 24–1). These results were shown even more dramatically for mouse oocytes, in which practically all mutations were healed if radiation intensity was lowered sufficiently.

There is no obvious answer to account for the differences in "mutation healing" between mice and *Drosophila.* However, as pointed out by Muller, at least two important differences between these two organisms should be taken into account: (1) the fact that induced mutation rates seem to be much higher in mammals than in flies, and (2) the longer generation time of mammals. Russell's findings had shown that the mutation rate per locus per roentgen was 22×10^{-8} for mice (seven loci) compared to 1.5×10^{-8} for *Drosophila* (eight loci), or an increase of 15 times. Such an extreme sensitivity to induced mutation (perhaps caused by increased amounts of DNA, see Fig. 24–8) would be extremely disadvantageous to mammals if there were no mechanism permitting "healing" under normal conditions of low radiation intensity (see "repair" enzymes, p. 568). Similarly the longer generation time of mammals would permit them to accumulate gene and chromosome mutations much more readily than flies if there were no healing mechanism for mutations. Muller has therefore proposed that healing mechanisms for mutations have been of survival value in animals and have consequently been selected through evo-

TABLE 24–1

Mutation frequencies in mice for the same amount of roentgens given at different radiation intensities

Radiation Intensity, r/Minute	*Mutation Frequencies*		
	Spermatozoa	Spermatogonia	Oocytes
90	high	high	very high
9	high	intermediate	—
.8	high	low	intermediate
.009	high	low	very low
.001	high	low	—

From Russell, 1963.

lutionary history because of the general presence of background mutation or of chemical substances in the cellular environment which cause mutation. In any event Russell's findings indicate that the principle of one hit–one mutation is not ruled out in mice, only that certain cellular conditions may permit repair of mutation damage.

Another consequence of the target theory is an expected relationship between the *quality* of radiation, in terms of the density of ions produced, and the rate of gene mutation and chromosome aberration. Thus, for example, radiation that produces a large number of ions along a path would be expected to have a greater chance of producing a mutation. This relationship, however, appears to be true for chromosome breaks but is reversed for recessive lethal mutations in *Drosophila* (Table 24–2). According to Lea and Catcheside, the explanation for this difference resides in the amount of ionization necessary for each type of effect. Recessive lethals can be caused by a single ionization; therefore, many scattered individual rays, each producing a small number of ions (e.g., soft x-rays), are more effective than relatively few rays, each producing a chain of many ions (e.g., α rays). Chromosome breaks, on the other hand, demand more ions per break (15 to 20 according to their calculation) and are therefore more effectively induced by rays in which many ionizations are placed closely together along their path.

On the assumption that a single ion may produce a single mutation, a number of attempts were made to estimate the size of the target. For sexually reproducing species the mutation target is of course located in the gonads, or the tissue that forms the gonads, since these provide the gametes. Within the gamete itself, the target is considerably smaller in size than the cell. According to calculations by Lea, the mutation target is on the order of about 50 angstroms in diameter. The validity of this value has been considerably debated, and most geneticists at present find it difficult to calculate any precise size of the mutation target through ionization damage. It is, nevertheless, of interest that there appears to be a correlation between chromosome breakage caused by irradiation and the size of the nucleus. Experiments performed by Sparrow and others (1963) indicate that plant species with larger nuclear volumes are more sensitive to radiation damage than those with smaller nuclei. Diploid cells in *Tradescantia* were found to have twice the frequency of mutation of haploid cells (Conger and Johnston), while tetraploid cells in *Vicia* had twice the mutation frequency of diploids (Evans). Evidence that the chromosome component of the nucleus was most instrumental in determining mutation rate was obtained by Ostergren in experiments with two species of hyacinth. Although each species had the same number of chromosomes, they were larger in one species than in the other. Equal radiation doses given to each of these two species resulted in a greater frequency of abnormalities for the larger chromosome species.

Further experiments in plants have generally supported the view that there is a direct correlation between radiation damage and the amount of DNA per chromosome or the number of chromosomes per given nuclear volume (see Sparrow et al., 1968). In mammals, however, the results are

TABLE 24–2

Relationship between number of ions produced by different types of radiation and their relative efficiencies in inducing chromosome breaks and lethal mutations

	Ionization Density Along Path	*Chromosome Breaks in Tradescantia*	*Recessive Lethals in Drosophila*
soft x-rays	least	1.0	.96
neutrons	more	3.7	.66
α rays	greatest	7.8	.29

After Wagner and Mitchell, 1964.

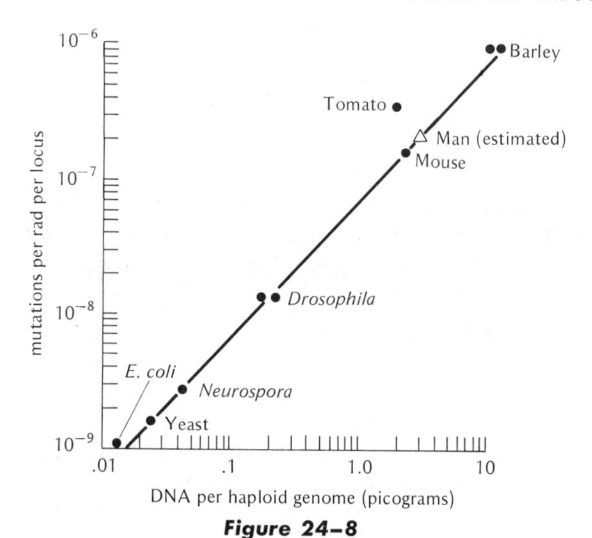

Figure 24–8

Relationship between the amount of DNA per haploid nucleus (picogram = 10^{-12} gram), and the mutation rate per locus per rad. The mutation rates are given only for forward mutations (e.g., prototroph → auxotroph, wild-type → mutant) and for acute rather than for chronic radiation dosages. Since both vertical and horizontal measurements are on a logarithmic scale, the straight-line relationship indicates that multiplying one measurement multiplies the other by the same amount. The point given for man is based only on DNA content, since the mutation rate per locus per rad is not known. (After Abrahamson, et al.)

somewhat more equivocal: Burki and Carrano have shown lower survival rates (i.e., greater mutational damage) for irradiated tetraploid Chinese hamster cells compared to diploid cells, but this relationship is reversed at low doses (below 600 rads).

Recently Abrahamson and co-workers have collected data for various species in which both the mutation rate per radiation dosage is known and the amount of DNA per nucleus. For forward mutations of specific loci under acute radiation, they have found a remarkably constant dependency of mutation rate on DNA nuclear content. As shown in Fig. 24–8, as the DNA content multiplies between species, the mutation rate per gene multiplies by almost exactly the same factor. This correlation may arise from an increase in the size of each gene as the DNA content increases. That is, each gene in organisms with greater amounts of DNA may contain additional nucleotides for regulating its activity compared to genes in organisms with smaller amounts of DNA. Whatever the explanation, there certainly seems to be a proportionate increase in the mutation target size with increased amounts of DNA.

Despite the attractiveness of the target theory, however, important evidence has accumulated that radiation alone may not be sufficient to cause mutation but that certain receptive cellular stages are also necessary. As shown in Table 24–3, mutational sensitivity in *Drosophila* depends strongly upon the maturation stages irradiated. In a detailed study of meiotic stages in the plant *Trillium*, Sparrow demonstrated that chromosome aberrations are induced about 60 times more frequently at metaphase than at interphase (Fig. 24–9). In general, when cell division is absent, such as in adult *Drosophila* somatic tissues, mutation sensitivity seems to be low and physiological tolerance high. Adult flies can therefore withstand very high irradiation dosages. Rapidly dividing tissues, however, are extremely sensitive to radiation damage, and this fact has been utilized to destroy fast-growing cancer cells. Thus, irradiation effect is, as a rule, highly correlated to cell-stage condition.

OXYGEN AND ENVIRONMENTAL EFFECTS

Other evidence opposed to accepting the target theory in simple form includes observations that ionization effects may also be modified by changing the internal environment of the cell experimentally. First among these studies were those of Thoday and Read, who found that lowering the oxygen concentration in *Vicia faba* during irradiation lowered the frequency of mutation. This relationship has since been confirmed in other organisms. In *Drosophila*, for example, the mutation rate of sex-linked lethals at 4000 *r* irradiation may vary from about 8 to 16 percent, depending on the oxygen concentration (Fig. 24–10). This increased radiation damage is also seen when different genetic stages themselves differ in O_2 content.

TABLE 24–3

General order of mutational sensitivity of different meiotic stages in *Drosophila*

high ↑	spermatids, late spermatocytes, early cleavage stages
	spermatozoa located in female seminal receptacles
mutation	early spermatocytes
sensitivity	first-day sperm, zygotes
	second-day sperm, late-stage oocytes
↓	early-stage oocytes
low	spermatogonia, oogonia

After Sobels, 1965.

Spermatids, for example, are generally higher in O_2 content than spermatozoa and also show higher mutational susceptibility to x-rays. Also, as shown by Sobels, the O_2 effect can be increased by pretreatment with cyanide, which inhibits cellular O_2 respiration and thereby permits the nonmetabolized O_2 to become involved in the mutation process.

It seems at present that the effect of oxygen on mutation arises from its utilization in the formation of peroxides during irradiation. Irradiation ionizes water molecules to produce hydrogen and hydroxyl radicals:

$$H_2O \xrightarrow{\text{irradiation}} H + OH$$

In the presence of oxygen, the hydrogen atoms may easily produce hydrogen peroxide:

$$\begin{aligned} H + O_2 &\rightarrow HO_2 \\ HO_2 + H &\rightarrow H_2O_2 \text{ (hydrogen peroxide)} \\ 2HO_2 &\rightarrow H_2O_2 + O_2 \end{aligned}$$

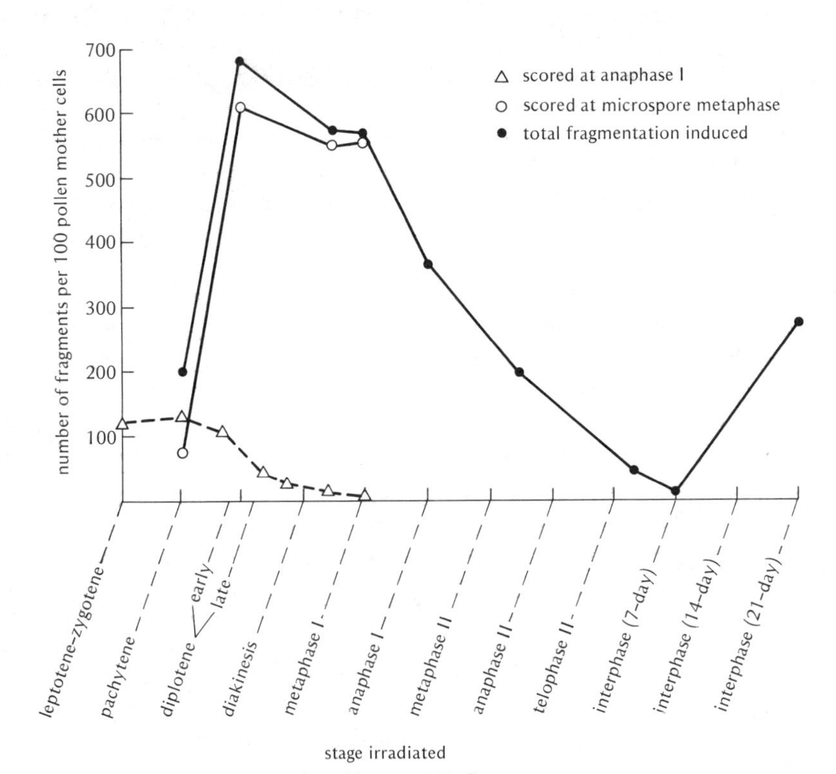

Figure 24–9

Relationship between the number of chromosomal fragments induced in meiotic cells of Trillium erectum at a specific dosage of radiation (50 r) and the stage at which the irradiation occurred. (From Sparrow.)

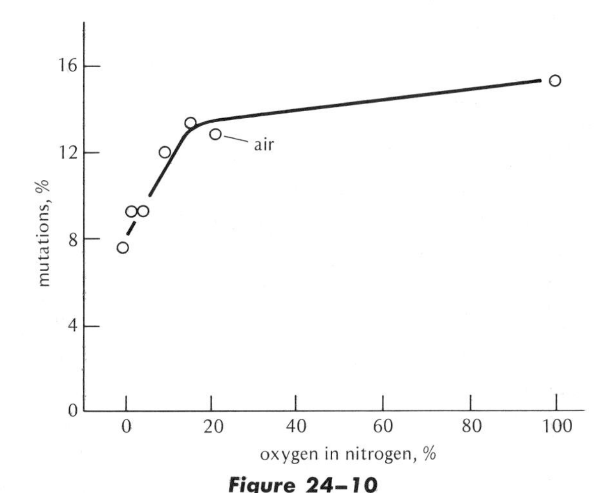

Figure 24–10

Frequency of sex-linked lethals produced in Drosophila melanogaster *males irradiated with 4000 r units of x-ray at different concentrations of oxygen. (After Hollaender, Baker, and Anderson.)*

Since peroxides are highly reactive molecules, they may be responsible for at least some of the observed mutations. It seems likely therefore that, aside from direct physical impact on the gene or chromosome, high-energy irradiation probably produces indirect effects that may themselves be mutagenic.

Indeed, such chemical effects may even occasionally arise in ordinary nonirradiated cells and may therefore help explain spontaneous mutations. The mutagenic activity of hydrogen peroxide, for example, has been shown to be affected by enzymes (catalases) produced within the cell, so that if enzyme poisons (potassium cyanide, sodium azide) are added, the mutation rate is increased. Since this increase occurs even in the absence of extracellular hydrogen peroxide, it is presumed that the normal cell is capable of producing peroxides of its own. The need for such chemical explanations arises because of the relatively inadequate amount of "background" radiation (cosmic rays, radioactive terrestrial material) that would be sufficient to account for observed mutation rates in most organisms. In *Drosophila*, for example, Muller calculated that the spontaneous mutation rate was about 1300 times larger than the rate expected on the basis of background irradiation. Thus the small part played by background effects has put the burden of explaining spontaneous mutation on internal cellular causes.

Some of these causes may be associated with temperature changes, and, in *Drosophila melanogaster*, very high or very low temperature shocks produce an increased frequency of recessive lethals. Some of the genes influenced by position effects are also known to be subject to temperature influence. For instance, the *a* gene in corn, in the presence of *Dt* (see p. 543), mutates much more rapidly at low temperatures.

In addition to temperature, aging effects are also known, so that seed and pollen stored for long periods show increased mutation rates. In the snapdragon, *Antirrhinum majus*, Stubbe showed that the frequency of recessive mutations increased to as high as 14 percent in some seeds aged 5 to 10 years. *Drosophila* sperm stored for 2 to 3 weeks, either in the male or female, have also shown significant increases in the mutation rate of sex-linked lethals. It is not clear at present whether such mutations arise because of the accumulation of toxic substances that lead to copy errors during the replication of DNA, or whether the DNA is directly changed by aging. In mammals various experiments have attempted to demonstrate that aging itself arises because of the accumulation of mutations in somatic tissues. Although inability to repair somatic mutation is undoubtedly one of the causes for mammalian aging (see p. 570), other factors are probably involved as well (see, for example, Orgel).

Among the most obvious cellular causes for mutation are genes themselves. So far, a considerable number of "mutator genes" have been shown to be responsible for mutation rate increases (see p. 543). In almost all cases that have been analyzed in detail, these genes appear to act by causing "copy errors" during DNA replication, and may be directly or indirectly involved in the function of the polymerase replicating enzymes, or the recombination and repair systems responsible for accurate genetic exchange and strand replacement (see Drake, 1973). In one instance Hershfield and Nossal have localized mutator gene action in bacteriophage T4 to a specific change in the structure of DNA polymerase leading to increased incorporation of the noncomplementary nucleotide bases, cytosine and guanine, during the replication of a template molecule of DNA which bears only adenine and thymine. These findings lead to a simple conclusion: if mutations arise primarily by DNA copy errors, one would expect a proportionate relationship between mutation rate and DNA replication.

The first serious attempt to determine the relationship between mutation rate and cell division rate was undertaken by Novick and Szilard. They utilized a bacterial population unable to synthesize tryptophan, whose growth conditions were controlled by the addition or deletion of tryptophan in the culture medium. As the experimenters changed the growth rate of their populations they found they were able to modify the generation time of each bacterial cell. As generation time increased, the mutation rate per generation increased as well. A most interesting observation, however, was that the mutation rate calculated on an hourly basis remained the same for an extended range of generation times greater than 80 minutes (Table 24-4).

The Novick and Szilard experiments therefore seemed to indicate that mutation was not necessarily connected to cell division; that is, the reproductive rate of the gene, during which period copying errors are believed to occur, was somehow different from the cell-division rate of bacteria. This seeming paradox has recently been resolved by Kubitschek, who has shown that the peculiarities of tryptophan-limited growth leads to intensive selection for the presence of mutant bacterial nuclei. This phenomenon leads, in turn, to what only appears to be a higher (and constant hourly) mutation rate, whereas, in reality, Kubitschek's experiments and others show that the mutation rate is strongly tied to the DNA replication rate. So far, therefore, the copy-error explanation for most point mutations seems to be valid.

TABLE 24-4

Mutation rate of *E. coli* to T5 phage resistance for different generation times

Generation Time, Hours	Mutation Rate per Generation	Mutation Rate per Hour
2	2.5×10^{-8}	1.25×10^{-8}
6	7.5×10^{-8}	1.25×10^{-8}
12	15.0×10^{-8}	1.25×10^{-8}

From Novick and Szilard, 1950.

However, not all gene mutations are caused by changes within the gene itself, nor are they restricted to simple "points" at which a copying error has occurred. Position effect, as discussed previously (p. 547), demonstrates that in some instances mutation can occur through even a small change in gene position. Also, very small deletions and inversions of chromosomal material may not be cytologically distinguishable from point mutations. One test that has been used to determine whether ionizing irradiation actually causes gene or point mutations has been to find instances where forward mutations produced by ionizing irradiation could also be reversed to wild type by ionizing irradiation. According to this argument, chromosomal changes cannot be reversed, since new irradiation cannot supply missing deletions or precisely reinvert inverted segments; only very small point mutations, in which genetic material has not been destroyed or misplaced, might therefore be subject to reversion. Although reversion of this type has not been indisputably demonstrated in *Drosophila*. such changes have been observed in *Neurospora*, bacteria, and yeast. These observations therefore indicate that irradiation may cause small reversible genic effects as well as the more obvious irreversible chromosomal changes.

Summing up our present findings on ionizing radiation, we can say that the target theory alone is no longer sufficient to explain radiation effects. The physiological state of the cell and the condition of the chromosomes, as well as temperature and oxygen pressure, are all instrumental in determining whether a mutation will or will not occur. In addition the precise chromosomal location of the "hit" is also important, since chromosome aberrations are more frequently noted in certain locations (e.g., the centromere region) than in others. Russell has also shown that there are significant

TABLE 24–5
Distribution of mutations induced by x-ray and gamma rays among seven loci in mice

	a	*b*	*c*	*p*	*d*	*se*	*d-se* (*Deletions*)	*s*	*Total*
spermatogonia	5	45	27	38	42	3	1	108	269
postspermatogonia	4	16	4	10	6	3	7	25	75

From Russell, 1965.

differences in the occurrence of induced mutation at the seven loci scored in mice (Table 24–5). The importance of an additional factor, mechanical stress, is reflected in the observation that treatment of the cells with centrifugation or sonic vibration during irradiation seems to prevent restitution and increases the number of breaks. Colchicine, on the other hand, reduces chromosome movement (because of impaired spindle formation) and consequently decreases the number of breaks during irradiation. Figure 24–11 shows some of the possible interactions that may take place during intermediary stages between the irradiation and the appearance of aberrations. The complexity of these relationships will become even more striking when we consider mutagenic effects caused by ultraviolet radiation and chemicals.

ULTRAVIOLET IRRADIATION

The mutagenic effect of ultraviolet light (UV) was first discovered by Altenburg in the irradiation of the polar cap cells (early gonadic tissue, see p. 255) of *Drosophila* eggs. The mutagenic potency of these rays has since been confirmed in many organisms in which germ tissue can be easily exposed to the low-penetrating UV. Although UV may also cause chromosomal aberrations, its effect is considerably milder than x-rays and it has therefore been mainly used to study "point" mutations. Its mutation effect, however, does not fit in with the target theory, since the relationship between mutation rate and UV dosage is not generally linear. As shown in Fig. 24–12, streptomycin resistance in *E. coli* is the only one of six mutation rates charted against increased amounts of UV dosage that follows the expected linear effect for one hit–one target. The other mutation rates follow various types of curves, which suggests that several UV "hits" are necessary for a mutation to occur. In *Aspergillus* the rate of morphological mutations seems to indicate a "healing" process occurring at high UV dosage.

Possessing a wavelength too long to produce ions, UV appears to act by affecting only those compounds that absorb it directly. In the cell direct absorption of UV is mainly confined to compounds with organic ring structures such as purines and pyrimidines (Chapter 4). Other compounds do not appear to absorb UV as greatly and are therefore not as directly affected by it. This close relationship between UV and DNA components also appears in comparisons between the UV absorption spectrum of DNA and the mutation rates caused by UV wavelengths. As shown previously in Fig. 4–12, the 2540-angstrom wavelengths at which DNA absorbs UV rays is also the most productive in increasing the mutation rate in corn pollen.

Support for a direct relationship between UV and DNA has been gained from *in vitro* studies showing that the pyrimidines, thymine and cytosine, are especially absorptive of UV wavelengths. Cytosine is hydrated by UV by the insertion of a water molecule into the C═C double bond (Fig. 24–13a). The double-carbon bonds of thymine are also disrupted, and two thymine bases may be connected to form a thymine dimer (Fig. 24–13b). *In vitro* studies indicate that thymine dimerization may be the primary mutagenic effect produced by

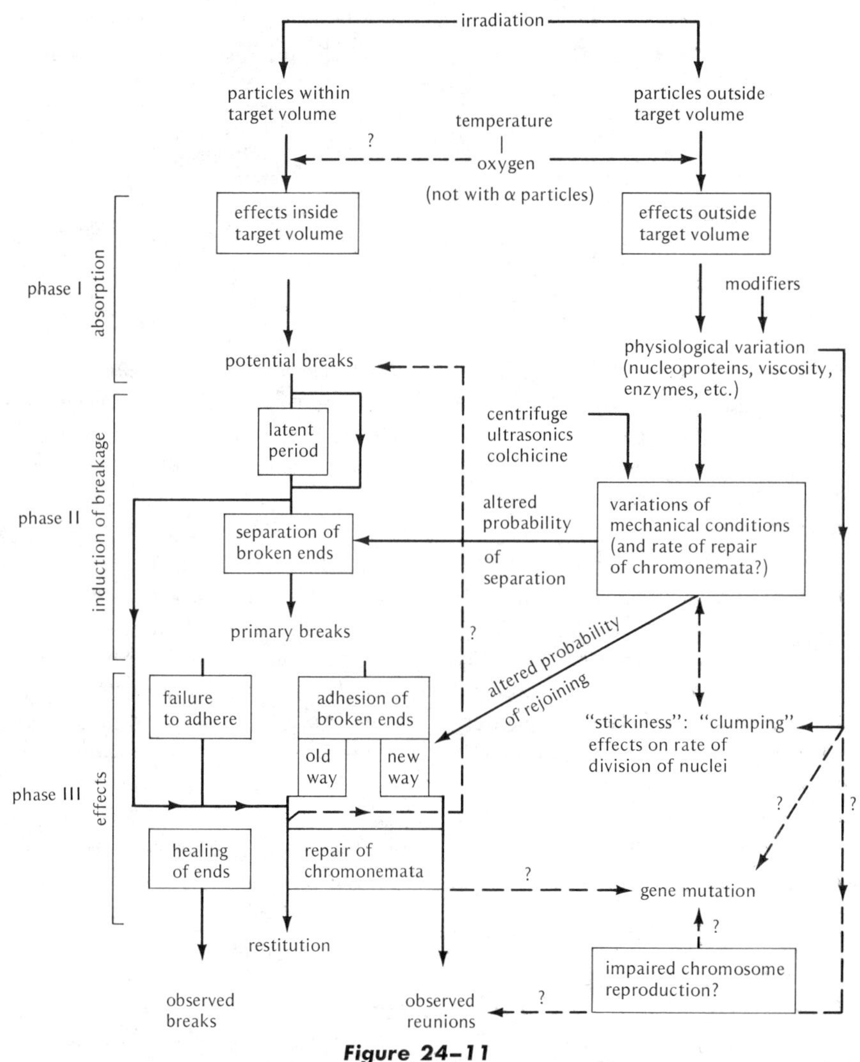

Figure 24–11

Different possible pathways and interactions causing genetic damage in organisms as a result of irradiation. (After Wagner and Mitchell, from Thoday.)

UV. Such dimers would distort the DNA helix and interfere with proper replication.

Despite these direct effects on DNA, indirect action by UV through absorption by intermediate compounds is also possible. For example, Stone and others have observed increased mutation rates in *Staphylococcus aureus* when the bacterial culture medium is irradiated by UV shortly *before* nonirradiated bacteria are placed upon it. Second, demonstrations by Witkin, and Doudney and Haas, have shown that pretreatment of bacterial cells with nucleic acid bases before UV irradiation increases the mutation rate. On the other hand, post-treatment of irradiated bacteria with compounds that inhibit protein synthesis (chloramphenicol) decreases the mutation rate. Taken together, these observations suggest the likelihood that UV acts also on various DNA precursors and enzymes which, in turn, affect mutation.

The UV effect on an organism, however, may

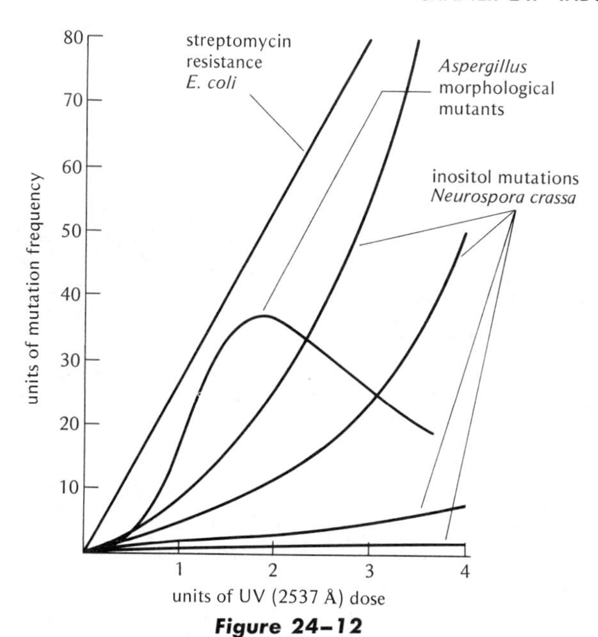

Figure 24–12

Relationships between different mutation rates and UV dosage. (After Swanson, 1957.)

be considerably influenced by whether other forms of radiation are also being used. In 1944 Swanson noted that irradiation of *Tradescantia* pollen tube nuclei with UV, either prior to or after x-ray exposure, had marked effects on the frequency of induced aberrations (Table 24–6). Depending upon the strength of the UV dose and the interval between UV and x-ray, the frequency of aberrations could be reduced to almost zero if x-ray doses were below 250 *r*. Similar but less striking results were obtained by Kaufmann and Hollaender in *Drosophila* and by Schultz in corn.

On the other hand, when dehydrated *Tradescantia* pollen was used (Kirby-Smith, Nicoletti, and Gwyn), a combined radiation effect was still noted, but the results were dramatically different. In this case combined radiation between a small normally undetectable dose of 1 *rad* of x-rays and a very small dose of UV produced an aberration frequency equal to that given by 100 *rad* of x-rays alone.

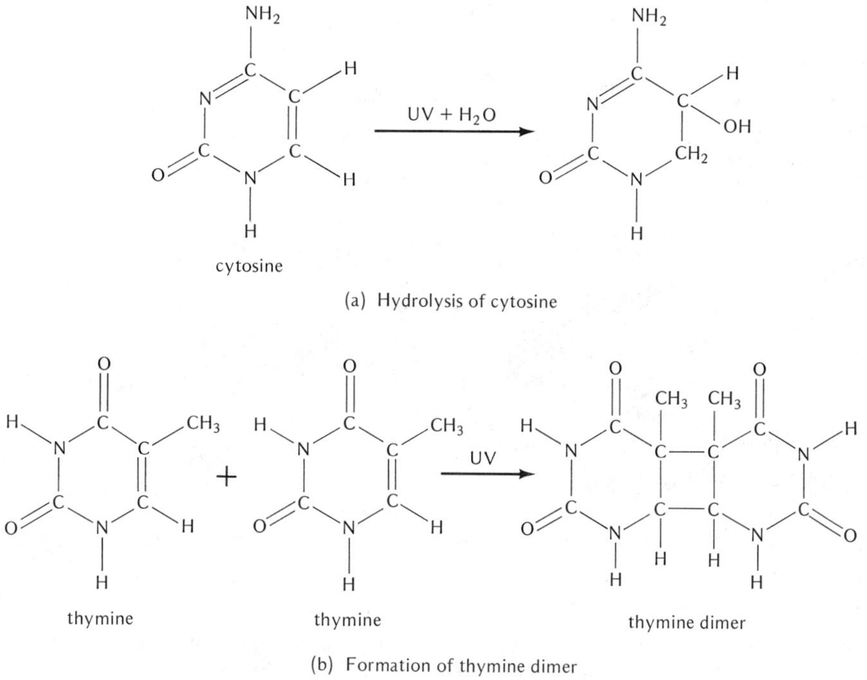

cytosine

$UV + H_2O$

(a) Hydrolysis of cytosine

thymine + thymine $\xrightarrow{UV}$ thymine dimer

(b) Formation of thymine dimer

Figure 24–13

Effect of UV on pyrimidines.

TABLE 24–6

Effect of UV and x-ray treatments, alone and combined, on the frequency of chromosome deletions in *Tradescantia* pollen tube nuclei

Treatment	*Percent Chromosome Deletions*
UV alone (30 sec)	2.25
x-ray alone (2 min)	3.51
UV first (30 sec) + x-ray (2 min)	1.73
x-ray first (2 min) + UV (30 sec)	3.29

Data of Swanson, 1944.

Interaction between infrared rays and x-rays has also been noted both for *Drosophila* and *Tradescantia*. In both cases infrared increased the efficiency of chromosomal aberrations, but some of these combined effects are also time-dependent or temperature-sensitive. These effects, as many others considered so far, are therefore associated with premutational and postmutational stages at which mutational resistance or repair can be distinctly modified.

REPAIR OF DNA

A most unusual facet of UV-induced cellular damage was the discovery by Kelner and others that this effect could be reversed by exposing cells to visible light containing wavelengths in the blue spectrum. This repair process, known as *photoreactivation*, indicates that the damage caused by UV can be reversed before the genetic material is permanently affected. It has now been observed in numerous organisms including bacteria, bacteriophages (Fig. 24–14), protozoans, algae, fungi, frogs, birds, and marsupials, but there are contradictory reports of its presence in mammals (see, however, Sutherland). As demonstrated by Setlow and Setlow, the photoreactivation effect primarily relies upon a specific enzyme that can split the pyrimidine dimers produced by irradiation (Fig. 24–15).

Photoreactivation, however, is only one of a number of possible repair mechanisms, since recovery from UV damage can also occur in the dark. Among the two "dark repair" mechanisms known so far is a system that involves enzyme recognition of the distorted section of a DNA strand, excision of the defective sequence, and replacement of this section by a new sequence of nucleotides formed as a complement to the undamaged DNA strand (Fig. 24–16). In *E. coli*, *uvr* mutations may affect different steps in this process, thus increasing the lethality of mutant bacteria that are exposed to UV. Interestingly the polymerase I enzyme, traditionally used by Kornberg and others for the *in vitro* replication of DNA (see p. 85), is also involved in the dark repair process; polymerase I mutants (*pol A*) are extremely sensitive to UV irradiation. There does seem to be a difference, however, between replication and UV repair enzyme systems, since cells that are deficient in ability to replicate can still perform repair functions.

A significant portion of DNA aberrations caused by irradiation are probably also repaired by enzyme systems involved in the recombination process. Evidence for this is the fact that irradiation-damaged bacteriophages can produce viable

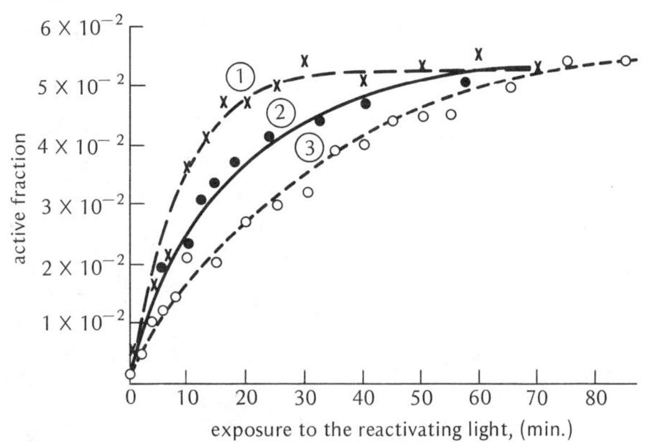

Figure 24–14

Effect of three different light intensities (1, 2, 3) on the infective activity of phage T2 particles irradiated by UV. In each case exposure to light increased the fraction of active phage particles far beyond that when light was absent. Also, the longer the exposure to light, the greater the increase in the size of the active fraction. (After Dulbecco.)

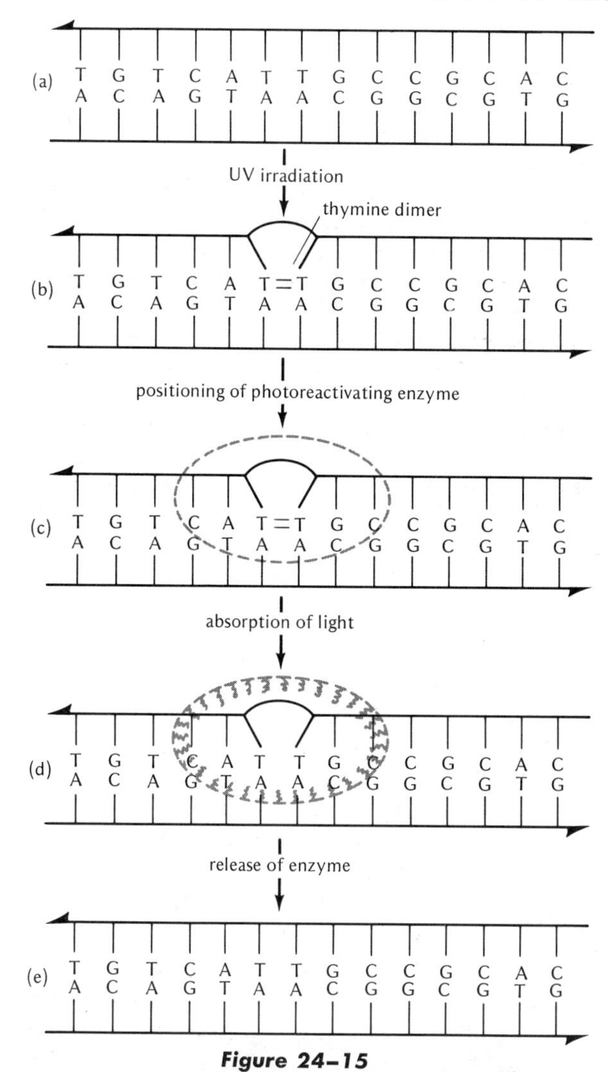

Figure 24–15

Diagram of the mechanism responsible for the photoreactivation process. (a) Section of intact DNA molecule showing complementary strands with opposite polarity (arrows). (b) Changed spatial configuration of the DNA molecule because of a thymine dimer formed by ultraviolet irradiation. (c) The photoreactivation enzyme locates itself on the DNA molecule at the point of distortion. (d) Absorption of visible light radiation in the blue end of the spectrum provides the energy which enables the enzyme to split the dimer. (e) The DNA molecule then resumes its normal shape and the enzyme is released. According to Setlow, Boling, and Bollum, this same photoreactivating enzyme can split dimers of the other pyrimidine, cytosine, as well as thymine-cytosine dimers.

progeny when many of them are permitted to infect a single cell. This procedure of "multiplicity reactivation" seems to depend upon recombinational events between the various intact portions of the damaged phages leading to the formation of recombinant viable phages. Furthermore *E. coli* cells defective in recombinational ability, such as *rec A* mutants (see p. 402), are also defective in their ability to withstand UV and x-rays.

Howard-Flanders has suggested that the recombinational repair process acts mostly on DNA strands carrying gaps caused either by intrastrand or interstrand crosslinks. As shown in Fig. 24–17, when such crosslinked strands replicate or are excised, gaps remain which then enable the "free" end of a single strand section to initiate an exchange with a similarly oriented strand on a homologous DNA molecule. According to this view, the exchange forms a new unbroken single strand which can then serve as the template for an enzyme complex to remove and replace the defective DNA portion.

The use of such recombinational systems in DNA repair may help to explain the fact that increased frequencies of recombination have been observed after irradiation in various organisms ranging from bacteriophage λ to *Drosophila* (see p. 347).* However, although recombinational repair may correct physical deformities in DNA, it can nevertheless lead to inaccuracies of its own in base pairing and base substitution. Witkin, Kondo, and others have shown that the mutation rate increases significantly in bacteria which depend upon recombination for the correction of DNA damage.

It is apparent that the various repair systems discussed are all maintained in order to prevent both somatic and genetic damage caused by pyrimidine dimers, intrastrand and interstrand crosslinks, etc. Such DNA aberrations may arise from

* In fact Manney and Mortimer have used x-ray irradiation to induce mitotic recombination in diploid yeast cells, and have then calculated the "x-ray distance" between various genes by the relationship between the amount of radiation and the recombinant frequency. In general the x-ray map distances follow the same rules of linearity and additivity as do the meiotic genetic maps.

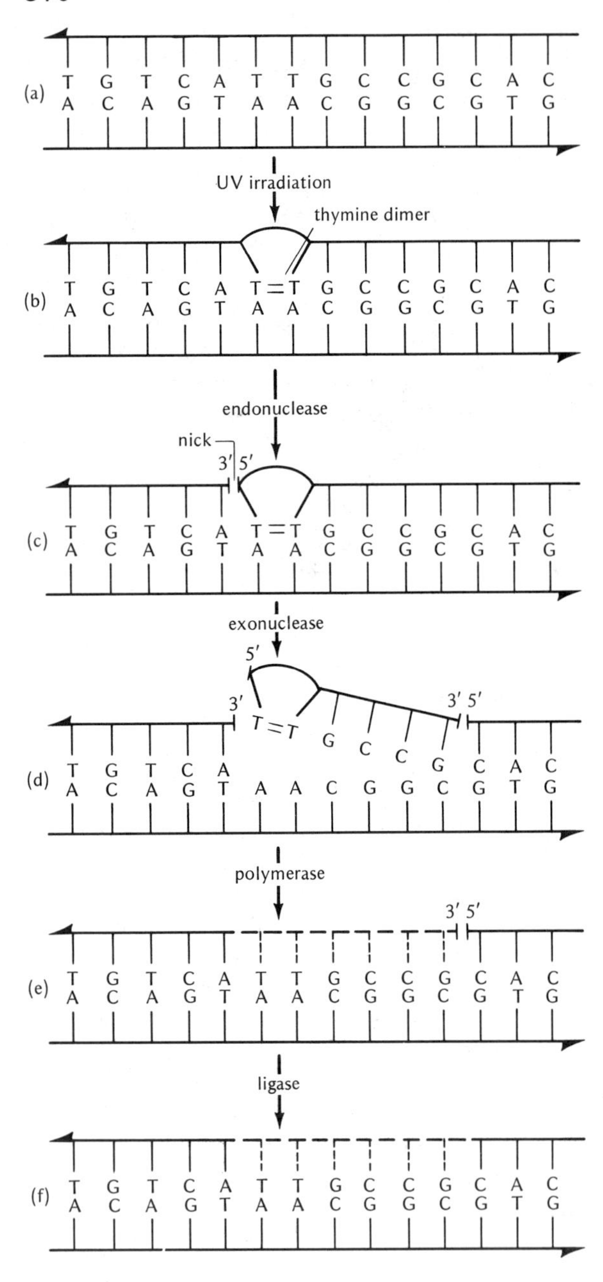

Figure 24–16

"Dark repair" mechanism for the excision and replacement of pyrimidine dimers. (a) Section of DNA double helix with each complementary strand oriented according to opposite polarity (arrows). (b) Formation of a thymine dimer as a result of exposure to UV irradiation. (c) The repair sequence begins with a special endonuclease enzyme producing a "cut" in the phosphate-sugar backbone at a nucleotide position adjacent to the thymine dimer. The 3′ phosphate group at this point is then removed by another enzyme (a phosphatase or 3′ exonuclease) leaving a nucleotide end bearing a 3′ hydroxyl group. (d) A 5′ exonuclease then removes a 6 or 7 nucleotide long section including the thymine dimer. In E. coli *this function is part of the activity of DNA polymerase I (see p. 85). (e) The deleted portion of the excised strand is then refilled by the DNA polymerase enzyme in the 3′ → 5′ direction using the complementary strand as a template. (f) The 3′–5′ gap that remains between adjacent nucleotides after polymerase activity is closed by polynucleotide ligase "joining" enzyme.*

external events such as irradiation, or from internal cellular events whose origins are as yet unknown. The prevalence of repair systems, however, indicates that all forms of life face common problems of DNA damage. Man too utilizes DNA repair systems, and evidence for their presence can be gained from the serious consequences caused by their malfunction. Thus the rare human autosomal recessive disease xeroderma pigmentosum is accompanied by extreme sensitivity of the skin to UV and increased incidence of skin cancer. Cleaver has shown that the cells of such affected individuals are unable to repair DNA damage induced by UV. The defect, in this case, appears to lie in the inability of an endonuclease enzyme to make the original "cut" that initiates the excision-repair process of pyrimidine dimers. In patients suffering from premature aging ("progeria"), a somewhat different repair enzyme appears to be defective, perhaps a ligase necessary to rejoin broken DNA strands (Epstein et al.).

CHEMICAL MUTAGENS

The discovery of mutagenic effects produced by the various types of radiation furnished, for the first time, an experimental probe into methods of changing gene structure and function. Irradiation, however, was a coarse instrument for this purpose since it was difficult to distinguish between direct and indirect effects of irradiation and to analyze

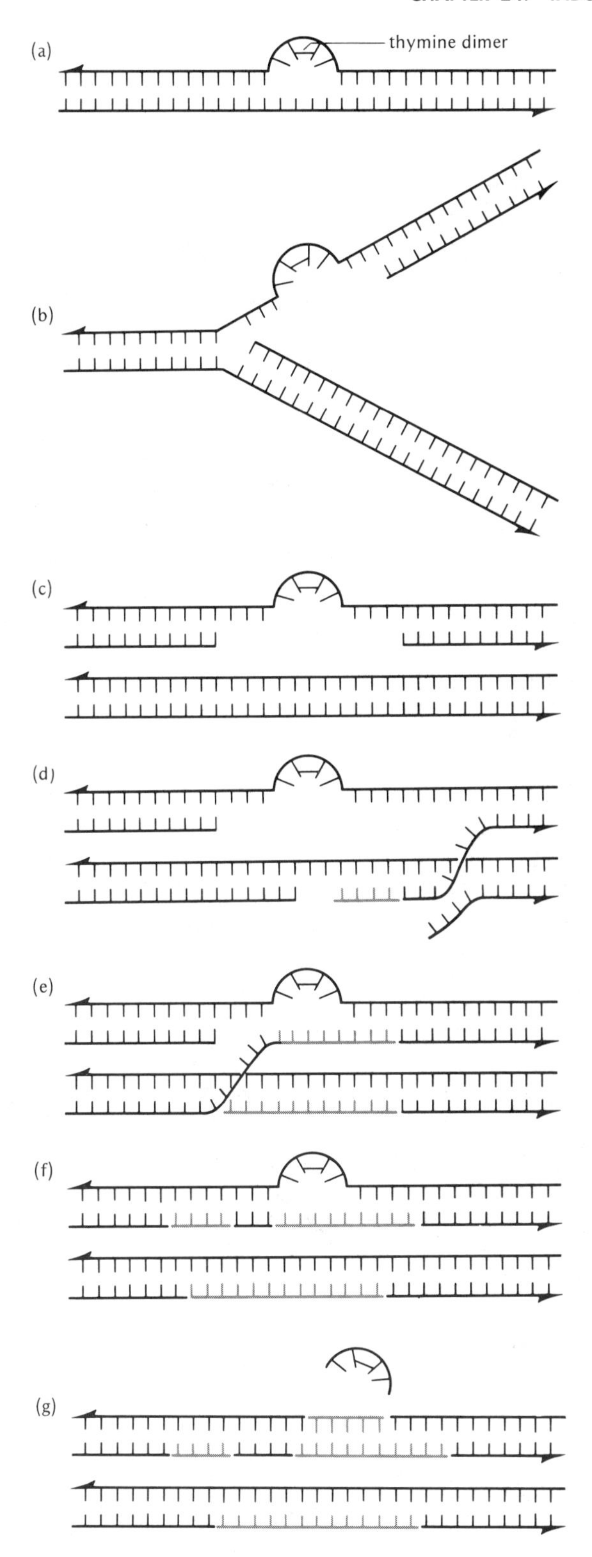

Figure 24–17

Model of a recombination mechanism which fills in the gaps caused by the replication of strands carrying pyrimidine dimers. (a) DNA molecule with thymine dimer on one strand. Arrows indicate polarities of the complementary strands. (b) Replication of the DNA molecule leading to a gap on the complementary strand opposite the thymine dimer. This gap arises because the polymerase enzyme cannot use the distorted dimer section as a template, and synthesis of a complementary strand is not reinitiated until an initiation point (see Fig. 5–18) is reached further down the DNA molecule. (c) Pairing between homologous DNA molecules. (d) and (e) An endonuclease cut is made in the intact DNA molecule enabling an exchange between the cut DNA strand and a free end of the similarly oriented but deficient homologous DNA strand. (The DNA strand bearing the thymine dimer can also be involved in exchange with its homologous strand but, for convenience, this is omitted). The gray sections represent newly synthesized nucleotide sequences made by the polymerase portion of the recombination enzyme complex. (f) and (g) The gap opposite the thymine dimer is now filled with a newly synthesized sequence that has been formed in the intact DNA molecule. The thymine dimer may then be excised as part of the final repair process. If not excised, the molecule may remain in the condition shown in (f), and the next round of replication will again produce a DNA molecule bearing a gap. The cycle (a) → (f) may then continue for a number of rounds of replication producing one complete and one defective DNA molecule at each round. Since the complete DNA molecule also replicates, the defective DNA molecule gradually diminishes in proportion, numbering 1 out of 2 molecules at the first replication, 1 out of 4 at the second, 1 out of 8 at the third, etc. The deleterious effect of dimerization may therefore also be diluted out by recombination. (Adapted from Howard-Flanders and also from Hanawalt.)

the exact nature of the biochemical compounds produced. More promising seemed the search for chemical mutagenic agents which could perhaps lead to a more precise characterization of the changes involved in mutation. After many attempts this search finally met with success during World War II, when a number of chemicals were found to be mutagenic in both plants and animals. Thom and Steinberg, for example, found that nitrous acid was effective in causing mutations in *Aspergillus*. In *Drosophila* Auerbach and Robson found that mutation could be induced by nitrogen and sulfur mustards. Rapaport and others discovered mutagenic activity in formaldehyde, diethylsulfate, diazomethane, and other compounds. Most of these chemicals also caused severe skin irritations in mammals and many were cancer-producing as well. In fact many chemicals were first tested for mutagenic activity simply on the basis of prior skin or cancer tests. Since then hundreds of chemical agents have been found to produce mutagenic activity in a variety of organisms.

In general, however, prediction of mutagenic activity cannot be made on the basis of chemical structure alone. Some compounds, for example, affect certain organisms but not others, while some are restricted in action to specific developmental stages or to a specific sex. Formaldehyde, for instance, acts only to produce mutations during larval stages in *Drosophila* male spermatogenesis but affects neither adults nor female larvae. Mustard gas and its related compounds usually produce mutations in *Drosophila* that may be delayed even as long as one generation before they appear.

Of special concern have been genes that respond differently to mutagenic agents. In *E. coli*, for example, Demerec found that manganese chloride produces a high frequency of reversion of arginine-requiring mutants to wild type ($1720/10^8$) and a low frequency of reversion of phenylalanine-requiring mutants ($11/10^8$). Surprisingly, when Demerec used x-rays as a mutagen in the above experiment, the mutation rates were reversed; the arginine mutants now had a low frequency of reversion ($54/10^8$), while the phenylalanine mutations reverted more frequently ($2460/10^8$). The fact that even greater differences in the action of a mutagen on particular genes can be observed (see, for example, Shukla), indicates that different mutagenic agents may produce different sets of conditions to which certain genes may respond, each in their unique way. This is also very obvious in *Chlamydomonas*, where streptomycin causes mutation of cytoplasmic (chloroplast) genes but has no detectable effect on nuclear genes. Nevertheless, despite the wide search, no mutagen was found that was specific for only one particular gene on a chromosome and affected no

adenine
(imino form)

cytosine

base-pair changes:

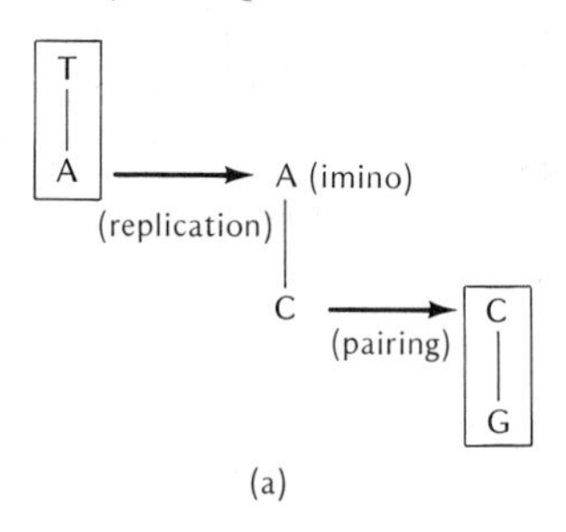

thymine
(enol form)

guanine

base-pair changes:

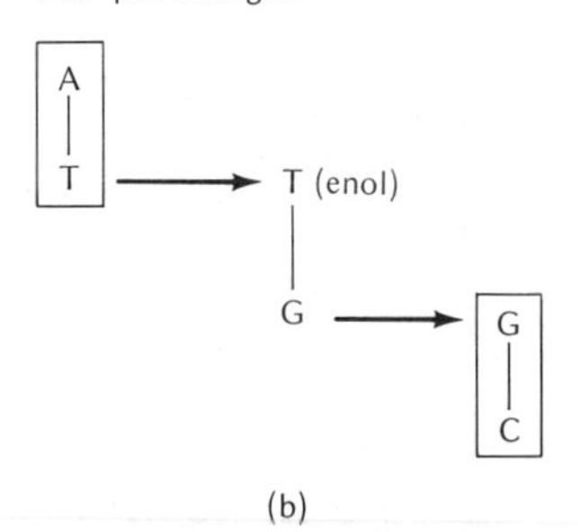

Figure 24–18

Illegitimate pairing between adenine and cytosine (a) and thymine and guanine (b), leading to base-pair substitutions.

others at all. Thus the wide array of possible chemical reactions caused by most compounds and the wide array of compounds that cause mutations made it difficult for a long period to assign any specific activity in the cell to a particular mutagen.

In the 1950s, however, when DNA was firmly identified as the genetic material, a search began for agents or mechanisms that directly affect nucleotide structures. Watson and Crick were perhaps the first to suggest such a mechanism, proposing that mutation could occur as a result of occasional changes in the hydrogen bonding of nucleotide bases. Adenine, for example, normally bears an NH_2 *amino* group that provides a hydrogen atom for bonding with the complementary *keto* (C═O) group of thymine (see Fig. 4–9). Occasionally, however, a *tautomeric shift* is believed to occur in which the amino form of adenine is changed to an *imino* (NH) form. As shown in Fig. 24–18a, this base can now bond in complementary fashion with cytosine. Similarly, a tautomeric shift may occur in thymine, changing it from the keto form to the rare *enol* (COH) form. In this form the base can now bond with guanine (Fig. 24–18b).

Note that the pairing errors arise through the transition of the pairing relationship of one purine base (e.g., adenine) into the pairing form of another purine (e.g., guanine) or of one pyrimidine (e.g., thymine) into that of another pyrimidine (e.g., cytosine). Because of this they have been called *transitional* errors. Freese has also proposed the possibility of *transversions*, in which a purine such as adenine, for example, may be changed into a pyrimidine such as cytosine (Fig. 24–19). Thus, for both transitions and transversions, a base substitution is produced which leads to a new nucleotide sequence. As we shall see in the following chapters, such simple changes may have far-reaching mutational consequences.

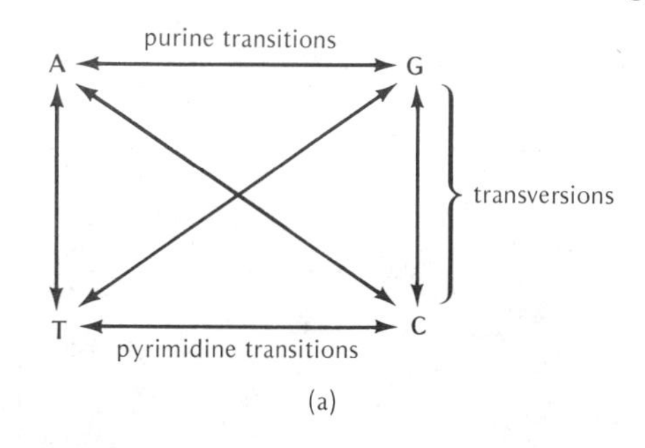

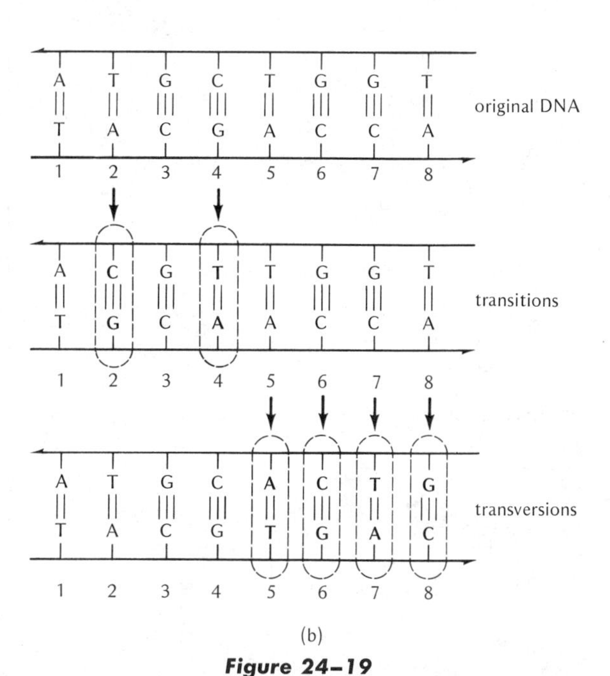

Figure 24–19

Types of possible nucleotide substitutions. (a) The mode of designating purine and pyrimidine changes. (b) Examples of specific base-pair changes in a section of double-stranded DNA. (Modified after Freese, 1971.)

Replication of the DNA is necessary if the errors produced by tautomeric shifts are to appear in DNA molecules, and they are therefore classified as *copy errors*. Similar mistakes in copying mechanisms can take place when base analogues are incorporated whose pairing relationships are occasionally ambiguous. Bromodeoxyuridine (BUdR) or 5-bromouracil (BU), for example, are analogues of thymine that are usually in the keto form, but may possibly undergo tautomeric shifts to the enol form that enable them to pair with guanine (Fig. 24–20). Thus BU may be incorporated as a pairing mate to adenine and then produce a G–C substitution for the original A–T, or it may occasionally be incorporated in the enol form as a pairing mate with guanine and then revert to its keto form to produce an A–T substitution for the original

(a) Pairing of 5-bromouracil with adenine:

5-bromouracil
(normal keto state)

adenine

(b) Pairing of 5-bromouracil with guanine:

5-bromouracil
(rare enol state)

guanine

(c) Mistake at BU replication (BU incorporated for T, but pairs with G)

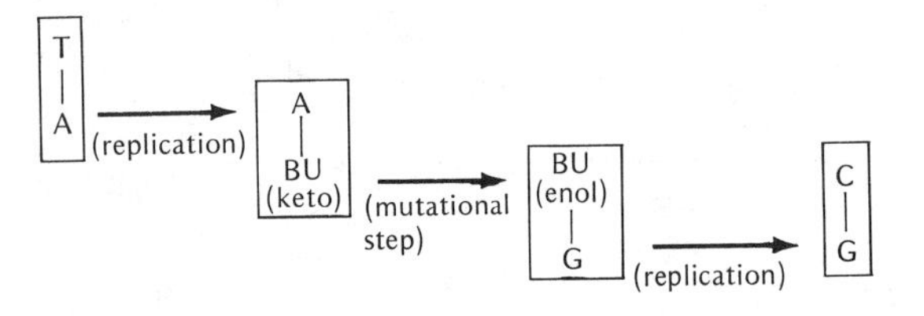

(d) Mistake at BU incorporation (BU incorporated for C, then pairs with A)

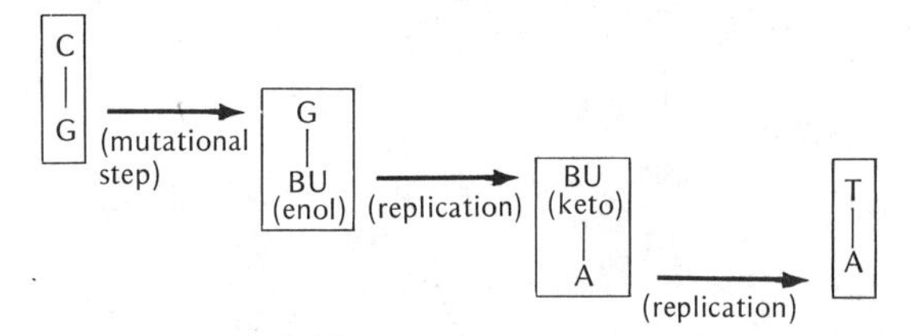

Figure 24–20

Pairing relationships of 5-bromouracil in its (a) normal keto and (b) rare enol states. The consequences of a tautomeric shift to the enol state may lead to base-pair changes in which (c) a G–C pair is substituted for T–A, or (d) A–T substituted for C–G.

deamination of adenine:

adenine

nitrous acid

hypoxanthine

pairing of hypoxanthine with cytosine:

cytosine

hypoxanthine

substitution of G–C pairs for A–T:

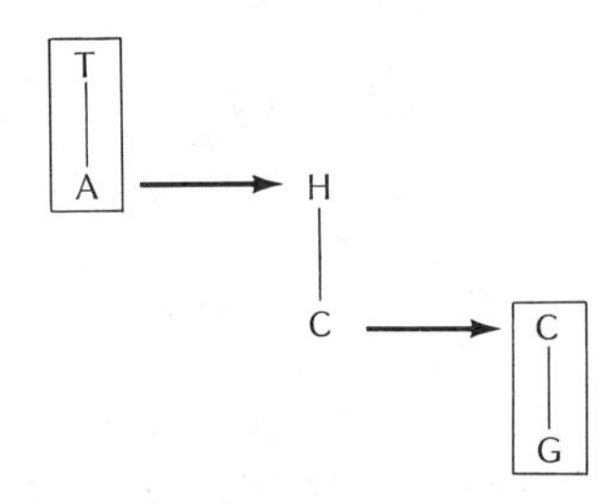

Figure 24–21

Action of nitrous acid on adenine resulting in the substitution of C–G pairs for A–T. Note that T is now changed to C; therefore the mutation is a transition.

G–C.* The base analogue 2-aminopurine (AP) shows similar mutational properties, enabling it to be incorporated as a substitute of adenine but to pair subsequently with cytosine, or to pair initially with cytosine and subsequently with thymine.

* Lawley and Brookes have suggested that mispairing may be caused by ionization of bases rather than by tautomeric shifts. In this mechanism a base such as BU loses the hydrogen normally associated with its 3′ nitrogen atom (see Fig. 24–20) and may now pair with guanine.

In contrast to copy errors, direct changes in nucleotide structure may be produced by agents such as nitrous acid and nitrogen mustard. According to Schuster and others, nitrous acid acts primarily through removal of the amino group (*deamination*) from adenine and cytosine, thereby converting these bases to hypoxanthine (H) and uracil (U), respectively (Figs. 24–21 and 24–22). Since H can now pair with C and U with A, further replications lead to transitions in which G is substituted for A, and T for C. Deamination of guanine also occurs with nitrous acid, but the purine product (xanthine) is unable to pair either with thymine or cytosine, and thus leads to inactivation of the DNA and often therefore to a lethal effect.

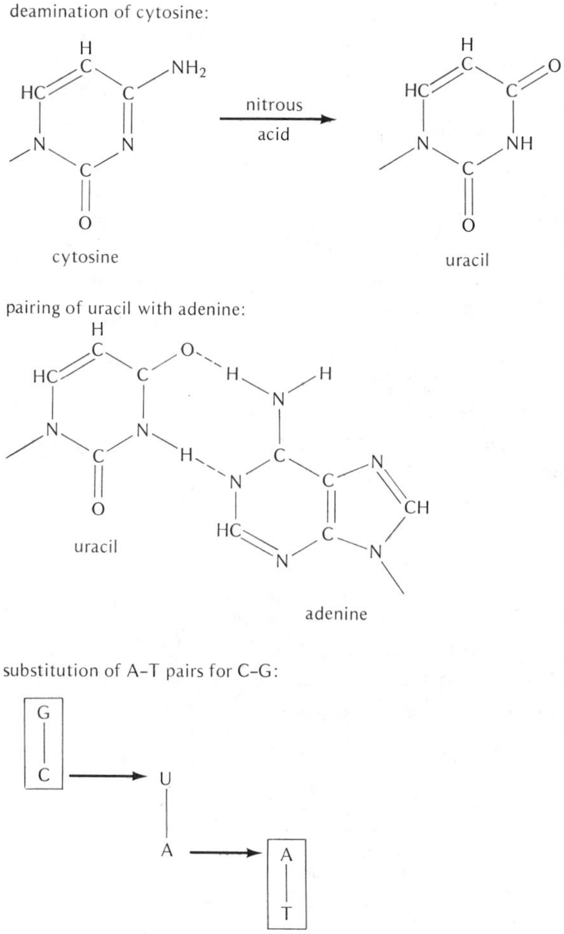

deamination of cytosine:

cytosine → nitrous acid → uracil

pairing of uracil with adenine:

uracil adenine

substitution of A–T pairs for C–G:

G–C → U–A → A–T

Figure 24–22

Action of nitrous acid on cytosine resulting in the substitution of A–T pairs for C–G. Note that G is now changed to A; therefore the mutation is a transition.

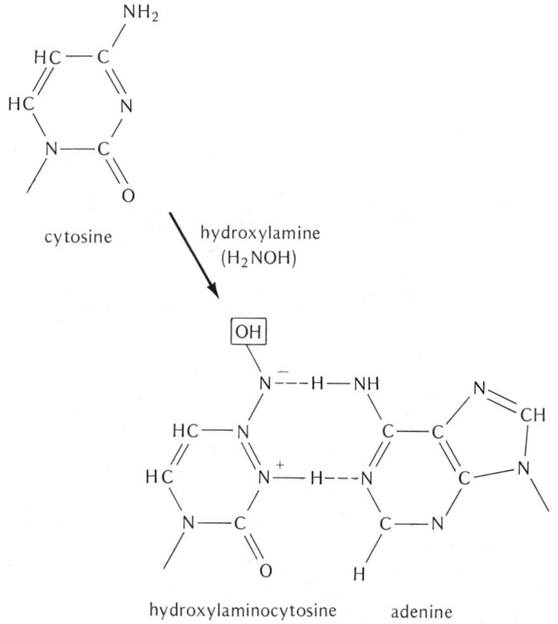

cytosine → hydroxylamine (H_2NOH)

hydroxylaminocytosine adenine

Figure 24–23

One possible effect of hydroxylamine: hydroxylation of the amino group of cytosine and subsequent pairing with adenine.

Other compounds that seem to cause mutations through transitions include hydroxylamine and other hydroxyl (OH) donors. According to Freese and co-workers, hydroxylamine is perhaps the most specific of all mutagens by adding a hydroxyl group to the amino group of cytosine, which then enables the base to undergo a tautomeric shift so that it can pair with adenine (Fig. 24–23).

Alkylating agents such as nitrogen mustard and ethyl ethanesulfonate may produce mutations in at least three ways (see Loveless): (1) addition of methyl or ethyl groups to guanine, causing it to behave as a base analogue of adenine and thereby produce pairing errors (Fig. 24–24); (2) loss of alkylated guanine bases (*depurination*), producing gaps in the DNA chain which may interfere with DNA replication or cause a shortened nucleotide sequence; and (3) crosslinkage between the strands

alkylation of guanine:

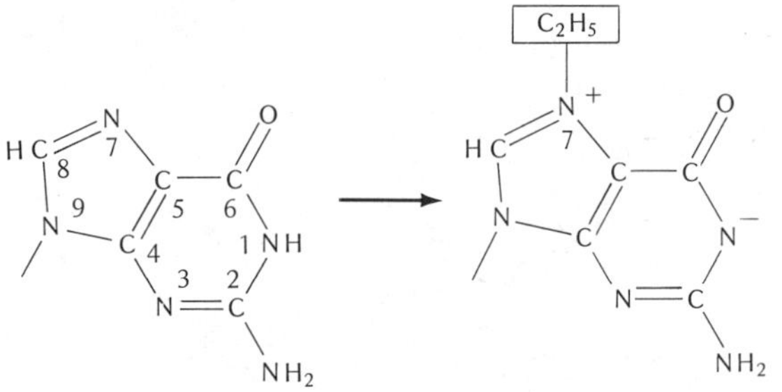

guanine → 7-ethylguanine

pairing of 7-ethylguanine with thymine:

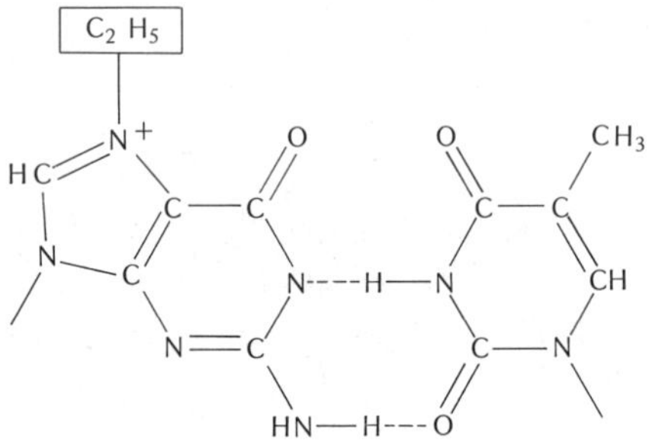

7-ethylguanine thymine

Figure 24–24

Proposed effect of treating guanine with the alkylating agents ethyl methanesulfonate or ethyl ethanesulfonate, according to Krieg.

of the same or different DNA molecules causing the loss or excision of nucleotides.

Acridine dyes such as proflavin and acridine orange are other mutagens that seem to produce direct effects on the DNA molecule. According to Lerman, acridine dyes act by inserting themselves between two neighboring purine bases in a single DNA strand. The consequence of such incorporation, according to Brenner and co-workers, is to cause either the insertion or deletion of a single nucleotide. Thus acridine mutations would not be expected to cause transitions as do base analogues, nitrous acid, hydroxylamine, and alkylating agents.

To distinguish differences in the mutation-inducing processes of these agents, Freese performed a test on *rII* mutations in phage T4. His procedure was to induce and collect *rII* mutations caused by all these agents and then test whether reverse mutation to normal could be induced by the 2-aminopurine and 5-bromouracil base analogues. As shown in Table 24–7, the base analogues had considerable success in reverting *rII* mutations that they themselves produced. Similarly they could revert high percentages of mutations caused by hydroxylamine, nitrous acid, and alkylating agents. Proflavin and some spontaneous mutations, on the other hand, could not generally be reverted by the transition-producing mechanisms

TABLE 24–7

Effect of base analogues (transition-producing mutagens) on the reversion of rII mutations in phage T4 that originally occurred either spontaneously or through the action of six different chemical mutagens

rII Mutations Induced by	*No. Mutants Tested*	*Induced to Revert by Base Analogues, Percent*	*Noninducible to Revert by Base Analogues, Percent*
2-aminopurine	98	98	2
5-bromouracil	64	95	5
hydroxylamine	36	94	6
nitrous acid	47	87	13
ethyl ethanesulfonate	47	70	30
proflavin	55	2	98
spontaneous	110	14	86

From Freese, 1963.

of base analogues. Therefore, at least two primary modes of mutation-inducing mechanisms seem to occur: (1) that caused by transitions and shared by the first five agents listed, and (2) the mechanisms involved in proflavin and certain spontaneous mutations—most likely caused by deletions and insertions of base pairs.

In those cases where mutator gene action can be identified, either transitions or transversions may occur. Thus, the Treffers mutator gene in *E. coli* (p. 543) appears to cause primarily A–T → C–G transversions (Yanofsky, Cox, and Horn), whereas mutators that arise from genetic changes in T4 DNA polymerase (p. 564) may cause both transitions and transversions. Surprisingly some of the T4 polymerase mutators act also as "antimutators"; that is, some of the genes that cause increased transversions cause also a reduction in A–T → G–C transitions. It is therefore apparent that appropriate balances between mutator and antimutator genes can supply considerable control and evolutionary flexibility over the mutation rate in any organism.

If we consider lysogenic viruses as cellular genetic factors, these too can be shown to exercise mutator activity, especially when they are inserted into host chromosome sites other than the one usually used. Thus a deletion of the λ attachment site in *E. coli* leads to the integration of λ into other sites which may provide opportunity for pairing and insertion. About one percent of such "abnormal" lysogens have been shown to produce mutant bacterial phenotypes as long as λ is present (Shimada, Weisberg, and Gottesman). As is also true of the lysogenic bacteriophage *mu* that occupies different positions on the bacterial chromosome (Taylor), the activity of λ as a mutagen appears to be confined to those bacterial genes in the immediate vicinity of the prophage. Once the λ prophage is excised, the bacterium reverts to wild type.

Viral mutagenesis can also extend to visible chromosome damage such as breaks and deletions. These are known to be caused in animals by a wide variety of viruses including herpes simplex, SV40, rubella (German measles), and chicken pox. The breaks and deletions may either be minute or extensive, and probably arise from the release of nucleases that attack host DNA or from the inhibition of DNA repair mechanisms. Should such breaks occur early in the development of an organism, as in fetuses carried by pregnant human females afflicted with rubella, serious abnormalities may ensue.

INDUCED MUTATION IN MAN

Questions about the extent and effects of induced mutation in humans have become of serious concern in recent years. Fallout from atomic and hydrogen bombs, medical x-rays, and other possible mutagens are now experienced by many persons to a greater degree than ever before.

In respect to their physiological or somatic effects, there is little question that such mutagens have serious effect. For example, the spontaneous appearance of cancerous leukemia is about .05 per thousand individuals. For the survivors of the Hiroshima and Nagasaki atomic bombs, the leukemia rate is much higher; persons who were close to the bomb when it exploded ("ground zero") show a leukemia rate of about 10 per thousand, or 200 times the spontaneous rate. Similarly the leukemia rate is high for persons x-rayed for spinal arthritis, with an incidence of about 20 times the spontaneous rate.

Data on the effect of mutagens on human heredity, however, have not yet been obtained. Some of the Hiroshima and Nagasaki victims who were exposed to more than 100 rads *in utero* do show an increase in the frequency of chromosomal aberrations, but this applies only to the somatic tissues tested. To date, no significant increase in mutation rate has been observed among the offspring of the atom bomb survivors. This does not mean that there were no mutations induced by the heavy doses of irradiation, but only that such mutations cannot yet be detected. The inability to distinguish mutation effects among the Japanese survivors may well be caused by the relatively low numbers of childbearing individuals who were studied. According to Neel, such studies had only a partial chance of detecting a 50 percent increase among deaths in newborn children of the irradiated groups.

The effects of mutagens have therefore been investigated in other mammals and the results have then been extrapolated to man. The mammal for which most mutation data is available at present is the mouse, which shows a spontaneous mutation rate of about 1×10^{-5} per gene per generation and an induced mutation rate per roentgen of 22×10^{-8} per gene per generation. In other words, $1 \times 10^{-5}/22 \times 10^{-8}$ or 45 *r* will produce an amount of mutation equivalent to the spontaneous mutation rate. Although other such estimates have also been offered (see, for example, Lüning and Searle), this approximate value can be called the "doubling dose" for mice. The difficulty lies in comparing mice to men. Brewen and co-workers suggest that human leukocyte cells are at least equally sensitive as those of mice to the induction of chromosomal deletions by radiation, and may even be more sensitive to the induction of translocations (see also Fig. 24–8). However, because man is a larger animal and may have a larger spontaneous mutation rate (see Table 23–1), the doubling dose may be somewhat higher than 45 *r*. Since it has been estimated (1972) that the average American receives less than 5 *r* from man-made sources of irradiation over a 30-year period, and that considerable "healing" probably occurs, it is hardly likely that such induced mutations add more than a few percent to those spontaneously present.

Nevertheless, even if the extent of induced mutation is relatively low, the question for man remains whether any preventable increase in mutagens, such as excessive medical x-rays or fallout, should be permitted. Mutation in man is accompanied not merely by a statistical increase in deleterious genes, but by personal suffering whose effects involve ethical and social concerns far wider than the immediate effect of the gene itself.

PROBLEMS

24–1. A dose of 4000 *r* of x-rays produced sex-linked lethal recessive mutations in 12 percent of treated *Drosophila* gametes. Assuming that almost no detectable mutations occur in the absence of radiation and that the relationship between mutation rate and dosage is strictly linear, state the expected lethal mutation frequency if the following dosages had been used: (a) 1000 *r*. (b) 2000 *r*. (c) 5000 *r*. (d) 6000 *r*.

24–2. Demerec irradiated males of a Swedish-B stock of *Drosophila melanogaster* at different x-ray dosages and found the following percentages of recessive sex-linked lethals:

X-ray Dosage, *r*	Percent Lethals
0	.18
1,000	1.97
3,000	6.01
4,000	6.77
5,000	10.73

(a) Graph the results. (b) Calculate the regression coefficient that best fits these data (Chapter 15). (c) What is the expected lethal frequency at 6000 *r*?

24–3. Léonard irradiated male mice with various dosages of x-rays and then bred them to virgin females. On the 17th day after mating, the female uteri and ovaries were dissected and the ratio of live embryos to erupted egg follicles (corpora lutea) recorded for each dosage. (Original data can be found in *Mutation Research*, **3**, 73–78, 1966.)

X-ray dosage, *r*	Relative survival of eggs	X-ray dosage, *r*	Relative survival of eggs
0	1.000	700	.424
100	.940	800	.399
200	.829	900	.366
300	.700	1,000	.319
400	.687	1,100	.277
500	.544	1,200	.255
600	.500	1,500	.195

(a) Graph the relationship that exists between dominant lethality and x-ray dosage. (Consider dominant lethality as 1 − relative survival of eggs.) (b) Using the techniques given in Chapter 15, calculate the correlation coefficients for this relationship.

24–4. How would you attempt to detect the induction of lethal mutations by x-ray in *Paramecium*?

24–5. A decrease in temperature appears to increase

the solubility of oxygen in the tissues and also leads to lower metabolism which permits oxygen in the cell to remain unused. Since a decrease in temperature is also known to increase the mutation rate in *Drosophila*, would you expect such a temperature effect on mutation rate when flies are irradiated in the absence of oxygen?

24-6. If ultraviolet irradiation causes mutation primarily through formation of thymine dimers along a single strand of DNA, would you expect an increase or decrease in the frequency of A-A (adenine-adenine) nearest-neighbor frequencies on the complementary strand of DNA?

24-7. Among the various kinds of mutations induced in *Drosophila* by mustard gas are lethals and translocations. This mutagen has been found to cause an increase in the frequency of translocations that is approximately proportional to the *square* of the frequency of lethals. Why this difference?

24-8. A DNA molecule which contains the tetranucleotide sequence

C-G
G-C
A-T
T-A

is subjected to the following mutagenic agents: (a) 5-bromouracil, (b) 2-aminopurine, (c) nitrous acid, (d) hydroxylamine, (e) ethyl ethanesulfonate. For each of these mutagenic agents, show the sequence of changes that leads to a mutational alteration in at least one nucleotide pair in replicates of this DNA chain.

24-9. As shown in Fig. 24-20, bromouracil may cause two types of transitions: A → G (A-T → G-C) or G → A (G-C → A-T). Which type would you say occurs when bromouracil has already been incorporated into DNA and is no longer present in the medium ("template transitions"), and which type can occur only when bromouracil is present in the medium ("substrate transitions")?

24-10. Tessman exposed two types of bacteriophage, ϕX174 and T4, to nitrous acid. Viable phages surviving this treatment were then cloned separately and each clone tested for the occurrence of mutation. For ϕX174 the mutation tested was that for *host range*, $h^+ \rightarrow h$, and the T4 test was for *rapid lysis*, $r^+ \rightarrow r$. The results showed that ϕX174 clones containing *h* mutations were all "pure" and did not contain h^+. Most of the T4 mutations, on the other hand, arose in mixed colonies containing *r* and r^+. How do these results reflect on the type of DNA (double-stranded or single-stranded) present in each of these phages?

24-11. Because of its specific pairing properties with cytosine, hydroxylamine is believed to produce transitions in only one direction, from a G-C base pair to an A-T base pair. Hydroxylamine also cannot produce reversions of mutations that were initially induced by itself nor does it cause reversions of mutations induced by 5-bromouracil. Hydroxylamine does, however, cause reversions of mutations induced by 2-aminopurine. What kinds of base-pair transitions would you suggest are primarily produced by 5-bromouracil and 2-aminopurine?

24-12. Smith and co-workers found that two *rII* mutations in bacteriophage T4 could revert spontaneously to wild type but could not be reverted either by base analogues such as 2-aminopurine, or by hydroxylamine, or by treatment with proflavin. These two mutations, however, could revert to wild type when the cells in which they were grown were deprived of thymine for a period of time. Since thymine deprivation probably causes mutation by incorporation of an incorrect base for thymine, explain what kinds of mutations—transitions or transversions—are probably responsible for the reversion of the two *rII* mutations.

REFERENCES

Abrahamson, S., M. A. Bender, A. D. Conger, and S. Wolff, 1973. Uniformity of radiation-induced mutation rates among different species. *Nature*, **245**, 460-462.

Altenburg, E., 1934. The artificial production of mutations by ultraviolet light. *Amer. Naturalist*, **68**, 491-507.

Auerbach, C., and J. M. Robson, 1946. Chemical production of mutations. *Nature*, **157**, 302.

Brenner, S., L. Barnett, F. H. C. Crick, and A. Orgel, 1961. The theory of mutagenesis. *Jour. Mol. Biol.*, **3**, 121-124. (The first research paper suggesting a cause for frame-shift mutations.)

Brewen, J. G., R. J. Preston, K. P. Jones, and D. G. Goslee, 1973. Genetic hazards of ionizing radiations: Cytogenetic extrapolations from mouse to man. *Mut. Res.*, **17**, 245-254.

Burki, H. J., and A. V. Carrano, 1973. Relative radio-sensitivities of tetraploid and diploid Chinese hamster cells in culture exposed to ionizing radiation. *Mut. Res.*, **17,** 277–282.

CATCHESIDE, D. G., 1948. Genetic effects of radiation. *Adv. in Genet.*, **2,** 271–358.

CHU, E. H. Y., N. H. GILES, and K. PASSANO, 1961. Types and frequencies of human chromosome aberrations induced by X-rays. *Proc. Nat. Acad. Sci.* **47,** 830–839.

CLEAVER, J. E., 1968. Defective repair replication of DNA in xeroderma pigmentosum. *Nature,* **218,** 652–656.

CONGER, A. D., and A. H. JOHNSTON, 1956. Polyploidy and radiosensitivity. *Nature,* **178,** 271.

DAVIDSON, G., 1974. *Genetic Control of Insect Pests.* Academic Press, London.

DEMEREC, M., 1953. Reactions of genes of *E. coli* to certain mutagens. *Symp. Soc. Exp. Biol.*, **7,** 43–54.

———, 1963. Selfer mutants of *Salmonella typhimurium. Genetics,* **48,** 1519–1531.

DOUDNEY, C. O., and F. L. HAAS, 1960. Some biochemical aspects of the post-irradiation modification of ultraviolet-induced mutation frequency in bacteria. *Genetics*, **45,** 1481–1502.

DRAKE, J. W., 1970. *The Molecular Basis of Mutation.* Holden-Day, San Francisco.

———, (ed.), 1973. *The Genetic Control of Mutation, Fogarty International Center Proceedings No. 17.* Published as a supplement to *Genetics*, Vol. **73,** by the Genetics Society of America, Austin, Texas.

DULBECCO, R., 1950. Experiments on photoreactivation of bacteriophages inactivated with ultraviolet radiation. *Jour. Bact.*, **59,** 329–347. (Reprinted in *Selected Papers in Biochemistry*, Vol. 4; see References, Chapter 1.)

EPSTEIN, J., J. R. WILLIAMS, and J. B. LITTLE, 1973. Deficient DNA repair in human progeroid cells. *Proc. Nat. Acad. Sci.,* **70,** 977–981.

EVANS, H. J., 1961. Chromatid aberrations induced by gamma irradiation. I. The structure and frequency of chromatid interchanges in diploid and tetraploid cells of *Vicia faba. Genetics*, **46,** 257–275.

FREESE, E., 1963. Molecular mechanism of mutations. In *Molecular Genetics,* Part 1, J. H. Taylor (ed.). Academic Press, New York, pp. 207–269.

———, 1971. Molecular mechanisms of mutations. In *Chemical Mutagens,* Vol. 1, A. Hollaender, ed. Plenum Press, New York, pp. 1–56.

GILES, N. H., 1954. Radiation induced chromosome aberrations in *Tradescantia.* In *Radiation Biology,* Vol. I, A. Hollaender (ed.). McGraw-Hill, New York, pp. 713–762.

HANAWALT, P. C., 1975. Molecular mechanisms involved in DNA repair. *Genetics*, **79,** 179–197. (Proc. XIII Intern. Congr. Genet.)

HEDDLE, J. A., D. WHISSELL, and D. J. BODYCOTE, 1969. Changes in chromosome structure induced by radiations: A test of the two chief hypotheses. *Nature*, **221,** 1158–1160.

HEIDENTHAL, G., 1945. The occurrence of X-ray induced dominant lethal mutations in *Habrobracon. Genetics*, **30,** 197–205.

HERSHFIELD, M. S., and N. G. NOSSAL, 1973. *In vitro* characterization of a mutator T4 DNA polymerase. *Genetics*, **73** suppl.: 131–136.

HOLLAENDER, A., (ed.), 1971–1973. *Chemical Mutagens, Principles and Methods for Their Detection,* Vols. 1–3, Plenum Press, New York.

HOLLAENDER, A., W. K. BAKER, and E. H. ANDERSON, 1951. Effect of oxygen tension and certain chemicals on the X-ray sensitivity of mutation production and survival. *Cold Sp. Harb. Symp.*, **16,** 315–325.

HOWARD-FLANDERS, P., 1973. DNA repair and recombination. *Brit. Med. Bull.*, **29,** 226–235.

KAUFMANN, B., and A. HOLLAENDER, 1946. Modification of the frequency of chromosomal rearrangements induced by X-rays in *Drosophila.* II. Use of ultraviolet radiation. *Genetics*, **31,** 368–375.

KELNER, A., 1949. Photoreactivation of ultraviolet-irradiated *Escherichia coli* with special reference to the dose-reduction principle and to ultraviolet-induced mutation. *Jour. Bact.*, **58,** 511–522.

KIRBY-SMITH, J. S., B. NICOLETTI, and M. L. GWYN, 1960. The induction of chromosomal aberrations in *Tradescantia* pollen by combined X-ray and ultraviolet treatment. *Genetics,* **45,** 996–997.

KONDO, S., 1973. Evidence that mutations are induced by errors in repair and replication. *Genetics*, *Suppl.*, **73,** 109–122.

KRIEG, D. R., 1963. Specificity of chemical mutagenesis. *Progr. Nuc. Acid Res.*, **2,** 125–168.

KUBITSCHEK, H. E., 1973. Intensive nuclear selection in tryptophan limited cultures of *Escherichia coli* B/r/1, *trp. Molec. Gen. Genet.,* **124,** 269–290.

LAWLEY, P. D., and P. BROOKES, 1962. Ionization of DNA bases or base analogues as a possible explanation of mutagenesis with special reference to 5-bromodeoxyuridine. *Jour. Mol. Biol.*, **4,** 216–219.

LEA, D. E., 1955. *Actions of Radiations on Living Cells*, 2nd ed. Cambridge Univ. Press, Cambridge.

Lerman, L. S., 1961. Structural considerations in the interaction of DNA and acridines. *Jour. Mol. Biol.*, **3,** 18–30.

Loveless, A., 1966. *Genetic and Allied Effects of Alkylating Agents.* Penn. State Univ. Press, University Park, Pa.

Lüning, K. G., and A. G. Searle, 1971. Estimates of the genetic risks from ionizing irradiation. *Mut. Res.*, **12,** 291–304.

Manney, T. R., and R. K. Mortimer, 1964. Allelic mapping in yeast using X-ray induced mitotic reversion. *Science*, **143,** 581–583.

Muller, H. J., 1927. Artificial transmutation of the gene. *Science*, **66,** 84–87. (Reprinted in the collections of Gabriel and Fogel, and Peters; see References, Chapter 1.)

———, 1954a. The nature of the genetic effects produced by radiation. In *Radiation Biology*, Vol. I, A. Hollaender (ed.). McGraw-Hill, New York, pp. 351–473.

———, 1954b. The manner of production of mutations by radiation. In *Radiation Biology,* Vol. I., A. Hollaender (ed.). McGraw-Hill, New York, pp. 475–626.

———, 1965. Radiation genetics: Synthesis. In *Genetics Today*, Vol. II, Proc. XI Intern. Congr. Genet., S. J. Geerts (ed.). Pergamon, Oxford, pp. 265–274.

Muller, H. J., and L. M. Mott-Smith, 1930. Evidence that natural radioactivity is inadequate to explain the frequency of "natural" mutations. *Proc. Nat. Acad. Sci.*, **16,** 277–285.

Neel, J. V., 1963. *Changing Perspectives on the Genetic Effects of Radiation.* C C Thomas, Springfield, Ill.

Novick, A., and L. Szilard, 1950. Experiments with the chemostat on spontaneous mutations of bacteria. *Proc. Nat. Acad. Sci.*, **36,** 708–719.

Orgel, L. E., 1973. Ageing of clones of mammalian cells. *Nature*, **243,** 441–445.

Ostergren, G., 1958. A study in hyacinthus on chromosome size and breakability by x-rays. *Hereditas*, **44,** 1–17.

Rapaport, J. A., 1946. Carbonyl compounds and the chemical mechanism of mutations. C. R. (*Dokl.*) *Acad. Sci.* URSS, N.S., **54,** 65–67.

Revell, S. H., 1959. The accurate estimation of chromatid breakage, and its relevance to a new interpretation of chromatid aberrations induced by ionizing radiations. *Proc. Roy. Soc. London*, B, **150,** 563–589.

Russell, W. L., 1956. Comparison of x-ray induced mutation rates in *Drosophila* and mice. *Amer. Naturalist*, **90,** 69–80.

———, 1963. Genetic hazards of radiation. *Proc. Amer. Phil. Soc.*, **107,** 11–17.

———, 1965. Evidence from mice concerning the nature of the mutation process. In *Genetics Today*, Vol. II, Proc. XI Intern. Congr. Genet., S. J. Geerts (ed.), Pergamon, Oxford, pp. 257–264.

Schultz, J., 1936. Radiation and the study of mutation in animals. In *Biological Effects of Radiation*, B. M. Duggar (ed.). McGraw-Hill, New York, pp. 1209–1261.

———, 1951. The effect of ultraviolet radiation on a ring chromosome in *Zea mays. Proc. Nat. Acad. Sci.*, **37,** 590–600.

Schuster, H., 1960. The reaction of nitrous acid with deoxyribonucleic acid. *Biochem. Biophys. Res. Commun.*, **2,** 320–323.

Setlow, J. K., M. E. Boling, and F. J. Bollum, 1965. The chemical nature of photoreactivable lesions in DNA. *Proc. Nat. Acad. Sci.*, **53,** 1430–1436.

Setlow, J. K., and R. B. Setlow, 1963. Nature of the photoreactivable ultraviolet lesion in deoxyribonucleic acid. *Nature*, **197,** 560–562.

Shimada, K., R. A. Weisberg, and M. E. Gottesman, 1973. Prophage λ at unusual chromosomal locations. II. Mutations induced by bacteriophage lambda in *Escherichia coli. Jour. Mol. Biol.*, **80,** 297–314.

Shukla, P. T., 1972. Analysis of mutagen specificity in *Drosophila melanogaster. Mut. Res.*, **16,** 363–371.

Smith, R. H., and R. C. von Borstel, 1972. Genetic control of insect populations. *Science*, **178,** 1164–1174.

Sobels, F. H. (ed.), 1963. *Repair from Genetic Radiation Damage and Differential Radiosensitivity in Germ Cells.* Macmillan, Inc., New York.

———, 1965. Radiosensitivity and repair in different germ cell stages of *Drosophila*. In *Genetics Today*, Vol. II, Proc. XI Intern. Congr. Genet., S. J. Geerts (ed.). Pergamon, Oxford, pp. 235–254.

Sparrow, A. H., 1951. Radiation sensitivity of cells during mitotic and meiotic cycles with emphasis on possible cytochemical changes. *Ann. N.Y. Acad. Sci.*, **51,** 1508–1540.

Sparrow, A. H., K. P. Baetcke, D. L. Shaver, and V. Pond, 1968. The relationship of mutation rate per roentgen to DNA content per chromosome and to interphase chromosome volume. *Genetics*, **59,** 65–78.

Sparrow, A. H., L. A. Schairer, and R. C. Sparrow, 1963. Relationship between nuclear volumes, chromosome numbers, and relative radio-sensitivities. *Science*, **141,** 163–166.

STADLER, L. J., 1928. Mutations in barley induced by X-rays and radium. *Science*, **68,** 186–187.

STONE, W. S., O. WYSS, and F. HAAS, 1947. The production of mutations in *Staphylococcus aureus* by irradiation of the substrate. *Proc. Nat. Acad. Sci.*, **33,** 59–66.

STUBBE, H., 1936. Die Erhöhung der Genmutationstrate in alternden Gonen von *Antirrhinum majus* L. *Biol. Zentralbl.*, **56,** 562–567.

SUTHERLAND, B. M., 1974. Photoreactivating enzyme from human leukocytes. *Nature*, **248,** 109–112.

SWANSON, C. P., 1944. X-ray and ultraviolet studies on pollen tube chromosomes. I. The effect of ultraviolet (2537 Å) on X-ray induced chromosomal aberrations. *Genetics*, **29,** 61–68.

———, 1957. *Cytology and Cytogenetics*. Prentice-Hall, Englewood Cliffs, N.J., Chaps. 10 and 12.

TAYLOR, A. L., 1963. Bacteriophage-induced mutation in *Escherichia coli*. *Proc. Nat. Acad. Sci.*, **50,** 1043–1051.

THODAY, J. M., 1952. Sister-union isolocus breaks in irradiated *Vicia faba*: The target theory and physiological variation. *Heredity*, **6,** (Supplement: *Symposium on Chromosome Breakage*) 299–309.

THODAY, J. M., and J. M. READ, 1947. Effect of oxygen on the frequency of chromosome aberrations produced by X-rays. *Nature*, **160,** 608.

THOM, C., and R. A. STEINBERG, 1939. The chemical induction of genetic changes in fungi. *Proc. Nat. Acad. Sci.*, **25,** 329–335.

TIMOFÉEFF-RESSOVSKY, N. W., K. G. ZIMMER, and M. DELBRUCK, 1935. Über die Natur der Genmutation und der Genstruktur. *Nach. Ges. Wiss. Göttingen*, **1,** 189–245.

WAGNER, R. P., and H. K. MITCHELL, 1964. *Genetics and Metabolism*. John Wiley, New York.

WALLACE, B., and Th. DOBZHANSKY, 1959. *Radiation, Genes and Man*. Holt, Rinehart and Winston, New York.

WHITTINGHILL, M., 1965. *Human Genetics and Its Foundations*. Reinhold, New York.

WITKIN, E. M., 1956. Time, temperature, and protein synthesis: A study of ultraviolet-induced mutation in bacteria. *Cold Sp. Harb. Symp.*, **21,** 123–138.

———, 1969. Ultraviolet-induced mutation and DNA repair. *Ann. Rev. Microbiol.*, **23,** 487–514.

WOLFF, S. (ed.), 1963. *Radiation-Induced Chromosome Aberrations*. Columbia Univ. Press, New York.

YANOFSKY, C., E. C. COX, and V. HORN, 1966. The unusual specificity of an *E. coli* mutator gene. *Proc. Nat. Acad. Sci.*, **55,** 274–281.

25
GENETIC FINE STRUCTURE

Previous chapters have discussed the basic unit of genetics, the gene, in terms of three main properties that can provisionally be described as follows:

1. The gene has a *functional* property, as evidenced by producing a particular phenotypic effect. It is because of this property that we have distinguished one allele from another. For example, the sex-linked gene for raspberry eye color in *Drosophila melanogaster, ras,* is an allele of the sex-linked wild-type gene ras^+, which produces the normal red eye color. The origin of *ras* is derived from the second property of the gene.
2. The gene has a *mutational* property, as evidenced by the presence and occurrence of alleles producing different phenotypes. On this basis the mutational property of the gene is reflected by the functional differences that arise. Our use of the gene concept, however, has also to distinguish between different genes that may produce similar phenotypes but are nevertheless not allelic. For example, the sex-linked gene *garnet* in *Drosophila melanogaster* produces a similar phenotypic effect as the gene for *raspberry*. Are *garnet* and *raspberry* mutations of the same gene or are they modifications of different genes? To distinguish between these alternatives, the third property of the gene was brought into play.
3. The gene has a *recombinational* property, as evidenced by the observation that it can separate itself from other genes through crossing over. For the *raspberry* and *garnet* genes, separation or recombination occurred about 12 percent of the time, and they were given appropriate positions on the X-chromosome linkage map of 32.8 and 44.4, respec-

tively. Genes separated from each other on a chromosome by recombinational analysis were considered to be distinctly different functional and mutational units with the proviso, long argued by Goldschmidt, that they bore certain essential relationships to each other by being located in a particular linear arrangement on the same chromosome or even by being in the same nucleus. That is, interactions between different genes may occur although physically separated and, as in position effects, these interactions may be modified by chromosomal structure and composition.

The relationships between genes, however, do not usually go so far as to enable distinction between phenotypes produced when mutant alleles of different genes are present on the same homologue or on opposite homologues. To illustrate this point we can make use of the stereochemical terminology introduced by Haldane and others—that two mutations located on the same homologous chromosome are in the *cis* position, but when located on opposite homologous chromosomes are in the *trans* position. A female fly heterozygous for *raspberry* and *garnet*, for example, may carry them either in the *cis* position, $ras\ g/+\ +$, or in the *trans* position, $ras\ +/+\ g$. In either case, whether *cis* or *trans*, the heterozygote for these mutations appears wild type. The mutant appearance only occurs when one or both of these genes is homozygous and present on both homologous chromosomes.

When new mutant alleles of a gene were discovered, each producing somewhat different phenotypic effects, they were tested on the basis of recombination with known genes. For example, the alleles $garnet^2$, $garnet^3$, etc., were so named because they produced variations of the garnet eye color, and recombined with other genes on the X chromosome but apparently not with *garnet*. On the basis of recombination it was therefore historically believed that different genes, i.e., different loci, could separate from each other by crossing over, while different alleles of a single gene were distinguished by their phenotypic effect but could not recombine with each other.

A further distinction between different genes and different alleles of the same gene was also made in terms of function. That is, nonallelic mutants in the *trans* position, i.e., on opposite homologues, supplement or *complement* each other's deficiencies to form wild type, while different allelic mutants in the *trans* position are *noncomplementary* and usually produce intermediate mutant phenotypes. *Raspberry* and *garnet*, as we have seen, are complementary, while the *trans* combination $garnet^1/garnet^2$ produces a mutant eye color.

Thus the concept of the gene was initially developed as that of a single unit that functioned as a unit, changed as a unit, and recombined as a unit. In visual terms the gene was often presented as a bead on a string that produced a particular product, could be transformed from one type of bead to another through mutation, and could only recombine with its neighboring beads. This view, as we shall see, was an oversimplification and received its first setback early in the history of modern genetics when Sturtevant and Morgan demonstrated that the *Bar* locus in *Drosophila melanogaster* contained more than one functional and recombinational "bead."

THE "BAR" LOCUS

Bar, one of the first semidominant mutations found, was a sex-linked gene that reduced the number of eye facets to about 68 in the B/B homozygote and to about 358 in the B/B^+ heterozygote, compared to the average of 779 facets in the wild-type B^+/B^+ homozygote. Within pure *Bar* strains, however, new types arose with a notable frequency (1/1600) which carried the normal allele (B^+), or a new allele with an even more drastic effect than *Bar*. The new allele was called *Ultrabar* (B^U) and in homozygous condition produced only about 24 facets.

Pure *Ultrabar* stocks were then established and were found to produce revertants to the wild-type allele or to the *Bar* allele with the same frequency in which they themselves had arisen. Since the required mutation rates necessary to explain those frequencies were much higher than any previously

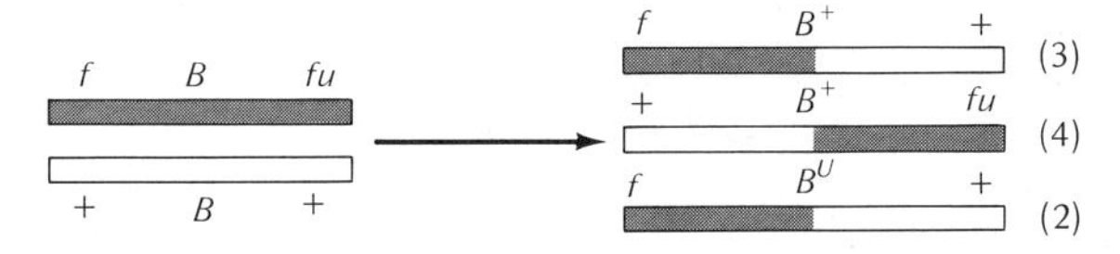

Figure 25–1

Technique used by Sturtevant and Morgan to show that the appearance of new Bar alleles (B^+ or B^U) is accompanied by crossing over of genes (f and fu) on either side of the Bar locus. On the left side are the two X chromosomes of the parental females tested; the right side shows the recombinant X chromosomes found among the progeny. Within parentheses are the numbers of each type observed.

known, it was difficult to account for the appearance of these new types on the basis of mutation alone. Instead, Sturtevant and Morgan hypothesized and also demonstrated that all these new types arose through recombination at the *Bar* locus.

The technique they used was to place other mutations at loci on either side of *Bar* or *Ultrabar* and then to observe whether these neighboring mutant alleles crossed over when the new types appeared at the *Bar* locus. The *Bar* locus was at 57.0, and the two neighboring loci used were *forked* (f) at 56.7 and *fused* (fu) at 59.5. Among the various crosses made, *Bar* females of the genetic constitution $f^+\ B\ fu^+/f\ B\ fu$ were mated to *forked fused* males (either $f\ B\ fu$ or $f\ B^+\ fu$) and their offspring observed. Out of about 20,000 offspring produced by these females, the *Bar* allele B reverted seven times to wild type B^+ and changed twice to *Ultrabar* B^U. In every one of these cases, however, there was also crossing over between *forked* and *fused* (Fig. 25–1); three of the new wild-type *Bar* alleles showed only *forked,* four showed only *fused,* and the two *Ultrabar* alleles showed only *forked.*

On the basis of these and other experiments, Sturtevant proposed that the new types arose through unequal crossing over. According to this theory, the wild-type allele at the *Bar* locus was deficient for the *Bar* gene. A *Bar* mutation therefore consisted of the insertion of a *Bar* gene at that particular locus. As shown in Fig. 25–2, unequal crossing over at this locus could produce wild-type (no *Bar* loci) and *Ultrabar* (two *Bar* loci) alleles. Similarly unequal crossing over between *Bar* and *Ultrabar* chromosomes should produce wild-type and "*triple Bar*" alleles. Although the first was observed, the presence of the second could not be tested, since many specimens with small eye size appeared to be sterile. Additional crosses were made and also found to be consistent with the interpretation of unequal crossing over.

Approximately 10 years later the cytological proof of Sturtevant's theory was found by Bridges, Muller, and others in analysis of the salivary banding of the X chromosome (Fig. 25–3). *Bar* was shown to be accompanied by a "repeat" of the 16*A* region which was "lost" when reversion to the wild-type allele occurred, and which appeared as two "repeats" in *Ultrabar.* The precise functioning of the *Bar* gene therefore seemed to be influenced by the quantity of genetic material at this locus. Quantity alone, however, was not sufficient to explain all the results. When Sturtevant compared facet numbers in homozygous *Bar* individuals (B/B) and *Ultrabar* heterozygotes ($BB/+$) he found that there was an average of 68 in the first case and 45 in the second. In other words, not only was the quantity, or dosage, of a gene important, but its position, whether on the same or on opposite chromosomes, could also influence the phenotype. Another mutant allele of *Bar,* called *Infrabar* (B^i), had a less drastic effect on eye size than *Bar,* but nevertheless caused similar position effects, as follows:

$$B/B^i = 73.5 \text{ facets} \qquad B^i/B^i = 292.6 \text{ facets}$$
$$BB^i/+ = 50.5 \text{ facets} \qquad B^iB^i/+ = 200.2 \text{ facets}$$

Among the important conclusions that can be drawn from this work are the following: (1) A gene does not necessarily represent a constant amount of genetic material and its alleles may therefore differ from each other in this regard; (2) certain sections or "suballeles" of a locus may recombine with each other, thereby producing new combinations of gene sections; and (3) when such recombined sections appear in the *cis* position, the phenotype produced may be quite different than when they are in the *trans* position.

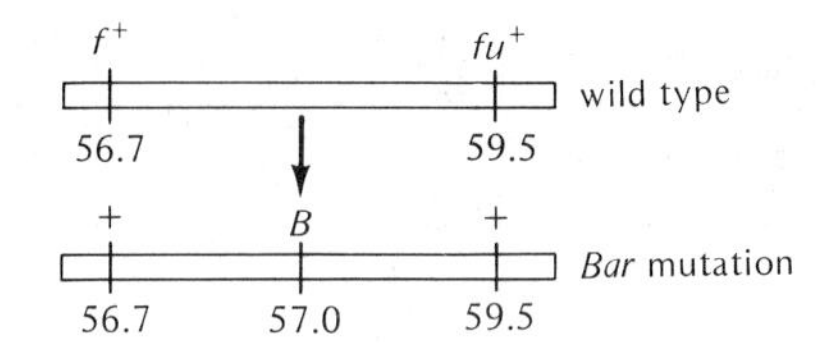

unequal crossing over in *Bar* females

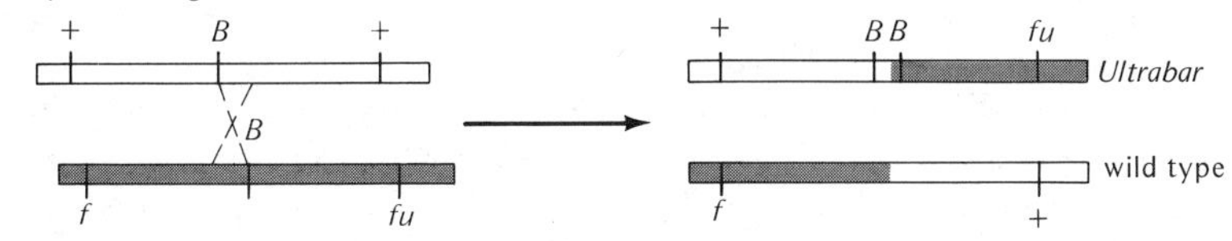

unequal crossing over in *Bar*/*Ultrabar* females

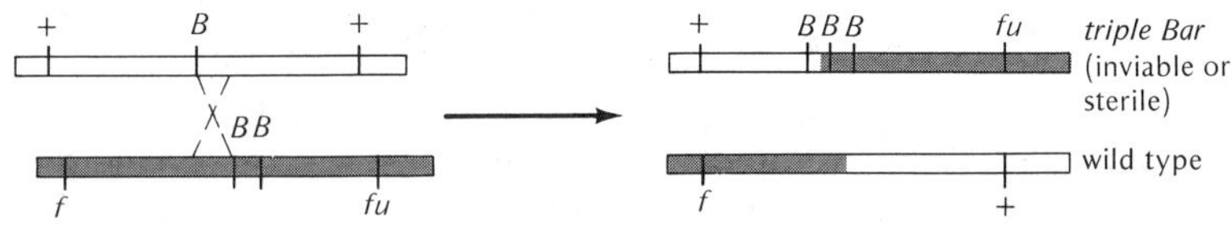

Figure 25–2

Explanation for the appearance of new Bar *alleles as a result of unequal crossing over. Note that chromosomal material is added to one of the crossover products (*Ultrabar*, or* triple Bar*) and removed from the other (wild type). Thus both gene duplications and gene fusions may arise from the same unequal crossing over event; the former caused by repeating the sequence of an entire gene such as* Bar*, and the latter caused by the removal of genetic material between two formerly separate genes (see also p. 841).*

PSEUDOALLELES AND COMPLEX LOCI

Variations on the theme discovered by Sturtevant and Morgan have been abundantly demonstrated for a variety of loci and organisms. At the *Drosophila garnet* locus Chovnick has shown that new wild-type chromosomes may occasionally arise through recombination among different *garnet* alleles. For example, recombination between *garnet*^{1} and *garnet*^{2} can be demonstrated by constructing female heterozygotes of the genotype *sable* (s) and *garnet* (g^1) on one X chromosome and *garnet*^{2} (g^2) and *pleated* (pl) on the other X. *Sable* is at locus 43.0 and lies to the left of *garnet* (locus 44.4), while *pleated* lies to the right of *garnet* at locus 47.9. Since these heterozygous females also carry the wild-type alleles for *sable* and *garnet*, their genotype can be symbolized $s\,g^1 +/+\,g^2\,pl$. If wild-type eye color is the result of recombination at the *garnet* locus, this process should also produce recombinants for the two outside markers at the same time, i.e., $s + pl$ chromosomes. Chovnick found two such chromosomes out of 68,000 examined, or a crossover frequency between g^1 and g^2 of about .003 percent.

Close-linked alleles of this type, which have similar phenotypic effects but can nevertheless still recombine with each other, have been named *pseudoalleles.* Such alleles are considered to occupy a *complex locus* which can, on this basis, be divided into *subloci* between which recombination can occur.* The complex locus *lozenge*, for example, consists of four groups, each containing one or more different lozenge alleles (Fig. 25–4). According to work by M. M. and K. C. Green, recombination distances between these groups are on the order of approximately .03 to .09 percent, or frequencies of less than one per thousand chromosomes. Since all these mutant alleles are recessive, they show no phenotypic effect when two that

* See the reviews of Carlson, 1959; Green, 1963; and Chovnick, 1966.

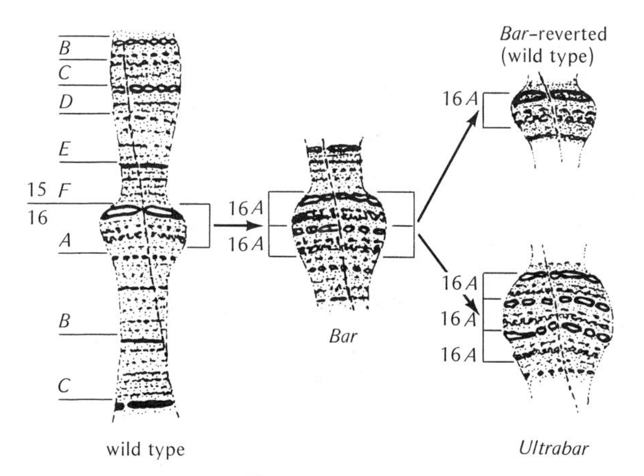

Figure 25–3

Salivary banding of the 16A region of the X chromosomes of Drosophila melanogaster *in several* Bar *alleles showing the* Bar *"repeats". (After Bridges.)*

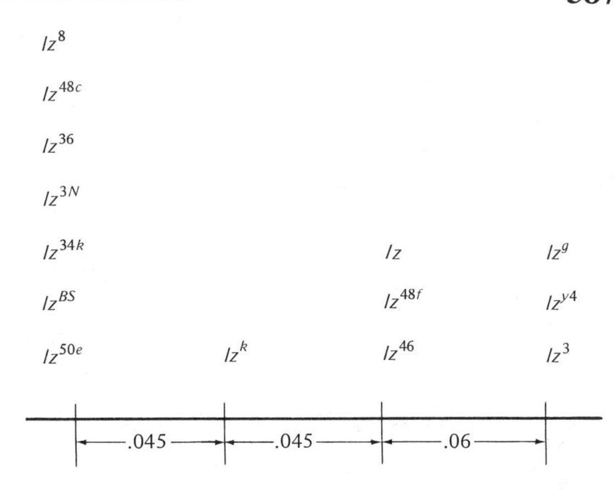

Figure 25–4

Some alleles of lozenge occupying different subloci within the lozenge locus of Drosophila melanogaster. *(Data of M. M. and K. C. Green; and M. M. Green, 1961.)*

are present in the *cis* position have their wild-type alleles on the other homologue (Fig. 25–5). However, contrary to nonallelic genes, when placed in the *trans* position many heterozygous combinations of different lozenge alleles do show mutant effects.

When this *cis-trans* position effect was first noted, Lewis hypothesized that it arose because of the sequential nature of gene-product synthesis. That is, a gene, being linear, produces a linear complex substance which can be interrupted at a specific step by a mutant allele. If the mutations are recessive and are present only on one chromosome, this frees the other chromosome to produce the necessary wild-type substance. However, pseudoallelic mutations within the same gene locus on both homologous chromosomes prevent either chromosome from producing the normal product. Lewis and others therefore envisaged the gene as an assembly line consisting of a number of subloci, each responsible for a different necessary step in the synthesis of the over-all product.

This sequential nature of pseudoallelic products may be dramatically observed when each of these products has somewhat different morphological effects. For example, in the *bithorax-bithoraxoid* gene complex in *Drosophila* all mutant alleles have some effect in modifying the small halteres or balancers into wing-shaped organs, but one group of alleles modifies the anterior part of the halteres into a wing (e.g., *abx/abx*), while a second group of alleles .03 percent recombinational units away has a similar effect on the posterior part (e.g., *pbx/pbx*). Any single bithorax gene that bears a mutation in one of these groups may therefore be written *abx* + or + *pbx*, each variety producing

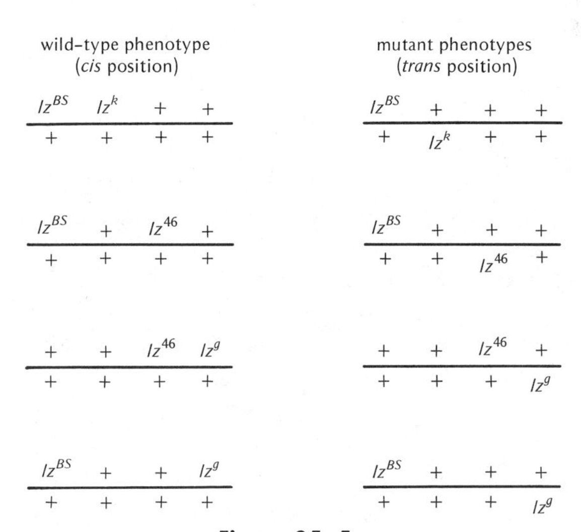

Figure 25–5

Phenotypes produced by cis *and* trans *combinations of different lozenge alleles.*

only a partial winglike structure in homozygous condition. Interestingly, Lewis found that the double mutant *abx pbx* (formed by recombination in the heterozygote *abx* +/+ *pbx*) may produce a fully developed second pair of wings. Thus, in accord with the "assembly-line" theory, each group of alleles in the bithorax complex seems to have a different function in the production of haltere substances.

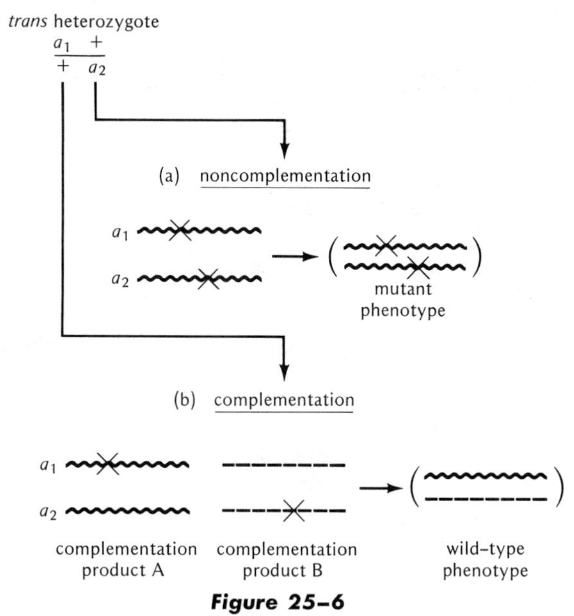

Figure 25–6

Interpretation of complementation and noncomplementation in the phenotype of a trans heterozygote. (a) Noncomplementation signifies that both mutant alleles, a_1 and a_2, affect the same gene product. (b) Complementation can signify that each of the mutant genes affects a different product; that is, a_1, for example, produces the two products A and B of which only one product is mutant (e.g., A), and a_2 produces these same products but the other is mutant (e.g., B). Assuming that A and B can interact together, complementation can therefore occur between nonmutant A from a_2 and nonmutant B from a_1 to form a wild-type phenotype.

COMPLEMENTATION MAPPING

The discovery of the *cis-trans* effect enabled a simple functional test for allelism between recessive mutations with similar effect: Two mutations in the *trans* heterozygote are usually allelic if they produce a mutant phenotype and nonallelic if they form a wild-type product. The ability of two recessive mutations to restore the wild-type phenotype either partially or completely in the *trans* position has been termed *complementation*. For example, the two *lozenge* pseudoalleles in *Drosophila*, lz^k and lz^{50e}, are complementary and represent mutations at different subloci since the *trans* combination lz^k/lz^{50e} is wild type. As shown in Fig. 25–6, complementation suggests that different products or portions of products are affected by each allele, while the lack of complementation indicates that both alleles affect the same gene product or same portion of the gene product.

The abundance of nutritional mutations in microorganisms and the ease with which they can be detected have made them especially suitable for complementation studies. In *Neurospora*, for example, many mutations produce strains that lack the ability to synthesize the amino acid histidine, and such strains will consequently be unable to grow on minimal medium. However, by introducing nuclei from each of two such different strains into a heterocaryon, complementation may be observed through the appearance of wild-type growth. Let us consider only five such histidine mutations that map in the same *his*-3 locus on the *Neurospora* first chromosome, *CD-16*, *245*, *261*, *D-566*, and *1438*. If we draw a complementation matrix as shown in Fig. 25–7a, we are not surprised to see that no complementation occurs in the heterocaryon between each of these mutations and itself. For *CD-16* the absence of complementation extends also to all other strains. Mutations *245* and *261*, however, complement with all other strains except *CD-16*, while *D-566* and *1438* complement with *245* and *261* but not with each other. Using these data we can then draw a complementation map in which the extent of noncomplementation is indicated as a straight line, and complementation is indicated by the absence of overlapping between these lines. Thus, as illustrated in Fig. 25–7b, *CD-16* is represented as a straight line overlapping (noncomplementing) all others, while *245* and *261* are represented as small lines that overlap only *CD-16*. *D-566* and *1438* overlap each other and *CD-16*, but do not overlap *245* and *261*. Note that on this basis three functional regions can be distinguished within which all mutations except *CD-16* affect only one region, while *CD-16* affects all three. The order in which the mutations should be placed within these regions cannot of course be determined from these data, but as more data are accumulated the complementation map becomes clearer. For example, another mutation, *430*, overlaps *245* and *261* but does not overlap *D-566* and *1438*. By similar reasoning, further complementation analysis with additional mutations indicates that the functional order is *245–261–D-566–1438*. Since *CD-16* affects all three functional regions, its complementation position cannot be determined from these data alone, but other data indicate that it extends to the left of *245*.

A comparison between the complementation map and the genetic map determined by recombinational analysis is of great interest. As shown in Fig. 25–7c, the genetic map for these mutations preserves the same order as the complementation map; that is, they are *colinear*. In *Drosophila melanogaster*, Judd and co-workers have recently shown that colinearity may also be detected between the order of a large number of mutations in the *zeste* region of the X chromosome, and their cytological positions derived from the use of deficiencies which cover different salivary chromosome bands (see p. 498). They conclude, in fact, that each chromomere or salivary chromosome band represents a single complementation region. However, colinearity between the genetic and complementation maps is not always true, and an allele that maps in one position genetically may affect another position functionally. The *CD-16* allele located in Fig. 25–7c on the left of the genetic map, for example, affects distant functional regions even at the extreme right. As will be discussed in Chapter 28 (pp. 666ff.), complexities of this kind arise because protein gene products assume complex forms and amino acids at one position on a protein may affect the function of others at distant positions.

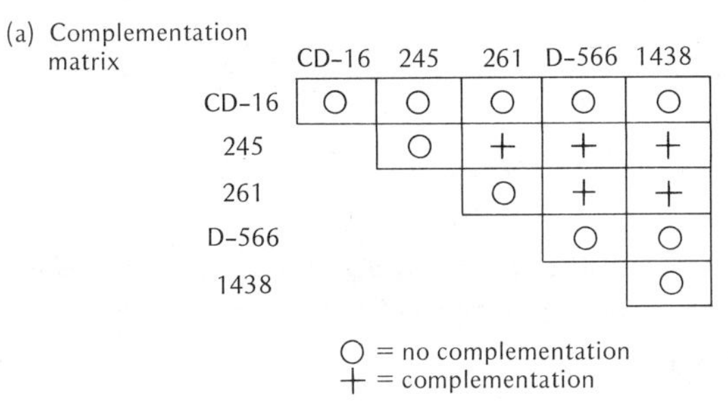

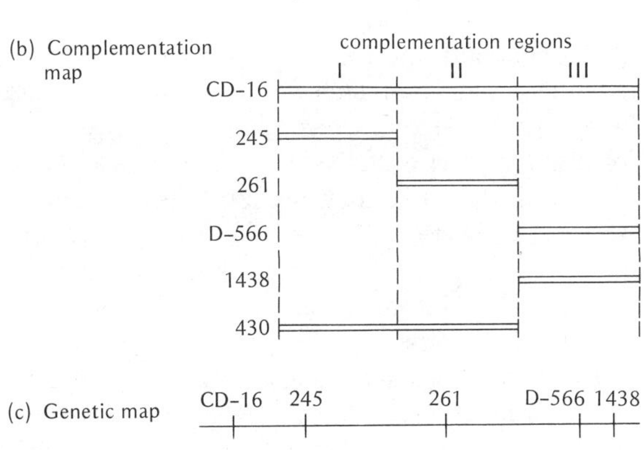

Figure 25–7

Complementation matrix, complementation map, and genetic map for some of the mutant genes at the his–3 *locus of* Neurospora crassa. *(Adapted from Ahmed, Case, and Giles.)*

THE *rII* LOCUS

The most refined analysis of a single genetic region to date is that undertaken in phage T4 for one of the loci affecting *rapid lysis* (p. 436). Benzer found that mutations at this locus, called *rII*, could be distinguished from mutations at other *r* loci by the inability of the *rII* mutants to grow on the lysogenic *E. coli* strain K that carried the λ prophage. That is, although *rII* mutants could infect the K bacterium, they could not destroy it. Only when strain K was simultaneously infected by different *rII* mutants that *complemented* each other in *trans* position could bacterial lysis and viral growth proceed. Through this procedure Benzer discovered that all *rII* mutants fell into two complementary groups, *A* and *B*, and mutations in one group were only complementary to mutations in the other group. Because these distinctions were made on the basis of a *cis-trans* test, Benzer called each of these two functional groups a *cistron.* It now seems likely that these cistrons produce proteins which bind to the cell membrane of the bacterial host and force it to break open more easily.

The importance of the *rII* mutations, however, lay in the analysis of their genetic fine structure. Through "mixed" infection of ordinary *E. coli* strain B bacteria with numerous viral particles from two different *rII* mutants, e.g. r_x and r_y, Benzer found that wild-type phage would occasionally be formed even when r_x and r_y were from the same cistron. That is, within cistron *A*, for example, r_x may occupy one sublocus while r_y may occupy another sublocus sufficiently far away to enable recombination to occur between them, $r_x r_y^+/r_x^+ r_y \rightarrow r_x^+ r_y^+$. These wild-type recombinants could then be detected by permitting phage derived from the lysis of the multiply infected *E. coli* B to infect *E. coli* K (Fig. 25–8). The frequency of appearance of normal r^+ plaques on plates of *E. coli* K would then indicate the frequency of recombination between the two original phage mutants. This method was extraordinarily sensitive to even low recombination frequencies, since millions of phage particles could be plated on cultures of strain K and rare recombinant products were easily observable. Benzer estimated that even recombination frequencies as low as .0001 percent could be detected.

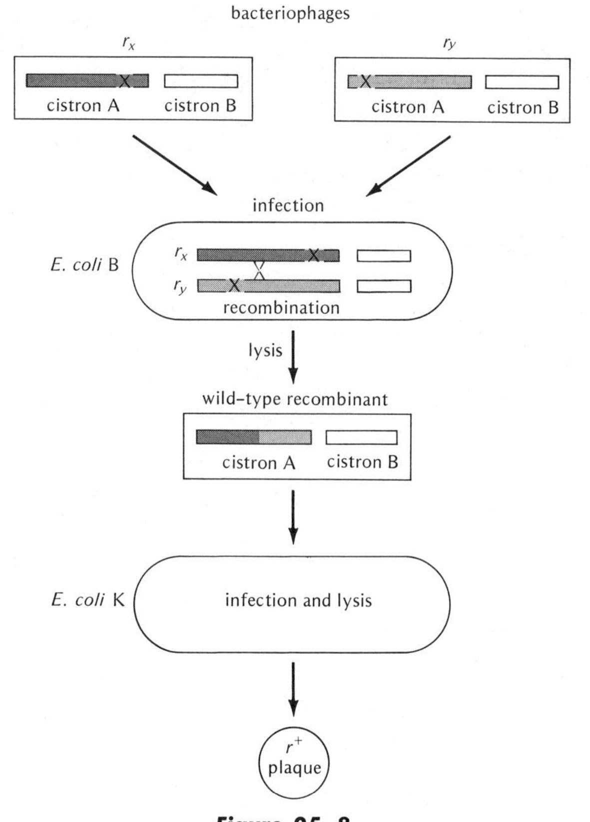

Figure 25–8

Technique used by Benzer to detect wild-type recombinants between two rII mutations affecting the same cistron.

Benzer also found that through recombination he could distinguish between "point" mutations affecting only one mutation site and longer "multisite" mutations. The distinction was made by observing whether r^+ recombinants were produced from a cross between an unknown *r* mutant and *r* mutants previously known to recombine with each other to produce r^+ recombinants. Were such r^+ recombinants formed, the mutation could be considered a point mutation. If r^+ recombinants were not formed, the mutation was of multisite nature. For example, *r* mutants *155* and *274*, known to recombine with each other to produce wild-type phages, do not produce r^+ recombinants when crossed to *r* mutant *164*. Mutant *r164* is therefore considered to be a multisite mutation that covers both the *155* and *274* regions. The fact that the presence of such multisite mutations can be shown to decrease recombination frequencies for mutations on either side of them (e.g., *228* and *196*) indicates that they are probably deletions. Also, as expected of deletions, such mutations never show spontaneous reversion to wild type, whereas the point mutations occasionally do revert (see pp. 539, 564).

These deletions served valuably in minimizing the labor necessary to locate new *rII* mutants. The procedure was to infect *E. coli* strain B cells simultaneously with an unknown *r* mutant and a known *r* deletion mutant, and then to observe whether r^+ recombinants were present in the progeny phage by testing them on *E. coli* strain K. If the area of one deletion included the unknown *r* mutant site, r^+ recombinants would be absent, since both mutants are defective at this same site. If the deletion did not include the unknown *r* site, however, r^+ recombinants would be produced and easily noted. The deletions used by Benzer covered known areas of the *A* and *B* cistrons, as shown in Fig. 25–9, so that a single test could locate an *rII* mutant within a definite segment. For example, mutant *r795* is located within the *A*5 segment since it only gives recombinants with deletions *rA105* and *r638* in set 1. When this same mutant was tested with the next set of deletions (set 2), it was found to produce recombinants only with deletion *rPB230*. This procedure localized *r795* within the *A*5c subsegment. Further localization (set 3) showed that *r795* was within the 2a2 subsegment of *A*5c. Then, through recombination mapping with other genes in this particular segment, *r795* could be further localized to a specific site.

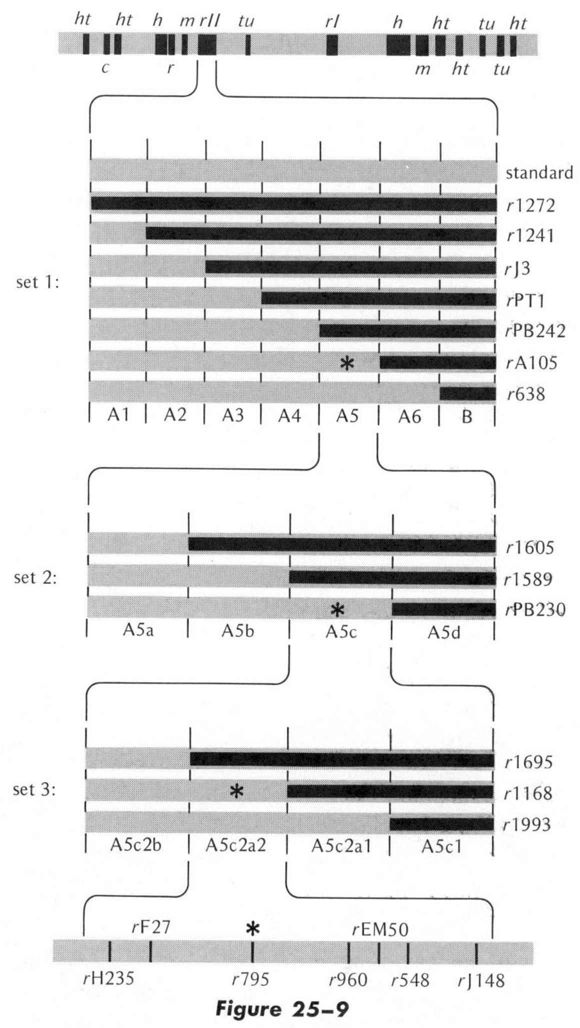

Figure 25–9

Method of deletion mapping for localizing an unknown mutation whose actual location is marked by an asterisk (i.e., r795). Black bars represent the relative lengths of the deletions. (Adapted from S. Benzer, "The fine structure of the gene." Copyright © 1962 by Scientific American, Inc. All rights reserved.)

The recombination mapping enabled the distances between genes to be calculated as well as their linkage order. For example, in Fig. 25–10 the recombination frequency between *155* and *274* is .014 percent and between *201* and *274* is .017 percent. The linkage order of mutations is therefore *201–155–274*. If this method is consistently followed, the expected recombination frequency between *155* and *201* should be the difference between .014 and .017, or .003 percent. However, despite the fact that Benzer's method was sensitive enough to detect recombination frequencies many times smaller, he found that wild-type recombinants were not produced at frequencies less than .01 percent. Since wild-type recombinants are only one of the two products of a recombination between two *r* mutants (the other is the nondetectable double *r* mutant recombinant), the actual limits of recombination are therefore on the order of .02 percent. The unit of genetic subdivision beyond which recombination does not occur Benzer called the *recon*.

Through recombinational analysis about 2000 spontaneous mutations in the *rII* region were mapped and found to occupy 308 different sites within the area. These sites are scattered fairly evenly throughout the area although some appear to be "hot spots" and contain numerous mutations as compared to other sites (Fig. 25–11). There are also, of course, sites within which mutations have not occurred ("zero sites") or have not yet been detected. In all, Benzer estimated that at least 400 and perhaps 500 different mutational sites are present in the *rII* region, and he referred to each

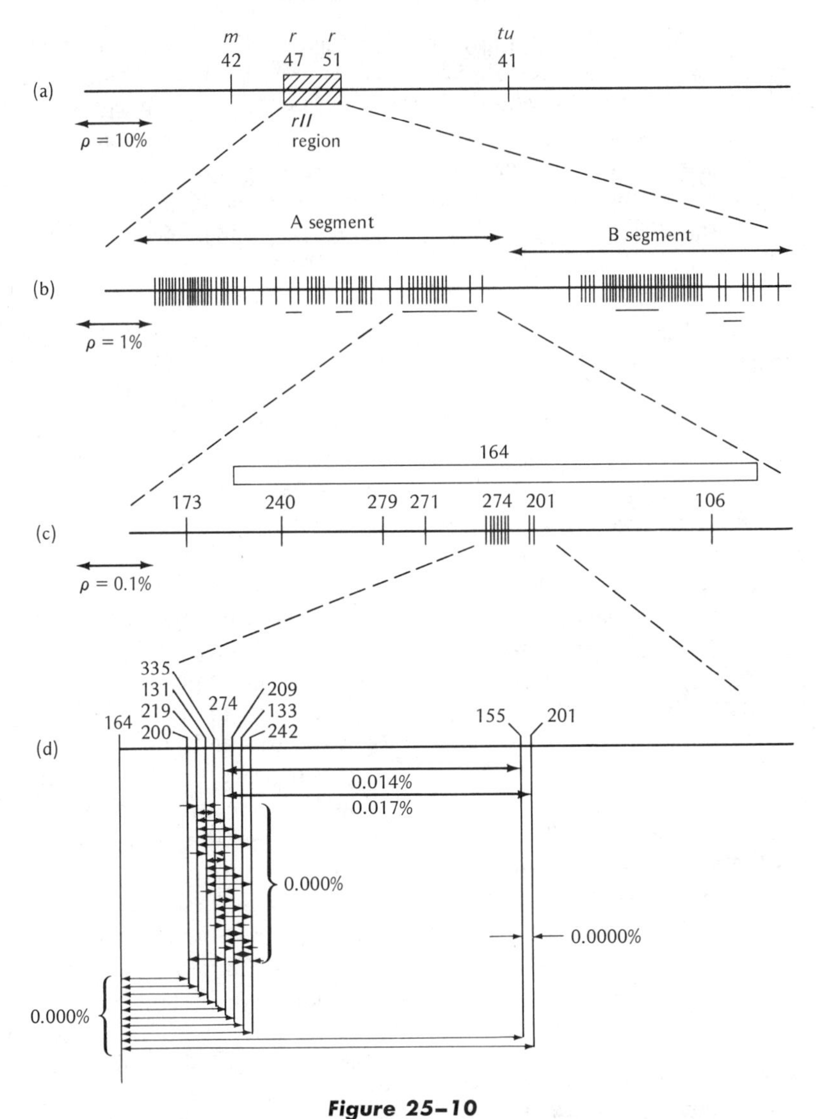

Figure 25–10

Levels of genetic mapping at the rII region of T4 bacteriophage given in percent recombination frequency. (After Benzer, 1956.)

of these as an elementary unit of mutation, or a *muton.*

Combining Benzer's results with some general estimates of the size of the phage chromosome, we can see that the recombinational and mutational units of a gene are extremely small. T4 double-helical DNA is presumed to have a linear length of approximately 200,000 nucleotides, constituting about 100 functional units or cistrons. Since the *rII* region contains two cistrons, it may be considered to have a length of about 1/50 of the total length, or 4000 nucleotides long. Within this length recombination values range from 10 percent between its extreme end points to the .02 percent values found by Benzer. In other words, there are 10/.02, or 500, recombinational units within the *rII* locus. This corresponds, as you will note, to the 400 to 500 mutational units. Interpreted in terms of nucleotides, 4000/500 gives a length of about 8 nucleotides for each of these units. There are, of course, many sources of error in this estimate, since the number of nucleotides available for recombination within the *rII* locus may be considerably less than 4000. Also, certain chromosomal regions may not carry any genetic information in the usual sense but may serve as "spacers" between cistrons. Each of these conditions would decrease the nucleotide length of the recombinational and mutational units.

More recently Tessman has shown that many artificially induced *rII* mutations may be separated by recombinational distances as low as .00001 percent. This would indicate that the number of recombinational units within the *rII* locus is as high as 10/.00001, or one million! Obviously, such recombinational distances would mean partitioning each *rII* nucleotide into 1,000,000/4000, or 250, parts, since only then would each recombinational unit correspond to a DNA unit. Partitioning of a nucleotide during recombination is difficult to visualize, however, and it seems more likely that the unit of recombination can be no smaller than the adjacent nucleotides between which exchange occurs, and that the unit of mutation is a single nucleotide. Tessman's results may reflect the peculiarities of recombinations that are influenced by specific mutations. That is, certain combinations of mutations affect the recombination frequency at a particular point and diminish it so that very low recombination frequencies are observed.

Such unique interactions may, according to Freese and others, also help explain the existence of mutational "hot spots." More simply, the neighboring nucleotides at a specific point may cause mutation to occur more frequently at that point than at others. Support for this view exists in the finding that enzymes which attach glucose to the DNA molecule act preferentially at certain sites which are determined by the neighboring nucleo-

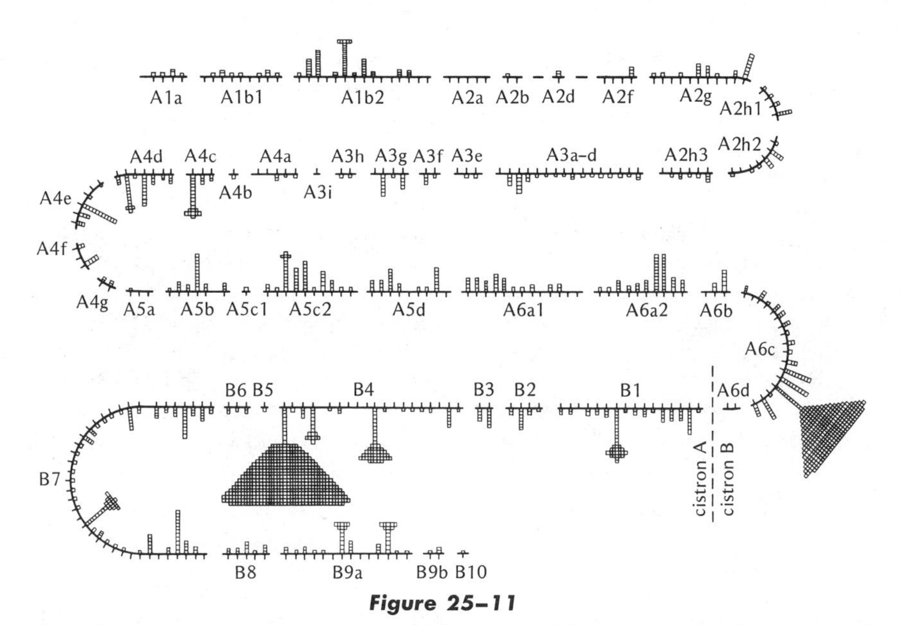

Figure 25–11

Map of spontaneous mutations arising at different loci of the two cistrons in the rII *region of bacteriophage T4. Each square symbolizes one mutation observed to arise at a particular locus. (After S. Benzer, 1961.* Proc. Nat. Acad. Sci., ***47:****403–415.)*

tides. Also, unique nucleotide bases are found along the DNA chain which may have special mutational (and recombinational) properties. In the T-even phages, for example, the substitution of 5-hydroxymethylcytosine for cytosine causes quite different mutational properties when this DNA is subjected to the mutagen hydroxylamine.

Perhaps the most impressive evidence of this kind is the observation by Koch that a change in one base pair in the *rII* locus (from A–T to G–C) causes a 23-fold increase in transitional mutations at the next base pair. Such phenomena may be caused by the defective ability of a DNA polymerase or repair enzyme to distinguish accurately a specific nucleotide base if the neighboring bases produce an unusual molecular configuration. We must also keep in mind that the DNA molecule is in folded condition, so that not all its length follows exact Watson-Crick base pairing. Certain unpaired loops may be formed at which one or more particular nucleotides can undergo preferential mutational activity. The present view therefore remains that the smallest recombinational unit is probably a single nucleotide, which, in turn, is probably the smallest mutational unit. These units taken individually are not, of course, genes. To understand the gene itself we must turn to its own special functions.

PROBLEMS

25-1. E. B. Lewis investigated two alleles at a particular locus (position 1.3) on the 2nd chromosome of *Drosophila melanogaster*, both causing the production of star eyes but one dominant (*S*) and one recessive (*s*; later called *asteroid*). Females heterozygous for these alleles were raised that also carried the nearby recessive markers *aristaless* (*al*: position 0.0) and *heldout* (*ho*: position 4.0) in the form *al S ho*/+ *s* +. Since these females were mated to star males, their progeny were all expected to be of the star phenotype, and this was true of more than 31,000 offspring examined. In four of the progeny, however, the eyes were normal, and these were shown to have developed from eggs of the constitution + + *ho*. (a) What factors indicate that these events were not caused by mutation? (b) Explain how the + genes at the *star* locus probably occurred. (c) In what linkage positions, relative to *al* and *ho*, would you map the two alleles (*S* and *s*) at the *star* locus?

25-2. If a heterozygote for two alleles at a complex locus (c^1 and c^2) carries the *a* and *b* markers on each side of this locus in the form $a\,c^1\,+/+\,c^2\,b$, what are the single, double, and triple crossover chromosomes that are wild type for *c* when the linkage order is: (a) a-c^1-c^2-b? (b) a-c^2-c^1-b?

25-3. Answer (a) and (b) of Problem 2 if the heterozygote has the constitution $a\,c^1\,b/\,+\,c^2\,+$.

25-4. In tests to investigate linkage relationships between different alleles at a *pantothenic* (*pan*) locus in *Neurospora crassa*, Case and Giles crossed *yellow* marker strains carrying one *pan* allele with *tryptophan* marker strains carrying another *pan* allele. Ascospores from these crosses were then plated on a medium devoid of pantothenic acid to detect pan^+ recombinants. These pan^+ recombinants were found in association either with the *yellow* markers (*ylo* +), the *tryptophan* markers (+ *tryp*), both markers (*ylo tryp*), or no markers (+ +). For example, the results of four crosses are given in the table at the foot of this page. What are the linkage relationships between *pan 18*, *pan 20*, and *pan 25* and the *ylo* and *tryp* markers?

25-5. In the synthesis of the amino acid histidine in *Salmonella* bacteria, a number of different mutations (*A* to *D*) affect one step in this process by interfering with the production of a particular enzyme. These mutant stocks may be "mated" together (through trans-

Cross	Percent pan^+ in the Four Classes			
	ylo +	+ *tryp*	*ylo tryp*	+ +
ylo pan 20 + × + *pan 18 tryp*	8.4	5.0	6.7	79.6
ylo pan 18 + × + *pan 20 tryp*	3.1	5.2	71.0	10.5
ylo pan 25 + × + *pan 20 tryp*	15.8	20.0	10.0	54.4
ylo pan 20 + × + *pan 25 tryp*	8.0	27.0	52.0	13.0

duction) to produce a wild-type recombinant as follows (+ means production of wild-type recombinant; 0 means lack of appearance of wild type):

Mutant Stock	Mated to Mutant Stock			
	A	*B*	*C*	*D*
A	0	+	0	+
B		0	0	0
C			0	+
D				0

(a) Considering that some or all of these mutations may be deletions, draw a possible topological map of this area. (b) If a point mutation produced wild-type recombinants with all the above mutations except *C*, at which position on this map would it most likely be located?

25-6. Using complementation tests, Catcheside has classified numerous *Neurospora* mutations that affect the synthesis of tryptophan (*tryp-1* locus) into nine groups, *A* to *I*, whose complementation patterns are as follows:

	A	*B*	*C*	*D*	*E*	*F*	*G*	*H*	*I*
A		0	0	0	0	0	0	0	0
B	0		+	+	+	+	+	+	+
C	0	+		0	0	0	+	+	+
D	0	+	0		0	0	0	+	+
E	0	+	0	0		0	0	0	+
F	0	+	0	0	0		0	0	0
G	0	+	+	0	0	0		0	0
H	0	+	+	+	0	0	0		0
I	0	+	+	+	+	0	0	0	

Draw a complementation map of this area.

25-7. Variations on the following type of circular complementation map have been found in a number of organisms. (Overlapping lines indicate noncomplementation.) On the basis of this map, make up a complementation matrix for mutations *1* to *6*, using the symbols 0 for noncomplementation and + for complementation.

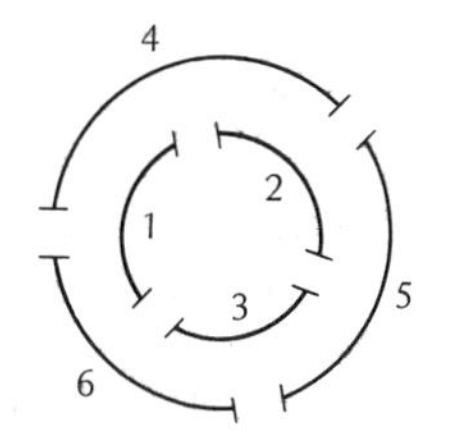

25-8. Various pairs of *rII* mutants of phage T4 were tested in both *cis* and *trans* positions in infection of *E. coli* and comparisons made as to the average number of phage particles produced per bacterium ("burst size"). One set of hypothetical results for six different *r* mutants *rU*, *rV*, *rW*, *rX*, *rY*, and *rZ* is as follows:

Cis	Burst size	Trans	Burst size
rU rV/wild type	250	*rU*/*rV*	258
rW rX/wild type	255	*rW*/*rX*	252
rY rZ/wild type	245	*rY*/*rZ*	0
rU rW/wild type	260	*rU*/*rW*	250
rU rX/wild type	270	*rU*/*rX*	0
rU rY/wild type	253	*rU*/*rY*	0
rU rZ/wild type	250	*rU*/*rZ*	0
rV rW/wild type	270	*rV*/*rW*	0
rV rX/wild type	263	*rV*/*rX*	270
rV rY/wild type	240	*rV*/*rY*	250
rV rZ/wild type	274	*rV*/*rZ*	260
rW rY/wild type	260	*rW*/*rY*	240
rW rZ/wild type	250	*rW*/*rZ*	255

If we assign mutation *rV* to the *A* cistron, what are the locations of the other five *rII* mutations in respect to the *A* and *B* cistrons?

REFERENCES

Ahmed, A., M. E. Case, and N. H. Giles, 1964. The nature of complementation among mutants in the *Histidine-3* region of *Neurospora crassa*. *Brookhaven Symp. Biol.*, **17,** 53–65.

Benzer, S., 1955. Fine structure of a genetic region in bacteriophage. *Proc. Nat. Acad. Sci.*, **41,** 344–354. (Reprinted in collections of Levine, of Peters, and of Stent, and in *Selected Papers in Biochemistry*, Vol. 2; see References, Chapter 1.)

———, 1956. Genetic fine structure and its relation to the DNA molecule. *Brookhaven Symp. Biol.*, **8,** 3–5.

———, 1957. The elementary units of heredity. In *A Symposium on the Chemical Basis of Heredity*, W. D. McElroy and B. Glass (eds.). Johns Hopkins Press,

Baltimore, pp. 70–93. (Reprinted in Taylor's collection; see References, Chapter 1.)

———, 1962. The fine structure of the gene. *Sci. American*, **206**, January issue, 70–84. (Reprinted in the collection of Srb, Owen, and Edgar; see References, Chapter 1.)

BRIDGES, C. B., 1936. The bar "gene," a duplication. *Science,* **83,** 210–211. (Reprinted in Peters' collection; see References, Chapter 1.)

CARLSON, E. A., 1959. Comparative genetics of complex loci. *Quart. Rev. Biol.*, **34,** 33–67.

———, 1966. The *Gene: A Critical History*. Saunders, Philadelphia. (Chaps. 13 and 21 through 23.)

CHOVNICK, A., 1961. The garnet locus in *Drosophila melanogaster.* I. Pseudoallelism. *Genetics,* **46,** 493–507.

———, 1966. Genetic organization in higher organisms. *Proc. Roy. Soc. Lond.* (B), **164,** 198–208.

FINCHAM, J. R. S., 1966. *Genetic Complementation.* W. A. Benjamin, New York.

FREESE, E., 1965. The influence of DNA structure and base composition on mutagenesis. In *Genetics Today*, Vol. II, Proc. XI Intern. Congr. Genet. S. J. Geerts (ed.). Pergamon, Oxford, pp. 297–305.

GREEN, M. M., 1961. Phenogenetics of the lozenge loci in *Drosophila melanogaster.* II. Genetics of lozenge-Krivshenko (lz^k). *Genetics,* **46,** 1169–1176.

———, 1963. Pseudoalleles and recombination in *Drosophila.* In *Methodology in Basic Genetics,* W. J. Burdette (ed.). Holden-Day, San Francisco, pp. 279–286.

GREEN, M. M., and K. C. GREEN, 1949. Crossing over between alleles at the lozenge locus in *Drosophila melanogaster. Proc. Nat. Acad. Sci.*, **35,** 586–591.

JUDD, B. H., M. W. SHEN, and T. C. KAUFMAN, 1972. The anatomy and function of a segment of the X chromosome of *Drosophila melanogaster. Genetics,* **71,** 139–156.

KOCH, R. E., 1971. The influence of neighboring base pairs upon base-pair substitution mutation rates. *Proc. Nat. Acad. Sci.*, **68,** 773–776.

LEWIS, E. B., 1951. Pseudoallelism and gene evolution. *Cold Sp. Harb. Symp.*, **16,** 159–172.

———, 1964. Genetic control and regulation of developmental pathways. In *The Role of Chromosomes in Development,* M. Locke (ed.). Academic Press, New York, pp. 231–252.

STURTEVANT, A. H., 1925. The effects of unequal crossing over at the Bar locus in *Drosophila. Genetics,* **10,** 117–147. (Reprinted in Peters' collection; see References, Chapter 1.)

STURTEVANT, A. H., and T. H. MORGAN, 1923. Reverse mutation of the bar gene correlated with crossing over. *Science,* **57,** 746–747.

TESSMAN, I., 1965. Genetic ultrafine structure in the T4 *rII* region. *Genetics,* **51,** 63–75.

PART

FUNCTION OF GENETIC MATERIAL

He [Anaxagoras] said that in the same semen there was contained both hair and nails and veins and arteries and nerves and bones and that they happened to be invisible because of their smallness, but as they developed little by little they were separated out. For how could hair be produced from non-hair and flesh from non-flesh?

ST. GREGORY OF CONSTANTINOPLE
Scholia

26
GENETIC CONTROL OF PROTEINS

The function of a gene within an individual is, in its ultimate sense, to control and influence the phenotype. Between the gene and the eventual phenotype, however, are many complex events which make it difficult to determine exactly how this control is exercised. Progress on this question was therefore slow until 1941 when, concurrent with their discovery of biochemical mutations in *Neurospora*, Beadle and Tatum announced an important and provoking concept called the *one gene–one enzyme* hypothesis. The main ideas of their proposal have since been elaborated in the following form (Tatum, 1959):

"1. All biochemical processes in all organisms are under genic control.

2. These overall biochemical processes are resolvable into a series of individual stepwise reactions.

3. Each single reaction is controlled in a primary fashion by a single gene, or in other terms, in every case a 1:1 correspondence of gene and biochemical reaction exists, such that

4. Mutation of a single gene results only in an alteration in the ability of the cell to carry out a single primary chemical reaction.

. . . The underlying hypothesis . . . is that each gene controls the reproduction, function, and specificity of a particular enzyme."

In its simplest form, this meant that the ultimate product of a metabolic process may be considered to be affected by a stepwise succession of enzymes, each produced by a particular gene, as diagrammatically illustrated in Fig. 26–1. Support for this fundamental hypothesis lay not only in the

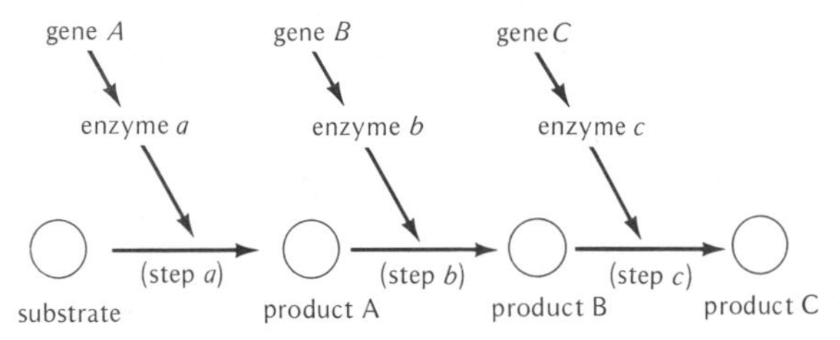

Figure 26–1

Diagrammatic representation of a metabolic sequence producing product C from a particular substrate in which three steps controlled by three enzymes are necessary. Each enzyme is, in turn, controlled by an individual gene. (After V. Grant, The Architecture of the Germplasm, John Wiley, New York, 1964; from Beadle.)

Neurospora results but, as pointed out by Beadle and Tatum, also derived from findings in man, plants, and *Drosophila*.

GARROD AND INBORN ERRORS OF METABOLISM IN MAN

Garrod, an English physician whose investigations began soon after the rediscovery of Mendel's experiments in 1900, made the important observation that certain hereditary human diseases caused particular biochemical defects. Alkaptonuria, for example, inherited as a mendelian recessive effect, causes the excretion of alkapton or homogentisic acid, which, in turn, causes urine to appear black upon exposure to air. In normal individuals alkapton is broken down by the catalytic action of an enzyme so that acetoacetic acid is excreted (Fig. 26–2). Although Garrod did not know all the steps of this process he surmised that some responsible enzyme was either absent or inactive in alkaptonuria. Furthermore, by feeding alkaptonuriacs compounds such as phenylalanine and tyrosine he was able to demonstrate that some of these were precursors of alkapton, since the excretion of homogentisic acid was consequently increased. This same metabolic pathway, which begins with the ingestion of phenylalanine, was later shown to be broken at various steps by genes which produce phenylketonuria, tyrosinosis, and albinism.

FLOWER PIGMENT MUTATIONS

After Garrod, a group of English investigators (Onslow, Basset, Scott-Moncrieff, and Lawrence, see the review of Lawrence, 1950), began the biochemical analysis of flower pigments and eventually showed a precise correlation between genetic and biochemical changes. For example, in the anthocyanin pigments of the cape primrose (*Streptocarpus*), three individual biochemical effects can be traced to separate genes. The three genes, called *R*, *O*, *D*, appear to produce their effects by adding or subtracting hydroxyl (OH) units, methoxyl ($O—CH_3$) units, or sugar molecules. The *rr oo dd* triple recessive is salmon-colored, containing mainly the pigment shown in Figure 26–3a. Presence of the dominant allele *R* (*R–oodd*) results in hydroxylation or methoxylation at the 3′ position and causes a change to rose color (Fig. 26–3b). The addition of *O* causes hydroxylation or methoxylation at both the 3′ and 5′ positions, and the color changes to mauve (Fig. 26–3c). In plants dominant for gene *D* (Fig. 26–3d, e, f), hexose (6 carbon) sugars are found at the 3 and 5 positions instead of the hexose-pentose combination at the 3 position, and the anthocyanin appears bluer.

Investigations of numerous other plants have shown similar findings; the anthocyanin pigments are usually bluer with the addition of sugar and hydroxyl groups, and somewhat redder when the hydroxyl groups are methylated. Acidity of the cell, another hereditary factor in some plants, is also of considerable importance, since dyes such as cyanin change from blue to red with increased acidity. Pigment inheritance therefore helped to show the effect of genes in controlling specific biochemical steps.

INSECT EYE PIGMENTS

The investigation of eye pigments in *Drosophila* by Beadle and Ephrussi in the thirties led to similar conclusions, and these, in turn, sparked the intensive investigations of biochemical mutations begun by Beadle and Tatum in the forties. *Dro-*

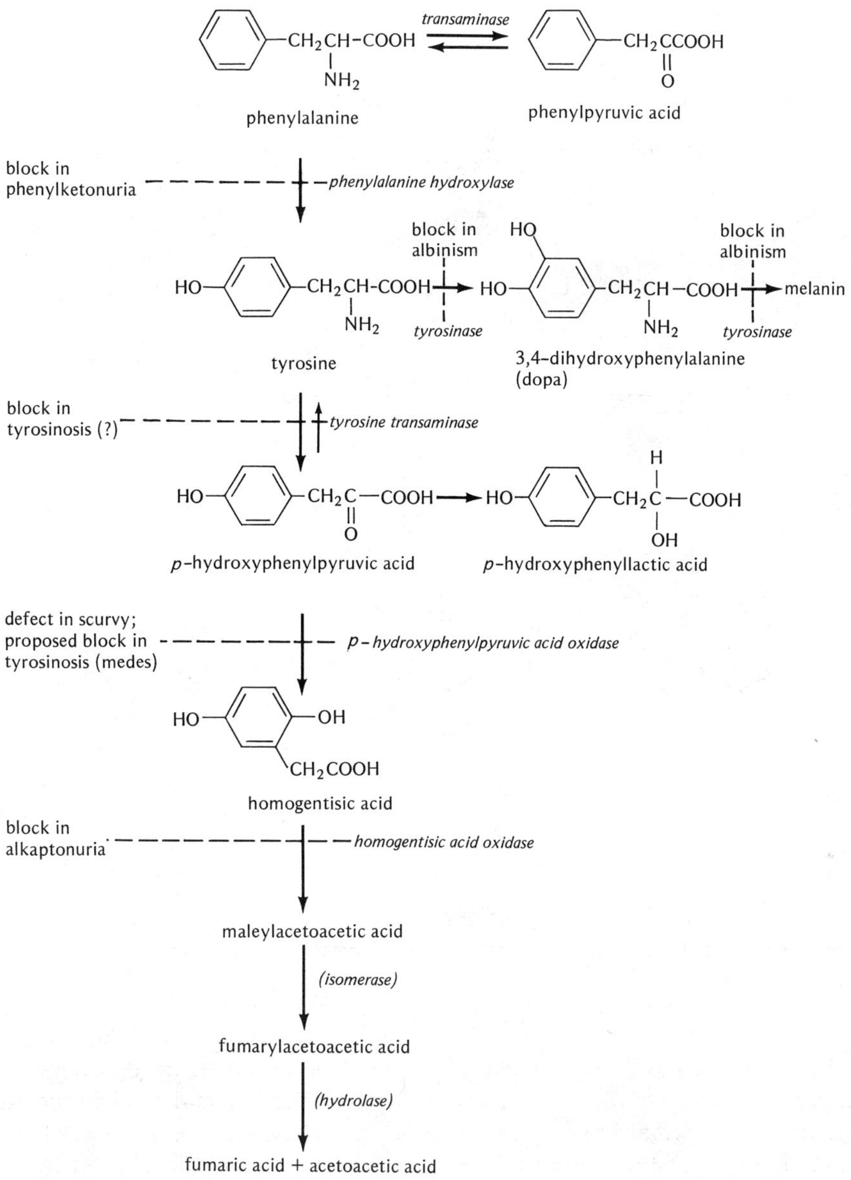

$C_6H_5-CH_2CH(NH_2)-COOH$ ⇌ (*transaminase*) $C_6H_5-CH_2C(=O)COOH$

phenylalanine — phenylpyruvic acid

block in phenylketonuria — *phenylalanine hydroxylase*

$HO-C_6H_4-CH_2CH(NH_2)-COOH$ → (*tyrosinase*; block in albinism) $(HO)_2C_6H_3-CH_2CH(NH_2)-COOH$ → (*tyrosinase*; block in albinism) melanin

tyrosine — 3,4-dihydroxyphenylalanine (dopa)

block in tyrosinosis (?) — *tyrosine transaminase*

$HO-C_6H_4-CH_2C(=O)-COOH$ → $HO-C_6H_4-CH_2CH(OH)-COOH$

p-hydroxyphenylpyruvic acid — *p*-hydroxyphenyllactic acid

defect in scurvy; proposed block in tyrosinosis (medes) — *p-hydroxyphenylpyruvic acid oxidase*

$HO-C_6H_3(OH)-CH_2COOH$

homogentisic acid

block in alkaptonuria — *homogentisic acid oxidase*

maleylacetoacetic acid

(isomerase)

fumarylacetoacetic acid

(hydrolase)

fumaric acid + acetoacetic acid

Figure 26–2

Sequence of some biochemical reactions in humans beginning with the metabolism of phenylalanine. In this sequence, various metabolic blocks are shown, each believed to be caused by the effect of a separately inherited genetic defect on a specific enzyme. For example, in phenylketonuria a defect in phenylalanine hydroxylase prevents the hydroxylation of phenylalanine thereby leading to the accumulation of phenylalanine and subsequent neurological damage caused by high concentrations of this amino acid. Some of this phenylalanine also undergoes transamination to phenylpyruvic acid which is then excreted in the urine. (After Stanbury, Wyngaarden, and Frederickson.)

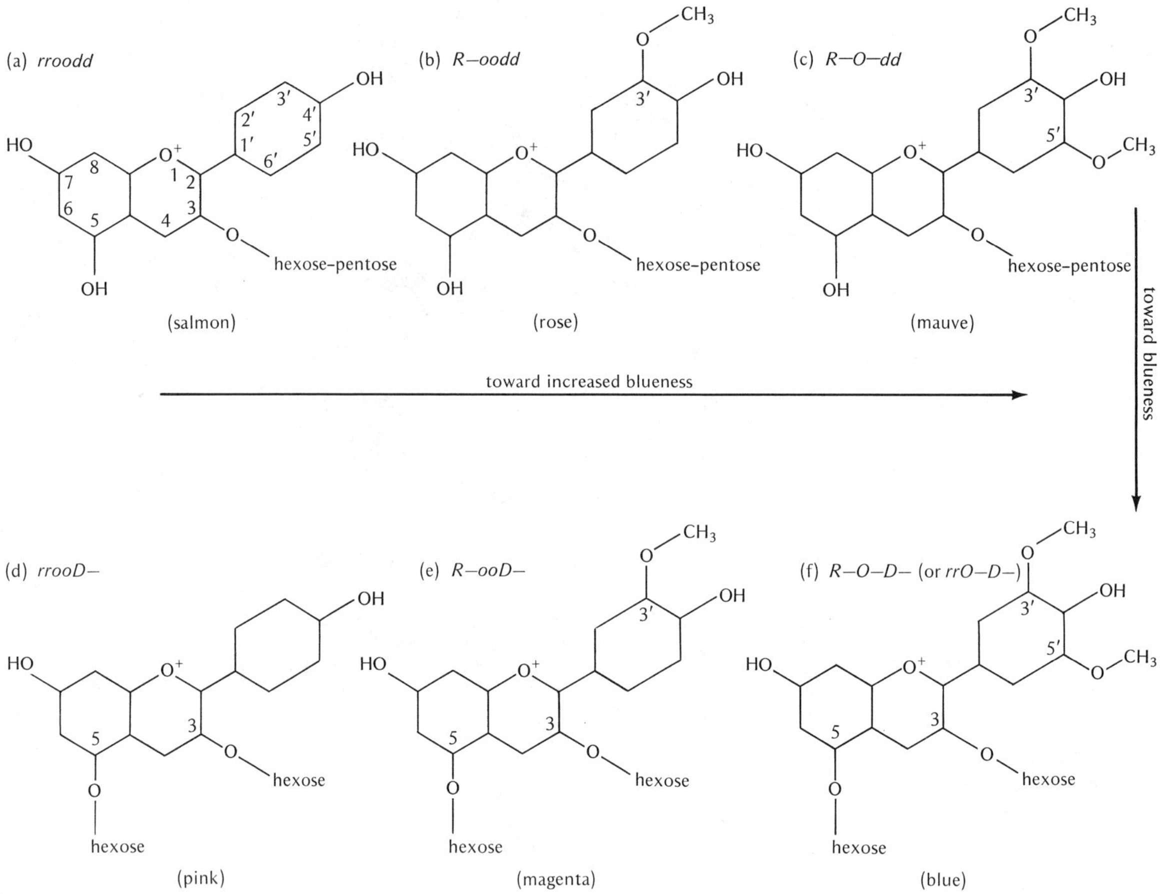

Figure 26–3

Differences in anthocyanin pigments produced by different genotypes among garden forms of Streptocarpus. *(After Lawrence and Sturgess.)*

sophila eye pigments had long been known to be affected by many genes, such as *vermilion*, *cinnabar*, *garnet*, *raspberry*, *white*, etc. Some of these caused the absence of the normal brown pigment and appeared bright red (e.g., *vermilion*, *cinnabar*); others lacked the bright red pigment and appeared ruby or brown (e.g., *garnet*, *raspberry*), while others lacked both pigments and were devoid of color (i.e., *white*).

Using 26 different eye-color mutations, Beadle and Ephrussi transplanted larval embryonic eye tissue (*imaginal discs*, see Chapter 30) from each one of these mutant stocks into individual wild-type larvae. The transplanted discs thus developed into an additional eye whose mature phenotype could be observed by dissecting it out after the wild-type larvae had reached the adult stage (Fig. 26–4). They found that most of these embryonic transplants were not affected by the larval environment in which they were placed but developed eye colors similar to those of the stock from which they were taken (*autonomous* development). Two exceptions existed, however, *vermilion* and *cinnabar*, each of which developed into a wild-type eye (*nonautonomous* development). It was evident that a diffusable substance entered into *vermilion* and *cinnabar* eye-tissue transplants and permitted full pigment development. Was this diffusible sub-

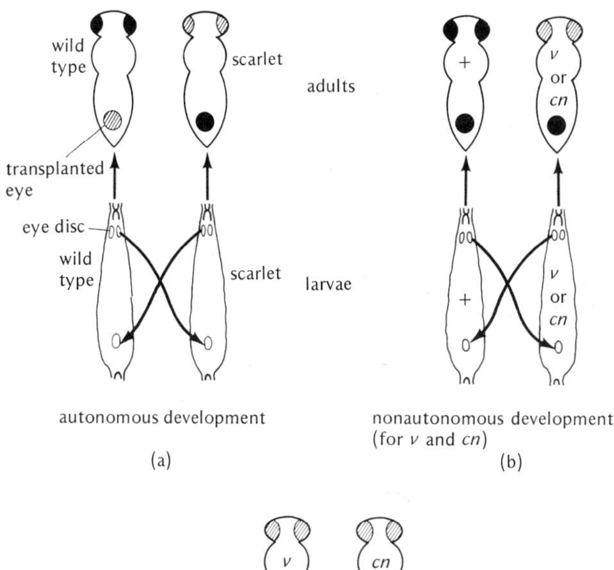

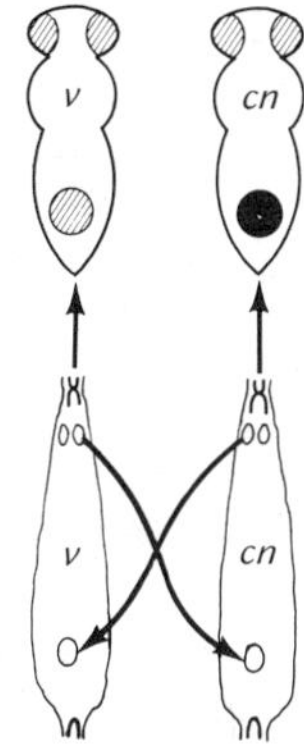

Figure 26–4

Scheme of transplantation experiments in Drosophila *showing autonomous and nonautonomous development for embryonic eye discs implanted into various genotypes (*v = vermilion, cn = cinnabar*). Black circles represent wild-type eyes; hatched circles represent mutant eye colors. (After Ephrussi.)*

stance identical in both cases or was it different? Beadle and Ephrussi felt this question could be answered by making transplants between *vermilion* and *cinnabar* larvae. If the same substance were missing in both mutations, eye-tissue transplants from one mutant stock to another would lack the necessary substance and therefore develop autonomously, i.e., be incapable of forming wild-type eyes.

They found, however, that only *cinnabar* transplants into *vermilion* larvae developed autonomously, but *vermilion* transplants into *cinnabar* larvae developed into wild-type eyes. To explain this result they proposed that there were two reaction steps involved, one of which was lacking in *vermilion* but was present in *cinnabar*, and another which was lacking in *cinnabar* but was present in wild type. Since each of these steps appeared to be represented by a diffusible substance, they called the first one the "v^+" substance (produced by the wild-type allele of *vermilion*) and the second the "cn^+" substance (produced by cn^+). These substances must be formed in the sequence $v^+ \rightarrow cn^+$, since *cinnabar* flies could supply v^+ substance to *vermilion*, but *vermilion* flies could not supply cn^+ substance to *cinnabar*. In other words, *vermilion* was blocked at v^+ but not at cn^+, and supplying v^+ to *vermilion* permitted these tissues to form cn^+ and the reaction to go to completion. The other 24 mutations which had developed autonomously in wild-type tissues were apparently lacking in additional substances which were either not available in the host or not diffusible.

At present two main types of pigments are known to be carried in *Drosophila* eyes, the *pterins* and the *ommochromes*, each affected by a different set of genes. Both pterins and ommochromes function by becoming attached to protein granules, in which form they are deposited in the eye cells, or *ommatidia*. The mutation *white* prevents the attachment of pigment to granules, and such mutants therefore appear white-eyed, as though they possessed no pigment at all. A fly deprived of its brown ommochromes will show only the bright pterin pigments, and its eyes therefore appear bright red, e.g., *vermilion*, *scarlet*, and *cinnabar*. On the other hand, a fly deprived of its pterins will have a darker eye color, such as *brown*, *raspberry*, or *garnet*. Flies deprived of both the ommochromes and the pterins, such as the double mutants *cinnabar brown* and *scarlet brown*, will have no eye color at all and will appear white-eyed.

Investigations by Caspari and Kühn in the flour

moth (*Ephestia*), and the biochemical analyses of Butenandt and others, have identified the v^+ and cn^+ substances as two of the precursors in the synthesis of one of the ommochrome pigments. As shown in Fig. 26–5, a metabolic block at either of two essential steps in this synthesis cannot be overcome without the addition of one of these substances.

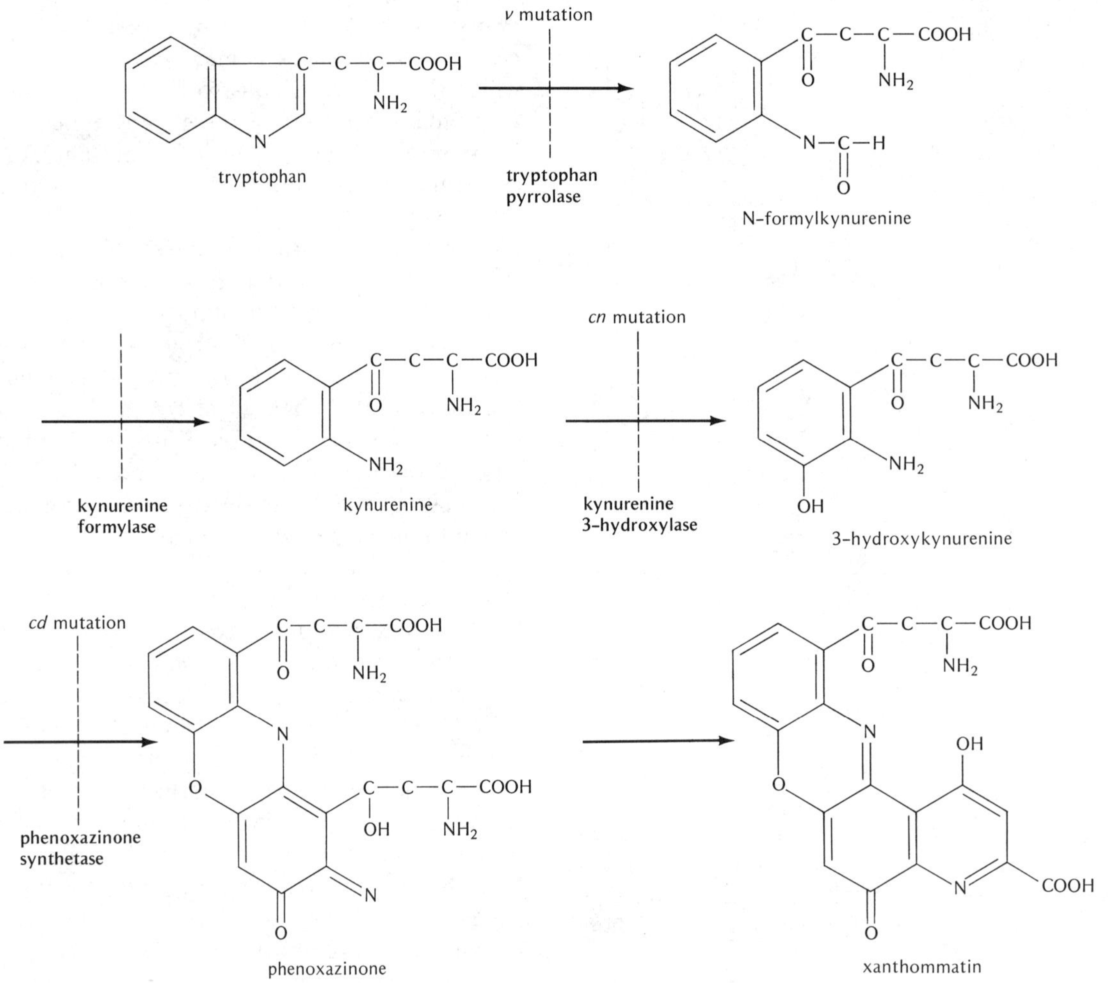

v mutation
tryptophan
tryptophan pyrrolase
N-formylkynurenine
cn mutation
kynurenine formylase
kynurenine
kynurenine 3-hydroxylase
3-hydroxykynurenine
cd mutation
phenoxazinone synthetase
phenoxazinone
xanthommatin

Figure 26–5

Abbreviated pathway for the synthesis of an ommochrome pigment from tryptophan, and the action of the genes vermilion (v) and cinnabar (cn) in blocking essential metabolic steps. Note that the addition of kynurenine (produced by cn flies) will overcome the block caused by vermilion. However, growth of cn flies in a kynurenine medium will have no effect since they will still be unable to synthesize 3-hydroxykynurenine. The enzyme that is lacking in vermilion mutants has been identified as tryptophan pyrrolase, and the enzyme lacking in cinnabar mutants is now known to be kynurenine hydroxylase. Enzymes that catalyze other steps in this pathway are indicated on the drawing. The metabolic block shown for the cardinal (cd) mutation is believed also to occur in scarlet and other mutations that can be distinguished by their different genetic loci (see Fig. 17–3). So far, however, no mutation is known which prevents the appearance of the enzyme kynurenine formylase. (From data of Phillips et al.)

BIOCHEMICAL MUTATIONS AND PATHWAYS IN MICROORGANISMS

Despite many advantages, functional and recombinational analysis with *Drosophila* and "higher" forms has serious shortcomings because of the large developmental gap between action at the gene level and the final morphological phenotype. In microorganisms, on the other hand, the steps involved in nutritional mutations are simpler to detect by biochemical methods and uncomplicated by dominance. For the purpose of analyzing the biochemical effect of genes, Beadle and Tatum were therefore led to *Neurospora,* since it is a simple organism which can be grown on a defined "minimal medium" and particular metabolic mutations can be detected by discovering which supplementary compounds are necessary to enable growth of the mutant (p. 536). In this fashion they discovered a series of mutations, each of which could be restored to full growth by addition of a *single* substance. Their first three mutant stocks, for example, called *pdx*, *thi*, and *pab*, were grown on a minimal medium to which were added pyridoxine, thiamine, and *p*-aminobenzoic acid, respectively.

As this work was extended by numerous others, it was discovered that mutations blocking the synthesis of a particular compound were not always caused by a breakdown at the same metabolic step. Some mutations, such as those affecting tryptophan synthesis, were shown by Bonner, Yanofsky, Brenner, and others to consist of at least four mutant subgroups. In *Salmonella typhimurium*, for example, tryptophan synthesis is impaired by the inability of some mutant stocks, such as *trp-8*, to synthesize anthranilic acid (Table 26–1), but can be restored to full growth by the addition of anthranilic acid, indole glycerol phosphate, indole, or tryptophan. Certain other mutations can grow upon the addition of either indole glycerol phosphate, indole, or tryptophan. Still others, such as *trp-3*, can grow either on indole or tryptophan, while certain mutants, such as *trp-1*, can only grow upon the addition of tryptophan itself. These results therefore indicate the presence of a metabolic pathway to the synthesis of tryptophan that can be interrupted at many steps along the way. Fig. 26–6 shows the known classes of *E. coli* mutations (*A-E*) that affect this pathway. As might be expected, a mutation blocking a particular step in such pathways will usually cause an accumulation of some of the preceding compounds in the pathway which are unable to be metabolized further. *Trp-D* mutations in *E. coli*, for example, accumulate anthranilic acid and *trp-A* mutations accumulate indole glycerol phosphate.

In analyzing the genetics of metabolic pathways, note that substances that enable growth of a mu-

TABLE 26–1

Effect of different media on the growth response of tryptophan mutations in *Salmonella typhimurium*

	Growth Response					
		Supplemented Medium				
Gene mutation	Minimal unsupplemented medium	Anthranilic acid (Ant)	Indole glycerol phosphate (IGP)	Indole (I)	Trypto-phan	Accumulated product
try-8	−	+	+	+	+	
try-2, -4	−	−	+	+	+	Ant
try-3	−	−	−	+	+	IGP
try-1, -6, -7, -9, -10, -11	−	−	−	−	+	IGP and I

Data of Brenner.

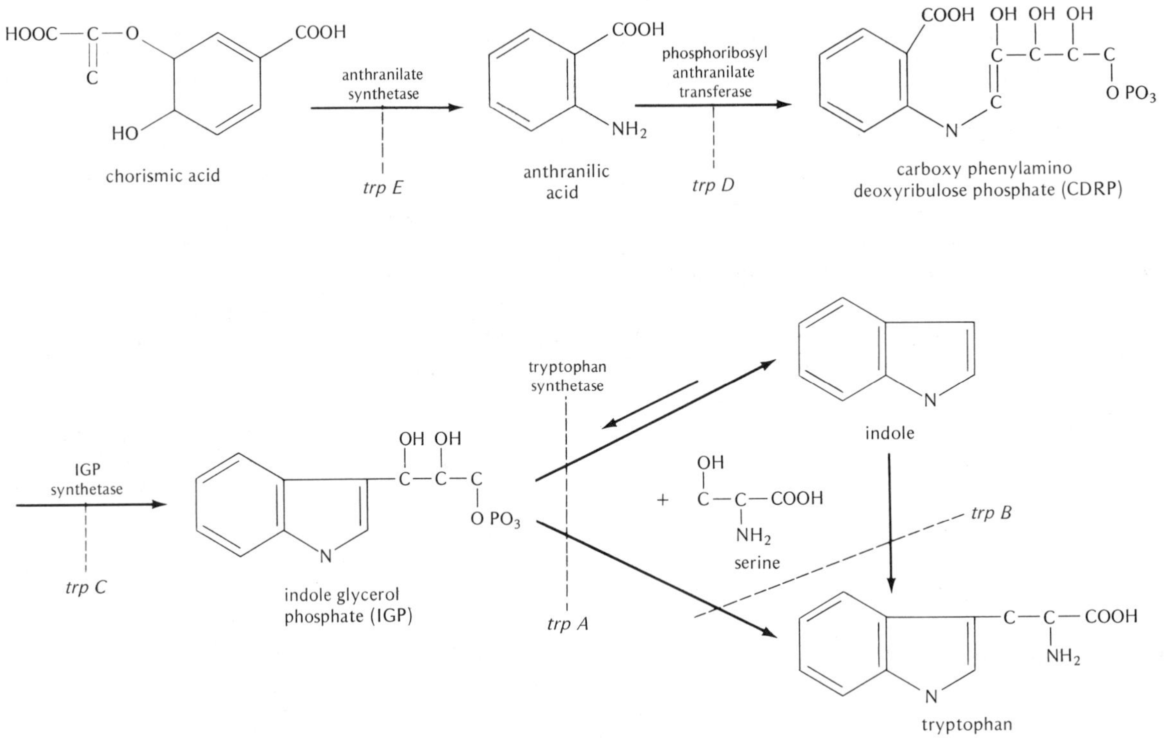

chorismic acid
anthranilate synthetase
trp E
anthranilic acid
phosphoribosyl anthranilate transferase
trp D
carboxy phenylamino deoxyribulose phosphate (CDRP)
IGP synthetase
trp C
indole glycerol phosphate (IGP)
tryptophan synthetase
trp A
indole
serine
trp B
tryptophan

Figure 26–6

Pathway of tryptophan synthesis, and the classes of trp gene mutations in E. coli *(A–E) that block various steps by affecting the function of the indicated enzymes. Similar enzymes function in the* Salmonella *tryptophan pathway, but the genes responsible are given different letters.*

tant strain are produced *after* the block caused by the mutation. For example, *E. coli trp E* mutants can be supplemented with anthranilic acid, IGP, indole and tryptophan. This means that the earlier a mutation acts in a metabolic pathway the more intermediary substances can be used for growth, whereas mutations that act later in the pathway can only utilize fewer substances produced at the end of the pathway (e.g., *trp B* mutations must be directly supplemented with the end product tryptophan, and nothing else will do). By the same reasoning we can distinguish whether a metabolic substance is formed at the beginning or end of a pathway by the numbers of different kinds of mutations that can utilize the substance for growth. That is, substances produced toward the end of the pathway can be utilized by more different mutations than substances produced at the beginning of the pathway. Thus, *all* of the various *trp* mutations can utilize tryptophan for growth, indicating of course that this compound is at the end of the pathway; *most* can utilize indole for growth, indicating that indole is also produced toward the end of the pathway and that many mutational blocks come before the indole-producing step; and *fewer* can utilize anthranilic acid for growth, indicating that this product is at the beginning of the pathway since these mutational blocks must occur even before anthranilic acid formation. In general, if two mutations in a particular pathway differ in the numbers of intermediary substances they can use, (e.g., mutant *#1* can use X, Y, Z, and mutant *#2* can use X and Y), the intermediary substance(s) that cannot be used by the mutation with the fewer numbers (e.g., substance Z), must come earlier in the pathway than the substances (X and Y) which can be used by both mutations. (If Z was produced *after* the synthesis of X and Y, both the *#1* and *#2* mutations that are blocked at the X and Y steps should be capable of utilizing Z to enable growth.) However, the determination of biochemical pathways is not always as simple as this method indicates, since some intermediary substances cannot be used experimentally because they are quite unstable or are not able to enter the cell from the external medium.

In most organisms the last two steps of tryptophan metabolism are affected by a single enzyme called *tryptophan synthetase*. Biochemical tests for the presence of this enzyme have generally shown that mutant stocks lacking the ability to utilize indole also have little or no tryptophan synthetase activity. In *Neurospora* this enzyme consists of only one main protein component, while in *E. coli* and *Salmonella typhimurium* the enzyme can be further separated into two components: (1) an *A* component, which affects next to the last step in Fig. 26–6, so that bacteria carrying such mutations can be supplemented by both indole and tryptophan; and (2) a *B* component, which affects the last step in Fig. 26–6, so that *trp-B* mutants must be supplemented directly with tryptophan in order to grow.

In both *Neurospora* and bacteria, however, the inability to form normal tryptophan synthetase does not necessarily mean that no protein at all is formed. Using immunological techniques, antibodies sensitive to normal tryptophan synthetase can be prepared in rabbit serum (see p. 167) and shown to react with extracts from many of the mutant tryptophan synthetase stocks. The reacting substances are called cross-reacting material, CRM, and tryptophan synthetase mutants producing CRM are designated CRM^+. Because of its sensitivity, the immunological reaction indicates that a protein similar to tryptophan synthetase is produced in normal or almost normal amounts by these mutants but nevertheless is unable to perform the necessary enzymatic role in tryptophan metabolism. This means that such mutations affect a critical portion of the tryptophan synthetase molecule. In contrast to CRM^+ mutants, more drastic effects on the tryptophan synthetase molecule seem to occur in CRM^- stocks, bacteria whose mutant product cannot be detected by the usual test for CRM.

In combination with other studies, these findings indicate that the action of a mutation appears as a metabolic "block," usually affecting only a single step of a chemical reaction. The cause of the block has been localized, in most investigated cases, to an impaired function of the enzyme that is usually active at that particular metabolic step (Table 26–2). The genetic control of phenotypes through rates of metabolic reaction had long been stressed by Goldschmidt but he possessed neither detailed biochemical information of the reactions nor of the enzymes involved. Beadle and Tatum's contribution, therefore, not only related genes to metabolism but also led to more precise investigations of the relationship between genes and enzymes. What are enzymes?

ENZYMES, PROTEINS, AND AMINO ACIDS

Interest in enzymes developed in the nineteenth century with the observation that certain cells or cellular products were capable of catalyzing reactions outside the cell. Among the first processes investigated was the fermentation of fruits and fruit juices caused by yeast cells. As demonstrated by Pasteur, yeast was a necessity for the conversion of the sugar and water in grape juice into alcohol and carbon dioxide. Somewhat later, in 1897, Büchner showed that the fermentive and catalyzing properties of yeast were not affected even when the cells were ground up with fine sand. Thus a biological catalyst or enzyme capable of existing both inside and outside of the cell was firmly identified with a particular metabolic process. Chemical identification of the enzyme itself, however, was difficult because of its presence in only small concentrations in nature. Finally, in 1926, Sumner developed a method of isolating a pure crystalline enzyme, urease, that functioned in the breakdown

TABLE 26–2

A sampling from the more than 100 human hereditary conditions known so far to show a loss of enzymatic activity

Condition	*Phenotypic Effect*	*Enzyme*	*Normal Enzyme Activity*
acatalasia	oral gangrene	catalase	acts in decomposition of hydrogen peroxide
alkaptonuria	darkening of urine, arthritis	homogentisic acid oxidase	catalyzes transformation of homogentisic acid into maleylacetoacetic acid
cretinism (goitrous)	thyroid malfunction, mental retardation	iodotyrosine deiodinase	catalyzes removal of iodine from iodotyrosine molecules
cystic fibrosis	excretion of highly viscous mucus; exocrine gland deficiencies	beta glucuronidase	metabolism of mucopolysaccharides
galactosemia	liver enlargement, galactose in urine and blood, cataracts	galactose-1-phosphate uridyl transferase	catalyzes transformation of galactose into glucose
Gaucher's disease	presence of Gaucher's cells causing enlargement and destruction of many tissues (spleen, liver, bone marrow, brain)	glucocerebrosidase	catalyzes the cleavage of glucose from glucocerebroside
glycogen storage disease	liver enlargement caused by accumulation of glycogen	glucose-6-phosphatase : brancher enzyme : muscle or liver phosphorylases : debrancher enzyme	catalyzes transformation of glycogen into glucose
Huntington's chorea	progressive mental deterioration	glutamine acid decarboxylase	catalyzes synthesis of γ-aminobutyric acid
hypophosphatasia	deficient calcification	alkaline phosphatase	probably catalyzes hydrolysis of phosphorylethanoalamine
jaundice (congenital nonhemolytic)	presence of bile in various tissues	glucouronyl transferase	catalyzes transfer of glucouronic acid from uridine diphosphate to bilirubin receptors
Lesch-Nyhan syndrome	mental retardation, self-mutilation, spasticity, increased uric acid production	hypoxanthine-guanine phosphoribosyl transferase	purine metabolism
methemoglobinemia	increased proportion of oxidized hemoglobin	diaphorase	catalyzes the reduction of methemoglobin into normal hemoglobin
Niemann-Pick disease	accumulation of lipids in reticuloendothelial cells and destructive changes in nervous system	sphingomyelinase	catalyzes hydrolysis of phosphorylcholine from sphingomyelin
phenylketonuria	mental retardation, excretion of phenylalanine	phenylalanine hydroxylase	converts phenylalanine into tyrosine
primaquine sensitivity ("favism")	hemolytic anemia	glucose-6-phosphate dehydrogenase	oxidizes glucose, permitting accumulation of protective levels of glutathione
Tay-Sachs disease	mental and motor retardation, death by ages 2–4 years	β-D-N-acetylhexosaminidase A	cleaves the terminal residue from a stored ganglioside (brain lipid)

of urea into carbon dioxide and water. This enzyme, as all others since then, proved upon chemical analysis to be a protein.

In a brief way, we have discussed proteins previously (p. 12) and pointed out that they are usually large molecules consisting of linked amino acids. It was the classical contribution of Emil Fischer to show that the amino acids are connected by *peptide linkages* that connect the carboxyl radical of one amino acid with the amino group of another through loss of a molecule of water (see Fig. 4–1). *Polypeptide chains* are thus formed which contain as high as a hundred or more amino acids. A protein, in turn, may consist of one or more polypeptide chains and reach a molecular weight as high as 6,000,000 or more. Given a molecule of a particular size, the possible kinds of protein that may exist therefore depend upon the kinds of amino acids that the protein contains, their respective order, relationship, and frequency. Surprisingly, although many kinds of amino acids are chemically possible, only about 20 different amino acids are found in most proteins.

In their basic structure, all amino acids contain an *alpha carbon* atom to which four subunits are attached:

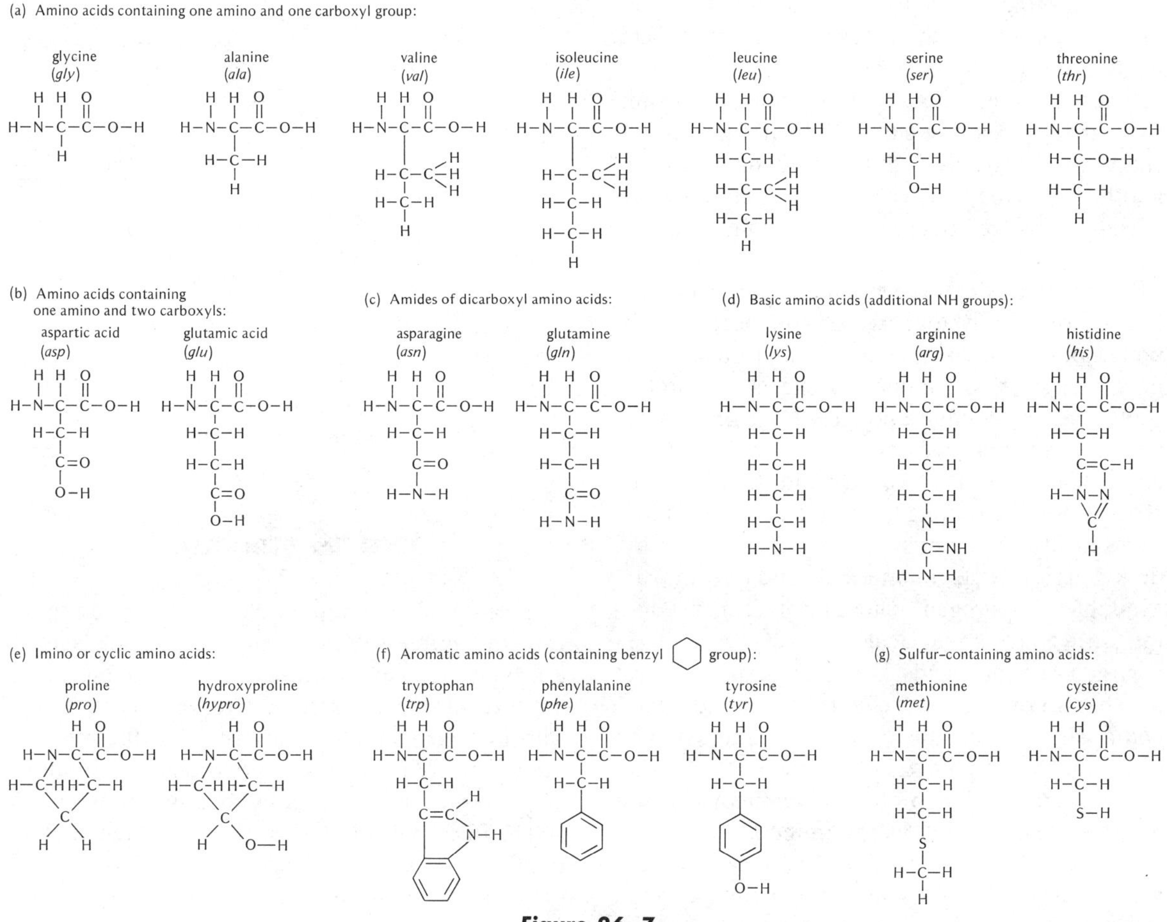

(a) Amino acids containing one amino and one carboxyl group:

glycine (*gly*) alanine (*ala*) valine (*val*) isoleucine (*ile*) leucine (*leu*) serine (*ser*) threonine (*thr*)

(b) Amino acids containing one amino and two carboxyls:

aspartic acid (*asp*) glutamic acid (*glu*)

(c) Amides of dicarboxyl amino acids:

asparagine (*asn*) glutamine (*gln*)

(d) Basic amino acids (additional NH groups):

lysine (*lys*) arginine (*arg*) histidine (*his*)

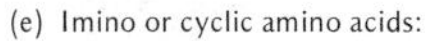

(e) Imino or cyclic amino acids:

proline (*pro*) hydroxyproline (*hypro*)

(f) Aromatic amino acids (containing benzyl ⬡ group):

tryptophan (*trp*) phenylalanine (*phe*) tyrosine (*tyr*)

(g) Sulfur-containing amino acids:

methionine (*met*) cysteine (*cys*)

Figure 26–7

Common amino acids found in proteins.

1. A carboxyl group C(=O)—OH with a potential negative charge (C(=O)—O^-).
2. An amino group NH_2 with a potential positive charge (NH_3^+).
3. An H atom.
4. A subunit called R that varies in structure from one amino acid to another.

As shown in Fig. 26–7, R may vary from a single H atom (glycine) to a benzene ring (phenylalanine). These differences are important not merely in establishing differently sized amino acids but also in determining the charge of the amino acid, its relationship to water molecules, and its interaction with other amino acids. Cysteine, for example, carries a reactive sulfhydryl (SH) group that can bond chemically with the sulfhydryl group of another cysteine molecule to produce a cystine "disulfide bridge" between two sections of a protein chain.

In many cases, however, the relationship between molecules is not based on chemical covalent bonds but on weaker noncovalent forces which, nevertheless, have important effects on protein properties and configuration. Hydrogen bonds, for example, are noncovalent, but are responsible to a large degree for the structure and stability of the DNA molecule. Among the amino acids the H ions exposed in the R units carrying carboxyl or hydroxyl groups (e.g., glutamic acid and tyrosine) are available as hydrogen donors for hydrogen bonding with other amino acids (Fig. 26–8b, c, d, e). Charged amino acids such as aspartic acid (− charge) or lysine (+ charge) are called *hydrophilic*, since their side chains can interact with water molecules, and they are most often found on the surface of a protein. *Hydrophobic bonds* may also be formed in which amino acids bearing methyl groups such as valine and isoleucine are much less soluble in water and therefore bond together in the interior of the protein (Fig. 26–8f, g).

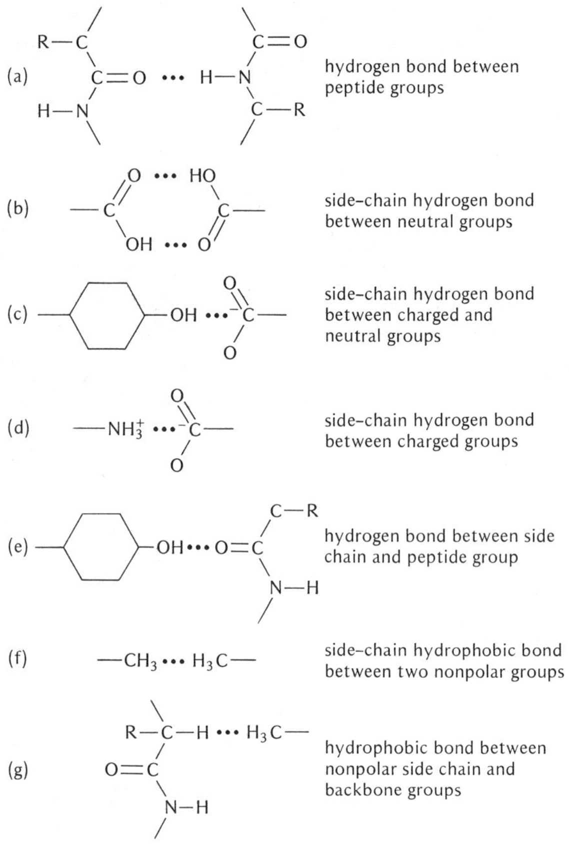

(a) hydrogen bond between peptide groups

(b) side-chain hydrogen bond between neutral groups

(c) side-chain hydrogen bond between charged and neutral groups

(d) side-chain hydrogen bond between charged groups

(e) hydrogen bond between side chain and peptide group

(f) —CH_3 ••• H_3C— side-chain hydrophobic bond between two nonpolar groups

(g) hydrophobic bond between nonpolar side chain and backbone groups

Figure 26–8

Types of noncovalent bonds formed between amino acids in proteins. (From Scheraga.)

PROTEIN STRUCTURE

Because of the unique qualities conferred upon proteins by the different amino acids, analysis of a protein depends upon learning the sequence of amino acids in its structure, or its *primary structure.* The first step in this analysis is to discover the kinds and frequencies of amino acids it contains. This was initially done by heating proteins to 100°C in strong acid (*hydrolysis*) and then analyzing the resultant amino acid mixture into its components by chemical tests. At present the amino acid mixture is dissolved in special solutions and then placed on filter paper. As the solvent moves along the filter paper, the amino acids move at different speeds along the path, depending upon their mobility. Usually the distinction between individual amino acids cannot be made at this stage, since many of them have similar mobilities in a single solvent. Further separation between amino acids is then made by turning the filter paper 90 degrees and subjecting it to a new solvent which then produces a two-dimensional array of spots separated both vertically and horizontally (Fig. 26–9). Staining by a special reagent (ninhydrin) colors each spot and gives the name *paper chromatography* to this process. Amino acids can also be separated into separate bands in columns containing ion-exchange resins in a process called *column chromatography.*

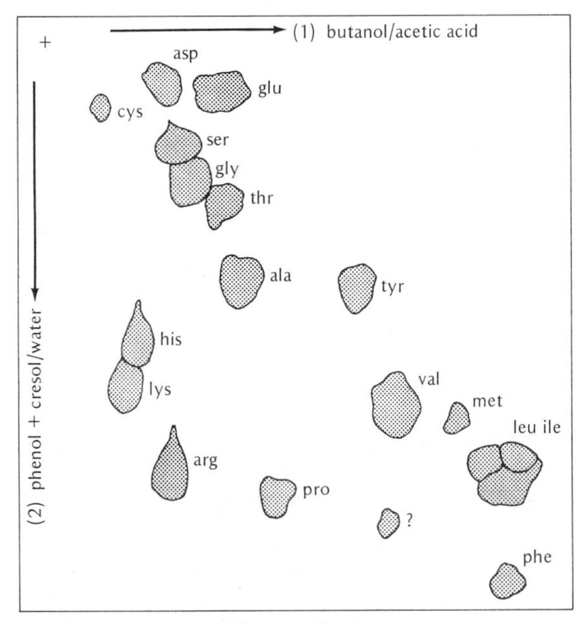

Figure 26–9

Paper chromatogram of a hydrolyzed protein. The hydrolyzed product was placed at the point marked + in the upper-left-hand corner and subjected to a mixture of acetic acid and butanol. The paper was then turned 90 degrees and subjected to a mixture of phenol, cresol, and water. After being sprayed by ninhydrin, the illustrated spots appeared, indicating the presence of specific amino acids. (After Haurowitz.)

Although these methods give information about the kinds of amino acids present and their relative concentrations (intensity of staining), they do not furnish their sequence in the polypeptide chain. Amino-acid-sequence analysis had to await a technique developed by Sanger for his study of the insulin molecule. This technique enables identification of the terminal amino acid of the peptide chain bearing the chemically unbonded amino group (the "N" terminal). The dye used for this purpose forms a yellow compound with the N-terminal amino acid that can be isolated after protein hydrolysis. Similar methods were then developed to identify the terminal amino acid chain that bears the carboxyl group (the "C" terminal). The breakdown of peptide chains into smaller differently sized fragments through use of acid and various enzymes (trypsin, chymotrypsin, subtilisin) enables both the terminal and interior amino acids in many of these shorter peptides to be identified. In the fashion of a puzzle, all these short sequences can then be fitted together until only a single amino acid sequence for the entire protein can satisfy the sequences in the subsidiary peptides (Fig. 26–10).

A second aspect of protein structure concerns the configuration imposed upon it by the peptide linkages. As shown in Fig. 26–11, this *secondary structure* takes the form of a screwlike alpha helix. If no other forces were involved, such a structure would impose upon a protein the appearance of a long coiled thread. In actuality, however, most proteins are more spherical than threadlike, indicating that the alpha helix has been bent and folded at various places to produce a new configuration called the *tertiary structure.*

As an illustration let us consider the protein myoglobin, in which primary, secondary, and tertiary structures have been investigated by Kendrew and co-workers in considerable detail. Myoglobin is a protein approximately 150 amino acids long that acts as a storehouse for oxygen within the cell. For this purpose it bears an iron-containing *heme* group capable of combining reversibly with oxygen that has been carried to the cells by the somewhat similar blood protein, hemoglobin.

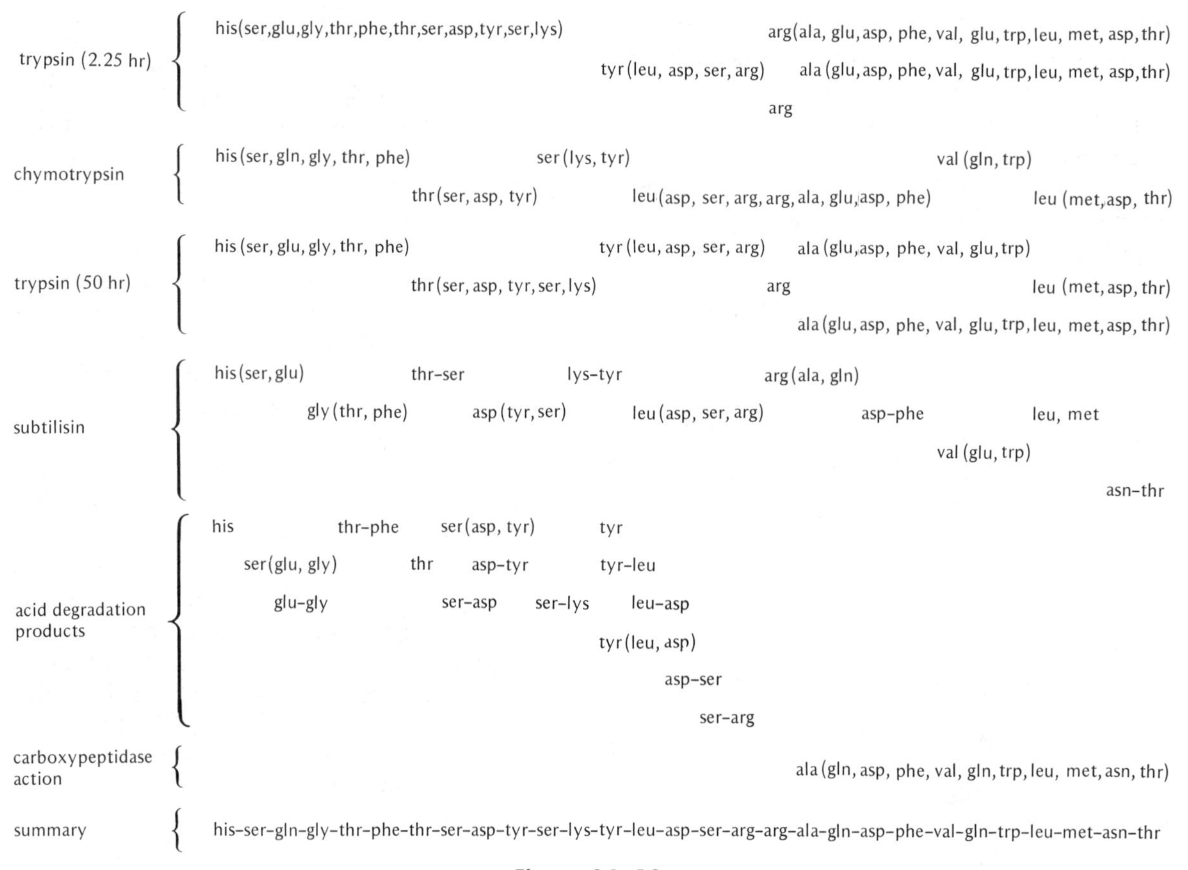

Figure 26–10

Structure of the polypeptide glucagon (bottom line) as determined by the peptide fragments arising from the action of acid and various enzymes. Amino acids within parentheses represent unordered sequences in the peptide fragment, while an amino acid to the left of the parenthesis represents the known N-terminal amino acid of the particular peptide. Thus histidine is at the N-terminal of the entire polypeptide and is then followed by serine, since serine is at the N-terminal of the acid degradation product that occurs after histidine. (After Canfield and Anfinsen.)

Table 26–3 shows the sequence of amino acids in myoglobin that has been determined to date by a combination of techniques. Determination of the secondary and tertiary structure has been solved by x-ray analysis in which the density of each portion of the molecule, as well as many single atoms within it, has been mapped in a three-dimensional fashion.

As shown in Fig. 26–12, the amino acid relationships in this molecule follow the alpha-helix pattern for straight sequences, but at the "corners" of the molecule the patterns are irregular. The formation of corners is produced by a variety of devices, one of them being the amino acid proline. In proline the R group forms a ring that includes the alpha amino N atom, thus changing the nature of the peptide bond and interfering with alpha-helix formation. Because of these interactions, the tertiary structure of the myoglobin molecule assumes the shape of a compact mass without channels going through it. This globule is stabilized by hydrophobic bonds between many of the methyl-containing amino acids that are pushed toward the center of the molecule, as well as by bonding of the heme group to various sections.

Of utmost interest is the finding that the secondary and tertiary structures of myoglobin are dictated by the primary amino acid structure. For example, loss of the original form of the myoglobin molecule by treatment with mild acid (denaturation) can be regained by reversing this treatment. Similarly Anfinsen and his collaborators have demonstrated that ribonuclease molecules will resume their normal shape and activity after having been denatured and their intramolecular sulfide bridges disrupted (Fig. 26–13). A specific sequence of amino acids thus produces a specific shape of protein.

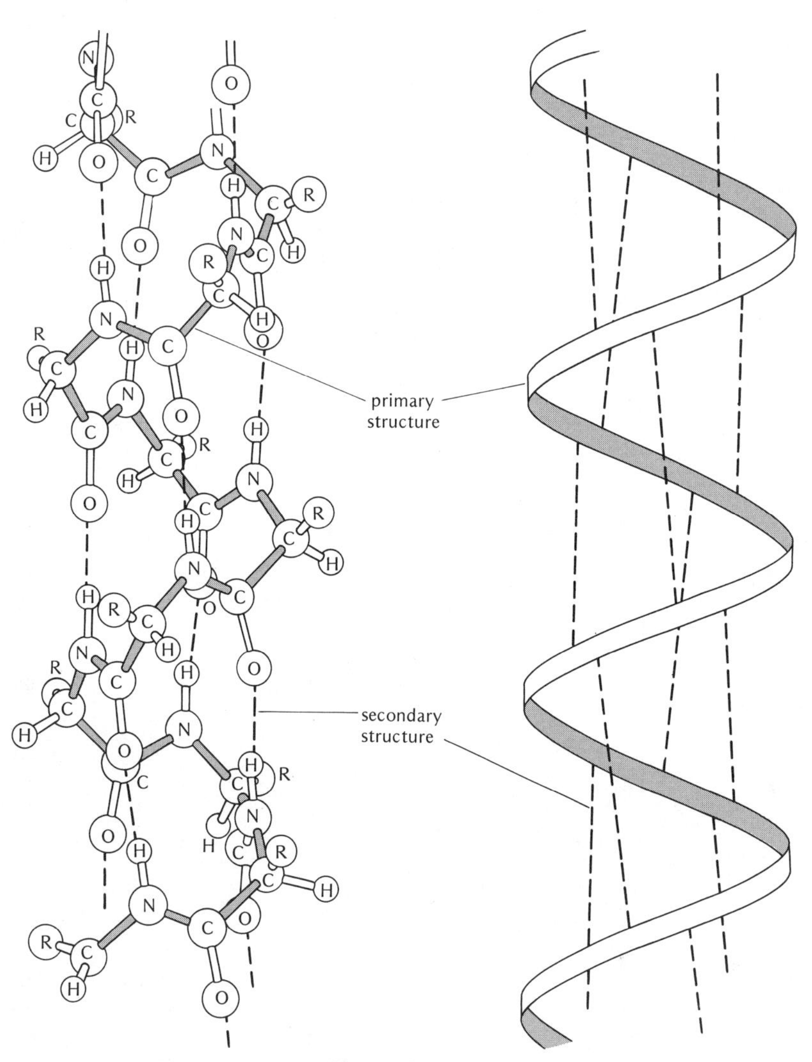

Figure 26–11

Relationship between primary structure (amino acid sequence) and secondary structure (alpha helix) in a protein. (From Corey, R. B., and Pauling, L.: Proc. Intern. Wool Textile Research Conf. Australia, 1955 B, 249, 1956.)

TABLE 26–3

Amino acid sequence in sperm whale myoglobin starting at the NH_2 terminal (NA1) of the chain and continuing to the carboxyl terminal (HC5)*

Position (Fig. 26–12)	Amino Acid	Position (Fig. 26–12)	Amino Acid	Position (Fig. 26–12)	Amino Acid
NA 1	val	2	glu	4	tyr
2	leu	3	ala	5	leu
A 1	ser	4	glu	6	glu
2	glu	5	met	7	phe
3	gly	6	lys	8	ile
4	glu	7	ala	9	ser
5	trp	E 1	ser	10	glu
6	gln	2	glu	11	ala
7	leu	3	asp	12	ile
8	val	4	leu	13	ile
9	leu	5	lys	14	his
10	his	6	lys	15	val
11	val	7	his	16	leu
12	trp	8	gly	17	his
13	ala	9	val	18	ser
14	lys	10	thr	19	arg
15	val	11	val	GH 1	his
16	glu	12	leu	2	pro
AB 1	ala	13	thr	3	gly
B 1	asp	14	ala	4	asn
2	val	15	leu	5	phe
3	ala	16	gly	H 1	gly
4	gly	17	ala	2	ala
5	his	18	ile	3	asp
6	gly	19	leu	4	ala
7	gln	20	lys	5	gln
8	asp	EF 1	lys	6	gly
9	ile	2	lys	7	ala
10	leu	3	gly	8	met
11	ile	4	his	9	asn
12	arg	5	his	10	lys
13	leu	6	glu	11	ala
14	phe	7	ala	12	leu
15	lys	8	glu	13	glu
16	ser	F 1	leu	14	leu
C 1	his	2	lys	15	phe
2	pro	3	pro	16	arg
3	glu	4	leu	17	lys
4	thr	5	ala	18	asp
5	leu	6	gln	19	ile
6	glu	7	ser	20	ala
7	lys	8	his	21	ala
CD 1	phe	9	ala	22	lys
2	asp	FG 1	thr	23	tyr
3	arg	2	lys	24	lys
4	phe	3	his	25	glu
5	lys	4	lys	HC 1	leu
6	his	5	ile	2	gly
7	leu	G 1	pro	3	tyr
8	lys	2	ile	4	gln
D 1	thr	3	lys	5	gly

After Perutz, 1965, from data of Kendrew and others.

* The positions of the amino acids in the various segments (A, AB, B, etc.) correspond to the positions seen in Fig. 26–12.

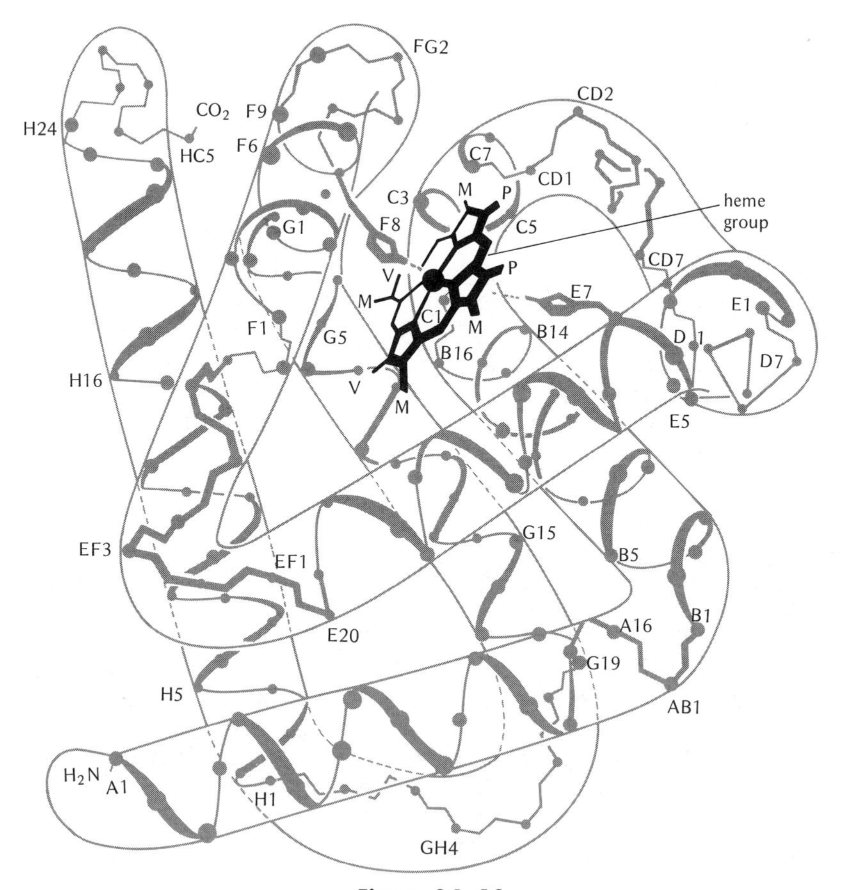

Figure 26–12

Structure of the sperm whale myoglobin molecule obtained by x-ray analysis. The alpha carbons of the amino acids are represented by large dots and the amino acids are identified in Table 26–3. (After Dickerson.)

For both myoglobin and ribonuclease, this shape is confined to a single folded polypeptide chain. Many proteins, however, have a *quaternary structure* in the sense that they exist as associations of two or more polypeptide chains bound together by noncovalent and occasionally covalent bonds. Among the most interesting of these are the blood hemoglobins—proteins that provided the first illus-

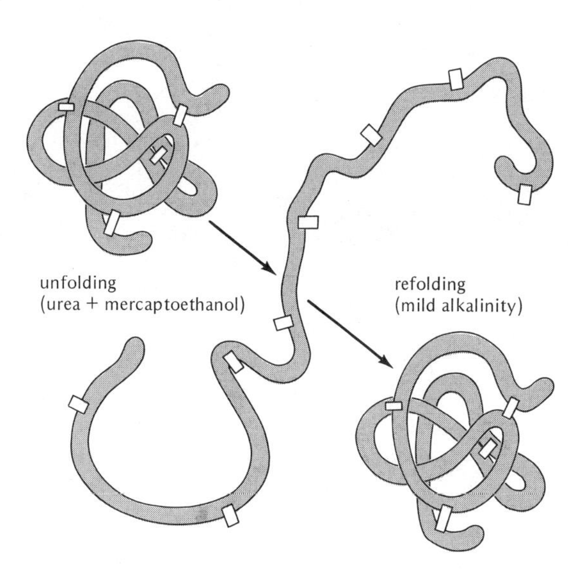

Figure 26–13

Diagrammatic representation of the unfolding of a native protein through the experimental action of urea and mercaptoethanol, and its eventual refolding into the same native configuration after the removal of these agents. The small rectangles represent locations where disulfide bonds are formed in the folded molecule. (After Epstein, Goldberger, and Anfinsen.)

tration of the effect of a mutation on an amino acid sequence.

ALTERED AMINO ACID SEQUENCES

Sickle-cell anemia, one of the blood diseases described early in this century, involves a marked change in the normal round shape of red blood cells to sickle-shaped. This change is especially apparent when oxygen concentration is lowered throughout the circulatory system at high altitudes or at even low altitudes in the peripheral circulatory system. Such affected individuals may develop various abnormalities, many of which lead to serious illness and death. However, not all individuals are affected to the same degree; some show only moderate sickling ("sickle-cell trait"), and others show considerable sickling ("sickle-cell anemia").

In 1949 Neel and Beet, independently, proposed the hypothesis that sickling was caused by a single mutant gene that was heterozygous in individuals with sickle-cell trait and homozygous in individuals with sickle-cell anemia. In the same year Pauling, Itano, Singer, and Wells observed that the

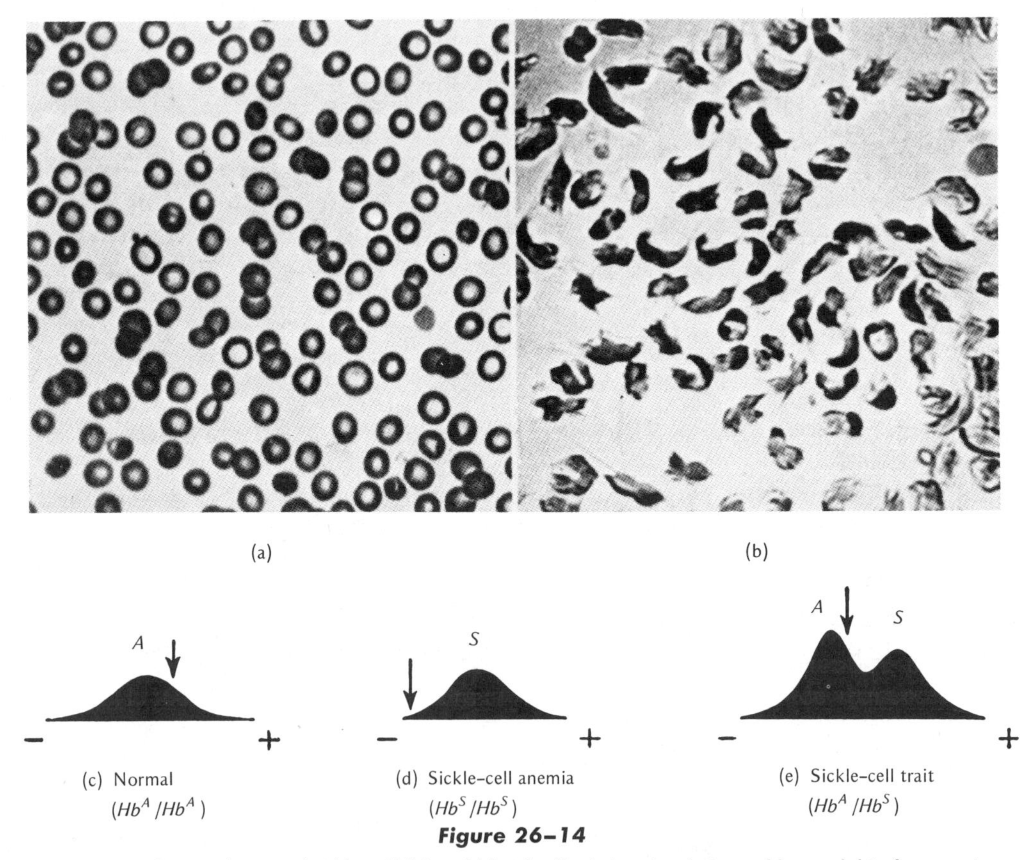

Figure 26–14

Appearance of normal (a) and sickle-cell (b) red blood cells. In (c–e) solutions of hemoglobin from various genotypes are placed at a particular point in an electrophoretic field indicated by the arrows. Normal hemoglobin migrates toward the cathode, while hemoglobin from sickle-cell anemics migrates toward the anode. Individuals with sickle-cell trait show both types of hemoglobin. (From C. Stern, Principles of Human Genetics, 3rd ed. W. H. Freeman, San Francisco, 1973; after Conley, and Pauling, Itano, Singer, and Wells.)

hemoglobins of normal and sickle-cell anemic individuals differ significantly when placed in a solution in an electrically charged (electrophoretic) field. Both types of hemoglobin migrate because of electric charges carried by each, but, at certain acidities of the solution, the hemoglobin of sickle-cell anemics migrates in an opposite direction to that of the normal. Also, as Fig. 26–14 shows, heterozygotes bearing the sickle-cell trait show a mixture of these two hemoglobin components.

The effect of sickle-cell hemoglobin on the red blood cell is believed to involve a change in the viscosity of the cell that probably helps prevent parasitical infection by falciparum malaria. The molecular mechanism proposed to account for this change in shape and viscosity is the "stacking" of the sickle-cell molecules in a uniform alignment (see Murayama). In the sickle-cell homozygote, ali the hemoglobin can be transformed into this crystallized type, and distortion of red-blood-cell shape can become sufficiently severe to cause hemolytic anemia. In the heterozygote enough normal hemoglobin is maintained to prevent anemia except under rare conditions. Here then, at last, was a "molecular" disease, whose modified protein could be isolated in significant amounts.

Normal and sickle-cell hemoglobins were named *A* and *S*, respectively, and the two alleles producing these hemoglobins were named Hb^A and Hb^S. A third abnormal hemoglobin was soon discovered, called *C*, which could be separated electrophoretically from the other two hemoglobins and was produced by the Hb^C allele. The fact that each of these hemoglobins had a different charge indicated that either the folding of the protein had changed and exposed or masked differently charged groups, or that the amino acid sequence had changed so that differently charged amino acids had been substituted for others.

The task of distinguishing between these possibilities was a difficult one. Hemoglobin is a fairly large molecule consisting, as does myoglobin, of an iron-containing "heme" portion which functions in oxygen respiration, and a protein "globin" portion which appears to be the active mutant element in the various inherited abnormalities. In a complete hemoglobin molecule there are four individual polypeptide chains divided into two identical α (alpha) chains and two identical β (beta) chains. The α's are approximately the same length as the β's but recognizably different in respect to chemical and electrophoretic properties.

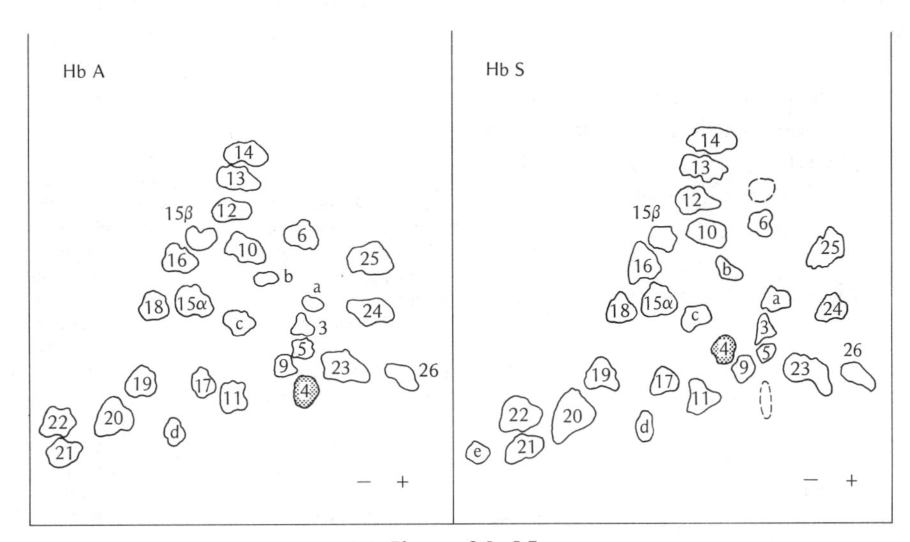

Figure 26–15

Chromatographic comparison of peptides produced by trypsin digestion of normal (Hb^A) *and sickle-cell* (Hb^S) *hemoglobins, showing shift in the position of peptide 4. (After Baglioni.)*

Since each chain contains about 140 amino acids, the task of analyzing hemoglobin abnormalities involves tracing sequences of about 280 amino acids in the two kinds of polypeptides.

For this purpose Ingram extended Sanger's method of protein analysis (p. 611) and devised a "fingerprinting" technique in which hemoglobin is subjected to trypsin digestion and the resulting shorter peptides are separated by a combination of paper electrophoresis and chromatography. This treatment produces about 30 uniquely identifiable spots, almost evenly divided between α- and β-chain peptide fractions. As Fig. 26–15 shows, the difference between the fingerprints of hemoglobins A and S resides entirely in the position of peptide 4. The position of this peptide, part of the β chain, was also found to be modified in hemoglobin C, and Ingram's task therefore resolved itself to analyzing the sequence of the eight amino acids in this changed peptide.

When this analysis was accomplished, there was found to be only a single amino acid difference between each of the three hemoglobins. As shown in Fig. 26–16, this amino acid difference occurred at amino acid position 6 in all three cases. In normal hemoglobin this position is occupied by glutamic acid, which is then replaced by valine in hemoglobin S. In this position valine does not ionize and therefore does not possess an electric charge. Since glutamic acid does ionize in the same position, an increased positive charge occurs in this hemoglobin S peptide, which may explain the electrophoretic pattern of the entire protein (see Fig. 26–14). The presence of lysine at this same position in hemoglobin C results in neutralization of the associated negatively charged glutamic acid, causing an even more rapid migration to the positive pole during electrophoresis.

Since the initial study by Ingram, the amino acid sequences of the entire α and β chains of normal hemoglobin have been determined (Fig. 26–17) and more than 100 variants of hemoglobin A have been discovered, each involving a different specific amino acid substitution on the α or β chains (Fig. 26–18). The chain designations of these mutant hemoglobins derive from symbolizing normal A hemoglobin as $\alpha_2^A\beta_2^A$, meaning two normal α and β chains. Thus sickle-cell hemoglobin is symbolized as $\alpha_2^A\beta_2^S$, and hemoglobin C is symbolized $\alpha_2^A\beta_2^C$, since only the β chains have been modified in each case. The independence of amino acid changes in one chain from amino acid changes in the other is evidence that the genes determining the α and β sequences are at different loci. Since the hemoglobin protein is so obviously composed of two independent gene products, each a single polypeptide, Ingram pointed out that it may be appropriate to modify the "one gene–one enzyme" hypothesis to "one gene–one polypeptide." Defining the gene as a section of DNA that determines the amino acid sequence of a polypeptide has the advantage that it relates a unique structural and functional component of a protein to a localized segment of genetic material (see also cistron, p. 590).*

	1	2	3	4	5	6	7	8
	+	+				–	–	+–
Hemoglobin A	val	his	leu	thr	pro	*glu*	glu	lys
	+	+					–	+–
Hemoglobin S	val	his	leu	thr	pro	*val*	glu	lys
	+	+				+–	+–	+–
Hemoglobin C	val	his	leu	thr	pro	*lys*	glu	lys

Figure 26–16

Sequences of amino acids in β-chain peptide 4 for hemoglobins A, S, and C. Presumed electrical charges on the amino acids are indicated by + and –. (After Ingram, 1963.)

*The advantage of defining a particular physical unit in terms of a functional unit may appear offset by the fact that different kinds of functional units in the cell are also determined by such physical units. Ribosomal RNA and transfer RNA molecules, for example, derive their sequences from DNA nucleotides, yet are not translated into proteins. Suggestions have therefore been made to broaden the definition of a gene to include any nucleotide sequence that determines a biologically functional nucleic acid, including messenger RNA. Such definitions, however, seem much too broad since it can be claimed that all or most nucleotides or nucleic acids in the cell provide some function whether they are transcribed (ribosomal and transfer RNA), translated (messenger RNA) or replicated (e.g., operators and promoters, Chapter 29). Because of these ambiguities it seems more appropriate to define the term *gene* at the particular functional level being considered with the proviso that the gene is a section of genetic material determining this function. In terms of proteins, therefore, it is useful and commonly accepted to consider a gene as the section of DNA that determines a polypeptide since it is the polypeptide that is the basic unit characterizing the function of a particular protein.

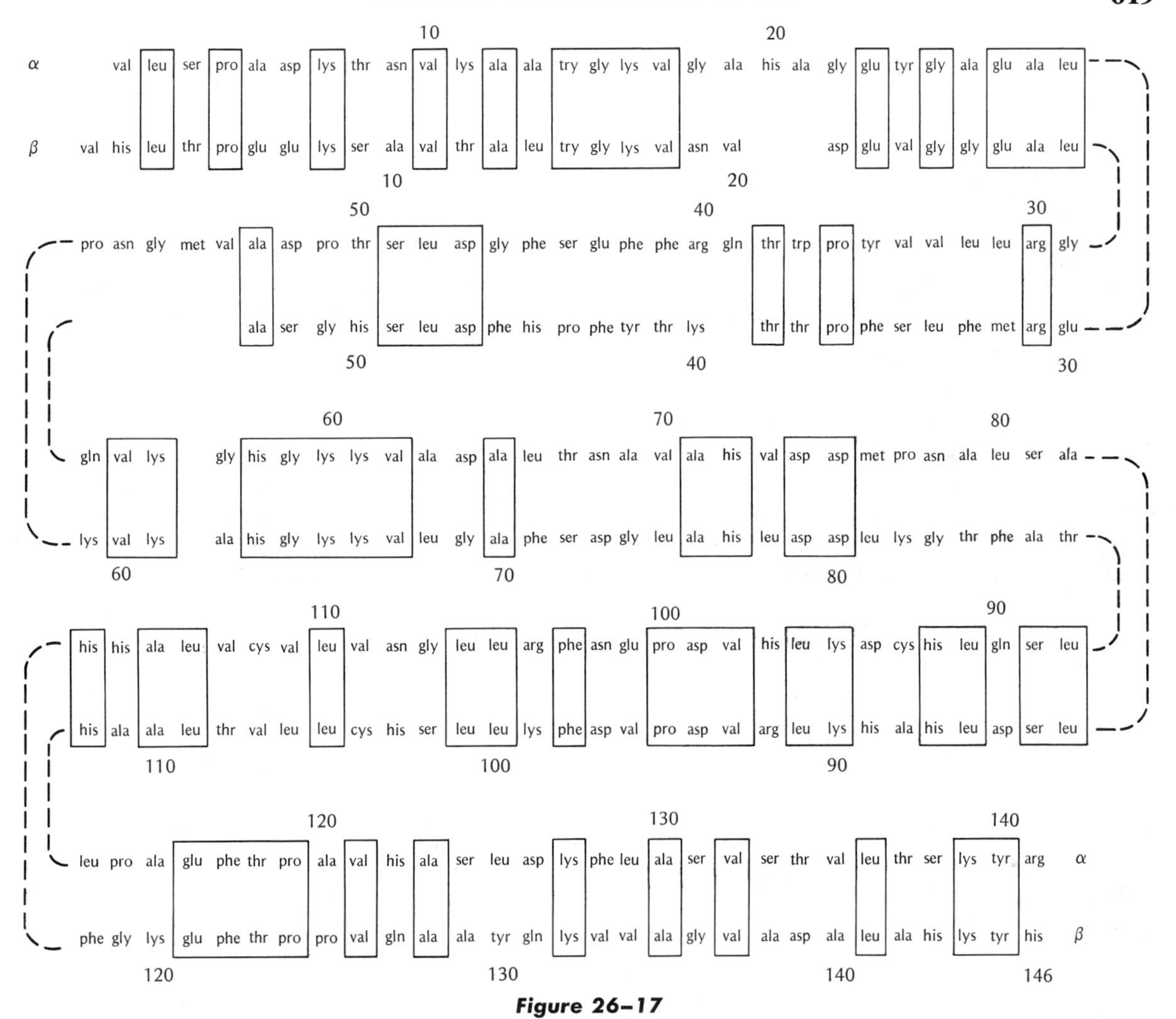

Figure 26–17

Amino acid sequence in the normal α and β chains of human hemoglobin. The positions of amino acids are numbered sequentially from the N-terminal, and those amino acids enclosed by solid lines occupy the same relative positions in both chains. (From Ingram, 1963; after Braunitzer, Konigsberg, Goldstein, and others.)

The separate genetic determination of α and β chains is accompanied in the cell by independent association between them. Heterozygotes for hemoglobin S, for example, produce both the $\alpha_2^A\beta_2^S$ and $\alpha_2^A\beta_2^A$ hemoglobins, as would be expected if both types of β chains could associate with the normal α chain. It is of interest, however, that this heterozygote does not commonly produce mixed hemoglobin molecules that are $\alpha_2^A\beta^A\beta^S$. Apparently, both α and β hemoglobin chains are formed mainly in pairs, or *dimers* (i.e., α_2 and β_2). These dimers, in association with heme groups, are then fitted together into a four-chained molecule (*tetramer*) whose structure has been carefully elucidated by the x-ray studies of Perutz and others. It seems clear so far that many aspects of the hemoglobin molecule, including its quaternary structure, are responsible for the efficiency with which oxygen molecules can be absorbed and released.

Hemoglobin A, however, is only one of the

α chain

	1	5	15	16	22	30	47	54	58	68	87	116
normal	*val*	*ala*	*gly*	*lys*	*gly*	*glu*	*asp*	*gln*	*his*	*asn*	*his*	*glu*

HbJ Toronto ----asp----
HbJ Oxford ----asp----
HbI ----asp----
HbJ Medellin ----asp----
HbG Chinese ----gln----
HbL Ferrara ----gly----
Hb Mexico ----glu----
Hb Shimonoseki ----arg----
HbM Boston ----tyr----
HbG Philadelphia ----lys----
HbM Iwate ----tyr----
HbO Indonesia ----lys----

β chain

	1	2	6	7	16	22	26	43	46	61	63	67	70	79	87	95	121	132	143
normal	*val*	*his*	*glu*	*glu*	*gly*	*glu*	*glu*	*glu*	*gly*	*lys*	*his*	*val*	*ala*	*asp*	*thr*	*lys*	*glu*	*lys*	*his*

Hb Tokuchi ---try---
HbC ----lys----
HbS ----val----
Hb San Jose ----gly----
Hb Siriraj ----lys----
Hb Baltimore ----asp----
HbG Coushatta ----ala----
HbE ----lys----
HbG Galveston ----ala----
HbK Ibadan ----glu----
Hb Hikari ----asn----
HbM Saskatoon ----try----
Hb Zurich ----arg----
Hb Milwaukee ----glu----
Hb Sydney ----ala----
Hb Seattle ----glu----
HbG Accra ----asn----
HbD Ibadan ----lys----
HbN ----glu----
HbO Arab ----lys----
HbK Woolwich -----gln----
Hb Kenwood ----asp----

Figure 26–18

A sampling of the known hemoglobin mutations in α and β chains. The numbers in the top line indicate the specific amino acid involved in each mutation according to Fig. 26–17. (Data from many sources; see Hunt, Sochard, and Dayhoff. Modeled after a diagram by Ingram.)

hemoglobins used in oxygen transport. Another type, hemoglobin F, is produced prenatally in the fetus and has higher affinity for oxygen than adult hemoglobin. As with hemoglobin A, hemoglobin F is also a tetramer composed of two pairs of polypeptide chains: two α chains and two unique γ (gamma) chains that bear some resemblance to β chains (formula $\alpha_2^A\gamma_2^F$). Usually hemoglobin F begins to diminish before birth and is at practically zero level at 6 months of age. Concurrent with this diminution of γ chains in the normal individual is an increase in β-chain production (Fig. 26–19). Apparently the onset of β-chain production either triggers or is triggered by the decrease in γ. This close relationship is also seen when individuals suffering from the inherited disease *thalassemia major* are unable to make β chains and continue to produce hemoglobin F in adult life. In cases where thalassemia seems to affect the production of α chains, a considerable amount of hemoglobin in infants may be composed only of γ^F chains. Such hemoglobin has the formula γ_4^F and is called *Hb-Bart's*.

Another chain, δ (delta), is found in one of the minor hemoglobin components, hemoglobin A_2,

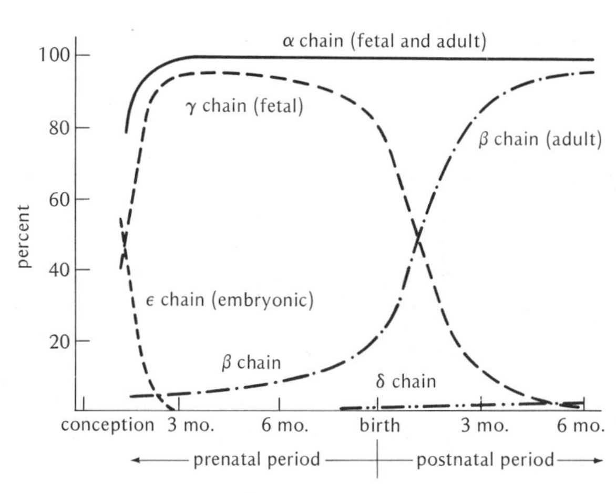

Figure 26–19

Changes in the amounts of various hemoglobin chains produced in human subjects from the embryonic period to the adult stage. These hemoglobins are normally found as tetramers, the earliest being an embryonic hemoglobin consisting of two α chains and two ϵ (epsilon) chains. (After Huehns et al.)

which can be symbolized as $\alpha_2^A\delta_2^A$. However, although both hemoglobins A and A_2 are produced in the same cell, the amount of hemoglobin A is normally about 40 times the amount of hemoglobin A_2. Since the only difference between these two types of hemoglobin lies in the presence of β chains in one and δ chains in the other, it can be presumed that the production of substances by β and δ genes are controlled, or "regulated." As we have seen, regulation also occurs for the β and γ genes during the birth period. Such regulatory phenomena are, of course, no different than we would expect, since the requirements of organisms change during development and some of the gene products involved in these changes must be turned on or off. Mechanisms to account for the process of regulation have been proposed and will be discussed in Chapters 29 and 30.

The hemoglobin studies, as well as similar analyses of tobacco mosaic virus protein, tryptophan synthetase enzyme, and other proteins, therefore indicate that most gene mutations cause single amino acid substitutions. Furthermore, the hemoglobin studies show that even single amino acid substitutions may have far-reaching effects. Apparently not only is the primary structure of a protein changed by such substitutions, but its secondary, tertiary, and quaternary properties may also be radically affected. For example, the removal of two methyl groups from normal contact with the heme group in the β chain mutation, hemoglobin Sydney (valine $\rightarrow$ alanine), causes the heme group to drop out of the molecule and leads to hemolytic anemia even in heterozygotes (Perutz and Lehmann).

In sickle-cell hemoglobin, the large nonpolar side chain of valine at the #6 β-chain position is extruded from the surface of the molecule causing a change in the behavior of deoxygenated hemoglobin. That is, sickle-cell hemoglobin remains in solution within the red blood cell as long as oxygen tension is high. As the oxygen tension falls, filamentous aggregations of these molecules become packed together into nonsoluble crystalline structures that precipitate out of solution (Finch et al.), distorting and stiffening the red blood cell into its characteristic sickle shape. As shown in Fig. 26–20 this single amino acid substitution causes marked pleiotropic effects (p. 215) on various biological functions leading to a large "pedigree of causes" for the disease.

Unfortunately, sickle cell disease is a common and undesirable relic among populations that have maintained sickle-cell genes during evolution because they protected heterozygotes against malaria. Even after the danger of malaria is removed, such populations may possess relatively high frequencies of such genes and are continually producing sickle-cell homozygotes from matings between heterozygotes. The past exposure of African populations carrying the sickle-cell gene to malaria (see Fig. 34–9) is thus responsible for the fact that sickle-cell anemia occurs in about 1 out of 500 American blacks.

However, not all amino acid substitutions in proteins lead to acute disease. So far a number of variants in hemoglobin and other proteins have been discovered which cause no observable functional effect. Insulin, for example, is known to exist in a number of forms which differ from each other by single amino acid substitutions but nevertheless perform equally well in diabetics in the removal of excess glucose from the bloodstream. Similar observations have been noted for the proteins produced by different strains of the tobacco mosaic

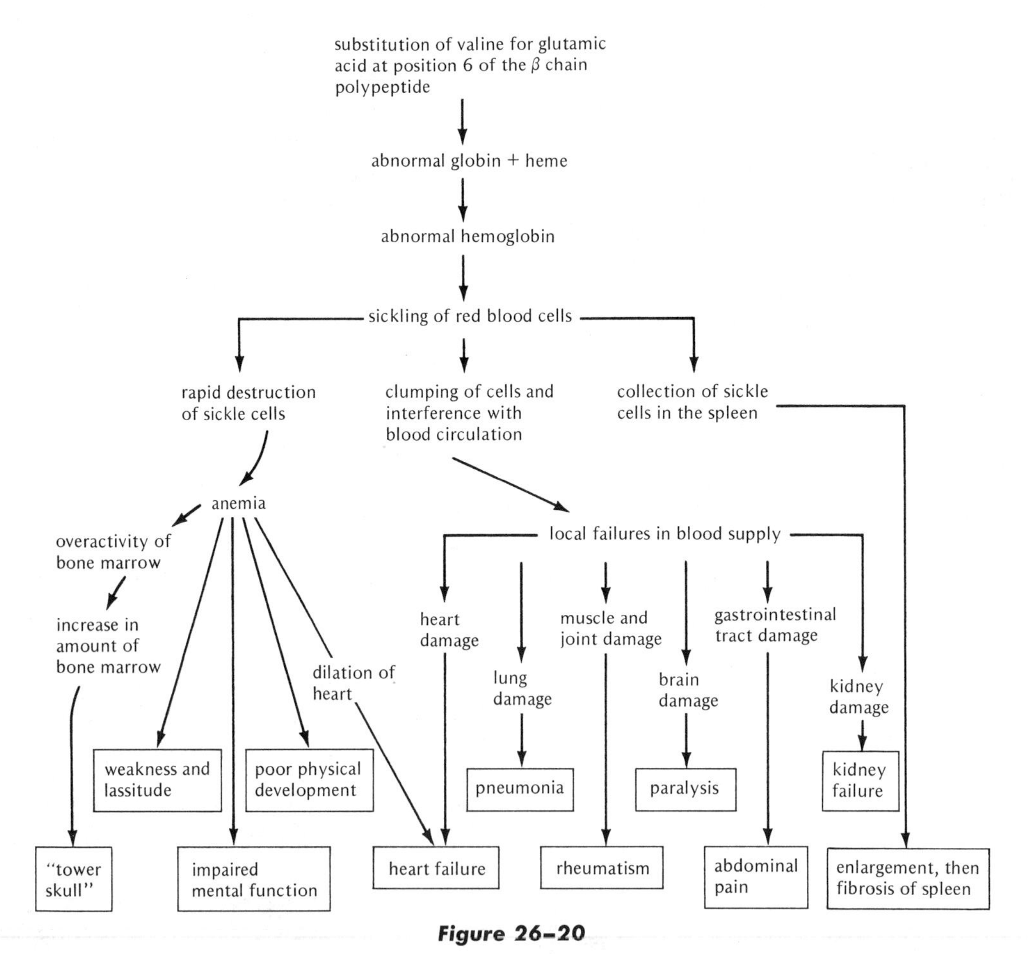

Figure 26–20

Possible consequences of the single amino acid substitution that occurs in hemoglobin in sickle-cell homozygotes. Among American blacks, the disease kills about one-half such homozygous individuals before the age of 20. (After Neel and Schull.)

virus that differ in amino acid composition yet retain their general mode of infection and duplication. Whether an amino acid change leads to a phenotypic change probably depends upon the particular protein involved as well as the particular point at which the amino acid substitution takes place.

GENE SEQUENCE AND ENZYME SEQUENCE

The separation of the *rII* locus of bacteriophage T4 into two cistrons, *A* and *B* (p. 589), indicates that two functional products affecting the same process (rapid lysis) are closely linked. The exact functional products of the *rII* gene, although identified, are still difficult to isolate, and the detailed relationship between the genes and enzymes involved in the same metabolic sequence has therefore been investigated most closely in those cases in which both genetic and protein analysis can more easily be accomplished. Tryptophan metabolism in bacteria, for example, is known to be affected by a number of enzymes involved in the sequential conversion of chorismic acid to tryptophan (Fig. 26–6). At the same time Demerec and Hartman have shown that the genetic loci affecting these enzymes in *Salmonella typhimurium* are "clustered" together and can be arranged in the same genetic order as their metabolic sequence.

An even more detailed correspondence between a sequence of genes and a sequence of organic synthesis was discovered by Hartman, Ames, and their collaborators for the histidine region of *Salmonella*. Through transduction techniques more than 1000 individual mutations have been mapped and found to be clustered into nine closely linked genetic loci, *A* to *I*, each affecting a different enzyme in the synthesis of histidine, and one gene affecting two different enzymes (Fig. 26–21). These genes map in the order *E-I-F-A-H-B-C-D-G*, that is, the gene sequence follows largely the histidine metabolic sequence, with the basic exception that the last gene, *G*, controls the first enzyme, pyrophosphorylase. Similar clustering of genes with related functions has been observed for proline, threonine, and isoleucine-valine metabolism in *Salmonella* as well as for tryptophan metabolism and arabinose fermentation in *E. coli*.

In bacteriophage the effect of mutant genes can be observed by breaking open the bacterial host and determining phage defects under the electron microscope or through the immunological reactions of phage proteins with specific antibodies. Some mutations will affect the ability of the phage to produce enzymes that are used in the synthesis of phage DNA and there will be no increase in DNA after phage infection. Others will affect enzymes involved in the synthesis of phage head proteins, phage tail proteins, etc. By using a variety of phage-T4 conditional temperature-sensitive mutants (can grow at 25°C but not at 42°C) as well as *amber* mutants (can grow on *E. coli* strain CR 63 but not on *E. coli* strain B) R. H. Epstein and his co-workers were able to classify each of these mutations phenotypically. When seen together with their linkage relationships, as in Fig. 26–22, genes with related functions show a marked clustering on the chromosome (see also Fig. 20–12).

Although striking and informative these correlations between gene and enzyme sequences have not been universally demonstrated. In *Neurospora*, for example, many mutations affecting the synthesis of the same amino acid are known to be on different linkage groups. The *Salmonella*, *E. coli*, and bacteriophage gene-enzyme sequences may therefore be indicative of a primitive evolutionary situation in which biochemical synthesis must be organized on the chromosome itself rather than through interaction among many cellular components in the cytoplasm.

COLINEARITY

In 1958 Crick proposed the hypothesis that DNA determines the sequence of amino acids in a polypeptide, and once this sequence is determined, the three-dimensional structure of the protein is determined. Fundamental to this relationship between the gene and the polypeptide is that they are both linear structures, in one case

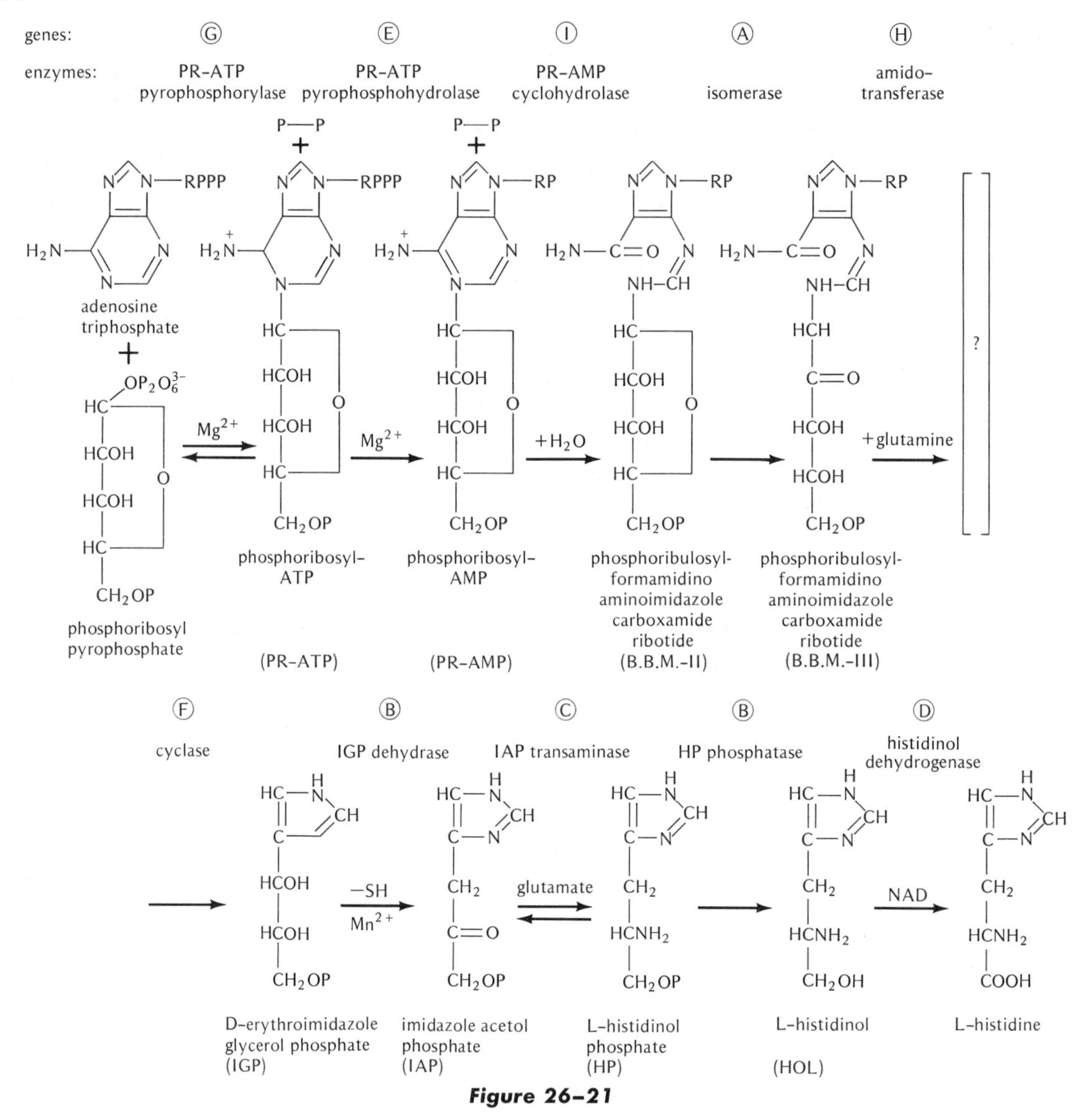

Figure 26–21

Biochemical pathway for the synthesis of histidine in Salmonella typhimurium. *Steps in the synthesis are governed by particular genes, beginning with Ⓖ and ending with Ⓓ. One gene, Ⓑ, affects two steps. (After Ames and Hartman, and other sources.)*

a sequence of nucleotides, in the other case a sequence of amino acids. If Crick's hypothesis were correct, it seems therefore obvious that two consequences should follow: (1) A linear sequence of nucleotides should determine a specific amino acid; as will be discussed later (Chapter 28), this relationship is specified by the genetic code, and this code is now almost entirely known. (2) A second most important consequence of the gene-polypeptide relationship is that a mutational change in a particular position of the nucleotide sequence should produce a change in a corresponding linear position in the amino acid sequence. That is, the nucleotide and amino acid sequences should be *colinear*; a genetic map of nucleotide changes should correspond with a mutational map of amino acid changes. For example, four nucleotide changes in a gene, *A*, *B*, *C*, and *D*, which are in the linkage order *A-B-C-D*, should affect the amino acid sequence in the same order. Thus, if each nucleotide change produces a specific amino acid change in the polypeptide (i.e., *A* nucleotide → *a* amino acid, *B* → *b*, etc.), the amino acid changes should be mapped on the polypeptide in the linear order *a-b-c-d*.

Evidence for colinearity therefore depends upon knowledge of both genetic and polypeptide sequences. In at least two important instances so far

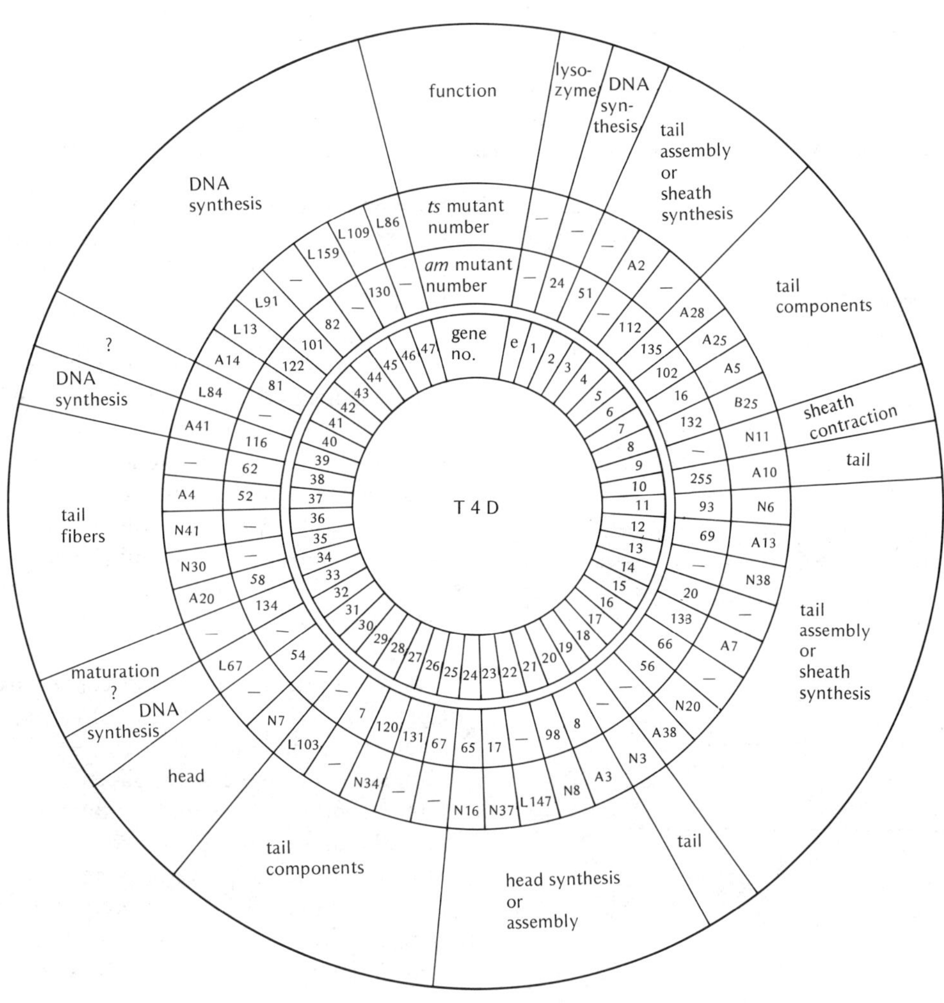

Figure 26–22

Diagrammatic presentation of the genetic (inside circles) and functional (outside circle) maps of bacteriophage T4 strain D. (From Jawetz, Melnick, and Adelberg, after various sources.) Further studies by Mosig and others have added considerable genetic detail to the T4 map (see Fig. 20–7), and continue to show close linkage between many genes with related function.

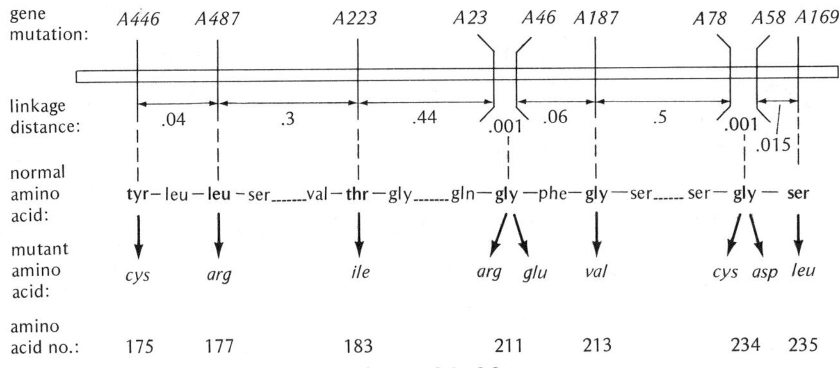

Figure 26–23

A partial genetic map of the tryptophan synthetase A cistron of E. coli (top) and the corresponding effects of these mutations on an amino acid sequence in the α polypeptide chain of the enzyme. This polypeptide has a total of 268 amino acids, and the numbers given in the figure refer to amino acid positions from one end (the N-terminal). (After Yanofsky et al.)

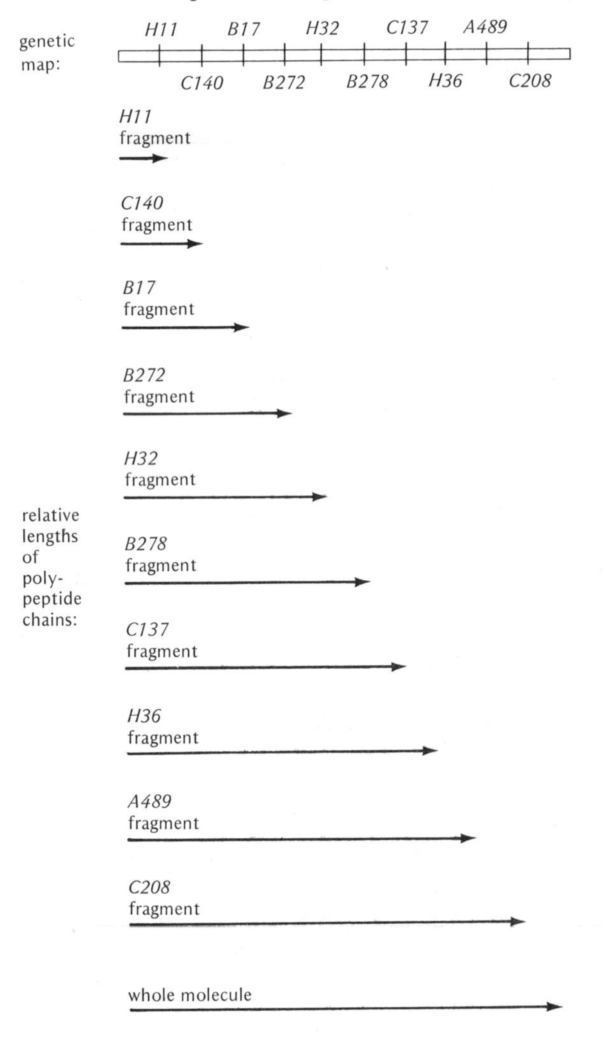

this knowledge has been demonstrated and is considered powerful evidence for colinearity. In one instance Yanofsky and co-workers utilized the *A* polypeptide chain of the tryptophan synthetase enzyme of *E. coli* (see p. 607). This chain, 267 amino acids long, shows a number of mutations whose genetic loci have been mapped in great detail. Corresponding to a section of this genetic map is a known amino acid sequence of protein A, in which single amino acid changes can be assigned to specific mutations. As Fig. 26–23 shows, the linear order of amino acid substitutions caused by the different mutations follows the genetic sequence in which these mutations occur on the bacterial chromosome.

A second demonstration of colinearity was performed by Sarabhai, Stretton, Brenner, and Bolle. Their study made use of a number of different T4 mutants, which have been collectively called *amber* mutants (p. 623). These phage mutants have in common the ability to grow on strains of *E. coli* that carry particular suppressor genes but are unable to grow on bacterial strains without these suppressors. According to various investigators (Benzer and Champe, and others) it is now clear that these

Figure 26–24

Correlation between length of a polypeptide chain (head protein) produced by different T4 amber mutants and the position of the mutation on the genetic map (above). (After Sarabhai et al.)

mutations produce only a partial polypeptide chain and that the action of the suppressor permits the chain to be completed (see p. 668). Sarabhai and co-workers found that the different *amber* mutants they used produced uniquely different lengths of the polypeptide chain involved in the head protein of phage T4. These different mutant polypeptide chains could then be ranked in order according to length. For example, the polypeptide product of the *A* mutant was longer than that of the *B* mutant, which was, in turn, longer than that of the *C* mutant, in the hierarchical order $A > B > C > D > E$, etc. Evidence for colinearity is seen in the fact that this order corresponds exactly to the position of the corresponding *amber* mutant on the chromosome; e.g., the linkage map is *A-B-C-D-E*. Thus each different *amber* gene can be considered to affect a point on the polypeptide chain that corresponds to its relative chromosomal position (Fig. 26–24).

PROBLEMS

26–1. Biochemical analysis shows that a particular protein molecule has four identical N-terminal amino acid residues and four identical C-terminal residues. What suggestions would you make as to its structure?

26–2. A biochemist analyzed a peptide containing only four amino acids, each of a different kind; valine, alanine, serine, and histidine (but not necessarily in that order). After degradation three kinds of dipeptides were recovered: val-ala, val-his, ser-his. What is the amino acid sequence in the peptide?

26–3. A peptide containing 10 amino acids is known to have valine at its N-terminal, but the sequence of the other amino acids is not known. This peptide is then treated with trypsin and chymotrypsin and each resulting fragment analyzed as to its N-terminal amino acid and remaining residues. In the case of trypsin, the carboxyl end produced by splitting is known to be either lysine or arginine. Determine the order of the 10 amino acids from the following fragments. (The N-terminal amino acid of each fragment is written to the left and connected to the remaining amino acids by a hyphen. Amino acids whose order is not known are enclosed in parentheses.)

trypsin treatment
  val-(his, leu) lys
  ala-(gly, tyr) arg
  pro-ser
chymotrypsin treatment
  val-leu
  gly-(ser, arg, pro)
  his-(tyr, ala, lys)

26–4. According to the studies of Srb, Horowitz, and others, a number of different nutritional mutation groups in *Neurospora* can be made to grow by the addition of arginine to minimal medium. Some of these mutations will also grow (+ sign) by the addition of other substances to the medium as shown below. (a) What metabolic pathway would you suggest for the synthesis of arginine? (b) What product might accumulate in *arg-12* mutations? (c) It has been found that the *arg-1* mutation listed below can also utilize argininosuccinic acid for growth, but the *arg-10* mutation can utilize only arginine. Which of these two mutations comes earlier in the arginine metabolic pathway, and at what step of the pathway is argininosuccinic acid synthesized?

Mutational Group	Growth Response: Minimal Unsupplemented Medium	Glutamic Semialdehyde	Ornithine	Citrulline	Arginine
arg-8, -9	−	+	+	+	+
arg-4, -5, -6, -7	−	−	+	+	+
arg-2, -3, -12	−	−	−	+	+
arg-1, -10	−	−	−	−	+

REFERENCES

AMES, B. N., and P. E. HARTMAN, 1974. The histidine operon of *Salmonella typhimurium*. In *Handbook of Genetics*, Vol. 1, R. C. King (ed.). Plenum Press, New York, pp. 223–235.

ANFINSEN, C. B., 1973. Principles that govern the folding of protein chains. *Science*, **181**, 223–230.

BAGLIONI, C., 1963. Correlations between genetics and chemistry of human hemoglobins. In *Molecular Genetics*, Part I, J. H. Taylor (ed.). Academic Press, New York, pp. 405–475.

BEADLE, G. W., and B. EPHRUSSI, 1937. Development of eye colors in *Drosophila*: Diffusable substances and their interrelations. *Genetics*, **22**, 76–86.

BEADLE, G. W., and E. L. TATUM, 1941. Genetic control of biochemical reactions in *Neurospora*. *Proc. Nat. Acad. Sci.*, **27**, 499–506. (Reprinted in the collections of Gabriel and Fogel, Levine, Peters, and Taylor; see References, Chapter 1.)

BEET, E. A., 1949. The genetics of the sickle-cell trait in a Bantu tribe. *Ann. Eugenics*, **14**, 279–284.

BONNER, D. M., 1965. Gene-enzyme relationships. In *Genetics Today*, Vol. II, Proc. XI Intern. Congr. Genet., S. J. Geerts (ed.). Pergamon, Oxford, pp. 141–148.

BRENNER, S., 1955. Tryptophan biosynthesis in *Salmonella typhimurium*. *Proc. Nat. Acad. Sci.*, **41**, 862–863.

BÜCHNER, E., 1897. Alkoholische Gährung ohne Hefezellen. *Ber. Dtsch. Chem. Ges.*, **30**, 117–124. (Translated into English in abridged form and reprinted in the collection of Gabriel and Fogel; see References, Chapter 1.)

BUTENANDT, A., 1953. Biochemie der Gene und Genwirkungen. *Naturwiss.*, **40**, 91–100.

CANFIELD, R. E., and C. B. ANFINSEN, 1963. Concepts and experimental approaches in the determination of the primary structures of proteins. In *The Proteins*, Vol. I, H. Neurath (ed.). Academic Press, New York, pp. 311–378.

CASPARI, E., 1949. Physiological action of eye color mutants in the moths *Ephestia kuhniella* and *Ptychopoda serrata*. *Quart. Rev. Biol.*, **24**, 185–199.

CRICK, F. H. C., 1958. On protein synthesis. *Symp. Soc. Exp. Biol.*, **12**, 138–163.

DEMEREC, M., and Z. HARTMAN, 1956. Tryptophan mutants in *Salmonella typhimurium*. *Carneg. Inst. Wash. Publ. No. 612*, Washington, D.C., pp. 5–33.

DICKERSON, R. E., 1964. X-ray analysis and protein structure. In *The Proteins*, Vol. II, H. Neurath (ed.). Academic Press, New York, pp. 603–778.

DICKERSON, R. E., and I. GEIS, 1969. *The Structure and Action of Proteins*. Harper & Row, New York.

EPHRUSSI, B., 1942. Chemistry of "eye color hormones" of *Drosophila*. *Quart. Rev. Biol.*, **17**, 327–338.

EPSTEIN, C. J., R. F. GOLDBERGER, and C. B. ANFINSEN, 1963. The genetic control of tertiary protein structure: Studies with model systems. *Cold. Sp. Harb. Symp.*, **28**, 439–449.

EPSTEIN, R. H., A. BOLLE, C. M. STEINBERG, E. KELLENBERGER, E. BOY DE LA TOUR, R. CHEVALLEY, R. S. EDGAR, M. SUSMAN, G. DENHARDT, and A. LIELAUSIS, 1963. Physiological studies of conditional lethal mutations of bacteriophage T4D. *Cold Sp. Harb, Symp.*, **28**, 375–392. (Reprinted in Stent's collection, and in *Selected Papers in Biochemistry*, Vol. 2; see References, Chapter 1.)

FINCH, J. T., M. F. PERUTZ, J. F. BERTLES, and J. DÖBLER, 1973. Structure of sickled erythrocytes and of sickle-cell hemoglobin fibers. *Proc. Nat. Acad. Sci.*, **70**, 718–722.

GARROD, A. E., 1902. The incidence of alkaptonuria: A study in chemical individuality. *Lancet*, **2**, 1616–1620. (Reprinted in the collections of Boyer and of Levine; see References, Chapter 1.)

———, 1909. *Inborn Errors of Metabolism*. Oxford Univ. Press, Oxford. (Second edition, 1923.)

HARRIS, H., 1975. *The Principles of Human Biochemical Genetics*, 2nd ed. Elsevier, New York.

HARTMAN, P. E., Z. HARTMAN, R. C. STAHL, and B. N. AMES, 1971. Classification and mapping of spontaneous and induced mutations in the histidine operon of *Salmonella*. *Adv. in Genet.*, **16**, 1–34.

HAUROWITZ, F., 1955. *Biochemistry, An Introductory Textbook*. John Wiley, New York.

HUEHNS, E. R., N. DANCE, G. H. BEAVEN, F. HECHT, and A. G. MOTULSKY, 1964. Human embryonic hemoglobins. *Cold Sp. Harb. Symp.*, **29**, 327–331.

HUNT, L. T., M. R. SOCHARD, and M. O. DAYHOFF, 1972. Mutations in human genes: Abnormal hemoglobins and myoglobins. In *Atlas of Protein Sequence and Structure*, Vol. 5, M. O. Dayhoff (ed.). National Biomed. Res. Found., Washington, D.C. pp. 67–87.

INGRAM, V. M., 1957. Gene mutations in human hemoglobin: The chemical difference between normal and sickle-cell hemoglobin. *Nature*, **180**, 326–328. (Reprinted in the collection of Boyer, and in *Selected Papers in Biochemistry*, Vol. 3; see References, Chapter 1.)

———, 1963. *The Hemoglobins in Genetics and Evolution*. Columbia Univ. Press, New York.

JAWETZ, E., J. L. MELNICK, and E. A. ADELBERG, 1974. *Review of Medical Microbiology*, 11th ed. Lange Medical Publishers, Los Altos, Calif.

KENDREW, J. C., 1962. Side-chain interactions in myoglobins. *Brookhaven Symp. Biol.*, **15**, 216–226.

KENDREW, J. C., R. E. DICKERSON, B. E. STRANDBERG, R. G. HART, D. R. DAVIES, D. C. PHILLIPS, and V. C. SHORE, 1960. Structure of myoglobin, a three-dimensional Fourier synthesis at 2 Å resolution. *Nature*, **185**, 422–427.

KÜHN, A., 1948. Über die Determination der Forum—Struktur—und Pigment bildung der Schuppen bei *Ephestia kuhniella*. *Zeit. Arch. Entwick.-Mech. Org.*, **143**, 408–487.

LAWRENCE, W. J. C., 1950. Genetic control of biochemical synthesis as exemplified by plant genetics—flower colors. *Biochem. Soc. Symp.*, **4**, 3–9.

LAWRENCE, W. J. C., and V. C. STURGESS, 1957. Studies on *Streptocarpus*. III. Genetics and chemistry of flower color in the garden forms, species, and hybrids. *Heredity*, **11**, 303–336.

MOSIG, G., 1970. Recombination in bacteriophage T4. *Adv. in Genet.*, **15**, 1–53.

MURAYAMA, M., 1966. Molecular mechanism of red cell "sickling." *Science*, **153**, 145–149.

NEEL, J. V., 1949. The inheritance of sickle-cell anemia. *Science*, **110**, 64–66. (Reprinted in Boyer's collection; see References, Chapter 1.)

NEEL, J. V., and W. J. SCHULL, 1954. *Human Heredity*. Univ. Chicago Press, Chicago.

PAULING, L., H. A. ITANO, S. J. SINGER, and I. C. WELLS, 1949. Sickle-cell anemia, a molecular disease. *Science*, **110**, 543–548. (Reprinted in the collections of Taylor and of Boyer; see References, Chapter 1.)

PERUTZ, M. F., 1964. The hemoglobin molecule. *Sci. American*, **211**, November issue, pp. 64–76.

———, 1965. Structure and function of hemoglobin. I. A tentative atomic model of horse oxyhemoglobin. *Jour. Mol. Biol.*, **13**, 646–668.

PERUTZ, M. F., and H. LEHMANN, 1968. Molecular pathology of human hemoglobin. *Nature*, **219**, 902–909.

PHILLIPS, J. P., H. S. FORREST, and A. D. KULKARNI, 1973. Terminal synthesis of xanthommatin in *Drosophila melanogaster*. III. Mutational pleiotropy and pigment granule association of phenoxazinone synthetase. *Genetics*, **73**, 45–56.

SANGER, F., 1955. The chemistry of simple proteins. *Symp. Soc. Exp. Biol.*, **9**, 10–31.

SARABHAI, A., A. O. W. STRETTON, S. BRENNER, and A. BOLLE, 1964. Colinearity of the gene with the polypeptide chain. *Nature*, **201**, 13–17. (Reprinted in *Selected Papers in Biochemistry*, Vol. 3; see References, Chapter 1.)

SCHERAGA, H. A., 1963. Intramolecular bonds in proteins. II. Noncovalent bonds. In *The Proteins*, Vol. I, H. Neurath (ed.). Academic Press, New York, pp. 477–594.

STANBURY, J. B., J. B. WYNGAARDEN, and D. S. FREDERICKSON (eds.), 1972. *The Metabolic Basis of Inherited Diseases*, 3rd ed., McGraw-Hill, New York.

SUMNER, J. B., 1926. The isolation and crystallization of the enzyme urease. *Jour. Biol. Chem.* **69**, 435–441. (Reprinted in the collection of Gabriel and Fogel; see References, Chapter 1.)

TATUM, E. L., 1959. A case history in biological research. *Science*, **129**, 1711–1715.

YANOFSKY, C., G. R. DRAPEAU, J. R. GUEST, and B. C. CARLTON, 1967. The complete amino-acid sequence of the tryptophan synthetase A protein (α subunit) and its colinear relationship with the genetic map of the *A* gene. *Proc. Nat. Acad. Sci.*, **57**, 296–298.

ZIEGLER, I., 1961. Genetic aspects of ommochrome and pterin pigments. *Adv. in Genet.*, **10**, 349–403.

27
PROTEIN SYNTHESIS

Early concepts of development proposed by Weismann and DeVries considered the different kinds of cytoplasm in an organism to arise from the presence of special determinants in the nuclei of each tissue that could pass into the cytoplasm at certain times. Although Weismann called these determinants "biophores," and DeVries called them "pangens," they had in common an inherited mode of transmission as well as the ability to produce different tissues depending upon their special molecular constitution. With the discovery and elucidation of enzymes and proteins, these ideas were expanded in the early 1900s by Driesch, Verworn, Wilson, and others who proposed that the nucleus itself was a storehouse for enzymes and served as the main center for protein activity. A primary task since then has therefore been to demonstrate the role of the nucleus in protein formation. Does the nucleus produce proteins or does it only produce the determinants of proteins?

One of the first experiments testing the physical role of the nucleus on cytoplasmic growth was performed by Hämmerling on the unicellular alga *Acetabularia.* This single-celled organism, shaped at one stage of its life cycle like a slender 1 to 2 inch umbrella, has feet called *rhizoids*, one of which contains the nucleus. Hämmerling found it a simple matter to remove the nuclear rhizoid or to cut the stem in half without damaging the organism in any obvious fashion. Such enucleated fragments, surprisingly, could survive for months. Moreover, tests performed by Brachet, Chantrenne, Baltus, Clauss, and others have shown that enucleated algae will incorporate both radioactive carbon dioxide and amino acids to form proteins and known enzymes such as phosphorylase and invertase. Protein synthesis, however, occurs only for approximately 2 weeks after enucleation. After that time, carbon dioxide incorporation drops off sharply and net protein synthesis stops completely. Analysis of enucleated fragments of the protozoa *Amoeba proteus* has led to similar conclusions; the nucleus is not necessary for the *immediate* functions of the cell, although the activity of enucleated cells will sooner or later decrease. Because of these findings, Mazia, in 1952, proposed that the nucleus was mainly concerned with *replacing* rather than directly producing cellular activity.

NUCLEIC ACID–PROTEIN RELATIONSHIPS

If not all proteins are made in the nucleus, nor the cytoplasm completely reliant on the nucleus for protein synthesis, how are genes involved in protein formation? In a more basic way, we can ask this question in terms of the activity and transfer of DNA itself; that is, does DNA partake directly in protein synthesis by increasing in amount and passing into the cytoplasm? As we have seen previously (Chapter 4), except for the expected reduction that occurs in the gametes of diploid organisms, the amount of DNA is approximately constant for all cells. Also, there is no indication that DNA is present in the cytoplasm (with the exception of isolated organelles such as mitochondria and plastids), nor is there evidence that it occurs there during protein formation. An intermediary between DNA and proteins was therefore sought that could be found both in the nucleus and cytoplasm in variable amounts, depending on the extent of protein synthesis.

In the early 1940s, through the efforts of Brachet and Caspersson, attention was called to the fact that both nuclear and cytoplasmic RNA existed, and that there appeared to be a correlation between the amount of RNA and the extent of protein production. High-protein-producing cells, such as those of the liver, pancreas, silk gland, and frog oocytes showed much larger amounts of RNA than low-protein producers, such as kidney, heart, and lung cells. It appeared that the high-protein producers had specific cytoplasmic areas that stained densely with basic dyes and absorbed ultraviolet radiation at a wavelength similar to the nucleic acids. The amount of this *basophilia* in a tissue and the amount of RNA were shown to be directly proportional. Also, the enzyme that breaks down RNA (ribonuclease) caused a cessation of protein production and acted specifically to remove the basophilic areas.

This relationship between RNA and protein was further clarified when methods were devised by Claude, Hogeboom, and others making it possible to break down, or *fractionate*, cells into various portions that could be isolated and separately analyzed. The procedure involves breaking many cells to form a homogenate and then spinning this mixture at various speeds in a centrifuge to separate different fractions. Each such fraction contains specific cellular organelles and particles that share common factors such as size and weight. The homogenate was usually separated into four fractions, containing respectively nuclei, mitochondria, microsomes (small particles), and a remainder called the supernatant. Chemical analysis of the microsomal fraction showed that it contained most of the cellular RNA.

Labeling experiments were then undertaken by injecting radioactive amino acids into animals and observing which fractions of liver or pancreatic cells were most rapidly "labeled." In all cases the fastest uptake of radioactivity *in vivo* was in the microsomal fraction. When such tests were made directly with homogenates (*in vitro*), the results were strikingly similar. It was then shown that the labeled amino acids were connected by peptide linkages and incorporated into proteins. In other words, although all sections of the cell could form proteins, such compounds were primarily produced by the microsomal fraction.

The microsomal fraction, however, was not a simple homogeneous solution. Under the electron microscope it usually appeared to be composed of two structures: small dense granules, associated with larger membranes. The granules, named *ribosomes*, were identified by Palade and Siekevitz as portions of the endoplasmic reticulum in normal cells (p. 11) and could be separated from the membranes by treatment with deoxycholate. Of great significance was the finding that *it was these ribosomes that contained most of the RNA and performed protein synthesis.* In fact, for the various animal tissues examined, the rule was that the intensity of protein synthesis depended on the number of ribosomes present.

These investigations led to experiments in which the protein-forming ability of ribosomes could be tested in noncellular or cell-free systems with various other components. The most efficient of these systems was found to contain ribosomes; the nucleosides adenosine triphosphate, and guanosine triphosphate (for energy transfer); some enzymes; and a fraction isolated from the supernatant solution which contained RNA in soluble rather than in granular form. It therefore seemed clear that the ribosome is a protein factory in itself and, if not the only source, at least the most productive source of protein synthesis.

The question of how the ribosomes obtained messages from the nucleus as to the kind of protein to produce was at first difficult to answer. One early hypothesis proposed that DNA is used as a template upon which RNA is copied, and this RNA molecule then proceeds to cross the nuclear membrane. Once in the cytoplasm it condenses to form an RNA ribosome which then begins to synthesize a protein determined by its own RNA composition. In other words, each ribosome can only produce a specific message, i.e., "one gene–one ribosome–one protein."

Although this theory seems to be a simple and direct account of gene-protein relationship, there are numerous indications that it does not truly reflect the situation in the cell. A basic objection is the fact that ribosomal products are not synthesized independently of the nucleus even after ribosomes are formed. Destruction of bacterial DNA, for example, results in a rapid drop of protein produced on the ribosome, indicating that additional "messages" are necessary for the ribosomes to continue protein production. Also, new protein synthesis seems, in some cases, far too rapid a process to await the synthesis of intermediary ribosomes. For example, in *E. coli* an F′ factor bearing the β-galactosidase gene (F′ lac^+, see p. 421) causes maximum enzyme synthesis within a few minutes of entry into a recipient female cell (F^- lac^-).

Furthermore, if ribosomes are simple RNA units producing different proteins, they should vary in size and shape depending on the kind of protein they produced. All measurements, however, indicate that ribosomes are fairly uniform in size and appearance. Under the electron microscope they are generally somewhat spherical structures, about 200 or so angstroms in diameter, and are constituted in *E. coli* of about 60 percent ribosomal RNA (rRNA) and a remainder consisting of about 40–50 proteins (see Fig. 27–10). They can also be uniformly characterized by their rate of sedimentation in the ultracentrifuge, measured in terms of the sedimentation constant, *S*. Such measurements can differentiate between two ribosomal subunits in *E. coli* and other procaryotic cells; the larger component having a 50*S* constant, and the smaller one showing a 30*S* sedimentation constant. Depending upon the concentration of magnesium ions and other factors, these subunits can be found separately or associated together in 70*S* units (Fig. 27–1). In eucaryotes cytoplasmic ribosome sizes are somewhat larger (40*S* and 60*S* components that associate into 80*S* units) and the proportions of chemical constituents are somewhat different (40% rRNA:60% protein) but, again, they all appear uniform in different tissues and different organisms. Only the ribosomes of mitochondria and chloroplasts differ from cytoplasmic ribosomes in

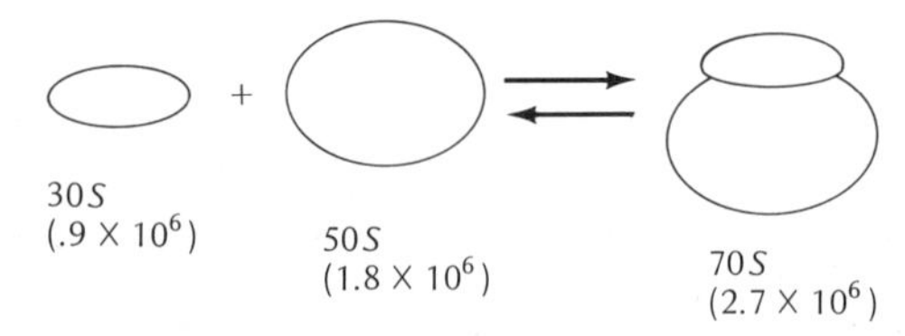

Figure 27–1

Diagrammatic appearance of an E. coli *ribosome based on electron microscope studies. The sedimentation coefficient S and approximate molecular weight (in parentheses) is given below each particle. The overall dimensions of the ribosome (70S) under hydrated conditions have been reported as about 135 × 220 × 400 angstroms. According to Tischendorf et al. the 30S subunit is bent and asymmetric, sitting like a "telephone receiver" on the 50S subunit.*

eucaryotes, with size and properties similar to procaryotic ribosomes.

MESSENGER RNA

To account for the involvement of ribosomes in protein synthesis and at the same time explain their nevertheless stable and uniform constitution, Jacob and Monod proposed that the ribosomes were not formed anew for each new protein synthesis, but were occupied by different transient RNA molecules at different times. According to their theory, an RNA "messenger" coming from the nucleus arranged itself in some way on an unoccupied ribosome and had its "message" translated into a protein. After a few translations the RNA "record" would wear out and the ribosome would then be free to pick up a new messenger.

The likelihood that messenger RNA (mRNA) was of a different type than rRNA had already appeared in a few findings. First, mRNA was usually considerably less stable than rRNA. Hershey, for example, had long noted that although phage infection of *E. coli* causes rapid synthesis of RNA molecules (mRNA), there is no net accumulation of RNA, since this same fraction undergoes a rapid breakdown. Davern and Meselson later labeled ribosomal RNA in *E. coli* with the "heavy" isotopes ^{13}C and ^{15}N and showed that rRNA remained stable for at least two generations, a much longer interval than the few minutes of stability shown by bacterial mRNA. Second, neither the base ratios of ribosomal nor soluble RNA had the same obvious relationship to DNA as did the base ratios of unstable mRNA (see the findings of Volkin and Astrachan, p. 100). As Fig. 27–2 shows, changes in guanine and cytosine content of DNA produce corresponding changes only in the GC content of mRNA. This finding would, of course, be expected if ribosomal and soluble RNA were formed from only small sections of the DNA molecule, whereas mRNA was formed from samples of all DNA sections.

Evidence demonstrating completely different

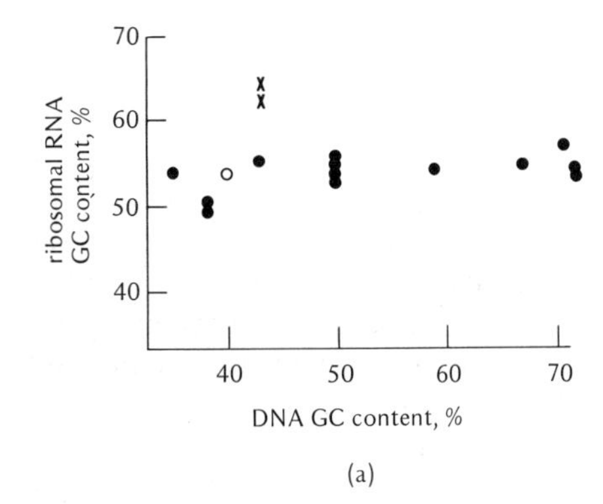

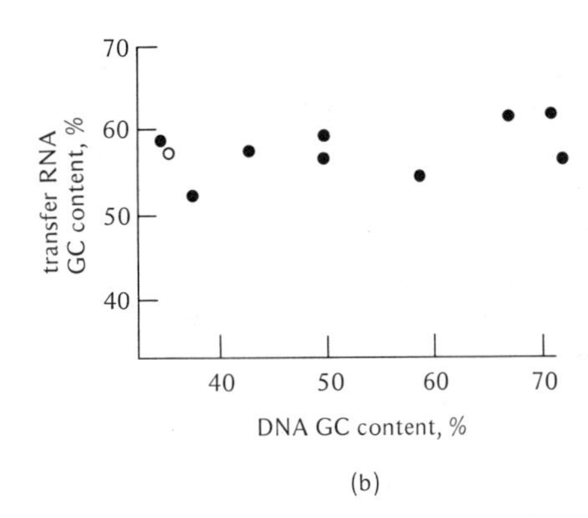

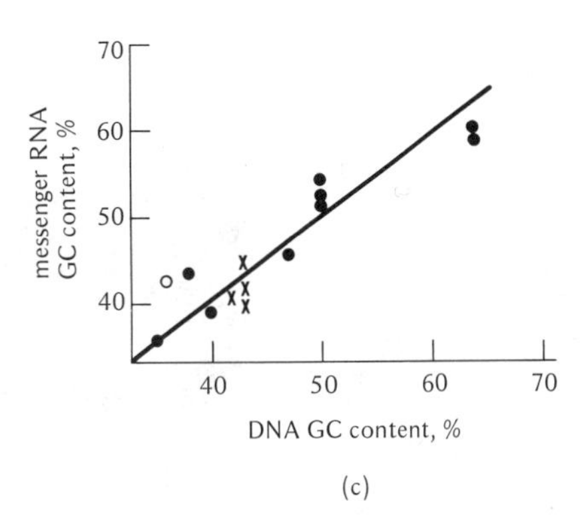

Figure 27–2

Relationship between guanine (G) and cytosine (C) content in different bacterial RNAs to the content of guanine and cytosine in DNA. Note that only in mRNA is there a proportional relationship; i.e., the greater the GC content in DNA, the greater the GC content of mRNA. (After Sueoka.)

types of activity for ribosomal and messenger RNA then came from two sources: phage-infected and uninfected bacterial cells.

The experiments of Brenner, Jacob, and Meselson were based on the knowledge that a bacteriophage, soon after infecting a cell, causes bacterial protein production to cease and viral protein production to commence. If the ribosomes of bacteria could be labeled before phage infection and then placed on an unlabeled medium after phage infection, the association of ribosome labeling with both mRNA and newly synthesized proteins would indicate a distinct role for mRNA as well as the use of old ribosomes in phage protein production. The ribosome labels used by Brenner and coworkers were the heavy isotopes ^{15}N and ^{13}C, which could be detected by their optical density when spun in a centrifuge. In addition they also labeled the new mRNA formed after viral infection by placing the bacteria on a radioactive phosphorous (^{32}P) medium. In this way they kept track both of changes in ribosome composition (heavy isotopes changing to light isotopes) and new RNA production (^{32}P labeling). Their results showed that after viral infection, new ^{32}P labeled RNA was produced as expected, representing RNA "messengers" for viral protein production. However, in all cases the RNA messenger and the newly synthesized proteins were found associated with ribosomes whose composition consisted solely of the previously labeled isotopes ^{15}N and ^{13}C. Viral proteins were therefore produced on "old" bacterial ribosomes, although with the help of "new" mRNA.

The mRNA of uninfected bacterial cells was investigated by Gros and co-workers through rapid pulse labeling of newly synthesized RNA. This mRNA was found to be associated with 70*S* and 100*S* ribosomes, but could be separated from the ribosomes as a 14*S* fraction when the magnesium concentration was lowered. As expected in this case the mRNA base ratios were similar to those of bacterial DNA.

Evidence that such messenger RNA was actually formed from DNA templates was presented by Hall, Spiegelman, Geiduschek, and others, in experiments performed both *in vivo* and *in vitro*. In general the experiments followed the procedure of extracting newly synthesized mRNA and demonstrating that it would "hybridize" with the DNA that produced it (see p. 101) but none other. For example, mRNA formed by the DNA of T2 bacteriophage upon infection of *E. coli* would hybridize only with T2 DNA and not with DNA of its *E. coli* host.

Experiments soon followed showing that cells briefly exposed to a radioactive precursor of RNA formed molecules of labeled messenger RNA that

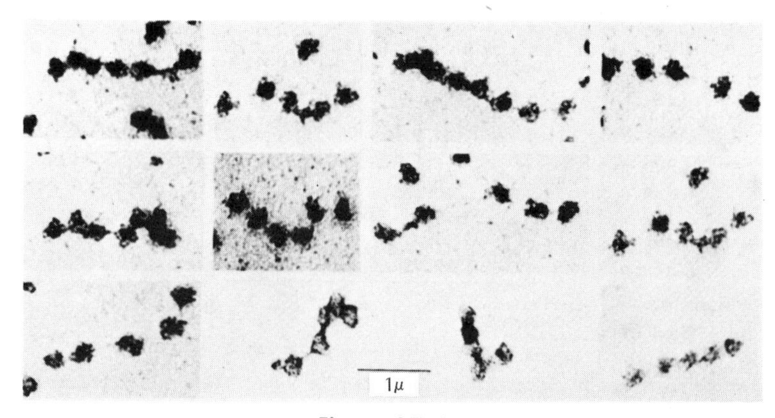

Figure 27–3

Electron micrographs of polysome complexes in reticulocytes producing hemoglobin. (From A. Rich, J. R. Warner, and H. M. Goodman, 1963. The structure and function of polyribosomes. Cold Sp. Harb. Symp., ***28,*** *269–285.)*

became reversibly joined to ribosomes. The average molecular weight of this messenger RNA was about 500,000 in *E. coli* bacteria, or of sufficient length to permit the attachment of numerous ribosomes. How many ribosomes were attached to an mRNA molecule? One of the first suggestions that mRNA carried more than one ribosome was the finding by Risebrough, Tissières, and Watson that mRNA-attached ribosomes appeared to have a higher sedimentation rate (i.e., were heavier) than unattached ribosomes. More precise evidence came from the findings of Barondes and Nirenberg, and others, utilizing synthetic mRNA (see Chapter 28) which could be shown by sedimentation data to attach groups of four or more ribosomes to each mRNA molecule. In mammalian reticulocyte cells Warner, Rich, and others have demonstrated by electron micrographs that as many as seven, eight, or more ribosomes align themselves linearly along the mRNA molecules to form a *polyribosome* or *polysome* (Fig. 27–3). Further experiments have shown that the polysome complexes are much more active in protein synthesis than single ribosomes; the polysomes are labeled faster by radioactive amino acids, and they also lose their labeling faster than do the single ribosomes. Thus the "turnover" of amino acids, a measure of protein synthesis, is greater in polysomes. However, since most mRNA persists only for a short time, polysome complexes are soon reduced to smaller independent ribosome components which, in turn, re-form into polysomes when a new strand of mRNA appears (Fig. 27–4).

The short-lived nature of most bacterial mRNA molecules appears, at first, puzzling and wasteful. In *E. coli* the stability of an mRNA molecule seems to last only as long as it is attached to a ribosome, and may have an average lifetime of only 2 minutes at 37°C. One important reason for

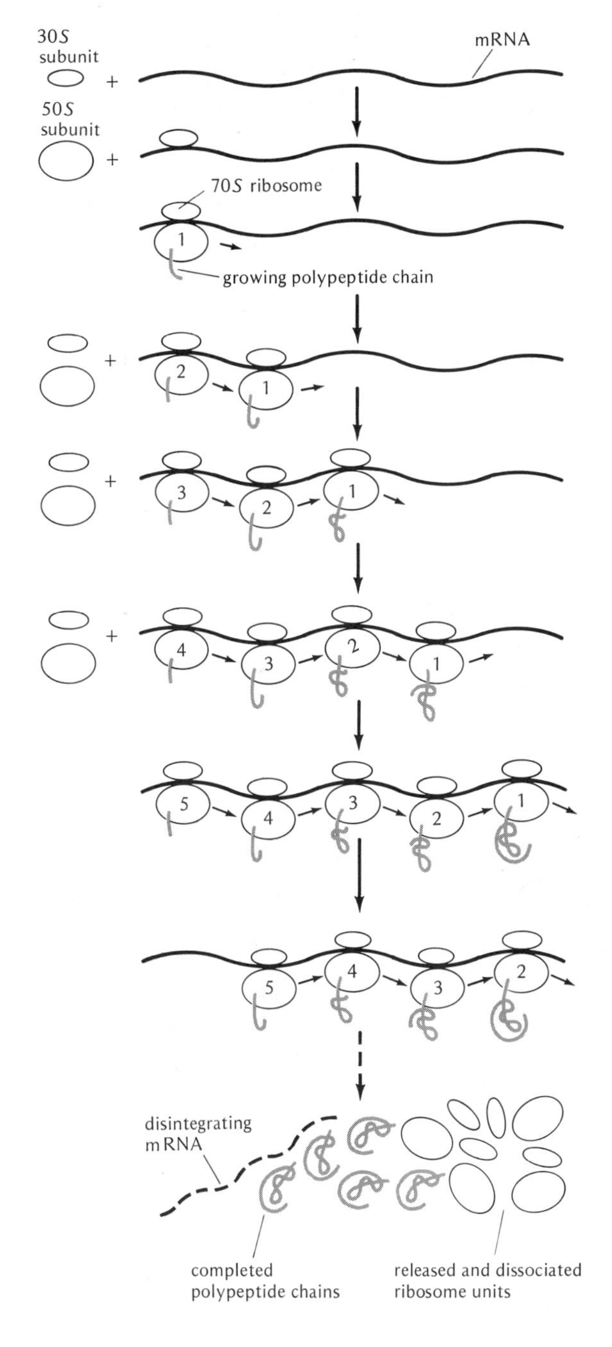

Figure 27–4

Mode of polysome formation in protein synthesis. Oncoming smaller ribosomal subunits attach at a specific initiation site on the mRNA molecule, forming a complex to which the larger subunit then attaches to form the functioning ribosome (further discussed on pages 641 ff.). Each ribosome then moves along the mRNA molecule synthesizing one continuous polypeptide chain. When it completes synthesis of a polypeptide, the ribosome separates from the mRNA and may then join a pool of 70S ribosome particles from which it is dissociated once more into 30S and 50S subunits by a specific dissociation factor.

this short mRNA life-span may be the obvious advantage it confers upon organisms that must produce a wide variety of proteins in the same cell, each such protein or group of proteins responding to the needs of the environment at a different time. Were mRNA molecules long-lived, it is easy to see that the bacterial cell would continue to produce all the proteins coded for by these molecules, many of which might be needless. Rapid mRNA turnover, however, is not universal. In the differentiated tissues of higher organisms the need for protein variety and a consequent variety of mRNA molecules may be restricted in any particular cell. Blood-cell reticulocytes, for example, produce hemoglobin throughout most of their life-span, and the mRNA responsible for this protein is stable enough to persist even though the nucleus of the mammalian blood cell disintegrates.

Essentially, two intermediary processes are involved in the transformation of DNA sequences into proteins:

1. A *transcription* process in which mRNA is copied or transcribed from a DNA strand through the enzymatic action of DNA-directed RNA polymerase (Chapter 5).
2. A *translational* process in which a particular nucleotide sequence on mRNA is translated into a particular amino acid sequence with the help of the ribosomes.

How is translation accomplished?

TRANSFER RNA AND PROTEIN SYNTHESIS

Discoveries made in the 1950s by Lipmann, Hoagland, Zamecnik, and others pointed to two important preliminary steps in the incorporation of amino acids into proteins. The first of these involves the "activation" of amino acids in cellular extracts through their attachment to adenosine triphosphate (Fig. 27–5). These attachments form highly reactive amino acid phosphate–adenyl groups, called *aminoacyl adenylates*. The enzymes that form these groups are specific under most circumstances for particular amino acids. Thus each of the 20 or so different amino acids has its own activating enzyme or enzymes. However, although activated, the aminoacyl adenylates are strongly bound to these enzymes and are not by themselves able to attach to each other or to the ribosomal particles.

The second step in amino acid incorporation involves the presence of an intermediary between the adenylates and the ribosomes in the form of the free-floating soluble RNA fraction of the supernatant solution (p. 631). This soluble RNA has within it molecules, called *transfer RNA* or tRNA, that become attached to radioactively labeled amino acids and then transfer these amino acids to ribosomes. Particular fractions of tRNA have been shown to be specific for particular varieties of amino acids, some of these attaching only to alanine, others to leucine, and so on. In this attachment process the activating enzymes (now called *aminoacyl-tRNA synthetases*) perform two roles: that of attaching amino acids to ATP, then that of transferring the amino acids without the adenyl group to tRNA (Fig. 27–6).

Interestingly, Yamane and Sueoka have found that some activating enzymes may be isolated from one bacterial species and perform the same role of attaching a particular amino acid to the tRNA used for that purpose in another species. Similar interactions have now been shown to exist between eucaryotes and procaryotes, although the optimal rate of amino acid attachment is usually between the activating enzymes and tRNA of similar organisms (see Jacobson). In general these reactions indicate that although some minor changes occurred in aminoacyl synthetases and tRNAs during evolution, the mutual recognition sites of many tRNA molecules and their activating enzymes have been conserved. Exactly where these recognition sites are located on tRNA is not yet fully known, and indications exist that the different aminoacyl tRNAs do not all use the same combinations of sites.

The role of tRNA in protein synthesis therefore seems to be twofold: on the one hand, to carry a specific amino acid, and, on the other hand, to attach itself to the ribosomes in accord with the sequence specified by mRNA. These findings thus

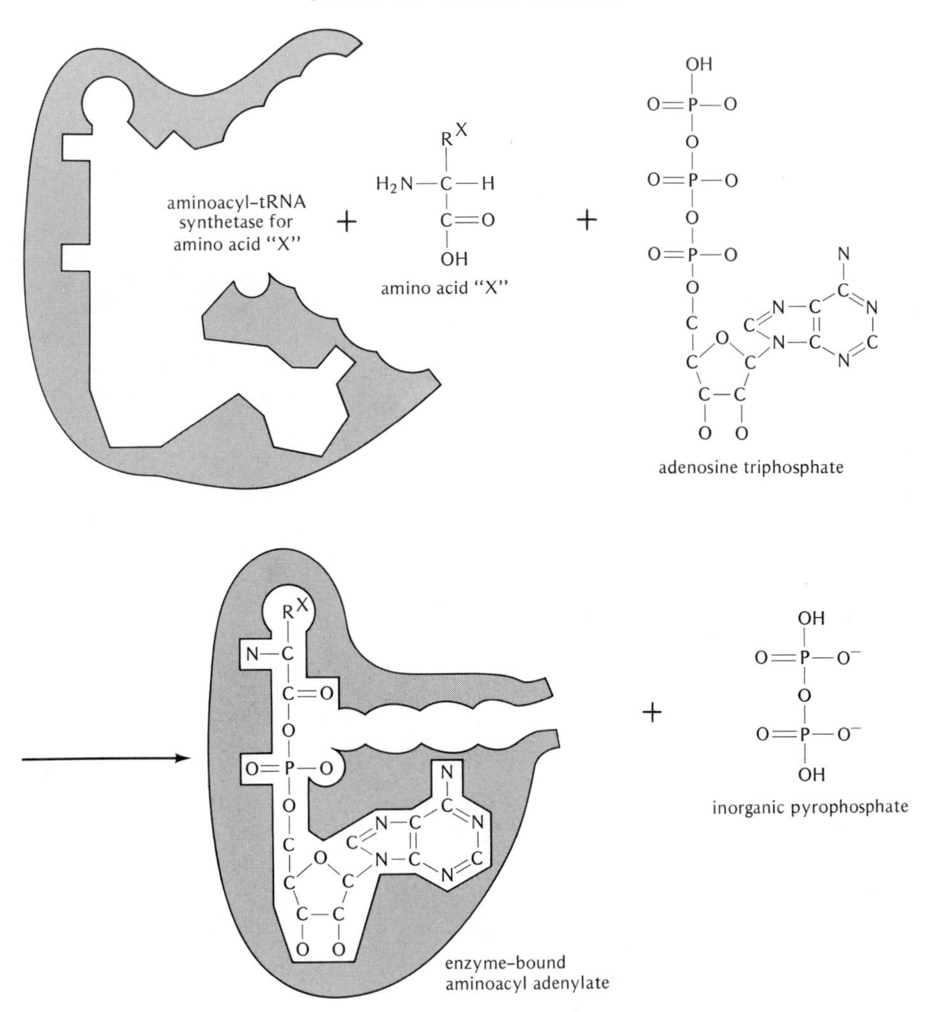

Figure 27–5

Attachment of amino acid "X" to adenosine triphosphate by an activating enzyme specific for that particular amino acid. Although portrayed in a highly schematic fashion in this illustration, the aminoacyl synthetase enzyme "X" has specific recognition sites for the ATP molecule and for the specific amino acid that it activates. The circle around the amino acid subunit R^X (see p. 610) indicates the specific enzyme-amino acid recognition site. (For simplicity, many of the H atoms in this and subsequent figures have been omitted.)

support a hypothesis previously presented by Crick in which each type of amino acid attaches to its own "adaptor," which then aligns itself at the ribosome on mRNA.*

The fact that it is the tRNA rather than the individual amino acid that attaches to mRNA was shown in two ingenious experiments. In one of these Chapeville, Lipmann, and others isolated tRNA labeled with radioactive (^{14}C) cysteine and then removed the cysteine sulfur group with a special catalyst (Raney nickel), thereby transforming the cysteine amino acid to alanine (Fig. 27–7). Since this newly formed alanine has not been removed from the original cysteine-tRNA, it can

* See discussion, 1957, *Biochem. Soc. Symp.*, **14**, 25–26.

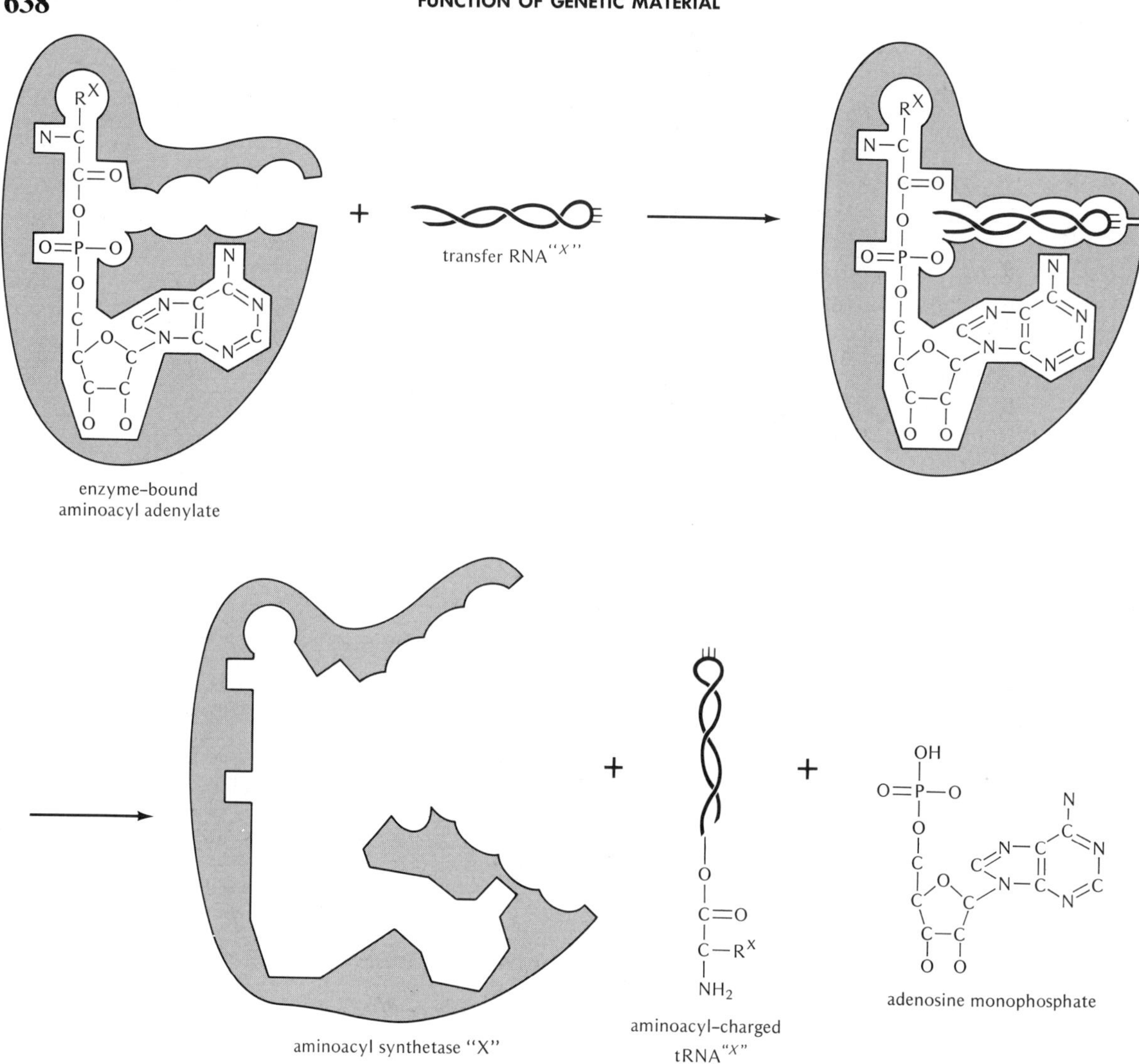

Figure 27–6

Transfer of enzyme-bound amino acid "X" to tRNA^X. Note that the aminoacyl-tRNA synthetase enzyme "X" has a specific site which recognizes only tRNA^X. Because of this specificity and that for amino acid "X", only amino acid "X" becomes bound to tRNA^X. (The drawing is not to scale: the enzyme and tRNA units are relatively much larger than pictured. The exact nucleotide sequence of a tRNA molecule is given in Fig. 27–8.)

be symbolized as ^{14}C alanyl-tRNA^{cys}. They also isolated radioactively labeled alanine associated with its normal tRNA, alanyl-tRNA^{ala}. Each of these were then added to separate solutions containing a synthetic messenger RNA that was known to form polypeptides *containing cysteine but not alanine* (see poly-UG, p. 659). The results showed that alanine was incorporated into the system *only when attached to the cysteine-tRNA.* Alanine attached to its own alanine-tRNA was not incorporated.

Using a similar technique, von Ehrenstein, Weisblum, and Benzer showed that ^{14}C alanyl-tRNA^{cys} placed in a system producing hemoglobin

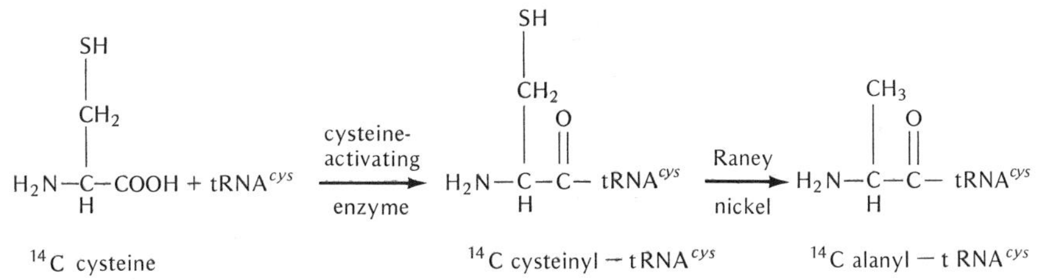

$$\underset{^{14}\text{C cysteine}}{H_2N{-}\overset{\overset{\displaystyle SH}{|}\atop \overset{\displaystyle CH_2}{|}}{\underset{H}{C}}{-}COOH} + tRNA^{cys} \xrightarrow[\text{enzyme}]{\text{cysteine-activating}} \underset{^{14}\text{C cysteinyl} - tRNA^{cys}}{H_2N{-}\overset{\overset{\displaystyle SH}{|}\atop \overset{\displaystyle CH_2}{|}}{\underset{H}{C}}{-}\overset{O}{\overset{\|}{C}}{-}\ tRNA^{cys}} \xrightarrow[\text{nickel}]{\text{Raney}} \underset{^{14}\text{C alanyl} - t\ RNA^{cys}}{H_2N{-}\overset{\overset{\displaystyle CH_3}{|}}{\underset{H}{C}}{-}\overset{O}{\overset{\|}{C}}{-}\ tRNA^{cys}}$$

Figure 27–7

Transformation of tRNA^{cys} carrying cysteine to tRNA^{cys} carrying alanine by the action of Raney nickel.

proteins was only incorporated into that peptide fraction that normally contains cysteine and not alanine. Peptides containing alanine but not cysteine received nonlabeled alanine and therefore showed no radioactivity. *In other words, the "message" for incorporation of amino acids into proteins resides solely in the nucleotide configuration of tRNA, not in its associated amino acid.*

In *E. coli* tRNA appears to comprise approximately 10 to 15 percent of all cellular RNA and has a molecular weight of about 25,000, equivalent to a chain 75 to 80 nucleotides long. Holley and co-workers were first to sequence the nucleotides of a tRNA molecule and considerable information rapidly accumulated. tRNA seems to have a large portion of its structure in the form of a double helix, and also contains a number of rare bases such as pseudouridine (Ψ) and inosine (I), as well as some normal bases to which methyl groups have been added. Some of these unusual nucleotides are unable to hydrogen-bond with other bases and therefore produce looped sections in which the double helical structure of tRNA is interrupted. All tRNA molecules appear to possess the four loop structures shown in Fig. 27–8, as well as an exposed unpaired sequence of three nucleotides, C–C–A, at the 3′ end. It is at this end that one of the hydroxyl groups (3′) in the ribose portion of the terminal adenine nucleotide serves as the "acceptor" point to which a single amino acid is covalently bonded by the particular amino-acid-activating enzyme.

Since the code is triplet (Chapter 28), three nucleotides of each tRNA molecule are used for coding purposes to pair with triplet sequences of mRNA. If we name the coding triplet of mRNA the *codon*, the complementary triplet on tRNA can be called the *anticodon*. It is now clear that the anticodon lies exposed in the loop indicated in Fig. 27–8, at a uniform distance of about 66 angstroms from the 3′ acceptor end. The constancy of this distance ensures that incoming amino acids brought to mRNA by their tRNA carriers will be positioned so as to allow a peptide bond to be formed most easily with another amino acid placed in close proximity by a previous tRNA molecule.

PROTEIN SYNTHESIS AND THE RIBOSOME

For tRNA to pair properly with mRNA, the ribosome serves the important intermediary function of stabilizing the trinucleotide attachment between mRNA and tRNA. Without the ribosome such an attachment would not be sufficiently strong or stable to permit the amino acids carried by tRNA to become linked together in peptide formation. Evidence for this is the fact that thymine trinucleotides (T–T–T) will not remain attached to polyadenylic acid (A–A–A—) for even a short period of time at temperatures above freezing. Various experiments indicate that the large 50*S* subunit of the ribosome serves as the stabilizing surface for the amino-acid-charged tRNA and the smaller 30*S* subunit serves as the stabilizing surface for the mRNA.

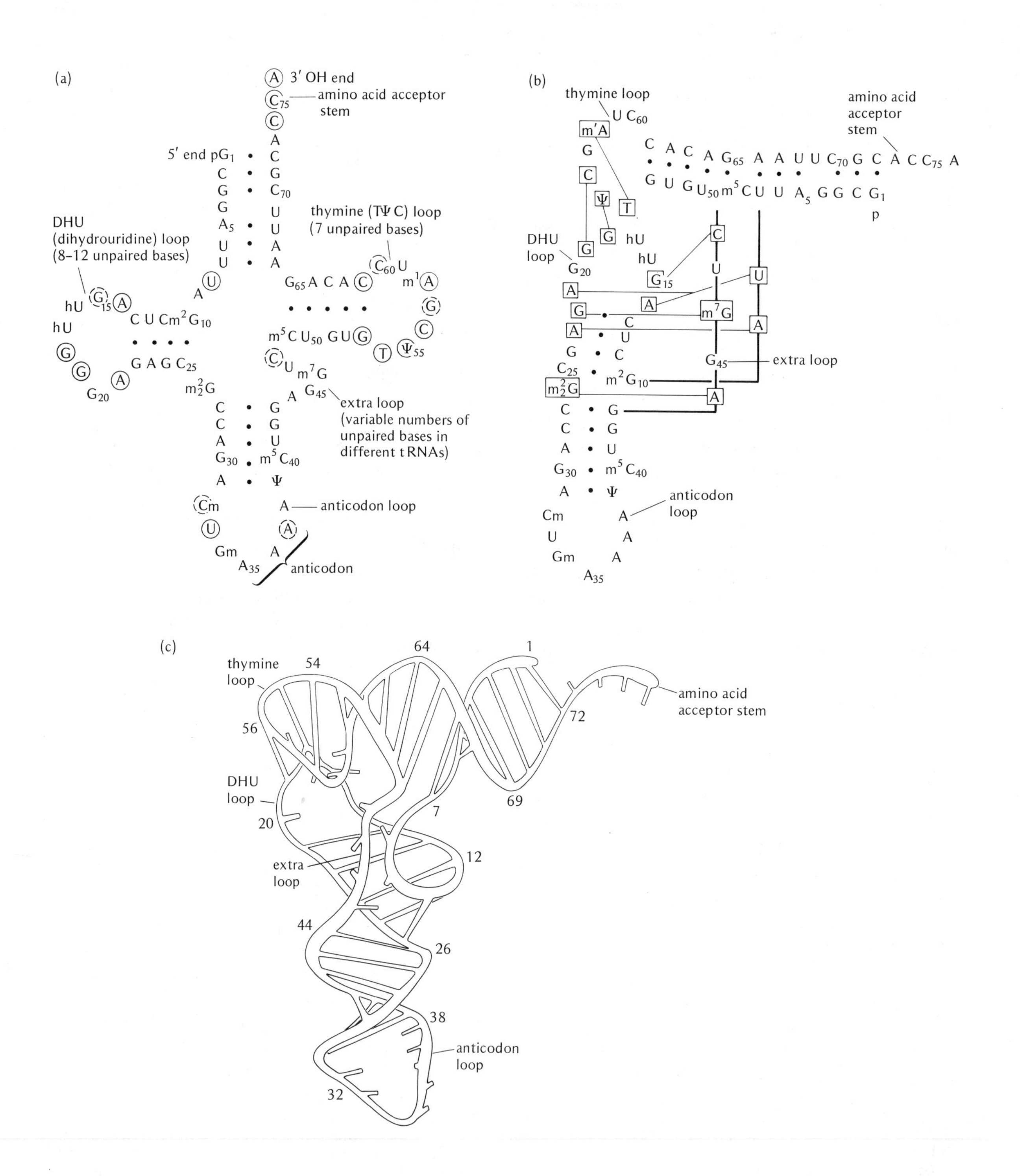

(a)
3′ OH end
amino acid acceptor stem
5′ end pG1
DHU (dihydrouridine) loop (8–12 unpaired bases)
thymine (TΨC) loop (7 unpaired bases)
extra loop (variable numbers of unpaired bases in different tRNAs)
anticodon loop
anticodon
(b)
thymine loop
amino acid acceptor stem
DHU loop
extra loop
anticodon loop
(c)
thymine loop
DHU loop
extra loop
anticodon loop
amino acid acceptor stem

Figure 27–8

(a) Sequence of the 76 nucleotides in phenylalanine transfer RNA of yeast shown in the commonly portrayed two-dimensional "cloverleaf" form. So far the more than fifty different kinds of tRNA molecules sequenced in a variety of organisms can be fitted into this same cloverleaf pattern, offering a maximum of pairing (dots) between complementary bases. The four major transfer RNA "loops" are indicated, including the anticodon loop which contains, in this case, the special sequence A-A-A that matches the messenger RNA codon for phenylalanine U-U-U. Transfer RNA molecules that are specific for other amino acids bear, of course, different anticodon sequences as well as different nucleotides in some of the other positions. In the present diagram, bases encircled with solid lines occupy the same positions in all the different tRNAs examined so far, whereas those with dashed circles are more variable, indicating base pair positions that are always occupied by either purines or pyrimidines. Numbering of the bases begins at the 5′ end. Unusual nucleotides found in tRNA include: hU = dihydrouridine (also designated DihU), ψ = pseudouridine, mx = methylated nucleotide, and T = thymine (i.e., a uracil methylated at the 5 position on the pyrimidine ring, see p. 50). (b) Schematic diagram showing the manner in which $tRNA^{phe}$ can be twisted to produce some of the physical relationships observed between various parts of the molecule. Thin lines indicate hydrogen bonds that are believed to arise between bases in the folded tertiary structure. (c) A three-dimensional representation of $tRNA^{phe}$ according to an x-ray diffraction study. The ribose-phosphate backbone of the nucleotide chain is shown as a coiled tube whose thickness indicates three-dimensional perspective: coils with greater thickness are closer to the reader than thinner coils. The crossbars indicate hydrogen bonds, including those that arise during folding of the molecule. (After Kim et al.)

Protein formation on ribosomes proceeds by a succession of steps that are practically unchanged in all organisms investigated so far. These steps are diagrammatically shown in Figure 27–9 and begin with the formation of an initiation complex between the smaller of the ribosomal subunits, mRNA, and a specific initiation transfer RNA molecule called $tRNA_f^{met}$.* The mRNA initiation site is always marked by a special codon, AUG, as well as by an, as yet, unknown secondary structure of the mRNA molecule which enables the initiation complex to form (Fig. 27–9b). In procaryotes some workers suggest that the initiation complex begins with the adhesion of a 30*S* ribosomal subunit to the initiation site on mRNA aided by the presence of protein *initiation factors.* To this complex the $tRNA_f^{met}$ molecule then attaches bearing the anticodon UAC (see, however, Noll and Noll). In eucaryotes there is evidence at present that the 40*S*-$tRNA_f^{met}$ combination is formed first, and the mRNA molecule then attaches with its AUG initiation codon. In both cases attachment of the larger ribosomal subunit immediately follows (Fig. 27–9c). When this occurs the completed ribosome now possesses two distinct attachment sites for tRNA molecules. One site, called the aminoacyl attachment site ("A"), is the point to which an entering tRNA molecule is bound, so that its anticodon can pair with the mRNA codon on the 30*S* particle, and the amino acid it carries can undergo peptide-bond formation on the 50*S* particle. The other site, called the peptidyl site ("P"), is extremely close, if not immediately parallel, to the A site, and serves as the site to which the entering tRNA molecule becomes "translocated" after peptide-bond formation.

In detail, a tRNA molecule bearing a specific amino acid enters at the A site (Fig. 27–9d), and then receives the amino acid(s) from the tRNA molecule at the P site through a peptide bond produced by the peptidyl transferase enzyme (Fig. 27–9e-g). The P-site tRNA molecule, having lost its amino acid(s), is now released from the ribosome (Fig. 27–9h), and the A-site tRNA molecule is translocated to the P site along with its newly attached chain of amino acids (Fig. 27–9i). Translocation of the tRNA molecule to the P site is

* In procaryotes this initiation tRNA molecule carries the amino acid methionine with a formyl group (CHO) attached to its amino nitrogen atom. The formyl group prevents peptide formation between the amino terminal of this methionine residue to the carboxyl terminal of any other amino acid. N-formylmethionine is thus restricted to function as an "N-terminal" amino acid, and is not ordinarily found at any other position on the nucleotide chain. In eucaryotes the $tRNA_f^{met}$ also carries methionine but it is not formylated. In both procaryotes and eucaryotes, peptidase enzymes usually cleave the initial methionine amino acid from the polypeptide chain after its formation. (Interestingly the fact that protein synthesis in mitochondria and chloroplasts begins with N-formylmethionine rather than methionine has been considered as evidence of a procaryotic origin for these eucaryotic organelles. Other interpretations, however, are also possible as indicated in papers cited in Chapter 13, p. 270.)

accompanied by a coordinated movement of the mRNA molecule along the ribosome, thus exposing a new mRNA codon at the A site.

For the entering tRNA to bind at the A site, and peptide-bond formation and elongation to occur, a number of protein *elongation factors* are necessary which use the ATP-like molecule guanosine triphosphate (GTP) as an energy source. The elongation cycle normally repeats itself until one of the termination or "nonsense" codons is reached: UAA, UAG, or UGA (Fig. 27–9j, k). Ordinarily no tRNA molecules bear anticodons to pair with these nonsense codons, so that the appearance of such codons in an mRNA message prevents elongation of the polypeptide chain. The termination codon, in fact, provides a signal to the ribosome for the attachment of one or more protein *release factors* which produce three immediate consequences: cleavage of the polypeptide chain from the tRNA molecule at the P site, release of the tRNA molecule from the ribosome, and dissociation of the ribosome into the two subunits (Fig. 27–9l, m).

In addition to the termination of polypeptide chain formation by nonsense codons, protein synthesis can also be stopped by chemical means. In bacteria, chloramphenicol blocks the attachment of the amino terminal on the entering amino acid to the 50*S* ribosomal subunit, thereby preventing peptide-bond formation. Erythromycin and the eucaryotic inhibitor cycloheximide interfere with translocation of the tRNA molecule from the P site, thus preventing further translation. Puromycin is also an effective antibiotic, since it has a chemical structure that enables it to "pose" as an amino-acid-charged tRNA molecule, and thus to accept the amino acid chain from the tRNA at the P site. However, since it carries neither a tRNA

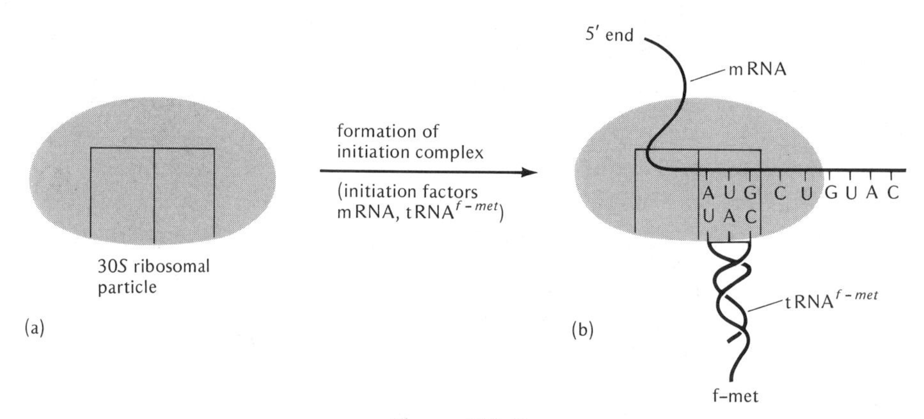

Figure 27–9

Scheme of protein synthesis beginning with the AUG initiation codon for tRNA$_f^{met}$ (also designated tRNA$^{f\text{-}met}$) and ending with a termination codon UAA. Although this illustration follows the sequence of events explained in the text, it is highly diagrammatic and the following should be noted: (1) The AUG codon does not ordinarily occur as the first set of nucleotides on an mRNA molecule. Various studies indicate that AUG is often preceded by a number of other nucleotides which probably enable the AUG initiator codon to occupy a specific position on the mRNA chain so that it can be recognized. (2) The UAA termination codon is not necessarily the last codon on mRNA, but may occur anywhere on the molecule. Thus a long mRNA molecule may possess a number of initiator and terminator codons, and may therefore produce a number of individual polypeptide chains (see p. 649). For E. coli mRNA specifying tryptophan synthetase proteins, Platt and Yanofsky have shown that the mRNA termination codon for the trp B gene is, in fact, also partially used as the initiation codon for the trp A gene: i.e., the mRNA nucleotide sequence is . . . trp B codons . . . UGAUG . . . trp A codons. . . . (3) Although the exact configuration of the tRNA molecule on the ribosome is not yet known, it probably changes in shape between the A and P sites. One purpose of this change is apparently to permit the amino acid(s) at the P site to come into close contact with the amino acid at the A site in order to allow the peptidyl transferase enzyme to function.

molecule nor a carboxyl terminal, it cannot attach its newly acquired peptide chain to the P site nor enable further peptide bonding to occur.

From all of the above considerations it is apparent that the crucial role played by the ribosome in protein synthesis depends upon its ability to

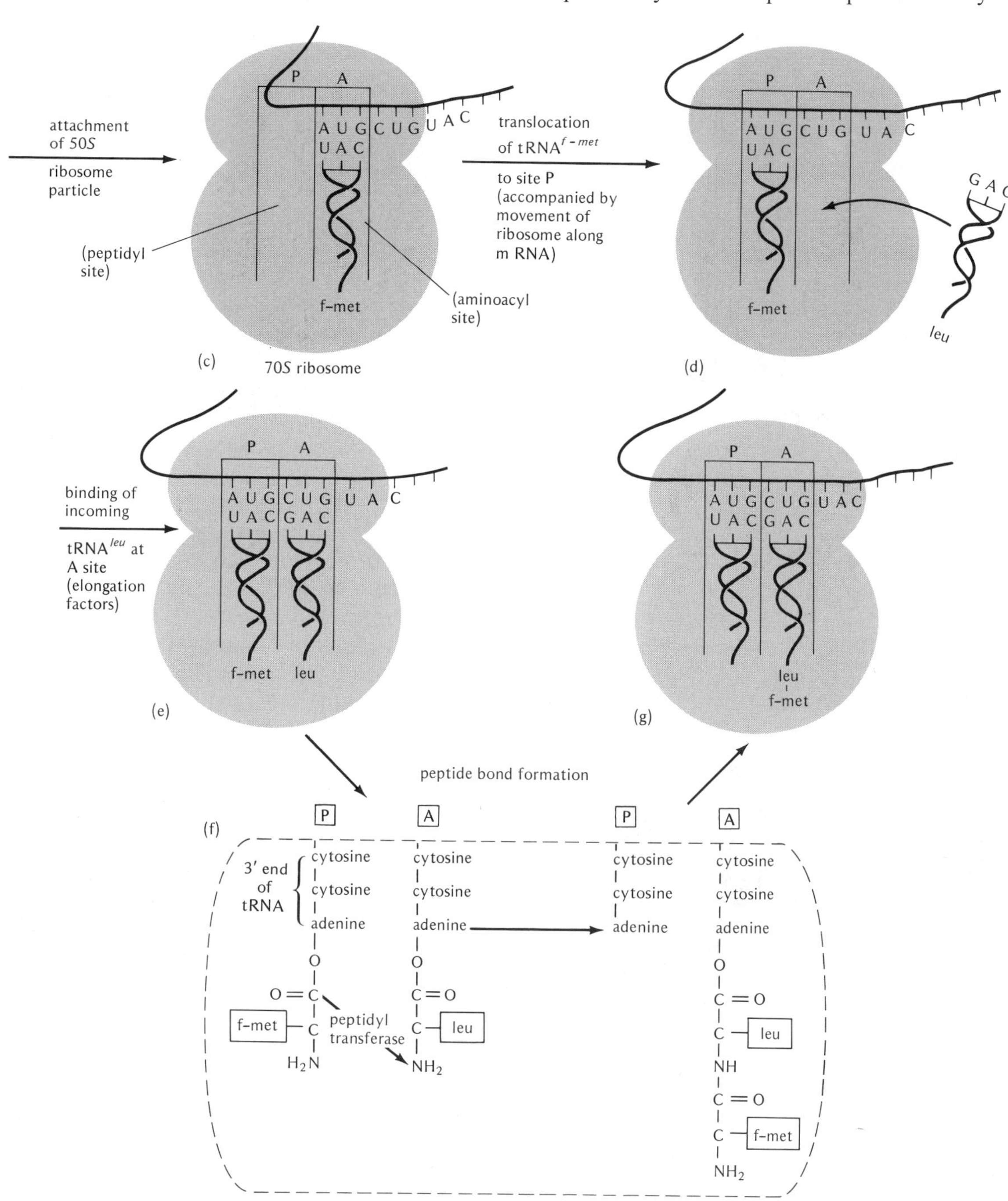

Figure 27–9. *Continued.*

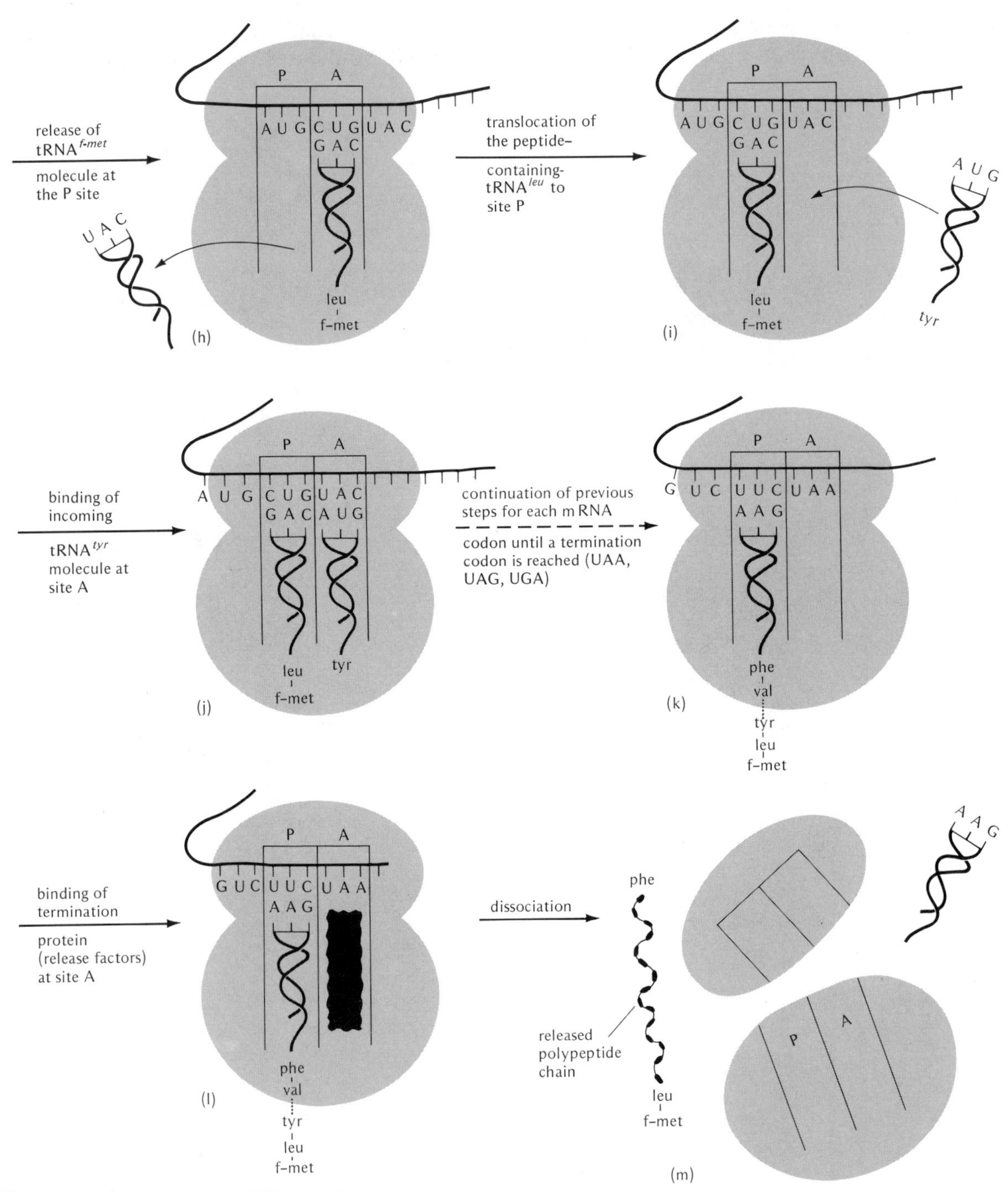

Figure 27–9. *Continued.*

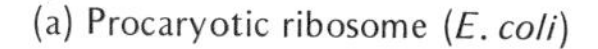

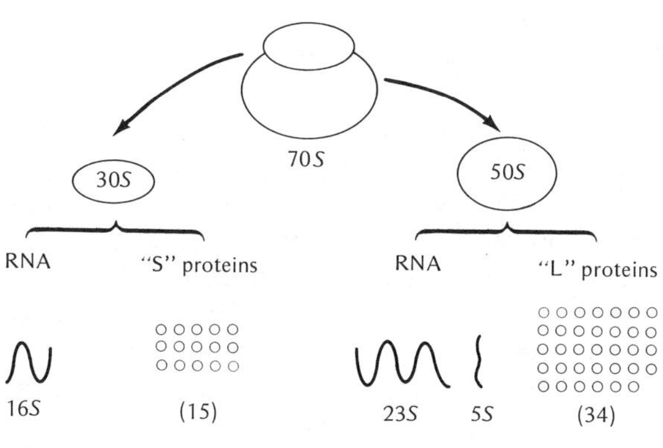

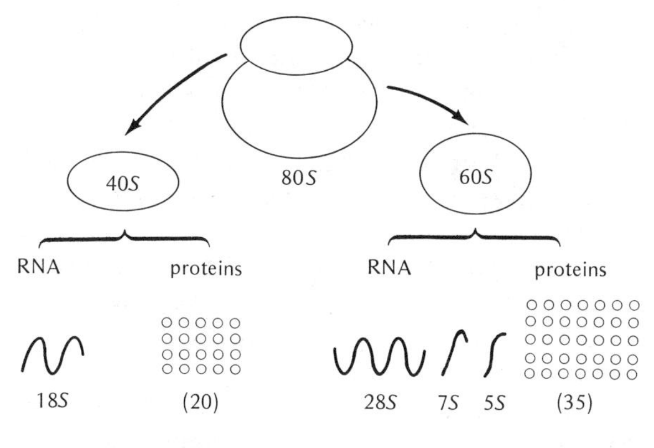

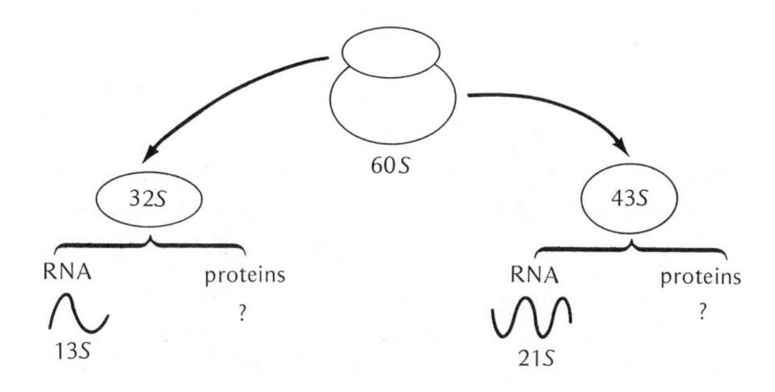

Figure 27–10

Diagram of ribosome components in (a) procaryotes, (b) eucaryotes, and (c) mitochondria. So far, all of the RNA components and most of the protein components in procaryotes have been found in a frequency of about one copy each per ribosome. However, a few of the protein components are not present in all of the ribosomes, but may perhaps become attached to them for specific protein synthesizing functions. In origin the ribosomal RNA molecules derive from precursors which are often considerably longer than the final products. Thus the 16S and 23S rRNA components of E. coli ribosomes derive from a long 30S RNA molecule which is "cleaved" after transcription (endonucleases) into 17.5S and 25S components and then degraded (exonucleases) to their final sizes. Similarly the 18S, 28S, and 7S, rRNA components of mammalian ribosomes derive from a precursor RNA with a sedimentation constant of 45S which undergoes a succession of cleavages and degradations in the order: $45S \rightarrow 41S \begin{matrix} \nearrow 32S \rightarrow 28S + 7S \\ \searrow 20S \rightarrow 18S. \end{matrix}$ *During this process about 1–2 percent of the larger ribosomal RNA components become methylated, carrying these additional* CH_3 *groups mostly on specifically localized ribose sugars, and relatively few on nucleotide bases. (Numbers of ribosome proteins in procaryotes and eucaryotes are derived from McConkey.)*

bind reversibly with both mRNA and tRNA. Considerable effort is therefore being expended to learn the exact mechanism of this relationship by analyzing the structure and function of various ribosomal components. It is, for example, now known that both the lengths and sequences of ribosomal RNA differ between the small and large ribosomal subunits (Fig. 27–10). In function, the 16*S* RNA chain in the smaller *E. coli* ribosomal subunit has been shown by Nomura to be responsible for the *in vitro* assembly of the entire subunit structure. Only when this RNA component is present will the addition of the various protein components enable reconstitution of a biologically active 30*S* particle.

However, neither this 16*S* RNA nor the 23*S* RNA molecule from the larger ribosomal subunit shows direct interactions with either mRNA or tRNA in the various *in vitro* studies performed so far. Only the small 5*S* RNA component of the larger subunit appears to interact with tRNA by binding it to the 50*S* ribosomal particle, but even this seems to depend upon aid of two 50*S* subunit proteins: L18 and L25.

Other ribosomal proteins have also been implicated in the binding of formylmethionyl $tRNA_f^{met}$, as well as in the processes of elongation (the peptidyl transferase is most likely a complex of some of the L proteins), translocation (two proteins associated with this event, L7 and L12, are believed to have "contractile" properties), and termination. The further importance of ribosomal protein function can be derived from the fact that the antibiotics mentioned above, as well as many other protein inhibitors, seem to affect polypeptide synthesis by direct interaction with ribosomal proteins. This is especially obvious in the case of streptomycin, which affects attachment of the tRNA molecule (as well as the 50*S* subunit) to the 30*S* subunit by interacting with a specific ribosomal protein, S12. Various streptomycin-resistant mutations have been analyzed and found to possess amino acid substitutions in this protein but not in any other ribosomal protein. On the other hand, the antibiotic kasugamycin also inhibits protein synthesis, but kasugamycin-resistant strains of *E. coli* have only their 16*S* rRNA component modified. Thus there is probably some function other than a purely structural one in ribosomal RNA.*

SEQUENCE AND RATE OF PROTEIN SYNTHESIS

Protein synthesis, as we have seen, should proceed sequentially from an initiation point to a termination point, rather than occur spontaneously at random points on a polypeptide chain. The first direct evidence for a sequential direction of protein synthesis has been derived from experiments of Dintzis and of Bishop, Schweet, and others showing that growth starts at the amino end (N-terminal) of the polypeptide chain and continues toward the carboxyl end (C-terminal). These experiments utilize the technique of labeling hemoglobin with a short pulse of radioactive isotopes as it is being synthesized in rabbit reticulocytes. Under these conditions the portion of the complete hemoglobin molecules which is always labeled is the C-terminal. Since the introduction of the radioactive label occurs when some hemoglobin molecules are partly completed but unlabeled, the finding of C-terminal labeling in all cases indicates that this terminal is the last portion of the hemoglobin chain that is synthesized.

By similar reasoning the sequence of synthesis can be traced through the entire polypeptide by observing which portions of the molecule are synthesized at which times. For this purpose a short pulse of tritium-labeled leucine is introduced at a certain time in the experiment, called T_2 in Fig. 27–11. The soluble hemoglobin products are then broken down by trypsin and analyzed for radioactivity. It is found that shortly after the introduction of labeled amino acid only the C-terminal peptides are labeled. As time continues, the opportunity for labeling enables a larger number of peptides to be labeled. These, however, are all labeled in sequential order, corresponding to their linear order in the polypeptide chain.

It now seems fairly clear, in procaryotes at least, that the sequential order of polypeptide synthesis follows the order of synthesis of the mRNA molecule itself. That is, the 5′ end of the mRNA molecule that is synthesized first (see p. 105) is also the end that is translated first (see p. 661). It is apparently unique to procaryotes that these processes occur together, as evidenced in the beautiful electron micrographs taken by Miller and co-workers. In Fig. 27–12, for example, translation is marked by the clustering of ribosomes along a linear mRNA molecule which is simultaneously being transcribed from the highly visible DNA double helix. As the DNA-directed RNA polymerase moves along the DNA template, the mRNA chain is lengthened, and this then furnishes one or more initiation sites for the attachment of ribosomes and the commencement of protein synthesis. However, the number of ribosomes is not constant; it increases during the lengthening of the mRNA molecule, and is then diminished as the RNA poly-

* Some polypeptide synthesis has also been found to occur outside the normal mRNA-ribosome translational process. In one such example a cell-free system from *Bacillus brevis* has been shown to synthesize the antibiotic gramicidin *S*, a circular chain of 10 amino acids consisting of two 5-amino acid repeats. There are no ribosomes in this system and it is not inhibited by agents which normally prevent translation such as chloramphenicol, puromycin, or RNase. According to Kleinkauf and others this polypeptide is made by a specific gramicidin enzyme complex to which amino acids are carried by activating enzymes. Interesting as such systems are, however, they are apparently quite rare, and it is unlikely that they produce more than a few polypeptides, or chains of any great length or complexity. On the contrary, the enzymes involved in such systems must themselves be made by the normal transcriptional-translational process.

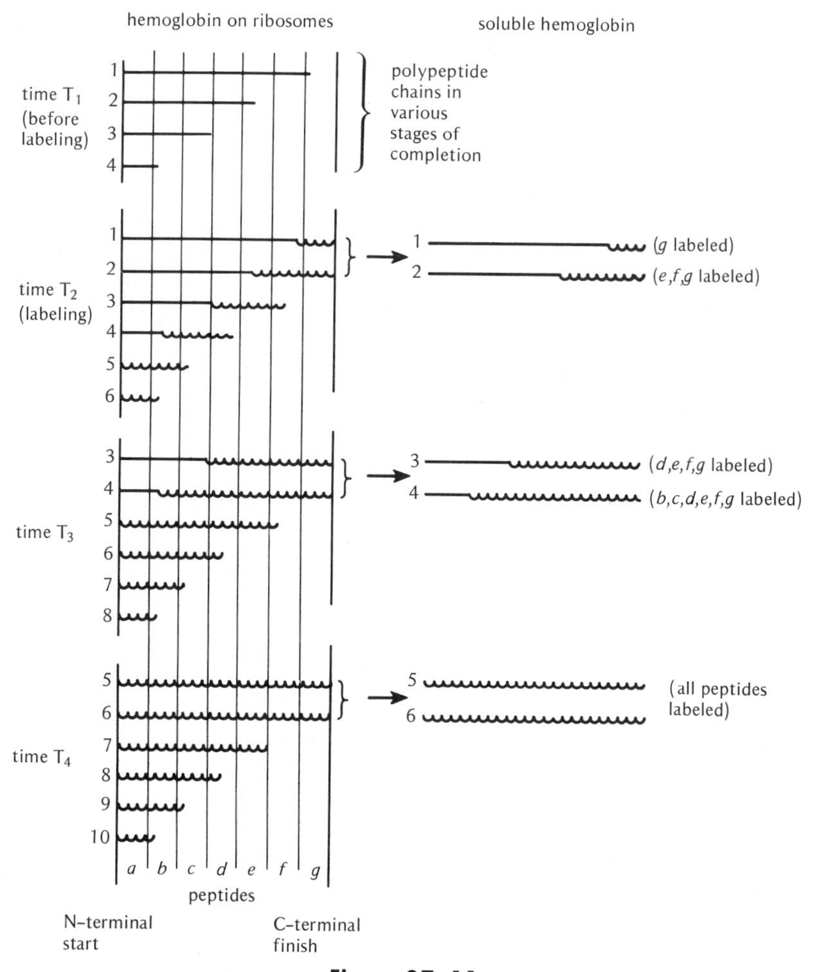

Figure 27–11

Sequence of growth of a protein determined by inserting a radioactive label into a protein-synthesizing system (time T_2) and then observing which peptides on the resulting protein are labeled at specific times after that. (After Dintzis.)

merase enzyme reaches its termination point (Fig. 27–12d), indicating a reduction in the length of the mRNA molecule. According to Kuwano, Schlessinger, and Apirion, and others, this reduction is probably accomplished by a specific exonuclease which degrades the mRNA molecule in the $5' \rightarrow 3'$ direction after a prescribed number of ribosomes have translated the message. Thus mRNA degradation in procaryotes begins even before transcription is completed.

The speed in which transcription and translation occur has now been determined for a number of genes. In *E. coli* the RNA messenger which codes for enzymes responsible for tryptophan synthesis (see Fig. 26–6) is transcribed from DNA at a rate of about 28 nucleotides per second at 37°C. Approximately 30 ribosomes then proceed to translate this messenger, each one producing protein at a rate of about 7 amino acids per second (Baker and Yanofsky). Somewhat less certainty has been

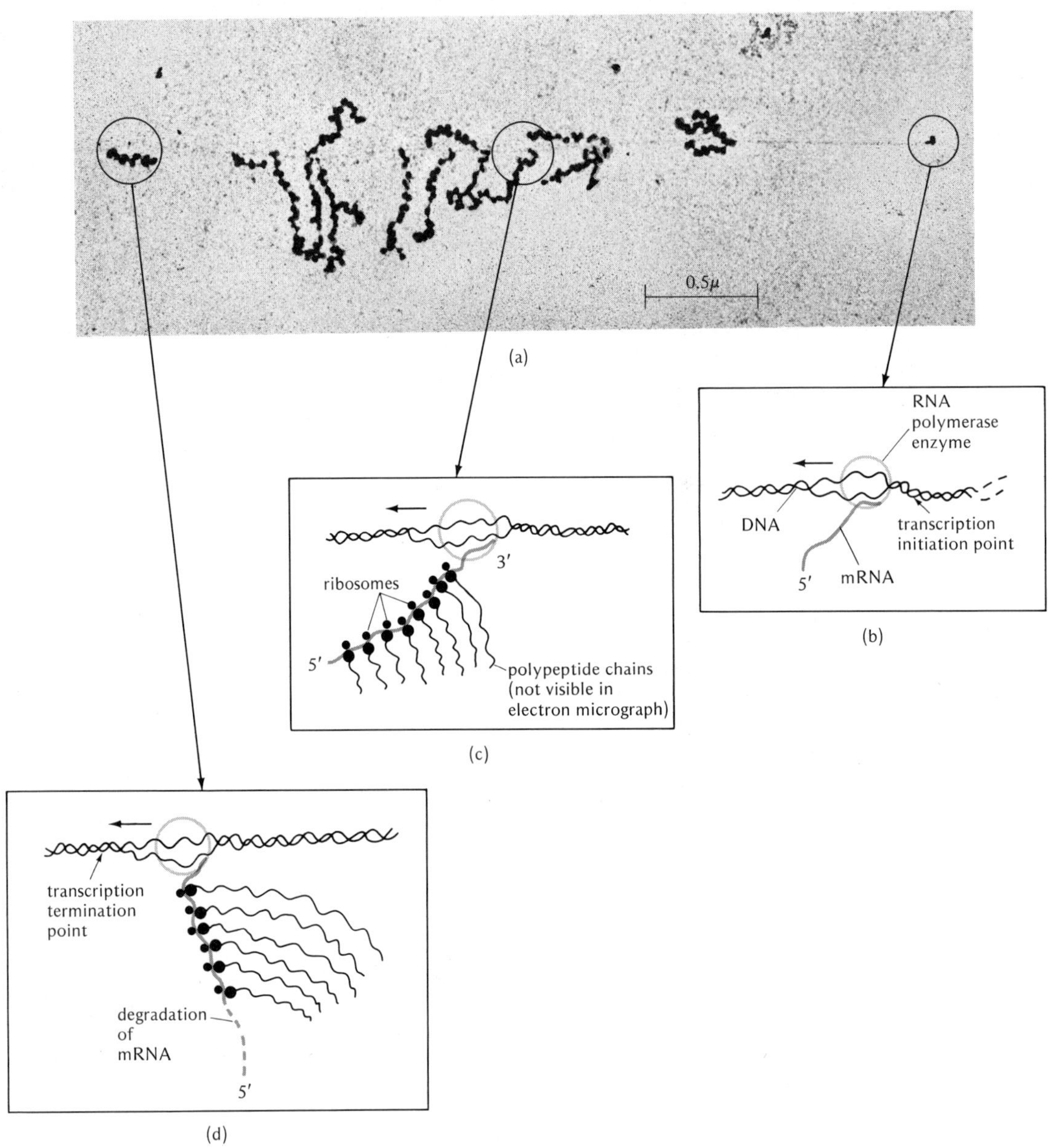

Figure 27–12

(a) Electron micrograph showing a horizontal thread of DNA in E. coli along with attached RNA polymerase enzymes and ribosomes. (b) Diagram of the transcription initiation area showing the RNA polymerase moving leftward and generating an mRNA molecule complementary to one of the DNA strands. (c) In its progress the RNA polymerase enzyme has produced a long mRNA molecule to which a number of ribosomes are now attached. Each ribosome attaches at the 5′ end of the mRNA molecule and translates the nucleotide sequence into a polypeptide chain as it moves toward the 3′ "growing" end of the messenger. (d) Diagram of mRNA degradation that is believed to account for the diminution in the number of ribosomes as the RNA polymerase enzyme reaches the transcription termination point. Once degradation has begun at the 5′ end of the mRNA, "new" ribosomes can no longer attach to initiate translation, and "old" ribosomes dissociate from the mRNA when they complete translation. (Electron micrograph from Miller, Hamkalo, and Thomas; courtesy of O. L. Miller, Jr.)

attached to estimates of eucaryotic transcriptional and translational speeds, but these are probably of the same order of magnitude as in procaryotes. Hunt and co-workers, for example, have calculated that each full-length hemoglobin mRNA molecule which has about 670 nucleotide bases (sedimentation constant 9*S*) is translated into a complete hemoglobin polypeptide chain every 35 seconds, or a translational speed of about 4 amino acids per second.

Although much longer mRNA molecules have been found, this does not necessarily mean that a single long polypeptide is synthesized. The histidine locus of *Salmonella* (p. 623), for example, produces an mRNA molecule containing about 13,000 nucleotides or of sufficient length to produce a protein containing 4000 amino acids. In actuality all 10 enzymes of the histidine locus are coded by this mRNA, and indications exist that each different enzyme (usually associated with one cistron) is separately synthesized on this *polycistronic* mRNA template. These indications include the fact that some of the histidine enzymes are synthesized more frequently than others. Interestingly the most frequently synthesized enzymes are those at the *G* end of the mRNA chain (pyrophosphorylase, etc.), whereas those at the opposite end (isomerase, etc.) are less frequently synthesized. Hartman and Ames termed this phenomenon *modulation*, and proposed that it occurs because of the unidirectional translation of this mRNA starting at the *G* end. Genes that are far from the *G* end may less likely be translated than those nearby, because of accidental mishaps to the ribosomes, or because of occasional breakage of the long mRNA molecule, or because of the presence of specific nucleotide structures (see p. 641) that govern the initiation of translation. Supporting evidence for modulation exists in the finding of numerous *polarity* mutations that not only produce a particular mutant enzyme but also slow down the rate of synthesis of other enzymes toward the *E* end of the histidine locus.

At the present time a considerable amount of information about procaryotic mRNA has accumulated because of the ready availability of mutations that affect the presence or absence of such molecules and their translation into protein (Chapter 29). Many of them, such as the mRNAs involved in the metabolism of tryptophan, lactose, arabinose, and others, show characteristics similar to those of histidine mRNA; that is, a fairly long length and a polycistronic message. Eucaryotic mRNA, on the other hand, cannot be isolated as readily, since genetic methods for mRNA detection are more difficult to apply. Nevertheless direct biochemical techniques have enabled some description of various eucaryotic mRNA molecules, and so far these are known to possess a number of unique features.

First of all, there are no definitely proven examples of polycistronic eucaryotic messengers. Most, if not all, eucaryotic mRNAs seem to produce only a single cistronic polypeptide chain, although this single chain may possibly be involved in more than one function. Second, many functional eucaryotic messengers seem to possess only a small fraction of the nucleotide length of longer mRNA precursors found in the nucleus. These long-length precursors are found among RNA molecules called "heterogenous nuclear RNA," or HnRNA, that are from 5000 to 50,000 nucleotides long. In the case of duck reticulocytes, for example, Imaizumi, Diggelmann, and Scherrer have shown that HnRNA precursors are from 10 to 100 times longer than cytoplasmic hemoglobin mRNA.* Obviously such eucaryotic mRNA molecules represent the sequences remaining after cleavage and degradation of their precursor HnRNA molecules have taken place. Although there is some evidence for cleavage and degradation in some of the procaryotic RNA precursors, such as those for ribosomal RNA (see Fig. 27–10), these do not seem as prevalent or as extensive as in eucaryotes.

A third distinctive feature of eucaryotic mRNA appears to lie in some of the nucleotide sequences attached to it. Various studies have shown that many eucaryotic mRNA molecules share with their HnRNA precursors a long sequence of 50 to

* Isolation of these long mRNA precursors was accomplished by using RNA-directed DNA polymerase (see p. 106) to produce DNA strands complementary to the hemoglobin mRNA. These DNA molecules were then permitted to anneal with HnRNA strands, and measurements were taken on those RNA strands that were "hybridized."

200 adenosine nucleotides at the 3′ end. Possession of this repeating adenine chain (poly-A) essentially marks these molecules as messenger RNA and differentiates them from both ribosomal and transfer RNA. According to Darnell and others, it seems likely that this poly-A section is added after transcription by enzymes that are independent of DNA. However, in addition to the poly-A section some organisms show other unique sequences. In *Xenopus* mRNA, Dina and coworkers have implicated the presence of repetitive sequences that are reminiscent of repetitive DNA sequences found in various eucaryotes (see p. 97). So far the function of the poly-A sections and other repeated sequences are not known, although it is possible that they enable the cell to identify and localize mRNA molecules, or prevent their degradation by various intracellular enzymes. Since there is evidence that two proteins of fairly constant molecular weight are bound to a wide variety of different mRNA molecules (Bryan and Hayashi), perhaps some of these repeated sections serve as attachment points for proteins that protect mRNA or enable it to be transported through the cell. The absence of poly-A in the mRNA of nuclear histone proteins may then be explained by the fact that histone mRNA exists for only part of the cell cycle and is then destroyed, or perhaps because such molecules are not transported to the cytoplasm.

A fourth unique feature has been mentioned previously (p. 636) and pertains to the long-lived nature of eucaryotic mRNA. Instead of lasting for only minutes, eucaryotic messengers continue the process of ribosomal attachment and translation for hours or days (see, for example, Singer and Penman). This longevity confers a more permanent protein complement upon the eucaryotic cell, and is undoubtedly essential in enabling differentiation between cells to occur. There are recent reports that mRNA longevity is enhanced by the presence of poly-A (see Huez et al.).

In summary, we can see that a considerable amount of control over protein synthesis is possible through regulation of (1) the rate at which ribosomes attach to the mRNA, (2) the stability of the mRNA molecule, (3) the amount of ribonucleotides present in the precursor pool from which mRNA is synthesized, (4) the size of the amino acid pool which contributes to the production of enzymes necessary for the synthesis of the different RNA's and their precursors, and (5) the relative concentrations of the different types of tRNA molecules necessary to bring required amino acids to the ribosomes. Also, as will be discussed in Chapter 29, more direct control over gene products occurs on the level of the gene itself; that is, mechanisms exist that permit or inhibit mRNA synthesis by particular sections of genetic material. All of these regulatory relationships derive ultimately from the coding properties of the nucleotides in DNA and RNA which specify the existence of particular amino acids in proteins, and which, in turn, affect the functioning of particular nucleotide sequences. The following chapter considers specifically which nucleotide sequences determine which amino acids.

PROBLEMS

27-1. If polypeptides were not synthesized sequentially from one end of the chain but at many points simultaneously, what results would you have expected to find in the labeling experiment performed by Dintzis (Fig. 27-11)?

27-2. If the DNA of two organisms differ markedly in their base ratios, would you also expect differences in their: (a) Transfer RNA's? (b) Ribosomal RNA's? (c) Messenger RNA's?

27-3. If two species of organisms are alike in their DNA base ratios, would you expect all three types of RNA to be alike as well?

27-4. In *Bacillus cereus*, synthesis of the enzyme penicillinase can begin in response to an inducer, ceph-

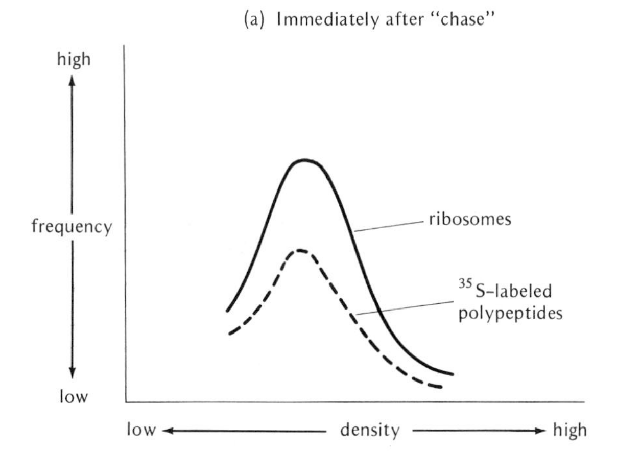

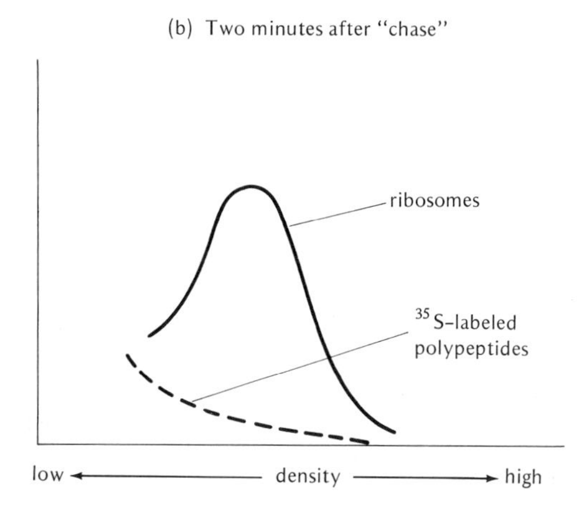

alosporin C. Harris and Sabath found that such induced cells treated with actinomycin D ceased producing RNA after 1 or 2 minutes but nevertheless continued producing penicillinase. What would you conclude about the longevity of penicillinase mRNA?

27-5. McQuillen, Roberts, and Britten added a compound containing radioactive sulfur (^{35}S is incorporated into the amino acids cysteine and methionine) for 15 seconds to a growing culture of *E. coli* cells, and then "chased" the ^{35}S out of the medium by adding large amounts of normal ^{32}S. They then sampled the *E. coli* culture for the association of ^{35}S radioactivity with the ribosomes at two different times: (a) immediately after the chase, and (b) two minutes after the chase. The sampling technique used, sucrose density centrifugation, enabled them to recognize the ribosomes by the gradient density at which they were found, and then to measure the extent of radioactivity of these ribosomes at each of the different densities. Fig. 27-13 shows their results in schematic fashion: the solid line represents the presence of ribosomes that are recognized both by their density and the degree of ultraviolet absorption, and the dashed line represents radioactive polypeptides containing ^{35}S-labeled cysteine and methionine. What would you say these findings indicate?

Figure 27–13

Diagram of some of the results obtained by McQuillen et al. as explained in Problem 27–5.

REFERENCES

Baker, R., and C. Yanofsky, 1972. Transcription initiation frequency and translational yield for the tryptophan operon of *Escherichia coli*. *Jour. Mol. Biol.*, **69,** 89–102.

Barondes, S. H., and M. W. Nirenberg, 1962. Fate of a synthetic polynucleotide directing cell-free protein synthesis. II. Association with ribosomes. *Science*, **138,** 813–816.

Bishop., J., J. Leahy, and R. Schweet, 1960. Formation of the peptide chain of hemoglobin. *Proc. Nat. Acad. Sci.*, **46,** 1030–1038.

Brachet, J., 1957. *Biochemical Cytology.* Academic Press, New York, Chap. VII.

Brenner, S., F. Jacob, and M. Meselson, 1961. An unstable intermediate carrying information from genes to ribosomes for protein synthesis. *Nature*, **190,** 576–581. (Reprinted in the collections of Taylor and of Stent, and in *Selected Papers in Biochemistry*, Vol. 7; see References, Chapter 1.)

Bryan, R. N., and M. Hayashi, 1973. Two proteins are bound to most species of polysomal mRNA. *Nature New Biol.*, **244,** 271–274.

Caspersson, T., 1941. Studien über den Eiweissumsatz der Zelle. *Naturwiss.*, **29,** 33–43.

Chapeville, F., F. Lipmann, G. von Ehrenstein, B. Weisblum, W. J. Roy, Jr., and S. Benzer, 1962. On the role of soluble ribonucleic acid in coding for amino acids. *Proc. Nat. Acad. Sci.*, **48,** 1086–1092. (Reprinted in Vols. 3 and 6 of *Selected Papers in Biochemistry*; see References, Chapter 1.)

Claude, A., 1948. Studies on cells: Morphology, chemical constitution, and distribution of biochemical functions. *Harvey Lectures*, **48,** 121–164.

DARNELL, J. E., W. R. JELINEK, and G. R. MOLLOY, 1973. Biogenesis of mRNA: Genetic regulation in mammalian cells. *Science*, **181,** 1215–1221.

DAVERN, C. I., and M. MESELSON, 1960. The molecular conservation of ribonucleic acid during bacterial growth. *Jour. Mol. Biol.*, **2,** 153–160.

DINA, D., I. MEZA, and M. CRIPPA, 1974. Relative positions of the 'repetitive,' 'unique' and poly(A) fragments of mRNA. *Nature*, **248,** 486–490.

DINTZIS, H. M., 1961. Assembly of the peptide chains of hemoglobin. *Proc. Nat. Acad. Sci.*, **47,** 247–261. (Reprinted in the collection of Zubay and Marmur, and in *Selected Papers in Biochemistry*, Vol. 3; see References, Chapter 1.)

EHRENSTEIN, G. von., B. WEISBLUM, and S. BENZER, 1963. The function of sRNA as amino acid adaptor in the synthesis of hemoglobin. *Proc. Nat. Acad. Sci.*, **49,** 669–675.

GEIDUSCHEK, E. P., T. NAKAMOTO, and S. B. WEISS, 1961. The enzymatic synthesis of RNA: Complementary interaction with DNA. *Proc. Nat. Acad. Sci.*, **47,** 1405–1415.

GROS, F., H. HIATT, W. GILBERT, C. G. KURLAND, R. W. RISEBROUGH, and J. D. WATSON, 1961. Unstable ribonucleic acid revealed by pulse labelling of *Escherichia coli*. *Nature*, **190,** 581–585. (Reprinted in *Selected Papers in Biochemistry*, Vol. 5; see References, Chapter 1.)

HALL, B. D., and S. SPIEGELMAN, 1961. Sequence complimentarity of T2-DNA and T2-specific RNA. *Proc. Nat. Acad. Sci.*, **47,** 137–146. (Reprinted in the collections of Levine and of Taylor; see References, Chapter 1.)

HÄMMERLING J., 1953. Nucleo-cytoplasmic relationships in the development of *Acetabularia. Intern. Rev. Cytol.,* **2,** 475–498.

HARTMAN, P. E., and S. R. SUSKIND, 1969. *Gene Action*, 2nd ed. Prentice-Hall, Englewood Cliffs, N.J.

HERSHEY, A. D., 1953. Nucleic acid economy in bacteria infected with bacteriophage T2. *Jour. Gen. Physiol.*, **37,** 1–23.

HOAGLAND, M. B., and M. L. STEPHENSON, J. F. SCOTT, L. I. HECHT, and P. C. ZAMECNIK, 1958. A soluble ribonucleic acid intermediate in protein synthesis. *Jour. Biol. Chem.*, **231,** 241–257. (Reprinted in the collections of Taylor and of Zubay and Marmur, and in Vols. 6 and 7 of *Selected Papers in Biochemistry;* see References, Chapter 1.)

HOAGLAND, M. B., P. C. ZAMECNIK, and M. L. STEPHENSON, 1957. Intermediate reactions in protein biosynthesis. *Biochim. Biophys. Acta*, **24,** 215–216.

HOGEBOOM, G. H., and W. C. SCHNEIDER, 1955. The Cytoplasm. In *Nucleic Acids*, Vol. 2, E. Chargaff and J. N. Davidson (eds.). Academic Press, New York, pp. 199–246.

HOLLEY, R. W., J. APGAR, G. A. EVERETT, J. T. MADISON, M. MARQUISEE, S. H. MERRILL, J. R. PENSWICK, and A. ZAMIR, 1965. Structure of a ribonucleic acid. *Science*, **147,** 1462–1465. (Reprinted in Vol. 6 of *Selected Papers in Biochemistry*; see References, Chapter 1.)

HUEZ, G., G. MARBAIX, E. HUBERT, M. LECLERQ, U. NUDEL, H. SOREQ, R. SALOMON, R. LEBLEU, M. REVEL, and U. Z. LITTAUER, 1974. Role of the polyadenylate segment in the translation of globin messenger RNA in *Xenopus* oocytes. *Proc. Nat. Acad. Sci.*, **71,** 3143–3146.

HUNT, T., T. HUNTER, and A. MUNRO, 1969. Control of hemoglobin synthesis: Rate of translation of the messenger RNA for the α and β chains. *Jour. Mol. Biol.*, **43,** 123–133.

IMAIZUMI, T., H. DIGGELMANN, and K. SCHERRER, 1973. Demonstration of globin messenger sequences in giant nuclear precursors of messenger RNA of avian erythroblasts. *Proc. Nat. Acad. Sci.*, **70,** 1122–1126.

INGRAM, V., 1972. *The Biosynthesis of Macromolecules*, 2nd ed. W. A. Benjamin, Menlo Park, Calif.

JACOBSON, K. B., 1971. Reaction of aminoacyl-tRNA synthetases with heterologous tRNA's. *Progr. Nuc. Acid Res. and Mol. Biol.*, **11,** 461–488.

KIM, S. H., F. L. SUDDATH, G. J. QUIGLEY, A. MCPHERSON, J. L. SUSSMAN, A. H. J. WANG, N. C. SEEMAN, and A. RICH, 1974. Three-dimensional tertiary structure of yeast phenylalanine transfer RNA. *Science*, **185,** 435–440.

KLEINKAUF, H., and W. GEVERS, 1969. Nonribosomal polypeptide synthesis: The biosynthesis of a cyclic peptide antibiotic, gramicidin *S. Cold Sp. Harb. Symp.*, **34,** 805–813.

KUWANO, M., D. SCHLESSINGER, and D. APIRION, 1970. Ribonuclease V of *Escherichia coli* IV. Exonucleolytic cleavage in the 5′ to 3′ direction with production of 5′-nucleotide monophosphates. *Jour. Mol. Biol.*, **51,** 75–82.

LEVITT, M., 1973. Orientation of double-helical segments in crystals of yeast phenylalanine transfer RNA. *Jour. Mol. Biol.*, **80,** 255–263.

LIPMANN, F., 1958. Chairman's introduction: Some facts and problems. (Symposium on amino acid activation.) *Proc. Nat. Acad. Sci.*, **44,** 67–73.

MAZIA, D., 1952. Physiology of the cell nucleus. In

Modern Trends in Physiology and Biochemistry, E. S. G. Barron (ed.). Academic Press, New York, pp. 77–122.

McConkey, E. H., 1974. Composition of mammalian ribosomal subunits: A re-evaluation. *Proc. Nat. Acad. Sci.*, **71,** 1379–1383.

Miller, O. L., B. A. Hamkalo, and C. A. Thomas, 1970. Visualization of bacterial genes in action. *Science*, **169,** 392–395.

Noll, M., and H. Noll, 1974. Translation of R17 RNA by *Escherichia coli* ribosomes. Initiator transfer RNA-directed binding of 30*S* subunits to the starting codon of the coat protein gene. *Jour. Mol. Biol.*, **89,** 477–494.

Nomura, M., 1973. Assembly of bacterial ribosomes. *Science*, **179,** 864–873.

Nomura, M., A. Tissières, and P. Lengyel (eds.), 1974. *Ribosomes.* Cold Spring Harbor Laboratory, Cold Spring Harbor. (Contains a large number of reviews on the activity and genetics of ribosomes in both procaryotes and eucaryotes.)

Palade, G. E., and P. Siekevitz, 1956. Liver microsomes: An integrated morphological and biochemical study. *Jour. Biophys. Biochem. Cytol.*, **2,** 171–198.

Platt, T., and C. Yanofsky, 1975. An intercistronic region and ribosome-binding site in bacterial messenger RNA. *Proc. Nat. Acad. Sci.,* **72,** 2399–2403.

Risebrough, R. W., A. Tissières, and J. D. Watson, 1962. Messenger RNA attachment to active ribosomes. *Proc. Nat. Acad. Sci.*, **48,** 430–436.

Roberts, J. D., J. E. Ladner, J. T. Finch, D. Rhodes, R. S. Brown, B. F. C. Clark, and A. Klug, 1974. Structure of yeast phenylalanine tRNA at 3 Å resolution. *Nature*, **250,** 546–551.

Singer, R. H., and S. Penman, 1973. Messenger RNA in HeLa cells: Kinetics of formation and decay. *Jour. Mol. Biol.*, **78,** 321–334.

Sueoka, N., 1964. Compositional variation and heterogeneity of nucleic acids and protein in bacteria. In *The Bacteria*, Vol. V, I. C. Gunsalus and R. Y. Stanier (eds.). Academic Press, New York, pp. 419–443.

Tischendorf, G. W., H. Zeichardt, and G. Stoffler, 1975. Architecture of the *Escherichia coli* ribosome as determined by immune electron microscopy. *Proc. Nat. Acad. Sci.,* **72,** 4820–4824.

Tissières, A., J. D. Watson, D. Schlessinger, B. R. Hollingsworth, 1959. Ribonucleoprotein particles from *Escherichia coli. Jour. Mol. Biol.*, **1,** 221–233.

Warner, J. R., and A. Rich, 1964a. The number of soluble RNA molecules on reticulocyte polyribosomes. *Proc. Nat. Acad. Sci.*, **51,** 1134–1141.

———, 1964b. The number of growing polypeptide chains on reticulocyte polyribosomes. *Jour. Mol. Biol.*, **10,** 202–211.

Watson, J. D., 1963. Involvement of RNA in the synthesis of proteins. *Science*, **140,** 17–26.

———, 1976. *Molecular Biology of the Gene,* 3rd ed. W. A. Benjamin, Menlo Park, Calif.

Yamane, T., and N. Sueoka, 1963. Conservation of specificity between amino acid acceptor RNA and amino acyl-sRNA synthetase. *Proc. Nat. Acad. Sci.*, **50,** 1093–1100.

28 NATURE OF THE GENETIC CODE

Since there are 20 different kinds of amino acids in proteins and only four kinds of nucleotides in DNA, the relationship between the gene and its most elementary functional product, i.e., between DNA and protein, can hardly be interpreted through a code of one nucleotide = one amino acid. However, if a sequence of more than one nucleotide codes one particular amino acid, we may ask how long this nucleotide sequence is: two nucleotides, three nucleotides, or more? A coding sequence of two nucleotides for one amino acid, or a "doublet code," would produce only 4^2, or 16, possible coding combinations, or *codons*. To code for 20 amino acids, four of these codons would have to serve the double function of coding for two amino acids each (or two codons for three amino acids each, etc.). Such a doublet code would therefore cause an *ambiguous* determination of a significant number of amino acids, a conclusion which hardly accords with the precise amino acid composition and sequence of almost any protein (Chapter 26).

A codon size of three nucleotides for one amino acid or triplet code seems more likely, since it produces $4^3 = 64$ possible codons (Fig. 28–1). Note, however, that since only 20 amino acids need to be coded, 44 codons in a triplet code seem to be superfluous. This excess of seemingly unnecessary codons would even be greater if the coding ratio were 4:1 (256 codons), 5:1 (1024 codons), etc. To account for the excess of codons beyond the necessary 20, we can suppose that more than one codon can code for a particular amino acid. For example, if each kind of amino acid were coded singlet code

A
G
C
T

doublet code

AA	AG	AC	AT
GA	GG	GC	GT
CA	CG	CC	CT
TA	TG	TC	TT

triplet code

AAA	AAG	AAC	AAT
AGA	AGG	AGC	AGT
ACA	ACG	ACC	ACT
ATA	ATG	ATC	ATT
GAA	GAG	GAC	GAT
GGA	GGG	GGC	GGT
GCA	GCG	GCC	GCT
GTA	GTG	GTC	GTT
CAA	CAG	CAC	CAT
CGA	CGG	CGC	CGT
CCA	CCG	CCC	CCT
CTA	CTG	CTC	CTT
TAA	TAG	TAC	TAT
TGA	TGG	TGC	TGT
TCA	TCG	TCC	TCT
TTA	TTG	TTC	TTT

Figure 28–1

Possible coding combinations using different-sized DNA nucleotide sequences. On the level of mRNA translation into protein, the mRNA codon sequences would be complementary, of course, to those shown here, e.g., ATC (DNA) = UAG (mRNA). (After Nirenberg, 1963.)

by three different possible codons, 60 possible codons would be accounted for. A code in which there is more than one codon for the same amino acid is called *degenerate*. It is also possible that some or all of the codons in excess of 20 do not code for *any* amino acid and are therefore *nonsense* codons.

Whatever the codon size, however, the form of the code may be *overlapping* or *nonoverlapping*. If, for example, the codons were triplets, this would mean that a sequence of six nucleotides could code either two amino acids, if it were nonoverlapping, or more amino acids, if it were overlapping (Fig. 28–2). A decision between these alternatives can be made by determining whether a single mutation produces a change in only a single amino acid in a protein product (nonoverlapping code) or in more than one amino acid (overlapping code). As has been discussed in Chapter 26, all evidence to date indicates that single nucleotide mutations affect only a single amino acid in a protein chain, meaning that the code is nonoverlapping.

For a nonoverlapping code two further possibilities exist: (1) The code has "commas," consisting of one or more nucleotides placed between the codons; such commas might serve the purpose of selecting the proper codons to be translated into amino acids. (2) The code is "commaless," without interspersed nucleotides among the codons. A commaless code would have to be "read" from a

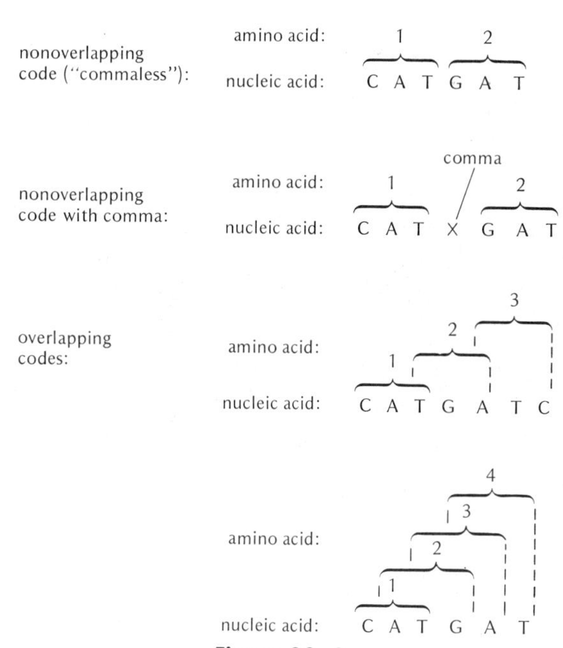

Figure 28–2

Nonoverlapping, and overlapping triplet codes.

particular starting point on the mRNA chain in a "reading frame" consisting of codons placed side by side. Thus a reading frame in a triplet code might begin on the DNA level with the nucleotide sequence CATGAT···, in which CAT and GAT are subsequently transcribed and translated into two particular amino acids. Removal of the cytidine nucleotide (C) from the beginning of this DNA sequence, e.g., ATGAT···, would produce triplets (ATG) that can translate into different amino acids than the originals ("missense" mutations) or yield no amino acids at all ("nonsense" mutations). These and other coding terms are briefly defined in Table 28–1.

TABLE 28–1

Definitions of some common terms used in describing the genetic code

Term	Meaning	Term	Meaning
code letter	nucleotide, e.g., A,U,G,C (in mRNA) or A,T,G,C (in DNA)	*ambiguous code*	when one codon can code for more than one amino acid, e.g., GGA = glycine, glutamic acid
code word, or *codon*	sequence of nucleotides specifying an amino acid, e.g., UUU = phenylalanine	*commaless code*	when there are no intermediary nucleotides (spacers) between words, e.g., UUUCCC = two amino acids in triplet nonoverlapping code
anticodon	sequence of nucleotides on tRNA that complements the codon, e.g., AAA = anticodon for phenylalanine	*reading frame*	the particular nucleotide sequence that starts at a specific point and is then partitioned into codons until the final code word of that sequence is reached.
genetic code or *coding dictionary*	a table of all the code words or codons that specify amino acids (see Table 28–3)	*frame-shift mutation*	a change in the reading frame because of the insertion or deletion of nucleotides in numbers other than multiples of the codon length. This modifies the previous partitioning of codons in the reading frame, and causes a new sequence of codons to be read
word size or *codon length*	the number of letters in a code word, e.g., three letters in a triplet code (these are the same as *coding ratio* in a nonoverlapping code)	*sense word*	a codon that specifies an amino acid normally present at that position in a protein
nonoverlapping code	when only as many amino acids are coded as there are code words in end-to-end sequence, e.g., (triplet code), UUUCCC = phenylalanine (UUU) + proline (CCC)	*missense mutation*	a change in nucleotide sequence, either by deletion, insertion, or substitution, resulting in the appearance of a codon that produces a different amino acid in a particular protein, e.g., UUU (phenylalanine) mutates to UGU (cysteine)
overlapping code	when more amino acids are coded for than there are code words present in end-to-end sequence, e.g., UUUCCC = phenylalanine (UUU) + phenylalanine (UUC) + serine (UCC) + proline (CCC)	*nonsense mutation*	a codon that does not produce an amino acid, e.g., UAG (also called a "stop" mutation or "termination" codon)
nondegenerate code	when there is only one codon for each amino acid, e.g., 20 different amino acids have a total of 20 codons	*universality*	utilization of the same genetic code in all organisms, e.g., UUU = phenylalanine in bacteria, mouse, man and tobacco
degenerate code	when there is more than one codon for a particular amino acid, e.g., UUU, UUC = phenylalanine, or 20 different amino acids have a total of more than 20 codons		
synonymous codons	different codons that specify the same amino acid in a degenerate code, e.g., UUU = UUC = phenylalanine		

THE TRIPLET CODE

Evidence on the length of the codon, the degeneracy of the code, and its commaless nature was presented in 1961 in an ingenious experiment by Crick, Barnett, Brenner, and Watts-Tobin. Taking advantage of the likelihood that mutations induced by acridine dyes such as proflavin and acridine orange, as well as certain spontaneous mutations, represent deletions or insertions (see p. 576), Crick and his group tested a number of such mutations located in the *B* cistron of the *rII* locus of T4. One of the mutations was arbitrarily designated as +, and, if this mutant effect was suppressed by a different mutation, the latter was then designated as −. *FC* 0, for example, was called +, and produced an *rII* mutant effect. *FC* 1, another *rII* mutation, suppressed the action of *FC* 0, producing wild type when they were both present in the *cis* configuration, and was therefore designated as −. *FC* 58, on the other hand, did not produce wild type when present in *cis* configuration with *FC* 0, but did produce wild type when present in *cis* configuration with *FC* 1. Thus *FC* 58 could be designated as + ("a suppressor of a suppressor"). In this fashion all the acridine mutations were classified as + or − and various combinations of them were then further tested in the *cis* configuration.

As shown in Table 28–2 for a sample group of mutations, these *cis* tests had most illuminating results: When two + mutations or two − mutations were combined, the mutant effect was not modified; wild type only appeared in the +− combination or in combinations of three +'s or three −'s. This meant that three changes of the same kind or a multiple of three were necessary to permit wild type to be formed. This relationship is visualized in Fig. 28–3 by considering that the *rII* effect of a single + mutation (e.g., insertion) can be corrected by either a − mutation (e.g., deletion) or by the presence of two further + mutations if the code is in triplets. Similarly a − mutant effect is restored to wild type by a + mutation or two further − mutations. Note that the restoration of wild type by these various com-

TABLE 28–2

Phenotypes produced by various *cis* combinations among a sample group of acridine-induced mutations at the *rII* locus of T4

Designations: +	Designations: −
FC 0	*FC* 1
FC 40	*FC* 21
FC 58	*FC* 23

Wild-type combinations	*rII combinations*
FC 0, *FC* 1, (+ −)	*FC* 0, *FC* 40, (+ +)
FC 0, *FC* 21, (+ −)	*FC* 0, *FC* 58, (+ +)
FC 40, *FC* 1, (+ −)	*FC* 1, *FC* 21, (− −)
FC 58, *FC* 1, (+ −)	*FC* 1, *FC* 23, (− −)
FC 0, *FC* 40, *FC* 58, (+ + +)	
FC 1, *FC* 21, *FC* 23, (− − −)	

From the data of Crick et al., 1961.

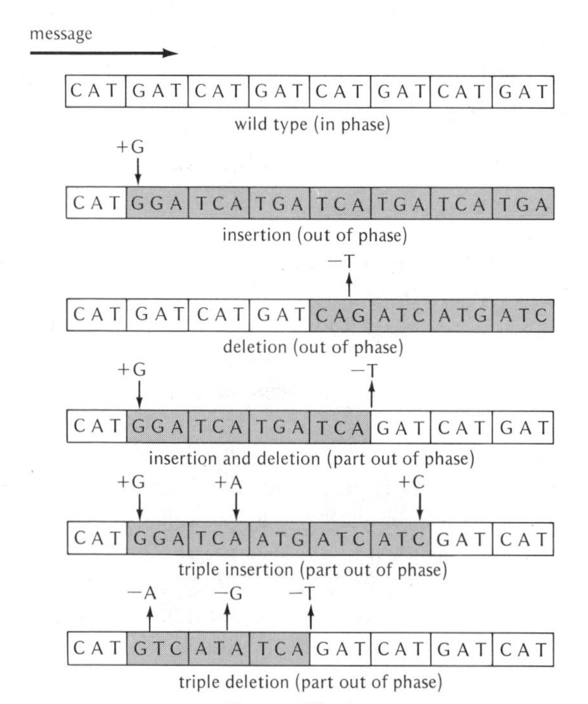

Figure 28–3

Effect of nucleotide insertions, deletions, and their combinations upon a nucleotide chain "read" as triplet codons in a commaless sequence ("reading frame") from left to right. Since such insertions or deletions affect a number of codons beyond the one carrying the mutation, these changes are considered as reading-frame or frame-shift mutations.

binations depends upon the commaless nature of this triplet code. That is, the wild-type message can be restored by changes further along the nucleotide chain because all the intervening nucleotides are involved in the reading frame (i.e., none are used as commas). Also, the fact that genetic mapping of the + and − mutants may show considerable distances between them indicates that the wild-type product formed by a +− combination must have missense mutations in the out-of-phase interval rather than nonsense mutations. That is, if no amino acid were coded because of a nonsense mutation, the protein formed would end abruptly at that point, because of the halt in the reading frame, and a wild-type product would hardly occur. The interval between a + and − combination (or between three +'s or three −'s) must therefore code for amino acids for the reading frame to continue, although these amino acids are probably not those originally present. Thus a +− combination produces a protein that is wild type for only part of its length, those sections before and after the missense interval. Therefore, if the codons in the missense interval code for amino acids, it seems likely that some of these newly produced codons are not restricted to the 20 types which would be found in a nondegenerate code. In other words, since the triplet code provides 64 possibilities for coding 20 amino acids, some amino acids must be coded by more than one codon for amino acids to be produced in a relatively long out-of-phase interval. These experiments therefore indicate that the code is commaless, degenerate, and most likely triplet. Since there are four possible nucleotide bases for each position in a codon, a triplet code would provide 4^3 different codons or code words in the *genetic code* or *coding dictionary*. Essentially the fundamental question of coding is then simply what codons specify what amino acids?

THE CODING DICTIONARY

In 1961 Nirenberg and Matthei reported the first experiment demonstrating that a particular RNA sequence produced a particular amino acid sequence. They used a cell-free system extracted from *E. coli* bacteria containing ribosomes, radioactively labeled amino acids, transfer RNA, and the necessary enzymes and energy sources (ATP) for protein synthesis. In the absence of messenger RNA no observable protein synthesis took place. They then added RNA from tobacco mosaic virus (TMV) and showed an increase in amino acid incorporation into proteins. In other words, mRNA was necessary for protein formation and an *E. coli* system could use TMV-RNA to synthesize TMV-like proteins. However, their most remarkable discovery occurred when they added the synthetic polyribonucleotide (see p. 101), poly-U, to this system and obtained a polypeptide whose amino acids were all phenylalanine in spite of the presence of other amino acids in the original mixture.

Nirenberg and Matthei's results meant that phenylalanine, or more precisely the *anticodon* of a phenylalanine-carrying tRNA, was coded by a sequence of uracil bases (UUU codon of mRNA = AAA anticodon of $tRNA^{phe}$). It was immediately evident that other synthetic RNA messengers introduced into this system could be used to discern the code for a variety of amino acids. Thus poly-A was found to be the code for lysine and poly-C for proline. Poly-G was also tried but could not attach to ribosomes, since it remained in solution as multistranded helixes caused by hydrogen bonding between its guanine bases. Aside from specifying the coding sequence of these three amino acids, such findings also indicated that the code was commaless, without special intervening nucleotides between codons. The argument for this is as follows: If the code has specific commas, the incorporation of phenylalanine into poly-U means that the commas are all uracil. However, the incorporation of lysine into poly-A and of proline into poly-C, in turn, means that the commas are made of adenine and of cytosine—explanations which are extremely inconsistent.

Within a short period of time after this discovery, synthetic ribonucleotides containing mixed bases were also tried, both by Nirenberg's group and by Ochoa's group. Poly-UC, for example, potentially contained only eight different codons:

UUU, UUC, UCU, UCC, CUU, CUC, CCU, CCC. Some codon assignments could then be made by varying the relative amounts of the two bases and observing which amino acids are preferentially incorporated. In this example poly-UC caused the polypeptide incorporation of leucine and serine in addition to the expected incorporation of phenylalanine (UUU) and proline (CCC). Modifying the proportion of bases in poly-UC so that it contains more uracil than cytosine causes more serine to be incorporated into the polypeptide, indicating that serine may either be UUC, CUU, or UCU. Further determinations with other nucleotide polymers enabled other amino acids to be characterized by their general code; e.g., poly-UG caused the polypeptide incorporation of cysteine, glycine, tryptophan, and valine, and poly-UA led to incorporation of asparagine, isoleucine, leucine, lysine, and tyrosine. As in serine, however, the precise nucleotide sequence within the codons of most amino acids could not be determined.

Since then, Nirenberg and Leder, as well as Khorana and co-workers, have obtained more exact characterizations of codon structure by using ribonucleotides whose base sequences are known. In the Nirenberg-Leder technique a "species" of mRNA molecules containing three nucleotides in known order is placed into a cell-free system containing ribosomes mixed with different activated tRNA molecules bearing amino acids. In this system the mRNA is too short to enable protein synthesis, but nevertheless causes the binding of the amino-acid-charged tRNA molecules to the ribosomes. The amino acid bound by a particular trinucleotide mRNA sequence is detected by radioactive labeling of only one kind of amino acid in the mixture and then observing whether there is pronounced incorporation of this radioactivity into the ribosomal fraction that is specially isolated on a separate filter. UUU trinucleotides, for example, will cause radioactivity of the ribosomal fraction only when phenylalanine is the radioactive amino acid in the mixture.

Evidence for the triplet nature of the code is obtained in these experiments from the fact that the binding of specific amino acids to ribosomes occurs preferentially in the presence of trinucleotide mRNA rather than in the presence of dinucleotide mRNA. Thus, $tRNA^{phe}$ bearing ^{14}C radioactive phenylalanine becomes bound to ribosomes when UUU or UUC trinucleotides are used. Radioactive leucine becomes ribosomal-bound by the trinucleotides UUA, UUG, CUC, and CUG. In the Khorana technique longer ribonucleotide sequences are formed by first synthesizing specific DNA sequences and then using DNA-directed RNA polymerase to make complementary RNA molecules. These precisely sequenced RNA messengers are then used in *in vitro* translational systems from which the amino acid sequences in the resultant polypeptides are analyzed. Such studies not only reaffirmed the triplet code and provided information on disputed codons but also demonstrated that the code was commaless and nonoverlapping. To date, all the 64 possible trinucleotides have been tested and found to incorporate among them all 20 amino acids. As shown in Table 28–3, there are numerous instances in which a single amino acid is coded by more than one codon; i.e., some of the different codons are *synonomous* and the code is degenerate. One cause for this degeneracy arises from the fact that the tRNA molecules to which a particular amino acid attaches may consist of more than one variety, each variety with its own anticodon. For example, different varieties of $tRNA^{leu}$ have been separated, each attaching to different mRNA nucleotide sequences (Bennett, Goldstein, and Lipmann). Another cause resides in the ability of some types of tRNA bearing a specific anticodon to pair with two or more synonymous codons of a particular amino acid (e.g., yeast $tRNA^{ala}$ bearing anticodon CGI can pair with alanine codons GCU, GCC, and GCA).*

According to Crick, the degeneracy caused by pairing behavior of a single tRNA with different synonomous codons can be explained by the "wobble hypothesis." That is, the first two positions of the triplet codon on mRNA pair precisely

* Inosinic acid (I) is a nucleotide that uses the purine, hypoxanthine, as a base. Hypoxanthine, in turn, derives from the loss of an amino group from adenine which thereby enables it to pair with cytosine, as shown in Fig. 24–21.

TABLE 28–3

Nucleotide sequence of RNA codons based on binding of amino-acid-charged tRNA molecules to ribosomes and on incorporation of amino acids into polypeptides*

Codon		Codon		Codon		Codon	
UUU	*phe*	UCU	*ser*	UAU	*tyr*	UGU	*cys*
UUC		UCC		UAC		UGC	
UUA	*leu*	UCA		UAA	nonsense or chain termination	UGA	nonsense
UUG		UCG		UAG		UGG	*trp*
CUU	*leu*	CCU	*pro*	CAU	*his*	CGU	*arg*
CUC		CCC		CAC		CGC	
CUA		CCA		CAA	*gln*	CGA	
CUG		CCG		CAG		CGG	
AUU	*ile*	ACU	*thr*	AAU	*asn*	AGU	*ser*
AUC		ACC		AAC		AGC	
AUA		ACA		AAA	*lys*	AGA	*arg*
AUG†	*met*	ACG		AAG		AGG	
GUU	*val*	GCU	*ala*	GAU	*asp*	GGU	*gly*
GUC		GCC		GAC		GGC	
GUA		GCA		GAA	*glu*	GGA	
GUG		GCG		GAG		GGG	

From data of Nirenberg and co-workers, Khorana and co-workers, and others.

* These codons are oriented on the mRNA molecule so that the nucleotide on the left is toward the 5′ end, and the nucleotide on the right is toward the 3′ end.

† AUG is the most common initiator codon (see Chapter 27).

with the first two nucleotides on the anticodon of tRNA, but pairing at the third position may be "wobbly," depending upon which nucleotide is present at the third position. Three types of wobbly pairing have been proposed: U at the third position of the tRNA anticodon may pair with A or G on tRNA, G may pair with U or C, and I may pair with U, C, or A. Thus, for example, codons that are alike at the first two positions but that differ only at the third position for U or C should code for the same amino acid, since the anticodon bearing G at the third position will pair with either of the two bases. As evidence for this, Söll and Rajbhandry found that a single type of $tRNA^{phe}$ from yeast (anticodon AAG) could translate the codons, UUU and UUC, with equal efficiency.

In only one case does the code seem *ambiguous*, in the sense that AUG specifies two kinds of tRNA: one kind ($tRNA_f^{met}$) at translation initiation sites on mRNA, and the other kind ($tRNA^{met}$) at interior mRNA sites. The AUG initiation codon is therefore believed to be identified differently than other AUG codons, probably because of a unique secondary structure of the mRNA initiation site.

The presence of triplets which do not experimentally bind any of the amino acids indicates the existence of "nonsense" codons. Evidence by Brenner and co-workers and Weigert and Garen suggests that UAA, UAG, and UGA probably fall into this category. As noted in Chapter 27, it is likely that at least one of the nonsense codons may serve the purpose of terminating the amino acid chain in protein synthesis. In the RNA virus MS2, two termination codons occur consecutively (see Fig. 28–5).

Using the known codon designations and the technique of RNA synthesis, the sequence of translation along the mRNA molecule itself can be determined. As discussed previously (p. 78ff.), a nucleotide sequence has two ends: the 5′ end, at which the phosphate group is usually attached to the 5′ carbon of the sugar molecule, and the 3′ end, which does not ordinarily contain a phosphate group. RNA with known nucleotide sequences can then be synthesized in which the 5′ and 3′ ends are occupied by specific triplets. Since the amino acid at the N-terminal of the polypeptide is synthesized first and the amino acid at the C-terminal is synthesized last (p. 646), observations of which amino acids coded by the synthetic RNA are at the C-terminal would indicate that translation is completed at this point and must therefore begin at the opposite end of the RNA molecule. For example, since the codon designation for asparagine is AAC, a synthetic RNA molecule with AAC at its 3′ end that produced C-terminal asparagine would indicate that the amino acid at the 3′ end is synthesized last. Such experiments have been performed by Smith and others, and the direction of translation along the mRNA molecule appears to be from the 5′ to the 3′ end. Interestingly this sequence of translation corresponds to the sequence in which the mRNA molecule itself appears to be synthesized (see p. 105 and also Fig. 27–12).

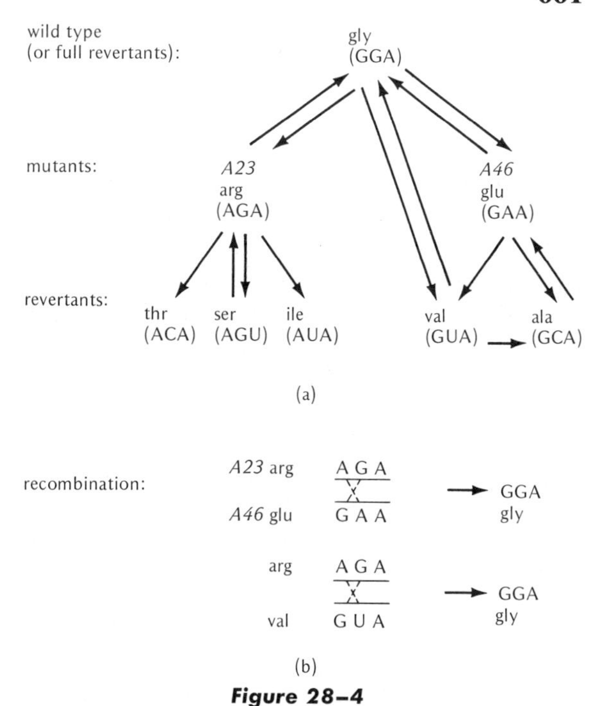

Figure 28–4

(a) Some of the observed amino acid replacements at the number 210 amino acid position on the protein A chain of E. coli tryptophan synthetase, and one interpretation of these replacements in terms of single nucleotide changes in the codons. (b) The production of wild type in crosses between A23 and A46 and between arginine and valine mutants is explained as resulting from a single crossover event. The reciprocal products of these crossovers (AAA, AUA) may produce mutant proteins that are not individually detectable because of the presence of the mutant parental strains.

EVIDENCE FOR THE CODE

In vivo evidence supporting the proposed genetic code has come from studies of proteins in bacteria and viruses in which amino acid positions have been carefully determined by biochemical means and then correlated with genetic data or exact nucleotide sequences. In *E. coli* Yanofsky and co-workers have shown that mutations at the tryptophan synthetase locus correlate closely with specific amino acid changes in the A protein of the tryptophan synthetase enzyme (see Fig. 26–23). In two instances, mutations *A23* and *A46*, only the same glycine amino acid is changed: to arginine in *A23* and to glutamic acid in *A46*. These mutant stocks are generally stable, although reverse mutations occasionally occur, either back to glycine (full revertants) or to other amino acids (partial revertants). Because of the pattern of reversion caused by base analogues and the frequency of occurrence of these mutations, it is believed that they are, in almost every case, caused by single nucleotide changes. As Fig. 28–4 shows, we can assign codons from Table 28–3 to these amino acids, and the hypothesis of single nucleotide changes for each mutation is consistent with the designated codons. Moreover, the observation that recombination between *A23* and *A46* produces wild-type recombinants means that recombinationally distinct mutant nucleotides are involved in the two codons. The recombinational event can therefore be explained as a single crossover consistent with the codon designations given.

A direct biochemical approach confirming most of the codon designations given in Table 28–3 was used in a study of the genes and proteins in the RNA bacteriophage MS2. As in other RNA phages of this type (f2, R17), MS2 is a small *E.*

coli phage containing about 3000 nucleotides which, according to complementation tests, are known to possess only three cistrons. The three different polypeptide products of these cistrons include a coat protein, an RNA-directed RNA replicase, and an assembly or maturation protein. It was the small size of this genotype, the limited number of products made by it, and the relative ease of labeling and isolating these materials that, for the first time, enabled the complete sequential analysis of a gene and its product.

The protein part of this analysis was accomplished by 1970 in experiments showing the exact sequence of the 129 amino acids in the coat protein. Using ribonucleotide-sequencing methods developed by Sanger and others, Fiers and coworkers then succeeded in analyzing all 387 nucleotides in this gene as well as a number of nucleotides at each end, including the beginning of the replicase gene. The information they derived has helped to clarify a number of important issues:

1. As shown in Fig. 28–5, there is an exact correspondence between each amino acid in the protein and one or more of its designated codons as given in Table 28–3. Furthermore, among the 49 different codons used in this gene there is, as expected, an obvious degeneracy; e.g., UUU, UUC = phenylalanine, UCU, UCC, UCA, UCG, AGC = serine, etc. The degeneracy, however, seems somewhat restricted since at least two of the amino acids appear to exclude the use of certain synonymous codons, although the use of such codons might have been expected on a chance basis. Thus isoleucine uses the codons AUU and

. . . (G)· AUA·GAG·CCC·UCA·ACC·GGA·GUU·UGA·AGC·AUG·

GCU·UCU·AAC·UUU·ACU·CAG·UUC·GUU·CUC·GUC·GAC·AAU·GGC·GGA·ACU·GGC·GAC·GUG·ACU·GUC·GCC·CCA·AGC·AAC·UUC·
Ala Ser Asn Phe Thr Gln Phe Val Leu Val Asp Asn Gly Gly Thr Gly Asp Val Thr Val Ala Pro Ser Asn Phe
1 5 10 15 20 25

GCU·AAC·GGG·GUC·GCU·GAA·UGG·AUC·AGC·UCU·AAC·UCG·CGU·UCA·CAG·GCU·UAC·AAA·GUA·ACC·UGU·AGC·GUU·CGU·CAG·
Ala Asn Gly Val Ala Glu Trp Ile Ser Ser Asn Ser Arg Ser Gln Ala Tyr Lys Val Thr Cys Ser Val Arg Gln
30 35 40 45 50

AGC·UCU·GCG·CAG·AAU·CGC·AAA·UAC·ACC·AUC·AAA·GUC·GAG·GUG·CCU·AAA·GUG·GCA·ACC·CAG·ACU·GUU·GGU·GGU·GUA·
Ser Ser Ala Gln Asn Arg Lys Tyr Thr Ile Lys Val Glu Val Pro Lys Val Ala Thr Gln Thr Val Gly Gly Val
55 60 65 70 75

GAG·CUU·CCU·GUA·GCC·GCA·UGG·CGU·UCG·UAC·UUA·AAU·AUG·GAA·CUA·ACC·AUU·CCA·AUU·UUC·GCU·ACG·AAU·UCC·GAC·
Glu Leu Pro Val Ala Ala Trp Arg Ser Tyr Leu Asn Met Glu Leu Thr Ile Pro Ile Phe Ala Thr Asn Ser Asp
80 85 90 95 100

UGC·GAG·CUU·AUU·GUU·AAG·GCA·AUG·CAA·GGU·CUC·CUA·AAA·GAU·GGA·AAC·CCG·AUU·CCC·UCA·GCA·AUC·GCA·GCA·AAC·
Cys Glu Leu Ile Val Lys Ala Met Gln Gly Leu Leu Lys Asp Gly Asn Pro Ile Pro Ser Ala Ile Ala Ala Asn
105 110 115 120 125

UCC·GGC·AUC·UAC·UAA·UAG·ACG·CCG·GCC·AUU·CAA·ACA·UGA·GGA·UUA·CCC·AUG·UCG·AAG·ACA·ACA·AAG·AAG·(U)
Ser Gly Ile Tyr Ser Lys Thr Thr Lys Lys
129 1 5

Figure 28–5

Nucleotide and amino acid sequences respectively for the coat protein gene of RNA phage MS2 and its protein product. As indicated, there are a total of 129 amino acids in the coat protein beginning with alanine (No. 1) and ending with tyrosine (No. 129), corresponding respectively to codons GCU and UAC. At either end of the gene are sequences of about 30 or more nucleotides which are not translated into amino acids. In the nucleotide sequence following the coat protein gene (lowermost line) there is, after a nontranslated interval, a codon sequence belonging to the replicase gene (No. 1–No. 6). Note that the initiation codon AUG precedes the translated sequences of each gene, and there are two consecutive termination codons at the end of the coat protein gene, UAA and UAG. (From Min Jou et al.)

AUC four times each, but there are no instances in which it uses AUA. Similarly UAC is used four times for tyrosine, but UAU is not used at all. It is suggested that the restricted use of such codons may arise from two causes: (a) The need of nucleotides in this phage to fulfill pairing requirements for maintaining a particular secondary structure of the RNA molecule (see below). (b) The need of phage codons to match the translational properties of the host *E. coli* cell (*E. coli* tRNA). These considerations show that, at least for these phages, the selection of which degenerate codons are to be used for an amino acid is probably not a random matter.

2. The initiation codon for translation, AUG, is exactly in its expected place at the beginning of the coat protein gene, and also starts the codon sequence for the replicase gene. However, in terms of the nucleotides around the initiation codon, there is no immediate clue to how this codon is distinguished from other AUG codons by the initiation tRNA molecule, $tRNA_f^{met}$. Interestingly the 30 or so nucleotides that precede the initiation codons of the coat protein and replicase genes are almost identical to those known for the same genes in phages f2 and R17. These nucleotides apparently designate ribosome attachment sites that have been conserved during evolution among this group of phages.

3. The entire nucleotide sequence of the coat protein gene can be arranged in a unique secondary structure based on the maximum amount of base pairing and the most molecularly stable configuration.* As shown in Fig. 28–6, this configuration assumes a flowerlike form with an essentially double-stranded stem leading into a number of petal-shaped hairpin loops. It is reasonable to suppose that this two-dimensional "flower" model is folded, in turn, into an as yet unknown three-dimensional structure. All this reaffirms the notion that at least some of the behavior of genetic material and mRNA is based on their secondary and tertiary structures as well as on their primary nucleotide sequences.

4. Questions such as the specificity of mutagens (Chapter 24) and the existence of mutational hot spots (Chapter 25) may well be resolved by these multidimensional considerations. For example, Fiers and co-workers point out that the nonsense codon UAG which is suppressed in *E. coli* "amber" suppressors (pp. 626 and 668), is known to arise in these RNA phages when a transitional mutation occurs (C → U) changing the glutamine codon CAG to UAG. However, such nonsense mutations are only mutagenically induced when a nucleotide in the glutamine codon is unpaired with a nucleotide on the opposite strand (codon positions 6, 50, 54, and 70). Thus the observation that the glutamine CAG codon at position 40 has not yet been found to mutate to a nonsense codon may be caused by the fact that all of its nucleotides are paired with those of the opposite strand. It is also of interest that differences exist with respect to which glutamine codon sites are affected by the mutagens used: nitrous acid causes CAG → UAG mutations at codon sites 6 and 50, fluorouracil at sites 6, 50, and 54, and hydroxylamine at sites 50 and 70.

5. A further consequence of MS2 molecular structure bears on the question of polarity mutations (p. 649). It was, for example, known that a nonsense mutation at the No. 6 codon of the coat protein terminated coat protein synthesis and prevented translation of the replicase gene. On the other hand, nonsense mutations at codons 50, 54, or 70 also terminated coat protein synthesis, but had no effect on replicase gene translation. This unusual polarity effect can now be explained as arising from the pairing between the AUG initiation codon for the replicase gene and parts of codons 27 and 28 of the coat protein gene (Fig. 28–6). Only when these latter two codons are translated by a ribosome on the coat protein gene is the AUG initiation codon of the replicase gene available for ribosomal attachment and translation. Thus the presence of a nonsense codon at position 6 prevents translation of the coat protein

*The existence of such a structure is supported by the fact that the various lengths of polynucleotide fragments produced by enzymatic breakdown were consistently similar in all of the MS2 RNA molecules. This indicated that these fragments derived from breaks at identical positions in a highly organized structure, since molecules with random secondary structures would have produced a much more varied population of nucleotide fragments when subjected to the ribonuclease enzyme used.

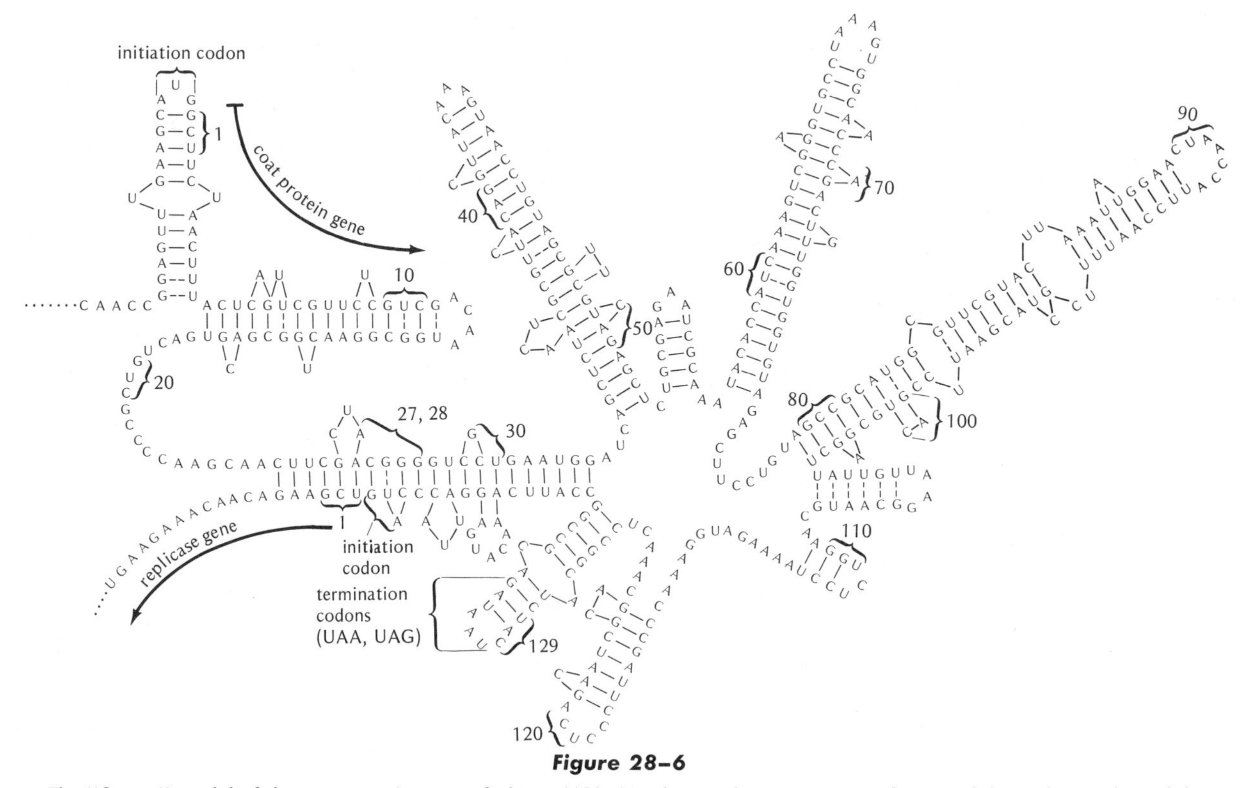

Figure 28–6

The "flower" model of the coat protein gene of phage MS2. Numbers indicate positions of some of the codon triplets of the gene as given in Fig. 28–5. (Modified after Min Jou et al.)

27–28 codon sequence, thereby preventing replicase initiation, whereas nonsense codons at positions 50 and thereafter enable translation of codons 27–28, and permit initiation and translation of the replicase gene.

UNIVERSAL CODE

One of the first indications that mRNA codons of one species could be translated into the correct amino acid sequence by the tRNA of another species were the *in vitro* experiments performed on hemoglobin synthesis by von Ehrenstein and Lipmann. They found that *E. coli* tRNA to which labeled amino acids were added would form hemoglobin when incubated with the mRNA and ribosomes of rabbit reticulocytes. The precision with which this interspecific attachment occurs was shown by converting cysteine into alanine in amino-acid-activated tRNA^{cys} (Fig. 27–7) and then observing that this alanine was now inserted into peptide positions ordinarily occupied by cysteine. In other words, the anticodon of the cysteine tRNA of a bacterial species recognized the cysteine codon of mammalian mRNA in spite of the fact that the tRNA was carrying an alanine amino acid! It is also indicative of universality that synthetic ribonucleotides such as poly-U, poly-UA, etc., will direct incorporation of the same amino acids into polypeptides in cell-free extracts derived from almost any organism (Table 28–4). Furthermore, trinucleotide binding experiments (p. 659) performed by Nirenberg and his colleagues have shown conclusively that the tRNAs of bacteria (*E. coli*), amphibians (*Xenopus laevis*), and mammals (guinea pig) respond similarly to the same triplet

TABLE 28–4

Stimulation of specific amino acid incorporation into polypeptides by synthetic ribonucleotides in cell-free extracts of different organisms

Source of Cell-free Extract	Amino Acid Incorporated Into Polypeptides					
	phe	*leu*	*ile*	*tyr*	*val*	*ser*
E. coli	poly-U, -UA, -UC, -UG	poly-U, -UA, -UC, -UG	poly-UA	poly-UA	poly-UG	poly-UC
Chlamydomonas	poly-U, -UA, -UC, -UG	poly-U, -UA, -UC, -UG	poly-UA	poly-UA	poly-UG	poly-UC
rat liver	poly-U, -UA, -UC, -UG	poly-U, -UA, -UC, -UG	poly-UA	poly-UA	poly-UG	poly-UC
mouse plasmocytoma	poly-U, -UA, -UC, -UG	poly-U, -UA, -UC, -UG	poly-UA	poly-UA	poly-UG	poly-UC
mouse leukemia	poly-U, -UA, -UC, -UG	poly-U, -UA, -UC, -UG	poly-UA	poly-UA	poly-UG	poly-UC

After Weinstein, 1963.

codons by binding the specific amino acids designated by the code (Table 28–3) to each of 50 different codons tested.* Even nonsense codons appear to be universal both in procaryotes and eucaryotes.

Code universality is also noted *in vivo* when transferring an F′ episome bearing the *E. coli lac*$^+$ factor into the bacterial species *Serratia marcescens*. The β galactosidase enzyme produced by such sexducted *Serratia* is found to be characteristic of that normally produced by *E. coli*. Similarly Gurdon and co-workers have shown that hemoglobin mRNA molecules from both mice and rabbits are efficiently translated into α and β polypeptide chains when injected into *Xenopus* eggs.

In respect to humans, perhaps the most striking example of universality arises from the claim made by Merril and co-workers that a bacterial enzyme catalyzing the metabolism of galactose sugars (α-D-galactose-1 phosphate uridyl transferase) is produced in human tissue culture cells after infection by a λ virus carrying the *E. coli gal*$^+$ gene. These cells, originating from a galactosemic patient, were unable to make the enzyme before the infection (see Table 26–2), nor could they be induced to make the enzyme by infection of λ virus carrying the *gal*$^-$ gene. Thus, Merril suggests that human cells are not only capable of translating bacterial mRNA codons into the correct amino acids but they can also transcribe bacterial DNA into mRNA. The common mechanisms of transcription and translation seen in all organisms may therefore enable the successful interspecific transfer and function of at least some genes.

The reason for code universality is not hard to find if we consider that a species must always form proteins no matter how many changes it undergoes during evolution. Since protein formation depends upon the precise positioning of the 20 different amino acids into numerous polypeptides, any mutation that will change the code will also change the positioning of amino acids and will thereby affect practically every protein formed by the cell. For example, if the code for alanine were changed to that for serine, then all or most tRNA bearing the amino acid alanine would now insert into serine positions. Such mutations would undoubtedly lead to drastic lethal effects, since many proteins are changed all at once. It is certainly much more

* The efficiency of amino acid binding to some of the triplets differed between the procaryotes and eucaryotes, indicating differences between some of their tRNAs. Along with further *in vitro* evidence, this suggests that not all 64 codons are used in every species, nor are identical choices always made of which codons are to be selected among the various synonymous codons that can specify a particular amino acid. Thus AUA is a good codon for isoleucine in eucaryotes but is apparently not ordinarily used for coding this amino acid in *E. coli*, which seems to rely instead on AUU and AUC. Restrictions in the use of certain codons is quite obvious in the MS2 virus described in the previous section. In spite of limitations of this kind, however, the code is still considered to be universal, since there are no instances in which a particular codon designates different amino acids in different species.

likely that protein changes, when they occur, take place one or a few at a time, and then only in respect to the positions of particular amino acids. For example, a change in one nucleotide sequence of mRNA might result in alanine being inserted into a former serine position in one of the proteins formed, or vice versa. Through analysis of the amino acid sequence in mutant proteins (Chapter 26) it is apparent that most observable mutations are the result of single amino acid substitutions. For these substitutions to take place, a change in nucleotide sequence must occur at a particular place on the mRNA. Thus, although the basic "code" remains the same in all organisms, it is the "messages" that change because of mutations.

The universality of the code, however, does not mean that DNA base ratios must be similar in different species for genes specifying similar proteins. The fact that the code is degenerate enables many bases to be changed by mutation in a sequence of mRNA, but this mRNA could still produce the same amino acid sequence. Thus the finding that DNA base ratios may differ greatly between organisms (see Table 4–1) does not detract from use of the same code in all species.

INTRAGENIC SUPPRESSION

Mutant changes in nucleotide sequences do not necessarily always lead to mutant effects. Some mutations are "silent" or "neutral," and may affect portions of the protein molecule that do not lead to functional change. Bonner's finding that there are sections of the tryptophan synthetase gene in *Neurospora* in which no known mutations occur is presumed to indicate areas of this kind. Such mutations are not detected simply because amino acid substitutions in these areas do not produce defective proteins. On the other hand, even areas that seem crucial for the functioning of a protein may show specific amino acid substitutions which produce no marked change in function. Thus the substitution of the serine and alanine codons for that of glycine at the *A23–A46* locus of the *E. coli* tryptophan synthetase gene does not appear to affect the function of the enzyme, although other changes at this locus have a marked effect (see Fig. 28–4).

Aside from "silent" mutations, however, mutant effects may be absent simply because they are suppressed by the presence of other mutations called *suppressors*. Two main types of suppressors can be distinguished: (1) *intragenic suppressors*, those that suppress the effect of a mutation in the same gene in which they are located; and (2) *intergenic suppressors*, those that suppress the effect of mutations in other genes.

One type of mutation causing intragenic suppression has already been discussed (p. 657) and it involves rectification of a change in reading frame of the nucleotide code by a further change, as, for example, when deletion of one nucleotide in the reading frame is followed by insertion of another nearby. A second type of intragenic suppression may occur when the suppressor mutation produces a new amino acid substitution whose effect on protein structure compensates for the change caused by an amino acid substitution at another point along the protein. For example, in the *A* gene of *E. coli* tryptophan synthetase, two mutations (*A46* and *A446*) mapping about 100 nucleotides apart cause a partial reversion of the mutant effect when they are present together (see Helinski and Yanofsky). Since a number of other mutations have now been found in these same areas which enable reversion of the double mutant, it has been suggested that these two *trp* sites, separated by about 35 amino acids, interact with each other in the three dimensional structure of the enzyme (Fig. 28–7). Analysis of transfer RNA mutations also shows such precisely localized interactions, but some of these are more readily explainable, since they directly affect base pairing. Thus mutations have been found in the tRNA molecule, shown in Fig. 28–9, which change G to A at position 31, or C to U at position 41, each of which interferes with base pairing in that double-stranded stem (A$\not-$C, G$\not-$U) and thereby prevents proper function of this tRNA. However, when the double mutant occurs, *A31–U41*, base pairing is restored (A–U) and the tRNA molecule can function.

In the diploid or heterocaryon, mutations may

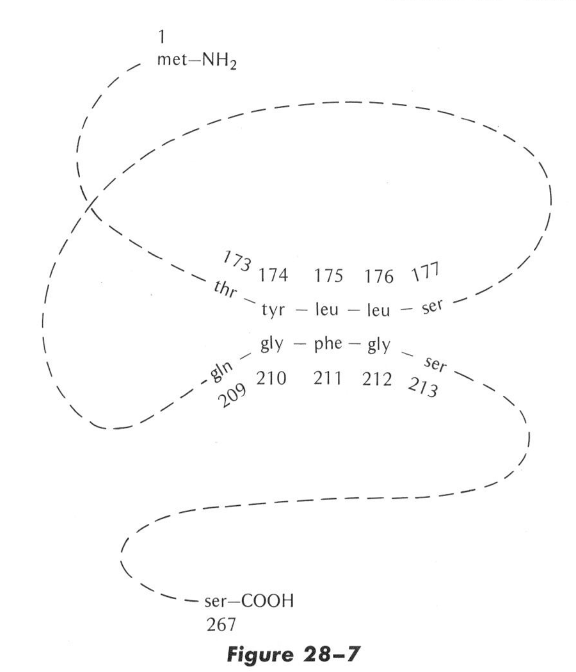

Figure 28–7

Amino acids found at the designated numbered positions on the tryptophan synthetase A protein. The diagram shows the close proximity that may exist in the folded polypeptide chain between amino acids on the 174–176 sequence and those on the 210–212 sequence. Evidence for this is the fact that certain amino acid substitutions at position 174 compensate for other substitutions at 210, and mutations at position 176 compensate for others at 212. (After Yanofsky, Horn, and Thorpe.)

be considered to be suppressors if they can "correct" through complementation the mutant phenotype caused by the presence of other mutant genes. We have seen that such complementation often occurs between mutations on two different cistrons as demonstrated by Benzer (Chapter 25) in the normal growth of two strains of T4 phage that infect the same bacterial cell but carry mutations in different *rII* cistrons. Other examples of this kind are now plentiful, showing intercistronic or interallelic complementation both *in vivo* and *in vitro* (e.g., see Glassman). Of special interest, however, is the observation that complementation can also occur intracistronically between the polypeptide products of two different mutations that affect the *same* cistron.

One clear example of such intracistronic complementation is the study of *E. coli* alkaline phosphatase mutations by Schlesinger, Levinthal, and others. Wild-type alkaline phosphatase is ordinarily a dimer composed of two identical polypeptide chains that may be replaced by mutant forms without enzyme activity. All these mutations map in the alkaline phosphatase cistron. Using immunological methods, some of these mutant proteins can be isolated (see cross-reacting material, p. 607) and shown to complement each other *in vitro*, producing active alkaline phosphatase. The product of these complementation reactions has been demonstrated to be a hybrid protein molecule in dimer form containing one polypeptide chain from each of the two mutations. Since each *in vitro* complementation between two mutations is paralleled by a similar *in vivo* complementation, it is presumed that the complementation mechanism is probably similar under both conditions. That is, a diploid organism or heterocaryon bearing two intracistronic complementary mutations produces polypeptide chains from each type of mRNA which may then aggregate to yield active proteins. The fact that most proteins contain two or more polypeptide chains and that their functional activity is based on the configuration of this multichained (*multimer*) structure allows for the possibility that polypeptide chains defective at different positions may nevertheless form a functional protein (Fig. 28–8). The advantages of heterozygosity in many diploid species (see Chapter 33) may therefore arise as a consequence of complementation reactions between polypeptides.

INTERGENIC SUPPRESSION

Among the intergenic mechanisms that may cause suppression are those that do not change the amino acid sequence in the mutant protein itself but rather modify the environment, thereby enabling the mutant protein to function. An example of a change in physiological conditions that acts to suppress a mutant effect is one discovered by Suskind and Kurek for the *td-24* tryptophan synthetase mutation in *Neurospora*. This mutation produces a protein that is extraordinarily sensitive

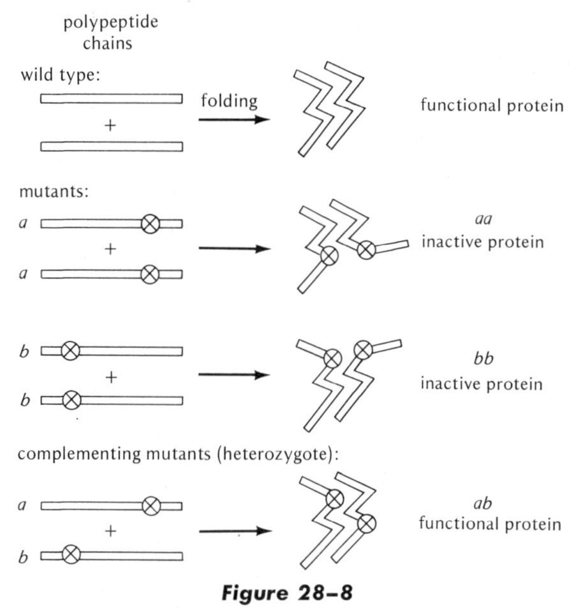

Figure 28–8

Intracistronic complementation between mutants that affect two polypeptide chains that combine to form one dipolypeptide (dimer) molecule.

to zinc ions but will function more normally when the zinc-ion concentration is lowered. A particular suppressor is known which has no effect on the *td-24* mutant protein itself but enables it to function in the cell, ostensibly by lowering the zinc-ion concentration. Other suppressors may overcome the effect of a mutant gene by permitting an alternate metabolic pathway to produce a necessary product or by metabolizing an inhibitory substance that has been accumulated because of the mutation.

Intergenic suppression, however, may also occur by actual changes in the amino acid sequence of the mutant protein itself. That is, a nonallelic suppressor gene may cause sufficient "errors" in the nucleotide–amino acid translation mechanism to enable a mutant nucleotide sequence to produce a few wild-type-like proteins. Possible intergenic suppressor effects of this kind include the following:

1. Appearance of a new kind of tRNA that enables an amino acid to be coded by a former "nonsense" codon. Examples of this mechanism include suppressors of the *amber* mutations that lead to the production of only partial polypeptide chains (see p. 627). These were the first nonsense mutations to be identified and have now been shown to be caused by the mutant appearance of the termination codon UAG on mRNA before the normal termination point. Since then, suppressors have also been found for the similarly acting nonsense mutations *ochre* (UAA) and *opal* (UGA). According to Brenner and co-workers, as well as Garen and others, the effect of such nonsense mutations can be overcome by suppressor genes that enable a particular tRNA to recognize the nonsense codon (Fig. 28–9). For each suppressor tRNA a specific amino acid is inserted into the polypeptide chain at the nonsense codon position, so that *amber* mutations in some suppressor strains, for example, have tyrosine in place of the former nonsense termination point, others serine, others glutamine, etc.* These suppressor tRNA molecules are therefore essentially mutations of normal tRNAs which continue to carry their former amino acid. Some of them have been mapped in *E. coli* and their loci show a fairly widespread distribution along the chromosome (see J. D. Smith).

---

*One can, of course, raise the question of the effect of such suppressors on normal chain termination. That is, if the suppressor enables the continuation of polypeptide synthesis past the mutant nonsense codon, it should also cause lethality by synthesizing long protein chains that are extended beyond their normal termination points. However, strains carrying nonsense suppressors are often viable, both in bacteria and yeast. So far the methods by which such suppressor strains protect themselves against simultaneous suppression of some or most of their normal termination signals is not fully known. There are apparently some instances in which normal polypeptide synthesis is terminated by *two different* consecutive nonsense codons (e.g., UAA UAG, see Fig. 28-5). In the phage $Q\beta$, on the other hand, there is apparently only one termination codon (UGA) for the coat protein gene, and an opal suppressor does seem to cause production of a nonfunctioning $Q\beta$ protein that is considerably longer than the normal coat protein. It is possible that some mechanism in addition to the nonsense codon itself ensures normal polypeptide termination at the end of a gene in nonsense suppressor strains; perhaps a special secondary structure on the mRNA molecule at the termination point, or a recognition device on the ribosome which "rejects" many of the mutant tRNAs carrying nonsense anticodons. Evidence for the latter alternative has been the discovery of *E. coli* ribosomal mutations called *ram* (ribosomal ambiguity) which greatly reduce the viability of cells carrying suppressor tRNA molecules. According to Biswas and Gorini, the *ram* mutation probably interferes with a ribosomal function that normally discriminates between different tRNAs and accepts only those that have not mutated.

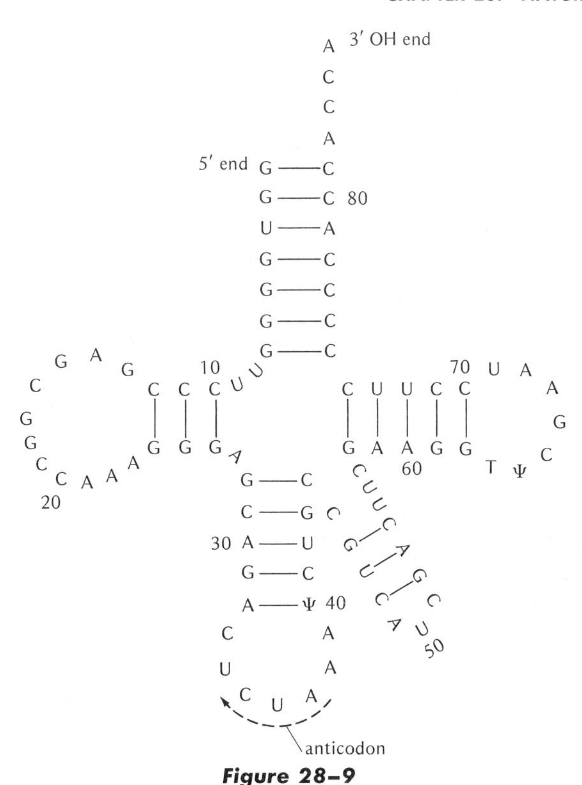

Figure 28–9

The nucleotide sequence of the tyrosine tRNA molecule in E. coli *responsible for the suppression of amber (UAG) nonsense mutations. Numbers indicate the position of each nucleotide from the 5′ end of the molecule. This nonsense suppressor tRNA has arisen from a fraction of* $tRNA^{tyr}$ *molecules which carry a mutation at their No. 35 position (G → C) giving them an anticodon of AUC (dashed arrow) rather than the normal AUG. This nucleotide substitution does not apparently affect the ability of the aminoacyl tRNA synthetase enzyme to attach tryosine to this tRNA molecule, and these tRNAs therefore insert tyrosine into polypeptides at amber UAG nonsense codon positions. The cells, however, still contain a major fraction of nonmutated* $tRNA^{tyr}$ *molecules transcribed at a different gene locus, which enables them to incorporate tyrosine into polypeptides at the usual UAC and UAU positions.*

2. Appearance of a new kind of tRNA that enables the "correct" amino acid to be coded by a former "missense" codon. Just as some suppressor tRNAs insert an amino acid into a polypeptide at a nonsense codon position, other suppressor tRNAs arise that can substitute the correct amino acid for a mutant type by recognizing a missense codon. Thus, for example, the *A36* mutation in *E. coli* tryptophan synthetase which results in the substitution of glycine (GGA) by the mutant amino acid arginine (AGA) can be suppressed by a particular mutant glycine tRNA which can pair with the arginine codon AGA but nevertheless carries the amino acid glycine. Of course, this suppressor also inserts glycine into some of the other nonmutant arginine positions, and growth in this strain is therefore somewhat diminished because of the presence of defective polypeptides.

3. Appearance of a new kind of tRNA that causes a change in the codon reading frame so that a frame-shift mutation can be corrected. Since frame-shift mutations arise from the insertion or deletion of any number of nucleotides in a reading frame other than a multiple of three (p. 657), suppressors of such mutations would have either to eliminate the reading of the added nucleotide(s) or compensate for the missing nucleotide(s). For example, a frame-shift mutation caused by a single added nucleotide to a reading frame could be suppressed by a tRNA molecule which would presumably "read" a single nucleotide by having only one nucleotide in its anticodon, or by a tRNA that has four nucleotides in its anticodon and thereby reads or occupies the space of one nucleotide in addition to the normal three. Similarly a deletion of one nucleotide in a reading frame could perhaps be suppressed by a tRNA that can read only two nucleotides. So far, various frame-shift mutation suppressors have been found and they all appear to be caused by mutant tRNA molecules. In *Salmonella typhimurium* a number of cases are known in which suppressors of frame-shift mutations in the histidine pathway (p. 624) are associated with altered $tRNA^{pro}$ or $tRNA^{gly}$. One of these suppressor $tRNA^{gly}$ molecules has been shown by Riddle and Carbon to differ from a normal $tRNA^{gly}$ (anticodon CCC) only in having four nucleotides in its anticodon (CCCC).

4. A change in ribosomal configuration that enables codon-anticodon misreadings to occur and thereby suppresses some mutations through misreading. According to Davies, Gilbert, and Gorini, the antibiotic streptomycin seems to function in this fashion by distorting ribosome structure and

changing the codon recognition system. In streptomycin-sensitive cells exposed to streptomycin, most proteins produced are nonfunctional, because of widespread misreading of the genetic code. However, in streptomycin-resistant cells many fewer proteins show changes upon streptomycin exposure, and, in such cells, streptomycin may actually function as a suppressor of some gene mutations. For example, in a cell-free system using poly-U mRNA, Pestka, Marshall, and Nirenberg have shown that streptomycin causes the occasional substitution of isoleucine tRNA (anticodon UAA) for that of phenylalanine (anticodon AAA). As discussed previously (p. 646), it is now known that streptomycin affects the ribosome directly, and streptomycin-resistant strains carry mutations in a 30*S* ribosomal protein.

5. A change in the amino-acid-activating enzyme that causes attachment of a different amino acid to a particular tRNA. A hypothetical example would be suppression of a mutation in a particular triplet, AUU(isoleucine) $\rightarrow$ UUU(phenylalanine), by the occasional attachment of isoleucine to $tRNA^{phe}$.

PROBLEMS

28-1. Given that the DNA-protein code is triplet and that the average protein is about 500 amino acids long, how many different proteins are possible in a cell having a haploid DNA complement of 2×10^{12} molecular weight? (One nucleotide pair in a DNA double helix has a molecular weight of 660.)

28-2. If DNA contained only the bases adenine and thymine, how long a code word would be necessary to enable coding for each of 20 different amino acids?

28-3. (a) Can you devise a nonoverlapping commaless triplet code for the 20 different amino acids in which the codons can be translated correctly without the necessity of a reading frame? (b) What theoretical arguments can you present against the existence of this type of code?

28-4. A particular DNA base sequence transcribed into messenger RNA is

TTATCTTCGGGAGAGAAAACA

(a) If reading begins at the left, what amino acids are coded by this sequence? (b) If proflavin treatment caused the deletion of the first adenine nucleotide on the left, what changes will occur in the first six amino acids coded by this sequence?

28-5. If a polyribonucleotide contained equal amounts of randomly positioned adenine and uracil bases, what proportion of triplets will code for: (a) Phenylalanine? (b) Isoleucine? (c) Leucine? (d) Tyrosine?

28-6. Using a polyribonucleotide that contained 47 percent adenine and 53 percent cytosine placed randomly, Jones and Nirenberg obtained the following frequencies of amino acids incorporated into proteins: 10.8% lysine, 11.6% asparagine, 9.3% glutamine, 9.4% histidine, 26.3% threonine, and 32.6% proline. Do the observed frequencies of amino acids accord with those that would be expected?

28-7. In addition to the primary amino acid structure of beef insulin discovered by Sanger, the structures of other animal insulins have since been determined. The only differences between these insulins lies in a small amino acid sequence adjacent to a cystine residue. Some of the observed differences for this sequence are:

beef: ala-ser-val
sheep: ala-gly-val
pig: thr-ser-ile
horse: thr-gly-ile

(a) Would you say this region is crucial for hormonal activity? (b) Would single nucleotide substitutions account for the amino acid substitutions?

28-8. Guest and Yanofsky observed the following amino acid substitutions at one particular site (position 233) on protein A of *E. coli* tryptophan synthetase:

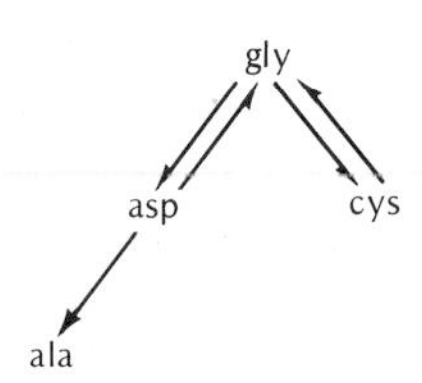

(a) Using the coding dictionary in Table 28-3, what codons would fit this set of events if each amino acid substitution is caused by only a single nucleotide change? (b) Which mutant combinations would you predict could recombine to form wild type (gly)? (c) Which mutant combinations could not recombine to form wild type?

28-9. Which of the above amino acid changes represent transitions and which represent transversions?

28-10. According to Beale and Lehmann, the following are some of the amino acid substitutions that occur in the α and β chains of human hemoglobin. (Dashed lines indicate amino acid differences between β and δ chains.)

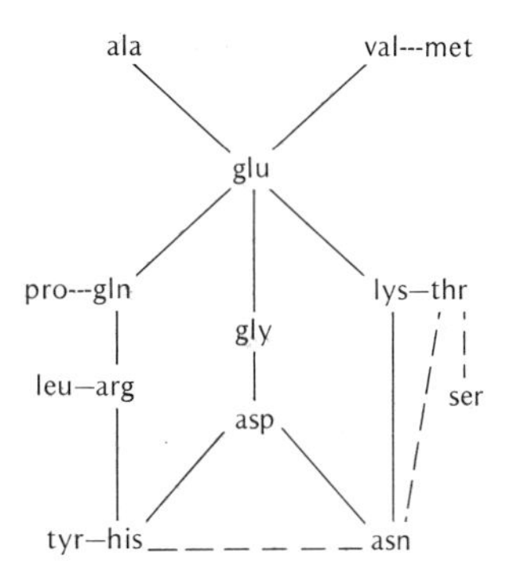

Using the coding dictionary in Table 28-3, propose codons for these amino acids which would most closely explain the observed substitutions in the form of single nucleotide changes.

28-11. Streisinger and co-workers studied amino acid sequences in the lysozyme protein produced by the T4 phage. One sequence is lys-ser-pro-ser-leu-asn-ala, but as a result of a deletion of a single nucleotide and subsequent insertion of another nucleotide this amino acid sequence was found to change to lys-val-his-his-leu-met-ala. (a) Using the codons in Table 28-3, determine the nucleotide sequences that produced the original amino acid sequence and determine the subsequent changes. (b) Which end of this amino acid sequence is toward the 5′ end of the mRNA molecule?

28-12. Sherman and co-workers analyzed four different mutations (*A-D*) that affect the synthesis of one of the cytochrome C proteins in yeast. All four of these mutations were localized to the same genetic locus, but could nevertheless be distinguished from each other by the fact that they produced different (partial) revertants to wild-type cytochrome C activity. The following comparison shows the sequence of amino acids at the N terminal of the normal gene, and the sequences determined for the revertants of the four mutations. The bottom line shows the N-terminal amino acid sequence for occasional revertants found in all of the four different mutant strains.

NORMAL SEQUENCE:	thr-glu-phe-lys-ala-gly-
Revertant of mutation *A*:	ile-thr-glu-phe-lys-ala-gly-
Revertant of mutation *B*:	leu-thr-glu-phe-lys-ala-gly-
Revertant of mutation *C*:	arg-thr-glu-phe-lys-ala-gly-
Revertant of mutation *D*:	val-thr-glu-phe-lys-ala-gly-
Revertant of mutations *A-D*:	ala-gly-

(For simplicity the initial methionine amino acid found in some of the above revertants is not indicated. However, as discussed on page 641, it should be understood that each mRNA translational sequence begins with an initiator codon that specifies methionine, although the methionine residue may be excised after incorporation into the polypeptide chain.)

Based on the above information: (a) What were the nucleotide changes caused originally by each of the mutations *A-D*? (b) What accounts for the fact that all four of the mutations can produce the same revertant as shown in the lower line?

28-13. Using Crick's wobble hypothesis (p. 659): (a) How would you explain the specificity of tRNA anticodons in distinguishing between the codons for cysteine (UGU, UGC), tryptophan (UGG) and the UGA nonsense codon? (b) Would you expect to find an aminoacyl tRNA that recognized only A at the third codon position? (c) How would you use the wobble hypothesis to account for the observation that suppressors of *ochre* nonsense mutations also suppress *amber*, whereas suppressors of *amber* nonsense mutations do not suppress *ochre*?

28-14. Explain whether you would expect intragenic complementation in both hemoglobin and myoglobin.

28-15. Brenner and Stretton found that *amber* or *ochre* mutations did not terminate polypeptide synthesis in the *rII* gene of T4 phage when these mutations were located within an interval in which a single nucleotide insertion had been made on one end and a single nucleotide deletion had been made on the other. How would you explain this finding?

28-16. A particular T4 phage nonsense mutation was found to affect the synthesis of a particular phage protein when grown in *E. coli* cells. This nonsense mutation, however, could undergo reverse mutation and thereby produce functional protein when the mutant phage T4 DNA was subjected to the transition-producing mutagen, 2-aminopurine (see p. 574). In

α Chain Positions

	138			141					146		
Hb A	. . . ser	lys	tyr	arg(COOH)							
Hb W1	. . . ser	asn	thr	val	lys	leu	glu	pro	arg(COOH)		
Hb CS	. . . ser	lys	tyr	arg	gln	ala	gly	ala	ser	val	ala . . .

some cases the reverse mutation led to the substitution of the arginine amino acid for the previous nonsense codon, and in other cases it caused the substitution of tryptophan. Furthermore, the phage strain carrying the nonsense mutation could be shown to recombine with another strain carrying an *amber* nonsense mutation leading to the production of wild-type recombinants. Explain these results by specifying the codon of the unknown nonsense mutation.

28-17. What tests would you perform to distinguish whether a particular protein defect in *E. coli* is caused by a nonsense mutation or by a frame-shift mutation?

28-18. Two human hemoglobin genetic variants are known to cause a lengthening of the α chain at its carboxyl (COOH) terminal. In one of the variants, called hemoglobin Constant Springs (Hb CS), no other amino acid changes are noted within the chain, whereas the other hemoglobin variant, Wayne 1 (Hb W1), also shows substituted amino acids beginning with amino acid position #139. Comparison between these variants and normal hemoglobin (Hb A) shows the above sequences. Assuming that only *single* nucleotide changes (either substitution, deletion, or insertion) are responsible for the appearance of each of these two genetic variants, how would you explain their amino-acid sequences in terms of nucleotide sequences?

REFERENCES

BENNETT, T. P., J. GOLDSTEIN, and F. LIPMANN, 1965. Coding and charging specificities of sRNA's isolated by countercurrent distribution. *Proc. Nat. Acad. Sci.*, **53,** 385–392.

BISWAS, D. K., and L. GORINI, 1972. Restriction, de-restriction, and mistranslation in missense suppression: Ribosomal discrimination of transfer RNA's. *Jour. Mol. Biol.*, **64,** 119–134.

BONNER, D. M., 1965. Gene-enzyme relationships. In *Genetics Today*, Vol. II, Proc. XI Intern. Congr. Genet., S. J. Geerts (ed.). Pergamon, Oxford, pp. 141–148.

BRENNER, S., A. O. W. STRETTON, and S. KAPLAN, 1965. Genetic code: The "nonsense" triplets for chain termination and their suppression. *Nature*, **206,** 994–998. (Reprinted in *Selected Papers in Biochemistry*, Vol. 3; see References, Chapter 1.)

CLARK, B. F. C., and K. A. MARCKER, 1966. The role of N-formyl-methionyl-sRNA in protein biosynthesis. *Jour. Mol. Biol.*, **17,** 394–406. (Reprinted in *Selected Papers in Biochemistry*, Vol. 3; see References, Chapter 1.)

CRICK, F. H. C., 1966. Codon-anticodon pairing: The wobble hypothesis. *Jour. Mol. Biol.*, **19,** 548–555. (Reprinted in the collections of Abou-Sabé and of Zubay and Marmur and in Vols. 3 and 6 of *Selected Papers in Biochemistry*; see References, Chapter 1.)

CRICK, F. H. C., L. BARNETT, S. BRENNER, and R. J. WATTS-TOBIN, 1961. General nature of the genetic code for proteins. *Nature*, **192,** 1227–1232. (Reprinted in the collections of Stent, of Taylor, and of Zubay and Marmur; see References Chapter 1.)

CRICK, F. H. C., and L. E. ORGEL, 1964. The theory of inter-allelic complementation. *Jour. Mol. Biol.*, **8,** 161–165.

DAVIES, J., W. GILBERT, and L. GORINI, 1964. Streptomycin, suppression, and the code. *Proc. Nat. Acad. Sci.*, **51,** 883–890. (Reprinted in the collection of Adelberg; see References, Chapter 1.)

EHRENSTEIN, G. VON, and F. LIPMANN, 1961. Experiments on hemoglobin biosynthesis. *Proc. Nat. Acad. Sci.*, **47,** 941–950.

GAREN, A., S. GAREN, and R. C. WILHELM, 1965. Suppressor genes for nonsense mutations. *Jour. Mol. Biol.*, **14,** 167–178.

GLASSMAN, E., 1962. In vitro complementation between nonallelic *Drosophila* mutants deficient in xanthine dehydrogenase. *Proc. Nat. Acad. Sci.*, **48,** 1491–1497.

GURDON, J. B., J. B. LINGREL, and G. MARBAIX, 1973. Message stability in injected frog oocytes: Long life of mammalian α and β globin messages. *Jour. Mol. Biol.*, **80,** 539–551.

HARTMAN, P. E., and J. R. ROTH, 1973. Mechanisms of suppression. *Adv. in Genet.*, **17,** 1–105.

HELINSKI, D. R., and C. YANOFSKY, 1963. A genetic and biochemical analysis of second-site reversion. *Jour. Biol. Chem.*, **238,** 1043–1048.

KHORANA, H. G., H. BÜCHI, H. GHOSH, N. GUPTA, T. M. JACOB, H. KÖSSEL, R. MORGAN, S. A. NARANG, E. OHTSUKA, and R. D. WELLS, 1967. Polynucleotide synthesis and the genetic code. *Cold Sp. Harb. Symp.,* **31,** 39–49.

MARSHALL, R. E., C. T. CASKEY, and M. NIRENBERG, 1967. Fine structure of RNA codewords recognized by bacterial, amphibian, and mammalian transfer RNA. *Science*, **155,** 820–826.

MARTIN, G., and F. JACOB, 1962. Transfer de l'episome sexuel d'*Escherichia coli à Pasteurella pestis. C. R. Acad. Sci., Paris*, **254,** 3589–3590. Cites demonstration of β galactosidase synthesis by *E. coli* F'*lac* factor in *Serratia.*

MERRIL, C. R., M. R. GEIER, and J. C. PETTRICIANI, 1971. Bacterial virus gene expression in human cells. *Nature*, **233,** 398–400.

MIN JOU, W., G. HAEGEMAN, M. YSEBAERT, and W. FIERS, 1972. Nucleotide sequence of the gene coding for the bacteriophage MS2 coat protein. *Nature*, **237,** 82–88.

NIRENBERG, M. W., 1963. The genetic code: II. *Sci. American*, **190,** March issue, pp. 80–94.

NIRENBERG, M. W., and P. LEDER, 1964. RNA codewords and protein synthesis: The effect of trinucleotides upon the binding of sRNA to ribosomes. *Science*, **145,** 1399–1407. (Reprinted in Zubay and Marmur's collection, and in *Selected Papers in Biochemistry*, Vol. 3; see References, Chapter 1.)

NIRENBERG, M. W., and J. H. MATTHEI, 1961. The dependence of cell-free protein synthesis in *E. coli* upon naturally occurring or synthetic polyribonucleotides. *Proc. Nat. Acad. Sci.*, **47,** 1588–1602. (Reprinted in the collections of Taylor and of Zubay and Marmur and in *Selected Papers in Biochemistry*, Vol. 3; see References, Chapter 1.)

PESTKA, S., R. MARSHALL, and M. NIRENBERG, 1965. RNA codewords and protein synthesis, V. Effect of streptomycin on the formation of ribosome-sRNA complexes. *Proc. Nat. Acad. Sci.*, **53,** 639–646.

RIDDLE, D. L., and J. CARBON, 1973. Frameshift suppression: A nucleotide addition in the anticodon of a glycine transfer RNA. *Nature New Biol.*, **242,** 230–234.

SANGER, F., G. G. BROWNLEE, and B. G. BARRELL, 1965. A two-dimensional fractionation procedure for radioactive nucleotides. *Jour. Mol. Biol.*, **13,** 373–398.

SCHLESINGER, M. J., and C. LEVINTHAL, 1963. Hybrid protein formation of *E. coli* alkaline phosphatase leading to *in vitro* complementation. *Jour. Mol. Biol.*, **7,** 1–12.

SIGNER, E. R., 1965. Gene expression in foreign cytoplasm. *Jour. Mol. Biol.*, **12,** 1–18.

SMITH, J. D., 1972. Genetics of transfer RNA. *Ann. Rev. Genet.*, **6,** 235–256.

SMITH, M. A., M. SALAS, W. M. STANLEY, Jr., A. J. WAHBA, and S. OCHOA, 1966. Direction of reading of the genetic message, II. *Proc. Nat. Acad. Sci.*, **55,** 141–147.

SÖLL, D., E. OHTSUKA, D. S. JONES, R. LOHRMANN, H. HAYATSU, S. NISHIMURA, and H. G. KHORANA, 1965. Studies on polynucleotides, XLIX. Stimulation of the binding of aminoacyl-sRNA's to ribosomes by ribotrinucleotides and a survey of codon assignments for 20 amino acids. *Proc. Nat. Acad. Sci.*, **54,** 1378–1385.

SÖLL, D., and U. L. RAJBHANDRY, 1967. Studies on polynucleotides. LXXVI. Specificity of transfer RNA for codon recognition as studied by amino acid incorporation. *Jour. Mol. Biol.*, **29,** 113–124.

SUSKIND, S. R., and L. I. KUREK, 1959. On a mechanism of suppressor gene regulation of tryptophan synthetase activity in *Neurospora crassa. Proc. Nat. Acad. Sci.*, **45,** 193–196.

WEIGERT, M. G., and A. GAREN, 1965. Base composition of nonsense codons in *E. coli.* Evidence from amino acid substitutions at a tryptophan site in alkaline phosphatase. *Nature*, **206,** 992–994. (Reprinted in *Selected Papers in Biochemistry*, Vol. 3; see References, Chapter 1.)

WEINSTEIN, I. B., 1963. Comparative studies on the genetic code. *Cold Sp. Harb. Symp.*, **28,** 279–280.

YANOFSKY, C., V. HORN, and D. THORPE, 1964. Protein structure relationships revealed by mutational analysis. *Science*, **146,** 1593–1594.

29
GENE REGULATION

Two basic properties of cells have, so far, been described: the transfer of information between cells by means of DNA, and the elaboration of this information into the amino acid sequences of proteins. The behavior of cells, however, depends greatly upon the time at which genetic information is expressed, the precise portion of information that is expressed, and the extent to which this expression occurs. In terms of cellular behavior, a number of general questions pose themselves: When is DNA to be synthesized? Which of the various kinds of RNA—messenger, ribosomal, transfer—are to be transcribed? Which of the messengers are to be translated? Which of the translated proteins are to function? How often are transcription and translation to occur? Where in the organism or cell are these events to be localized?

The answers to these questions derive from phenomena that can broadly be called *regulatory*; that is, their solution depends upon mechanisms that enable a living organism to selectively control or regulate the various processes under its command. As implied in the above questions, controls must exist on the levels of DNA synthesis, RNA transcription, RNA translation, and protein function. How are these controls exercised?

REGULATOR GENES

The first clear example of the existence of precise control by one gene over the function of another involved regulation of mRNA transcription for a particular group of enzymes concerned with lactose metabolism in

E. coli. As we know, bacterial cells are capable of making many kinds of enzymes in relatively large quantities, but the varieties and amounts of enzymes actually produced at any given time are usually only those necessary for the task called for by their environment. For example, a broad category of enzymes, called *inducible enzymes*, are ordinarily absent or only present in very small quantities unless a specific compound, called an *inducer*, is present. β galactosidase, an enzyme that splits lactose into galactose and glucose (Fig. 29–1a) has rarely more than one or two molecules present per cell in *E. coli* raised on nonlactose medium (e.g., glycerol). Lactose, however, is converted to a product (allolactose, Fig. 29–1b) which acts as an inducer for β galactosidase, and within 2 to 3 minutes after lactose medium is introduced *E. coli* cells will start producing the enzyme until about 3000 molecules of β galactosidase are present per cell.

Along with the induction of β galactosidase, a second protein called (galactoside) *permease* is produced which facilitates lactose entering the cell and becoming hydrolyzed by β galactosidase into its component sugars. Another enzyme (thiogalactoside) *acetylase*, also appears upon lactose induction, although in small quantities. All three of these proteins are affected by genes that map within the *lac* locus, in the sequence *z* (β galactosidase)-*y*(permease)-*a*(acetylase). Among the various lac^- mutations some, called *Cz*, are like CRM mutations (p. 607) in producing an enzymatically abnormal but antigenically detectable cross-reacting protein.

In 1956 Monod and others discovered a new class of mutations that affected the production of the three proteins by changing their synthesis from inducible to *constitutive*. That is, these proteins were now produced in large quantities in the absence of lactose inducer, but their structure was not the least bit affected. Such constitutive mutations, called i^-, generally mapped at one end of the *lac* locus not far from *z*. The fact that *i* determined the inducibility of the enzymes and not

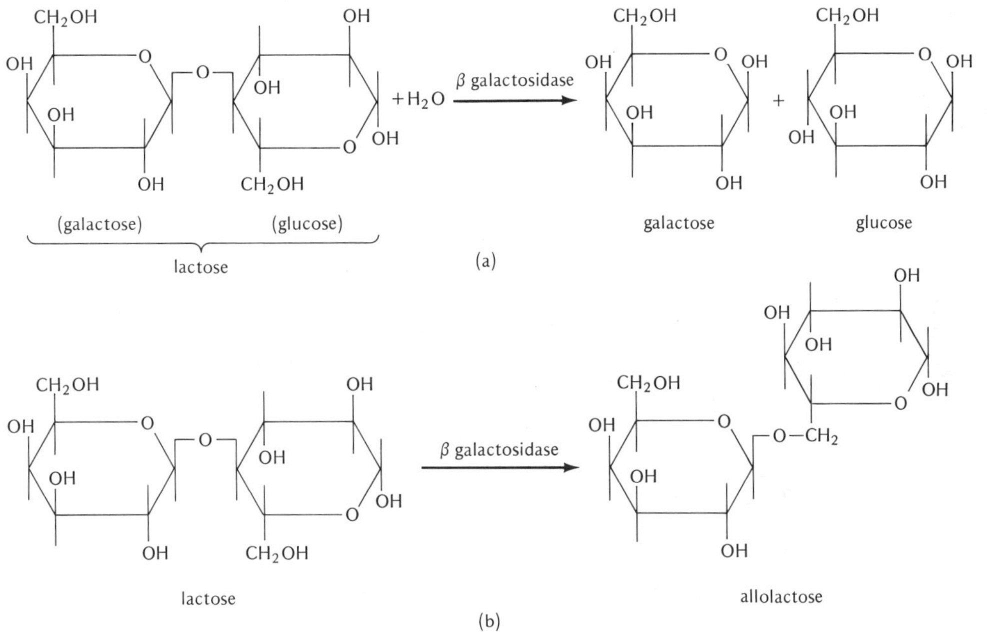

(a)

(b)

Figure 29–1

(a) Hydrolysis of the sugar lactose into galactose and glucose through action of the enzyme β galactosidase. (b) Some of the galactose and glucose molecules hydrolyzed by the enzyme are converted into allolactose, the inducer for β galactosidase synthesis.

their structure led to the description of *i* as a *regulator gene* (see also "controlling elements" in corn, p. 545) and of the *z, y,* and *a* genes determining the amino acid sequences of the enzymes as *structural genes*.

Further experiments showed that the wild-type regulator i^+ had two important features. One was its dominance over i^- in partial diploids (i^+/i^-). For example, a mating of i^- constitutive Hfr donors to i^+ inducible F^- recipients produced partially diploid *merozygotes* carrying both genes for a period of time whose production of β galactosidase was inducible and not constitutive. A second feature was that the wild-type effect of i^+ appeared to accumulate in the cytoplasm and cause enzyme inducibility (i.e., prevented constitutive synthesis). This was demonstrated in an experiment by Pardee, Jacob, and Monod in which they mated wild-type $i^+ z^+$ Hfr sensitive to T6 phage and streptomycin with F^- recipients that were $i^- z^-$ and resistant to both agents. The mutant recipients, although carrying the constitutive regulator i^-, were unable to synthesize wild-type galactosidase until the z^+ gene of the donor entered. At that point, constitutive synthesis of β galactosidase began (Fig. 29-2) and the Hfr donors were then destroyed by the addition of T6 phage and streptomycin. However, further constitutive synthesis of galactosidase in these merozygotes stopped at about 60 to 80 minutes after mating, and enzyme synthesis then became inducible. In other words, entry of the wild-type i^+ gene into the constitutive mutant causes production of a regulatory substance that prevents constitutive synthesis in spite of the presence of the i^- gene. As shown in the last line of Table 29-1, this substance prevents constitutive synthesis of β galactosidase even when the wild-type z^+ gene for the enzyme is on the recipient chromosome. Moreover, the control exercised by the regulatory i^+ substance occurs even over the inactive but antigenically detectable cross-reacting material (CRM) produced by *Cz* mutations which are probably of the "missense" variety.

The regulatory substance produced by the i^+ gene has been called a *repressor*, since it represses the function of the *lac* structural genes in the ab-

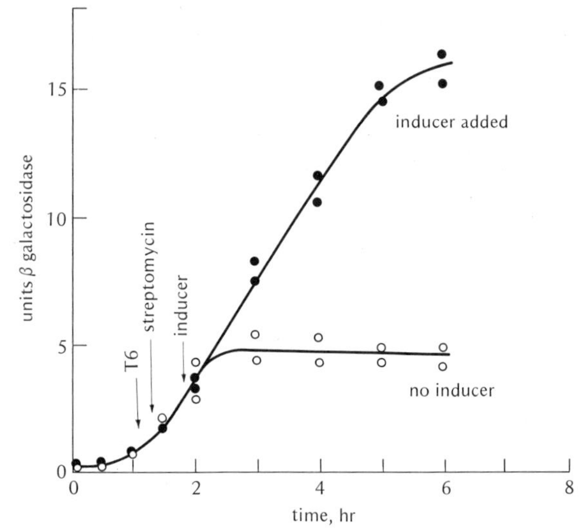

Figure 29-2

Synthesis of β galactosidase in E. coli cultures containing $i^+ z^+$ T6^s Sm^s Hfr donors and $i^- z^-$ T6^r Sm^r recipients. The Hfr donors are destroyed by addition of T6 and streptomycin, and the only galactosidase produced is in z^- recipients carrying the donor z^+ gene. Synthesis of β galactosidase is then constitutive for a period of time but levels off at about 1 to 2 hours. At that point sufficient i^+ substance has been produced to enable enzyme synthesis to become inducible. (After Pardee, Jacob, and Monod.)

TABLE 29-1

Production of β galactosidase in different *E. coli* genotypes, some of which are heterogenotes bearing F′ *lac* factors (see p. 421)

Genotype	β Galactosidase Production: Noninduced Bacteria	Induced Bacteria
$i^- z^-$	−	−
$i^- z^+$	+	+
$i^+ z^+$	−	+
$i^+ z^-$	−	−
$i^- z^+/F' i^- z^+$	+	+
$i^+ z^+/F' i^- z^-$	−	+
$i^+ z^-/F' i^- z^+$	−	+
$i^- z^-/F' i^+ z^+$	−	+
$i^- z^+/F' i^+ z^-$	−	+

After Jacob and Monod, and other sources.

sence of inducers. Through the efforts of Gilbert and Müller-Hill it is now clear that the repressor is a protein that binds directly to *lac* DNA. According to their experiments, the DNA of a particular defective strain of λ which has incorporated the *E. coli lac* region can be shown to carry the repressor protein along with it during sedimentation in the ultracentrifuge. Induction by a lactose-like sugar causes the repressor to be released from this DNA. Moreover, λ DNA which does not carry the bacterial *lac* region will not bind the repressor.

Although the amino acid sequence of the repressor has now been determined (see Beyreuther et al.), the exact mechanism by which this protein functions is not yet known. Nevertheless both *in vivo* and *in vitro* studies indicate that the wild-type repressor protein is capable of adopting one of two alternate states: (1) to bind with *lac* DNA and thereby prevent constitutive function of the *lactose* genes in the absence of inducer; (2) to become inactive as a repressor upon the presence of inducer. The existence of these two alternatives has suggested to Monod and others that inducible repressors are members of a large class of *allosteric proteins* that can change their shape by interacting with a particular molecule. In the present case the wild-type repressor is presumed to have two sites: one that binds to *lac* DNA, and the other that binds to inducer. When inducer is absent, the DNA binding site of the repressor is functional, and *lac* enzyme synthesis is prevented. As will be discussed later (p. 679), this control is exercised through the prevention of mRNA transcription. However, when inducer is present, it binds to the repressor and transforms its DNA-binding site into a nonfunctional state, thereby permitting mRNA transcription and subsequent enzyme synthesis.

The effect of repressor mutations is therefore easily understood. The i^- regulatory mutant produces either a nonactive repressor or no repressor, and therefore allows constitutive enzyme synthesis. Of the two repressor binding sites, the mutation probably alters the site that attaches to *lac* DNA. Other *i* mutations may occur, and one type, known as i^s, or the *superrepressor* mutation, prevents the synthesis of lactose enzymes even in the presence of inducer. Furthermore, the i^s superrepressor substance is dominant over the ordinary repressor substance produced by the i^+ gene. Thus the inducer can be considered capable of inactivating the effect of normal i^+ repressor but incapable of inactivating the i^s superrepressor. Apparently the i^s superrepressor can continue repression because it has lost the binding site for inducer attachment, and *lac* enzyme synthesis therefore remains *noninducible*.

REPRESSIBLE SYSTEMS

In some enzyme systems, such as those involving the synthesis of histidine, tryptophan, and many other amino acids as well as purines and pyrimidines, enzymes are not produced by induction. Instead such enzymes are *repressible* in the sense that their production is repressed by a specific substance that is activated during the course of metabolism. Tryptophan production, for example, is reduced by the accumulation of tryptophan, and the synthesis of each of the five enzymes in the tryptophan pathway (Fig. 26–6) is reduced to a similar extent when a certain concentration of tryptophan is reached (*coordinate repression*). This mechanism obviously serves to prevent synthesis of materials that are already present in sufficient concentration.* Both repressible systems and in-

*The excessive synthesis of histidine and other small molecules may also be reduced by another mechanism, called *feedback* or *end-product inhibition*. In this case control is exercised over the *activity* of enzymes but not over the *synthesis* of enzymes. For example, Gorini has shown that at certain concentrations of arginine supplied to bacterial cells, intracellular arginine synthesis is reduced (feedback inhibition) but the quantitative level of arginine-synthesizing enzymes remains the same as in minimal medium; i.e., repression is absent. It is only when relatively high concentrations of arginine are supplied that the level of arginine-synthesizing enzymes is repressed. Feedback control has been shown by Umbarger and others to be caused by end-product inhibition of the initial enzyme in many metabolic sequences that produce small molecules. Since the enzymes that initiate such metabolic pathways interact with their own metabolic substrates as well as with the much differently shaped end product, it is believed that these enzymes are allosteric proteins with two combining sites, one that can combine reversibly with the normal substrates to perform the normal metabolic function, and another that can combine reversibly with the end product. When combined with the end product (i.e., at high end-product concentrations), such an enzyme is unable to function metabolically because it has undergone an allosteric change, and the metabolic pathway leading to the end product is therefore inhibited.

ducible systems can therefore control gene expression through repressors (*negative control*), the former producing an inactive repressor (occasionally called an "aporepressor") that is activated by a substance called a *corepressor*, the latter producing an active repressor that is inactivated by an inducer.

In both systems, inducers and corepressors are usually small molecules described as *effectors* whose presence lead to changes in the function of regulatory repressor proteins that control transcription. In the case of the *lac* system the *actual effector* (allolactose) is different from the *apparent effector* (lactose), and this can also be true for repressible systems as we shall see below. Briefly, the two systems can be characterized as follows:

> *Inducible system:* active repressor + inducer → inactive repressor → permits transcription of structural gene to produce mRNA
> *Repressible system:* inactive repressor + corepressor → active repressor → prevents transcription of structural gene from producing mRNA

As in inducible systems, mutations of regulatory genes can take place in repressible systems leading to constitutive enzyme synthesis. Some mutations in *Salmonella*, for example, map at loci distant from the *histidine* region and affect regulation so that histidine enzyme levels in the presence of histidine are "derepressed" or much higher than normal. This indicates that histidine itself cannot be the sole repressor but must interact with the normal repressor product of a regulator gene. Mutations causing derepression in the presence of histidine therefore probably involve a change in the repressor so that it cannot be activated by the histidine corepressor.

In *Aerobacter*, Schlesinger and Magasanik have shown that the repression in the histidine system is closely associated with transfer RNA activated with histidine molecules. They found that α-methylhistidine, an analogue of histidine which inhibits the activity of histidine-activating enzyme (histidyl-tRNA synthetase), causes derepression of histidine enzyme synthesis. Similarly, Roth and others have shown in *Salmonella* that histidine enzyme derepression may arise from mutational sources affecting the function of histidine-activating enzyme (*his S*), or affecting the synthesis and structure of histidyl-tRNA (e.g., *his R*, *his T*). Recent observations suggest that the regulator gene product with which histidyl-tRNA interacts is the first enzyme in the histidine biosynthetic pathway, PR-ATP pyrophosphorylase (see Fig. 26–21). This enzyme may thus be an example of autogenous regulation in which a protein directly controls the expression of its own structural gene (see Goldberger; but see also Scott et al.)

The use of activated amino acids as part of the regulatory system is also seen in *E. coli* where Eidlic and Neidhart have demonstrated that valine cannot regulate enzyme synthesis unless the valine-activating enzyme is functional. When the temperature is raised in certain temperature-sensitive mutants, the valine-activating enzyme becomes nonfunctional, and an excess of valine will no longer repress the synthesis of enzymes in the valine metabolic pathway. Thus the likelihood exists in these cases that the actual effector is a transfer RNA, or a transfer RNA charged with amino acid, or some other substance which interacts with these molecules. On the other hand, neither tRNA nor tRNA synthetases seem to be necessary for the repression of enzymes used in the synthesis of tryptophan, arginine, or methionine. Apparently these amino acids can themselves serve as corepressors.

THE OPERON

The discovery of the regulator gene *i* affecting the synthesis of lactose enzymes was soon followed by the discovery of mutations that produced β galactosidase constitutively in spite of the presence of i^+ genes. Such mutations, designated as operator constitutive, or o^c mutations, mapped in the *lac* region close to the *z* locus. They were unique in only affecting those genes to which they were linked in the *cis* position. For example, an F′-*lac* particle bearing $i^+ \ o^c \ z^+$ genes in an $i^+ \ o^+ \ z^-$ recipient produced β galactosidase constitutively, while an F′ $i^+ \ o^+ \ z^+/i^+ \ o^c \ z^-$ heterogenote produced β galactosidase only when induced with lactose (Table 29–2). When *Cz* mutations were used, serological cross-reacting material (CRM) was produced constitutively when *Cz* was on the o^c chromosome, and was produced by induction with lactose when *Cz* was on the o^+ chromosome.

These findings indicated that there was at least one intermediary step between the repressor substance produced by the regulator *i* locus and the consequent repression of the enzymes produced by the structural genes. This intermediary step, controlled by the *operator* locus, enabled (o^+) or prevented (o^c) repression of structural gene function. The fact that a single operator-gene mutation controlled all three enzymes in the lactose region indicated that the linkage between operator and structural genes was not only physical but also functional. To this integrated unit Jacob and Monod gave the name *operon* and pictured its activity as the controlled synthesis of mRNA by RNA polymerase enzyme (Fig. 29–3a, b). Thus presence of a repressor on the operator locus in the *lac* system acts as a block to transcription of the operon by RNA polymerase (Fig. 29–3c, d), and presence of an appropriate inducer removes the repressor from the operator to allow transcription (Fig. 29–3e, f). Constitutive enzyme synthesis, meaning continued transcription of structural genes uninhibited by repressor, could then occur in the following ways: (1) A mutation of the regulator gene produces an inactive repressor that is unable to bind to the operator (Fig. 29–3g), and (2) a mutation of the operator makes it incapable of binding the repressor (Fig. 29–3h). Because of its simplicity, this notion of the operon, as a set of structural genes controlled by an adjacent site, has rapidly become one of the central concepts of genetics and has led to widespread investigation of transcriptional controls.

TABLE 29–2

Production of β galactosidase (and cross-reacting material CRM) in different *E. coli* genotypes

Genotype	*β Galactosidase Production*	
	Noninduced Bacteria	Induced Bacteria
$i^+ o^+ z^+/\text{F}' \, i^- o^+ z^+$	−	+
$i^+ o^+ z^+/\text{F}' \, i^- o^+ z^-$	−	+
$i^+ o^c z^+/\text{F}' \, i^- o^+ z^-$	+	+*
$i^- o^c z^+/\text{F}' \, i^+ o^+ z^-$	+	+*
$i^+ o^+ z^-/\text{F}' \, i^- o^c z^+$	+	+*
$i^- o^+ z^-/\text{F}' \, i^- o^c z^+$	+	+
$i^+ o^+ z^+/\text{F}' \, i^- o^c z^-$	−	+
$i^+ o^+ z^-/\text{F}' \, i^+ o^c z^+$	+	+*
$i^+ o^c z^-/\text{F}' \, i^+ o^+ z^+$	−	+
$i^+ o^c z^+/\text{F}' \, i^+ o^+ Cz$	+	+ and CRM*
$i^+ o^+ z^+/\text{F}' \, i^+ o^c Cz$	CRM	+ and CRM*

After Jacob and Monod, and other sources.

* o^c mutations present in i^+ cells are not fully constitutive and will generally respond to inducer, producing more enzyme than in noninduced cells. However, induced o^c mutants will not ordinarily produce more enzyme than induced wild type.

Transcription, however, depends not only on the operator but also on the attachment sites for RNA polymerase: the easier it is for polymerase enzymes to attach to this site, the greater the number of mRNA molecules that can be made before repression recurs. The first known evidence that an attachment site could vary in its ability to attract or bind polymerase enzymes was found for the polymerase attachment of the *lac* operon, called the *promoter* (*p*). As expected, mutation or deletion of the promoter does not cause constitutive enzyme synthesis, but instead such aberrations may prevent any *lac* enzyme synthesis at all (Fig. 29–3i), or change the rate of such synthesis (Fig. 29–3j). Interestingly polymerase attachment to the promoter is itself controlled by a second site on the promoter that serves to bind a protein called CRP (cyclic AMP receptor protein; also called CAP, catabolite activator protein). Since RNA polymerase will not attach to the promoter without CRP, this protein can be described as a *positive* regulator because it enhances function rather than represses it. However, even attachment of CRP to the promoter is not an independent event since it relies on the presence of cyclic AMP (adenosine cyclic monophosphate). The concentration of this compound is, in turn, dependent on

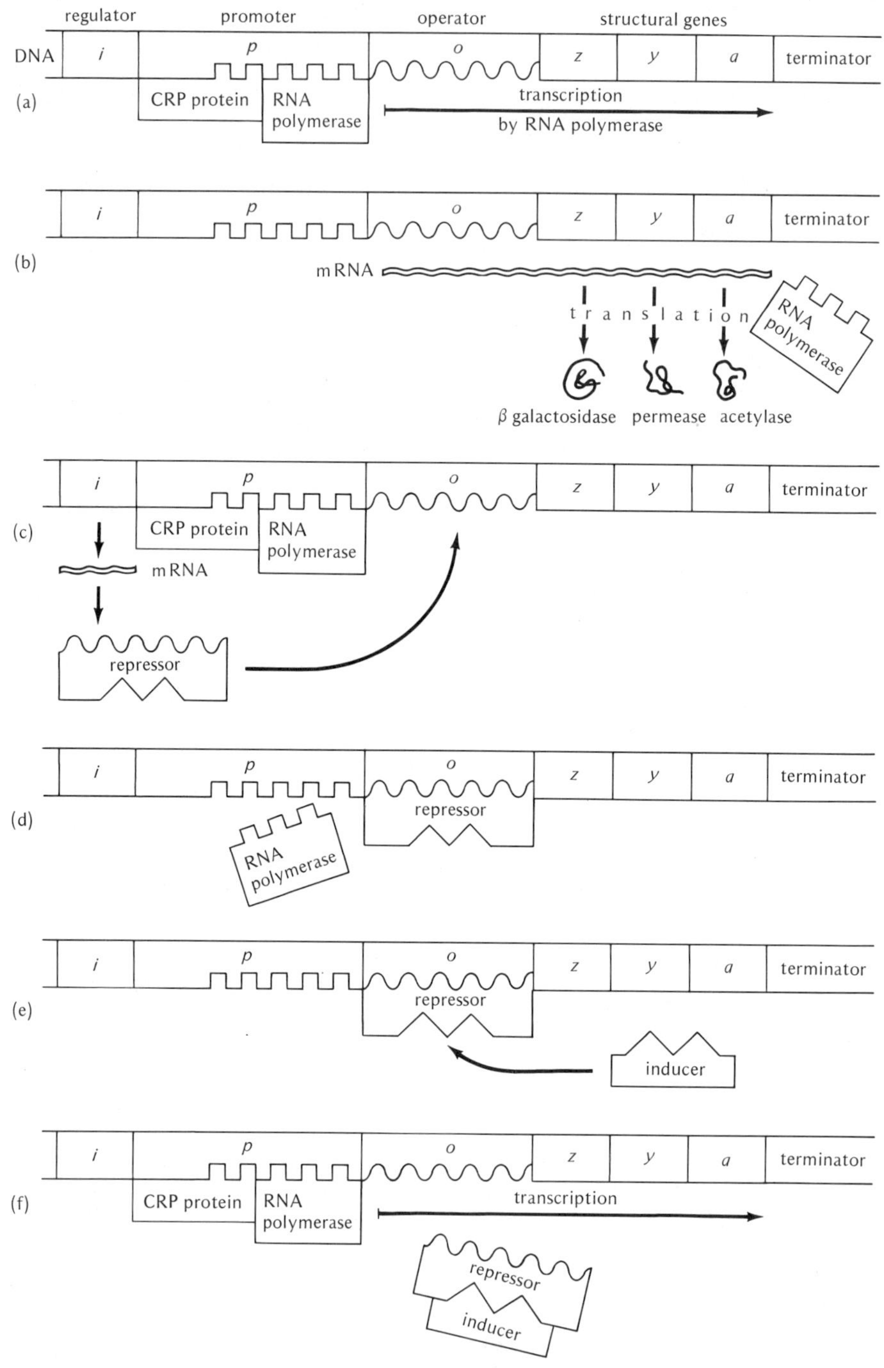

Figure 29–3

Scheme for the regulation of lac enzyme synthesis through transcription control as explained on previous page. (a) In the absence of repressor, transcription of the lac operon is accomplished by the RNA polymerase enzyme which attaches at the promoter site (p) in the presence of CRP protein, and then proceeds to transcribe the operator until it reaches the terminator sequence. (b) Subsequent translation of the lac mRNA leads to the formation of lac enzymes. (c) Synthesis of the lac repressor protein occurs via translation of an mRNA which is separately transcribed at the i regulator gene locus.

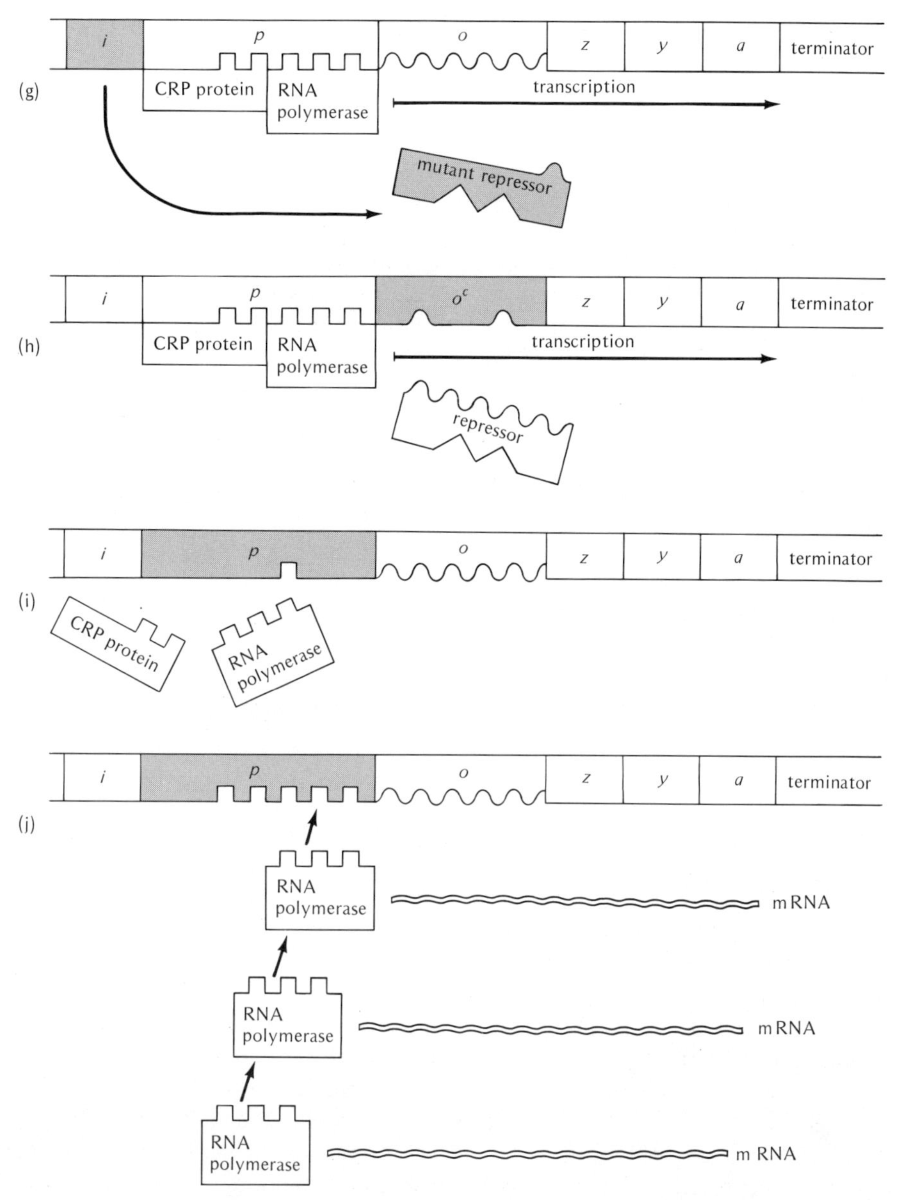

(d) Attachment of the repressor protein to the operator prevents the polymerase enzyme from transcribing the operon. Some studies suggest that the repressor will also prevent the RNA polymerase from binding to the promoter. (e) An appropriate inducer attaches to a specific site on the allosteric repressor (see p. 677) causing it to change its shape and detach from the operator, thereby permitting transcription (f). In g-j the effects of various regulatory mutations (shaded sections) are shown: (g) A mutation of the regulator gene produces a repressor which is unable to bind to the operator thereby causing constitutive lac *enzyme synthesis. (h) A mutation in the operator prevents binding of the repressor, and therefore also leads to constitutive enzyme synthesis. (i) A mutation or deletion in the promoter interferes with the attachment of RNA polymerase thus preventing transcription of the* lac *operon. (j) A mutation in the promoter ("up-promoter" mutation, see p. 687) enables more rapid attachment of the RNA polymerase, thereby increasing the number of mRNA molecules and subsequent rate of* lac *enzyme synthesis. "Down-promoter" mutations also exist which diminish the attachment rate of RNA polymerase, and some promoter mutations may reduce the requirement for cyclic AMP.*

occurrence of some of the metabolic activities of the cell.*

The exquisite sensitivity of *lac* enzyme synthesis to various regulatory compounds has placed considerable emphasis on unraveling the precise molecular structures involved in these interactions. As shown in Fig. 29–4, recent studies by Gilbert and Maxam, Maizels, and Dickson et al., have now provided the first complete nucleotide sequence known for the control region of an operon. This sequence extends from the final codons in the regulator gene *i*, to the beginning codons of the structural gene, *z*, as indicated by the presence of the termination codon of the former (ACT on DNA = UGA on mRNA) and the initiator codon of the latter (TAC = AUG).

Among the unusual features of the *lac* promoter and operator regions are sequences of "twofold symmetry" in which sections of DNA to the left of a particular point are exactly symmetrical to sections on the right.† It has been suggested that these symmetrical sections may be recognized by proteins which themselves possess symmetrical configurations that can fit these sections. Evidence supporting this view may be derived from the observation that the *lac* repressor is composed of four subunits which join to form two symmetrical grooves, each of which may function to recognize one-half of the bisymmetrical operator. According to Steitz et al., "the repressor-operator complex might bear some similarities to a hotdog [operator] in a hotdog bun [repressor]." Both Maniatis and co-workers as well as Pirotta have also detected such symmetry in the operator regions of phage λ, and they too suggest that these DNA sequences probably interact with symmetrically arranged subunits in the repressor protein. Since the CRP site also possesses a section of twofold symmetry, it is possible that the CRP protein may function in a similar manner.

On the other hand, the attachment of RNA polymerase to its specific site in the *lac* region is apparently not dependent on symmetrical nucleotide sequences, although this site does possess a small region of A-T pairs surrounded by regions of numerous G-C pairs. Dickson et al. suggest the presence of CRP protein reduces the stability of hydrogen bond pairing in the high G-C regions enabling the A-T region with fewer hydrogen bonds to "open" and the RNA polymerase to attach. It is, however, still difficult to visualize why transcription begins in the operator rather than in the polymerase attachment site. In any case, once transcription begins, the presence of repressor can no longer prevent it, since the operator binding site is not available to the repressor until the RNA polymerase has moved beyond it. As shown by Maizels, transcription of the *lac* operon then proceeds by forming an RNA complement to that DNA strand oriented in the 3′ to 5′ direction from the starting point. This strand, called the "sense" strand, is the outer strand of Fig. 29–4.

The sequential transcription of genes along the operon in the operator → terminator direction is further reflected in the sequential reduction in enzyme synthesis caused by polarity mutations (pp. 649 and 663). As shown by Beckwith and others, polarity mutations in the *z* locus, for example, not only affect β galactosidase synthesis but also reduce the quantities of permease and acetylase. That is, a polar mutation reduces the amount of protein synthesized by genes on the side of the mutation that is distal to the operator. Since these mutations are now known to be caused by nonsense codons, they apparently lead to the detachment of the ribosome from mRNA, and thereby prevent further translation. Surprisingly, however, the polarity effect also seems to reduce the amount of mRNA that is formed from genes on the distal side of the mutation. According to Imamoto and co-workers, the reason for this reduction in mRNA may lie in the tight coupling of transcription to translation in bacteria; that is, the cessation of ribosome movement along the mRNA molecules caused by the polarity nonsense codon prevents

*A number of instances have now been found in which enzyme synthesis by an operon is repressed by the accumulation of products that result from the breakdown (catabolism) of various carbon compounds such as glucose. This process has been named *catabolite repression*, and is caused by a reduction in the amount of cyclic AMP, which then reduces the frequency of attachment of RNA polymerase to the promoters of "catabolite-sensitive" operons (e.g., those involved in producing enzymes for the degradation of arabinose and galactose in *E. coli*). On the other hand, catabolite-insensitive operons are also known (e.g., leucine biosynthesis in *Salmonella*).

†Also demonstrated in the genetic studies of Sadler and Smith.

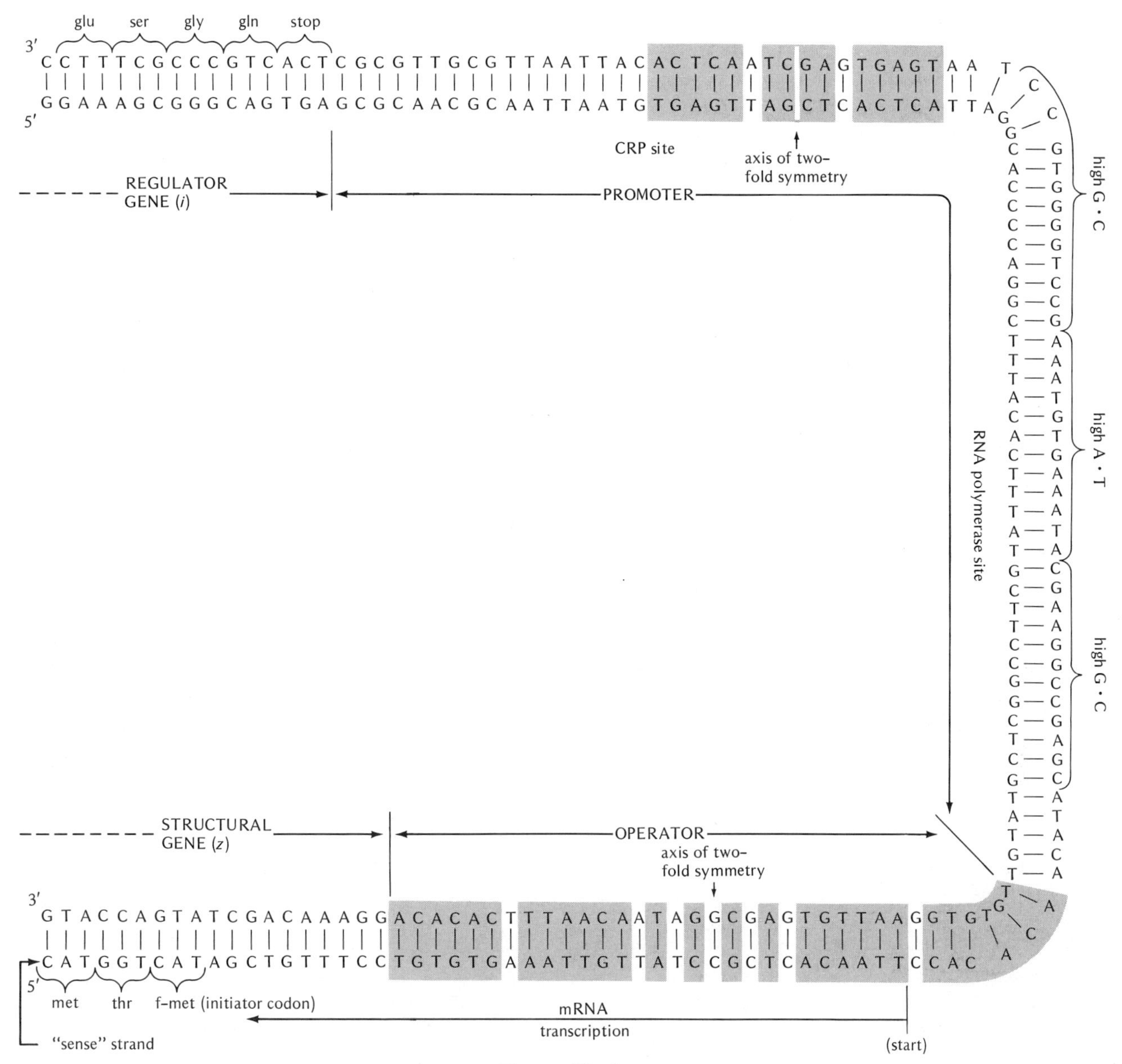

Figure 29–4

DNA nucleotide sequences at the lac operator and promoter sites of E. coli, including the end of the regulator gene (i) and the beginning of the first structural gene of the operon (z). The base pairs enclosed in rectangular outlines show twofold symmetry, with base pair sequences to the right of each arrow symmetrical to those on the left. These sequences are thus "palindromic" in the sense that the nucleotides in the upper strand going in one direction are the same as the nucleotides in the lower strand going in the other direction (see also Fig. 19–22). (After Dickson et al.)

further transcription of DNA. Morse and co-workers, however, suggest that mRNA is actually produced distal to the polar mutation but is rapidly degraded by ribonucleases that attack mRNA unprotected by ribosomes.

Some light has been shed on this dispute by the findings of Adhya et al. and of Franklin that polarity effects vanish when genes bearing polarity mutations are transcribed by a particular RNA polymerase ("juggernaut enzyme") that does not halt transcription when it reaches ordinary "rho" termination signals (p. 105). This indicates that the juggernaut polymerase is not sensitive to termination sites induced by the presence of polarity mutations that would have been recognized by ordinary RNA polymerase. A polarity mutation may therefore cause reduction in the length of mRNA produced by an operon by increasing the sensitivity of RNA polymerase to termination sites that are not usually used to halt transcription. As yet, the mechanism responsible for this increased sensitivity is not known.

Whatever the cause, polarity effects are obviously dependent upon the fact that transcription in bacteria can produce long "polycistronic" mRNA molecules in which each structural gene bears its own initiator (AUG) codon. Thus genes distal to a polar mutation can nevertheless occasionally be translated into protein when long mRNA molecules can somehow manage to be synthesized or escape degradation. In eucaryotes, however, mRNA appears to be mostly monocistronic (see p. 649), and polarity effects on adjacent structural genes would not ordinarily be expected.

In repressible enzyme systems in which concentrations of an end product such as histidine and tryptophan usually prevent further enzyme synthesis, the function of the operon is similar to inducible systems except that the regulator gene now produces an *inactive* repressor which is activated by the end product. As may be anticipated, mutations that cause insensitivity to the activated repressor also occur and, in most cases, map at operator sites at one end of the repressible operon. Similarities between inducible and repressible systems also extend to the occurrence of both polar and promoter mutations, as has been shown for the repressible *trp* operon in *E. coli*. In fact there are instances in which the seemingly large differences between inducible and repressible systems can be bridged. For example, Myers and Sadler have demonstrated the presence of a regulator gene mutation, called i^{rc}, which completely transforms the role of the *lac* repressor in *E. coli*, so that it now acts as though it were a repressor in a repressible system rather than in an inducible system. That is, the *lac* genes function constitutively in i^{rc} cells, but are repressed by the presence of lactose! Apparently the mutant repressor molecule can now only bind to the operator when it is carrying a galactoside. Thus a mutational change can lead to the transformation of an inducer into a corepressor. On the other hand, the conversion of a corepressor into an inducer can also occur as shown by Jacoby and Gorini for arginine in *E. coli*. Although arginine is normally a corepressor, it acts quite differently in one regulator mutant strain, *arg* R^{B}. There the normally inactive repressor is now active, and it only becomes inactivated in the presence of arginine.

Operon function may also be strikingly changed by fusion to a new operator as was first shown by Jacob, Ullman, and Monod for the structural genes of the *lac* operon of *E. coli*. In their experiment, F′ particles carrying parts of the *lac-pur* region but with a long intercalary deletion that eliminated the *lac* operator (and part of the *z* gene) were inserted into F^- bacteria unable to synthesize permease and acetylase (y^-). The functional *lac* enzymes produced by the F′ particle (permease and acetylase) were no longer induced by lactose. Instead such enzymes were now under control of the purine operator. That is, an excess of purine causes repression of galactoside permease and acetylase!

Chromosomal changes can therefore connect the structural genes of operons to new operator genes affected by different repressors, or, conversely, an operator gene affected by the same repressor may be duplicated in various operons throughout the genome. For example, in arginine synthesis in certain strains of *E. coli*, practically all the different enzymes involved show repression by certain concentrations of arginine yet nevertheless map at different loci. However, since the regulator gene that affects all of them maps at a single locus, the conclusion has been drawn that only one type of activated repressor is responsible for controlling each of these genes; i.e., each different gene locus involved in arginine synthesis has a similar operator. Such duplicated operators, found also for *trp* and *ara* genes, would help explain the observed dispersion of many genes that affect the same pathway and are repressed or induced as a unit among one or more chromosomes.

THE ISOLATION OF A GENE

In addition to being the first genetically identified operon, the *lac* system also offers an advantage in genetic manipulation by enabling the use of F′ *lac* factors (see p. 421). Such *lac*-carrying episomes can be incorporated into various sites on the bacterial chromosome including loci close to the attachment points of bacterial prophages such as λ or φ80. As shown by Beckwith, Signer and Epstein, this allows transducing phages to be produced that carry primarily the bacterial *lac* region. Isolation of such transducing phages enabled, for the first time, the identification of a small DNA molecule belonging exclusively to one particular operon.

To accomplish this, a strain of φ80 carrying part of the *lac* operon in inverted position relative to the usual flanking bacteriophage genes, and a strain of λ carrying *lac* in the normal noninverted sequence were isolated. From Fig. 29–5a one can see that this difference in orientation of the *lac* genes determines which of the DNA strands in the two phages can pair together: the λ DNA strand with the high purine content labeled H (heavy) has a *lac* operon section that is *not* complementary to the L (light) DNA strand of φ80 *lac*. That is, if the λ H strand had the *lac* nucleotide base sequence of TCAAGC. . . , the L strand of *lac* in the φ80 strain would have the same sequence in reverse, whereas the φ80 H strand would have the complementary sequence.

Based on this reasoning, Shapiro and co-workers denatured double-stranded DNA of each of these transducing phages, isolated their heavy strands by centrifugation (Fig. 29–5b), and then permitted the heavy strands of the two phages to anneal together (Fig. 29–5c). The bacteriophage nucleotide sequences flanking the *lac* operon are not complementary and therefore cannot hybridize. Only the *lac* operon genes carried by these phages bear perfectly complementary nucleotide sequences. Thus treatment of the annealed hybrid λ-φ80 DNA with a nuclease specific for single-stranded DNA leaves only a small double-stranded DNA complex that consists entirely of *lac* genes (Fig. 29–5d). To ensure that no flanking bacterial genes were carried in these transducing phages, deletions were used which eliminated everything but the *lac* promoter, operator, and *z* gene (and parts of *i* and *y*). Under the electron microscope the isolated double-stranded *lac* region is about 1.4 microns long or somewhat more than 4200 base pairs. This corresponds fairly closely to the estimated 4700 base pairs for the complete *i p o z y a* *lac* region. Further evidence that the isolated DNA is the *lac* operon comes from its ability to hybridize with *lac* mRNA.

The success of this experiment indicates that techniques are becoming available which will permit the isolation and proliferation of specific operons or parts of operons. Even in eucaryotes, one can now isolate hemoglobin mRNA, and, by means of reverse transcriptase (p. 106), make a DNA strand complementary to this messenger.* The potential for genetic manipulation provided by these methods raises therefore the possibility of manipulation of human genes as well. Shapiro, Eron, and Beckwith expressed their social concerns in a letter to *Nature* in the following form: "Let us simply point out to those who feel we have ample time to deal with these problems that less than 50 years elapsed between Becquerel's discovery of radioactivity in 1896 and the use of an atomic weapon against humans in 1945. As to the specific issue of genetic engineering, we cannot

*Although these isolated nucleotide sequences have not yet been used to make proteins, cell-free systems have been discovered which enable the *in vitro* transcription of larger viral DNA molecules such as λ *lac* transducing DNA, and their translation into active enzymes such as β galactosidase (see Zubay).

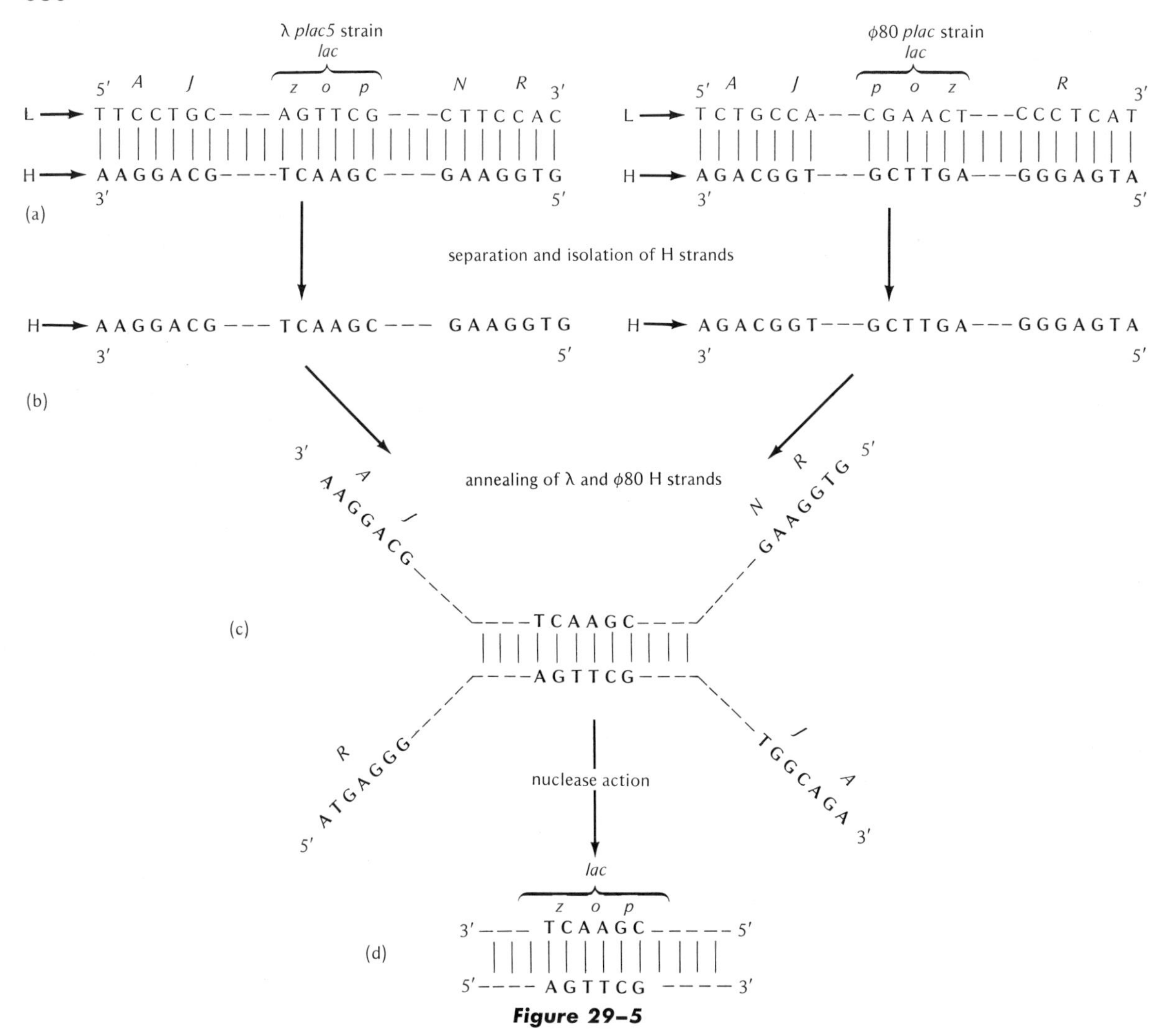

Figure 29–5

Technique used by Shapiro et al. for the isolation of genes in the lac *operon. (a) DNA double strands showing short hypothetical nucleotide sequences in the strains of λ and φ80 transducing phages that were used. (As shown in Fig. 29–4, the actual sequences encompass, of course, large numbers of nucleotides.) Note the change in order of the* lac *operon genes (*z o p*) in the two phages relative to the flanking markers* A *and* R. *(b–c) Denaturation of the double helices and isolation of the heavy "H" strands by differential centrifugation is followed by mixing the two different kinds of H strands together. (d) Attack by an endonuclease specific for single-stranded DNA removes the single-stranded sections and leaves only the double-stranded* lac *operon genes.*

predict the future. But who in 1896 could have foreseen the weapons of mass destruction which now threaten us all?" Some aspects of these matters will be further discussed in Chapter 36.

REGULATING THE REGULATORS

Since the initiation of operon function depends upon the presence of regulator genes producing active or inactive repressor, one may ask what regulates the regulator, what regulates the regulator's regulator, and so on. The number of different kinds of regulating molecules in a cell must eventually reach a limit, and regulatory genes must, at some point, produce regulatory products either by constitutive means or by some self-regulating mechanism.

An example of how the regulator itself can be controlled concerns the gene *i* which produces the *lac* repressor. Normally the cell contains no more than 5–10 repressor molecules which regulate *lac* enzyme synthesis. This means that the *i* gene must be transcribed into mRNA only rarely during the life of a cell, since a single mRNA molecule translated by 5–10 ribosomes would fulfill the cell's entire need for repressor protein. This restricted transcription of the *i* gene may well be caused by its having a relatively inactive promoter unable to complex with RNA polymerase more than once or twice during the cell cycle.

Evidence for this view comes from the finding that a number of *i* promoter mutations control the repressor transcription rate. Some, called i^Q, are "up-promoter" mutations that increase the amount of repressor produced, and in one extreme mutant strain, $i^{super\,Q}$, the repressor concentration is 50 times normal. Since "down-promoter" mutations undoubtedly also occur in repressor genes, it is clear that the products of some regulator genes may be modulated by their promoters.

A mechanism in which regulated gene activity can be initiated by inducers is shown in Fig. 29–6. Two operons are involved in this model (I and II), each of which contains a regulator gene (RG) as well as structural genes (SG). The regulator gene of each operon produces a repressor that affects the operator gene (O) of the other operon. Thus activity of RG_{I} blocks O_{II} and consequent synthesis of the proteins produced by $SG^1_{\mathrm{II}}SG^2_{\mathrm{II}}SG^3_{\mathrm{II}}$. Similarly, RG_{II} activity blocks O_{I} and prevents protein synthesis by SG^1_{I} and SG^2_{I}. The two repressors produced by these regulators, however, are each affected by different inducers (I); RG_{I} repressor is made inactive by inducer I_{I}, and RG_{II} repressor is inactivated by I_{II}. The first inducer to which this system is exposed will therefore determine which set of structural genes will subsequently function. For example, the introduction of I_{II} would permit O_{I} to function and lead to continued repression of operon II even after the inducer is removed.

By means such as this, cell differences in proteins and other products in multicellular organisms may be established; i.e., introduction of a particular inducer may cause a particular set of genes to begin functioning in a cell and repress the activity of other genes, while other cells subject to other

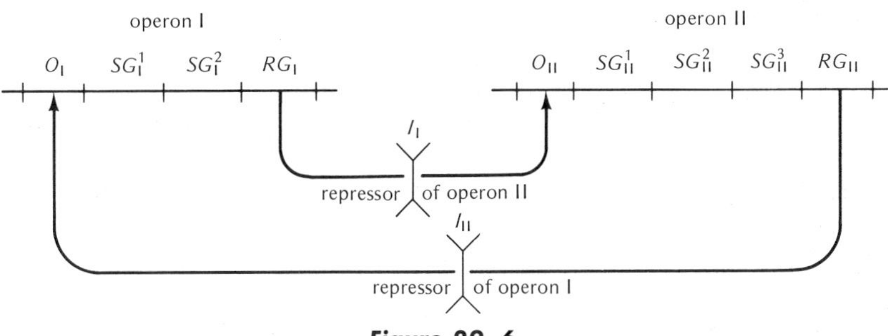

Figure 29–6

Mutual control of gene activity in two operons whose regulator gene products are affected by different inducers. Inducer I_I inactivates the repressor of operon II and thereby permits operon II to begin functioning and subsequently inhibits operon I. Inducer I_{II} permits operon I to begin functioning and inhibits operon II. (After Monod and Jacob.)

inducers develop differently. So far, both operator and promoter sites have been identified by Arst and co-workers for some genes in the fungus, *Aspergillus nidulans*; and similar regulatory loci must also exist in other eucaryotes. As discussed previously (p. 544), the *Activator* gene in corn shows obvious regulatory relationships in the sense that it can cause mutational changes in the *Dissociation* gene even at a distance.

Among humans, an example of eucaryotic regulation is believed to occur in the β thalassemia condition (p. 621) in which the amount of β hemoglobin chains is significantly reduced. Since some β thalassemics have also been shown to produce a small percentage of normal β hemoglobin chains, this condition is probably not caused by a deletion or by a missense or nonsense mutation. Instead a reduction in the quantity of β hemoglobin mRNA has been suggested in such individuals, either by a "down-promoter" transcription mutation or by the decreased stability of mRNA.

Of course, regulation need not always be exercised through transcriptional controls. Since transcription depends upon the presence of DNA, the numbers of available DNA copies of a particular gene may readily control the number of RNA molecules transcribed. Perhaps the most prominent example of this occurs in certain eucaryotes where large numbers of ribosomal RNA molecules are formed in the nucleolus. These are now known to be transcribed from numerous DNA replicates of ribosomal genes that have arisen through specific "gene amplification" processes (see p. 710). It has also been suggested that genes producing ribosomal RNA are transcribed more frequently in *Bacillus subtilis* because these genes are located near the origin of replication. That is, the replication forks that arise at this point in the chromosomes of rapidly growing cells provide additional copies of ribosomal genes during this stage, and permit their relatively increased transcription rate.

However, even though mRNA is transcribed, its translation need not follow. At the translational level various controls are possible including those that affect (1) the presence of appropriate initiation, elongation, and termination factors (Chapter 27); (2) the ability of ribosomes to attach to particular mRNA molecules or to particular sections of such molecules (e.g., p. 663); (3) the longevity of the mRNA templates (p. 636); and (4) the kinds and amounts of tRNA molecules carrying the necessary anticodons to enable translation (pp. 662–663).

Evidence for the presence of translational controls includes the finding that T4 infection of *E. coli* modifies the translational ability of host ribosomes so that they preferentially translate T4 mRNA. Such phage infection also inhibits translation of phage R17 or MS2 RNAs because of a specific change in structure of one of the initiation factors. Furthermore, it is known that some viruses produce nucleases that destroy specific kinds of host tRNA molecules, thereby helping to shut down host mRNA translation. At the same time such viruses may produce unique tRNA species of their own which ensure viral mRNA translation.

Even posttranslational control is quite possible through gene products (e.g., proteases) which can affect the longevity of specific proteins, or which can postpone or modify their function. Some enzymes, such as trypsinogen, are known to be activated (e.g., into trypsin) by the selective deletion of a portion of their polypeptide chain. On the other hand, the function of an enzyme, such as alkaline phosphatase, appears to be modified by the addition of sialic acid residues to the polypeptide chain. Certainly the feedback inhibition discussed on page 677 is a good example of posttranslational control, since the enzymes in such metabolic pathways have already been synthesized, and only their function is regulated by the end products. Because controls are similarly exercised in the relationships between cells, between organs, and between organisms, all present living phenomena can be characterized as a series of highly regulated responses and interactions.

POSITIVE CONTROL

In addition to the examples of negative control considered above, mechanisms of positive control over gene function also exist. On the transcriptional level a regulatory substance in such systems would act on the operon as an *activator* that enhances enzyme synthesis, rather than as a repressor that inhibits it. For example, we have seen that the σ factor enables the RNA polymerase enzyme to begin mRNA transcription at specific initiation sites on DNA (p. 104). Similarly presence of the CRP protein is necessary to activate transcription of "catabolite-sensitive" operons which are dependent on the presence of cyclic AMP (p. 679). Positive control by activator proteins is also implicated in the appearance of enzymes necessary for the metabolism of the sugars arabinose, maltose, and rhamnose.

In the case of arabinose, Englesberg and others have suggested that the regulatory protein produced by the *ara C* gene in *E. coli* may act as *both* repressor and activator. Normally, in the absence of arabinose the *C* protein acts as a repressor that binds to the operator locus of the *ara* operon, thereby helping to prevent transcription. However, upon induction by arabinose, the *C* protein-arabinose combination is released from the operator site, and now becomes an activator which attaches to a separate *initiator* site on the operon. Although the exact mechanism is not known, the *C* protein, in its activator role, probably imposes a particular secondary structure on the initiator site which allows the RNA polymerase to transcribe the operon. Positive controls, as observed previously for negative controls, may also be inducible or repressible (Table 29-3).

TABLE 29-3
Comparisons between negative and positive control systems

	Negative Control		*Positive Control*	
Regulator protein	Repressor		Activator	
Controlling site on DNA	Operator		Initiator	
Role of regulator protein	Prevents gene transcription		Enables gene transcription	
	Inducible System	Repressible System	Inducible System	Repressible System
Effector molecule	Inducer	Corepressor	Inducer	Corepressor
Role of effector	Prevents repressor function	Stimulates repressor function	Stimulates activator function	Prevents activator function
Presence of effector	Enables gene transcription	Prevents gene transcription	Enables gene transcription	Prevents gene transcription
Effect of mutation in regulator protein causing loss of effector attachment site	Noninducible synthesis (dominant)*	Derepressed synthesis (recessive)	Noninducible synthesis (recessive)	Derepressed synthesis (dominant)
Effect of mutation preventing attachment of regulator protein to DNA (When mutation is in protein: recessive. When mutation is in DNA attachment site: dominant in *cis* position, recessive in *trans* position.)	Structural genes are transcribed		Structural genes are not transcribed	
	Constitutive synthesis	Derepressed synthesis	Noninducible	Superrepressed
Effect of mutation that prevents regulator protein from leaving DNA (dominant)	Structural genes are not transcribed (repression continues)		Structural genes are transcribed (activation continues)	
	Noninducible	Superrepressed	Constitutive synthesis	Derepressed synthesis

* Mutations are classified as dominants or recessives depending upon their behavior in the presence of a wild-type gene in a merozygote.

One of the most important examples of positive control is that of the synthesis of DNA itself. According to Jacob and others, the bacterial chromosome, as well as other independent genetic particles in the cell, such as F factors, can be considered genetic units capable of independent replication, or *replicons*. Replicons control their own replication, as evidenced by the discovery of certain mutations that affect only the replication of the mutated replicon but not of other replicons in the same cell. It has been proposed that one of the structural genes of the replicon produces a cytoplasmic product, or activator, which is subject to various cytoplasmic signals. When cellular growth has proceeded to a certain point, the activator interacts with a particular section of replicon DNA, called the *replicator*, and replication begins at this point (see Fig. 19–6). Recent evidence indicates that both the bacterial chromosome and independent F particle are associated with the bacterial membrane, and that the replicator is probably at or near the membrane attachment point (see Jacob, Ryter, and Cuzin, 1966).

In higher organisms the likelihood of numerous independent replicating units along the chromosome has been previously discussed (p. 94), and has also been pointed out by Plaut, Pelling, and others for *Diptera* polytene chromosomes, humans, Chinese hamsters, etc.

THE LOGIC OF LAMBDA

Among the most elegant examples of gene regulation known so far are the many precisely coordinated interactions between λ genes. We have seen that when sensitive *E. coli* strains are infected by the temperate λ virus, a choice exists between two alternative pathways: lysis or lysogeny (Chapter 19). If the lytic pathway is followed, approximately 100 viral particles are produced within one hour by an infected bacterial cell, and the cell is destroyed. On the other hand, the lysogenic pathway leads to the integration of the λ chromosome into the bacterial chromosome, and the cell can now continue to live and replicate, immune from lysis by λ until induction occurs. The choice of which pathway to follow is dependent upon as yet undetermined factors in both host and virus. However, once a pathway is selected by λ, its course is then regulated by various λ gene products. Through genetic and biochemical techniques, these products can be shown to include both repressors and activators, that is, both negative and positive controls. Some of the basic regulatory steps are diagrammed in Fig. 29–7 and can be described as follows:

1. Upon infection of the bacterial cell, the λ chromosome circularizes (see p. 415), and tran-

Figure 29–7

Diagram of some of the regulatory steps known to occur in λ development. (a) "Immediate-early" stage in which the E. coli *RNA polymerase transcribes the λ N gene and probably the cro gene. Operator and promoter sites for transcription of the left and right strands are designated respectively by o_L, p_L, and o_R, p_R. (b) "Delayed-early" stage in which further transcription of λ genes is enhanced by the N gene product which prevents the RNA polymerase from terminating at the indicated termination sites. (c, d) Lytic pathway showing the repression of the cI gene by the cro gene product and activation of the genes necessary for lysis by the Q gene product. In (d) the λ chromosome is shown as a circle whose DNA ends have been covalently joined at a point between genes A and R. This bonding site represents the m and m′ loci described in Chapter 19 which are usually at opposite ends of the λ prophage before it is circularized. Because λ has assumed this circular form, activation by the Q gene product enables continuous polycistronic transcription of the right strand from gene S to gene J ("late mRNA"). (The att site in this figure designates the previously described P and P′ attachment sites that allow homologous pairing with the B and B′ sites of the bacterial chromosome.) (e-j) Steps in the lysogenic pathway beginning with transcription of the cI gene activated by cII and cIII. Note that although the pre promotor transcribes the cro gene this transcription is only of the left strand whereas the cro gene repressor product is only formed by transcription of the right strand. Thus, transcription from pre onward produces only mRNA for the cI protein which then represses further λ transcription (f), and appears to activate its own transcription (g). Further details are described in the text.*

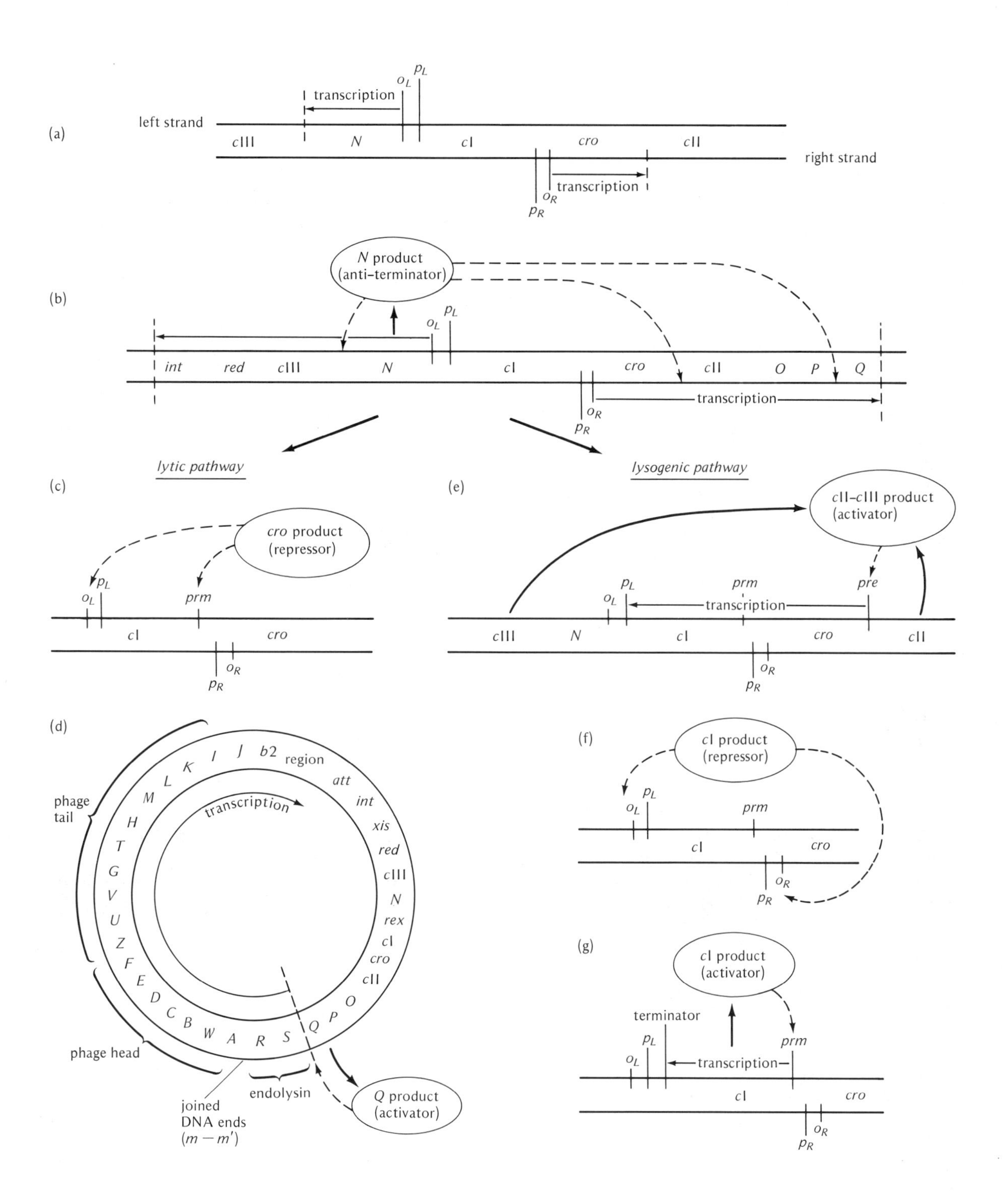
(a)
left strand
right strand
transcription
transcription
cIII
N
cI
cro
cII
(b)
N product
(anti-terminator)
int
red
cIII
N
cI
cro
cII
O
P
Q
transcription
lytic pathway
lysogenic pathway
(c)
cro product
(repressor)
prm
cI
cro
(e)
cII–cIII product
(activator)
prm
pre
transcription
cIII
N
cI
cro
cII
(d)
phage tail
phage head
joined
DNA ends
endolysin
Q product
(activator)
transcription
b2 region
att
int
xis
red
cIII
N
rex
cI
cro
cII
O
P
Q
S
R
A
W
B
C
D
E
F
Z
U
V
G
T
H
M
L
K
I
J
(f)
cI product
(repressor)
prm
cI
cro
(g)
cI product
(activator)
terminator
prm
transcription
cI
cro

scription of λ begins by means of the host RNA polymerase. This is known as the "immediate-early" stage, and mRNA is produced for the *N* gene locus and to some extent for the *cro* gene (also called *tof*) as shown in Fig. 27–7a.

2. The *N* gene product is a protein that enhances the transcription of genes that are necessary for recombination (*red*) and integration (*int*), as well as those needed for DNA replication (*O*, *P*) and further regulation (*cII*, *cIII*, *Q*). It does this by changing the host RNA polymerase into a "juggernaut enzyme" (p. 684) that continues transcription beyond the three termination sites indicated in Fig. 29–7b. This is called the "delayed-early" stage, and λ apparently still maintains the option at this point of embarking upon either the lytic or lysogenic pathways. Interestingly the transcription process in λ is not confined to only one of the two DNA strands. As shown in Fig. 29–7, promoter and operator sites exist on both the "left" strand (e.g., p_L,o_L) and "right" strand (e.g., p_R,o_R). Thus, only when *N* is transcribed from o_L leftwards is the *N* gene product made.

3. If the lytic pathway is adopted, synthesis of the λ repressor produced by the *cI* gene must be inhibited, since the *cI* repressor would prevent transcription of all the remaining λ genes. Control over *cI* transcription during the lytic pathway is exercised by a repressor produced by the *cro* gene which affects the *cI* operon as shown in Fig. 29–7c. The *cro* product also represses transcription of the "early" genes such as those involved in recombination and integration, as well as the *cII* and *cIII* genes which would ordinarily activate transcription of the *cI* repressor. The ratio between the *cro* and *cI* repressors has been suggested as a primary factor in the decision for lysis or lysogeny (see Herskowitz).

4. Continued repression of the *cI* gene enables the "late" stage of lytic development to proceed, activated by the *Q* gene product (Fig. 29–7d). This protein enhances transcription of the genes needed for cell lysis (*S* and *R*), and for components of viral heads and tails (*A-J*).

5. Adoption of the lysogenic pathway for λ depends upon integration of λ DNA into the bacterial chromosome followed by the repression of those λ genes necessary for lysis. The integration step is achieved by the *int* gene product formed during the "delayed-early" phase of development. As explained in Chapter 19, this *int* product enables a crossover event to occur between λ and the host chromosome at homologous attachment sites (see Fig. 19–18). Repression, on the other hand, is achieved by the *cI* repressor whose synthesis first occurs during the "establishment" phase of lysogeny. At that stage a specific promoter site, *pre*, enabling transcription of the *cI* gene is activated by the combined products of the *cII* and *cIII* genes (Fig. 29–7e). Bacterial host gene products also exist which interact at the *pre* site to control whether or not transcription is to proceed. For example, *E. coli hfl* mutants cause a "*h*igh *f*requency of *l*ysogenization" even when infecting λ phage bearing a *cIII* mutation (Belfort and Wulff).

6. When finally transcribed and translated, the *cI* repressor protein prevents transcription of all λ "early" genes and also inhibits the activator functions of the *N* gene product, as well as preventing λ DNA replication by the *O* and *P* products (Fig. 29–7f). Once established, the "maintenance" phase of lysogeny is continued by the apparent ability of the *cI* protein to activate its own synthesis at a special promoter site, *prm*. The *cI* gene can therefore be considered as an example of autogenous regulation (p. 678).

PROBLEMS

29-1. What results would you have expected to find in the experiment of Pardee, Jacob, and Monod (p. 676) if the i^- mutant was constitutive because it produced its own inducer rather than because it lacked an active repressor?

29-2. It has been suggested that the polypeptide chains of the enzyme β galactosidase are present in a bacterial cell even in the absence of an inducer, and the function of the inducer is merely to convert the enzyme molecules into a more active form. In an experiment performed to test this hypothesis, bacterial cells were labeled for a number of generations with radioactive ^{35}S (incorporated into protein via the amino acids methionine and cysteine) in the absence of inducer. These cells were then grown in a medium containing normal ^{32}S in the presence of inducer. The β galactosidase isolated from this experiment did not show the presence of ^{35}S. Explain how this finding reflects upon whether the enzyme or its precursors are the direct target of the inducer.

29-3. Assume that the three alternatives for the synthesis of the acetylase enzyme in the *lac* operon of *E. coli* are either inducible synthesis, constitutive synthesis, or no synthesis at all (caused either by the superrepression of a normal operator or the absence of an ac^+ gene). Explain which of these three alternatives you would expect to find in the following merozygotes:

(a) $i^- \ o^+ \ z^- \ ac^+$ / F′ $i^+ \ o^c \ z^+ \ ac^-$
(b) $i^s \ o^+ \ z^- \ ac^+$ / F′ $i^- \ o^c \ z^- \ ac^+$
(c) $i^s \ o^c \ z^+ \ ac^-$ / F′ $i^- \ o^+ \ z^+ \ ac^+$

29-4. Explain whether you would expect to find a promoter mutation that causes constitutive enzyme synthesis by the *lac* operon.

29-5. In Fig. 29-6, once the function of one of the operons has begun, the other becomes permanently inactive. Can you devise a system in which two operons, each blocked by the products of separate regulator genes, can be regulated so that they function simultaneously?

29-6. What would be the effect of an operator constitutive mutation at the o_L site on the left strand of λ DNA (Fig. 29-7)?

29-7. In an experiment to test which genes were sensitive to catabolite repression (p. 682), Silverstone and Magasanik used a strain of *E. coli* that carried a deletion for the permease and acetylase genes in the *lac* operon (*lac* $i^+ \ p^+ \ o^+ \ z^+ \ [y \ a]^{\text{deletion}}$). The cells of this strain, however, also had an F′ episome that carried a section of the bacterial *lac* locus which was normal for permease and acetylase, along with a deletion of the *lac* operon (F′ $[pur \ E - lac \ z]^{\text{deletion}} \ lac \ y^+ \ a^+$) to the y^+ gene. This subjected transcription of the galactoside permease and acetylase genes to control by the purine operon (see p. 684). In their tests they used both a glycerol medium which ordinarily does not lead to catabolite repression, and a glucose medium which does lead to repression of catabolite-sensitive operons. Their results were as follows:

Medium	Enzyme Units Produced per mg. Protein	
	β Galactosidase	Acetylase
glycerol	18,000	44.6
glycerol and purine (adenine)	16,400	1.5
glucose	6,880	51.1

On the basis of these results, explain whether the *lac* operon, *pur E* operon, or both are sensitive to catabolite repression.

REFERENCES

ADHYA, S., M. GOTTESMAN, and B. DE CROMBUGGHE, 1974. Release of polarity in *Escherichia coli* by gene *N* of phage λ: Termination and antitermination of transcription. *Proc. Nat. Acad. Sci.*, **71,** 2534–2538.

ARST, H. N., JR., and D. W. MACDONALD, 1975. A gene cluster in *Aspergillus nidulans* with an internally located *cis*-acting regulatory region. *Nature*, **254,** 26–31.

ARST, H. N., JR., and C. SCAZZOCCHIO, 1975. Initiator constitutive mutation with an 'up-promoter' effect in *Aspergillus nidulans*. *Nature*, **254,** 31–34.

BAUTZ, E. K. F., 1972. Regulation of RNA synthesis. *Prog. Nuc. Acid Res. Mol. Biol.*, **12,** 129–160.

BECKWITH, J. R., 1964. A deletion analysis of the *lac* operator region in *Escherichia coli*. *Jour. Mol. Biol.*, **8,** 427–430.

BECKWITH, J. R., E. R. SIGNER, and W. EPSTEIN, 1966. Transposition of the *lac* region of *E. coli*. *Cold Sp. Harb. Symp.*, **21,** 393–401.

BECKWITH, J. R., and D. ZIPSER (eds.), 1970. *The Lactose Operon*. Cold Spring Harbor Laboratory, Cold Spring Harbor.

BELFORT, M., and D. WULFF, 1974. The role of the Lambda *cIII* gene and the *Escherichia coli* catabolite gene activation system in the establishment of lysogeny by bacteriophage Lambda. *Proc. Nat. Acad. Sci.*, **71,** 779–782.

BEYREUTHER, K., K. ADLER, N. GEISLER, and A. KLEMM, 1973. The amino-acid sequence of *lac* repressor. *Proc. Nat. Acad. Sci.*, **70,** 3576–3580.

DAVISON, J., 1973. Positive and negative control of

transcription in bacteriophage λ. *Brit. Med. Bull.*, **29,** 208–213.

DICKSON, R. C., J. ABELSON, W. M. BARNES, and W. S. REZNIKOFF, 1975. Genetic regulation: The *lac* control region. *Science*, **187,** 27–35.

EIDLIC, L., and F. C. NEIDHARDT, 1965. Role of valyl-sRNA synthetase in enzyme repression. *Proc. Nat. Acad. Sci.*, **53,** 539–543.

ENGLESBERG, E., 1971. Regulation in the L-arabinose system. In *Metabolic Pathways*, 3rd edition, Vol. V, (Metabolic Regulation), H. J. Vogel (ed.). Academic Press, New York, pp. 257–296.

ENGLESBERG, E., J. IRR, J. POWER, and N. LEE, 1965. Positive control of enzyme synthesis by gene C in the L-arabinose system. *Jour. Bact.*, **90,** 946–957.

ENGLESBERG, E., and G. WILCOX, 1974. Regulation: Positive control. *Ann. Rev. Genet.*, **8,** 219–242.

FRANKLIN, N. C., 1974. Altered reading of genetic signals fused to the operon of bacteriophage: Genetic evidence for modification of polymerase by the protein product of the *N* gene. *Jour. Mol. Biol.*, **89,** 33–48.

GILBERT, W., and A. MAXAM, 1973. The nucleotide sequence of the *lac* operator. *Proc. Nat. Acad. Sci.*, **70,** 3581–3584.

GILBERT, W., and B. MÜLLER-HILL, 1966. Isolation of the *Lac* repressor. *Proc. Nat. Acad. Sci.*, **56,** 1891–1898. (Reprinted in the collection of Zubay and Marmur; see References, Chapter 1.)

GILBERT, W., and B. MÜLLER-HILL, 1967. The *lac* operator is DNA. *Proc. Nat. Acad. Sci.*, **58,** 2415–2421. (Reprinted in the collection of Zubay and Marmur; see References, Chapter 1.)

GOLDBERGER, R. F., 1974. Autogenous regulation of gene expression. *Science*, **183,** 810–816.

GORINI, L., 1959. Régulation en retour (feedback control) de la synthese de l'arginine chez *Escherichia coli*. *Bull. Soc. Chim. Biol. Paris*, **40,** 1939–1952.

GORINI, L., W. GUNDERSEN, and M. BURGER, 1961. Genetics of regulation of enzyme synthesis in the arginine biosynthetic pathway of *Escherichia coli*. *Cold Sp. Harb. Symp.*, **26,** 173–182.

HERSKOWITZ, I., 1973. Control of gene expression in bacteriophage lambda. *Ann. Rev. Genet.*, **7,** 289–324.

IMAMOTO, F., 1973. Diversity of regulation of genetic transcription. I. Effect of antibiotics which inhibit the process of translation on RNA metabolism in *Escherichia coli*. *Jour. Mol. Biol.*, **74,** 113–136.

IMAMOTO, F., Y. KANO, and S. TANI, 1970. Transcription of the tryptophan operon in nonsense mutants of *Escherichia coli*. *Cold. Sp. Harb. Symp.*, **35,** 471–490.

JACOB, F., and J. MONOD, 1961. Genetic regulatory mechanisms in the synthesis of proteins. *Jour. Mol. Biol.,* **3,** 318–356. (Reprinted in the collection of Taylor, and in *Selected Papers in Biochemistry*. Vol. 1 and Vol. 5; see References, Chapter 1.)

JACOB, F., A. RYTER, and F. CUZIN, 1966. On the association between DNA and membrane in bacteria. *Proc. Roy. Soc. Lond.* (B), **164,** 267–278.

JACOB, F., A. ULLMANN, and J. MONOD, 1965. Délétions fusionnant l'opéron lactose et un opéron purine chez *Escherichia coli*. *Jour. Mol. Biol.*, **13,** 704–719.

JACOBY, G. A., and L. GORINI, 1969. A unitary account of the repression mechanism of arginine biosynthesis in *Escherichia coli*. I. The genetic evidence. *Jour. Mol. Biol.*, **39,** 73–87.

MAIZELS, N. M., 1973. The nucleotide sequence of the lactose messenger ribonucleic acid transcribed from the UV5 promoter mutant of *Escherichia coli*. *Proc. Nat. Acad. Sci.*, **70,** 3585–3589.

MANIATIS, T., M. PTASHNE, B. G. BARRELL, and J. DONELSON, 1974. Sequence of a repressor-binding site in the DNA of bacteriophage λ. *Nature*, **250,** 394–397.

MCCLINTOCK, B., 1961. Some parallels between gene control systems in maize and in bacteria. *Amer. Naturalist*, **95,** 265–277. (Reprinted in Taylor's collection; see References, Chapter 1.)

MONOD, J. 1956. Remarks on the mechanism of enzyme induction. In *Enzymes: Units of Biological Structure and Function*, O. H. Graebler (ed.). Academic Press, New York, pp. 7–28.

MONOD, J., J. P. CHANGEUX, and F. JACOB, 1963. Allosteric proteins and cellular control systems. *Jour. Mol. Biol.*, **6,** 306–329. (Reprinted in Vol. 8 of *Selected Papers in Biochemistry*; see References, Chapter 1.)

MONOD, J., and F. JACOB, 1961. Teleonomic mechanism in cellular metabolism, growth, and differentiation. *Cold Sp. Harb. Symp.*, **26,** 389–401.

MORSE, D. E., and C. YANOFSKY, 1969. Polarity and the degradation of mRNA. *Nature*, **224,** 329–331.

MYERS, G. L., and J. R. SADLER, 1971. Mutational inversion of control of the lactose operon of *Escherichia coli*. *Jour. Mol. Biol.*, **58,** 1–28.

PARDEE, A. B., F. JACOB, and J. MONOD, 1959. The genetic control and cytoplasmic expression of inducibility in the synthesis of β galactosidase by *E. coli*. *Jour. Mol. Biol.*, **1,** 165–178. (Reprinted in Adelberg's collection, and in *Selected Papers in Biochemistry*, Vol. 1; see References, Chapter 1.)

PELLING, C., 1966. A replicative and synthetic chromosomal unit—the modern concept of the chromomere. *Proc. Roy. Soc. Lond.* (B), **164,** 279–289.

PIROTTA, V., 1975. Sequence of the O_R operator of phage λ. *Nature*, **254**, 114–117.

PLAUT, W., D. NASH, and T. FANNING, 1966. Ordered replication of DNA in polytene chromosomes of *Drosophila melanogaster*. *Jour. Mol. Biol.*, **16**, 85–93.

REZNIKOFF, W. S., 1972. The operon revisited. *Ann. Rev. Genet.*, **6**, 133–156.

ROTH, J., 1965. Regulatory mutants of *Salmonella typhimurium* selected by resistance to a histidine analogue, triazole alanine (tra). *Fed. Proc.*, **24**, 416.

SADLER, J. R., and T. F. SMITH, 1971. Mapping of the lactose operator. *Jour. Mol. Biol.*, **62**, 139–169. (Reprinted in Abou-Sabé's collection; see References, Chapter 1.)

SCHLESINGER, S., and B. MAGASANIK, 1964. Effect of α-methylhistidine on the control of histidine synthesis. *Jour. Mol. Biol.*, **9**, 670–682.

SCOTT, J. F., J. R. ROTH, and S. W. ARTZ, 1975. Regulation of histidine operon does not require *his G* enzyme. *Proc. Nat. Acad. Sci.*, **72**, 5021–5025.

SHAPIRO, J., L. MACHATTIE, L. ERON, G. IHLER, K. IPPEN, and J. BECKWITH, 1969. Isolation of pure *lac* operon DNA. *Nature*, **224**, 768–774.

STEITZ, T. A., T. J. RICHMOND, D. WISE, and D. ENGELMAN, 1974. The *lac* repressor protein: Molecular shape, subunit structure, and proposed model for operator interaction based on structural studies of microcrystals. *Proc. Nat. Acad. Sci.*, **71**, 593–597.

UMBARGER, H. E., 1961. Endproduct inhibition of the initial enzyme in a biosynthetic sequence as a mechanism of feedback control. In *Control Mechanisms in Cellular Processes*, D. M. Bonner (ed.), Ronald Press, New York, pp. 67–85.

ZUBAY, G., 1973. In vitro synthesis of protein in microbial systems. *Ann. Rev. Genet.*, **7**, 267–287.

30
DIFFERENTIATION AND PATTERN

Differentiation is a term often used to describe the many developmental processes which cause distinction between parts of an organism. In single-celled organisms such developmental processes affect, by definition, only one cell or parts of a cell. In multicellular organisms large groups of cells such as tissues and organs come to differ from each other. Differentiation in both types of organisms have in common the fact that changes generally arise as a result of the appearance of gene products differing in respect to quality or quantity from those in other parts of the organism or because of differences in the efficiency with which a gene product functions in different parts of the organism. However, unlike temporary cellular changes, such as the many reversible modifications in enzyme synthesis described in the previous chapter, differentiation is used mostly to define stable or irreversible changes in structure or function, that is, "permanent" activation or repression. Some of the causes that induce these changes, their genetic control, and their effect on the phenotypes that result are the subjects of this chapter.

DIFFERENTIATION IN VIRUSES AND BACTERIA

Development at the level of procaryotes involves, as in higher organisms, the production of proteins that provide specific structural and functional phenotypes to their carriers. Although the small size of these organisms makes developmental changes difficult to observe, procaryotes often have the advantage of being relatively simple systems. For example, rod-shaped tobacco mosaic virus (TMV) consists of only two components, RNA genetic material and coat protein, which can be separated and reconstitute themselves into infectious particles under the proper conditions (p. 67). During this *self-assembly* process approximately 2100 identical protein subunits interact with each other independently beginning at the 5′ end of a single TMV RNA molecule to form one rod-shaped virus. Beyond these RNA and coat protein components, no other information appears necessary to erect this simple structure.

However, the morphology of most organisms, including many procaryotes, is far more intricate. In general, complex assembly systems prevail which depend upon the timed arrival of many different kinds of proteins as well as other molecules, and upon their sequential incorporation into a relatively elaborate morphological structure. The complexity of most developmental processes has therefore been quite difficult to analyze, but some are now being successfully approached by a variety of genetic and biochemical techniques. Among the most important of analytical methods is the search for mutations that enable the experimenter to determine the normal sequence of events that would occur in the absence of mutation. A simple example of this would be the discovery of mutations that affect particular products which, in turn, prevent development beyond particular observable stages. If a mutation in protein X prevents the development of step K, we can surmise that this protein is ordinarily necessary for the K stage to occur. Knowledge of each cellular product affected in a stepwise series can then reveal the normal course of developmental events. This technique has proved of great value in understanding the biochemistry of metabolic pathways (see p. 605) and, as can be seen in Fig. 30–1, has been used with considerable success in analyzing developmental pathways of T4 viruses.

According to Cohen and others, the T-even phage DNA utilizes the bacterial host DNA-dependent RNA polymerase to synthesize phage mRNA. This phage-induced mRNA is then translated on host ribosomes to produce a number of "early" proteins, many of which are necessary for the subsequent synthesis of phage DNA. These proteins are for the most part unique enzymes such as those involved in the production of hydroxymethylcytosine nucleotides (HMC) as well as those that induce the substitution of glucose for the deoxyribose sugar in some sections of phage DNA. Within 5 to 7 minutes of infection these early enzymes lead to formation of a pool of "vegetative" phage DNA fibrils. If a protein synthesis-inhibiting agent such as chloramphenicol is applied to the cultures at this point, the phage DNA remains in its loose fibrillar hydrated state. If protein synthesis is permitted to continue, a *condensing protein* is formed that causes the vegetative DNA fibrils to lose water and compact into condensed polyhedric-shaped particles (see Fig. 3–2).

After early protein synthesis has ceased, a number of "late" proteins appear. According to Kellenberger and others, the molecules of *head protein* polymerize around the condensed DNA to form the DNA-containing head capsule. This induction seems to be caused by another protein, perhaps the condensing protein. In certain mutations (gene *22*) *polyheads* are formed which are long tubes of polymerized head protein containing a material that may be the inducing protein for head formation. It has been suggested that the polyhead mutations cause an unregulated production of this inducer protein or cause the inability of the inducer protein to bind with DNA particles. In any case the fact that cylindrical polyheads appear indicates that the shape of the head is, to some degree, a self-assembly property of the polymerizing head protein itself. Other late proteins include those involved in the various tail structures as well as the lysozyme that is used to rupture the host cell wall. Altogether, a completed phage particle is composed of about 30 different components. As shown in Fig. 30–1, these components derive from three main subassembly processes involving the tail, the head, and the tail fibers, each produced by polymerization and attachment of particular proteins to previously existing structures. Since mutations that block one process do not block the others, these three subassemblies are independent of each other until they gather to-

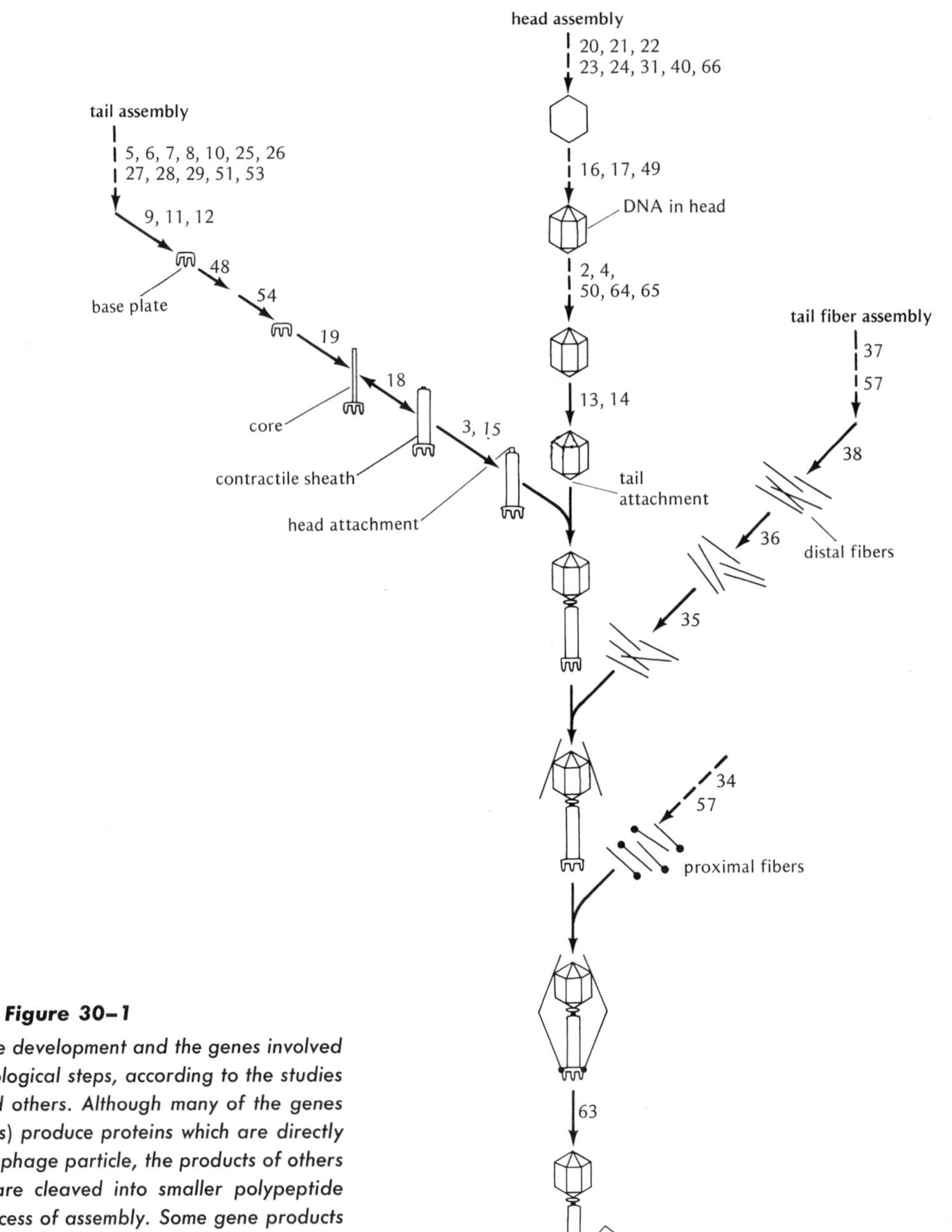

Figure 30–1

Sequence of T4 phage development and the genes involved in the various morphological steps, according to the studies of Edgar, Wood, and others. Although many of the genes (indicated by numbers) produce proteins which are directly incorporated into the phage particle, the products of others (e.g., 22, 23, 24) are cleaved into smaller polypeptide chains during the process of assembly. Some gene products (e.g., 63) appear to be necessary for assembly, but are not themselves incorporated into the final structure. Among gene products that are incorporated, some can also serve as "jigs" for the assembly of later products. Thus, Terzaghi suggests that the distal half-fibers of the tail are first attached to sites on the phage head, and the proximal half-fibers attach to the base plate. Only when head, tail, and base plate are connected together, can complete tail fibers be formed by attachment of the distal half-fibers to the proximal half-fibers. (After Wood, modified.)

gether in the final viral particle. As previously noted in the phage T4 linkage map (Fig. 26–22), there is often a close relationship between some of the genes involved in these subassemblies and their distribution on the chromosome.

In some bacteria the process of spore formation has received considerable attention because it involves marked changes in bacterial phenotype and some of the biochemical and genetic steps can be analyzed in detail. Spores in bacteria are small dormant structures that neither replicate nor show metabolic activity and may survive for hundreds of years under adverse conditions. However, when exposed to a proper environment, they will germinate and resume vegetative growth and replication. The decision made in a bacterial cell to form a spore (sporulation) is usually caused by a limitation in available nutrients, especially carbon and nitrogen. In some *Bacilli* and *Clostridia* species a *prespore* or *forespore* structure first appears, enclosing in a separate compartment half the cellular DNA and a portion of cytoplasm. Accompanying these developments are marked enzymatic changes that eventually lead to the mature spore (*endospore*). Some of the usual bacterial enzymes increase in amount during sporulation, other decrease, and still others occur which are unique to this type of development.

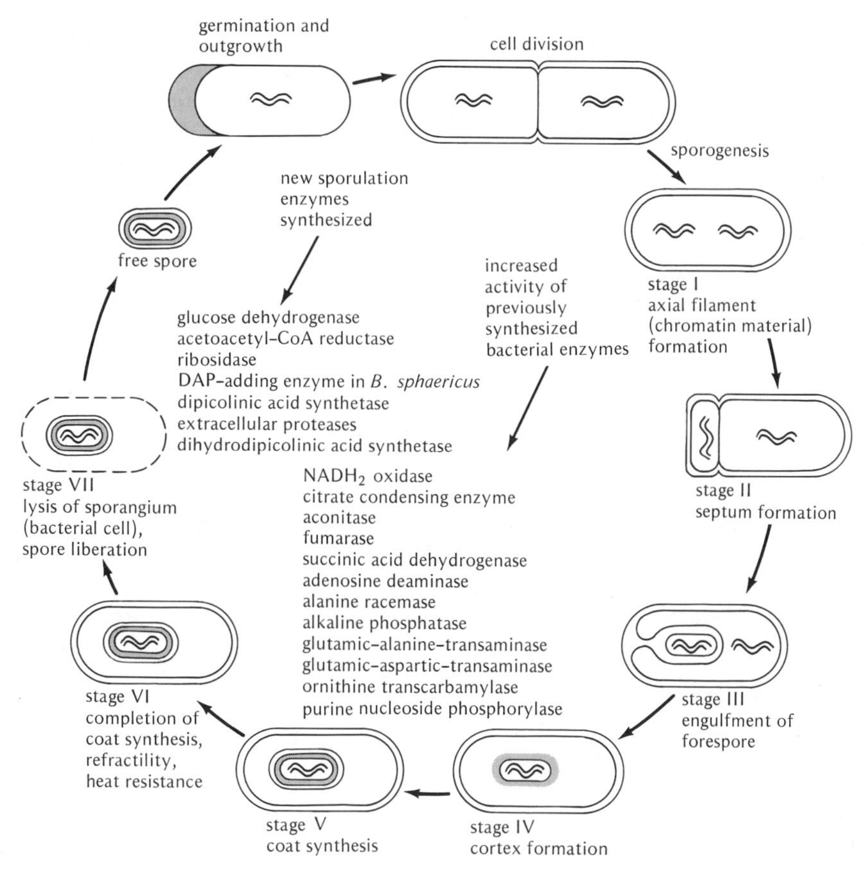

Figure 30–2

Sequence of morphological steps in sporogenesis and some of the enzymatic events involved. From the onset of sporulation to the production of a finished spore is about 6 to 8 hours. The final spore is coated by layers of proteinaceous materials that are not found in vegetative cells and is subsequently liberated by cell lysis. Spores are resistant to heat, ultraviolet radiation, organic solvents, enzymes, and dessication. (After Halvorson and Szulmajster.)

Experiments by Linn and co-workers have suggested that sporulation in *Bacillus subtilis* is associated with loss of activity of the σ subunit in DNA-directed RNA polymerase. This loss may signify changes that enable the polymerase to recognize DNA templates different from those recognized by the vegetative form of the enzyme. The view that transcriptional control accounts for the onset of sporulation is also supported by the fact that the transcription-inhibitor, actinomycin D (see p. 103), will prevent the appearance of specific stages in the sporogenesis sequence if given early in the process. In general the mRNA involved in sporogenesis events is often produced about one hour prior to its translation into protein, i.e., this mRNA has a lifetime many times longer than the approximate 2 to 3 minute lifetime of vegetative mRNA (p. 635). Once initiated, the morphological and enzymatic events in sporogenesis then follow an exact pattern that appears to be under precise genetic control (Fig. 30–2). Some of these genes have been mapped in *B. subtilis* by transduction techniques, and are apparently distributed throughout the genome.

DIFFERENTIATION IN MULTICELLULAR ORGANISMS

In eucaryotes interest in differentiation had its origin in the observations of early embryologists as far back as Aristotle, who noted that embryos with few or no observable structural differences, such as a chick egg, give rise to complex differentiated organisms. The fact that development produces an exact replica of the parents was long believed to be caused by the transmission of adult structures in miniature form. As we discussed in Chapter 1, this *preformationist* doctrine was supplanted in the eighteenth and nineteenth centuries by the *epigenetic* view that adult structures were absent from the early embryo but appeared *de novo* during embryonic development. However, although epigenesis more readily explained the superficial similarities of all cells of an early embryo, it left unanswered the question of how and why cellular differences arise.

The notion that cellular differentiation was caused by differences in cell nuclei between tissues was first expounded by Weismann and by Roux, who proposed that the products of embryonic mitoses were dissimilar, and that different cells obtained different hereditary determinants. A muscle cell, for example, lost all but muscle determinants during the divisions that led to its formation; a nerve cell retained only nerve determinants, etc. Support for this view existed in the finding of chromosomal loss from the somatic tissues of *Ascaris megalocephala* (see p. 26) as well as in the *mosaic* development shown by many annelids and molluscs. In these latter organisms each early embryonic cell appears to be capable of forming only a restricted range of tissues or organs. The embryo was consequently conceived to be a mosaic of separately determined cells that develop independently of each other into adult structures.

The Weismann-Roux theories, however, were soon cast into doubt by the findings of Driesch, Boveri, and others that development was not "determined" or mosaic in numerous other species of embryos. The cells of early sea urchin embryos, for example, can be isolated from each other at the two- and four-cell stages and nevertheless develop into complete embryos. In salamanders, Spemann showed that a single cell at the embryonic 16-cell stage could produce an entire embryo. More recent experiments by Briggs and King have shown that some nuclei from blastula and gastrula stages of frog embryos (*Rana pipiens*) are still sufficiently potent to produce a complete embryo when transplanted into an enucleated egg (Fig. 30–3). In *Xenopus laevis*, the African clawed frog, Gurdon has shown that at least 20 percent of the intestinal cells of feeding tadpoles can be transplanted to enucleated eggs and produce embryos with functional muscle and nerve cells. Furthermore, some of these intestinal cells may even produce a completely viable and fertile embryo. In plants Steward has found that individual carrot root cells can, with proper nourishment, be made to differentiate into complete carrot plants. In *Drosophila*, Ilmensee has demonstrated that nuclei derived from various regions of the early gastrula embryo can reach the larval stage when trans-

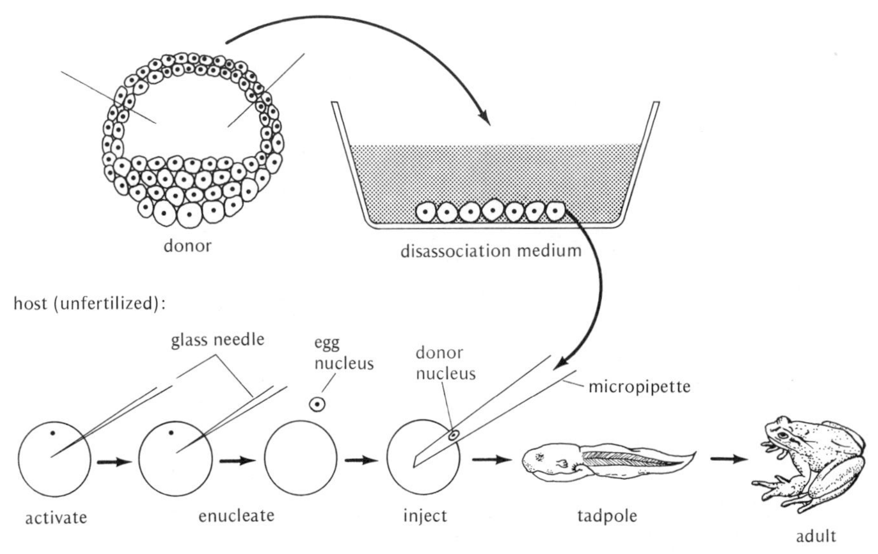

Figure 30–3

Transplantation of nuclei from a donor in the blastula stage to an unfertilized host. The host egg is "activated" by pricking with a glass needle to stimulate development and the female egg nucleus removed. (In many experiments, the egg nucleus is destroyed by ultraviolet irradiation.) A diploid donor cell is then obtained by dissociating donor tissue into single cells using versene or trypsin. One of these cells is then sucked up by a very thin micropipette whose diameter is small enough to rupture the cell wall. This nucleus, along with a very small amount of cytoplasm, is then injected into the enucleated host egg. Development in such cases can produce completely normal individuals. (After Moore, 1972, with additions.)

planted into enucleated eggs. Although these larvae do not survive to the adult stage (probably because of damage to the egg in the transplantation procedure), their cells can be further transplanted to viable hosts and will continue to grow and divide. In fact, descendants of these gastrula nuclei will form gametes that produce fertile adults when they can be placed into developing gonadal tissues of normal embryos and larvae. In mice, nucleic acid hybridization studies (p. 101) show no detectable differences between DNA extracted from embryonic and adult tissues such as brain, kidney, liver, spleen, and even cancer cells.

Taken as a whole, these findings indicate that many embryonic cells are not unalterably determined to produce a particular tissue but maintain the potentiality for full development. In this type of *regulative development*, differentiation is therefore the result of mutual interaction or regulation between fully competent cells, each maintaining an intact set of genes. That is, the adoption of a particular phenotype by a cell is caused by its relationship with other cells or with the cytoplasm produced by other cells rather than by an irreversible change in the amount of its genetic complement. Even "mosaic" development can therefore also be explained in a regulative way. For example, nuclei of developing *Drosophila* eggs that enter the posterior "polar cap" region of the egg are destined to become the gonadal tissues of the adult (p. 255). However, should the egg cytoplasm in this region be transplanted to the anterior pole of an egg, nuclei that enter this pole now develop into gonadal cells (Ilmensee and Mahowald). It

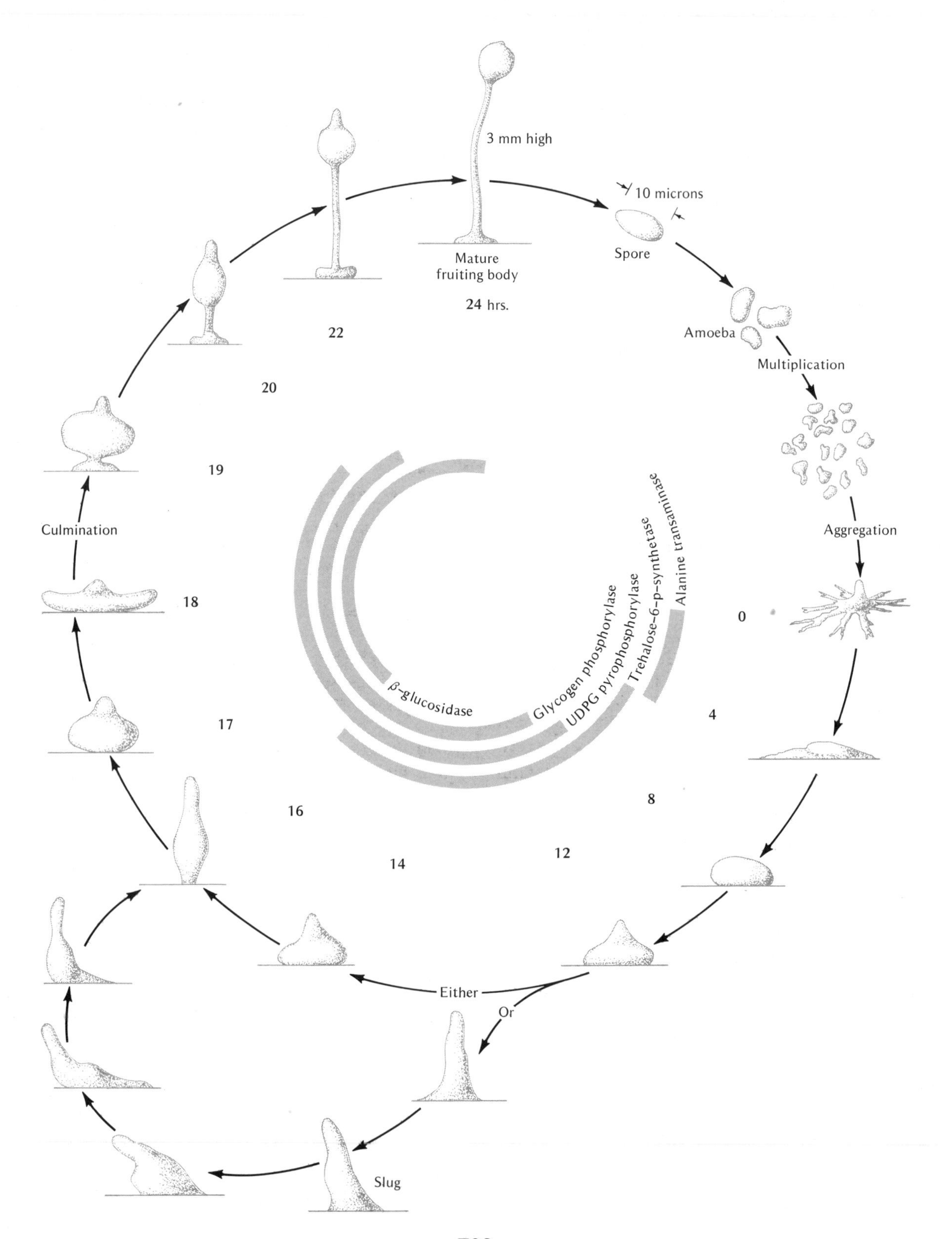
3 mm high
10 microns
Mature
fruiting body
24 hrs.
Spore
Amoeba
Multiplication
Aggregation
22
20
19
Culmination
18
17
16
14
12
8
4
0
β-glucosidase
Glycogen phosphorylase
UDPG pyrophosphorylase
Trehalose-6-p-synthetase
Alanine transaminase
Either
Or
Slug

Figure 30–4

Life cycle of the cellular slime mold, Dictyostelium discoideum. *In its multicellular phase, the transition from an aggregate to a mature fruiting body usually takes about 24 hours (numbers indicated on the figure). During one part of the life cycle, the organism may adopt a migratory habit (lower left) which can last for a few hours or a few days. The time spans during which given enzymes appear are in the center of the figure. Transcription of mRNA specifying these enzymes was previously believed to occur 2–4 hours earlier, but Firtel, Baxter, and Lodish now suggest that translation probably occurs soon after transcription. Some of the enzymes are specifically associated with steps necessary for the synthesis of cellulose in stalk cells, or the sugar trehalose in spore cells. (After Sussman, with modifications.)*

seems therefore obvious that the mosaic development observed in insects, molluscs (Fig. 13–2), and annelids, is primarily caused by maternally produced cytoplasmic differences leading to nuclei with functional but not genetic differences. The fact that mitosis in practically all organisms produces constant chromosomal numbers in daughter cells, no matter what their developmental fate, may be considered as firm support for regulative development.

DIFFERENTIATION IN SLIME MOLDS

Regulatory relationships between cells are especially obvious and are being extensively investigated in the cellular slime mold *Dictyostelium discoideum.* In this simple haploid eucaryote only three basic types of cells are found: the amoeboid migratory cell, the spore cell, and the stalk cell. As shown in Fig. 30–4, large numbers of *Dictyostelium* amoeboid cells will ordinarily aggregate together when their food supply of bacteria is exhausted. Depending on environmental conditions (light, acidity, etc.), this aggregate may then form a migrating pseudoplasmodium or *slug.* However, regardless of whether this multicellular aggregate migrates or remains in place, a *fruiting body* is eventually erected consisting of a slender *stalk* of vacuolated cells bearing on top thousands of *spores* contained within a relatively large spherical sporangium. Under appropriate conditions the spores germinate, each producing one amoeboid cell which then feeds and divides mitotically to begin the cycle once again. Only two major developmental alternatives therefore exist for a *Dictyostelium* amoeba: stalk cell or spore cell. How is this choice made?

According to Bonner, Shaffer, Sussman, and many others, the formation of an aggregate is initiated by the presence of a chemical signal, cyclic AMP (see p. 679), whose highest concentration emanates from one or a few cells at the aggregate center. As this chemical diffuses through the medium it creates a gradient that causes the movement of other amoeboid cells toward the region of high concentration. Aggregating cells then produce more cyclic AMP, and also develop changes in membrane structures that enable them to adhere together. The eventual role adopted by an amoeboid cell, as stalk cell or spore cell, appears to be regulated primarily by its relative position within the aggregate. Cells entering the aggregate first usually give rise to the tip of the slug, while cells entering later give rise to the base of the slug. According to some workers, it is the concentration of cyclic AMP that acts to differentiate the two kinds of cells. Other workers, such as Bonner, suggest that the aggregation process merely sorts out cells that have already differentiated.

Whatever the cause for the initial differentiation between the two kinds of cells, their fate becomes firmly decided by the time migration of the slug ceases. Cells that are at the tip move downward toward the base of the slug to form the cellulose-containing stalk cells, whereas cells at the base move upwards to form the spores (Fig. 30–5). The pattern of differentiation of an aggregate into stalk cells or spore cells does not seem to be affected by the numbers of cells in such aggregates: small aggregates form small fruiting bodies, and large aggregates form large ones. In the "fruity" mutant strain of *Dictyostelium* a miniature but recognizable fruiting body is formed from only 12 cells! Obviously the cells can sense their positions relative to each other and differentiate themselves

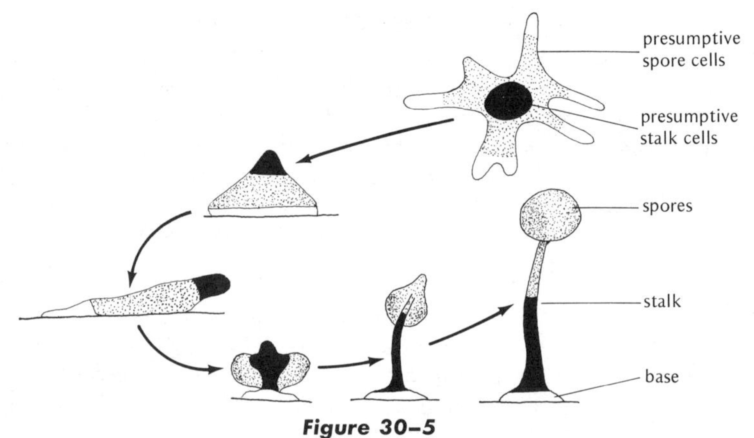

Figure 30–5

Diagram showing the roles of cells that occupy different positions in the aggregate (top) and slug (left) during the formation of spores, stalk, and base of the Dictyostelium *fruiting body. (After Sussman.)*

according to an inherent fruiting-body pattern. Such precisely regulated differentiation even occurs when a slug is cut into independent pieces, and groups of cells that would formerly have differentiated only into stalk tissue or spore tissue now differentiate in isolation from each other into complete fruiting bodies.

Although the exact regulatory mechanisms that cause the differentiation of cells are still undiscovered, there is little question of their genetic basis. A large variety of gene mutations are now known in *Dictyostelium* that do not affect growth at the amoeboid stage but do affect aggregation, or fruiting body formation, or the rate of development.

CELL REGULATION

Gene regulation must certainly be in force in those instances where a particular cell produces more of one protein than the other, although the genes for both proteins are present. We have seen this phenomenon in the large differences between the amounts of β and δ chains found in adult hemoglobin, and also noted the relative changes in the rates of production of different hemoglobin chains during birth (p. 621). Similar observations have been made by Markert and others for the enzyme lactate dehydrogenase (LDH). This enzyme catalyzes particular steps during carbohydrate metabolism that affect the concentrations of the intermediary compounds lactate and pyruvate.

LDH has been found to exist in animal tissues in at least five forms, or *isozymes*, which differ from each other sufficiently to be separated by electrophoresis on starch strips. These five isozymes have been named LDH-1, -2, -3, -4, and -5. Examples of their distribution in 10 different tissues of the same organism can be seen in Fig. 30–6. Note that one or more isozymes may be almost completely absent in some tissues or their relative amounts markedly changed between different tissues. Accounting for some of these differences are observed differences in the function of certain of the isozymes during oxygen starvation (anaerobic conditions). LDH-5 permits more anaerobic metabolism and a greater accumulation of lactate, while the lactate-forming activity of LDH-1 is more easily inhibited under anaerobic conditions. Thus skeletal muscle which can tolerate some degree of temporary muscle impairment caused by accumulation of lactate, possesses LDH-5 primarily, while heart muscle, which cannot tolerate any temporary impairment, has only minor amounts of LDH-5 and larger amounts of LDH-1. The synthesis of the LDH isozymes,

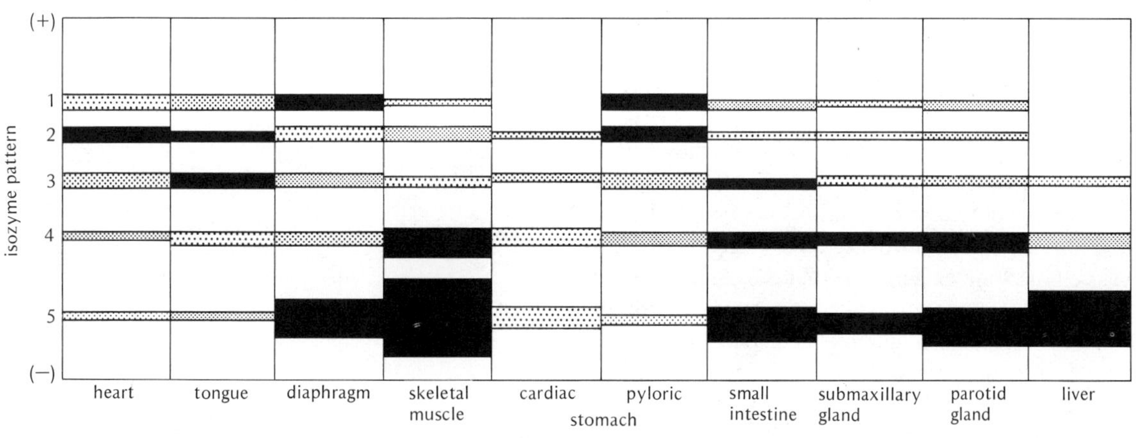

Figure 30–6

Distribution of the five isozymes of lactate dehydrogenase (LDH–1 to LDH–5) in 10 different tissues of the adult mouse. The (+) and (−) indicate positive and negative poles in the electrophoretic field. (After Markert.)

therefore, seems to be regulated in accordance with the needs of a tissue. However, even the needs of a tissue may change at different times, and Markert and Ursprung have shown that the relative proportions of the LDH isozymes undergo significant changes during development in any one tissue (Fig. 30–7).

On the molecular level isozyme differences arise from the fact that numerous enzymes are in multimer form (see p. 667) composed of two or more

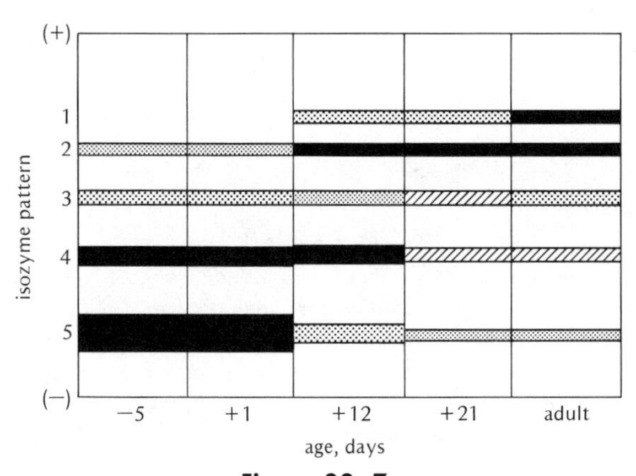

Figure 30–7

Developmental changes in the distribution of LDH isozymes in the stomach of the mouse from 5 days before birth (− 5) to postnatal 1, 12, 21 days, and the adult stage. (After Markert.)

polypeptide subunits which associate together to form the functioning enzyme. For example, LDH is a four-chained tetramer derived from polypeptides that can be produced by two different gene loci, LDH^{α} and LDH^{β}. When only one of these two genes is activated in a cell, only one form of LDH is produced. Thus four α chains constitute the LDH-1 tetramer or isozyme, and four β chains constitute the LDH-5 isozyme. Activation of both genes in the same cell then permits the "hybrid" isozymes to be produced: LDH-2 ($3\alpha:1\beta$), LDH-3 ($2\alpha:2\beta$), and LDH-4 ($1\alpha:3\beta$). Since the α and β subunits seem to associate together randomly to form LDH enzyme, it is likely that the proportion of each of the different isozymes found in a particular tissue at a particular time depends upon the relative amounts of the α and β chains present. That is, isozyme levels appear to be controlled by the extent to which each of the two LDH genes are activated.*

*Additional gene loci are known to be involved in the formation of LDH isozymes; and in fish, a third gene locus (LDH^{γ}) can contribute polypeptide chains that combine with those of LDH^{α} and LDH^{β}, e.g., $3\alpha:1\gamma$, $2\beta:2\gamma$, etc. Although the relative amount of LDH mRNA productivity for each of these genes has not yet been determined, it seems likely that production of these proteins is probably primarily regulated by transcriptional controls. In addition some evidence exists that controls may also be exercised through posttranslational mechanisms: in some tissues, such as heart muscle, the rate of degradation of the LDH-5 isozyme is much higher than in other tissues, e.g., skeletal muscle.

A system that may be considered perhaps the most extreme form of cell differentiation in eucaryotes is that for antibody production found in many vertebrates. The antibody system seems to be considerably unlike the LDH isozyme system in the sense that many *thousands* of different kinds of antibody polypeptide chains can be produced, each normally the product of only a single lymphocyte cell or clone of lymphocyte cells. As explained in Chapter 9, the production of a specific type of antibody molecule is elicited in a particular lymphocyte by the presence of a particular antigen. Such antibody molecules are composed of four polypeptide chains (Fig. 30–8a): two "heavy" (each 440 amino acids long) and two "light" (each 214 amino acids long) held together by disulfide bonds.

Antibody specificity toward a particular antigen lies in what is called the "variable" regions of the antibody molecule (Fig. 30–8b) at the amino ends of the polypeptide chains. These variable regions have different amino acid sequences in each of the different antibodies analyzed so far, but are of approximately the same length (108–118 amino acids long) in both light and heavy chains. The remainder of each antibody molecule, called the "constant" region, does not change along with the variable region, but maintains mostly a common amino acid sequence in different antibody polypeptide chains. There are nevertheless different classes of heavy and light chains whose constant regions differ slightly, and whose functions also differ in respect to the kinds of cells they can enter, or membranes they can cross, or activities they can perform after binding the antigen.

The existence of what seems to be enormous antibody diversity, of a magnitude that may perhaps reach a potential level of 100,000 or more different kinds of antibody proteins produced by an individual, raises many questions of interest to developmental biologists. Are there separate genes for the constant and variable regions of an antibody polypeptide chain? How are these genes or their products combined to form one polypeptide chain? Are there separate genes for each of the many different possible variable regions that an organism can produce (see, for example, Hood and Talmadge)? Or, on the other hand, does the diver-

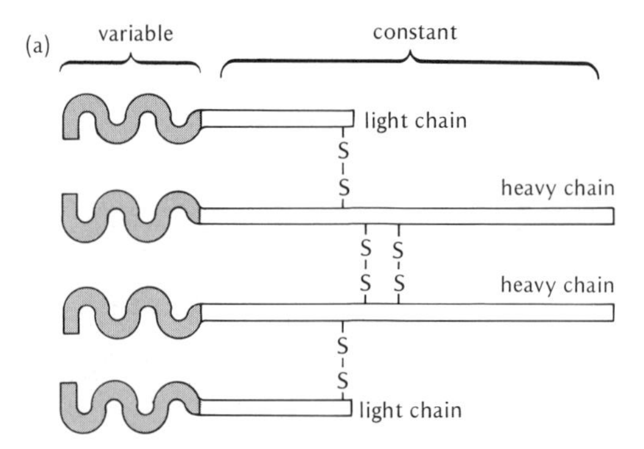

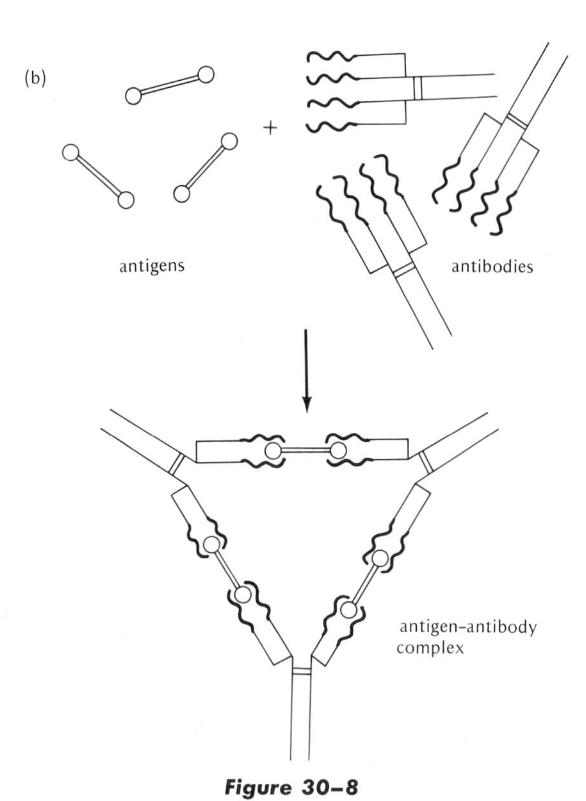

Figure 30–8

Schematic representation of an antibody molecule, and one of the ways it can bind a particular antigen. (a) Relative positions of the heavy and light chains, and the variable and constant regions. (b) The formation of an antigen-antibody complex by binding a two-pole divalent antigen into a trimer arrangement.

sity of variable regions arise from only a few genes by a process that generates gene differences within the organism itself; that is, by means of somatic mutation (see Cohn) or somatic recombination (see Smithies) or somatic translocation (see Gally and Edelman)?

Although no clear-cut answer to these questions has yet been demonstrated, most developmental biologists suggest that cellular differences within an organism primarily arise from mechanisms that produce changes in cellular function without necessarily accompanying changes in the number of genes. We have seen such mechanisms in microorganisms with regard to the action of operator and regulator genes and in the phenomena of induction and repression. In higher organisms similar regulatory mechanisms must exist, although they are undoubtedly further complicated by the relationship between cells and tissues. That is, the regulation of development in higher organisms must be such that some tissues are subject to different stimuli than others and some tissues respond differently to the same stimulus. Among these various stimuli are initial differences between various parts of the zygote, the influence that other cells and tissues exert by contact, and the products of cells and tissues in other parts of the organism. Since development proceeds in time as well as in space, these stimuli and their responses must not only be spatially regulated but also temporally regulated.

THE EGG

At the base of differentiation in most multicellular organisms stands the fertilized zygote, which, through successive cleavages, produces the many stages of the embryo (Fig. 30–9). The finding by Hultin and others that protein synthesis begins with fertilization was at first thought to imply that this was caused by mRNA templates sent from both the zygotic nucleus and the daughter nuclei produced by early cleavage. This interpretation, however, has been contradicted by the following facts:

1. In spite of many attempts, no new mRNA can be observed immediately after fertilization in sea urchin eggs.
2. When enucleated parts of an unfertilized sea urchin egg are stimulated by parthenogenetic treatment (hypertonic sea water), they produce protein at the same rate as normally fertilized eggs with nuclei. Nuclear mRNA produced upon fertilization is therefore certainly not the cause of protein production in the enucleated fragments.
3. Harvey has found that these enucleated egg halves can divide and produce structures resembling the normal mitotic apparatus.
4. Actinomycin D, which inhibits RNA synthesis from DNA templates, does not greatly affect protein synthesis when applied to eggs soon after fertilization. For example, Gross has shown that protein synthesis of actinomycin-treated eggs of *Strongylocentratus purpuratus* is not affected until the blastula stage. After that point protein synthesis declines in the actinomycin-treated eggs, suggesting that further mRNA production were then necessary.

Coupled with observations by Monroy and others that *unfertilized* sea urchin eggs contain protein-forming mRNA, these findings indicate that the protein synthesis that occurs upon fertilization utilizes mRNA that has been stored in the egg during its maternal maturation. The mechanism that activates the cytoplasmically stored RNA to act as a template is unknown, but it may involve a change in ribosomal structure since the subjection of ribosomes from unfertilized eggs to a short treatment of trypsin causes them to become active in protein synthesis. Because trypsin degrades protein, it has been suggested that prefertilization ribosomes are prevented from functioning because of the presence of a "masking" protein.

According to some workers, masking proteins are often associated with mRNA in eucaryotes to protect mRNA molecules from degradation during their transport from nucleus to cytoplasm (see also

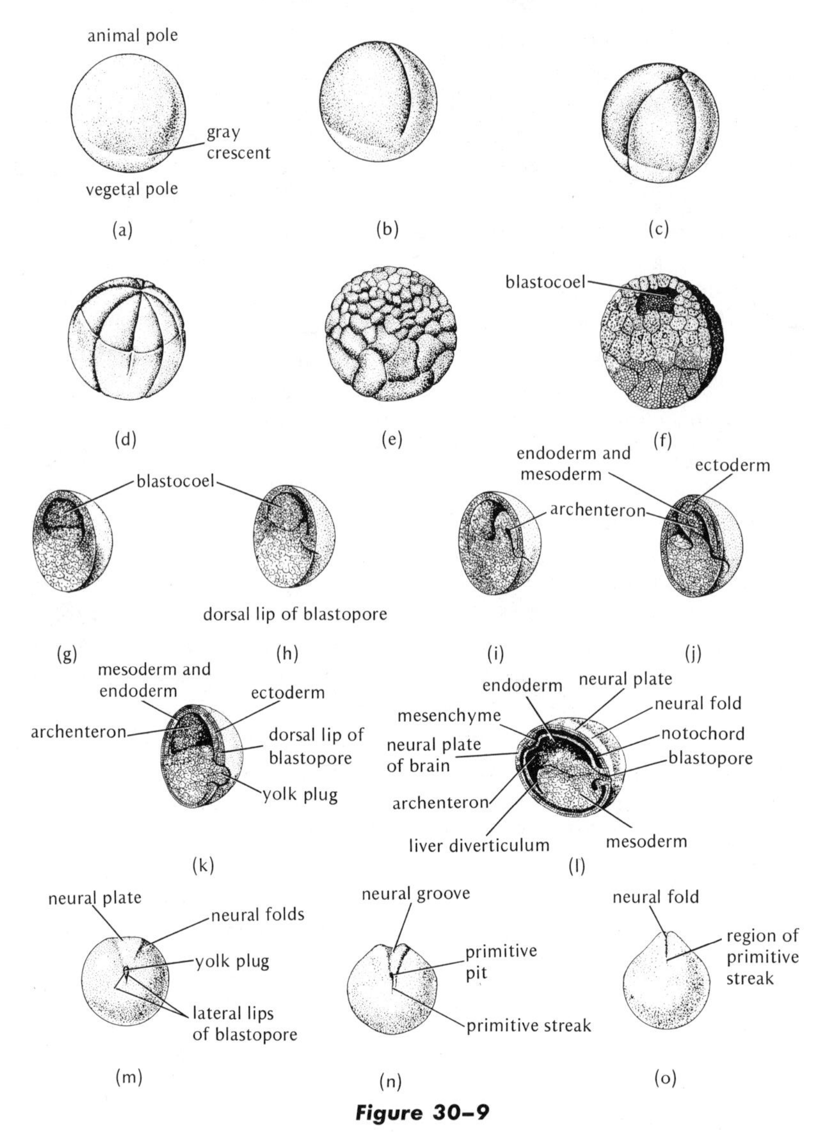

Figure 30–9

Some early stages in development of the frog embryo: (a–d) early cleavage; (e–g) blastula; (h–k) gastrula; (l–o) neurula. (After Huettner.)

p. 650). In the present case, masked maternal messenger RNA may serve the purpose of enabling cell division to occur rapidly in early embryogenesis without the need of synthesizing new mRNA by zygotic nuclei. In *Drosophila melanogaster* development, for example, the length of early nuclear replication cycles is only about 10 minutes—hardly enough time to selectively transcribe significant amounts of all the various kinds of cellular mRNA. Cytoplasmic particles containing protein and RNA, perhaps masked mRNA, have been called *informosomes*. The RNA in such complexes presumably remains inactive until it becomes attached to functioning ribosomes.

Whatever the mechanism that activates preexisting mRNA in the egg, the nuclear mitotic activity induced upon fertilization involves DNA replication and consequent multiplication of cell nuclei but is not necessarily accompanied by significant nuclear mRNA production until a large number of cells has been formed. Differentiation at early stages is therefore probably caused by maternally produced differences in the egg cytoplasm or egg cortex (p. 701). In amphibians this view is supported by the finding of Curtis, who grafted the enucleated dorsal section of the egg cortex ("gray crescent" region) from one fertilized egg of *Xenopus* to the ventral surface of another fertilized egg and found that the product was a double embryo (Fig. 30–10). It is likely, therefore, that the gray crescent region of the egg contains surface determinants (see also p. 265) which, when invaded by functioning nuclei, will produce an embryo.

On a more biochemical level Gurdon and others have shown that both RNA and DNA synthesis is very firmly controlled by the egg cytoplasm in *Xenopus* eggs. A nucleus taken from differentiating embryonic cells in which considerable RNA synthesis is being carried out will cease RNA synthesis when transplanted into enucleated eggs. Instead, these nuclei then begin DNA synthesis as normally occurs in fertilized eggs. Only after the nuclei in such eggs have undergone a successive series of five cell divisions is RNA synthesis resumed. At that point heterogeneous nuclear RNA is made (see p. 649), and smaller mRNA-like molecules appear shortly thereafter.

This theme of cytoplasmic control over cellular differentiation is not new, and extends back to Boveri, who showed that the chromosome diminution observed in *Ascaris* is controlled by the nature of the cytoplasm; the movement of nuclei into certain sections of the cytoplasm by centrifugation causes chromosomal loss. Support for this view now derives from various sources. Cytoplasmic effects obviously operate on the paternal chromosome set of coccids which becomes heterochromatinized in male somatic cells (p. 545). In *Acetabularia* information about the kind of crown or "cap" produced by this single-celled alga has been shown to be stably maintained in the cytoplasm for considerable lengths of time after the nucleus has been removed (see review by Harris). The phenomenon of maternal effects such as kynurenine distribution in *Ephestia* and cleavage pattern in *Limnaea* (Chapter 13) also points strongly to the existence of cytoplasmic determinants before fertilization, among which may be maternal mRNA templates. In fact Gross and co-workers have now demonstrated the presence of at least one form of maternal mRNA in unfertilized sea urchin eggs—that for histone proteins.

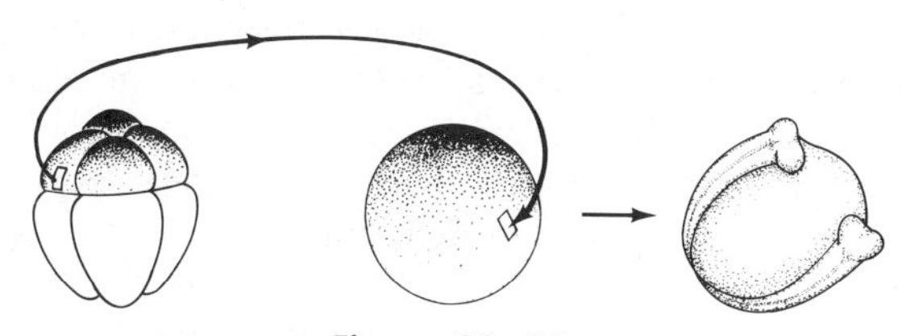

Figure 30–10

Grafting of a section of the cortex of the gray crescent area from an early-cleavage egg to an uncleaved fertilized egg produces a pair of twin embryos. (After Curtis.)

GENE ACTION IN DEVELOPMENT

The involvement of the zygotic nucleus and its daughter cells in new mRNA synthesis seems postponed, at least in some cases, to a developmental stage beyond the early mitoses or cleavage stages. The precise point at which RNA production by the fertilized egg nucleus affects development, however, is difficult to determine and must vary between different kinds of mRNA. Perhaps the earliest of gene actions observed so far in multicellular organisms is believed to be that found by Counce in *Drosophila* matings involving homozygous *deep orange* females. Normally such females produce eggs that show malfunctioning within less than an hour after fertilization if they are fertilized by sperm containing the sex-linked *deep orange* allele (*dor*) or by sperm containing the Y chromosome. Only fertilization with the wild-type allele dor^{+} can produce viable embryos. Although somewhat different results have been obtained by Hildreth and Lucchesi (see discussion by Wright), the effect of this allele on development seems to occur immediately after its entry into the egg or the entry of a *dor*$^+$ substance produced by it.

A second gene mutation whose effect appears quite early in development is that associated with the nucleoli of *Xenopus*. Normally, *Xenopus* nuclei have two nucleoli (2-*nu*) but these can be diminished by the *anucleolate* mutation, which, in heterozygous condition, produces nuclei with only one nucleolus (1-*nu*). The *anucleolate* homozygote has no distinct nucleoli (0-*nu*) and usually dies in the tadpole stage before being able to feed. Interestingly Brown and Gurdon have shown that this mutation in homozygous condition specifically affects production of ribosomal RNA; neither DNA, messenger RNA, or transfer RNA are affected. Although such *anucleolate* homozygotes maintain the original rRNA of the egg, death nevertheless ensues because they are unable to synthesize any of the new rRNA that normally appears at the gastrula stage.

According to Wallace, Birnstiel, Brown, and others, the cause responsible for *anucleolate* behavior is a deficiency of that portion of *Xenopus* DNA (rDNA) that normally hybridizes with ribosomal RNA (rRNA). In normal *Xenopus* there are approximately 450 copies of the same ribosomal gene in each nucleolar organizer region. Each of these repeated or "reiterated" genes is transcribed into a 45 *S*-40 *S* molecule of precursor RNA which is then cleaved into 18 *S* and 23 *S* rRNA components (see Fig. 27–10). As shown in the electron micrographs of Miller and Beatty (Fig. 30–11), the reiterated ribosomal genes are arranged in tandem, separated from each other by "spacer" DNA which is not transcribed. The defect in *anucleolate* consists in a loss of almost the entire rDNA section, and parallels according to Ritossa, Atwood, and Spiegelman, the loss of rDNA in the *bobbed* mutations in *Drosophila* (p. 243). Since each nucleolar organizer usually produces its own nucleolus, the absence of rDNA causes loss of the nucleolus.

In the amphibian oocyte the increase in amount of rDNA responsible for the production of rRNA is even greater than in normal cells since a mature oocyte contains about a thousand nucleoli, each with a complement of rDNA genes normally found in the nucleolar organizer of only one somatic cell! Although the purpose for the amplification of rDNA genes in the oocyte seems clear—to enable a large "unicellular" mass of cytoplasm to possess sufficient ribosomal RNA so that protein synthesis can continue in the interval between fertilization and gastrulation—the molecular mechanism responsible for such *gene amplification* is not yet fully determined.

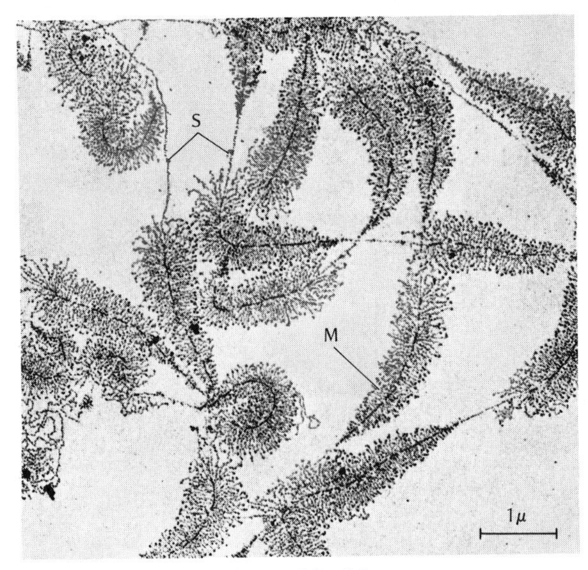

Figure 30–11

Electron micrograph of about 25 genes, all producing ribosomal RNA, isolated from the nucleolar organizer region of a Triturus viridescens *oocyte. As previously explained (see Figs. 5–24 and 27–12), each "bush" or matrix of molecules (M) arising from the thin DNA axis represents RNA molecules in various stages of transcription: the smaller RNA molecules are near the initiation point for transcription, and the longer RNA molecules are near the terminating point. Note that the transcription regions are not continuous, but there are spacer intervals (S) between the genes that are not transcribed. (From O. L. Miller, Jr., B. R. Beatty, B. A. Hamkalo, and C. A. Thomas, Jr., 1970. Electron microscopic visualization of transcription.* Cold Sp. Harb. Symp., **35,** *505–512.)*

According to Hourcade, Dressler, and Wolfson, amplification of the rDNA genes occurs in each oocyte nucleolus through a "rolling circle" intermediate. This mechanism, shown previously to exist in some procaryotes (p. 88), could "crank out" long multigene concatemers of rDNA from a single circular rDNA gene. However, the source for the initial circular rDNA template is not known, and attempts have not as yet been successful in finding similar amplification mechanisms to explain the increased productivity of specific genes in specialized tissues.

For example, there are relatively few copies of both hemoglobin genes in developing vertebrate erythrocytes and silk fiber genes in *Bombyx mori* silk glands. Increased protein productivity in these tissues apparently relies on increased amounts of mRNA derived from a small number of genes (perhaps one per haploid genome) through increased transcription rates and increased messenger stability. Even when many copies of a particular gene are found, such as the thousands of genes that produce histone proteins in sea urchins, such genes are already present in reiterated form in the normal gene complement, and their numbers are not further increased through gene amplification. The process of gene amplification may therefore be limited to the large 18*S* and 23*S* ribosomal genes which would probably take up more than half of the DNA in all amphibian cells if they could not be specifically amplified in the additional nucleolar rDNA of the oocyte.

A gene in mice that also affects the nucleolus is one of the *tailless* (t) alleles. This allele, t^{12}, as other t alleles, produces a normal tail as a heterozygote with the wild-type allele +, but produces a tailless mouse when combined with the dominant *Brachyury* allele, T. It can be kept as a balanced lethal stock T/t^{12}, since both T/T and t^{12}/t^{12} homozygotes are lethal. In the case of the lethal t^{12} homozygote, development stops at the morula stage, approximately 4 days after fertilization, and shows, among its main characteristics, a change in shape of the nucleoli and a relative decrease in RNA synthesis. Whether this defect in RNA synthesis is caused directly by the t^{12} gene or whether it is a secondary effect of some other disturbance is unknown. As shown in Table 30–1, other recessive t alleles also have specific times of action, although these are generally later than the t^{12}/t^{12} homozygote. They all bear in common their chromosomal location and their effect on the differentiation of ectodermal and neural structures, but again the mechanism involved is unknown.

TABLE 30–1

Effects of different genotypes of *tailless* alleles in mice (in days after fertilization)

Genotype	Effect
t^{12}/t^{12}	lethal at morula stage (4 days); affects ability to produce blastocyst, nucleolar differentiation, and RNA formation (probably ribosomal RNA)
t^{12}/T	tailless but viable; a constriction appears at base of tail at about 11 days that affects the tail notochord and neural tube and is accompanied by resorption of the complete tail
$t^{12}/+$	normal tail
t^{0}/t^{0}	lethal at early egg cylinder stage (6–7 days); affects division of inner cell mass into embryonic and extraembryonic ectoderm
t^{0}/T	as t^{12}/T
$t^{0}/+$	normal tail
t^{w5}/t^{w5}	lethal at egg cylinder stage (7–10 days); affects growth and maintenance of embryonic ectoderm
t^{w5}/T	as t^{12}/T
$t^{w5}/+$	normal tail
t^{w18}/t^{w18}	lethal at late egg cylinder stage (8–10 days); affects growth of primitive streak causing neural-tube complications
t^{w18}/T	as t^{12}/T
$t^{w18}/+$	normal tail
t^{w1}/t^{w1}	lethal at embryonic shield stage and later (9 days–birth); affects growth and maintenance of neural tube and brain, causing microcephaly and abnormal neural tube
t^{w1}/T	as t^{12}/T
$t^{w1}/+$	normal tail
T/T	lethal at 11th day; affects the notochord and allantois, causing the absence of the posterior half of the embryo and lack of an umbilical system
$T/+$	short tail; a constriction appears midway along the tail at about 11 days and is accompanied by notochord abnormalities and consequent resorption of part of the tail structure

After Bennett, and after E. W. Sinnott, L. C. Dunn, and T. Dobzhansky, 1958. *Principles of Genetics*, 5th ed. McGraw-Hill, New York.

In the T/T homozygote more information is available, and the mutant effect seems to lie in the inability of T/T tissues to respond to specific stimuli known as "organizers" or "inducers." Normally the dorsal mesoderm tissue of the embryo becomes segmented into paired structures called "somites," which produce the cartilage and bone of the spinal column. As shown by Grobstein and Holtzer in tissue-culture experiments, T/T somites are unable to respond to spinal cord tissue and form cartilage. T/T spinal cord, however, can induce the production of cartilage in normal somites.

In some cases neither the inductive nor responsive capacity of the tissues is affected; rather, the mutation acts by preventing proper contact between tissues that would otherwise normally interact. For example, *Danforth's short-tail* mutation in mice (*Sd*) will, when homozygous, cause the appearance of individuals lacking ureters and kidneys among other defects. The absence of kidneys, however, is not caused by the inability of the kidney-forming tissues to respond to the action of the "organizer" present in the ureters. As shown by Gluecksohn-Schoenheimer, kidneys are generally formed in such individuals if the ureters can reach the kidney area. The *Sd* mutation therefore seems to prevent kidney formation by interfering with the growth of ureters but does not affect their inductive powers.

These examples represent only a few of the many developmental effects known to be caused by genes. Indeed, the literature of developmental genetics is large, and summaries can be found in the reviews of Hadorn (1961), Wagner and Mitchell (1964), Markert and Ursprung (1971), and others. In general the effects of mutant genes indicate that development is normally a complex but ordered sequence of interactions between cell, tissues, and organs. Since many of these interactions seem to be subject to genic control, the question now turns from the effects and products of genes during development to the regulation of genes. Is there direct evidence that genes are turned on and off during development?

REGULATION OF CHROMOSOME DIFFERENCES

Among the most graphic demonstrations correlating the onset of gene activity with chromosome differentiation are those that have arisen from studies of chromosome "puffs" in the polytene nuclei of *Diptera* insects. These puffs, or Balbiani rings, arise at different points on polytene chromosomes and, as discovered by Beermann in *Chironomus*, and by Breuer and Pavan in *Rhynchosciara*, many of them are distinctive for specific tissues, and also vary within a tissue at different times (Fig. 30–12). The present view is that the puff signifies RNA synthesis, often localized to an expanded portion of a single DNA band (Fig. 2–13d,e). This view is supported by experiments that stain RNA differentially and show its localization in the puff and by labeling experiments with radioactive RNA precursors that also show marked RNA synthesis in the puff region. Furthermore, drugs such as actinomycin, which inhibit DNA-directed RNA synthesis, also inhibit RNA synthesis in the puff and puff formation. The pres-

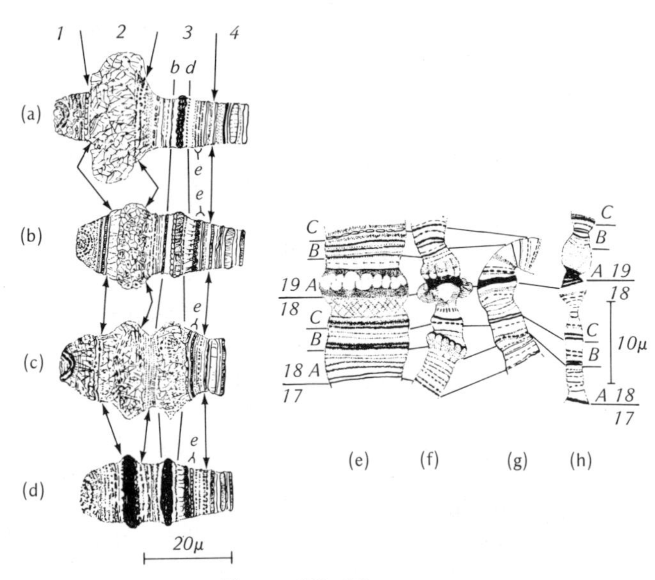

Figure 30–12

Polytene chromosome puffing and banding patterns in Diptera. *(a–d) Changes in puffing patterns in the salivary chromosomes of* Rhynchosciara angelae *at different stages in larval development (after Breuer and Pavan). (e–h) Different puffing patterns in the same chromosome of different tissues in a single individual of* Chironomus tentans*: (e) salivary gland, (f) malpighian tubule, (g) rectum, (h) midgut. (After Beermann, 1952.)*

ence of giant lampbrush chromosomes in amphibian oocytes (p. 26) is also correlated with RNA production, as shown by Gall and Callan, Izawa, and others.

In the salivary chromosomes the RNA products of the puff become bound to protein molecules and appear under the electron microscope as round particles approximately 300 angstroms in diameter. It is of great interest that these particles seem to approach the nuclear membrane, penetrate the membrane pores, and subsequently undergo a change in shape. On the possibility that these observations reflect the true sequence of events, Beermann has proposed that the RNA synthesized in the puff may be protected from the enzymatic action of ribonuclease by a transport protein that carries it to the nuclear membrane. The RNA then enters the cytoplasm as single-stranded mRNA and is picked up by ribosomes to form polysome complexes.

Whatever the mechanism, there seems to be a direct relationship between puffing and gene action in these tissues. As evidence, Beermann has shown that a trait inherited as a single gene difference in chromosomes is associated with the appearance of a single puff. The two species of *Chironomus* involved, *tentans* and *pallidivittatus*, differ in the kind of proteins secreted by a special portion of the salivary gland and differ also in the presence of a puff on the 4th chromosome. In *tentans* the protein secretion is clear in this part of the gland, and the tip of the 4th chromosome shows no puff. In *pallidivittatus* the secretion is granular, and the tip of the 4th chromosome shows puffing. This puffing appears to be directly associated with the granular secretion, since it is the only distinctive difference between the special cells producing this substance and other nongranular salivary cells in the same gland. Crosses made between the two species and their fertile hybrids have enabled the gene for granular secretions (*SZ*) to be mapped as a single gene in the 4th chromosome in the region of the puff. Furthermore, heterozygotes for this gene (*SZ/sz*) show salivary chromosomes which are also heterozygous for the puff (Fig. 30–13). Since the only observable difference between the *SZ* and *sz* gene regions lies in the puff but not in the number of chromosome bands, the lack of puffing ability in *tentans* is not apparently caused by a deletion of the gene for granular secretion. Beermann has therefore proposed that the *SZ* mutation is probably of the "regulator" or "operator" type, which controls gene productivity, rather than of the "structural" type, which produces mutant proteins.

The association between a single salivary chromosome puff or band and a single gene is also supported by genetic and cytological studies in *Drosophila melanogaster*. As previously mentioned (p. 589), Judd and co-workers have shown that single complementation groups (cistrons) of the *zeste* gene can be localized to single bands. (Hochman

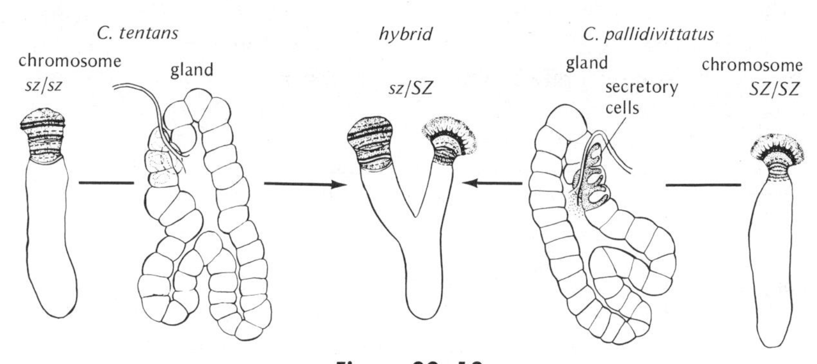

Figure 30–13

Fourth-chromosome puff in the salivary cells associated with granular secretion in pallidivittatus *and nonsecretion in* tentans *and in the hybrid between them. (After Beermann, 1961.)*

has also pointed to the close correspondence between the number of complementation groups and the number of bands on the small *Drosophila* fourth chromosome.) This one-to-one relationship between a "gene" and an observable cytological unit, according to Callan and Lloyd, is also true for each individual loop in the lampbrush chromosomes of amphibian oocytes. However, a paradox exists: if each gene occupies one puff or one loop, then the size of these genes is far greater than the 1000 nucleotide gene length ($\cong$300 amino acid polypeptide chain) normally expected. For example, a *Drosophila* band, out of which a puff may develop, contains about 30,000 nucleotide pairs or more, and each amphibian lampbrush loop contains DNA lengths that are even many times longer.

In answer to this puzzle Callan has proposed that an amphibian lampbrush loop represents a number of identical "slave" copies of a "master" gene. The slave genes are connected in end-to-end tandem sequence extending out along the loop, whereas the master gene remains in the chromomere section of the lampbrush chromosome. To ensure that the slave genes are exact copies of the master, Callan's hypothesis demands a complex DNA repair system that would match the nucleotide sequence of each slave against the master during meiosis of every generation. Although repair systems exist (Chapter 24), evidence for a mechanism that could achieve such continuous sequential matching is unknown.

Nevertheless various studies show that the organization of small chromosome sections, that is, chromomeres or bands, may well contain a number of tandem DNA sequences of "gene" size, each 700–800 nucleotides long. Some of these studies depend upon DNA reassociation techniques (see p. 97), whereas experiments by Thomas and coworkers are based on the detection of "rings" that result from the action of exonucleases on DNA fragments with redundant sequences (see p. 448). In practically all cases, from *Xenopus* to *Drosophila* to rats, these lengths of DNA, which are sufficient to code for polypeptide chains of 200 amino acids or more, seem to be linearly separated from each other by small repeated sequences perhaps 100–200 nucleotides long.

According to Bonner and Wu, there are about 4500 "families" of these small repeated sequences in *Drosophila melanogaster*. Since there are about 5000 chromomeres or bands visible in the chromosomes of this species, this correspondence may be meaningful in that each family is perhaps localized to a single chromomere. They suggest that a *Drosophila* chromomere consists of 30 to 35 sequences that are about 750 nucleotides long, each separated from the other by the smaller repetitive sequences that are identical within a single band (Fig. 30–14). Assuming this view to be correct, it is still not known whether the 750 nucleotide-long sequences in a chromomere are identical slave genes, or different structural genes comprising one specially regulated chromosome unit. Davidson and Britten suggest a regulatory role for the shorter repetitive sequences, which may function perhaps as promoters or operators in transcription (see also Crick, 1971, and Beermann, 1972). Other conjectured roles for these repetitive sequences are specialized intervals that enable crossing over;

Figure 30–14

Possible chromosome organization in which DNA sequences about 750 nucleotides long (s_1, s_2 . . . s_n) alternate with smaller repetitive sequences that are about 100–200 nucleotides long (r_1, r_2, etc.). A section of about 30 of the longer sequences are organized into a single chromomere or band (e.g., chromomere 1) in which the shorter sequences are all identical (e.g., r_1). It is not yet clear whether the 750 nucleotide-length sequences within a chromomere differ from each other as do the structural genes in a procaryotic operon, or whether they are all identical "slaves" of one "master." (After Bonner and Wu.)

structural sequences necessary for the tight physical packing of chromomere bands; and sequences that permit the long heterogenous nuclear mRNA molecules (see p. 649) to be cut into shorter cytoplasmic mRNA fragments.

Experiments that bear on the environmental activation of chromosome puffs and consequently on the induction of gene action during differentiation have been performed by Clever and others using *Chironomus tentans*. In this organism two of the puffs that appear during molting of the third and fourth instars (larval stages accompanied by shedding of skin) are called A and B, and are located on the 1st and 4th chromosomes, respectively. Concurrent with the appearance of puffs at these points is a rise in the level of molting hormone, *ecdysone*. This relationship is more than accidental, since Clever has found that ecdysone injected into larvae that are between molts nevertheless produces puffing effects similar to normal molting. The A and B puffs that are produced by ecdysone are not restricted to salivary-gland cells but are also found in other tissues with polytene chromosomes. To use an analogy with microorganisms, ecdysone seems to act as an "inducer" for the synthesis of whatever proteins are produced by the genes at the A and B puffs.

In *Drosophila* ecdysone has similar effects on puffing (Fig. 30–15), and Poels has shown that the hormone causes the release of a specific kind of mucopolysaccharide by the salivary cells of larvae. However, since the puffing patterns of chromosomes in different tissues of the larvae are similar, the exact relationship between ecdysone, puffs, and subsequent proteins is still not clear. What accounts for differences and what accounts for similarities? Moreover, ecdysone is not the only external "inducer": the experimental modification of puffing patterns has been achieved through changes in temperature, oxygen pressure, nutrition, etc. (see Ashburner).

An important question we can now ask is related to the previously discussed potentiality of nuclei to remain competent to produce full development. That is, are the chromosomal changes induced by the cellular environment reversible or irreversible?

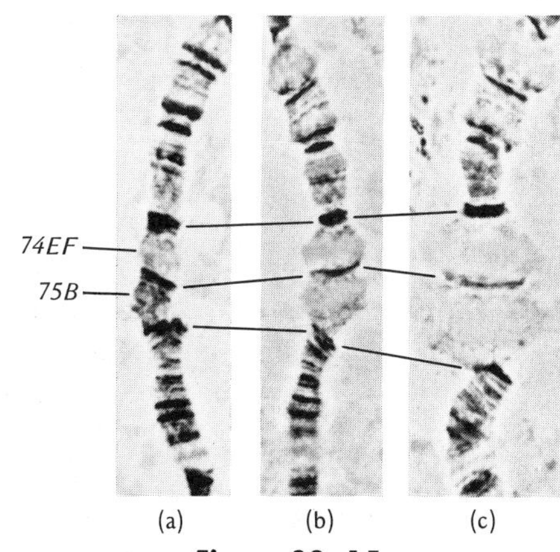

Figure 30–15

Photomicrograph showing different degrees of puffing in the left arm of chromosome 3 in the nuclei of larval salivary glands of Drosophila melanogaster. *(a) Before exposure to ecdysone. (b) Thirty minutes after injection of ecdysone. (c) Three hours after ecdysone injection. (Courtesy of Michael Ashburner.)*

In the case of puffing, the changes seem to be reversible, and puffs have been shown to expand and contract without any permanent alteration in their puffing ability. On the cellular level, however, once a cell has differentiated into a specific type, cell division appears to be necessary for such a cell to change into a new form. It has therefore been suggested that it is primarily through cell division that the regulatory mechanisms can be reshuffled which ordinarily restrict the function of major chromosome sections. However, exceptions to this rule are known: In the coelenterate *Hydra*, digestive cells have been shown to differentiate into epidermal cells without cell division. In *Xenopus*, Cooke found that embryonic cells can be made to differentiate into more specialized tissue although cell division has been prevented by colcemid treatment.

Since it seems unlikely that the quantity of DNA is affected by differentiation, what then restricts the developmental course of a cell and allows functional discrimination between different DNA segments? Examined in terms of controlling agents such as hormones, we can therefore ask what functions as the repressor. That is, what prevents all genes from being turned on all the time?

HISTONES

In 1950 Edgar and Ellen Stedman proposed the hypothesis that the histone constituents of chromosomes in higher organisms play a role in controlling gene action. Although this view was not supported at that time by experimental evidence, recent work by Huang and Bonner, Allfrey, Paul and Gilmour, and others has pointed strongly to the involvement of histones and other nuclear proteins in regulating gene action.

Histones are positively charged, or "basic," proteins that form complexes with the negatively charged phosphate groups of DNA. The positive charges are furnished by a high proportion of lysine and arginine amino acids, some histones bearing the description "lysine-rich" and others "arginine-rich." In general, histones are found only in organisms in which cell differentiation occurs and are relatively limited in variety to five major types, mostly distinguished by their lysine/arginine ratios. The close complexing of histones with DNA causes an increase in diameter of the ordinary DNA double helix from 20 angstroms to 35 angstroms. This close complexing also changes the properties of DNA, so that the temperature at which DNA strands change from regular double helix form to loose single-stranded form (the *melting temperature*) is generally increased.

These and other distinctions between histone-complexed DNA and noncomplexed DNA serve the purpose of estimating the extent of histone DNA in various cells. In young pea plants, for example, Bonner and collaborators have found that the cotyledons (first leaves) contain about 95 percent histone-complexed DNA, while the embryonic cells contain about 80 percent of such DNA. Associated with these findings are important differences in the relative efficiencies with which the two kinds of chromatin can produce new RNA. When these two chromatins are extracted and incubated with DNA-directed RNA polymerase, the rate of RNA synthesis approximately follows the extent to which non-histone-complexed DNA is present; i.e., pea-embryo chromosomes produce more RNA than cotyledon chromosomes. If the histones are removed from the pea-embryo chromosomes, a further marked increase in RNA synthesis occurs. On the other hand, in cells in which almost all the DNA is complexed with histones, such as duck erythrocytes, very little RNA is produced. General repression of RNA synthesis seems therefore to be a function of the presence of histones.

If histones can serve as repressors of transcription, the question arises of how such repression is exercised on specific gene loci. There are certainly too few different histones to account for the wide variety of specific regulators needed to control the many different genes in a eucaryote. In fact most histones are amazingly constant even in different organisms. Histone IV, for example, is a polypeptide chain 102 amino acids long that shows differences at only two amino acid positions between cattle and peas! It is difficult to conceive that a regulator with any specificity at all would change so little over the more than one billion years of evolution that have separated plants from animals.

The constancy of histones therefore indicates that they perform the same functions in different eucaryotes, and these functions are probably tied to the general structure of DNA, but unrelated to genetic differences between organisms. From x-ray studies and other evidence, R. D. Kornberg has recently suggested that chromatin is composed of repeating DNA subunits 200 base pairs long associated with two molecules of each of four different histones. These subunits are connected end-to-end to form a jointed flexible chain. A function of this kind may, of course, be regulatory—the coiling and uncoiling of chromosomes—but histones are obviously not the sole regulators of individual genes.

Instead of histones, recent work has placed a more specific regulatory role on the nonhistone nuclear proteins, which are generally more acidic, containing increased amounts of aspartic and glutamic acids. Although less plentiful than histones, the acidic nuclear proteins possess much greater variety and usually vary considerably in amount and composition between different cell types. They are also synthesized quite rapidly, showing high rates of "turnover" in metabolically active tissue, as might be expected in cells where changes in gene activity arise from changes in regulatory proteins.

Combining the results of various investigations involving nonhistone proteins (see Stein, Spelsberg, and Kleinsmith), a hypothetical regulatory sequence is shown in Fig. 30–16, beginning with the entry of a steroid hormone (e.g., progesterone) into a eucaryotic cell. The hormone (H) attaches to a specific cytoplasmic receptor (R) which is then further processed into a form (R′) enabling the H-R′ complex to cross the nuclear membrane. In the nucleus the H-R′ segment is bound to a nonhistone protein acceptor to which phosphate groups are then added by specific phosphorylating kinase enzymes.* Some evidence exists that the negatively charged phosphorylated nonhistone proteins can combine with the positively charged basic histones to cause removal of histones from DNA, and thereby allow transcription to proceed. Some phosphorylated nonhistones may also act as "sigma factor" components of RNA polymerase (see p. 104), enabling the enzyme to recognize promoters and transcribe specific DNA sections. In any case the specificity of nonhistone proteins is now presumed to dictate which genes are to be transcribed. The presence of different kinds of histones in a cell indicates only that histones are involved in different nonspecific functions and different generalized levels of repression, such as DNA complexing, strand-to-strand complexing, chromosome coiling, etc. A particular gene may therefore be subject to different degrees of repression in different tissues, accounting for the fact that

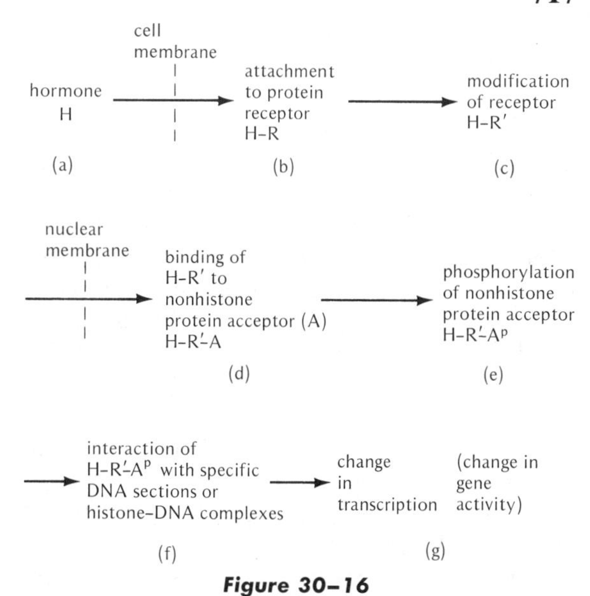

Figure 30–16

Possible sequence of the regulatory events that can occur in a eucaryote by means of hormones, nonhistone proteins, and histones. The reverse sequence leading to the secretion of a hormone would, of course, also be possible. (Based on Stein, Spelsberg, and Kleinsmith.)

an inducer does not cause the same gene action in all cells. Although these theories are attractive, many of their molecular details remain to be demonstrated.

PATTERN OF DEVELOPMENT

One of the most obvious features of multicellular organisms is the spatial arrangement of their tissues and organs. That is, not only are these body parts different from each other, but they occupy localized positions that are fairly uniform to all individuals of the species. It is evident that for such precise and repeatable patterns to occur, they must be preceded by a precisely ordered pattern of development. Unfortunately patterns of development are still quite difficult to discern in detail because neither the embryologist nor the geneticist can as yet trace the precise sequence of activity and interaction between all the cells involved in such processes.

*Various kinds of kinase enzymes have been demonstrated, some activated and others inactivated by the presence of cyclic AMP (see p. 679). Thus, in addition to control by hormones such as steroids, gene activity may also be controlled by the phosphorylation of nonhistone proteins. Histones can also be phosphorylated but the result is probably a generalized response involving the configuration of large sections of chromosome rather than a specific localized response. For example, Bradbury, Inglis, and Matthews have recently shown that the phosphorylation of a particular histone fraction causes the chromosome condensation and coiling that lead to mitosis.

In a general way the initiation of pattern begins with the egg. Differences, such as those imposed by the uneven deposition of maternal substances in the egg cytoplasm or by external factors such as gravity, undoubtedly help to initiate localized differences in development. Superimposed upon these differences are new localized influences that occur once development is initiated, such as the effect of substances diffusing from particular points, or the effect of differentiated cells migrating from special areas. These cause further differentiation which, in turn, causes the production of other inducer substances and still further differentiation. A particular pattern seen late in the development of an individual therefore does not arise *de novo* but is based upon previous patterns. It is because the previously existing patterns are not identical to those that arise later that development is considered epigenetic.

A diagrammatic view of epigenetic differentiation is shown in Fig. 30–17, in which development begins with three differentiated areas, A, B, C, present in the egg cytoplasm. Fertilization provides an inducer I^1 which leads to differentiation within the three previously existing areas and initiation of a new differentiated area, D. The inducer may act within these areas by establishing a gradient (more of it present on one side than the other) or by influencing gradients that are already present. Further inducers then arise (I^2, I^3, I^4), causing further differentiation within previously differentiated areas (E, F) or new areas (G).

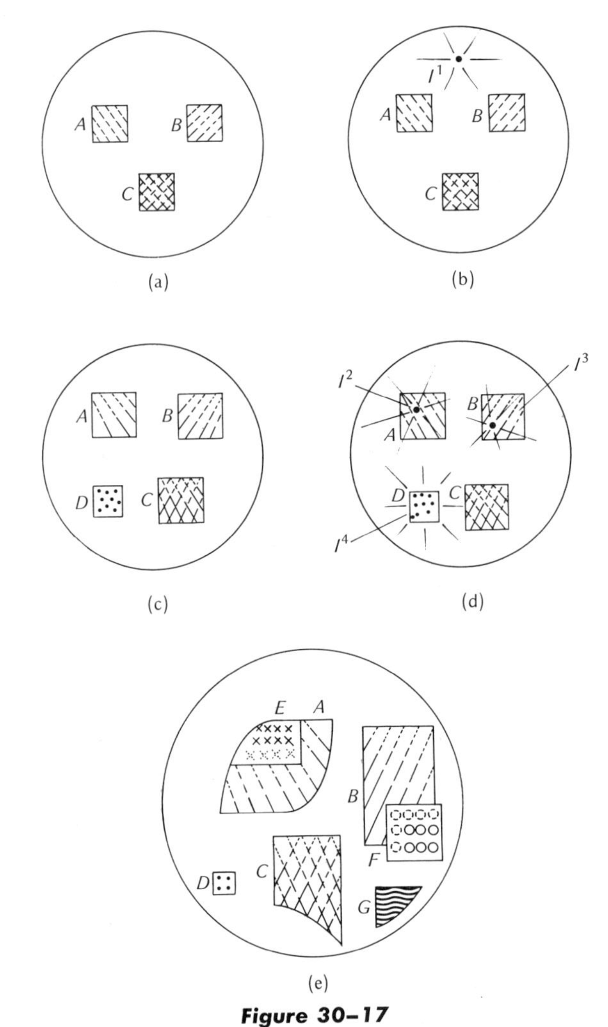

Figure 30–17

Diagrammatic representation of differentiation in a tissue that grows more and more complex because of the presence of gradients and the localized origin of inducers. Further explanation in the text.

These gradients and their interactions thus provide "positional information" to individual cells enabling them to "know" where they are located relative to other cells. Such differentiated groups will then respond to new gradients according to regulatory pathways now determined by their previous histories. The uniformity of the developmental pattern in different individuals lies in the fact that the succession of regulatory events, or pattern history, is identical in each individual. When the history of the pattern is changed, such as a change in the relative positions of A, B, and C, or a change in the effect of one of the inducers, a changed phenotype results.

Some genetic aspects of these relationships are illustrated by the wing vein pattern in *Drosophila*. This pattern, that of five longitudinal veins and two crossveins, varies little from one individual to another and from one generation to another in any *Drosophila* species. As shown in Fig. 30 18 for *Drosophila melanogaster*, it has its origin in an embryonic disc which begins to proliferate and

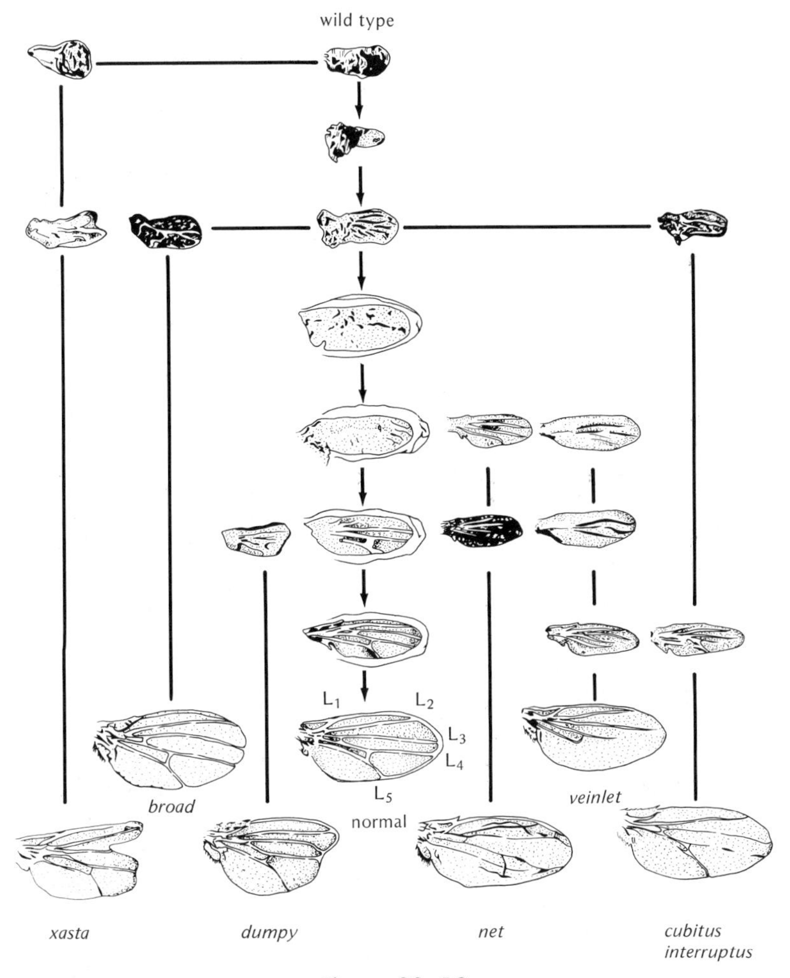

Figure 30–18

Comparison of stages in normal wing development in Drosophila melanogaster *(center column) and mutant wing development (sides) beginning with the imaginal (embryonic) wing-disc stage shown at the top. (After Waddington, 1956.)*

expand in that stage of metamorphosis when the pupa is first formed (prepupa). The two sheets of tissues which represent the two surfaces of the wing are later pushed apart by the body fluid of the developing pupa, forming a wing sac. At later stages this fluid is expelled from the wing and the two surfaces come together again, leaving narrow cavities, which are the adult veins.

The occurrence of mutation affecting each of these steps has been shown in numerous investigations by Goldschmidt, Waddington, and others, some of which are diagrammed in Fig. 30–18. *Cubitus interruptus*, for example, acts early in development and causes gaps near the distal tip of vein L_4. Interestingly the general pattern produced by this mutation is such that vein L_5 moves in the direction of the defective vein. Genes with later-acting effects, such as *veinlet*, also cause compensating changes in pattern, such as movement of the posterior crossvein toward the base of the wing. Some mutations, such as *shifted*, "squeeze" the veins together more closely without any change in the shape of the wing. In others, such as *dachs*, the veins diverge at a greater angle than normal. Mutations that enlarge the wing mass, such as *blot*, may cause a small, mirror-image wing to be formed. Wing formation is therefore subject to numerous forces such as hydrostatic pressure and cellular tension, each of which may affect the presence, absence, and relationship between wing veins.

In spite of these mutant effects on development, it is surprising that the pattern of the *Drosophila* wing is rarely basically changed by these mutations. Also surprising is the constancy of this pattern in wild-type populations in the face of numerous modifying genes that undoubtedly exist. As will be discussed more fully (pp. 808–811), the reason for constancy in many such patterns lies in the fact that their development is caused by interactions between many genes, so that a minor change in one gene usually has little or no effect on the basic pattern. The developmental system for an essential pattern is therefore said to be *buffered* or *canalized*, terms signifying protection against variability. Using such concepts, Waddington has likened development to balls rolling down a succession of valleys (called the *epigenetic landscape*) whose depths and directions are controlled by genes (Fig. 30–19). The end points of the valleys are the adult structures or patterns. Pattern constancy is therefore a reflection of the depth of the "grooved" pathway followed by development; the deeper the groove, the more difficult for the pattern to be changed.

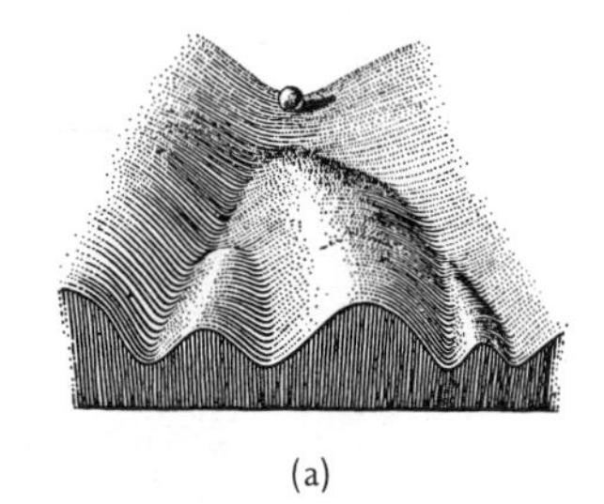

(a)

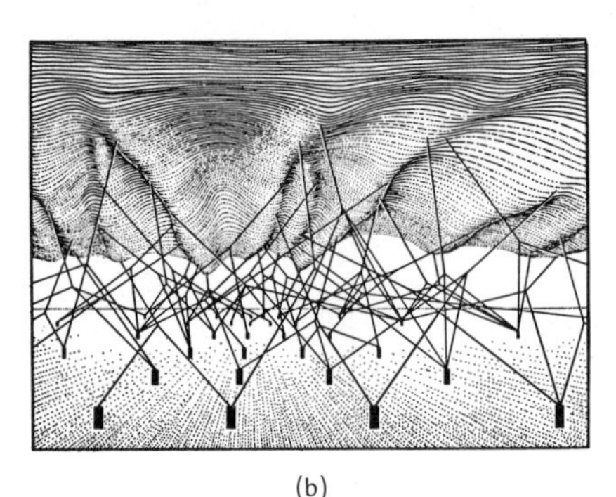

(b)

Figure 30–19

Diagrammatic representation of development as a three-dimensional epigenetic landscape. In (a) development of a particular part of the individual is pictured as a ball rolling down a succession of developmental pathways toward an adult structure. The different valleys represent alternative courses of development. Selection of a particular pathway depends upon numerous factors but most importantly upon the control of the epigenetic landscape by the underlying genes and gene products pictured in (b). In (b) the pegs represent genes and the strings represent gene products and their various interactions. The "pull" (or effect of these many interactions upon the strings) produces the surface configurations. (Courtesy C. H. Waddington, 1957. Strategy of the Genes. George Allen & Unwin Ltd., London.)

In this sense a pattern observed in an adult structure represents only the end point of a succession of previous patterns. That is, the observable pattern is preceded by a "prepattern," in turn preceded by other prepatterns, etc. A graphic example of a prepattern was demonstrated by Stern (1954) for some of the large bristles found on the dorsal surface of *Drosophila melanogaster*. Normally the number and position of these bristles on the fly's body are constant, since the bristles probably serve the essential purpose of detecting differences in air pressure during flight. In some mutations, however, bristles are missing, and in one such mutation, *achaete* (*ac*), one or more of three thoracic bristles are characteristically absent, thus modifying the usual bristle pattern (Fig. 30–20).

Stern asked whether the underlying bristle pattern or prepattern is actually modified by *achaete*—i.e., whether two patterns exist, wild type and *achaete*—or whether the *achaete* fly has a

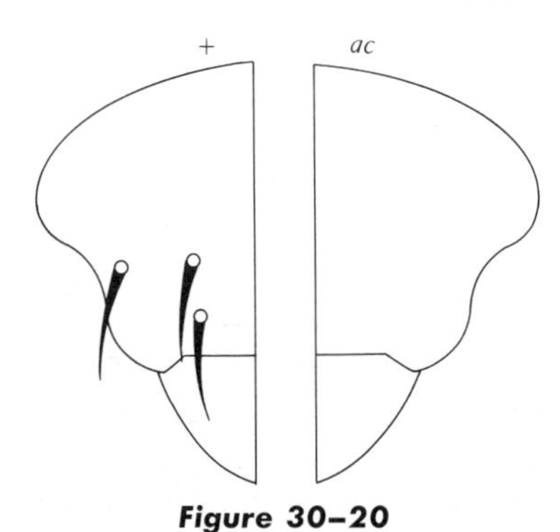

Figure 30–20

Positions of the three thoracic bristles in Drosophila melanogaster *affected by the achaete (ac) gene. A half thorax showing the wild-type phenotype is on the left. A half thorax showing an extreme achaete phenotype is on the right. (After Stern, 1954.)*

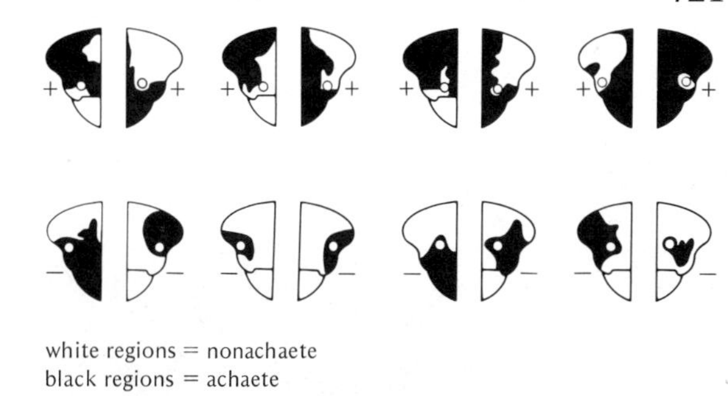

Figure 30–21

Top row: half thoraxes of different mosaic individuals in which wild-type tissue (white regions) has produced one of the three thoracic bristles although surrounded by achaete tissue (black regions). Bottom row: half thoraxes of mosaics in which bristles are absent in achaete tissue although surrounded by wild-type tissue. + indicates presence of the circled thoracic bristle, – indicates the absence of the bristle. (After Stern, 1954.)

basically normal bristle prepattern but is simply unable to respond to it. To distinguish between these two possibilities Stern designed experiments creating "mosaics" (p. 482) in which part of the thorax was of one kind of tissue and part of the other. For a sex-linked gene such as *achaete*, mosaics could be made by obtaining *achaete*/+ female heterozygotes whose wild-type X chromosome is ring-shaped. Under certain conditions (increased maternal age) such heterozygotes show frequent loss of their unstable ring X chromosome during the embryonic stages and appear as male/female gynandromorphs. Recognition of *achaete* tissue in the experiment was aided by the fact that the X chromosome carrying *achaete* also carried *yellow* (color) and *singed* (bristles). Thus *achaete* tissue was clearly distinguished from wild-type tissue, the latter being nonyellow and nonsinged.

Diagrams of the thoraxes of some of the gynandromorphs produced in this experiment are shown in Fig. 30–21. If we consider those instances in which a bristle appears in wild-type tissue that is completely surrounded by *achaete* tissue, it is obvious that the bristle-forming pattern of the thorax is not the least affected by the large amounts of *achaete* tissue. In other words, the underlying pattern producing bristles in *achaete* is identical to that of wild type with the exception that wild-type tissue can respond to this prepattern while *achaete* tissue usually cannot. The *achaete* areas generally lacked bristles, no matter in how large an area of wild-type tissue they were contained. These results therefore indicated the existence of an underlying prepattern for bristle formation.

Use of X-chromosome mosaics has also demonstrated the presence of a prepattern for male sex combs in *Drosophila*. In this instance the presence of male tissue in the sex-comb area produces the unique sex-comb bristles (Fig. 30–22) and the presence of female tissue produces the normal female bristles. Since the size and extent of the surrounding tissue do not matter in either case, the underlying prepattern is obviously the same for both males and females, but only male tissue can respond to this prepattern by forming sex combs.

Surprisingly, in some cases prepatterns may exist to which only mutant tissues can respond. Tokunaga, for example, found that the secondary sex comb produced by the *Drosophila* recessive gene *engrailed* (*en*) apparently arises as a response to a prepattern that is actually present in wild-type tissue. This was demonstrated by producing mosaics through somatic crossing over in *en*/+ heterozygotes (p. 378), thereby causing the presence of *en*/*en* tissue that was also distinguished by

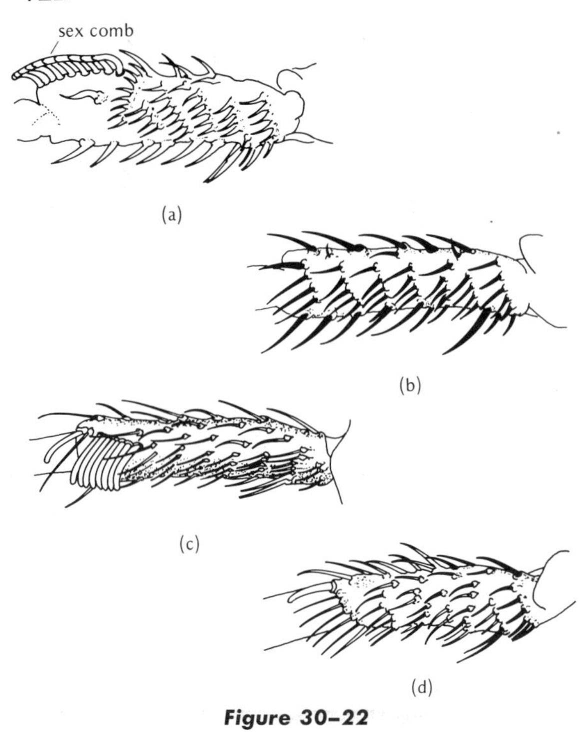

Figure 30–22

First tarsal segments of the foreleg of Drosophila melanogaster. *(a) Male tarsus showing special bristles of male sex comb, with all bristles that are drawn in outline signifying yellow and singed phenotype. (b) Female tarsus with normal bristles. (c) Mosaic containing mostly male tissue but showing a female bristle in a gap in the sex comb. (d) Mosaic containing mostly female tissue but showing the unique male sex-comb bristle for the male tissue located in the sex-comb area. (From Stern.)*

a color marker in the wild-type background. Tokunaga found that whenever the secondary sex-comb area is occupied by *en/en* tissue, secondary sex-comb bristles are differentiated no matter how large the surrounding wild-type area. Thus the absence of secondary sex combs in wild-type tissue is probably caused by a failure of wild-type tissue to respond to the secondary sex-comb pattern.

These findings raise the question of the origin of prepatterns and their relative persistence. Of what value is a secondary sex-comb prepattern to a fly that has only one sex comb? Certainly some species of *Drosophila* have secondary sex combs, and it might therefore be possible that such a structure was passed on from an ancestor of *Drosophila melanogaster*. In a set of experiments performed by Maynard Smith and Sondhi, and later by Sondhi alone, a prepattern was found to exist for two head bristles of *Drosophila subobscura* that are ordinarily not found in the *Drosophila* family but occur in a closely related family of flies, the *Aulicigastridae*. Prepatterns may therefore represent part of an evolutionary accumulation of patterns that are condensed during development. A normal individual will respond to some of these but not others.

However, prepatterns themselves may be changed since some mutations can produce new underlying patterns that cause even wild-type tissues to respond in an abnormal fashion. One such example, investigated by Stern and Tokunaga, is believed to be the effect caused by the dominant mutant allele for *eyeless*, ey^D. Among other effects this gene enlarges certain portions of the tarsus, leading to the appearance of multiple sex combs in the *Drosophila* male foreleg. When mosaics were composed of both non-ey^D and ey^D tissues, the non-ey^D tissue produced multiple sex combs as though it were responding to a new prepattern. As will be discussed in Chapter 34, Rendel and others have also shown that prepatterns may be changed by gene selection.

IMAGINAL DISCS AND FATE MAPPING

The large numbers of identified mutations in *Drosophila*, their localization on gene linkage maps, and the extensive work already done on *Drosophila* development from the time of T. H. Morgan on, have made this insect a primary source for understanding the genetic control of development in eucaryotes. *Drosophila* has the advantage that changes in adult morphological patterns can often be traced to small discs of tissue, called *imaginal discs*, or *anlage*, already present in the developing embryo or early larva. These discs begin as isolated groups of about 20 cells that grow in the form of folded epithelial sacs throughout the larval period. As far as is known, the imaginal discs perform no function in the larva, but are first prominent in the pupal stage. In the pupa, under the influence of metamorphosing hormones such as ecdysone, the imaginal discs evert, unfold, and then undergo marked morphogenetic changes leading to the adult structures shown in Fig. 30–23.

Transplantation experiments have shown that many mutations affecting a particular body part of the adult are already determined in the specific imaginal disc that develops into the adult mutant structure. For example, a leg disc from a *yellow* mutant genotype will develop into leg structures showing the yellow phenotype when it is transplanted into the abdomen of a wild-type larva that undergoes metamorphosis. Mixing of cells from the leg discs of both *yellow* and *ebony* mutant strains into a single transplant produces, as expected, leg parts that are mosaic for both *yellow* and *ebony* tissue. Also, there is now considerable evidence that specific areas within an imaginal disc are, if not completely determined, certainly strongly inclined to form specific substructures within an adult organ. According to Garcia-Bellido and co-workers (see review of Crick and Lawrence), cells from the imaginal wing discs of *Drosophila* produce clones of descendants that become confined to fixed-boundary "compartments" in later development.

A finding with widespread implications has been the demonstration by Hadorn and co-workers that the developmental pattern of an imaginal disc can be maintained even after its cellular descendants have been succesively transplanted into different hosts for many generations. In the Hadorn technique a disc is removed from a larva and inserted into an adult fly that serves as a "tissue culture" host for the transplant. The disc grows in the abdomen of the fly but does not undergo metamorphosis unless it is removed and inserted into a larva that eventually undergoes metamorphosis. Thus discs or portions of discs may be transferred each generation from adult to adult, and tested

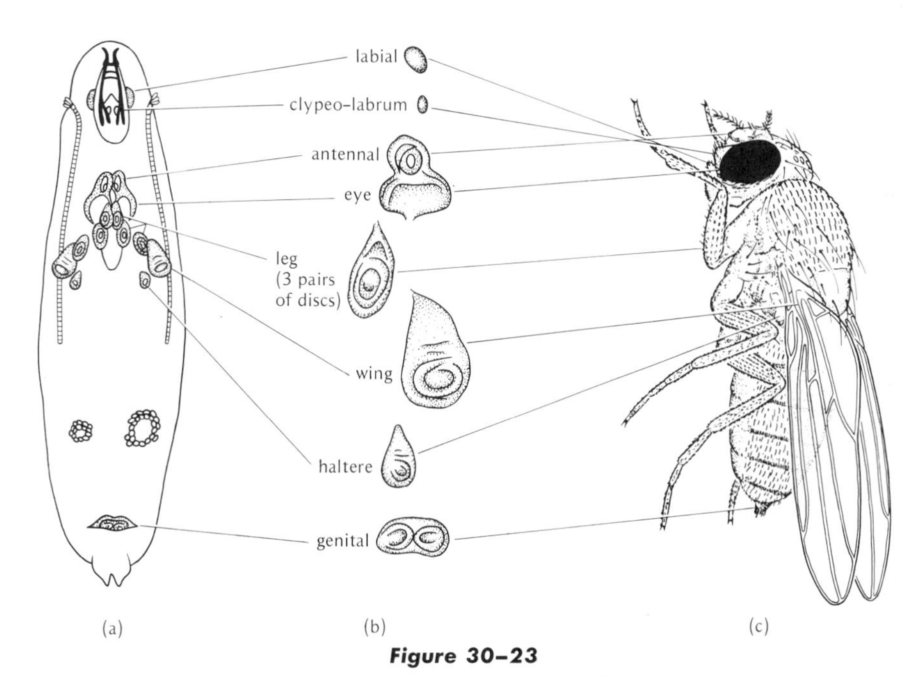

Figure 30–23

(a) Location of the various imaginal discs in a Drosophila *larva. There are a total of nine pairs plus the single genital disc. (b) General appearance of some of the discs during the larval stage. (Many of their developmental aspects are reviewed by Gehring and Nöthiger.) (c) Adult structures derived from the designated discs. (a and b after Fristrom et al.; c after Stern, 1968.)*

at any time for their ability to differentiate into the originally destined adult tissue. With only rare exceptions, the imaginal discs do not change their developmental destinies. Occasionally, however, the tissue of a disc does change its developmental role to produce what seems like an unexpected structure; e.g., a leg disc produces a wing structure, etc. These changes, known as *transdeterminations*, seem to follow particular patterns (leg → wing, proboscis → antenna, genital → antenna, etc.), although their causes are not yet known. Other experiments also suggest that imaginal disc cells, "determined" though they may be, still possess the potentiality to change their fate within some general limits (see review of Postlethwait and Schneiderman).

Interestingly the origin of many of the discs themselves can be traced to an even earlier stage called the *blastoderm* (also blastula). As shown previously in Fig. 21–13, *Drosophila* blastoderm arises from the outward movement of the many rapidly dividing nuclei formed after fertilization in the interior of the egg. These nuclei reach the egg surface, where they undergo a few further divisions, forming a surface monolayer of about 4000 cells. Chan and Gehring have shown that such blastoderm cells are already differentiated in respect to their roles of forming anterior or posterior organs. When each half of a blastoderm embryo is cultured so that the cells can complete their development, the anterior half produces only head and thorax structures, whereas the posterior half produces thoracic and abdominal structures

More precise "fate mapping" of blastoderm cells is now being performed by a technique originally invented by Sturtevant. This technique depends upon the occurrence of mosaics in which one organ or structure is genetically different from the other because of a chromosomal loss during one of the early mitotic divisions (Fig. 21–13a). Fate mapping is then performed by noting a simple correlation: the greater the physical distance between two organs whose cells derive from the blastoderm, the greater the chances that a mosaic boundary will pass between these organs. Thus, as expected, gynandromorphs show greater mosaicism between head and genitals (whose cells derive from opposite ends of the egg) than between head and eye.

Fate map units are therefore measured in "sturts" (named after Sturtevant) based on the frequency in which a mosaic boundary arises between two given structures. For example, gynandromorphs can be produced in which the male and female tissues are recognizably different because they carry specific sex-linked genetic markers; e.g., XX female tissue is wild-type ($y/+$) and XO male tissue is yellow ($y/0$). If 100 such gynandromorphs are scored, and there are 10 flies that show a mosaic difference between the first and second legs on one side, then the blastoderm sites for these legs can be said to be separated by 10 sturts. The analyses of numerous such gynandromorphs have enabled various workers (see Garcia-Bellido and Merriam; Hotta and Benzer) to establish fate maps in which many adult structures are localized to their relative positions on the blastoderm surface (Fig. 30–24). Note that, in contrast to the one-dimensional linear linkage map derived from the linear chromosome, the fate map is two-dimensional since it derives from the surface of the blastoderm embryo. Thus "cross distances" between different structures are also part of the map.

BEHAVIOR AND DEVELOPMENT

Of considerable interest has been the finding that mutations that affect behavior can also be localized in *Drosophila* by fate mapping. The technique in such cases is to observe the relationship between a particular behavioral trait and the tissues that are separated by a mosaic boundary. For example, the sex-linked semidominant *Hyperkinetic* gene (*Hk*) causes the shaking of flies' legs when they are anesthetized by ether. Gynandromorphs are then constructed which contain wild-type alleles for both *yellow* and *Hyperkinetic* on one X chromosome and the mutant alleles for these genes on the other. If the developmental blastoderm locus of *Hyperkinetic* is very close to that of a particular organ, then the hyperkinetic activity of a leg should be correlated with the color pheno-

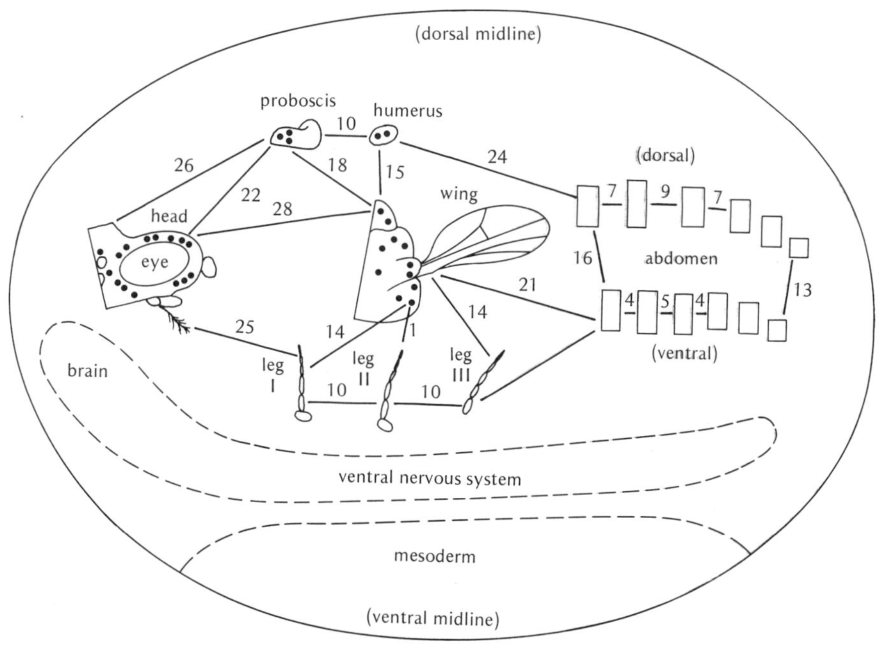

Figure 30–24

Fate map of some of the external parts of the right side of an adult Drosophila melanogaster *projected onto the right half of the blastoderm surface of the egg as seen from the inside. The numbers represent distances (in "sturts") between the blastoderm sites for some of these adult structures. (Similar maps have been made for larval structures.) As may be expected because of the symmetry of the fly, the left half of the blastoderm surface is a mirror image of the right half. Small black circles on some of the structures represent bristles, many of which have also been localized on the fate map. (Modified after Hotta and Benzer.)*

type of the organ; that is, the leg should be markedly hyperkinetic (*Hk*/0) when the organ is yellow (*y*/0), and the leg should be less hyperkinetic (*Hk*/+) when the organ is normal-colored (*y*/+). Those instances where the leg is mutant (for shaking activity) and the organ is normal (for color), or vice versa, indicate that the mosaic boundary has passed between them.

Thus, Hotta and Benzer showed that among 300 gynandromorphs the mosaic boundary passed between the antenna and shaking of the first leg in 83 flies, meaning that there are approximately $83/300 = 28$ sturts distance between these two developmental sites on the blastoderm. These studies also indicate that each leg has its own independent hyperkinetic site, and these sites fall within those blastoderm cells that give rise to the ventral nervous system (see also Ikeda and Kaplan). Similar analyses have been made of some of the many mutations that affect visual sensitivity, and others such as *drop-dead* (precocious sudden death) and *wings-up* (permanent vertical wing position). In some cases the behavioral effect seems to originate from only a single blastoderm site, whereas others derive from interactions between two or more sites.

Analysis of behavior patterns has also been pursued through various other techniques and on other levels. *Drosophila* courtship, for example, is one area where there is now considerable genetic information. A classic study by Bastock has shown that the *yellow* gene in *Drosophila* males affects their courtship pattern (wing vibrations, genital licking, etc.), so that such males are less attractive to females than are normal males. Strains of *Drosophila* showing homosexual behavior have also been discovered in which males court other males or females court females.

Methods are continually being invented to identify mutations that affect almost any normally accepted pattern of behavior. For example, the direction of movement in *Drosophila*, toward gravity or away from it, has been shown by Hirsch and others to be caused by genetic determinants that can be localized to specific chromosomes. The inheritance of sharp differences in olfactory behavior has been demonstrated by Kikuchi who has found a dominant *Drosophila* mutation that causes its carriers to be attracted to odors that repel normal flies. Konopka and Benzer have shown the existence of a group of mutant genes or alleles that change those rhythmic patterns ("circadian rhythm" or "biological clock") that enable developing flies to emerge from their pupae at specific times of day. Perhaps one of the most dramatic of behavioral genes is the temperature-sensitive mutant described by Suzuki, Grigliatti, and Williamson, in which paralysis of the fly occurs within 5 seconds of a shift from 22°C to 29°C, and can then be reversed by a shift to the lower temperature.

Organisms other than *Drosophila* are also being used to yield information on the developmental source of behavior.* Among the simplest of these is the nematode *Caenorhabditis elegans* investigated by Brenner in a search for specific inherited physical defects in nerve connections. By means of electron micrograph analysis he has shown that some behavioral mutations cause small but consistent changes in the "wiring patterns" of neurone cells. Since there are—even in this organism—many billions or trillions of possible wiring patterns, it seems unlikely that each individual neuronal synapse can be caused by a separate gene. What, then, is the relationship between the gene and the behavior? This question remains unanswered since the complexity of developmental patterns in behavior is still far from unraveled. Nevertheless there is now little doubt that genes exert effects on all patterns and at all developmental levels.

PROBLEMS

30-1. Werz has reported that normal nucleated cells of the unicellular alga *Polyphysa cliftonii* will produce "caps" when their stalks have reached a length of 10 centimeters but that young enucleated cells can produce caps even though they are only 3 centimeters long. How would you explain this?

30-2. The gene *lozenge-clawless* (lz^{cl}) affects the female genitalia of *Drosophila melanogaster*, so that the sperm-storage organs (spermathecae) and ovarian glands (parovaria) do not develop, but the gene does not affect male genitalia. The *transformer* (*tra*) gene (p. 229) converts XX individuals into phenotypic males. Anders studied the interaction between both of these genes and found that XX individuals homozygous for lz^{cl} and *tra* were phenotypically males with normal genitalia. Explain which of these genes acts first in development.

30-3. In the *o* mutation discovered by Humphrey in the axolotl (see Problem 13, Chapter 13), a maternal effect produced by *o/o* mothers prevents normal development of the embryo, no matter what its genotype. Briggs and Cassens have shown that development can be improved by injecting the eggs of *o/o* mothers with material from normal eggs. Interestingly the corrective material is found to be more highly concentrated in the nuclear sap than in the cytoplasm. Interpret this finding.

30-4. A maternal effect in *Drosophila* is noted when phenotypically wild-type females heterozygous for the sex-linked recessive eye color gene *maroon-like* (*mal*) are mated to *mal* males and produce all wild-type offspring, male and female. Since the *mal* gene is involved in the synthesis of the enzyme xanthine dehydrogenase, we can assume that the genotype of the zygote has no affect on the presence or absence of the enzyme, but that the enzyme itself or a precursor is transmitted to the progeny in each egg derived from

* Recent books on various aspects of behavior genetics include those of Hirsch (1967), Thiessen (1972), Parsons (1973), McClearn and DeFries (1973), and Wilson (1973).

the wild-type female. On the other hand, an autosomal recessive gene, *rosy* (*ry*), also affects xanthine dehydrogenase, and flies homozygous for this gene do not seem to produce the enzyme at all.

When females heterozygous for *maroon-like* and homozygous for *rosy* (mal^+/mal; ry/ry) are mated to *maroon-like* males homozygous for the wild-type alleles for *rosy* (mal; ry^+/ry^+), the progeny again are *all* wild type in respect to xanthine dehydrogenase. Since the maternal cytoplasm in this case is from a homozygous *ry*/*ry* female, what kind of mechanism may possibly be responsible for the wild-type appearance of the offspring?

30-5. Using only two gene loci (α and β), the tetrameric LDH enzyme can assume five different isozyme phenotypes (p. 704). How many different isozyme phenotypes would you expect to find: (a) If LDH were a trimer rather than a tetramer? (b) If LDH were a tetramer, but three gene loci were involved in producing the different component subunits (e.g., α, β, and γ)?

30-6. The appearance of two different enzymes in a tissue during development appears to be synchronously controlled: both of them increase or decrease in amount at the same time. Explain whether this necessarily means that the genes for these two enzymes are associated together in the same operon.

30-7. Fig. 30-25 represents the RNA obtained by Attardi and coworkers when various preparations of duck erythroblasts (red cell progenitors) were incubated with radioactive (^{14}C) uridine medium. One curve (—○—○—○—) shows the amount of RNA synthesized (in terms of counts per minute of radioactive uridine incorporated) over a 13 hour period when the cells are subjected to no further treatment. The maximum RNA value remains at approximately the same level even when the incubation time is increased to about 40 hours, indicating that a "plateau" of RNA synthesis is reached at 600–800 counts per minute. Another curve (—△—△—△—) shows the effect of actinomycin D on RNA synthesis when a sample of these cells are subjected to the transcription-inhibitor at about three hours. Actinomycin causes RNA synthesis to fall to about 20 percent of its previously noted maximum level. A third curve (—□—□—□—) shows the RNA obtained when a sample of erythroblasts are allowed to synthesize RNA for a period of three hours in the presence of nonradioactive uridine, and the ^{14}C-uridine is only added at this point. In this

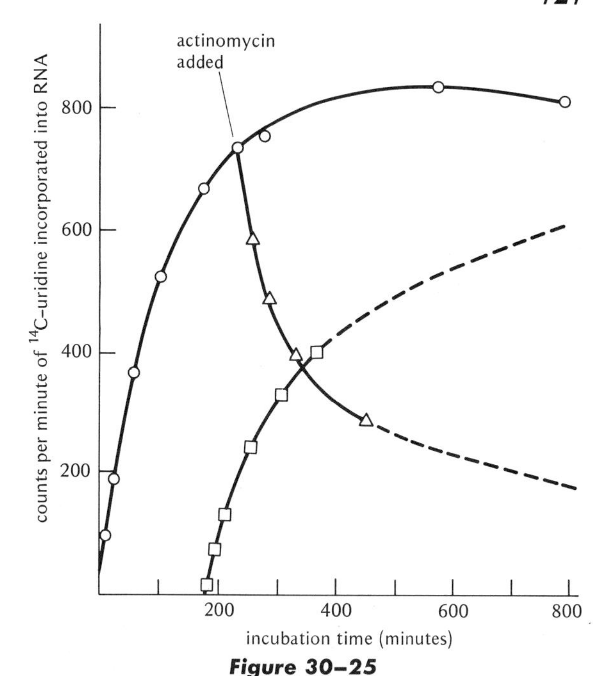

Figure 30–25

The amounts of tritium-labeled RNA synthesized by duck erythroblasts after various treatments. See Problem 30–7. (After G. Attardi et al., 1966. Jour. Mol. Biol., ***20,*** *145–182.)*

case RNA synthesis increases. Since RNA is apparently being continually synthesized by duck erythroblasts through the entire incubation period, what might account for the fact that the amount of RNA synthesized does not increase beyond the plateau level given in the figure?

30-8. Using some of the hypotheses about regulation in higher organisms, how would you explain the fact that the simpler protamines replace the complex histones in some animal sperm chromosomes?

30-9. Hotta and Benzer scored *Drosophila* gynandromorphs carrying one X chromosome that was mutant for both the sex-linked recessive *drop-dead* gene (see p. 725) and for recessive body surface genes such as *yellow* and *forked.* Since the other X chromosome had the wild-type alleles for these genes, these gynandromorphs were recognized by the fact that their XX female tissue was wild type, and their XO male tissue showed the mutant surface markers. A mosaic fly could therefore be of the drop-dead phenotype if the appropriate tissue causing the drop-dead effect was of the XO type. However, if this tissue was not of the XO type, then the gynandromorph was normal in this respect, but was nevertheless a mosaic for the surface markers. In their experiment Hotta and Benzer obtained the following data for 136 gynandromorphs:

54 were normal for drop-dead and normal for the abdominal surface
28 were normal for drop-dead and mutant (yellow and forked) for the abdominal surface
23 showed the drop-dead phenotype but were normal for the abdominal surface
31 showed the drop-dead phenotype and were mutant for the abdominal surface

Using these data, calculate the "fate-map" distance between the developmental site for the drop-dead effect and the site for the abdomen.

REFERENCES

ALLFREY, V. G., V. C. LITTAU, and A. E. MIRSKY, 1963. On the role of histones in regulating ribonucleic acid synthesis in the cell nucleus, *Proc. Nat. Acad. Sci.*, **49**, 414–421.

ASHBURNER, M., 1972. Puffing patterns in *Drosophila melanogaster* and related species. In *Developmental Studies on Giant Chromosomes*, W. Beermann, (ed.), Springer-Verlag, New York, pp. 101–151.

BASTOCK, M., 1956. A gene mutation which changes a behavior pattern. *Evolution,* **10**, 421–439.

BEERMANN, W., 1952. Chromomerenkonstanz und spezifische Modifikationen der Chromosomenstruktur in der Entwicklung und Organdifferenzierung von *Chironomus tentans. Chromosoma,* **5**, 139–198.

———, 1961. Ein Balbiani-Ring als Locus einer Speicheldrüsen-Mutation. *Chromosoma*, **12**, 1–25.

———, 1972. Chromomeres and genes. In *Developmental Studies on Giant Chromosomes*, W. Beermann (ed.). Springer-Verlag, New York, pp. 1–33.

BENNETT, D. 1964. Abnormalities associated with a chromosome region in the mouse. II. Embryological effects of lethal alleles in the t-region. *Science*, **144**, 263–267.

BONNER, J., 1965. *The Molecular Biology of Development.* Oxford Univ. Press, London.

BONNER, J., and J.-R. WU, 1973. A proposal for the structure of the *Drosophila* genome. *Proc. Nat. Acad. Sci.*, **70**, 535–537.

BRADBURY, E. M., R. J. INGLIS, and H. R. MATTHEWS, 1974. Control of cell division by very lysine rich histone (F1) phosphorylation. *Nature*, **247**, 257–261.

BRENNER, S., 1973. The genetics of behavior. *Brit. Med. Bull.*, **29**, 269–271.

———, 1974. The genetics of *Caenorhabditis elegans. Genetics*, **77**, 71–94.

BREUER, M. E., and C. PAVAN, 1955. Behavior of polytene chromosomes of *Rhynchosciara angelae* at different stages of larval development. *Chromosoma*, **7**, 371–386.

BRIGGS, R., and T. J. KING, 1959. Nucleocytoplasmic interactions in eggs and embryos. In *The Cell*, Vol. I., J. Brachet and A. E. Mirsky (eds.). Academic Press, New York, pp. 537–617.

BROWN, D. D., and J. B. GURDON, 1964. Absence of ribosomal RNA synthesis in the anucleolate mutant of *Xenopus laevis. Proc. Nat. Acad. Sci.*, **51**, 139–146.

CALLAN, H. G., 1967. On the organization of genetic units in chromosomes. *Jour. Cell Sci.*, **2**, 1–7.

CALLAN, H. G., and L. LLOYD, 1960. Lampbrush chromosomes of crested newts *Triturus cristatus* (Laurenti). *Phil. Trans. Roy. Soc. Lond.* (*B*), **243**, 135–219.

CHAN, L.-N., and W. GEHRING, 1971. Determination of blastoderm cells in *Drosophila melanogaster. Proc. Nat. Acad. Sci.*, **68**, 2217–2221.

CLEVER, U., 1961. Genaktivitäten in den Riesenchromosomen von *Chironomus tentans* und ihre Beziehungen zur Entwicklung. I. Genaktivierungen durch Ecdyson. *Chromosoma*, **12**, 607–675.

———, 1963. Genaktivitäten in den Riesenchromosomen von *Chironomus tentans* und ihre Beziehungen zur Entwicklung. IV. Das Verhalten der Puffs in der Larvenhäutung. *Chromosoma*, **14**, 651–675.

COHEN, S., 1965. Enzyme formation in virus-infected cells. In *Genetics Today*, Vol. II. Proc. XI Intern. Congr. Genet., S. J. Geerts (ed.). Pergamon, Oxford, pp. 151–164.

COHN, M., 1973. Antibody diversification: The somatic mutation model revisited. In *The Biochemistry of Gene Expression in Higher Organisms*, J. K. Pollak, and J. W. Lee (eds.). D. Reidel Publ. Co., Dordrecht, Holland, pp. 574–592.

COOKE, J., 1973. Morphogenesis and regulation in spite of continued mitotic inhibition in *Xenopus* embryos. *Nature*, **242**, 55–57.

COUNCE, S. J., 1956. Studies of female-sterility genes in *Drosophila melanogaster. Zeit. Abst. Induk. u. Vererbung.*, **87**, 443–492.

CRICK, F., 1971. General model for the chromosomes of higher organisms. *Nature*, **234**, 25–27.

CRICK, F. H. C., and P. A. LAWRENCE, 1975. Compartments and polyclones in insect development. *Science*, **189**, 340–347.

CURTIS, A. S. G., 1963. The cell cortex. *Endeavour*, **22**, 134–137.

DAVIDSON, E. H., and R. J. BRITTEN, 1973. Organization, transcription, and regulation in the animal genome. *Quart. Rev. Biol.*, **48**, 565–613.

ELGIN, S. C. R., and J. BONNER, 1973. Isolated chromatin in the study of gene expression. In *The Biochemistry of Gene Expression in Higher Organisms*. J. K. Pollak, and J. W. Lee (eds.). D. Reidel Publ. Co., Dordrecht, Holland, pp. 142–163.

FAVRE, R., E. BOY DE LA TOUR, N. SEGRE, and E. KELLENBERGER, 1965. Studies on the morphopoiesis of the head of phage T-even. 1. Morphological, immunological and genetic characterization of polyheads. *Journal Ultrastructure Research,* **13,** 318–342.

FIRTEL, R. A., L. BAXTER, and H. F. LODISH, 1973. Actinomycin D and the regulation of enzyme biosynthesis during development of *Dictyostelium discoideum. Jour. Mol. Biol.*, **79**, 315–328.

FRISTROM, J. W., R. RAIKOW, W. PETRI, and D. STEWART, 1969. *In vitro* evagination and RNA synthesis in imaginal discs of *Drosophila melanogaster*. In *Park City Symposium on Problems in Biology*, E. W. Hanly (ed.). Univ. Utah Press, Salt Lake City, pp. 381–401.

GALL, J. G., and H. G. CALLAN, 1962. H^3 uridine incorporation in lampbrush chromosomes. *Proc. Nat. Acad. Sci.*, **48**, 562–570.

GALLY, J. A., and G. M. EDELMAN, 1972. The genetic control of immunoglobin synthesis. *Ann. Rev. Genet.*, **6**, 1–46.

GARCIA-BELLIDO, A., and J. R. MERRIAM, 1969. Cell lineage of the imaginal discs in *Drosophila melanogaster. Jour. Exp. Zool.*, **170**, 61–76.

GEHRING, W. J., and R. NÖTHIGER, 1973. The imaginal discs of *Drosophila*. In *Developmental Systems: Insects,* Vol. 2, S. J. Counce and C. H. Waddington (eds.). Academic Press, London, pp. 211–290.

GLUECKSOHN-SCHOENHEIMER, S., 1943. The morphological manifestations of a dominant mutation in mice affecting tail and urogenital system. *Genetics*, **28**, 341–348.

GOLDSCHMIDT, R., 1955. *Theoretical Genetics*. Univ. of California Press, Berkeley, Part III.

GROSS, K. W., M. JACOBS-LORENA, C. BAGLIONI, and P. R. GROSS, 1973. Cell-free translation of maternal messenger RNA from sea urchin eggs. *Proc. Nat. Acad. Sci.*, **70**, 2614–2618.

GROSS, P. R., 1967. The control of protein synthesis in embryonic development and differentiation. *Curr. Top. in Devel. Biol.*, **2**, 1–47.

GURDON, J. B., 1974. *The Control of Gene Expression in Animal Development*. Clarendon Press, Oxford.

HADORN, E., 1961. *Developmental Genetics and Lethal Factors*. John Wiley, New York.

———, 1965. Problems of determination and transdetermination. *Brookhaven Symp. Biol.*, **18**, 148–159.

HALVORSON, H., and J. SZULMAJSTER, 1973. Differentiation: Sporogenesis and germination. In *Biochemistry of Bacterial Growth*, J. Mandelstam and K. McQuillen (eds.). John Wiley, New York, pp. 494–516.

HARRIS, H., 1974. *Nucleus and Cytoplasm*, 3rd ed. Clarendon Press, Oxford.

HARVEY, E. B., 1956. *The American Arbacia and Other Sea Urchins*. Princeton Univ. Press, Princeton, N.J.

HILDRETH, P. E., and J. C. LUCCHESI, 1967. Fertilization in *Drosophila*. III. A reevaluation of the role of polyspermy in development in the mutant *deep orange*. *Dev. Biol.*, **15**, 536–552.

HIRSCH, J., 1962. Individual differences in behavior and their genetic basis. In *Roots of Behavior*, E. Bliss (ed.). Paul B. Hoeber, New York, pp. 3–23.

HIRSCH, J. (ed.), 1967. *Behavior-Genetic Analysis*. McGraw-Hill, New York.

HOCHMAN, B., 1973. Analysis of a whole chromosome in *Drosophila. Cold Sp. Harb. Symp.*, **38**, 581–589.

HOOD, L., and D. W. TALMADGE, 1970. Mechanism of antibody diversity: germ line basis for variability. *Science*, **168**, 325–334.

HOTTA, Y., and S. BENZER, 1973. Mapping of behavior in *Drosophila* mosaics. In *Genetic Mechanisms in Development*, F. H. Ruddle (ed.). Academic Press, New York, pp. 129–167.

HOURCADE, D., D. DRESSLER, and J. WOLFSON, 1973. The amplification of ribosomal RNA genes involves a rolling circle intermediate. *Proc. Nat. Acad. Sci.*, **70**, 2926–2930.

HUANG, R. C., and J. BONNER, 1962. Histone, a suppressor of chromosomal RNA synthesis. *Proc. Nat. Acad. Sci.*, **48**, 1216–1222.

HUETTNER, A. F., 1949. *Comparative Embryology of the Vertebrates*, 2nd ed., Macmillan, Inc., New York.

HULTIN, T., 1950. The protein metabolism of sea urchin eggs studied by means of N^{15} labeled ammonia. *Exp. Cell Res.*, **4,** 599–602.

———, 1961. Activation of ribosomes in sea urchin eggs in response to fertilization. *Exp. Cell Res.*, **25,** 405–417.

IKEDA, K., and W. D. KAPLAN, 1970. Patterned neural activity of a mutant *Drosophila melanogaster. Proc. Nat. Acad. Sci.*, **66,** 765–772.

ILLMENSEE, K., 1973. The potentialities of transplanted early gastrula nuclei of *Drosophila melanogaster.* Production of their imago descendants by germ-line transplantation. *Wilhelm Roux' Arch.*, **171,** 331–343.

ILLMENSEE, K., and A. P. MAHOWALD, 1974. Transplantation of posterior polar plasm in *Drosophila.* Induction of germ cells at the anterior pole of the egg. *Proc. Nat. Acad. Sci.*, **71,** 1016–1020.

IZAWA, M., V. G. ALLFREY, and A. E. MIRSKY, 1963. Composition of the nucleus and the chromosomes in the lampbrush stage of the new oocyte. *Proc. Nat. Acad. Sci.*, **50,** 811–817.

JOHNSON, J. D., A. S. DOUVAS, and J. BONNER, 1974. Chromosomal proteins. *Intern. Rev. Cytol.*, **Suppl. 4,** 273–361.

KIKUCHI, T., 1973. Specificity and molecular features of an insect attractant in a *Drosophila* mutant. *Nature*, **243,** 36–38.

KONOPKA, R. J., and S. BENZER, 1971. Clock mutants of *Drosophila melanogaster. Proc. Nat. Acad. Sci.*, **68,** 2112–2116.

KORNBERG, R. D., 1974. Chromatin structure: A repeating unit of histones and DNA. *Science*, **184,** 868–871.

LINN, T. G., A. L. GREENLEAF, R. G. SHORENSTEIN, and R. LOSICK, 1973. Loss of the sigma activity of RNA polymerase of *Bacillus subtilis* during sporulation. *Proc. Nat. Acad. Sci.*, **70,** 1865–1869.

LOSICK, R., 1973. The question of gene regulation in sporulating bacteria. In *Genetic Mechanisms of Development*, F. H. Ruddle (ed.). Academic Press, New York, pp. 15–27.

MARKERT, C. L., 1963. Epigenetic control of specific protein synthesis in differentiating cells. In *Cytodifferentiation and Macromolecular Synthesis*, M. Locke (ed.). Academic Press, New York, pp. 65–84.

MARKERT, C. L., and H. URSPRUNG, 1971. *Developmental Genetics.* Prentice-Hall, Englewood Cliffs, N.J.

MAYNARD SMITH, J., and K. C. SONDHI, 1960. The genetics of a pattern. *Genetics*, **45,** 1039–1050.

MCCLEARN, G. E., and J. C. DEFRIES, 1973. *Introduction to Behavioral Genetics.* W. H. Freeman, San Francisco.

MONROY, A., R. MAGGIO, and A. M. RINALDI, 1965. Experimentally induced activation of the ribosomes of the unfertilized sea urchin egg. *Proc. Nat. Acad. Sci.*, **54,** 104–111.

MOORE, J. A., 1972. *Heredity and Development*, 2nd ed. Oxford Univ. Press, London.

PARSONS, P. A., 1973. *Behavioural and Ecological Genetics.* Clarendon Press, Oxford.

PAUL, J., and R. S. GILMOUR, 1968. Organ-specific restriction of transcription in mammalian chromatin. *Jour. Mol. Biol.*, **34,** 305–316.

POELS, C. L. M., 1970. Time sequence in the expression of various developmental characters induced by ecdysterone in *Drosophila hydei. Devel. Biol.*, **23,** 210–225.

POSTLETHWAIT, J. H., and H. A. SCHNEIDERMAN, 1973. Developmental genetics of *Drosophila* imaginal discs. *Ann. Rev. Genet.*, **7,** 381–433.

RITOSSA, F. M., K. C. ATWOOD, and S. SPIEGELMAN, 1966. A molecular explanation of the *bobbed* mutants of *Drosophila* as partial deficiencies of "ribosomal" DNA. *Genetics*, **54,** 819–834.

RUDDLE, F. H. (ed.), 1973. *Genetic Mechanisms of Development.* Academic Press, New York.

SMITHIES, O., 1973. Immunoglobin genes: Arranged in tandem or in parallel? *Cold Sp. Harb. Symp.*, **38,** 725–737.

SONDHI, K. C., 1962. The evolution of a pattern. *Evolution*, **16,** 186–191.

STEARNS, L. W., 1974. *Sea Urchin Development: Cellular and Molecular Aspects.* Dowden, Hutchinson & Ross, Stroudsburg, Pa.

STEDMAN, E., and E. STEDMAN, 1950. Cell specificity of histones. *Nature*, **166,** 780–781.

STEIN, G. S., T. C. SPELSBERG, and L. J. KLEINSMITH, 1974. Nonhistone chromosomal proteins and gene regulation. *Science*, **183,** 817–824.

STERN, C., 1954. Two or three bristles. *Amer. Sci.*, **42,** 213–247.

———, 1968. *Genetic Mosaics and Other Essays.* Harvard Univ. Press, Cambridge, Mass.

STERN, C., and C. TOKUNAGA, 1967. Nonautonomy in differentiation of pattern-determining genes in *Drosophila.* I. The sex comb of *eyeless-Dominant. Proc. Nat. Acad. Sci.*, **57,** 658–664.

STEWARD, F. C., M. O. MAPES, and K. MEARS, 1958. Growth and organized development of cultured cells, II: Organization in cultures grown from freely suspended cells. *Amer. Jour. Bot.*, **45,** 705–708.

SUSSMAN, M., 1973. *Developmental Biology: Its Cellular and Molecular Foundations*. Prentice-Hall, Englewood Cliffs, N.J.

SUZUKI, D. T., T. GRIGLIATTI, and R. WILLIAMSON, 1971. Temperature-sensitive mutations in *Drosophila melanogaster*. VII. A mutation (*para^{ts}*) causing reversible adult paralysis. *Proc. Nat. Acad. Sci.*, **68**, 890–893.

TERZAGHI, E., 1971. Alternative pathways of tail fiber assembly in bacteriophage T4? *Jour. Mol. Biol.*, **59**, 319–327.

THIESSEN, D. D., 1972. *Gene Organization and Behavior*. Random House, New York.

TOKUNAGA, C., 1961. The differentiation of a secondary sex comb under the influence of the gene *engrailed* in *Drosophila melanogaster*. *Genetics*, **46**, 157–176.

WADDINGTON, C. H., 1956. *Principles of Embryology*. Allen and Unwin, London.

———, 1973. The morphogenesis of patterns in *Drosophila*. In *Developmental Systems: Insects*, Vol. 2, S. J. Counce and C. H. Waddington (eds.). Academic Press, London, pp. 499–535.

WAGNER, R. P., and H. K. MITCHELL, 1964. *Genetics and Metabolism*, 2nd ed. John Wiley, New York.

WALLACE, H., and M. L. BIRNSTIEL, 1966. Ribosomal cistrons and the nucleolar organizer. *Biochem. Biophys. Acta*, **114**, 296–310.

WILSON, J. R., (ed.), 1973. *Behavioral Genetics: Simple Systems*. Colorado Associated University Press, Boulder.

WOOD, W. B., 1973. Genetic control of bacteriophage T4 morphogenesis. In *Genetic Mechanisms of Development*, F. H. Ruddle (ed.). Academic Press, New York, pp. 29–46.

WRIGHT, T. R. F., 1970. The genetics of embryogenesis in *Drosophila*. *Adv. in Genet.*, **15**, 261–395.

PART VI

COURSE OF GENETIC MATERIAL IN POPULATIONS

It is argued [by Empedocles] that where all things happened as if they were made for some purpose, being aptly united by chance, these were preserved, but such as were not aptly made, these were lost and still perish.

ARISTOTLE
Parts of Animals, Book I

31
GENE FREQUENCIES AND EQUILIBRIUM

The sections dealing with gene transmission, gene structure, and gene function have laid special stress on studies of individual organisms and their cells. However, to predict the fate of genes one must know things about populations as well as about single organisms, since it is in populations of individuals that the future of many genes is decided. For example, the reproductive ability of individuals carrying a specific gene may hinge upon the frequency of this gene in the population, the size of the population, the genotypes of other individuals in the population, and factors that may involve the relationship between the population itself and various aspects of its environment including other populations. Furthermore, the biological continuity of life occurs in space and time. In both of these dimensions the individual organism plays but a small role; rather, it is the population that is the agent for the distribution of organisms through space, and for the persistence of organisms through time. Thus, although genes exist within individuals, the fate of individuals and consequently the fate of their genes are strongly tied to factors concerning the population as a whole. What is a population, and how do we determine its genetic constitution?

Among geneticists a population is usually defined as a community of sexually interbreeding or potentially interbreeding individuals. Since mendelian laws apply to the transmission of genes among these individuals, such a community has been termed by Wright a *mendelian population*. The size of the population may vary, but it is usually considered to be a local group (also called "deme"), each member of which has an equal chance of mating with any other member of the opposite sex. Most of the theoretical and experimental emphasis has been laid so far on populations of diploid organisms, and the chapters that follow deal mostly with these cases. Whether diploid or haploid, however, populations can be said to have two important attributes: *gene frequencies* and a *gene pool*.

GENE FREQUENCIES

Gene frequencies are simply the proportions of the different alleles of a gene in a population. To obtain these proportions we count the total number of organisms with various genotypes in the population and estimate the relative frequencies of the alleles involved. Of course, except for occasional mutation, the genetic complements of all the cells of a multicellular organism are the same. One may therefore adopt the convention that a haploid organism has only one gene at any one locus, a diploid has two, a triploid three, etc. For example, if we consider the human MN blood group, there are a total of 200 genes in a population that contains 50 *MM*, 20 *MN*, and 30 *NN* individuals. Of these, 100 + 20, or .6 of the total, are *M* and 20 + 60, or .4 of the total, are *N*. The same gene frequencies can also be calculated from the frequencies of the three *genotypes, MM, MN, NN,* according to the formula: frequency of a gene = frequency of homozygotes for that gene + 1/2 frequency of heterozygotes (who each contain one such gene out of two). Thus frequency $M = .5MM + (1/2).2MN = .6$, and frequency $N = .3NN + (1/2).2MN = .4$.

GENE POOL

The gene pool is the sum total of genes in the reproductive gametes of a population. It can be considered as a gametic pool from which samples are drawn at random to form the zygotes of the next generation. Thus the genetic relationship between an entire generation and the subsequent generation is very similar to the genetic relationship between a parent and its offspring. Since the frequencies of genes in the new generation will hinge, to some degree at least, upon their frequencies in the old, one might say that gene frequencies rather than genes are "inherited" in populations. In what form can these gene-frequency relationships between generations be expressed and analyzed?

One of the first attempts at utilizing the concept of gene frequencies occurred in criticisms by Yule and others against the acceptance of particulate inheritance (see the discussion of biometricians and mendelians in Chapter 14). They argued that dominant alleles, no matter what their initial frequency, would be expected to reach a stable equilibrium frequency of three dominant individuals to one recessive, since this was the mendelian segregation pattern for these genes. The fact that many dominant alleles such as brachydactyly (short fingers) were present in very low frequency was therefore evidence that mendelian dominants and recessives were not segregating properly in populations. Although widely accepted at first, this argument was disproved in 1908 by both Hardy in England and Weinberg in Germany. Hardy and Weinberg demonstrated that gene frequencies are not dependent upon dominance or recessiveness but may remain essentially unchanged or be conserved from one generation to the next under certain conditions. The conservation of gene frequencies will be discussed in this chapter. Chapter 32 will deal with the forces that can change gene frequencies. Only simple algebra is needed to follow the derivation of the formulas used in the following sections.

CONSERVATION OF GENE FREQUENCIES

The principle discovered by Hardy and Weinberg may be simply illustrated. It may be presumed for present purposes that among humans the difference between those who can and those who cannot taste the chemical phenylthiocarbamide (PTC) resides in a single gene difference with two alleles, *T* and *t*. The allele for tasting,

T, is dominant over *t*, so that heterozygotes, *Tt*, are tasters, and the only nontasters are *tt*. If we were to choose an initial population composed of an arbitrary number of each genotype, we may ask what will be the frequency of these genes after many generations. Let us, for example, place upon an island a group of children in the ratio .40 *TT*:.40 *Tt*:.20 *tt*. The gene frequencies in this newly formed population are therefore .40 + .20 = .60 *T*, and .20 + .20 = .40 *t*. Let us also assume that the number of individuals in the population is large, and that tasting or nontasting has no effect upon survival (viability), fertility, or attraction between the sexes.

As these children mature they will choose their mates at random from those of the opposite sex regardless of their tasting abilities. Matings between any two genotypes can then be predicted solely on the basis of the frequency of those genotypes in the population. As shown in Table 31–1, nine different types of matings can occur, of which three matings are reciprocals of others (e.g., *TT* × *tt* = *tt* × *TT*). In all, therefore, there are six different mating combinations between these genotypes which will produce offspring in the ratios shown in Table 31–2.

Note that although the frequencies of genotypes have been altered by random mating, the *gene* frequencies have not changed. For *T* the gene frequency is equal to .36 + 1/2(.48) = .60, and the frequency of *t* is .16 + 1/2(.48) = .40, exactly the same as before. Under these conditions, no matter what the initial frequencies of the three genotypes, the gene frequencies of the next generation will be the same as those of the parental generation. For example, if the founding population of this island contained .25 *TT*, .70 *Tt*, and .05 *tt*, the gene frequency for *T* would be .25 + 1/2 (.70) = .60, and .05 + 1/2 (.70) = .40 for *t* (the same as

TABLE 31–1

Types of random-mating combinations and their relative frequencies in a population containing .40*TT*, .40*Tt*, and .20*tt* genotypes

		Males		
		TT .40	*Tt* .40	*tt* .20
Females	*TT* .40	.16 ①	.16 ②	.08 ③
	Tt .40	.16 ④	.16 ⑤	.08 ⑥
	tt .20	.08 ⑦	.08 ⑧	.04 ⑨

TT × *TT* (①) = .16
TT × *Tt* (② + ④) = .32
TT × *tt* (③ + ⑦) = .16
Tt × *Tt* (⑤) = .16
Tt × *tt* (⑥ + ⑧) = .16
tt × *tt* (⑨) = .04
1.00

TABLE 31–2

Relative frequencies of the different kinds of offspring produced by the matings shown in Table 31–1

Parents		Offspring Ratios			Offspring Frequencies		
Type of mating	Frequency of mating	*TT*	*Tt*	*tt*	*TT*	*Tt*	*tt*
TT × *TT*	.16	all (.16)			.16		
TT × *Tt*	.32	1/2 (.32)	+ 1/2 (.32)		.16	.16	
TT × *tt*	.16		all (.16)			.16	
Tt × *Tt*	.16	1/4 (.16)	+ 1/2 (.16)	+ 1/4 (.16)	.04	.08	.04
Tt × *tt*	.16		1/2 (.16)	+ 1/2 (.16)		.08	.08
tt × *tt*	.04			all (.04)			.04
					.36	.48	.16

TABLE 31–3

Frequencies of offspring produced by random mating in a population containing .25*TT*, .70*Tt*, and .05*tt*

Parents		Offspring		
Mating	Frequency	*TT*	*Tt*	*tt*
TT × *TT*	.25 × .25 = .0625	.0625		
* *TT* × *Tt*	.25 × .70 × 2 = .3500	.1750	.1750	
* *TT* × *tt*	.25 × .05 × 2 = .0250		.0250	
Tt × *Tt*	.70 × .70 = .4900	.1225	.2450	.1225
* *Tt* × *tt*	.70 × .05 × 2 = .0700		.0350	.0350
tt × *tt*	.05 × .05 = .0025			.0025
	1.0000	.3600	.4800	.1600

* These matings can occur in two ways, e.g., *Tt* × *Tt* matings occur in frequencies of both .25 ♂ × .70 ♀ and .25 ♀ × .70 ♂.

above). However, despite the new frequencies of genotypes, the offspring are again produced in the ratio .36 *TT*:.48 *Tt*:.16 *tt* (Table 31–3), or a gene frequency .60 *T*:.40 *t*.

There are two important conclusions therefore that follow:

1. Under conditions of random mating (*panmixis*) in a large population where all genotypes are equally viable, gene frequencies of a particular generation depend upon the *gene* frequencies of the previous generation and not upon the *genotype* frequencies.
2. The frequencies of different genotypes produced through random mating depend only upon the gene frequencies.

Both these points mean that by confining our attention to the genes rather than to the genotypes we can predict both gene and genotype frequencies in future generations (providing outside forces are not acting to change their frequency and random mating occurs between all genotypes). To continue our previous illustration we may predict that the initial gene frequencies in the taster-nontaster populations will not change in the next or succeeding generations. Also, after the first generation the genotype frequencies will also remain stable, i.e., at *equilibrium*. This genotypic equilibrium, based on stable gene frequencies and random mating, is known as the *Hardy-Weinberg principle* (or law) and has served as the founding theorem of population genetics.

The general relationship between gene frequencies and genotype frequencies can be described in algebraic terms by means of the Hardy-Weinberg principle as follows: if p is the frequency of a certain gene in a panmictic population (e.g., *T*) and q the frequency of its allele (e.g., *t*), so that $p + q = 1$ (i.e., there are no other alleles), the equilibrium frequencies of the genotypes are given by the terms $p^2(TT)$, $2pq(Tt)$ and $q^2(tt)$. If the gene frequencies of *T* and *t* are $p = .6$ and $q = .4$, respectively, the equilibrium genotype frequencies will then be $(.6)^2(TT) + 2(.6)(.4)(Tt) + (.4)^2(tt) = .36\ TT + .48\ Tt + .16\ tt$.

This relationship can be visualized by drawing a checkerboard in which the genotype frequencies are the result of random union between alleles that are in the frequencies of p and q (Fig. 31–1). The same results are also produced by the expansion of the binomial $(p + q)^2 = p^2 + 2pq + q^2$. Therefore, with any given p and q and random mating among genotypes, one generation is sufficient to establish an equilibrium condition for the frequencies of genes and genotypes. Once established, the equilibrium condition will persist until the *gene* frequencies are changed. An algebraic proof of equilibrium for the general case of random mating between p^2TT, $2\ pqTt$, and q^2tt is given in Table 31–4, on the condition that the total frequency of both alleles $(p + q)$ equals 1, and consequently $(p + q)^2 = (p^2 + 2pq + q^2) =$ sum of genotypic frequencies $= 1$.

The consideration of two alleles at a locus is only one example for which information may be desired. For genes in which there are more than two alleles at a locus, we can adopt one of three approaches:

1. If we are interested in the genotypic frequencies determined by only one of the alleles, e.g., A_1, we can consider the frequency of A_1 as p and lump the frequencies of all other alleles at that locus (e.g., $A_2, A_3, \ldots, A_n$, etc.) into a single frequency q. The equilibrium frequency is then similar to that of two alleles; $p^2\ A_1A_1 + 2pq\ A_1A_{2\ldots n} + q^2\ A_{2\ldots n}A_{2\ldots n}$. The last term will consist of numerous heterozygotes, but since we are interested only in the genotypes of A_1 relative to all the others, the precise composition of this is not essential for our purpose.

2. If we are interested in the equilibrium values for the genotypes of only two of the alleles, e.g., A_1 and A_2, with respective frequencies q_1 and q_2, the genotypic equilibrium values are $q_1^2\ A_1A_1 + 2q_1q_2\ A_1A_2 + q_2^2\ A_2A_2$. (Since $q_1 + q_2$ does not equal 1 because of the presence of other alleles, the total frequencies of the three genotypes will also be unequal to 1.)

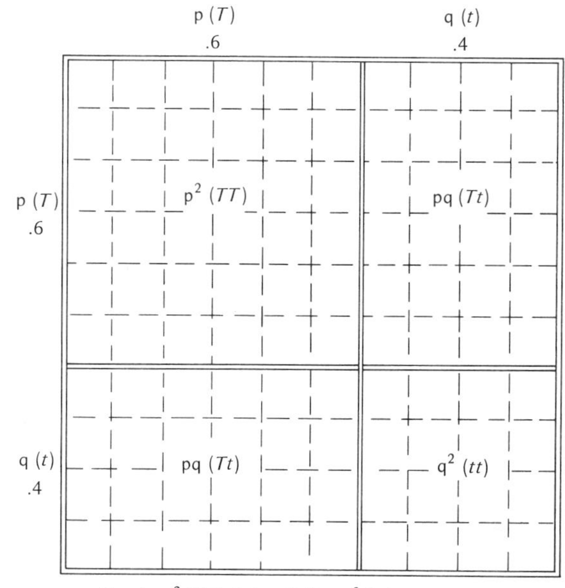

Figure 31–1

Genotypic frequencies generated under conditions of random mating for two alleles, T and t, at a locus when their respective frequencies are p = .6 and q = .4.

3. If our purpose is to find equilibrium values for the genotypes of three or more alleles, we must consider each allelic frequency as an element in a multinomial expansion. For example, if there are only three

TABLE 31–4

Mating combinations and frequencies of offspring produced under conditions of random mating when genotypic frequencies are p^2TT, $2pqTt$, and q^2tt

Parents		Offspring		
Matings	Frequencies	*TT*	*Tt*	*tt*
TT × *TT*	$p^2 \times p^2 = p^4$	p^4		
(2) *TT* × *Tt*	$2 \times p^2 \times 2pq = 4p^3q$	$2p^3q$	$2p^3q$	
(2) *TT* × *tt*	$2 \times p^2 \times q^2 = 2p^2q^2$		$2p^2q^2$	
Tt × *Tt*	$2pq \times 2pq = 4p^2q^2$	p^2q^2	$2p^2q^2$	p^2q^2
(2) *Tt* × *tt*	$2 \times 2pq \times q^2 = 4pq^3$		$2pq^3$	$2pq^3$
tt × *tt*	$q^2 \times q^2 = q^4$			q^4
	$p^2(p^2 + 2pq + q^2)$ $+ 2pq(p^2 + 2pq + q^2)$ $+ q^2(p^2 + 2pq + q^2)$ $= p^2 + 2pq + q^2 = (p + q)^2 = 1$	$p^4 + 2p^3q + p^2q^2$ $= p^2(p^2 + 2pq + q^2)$ $= p^2(1) = p^2$	$2p^3q + 4p^2q^2 + 2pq^3$ $= 2pq(p^2 + 2pq + q^2)$ $= 2pq(1) = 2pq$	$p^2q^2 + 2pq^3 + q^4$ $= q^2(p^2 + 2pq + q^2)$ $= q^2(1) = q^2$

possible alleles at a locus, A_1, A_2, and A_3, with respective frequencies p, q, and r, so that $p + q + r = 1$, the genotypic equilibrium frequencies are determined from the trinomial expansion $(p + q + r)^2$. The genotypic values are then $p^2A_1A_1 + 2pqA_1A_2 + 2prA_1A_3 + q^2 A_2A_2 + 2qr A_2A_3 + r^2 A_3A_3$. Thus, since each haploid gamete contains only a single allele for any one gene locus, zygotic combinations between such haploid gametes will then depend only upon the frequencies of the alleles (Fig. 31–2). In all three of the above conditions, therefore, equilibrium is established in a single generation of random mating.

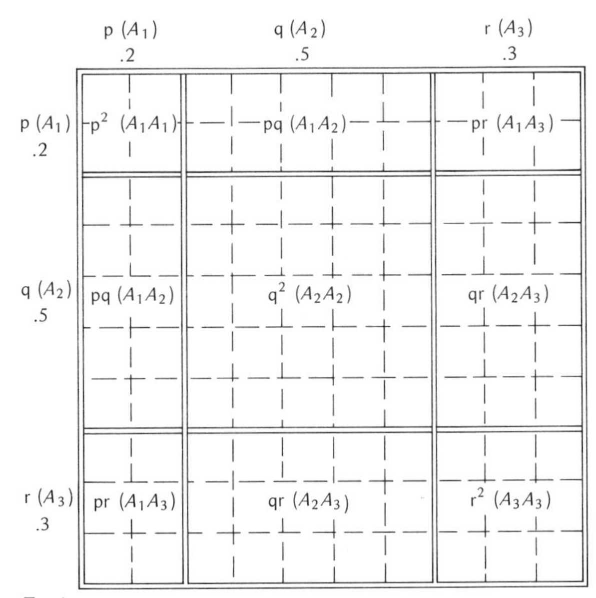

Total:
$p^2 A_1A_1 + 2pq A_1A_2 + 2pr A_1A_3 + q^2 A_2A_2 + 2qr A_2A_3 + r^2 A_3A_3$

Figure 31–2

Genotypic frequencies generated under conditions of random mating when there are three alleles, A_1, A_2, and A_3, present at a locus. For purposes of illustration the respective gene frequencies of these alleles have been given as $p = .2$, $q = .5$, and $r = .3$. Genotypic frequencies are therefore $.04A_1A_1$, $.20A_1A_2$, $.12A_1A_3$, $.25A_2A_2$, $.30A_2A_3$, and $.09A_3A_3$.

ATTAINMENT OF EQUILIBRIUM AT TWO OR MORE LOCI

The establishment of equilibrium in one generation holds true as long as we consider each single gene locus separately without being concerned about what is happening at other gene loci. If, however, we consider the products of two independently assorting gene-pair differences simultaneously, e.g., *Aa* and *Bb*, the number of possible genotypes increases to 3^2 (i.e., *AABB*, *AABb*, *AaBB*, *AaBb*, etc.). As expected, more terms are now involved in the multinomial expansion, so that if we call p, q, r, and s the gene frequencies of *A*, *a*, *B*, and *b*, respectively, the equilibrium ratios of their genotypes are expressed as $(pr + ps + qr + qs)^2$, or p^2r^2 *AABB*, $2p^2rs$ *AABb*, $2p^2s^2$ *AAbb*, $2pqr^2$ *AaBB*, . . . , q^2s^2 *aabb*.

This equilibrium formula depends on the terms pr, ps, qr, and qs, which are the equilibrium frequencies of the gametes *AB*, *Ab*, *aB*, and *ab*, respectively. Once the gametic frequencies have reached those equilibrium values, the equilibrium genotypic frequencies will also have been reached. The problem of attainment of equilibrium therefore resolves itself to the time that it takes for the gametic frequencies to reach these values. If we begin only with heterozygotes (*AaBb* × *AaBb*) in which the frequencies of all genes are the same (i.e., $p = q = r = s = .5$), all four types of gametes (*AB*, *Ab*, *aB*, *ab*) are immediately produced at equilibrium frequencies (.25), and genotypic equilibrium is reached within one generation. However, this is the only situation in which equilibrium is reached so rapidly. To take an extreme case, if we begin a population with the genotypes *AABB* and *aabb*, only two types of gametes are produced (*AB* and *ab*) and equilibrium for all genotypes cannot be reached in the next generation since numerous genotypes are missing (i.e., *AAbb*, *aaBB*, etc.). In general, two questions may be asked: (1) What are the expected equilibrium frequencies of gametes? (2) How rapidly are these frequencies achieved?

Let us first divide the gametes into those in "repulsion" (*Ab* and *aB*) and those in "coupling" (*AB* and *ab*). Since the frequencies of *genes* in gametes in repulsion are equal to the frequencies of *genes* in gametes in coupling, we would expect the products of the frequencies of both types of gametes to be equal at equilibrium: $(Ab) \times (aB) = (AB) \times (ab)$. For example, if the frequencies of A and B are .6 each, and the frequencies of a and b are .4 each, then at equilibrium $(.24)(.24) = (.36)(.16)$, or both products equal .0576. If there is a difference between the coupling and repulsion products in the initial population, this difference therefore represents the change in gametic frequencies that must occur for equilibrium value to be reached. If we call this difference, *disequilibrium,* or d, and it is positive, so that $(Ab)(aB) - (AB)(ab) = (+)d$, at equilibrium this fraction will have been added to each of the coupling gametes and subtracted from each of the repulsion gametes. If d is negative, the reverse operation will occur. In both cases, disequilibrium will have diminished to zero.

Until the final gametic ratios are reached, one-half of the difference from equilibrium is reduced each generation, so that within four to five generations more than 90 percent of this difference from equilibrium frequency has been attained by all gametes or less than 10 percent of disequilibrium value remains. Table 31–5 shows how d is calculated and how the changes in gametic frequencies occur until equilibrium is attained. For three gene pairs the speed of approach to equilibrium is even further diminished, and it becomes slower still as more gene pairs are involved.

If the two segregating loci in repulsion are linked, the chances that all the different types of dihybrid gametes will be found depends on crossover frequencies between the two loci. The closer the linkage, the longer it will take for the frequency of coupling-type gametes to equal the frequency of repulsion-type gametes, the greater *linkage disequilibrium* will remain, and equilibrium

TABLE 31–5

Calculation of d and the equilibrium frequencies of gametes for a population in which the frequencies of two separate gene pairs Aa and Bb are $A = B = .6$ and $a = b = .4$, and the initial genotypic frequencies are $AABB = AAbb = aaBB = .30$ and $aabb = .10$

Equilibrium Frequency of Gametes

Initial population	Gametes: Type	Initial frequency	Equilibrium frequency
30% *AABB*	*AB*	.3	.3 + *d*
30% *AAbb*	*Ab*	.3	.3 − *d*
30% *aaBB*	*aB*	.3	.3 − *d*
10% *aabb*	*ab*	.1	.1 + *d*

$d = (Ab)(aB) - (AB)(ab) = (.3)(.3) - (.3)(.1) = .06$

Attainment of Equilibrium

Generation	Amount added (AB, ab) or subtracted (Ab, aB)	Proportion of disequilibrium remaining	Gametes: AB	Ab	aB	ab
1		1.0*d*	.3	.3	.3	.1
2	.5*d*	.5*d*	.33	.27	.27	.13
3	.75*d*	.25*d*	.345	.255	.255	.145
4	.875*d*	.125*d*	.3525	.2475	.2475	.1525
5	.9375*d*	.0625*d*	.35625	.24375	.24375	.15625
·						
·						
·						
equilibrium	*d*	0*d*	.36	.24	.24	.16

TABLE 31–6

Attainment of the difference between initial and equilibrium value for the gametes formed by two loci under different degrees of recombination in terms of the proportion of initial disequilibrium added or subtracted*

	Recombination, Percent				
Generation	50 (Independent Assortment)	40	30	20	10
1					
2	.5*d*	.4*d*	.3*d*	.2*d*	.1*d*
3	.75*d*	.64*d*	.51*d*	.36*d*	.19*d*
4	.875*d*	.784*d*	.657*d*	.488*d*	.271*d*
5	.9375*d*	.8704*d*	.7599*d*	.5904*d*	.344*d*
·					
·					
·					
equilibrium	*d*	*d*	*d*	*d*	*d*

* The proportion of linkage disequilibrium that remains each generation is 1 — the values given, as shown in the previous table.

will be delayed accordingly (Table 31–6). This does not mean that the eventual equilibrium values for linked genes will be any different from those attained in the absence of linkage; *d* is dependent on gametic frequencies and not on linkage. Thus, once equilibrium is attained there is no way of distinguishing linked or unlinked genes except through tests for departures from independent assortment (Chapter 16).

In spite of these theoretical considerations, not all gametes of linked loci in natural populations reach equilibrium frequencies. A number of cases have, so far, been discovered in which linkage disequilibrium appears to be maintained because of advantages conferred upon certain linked allelic combinations. Some of these advantages will be discussed in later chapters (pp. 807, 817, and 820).

SEX LINKAGE

For sex-linked genes the number of possible genotypes is increased, since there is a difference in number of sex chromosomes between the homogametic and heterogametic sex. If females are XX and males XY, five genotypes can occur for a sex-linked pair of alleles *A* and *a*: three in females (*AA*, *Aa*, *aa*) and two in males (*A* and *a*). If we assign the frequencies p and q to *A* and *a*, respectively, the equilibrium genotypic values in females are the same as for an autosomal gene, p^2 *AA*, 2pq *Aa* and q^2 *aa*. In male "hemizygotes" the gene frequencies are expressed directly, and at equilibrium there will be p *A* and q *a* genotypes. As shown in Table 31–7, given random mating, stable genotypic frequencies for both males and females can be proved algebraically.

These equilibrium values are based on the fact that the gene frequencies for *A* and *a* (p and q) are identical in both sexes. Should a difference in gene frequencies be found between males and females, it would indicate that the population is not at equilibrium, assuming all genotypes are equally viable. For example, if we were to begin a population with the proportions .20 *A*:.80 *a* in males and .20 *AA*:.60 *Aa*:.20 *aa* in females, the frequency of *A* is .2 in males and .5 in females. We may then ask: What are the equilibrium frequencies of all five genotypes and how long will it take for these frequencies to be attained? Since there is only one X chromosome in males and two in females, the average frequency of a sex-linked gene in a breeding population with equal numbers of males and females will be the sum of one-third

TABLE 31–7

Frequencies of random-mating combinations and frequencies of their offspring for a sex-linked gene at which the allele *A* has the frequency p and the allele *a* has the frequency q

		Mothers		
		p^2 *AA*	2pq *Aa*	q^2 *aa*
Fathers	p *A*	p^3	$2p^2q$	pq^2
	q *a*	p^2q	$2pq^2$	q^3

Offspring

Matings		Females			Males	
Type	Frequency	*AA*	*Aa*	*aa*	*A*	*a*
AA × *A*	p^3	p^3			p^3	
Aa × *A*	$2p^2q$	p^2q	p^2q		p^2q	p^2q
aa × *A*	pq^2		pq^2			pq^2
AA × *a*	p^2q		p^2q		p^2q	
Aa × *a*	$2pq^2$		pq^2	pq^2	pq^2	pq^2
aa × *a*	q^3			q^3		q^3
		$p^3 + p^2q$ $= p^2(p + q)$ $= p^2(1) = p^2$	$2p^2q + 2pq^2$ $= 2pq(p + q)$ $= 2pq(1) = 2pq$	$pq^2 + q^3$ $= q^2(p + q)$ $= q^2(1) = q^2$	$p^3 + 2p^2q + pq^2$ $= p(p^2 + 2pq + q^2)$ $= p(1) = p$	$p^2q + 2pq^2 + q^3$ $= q(p^2 + 2pq + q^2)$ $= q(1) = q$

of its frequency in males plus two-thirds of its frequency in females, or $p = 1/3\,(p_{males}) + 2/3\,(p_{females}) = (p_{males} + 2p_{females})/3$. In the example cited, the frequency of *A* is $[.2 + 2(.5)]/3 = 1.2/3 = .4$, and the frequency of *a* is .6. Thus, according to our methods, the equilibrium values expected with p equal to .4 and q to .6 are .4 *A* : .6 *a* in males and .16 *AA* : .48 *Aa* : .36 *aa* in females.

However, in contrast to single autosomal loci, equilibrium will not be reached in a single generation. Since males inherit their X chromosomes only from their mothers, the frequency of a sex-linked gene among them is the same as its maternal frequency. Therefore, a founding population with a maternal frequency of *A* equal to .5 will produce male offspring with this same frequency of *A*. The daughters, however, inherit one paternal and maternal chromosome each, so that the frequency of a sex-linked gene among them will be the average of the parental frequencies. If the females in the founding population had an *A* frequency of .5, but the males had an *A* frequency of only .2, the daughters would have an *A* frequency of $(.2 + .5)/2 = .35$, while their brothers would have the .5 frequency of their mothers. Thus, in the first generation of random mating the *A* equilibrium value of .4 will not be reached by the daughters and will be exceeded by the sons. In the offspring

TABLE 31–8

Frequencies of the sex-linked gene *A* in males and females in successive generations under conditions of random mating when the initial frequency of *A* is .2 in males and .5 in females

	Generations						
	0	*1*	*2*	*3*	*4*	*5*	*6*
males	.2	.5	.35	.425	.3875	.40625	.396875
females	.5	.35	.425	.3875	.40625	.396875	.4015625

of the second generation, the difference from equilibrium values will be diminished, but this time the sons will be below equilibrium ($A = .35$) and the daughters above [$A = (.35 + .5)/2 = .425$]. Each succeeding generation will then show a similar reversal but nevertheless achieve a successively closer approximation to the final equilibrium values (Table 31–8).

ESTIMATION OF EQUILIBRIUM FREQUENCIES IN NATURAL POPULATIONS

The situations discussed above hinge either upon prior knowledge of the gene frequencies or a full identification of all genotypes, or both. Natural populations, however, do not always fulfill these conditions, and it may be difficult to discern just what the gene frequencies are and whether a population has achieved the expected equilibrium values. To analyze natural populations, therefore, much depends upon the extent of the data and on the kinds of genes involved.

CODOMINANCE IN NATURAL POPULATIONS

When all segregants of the gene differences at a single locus can be scored, the gene frequencies can be reliably estimated and the observed genotype frequencies can then be easily compared to their expected equilibrium values. Codominance at the *MN* blood group locus, for example, enabled Boyd to classify 104 American Ute Indians into the genotype frequencies .59 *MM*, .34 *MN*, and .07 *NN*. The gene frequency of *M* is therefore .59 + .17 = .76 and for *N* it is .07 + .17 = .24. At equilibrium, therefore, expected genotype frequencies should be $(.76)^2$ *MM*, 2(.76)(.24) *MN*, and $(.24)^2$ *NN*; or .58 *MM*, .36 *MN*, and .06 *NN*.

The difference between observed and expected genotypic numbers can then be tested by the chi-square method shown in Table 31–9. For the one degree of freedom in the present data,* the chance probability of the observed chi-square is sufficiently high that the hypothesis of equilibrium in this population can be accepted even without Yates' correction.

When the genotypes of the mated couples in the population are known, the assumption of random mating can also be tested. Under random mating the frequencies of the different mating combinations should depend only upon the frequencies of their genotypes. An actual set of data collected by Matsunaga and Itoh tested these expectations for the *MN* locus. They analyzed the blood types of 741 couples (or 1482 individuals) in a Japanese town and found genotypic frequencies of .274 *MM*, .502 *MN*, and .224 *NN*. Since the gene frequencies are .525 *M* and .475 *N*, the expected equilibrium genotypic frequencies were .276 *MM*, .499 *MN*, and .225 *NN*. It is obvious that the genotypes of the population are in equilibrium for this locus. We may now ask whether the mating combinations are also as expected.

TABLE 31–9

Chi-square test for goodness of fit between MN-blood-type data given in the text and Hardy-Weinberg equilibrium

	MM	*MN*	*NN*	*Total*
observed	61	36	7	104
expected	60	38	6	104
obs. − exp.	1	−2	1	
$(\text{obs.} - \text{exp.})^2$	1	4	1	
$\frac{(\text{obs.} - \text{exp.})^2}{\text{expected}}$	.0167	.11	.167	$\chi^2_{1\,\text{df}} = .294$

*Note that in calculating the degrees of freedom in a test of gene frequencies, the "free" variables are not equal to $k - 1$ phenotypic classes but are even further restricted. In the present case once the frequency of one allele and the total number of individuals is given, the only degree of freedom is the frequency of one of the genotypes, since these three values will now furnish the frequencies of all three genotypes. Thus, for most instances involving gene frequencies the number of degrees of freedom is equal to the number of phenotypes minus the number of alleles.

The number of observed matings of different combinations are given in the last column of Table 31–10. To the left of this column the expected mating combination frequencies are calculated on the basis of the gene frequencies, p for *M* and q for *N*. For example, since the frequency of the genotype *MM* is p^2 at equilibrium, the random-mating combination *MM* × *MM* should be p^4. When the frequencies of all expected mating combinations are calculated this way (see also Table 31–4), a comparison between observed and expected shows excellent agreement with the assumption of random mating.

TABLE 31–10

Comparison of mating combinations expected according to random mating and those observed in 741 couples by Matsunaga and Itoh (p = .525, q = .475)

	Expected frequency	Expected number	Observed number
MM × *MM*	$p^4 = .0760$	56.3	58
MM × *MN*	$4p^3q = .2749$	203.7	202
MM × *NN*	$2p^2q^2 = .1244$	92.2	88
MN × *MN*	$4p^2q^2 = .2487$	184.3	190
MN × *NN*	$4pq^3 = .2251$	166.8	162
NN × *NN*	$q^4 = .0509$	37.7	41
	1.0000	741	741

DOMINANCE IN NATURAL POPULATIONS

When the effect of one allele at a locus is completely dominant over that of another, the heterozygous genotype (e.g., *Aa*) cannot be phenotypically distinguished from the homozygous dominant (e.g., *AA*). Under these circumstances gene frequencies cannot be obtained directly, as in codominance, since two of the genotypic frequencies are unknown.

One method used to estimate gene frequencies in such cases relies upon the only genotype whose frequency is definitely known, the recessive homozygote. We have seen that under conditions of random mating, the genotype frequencies will be p^2 *AA*, $2pq$ *Aa*, and q^2 *aa*. The recessive homozygotes are therefore present in a frequency q^2 equal to the square of the recessive gene frequency q. If, let us say, q^2 is .49, q is $\sqrt{.49} = .70$, and the frequency of the dominant allele p is $1 - q$, or .30. (Note that when dealing with fractions, the square root of a fraction is a number larger than the fraction.) The homozygous dominants, therefore, have the frequency $p^2 = (.30)^2 = .09$, and the heterozygotes have the frequency $2pq = 2(.30)(.70) = .42$.

When recessive phenotypes are rare, it is surprisingly common (but mathematically inevitable) to find that the "carrier" heterozygotes are present in relatively high frequency. Albinism, for example, affects only about 1 out of 20,000 humans in some populations, or $q^2 = 1/20{,}000 = .00005$. The gene frequency, q, of the albino gene is therefore .007, and the frequency, p, of the nonalbino allele is .993. The frequency of albino "carriers" is therefore $2(.993)(.007) = .014$, or approximately 1 out of 70 individuals. There are thus $.014/.00005 = 280$ times as many heterozygotes for this trait as homozygotes. Similar high frequencies for carriers of recessive traits such as cystic fibrosis, Tay-Sachs disease, and alkaptonuria point to the difficulty of eliminating deleterious recessive characters, since they are carried mostly in the unexpressed heterozygous condition (see also Table 32–7).

Since genotypic differences cannot be detected when dominance is complete, calculation of q as the square root of the frequency of homozygous recessives assumes that Hardy-Weinberg equilibrium has been reached, but in no way provides supporting evidence for this assumption. A given frequency of one genotype (e.g., *aa*) need not mean that the frequency of other genotypes (*AA*, *Aa*) are in equilibrium. For such cases Snyder has devised a test for the presence of genetic equilibrium when family data are available. The test is based on the expected frequencies of recessives produced by the parental mating combinations. For example, a mating between two dominant phenotypes will produce recessives only if both are heterozygotes. If equilibrium exists, the frequency

TABLE 31–11

Frequencies of random-mating combinations for a gene at which the allele *A* has the frequency p and *a* has the frequency q*

			Female Parent		
			Dominant		Recessive
			p^2 *AA*	$2pq$ *Aa*	q^2 *aa*
Male Parent	Dominant	p^2 *AA*	p^4	$2p^3q$	p^2q^2
		$2pq$ *Aa*	$2p^3q$	$4p^2q^2$	$2pq^3$
	Recessive	q^2 *aa*	p^2q^2	$2pq^3$	q^4

dominant parent × dominant parent $= p^4 + 4p^3q + 4p^2q^2$
frequency of recessives produced $= (1/4)(4p^2q^2) = p^2q^2$
dominant parent × recessive parent $= 2p^2q^2 + 4pq^3$
frequency of recessives produced $= (1/2)(2pq^3 + 2pq^3) = 2pq^3$
recessive parent × recessive parent $= q^4$
frequency of recessives produced $= q^4$

* Mating combinations included within the boxed area are those that produce recessive phenotypes. The frequency of recessives produced by each such mating combination is shown below the table.

of such matings and the proportion of recessive offspring expected can then be calculated. Table 31–11 shows the various mating combinations and their total frequencies for a case where *A* is dominant over *a*. Note that the frequency of recessive offspring produced from the heterozygous dominant × dominant matings will be one fourth of $4p^2q^2$, or p^2q^2. This frequency expressed as a fraction of all offspring of this type of mating can then be formulated as

$$\frac{p^2q^2}{p^4 + 4p^3q + 4p^2q^2} = \frac{p^2q^2}{p^2(p^2 + 4pq + 4q^2)} = \frac{q^2}{p^2 + 4pq + 4q^2} = \frac{q^2}{[(p + q) + q]^2} = \frac{q^2}{(1 + q)^2}$$

Similarly rcccssivcs will only be produced in dominant × recessive matings when the dominant parent is heterozygous. In such cases, one half the offspring are recessive and have the expected frequency $2pq^3$. Compared to all offspring of dominant × recessive matings, this frequency is equal to

$$\frac{2pq^3}{2p^2q^2 + 4pq^3} = \frac{2pq^3}{2pq^2(p + 2q)} = \frac{q}{p + 2q} = \frac{q}{p + q + q} = \frac{q}{1 + q}$$

To use Snyder's terms, we now have two *population ratios* whose theoretical expectations can be compared to the observed frequencies. On this basis Snyder performed a test for equilibrium of the alleles for the ability to taste phenylthiocarbamide (Table 31–12). The frequency of the non-tasting allele was derived from the frequency of homozygous recessives (289 + 86 + 86) among 1600 parents, or $q = \sqrt{461/1600} = \sqrt{.288} = .537$. The expected frequency of recessive offspring from taster × taster parents should therefore be $q^2/(1 + q)^2 = (.537)^2/(1.537)^2 = .122$. Note that the observed proportion of recessives from these matings fits this frequency quite closely. Similarly the expected frequency of recessives from taster × nontaster combinations [$q/(1+q) = .349$] also fits the observed proportion closely. Thus there is little question that the population is at equilibrium for this gene. The only apparent anomaly is the occurrence of five tasters in the nontaster × nontaster combinations, but this has been ascribed to a variety of causes, including illegitimacy and wrong diagnosis.

SEX LINKAGE IN NATURAL POPULATIONS

When the two alleles of a sex-linked gene are codominant, each of the five different male and female genotypes has a unique phenotype. If the numbers of males and females are equal, the gene frequencies can then be easily estimated from the genotype frequencies according to the formula previously used $(p_{males} + 2p_{females})/3$, where p is the frequency of one codominant allele, e.g., *A*. If male and female numbers are not equal, the following estimate of p can be used:

$$p = \frac{2n_{AA} + n_{Aa} + n_{A(Y)}}{2N_f + N_m}$$

where N_f = total no. females
N_m = total no. males
n_{AA} = no. of female *AA* genotype
n_{Aa} = no. of female *Aa* genotype
$n_{A(Y)}$ = no. of male *A* genotype

To determine whether the population is at equilibrium, two tests can then be performed. The first test is simply to observe whether the frequency of p is the same in both sexes, by use of a contingency chi-square (p. 148). If this important requirement for equilibrium is fulfilled (p. 743), the observed and expected genotypic frequencies in females can then be compared.

The procedure to be used can be illustrated by a sample of 281 Boston cats scored by Todd for the frequency of the sex-linked gene *yellow* (*y*). In the heterogametic XY males, the presence of the *yellow* allele produces yellow fur, while its normal allele (y^+, or simply +) produces a darker color whose particular shade and pattern depend on the presence of other genes. In the female, three phenotypically distinct genotypes occur: the homozygous yellow (*y*/*y*), homozygous normal (+/+), and the "tortoise shell," or "calico," heterozygote (+/*y*).

A gene count of Todd's data (Table 31–13) gives 56 *yellow* alleles and 252 wild-type alleles among the X chromosomes of females and 28 *yellow* and 99 wild-type alleles among the X chromosomes of males. A chi-square test for independence based on these numbers shows no significant difference in gene frequencies between the sexes. Using the formula for sex-linked genes, the overall gene frequency for *yellow* is

$$\frac{2(4) + 48 + 28}{2(154) + 127} = \frac{84}{435} = .193$$

This value can now be substituted in the ex-

TABLE 31–12

Comparison of observed and expected frequencies of recessive phenotypes produced by a population of 800 couples

Parents		Offspring			Frequency of Nontasters Among Offspring	
Mating	No. Couples	Tasters	Nontasters	Total	Observed	Expected
taster × taster	425	929	130	1,059	.123	.122
taster × nontaster	289	483	278	761	.365	.349
nontaster × nontaster	86	(5)	218	223		
	800	1,417	626	2,043		

From data of Snyder.

TABLE 31-13

Genotype frequencies, gene counts, and chi-square test for independence between gene count and sex for 281 Boston cats

	No. Genotypes			No. Genes	
	+/+	+/*y*	*y*/*y*	+	*y*
females	102	48	4	(a) 252	(c) 56
males (hemizygotes)	99		28	(b) 99	(d) 28

$$\chi^2_{1\,df} = \frac{[|ad - bc| - 1/2\,N]^2\,N}{(a+b)(a+c)(b+d)(c+d)}$$

$$= \frac{[|(252)(28) - (99)(56)| - 1/2\,(435)]^2(435)}{(252+99)(252+56)(99+28)(56+28)} = .885$$

Data from Todd, 1964.

pected equilibrium genotypic frequencies for females and chi-square calculated. As shown in Table 31-14, the chi-square difference is .70, which, at one degree of freedom, is not significant. Thus there is little reason to doubt the existence of Hardy-Weinberg equilibrium for these genotypes.

When one sex-linked allele is dominant, two of the female genotypes can no longer be separately identified. However, the two different genotypes in males still remain phenotypically distinct. The gene-frequency estimates in males can then be used to calculate the expected frequencies of the female genotypes. If + were completely dominant in the example just considered, so that "tortoise shell" phenotypes could not be identified, the frequency of *y* calculated from the data on males would be 28/127 = .220 and + = .780. The expected genotypic frequencies in females would then be $(.780)^2$ +/+, 2(.780)(.220) +/*y*, and $(.220)^2$ *y*/*y*. Since the first two genotypes are now presumed to be phenotypically identical, the frequencies of female phenotypes should be $[1 - (.220)^2]$ + and $(.220)^2$ *y* or approximately 147 + and 7 *y*. A comparison of these values with those observed (150 + and 4 *y*) yields a chi-square of .94, which, for one degree of freedom, supports the hypothesis that genetic equilibrium has been attained.

Note that the relationship between the frequency of the sex-linked recessive phenotype in males and females is $q:q^2$ or $1:q$. When q is .1, the frequency of male recessives is 10 times the frequency of female recessive homozygotes. Thus the smaller q is, the relatively greater is the frequency of the recessive phenotype among males. Sex-linked hemophilia, for example, has a phenotypic frequency among males of less than 1 in 10,000. By contrast, the frequency of female homozygotes

TABLE 31-14

Calculation of chi-square for the hypothesis that the Hardy-Weinberg equilibrium exists among the female genotypes observed in Table 31-13 (y = .193, + = .807)

	Genotype			Total
	+/+	+/*y*	*y*/*y*	
observed	102	48	4	154
expected (× 154)	$(.807)^2 = 100$	$2(.193)(.807) = 48$	$(.193)^2 = 6$	154
		$\chi^2_{1\,df} = .70$		

would be $(.0001)^2$, or one in one hundred million, an expectation that accords with the very few observed cases of female "bleeders."

MULTIPLE ALLELES

When more than two alleles are present at a locus, the equilibrium condition is described by the multinomial expansion, $(p + q + r + \ldots)^2$, in which each letter represents the gene frequency of one allele. For example, for three alleles, A_1, A_2, and A_3, with respective frequencies p, q, and r, we would expect six genotypes at Hardy-Weinberg equilibrium in the proportion $p^2\ A_1A_1$, $2pq\ A_1A_2$, $2pr\ A_1A_3$, $q^2\ A_2A_2$, $2qr\ A_2A_3$, and $r^2\ A_3A_3$. If all the alleles are codominant with each other, each genotype has its own distinct phenotype, and the genotypic frequencies are easily scored. The gene frequencies are then calculated according to the following equations:

$$p = \frac{2(A_1A_1) + (A_1A_2) + (A_1A_3)}{2N}$$

$$q = \frac{2(A_2A_2) + (A_1A_2) + (A_2A_3)}{2N}$$

$$r = \frac{2(A_3A_3) + (A_1A_3) + (A_2A_3)}{2N}$$

where A_1A_1, A_1A_2, A_1A_3, etc., refers to the *numbers* of genotypes in each category, and N refers to the total number of individuals scored.

Such a system is found in human populations bearing different forms of the red-blood-cell enzyme acid phosphatase. According to Hopkinson, Spencer, and Harris, this enzyme may appear in six different phenotypic patterns, AA, BB, CC, AB, BC, or AC, determined by all possible combinations of the alleles *A*, *B*, and *C* at a single locus. Investigations of a Brazilian population by Lai, Nevo, and Steinberg have shown that the observed phenotypic frequencies of the acid phosphatase combinations conform closely to those expected according to the Hardy-Weinberg equilibrium (Table 31–15).

When there is a hierarchy of dominance so that A_1 is dominant over A_2 and both A_1 and A_2 are dominant over A_3, the A_3 phenotype is the only known homozygous genotype. Since the A_3A_3 frequency is r^2, the frequency of the A_3 allele (r) is the square root of its phenotypic frequency. The frequency of the other alleles can then be derived by first noting that the A_2 and A_3 phenotypes (A_2A_2, A_2A_3, A_3A_3) have a combined frequency of $q^2 + 2qr + r^2 = (q + r)^2$. Since the square root of this value, $q + r$, will account for the frequencies of two of the three alleles, the frequency p of the remaining allele A_1 must then be equal to $1 - (q + r)$. The values of p and r are now known, and q can be estimated as $1 - (p + r)$. Although these gene-frequency estimates are simple to calculate, their reliability is reduced because only a few genotypes can actually be distinguished. Consequently the attainment of equilibrium cannot be reliably tested.

TABLE 31–15

Comparison of observed acid phosphatase phenotypes and those expected according to Hardy-Weinberg equilibrium in a sample of 369 Brazilian individuals

	AA	BB	CC	AB	AC	BC
observed	15	220	0	111	4	19
expected	14.4	219.9	.4	112.2	4.4	17.7

From Lai, Nevo, and Steinberg, 1964.

When A_1 and A_2 are codominant in relationship to each other, yet dominant in relationship to A_3, the occurrence of an additional phenotype (A_1A_2) permits a better gene-frequency estimate and a reliable test for equilibrium. For example, the ABO blood group in humans can be classified into four phenotypes that arise from combinations of the alleles $A(I^A)$, $B(I^B)$, and $O(i)$ according to the following frequencies ($A = p$, $B = q$, $O = r$):

	A		AB	B		O
genotype	*AA*	*AO*	*AB*	*BB*	*BO*	*OO*
frequency	p^2	$2pr$	$2pq$	q^2	$2qr$	r^2

Under the assumption of equilibrium, the combined frequency of B and O phenotypes is $q^2 + 2qr + r^2 = (q + r)^2$. The square root of this

TABLE 31–16

Calculation of gene frequencies for A(p), B(q), and O(r) in a sample of 192 individuals bearing the phenotypes given in the text*

Phenotypes	Number Observed	Frequency	$\sqrt{\text{Frequency}}$	Estimate	Correction Formula	Corrected Values
B + O	123	.6407	.8004	p = .1996	p(1 + 1/2 *d*)	p = .2003
A + O	155	.8073	.8985	q = .1015	q(1 + 1/2 *d*)	q = .1018
O	92	.4792	.6923	r = .6923	(r + 1/2 *d*)(1 + 1/2 *d*)	r = .6979
				.9934		1.0000
				d = 1 − .9934 = .0066		

From data of Boyd.

* The Bernstein correction formulas utilize the value of *d* shown below the estimate column and produce the corrected values of p, q, and r given in the last column.

TABLE 31–17

Comparison of observed and expected frequencies for the ABO sample given in the text, utilizing the corrected values of p, q, and r in Table 31–16

	Phenotypes				
	A	B	AB	O	**Total**
observed	63	31	6	92	192
equilibrium frequencies	$(p^2 + 2pr)N$	$(q^2 + 2qr)N$	$(2pq)N$	$(r^2)N$	N
expected	61	29	8	94	192
χ^2 (with Yates' correction)	.037	.078	.282	.024	.421

value subtracted from 1 will then give the frequency of $A: 1 - (q + r) = p$. Similarly the square root of the combined A and O frequencies ($\sqrt{p^2 + 2pr + r^2} = p + r$) subtracted from 1 will give the frequency of $B: 1 - (p + r) = q$. The frequency of *O* can then be estimated as the square root of the O phenotype, $\sqrt{r^2} = r$.* An illustration of the calculation of all three frequencies can be taken from the following sample of 192 individuals from Wales. The observed phenotypes were:

	A	**B**	**AB**	**O**
number	63	31	6	92
phenotypic frequency	.3281	.1615	.0312	.4792

The fifth column of Table 31–16 shows the frequencies of p, q, and r calculated from these data.

* Although other procedures for gene-frequency estimation can be used, those given above are generally considered statistically superior.

Note that the total frequencies do not add up to 1.0000, although the difference from unity is slight. To eliminate this difference Bernstein proposed correction formulas that provide truer estimates of the gene frequencies. These corrected frequencies are calculated in the sixth column based on the difference (*d* = .0066) between the total observed frequency and 1.0000. For example, the corrected value of p is .1996(1.0033) = .2003 (last column).

Using these corrected gene frequencies, the expected genotypic equilibrium values can now be calculated and compared to those observed (Table 31–17). Since there is only one degree of freedom (four phenotypes minus three alleles), the chi-square of .421 indicates that the discrepancy from equilibrium is not significant and has a high probability of occurring in normal sampling of an equilibrium population.

For two pairs of alleles at different loci, *A*, *a* (frequencies p, q), and *B*, *b*, (frequencies r, s), nine genotypes are possible, *AABB*, *AABb*, . . ., with equilibrium frequencies p^2r^2 *AABB*, $2p^2rs$ *AABb*, . . . (see p. 740). When each genotype is distinctly recognizable because of codominance, the gene frequencies are easily estimated and the assumption of equilibrium can be tested directly. If there is dominance in the form $A > a$, $B > b$, the equilibrium frequency of the *aa* phenotype is q^2 and the frequency of all *A–* phenotypes is $1 - q^2$. Similarly the *bb* phenotype appears in the frequency of s^2 while the *B–* phenotypes have a frequency $1 - s^2$. Thus, under these conditions, the equilibrium frequency of the four phenotypes, *A–B–*, *A–bb*, *aaB–*, and *aabb*, have the values in Table 31–18. Gene frequencies can then be estimated as

$$q = \sqrt{q^2 \text{ phenotypes}} = \sqrt{(aaB\text{–}) + (aabb)}$$
$$s = \sqrt{s^2 \text{ phenotypes}} = \sqrt{(A\text{–}bb) + (aabb)}$$

When interaction occurs between two gene pairs (Chapter 11), some phenotypes can no longer be separately identified. For example, if a quantitative trait is influenced by two gene pairs so that a dominant allele at either gene pair produces the same effect, only three phenotypes can be distinguished: (1) *A–B–*, (2) *A–bb* or *B–aa*, and (3) *aabb*. The frequencies of these phenotypes (Table 31–19) are (1) $(1 - q^2)(1 - s^2)$, (2) $(1 - q^2)s^2 + q^2(1 - s^2)$, and (3) q^2s^2. Gene-frequency estimates are now difficult to make, since the two types of homozygous recessive genotypes (*aa–*, *bb–*) are not separately distinguishable.

A further complication in this case arises if both *a* and *b* have the same frequencies, so that $q = s$. The three frequencies of the phenotypes then reduce to (1) $(1 - q^2)(1 - q^2)$, (2) $2(1 - q^2)(q^2)$, and (3) q^2q^2. If we were to call $(1 - q^2)$, p, and (q^2), q, it is easy to see that the frequencies of these phenotypes at equilibrium would appear to be in the ratio $p^2{:}2pq{:}q^2$. In other words, equilibrium values for two gene pairs under these conditions are indistinguishable from a single gene pair. Similar problems occur when other types of interaction are present between gene pairs and point to the difficulty of estimating gene frequencies in a population when the mode of inheritance of the genes involved has not been established through breeding tests.

TABLE 31–18

Equilibrium frequencies produced by combinations between genes at two loci when complete dominance exists at each locus ($A > a$, $B > b$)

	B– $(1 - s^2)$	*bb* (s^2)
A– $(1 - q^2)$	*A–B–* $(1 - q^2)(1 - s^2)$	*A–bb* $(1 - q^2)s^2$
aa (q^2)	*aaB–* $q^2(1 - s^2)$	*aabb* q^2s^2

TABLE 31–19

Equilibrium frequencies produced under conditions similar to those of Table 31–18 except that two genotypes (*A–bb* and *aaB–*) are now phenotypically identical

	B– $(1 - s^2)$	*bb* (s^2)
A– $(1 - q^2)$	*A–B–* (1) $(1 - q^2)(1 - s^2)$	*A–bb* (2) $(1 - q^2)s^2$
aa (q^2)	*aaB–* (2) $q^2(1 - s^2)$	*aabb* (3) q^2s^2

PROBLEMS

31-1. In a large randomly mating population of snakes, 75 percent of individuals at birth have black stripes and 25 percent lack such stripes. Explain whether this indicates that black stripes are dominant.

31-2. What is the frequency of heterozygotes *Aa* in a randomly mating population if the frequency of recessive phenotypes (*aa*) is .09?

31-3. What is the frequency of heterozygotes *Aa* in a random-mating population in which the frequency of all dominant phenotypes is .19?

31-4. In a particular population of humans that has presumably reached Hardy-Weinberg equilibrium, the frequency of alkaptonuria (see p. 600), caused by homozygosity for a recessive gene, is one per million.

What is the probability that an affected offspring will be produced by: (a) A mating between two normal nonrelated individuals? (b) A mating between a person who has alkaptonuria and a normal nonrelated individual? (c) A mating between a normal individual with normal parents, who has an alkaptonuric brother, and a normal nonrelated individual?

31-5. The gamma-globulin portion of human blood serum may be classified into two types, Gm^{a+} and Gm^{a-}, the former caused by an allele whose effect is dominant over that of the latter. Assuming that Hardy-Weinberg equilibrium has been established, what are the frequencies of heterozygotes (Gm^{a+}/Gm^{a-}) in the following regional Swedish populations?

		No. Phenotypes	
Region	No. Tested	Gm^{a+}	Gm^{a-}
(a)	293	161	132
(b)	253	141	112
(c)	142	77	65
(d)	233	142	91
(e)	160	108	52

31-6. Two populations begin with the following genotypic frequencies:

population I: .24 *AA* .32 *Aa* .44 *aa*
population II: .33 *AA* .14 *Aa* .53 *aa*

If random mating occurs, what will be their genotypic frequencies in the next generation?

31-7. (a) Which of the following populations are in Hardy-Weinberg equilibrium:

	Genotypes		
	AA	*Aa*	*aa*
population I	.430	.481	.089
population II	.64	.32	.04
population III	.4225	.4550	.1225
population IV	.0025	.1970	.8005
population V	.0081	.0828	.9091

(b) What are the expected equilibrium frequencies for those of the above populations that are not in equilibrium? (c) How long will it take for these equilibrium values to be reached under random mating?

31-8. In testing the blood types of 2047 Guernsey cattle, Stormont found genotypes for the codominant alleles, *Z* and *z*, in the following frequencies: 542 *ZZ*, 1043 *Zz*, and 462 *zz*. Using the chi-square method, calculate how far the observed frequencies depart from the expected equilibrium frequencies.

31-9. Among 361 Navaho Indians tested in New Mexico, Boyd reported 305 of blood type M, 52 MN, and 4 N. (a) Calculate the departure from genotypic equilibrium frequencies and its significance (at the 5 percent level) using chi-square. (b) What proportion of children produced by women of N phenotype are expected to have the maternal phenotype? (c) What proportion of children produced by heterozygous MN females are expected to have the maternal phenotype?

31-10. As in Problem 9, 140 Pueblo Indians were tested and the following phenotypes found: 83 M, 46 MN, and 11 N. Is there a significant difference between the Navaho and Pueblo Indian population?

31-11. Among the same group of Navaho Indians tested in Problem 9, further tests for ABO blood types showed 77.7 percent frequency of O blood and 22.3 percent frequency of A blood. Using the gene-frequency estimates from these data and those of Problem 9: (a) Find the equilibrium values for the blood-group phenotypes NO, MNO, and MA. (b) Would such equilibrium values be different if 10 percent linkage existed between the *MN* locus and the *ABO* locus? (c) What would happen at 20 percent linkage? (d) 50 percent linkage?

31-12. If the gene frequency of the *M* blood-type allele is p, and the gene frequency of *N* is q, and $p + q = 1$, what is the probability that a man wrongly accused of being the father of a group N child will be cleared by MN-blood-group tests?

31-13. *A* and *B* are independently assorting genes, each having one recessive allele, respectively, *a* and *b*. What are the equilibrium frequencies for the homozygous recessive phenotype *aabb* in the following populations?

	Initial Genotype Frequency			
Population	*AABB*	*AAbb*	*aaBB*	*aabb*
(a)	.7	—	—	.3
(b)	—	.7	.3	—
(c)	.5	—	—	.5
(d)	—	.8	.2	—

31-14. In which of the populations in the above problem will it take longest for genotypic equilibrium to be achieved given the initial genotype frequencies?

31-15. Among 569 Egyptians tested for the ability to taste phenylthiocarbamide, 442 tasters were found.

If the allele for tasting has a dominant effect over the nontasting allele: (a) What proportion of matings will be of the type taster × taster, assuming that mating is purely random? (b) What is the ratio of taster to nontaster children expected from these taster × taster matings? (c) What is the proportion of taster to nontaster children expected from taster × nontaster matings in this population?

31-16. For the same Egyptian population as above, ABO blood tests on a sample of 502 individuals showed 137 O, 193 A, 128 B, and 44 AB. (a) Using chi-square and *corrected* gene estimates, test the departure of the observed proportions from those expected under equilibrium. (b) What frequency of tasters in this population is expected to have the blood type AB? (c) The blood type O?

31-17. Just as in humans (see Table 31-15), there are 6 possible phenotypes among the red cell acid phosphatase enzymes in macaque monkeys: A, B, C, AB, AC, and BC. In a population sample of 179 such monkeys, Lai observed 134 A, 1 B, 11 AB, 33 AC. Assuming that these phenotypes are caused by combinations of three codominant alleles at a single locus, aa = A, ab = AB, ac = AC: (a) What are the frequencies of the alleles a, b, and c in this population? (b) What are expected frequencies for each of the 6 possible phenotypes?

31-18. Azen and Smithies investigated a population of humans for one of the protein components of blood serum and found that the inheritance pattern of this serum protein was caused by a single gene with *four* codominant alleles, A, B, C, and D. Among 113 adults the following phenotypes were observed: 4 A, 38 AB, 68 B, 1 AC, 1 BC, and 1 AD. Compare these data with those you would expect to find under Hardy-Weinberg equilibrium.

31-19. The following are the observed ABO blood group phenotypes among a group of 600 Canadian Indians:

O	A	B	AB
288	280	19	13

(a) Calculate the *corrected* estimates of the A, B, and O gene frequencies. (b) How well does the observed distribution of phenotypes agree with those expected on the basis of the calculated gene frequencies?

31-20. If the frequency of the O gene in the ABO blood group system is .70 in a particular European population, what would be the expected frequency of the O phenotype among individuals who show Down's syndrome (trisomy 21) if the ABO locus is located on the 21 chromosome?

31-21. Let us assume that in humans a single autosomal gene, B, determines pattern baldness, behaving as a dominant in males and as a recessive in females. If an equilibrium human population has 51 percent bald men: (a) What is the frequency of bald women? (b) Assuming random mating, what proportion of marriages should be between a bald man and a nonbald woman? (c) In what proportion of the marriages in the population should the first child eventually develop pattern baldness: If male? If female? (d) If a nonbald couple produces a son, what is the probability he will become bald? (e) A bald woman has a daughter, but nothing is known about the father. What is the probability that the daughter will eventually become bald?

31-22. Heiken analyzed 174 Swedish children for MNSs antigens and found the following genotypes:

$$MS/MS = 10 \qquad Ms/Ns = 34$$
$$MS/Ms = 25 \qquad NS/NS = 0$$
$$Ms/Ms = 20 \qquad NS/Ns = 7$$
$$MS/NS = 9 \qquad Ns/Ns = 26$$
$$MS/Ns \text{ (or } Ms/NS) = 43$$

Do the frequencies of these genotypes accord with those expected in Hardy-Weinberg equilibrium?

31-23. Among the parents of the children tested above, Heiken noted the following matings between individuals tested for the secretor gene (alleles Se and se).

Type	No. Matings
Se Se × *Se Se*	3
Se Se × *Se se*	6
Se se × *Se se*	13
Se Se × *se se*	8
Se se × *se se*	33
se se × *se se*	18

Do these matings accord with those expected at Hardy-Weinberg equilibrium?

31-24. Histocompatibility reactions involve the response of recipients to grafted tissues from other individuals (see p. 175). If the donor bears a histocompatability antigen which the recipient does not possess, the donor tissue will be incompatible and subsequently rejected. Assume that a random breeding population has only one histocompatibility locus with two alleles,

A^1 and A^2, and that each allele has .5 frequency. What proportion of tissue grafts between random individuals will be incompatible?

31-25. For the same population as in Problem 31-24 above, what is the frequency of incompatible parent-child combinations; that is, where the child is the recipient of tissue from a parent?

31-26. Explain whether you would expect the frequency of incompatibility in tissue grafts between random individuals and between parents and children to increase or decrease as more alleles are involved in the histocompatibility reaction.

31-27. If hemophilia (type A, sex-linked recessive) has a frequency of 1 per 5000 males in a certain equilibrium population, what is the frequency of female hemophiliacs?

31-28. A population containing two sex-linked gene pairs, *A*, *a* and *B*, *b* (the effect of *A* is dominant to *a*, and *B* is dominant to *b*), shows the following *phenotypic* proportions among males: $AB = 40$ percent, $Ab = 20$ percent, $aB = 30$ percent, $ab = 10$ percent. (a) What are the expected equilibrium frequencies for these four types of males? (b) Will equilibrium be reached in one generation? (c) What is the expected equilibrium frequency for the *ab* phenotype among females?

31-29. In a London population of cats, Searle scored both males and females for the *yellow* genotype (see p. 747) and found the following:

	Number		
	$+/+$	$+/y$	y/y
females	277	54	7
males (hemizygotes)	311		42

(a) What are the gene frequencies in this population? (b) Is this population in Hardy-Weinberg equilibrium?

31-30. If the previous observations were unable to distinguish between $+/y$ and $+/+$ females, what results would you now obtain from a test for Hardy-Weinberg equilibrium?

31-31. Assuming red-green colorblindness is caused by a single sex-linked recessive allele: (a) What are the expected genotypic frequencies among the children produced by a population of 10 colorblind and 5 normal men mated to normal homozygous women, each mating producing the same size family (e.g., two children, one male and one female)? (b) What are the genotypic frequencies at equilibrium for this population?

31-32. Waaler tested 9049 school boys in Oslo for red-green colorblindness and found 8324 of them to be normal and 725 colorblind. He also tested 9072 school girls and found 9032 normal and 40 colorblind. Assuming that all forms of red-green colorblindness are caused by effects of the same sex-linked recessive allele, *c*, estimate whether this sample demonstrates equilibrium for *Cc*.

31-33. On closer analysis of the population in Problem 32, Waaler found that there was actually more than one *c* allele causing colorblindness in his sample, one kind for the "prot" type (c^p) and one for the "deuter" type (c^d), and that some of the "normal" females were probably of genotype c^pc^d. Through further analysis of the 40 colorblind females, he found that 3 were prot, c^pc^p, and 37 were deuter, c^dc^d. How does this finding affect your previous answer?

31-34. Hutt collected the following data from the families of Cornell University veterinary students:

Mating	Offspring	
	Tasters	Nontasters
taster × taster	654	76
taster × nontaster	354	205
nontaster × nontaster	7(?)	98

Calculate whether this population is in equilibrium with respect to genotypic frequencies.

REFERENCES

BOYD, W. C., 1950. *Genetics and the Races of Man*. Little, Brown, Boston.

DOBZHANSKY, TH., 1955. A review of some fundamental concepts and problems of population genetics. *Cold Sp. Harb. Symp.*, **20**, 1-15.

HARDY, G. H., 1908. Mendelian proportions in a mixed population. *Science*, **28**, 49-50. (Reprinted in the collections of Gabriel and Fogel and of Peters; see References, Chapter 1. Also found in the collections of Brosseau and of Morris, see below.)

HOPKINSON, D. A., N. SPENCER, and H. HARRIS, 1963. Red cell acid phosphatase variants: A new human polymorphism. *Nature*, **199**, 969-971.

LAI, L., S. NEVO, and A. G. STEINBERG, 1964. Acid

phosphatases of human red cells: Predicted phenotype conforms to a genetic hypothesis. *Science*, **145,** 1187–1188.

MATSUNAGA, E., and S. ITOH, 1958. Blood groups and fertility in a Japanese population, with special reference to intrauterine selection due to maternal-fetal incompatibility. *Ann. Hum. Genet.*, **22,** 111–131.

SNYDER, L. H., 1934. Studies in human inheritance. X. A table to determine the proportion of recessives to be expected in various matings involving a unit character. *Genetics*, **19,** 1–17.

TODD, N. B., 1964. Gene frequencies in Boston's cats. *Heredity*, **19,** 47–51.

WEINBERG, W., 1908. Über den nachweis der Vererbung beim Menschen. (Translated into English and reprinted in Boyer's collection; see References, Chapter 1.)

The following books provide derivations and discussions of many of the basic theoretical principles of population genetics covered in this chapter and in the chapters that follow.

CAVALLI-SFORZA, L. L., and W. F. BODMER, 1971. *The Genetics of Human Populations*. W. H. Freeman, San Francisco.

CROW, J. F., and M. KIMURA, 1970. *An Introduction to Population Genetics Theory*. Harper & Row, New York.

FALCONER, D. S., 1960. *Introduction to Quantitative Genetics*. Oliver and Boyd, Edinburgh.

JACQUARD, A., 1974. *The Genetic Structure of Populations*. Springer-Verlag, Berlin. (Translated from the French by D. and B. Charlesworth.)

KEMPTHORNE, O., 1957. *An Introduction to Genetic Statistics*, John Wiley, New York.

LI, C. C., 1955. *Population Genetics*. Univ. of Chicago Press, Chicago, Chaps. 1–10.

MATHER, W. B., 1964. *Principles of Quantitative Genetics*. Burgess, Minneapolis.

PIRCHNER, F., 1969. *Population Genetics in Animal Breeding*. W. H. Freeman, San Francisco.

RASMUSON, M., 1961. *Genetics on the Population Level*. Heinemann, London.

WRIGHT, S., 1969. *Evolution and the Genetics of Populations,* 2 vols. Univ. of Chicago Press, Chicago.

Numerous important papers in population genetics have been reprinted in:

SPIESS, E. B. (ed.), 1962. *Papers on Animal Population Genetics*. Little, Brown, Boston.

Other research papers relating to some of the topics on evolutionary genetics discussed in these chapters have been reprinted in:

BROSSEAU, G. E., Jr. (ed.), 1967. *Evolution: A Book of Readings*. William C. Brown, Dubuque, Iowa.

DAWSON, P. S., and C. E. KING (eds.), 1971. *Readings in Population Biology*. Prentice-Hall, Englewood Cliffs, N.J.

EHRLICH, P. R., R. W. HOLM, and P. H. RAVEN (eds.), 1969. *Papers on Evolution*. Little, Brown, Boston.

MORRIS, L. N. (ed.), 1971. *Human Populations, Genetic Variation, and Evolution*. Chandler, San Francisco.

32
CHANGES IN GENE FREQUENCIES

For gene frequencies to change, mutation must first introduce the innovation that leads to genetic differences. The mere appearance of new genes, however, is no guarantee that they will either persist or prevail over other genes. In fact, by very simple reckoning, a newly mutated gene will ordinarily have a very small chance of survival. Let us assume, for example, that a newly mutated gene, *a*, appears in a diploid "carrier" *Aa* that must mate in a population of homozygotes (*AA*). If this population is fairly constant in size, each mating between two individuals produces, on the average, two surviving offspring to compensate for the eventual loss of the parents. Of course, some matings are less and some are more fruitful, so that the number of offspring in each family ranges from zero to many.

The frequencies of the different family sizes can then be statistically described by a Poisson distribution which has a mean of 2 (Table 32–1). Note, however, that not all families have the same probability of preserving the mutant gene. For instance, a mating of the mutant carrier ($Aa \times AA$) that does not produce any children will result in the loss of the mutant gene. If only one child is produced from such a mating, the chance of losing the mutant gene is 1/2. In families with two children, the gene can be lost if both children are *AA*, giving a probability of loss of 1/4. As more surviving children are produced, the chances for the mutant gene to be lost decrease. Thus the chances for elimination of the mutant gene (frequency = 0) and "fixation" of the wild-type gene (frequency = 1) depend upon the frequency of each family size. As shown in Table 32–1, the total

TABLE 32–1

Calculation of the probability of loss in one generation of a mutant gene in a population whose families produce offspring according to a Poisson distribution with a mean of 2 (see Chapter 8)

number of offspring per family	0	1	2	3	4	5	
frequency of family	e^{-2}	$2e^{-2}$	$\frac{2^2}{2!}e^{-2}$	$\frac{2^3}{3!}e^{-2}$	$\frac{2^4}{4!}e^{-2}$	$\frac{2^5}{5!}e^{-2}$	
probability that mutant gene will not occur among offspring (probability of elimination)	1	1/2	1/4	1/8	1/16	1/32	
frequency of family × probability of elimination	e^{-2}	e^{-2}	$\frac{1}{2!}e^{-2}$	$\frac{1}{3!}e^{-2}$	$\frac{1}{4!}e^{-2}$	$\frac{1}{5!}e^{-2}$	
total probability of elimination	$e^{-2}\Big(1 +$	$1 +$	$\frac{1}{2!} +$	$\frac{1}{3!} +$	$\frac{1}{4!} +$	$\frac{1}{5!} +$	$\cdots\Big)$
			$= e^{-2}(e) = e^{-1} = \frac{1}{2.718} = .3679$				

probability for elimination in this first generation is .3679. The second generation will therefore begin with approximately a one-third likelihood that the mutant gene is no longer present among its parents. If we now add to this value the probability that this gene, even if present, might not be transmitted to the third generation, the likelihood of elimination rises to more than half ($e^{-(1-.3679)} = e^{-.6321} = .5315$). At the end of the third generation the probability of elimination is almost two-thirds ($e^{-(1-.5315)} = .6259$). By the time 30 generations have been reached, the probability of elimination has risen to almost 95 percent. Thus the chances that a new mutant gene will be lost is increased with each generation from the time of the initial mutation. After n generations, Fisher has shown that the probability that a gene will survive is approximately 2/n, when n is large. How then can we explain not only the persistence, but also the increase in frequency of many genes?

MUTATION RATES

One of the causes responsible for a change in gene frequency is the frequency of mutation itself. Obviously if gene *A* continually mutates to *a*, and the reverse mutation never occurs, the chances improve that *a* will increase in frequency with each generation. Given a long enough period of time and a persistent mutation rate in a population of constant size, *a* can eventually replace *A*. This is easy to see in a quantitative fashion if we begin by calling p_0 the initial frequency of *A*, and u the mutation rate of *A* to *a*. The first generation of mutation will cause the appearance of *a* in a frequency of $u \times p_0$. In other words, the frequency of *A* is now reduced to $p_0 - up_0 = p_0(1 - u)$. In the next generation new *a* genes arise in the frequency $u \times [p_0(1 - u)] = p_0(u - u^2)$, so the frequency of *A* is now $[p_0(1 - u)] - [p_0(u - u^2)] = p_0 - 2p_0u + p_0u^2 = p_0(1 - u)^2$. As the number of generations increases to n, the frequency of *A* becomes equal to $p_0(1 - u)^n$. Thus, even if the mutation rate of *A* to *a* is small, after a large enough number of generations the term $(1 - u)^n$ will approach zero and *A* will disappear.

Of course, the mutation rate does not always occur in only one direction. The allele *a*, for example, may mutate back to *A* with frequency v. If we call the initial frequencies of alleles *A* and *a* p_0 and q_0, respectively, a single generation of mutation will produce a frequency of *A* equal to $p_0 + vq_0$ and a frequency of *a* equal to $q_0 + up_0$. Let us confine our attention to only one of the alleles, *a*. We note that *a* has gained the fraction

up_0 but lost the fraction vq_0; in other words the *change* in the frequency of *a*, called *delta* q (Δq), can be expressed $\Delta q = up_0 - vq_0$. A simple examination of this equation shows that if p were relatively large and q small, Δq would be large and q would increase rapidly. As q then became larger and p became smaller, Δq would diminish in size. At a certain point, p and q would be "balanced" in relationship to their mutation frequencies, so that Δq would be zero. This point is known as the *mutational equilibrium* ($\hat{q}$) and can be expressed

$$(\text{zero } \Delta q) = 0 = up - vq$$
$$up = vq$$

However, since

$$p = 1 - q,$$

$$u(1 - q) = vq$$

or

$$u - uq = vq$$
$$u = uq + vq = q(u + v)$$

or

$$\hat{q} = \frac{u}{u + v}$$

When we apply the same procedure to the gene frequency of *A*, we obtain the equilibrium frequency $\hat{p} = v/(u + v)$, or $\hat{p}/\hat{q} = [v/(u + v)]/[u/(u + v)] = v/u$. It is easy to see that when there is equality between the mutation rates, that is, $u = v$, the equilibrium gene frequencies $\hat{p}$ and $\hat{q}$ will be identical. If the mutation rates differ, the equilibrium frequencies will also differ. For example, if $u = .00005$ and $v = .00003$, the equilibrium frequency $\hat{q}$ equals $5/8 = .625$, and $\hat{p} = 3/8 = .375$. The rate at which this equilbrium frequency is reached by mutation, however, is usually quite slow. It is derived through calculus from Δq and is given by the formula $(u + v)n = \log_e [(q_0 - \hat{q})/(q_n - \hat{q})]$*, where n is the number of generations required to reach a frequency q_n when starting with a frequency q_0. For the example just considered, the number of generations necessary for q to increase from a frequency of one-eighth to three-eighths is

$$(.00008)n = \log_e \frac{.125 - .625}{.375 - .625}$$
$$= \log_e 2.00 = .69315$$
$$n = \frac{.69315}{.00008} = 8664 \text{ generations}$$

Thus the approach to equilibrium based on the usually observed mutation rates of 5×10^{-5} or less is very slow, and equilibrium is probably rarely, if ever, reached, especially since mutation rates are subject to a variety of modifying genes (see p. 577) and are probably not constant. As a rule, therefore, the attainment of mutational equilibrium does not appear to be the sole cause that accounts for existing gene frequencies.†

SELECTION

One of the most important causes for the change in frequency of a gene lies in the ability of its carriers to produce surviving offspring. It is obvious that if individuals carrying the gene *A* are more successful in reproduction than individuals carrying its allele *a*, the frequency of the former will tend to be greater than that of the latter. The wide variety of mechanisms responsible for modifying the reproductive success of a genotype is known collectively as *selection*. Knowledge of certain aspects of selection dates back to antiquity and to the attempts of plant and animal breeders to modify the nature of organisms upon which

* $\log_e$ is the "natural" logarithm with a base $e = 2.718$. If one wishes to use "common" logarithms to the base 10, the equation can be written $(u + v)n = 2.303 \log_{10} [(q_0 - \hat{q})/(q_n - \hat{q})]$.

† On the molecular level, Kimura and others have pointed out that the chances for a reverse mutation to occur at a specific amino acid site on a protein is probably very small, and mutational equilibrium should not be expected. For example, a gene coding for 200 amino acids which has undergone one nucleotide substitution (e.g., A → C at position 20) has the potential for further mutation by substitution of three different alternative nucleotides for each of its 600 nucleotide positions. However, only one out of these 1800 possible mutations would provide an exact reversion to the previously existing codon (C → A at position 20). Since the mutation rate in most genes is already quite small (see Chapter 23), the chances of an exact nucleotide reversion is more than a thousand times smaller, and can safely be ignored as a factor enabling mutation rate equilibria to be reached for single nucleotides or single amino acids.

man is dependent. Since the reproductive success of these domesticated organisms is determined because the parents of each generation have been consciously chosen, this mechanism has been named *artificial selection*. By contrast, organisms whose reproductive success is not determined by human choice are considered to be subject to *natural selection*.

The mechanisms of selection may include such obvious factors as temperature and humidity fluctuations, changes in the availability of food, differences in sexual attraction, and also many more subtle influences that are quite difficult to detect. Whether its mechanisms are obvious or not, the widespread operation of selection can be observed by noting the many differences in reproductive success between varieties of an organism. For example, a mixture of barley seeds composed of many distinguishable varieties was sown in a number of locations throughout the United States. Each year these fields were harvested and seeds resown for the following year. After periods of 4 to 12 years, a census was taken of 500 randomly chosen plants from each location which were then classified according to variety. As shown in Table 32–2, the relative success of many varieties differed considerably in any one location. Furthermore, the relative success of any one variety showed marked changes when it was grown in different locations. Selection, as evidenced by the different frequencies of genotypes that survived, can thus be seen to have operated in every location, although differently in each. According to Darwin's usage, we may thus regard selection as a "scrutinizing" process that rewards vigorous varieties with reproductive success and helps explain how new varieties and species survive and increase.

FITNESS

Fitness has come to have a variety of applied meanings, such as fitness for sports, intellectual pursuits, business success—terms that generally refer to ability or aptitude. In its genetic sense, however, it is far more restricted, and refers only to *relative reproductive success*. Whether a genotype appears superficially "weak" or "strong," "beautiful" or "ugly," may be of no matter; only when a genotype can produce more offspring than another in the same environment, is its fitness superior. In this sense genotypes can also be described as having *adaptive* or *selective value*, which are merely other terms for reproductive success. How are these qualities measured?

In its simplest form this measurement consists of counting the number of offspring produced by one genotype as compared to those produced by

TABLE 32–2

Percentages of different barley varieties found in samples of 500 plants after 4 to 12 years of growth in five locations*

Variety	*Arlington, Virginia*	*St. Paul, Minnesota*	*Moccasin, Montana*	*Pullman, Washington*	*Davis, California*
A	89.2	16.6	17.4	30.0	72.4
B	2.6	3.0	11.6	.2	.2
C	1.2	2.8	5.0	1.0	0
D	2.2	5.4	7.4	.6	1.6
E	.8	0	.8	1.2	5.4
F	.8	.8	48.2	55.2	13.0
G	.8	61.0	3.8	6.0	6.8
H	2.2	10.0	1.6	4.6	.4
I	0	0	0	1.0	.2
J	.2	.4	4.2	.2	0

From data of Harlan and Martini, 1938.
*The initial population in each locale was founded from the same mixture of seeds of all varieties.

another. For example, if individuals of genotype *A* produce an average of 100 offspring that reach maturity while genotype *a* individuals produce only 90 in the same environment, the reproductive success of *a* is reduced by 10 offspring, or by the fraction 10/100 = .1.

Looked at in terms of fitness or adaptive value, *A* may therefore be considered to possess a superior adaptive value relative to *a*, which is now considered relatively "deleterious." If we designate the adaptive value of a genotype, *W*, as the proportion of offspring that survive to maturity relative to those of other genotypes, *W* can be chosen so as to fall in a range between 1.00 for the most productive genotypes and 0 for lethals. In the present case, therefore, the adaptive value of $A = W_A = 1.00$, and that for $a = W_a = .90$. The force acting on each genotype to reduce its adaptive value is defined by the *selection coefficient*, s. For the above example, s = 0 for *A*, and s = .1 for *a*. Thus the relationship between *W* and s is simply, $W = 1 - s$, or $s = 1 - W$. In many instances one genotype may be arbitrarily designated at the beginning of a study as having a fitness of 1.00 and the fitness of competing genotypes evaluated relatively. In such cases adaptive values may exceed the 1.00 value, as shown for certain of the human genotypes in Table 32–3.

GAMETIC SELECTION

Selection against a genotype may occur either in gametes (haploids) or zygotes (diploids). When selection occurs in gametes,* there is of course no difference between dominant and recessive genes, since both genotypes are phenotypically expressed. The population can then be considered to consist of *A* genotypes with a frequency p, and *a* genotypes with a frequency q. If the fitness of *A* is considered 1, and the fitness of its relatively deleterious allele *a* is reduced by a selection coefficient s, one generation of selection will cause the following changes:

	Genotypes		Total
	A	a	
initial frequency	p_0	q_0	1
fitness	1	$1 - s$	
after selection	p	$q(1 - s)$	$p + q - sq = 1 - sq$

If s is 1 (i.e., *a* is lethal), the frequency of *a* after selection is $q(1 - s) = q(0) = 0$, and *a* is thus eliminated in a single generation. If s is less than 1, *a* will persist, but its frequency will now be based on the new total that has been reduced by the amount sq. Therefore, the selected frequencies of *A* and *a* are now fractions of this new total, or $A = p/(1 - sq)$ and $a = q(1 - s)/(1 - sq)$. The change in the frequency of *a* (Δq) can then be calculated as q after selection (q_1) − q before selection (q_0), or

$$\Delta q = q_1 - q_0 = \frac{q(1-s)}{1-sq} - q$$
$$= \frac{q(1-s)}{1-sq} - \frac{q(1-sq)}{1-sq}$$
$$= \frac{q - sq - q + sq^2}{1-sq}$$
$$= \frac{-sq + sq^2}{1-sq} = \frac{-sq(1-q)}{1-sq}$$

If selection is proceeding slowly, s is small and sq in the denominator can be safely ignored. In other words, $\Delta q = -sq(1 - q)$, under these conditions. This relationship can then be summed for n generations by the methods of calculus to give

$$sn = \log_e \frac{q_0(1-q_n)}{q_n(1-q_0)} = 2.303 \log_{10} \frac{q_0(1-q_n)}{q_n(1-q_0)}$$

where q_0 is the initial frequency of the gene and q_n its frequency after n generations. For example, if a gene present in a haploid population with frequency .25 has a selection coefficient of .01

* Meiotic drive or segregation distortion (p. 219) may be considered as one form of gametic selection in which gametes carrying a particular gene show preferential survival. However, not too many instances of meiotic drive have yet been found and, when found, it is usually limited to gametes of only one sex.

TABLE 32–3

Estimates of relative fitnesses of certain human genotypes compared to normal homozygotes having a designated fitness of 1.00

Trait	*Population*	*Relative Fitness*
retinoblastoma (heterozygotes)		0
Tay-Sachs disease (homozygote)		0
achondroplasia (heterozygotes)	Denmark	.20
hemophilia (males)	Europe (average of 3 populations)	.29
neurofibromatosis (heterozygotes)	Michigan	
	males	.41
	females	.75
Huntington's chorea (heterozygotes)	Michigan	
	males	.82
	females	1.25
schizophrenia (genotype unknown)	U.S.	
	males	.49
	females	.78
sickle-cell anemia (heterozygotes)	East Africa (malarial regions)	1.26

From Spuhler, in Schull, 1963, and other sources.

acting against it, the number of generations necessary to reduce its frequency to .10 is

$$.01n = 2.303 \log_{10} \frac{.25(1 - .10)}{.10(1 - .25)}$$

$$= 2.303 \log_{10} \frac{.25(.90)}{.10(.75)}$$

$$= 2.303 \log_{10} 3 = 2.303(.47712) = 1.09861$$

$$n = \frac{1.09861}{.01} = 110 \text{ generations}$$

ZYGOTIC SELECTION

In most higher animals and plants selection takes place primarily in the diploid or zygotic stage. In diploids, however, there are three possible genotypes for a single gene difference (e.g., *AA*, *Aa*, *aa*), so that the effectiveness of selection depends, among other things, upon the degree of dominance. If complete dominance exists, and selection occurs only against the recessive *aa*, the heterozygote is shielded from selection, and harmful or even lethal recessive genes may persist for many generations.

For example, if we begin with a population in which *aa* is lethal but *AA* and *Aa* are not, the gene-frequency change (Δq) for *a* after one generation of selection is $-q^2/(1 + q)$ according to the algebraic calculations given in Table 32–4.* If q is initially large, its value will decrease rapidly. However, note that as q decreases to small fractions, the change in each generation is approximately equal to q^2, since the denominator $(1 + q)$ is not much different from 1. Thus, at low frequencies of *a* such as $q = .01$, the change in q per generation is only .0001, or a loss of only 1 out of 10,000 *a* genes. Selection against recessive genes with lethal effect is therefore quite inefficient when these genes are present in low frequency, since only very few recessive homozygotes are formed.

When the recessive homozygote is deleterious but not lethal, i.e., the selection coefficient is less than 1, the gene-frequency change in one genera-

* Algebraic proofs for obtaining the terms for various other gene frequency changes are provided in subsequent tables but, as in the present case, they are not essential to the main line of reasoning.

TABLE 32–4

Calculation of the change in frequency (Δq) of a gene (*a*) that is lethal in homozygous condition (s = 1)

	Genotypes			Total	Frequency of *a*
	AA	*Aa*	*aa*		
initial frequency	p^2	$2pq$	q^2	1	q
adaptive value	1	1	$1 - s = 0$		
after selection	p^2	$2pq$	0	$p(p + 2q) = p[(p + q) + q] = p(1 + q)$	$\frac{1}{2}\left[\frac{2pq}{p(1 + q)}\right] = \frac{pq}{p(1 + q)}$
relative frequency	$\frac{p^2}{p(1 + q)} = \frac{p}{1 + q}$	$\frac{2pq}{p(1 + q)} = \frac{2q}{1 + q}$	0		$= \frac{q}{1 + q}$

$$\Delta q = \frac{q}{1 + q} - q = \frac{q}{1 + q} - \frac{q(1 + q)}{1 + q} = \frac{q}{1 + q} - \frac{q + q^2}{1 + q} = -\frac{q^2}{1 + q}$$

tion is derived according to the calculations in Table 32–5. When *a* is rare, the product sq^2 is a very small number and the denominator for Δq can be considered 1. Under these conditions, sq^2 in the numerator is much larger than sq^3 and Δq is therefore essentially $-sq^2$. For a small q, therefore, the change in gene frequency per generation is quite small. For larger values of q (i.e., if s is small), the change in q is larger and reaches a maximum rate when $q = 2/3$.

These relationships are shown in Table 32–6 for a variety of selection coefficients acting on the recessive homozygote. The efficiency of selection is expressed in the form of the number of generations necessary to effect a change in gene frequency or to produce a corresponding change in the frequency of recessive homozygotes. Note that the initial change in gene frequency from .99 to .10 is relatively rapid for practically all selection coefficients. Further reductions in gene frequency

TABLE 32–5

Calculation of Δq for a gene (*a*) that is deleterious but not lethal in homozygous condition ($0 < s < 1$)

	AA	*Aa*	*aa*	Total	Frequency *a*
initial frequency	p^2	$2pq$	q^2	1	q
adaptive value	1	1	$1 - s$		
frequency after selection	p^2	$2pq$	$q^2(1 - s)$	$p^2 + 2pq + q^2 - sq^2 = 1 - sq^2$	
relative frequency	$\frac{p^2}{1 - sq^2}$	$\frac{2pq}{1 - sq^2}$	$\frac{q^2(1 - s)}{1 - sq^2}$		$\frac{pq + q^2(1 - s)}{1 - sq^2} = \frac{pq + q^2 - sq^2}{1 - sq^2} = \frac{q(p + q - sq)}{1 - sq^2} = \frac{q(1 - sq)}{1 - sq^2}$

$$\Delta q = \frac{q(1 - sq)}{1 - sq^2} - q = \frac{q(1 - sq)}{1 - sq^2} - \frac{q(1 - sq^2)}{1 - sq^2} = \frac{q - sq^2 - q + sq^3}{1 - sq^2} = \frac{-sq^2 + sq^3}{1 - sq^2} = \frac{-sq^2(1 - q)}{1 - sq^2}$$

TABLE 32–6

Number of generations necessary for a given change in q of a deleterious recessive gene under different selection coefficients

Change in Gene Frequency		Change in Frequency of Homozygotes		No. Generations for Different s Values						
From (q_0)	*To* (q_n)	*From* (q_0^2)	*To* (q_n^2)	s = 1 (*lethal*)	s = .80	s = .50	s = .20	s = .10	s = .01	s = .001
.99	.75	.980	.562	1	5	8	21	38	382	3,820
.75*	.50	.562	.250		2	3	9	18	176	1,765
.50	.25	.250	.062	2	4	6	15	31	310	3,099
.25	.10	.062	.010	6	9	14	35	71	710	7,099
.10	.01	.010	.0001	90	115	185	462	924	9,240	92,398
.01	.001	.0001	.000001	900	1,128	1,805	4,512	9,023	90,231	902,314
.001	.0001	.000001	.00000001	9,000	11,515	18,005	45,011	90,023	900,230	9,002,304

*The change in gene frequency is most rapid when $q_0 = .67$.

are considerably slower; to reduce the gene frequency below .01 may take thousands of generations, even when the selection coefficient is relatively high.

The reason for the relative inefficiency of selection against rare recessives is simply that most recessive genes are then present in heterozygotes where they are protected from selection. For example, when q equals .01, the frequency of homozygotes (q^2) equals .0001, but the frequency of heterozygotes (2pq) is approximately .02. This means that although only 1 per 10,000 offspring show the recessive trait, about 1 per 50 individuals are carrying it; i.e., the gene is almost entirely present in heterozygous condition. As shown in Table 32–7, the more rarely a gene is found in a population, the more frequently does it occur in heterozygotes compared to homozygotes.

TABLE 32–7

Genotype frequencies for different human diseases caused by recessive genes

Disease	Population	Gene Frequency (q)	Frequency of Homozygotes (q^2)	Frequency of Heterozygous Carriers (2pq)	Ratio of Heterozygous Carriers to Homozygotes ($2pq{:}q^2 = 2p{:}q$)
Congenital blindness	Pingalap (Caroline Islands)	.3	1 in 11	1 in 2.4	5:1
Sickle-cell anemia	Africa (some areas)	.2	1 in 25	1 in 3	8:1
Albinism	Panama (San Blas Indians)	.09	1 in 132	1 in 6	21:1
Ellis-van Creveld syndrome	Old Order Amish (Lancaster County, Pa.)	.07	1 in 200	1 in 8	26:1
Cystic fibrosis	U.S. (Conn.)	.032	1 in 1000	1 in 16	60:1
Tay-Sachs disease	Ashkenazi Jews (Northeastern Europe)	.018	1 in 3000	1 in 28	108:1
Albinism	Norway	.010	1 in 10,000	1 in 50	198:1
Phenylketonuria	U.S.	.0063	1 in 25,000	1 in 80	314:1
Cystinuria	England	.005	1 in 40,000	1 in 100	400:1
Galactosemia	U.S.	.0032	1 in 100,000	1 in 159	630:1
Alkaptonuria	England	.001	1 in 1,000,000	1 in 500	2,000:1

The selective situation can of course be reversed so that the dominant allele is selected against and the recessive is now favored. When this occurs selection will obviously be more effective, since the dominant gene is subject to selection in all genotypes in which it occurs. For example, should a dominant allele become lethal, its frequency is reduced to zero in a single generation. However, as the selection coefficient against the dominant allele decreases, replacement by the recessive is considerably slower. For the general case, selection against a dominant allele of gene frequency p results in a change of $-sp(1-p)^2/[1-sp(2-p)]$, as shown in Table 32–8. Note that if s is small, Δp is effectively equal to $-sp(1-p)^2$. Since $1-p$ is q and p is $1-q$, this means that Δp is now $-sq^2(1-q)$, or Δp is identical to Δq (Table 32–5) at low selection coefficients. Under these conditions we may apply the values of Table 32–6 but in reverse order. That is, for a selection coefficient of .10 in *favor* of a recessive allele, 90,023 generations are necessary to *increase* its frequency from .0001 to .001, or to reduce the frequency of the dominant allele from .9999 to .9990. Subsequent changes in frequency are more rapid as the favored recessive homozygote becomes more frequent.

When dominance of the advantageous allele is incomplete, heterozygotes will show the effect of a deleterious gene since the heterozygous phenotype is at least partially deleterious. If dominance is absent completely and the heterozygote has a phenotype exactly intermediate to that of the two homozygotes, its selection coefficient will be exactly half that of deleterious homozygotes. As shown in Table 32–9, the resultant change in gene frequency in one generation is almost identical to that for gametic selection (p. 760). In other words, the absence of dominance "uncovers" the genes and makes all of them available for selection, but not all genotypes are affected to the same degree. For selection coefficients of 20 percent or more, the gene-frequency changes are quite rapid and the frequency of the more deleterious gene is reduced to almost zero in relatively few generations. Lower selection coefficients shown in Table 32–10 do not have this drastic effect, but they do cause a rapid fall in gene frequencies in relatively few generations. Note that the values in Table 32–10 also indicate the number of generations necessary for increase in the frequency of the *favored* allele if dominance is absent. For example, 23 generations will produce a change from .9990 to .9999 in the frequency of a favored allele when the selection

TABLE 32–8

Calculation of the change in frequency (Δp) for one generation of selection against a deleterious gene (*A*) that is completely dominant

	AA	*Aa*	*aa*	*Total*	*Frequency A*
initial frequency	p^2	$2pq$	q^2	1	p
adaptive value	$1-s$	$1-s$	1		
frequency after selection	$p^2(1-s)$	$2pq(1-s)$	q^2	$1-sp(2-p)$*	
relative frequency	$\frac{p^2(1-s)}{1-sp(2-p)}$	$\frac{2pq(1-s)}{1-sp(2-p)}$	$\frac{q^2}{1-sp(2-p)}$		$\frac{p-sp\dagger}{1-sp(2-p)}$

$$\Delta p = \frac{p-sp}{1-sp(2-p)} - p = \frac{p-sp-p[1-sp(2-p)]}{1-sp(2-p)} = \frac{p-sp-p+sp^2(2-p)}{1-sp(2-p)}$$

$$= \frac{-sp+2sp^2-sp^3}{1-sp(2-p)} = \frac{-sp(1-2p+p^2)}{1-sp(2-p)} = \frac{-sp(1-p)^2}{1-sp(2-p)}$$

* Derived from adding the three postselection frequencies and equating $p^2+2pq+q^2=1$ and $p=1-q$ and $q=1-p$: $p^2(1-s)+2pq(1-s)+q^2 = p^2-p^2s+2pq-2pqs+q^2 = p^2+2pq+q^2-p^2s-2pqs = 1-p^2s-2p(1-p)s = 1-p^2s-2ps+2p^2s = 1-2ps+p^2s = 1-sp(2-p)$.

† The numerator is derived from adding the numerator of the *AA* frequency to half that of the *Aa* frequency as follows: $p^2(1-s)+pq(1-s) = p^2-p^2s+pq-pqs = p^2-p^2s+p(1-p)-sp(1-p) = p^2-p^2s+p-p^2-sp+sp^2 = p-sp$.

TABLE 32–9

Calculation of the change in frequency (Δq) in one generation of selection against a deleterious gene (*a*) that shows no dominance (i.e., its effect is additive)

	AA	*Aa*	*aa*	*Total*	*Frequency a*
initial frequency	p^2	$2pq$	q^2	1	q
adaptive value*	1	$1 - s$	$1 - 2s$		
frequency after selection	p^2	$2pq(1 - s)$	$q^2(1 - 2s)$	$1 - 2sq$†	$\frac{q - sq(1 + q)}{1 - 2sq}$‡

$$\Delta q = \frac{q - sq(1 + q)}{1 - 2sq} - q = \frac{q - sq(1 + q)}{1 - 2sq} - \frac{q(1 - 2sq)}{(1 - 2sq)} = \frac{q - sq - sq^2 - q + 2sq^2}{1 - 2sq} = \frac{-sq + sq^2}{1 - 2sq} = \frac{-sq(1 - q)}{1 - 2sq}$$

* The adaptive values can also be written as $AA = 1$, $Aa = 1 - hs$, $aa = 1 - s$, where h can have values from 1 (deleterious gene is completely dominant) to .5 (effect of deleterious gene is intermediate in heterozygotes, as above) to 0 (deleterious gene is fully recessive). The change in gene frequency according to this notation would then be:

$$\Delta q = \frac{-sq[h(1 - 2q)(1 - q) + q(1 - q)]}{1 - sq[2h(1 - q) + q]}$$

Note that Δq reduces to the values given in Tables 32–5 and 32–8 (p interchanges with q) when h is, respectively, 0 and 1.

† Derived from adding all genotypic frequencies and equating $p^2 + 2pq + q^2$ and $p + q$ to 1:

$$(p^2 + 2pq + q^2) - 2pqs - 2sq^2 = 1 - 2sq(p + q) = 1 - 2sq.$$

‡ Derived from adding the frequency of *aa* to .5 the frequency of *Aa*:

$$\frac{pq(1 - s)}{1 - 2sq} + \frac{q^2(1 - 2s)}{1 - 2sq} = \frac{pq(1 - s) + q^2 - 2q^2s}{1 - 2sq} = \frac{pq + q^2 - pqs - q^2s - q^2s}{1 - 2sq}$$

$$= \frac{q[(p + q) - s(p + q) - sq]}{1 - 2sq} = \frac{q - sq - sq^2}{1 - 2sq} = \frac{q - sq(1 + q)}{1 - 2sq}$$

coefficient against the deleterious allele is .10. Compare this to the 90,023 generations necessary for a similar change under conditions of complete dominance. The effectiveness of selection is therefore strongly dependent upon the degree to which the deleterious gene is expressed in the heterozygote. Since most recessive genes are believed to have some heterozygous expression, selection efficiency

TABLE 32–10

Approximate number of generations necessary for gene-frequency changes under different selection coefficients when dominance is absent and heterozygote fitness is exactly intermediate to the two homozygotes

Frequency Change				*No. Generations*						
For Deleterious Allele		For Favored Allele								
From (q_0)	*To* (q_n)	*From* (q_0)	*To* (q_n)	$s = 1$ (*lethal*)	$s = .80$	$s = .50$	$s = .20$	$s = .10$	$s = .01$	$s = .001$
.99	.75	.01	.25	3	4	7	17	35	350	3,496
.75	.50	.25	.50	1	1	2	5	11	110	1,099
.50	.25	.50	.75	1	1	2	5	11	110	1,099
.25	.10	.75	.90	1	1	2	5	11	110	1,099
.10	.01	.90	.99	2	3	5	12	24	240	2,398
.01	.001	.990	.999	2	3	5	12	23	231	2,314
.001	.0001	.9990	.9999	2	3	5	12	23	230	2,304

for or against them probably falls between the extremes of slow progress for complete dominance and rapid progress for absence of dominance.

HETEROZYGOUS ADVANTAGE

The examples of selection just considered always go in one direction, toward elimination of the deleterious allele and establishment or fixation of the favored allele. As long as the selection coefficient does not change, an equilibrium between favored and unfavored alleles is impossible without the occurrence of new mutations. Various conditions, however, permit the establishment of an equilibrium through which both alleles may remain indefinitely within the population provided the selection coefficients remain constant. One such condition is known as *overdominance*, and occurs when the heterozygote has superior reproductive fitness to both homozygotes. In general, if the heterozygote *Aa* has an adaptive value of 1 while the fitnesses of the homozygotes *AA* and *aa* are reduced by the selective coefficients s and t, respectively, the change in frequency of *a* in a single generation is calculated as in Table 32–11. When Δq is zero, equilibrium has been reached and there will be no further change in gene frequency. Note that there are three possible conditions that will cause the numerator $[pq(ps - qt)]$ to be equal to zero and therefore Δq to equal zero. The first two conditions occur when either p or q are zero. Under these conditions, however, both genes will not be present in the population at the same time, and "balance" or equilibrium will be absent. The third condition occurs when $ps = qt$, so that the numerator of Δq is $pq(0) = 0$. When this happens, the following relationships can be derived:

$$ps = qt$$

add qs *to both sides*	*add* pt *to both sides*
$ps + qs = qt + qs$	$ps + pt = qt + pt$
$s(p + q) = q(s + t)$	$p(s + t) = t(p + q)$

Now, since $p + q = 1$,

$$q = \frac{s}{s + t} \qquad p = \frac{t}{s + t}$$

It is easy to see that if s and t are constant values, both p and q will reach a stable equilibrium; i.e., if q departs from the equilibrium value, selec-

TABLE 32–11

Calculation of the change in frequency (Δq) of a gene (*a*) in one generation of selection under conditions in which the heterozygote has superior fitness to both homozygotes

	AA	*Aa*	*aa*	*Total*	*Frequency a*
initial frequency	p^2	$2pq$	q^2	1	q
adaptive value	$(1 - s)$	1	$(1 - t)$		
frequency after selection	$p^2(1 - s)$	$2pq$	$q^2(1 - t)$	$1 - p^2s - q^2t$*	$\frac{q(1 - qt)}{1 - p^2s - q^2t}$†

$$\Delta q = \frac{q(1 - qt)}{1 - p^2s - q^2t} - q = \frac{q(1 - qt) - q(1 - p^2s - q^2t)}{1 - p^2s - q^2t} = \frac{q - q^2t - q + qp^2s + q^3t}{1 - p^2s - q^2t}$$

$$= \frac{qp^2s + q^3t - q^2t}{1 - p^2s - q^2t} = \frac{qp^2s + q^2t(1 - p) - q^2t}{1 - p^2s - q^2t} = \frac{qp^2s + q^2t - pq^2t - q^2t}{1 - p^2s - q^2t}$$

$$= \frac{qp^2s - pq^2t}{1 - p^2s - q^2t} = \frac{pq(ps - qt)}{1 - p^2s - q^2t}$$

* $p^2 - p^2s + 2pq + q^2 - q^2t = (p^2 + 2pq + q^2) - p^2s - q^2t = 1 - p^2s - q^2t.$

† $\frac{pq + q^2(1 - t)}{1 - p^2s - q^2t} = \frac{pq + q^2 - q^2t}{1 - p^2s - q^2t} = \frac{q[(p + q) - qt]}{1 - p^2s - q^2t} = \frac{q(1 - qt)}{1 - p^2s - q^2t}$

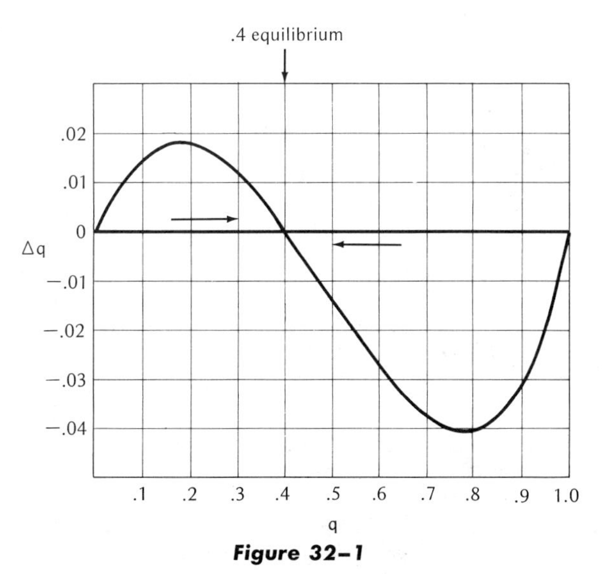

Figure 32–1

Change in the frequency of allele a (Δq) when the adaptive values are AA = .80, Aa = 1.00, aa = .70, and population size is infinite. These values provide a stable balanced polymorphism ($\Delta q = 0$) at q = .4. That is, Δq is positive (q increases) if q is less than .4 and negative (q decreases) if q is more than .4. If one allele is accidentally eliminated, i.e., q equals 0 or 1, Δq is of course zero, but polymorphism is lost. (After Li.)

tion pressure will force it back. This relationship arises from the change in q each generation. If Δq is positive, the gene frequency q increases. If Δq is negative, q decreases. The negative or positive sign of Δq depends on whether q is above or below its equilibrium value. For example, the equilibrium value for q when s = .20 and t = .30 is q = .20/(.20 + .30) = .40. When q is below this value, Δq is positive and acts to increase the value of q so that equilibrium is reached (Fig. 32–1). On the other hand, when q is above the equilibrium value, Δq is negative and thus acts to decrease the value of q toward the same equilibrium point. Both alleles therefore remain in the population at frequencies $\hat{q}$ and $\hat{p}$ so long as the selection coefficients confer superior fitness on the heterozygotes.

The maintenance of different genotypes in this fashion is an example of "balanced polymorphism," a term invented by Ford to describe the preservation of genetic variability (polymorphism) through selection. Experiments performed with *Drosophila* have extensively demonstrated balanced polymorphism for the presence of different allelic genes as well as for the presence of different chromosome arrangements. In *Drosophila melanogaster*, for example, L'Héritier and Teissier demonstrated the persistence of the mutant gene *ebony* in laboratory populations kept continuously for long periods of time. They suggested that it was the relative superiority of *ebony*/wild-type heterozygotes that accounted for balanced polymorphism in this case. In *Drosophila pseudoobscura*, Dobzhansky and Pavlovsky have shown that the frequencies of the Standard (ST) and Chiricahua (CH) 3rd-chromosome arrangements (see Fig. 22–16) will come to a stable equilibrium when flies carrying these arrangements are placed togther in a population cage that is kept continuously for a year or longer (Fig. 32–2). The superiority of the heterozygote is seen in the relative adaptive values calculated for this experiment: ST/ST = 0.895, ST/CH = 1.000, CH/CH = 0.413.

Even a small heterozygous advantage may enable genes that are otherwise lethal to remain in the population. For example, gene *a* lethal in homozygous condition but conferring a 1 percent advantage on the heterozygote compared to the *AA* homozygote would reach a frequency of approximately 1 percent at equilibrium:

	AA	**Aa**	**aa**
adaptive value	.99	1.00	0
selection coefficient	s = .01		t = 1

$$\hat{q}\ (\text{freq. of } a) = \frac{s}{s+t} = \frac{.01}{1.01} = .0099$$

Other conditions responsible for polymorphism may include a change in selection coefficients so that genes detrimental at one time are advantageous at another. Also, selection against a gene may depend upon its frequency and be reversed when it is at low frequency, before it can be eliminated. In some *Drosophila* populations a device that ensures such "frequency-dependent selection" is the increased sexual success of males of the rare genotype (see Petit and Ehrman). In plants the mechan-

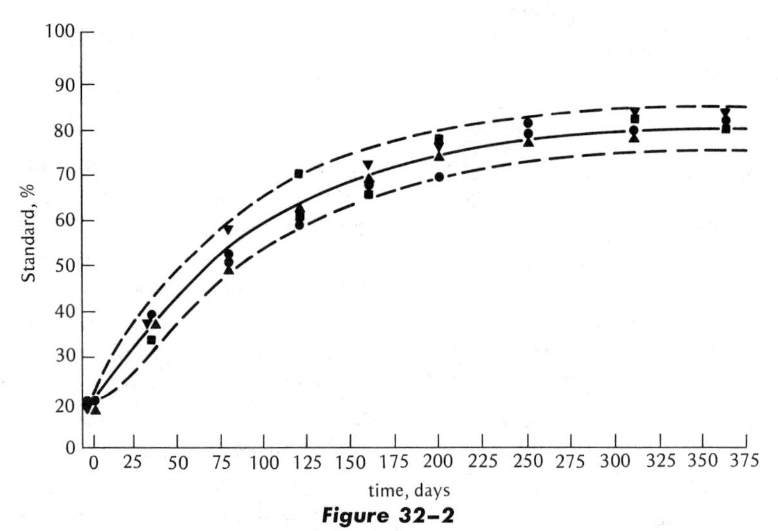

Figure 32–2

Results of four population-cage experiments with competing chromosomal arrangements of Drosophila pseudoobscura *in which each population was begun with 80% CH and 20% ST. Equilibrium values of 80 to 85% ST are reached after approximately 1 year. (After Dobzhansky and Pavlovsky, 1953.)*

ism that increases the number of self-sterility alleles in a population can also be considered frequency-dependent, since alleles that are rare can spread more rapidly than alleles that are common (p. 169). In addition polymorphism may also appear when selection coefficients vary from one environment to another. A population sufficiently widespread to occupy many environments may therefore maintain a variety of genotypes each of which is superior in a particular habitat. As pointed out by Levins, both the spatial and temporal organization of the environment may have significant effects on the extent to which a population will rely upon genetic polymorphism as an adaptive strategy. Further polymorphisms can arise because of interaction between gene loci so that all four alleles of two gene pairs (e.g., *Aa* and *Bb*) are maintained when the repulsion heterozygotes are superior to the coupling homozygotes (e.g., fitness is in the order $Ab = aB > AB = ab$). The maintenance of genetic variability may also result from differences in selection coefficients between the two sexes. The calculation of equilibrium and gene-frequency changes under some of these conditions is given by Li (see references Chapter 31) and derived from the work of Haldane, Fisher, and Wright.

In many of these instances selection will produce a departure from Hardy-Weinberg equilibrium by elimination of homozygotes. However, departures from Hardy-Weinberg equilibrium are not always caused by such factors, and an excess of heterozygotes does not necessarily mean that the heterozygote is of superior fitness in the population. As shown by Wallace (Table 32–12), a population beginning with 70 percent *A* and 30 percent *a*, for example, may have an intermediate adaptive value for the heterozygote yet produce an apparent excess of heterozygotes if a sample is only taken after selection has acted. Lewontin and Cockerham have also pointed out that observed genotypic values for a single generation will seem to correspond perfectly to expected Hardy-Weinberg equilibrium values when the adaptive values of *Aa*, *AA*, and *aa* are in the relationship $(W_{Aa})^2 = (W_{AA})(W_{aa})$ (e.g., $W_{Aa} = .5$, $W_{AA} = 1.0$, $W_{aa} = .25$). This apparent correspondence will occur in spite of the fact that selection is operating on these genotypes, and will therefore signify an erroneous equilibrium. The reason for these discrepancies between calculated adaptive values and real adaptive values is simply that final zygotic frequencies alone do not indicate the wide adaptive differences that may exist between genotypes. To obtain information on adaptive values it is also necessary to know the gametic or zygotic frequencies *before* selection. Only when gene frequencies are no longer changing and have reached equilibrium values can an accurate estimate of adaptive values be made from adult frequencies alone.

UNSTABLE EQUILIBRIUM

Not all polymorphisms, however, are stably balanced or at permanent equilibrium. They are considered "unstable" if any disturbance of equilibrium frequencies causes one of the genes to go to fixation. One type of unstable equilibrium is possible when selection acts against the heterozygotes at a gene locus with two alleles. If both homozygotes have equal adaptive value and the heterozygote is inferior, an equilibrium will be produced only when the frequency of each of the two alleles is exactly equal to .5. At this value the two genes are perfectly balanced, since the same proportion of each of the two genes is being removed in the heterozygote, i.e., .25 *AA*, .50 *Aa*, and .25 *aa*. However, any slight departure from this value will cause the less frequent allele to have proportionately more of its genes in heterozygotes than the more frequent allele. It will begin to lose proportionately more genes, and the other allele will go to fixation. For example, if the heterozygote were lethal and the gametic frequency of *A* rose accidentally to .6 and *a* fell to .4, random mating would yield gene frequencies of .69 *A* and .31 *a* (Table 32–13a). In the next generation the frequency of *A* would increase to .83, and the *a* frequency would fall to .17 (Table 32–13b). Within a relatively short time, the *A* gene would go to fixation and the *a* gene to elimination.

One example of selection against heterozygotes occurs when an Rh negative mother (*rr*) bears an Rh positive child (*Rr*) who then dies as a result of hemolytic disease (p. 172). As a consequence, we might expect an unstable equilibrium between these two Rh genes and the eventual elimination of one or the other. Actually this situation is far from simple, since selection against heterozygotes

TABLE 32–12

Production of an apparent excess of heterozygotes when observations are restricted to postselection genotypes in a population in which the heterozygote adaptive value is actually intermediate to both heterozygotes

	Gametes		Genotypes			
	A	*a*	*AA*	*Aa*	*aa*	Total
actual initial frequency	.7	.3	.49	.42	.09	1.00
actual adaptive value			.2	.6	1.0	
frequency after selection			.098	.252	.09	.44
relative frequency observed after selection	.509	.491	.223	.573	.204	1.00
expected frequency according to Hardy-Weinberg			.259	.500	.241	1.00
apparent excess of heterozygotes				.073		
apparent adaptive value if heterozygotes = 1.00 $\left(\frac{\text{obs.}}{\text{exp.}} \div \frac{\text{obs. } Aa}{\text{exp. } Aa}\right)$			.75	1.00	.74	

After Wallace.

TABLE 32–13

Change in gene frequency as a result of heterozygote lethality when initial gene frequencies are not equal (e.g., A = .60, a = .40)

a. First Generation ($AA = .60$, $aa = .40$)

	.60 *AA*	.40 *aa*
.60 *AA*	.36 *AA*	lethal *Aa*
.40 *aa*	lethal *Aa*	.16 *aa*

$$\text{frequency } A = \frac{.36}{.52} = .69$$

$$\text{frequency } a = \frac{.16}{.52} = .31$$

b. Second Generation ($AA = .69$, $aa = .31$)

	.69 *AA*	.31 *aa*
.69 *AA*	.476 *AA*	lethal *Aa*
.31 *aa*	lethal *Aa*	.096 *aa*

$$\text{frequency } A = \frac{.476}{.572} = .83$$

$$\text{frequency } a = \frac{.096}{.572} = .17$$

occurs only in Rh negative mothers and is inversely related to the frequency of the Rh negative gene. For example, if the Rh negative gene is common, we would expect the selection coefficient against the heterozygote to be high, since many mothers will be Rh negative. On the other hand, if the Rh negative gene is infrequent, few Rh negative mothers are found and the heterozygote has a lower selection coefficient. Furthermore, offspring of Rh negative mothers may be protected when their ABO blood types are incompatible (p. 174). Thus, although different human populations have different frequencies of these genes, the reasons for this may be quite numerous and have not yet been fully determined.

EQUILIBRIUM BETWEEN MUTATION AND SELECTION

Until now, changes in gene frequency have been considered to be caused by either mutation or selection acting separately. In nature, however, mutation and selection are going on simultaneously, and gene-frequency values are influenced by both factors. Predictions on the basis of one factor alone may therefore be quite misleading. For example, even though a recessive gene is detrimental in homozygous condition, it may nevertheless persist in a population because of its mutation frequency. That is, a certain equilibrium point is reached at which the number of genes being removed by loss of homozygotes is replaced by the same number of genes introduced into heterozygotes through mutation. We may determine this equilibrium frequency by the following argument.

We have seen that the change in gene frequency per generation for a deleterious recessive *a* with frequency q is equal to a loss of $sq^2(1-q)/(1-sq^2)$ per generation. If s is small, the denominator can be considered 1, and the loss in frequency is then $sq^2(1-q)$.* The frequency of newly mutated *a* genes, however, is equal to the mutation rate (u) of $A \rightarrow a$ multiplied by the *A* frequency, which is $1 - q$. Thus the loss of *a* genes through selection is exactly balanced by the gain of *a* genes through mutation when

$$sq^2(1-q) = u(1-q)$$
$$sq^2 = u$$
$$q^2 = \frac{u}{s}$$
$$q = \sqrt{\frac{u}{s}}$$

The equilibrium frequency of a mutant gene in a population is thus a function of both the mutation frequency and the selection coefficient. As would be expected, the formula shows that when the mutation rate increases, the gene frequency also will increase. In addition, when the selection coefficient increases, the gene frequency decreases. For a recessive that is lethal in homozygous condition ($s = 1$), the above expression reduces to $q = \sqrt{u}$, or $q^2 = u$. In other words, the frequency

*Because of the low frequency of *a*, the loss of *a* genes by back mutation of $a \rightarrow A$ is very small and can be ignored.

of homozygous lethals at equilibrium is about equal to the frequency of new genes introduced by mutation. (q will of course be larger than q^2 since most recessive genes under these conditions are in heterozygous rather than in homozygous form.)

For a deleterious dominant allele with frequency p, we have seen (p. 764) that the reduction in p under selection can be simplified to $sp(1 - p)^2$. Since q or $1 - p$ is the frequency of the recessive allele, the mutation rate of recessive to dominant, u, produces new dominant alleles at a frequency $u(1 - p)$. Equilibrium between selection and mutation then occurs when

$$sp(1 - p)^2 = u(1 - p)$$

$$p(1 - p) = \frac{u}{s}$$

Since small values of p can usually be expected when the dominant gene is selected against, $1 - p$ can be considered 1, and the equilibrium value of p is therefore equal to u/s. Note that when the dominant is lethal (i.e., $s = 1$) the frequency of the dominant in a population is equal to its mutation rate. As the selection coefficient decreases, the dominant frequency increases.

ESTIMATES OF MUTATION RATES AND EQUILIBRIUM FREQUENCIES

The mutation rate, if unknown, can be estimated for dominant genes from the relationship $p = u/s$, $u = ps$ if both p and s are known. For example, a Danish study of the dominant trait chondrodystrophic dwarfism showed a frequency of this gene of approximately 10 per 94,000 infants. Since each child has two genes and practically no homozygous dominants are expected at such low frequencies, the gene frequency p is $10/188{,}000 = .000053$. The selection coefficient was then calculated by comparing the number of children produced by chondrodystrophic dwarfs and those produced by their normal siblings. The data showed that 108 dwarfs produced 27 children, or the reproductive success of the heterozygous genotype was $27/108 = .250$. On the other hand, their 457 normal siblings produced 582 children, or the reproductive success of the normal genotype was $582/457 = 1.274$. Thus the fitness of the chondrodystrophic carrier relative to normal is $.250/1.274 = .196$, or the selection coefficient reducing its fitness to this value is $1 - .196 = .804$. The estimated mutation rate for chondrodystrophy is then $ps = (.000053)(.804) = .000043$. Not all cases of chondrodystrophy, however, are caused by the same dominant gene, and the actual mutation rates for the individual genes involved are probably lower.

When dominance is lacking, the reduction in gene frequency per generation for low values of q is very close to $sq(1 - q)$ (Table 32–9). Since the mutation frequency to the deleterious allele is $u(1 - q)$, the selection-mutation equilibrium is

$$sq(1 - q) = u(1 - q)$$

$$q = \frac{u}{s}$$

In other words, the equilibrium frequency of a deleterious gene in the absence of dominance is about the same as that for a deleterious dominant ($p = u/s$). Since such dominant or partially dominant genes are of considerable disadvantage to heterozygotes, Fisher has proposed that their deleterious effect is probably diminished in most organisms by selection of modifiers that change the degree of dominance. For example, mutant genes at a particular locus, A (e.g., $A^1, A^2 \cdots$), may act as partial dominants in the presence of the wild-type allele A^+. Since these mutant genes are mostly deleterious, modifiers that increase the dominance of A^+ will be selected until the effects of the mutations at this locus are relatively recessive. Evidence for this is seen in the successful selection for dominant and recessive modifiers demonstrated by Ford in the currant moth *Abraxas* (p. 215). Haldane has suggested that instead of modifiers, special wild-type alleles are selected (e.g., A^{X+}, A^{Y+}, A^{Z+}) that act as dominants in the presence of a mutant allele (e.g., $A^1, A^2, A^3, \cdots$). Further discussions of the evolution of dominance can be found in Merrell; Sved and Mayo; Murray; and others.

In general both mechanisms for producing dominance probably occur, as has been aptly demonstrated by Harland and others in two species of cotton, *Gossypium barbadense* and *G. hirsutum*. In these plants certain alleles show simple dominance when crosses are made between variants of the same species. Interspecific crosses, however, show the effect of numerous modifying genes on these traits, as well as differences in the degree of dominance of particular alleles. However, in spite of these dominance-producing mechanisms, many deleterious genes are probably still not completely recessive and seem to have some effect in heterozygous condition. Thus, in natural populations the equilibrium frequencies of deleterious genes are probably higher than for dominants but lower than for pure recessives.

MIGRATION

Mutation is not the only mechanism for introducing new genes into a population. A population may receive alleles by *migration* from a nearby population that maintains an entirely different gene frequency. When this occurs, two factors are of importance to the recipient population: the difference in frequencies between the two populations and the proportion of migrant genes that are incorporated each generation. The relationship between these factors can be expressed mathematically by designating q_0 as the initial gene frequency in the recipient or "hybrid" population, Q the frequency of the same allele in the migrant population, and m the proportion of newly introduced genes each generation. Since the proportion of replacement genes is m, the gene frequency in the hybrid population will suffer a loss of q_0 equal to mq_0 and a gain of Q equal to mQ. The first-generation hybrid frequency will therefore be $q_0(1 - m) + mQ$. As shown in Table 32–14, the difference in gene frequency between both populations is $(1 - m)(q_0 - Q)$. Each generation of migration then causes the preceding q of the hybrid population to be multiplied by the factor $(1 - m)$ and adds an additional increment of mQ. When the gene frequency of the hybrid population becomes q_n, where n is the number of generations of migration, the following relationship can be noted:

$$q_n - Q = (1 - m)^n(q_0 - Q)$$

$$(1 - m)^n = \frac{q_n - Q}{q_0 - Q}$$

Since this equation depends upon five factors (q_0, q_n, Q, m, n), knowledge of four factors enables us to derive the fifth one arithmetically. For example, one of the Rh alleles (R^0) which was used in the calculations of Glass and Li (1953) showed a frequency of .028 for American whites and .446 for American blacks. Can these values, together

TABLE 32–14

Effect of migration on the gene-frequency differences between the recipient (hybrid) population and the donor (migrant) population

	Gene Frequency in Population		
Generation	Hybrid	Migrant	**Gene Frequency Difference Between Hybrid and Migrant**
0	q_0	Q	$q_0 - Q$
1	$q_0(1 - m) + mQ = q_0 - mq + mQ$	Q	$q_1 - Q = q_0 - mq_0 + mQ - Q$ $= (1 - m)(q_0 - Q)$
2	$(q_0 - mq_0 + mQ)(1 - m) + mQ$	Q	$q_2 - Q = q_0 - 2mq_0 + m^2q_0 + 2mQ - m^2Q - Q$ $= (1 - m)^2(q_0 - Q)$
⋮			
n	$(q_{n-1})(1 - m) + mQ$	Q	$q_n - Q = (1 - m)^n(q_0 - Q)$

with others to be determined, be used in deriving the degree of gene exchange (m) between the two populations? Although some black genes undoubtedly enter the white population, the white population is so large that the introduction of black genes probably makes very little difference in gene frequency. On the other hand, the black population is much smaller and has remained isolated from its African origin for two or more centuries. On this basis the white population can be considered as the gene donor or migrant population (Q) and the present black population as the hybrid (q_n). To obtain the original gene frequency of R^0 in the black population (q_0), data of present East African blacks were used on the assumption that these data may reflect the original gene frequencies of 200 to 300 years ago. Among the East Africans, R^0 showed a frequency of .630, indicating a reduction in the frequency of this gene in American blacks because of interbreeding with the white population. Glass and Li estimated that this reduction had begun at the time of the initial introduction of blacks into the American colonies 300 years ago and continued throughout the 10 generations since. The formula derived above may then be written

$$(1 - m)^{10} = \frac{q_{10} - Q}{q_0 - Q} = \frac{.446 - .028}{.630 - .028} = .694$$

$$1 - m = \sqrt[10]{.694}$$

$$1 - m = .964^*$$

$$m = .036$$

This value of m means that, excluding all other causes such as mutation, 36 genes per 1000, or 3.6 percent of genes in the black population, were introduced from the white population each generation. Since 1 − m represents the proportion of nonintroduced genes, $(1 - m)^{10} = .694$ is the proportion of genes that have remained of African origin over the 10-generation period. Thus, according to these estimates, the American black population is genetically about 70 percent African and 30 percent white. Migration rates that somewhat approximate these values have been obtained from more recent data and, as shown in Table 32–15, these data also indicate that migration rates probably vary in different American localities.

*The tenth root is obtained by dividing the log of the number by 10 and then obtaining the antilog; i.e., $10\sqrt{9.84136 - 10} = \bar{1}.98414$, antilog = .964.

TABLE 32–15

Gene frequencies found in American blacks in two localities (one south, one north), American whites in one locality, and African blacks living in areas from which slaves were imported into the United States.

Gene	Blacks (Africa)	Blacks (Claxton, Georgia)	Blacks (Oakland, Calif.)	Whites (Claxton, Georgia)
R^0	.617	.533	.486	.022
R^1	.066	.109	.161	.429
R^2	.061	.109	.071	.137
r	.248	.230	.253	.374
A	.156	.145	.175	.241
B	.136	.113	.125	.038
M	.474	.484	.486	.507
S	.172	.157	.161	.279
Fy^a	.000	.045	.094	.422
P	.723	.757	.737	.525
Jk^a	.693	.743		.536
Js^a	.117	.123		.002
T	.631	.670		.527
Hp^1	.684	.518		.413
$G6PD$	.176	.118		.000
Hb^S	.090	.043		.000

From Adams and Ward.

RANDOM DRIFT

The three forces considered up to now, mutation, selection, and migration, have one important quality in common; they usually act in a *directional* fashion to change gene frequencies progressively from one value to another. When unopposed, these forces can lead to fixation of one allele and elimination of all others, or, when balanced, they can lead to equilibrium between two or more alleles. However, in addition to these directional forces, there are also changes that have no predictable constancy from generation to generation. One of the most important of such *nondirectional* forces arises from variable sampling of the gene pool each generation and is known as *random genetic drift*.

Genetic drift is caused by the fact that real populations are limited in size rather than infinite, so that gene-frequency changes occur because of sampling errors. This is easy to see if we consider that when the number of parents of a population is consistently large each generation, there is always a strong likelihood of obtaining a good sample of the genes of the previous generation as long as the directional forces are not acting to change them. On the other hand, if only a few parents are chosen to begin a new generation, such a small sample of genes may deviate widely from the gene frequency of the previous generation.

The extent of the deviation in both cases can be measured mathematically by the standard deviation of a proportion $\sigma = \sqrt{pq/N}$. Here p is the frequency of one allele, q of the other, and N the number of genes sampled. For diploid parents, each carrying two genes, $\sigma = \sqrt{pq/2N}$, where N is the number of actual parents. For example, if we begin with a large diploid population, where $p = q = .5$, and continue this population each generation by using 5000 parents, then $\sigma = \sqrt{(.5)(.5)/10{,}000} = \sqrt{.000025} = .005$. The values of such populations will therefore fluctuate mostly (68 percent of the time, see Fig. 8–7) around $.5 \pm .005$, or between .495 and .505. On the other hand, a choice of only two parents as "founders" will produce a standard deviation of $\sqrt{(.5)(.5)/4} = \sqrt{.0625} = .79$, or values of $.5 \pm .79$ ($-.29$ to 1.29). In other words, sampling accidents because of smaller population size will easily yield gene frequencies that are either 0 or 1. Since gene-frequency limits are also 0 or 1, sampling accidents because of small population size may easily result in fixation of one or the other of the alleles. If such small sizes are continued each generation, the likelihood increases that such a population will eventually reach fixation for an allele. This change in gene frequency can thus arise in the absence of any of the directional forces previously considered.

The attainment of fixation, however, does not mean that every small population will be fixed for the same allele. If many such small populations are considered together, the average gene frequency for all combined may remain fairly constant, although the individual populations involved have reached fixation. This can be illustrated by considering 96 such populations, each begun with two heterozygotes ($Aa \times Aa$) and then continued separately each generation through a random choice of only two parents. The offspring of the first generation in each population will be *AA*, *Aa*, *aa*, in the proportions 1:2:1. Since only two parents are chosen for the next generation, the probability of any particular mating combination depends upon the frequencies of the genotypes. As

TABLE 32–16

Probabilities of selecting given mating combinations in a population in which the three genotypes *AA*, *Aa*, and *aa* are in the ratio 1:2:1

Mating	Probability	Gene Frequencies in Mating Parents	
		A	*a*
$AA \times AA$	$1/4 \times 1/4 = 1/16$	1	0
(2) $AA \times Aa$	$2 \times 1/4 \times 1/2 = 1/4$	.75	.25
(2) $AA \times aa$	$2 \times 1/4 \times 1/4 = 1/8$	.50	.50
$Aa \times Aa$	$1/2 \times 1/2 = 1/4$	.50	.50
(2) $Aa \times aa$	$2 \times 1/2 \times 1/4 = 1/4$	.25	.75
$aa \times aa$	$1/4 \times 1/4 = 1/16$	0	1

shown in Table 32–16, there is a 1/16 chance of choosing two parents both *AA* and a 1/16 chance of choosing two parents both *aa*. Thus, for 96 populations, approximately 2/16, or 12, will probably reach fixation for one of the two alleles in the first generation, some for *A* and some for *a*. Among the remaining 84 populations there will be four mating combinations, *AA* × *Aa*, *AA* × *aa*, *Aa* × *Aa*, and *Aa* × *aa*. Three of these combinations may again produce new parental mates which are both homozygous for the same allele. In all, 15 such homozygous populations will probably appear in the second generation, or a frequency of 15/84 = 18 percent. In the next generation the drifting process will continue further among the remaining heterogeneous populations and a greater proportion of these will be added to those that have reached fixation. The proportion of populations attaining fixation, i.e., the "rate of fixation," will eventually become stable at 1/2N, where N is the actual number of parents used to begin a generation. Eventually all 96 populations will consist entirely of groups that have reached fixation, but their frequencies will be approximately evenly divided between groups of *AA* and groups of *aa*.

From the formula for the rate of fixation, 1/2N, it is obvious that if N is large, fixation proceeds very slowly. Given a long enough period of time, however, even relatively large populations will show some degree of drift. Wright has calculated the degree of fixation for different-sized populations under a variety of conditions. One set of relationships is shown in Fig. 32–3 for different population sizes all with an initial gene frequency of $p = .5$ when a small percentage of genes at this same frequency is introduced by migration ($m = .0001$). When the population size is very large ($N = 100{,}000$) there is very little drift over a long period of time and most frequencies cluster around the initial value $q = .5$. As population sizes decrease, values further from $q = .5$ are found, until when $N = 1000$, many populations have reached fixation for one of the two alleles.

The tendency toward fixation of an allele by random drift may of course be counteracted by selective forces acting to eliminate it. However, for alleles that are neutral in their effect, the chances of fixation seem to depend mostly on their mutation rate. This has been pointed out by Kimura and co-workers who reason that a population of N individuals bearing 2N genes will have $2N\mu$ newly arisen mutations if the chances for each gene to mutate is μ. If all or most of these mutant genes are of equal neutral effect, each will be present in about 1/2N frequency and will probably persist in this frequency because one is no better than the other. Thus there are a total of $2N\mu$ neutral mutations, each with a 1/2N chance of fixation (and a $1 - 1/2N$ chance of elimination), or the chance of fixation for a particular neutral gene is $2N\mu \times 1/2N = \mu$.

That is, the chances for fixation of neutral alleles in populations appear to be independent of population size. Kimura therefore suggests that if their collective mutation rate is substantial, neutral alleles may be widely prevalent in all populations and account for the large amount of genetic variability that is often present. (There is considerable controversy over this matter, and it will be further discussed in Chapter 34.)

However, on an individual basis the *time* necessary to attain fixation for any particular neutral allele will depend on population size, since the random drift process in small populations greatly speeds up the fixation of these genes, although it does not change their probability of fixation. Kimura and Ohta have calculated that the average number of generations necessary for the fixation of a new neutral mutation is approximately four times the number of parents in each generation.

The number of parents is therefore an important value for the determination of drift and is called *effective population size* (N_e). It differs from the observed population size because not all members of a population are necessarily parents. If, out of a total population of 1000, there are 300 mating pairs that contribute equal amounts of offspring to the next generation, the effective size is only 600. Population size is also reduced if one of the sexes is present in small numbers, so that the parentage of the next generation is limited by this sex. For example, a population of 3 males mated to 300 females will have an effective size somewhat

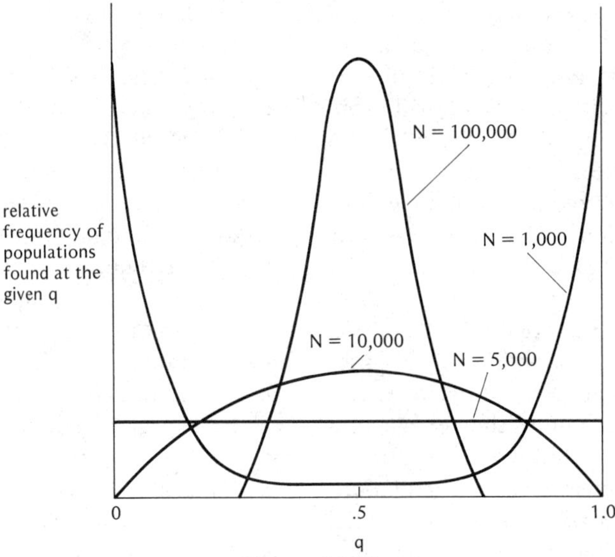

Figure 32–3

Distribution of equilibrium gene frequencies for populations of different sizes when selection is zero and a small amount of migration occurs into the population (m = .0001) from a population whose gene frequency is q = .5. In spite of this migration, populations of relatively small size (N = 1000, N = 5000) show a considerable amount of random drift, many reaching elimination (q = 0.0) or fixation (q = 1.0). Only populations of large size (N = 10,000, N = 100,000) maintain the initial gene frequency q = .5 in appreciable proportions. (After Wright.)

more than 3 but considerably less than 300. The relationship has been expressed by Wright as $N_e = 4N_fN_m/(N_f + N_m)$, where N_f is the number of parental females and N_m the number of parental males. In the above case N_e would be $4(300)(3)/303 = 11$. Inequalities between the contributions of different parents to the next generation will also reduce the effective population size.

As a result of these considerations, Wright has proposed that, under certain circumstances, genetic drift in populations of small effective size may produce notable changes in gene frequency between one population and another. At times such drift may take place even in the face of selection or of migration pressures from other populations (Fig. 32–3). Drift has been used to explain differences found among snails in Pacific Islands. As indicated by Gulick in studies beginning in 1872, almost every valley or even parts of a valley on some islands may contain snails of the same species that diverge widely in respect to such shell characteristics as color, size, shape, and type of coiling. Environmental differences between these valleys are not obvious, and it is difficult to explain all these divergences because of selection. The geographical breakdown of a slow-moving snail species into small populations therefore makes it likely that in addition to other factors these populations have diverged as a result of random genetic drift. Support for this point of view has been gained from studies of European *Cepaea* snail populations by Lamotte and by Goodhart, although other interpretations of such phenomena have also been presented (see Cain and Sheppard, and Clarke).

In addition to the persistence of small population size over many generations, genetic drift can also be caused only by occasional changes in size. At the extreme of such reductions, termed the *founder principle* by Mayr, a population may occasionally send forth only a few founders to begin a new population. Whatever genes or chromosome arrangements these founders take with them, detrimental or beneficial, all stand a good chance of becoming established in the new population because of this sudden sampling accident. Thus Carson, by careful analysis of salivary chromosome banding patterns, has recently shown that native *Drosophila* species on the island of Hawaii, the youngest of the Hawaiian islands, must have originated from one or few individual founders emigrating from the nearby island of Maui. These single founders provided unique chromosome arrangements that could be traced in the descendant species.

In experimental populations, genetic drift via the founder principle was demonstrated by Dobzhansky and Pavlovsky for *Drosophila pseudoobscura* carrying different 3rd-chromosome gene arrangements. They showed that when a number of populations are begun with the same frequency of a particular arrangement (PP) but differing in the number of founding individuals, the variability between the resultant populations for this chromosome arrangement depends upon the number of founders. As seen in Fig. 32–4, when populations begin with 4000 individuals, the frequency of PP progresses toward equilibrium in a fairly uniform fashion among the populations. On the other hand, populations begun with only 20 founding individuals show a considerably greater amount of variability in PP frequencies.

Among human populations historical instances of such unique sampling events may occasionally be traced to small groups of settlers who depart from a large population and begin their own. For example, Roberts has shown that the 267 individuals who were living on the Atlantic Ocean island of Tristan da Cunha in 1961 can be traced through records and pedigrees to relatively few ancestors. As shown in Table 32–17, the two initial founders of the population who settled there in 1816 (W. G. and his wife, M. L.) contributed more than 25 percent of the genes that were present in the popu-

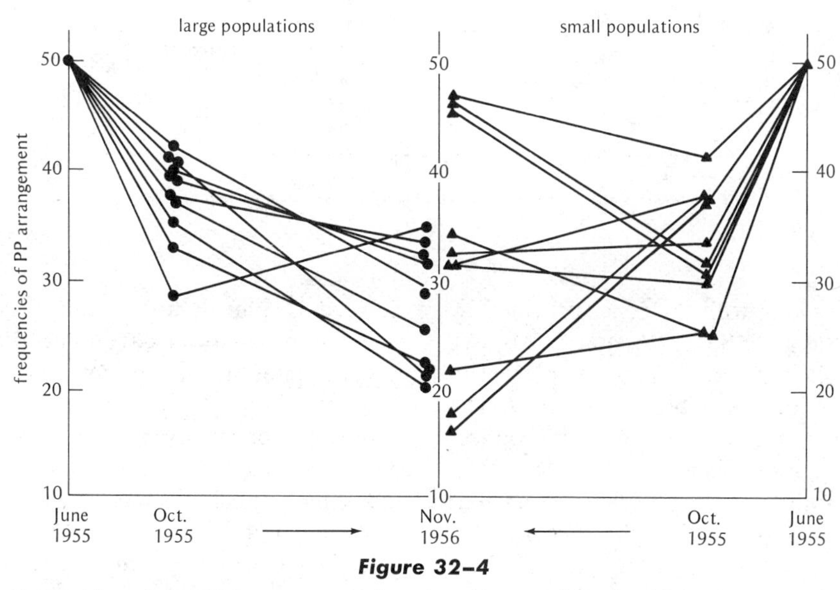

Figure 32–4

Frequencies of the PP arrangement in competition against the AR arrangement in Drosophila pseudoobscura *population cages begun with large population samples (left) and small population samples (right). (After Dobzhansky and Pavlovsky, 1957.)*

TABLE 32–17

Fractions of the human gene pool in Tristan da Cunha believed to have been contributed by 26 ancestors in designated years. Below each year, the size of the population is given in parentheses*

Ancestor	*Year*				
	1855 (103)	*1857 (33)*	*1884 (106)*	*1891 (59)*	*1961 (267)*
W. G.	0.1275	0.0625	0.0548	0.0657	0.0691
M. L.	0.1373	0.0625	0.0548	0.0657	0.0691
T. S.	0.0662	0.1389	0.0943	0.1441	0.1339
S. W.	0.0907	0.1910	0.1392	0.1864	0.1602
F. M. W.	0.0245	0.0521	0.0448	0.0424	0.0263
R. R.	0.0637	—	—	—	—
S. K.	0.0858	—	—	—	—
T. R.	0.0049	0.0139	0.0177	0.0424	0.0382
M. W.	0.0637	0.1389	0.1380	0.0805	0.0424
A. C.	0.0637	0.1389	0.1285	0.0636	0.0424
P. G.	0.0490	0.1042	0.0896	0.0847	0.0526
C. T.	0.0490	—	—	—	—
P. M.	0.0441	—	—	—	—
W. D.	0.0588	—	—	—	—
M. F.	0.0049	—	—	—	—
G.	0.0049	—	—	—	—
A. H.	0.0245	0.0833	0.0684	0.0890	0.0365
F. R. C.	0.0049	0.0139	—	—	—
F. F. K.	0.0221	—	—	—	—
B.	0.0098	—	—	—	—
J. B.	—	—	0.0472	—	—
S. P.	—	—	0.0613	0.1186	0.1045
F. R.	—	—	0.0283	0.0170	—
R. A. B.	—	—	0.0142	—	—
M. J.	—	—	0.0142	—	—
F. S. G.	—	—	0.0047	—	—

After Roberts.
* The contributions of 8 settlers who arrived after 1891 have been omitted.

lation by 1855. Along with the genes provided by three other settlers (T. S., S. W., and F. M. W.), approximately 45 percent of the genes present in 1961 can be ascribed to only five ancestors. The lack of uniformity in the amount of genes contributed by different ancestors is even further emphasized if we take into account the sharp reductions in the size of this population that occurred by chance in 1856 and 1885. After each of these "bottlenecks," there were marked effects on the composition of the gene pool caused by relative changes in the contributions of certain ancestors. Since it is difficult to attribute these changes in population size to selective forces, their effects on the gene pool can be considered mostly or entirely random.

In the prior history of man, as well as in many other migrating organisms, it is therefore likely that some populations were begun with only a few "Adams" and "Eves" carrying unique genes or gene frequencies. Certainly the relatively high incidences of some genes listed previously in Table 32–7, such as congenital blindness among the Pingalapese, and Ellis-van Creveld syndrome (polydactylous dwarfism) among the Lancaster County Amish are difficult to justify except as founding accidents, since they appear to confer no advantage on either their homozygous or heterozygous carriers.

However, the occurrence of such rare founding events is not as predictable as the genetic drift occurring in populations that remain constant in size. In other words, variability in gene frequency (i.e., σ) can be estimated more easily if population size is constant than when sudden temporary reductions occur. Other factors that cause the variability of gene frequencies to be unpredictable are unique historical events such as a change in the direction or intensity of selection because of a change in environment, an unusual favorable mutation, rare hybridization, and an unusual swamping by mass immigration. Although the effect of some of these factors on evolutionary changes may be of considerable importance, they are quite difficult to evaluate unless data are gathered at the time of their occurrence.

PROBLEMS*

32-1. If a gene mutates at a rate of 1×10^{-6} for only a single generation in a population of 10 million diploid individuals, how many such mutant genes will probably be passed on to the next generation if the population size remains constant?

32-2. Is the probability for the elimination of a single new mutant gene increased or decreased when the population size is diminishing? (Hint: Assume a Poisson distribution with a mean productivity per family of one child rather than two.)

32-3. If a gene has a mutation rate, $A \rightarrow a$, of 1×10^{-6} and back-mutation is absent, what will be the A frequency after (a) 10, (b) 100, (c) 1000, (d) 10,000, (e) 1,000,000 generations of mutation?

32-4. How many generations will it take for the a gene in Problem 3 to rise in frequency from .1 to .5?

32-5. A certain stock of *Drosophila* shows a mutation rate for normal (w^+) to eosin eye (w^e) of 1.3×10^{-4}, and a reverse mutation rate $w^e \rightarrow w^+$ of 4.2×10^{-5}. (a) What is the equilibrium value of w^e? (b) How many generations would it take for w^e to increase from .1 to .5?

32-6. Let us assume that the dominant mutation brachydactyly (short fingers) is not selected against, and that new brachydactylous genes arise with frequency 1×10^{-6}. On the other hand, the mutation rate in the reverse direction, brachydactyly $\rightarrow$ normal, is also 1×10^{-6}. What is the equilibrium frequency of this mutant gene?

32-7. Calculate the change in gene frequencies for a single generation of selection (Δq) when a gene has the following initial frequencies: .01, .10, .30, .50, .70, .90, .99 for each of the following three conditions: (a) The population is haploid and the selection coefficient against the gene is .30. (b) The population is diploid and there is a selection coefficient of .30 against the homozygous recessive genotype. (c) The population is diploid, but dominance is completely lacking and the selection coefficient is again .30. (d) Plot all your results on a single graph.

32-8. Dobzhansky and Pavlovsky found that a population cage of *Drosophila pseudoobscura* carrying two 3rd-chromosome gene arrangements, Arrowhead (AR) and Chiricahua (CH), showed the following fitness values for the combinations AR/AR, .75; AR/CH, 1.00; CH/CH, .42. (a) If they began this population with .20 AR and .80 CH, what are the expected frequencies of these arrangements in the next generation? (b) What are the expected equilibrium frequencies of AR and CH? (c) Assuming that selection occurs only among the larvae, what equilibrium frequency of heterozygotes (AR/CH) and homozygotes (AR/AR, CH/CH) would you expect to find in the population among the eggs? Among the adults?

32-9. In a certain locality a mutant gene is found to be lethal when homozygous but nevertheless enhances the fitness of the heterozygotes relative to the nonmutant homozygotes. In fitness comparisons the heterozygotes for this gene produce, on the average, twice as many offspring as the nonmutant homozygotes. (a) If the frequency of the mutant gene is .20 in a particular generation, what will its frequency be in the next? (b) If its frequency is .33, what will be the new frequency in the next generation and in the one following that?

* In these problems assume that only the stated factors are operating to change gene frequencies.

32-10. Da Cunha crossed homozygous *EE Drosophila polymorpha* (dark-colored abdomen) to a homozygous *ee* strain (light-colored abdomen) and obtained an F_1 generation that was phenotypically intermediate to both parents. A cross of $F_1 \times F_1$ produced 1605 *EE*, 3767 *Ee*, and 1310 *ee*. (a) Calculate the relative adaptive values of each genotype. (b) If the selection coefficients remained unchanged over many generations, what would be the equilibrium frequency of *e*?

32-11. In three different populations the relative fitnesses of genotypes differing in respect to one pair of genes, *A* and *a*, was found to be as follows:

Population	Relative Fitnesses AA	Aa	aa
1	.50	1.00	.00
2	.75	1.00	.50
3	.90	1.00	.80

(a) Assuming that only selection is operating on these genotypes, what are the equilibrium frequencies for the genes *A* and *a* in each of these populations? (b) If all three of these populations began with exactly the same gene frequencies, $A = .400$, $a = .600$, explain which of these populations would approach their equilibrium gene frequencies fastest?

32-12. (a) Assuming random mating, calculate the frequency of Rh negative mothers bearing Rh positive children if the frequency of the rh^- gene (r) is .1, .5, .9. (b) If all Rh positive children born to such mothers are eliminated because of the incompatibility, what would be the new frequency of the *r* gene in each of these three cases? (c) At which of these three frequencies would there be a stable equilibrium?

32-13. A randomly breeding population has the folloing ABO and Rh gene frequencies:

I^A	I^B	i^O	R	r
.30	.10	.60	.80	.20

(a) What proportion of the pregnancies in such a population would you expect to involve an rh^- mother and an Rh^+ fetus? (b) In what proportion of those cases in which the mother is rh^- and the fetus Rh^+ would you expect that some protection against Rh disease would be afforded the fetus by the ABO system (see p. 174)?

32-14. The following two genetically caused diseases are considered lethal since affected individuals do not reach reproductive age: retinoblastoma, caused by a dominant gene; infantile amaurotic idiocy, caused by a recessive gene in homozygous condition. (a) Assuming that both genes arise at mutation rates of about 3×10^{-6}, what are their respective equilibrium frequencies? (b) What frequency of homozygotes is expected for each gene?

32-15. Chung, Robison, and Morton analyzed data of deaf mutes in Northern Ireland and found two primary genetic causes for this disease; autosomal recessives, arising with a mutation frequency of about 3×10^{-5} per gamete, and autosomal dominants, arising with a mutation frequency of about 5×10^{-5} per gamete. Assuming that the reproductive success of these deaf mute genes is only one third that of the normal genes, what is the equilibrium value for each of these two types of gene?

32-16. Brachydactyly (Problem 6) is found to occur in a certain population with a frequency of .001. Assuming that this represents the equilibrium value of the gene, what is its adaptive value?

32-17. In some Michigan populations, Huntington's chorea (caused by a dominant gene) has a mutation rate of about 1×10^{-6} and is found in approximately the same frequency. (a) Assuming that equilibrium has been reached, what is the selection coefficient against this gene? (b) If the mutation rate is doubled in this population because of irradiation, what will be the new equilibrium frequency of this gene? (c) What are the equilibrium frequencies of a recessive gene with the same selection coefficient as Huntington's chorea and the same mutation rates used in (a) and (b)?

32-18. If two diseases, A and B, are caused, respectively, by two different genes each at different loci, one dominant and one recessive, which of these genes would increase in frequency most rapidly or decrease in frequency least rapidly if diseased individuals can be completely cured? Partially cured? Why?

32-19. In an Israeli population Brand tested for PTC tasting and nontasting among polio patients and nonpolio patients and found the following:

	Tasters	Nontasters
polio patients	79	36
nonpolio patients	100	20

Is there any association between PTC tasting and polio?

32-20. Reed and others have criticized the Glass and Li estimates of the migration rate of white genes into the American black population (p. 773) on the grounds that the estimate of the R^0 gene frequency is probably inaccurate for many of the African populations that may have been ancestral to American blacks. Reed has instead suggested that the Duffy blood group gene (Fy^a) would probably provide a better estimate of migration rate since this allele is generally absent in most black African populations. Based on the data for the Duffy blood group gene given in Table 32–15, offer an estimate of the proportion of white genes now present in the American black populations of: (a) Georgia; (b) California.

32–21. Two small separated human populations, A and B, have respective frequencies of phenylthiocarbamide tasters (caused by a dominant) of .85 and .25. If 5 percent of population B comes from population A each generation, what will be the frequency of the tasting gene in population B after (a) 1, (b) 5, (c) 10 generations?

32–22. Let us assume that three generations ago a certain tribe of South American Indians had only blood type O (gene i) of the ABO blood group. A recent sampling of this population now shows approximately 3 percent of the A gene (I^A). A neighboring population, many times the size of this tribe, has fairly constant frequencies of $i = .90$ and $I^A = .10$. Assuming no other factors are operative, what percentage of genes are introduced each generation into this tribe through migration?

32–23. In Problem 22, if you assume that migration is minimal and that selection for blood-group genes is absent in this tribal population, how would you attempt to explain the change in gene frequency? What sort of data would you look for?

32–24. (a) If 1000 diploid populations are each founded with a gene frequency of $q = .5$, and each continued for another generation by 10 male and 40 female parents, what average range of gene frequencies (standard deviation) can be expected? (b) What will be the standard deviation if only 10 mated couples are used?

32–25. If many generations of breeding continue under the conditions given in Problem 24, some of the 1000 populations will have reached fixation for one or the other allele. (a) What number of populations will reach fixation each generation for condition (a) above? (b) For condition (b)?

32–26. What data would you need to differentiate between gene frequency changes caused by selection and those caused by genetic drift?

32–27. Glass studied the ABO blood-group frequencies of the Dunkers, a religious sect that originated in the Rhineland region of Germany and settled in Pennsylvania during the eighteenth century. Since that time, the sect has remained relatively isolated from the surrounding population both in custom and mating. The frequency of ABO blood types found in the Dunkers compared to the surrounding American population and to a population of present-day Rhineland Germans is shown in the table below. What explanation would you offer to account for the blood-group frequencies among the Dunkers?

	No. Persons Tested	Blood-group Frequencies			
		A	AB	B	O
Dunkers	228	.593	.022	.031	.355
Rhineland Germans	5,036	.446	.047	.100	.407
Eastern U.S.A.	30,000	.395	.042	.112	.452

REFERENCES

ADAMS, J., and R. H. WARD, 1973. Admixture studies and the detection of selection. *Science*, **180,** 1137–1143.

CAIN, A. J., and P. M. SHEPPARD, 1954. Natural selection in *Cepaea. Genetics*, **39,** 89–116.

CARSON, H. L., 1970. Chromosome tracers of the origin of species. *Science*, **168,** 1414–1418.

CHARLESWORTH, B., and D. CHARLESWORTH, 1973. The measurement of fitness and mutation rate in human populations. *Ann. Hum. Genet.*, **37,** 175–187.

CLARKE, B. C., 1968. Balanced polymorphism and regional differentiation in land snails. In *Evolution and Environment*, E. T. Drake (ed.). Yale Univ. Press, New Haven, pp. 351–368.

DOBZHANSKY, Th., and O. PAVLOVSKY, 1953. Indeterminate outcome of certain experiments on *Drosophila* populations. *Evolution*, **7,** 198–210.

———, 1957. An experimental study of interaction between genetic drift and natural selection. *Evolution*, **11,** 311–319. (Reprinted in Spiess's collection; see References, Chapter 30.)

FISHER, R. A., 1930. *The Genetical Theory of Natural*

Selection. Clarendon, Oxford. (Reprinted 1958, Dover, New York.)

FORD, E. B., 1940. Polymorphism and taxonomy. In *The New Systematics*, J. Huxley (ed.). Clarendon, Oxford, pp. 493–513.

GLASS, H. B., and C. C. LI, 1953. The dynamics of racial intermixture: An analysis based on the American Negro. *Amer. Jour. Hum. Genet.*, **5,** 1–20.

GOODHART, C. B., 1962. Variation in a colony of the snail *C. nemoralis. Jour. Anim. Ecol.*, **31,** 207–237.

GULICK, J. T., 1905. Evolution, racial and habitudinal. *Carneg. Inst. Wash. Publ. No. 25*, Washington, D.C., pp. 1–265.

HARLAN, H. V., and M. L. MARTINI, 1938. The effect of natural selection on a mixture of barley varieties. *Jour. Agric. Res.*, **57,** 189–199.

HARLAND, S. C., 1936. The genetic conception of the species. *Biol. Rev.*, **11,** 83–112.

LAMOTTE, M., 1959. Polymorphism in natural populations of *C. nemoralis. Cold Sp. Harb. Symp.*, **24,** 65–84.

LEVINS, R., 1968. *Evolution in Changing Environments.* Princeton Univ. Press, Princeton, N.J.

LEWONTIN, R. C., and C. C. COCKERHAM, 1959. The goodness-of-fit test for detecting natural selection in random mating populations. *Evolution*, **13,** 561–564.

L'HÉRITIER, P., and G. TEISSIER, 1937. Elimination des formes mutants dans les populationes de *Drosophiles.* Cas de *Drosophiles* "ebony." C. R. *Acad. Sci., Paris*, **124,** 882–884.

LI, C. C., 1955. The stability of an equilibrium and the average fitness of a population. *Amer. Naturalist*, **89,** 281–295. (Reprinted in Spiess's collection; see References, Chapter 30.)

MAYR, E., 1942. *Systematics and the Origin of Species.* Columbia Univ. Press, New York.

MERRELL, D. J., 1969. The evolutionary role of dominant genes. In *Genetics Lectures*, Vol. 1, C. R. Bogart (ed.). Oregon State Univ. Press, Corvallis, pp. 167–194.

MURRAY, J., 1972. *Genetic Diversity and Natural Selection.* Oliver and Boyd, Edinburgh.

PETIT, C., and L. EHRMAN, 1969. Sexual selection in *Drosophila. Evol. Biol.*, **3,** 177–223.

ROBERTS, D. F., 1968. Genetic effects of population size reduction. *Nature*, **220,** 1084–1088.

KIMURA, M., and T. OHTA, 1971. *Theoretical Aspects of Population Genetics.* Princeton Univ. Press, Princeton, N.J.

SCHULL, W. J. (ed.), 1963. *Genetic Selection in Man.* Univ. of Michigan Press, Ann Arbor.

SVED, J. A., and O. MAYO, 1970. The evolution of dominance. In *Mathematical Topics in Population Biology*, K. Kojima (ed.). Springer-Verlag, New York, pp. 289–316.

WALLACE, B., 1958. The comparison of observed and calculated zygotic distributions. *Evolution,* **12,** 113–115.

WRIGHT, S., 1951. The genetic structure of populations. *Ann. Eugenics*, **15,** 323–354.

33
INBREEDING AND HETEROSIS

In most populations the mobility of individuals and the consequent mobility of their gametes are usually restricted. That is, individuals tend to mate with others that are close at hand. In this sense, mating between all members of a population is usually far from random, and a gene pool may consist of many small subpools, each of which may depart to some extent from the overall characteristics of the entire pool. We have seen that one such departure may occur when population size is restricted and new gene frequencies establish themselves in small subpools that become separate from the main population. Another type of event that can easily occur because of restrictions in population size or area is mating between relatives, or *inbreeding*. As we shall see, when such matings occur often, they may have little effect in changing the overall gene frequencies but important effects in increasing the frequencies of homozygotes. As a rule, therefore, if a recessive gene is rare, inbreeding will cause it to appear in greater homozygous frequency than under random mating. When this occurs, selection will then be provided with an increased opportunity to act upon rare recessives. The consequences of inbreeding may therefore have wide application for both natural and artificial selection.

MEASUREMENT OF INBREEDING

To measure inbreeding and its effects we must determine the extent to which any two homologous alleles in a zygote are descended from the same gene in the zygote's ancestry. Common descent may, of course, be claimed for any pair of genes when they are traced back far enough, since the chain of ancestry usually narrows down to very few individuals. In the case of inbreeding, however, descent from a common ancestor means only those common ancestors specified within a particular pedigree or a particular set of generations. Thus, although all alleles in a pedigree may ultimately trace their ancestry to a single gene, the measurement of inbreeding among them refers only to those relationships that are known.

In a diploid, when two alleles at a locus are descended from the same gene, they are called *identical*. If mutation has not changed either of these alleles during the pedigree interval, the organism is homozygous at that locus, e.g., *AA*. However, homozygotes may also occur when each of the two alleles are not descended from the same gene. For example, a mating between first cousins may produce a homozygote containing genes of identical origin and, at the same time, also produce an individual homozygous for genes of separate origin (Fig. 33–1). The two latter genes may be called *similar* but not identical. In inbreeding, however, our interest is primarily in identical genes,

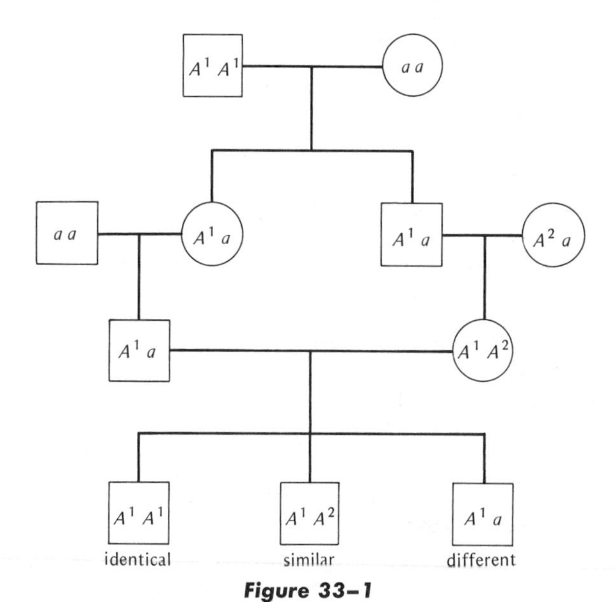

Figure 33–1

Three of the possible combinations of alleles in a diploid as a result of a first-cousin mating.

and the measurement used for the probability that two genes in a zygote are identical is given by the *inbreeding coefficient, F.*

To calculate the inbreeding coefficient, it is helpful if the extent to which mating occurs between relatives is known. If it is not known, it can nevertheless be estimated on the basis of population size. For example, let us assume that a population is begun with 50 diploid individuals, none of which carry the same gene at a particular locus; i.e., there are 100 alleles in the population: A^1, A^2, A^3, . . . A^{100}. Let us also assume that the gametes of each individual are of both male and female types and unite with each other randomly during a common mating season, as among some marine hermaphrodites. If we restrict our attention to this one locus, e.g., *A*, there are then 100 kinds of gametes, each different from the other but identical with all gametes of its own kind. In other words, 1/100 of the gametes are identical, and the probability of picking up a gamete identical to one already chosen is 1/100, or 1/2N, where N is the number of breeding diploid individuals. One may generalize from this case and say that the probability that an individual will be formed of gametes containing two identical genes is 1/2N. According to our previous definition, it is this quantity that is now the coefficient of inbreeding for a single generation of random union between gametes in population of size N.

In the next generation there will again be 2N different kinds of gametes produced by the new parents, and the probability of inbreeding is again 1/2N. In addition to gametes produced by heterozygous individuals, however, some gametes will have arisen from the identical homozygous individuals that are now among the parents. This extra proportion of identical gametes will increase the chances of forming new identical homozygotes. In mathematical terms, if the probability of newly arisen identical homozygotes is 1/2N for any generation, the probability that the remaining zygotes, 1 – (1/2N), will have identical genes is the inbreeding coefficient of the previous generation. Thus the inbreeding coefficient (F) of generation 2 is $F_2 = (1/2N) + [1 - (1/2N)]F_1$, where F_1 is the inbreeding coefficient of generation 1. Calculation of the inbreeding coefficient for succeeding generations follows this same pattern:

$$F_0 = 0$$

$$F_1 = \frac{1}{2N}$$

$$F_2 = \frac{1}{2N} + \left(1 - \frac{1}{2N}\right)F_1$$

$$F_3 = \frac{1}{2N} + \left(1 - \frac{1}{2N}\right)F_2$$

so that for any generation n,

$$F_n = \frac{1}{2N} + \left(1 - \frac{1}{2N}\right)F_{n-1}$$

PANMICTIC INDEX

Knowledge of the inbreeding coefficient permits us to measure the rate at which homozygosity should be attained. In other words, we can now estimate the extent to which mating between gametes in a population of limited size departs from that ideal *panmictic* or outbred state in which no two alleles of an individual are related. If we call F the inbreeding, or fixation, index, 1 − F can be called the panmictic index, or P. P is therefore a measure of the relative amount of random-mating heterozygosity (to which we give the initial value 1) that is diminished by inbreeding (F). Since F is also 1 − P, the last equation in the previous section, $F_n = (1/2N) + [1 - (1/2N)]F_{n-1}$, can be written

$$1 - P_n = \frac{1}{2N} + \left(1 - \frac{1}{2N}\right)(1 - P_{n-1})$$

$$-P_n = -1 + \frac{1}{2N} + 1 - \frac{1}{2N} - P_{n-1} + \frac{1}{2N}P_{n-1}$$

$$-P_n = -P_{n-1} + \frac{1}{2N}P_{n-1}$$

$$= -P_{n-1}\left(1 + \frac{1}{2N}\right)$$

$$P_n = P_{n-1}\left(1 - \frac{1}{2N}\right)$$

The loss of heterozygosity by random mating for any generation (also known as "the rate of disintegration") is therefore equal to the factor 1/2N. For a population of dioecious organisms in which self-fertilization is impossible, this rate is reduced to 1/(2N + 1), but if the population size is fairly large, the disintegration rate can be approximated by 1/2N. For large populations, therefore, the panmictic index, P_n, can be written in terms of previous generations, such as, P_{n-2}:

$$P_n = \left[P_{n-2}\left(1 - \frac{1}{2N}\right)\right]^{*}\left(1 - \frac{1}{2N}\right)$$

$$P_n = P_{n-2}\left(1 - \frac{1}{2N}\right)^2$$

When this is extended back to the initial panmictic index of the base population, P_0,

$$P_n = P_0\left(1 - \frac{1}{2N}\right)^{n\dagger}$$

If we begin with a base population without inbreeding, F = 0 or P = 1, values of P for different population sizes can be graphed according to Fig. 33–2. It is obvious that as the population size increases, the loss of heterozygosity (rate of disintegration) decreases. Despite this decrease, however, given a long enough period of time, most populations would be expected to show a fall in heterozygosity, provided that selection, mutation, and migration are not changing allelic frequencies. Inbreeding effects and sampling errors induced by random drift (p. 774) are therefore both functions of population size. (Note that population size refers only to the effective population number, N_e, which is usually less than the observed number of individuals.)

Since inbreeding reduces heterozygosity, let us consider the consequences of inbreeding on the

* Since $P_{n-1} = P_{n-2}\left(1 - \frac{1}{2N}\right)$.

† In terms of F, P = 1 − F, therefore $1 - F_n = (1 - F_0)\left(1 - \frac{1}{2N}\right)^n$. If no inbreeding was present in the initial population ($F_0 = 0$), then $1 - F_n = \left(1 - \frac{1}{2N}\right)^n$ or $F_n = 1 - \left(1 - \frac{1}{2N}\right)^n$.

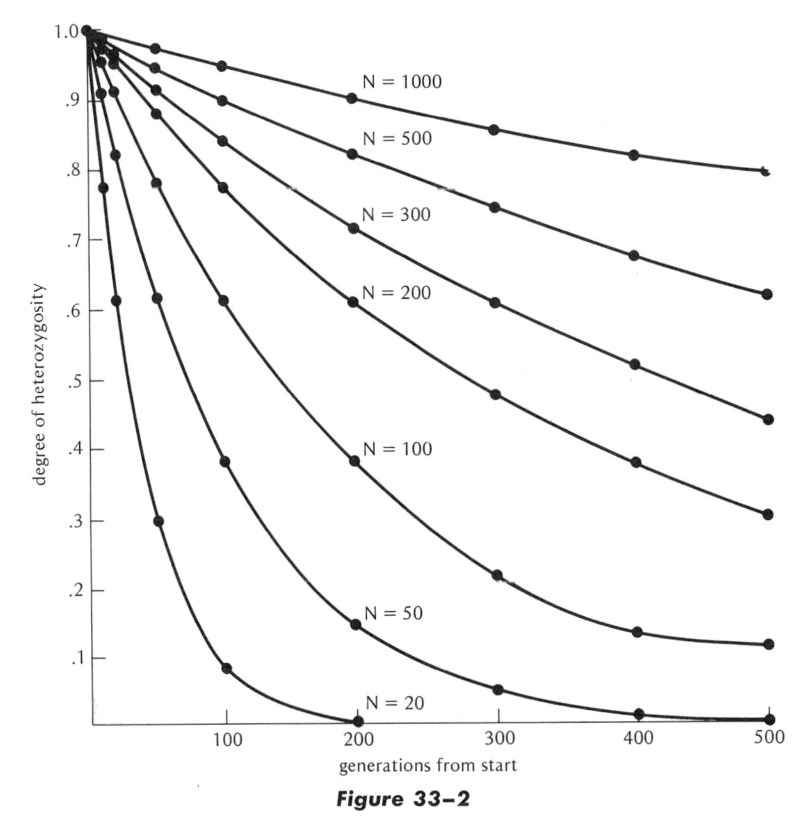

Figure 33–2

Degree of heterozygosity remaining in populations of different sizes after given generations of random union between gametes. Calculations are based on 1.00 as the initial degree of heterozygosity. (Under these conditions, the panmictic index P and the degree of heterozygosity are identical.)

frequencies of the different genotypes. If we confine our attention to a particular locus with only two alleles, *A* and *a*, there will be a proportion of F identical homozygotes according to our definition of the inbreeding coefficient. Of this inbred proportion, some will be *AA* and some *aa*, the frequencies of each depending upon their respective population gene frequencies p and q. Thus there will be p*FAA* and q*Faa* genotypes produced by inbreeding. In addition to these, however, the remaining individuals (1 − F) will bear genotypes whose frequencies are determined according to the Hardy-Weinberg equilibrium of p^2AA, $2pqAa$, and q^2aa. The three genotypes will therefore have the following frequencies:

$$AA = p^2(1 - F) + pF = p^2 - p^2F + pF$$
$$= p^2 + pF(1 - p) = p^2 + pqF$$

$$Aa = 2pq(1 - F) = 2pq - 2pqF$$

$$aa = q^2(1 - F) + qF = q^2 - q^2F + qF$$
$$= q^2 + qF(1 - q) = q^2 + pqF$$

It is now easy to see that the increase in the frequency of each type of homozygote by a factor of pqF comes from an equivalent reduction in the frequency of heterozygotes. Note also that this reduction in heterozygotes affects the gene frequencies p and q equally, so that only the genotypic frequencies are changed. When inbreeding is absent, F = 0, the above equations reduce to the Hardy-Weinberg frequencies p^2AA, $2pqAa$, and q^2aa. When inbreeding is complete, F = 1, 2pq − 2pqF equals zero, and the only remaining genotypes are pAA and qaa.

INBREEDING PEDIGREES

When exact family pedigrees are known, inbreeding coefficients can be more precisely calculated for the offspring of mates that have common ancestors. One method commonly used is based on the probability that identical alleles will be inherited at each stage of transmission. This is illustrated in Fig. 33–3 in a pedigree of an individual Z who is the offspring of a brother-sister mating which has two common ancestors, V and W. If we first confine our attention to V, we can see that the probability that its offspring X and Y inherit identical genes is 1/2. This is because V, being diploid, may produce two types of gametes, of which X receives one; so that the probability that Y receives an identical one is 1/2. To be more specific, V may have the genotype A^1A^2, so that X and Y may receive either of the two gametes, A^1 and A^2. There is a 1/4 probability that X will have A^1 when Y has A^1 and a 1/4 probability that X will have A^2 when Y has A^2, i.e., a 1/2 probability that both X and Y will have identical gametes from V. The remaining probability of 1/2 covers occasions when X and Y will not receive identical alleles from V; that is, one will receive A^1 and the other A^2. However, if V is itself inbred, so that it has an inbreeding coefficient F_V, the chance for identical alleles to be passed on to X and Y will be increased by this factor multiplied by the remaining 1/2.* In other words, the total probability that X and Y receive identical alleles from V is $1/2 + 1/2\ F_V = 1/2(1 + F_V)$.

However, even if X and Y obtain identical alleles from V, there is still a probability of 1/2 that X will not pass on this particular allele to Z (since Z contains only one gene from X), and for the same reason, a probability of 1/2 that Y will not pass on this same allele to Z. In all, therefore, the probability that Z will contain identical genes from ancestor V is $1/2\ (1 + F_V) \times 1/2 \times 1/2 = (1/2)^3(1 + F_V) = 1/8 + 1/8\ F_V$. Similarly, the chances that Z will contain identical genes from W is $(1/2)^3(1 + F_W) = 1/8 + 1/8\ F_W$, since each path is an independent route by which identical genes may have been transmitted.

In this scheme it is important to note that the inbreeding coefficient of Z can only be affected by the inbreeding coefficients of the two common ancestors V and W. If the inbreeding coefficients of V and W are 0, the inbreeding coefficient of Z becomes $1/8 + 1/8 = 1/4 = F_Z$. Thus, if we consider that an individual receives an inbreeding factor $1/2\ (1 + F)$ from an ancestor common to both its parents, its inbreeding coefficient from that source is $(1/2)^n 1/2\ (1 + F_A)$, where n is the number of steps leading from *both* parents of the individual to the common ancestor and F_A the inbreeding coefficient of the common ancestor. [If F_A is zero, this reduces to $(1/2)^n 1/2$. If F_A is one, the inbreeding coefficient contributed by this pathway is $(1/2)^n$.]

The number of steps between the parents of an individual and its common ancestor may be few or many. As shown in Fig. 33–4, first-cousin matings have one additional pair of steps leading to the common ancestor as compared to full-sib matings. To calculate the inbreeding coefficient of the progeny of such matings, we may assume that the common ancestors have not been inbred unless

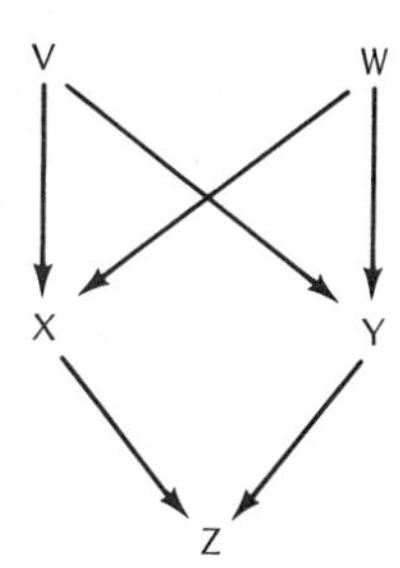

Figure 33–3

Pedigree of a brother–sister full–sib mating (X and Y) producing an offspring Z. The arrows indicate contribution of a gamete. Thus the two initial parents, V and W, each contribute gametes to both their offspring, X and Y.

*Remember that the genotype of V, as well as other individuals in the pedigree, is not known.

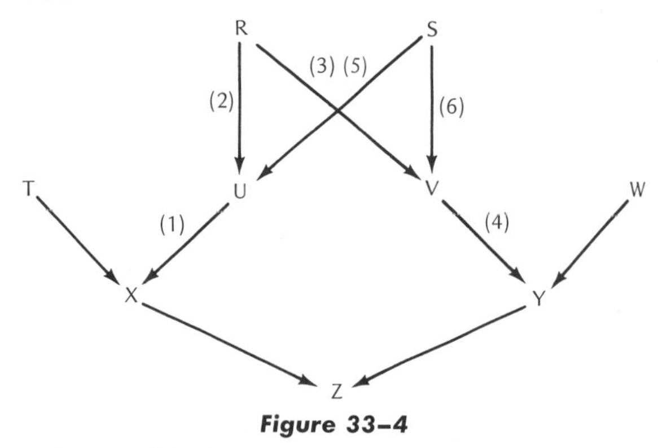

Figure 33–4

Pedigree of a first-cousin mating showing the various gametic paths (numbered in parentheses) from the common ancestors R and S to Z.

further information is given. The inbreeding coefficient of Z is then the sum of two paths 1–2–3–4 and 1–5–6–4, or $(1/2)^4\ (1/2) + (1/2)^4\ (1/2) = 1/16$.

If a few common ancestors occur in a pedigree, or more than one path of transmission for identical genes is possible, the inbreeding coefficient of the individual is the sum of all such paths. For example, Fig. 33–5 shows a possible pedigree in cattle breeding in which there are three common ancestors, U, V, and W. (In this pedigree, U has mated with its grandchild W to produce X, and V has mated with its offspring W to produce Y.) Since

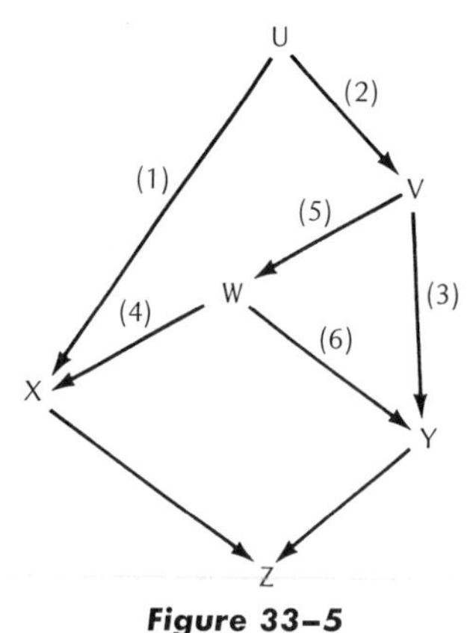

Figure 33–5

Highly inbred pedigree showing the gametic paths from the various common ancestors to Z.

the pedigree does not show inbreeding for these three common ancestors, we may assume that their inbreeding coefficients are zero and their contributions to the inbreeding of Z are a factor of 1/2. The inbreeding coefficient of Z then consists of the following contributions:

common ancestor U: path 1–2–3

$$(1/2)^3\,(1/2) = (1/2)^4$$

path 1–2–5–6

$$(1/2)^4\,(1/2) = (1/2)^5$$

common ancestor V: path 4–5–3

$$(1/2)^3\,(1/2) = (1/2)^4$$

common ancestor W: path 4–6

$$(1/2)^2\,(1/2) = (1/2)^3$$

The total inbreeding coefficient of Z is then $1/8 + 2/16 + 1/32 = .28125$. Note that each individual occurs only once along a particular path and that the path goes directly from one parent (X) to previous generations until it reaches the common ancestor and then returns to the other parent (Y) without reversing itself between generations (e.g., a path such as 1–2–3–6–4 is not included).

SYSTEMATIC INBREEDING

Inbreeding systems can be constructed so that the mating procedure carried out each generation produces individuals all of whom have the same coefficient of inbreeding. The choice of rates at which inbreeding is achieved under such systems is usually of value when the precise control of homozygosity is necessary or desirable. A system of self-fertilization each generation would be expected to produce homozygosity at a most rapid rate. Since the chances that two gametes of a diploid parent will be identical for a particular allele is 1/2, the inbreeding coefficient of the progeny of a self-fertilized individual will be $F = 1/2(1 + F_P)$, where F_P is the inbreeding coefficient of the parent. As shown in Table 33–1, if we begin with a population completely heterozygous, there will remain a proportion of only $1/2^n$

TABLE 33–1

Results of "selfing" when beginning with a population consisting entirely of heterozygotes, *Aa*

Generation	Genotypes *AA*	*Aa*	*aa*	Proportion of Heterozygotes Relative to Those in the Initial Population
0	—	1	—	1
1	1/4	2/4	1/4	1/2
2	3/8	2/8	3/8	1/4
3	7/16	2/16	7/16	1/8
4	15/32	2/32	15/32	1/16
5	31/64	2/64	31/64	1/32
10	1023/2048	2/2048	1023/2048	1/1024
⋮				
n	$(2^n - 1)/2^{n+1}$	$2/2^{n+1} = 1/2^n$	$(2^n - 1)/2^{n+1}$	$1/2^n$*

* The proportion of homozygotes remaining in any particular generation, n, is then $1 - 1/2^n = (2^n - 1)/2^n$, or adding the values of *AA* and *aa*

$$2 \times \left(\frac{2^n - 1}{2^{n+1}}\right) = \frac{2^n - 1}{2^n}$$

heterozygotes after n generations. This attempt at rapid inbreeding is, of course, only possible for hermaphroditic organisms.

For sexual organisms, the most rapid inbreeding system is that between brothers and sisters who share both parents in common. We have seen that such "full-sib" mating produces an inbreeding coefficient of 25 percent in the first generation of inbreeding. In succeeding generations this rate is diminished, since some of the genes made identical by sib mating were already previously identical. Nevertheless, within 10 generations full sib-mating can produce an inbreeding coefficient of about 90 percent. A first-cousin mating system uses two pairs of parents each generation (see Fig. 33–6), and the rate of inbreeding is therefore slower than for sib mating. A more elaborate system that does not increase the inbreeding coefficient as rapidly as other systems is second-cousin mating, in which four pairs of parents are used each generation. The rate of inbreeding for all four systems is compared in Fig. 33–7. There are, of course, other inbreeding systems, such as half-sib mating, parent-offspring mating, third-cousin mating, and mixtures of various kinds. Formulas for calculating these inbreeding coefficients can be obtained or derived from the basic work in this field by Wright (1921).

ASSORTATIVE AND DISASSORTATIVE MATING

The systems considered above are of the type called *genetic assortative mating*, since the parents are sorted and mated together on the basis of their genetic relationship. The main effect of such assortative mating is to increase or control the inbreeding coefficient. However, in addition to this, assortative mating may also occur on the basis of *phenotypic* similarity. For example, a population segregating for genes at a particular locus may be restricted to matings between dominant phenotypes (*AA* × *AA*, *Aa* × *AA*, *AA* × *Aa*) and between recessives (*aa* × *aa*). If continued long enough, heterozygotes are eliminated and the population becomes polarized into two groups: one homozygous dominant (*AA*) and the other homozygous recessive (*aa*). However, contrary to inbreeding, genes at other segregating loci, e.g., *Bb*, *Cc*, etc., will not necessarily become homozygous if phenotypic assortative mating can be confined to nonrelated individuals.

If phenotypic assortative mating is determined by more than one locus (e.g., *A–B–* × *A–B–* and *aabb* × *aabb*), the progress to homozygosity for these loci will evidently be slower than for one

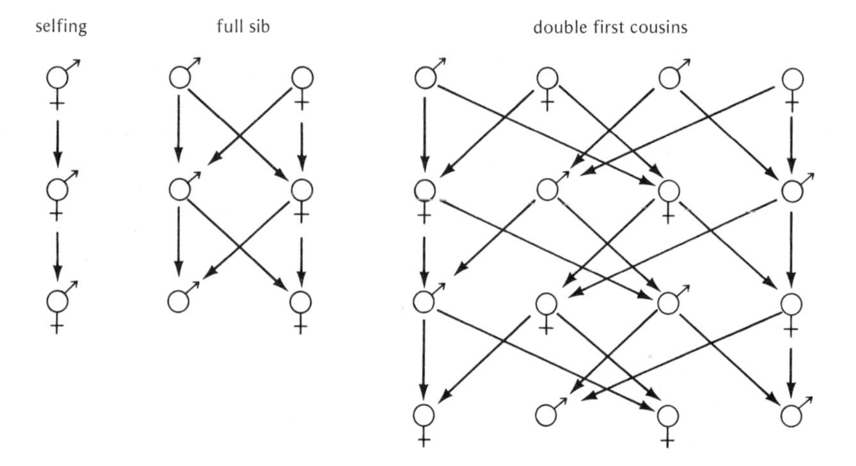

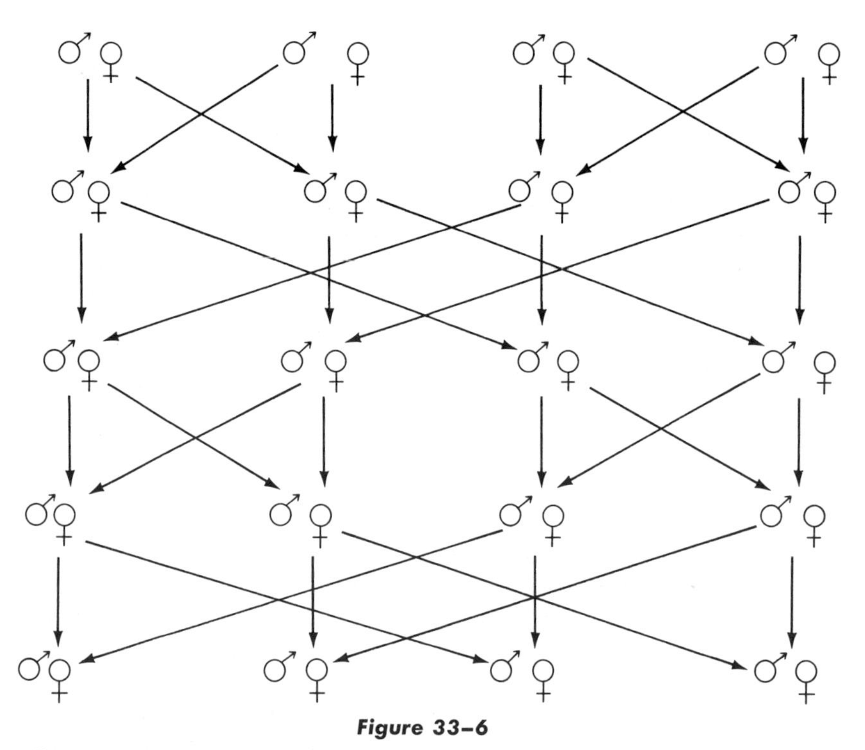

Figure 33–6

Different mating schemes showing different numbers of individuals used to perpetuate the population each generation: selfing = 1 individual, sib mating = 1 pair, double first cousin = 2 pairs, quadruple second cousin = 4 pairs. Common ancestry of an individual is in the parent generation for the selfing system, in the grandparent generation for full-sib mating, in the great-grandparent generation for first-cousin mating, and in the great-great-grandparent generation for second-cousin mating.

locus because more than one kind of heterozygote is now possible (e.g., $AaBB$, $AABb$, $AaBb$) and their rate of elimination is slower. Comparison between the rates of homozygosity for phenotypic assortative mating based on different numbers of loci is shown in the upper half of Fig. 33–8. Mating between like phenotypes without inbreeding, therefore, may be relatively ineffective in establishing homozygous strains if the phenotypic characteristics are based on numerous loci.

Phenotypic disassortative mating, or the mating of unlike phenotypes, has an effect opposite to assortative mating in that it tends to maintain heterozygosity (lower half, Fig. 33–8). We see evidence of this effect in mating between unlike sexes, which preserves both genetic and phenotypic dissimilarities. Primitively such sexual differences may have arisen at a single gene locus at which one sex was homozygous and the other heterozygous. If we call this locus A, disassortative mating restricted to dominant × recessive phenotypes $A- \times aa$ will immediately produce an equilibrium condition $.5\ Aa : .5\ aa$. Other examples of this kind can be found among some plants in which the female organs (styles) are at levels different from the male organs (anthers). Some flowers in *Primula* species, for example, are of two main types, *pin* and *thrum* (Fig. 33–9), which help prevent self-fertilization.

Disassortative mating may also be enforced on a *genetic* basis. Self-sterility alleles in plants provide an example. In such systems, pollen of a particular genotype (e.g., S^1) can fertilize only plants with different genetic constitutions (e.g., S^2S^3). Thus at least three sterility alleles are maintained in such populations, and usually more (see p. 169). These systems prevent self-fertilization and help maintain heterozygosity within a breeding population.

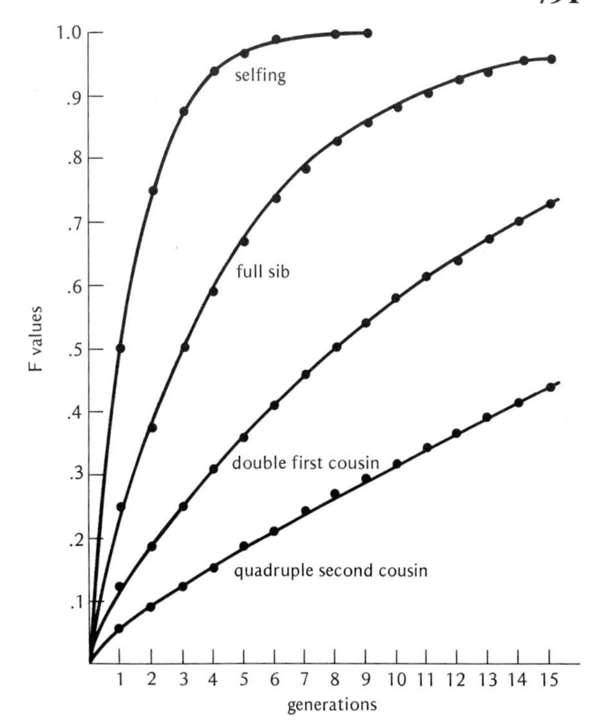

Figure 33–7

Inbreeding coefficients at generations 1 to 15 for different systems of inbreeding.

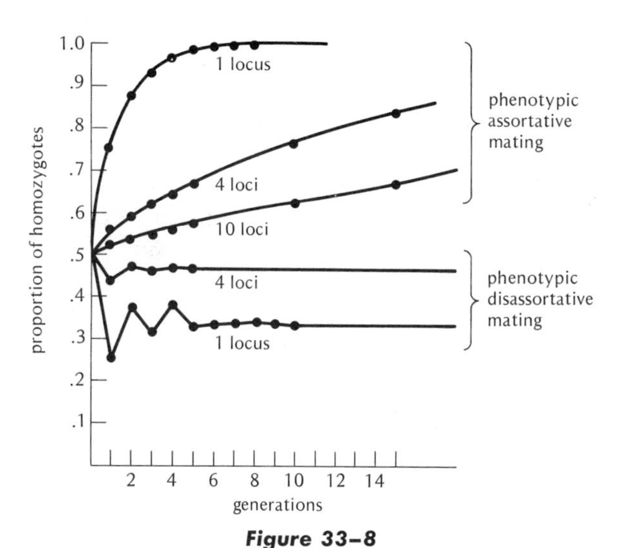

Figure 33–8

Comparison of the proportion of homozygotes achieved in different generations of phenotypic assortative and disassortative matings for different numbers of loci.

INBREEDING DEPRESSION

The consequences of inbreeding are of profound interest to geneticists, since there is considerable evidence that the efficiency of selection is closely related to the breeding system used. For the estab-

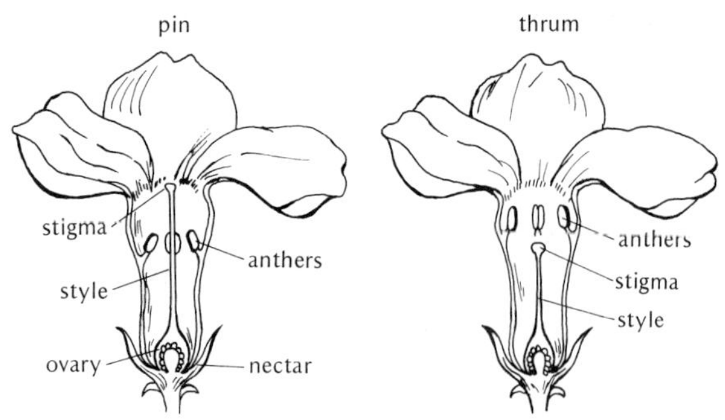

Figure 33–9

Two types of flowers found in Primula vulgaris. *The mechanical aspects of cross-fertilization between these flowers was pointed out by Darwin, who noted that the plants were pollinated by bees or moths with a long proboscis. An animal obtaining nectar from the thrum type of flower would pick up pollen around the base of its proboscis which, upon further feedings, would be deposited on the high stigma of a pin flower. Similarly, the feeding insect would pick up pollen from the pin flower near the tip of its proboscis which might later be deposited on the low stigma of a thrum flower. In addition to this mechanism, physiological incompatibilities have been found to prevent self-fertilization in* Primula.

lishment of superior homozygous genotypes it is apparent that selection will be more effective when the breeding system promotes homozygosity than when it does not. At the same time, we have seen that there are numerous devices ensuring cross-fertilization and the maintenance of various levels of heterozygosity. When normal crossbreeding patterns are obstructed by enforced inbreeding, one of the primary results is *inbreeding depression*. This can be understood by comparing the frequencies of homozygous recessives in inbred and outbred populations.

From previous calculations (p. 786) we have seen that the effect of inbreeding is to change the frequencies of genotypes to

$$AA = p^2 + pqF$$

$$Aa = 2pq - 2pqF$$

$$aa = q^2 + pqF$$

Thus, if a recessive disease with genotype *aa* occurs with frequency q^2 in a random outbred population, its frequency will be increased by pqF in an inbred population. The ratio of inbred to outbred frequency of the homozygous recessive will therefore be

$$\frac{q^2 + pqF}{q^2} = \frac{q(q + pF)}{q^2} = \frac{q + pF}{q}$$

Obviously, if q is large and F is small, the inbreeding increment pF will be relatively small, and the increased frequency of homozygous recessives will hardly be noticeable. However, if q is very small, p is large, and pF provides a notable increase in recessives even when F is fairly small. For example, if q is .5, first-cousin mating (F = .0625) will produce the following inbred to outbred ratio of homozygotes:

$$\frac{.5 + (.5)(.0625)}{.5} = \frac{.53125}{.5} = 1.06$$

However, if q is .005, this ratio increases to .067/.005 = 13.4. When q = .0005, the increase of homozygotes because of first-cousin mating is .0630/.0005, or 126 times that of the randomly bred populations.

This increase in frequency of recessive homozygotes upon inbreeding occurs, of course, only in those inbred families carrying the recessive gene. One might therefore argue that since the frequency of the gene is small, the frequency of such families will also be small. It should be remembered, however, that although each kind of deleterious gene might be present in only a few families, there are many different kinds of deleterious genes, and many families might be affected by inbreeding. Since different inbred families will often produce homozygotes for different genes, one consequence of inbreeding is therefore to increase the variability between inbred families. Thus, paradoxically, an increase in homozygosity can lead to an increase in variability.

Accompanying the increase in homozygosity under inbreeding is also a change in the mean of quantitative traits toward that of the homozygous recessive. If a gene has a quantitative effect equal to v, the dominant genotype may be considered

TABLE 33–2

Average value of genotypes when the homozygous recessive (*aa*) has a lower quantitative value than the dominant genotypes

Genotype	Frequency Before Inbreeding	Quantitative Value	Frequency × Value
AA	p^2	v	p^2v
Aa	2pq	v	2pqv
aa	q^2	−v	$-q^2v$

average value before inbreeding $= v(p^2 - q^2) + 2pqv = v(p + q)(p - q) + 2pqv = v(p - q) + 2pqv$

TABLE 33–3

Average value of the genotypes in Table 33–2 produced when the inbreeding coefficient is F

Genotype	Frequency After Inbreeding	Quantitative Value	Frequency × Value
AA	$p^2 + pqF$	v	$p^2v + pqvF$
Aa	$2pq - 2pqF$	v	$2pqv - 2pqvF$
aa	$q^2 + pqF$	−v	$-q^2v - pqvF$

average value after inbreeding $= v(p^2 - q^2) + 2pqv - 2pqvF = v(p - q) + 2pqv - 2pqvF =$ average value before inbreeding $- 2pqvF$

+v and the homozygous recessive −v. Under these conditions, Table 33–2 shows the average value of all genotypes produced by a population under random mating. If, however, inbreeding occurs with coefficient F, the mean value changes to that shown in Table 33–3.

Note that the reduction in quantitative value upon inbreeding, 2pqvF, arises from the exact reduction in quantitative value of the heterozygotes. Also, this change in quantitative value occurs in the direction of the recessive value, −v. If dominance is incomplete, the change will be somewhat diminished but nevertheless in the same direction. If dominance is completely absent, the quantitative value of the heterozygote is exactly midway between both homozygotes, or zero, and the average quantitative change or inbreeding effect reduces to zero.

The four primary features of inbreeding, therefore, are the increase in frequency of homozygotes, the increase in variability between different inbred families, the reduction in value of quantitative characters in the direction of recessive values, and the dependence of this reduction in value upon dominance. If this inbreeding effect is multiplied for many genes at many loci, there may be a large reduction in value for many traits, including those that affect fitness and survival. In corn, for example, inbreeding has been shown to seriously depress height and yield. A similar effect on fertility and mortality in rats is shown in Table 33–4. In some organisms inbreeding depression may be so severe that the strain can no longer reproduce. In alfalfa and some other plants very few strains can survive beyond two or three generations of inbreeding.

The important role played by homozygous recessives in decreasing both fitness and quantitative values is not unexpected. From our previous con-

TABLE 33–4

Effect of about 30 generations of inbreeding (parent × offspring, brother × sister) on fertility and mortality in rats

Year	Nonproductive Matings, Percent	Average Litter Size	Mortality from Birth to 4 Weeks, Percent
1887	0	7.50	3.9
1888	2.6	7.14	4.4
1889	5.6	7.71	5.0
1890	17.4	6.58	8.7
1891	50.0	4.58	36.4
1892	41.2	3.20	45.5

From Lerner, 1954, after Ritzema Bos.

sideration of selection (p. 764) we have seen that deleterious genes with dominant effects or deleterious genes lacking dominance will tend to be eliminated much more quickly than recessives. Populations can therefore be expected to contain a greater number of deleterious recessives than deleterious dominants or partial dominants. Thus it is probably these recessive genes, "uncovered" by inbreeding, that are primarily responsible for inbreeding depression.

Inbreeding depression, however, is not a universal phenomenon in all species, and certainly not in many species that are normally self-fertilized. An example of escape from inbreeding depression is believed to exist in some families of the ancient Egyptian pharaohs, in which brother-sister matings were practiced for many generations without obvious known ill effects. On the whole, however, there is ample evidence that most normally cross-fertilizing species will show deterioration upon consistent inbreeding although some strains may escape.

HETEROSIS

In contrast to inbreeding depression, the crossbreeding of two different strains will usually show more vigorous hybrid offspring than either of the parent strains considered separately. The superiority of the hybrid, known as *heterosis*, may show itself in improved general fitness characteristics such as longevity and resistance to disease. Such improvements are also called "hybrid vigor" and are considered different from the oft-noted increases in "luxuriant" factors such as size which do not necessarily improve fitness. The distinction between hybrid vigor and luxuriance is not always easy to make, since improvement in some characteristics may be accompanied by a deterioration in others. This is nowhere more obvious than in the hybrid between the female donkey and male horse. The offspring of this union, the mule, is superior to both parents in fitness in many important physical respects, yet it is almost always sterile.

Heterosis in corn has received considerable attention because of its marked effect on yield improvement. This increased productivity on crossing different strains of corn was first noted late in the nineteenth century and was then developed according to systematic genetic procedures by Shull, Jones, East, and others. Together with improved farming methods, hybrid corn has helped to treble the average corn yield in the United States from about 25 to 30 bushels per acre (1923 to 1932) to more than 80 bushels per acre since 1965. At present, hybrid corn accounts for almost the entire corn planting and adds an extra 1 billion

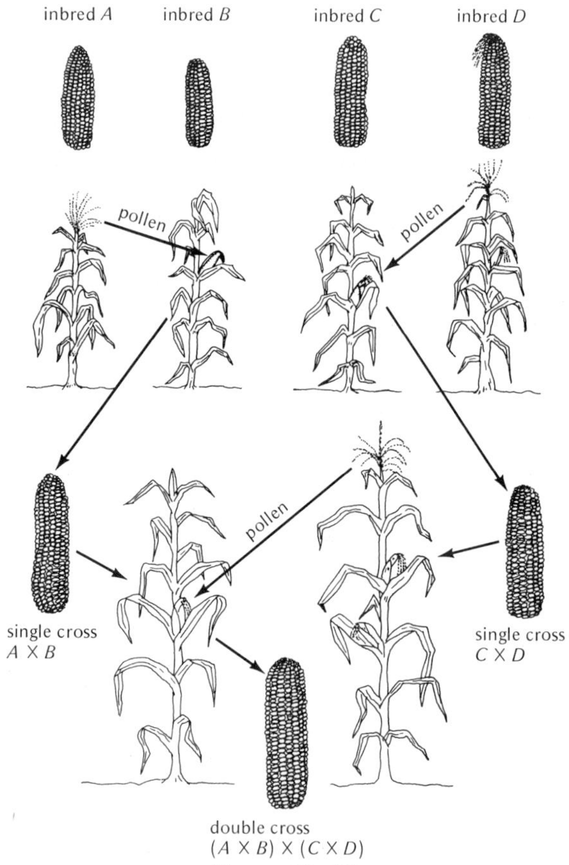

Figure 33–10

Crosses showing the production of hybrid corn from four inbred lines A, B, C, and D. Paired crosses between the initial inbred lines produce two vigorous hybrid plants, AB and CD, which are then intercrossed to yield the double-cross hybrid, ABCD. (After Dobzhansky.)

bushels of corn a year to the American agricultural economy. This increase has occurred in other countries as well; in Northern Italy, for example, hybrid corn accounts for increases of more than one million tons annually.

The usual method for raising hybrid corn is to establish many inbred lines, make intercrosses, and determine which hybrids are most productive in a given locality (have the best "combining ability"). Since more hybrid seed can be obtained from vigorous plants than from weak inbred lines, a "double cross" is made (Fig. 33–10) in which two pairs of selected hybrids are crossed to furnish the ultimately harvested corn. To eliminate the laborious detasseling of corn that is necessary to prevent self-fertilization in hybrid crosses, cytoplasmic factors which produce male sterility have been used in conjunction with "restorer" genes. As Fig. 33–11 shows, a cytoplasmic male sterility factor (S) prevents one of the strains from acting as a male parent. A cross between this female and a pollinating plant bearing restorer genes (R) can produce offspring, all or half of which are capable of producing pollen. The commercial hybrid seed that is planted by farmers using these strains contains an appreciable number of such restored plants, which are included to ensure full pollination of all corn ears. Unfortunately the cytoplasmic factors which confer pollen sterility have been found to be linked to factors that confer susceptibility to southern corn leaf blight, and the use of cytoplasmic sterility in corn is now limited.

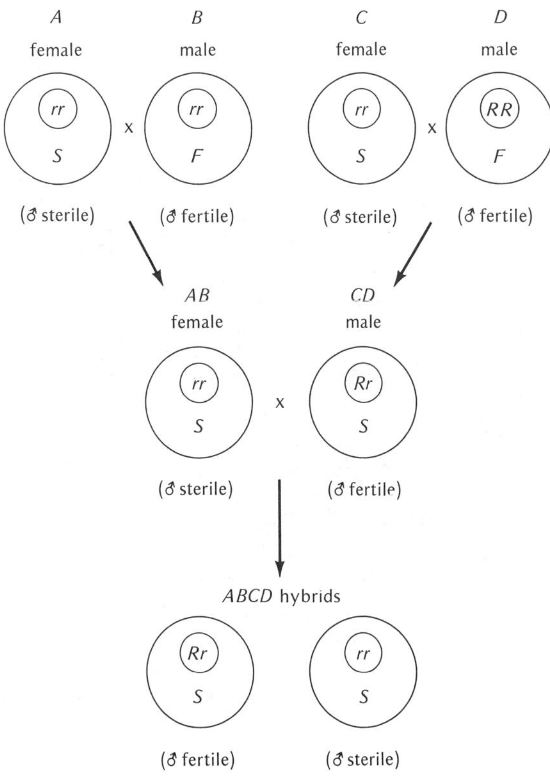

Figure 33–11

One possible scheme in which normal detassling in corn can be eliminated is through the use of cytoplasmic factors (S) which produce male sterility and restorer genes (R) which restore male fertility. (F designates the cytoplasmic factor for male fertility.) In this scheme, crosses between the inbred plants A × B and C × D produce a ♂ sterile AB and a CD pollinator. The progeny of AB × CD are ♂ sterile and ♂ fertile in the ratio 1 : 1. Thus planting of the double-hybrid ABCD seed will ensure sufficient pollen to enable fertilization of all the ABCD ears.

The vigor of hybrid corn, however, is not self-perpetuating, and pure breeding heterotic strains of corn are unknown. In fact all attempts to inbreed strains of hybrid corn in order to make them exclusively homozygous for beneficial genes have failed. This is graphically demonstrated in the "F_2 breakdown" that occurs in the generations following heterosis (Fig. 33–12).

Other plant species, such as sorghum, may also produce hybrid vigor when different strains are crossed, but hybrid plantings are not too common, because of the difficulty of ensuring cross-pollination. Cytoplasmic male sterility factors that prevent inbreeding have, however, been used to produce a large number of hybrid varieties of onions, carrots, and beets that are presently used in commercial plantings.

THEORIES OF HETEROSIS

The question of whether heterosis is a result of the superiority of the heterozygotes, or whether it arises from other causes, has been long debated. We have previously seen (p. 766) that a hetero-

Figure 33–12

Heterosis (F_1) arising from a cross between two inbred lines of corn (P_1 and P_2) followed by inbreeding depression in succeeding generations (F_2 to F_8). (From D. F. Jones, 1924. The attainment of homozygosity in inbred strains of maize. Genetics, 9, 405–418.)

zygote for a single gene difference may, at least theoretically, be superior to either of the two homozygotes. One reason for this superiority has been proposed to be the metabolic advantage of the gene products produced by two different kinds of alleles rather than by only one kind. For example, a heterozygote may produce a unique "hybrid substance" completely different from either of the homozygous products. Hybrid enzymes (Fig. 33–13), the AB antigen of human blood groups, and other hybrid compounds are examples of substances that show chemical differences from the substances produced by homozygotes for these genes. As shown for alkaline phosphatase mutations in *E. coli* (p. 667), mutant polypeptides produced by two different alleles may complement each other within a single multichained protein to produce a more effective product. In corn, McDaniel and Sarkissian have demonstrated that a hybrid possesses mitochondrial enzymes which significantly improve respiratory activities compared to the mitochondria of the parental strains.

Another kind of hybrid metabolism occurs when the heterozygote produces an intermediate amount of the same substance that is produced in different quantities by the two homozygotes. For example, Emerson found that a culture of *Neurospora* containing two sets of different nuclei (heterocaryon), one with *pab* and the other pab^+, produced a more optimum amount of *p*-aminobenzoic acid for growth than did either homocaryon. Also, if each homozygote produces a different metabolic sub-

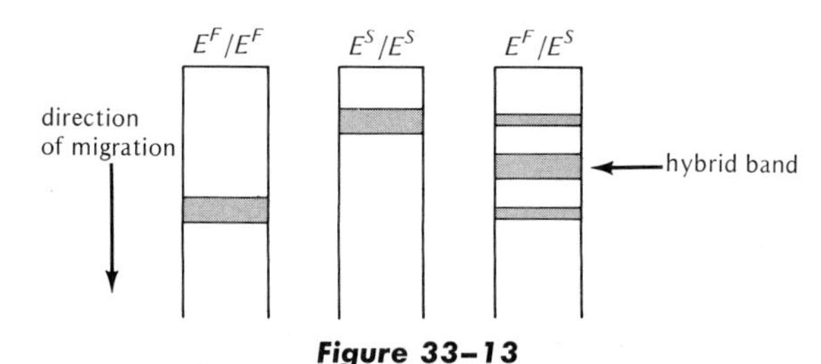

Figure 33–13

Electrophoretic zymograms formed from extracts of corn plants homozygous and heterozygous for a gene controlling the synthesis of a particular enzyme, the E_3 esterase. As shown above, the heterozygote (E^F/E^S) forms a band intermediate to the two homozygotes, indicating the presence of a hybrid enzyme or isozyme bearing new properties (see also p. 704). According to some terminologies, a unique electrophoretic form of an enzyme or protein determined by a specific allele is called an allozyme (see also p. 808), and the more general term, isozyme, is used for different molecular forms of a protein arising from any cause whether genetic or nongenetic. (After Schwartz.)

stance, the heterozygote may produce both substances and thereby reap whatever advantages may arise from the simultaneous possession of two different metabolites. We have seen this possibility in the sickle-cell heterozygote who possesses both sickle hemoglobin that is believed to prevent malarial infection and normal hemoglobin that prevents hemolytic anemia (p. 617).

Whatever the physiological causes, it has been presumed that the heterozygote can be better protected or "buffered" so that environmental disturbances in development do not throw the individual "off course." Evidence for this view exists in data showing much less variability for hybrids than for their inbred parents (Table 33–5). In other words, hybrids are superior in their ability to produce a uniform phenotype under conditions in which variable phenotypes are produced by homozygotes. The explanation of heterosis in terms of the superiority of heterozygous genotypes relative to homozygotes has been called the *overdominance hypothesis* and dates from proposals by Shull and by East in 1908.

An alternative explanation for heterosis, based on the superiority of dominant alleles when the recessive alleles are deleterious, is known as the *dominance hypothesis*, and has been espoused by Jones, Fisher, Mather, and others. Essentially the dominance hypothesis proposes that inbreeding in a particular line produces homozygosity for particular recessive deleterious genes. When crosses are made between such inbred lines, deleterious recessives of one line are hidden by the dominants of

TABLE 33–5

Variance comparisons between inbreds and hybrids for different characters*

	Inbreds	***Hybrids***
Drosophila melanogaster		
lifetime egg production (Gowen and Johnson) C.V.	48	31
wing length (Robertson and Reeve) C.V.^{2}	2.35	1.24
abdominal bristles (Rasmuson) M.S.	.1466	.0857
Chickens (Schultz) V.C.		
egg shape	4.235	3.585
egg weight	8.347	4.610
shank length	.0589	.0351
Mice (Chai) C.V.^{2}		
birth weight	119	59
weight at 3 weeks	98	47
weight at 60 days	24	19

From Lerner, 1954, and Falconer, 1960, after sources whose names are given in parentheses.

* The values given are derived from averages of two or more inbred lines and the hybrids between them. The units used are measures of variation, either mean squares (M.S.), variance components calculated from mean squares (V.C.), or coefficients of variation (C.V.).

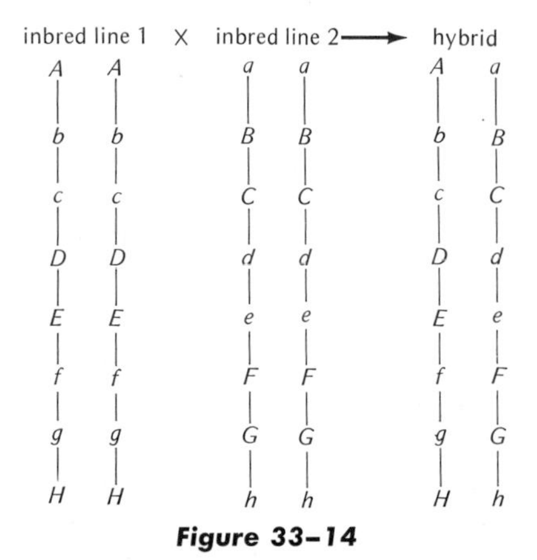

Figure 33–14

Cross between inbred lines homozygous for different deleterious recessives on a particular chromosome leading to a heterotic hybrid. The chromosome pair in the hybrid has all the deleterious recessive genes of both parents but also bears the normal dominant genes for each recessive.

the other line and the hybrid therefore appears to be heterotic (Fig. 33–14). Note, however, that it is the presence of dominant genes rather than heterozygous loci that produces heterosis.

If the dominance hypothesis is true, we might expect that pure heterotic lines can ultimately be derived that are homozygous for dominant alleles at all loci. However, although this had been attempted many times it has not yet been attained. One reason for this failure may be seen by studying Fig. 33–14 more closely. Note that if the genes *A* to *H* are considered to be on a single chromosome, crossing over in the hybrid would have to occur at a number of prescribed points simultaneously in order to produce a purely dominant chromosome. That is, crossing over must take place between *A* and *B*, *C* and *D*, *E* and *F*, and *G* and *H*. Crossing over at less than all of these points or at different points along the chromosome (i.e., *B* and *C*, *D* and *E*, *F* and *G*) would be inadequate for the purpose of obtaining only dominant alleles. The rarity or absence of such fortuitous crossing-over events among linked genes is therefore used by proponents of the dominance hypothesis to account for the absence of purely dominant heterotic lines.

According to the overdominance hypothesis, however, the failure to achieve pure breeding heterotic lines lies simply in the association between heterosis and heterozygosis. Because the F_1 hybrid is most heterozygous, further inbreeding produces homozygous strains with less fitness and productivity.

At present, clear-cut evidence has not yet been presented as to which of these two mechanisms is responsible for heterosis. It seems likely that in some cases dominance is sufficient to explain the observed superiority of hybrids. Other cases, in which superior fitness seems to reside in only one or a few gene differences, point strongly to the mechanisms of overdominant heterozygotes in certain environments (see, for example, Wills and Nichols). Suggestions have been made by Crow and by Kimura and Ohta that both mechanisms may operate, some loci being dominant, some overdominant. Discussions on this subject, pro and con, can be found in Falconer (1960), Lerner (1954, 1958), Mather (1955) and Wallace (1968). Experiments that have attempted to distinguish between these alternatives are reported by Wallace (1958), Falk (1961), Mukai et al. (1966), and others.

PROBLEMS

33-1. (a) If a small population of 75 hermaphroditic marine invertebrates are all diploid and unrelated to each other, what would be the inbreeding coefficient of the next generation if the parents all released their gametes into the water at the same time so that random fertilization could occur between all gametes? (b) If the population size remained constant and there were no outside forces acting to change the frequencies of genes or genotypes, what would be the inbreeding coefficient after five generations? (c) If self-fertilization were impossible in this group, what would be the inbreeding coefficient after one generation? (d) After five generations?

33-2. (a) What general change in inbreeding coefficients would you expect if the population size in the above example were 150 instead of 75? (b) If it were 25 instead of 75?

33-3. What inbreeding coefficient would result after five generations of random mating in a population of dioecious individuals that always consisted of 40 females and 10 males?

33-4. (a) If a population of 1000 cross-fertilizing sexual diploids remained constant in size for 1000 generations, what degree of heterozygosity would remain? (b) What would be the inbreeding coefficient after this length of time?

33-5. If the initial frequencies of a pair of alleles, *A* and *a*, in Problem 4 were .6 and .4, respectively, what genotypic frequencies would you expect after the 1000 generations of random mating?

33-6. A population has an initial frequency of allele *A* at .7 and allele *a* at .3. (a) What are the frequencies of genotypes *AA*, *Aa*, and *aa* when the inbreeding coefficient is .2? (b) When the inbreeding coefficient is .5? (c) When inbreeding is complete (F = 1).

33-7. Calculate the inbreeding coefficient of F in the following pedigree:

A B C
D E
F

33-8. Calculate the inbreeding coefficient of F in Problem 7 if the inbreeding coefficient of B is .4.

33-9. Calculate the inbreeding coefficient of H in the following pedigree:

A B C
D E F
G
H

33-10. Calculate the inbreeding coefficient of H in Problem 9 if the inbreeding coefficients of A and F are .2.

33-11. Calculate the inbreeding coefficient of the shorthorn bull, Comet, from the following pedigree. [Note that one of the parents of Comet, Favorite (bull), is also inbred, and therefore its inbreeding coefficient must be calculated and applied to the inbreeding coefficient that it contributes to Comet.]

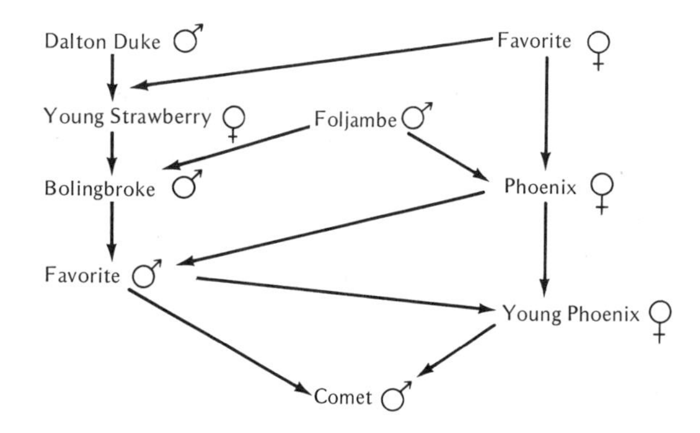

33-12. In the preceeding pedigree, what is the inbreeding coefficient for Comet for sex-linked genes?

33-13. If a continuously self-fertilizing population began with a single heterozygote, how many generations would it take to achieve approximately 97 percent homozygotes?

33-14. If a diploid population began with one individual that was heterozygous for a single gene difference and three homozygotes, what percentage of the population would be heterozygous after five generations of self-fertilization?

33-15. If the frequency of autosomal alleles *A* and *a* in a random-mating population is .8 and .2, respectively, what will be the frequency of genotypes after: (a) One generation of self-fertilization? (b) Five generations of self-fertilization? (c) One generation of full-sib mating? (d) One generation of first-cousin mating?

33-16. If the alleles *A* and *a* in Problem 15 were on the X chromosome (males XY, females XX), what would be the genotypic frequencies in males and females after one generation of full-sib mating?

33-17. In a small population the following numbers of genotypes were observed: 8 *AA*:24 *Aa*:68 *aa*. Assuming that the only forces acting to change genotypic frequencies are those caused by inbreeding, what inbreeding coefficient would you suggest for this population?

33-18. A population began with a gene frequency $A = .6$ and $a = .4$ and existed for a number of generations without any forces acting to change genotypic frequencies other than inbreeding. (a) What would be the genotypic frequencies (*AA*, *Aa*, and *aa*) if the total inbreeding coefficient attained was .4? (b) What would be the total inbreeding coefficient if the genotypic frequencies after inbreeding were $AA = .408$, $Aa = .384$, and $aa = .208$?

33-19. Assuming that the gene frequency of alkaptonuria is the same in all populations, what inbreeding coefficient would you assign to a small population in which alkaptonuric individuals (homozygous recessives) appear in a frequency of .0005 as compared to a large random-mating population (F = 0) in which they appear in frequency .000001?

33-20. The following is a pedigree of a female albino rat (homozygous recessive, *aa*, shaded figure no. 1) with given genotypes for certain individuals. (a) What are the chances that an offspring produced from a mating between the albino and its sibling (1 × 2) will be albino? (b) What are the chances that an offspring produced from the first-cousin mating (1 × 3) will be albino?

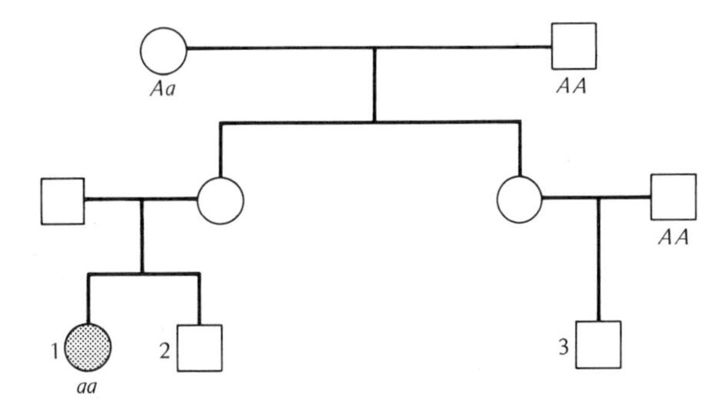

33-21. A large population carries two different gene pairs, *Aa* and *Bb*, in frequencies .6 *A*:.4 *a* and .999 *B*:.001 *b*. (a) What frequencies of homozygous recessives are expected for each gene if only first cousins were to mate? (b) How many times greater are these frequencies than the frequencies of homozygotes expected without inbreeding?

33-22. In some human populations the gene frequency for infantile amaurotic idiocy (recessive) is about .01. (a) What is the probability for this disease to occur in a child from a mating between two nonrelated normal individuals? (b) Will this probability increase or decrease in a mating between two *normal* first cousins?

33-23. In a large randomly bred population, the frequencies of a pair of alleles are .7 *A* and .3 *a*. (a) If *A* is completely dominant, and a quantitative character is affected by these genes so that its measurements are 1.0 for *AA* and *Aa*, and 0.0 for *aa*, what is the average value of this character? (b) What average value would this character have if this entire population were the result of one generation of full-sib mating?

33-24. (a) If a quantitative character were depressed by inbreeding to the extent calculated in (b) of Problem 23, what measurable effect would be produced if the population was then completely outbred (F = 0)? (b) How would this indicate, in the present instance, whether heterosis was caused by dominance or overdominance?

33-25. In a plant-breeding experiment one generation of self-fertilization caused a marked average reduction in yield when compared to the yield of a completely outbred population. On the other hand, crosses between certain of these self-fertilized lines produced an increase in yield greater than that of the outbred population. How would you explain these results?

33-26. (a) What is the probability that a self-fertilized corn plant heterozygous for four closely linked genes *AbCd/aBcD* will produce a homozygote if there is 1 percent crossing over between each locus and interference is absent? (b) How many generations are required to produce the homozygote *ABCD/ABCD*, if multiple crossing over is prevented in the region *A–D*?

REFERENCES

CROW, J. F., 1952. Dominance and overdominance. In *Heterosis*, J. W. Gowen (ed.). Iowa State Coll. Press, Ames, pp. 282–297.

DOBZHANSKY, TH., 1955. *Evolution, Genetics, and Man*, John Wiley, New York.

DUVICK, D. N., 1965. Cytoplasmic pollen sterility in corn. *Adv. in Genet.*, **13**, 1–56.

EAST, E. M., and D. F. JONES, 1919. *Inbreeding and Outbreeding*. Lippincott, Philadelphia.

EMERSON, S., 1948. A physiological basis for some suppressor mutations and possibly for one gene heterosis. *Proc. Nat. Acad. Sci.*, **34**, 72–74.

FALCONER, D. S., 1960. *An Introduction to Quantitative Genetics*. Oliver and Boyd, Edinburgh.

FALK, R., 1961. Are induced mutations in *Drosophila* overdominant? II. Experimental results. *Genetics*, **46,** 737–757.

FISHER, R. A., 1965. *The Theory of Inbreeding*, 2nd ed. Oliver and Boyd, Edinburgh.

GOWEN, J. W. (ed.), 1952. *Heterosis.* Iowa State Coll. Press. Ames.

JONES, D. F., 1917. Dominance of linked factors as a means of accounting for heterosis. *Genetics*, **2,** 466–479.

———, 1924. The attainment of homozygosity in inbred strains of maize. *Genetics*, **9,** 405–418.

———, 1958. Heterosis and homeostasis in evolution and in applied genetics. *Amer. Naturalist*, **92,** 321–328.

KIMURA, M., and T. OHTA, 1971. *Theoretical Aspects of Population Genetics.* Princeton Univ. Press, Princeton, N.J.

LERNER, I. M., 1954. *Genetic Homeostasis.* John Wiley, New York.

———, 1958. *The Genetic Basis of Selection.* John Wiley, New York.

MATHER, K., 1955. The genetical basis of heterosis. *Proc. Roy. Soc. Lond.* (*B*), **144,** 143–150.

———, 1973. *Genetical Structure of Populations.* Chapman and Hall, London.

MCDANIEL, R. G., and I. V. SARKISSIAN, 1968. Mitochondrial heterosis in maize. *Genetics,* **59,** 465–475.

MUKAI, T., I. YOSHIKAWA, and K. SANO, 1966. The genetic structure of natural populations of *Drosophila melanogaster.* IV. Heterozygous effects of radiation-induced mutations on viability in various genetic backgrounds. *Genetics*, **53,** 513–527.

SCHWARTZ, D., 1964. A second hybrid enzyme in maize. *Proc. Nat. Acad. Sci.*, **51,** 602–605.

SHULL, G. H., 1952. Beginning of the heterosis concept. In *Heterosis,* J. W. Gowen (ed.). Iowa State Coll. Press, Ames, pp. 14–48.

SPRAGUE, G. F., 1972. The genetics of corn breeding. *Stadler Genetics Symp.*, **4,** 69–81.

WALLACE, B., 1958. The average effect of radiation-induced mutations on viability in *Drosophila melanogaster. Evolution*, **12,** 532–552.

———, 1968. *Topics in Population Genetics.* Norton, New York.

WILLS, C. J., and L. NICHOLS, 1971. Single gene heterosis in *Drosophila* revealed by inbreeding. *Nature*, **233,** 123–125.

WRIGHT, S., 1921. Systems of mating. *Genetics*, **6,** 111–178.

34
GENETIC STRUCTURE OF POPULATIONS

The genetic structure of natural populations obviously departs from many of the ideal conditions considered in previous chapters; that is, populations are not of constant size, nor always of the same mating pattern, nor uniformly subject to constant conditions of mutation, migration, and selection. It is also obvious that because of multiple alleles there are usually more than three diploid genotypes for any one locus and that, through interaction, the fitness of these genotypes must depend upon genes at other loci. Because of these complexities it is therefore not surprising that the genetic structure of populations is quite difficult, if not impossible, to predict mathematically in detail, even by the most elaborate techniques. A more popular approach—and one that is perhaps more useful—has been to make measurements of various characteristics present in natural and experimental populations and then, in combination with mathematical analysis, use these observations to derive a concept of the genetic structure of populations. This approach began with a few scattered observations at the turn of the century, and has received special impetus in the last few decades through the work of Chetverikoff, Dobzhansky, Dubinin, Fisher, Ford, Haldane, Wright, and many others.

OPTIMUM PHENOTYPES AND SELECTION PRESSURE

Because of selection over long periods of time, most populations may be considered to have achieved phenotypes that are optimally adapted to their surroundings. That is to say, many phenotypes will tend to cluster around some value at which fitness is highest. Individuals that depart from these optimum phenotypes may therefore be expected to possess lower fitness than those closer to the optimal values. A classic study by Bumpus in 1899 was the measurement of nine different characters of sparrows that had died in a storm and of sparrows that had survived. For eight out of nine of these characteristics Bumpus found that the surviving sparrows had measurements that tended to cluster around intermediate phenotypic values, while the dead sparrows showed much greater variability. In Bumpus' terms, "It is quite as dangerous to be conspicuously above a certain standard of organic excellence as it is to be conspicuously below the standard." This view has since been supported by many studies on a variety of organisms including snails, lizards, ducks, chickens, and numerous other organisms.

Humans, too, are not excluded from selection for optimum values, and an actual example concerns measurements of the birth weight of babies in a London hospital. Data collected by Karn and analyzed by Karn and Penrose showed that of 6693 female births during a certain period, there were 6419 survivors 1 month later. The proportions of survivors at various birth weights are shown in Fig. 34–1. Note that the optimal birth weight, that which has the highest proportion of survivors, is 8 pounds. The further birth weight departs from this value, either more or less, the smaller the frequency of survivors. Data of this kind therefore enable us to measure an important attribute of a population, the degree to which selection acts upon a population because not all its members have achieved an optimum phenotype. In the present case the overall mortality was 274/6693 = 4.1 percent, and the mortality of the optimum 8-pound class was only 1.2 percent. In other words, 4.1 − 1.2 = 2.9 percent additional mortality occurred in the population among the phenotypes which were not of the optimal 8-pound class. This 2.9 percent value may therefore be used as a measure of the intensity or pressure of selection for an optimum birth weight.

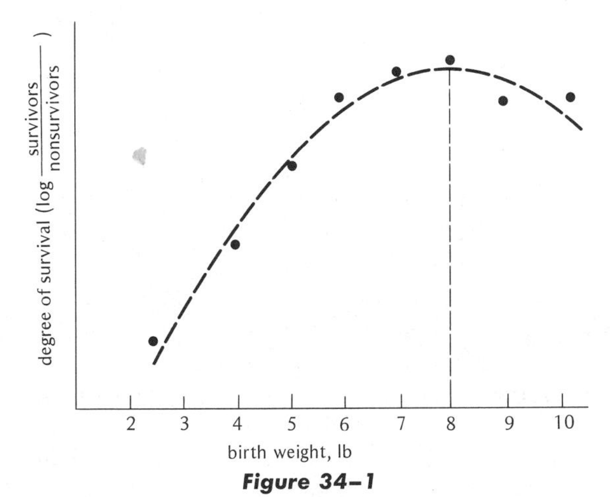

Figure 34–1

Relationship between birth weight and the degree of survival among 6693 female births in a London obstetric hospital. (After Karn and Penrose.)

In more exact terms, *selection intensity* or *pressure* (I) can be defined as the difference in survival rates between optimal (S_O) and suboptimal (S_S) phenotypes, multiplied by the frequency of suboptimal phenotypes [f(s)] in the population, $I = (S_O - S_S) \times f(s)$.* When selection pressure is zero, all phenotypes are optimal [$f(s) = 0$], and when selection pressure is 1, all phenotypes are suboptimal, with a survival rate zero ($S_O - S_S = 1$). Obviously existing natural populations that show any selective differences among

*Haldane, who first analyzed the data of Karn and Penrose from the viewpoint of selection intensity, calculated its value as $I = -\log_e (S_o/S)$, where S_o is the survival frequency of the optimal and S is the survival frequency of all phenotypes. According to Haldane's formula, however, the frequency of optimal phenotypes is dependent upon the intensity of selection and can never be more than $1/e = .368$. The present formula was derived in order to take into account the many possible differences in frequency between optimal and suboptimal phenotypes. This formula is restricted to estimating mortality caused by selective differences and does not take into account mortality that affects all phenotypes randomly. For another type of selection intensity formula see Van Valen (1965).

them will have selection pressures between these two extremes. The data of Karn and Penrose showed that 718 babies out of 727, or 98.8 percent, survived in the optimal 8-pound class and 5701 babies out of 5966, or 95.6 percent, survived in the other classes. Since the frequency of the suboptimal phenotypes was 5966 out of a total of 6693 babies, or 89.2 percent, selection pressure at this stage is $I = (.988 - .956) \times .892 = .029$.

The dependence of I upon the frequency of suboptimal phenotypes (s), as well as upon the departure of their survival rate from the optimum ($S_O - S_S$), has been graphed in Fig. 34-2. This relationship shows the expected increase in I with increased values of s and $S_O - S_S$.

The values of $S_O - S_S$ and of s vary between different populations and undoubtedly vary at different times in the life cycle of any population. It is nevertheless unlikely that high selective intensities can persist for many generations, since any increase in I because of a new departure from an optimum phenotype is an additional burden on a population that is already suffering from mortality caused by the usual nonselective factors that affect all phenotypes equally. However, even though small in value, I may nevertheless have relatively significant effects at certain stages. In the present human data, for example, there was a total of 4.1 percent deaths, of which 2.9 percent were caused by selection against suboptimal phenotypes, or 2.9/4.1 = 70.8 percent deaths were caused by selection against these phenotypes at birth. Furthermore, the data mentioned here cover only a small part of the life cycle, that between birth and 1 month of age. Selection undoubtedly acts at many intervals throughout the life cycle, although with varying pressure, so that a large number of deaths may arise because of departure from optimal values.

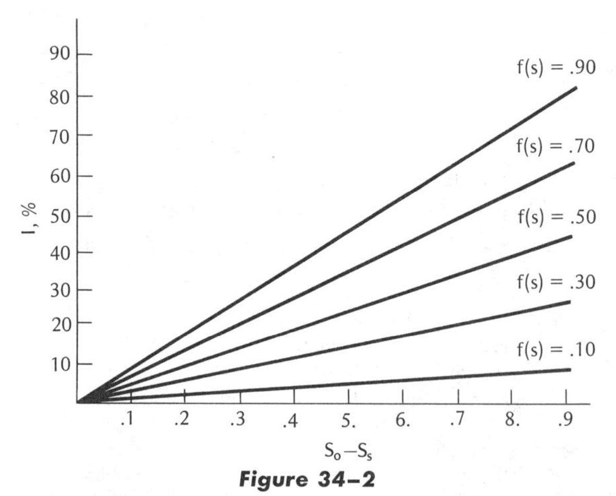

Figure 34-2

Selection pressures on a population (I) under different frequencies of suboptimal phenotypes [f(s)] and different departures of the suboptimal survival rate from the optimal ($S_o - S_s$).

KINDS OF SELECTION

In the data of Karn and Penrose the departure from optimum values occurs both at high and at low birth weight. The primary effect of selection pressure is therefore to reduce the variability of birth weight rather than to change its average value. The average birth weight in Karn and Penrose's sample was 7.06 ± 1.22 before selection and 7.13 ± 1.10 after selection. In other words, there was a change in average birth weight of $(7.13 - 7.06)/7.06$, or 1 percent, but a much greater reduction in the standard deviation, or variability, of $(1.22 - 1.10)/1.22$, or 10 percent. This reduction in frequency of extreme phenotypes has been termed *centripetal* or *stabilizing selection.* It signifies selection for an intermediate "stable" value. Not all selection, however, is of this type, since selection may well favor an extreme phenotype; this is known as *directional selection.* Much of the selection practiced by animal and plant breeders is of this kind, since they select for extremes of yield, productivity, resistance to disease, etc. The role of directional selection in evolution is also of importance when the environment of a population is changing and only extreme phenotypes happen to be adapted for new conditions.

Selection, whether stabilizing or directional, may be constant from one generation to the next if the selective environment is not fluctuating. However, when the environment is not stable between generations or between seasons, the optimum phenotype and consequently the optimum genotype may shift accordingly. Such shifts may

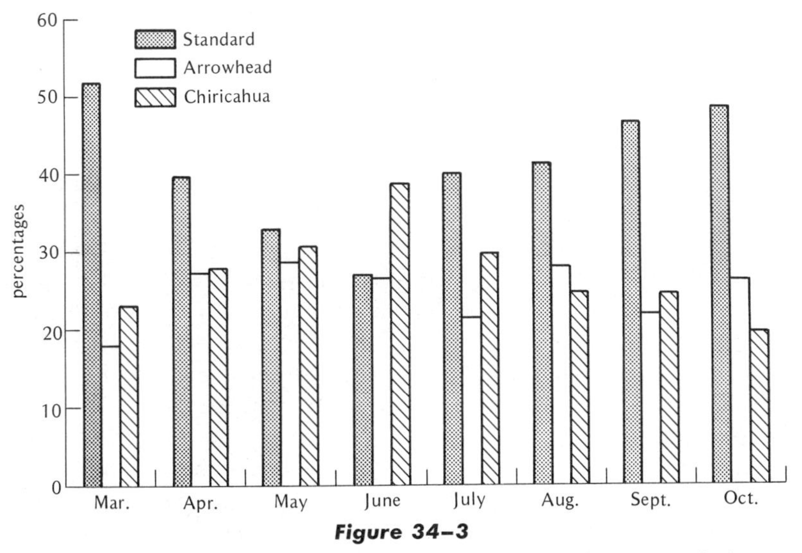

Figure 34–3

Percentages of different 3rd-chromosome arrangements in Drosophila pseudoobscura *at different months during the year in Pinon Flats, California. (After Dobzhansky, 1947.)*

result in selection in one direction for one generation or season, and selection in the opposite direction for the next (Fig. 34–3). This type of selection, called *cyclical selection*, will help to maintain genetic differences in a population, since different traits will be advantageous at different times (see Haldane and Jayakar).

In addition a population may be subjected within a single generation to various environments to which different genotypes among its members are most suited. This has been termed *disruptive selection*, and experiments by Thoday and others have shown that considerable genetic differences may be maintained in such populations in spite of cross-breeding between individuals from different environments. A diagrammatic comparison of these forms of selection is shown in Fig. 34–4 for a trait that is normally distributed. If this distribu-

stabilizing selection (selection for the mean):

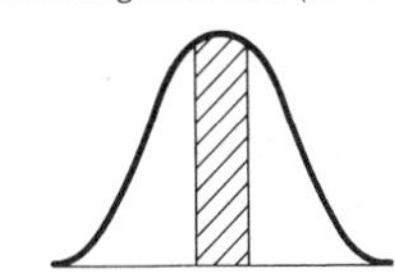

directional section (selection for one extreme):

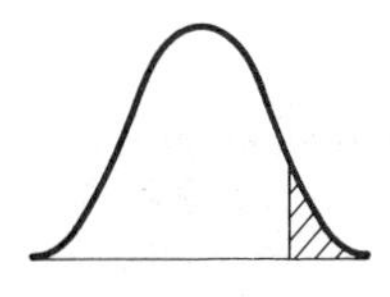

cyclical selection (selection favoring different phenotypes alternately):

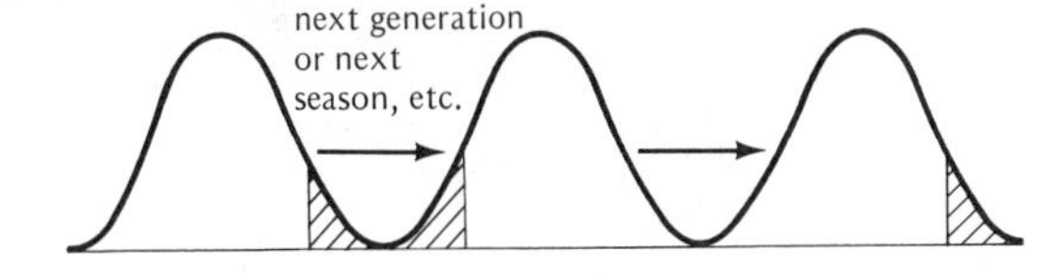

disruptive selection (selection for both extreme phenotypes at the same time):

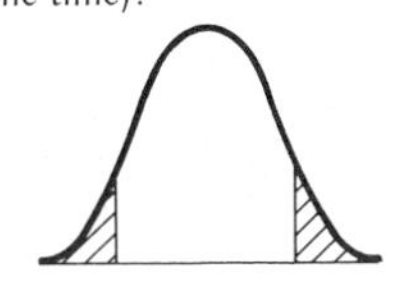

Figure 34–4

Comparison of different types of selection in terms of that section of the population selected as parents for the next generation (hatched areas). The population distribution is considered to be normal with respect to the selected trait; most individuals are at the mean, fewer at the extremes.

tion is caused by a number of different genotypes, all forms of selection except directional will cause the preservation of genetic variability. Thus selection may help sustain genetic differences, as do the factors of mutation and heterozygous advantage.

FISHER'S FUNDAMENTAL THEOREM OF NATURAL SELECTION

To the geneticist, selection among different phenotypes is usually of little consequence to the future evolution of a population unless it produces genetic change. By the same reasoning, the absence of genetic differences between phenotypes in homozygous "pure lines" offers little opportunity for selection to produce any noticeable effect. For example, in each of two pure lines of beans, Johannsen selected subsidiary lines for both light and heavy seeds each generation for six generations (Table 34–1). Although the difference between the seeds selected for these subsidiary lines was never less than 10 centigrams and occasionally as high as 40 centigrams, the difference between the seeds produced by the offspring of these selected lines never rose to more than 2 centigrams. In fact the weights among the offspring were occasionally reversed in direction so that the heavy-selected lines weighed less than the light-selected lines (negative values in Table 34–1)! Since each of Johannsen's pure lines had been long inbred and was probably homozygous for most or all genes, the absence of selection progress can be ascribed to the lack of genetic variability upon which selection could act.

In the same way as we think of selection for weight in Johannsen's lines, we can also think of selection for fitness in any population: *The greater the genetic variability upon which selection for fitness may act, the greater the improvement in fitness.* This principle is the basis of Fisher's *fundamental theorem of natural selection*, which states in mathematical terms that the fitness of a population increases at a rate that is proportional to the genetic variability or genetic differences in fitness present in the population. It is easy to see that if a population in a particular environment were completely homozygous for all genes, selection for fitness in a changed environment would have little effect, and produce no genetic improvement. Since populations do not always face constant environmental conditions, selection is not always for only one optimal genotype, and consequently genetic variability must be maintained for the population to survive. To what extent is genetic variability present?

TABLE 34–1

Comparison of average parental differences in weight for lines selected for light and heavy seeds in princess beans and resultant average differences in weight between their offspring*

	Line No. 1		*Line No. 19*	
Year	Differences Between Parental Beans	Average Differences Between Offspring	Differences Between Parental Beans	Average Differences Between Offspring
1902	10	1.70	10	−1.05
1903	25	−4.31	17	.81
1904	37	2.09	12	1.25
1905	40	.09	12	.89
1906	38	−1.38	16	1.95
1907	25	−1.41	23	− .41

After Babcock and Clausen, from Johannsen.

* The average differences in weight between offspring are determined by subtracting the average weight of offspring produced by the lighter parental beans from the average weight of offspring produced by the heavier parental beans.

GENETIC VARIABILITY IN NATURAL POPULATIONS

Among the first investigations of genetic variability or polymorphism in natural populations were those carried out on *Drosophila* species by the Russian workers Chetverikoff, Timoféeff-Ressovsky, Dubinin, Dobzhansky, and others. Because of relatively simple techniques for making chromosomes homozygous (i.e., inbreeding and utilizing marker chromosomes, see p. 534), many recessive genes normally heterozygous in the wild could be observed in homozygous condition in the laboratory. Surprisingly, a large variety of gene mutations were shown to exist in wild populations, although in low frequencies. When these tests were extended to detect the presence of recessive genes causing lethality, subvitality, and sterility, the results appeared even more impressive. Wild chromosomes made homozygous showed a distribution of viability ranging from zero to above normal, but were mainly in the below-normal range (Fig. 34–5). If we consider that these flies are diploid, the chances for each individual fly to carry such genes are also quite high. For example, the data on *Drosophila persimilis* in Table 34–2 show that .745 of the 2nd chromosomes, .773 of 3rd chromosomes, and .719 of 4th chromosomes in this sample are free of lethals and semilethals. Since the fly carries two of each chromosome, $.745 \times .745 \times .773 \times .773 \times .719 \times .719$, or only 17 percent, of all flies are free of lethals. If we consider the high frequency of all subvitals and also include genes causing sterility (last two columns of Table 34–2), it is almost a certainty that a fly will carry more than one deleterious recessive gene. However, since many loci are involved, not all such genes are allelic to each other, and a fly may carry a number of such genes without showing ill effect, as long as they are completely recessive.

When genetic material assumes a number of different recognizable arrangements on a chromosome, e.g., inversions, the degree of heterogeneity can easily be noted by cytological examination. In *Drosophila pseudoobscura*, for example, populations are known which maintain many such different arrangements throughout the year (see Fig. 34–3, for example). It has been suggested that the restricted amount of recombination conferred upon these inversions when present with other gene arrangements (see pp. 502 ff.) enables them to maintain linked groups of genes that are advantageous under particular environmental conditions. Such polymorphisms are widespread and include the maintenance of translocations (grasshoppers, snails, *Oenothera*, *Datura*), extra chromosomes (platyhelminthes, shrews, many insects, corn, rye, and other plants), and other cytological differences between individuals.

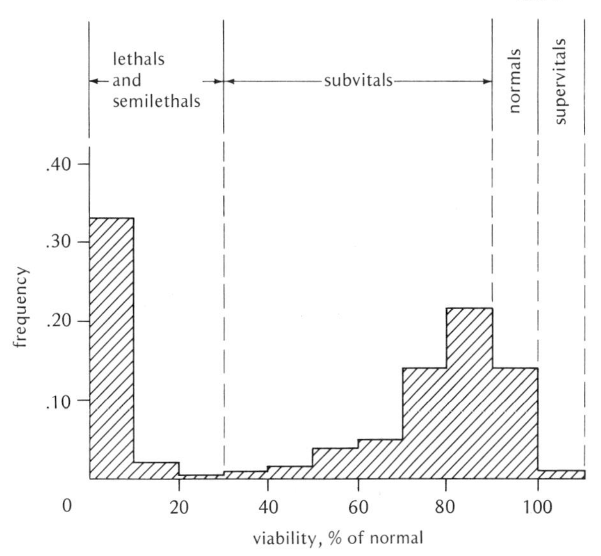

Figure 34–5

Frequency distribution of viabilities among 177 second chromosomes taken from a wild population of Drosophila willistoni *and made homozygous. These viabilities are based on the relative number of wild-type homozygous flies that emerge from cultures in which the expected frequency of such homozygotes is 33.3 percent (see Fig. 23–3). (After Krimbas.)*

Chromosomal differences and deleterious genes, however, are only two indications of the large amount of concealed genetic variability. In recent years a number of investigations have shown that many genes in natural populations are polymor-

TABLE 34–2

Polymorphism as reflected in percentages of chromosomes from natural populations of *Drosophila* species giving various effects when in homozygous condition

Species	*Chromosome*	*Lethals and Semilethals*	*Subvitals*	*Female Steriles*	*Male Steriles*
melanogaster	2nd	12.6–61.3	26.1–66.6		
persimilis	2nd	25.5	84.4	18.3	13.2
	3rd	22.7	74.2	14.3	15.7
	4th	28.1	98.4	18.3	8.4
prosaltans	2nd	32.6	33.4	9.2	11.0
	3rd	9.5	14.5	6.6	4.2
pseudoobscura	2nd	21.3–33.0	93.5	10.6	8.3
	3rd	25.0	41.3	13.6	10.5
	4th	25.9	95.4	4.3	11.8
willistoni	2nd	28.4–41.2	71.9	40.5	64.8
	3rd	25.6–32.8		40.5	66.7

Data from various sources; Dobzhansky, Spassky, Dubinin, Krimbas, Ives, Pavan, Cordeiro, Cavalcanti, Townsend, and others (see review of Dobzhansky 1959).

phic in respect to alleles that determine slight enzyme or protein differences. These differences have been mostly detected in studies that measure the mobility of an enzyme or protein in an electric field; different forms of the protein may migrate different distances from one of the poles (see Fig. 33–13). Since each different electrophoretic form may be caused by a different amino acid sequence in one or more of the polypeptide chains that compose the protein, they may each be considered to signify an allelic difference or *allozyme*.* The polymorphism discovered in this fashion provides, according to Lewontin and Hubby, a means of estimating genetic variability.

As shown in Table 34–3, this estimate is astonishingly high for quite a number of species. In man, for example, almost 30 percent of the 71 different proteins tested showed the presence of polymorphic alleles, with a 7 percent average heterozygosity per individual. If we consider that perhaps only one out of three amino acid changes in a protein cause charge differences that can be detected by electrophoretic techniques, there are undoubtedly many additional loci at which polymorphism exists. For various animal species including those listed in Table 34–3, Lewontin estimates that in general about two-thirds of all loci are polymorphic, and that an individual is heterozygous for about a third of all its loci! Considering the thousands of genes that each of these species possesses, attempts are therefore being made to explain the source for this extensive polymorphism. As will be discussed later (p. 816), some of these explanations rely primarily on selection, whereas others emphasize the prevalence of neutral mutations. Before pursuing such explanations, however, it would be valuable to examine some additional evidence for genetic variability and note some of the consequences.

*To ensure that these electrophoretic differences are caused by allelic differences rather than by posttranscriptional modifications (p. 688), or interactions with buffer solutions used in the procedure, etc., breeding tests should be performed (Chapter 9).

CANALIZATION

It is a common attribute of populations that they show a high degree of uniformity for various characters that are assumed to be strongly associated with fitness. For example, wing shape and body proportions in insects appear to be of utmost importance for survival in nature, and the variability of such phenotypic characters is usually small.

Since these phenotypes are genetically controlled, it might therefore seem as though the absence of phenotypic variability is caused by an absence of genetic variability. In 1953, however, Waddington showed that the high production of optimum phenotypes in a population does not necessarily mean that genetic variation for such characters is lacking.

The experiment he performed was to subject a stock of wild-type *Drosophila melanogaster* to environmental shocks that caused the production of occasional flies lacking wing crossveins. This character is usually caused by a specific sex-linked mutation, *crossveinless*, but in the present experiment, as in many others, it was produced as a noninherited phenocopy (see p. 188). Waddington, however, carried his experiment a number of important steps beyond the usual phenocopy treatment. He subjected the flies to environmental shock each generation and bred a separate line only from those showing the crossveinless effect. Within a short time the crossveinless character in this selected line began to appear in high frequency as a result of this continued phenocopy treatment. Even more important, however, some of these flies were now crossveinless *without* treatment. Such flies could then be selected, and a new stock established. The crossveinless effect in this stock was shown to be caused by numerous genes independent of the single *crossveinless* sex-linked gene. Phenocopy treatment had thus "exposed" these hidden *crossveinless* polygenes (see p. 280) to selection so that their small individual contributions could now be selected to produce the crossveinless phenotype.

To account for these results Waddington proposed that phenotypic characters such as crossveins are *canalized* (see p. 720), so that their development is normally unaffected by environmental stresses or by the underlying genetic variability that he demonstrated existed in his experiments. Only rare events such as heat shock, or the *crossveinless* mutation, will upset canalization and cause a change in the normal crossvein phenotype.

The existence of canalization for other phenotypes was soon found by the same method of disrupting development with environmental shocks and then selecting those genotypes showing marked divergence from the canalized developmental pathways. In 1958 Dun and Fraser showed that a genetic shock such as the presence of a mutant gene in mice may also upset the development of a character and expose considerable underlying genetic variability. This same approach was used by Rendel in the following experiments with the *scute* mutation in *Drosophila melanogaster*.

Normally wild-type *Drosophila* have four bristles on the scutellum, the posterior dorsal part of the thorax. However, in stocks homozygous for the recessive sex-linked *scute* gene (*sc*), the number of scutellar bristles ranges from zero to three. Rendel

TABLE 34–3
Estimates of genetic variability found in natural populations of a number of organisms

Species	Number of Populations Examined	Number of Loci (Proteins) Studied	Proportion of Loci Polymorphic per Population	Heterozygosity per Locus
Homo sapiens (man)	1	71	.28	.067
Mus musculus musculus (house mouse)	4	41	.29	.091
Peromyscus polionotus (old-field mouse)	7	32	.23	.057
Drosophila melanogaster (fruit fly)	1	19	.42	.119
Drosophila persimilis	1	24	.25	.106
Drosophila pseudoobscura	10	24	.43	.128
Drosophila simulans	1	18	.61	.160
Drosophila willistoni	10	20	.81	.175
Limulus polyphemus (horseshoe crab)	4	25	.25	.061

Abridged from Lewontin, 1974.

practiced selection on the *scute* stocks in order to obtain both low and high scutellar bristle lines. Selection proved to be highly successful, and the average scutellar number in the high lines, for example, approximated the normal number of four. Substitution of the wild type (sc^+) for the *scute* gene in this high line now produced flies with a mean of five scutellar bristles and some with as many as six. Surprisingly the selected *sc* high lines with a mean of four bristles showed little variability, but the wild-type sc^+ flies with a mean of five bristles showed greater variability. It appeared as though a zone of canalization existed at four scutellar bristles, and flies with genotypes outside this zone were consequently more variable.

According to Rendel, the relationship between genotype and phenotype can be expressed in the form of a curve. This curve can be envisaged as being initially in the form of a straight line (Fig. 34–6a), so that a wide variety of phenotypes may be caused by a wide variety of genotypes. If an optimal phenotype is being selected, most genotypes will be distributed around a mean optimal genotype (normal distribution curve shown on ordinate of Fig. 34–6). It will obviously be advantageous if the fewest possible members of this optimal genotypic class depart from the optimal phenotype. Similarly the phenotypic expression of suboptimal genotypes will be selected so that it is progressively more optimal. This new relationship can easily be visualized by changing the genotype-phenotype curve so that the portion determining the optimal phenotype covers a larger variety of genotypes (Fig. 34–6b). As selection

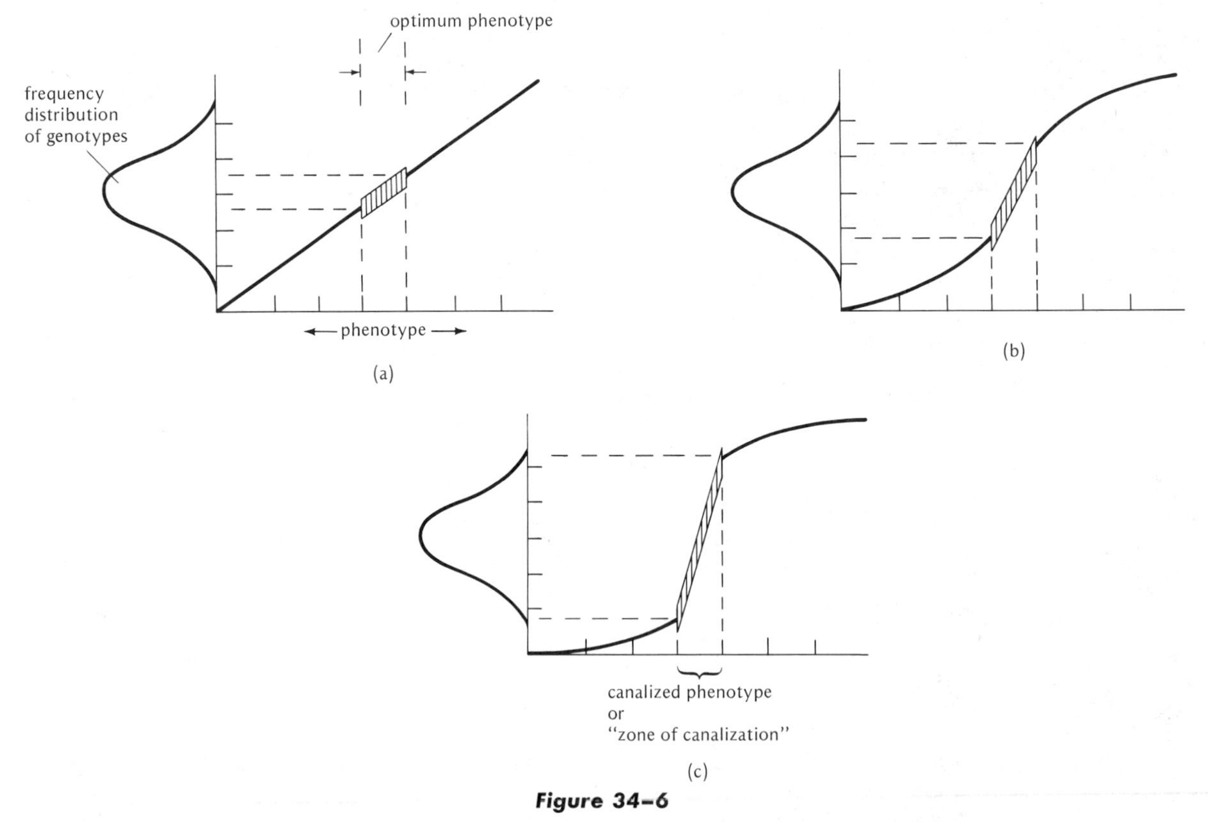

Figure 34–6

Sequence of selection for a canalized phenotype. (a) Only a small section of the genotypic distribution produces the optimum phenotype. (b and c) Larger sections of the genotypic distribution produce the optimum phenotype. In this process selection occurs for genotypes that can produce the same phenotype in spite of genotypic variability.

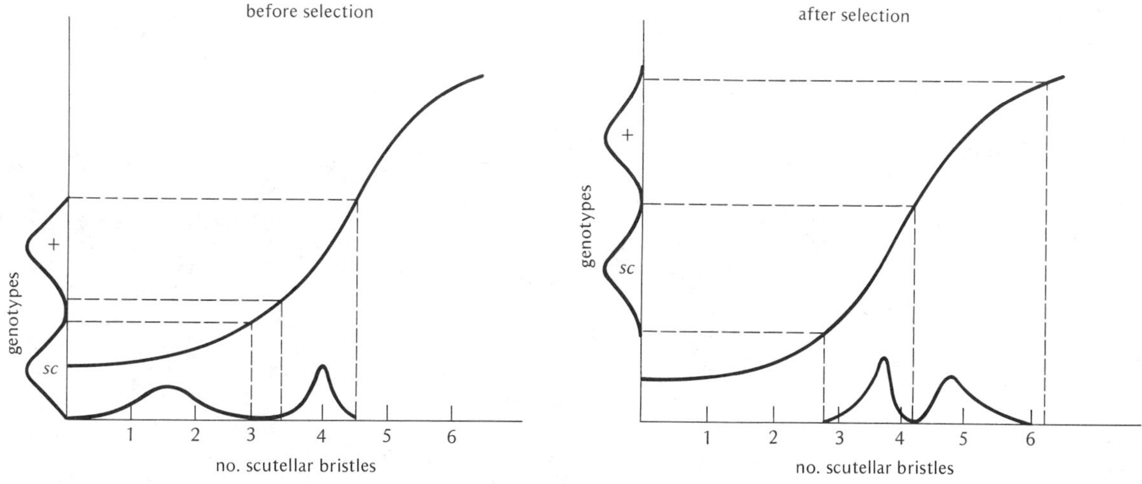

Figure 34–7

Explanation of Rendel's results in terms of canalization. The canalized phenotype is between three and four bristles and is normally produced by the wild-type genotype. Genotypes outside the canalized phenotypes (e.g., sc) ordinarily produce greater variability. After selection for increased bristle number in scute, the mutant genotype moved into the zone of canalization and its variability decreased. Substitution of the wild-type gene in the selected stock then increased the number of scutellar bristles beyond those in the canalized phenotype and therefore led to greater variability of wild type. (After Waddington, 1962.)

continues, the curve assumes an S-shaped configuration in which the steep portion now covers most of the different possible genotypes and produces a canalized phenotype (Fig. 34–6c).

Rendel's results with scutellar bristles can therefore be explained on the basis of two genotypic distributions: one containing the *scute* gene that produces an effect outside the zone of canalization, and the other for the wild-type gene that normally acts within the zone of canalization (left-hand side of Fig. 34–7). Selection among the variable genotypes revealed by the *scute* gene results in the genotypic distributions of *scute* and $scute^+$ being moved up higher along the curve, so that their relationship is now reversed: *scute* genotypes are within the zone of canalization, and wild-type genotypes are outside this zone (right-hand side of Fig. 34–7).

The effect of selection has therefore been to change the expression of the genotypes so that more scutellar-producing substance is produced by both *sc* and sc^+. However, no matter how much scutellar-producing substance is produced, the genes controlling the zone of canalization have apparently not been affected by selection in these experiments since variability always remains less in the four-bristle genotypes regardless of whether they are *scute* or wild type.

Selection thus appears to occur on two levels: first, it changes the primary expression of the genotype, and, second, it changes the zone of canalization of the phenotype. The first type of selection is probably the more common, since it involves little more than selecting those genotypes with extreme phenotypic expression. The second type of selection involves a more complex process of changing canalization from one phenotype to another. In other words, "canalizing selection" must not only proceed to a new phenotype but must also reduce the phenotypic variability of those genotypes whose expression falls within the range of this new phenotype. That canalizing selection is also possible has been shown by Rendel and Sheldon, who selected for reduced variability at a scutellar bristle number other than four. They found they could successfully canalize the number of scutellar bristles at two in a stock homozygous for *scute*.

GENETIC HOMEOSTASIS

The canalization of many characters to produce similar phenotypes in genetically different individuals indicates again a wide variety of genetic differences normally maintained in a population. Another source testifying to the prevalence of genetic variability arises from the common observation that artificial selection for most phenotypes reaches limits beyond which selection has no further effect, although gene differences persist.

For example, Robertson and Reeve selected for large and small body size in three strains of *Drosophila melanogaster* and were able to effect a significant change within 10 to 15 generations. As shown in Fig. 34–8, the high lines reached maxima of about 10 units above the control level and the low lines reached minima of about 15 units below the control level (1 unit = 1/100 mm). Beyond these points, continued selection had little effect, and the selected populations remained at the same level until selection ended. Since further selection was ineffective at these "plateaus," it might, therefore, appear as though homozygosity had been reached for all genes determining size in each of the selected lines. The inadequacy of this explanation, however, is quickly revealed when selection is reversed (dashed line), so that the large line is selected for small size, and the small line for large size. Under these circumstances all selected lines reverse rather quickly to the control level. Only the small lines after 25 generations of selection appear to have reached genetic uniformity, and reverse selection is ineffective at that point. Thus, for the large lines (generations 6 and 17) as well as for the small lines at early stages (generation 6), considerable genetic heterogeneity must have remained for reverse selection to be effective.

Similarly other experiments demonstrate that continued inbreeding in many populations may reduce heterozygosity only to a certain level beyond which further reduction does not occur. This maintenance of genetic variability in a population in the face of all the forces acting to reduce it is an illustration of a phenomenon termed by Lerner *genetic homeostasis*.*

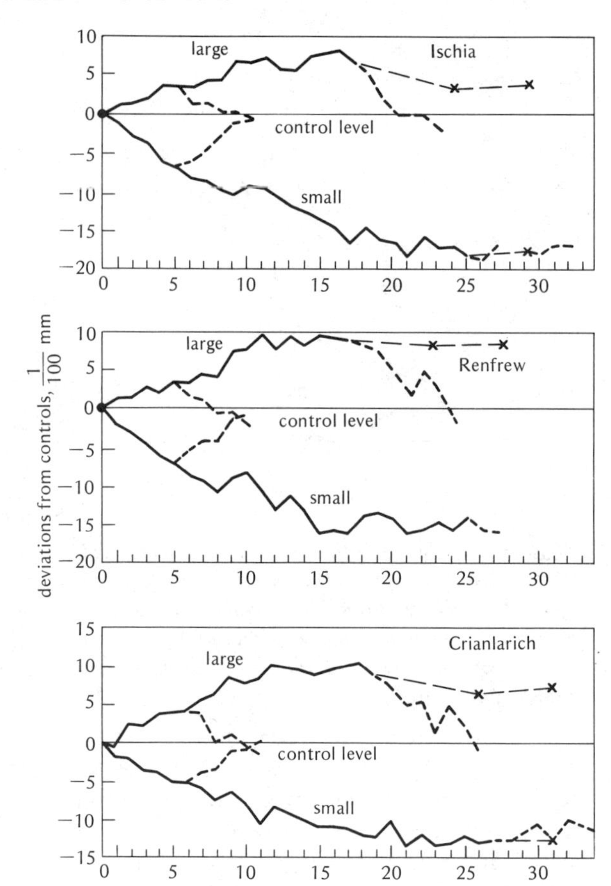

Figure 34–8

Thorax length of Drosophila melanogaster *stocks selected for large and small size. Average measurements for each generation are graphed in terms of deviation from the nonselected controls. The dashed lines represent back-selection toward the control level, and the lines marked with x's (e.g., generation 17 Renfrew strain, large line) represent relaxed selection in which no changes are attempted. (After Robertson.)*

*Homeostasis is a term devised by Walter B. Cannon to denote the tendency of a physiological system to react to an external disturbance in such a way that the system is not displaced from its normal values.

In general, genetic homeostasis depends upon the particular array of gene frequencies built up by a population over the long period of its evolution in a particular environment. Since these established gene frequencies have been selected to confer a high degree of fitness upon a population, any rapid departure from these frequencies may be expected to reduce fitness. According to Lerner, homeostasis may also arise from the necessity for the maintenance of certain levels of heterozygosity to ensure normal development. The number and location of the genes in an organism that must be heterozygous cannot be determined as yet, but it is possible that many such genes are inadvertently made homozygous during selection. That is, selection for a trait such as bristle number, when carried beyond certain limits, may result in homozygosity for a number of genes that have pleiotropic (secondary) effects on fitness and reduce viability. Such gene-frequency changes will therefore be resisted, and the population will continue to maintain genetic variability simply because heterozygous genotypes showing wild-type characteristics have superior fitness over the homozygotes favored by artificial selection. In any event, the notion that increased homozygosity does cause abnormal development is well established by the many observed cases of inbreeding degeneration (pp. 791 ff).

GENETIC LOAD AND GENETIC DEATH

In spite of the many advantages of genetic variability, many of the genes maintained by natural populations may be disadvantageous to their carriers either in certain combinations or in homozygous condition. As we have seen previously (p. 803), selection can account for a significant loss of individuals that lack optimal fitness even in long-standing populations. Thus, if we consider genetic perfection as the elimination of all deleterious inferior gene combinations, there is little doubt that most, if not all, populations are genetically imperfect.

The extent to which a population departs from a perfect genetic constitution is called the *genetic load* and is accompanied by the loss of a portion of its individuals through their *genetic death*. Genetic death is not necessarily an actual death before reproductive age but can also be expressed through sterility, inability to find a mate, or by any means that reduces reproductive ability relative to the optimum genotype. These values can be estimated in terms of the frequency and number of individuals eliminated because of selection against them. For example, if a gene is deleterious in homozygous condition, the frequency of homozygotes before and after selection will be as follows:

	AA	**Aa**	**aa**
frequency at fertilization	p^2	$2pq$	q^2
relative adaptive value	1	1	$1 - s$
frequency after selection	p^2	$2pq$	$q^2 - sq^2$

Thus, the loss in frequency of individuals, or incurred genetic load, is equal to sq^2. Therefore, if there were N individuals in the population before selection, sq^2N are now eliminated because of genetic imperfection.

This value of sq^2, however, also equals the mutation rate at equilibrium for $A \rightarrow a$ (p. 770). In other words, the genetic load caused by a deleterious homozygous recessive is equal to its mutation rate. An important feature of this relationship pointed out by Haldane is that if the mutation rate is constant, it will make little difference to the genetic load whether s is small or large. That is, if s is large, q will be small at equilibrium, and if s is small, q will be large. High selection coefficients will therefore cause the gene to be eliminated more rapidly (low q), and low selection coefficients will permit the gene to remain longer in the population (high q). In either case, however, the genetic load is still $sq^2 = u$ and the number of genetic deaths remain at sq^2N. Insofar as deleterious recessives are produced by mutation, therefore, any increase in mutation rate will cause a corresponding increase in genetic load and consequent increase in genetic death.

MUTATIONAL AND SEGREGATIONAL LOADS

The ideal of genetic perfection with its absence of genetic death is probably never reached in any species, since environments usually change with time and the advantages of different gentoypes change accordingly. It is conceivable that a population so perfectly adjusted to its environment—that is, has little or no genetic load—may become extinct within a short period because of a rapid environmental change. On the other hand, a population with a relatively large genetic load may be subjected to a new environment in which formerly deleterious genes now enable it to survive. The absence of genetic load may therefore be more detrimental to a population than its presence (see Li, and also Brues). Therefore, although the genetic load can be measured in terms of departure from the optimum genotype, the evolutionary value of a particular optimum genotype may be a very limited one; the optimum genotype may change in time, or may change from place to place, or may even differ in the same place, if, for example, a division of labor occurs between the individuals of a population (e.g., male and female).

Keeping these limitations in mind, the measurement of genetic load may nevertheless offer essential information concerning the structure of populations. According to Crow, various genetic components contribute to the genetic load, and techniques have been devised by him and others to evaluate their relative importance. Two of the most essential factors that have been considered responsible for genetic load are those caused by mutation and by segregation.

In Crow's terms, the mutational load is "the extent to which the population is impaired by recurrent mutation." Since the raw material of evolution is mutation, the mutational load is undoubtedly part of the genetic load of every species. This load has two components: one produced directly by deleterious mutations and one produced indirectly by beneficial mutations. Both components, according to Muller and others, arise because an ideal population consists only of individuals homozygous for all beneficial genes. Deleterious genes, the most common type produced by mutation, must eventually be eliminated, and therefore constitute an important part of the load. The few advantageous genes introduced by mutation will also produce a genetic load, since they will eventually replace the older genes, which now, by contrast, have become "transitional" and deleterious.

The segregational or "balanced" load, on the other hand, is restricted to those instances in which a heterozygous genotype is superior to both types of homozygotes. For a gene with two alleles, this load amounts to p^2s for the *AA* homozygotes plus q^2t for the *aa* homozygotes, as can be seen from the following calculations:

	AA	*Aa*	*aa*
frequency at fertilization	p^2	2pq	q^2
relative adaptive value	$1 - s$	1	$1 - t$
frequency after selection	$p^2 - p^2s$	2pq	$q^2 - q^2t$
reduction in frequency (load)	p^2s		q^2t

If we substitute the equilibrium frequencies for p and q (p. 766) into $p^2s + q^2t$ we obtain

$$\left(\frac{t}{s+t}\right)^2 s + \left(\frac{s}{s+t}\right)^2 t = \frac{st^2 + ts^2}{(s+t)^2}$$
$$= \frac{st(s+t)}{(s+t)^2} = \frac{st}{s+t}$$

Thus, if s and t are both about .1, the segregational load will be .01/.2, or .05. This value is considerably higher than most mutation rates and demonstrates the increased genetic load that may be expected to be caused in randomly breeding populations by segregation as compared with the load caused by mutation (p. 813). Inbreeding, however, has been shown by Crow to cause a reversal of the effects of the two types of load, so that there is greater depression of fitness if the load is mutational than if it is segregational.

Let us consider, for example, complete inbreeding which will eliminate all heterozygotes and permit only homozygotes to remain in the proportions $pAA:qaa$ (see p. 787). Since the selection coefficient against *aa* is s, the load caused by a deleterious recessive mutation in a completely inbred population is therefore qs. This value can now be used to obtain a ratio between inbred to outbred mutational loads of qs/q^2s, or $1/q$. If q is small, the increase in load upon inbreeding will therefore be a very large number. On the other hand, if the load has been exclusively segregational, complete inbreeding will also produce p*AA* and q*aa*, but the increase in load will now depend upon the selection coefficients for each type of homozygote, s and t. The total segregational genetic load upon inbreeding is therefore ps + qt. Substituting equilibrium values of p and q into this quantity gives a segregational inbred load of

$$ps + qt = \frac{ts}{s+t} + \frac{st}{s+t} = \frac{2st}{s+t}$$

The ratio of inbred to outbred segregational loads is then

$$\frac{2st}{s+t} \Big/ \frac{st}{s+t} = 2$$

This value is evidently quite small compared to the $1/q$ ratio produced by the mutational load upon inbreeding.

Of course, not all genes are fully recessive, and the effect of the mutational load upon inbreeding may cause a change considerably less than $1/q$ if some heterozygous expression of the gene appears under random mating. Also, the segregational load may arise in a multiple-allelic system in which many kinds of superior heterozygotes and many kinds of deleterious homozygotes are possible at a single locus. These deleterious homozygotes will be present in low frequency under random mating but will greatly increase in frequency upon inbreeding yielding inbred/outbred ratios of more than 2. Thus both the heterozygous expression of the mutational load and multiple allelism of the segregational load make it difficult to distinguish easily between the relative importance of these two types of load. Nevertheless the determination of these loads bears heavily on a number of important questions.

For example, whether a population follows mutational or segregational patterns will determine, to some extent, the genetic effects caused by added mutation, such as that produced by radioactive fallout. If the mutational load plays a minor role, added mutation may have a relatively smaller effect on the total genetic load than when it plays a major role. For purposes of plant and animal improvement, it is also important to know whether selection for heterozygotes or for homozygotes should be practiced. As yet, the relative value of each type of load has not been fully determined, although it seems likely that both types of load occur in most natural populations.

THE COST OF EVOLUTION: SELECTIONISTS VS. NEUTRALISTS

Whatever type of load a population bears, natural selection will cause new additions to the load by favoring some genes and discarding others. Since this process of gene replacement is an essential one for any species, it is reasonable to ask how many individuals a population must lose, in terms of genetic death, to replace a single gene. If a dominant mutation arises that is advantageous in comparison to the more frequent recessive, we have seen that the number of genetic deaths will be sq^2N for *one* generation of selection, where N is the number of individuals in the population. For *complete* replacement of the recessive gene to occur, Haldane has calculated that the total number of deaths will be determined by a factor D, multiplied by the population number N, such that $D = \Sigma\, sq^2 = -\ln p_0 = -2.303 \log_{10} p_0$, where p_0 is the initial frequency of the dominant. For example, if $p_0 = .00005$, $D = 9.9$, or the cost of evolution for this genetic change is about 10 times the average number of individuals in a single generation.* This cost (DN) will, of course, be spread over many generations, depending upon the magnitude of selection pressure (p. 803).

*The $\log_{10}$ of .00005 is 5.69897 − 10, and the negative log of this number is 10.00000 − 5.69897 = 4.30103. This is then multiplied by 2.303 to give 9.9.

If the favored gene (e.g., *A*) is not completely dominant, DN is increased, since the superior *AA* phenotypes are continually rarer and the unfavored *Aa* and *aa* phenotypes more plentiful than when *A* is dominant. When the favored gene is recessive, DN rises to approximately 100N. Since many beneficial mutants are not completely dominant, Haldane proposes an average death value of 30N for the replacement of a single gene. Because it is based only on N and p_0, DN is not affected by a change in s as long as s is small. That is, the value of 30N genetic deaths holds for any gene substitution, no matter what phenotypic effect is involved. When s is large, gene replacement becomes more rapid, although the population now suffers the danger of extinction because insufficient numbers may be present to ensure mating partners or survival in the face of accident. In the extreme case, when $s = 1$, replacement may occur in a single generation if the unfavored gene is dominant (p. 764). Under this circumstance, however, extinction is fairly sure to occur if the favored gene is present in very low frequency, since there will be very few favorable homozygous recessives available for survival. In general, Haldane suggests that a population is probably capable of sacrificing about one-tenth of its reproductive surplus for genetically selective purposes; that is, a gene substitution can occur at a rate of $1/10 \times 30N$, or every 300 generations.

However, not all gene substitutions may be caused by selection, and Kimura has proposed that the rate of evolution at the molecular level is actually far more rapid than Haldane suggests. Kimura notes, for example, that most amino acids in the hemoglobin proteins of vertebrates are replaced at a rate of approximately one amino acid change per 10^9 years (Chapter 35), and the amount of DNA in such organisms is usually more than necessary to code for 10^9 amino acids. If we limit our argument to these observations and to the assumption that amino acid substitution rates are about the same in all proteins, it is obvious that each vertebrate species would be expected to undergo at least one complete amino acid substitution per year. Since the cost for a single gene substitution by selection is 30N, and there is an average of about 3 years per vertebrate generation, then each vertebrate population must expend an enormous number of genetic deaths in order to maintain its size and escape extinction. In the present example, the population would have to devote 90 times its number in each generation were selection the primary cause for gene frequency changes.

Because of this presumably high cost of selection, Kimura and co-workers have instead proposed that most amino acid changes are neutral in their effect. Since selection among neutral alleles is absent, the fixation of such an allele obviously incurs no genetic load, and would depend only on its mutation rate and on random drift (p. 775). This "neutralist" hypothesis (also called by some "non-Darwinian evolution") appears to be supported by the almost universal finding of large amounts of enzyme and protein polymorphism. As discussed previously (p. 808), many species may be polymorphic for as much as 60 or 70 percent of their proteins, and even minimum estimates of 20 to 30 percent polymorphism indicate that allelic differences are being maintained at many thousands of loci. One could of course claim that allelic differences can be maintained without fixation and, since they will never be fixed, we can therefore exclude the high cost of evolution that Haldane has shown to be necessary for gene substitution. However, the maintenance of polymorphism by selection may itself entail a genetic load that can also be enormous and intolerable. It can be calculated by means of the following arguments:

We know that the segregational load accompanying the preservation of polymorphism through the superiority of heterozygotes is $st/(s + t)$ (p. 814). If such a load was, for example, caused by a balanced lethal system at one gene pair in which the selection coefficients acting against both homozygotes are 1, the load would be 1/2, and the remaining fitness of the population would be $1 - 1/2 = 1/2$. For two pairs of balanced lethal genes acting independently of each other, the fitness of the population is reduced to $(1 - 1/2)^2 = 1/4$, as shown in Fig. 11-7. In general, no matter what the value of the selection coefficients, the average fitness of a population bearing a balanced or segregational load is therefore $[1 - st/(s + t)]^n$ where n designates the number of gene pairs at which heterozygote superiority is being maintained. Kimura and Crow have calculated that the fitness of such a population is approximately equal to $e^{-\Sigma L}$ where e is the base of natural logarithms (2.718) and ΣL designates the sum of individual loads for each gene pair involved.

Not surprisingly this load can be quite large, even if the selection coefficients acting against homozygotes are small, as long as there are many gene pairs involved in maintaining superior heterozygotes. For example, if superior heterozygotes are being maintained at 100 loci, each bearing a genetic load of .01 (e.g., $s = .02$, $t = .02$), the average fitness of the population is reduced to about $e^{-100(.01)} = e^{-1} = .37$. For 500 gene pairs acting similarly, the fitness is approximately $e^{-5} = .007$, and it is $e^{-10} = .00005$ for 1,000 such gene pairs. Thus, to maintain polymorphism at 1,000 loci even with only a very small selective advantage, only one out of 20,000 offspring would survive. To prevent extinction of the population each female would then have to produce about 40,000 young for this selective purpose alone! Most polymorphic systems must therefore consist of neutral mutations according to Kimura and co-workers (see also Nei), and they support this argument with evidence that includes high frequencies of neutral mutations in a mutator strain of *E. coli* (Gibson, Schleppe, and Cox).

Nevertheless some selection schemes have been proposed which attempt to explain the persistence of numerous polymorphisms but confer minimal genetic loads. Frequency-dependent selection (p. 767) is one mechanism that incurs genetic loads only when a gene is relatively rare, but produces no genetic load when the gene has reached equilibrium (Kojima and Yarbrough). However, since polymorphism seems to exist at thousands of loci, it may be questioned whether there are also thousands of individual frequency-dependent mechanisms in the environment.

Sved, King, and others have suggested that selection in a natural population probably does not act separately on each gene locus, but "lumps" the effect of many genotypes together into two main groups: the fit and the unfit. That is, there are "threshold" numbers of polymorphic loci in a population which serve to distinguish these two *groups* of genotypes but do not usually distinguish *individual* genotypes within each group. In comparisons between genotypes, heterozygotes for more gene loci than the threshold number show no increased heterotic effect, and heterozygotes for fewer genes than the threshold number are presumably all equally deleterious. Although this threshold may shift, depending as it does on the environmental stresses placed on the population (see Milkman, 1967, and also Wallace), the genetic loads incurred by such populations may be relatively small since differences between each of the many genotypes above the threshold or below the threshold do not add very much to the load. The likelihood that genes are not individually replaced in evolution but are selected as linkage blocks (see Franklin and Lewontin) supports the view that there may be considerable interaction between genes to produce threshold effects, and also indicates that theoretical genetic loads may be erroneously calculated if it is assumed that each locus acts independently of all others. There is, however, little direct evidence as yet that thresholds exist, although genetic homeostasis (p. 813) may be one such indication.

At present considerable emphasis is being placed by "selectionists" on the following findings:

1. *There is an association between the patterns of some enzyme polymorphisms and ecological conditions.* The argument here is simply that a strong correlation between particular enzymatic alleles and particular environmental conditions might indicate that the allozyme frequencies are being maintained through selection. One well-known example of such a correlation is the relationship between the gene for sickle-cell anemia and malaria. As shown in Fig. 34–9, the geographic distribution of this gene, as well as some others that may offer protection against malaria, is mostly localized to areas in which malarial disease is prevalent. In these cases, however, the maintenance of polymorphism by selection obviously

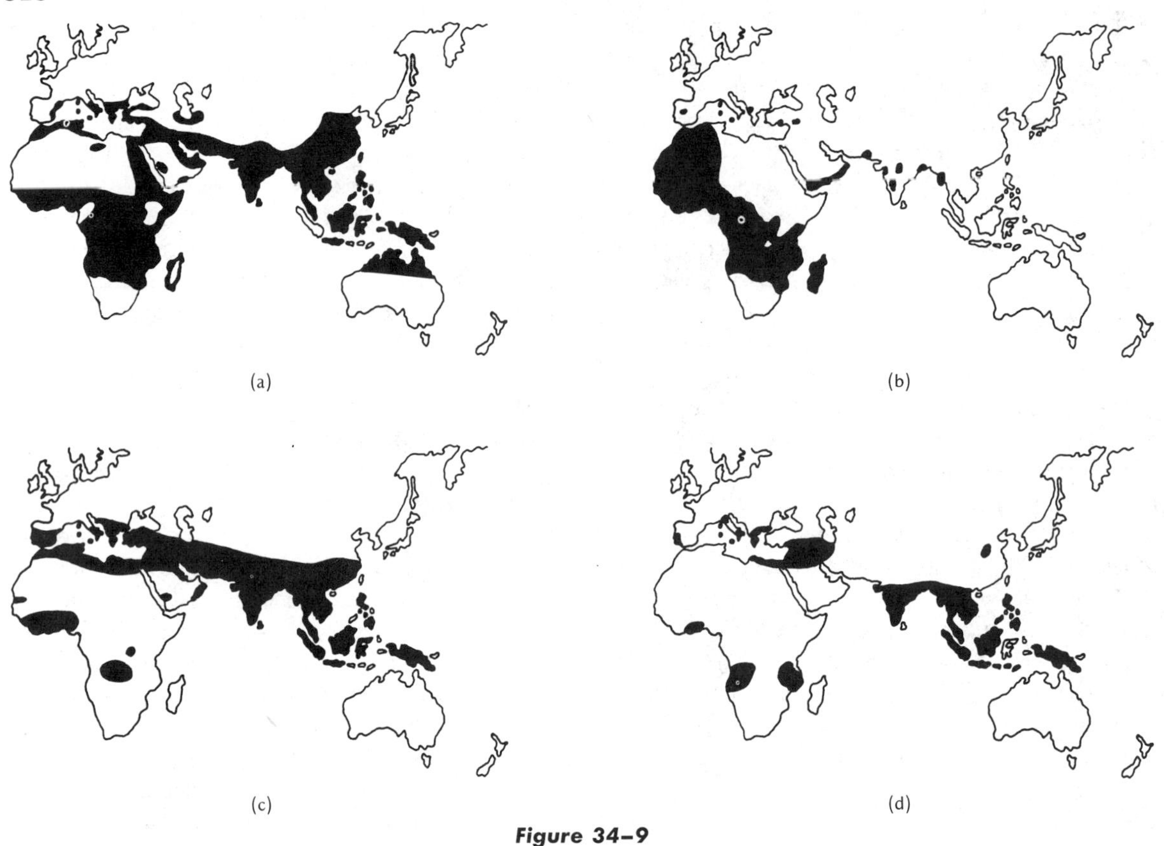

Figure 34–9

Relationship between the geographic distributions of malaria and genes believed to confer resistance against the disease. (a) Distribution of falciparum malaria in the Old World before 1930. (b) Distribution of the gene for sickle-cell anemia. (c) Distribution of the gene for thalassemia. (d) Distribution of the gene for glucose-6-phosphate dehydrogenase deficiency. (After Motulsky).

incurs a significant expense in the loss of homozygotes. A system that may have milder selective effects is the polymorphism discovered by Koehn in the freshwater fish *Catostomus clarkii*. The distribution of two alleles of an esterase enzyme in this fish seems to follow a temperature cline along the Colorado River basin. The homozygote for the allele that is most frequent in the more southern (and warmer) latitude produces an esterase enzyme that becomes more active as temperature increases, whereas the allele that is more frequent in the northern (and colder) latitude forms an enzyme that is more active as temperature decreases. Not unexpectedly the heterozygote for the two alleles forms an enzyme that is most active at intermediate temperatures. However, although other such correlations between alleles and particular environments have been shown in plants as well as in animals (e.g., see Allard and Kahler), there are numerous instances where no obvious correlation exists. It is therefore difficult to prove that such correlations are always necessarily causal and may not be accidental. Also, some enzyme loci being scored for polymorphism may be strongly linked to a gene locus at which selection is operating, and the protein polymorphism observed would only be the effect of such "hitchhiking."

2. *There are a number of enzyme polymorphisms*

whose advantages are unknown but whose frequencies are difficult to explain on a purely random basis. An early example of this kind was the observation by Prakash, Lewontin, and Hubby, that populations of *Drosophila pseudoobscura* ranging from California to Texas show remarkably similar allozyme frequencies for a number of proteins. Although it has been suggested that such similarities arise because of migration between different populations rather than through common selective factors, migration could hardly explain similarities in gene frequencies between *different species.* As Ayala and Tracey have recently pointed out, genetically isolated species of the *Drosophila willistoni* group share common gene frequencies for the alleles of many different enzymes. Furthermore, the pattern of similarity between them is not constant, and the data seem to show that different species share common selective factors for some enzymes but not for others. Perhaps even more striking is the finding by Milkman (1973) that clones of *E. coli* isolated from the intestinal tracts of animals as diverse as lizards and humans, and from localities as widespread as New Guinea and Iowa, appear to share common allozyme frequencies. For each of five different enzymes, Milkman found that one particular electrophoretic band was very frequent in almost all samples. Since other allozymes of these proteins exist, the finding of such a narrow distribution of allozymes can hardly be explained on the basis of neutral mutations and random drift.

3. *An association may exist between the function of an enzyme and the degree of polymorphism that it shows.* This was first suggested by Gillespie and Kojima, who grouped enzymes into two classes, namely those involved in restricted pathways of energy metabolism such as glycolysis (e.g., aldolase) and those that can use a variety of substrates (e.g., esterases, acid phosphatases, etc.). Their findings and others have since shown that the enzymes with more restricted uses show significantly less polymorphism than enzymes whose substrates are more variable. Johnson, extending this notion further, has shown that enzymes involved in regulating metabolic pathways (e.g., glucose-6-phosphate dehydrogenase, phosphoglucomutase, etc.) are generally more polymorphic than enzymes whose functions are not primarily regulatory (e.g., malate dehydrogenase, fumarase, etc.). Although the biochemical causes that sustain these differences in polymorphism are not yet known, it seems clear that they are not random, and many of the polymorphic alleles are consequently not neutral.

In summary, selective forces unquestionably operate to help maintain polymorphism. However, the number of genes on which selection operates at any one time, the linkage relationship between these genes, the kinds of selection that operate, and the size of the selection coefficients are all unknown. Because of this absence of information, neutral mutation and random drift cannot be excluded as an important cause for polymorphism on the biochemical level. Certainly some genetic variants may be neutral at certain times or under certain conditions, but have selective value when circumstances change.

POPULATION STRUCTURE IN EVOLUTION

If we define evolution as changes in gene frequencies, we can consider the population as the primary unit in evolution, since it is there that gene frequency changes take place. Individuals are, of course, the carriers of genes, but an individual lasts for only a single generation and contains only a small portion of the gene pool upon which evolutionary forces act. A population, on the other hand, has continuity between generations and possesses all the genetic material of the individuals within it. On this basis the genetic laws by which we measure evolutionary changes can only be expressed in terms of populations, although the many component measurements are made in terms of its individuals.

Populations, therefore, have a unique evolutionary distinction, and many genetic factors that affect the evolution of populations may act differently from the way we expect if individuals alone were considered. Sex, for example, may be of little value to an individual (or even a disadvantage) but can be a distinct advantage to a population (see Maynard Smith). The role of mutation can also be misunderstood if it is examined too narrowly. That is, since mutation is more or less random, one might expect as many adaptive mutations to occur as those that are deleterious. It would therefore appear as though evolution merely awaits the occurrence of superior adaptive individuals before it proceeds. In practically all populations, however, the role of new mutations is not of immediate significance. A population that has long been established in a particular environment will have many of its genes adapted for these conditions. New mutations that arise, if not neutral in their effect, will therefore rarely be better and will very likely be worse than the genes already present. However, even if some mutations are better, it is unlikely that they will exceed the fitness of the established genes by a large degree. Assuming these beneficial mutations are not lost by chance (pp. 756–757), their increase in frequency will take many generations (p. 764).

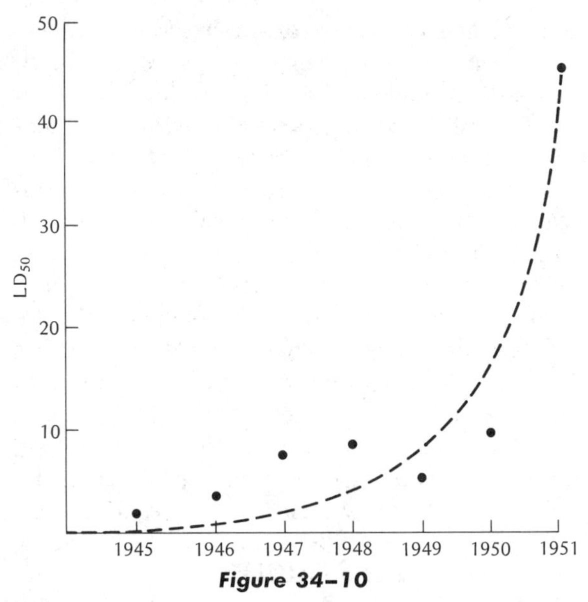

Figure 34–10

Resistance to DDT in houseflies collected from Illinois farms measured in terms of the lethal dose necessary to kill 50 percent of the flies (LD_{50}). (Data of Decker and Bruce.)

On the other hand, a change in environmental conditions may have a more important evolutionary effect, since many genes formerly in low frequency may suddenly possess high adaptive values. This can be seen in the rapid genetic changes that have occurred in many insect populations exposed to pesticides such as Dieldrin and DDT (Fig. 34–10). Mechanisms that confer DDT resistance may include (1) an increase in lipid content that enables the fat-soluble DDT to be separated from other parts of the organism, (2) the presence of enzymes that break DDT down into relatively less toxic products, (3) a reduction in toxic response of the nervous system to DDT, (4) changes in the permeability of the insect cuticle to DDT absorption, and (5) a behavioral response that reduces contact with DDT. It is therefore not surprising that insecticide resistance has been shown to be associated with numerous genes. The genes responsible for DDT resistance in *Drosophila*, for example, are located on all major chromosomes, each gene acting as a polygene with a small incremental effect (Fig. 34–11).

Selection among individual mutant genes on separate chromosomes, however, is not the only method by which genetic progress is achieved. As pointed out by Mather and others (see p. 817), linkage and recombination between genes affecting a selected character may have a marked effect on the response to selection. To illustrate this point let us assume that each of four loci, *Aa*, *Bb*, *Cc*, *Dd*, influence a character quantitatively, and that all capital-letter genes have a plus effect on the character and all small-letter genes have a minus effect. If the phenotypic optimum is an intermediate one, as it is for many characters, it would be advantageous that the genotype also be intermediate, e.g., *AaBbCcDd*. Under these circumstances the absence of linkage is disadvantageous, since other nonoptimum genotypes are quite easily formed through independent assortment.

A much more advantageous scheme for such a population is for tight linkage to be present between these four loci in the fashion *AbCd* on one homologue and *aBcD* on the other. Thus any combination of chromosomes will always have four plus genes and four minus genes, yet the population still retains the variability of all alleles.

Note, however, that a linked combination in this form must have three crossovers for chromosomes to be formed containing all plus or all minus genes. Thus if selection changes from an intermediate phenotype to an extreme phenotype (either all plus or all minus), a significant amount of time may elapse before the appropriate crossovers can furnish the most adaptive combinations of genes. In other words, recombination as well as mutation dictates the progress of selection (see Strickberger, 1969).

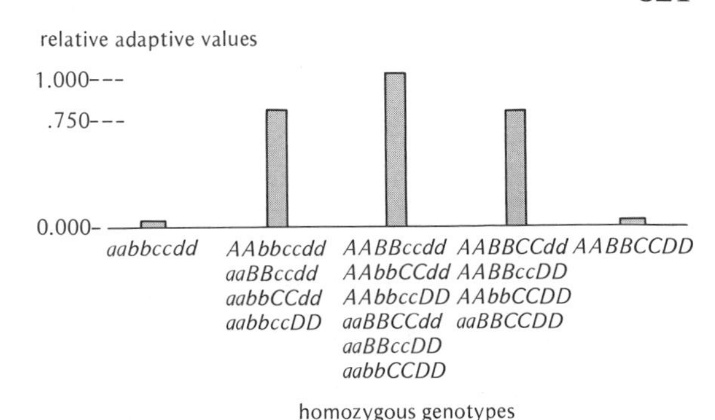

Figure 34–12

One possible distribution of homozygous genotypes if their relative adaptive values are arbitrarily fixed. The optimum genotype is considered to be intermediate with four capital-letter alleles at these four loci. Since six different homozygous genotypes may be at the optimum, six possible adaptive peaks are present at this adaptive value. (After Wright.)

ADAPTIVE PEAKS

The fact that more than one locus is involved in the evolution of fitness of a population provides many possible ways in which increased fitness can be achieved. For simplicity, we can consider a population containing only homozygous genotypes, where the same four loci as above affect a character, so that the optimum phenotype is determined by equal numbers of capital-letter and small-letter genes. Then, as shown in Fig. 34–12, a variety of six optimum genotypes (*AABBccdd*, *AAbbCCdd*, etc.) are possible. According to Wright each of these genotypes may be considered to occupy an "adaptive peak," and, as long as no other factors change the fitness of these genotypes, each of these six peaks is of equal height. A population consisting entirely of one of these genotypes would therefore achieve maximum fitness for this phenotype.

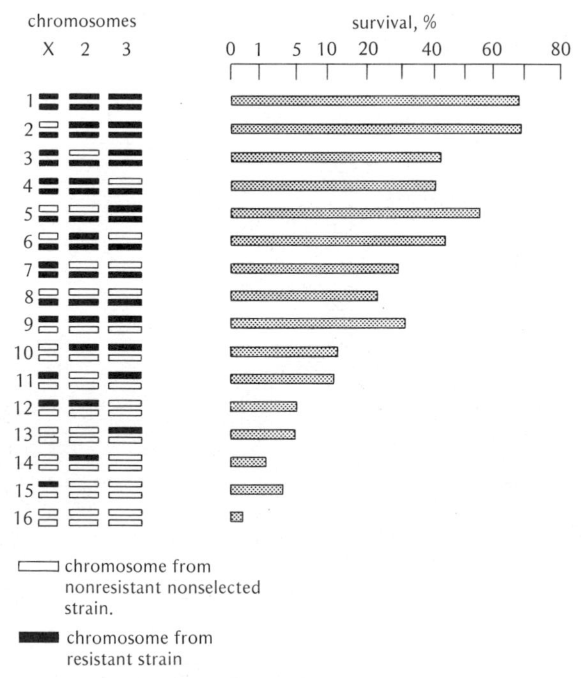

Figure 34–11

Percent survival of 16 different types of Drosophila melanogaster *flies exposed to a uniform dose of DDT. Each type of fly carries a unique set of chromosomes derived from DDT-resistant and DDT-nonresistant strains. (After Crow, 1957.)*

When more than four loci are involved in fitness with more than two alleles at any locus, the number of possible adaptive peaks increases astronomically. For a locus with only four alleles, there are 10 possible diploid gene combinations, and for 100 loci with four alleles each there are 10^{100} possible gene combinations. Even limited to this small number of loci, the number of possible combinations far exceeds the number of individuals in any species and even the estimated number of protons and neutrons in the universe (2.4×10^{70}). Even if

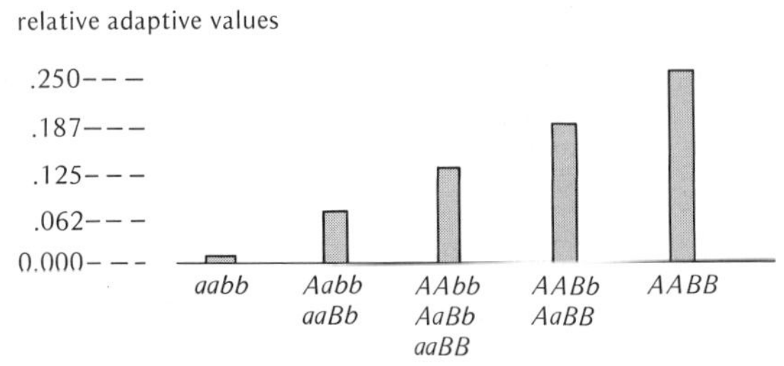

Figure 34–13

Adaptive values assigned to genotypes at the two gene pairs Aa and Bb, based on additive effects on fitness caused by capital-letter alleles. (After Wright.)

only a small portion of these gene combinations are adaptive, there are undoubtedly more possible adaptive peaks than can be occupied by a species at any one time.

Again, however, we must clearly differentiate between populations and individuals, i.e., between adaptive peaks for populations in this case and those for genotypes. High adaptive peaks for a population do not necessarily coincide with high selective peaks for a genotype, since the selective values of genotypes refer to competition with other genotypes but say nothing about their effect on

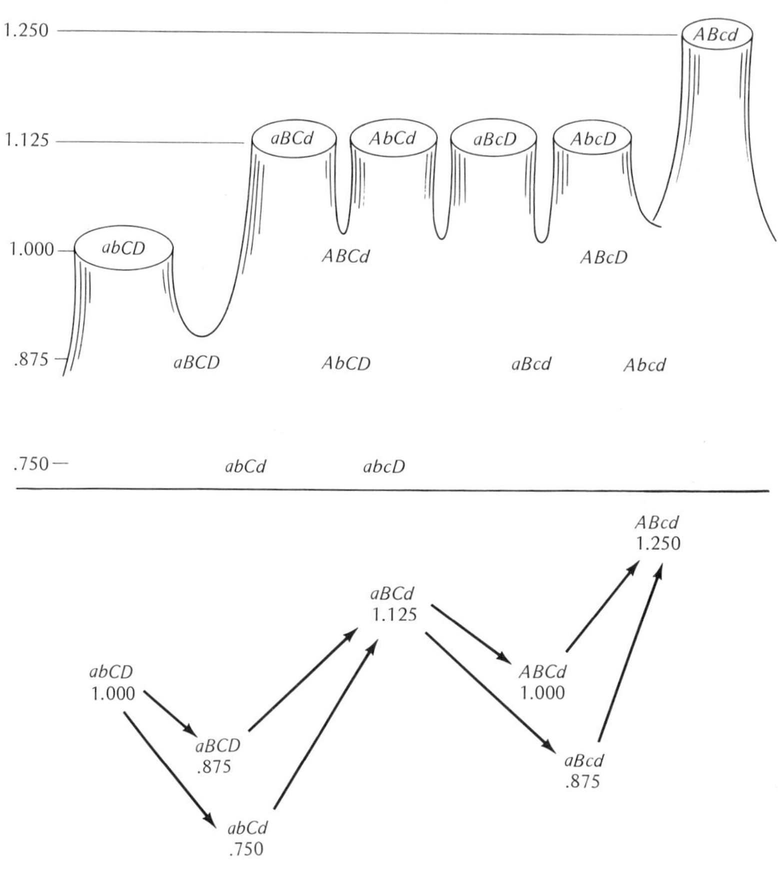

Figure 34–14

Above: adaptive landscape based on the assigned adaptive values for the various genotypic combinations considered in Figs. 34–12 and 34–13. Below: some of the possible paths in the progress of a population from a lower adaptive peak at aabbCCDD to the highest adaptive peak at AABBccdd. The ABCD (W = .25) and abcd (W = 0.00) populations are omitted. (After Wright.)

the population. For example, it has been pointed out by Haldane, Wright, and others that "altruistic" genotypes that sacrifice themselves for the good of the population may have low selective value, although a population bearing such genotypes may have higher reproductive values than one without them. Conversely "social parasites" that increase at the expense of other genotypes in a population may have high selective value, although they depress the reproductive fitness of the population. Illustrations of the latter type are genes that modify segregation ratios in their favor (segregation distorters, p. 219) but have a deleterious effect, such as *tailless* genes in mice (p. 711).

For simplicity, however, let us assume that such complications are not involved in the present example of four gene pairs. Even so, the concept of many adaptive peaks with uniform height undoubtedly departs from real conditions. It is likely that the effects of each of these four pairs of genes may differ considerably. In Fig. 34–13, Wright has therefore assigned adaptive values to the effects of the *A* and *B* genes on the basis that the greater the number of capital-letter genes in these two pairs, the greater the fitness. When combined with the previously given adaptive values, a "landscape" of peaks can now be constructed showing the adaptive heights of different possible genotypes (Fig. 34–14). Note that there is now one peak (*AABBccdd*) superior to all, and also a number of possible intermediate peaks each surrounded by genotypes that are relatively inferior. A population may therefore increase in fitness during evolution but nevertheless reach an intermediate peak which is not necessarily the most adaptive. To move from peak to peak until a population finds the highest one demands that it travel through inferior genotypes that occupy the "valleys" of this landscape. This reduction in fitness can be seen in the lower half of Fig. 34–14, which shows the general course of travel that a population located at *aabbCCDD* might take to reach the highest peak at *AABBccdd*. There are thus at least two nonadaptive stages at which the population will suffer in this illustration.

Once a population has reached an adaptive peak, further evolution will depend upon the origin of a new selective environment and the creation of new adaptive peaks. However, if conditions are not changing rapidly, the same set of adaptive peaks will remain for long periods of time. A population on one adaptive peak can then no longer reach a higher peak without going through a "nonadaptive" value. Since selection can hardly occur for nonadaptation, how can a population located on a relatively low adaptive peak evolve so that the highest or near-highest peaks on the adaptive landscape are occupied?

In answer to this problem Wright has proposed that many populations are broken into small groups of subpopulations. These local populations,

TABLE 34–4

Comparison of evolutionary processes in a single homogeneous population and in a subdivided population

	Homogeneous Population	*Population Subdivided into Demes*
What is Selected	a gene, differing from its alleles in net selective value	different gene frequencies
Source of Variation	gene mutation	random drift between demes and selection toward new adaptive peaks
Process of Selection	selection among individuals	selection among demes
Evolution Under Static Conditions	progress restricted to single peak	continued shifts as new adaptive peaks are encountered
Evolution Under Changing Conditions	progress up nearest adaptive peak	selection between different demes for occupancy of all available peaks

From Wright, 1963.

or "demes," are small enough to enable genetic differences to occur between them through the nonselective process of random drift (p. 773), but are nevertheless not so widely separated as to completely prevent gene exchange and the introduction of new genetic variability. The adaptive landscape is therefore occupied by a network of demes, some at higher peaks than others. Thus selection takes place not only between genotypes competing within demes, but also between demes competing within a general environment. According to Wright (Table 34–4), this scheme has many advantages over the evolution of a single homogeneous population.

PROBLEMS

34–1. Jayant recorded 1814 female births in a private hospital in India of which 1714 survived beyond 1 week of age. The birth weight showing the lowest mortality was 6 1/2 pounds, at which 622 females were born and 616 survived. What is the selection intensity associated with birth weight at this stage?

34–2. Ten percent mortality is caused by selection in a population in which the frequency of suboptimal phenotypes is 75 percent. What difference in survival rate exists between optimal and suboptimal phenotypes?

34–3. For a given frequency of suboptimal phenotypes, what selection effect is produced by a relative increase in survival rate of the optimal phenotype?

34–4. Explain your answers to the following questions: (a) Since selection for increased fitness depends upon available genetic variability, could an increase in fitness occur in the absence of new mutation? (b) Is the most fit population in a perfectly stable environment that which always has the least genetic variability remaining in it?

34–5. The frequency of recessive genes causing female sterility in a population of *Drosophila persimilis* is 18.3 percent on the 2nd and 4th chromosomes and 14.3 percent on the 3rd. (a) What percentage of *persimilis* flies carry female sterile genes on both of their 2nd chromosomes? (b) What percentage of flies are completely free of female sterile genes on these three pairs of chromosomes?

34–6. If a population were selected to produce an increased phenotypic value for a quantitative character, how could you distinguish if the genotypes in the population had changed so they produced more phenotypic substance, or if the zone of canalization had changed so that the new phenotype could be caused by genotypes producing the same amount of phenotypic substance as before?

34–7. Mather and Harrison selected for increased abdominal bristle number in a stock of *Drosophila melanogaster* and found excellent response to selection until generation 20. Near this point the fertility of the stock decreased considerably and selection had to be discontinued or "relaxed." Under relaxed selection the bristle number decreased and almost reached the initial preselection value. When selection began again on this same line of flies, bristle number increased to the previous high value but fertility was no longer affected. Moreover, when selection was relaxed at generation 50, the bristle number remained stable at a high value. What explanations can be offered for these events?

34–8. In a certain random breeding population, individuals suffering from phenylketonuria (homozygous recessives) occur with a frequency of .001 and may be considered lethal. (a) What is the genetic load incurred by the population because of phenylketonuria? (b) If there are two hundred million people in this population, what number of genetic deaths can be ascribed to this gene?

34–9. When phenylketonuria can be detected early enough, it can be treated by the removal of phenylalanine from the diet. Such individuals may live to reproductive age and reproduce normally. For purposes of this problem we may assume that, as a result of this treatment, selection against phenylketonurics is now one half its former value. (a) If this recently discovered treatment were applied to a similar population as in Problem 8, what number of genetic deaths caused by phenylketonuria would be expected in the present generation? (b) If the mutation rate is constant, what will be the genetic load caused by phenylketonuria in this treated population when it reaches equilibrium? (c) If the mutation rate is doubled in both the untreated population of Problem 8 and in the present treated population, and they both reach equilibrium, what would be the difference between them in the genetic load caused by phenylketonuria?

34-10. If induced recessive deleterious mutations are produced at a gene locus at a rate of 2×10^{-7} for each r of irradiation, what increase in genetic load would you expect from this source in a population which receives an average skin dose of 20 r each generation as a result of x-ray treatment, fallout, etc.? (Calculate your answer on the basis that one half the skin dose penetrates to the gonads and the haploid number of genes in this diploid population is 10,000.)

34-11. In a large randomly breeding population of *Drosophila pseudoobscura*, Dobzhansky observed the following relative adaptive values for genotypes carrying the 3rd-chromosome arrangements Standard (ST) and Chiricahua (CH): ST/ST = .90, ST/CH = 1.00, CH/CH = .41. (a) Calculate the genetic load produced by this population because of these genotypic differences. (b) How many genetic deaths will occur because of this load if the number of eggs laid each generation is 100,000?

34-12. (a) What change in genetic load of a population would you expect because of increased inbreeding? (b) Explain whether you expect a greater difference in genetic load upon inbreeding at a locus where heterozygotes are superior or at a locus where homozygotes are superior.

34-13. In humans a dominant gene A occurs through mutation in about 1 per 100,000 gametes. One hundred individuals affected by A were found to produce 80 offspring, half of which had the A phenotype. Their normal sibs, on the other hand, numbering 200, produced 400 normal offspring. (a) What is the equilibrium frequency for this gene? (b) What is the genetic load caused by this gene? (c) If the U.S. population is considered at 200 million, how many genetic deaths would be caused by this gene each generation?

34-14. If no further mutation occurred at the phenylketonuria locus, how many genetic deaths are necessary to completely replace the phenylketonuria allele by the normal allele: (a) If phenylketonuria is not treated (Problem 8)? (b) If phenylketonuria is treated (Problem 9)?

34-15. If a dominant gene having a similar frequency to that of phenylketonuria (Problem 8) were suddenly advantageous, how many genetic deaths would be necessary to completely replace its recessive allele?

34-16. In a population in which the Aa heterozygote is superior ($W = 1.00$), the aa homozygote is lethal and the AA homozygote has an adaptive value of .8: (a) Demonstrate which of these homozygotes will be responsible for the major portion of genetic deaths at equilibrium. (b) Would you say that the genetic loads produced by each of two different deleterious homozygotes at equilibrium depends upon their respective frequencies rather than upon their selection coefficients?

34-17. If one billionth of all possible gene combinations occupy adaptive peaks, how many different adaptive peaks are possible in a species that has 1000 genes with four alleles at each?

34-18. (a) Why do not all populations occupy all the adaptive peaks or fittest genotypic combinations that are available to them? (b) Why do not some populations occupy any adaptive peak at all?

34-19. The diagrams in Fig. 34-15 represent frequencies of a single gene in a number of populations under the same conditions of selection, mutation, and migration. (a) What is responsible for the different distributions of populations in these diagrams? (b) Explain which of these diagrams represents increased evolutionary possibilities.

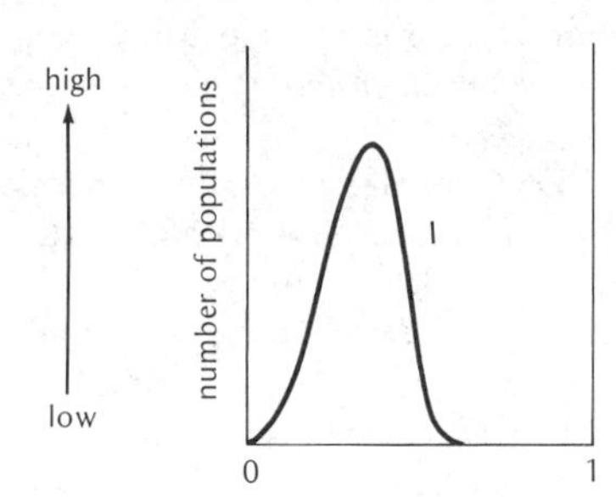

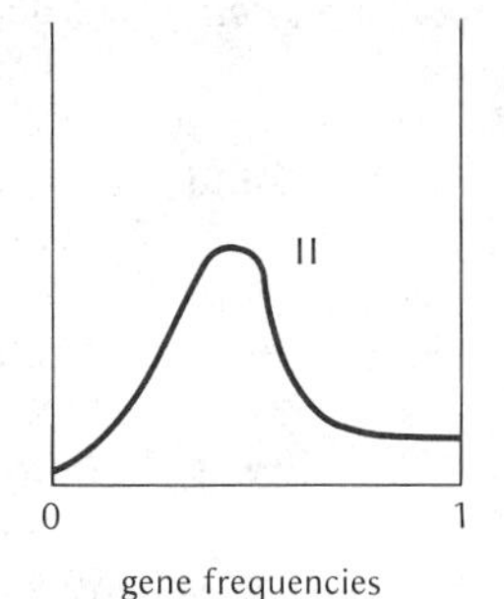

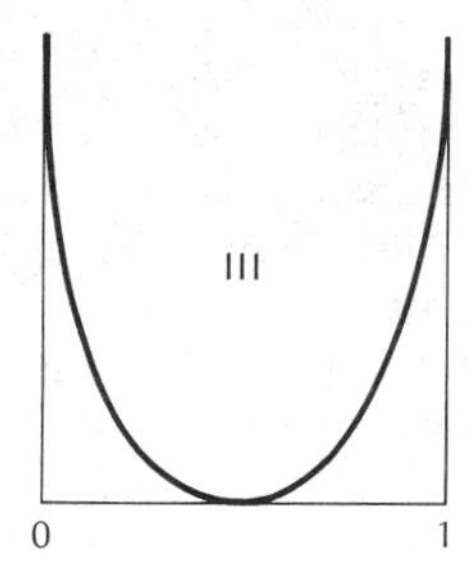

Figure 34-15

Distributions of gene frequencies in three collections of populations, each collection having been subject to the same conditions of selection, mutation and migration. See Problem 19.

REFERENCES

ALLARD, R. W., and A. L. KAHLER, 1972. Patterns of molecular variation in plant populations. In *Proceedings of the Sixth Berkeley Symposium on Mathematical Statistics and Probability*, Vol. 5. Univ. of Calif. Press, Berkeley, pp. 237–254.

AYALA, F. J., and M. L. TRACEY, 1974. Genetic differentiation within and between species of the *Drosophila willistoni* group. *Proc. Nat. Acad. Sci.*, **71,** 999–1003.

BABCOCK, E. B., and R. E. CLAUSEN, 1927. *Genetics in Relation to Agriculture*. McGraw-Hill, New York.

BRUES, A. M., 1964. The cost of evolution vs. the cost of not evolving. *Evolution*, **18,** 379–383.

BUMPUS, H. C., 1898. The elimination of the unfit as illustrated by the introduced sparrow. *Biol. Lect. Woods Hole*, pp. 209–226.

CHETVERIKOFF, S. S., 1926. On certain aspects of the evolutionary process from the standpoint of modern genetics. *Zhur. Eks. Biol.*, **2,** 3–54 (in Russian). (Translated into English by I. M. Lerner, and reprinted 1961 in *Proc. Amer. Phil. Soc.*, **105,** 167–195.)

CROW, J. F., 1957. Genetics of insect resistance to chemicals. *Ann. Rev. Entomol.*, **2,** 227–246.

———, 1958. Some possibilities for measuring selection intensities in man. *Hum. Biol.*, **30,** 1–13.

———, 1962. Population genetics: Selection. In *Methodology in Human Genetics,* W. J. Burdette (ed.). Holden-Day, San Francisco, pp. 53–75.

DECKER, G. E., and W. N. BRUCE, 1952. House fly resistance to chemicals. *Amer. Jour. Trop. Med. Hygiene*, **1,** 395–403.

DOBZHANSKY, TH., 1947. A directional change in the genetic constitution of a natural population of *Drosophila pseudoobscura*. *Heredity*, **1,** 53–64.

———, 1959. Variation and evolution. *Proc. Amer. Phil. Soc.*, **103,** 252–263.

DUN, R. B., and A. S. FRASER, 1958. Selection for an invariant character—vibrissa number—in the house mouse. *Nature*, **181,** 1018.

FISHER, R. A., 1930. *The Genetical Theory of Natural Selection*. Clarendon, Oxford. (Reprinted 1958, Dover, New York.)

FRANKLIN, I., and R. C. LEWONTIN, 1970. Is the gene the unit of selection? *Genetics*, **65,** 707–734.

GIBSON, T. C., M. L. SCHILEPPE, and E. C. COX, 1970. On fitness of an *E. coli* mutation gene. *Science*, **169,** 686–690.

GILLESPIE, J. H., and K. KOJIMA, 1968. The degree of polymorphisms in enzymes involved in energy production compared to that in nonspecific enzymes in two *Drosophila ananassae* populations. *Proc. Nat. Acad. Sci.*, **61,** 582–585.

HALDANE, J. B. S., 1954. The measurement of natural selection. *Proc. IX Intern. Congr. Genet.*, *Caryologia* (*Suppl.*): 480–487.

———, 1957. The cost of natural selection. *Jour. Genet.*, **55,** 511–524. (Reprinted in Spiess's collection; see References, Chapter 31.)

———, 1960. More precise expressions for the cost of natural selection. *Jour. Genet.*, **57,** 351–360.

HALDANE, J. B. S., and S. D. JAYAKAR, 1963. Polymorphism due to selection of varying direction. *Jour. Genet.*, **58,** 237–242.

JOHNSON, G. B., 1974. Enzyme polymorphism and metabolism. *Science*, **184,** 28–37.

KARN, M. N., and L. S. PENROSE, 1951. Birth weight and gestation time in relation to maternal age, parity, and infant survival. *Ann. Eugenics*, **161,** 147–164.

KIMURA, M., and J. F. CROW, 1964. The number of alleles that can be maintained in a finite population. *Genetics,* **49,** 725–738.

KIMURA, M., and T. OHTA, 1972. Population genetics, molecular biometry, and evolution. In *Proceedings of the Sixth Berkeley Symposium on Mathematical Statistics and Probability*, Vol. 5. Univ. of Calif. Press, Berkeley, pp. 43–68.

KING, J. L., 1967. Continuously distributed factors affecting fitness. *Genetics*, **55,** 483–492.

KOEHN, R. K., 1969. Esterase heterogeneity: Dynamics of a polymorphism. *Science*, **163,** 943–944.

KOJIMA, K., and K. M. YARBROUGH, 1967. Frequency-dependent selection at the esterase 6 locus in a population of *Drosophila melanogaster*. *Proc. Nat. Acad. Sci.*, **57,** 645–649.

KRIMBAS, C. B., 1959. Comparison of the concealed variability in *Drosophila willistoni* with that in *D. prosaltans*. *Genetics*, **44,** 1359–1369.

LERNER, I. M., 1954. *Genetic Homeostasis*. John Wiley, New York.

LEWONTIN, R. C., 1974. *The Genetic Basis of Evolutionary Change*. Columbia Univ. Press, New York.

LEWONTIN, R. C., and J. L. HUBBY, 1966. A molecular approach to the study of genic heterozygosity in natural populations. II. Amount of variation and degree of heterozygosity in natural populations of *Drosophila pseudoobscura*. *Genetics*, **54**, 595–609.

Li, C. C., 1963. The way the load ratio works. *Amer. Jour. Hum. Genet.*, **15**, 316–321.

Mather, K., 1953. The genetical structure of populations. *Symp. Soc. Exp. Biol.*, **7**, 66–95.

Maynard Smith, J., 1971. The origin and maintenance of sex. In *Group Selection*, G. C. Williams (ed.). Aldine-Atherton, Chicago, pp. 163–175.

Milkman, R. D., 1967. Heterosis as a major cause of heterozygosity in nature. *Genetics*, **55**, 493–495.

———, 1970. The genetic basis of natural variation in *Drosophila melanogaster*. *Adv. in Genet.*, **15**, 55–114.

———, 1973. Electrophoretic variation in *Escherichia coli* from natural sources. *Science*, **182**, 1024–1026.

Motulsky, A. G., 1960. Metabolic polymorphisms and the role of infectious diseases in human evolution. *Hum. Biol.*, **1**, 28–62.

Nei, M., 1975. *Molecular Population Genetics and Evolution*. North-Holland Publ. Co., Amsterdam.

Prakash, S., R. C. Lewontin, and J. L. Hubby, 1969. A molecular approach to the study of genic heterozygosity in natural populations. IV. Patterns of genic variation in central, marginal and isolated populations of *Drosophila pseudoobscura*. *Genetics*, **61**, 841–858.

Rendel, J. M., 1959. Canalization of the scute phenotype. *Evolution*, **13**, 425–439.

———, 1967. *Canalization and Gene Control*. Logos Press, London.

Rendel, J. M., and B. L. Sheldon, 1960. Selection for canalization of the scute phenotype in *Drosophila melanogaster*. *Austral. Jour. Biol. Sci.*, **13**, 36–47.

Robertson, F. W., 1955. Selection response and the properties of genetic variation. *Cold Sp. Harb. Symp.*, **20**, 166–177.

Sheppard, P. M., 1975. *Natural Selection and Heredity*, 4th ed. Hutchinson Univ. Library, London.

Strickberger, M. W., 1969. Factors determining rates of the evolution of fitness in laboratory populations of *Drosophila pseudoobscura*. *Genetics*, **62**, 639–651.

Sved, J. A., T. E. Reed, and W. F. Bodmer, 1967. The number of balanced polymorphisms that can be maintained in a natural population. *Genetics*, **55**, 469–481.

Thoday, J. M., 1972. Disruptive selection. *Proc. Roy. Soc. Lond.* (*B*), **182**, 109–143.

Van Valen, L., 1965. Selection in natural populations. III. Measurement and estimation. *Evolution*, **19**, 514–528.

Waddington, C. H., 1953. Genetic assimilation of an acquired character. *Evolution*, **7**, 118–126.

———, 1962. *New Patterns in Genetics and Development*. Columbia Univ. Press, New York.

Wallace, B., 1970. *Genetic Load: Its Biological and Conceptual Aspects*. Prentice-Hall, Englewood Cliffs, N.J.

Wright, S., 1963. Genic interaction. In *Methodology in Mammalian Genetics*, W. J. Burdette (ed.). Holden-Day, San Francisco, pp. 159–192.

35

SPECIATION AND EVOLUTION

Interest in organic evolution had its origins among the early Greeks and Romans in observations that indicated structural and functional relatedness between many living organisms. Although most explanations for these relationships were quite fanciful, some notions, such as those of Empedocles (p. 734), foreshadowed more serious attempts to understand biological change. Until the middle of the nineteenth century, however, numerous obstacles stood in the way of a rational scientific appreciation of evolution. Foremost among these obstacles was the widely accepted notion of the "fixity of species," which pictured all living forms as static entities, both unchanged in the past and unchangeable in the future. This notion, promulgated by numerous philosophers and supported by important religious institutions, was also espoused by many biologists, including those, such as Linnaeus, who were directly involved in the study and classification of species (see p. 4). To the biologists, belief in the fixity of species had its basis in the special morphological qualities that all members of a species appeared to share. That is, a species was ideally represented as a particular morphological "type" distinctly different from the morphological types of other species. The fact that such "typological" descriptions were of value in separating one group of organisms from another seemed, at the time, more important than the fact that many members of most species departed from the idealized types to various degrees. In addition many biologists felt it impossible to answer the question as to how species could differ from each other in such typological fashion if the boundaries between them were truly changeable and fluid. Why were there no intermediates? Classification, by its nature, therefore emphasized two notions that made it difficult to accept evolutionary change: (1) that species could be described as types and (2) that these types were sharply distinct and therefore could not have arisen from each other.

The marshaling of evidence by Charles Darwin in 1859 in support of evolution broke the hold of the doctrine of fixity of species in biological science. Darwin argued in *The Origin of Species* and in later works that although there were morphological species types, they arose by natural selection among the variable members of previous species. The tendency for population numbers to increase by normal methods of reproduction, combined with the fact that most populations were stationary in size, led to the condition called by Darwin "the struggle for existence." This struggle was not necessarily a battle between organisms ("nature red in tooth and claw") but symbolized more simply that death could occur by any means in preventing all offspring of a species from surviving to reproductive age. Given the many observed variations carried by most species, the struggle for existence led consequently to the selection and proliferation of only those organisms that were most adapted to their environment and those most successful in mating. The result of selection was described as "survival of the fittest," a term borrowed from Herbert Spencer, unfortunate because of its moral connotations. In modern terms, of course, "fittest" means only individuals with most reproductive success in a particular environment, whether such success is achieved by "strength" or "weakness" (p. 759). Schematically, these concepts are diagrammed in Fig. 35–1.

The source of variability and many of the mechanisms of evolution, however, remained unclear until postmendelian population concepts were developed by Haldane, Fisher, and Wright. At present we understand that evolution has as its base a change in the genetic constitution of a population. Far from one idealized genotype, considerable genetic variability may exist within a population as well as between different populations of the same species. It is these genetic differences, caused by mutation and recombination, and acted upon by selection and other forces discussed in previous chapters, that contribute to evolution. Let us now consider two primary steps that affect populations and lead to observable evolutionary changes: race formation, and speciation.

RACE FORMATION

We have learned that the interbreeding nature of a population serves as the important cohesive force that holds it together and enables it to share a common gene pool. At the same time, we understand that a species may consist of numerous individual populations with various degrees of interbreeding between them. Widely separated populations, for example, will have less opportu-

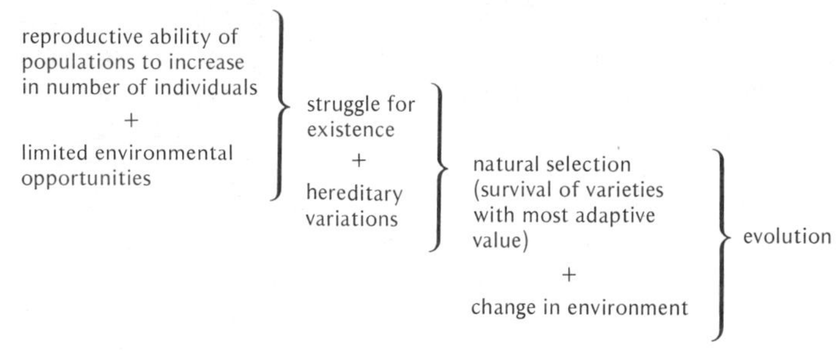

Figure 35–1

Schematic presentation of the main arguments for evolution by natural selection given by Charles Darwin and Alfred Russell Wallace. (After a table by Wallace.)

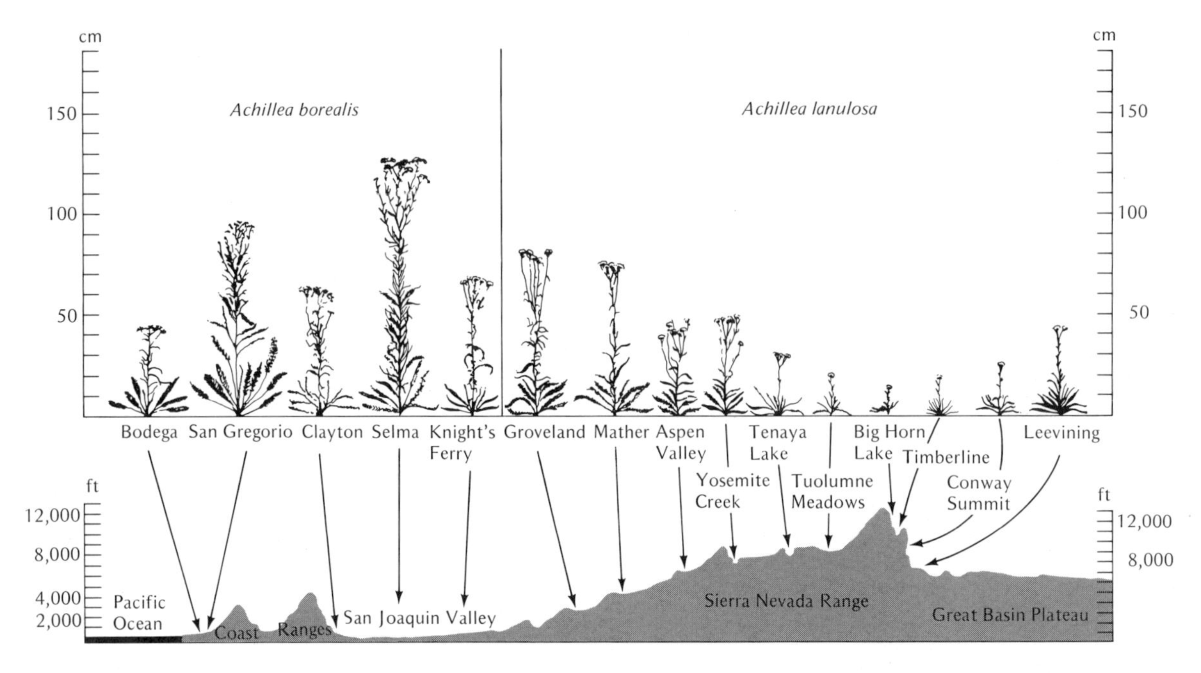

Figure 35–2

Representative plants from different populations of Achillea gathered from designated localities along a transect across central California and grown in a uniform garden at Stanford, California. (After Clausen, Keck, and Hiesey.)

nity to share their gene pools than those closer at hand. The structure of a species is therefore broken into various geographical subunits. Since the forces acting upon these subunits may vary in different localities, it will come as no surprise to find observable differences between populations. In the yarrow plant *Achillea* a transect across central California shows populations differing significantly in factors such as height and growing season (Fig. 35–2). The adaptive nature of most of these differences is seen in the different responses of these populations when grown in different localities. Thus the coastland plants are weak when grown at higher altitudes, and the high-altitude forms grow poorly at much lower altitudes (Fig. 35–3).

The adaptive shifting of gene frequencies between different localities and between different time intervals has been well-documented in numerous instances. In *Drosophila pseudoobscura*, Dobzhansky and others have shown that the frequencies of the 3rd-chromosome arrangements differ notably in a range of environments across the American Southwest (Fig. 35–4) and also undergo significant seasonal changes (see Fig. 34–3). Some genetic changes in this species may also extend over longer periods of time, such as the significant increase in the frequency of the Pikes Peak arrangement between 1940 and 1957 in many California populations from almost zero to as high as 10 percent.

More spectacular examples of adaptive changes in gene frequencies have occurred in association with the phenomenon known as industrial melanism. Industrial areas in England and Western Europe have seen a marked change in the appearance of certain moths and butterflies during the past century from light-colored forms to dark-colored melanic forms. The genetic basis of these differences generally involves a single, usually dominant, gene determining melanism, as well as a number of modifier genes that affect the dominance of the melanic gene.

In England investigations by Kettlewell and others on the British peppered moth, *Biston betularia*, have shown that the darkening of trees caused by air pollution in industrial areas confers a selective advantage upon melanic forms. Near the industrial city of Birmingham, for example, Kettlewell released known numbers of light and melanic forms and recaptured a significantly greater proportion of melanic forms. In other words, these sooty areas offer greater protection to the melanic forms than to the light-colored forms, since more of the former survive to be recaptured. As Kettlewell showed, the adaptive value of the melanic types probably lies in their ability to remain concealed on darkened tree trunks from bird predators. In nonindustrial areas, on the other hand, trees covered with normal gray lichens offer decided advantages to the light-colored moths. Thus, there are various populations of peppered moths in England, some of which bear

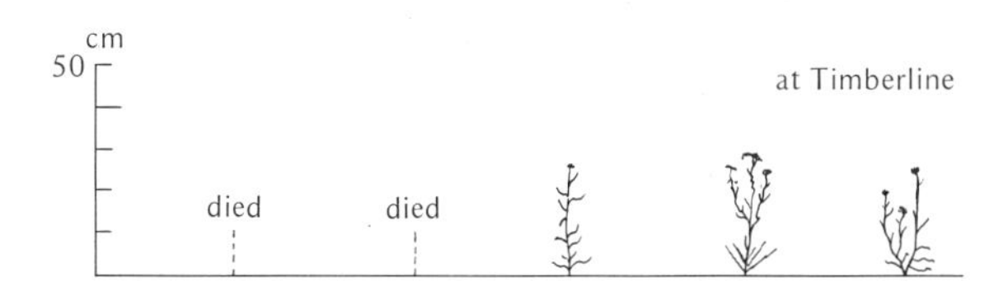

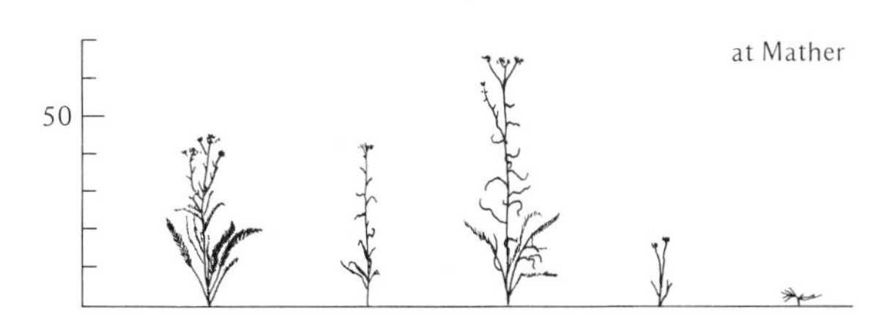

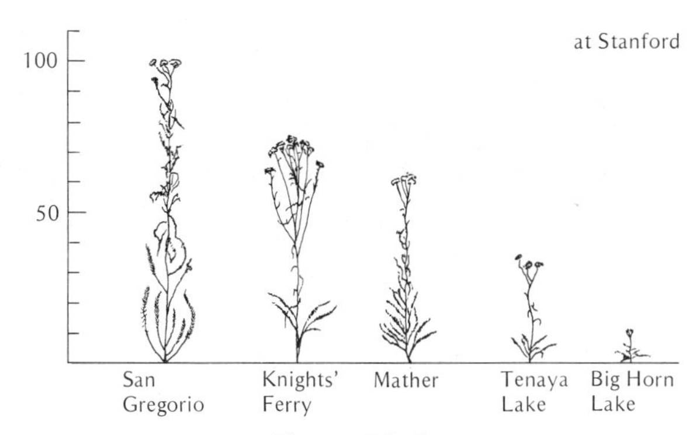

Figure 35–3

Responses of clones from representative Achillea *plants originating from five different localities in California and grown at three different altitudes; sea level* (Stanford), *4600 feet* (Mather), *and 10,000 feet* (Timberline). *(After Clausen, Keck, and Hiesey.)*

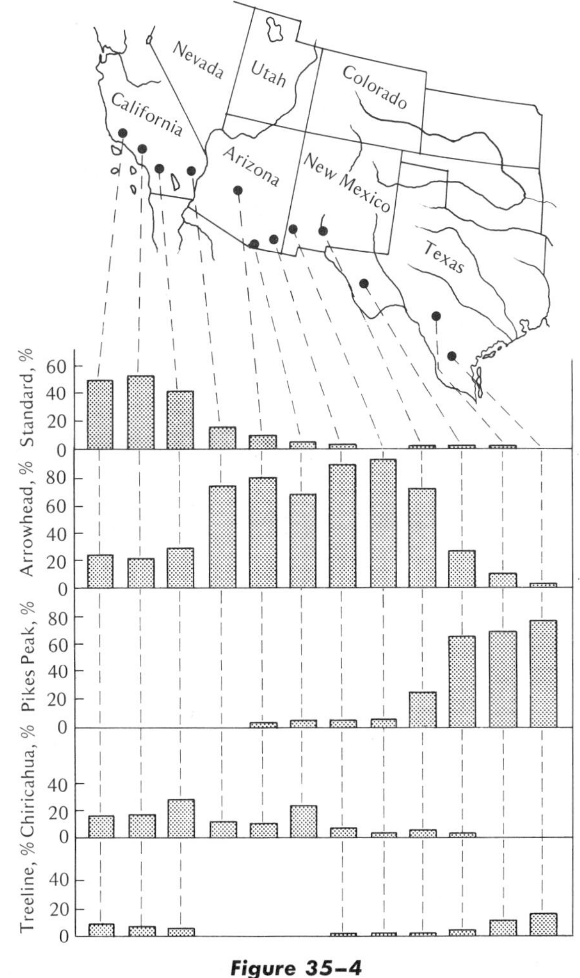

Figure 35–4

Frequencies of five different 3rd-chromosome gene arrangements in Drosophila pseudoobscura *in 12 localities on an east-west transect along the United States–Mexican border. (After Dobzhansky, 1944.)*

frequencies of the melanic gene almost as high as 100 percent (industrial areas), while in others it is almost completely absent (rural areas).

In the main, populations of the same species that differ markedly from each other have been characterized as *races*. Races share the possibility of participating in the gene pool of the entire species, although they are sufficiently separated to exhibit individually unique gene frequencies. The distinction between races is, therefore, not absolute; races may differ in the relative frequency of a particular gene, but these differences do not prohibit gene exchange.

Perhaps one of the most studied cases of racial differences has been in humans. On the basis of morphological appearance and gene frequencies, Boyd and others have proposed the establishment of five main racial groups, all of which can be further subdivided into smaller populations:

1. European group: includes a variety of populations, known as the White Caucasians, ranging from the Lapps of Scandinavia to the Mediterranean peoples of Southern Europe and North Africa.
2. African group: includes the African Blacks.
3. Asian group: includes the Mongoloid peoples and the population on the India-Pakistan subcontinent. (According to most authorities, the latter population is considered among the extra-European Caucasoids.)
4. American group: includes all aboriginal populations found on the American continents. (Many authorities consider American Indians as a Mongoloid subgroup.)
5. Pacific group: includes populations such as the Melanesians and Polynesians, and the race of Australian aborigines.

If we use genes whose frequencies can be detected and scored, the distinction between racial groups is usually not simply discerned by the presence or absence of particular genes but is, in many instances, a matter of gene frequencies. Table

TABLE 35–1

Frequencies (in percent) of various blood-group alleles in five broadly defined human races

Blood Group Allele	Caucasians	Africans	Asians (Mongolian)	American Indians	Australians
A_1	20–30	10–20	15–25	0–55	20–45
A_2	4–8	5	0	0	0
B	5–20+	10–20	15–30	0	0
L^{MS}	20–30	7–20	4	15–30	0
L^{Ms}	30	30–50	56	50–70	26
L^{NS}	5–10	2–12	1	2–6	0
L^{Ns}	30–40	30–50	38	5–20	74
R^0	1–5	40–70	0–5	0–30	9
R^1	30–50	5–15	60–76	30–45	56
R^2	10–15	6–20	20–30	30–60	20
R^Z	0–1	0	0–0.5	1.6	2
r	30–40	10–20	0–7	0	0
r'	0–2	0–6	0	0–17	13
r''	0–2	0–1	0–3	0–3	0
P	40–60	50–80	15–20	20–60	?
Fy^a	40	<10	90	0–90	?
Lu^a	2–5	0–4	?	0–10	0
Di^a	0+	0	1–12	0–25	0

Adapted from Stern, 1973.

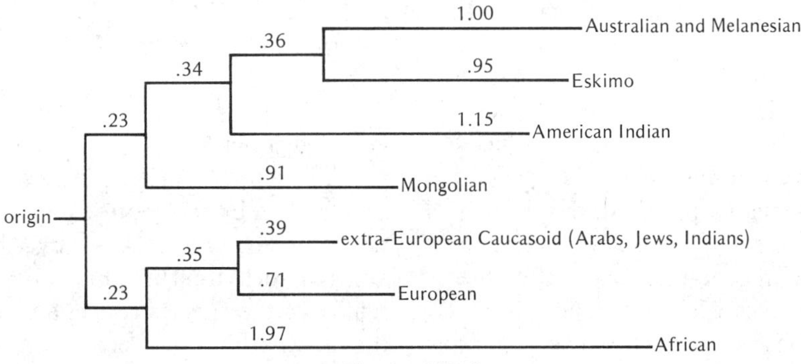

Figure 35–5

Phylogenetic relationship between seven different human groups based on some of the available genetic data. The numerical values are the average numbers of gene substitutions or "genetic distances" between the given points. Thus the African to Mongolian genetic distance is 1.97 + .23 + .23 + .91 = 3.34. (After Cavalli-Sforza).

35–1, for example, shows a comparison of frequencies for a variety of gene systems in five racial groups. In practically all these gene systems knowledge of a particular genotype alone is not by itself sufficient to indicate to which race an individual belongs. An individual of O blood type who is also Rh positive may, for example, belong to any of the races considered if our attention is confined only to these genes.

It is interesting to note that differences between human populations have not reached the point where one population is fixed for one allele at a particular locus and another population is fixed for a different allele: when a population shows fixation for one allele, other populations are always polymorphic for it. This has indicated to many population geneticists that the differentiation between human populations or races has not had the time to proceed very far. According to Cavalli-Sforza and others who have proposed phylogenetic trees of human populations based on the study of many blood groups (Fig. 35–5), the time of separation between the two most genetically differentiated groups, Africans and Mongoloids, may have occurred 25,000 to 40,000 years ago. (This time scale is based on an estimated 15,000-year-old separation of American Indians from the Asiatic mainland.)

The presence of so much genetic variability in all human races (see also Table 34–3) indicates the fictional nature of concepts such as "pure" races. Members of a race are not genetically "pure" in the sense of sharing a uniform genetic identity, nor does genetic uniformity even apply to members of the same family. Lewontin, in fact, points out that more than 90 percent of the genetic variability between humans comes from differences between individuals and groups of the same race, and only 6 percent comes from differences between races. From a genetic point of view "purity" can only be ascribed to asexual clones derived from a single individual. In clonal reproduction, however, it is debatable whether the terms races or species are appropriate, and other terms have been devised to describe populations among microorganisms (see Sonneborn, 1957).

Just as genetic differences at particular loci are not sufficient to indicate racial differences, similarities for particular gene frequencies between two populations do not necessarily indicate racial identity. An apt illustration of this is found in the tests performed by Fisher, Ford, and Huxley to detect PTC tasting among the chimpanzees of the London zoo. As with humans, this ability in chimpanzees is a genetically determined characteristic, and can be measured by observing their reaction to a PTC solution; the nontasters swallow the solution and the tasters spit it out. Remarkably the frequencies of nontasters among the chimpanzees were found to be similar to the frequencies of nontasters among Englishmen. Of course, Englishmen and chimpanzees differ in other respects!

Our criterion for evaluating the differences between populations of a single species is, therefore, based essentially upon gene-frequency differences. When these differences are numerous and it is advantageous to consider populations as separate entities, we may categorize them broadly as races. As we have seen, the forces producing racial differences are primarily adaptive; that is, gene-frequency changes are usually the response of a population to the selective forces operating within a particular environment. At times these racial differences are accompanied by observable morphological differences such as those between some human populations. At other times observed racial differences extend only to gene or chromosomal differences such as those between Texas and California populations of *Drosophila pseudoobscura*. In both cases race formation is a potentially reversible process, since different races may interbreed and combine again into a single populational unit. Thus, when the extent of migratory activity between individuals of a species is great, race formation may be considerably impeded. Rensch, for example, has calculated that migratory species of birds have an average of less than half the number of races of nonmigratory species: the greater the gene flow, the fewer the differences.

As a rule, therefore, race formation is accelerated by barriers that reduce gene exchange between populations. Initially such barriers are primarily geographical and occur when populations bud off from one another and occupy different areas. The potential for gene exchange, however, enables all these different populations to be considered as members of a single species. It is only when populations have achieved sufficient differences to inhibit any gene exchange at all between them that they may be considered to have diverged sufficiently to have reached the level of separate species.

THE SPECIES

The concept of a species as an interbreeding group distinct from other such groups arises in sexually reproducing organisms from the fact that such groups exist in nature and are mutually separated in many instances by "bridgeless gaps" across which interbreeding does not occur. The reality of the species concept is also supported by the fact that they are recognized as distinct groups both by man and other forms of life to whom such discrimination is essential. Predators of all kinds, for example, learn early to discriminate among different possible varieties of prey and to select those that are palatable and can be used for food. Where the recognition of species can be verbalized, such as in primitive human societies, the distinctions made are, in many cases, strikingly similar to the species classifications based on more sophisticated biological criteria. Thus a tribe of New Guinea islanders have distinct names for 136 species of birds found in this region, almost the exact number of species recognized by ornithologists (137). The transition of racial differences to species differences is, therefore, usually marked by a qualitative change accompanied by reproductive separation or isolation. What mechanisms prevent gene exchange between populations and how do such mechanisms originate?

Brought together under one heading, mechanisms that prevent gene exchange have been broadly termed *isolating mechanisms*. Some authors include in this category all factors that prevent gene exchange, even geographical and spatial isolation. Such geographically separated populations, also called *allopatric* populations, obviously do not have the opportunity for gene exchange, and it has, therefore, been debated whether, given the opportunity, many of them would still remain reproductively isolated. Other authors have, therefore, proposed that isolating mechanisms be restricted to those that prevent gene exchange between populations in the same geographic locality, i.e., mechanisms that isolate *sympatric* populations.

Mayr has classified sympatric isolating mechanisms into two broad categories: those that operate before fertilization can occur (premating), and those that operate afterward (postmating). Among the premating isolating mechanisms are:

1. *Seasonal or habitat isolation.* Potential mates do not meet because they flourish in different seasons or in different habitats. Some plant species, such as *Tradescantia canaliculata* and *T. subaspera*, for example, are sympatric throughout their geographical distribution, yet are isolated by the fact that their flowers bloom at different seasons. Also, one species grows in the sunlight and the other in deep shade.

2. *Behavioral or sexual isolation.* The sexes of two species of animals may be found in the same locality at the same time, but their courtship patterns are sufficiently different to prevent mating. The distinctive songs of many birds, the special mating calls of certain frogs, and the sexual displays of most animals are generally attractive only to mates of the same species. Numerous plants have floral displays that attract only certain insect pollinators. Even where the morphological differences between two species is minimal, behavior differences may suffice to prevent cross-fertilization. Thus *Drosophila melanogaster* and *D. simulans*, called *sibling species* because of their morphological similarity, will normally not mate with each other although kept together in a single population cage.

3. *Mechanical isolation.* Mating is attempted, but fertilization cannot be achieved because the genitalia do not fit together. This type of incompatibility was long thought to be a primary isolating mechanism in animals. At present, however, there is little evidence that matings in which the genitalia are markedly different are ever seriously attempted.

Among the postmating mechanisms that prevent the successs of an interpopulational cross even though mating has taken place are:

1. *Gametic mortality.* In this mechanism either sperm or egg is destroyed because of the interspecific cross. Pollen grains in plants, for example, may be unable to grow pollen tubes in the styles of foreign species. In some *Drosophila* crosses Patterson and others have shown that an insemination reaction takes place in the vagina of the female that causes swelling of this organ and prevents successful fertilization of the egg.

2. *Zygotic mortality and hybrid inviability.* The egg is fertilized, but the zygote either does not develop or develops into an organism with reduced viability. Numerous examples of this type of incompatibility have been observed in both plants and animals. For example, Moore made crosses between 12 frog species of the genus *Rana* and found a wide range of inviability. In some crosses, no egg cleavage could be observed; in others, the cleavage and blastula stages were normal but gastrulation failed; and in others, early development was normal but later stages failed to develop. The effect of nuclear-cytoplasmic interactions in preventing gene exchange is seen in the findings of Laven in crosses between certain European strains of *Culex pipiens.* When a cross is made using cytoplasm from one source, e.g., A♀ × B♂, the matings are fertile. However, when the reciprocal cross is made, B♀ × A♂, viable hybrids do not appear and the matings are sterile.

3. *Hybrid sterility.* The hybrid has normal viability but is reproductively deficient or sterile. This is exemplified in the mule (see p. 794) and many other hybrids. Sterility in such cases may be caused by interaction between genes from different sources (as in *D. pseudoobscura-persimilis* hybrids, below) or (as in *Culex pipiens* above) by interaction between the cytoplasm from one source and the chromosomes from another.

In general the barriers separating species are not confined to a single mechanism. The *Drosophila* sibling species *D. pseudoobscura* and *D. persimilis*

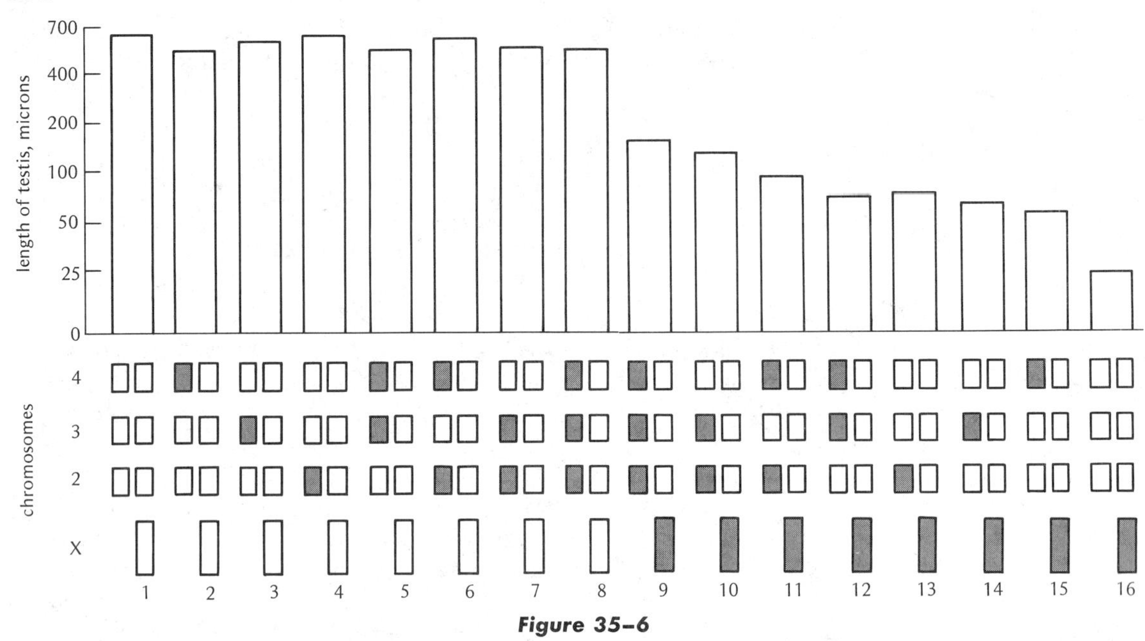

Figure 35–6

Length of testis in hybrids that contain different combinations of Drosophila pseudoobscura (*white*) *and* D. persimilis (*shaded*) *chromosomes. The smallest testis size occurs when the X chromosome is from one species and all the autosomes are from another.* (*After Dobzhansky, 1951.*)

are isolated from each other by habitat (*persimilis* usually lives in cooler regions and at higher elevations), courtship time (*persimilis* is usually more active in the morning, *pseudoobscura* in the evening), and mating behavior (the females of both species prefer males of their own species). Thus, although the distribution ranges of these two species overlap throughout large areas of western United States, these isolating mechanisms are sufficient to keep the two species apart. To date, only a few cross-fertilized females have been found in nature among many thousands of flies that have been examined. However, even when cross-fertilization occurs between these two species, gene exchange is still impeded, since the F_1 hybrid male is completely sterile. By using chromosomes with mutant markers and examining the progenies of fertile F_1 hybrid females backcrossed to males of either species, Dobzhansky has shown that the genetic factors causing this sterility are located on all chromosomes. A chromosomal interaction takes place in hybrid males, and the greater the difference in species origin between sex chromosomes and autosomes, the greater the degree of sterility (Fig. 35–6). Even the occasional presence of tetraploid tissue in the testes of these hybrids fails to improve chromosome pairing and restore fertility (see p. 470). Furthermore, in contrast to the vigorous but male-sterile F_1 hybrids, the progenies of the F_1 backcrosses show markedly lower viability (*hybrid breakdown*), further impeding gene exchange between the two species.

In plants both pre- and postmating isolating mechanisms may be operative in the same interspecific cross, the postmating barriers functioning when the premating barriers have been bypassed. For example, Latimer* has shown that the cross *Gilia australis* ♂ × *Gilia splendens* ♀ usually fails because of the retarded growth of the *australis* pollen tubes. On the other hand, *splendens* pollen tubes may reach *australis* ovules in the reciprocal cross, but the hybrid embryo, nevertheless, fails to

*Cited by Grant, 1963.

develop because of the degeneration of the endosperm. However, in some other species, even when the endosperm is functional, a further barrier may exist; in the cross *Linum perenne* ♀ × *Linum austriacum* ♂, the hybrid embryo is incapable of sprouting through the maternal seed coat. Only when the embryo is artificially dissected out of its seed can it develop into a full grown plant.

MODES OF SPECIATION

In 1889 A. R. Wallace proposed that natural selection might favor the establishment of mating barriers between populations if the hybrids were adaptively inferior. That is, genotypes that did not mate to produce inferior hybrids would be selected over genotypes that did. According to this hypothesis, selection for sexual isolation arises because most races and species are strongly adapted to specific environments. The "genetic dilution" of the parental gene complexes which occurs in hybrids may therefore be of considerable disadvantage in the original environments. Thus genotypes that have incorporated premating isolating mechanisms would have the advantage of not having wasted their gametes in the production of deleterious offspring.

Full utilization of this mode of speciation demands, of course, that the different populations producing deleterious hybrids be exposed to each other in the same locality; only then could the more sexually isolated genotypes be specifically selected. Speciation should, therefore, occur in the following sequence: (1) genetic differentiation between allopatric populations, (2) overlap of differentiated populations in a sympatric area, and (3) selection for sexual isolating mechanisms. Demonstration of this sequence among natural populations has, therefore, been attempted by comparing the degree of sexual isolation between different sympatric and allopatric populations; sexual isolation should be strongest among the sympatric populations, since they are sufficiently close to produce deleterious hybrids, and weakest among allopatric populations that are too distant to produce such hybrids. In one such comparison, Grant has reported that of nine species in *Gilia*, those that are most difficult to cross are the sympatric ones. The allopatric species, by contrast, show no barriers against hybridization although all the crosses are sterile.

In *Drosophila paulistorum*, Dobzhansky and co-workers have demonstrated the existence of five morphologically identical races in Central and South America that show complete isolation from each other when they are found in the same locality. The fact that these races can still exchange genes with various strains of a "transitional" race provides, in theory, a common gene pool for all of them and prevents them from being considered as full-fledged species. They have, therefore, been described as "incipient species" or "semispecies," and *Drosophila paulistorum* has been called a *superspecies*. Investigations by Ehrman have shown that the genetic basis for sexual isolation between these populations is polygenic, located on all chromosomes. However, even when this sexual premating mechanism is bypassed, sterility of the male hybrids will help prevent gene exchange. The source of this sterility may lie in a symbiont carried in the egg cytoplasm that affects the reproductive tract of the weakened hybrid males.

An experimental demonstration of selection for premating isolating mechanisms has been clearly presented by Koopman, who made use of the normally isolated sibling species, *Drosophila pseudoobscura* and *D. persimilis*. Although sexual isolation exists between these two species in nature and at normal temperatures in the laboratory, cold temperatures will cause an increase in interspecific mating and may produce as high as 50 percent hybrids. By marking each of the two species with different homozygous recessive markers, Koopman was able to recognize hybrids formed under these low-temperature conditions and remove them from interspecific population cages. He performed this operation each generation and found that, as time went on, fewer and fewer hybrids were produced. For example, after five generations the amount of hybrids produced in the mixed populations was generally 5 percent compared to values that were initially as high as 50 percent. This was striking evidence that selection against hybrids had caused rapid selection for sexual isolation and the consequent reduction in hybrid formation.

One may, of course, argue that, given sufficient time, even allopatric populations will accumulate a sufficient number of genetic differences to show sexual isolation when they are brought together in the same locality. In the *virilis* group of *Drosophila* species, Patterson and Stone have observed that the European *D. littoralis* is much more isolated from the American populations of *americana*, *texana*, and *novamexicana* than are American species in the same group. It is even possible that allopatric isolation may proceed quite rapidly, and Carson has suggested this mechanism to explain rapid speciation noted among the Hawaiian *Drosophilidae*. He puts emphasis on "founding accidents" (p. 777) that lead to large genetic differences between neighboring populations as one of the immediate causes for sexual isolation. In many cases, however, selection for sexual isolation because of hybrid sterility would seem to be much more effective in erecting species barriers than the accidental processes occurring in allopatric populations.

Where species barriers break down to produce viable and fertile hybrids—and there are such instances—*zones of hybridization* or *hybrid swarms* may occur whose genotypes and phenotypes are intermediate to both parental species. If a unique and discrete habitat exists to which the hybrids are better adapted than the parents, it is conceivable that the new population may eventually become isolated from the parental populations.

In some cases, especially plants, fertile hybrids can introduce genes from one species into the other, producing a phenomenon which Anderson has termed *introgressive hybridization*. Repeated hybrid backcrosses of this type are believed by Mangelsdorf to have introduced genes from a grass called *Tripsacum* into the ancestral Central American stock of modern corn, *Zea mays*. In plants, furthermore, even if the hybrid is sterile and must propagate asexually, polyploidy may arise enabling the hybrid to produce fertile gametes (allopolyploids; see p. 468). Since these gametes are diploid relative to the haploid gametes of the parental species, a new species is born at one stroke, fertile with itself or other such polyploid hybrids but sterile in crosses with either parental species.

The sequence of evolutionary events in speciation seems, therefore, to begin with race formation and end with reproductive isolation. In this sequence one disputed point among population geneticists is the degree to which spatial separation between populations is necessary to accumulate the initial genetic differences. Many workers in this field believe that populations can only accumulate genetic differences when they are sufficiently separated by space to prevent gene exchange which might eradicate these differences. Only after this important early period of geographical separation takes place can the speciation process take hold, either by the accidental origin of isolating mechanisms or by selection of isolating mechanisms because of defective hybrids.

Other workers, especially Mather and Thoday in recent years, have proposed that a population in a single locality selected for adaptation to different habitats within that locality could produce an increase in genetic variability (see disruptive selection, p. 805) that would lead to polymorphism. One such example is the polymorphism now found in the British peppered moth (p. 831), and a further important example is the polymorphism of mimicry in the butterfly *Papilio dardanus* (see Sheppard, 1961). It has also been proposed that under some circumstances, especially if the selected forms can exist independently of each other, isolation between the selected groups might result. Evidence for this view has been presented by Thoday and Gibson in selection experiments on bristle number in *Drosophila melanogaster*. They selected flies each generation for high and low bristle number and found that, although random mating was permitted, mating preferences of these flies went rapidly in the direction of positive assortative mating, high × high and low × low (Table 35–2). Increased isolation resulting from disruptive selection between populations in the same locality has since been achieved in other experiments (see Coyne and Grant, and Soans et al.). However, in spite of many attempts, some

TABLE 35–2

Results of tests for mating preferences among *Drosophila melanogaster* that have been selected for high bristle number (H) and low bristle number (L), and in which males and females are given a free choice of mates

Generation of Selection	Number of Matings			
	H × H	H × L	L × H	L × L
7	12	3	4	12
8	14	2	6	10
9	10	4	6	7
10	8	4	3	13
19	27	2	8	20
	71	15	27	62

From Thoday, 1972.

results have not been replicated (see Scharloo), and it has also been questioned whether any single locality in nature could consistently maintain divergent selective conditions for a long enough period of time to produce speciation (see Mayr). Even if such *sympatric speciation* has occurred, it is most likely uncommon.

THE CREATIVE ROLE OF NATURAL SELECTION

The genetic changes by which species adapt to their environments are the underlying structure of evolutionary progress. However, genetic opportunity at any one time is limited. The reason for this is that the cost of evolution to a population is probably high (pp. 815ff.), especially if many gene changes are being selected simultaneously. It is, therefore, hardly surprising to find that many species become extinct because of changing environments to which they are unable to respond effectively. These evolutionary limitations extend also to the direction toward which a species is capable of evolving. For example, given their genetic architecture and the conditions under which they live, it is most unlikely that vertebrates could evolve insect forms or vice versa. The genetic endowment of a species produced by its past evolutionary history thus closes off certain evolutionary pathways and opens others which appear to be unique and "creative." In this sense, the creativity we observe in evolution is restrictive; that is, it is guided by all of its many prior historical interactions.

The creativity of evolution, however, does not mean "purposefulness" in the human sense. Except for artificial selection, there are no observable agents either within or without the organism that are consciously capable of directing evolution toward any particular stage. Theories, such as that of Lamarck (p. 5) and that of postadaptation (p. 548), which laid emphasis on the ability of an organism to respond directly to its environment by sensing its needs and evolving necessary structures, have never been demonstrated. According to the modern view, evolutionary creativity is primarily caused by the important role played by natural selection as it acts upon genetic variability and thereby exposes a species to further selection for the same or closely connected environment.

These considerations lead to the following important distinction. If we were artificially to partition the causation of evolution into two forces, selection and mutation, the argument that mutation *alone* is insufficient to produce most of the observed complex biological structures would be quite true. Without the creativity of selection, millions of possible structures may be produced at random, but it is hardly likely that any of them will show the remarkably precise functional relationships of organs such as the vertebrate eye or the human brain. It is primarily because of the guiding role of selection in choosing only those mutations that increase vision in animals and intelligence in humans that structures such as the eye and brain have evolved. In other words, the random process of mutation produces the variability which natural selection then molds into structures which could not have arisen all at once by themselves. As aptly phrased by Fisher, "natural selection is a mechanism for generating an exceedingly high degree of improbability."*

Restriction of future genetic change because of

*Quoted by Huxley, 1943, and others.

natural selection, however, does not mean that each type of organism must evolve a unique set of structures, different from those evolved in all other evolutionary lines. Many different organisms have similar phenotypic adaptations which have evolved separately, such as eyes bearing retinal pigments, lenses, and focusing devices. In such instances, called "parallelism" or "convergence," different genes in different organisms act to produce the same phenotypic result. Thus, on the one hand, organisms are different because of their unique histories and adaptations, and, on the other hand, they bear many similar features because they have faced many similar adaptive problems. Further use of genetic principles in understanding evolutionary phenomena has been set forth in the books of Dobzhansky, Ford, Grant, Huxley, Mayr, Rensch, Simpson, Stebbins, and others.

EVOLUTION OF PROTEINS

Our application of genetics to evolution has so far been limited to some of its grosser and more obvious features. On the molecular level evolution is, of course, reflected in protein differences, and a large number of investigations are now directed toward discerning and evaluating such differences. Proteins have the important evolutionary advantage that they are, in a sense, "chemical fingerprints" of evolutionary history, bearing amino acid sequences that have changed only as a result of genetic changes. Organisms that bear large numbers of amino acid sequences in common may, therefore, be considered to be more directly related than organisms which differ greatly in amino acid structures. Among the more than 100 proteins with known amino acid sequences (see Dayhoff), the hemoglobins are especially prominent.

In comparisons between the hemoglobins of vertebrates, the polypeptide chains are so remarkably similar both in respect to length and to general amino acid composition that the amino acid differences between them stand out quite distinctly. Zuckerkandl, for example, has calculated an average of 22 differences between human hemoglobin chains (α and β) and the similar hemoglobins of the horse, pig, cattle, and rabbit. Assuming from fossil evidence that a common ancestor existed for these groups approximately 80 million years ago, this finding can be interpreted to mean that the human hemoglobins diverged from those of these animals by about 11 new mutations in each hemoglobin line, or an average of about one amino acid change per 7 million years.

However, not all proteins share the same amino acid substitution rate. Relatively few changes, for example, have been observed in the evolution of insulin, cytochrome C, and histones, whereas the fibrinopeptides formed during blood clotting have changed more rapidly than any other known protein. This wide divergence in evolutionary rates indicates that there are selective constraints that limit the amount of mutational changes that a particular protein can undergo. That is, most of the insulin molecule is apparently involved in its endocrine functions, and most changes would therefore, be deleterious. The fibrinopeptides, on the other hand, primarily serve as a "filler" in the production of fibrin, so that most amino acid positions are not crucial for proper function. But note that although selection may help determine which and how many amino acid positions can be changed, it need not severely limit the kinds of amino acids that may be substituted at "permissible" positions.* That is, some amino acid changes may be neutral in their effect on the function of the protein.

The proponents of neutral mutation and random drift (see p. 817) have therefore used these findings in attempts to explain the constancy of evolutionary rates in particular proteins. Kimura and Ohta, for example, point to the fact that the same number of changes in the α hemoglobin chain has occurred, relative to amino acids in β hemoglobin, whether the α chain comes from the same species or different species. Compared to the human β chain, the human α chain shows 75

*Amino acid positions on a protein that show changes during the same particular evolutionary period have been called concomitantly variable codons, or *covarions*. According to Fitch, there are different numbers of covarions in different proteins at any particular time (e.g., 10 in cytochrome C, 39 in the β hemoglobin chain, etc.), but the rate at which mutations are fixed in these covarions seems to be similar.

differences, the horse α chain shows 77 differences, and the carp α chain shows 77 differences. They ask: Why should the α hemoglobin chains of humans, horses, and fish, each with a different selective history, have diverged from the human β chain at almost exactly the same rate? According to the neutralists, this uniformity can most easily be explained by a common rate of neutral mutation and drift rather than common selective conditions. Striking as this evidence is, there are exceptions to constant evolutionary rates in some proteins, for example, cytochrome C (Jukes and Holmquist), and different rates of nucleotide substitution are noted in comparisons between different proteins (Langley and Fitch).

Assuming, however, that the data for the hemoglobins really reflect a constant rate of evolution for this protein, the number of amino acid differences between various hemoglobinlike chains would then reflect their relative age. It has therefore been proposed that evolutionary relationships between myoglobin and the various hemoglobins follow the hypothetical path shown in Fig. 35–7. Thus myoglobin which differs from the β chain of human hemoglobin in 86 percent of the amino acid sequence can be assumed to have differentiated from a molecular ancestor common to both chains about 650 million years ago. On the evolutionary time scale this period was in the pre-Cambrian era, and the divergence probably occurred in an invertebrate or prevertebrate progenitor. The split between α and β chains involves differences in 52 percent of the amino acid sequences and can therefore be ascribed to an evolutionary divergence during the Devonian period at the time of appearance of the first amphibians. If we follow this scheme further, the most recent split between these hemoglobin chains occurred approximately 35 million years ago and accounts for the 7 percent difference in amino acids between β and δ chains.

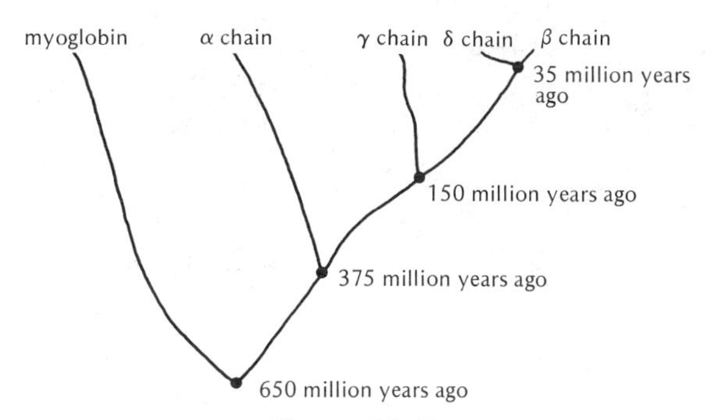

Figure 35–7

Conjectured phylogenetic relationships among myoglobin and four hemoglobin chains. (After Zuckerkandl.)

The evolutionary picture emerging from these structures indicates that at least five types of polypeptide chains in humans, and more types if other animals are considered, can be traced back to a common ancestry. How can new hemoglobin types appear and persist in an organism in the presence of old ones? It seems most probable at present that each new hemoglobin polypeptide, i.e., each step in the branching process shown in Fig. 35–7, originated by duplication of the gene for a hemoglobin chain. This duplication was then followed by the appearance of a duplicate polypeptide and its consequent evolutionary differentiation through mutation.

The gene duplication process, leading to the presence of a pair of similarly functioning genes in sidy-by-side position along the chromosome, can be partially reversed by a *gene fusion* process that unites them into a single "hybrid" gene. One of the mechanisms responsible for this union may be the same unequal crossing-over event that causes gene duplication by changing the amount of chromosomal material in a recombinant product (see Fig. 25–2). Striking examples of gene fusion are hemoglobin abnormalities in which the β and δ hemoglobin genes, lying in tandem, are fused to produce a combined gene product composed of portions of the two formerly separate polypeptide chains. Thus, individuals affected by "Hb Lepore" have hybrid polypeptide chains 146 amino acids long, with the δ portion at the amino end and the β portion at the carboxyl end. Homozygotes for this fused gene do not produce normal β or δ chains, but continue to produce relatively large amounts of γ chains as though suffering from thalassemia (p. 621). The Lepore chromosome may therefore be represented as --- $\widehat{\delta\beta}$ --- in contrast to the normal chromosome --- δ - β ---. Individuals representing the reciprocal recombinant product of the crossover event that produced Lepore hemoglobin have now also been identified, and these are called "anti-Lepore." They produce a hybrid polypeptide chain with a section of the β sequence at the amino end and a section of the δ sequence at the carboxyl end. Presumably, these individuals carry a new gene duplication in the form --- δ - $\widehat{\beta\delta}$ - β ---.

Evolution by means of gene duplication has now been shown to have occurred for a variety of proteins. In many of these instances, such as the hemoglobins described above and the different serine proteases (chymotrypsin and trypsin), the duplicated polypeptide chains have preserved similar functions. In other instances duplicated genes have evolved in different functional directions, although they still share enough common amino acid sequences to indicate their common origin. Thus the vertebrate "nerve growth factor" protein that enhances the outgrowth of neural cells from sympathetic and sensory ganglia, has amino acid sequences similar to insulin, and may share some functional similarities as well (Frazier et al.). Also, the α-lactalbumin protein, which is part of an enzyme used in the synthesis of lactose in mammalian milk, has an amino acid sequence remarkably similar to that of the lysozyme enzyme found in tears that is used to degrade the mucopolysaccharides of bacterial cell walls (Hill et al.). This similarity is further reflected in the fact that both proteins are the products of tissues (mammary, tear duct) that were at one time probably sebaceous glands, and they both use sugar molecules as their substrates.

Within genes duplications also exist, as shown in the presence of three homologous amino acid regions within the heavy chain of the gamma G immunoglobin antibody (p. 706). In the human haptoglobin α-2 protein found in blood serum, a segment of 59 amino acids (positions 13 to 71) is an almost exact repetition of an adjacent segment (positions 72 to 130). Repeated amino acid sequences have also been discerned within the ferredoxin protein (used as an electron carrier in various biochemical processes), in the glutamate dehydrogenase enzyme, and various other proteins (see Jukes, 1972).

However, attempts to determine phylogenetic relationships exclusively by comparisons of amino acid sequences in a polypeptide cause some obvious inconsistencies. For example, the cytochrome C protein of man differs from that of horses by 12 amino acid substitutions, but it differs from kangaroos by only 8 amino acid substitutions. Another protein, the β chain of human hemoglobin, differs from that of distantly related primates (lemurs) by more than 20 amino acid substitutions but differs from the β chain of pigs by only 14 amino acid changes. Without further information the comparisons here would lead to a strange phylogeny: man is more related to marsupials than to some placentals and, among mammals, more related to pigs than to primitive living primates.

The fact that one type of protein alone is an insufficient base upon which to declare phylogenetic relationships is also seen in comparisons made by Zuckerkandl and Schroeder between the hemoglobins of man and gorilla; only a single amino acid difference exists between these two forms. The extent of this difference is, therefore, no greater than that found in some of the mutant variants in human populations. Obviously more than one type of protein must be compared between different organisms before reliable relationships can be determined. As pointed out by Mayr, since most organisms contain thousands of cistrons, to select only one or a few for examination may give a rather limited picture of how the organism really looks or looked like phenotypically. Furthermore, although selection acts ultimately upon genes and their immediate products, the proteins, the action of selection in many organisms is more directly upon the entire phenotype, i.e., upon interactions between numerous proteins. In the words of Simpson, "On an average the farther we are from genes the nearer we are to the action of selection, and thus the better able we are to interpret the adaptive processes involved." Viewed, therefore, as an aid in understanding evolutionary relationships rather than as the sole criterion for such relationships, comparisons between amino acid sequences will undoubtedly offer valuable contributions.

One evolutionary area in which a molecular and biochemical approach offers unique advantages is in determining relationships between basic metabolic pathways. Haldane at first, and more recently Horowitz, proposed that the evolution of metabolic pathways began with the various modes of absorption and utilization of different organic compounds in the primeval "soup" four to five billion years ago. These early waters are presently believed to have contained a large amount of organic material produced by interactions among the primitive atmosphere components, hydrogen, methane (CH_4), ammonia, and water. Experiments by Miller and Urey have demonstrated that such mixtures are capable of producing a wide variety of important organic compounds, including amino acids. Since the concentration of organic material in these primeval seas may have ranged as high as 10 percent, the first living forms would have been able to utilize large amounts and varieties of compounds necessary for subsistence and growth.

Within a relatively short time after the appearance of living organisms, however, some or many of these organic compounds must have diminished to a short supply. An amino acid such as histidine, for example, may have become relatively rare, but some of its molecular relatives, such as histidinol (see Fig. 26–21), may have remained plentiful. Adaptive value would, therefore, have been conferred upon organisms carrying enzymes that were capable of catalyzing the reaction histidinol → histidine. As the supply of histidinol was depleted, in turn, selection may then have operated to confer adaptive value upon an organism capable of catalyzing a histidinol precursor into histidinol. In this fashion a metabolic pathway would be established, beginning with the final compound, e.g., histidine, and leading in a descending stepwise fashion to compounds which could be used as precursors. Thus the more primitive enzymes may be those which are at the final steps of a metabolic pathway rather than those at earlier steps.

EVOLUTION OF NUCLEOTIDE SEQUENCES

A molecular approach to evolution more directly concerned with genes rather than gene products has been to analyze the genetic material itself. However, except for very small viruses (e.g., MS2, see pp. 661ff.), complete nucleotide sequences are still unknown, and direct comparisons between the primary structures of nucleic acids cannot be made. More indirect methods have therefore been developed based on measuring the degree of *in vitro* hybridization between nucleic acids extracted from different organisms. In one of the techniques presently used, DNA is extracted from two organisms, e.g., X and Y, dissociated into single strands, and then given the opportunity to form X-Y "hybrid" double-stranded DNA by incubating the different DNA molecules together at appropriate temperatures. To enable the separation of interspecific X-Y DNA from intraspecific X-X or Y-Y DNA, the DNA of one species, e.g., X, is radioactively labeled, and only relatively small amounts of it are used in the incubation mixture. Because of its rarity the DNA of X will therefore have very little chance of forming X-X double strands, and all radioactively labeled double-stranded DNA can be assumed to be X-Y. This double-stranded DNA can then be extracted (on hydroxyapatite crystals) and its properties examined.

If the DNA from X and Y are perfectly homologous and possess no nucleotide differences at all (i.e., they are of the same species), then the melting temperature at which the hybrid X-Y DNA dissociates into single strands (see p. 75) will, of course, be the same as either X-X or Y-Y. However, should there be nucleotide differences between X and Y, then the X-Y hybrid will dissociate more easily because of nucleotide mismatching; that is, its stability is reduced and its melting temperature will be lowered. Experiments by McCarthy and others show that for each 1 percent difference in nucleotide composition between X and Y, the thermal stability of the X-Y hybrid DNA molecule is lowered by 1.6 °C.

An illustration of results that have been obtained by this method is shown in Fig. 35–8, along with corresponding estimates derived from amino acid analysis of various proteins. In this example the nonrepetitive or "unique" portion of cattle DNA (see Fig. 5–20) is hybridized to nonrepetitive DNA from sheep and pig. The approximate reduction in thermal stability for these two hybrid DNAs turns out to be respectively 6 °C (3.7 percent mismatching) and 12 °C (7.5 percent mismatching), indicating a significantly further evolutionary distance between cow and pig than between cow and sheep. Note also the number of nucleotide substitutions estimated by DNA hybridization is higher than the number of nucleotide substitutions that have occurred in the hemoglobin and insulin genes (see also Gummerson and Williamson), and less than that for the rapidly evolving fibrinopeptides. Obviously, different portions of nonrepetitive DNA can change at different rates, and some portions of this DNA may not even code for proteins.

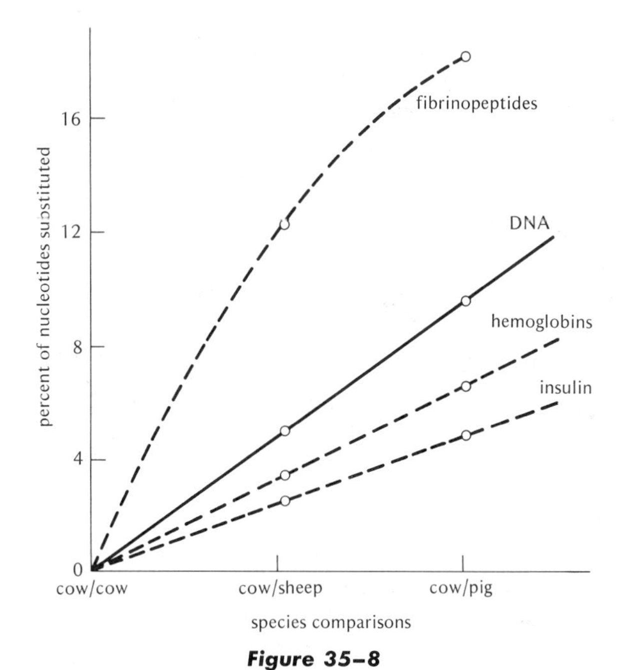

Figure 35–8

Nucleotide differences observed among three species of artiodactyls using the DNA hybridization technique described in the text (solid line) and estimates of nucleotide substitutions derived from amino acid analysis of various proteins (dashed lines). (After McCarthy and Farqhar.)

In general, therefore, DNA-DNA comparisons enable a broad determination of the evolutionary relationship between species which reflects, so far, many of the phylogenetic relationships determined by other taxonomic methods. To obtain more precise estimates of the rate of nucleotide substitution during evolution, called by some the "evolutionary clock," it would be necessary to know the exact nucleotide mutation rates and the exact proportion of nucleotide mutations that have undergone selection. We could then ask: Is there a single evolutionary clock? Many evolutionary clocks? Changing evolutionary clocks? At present, however, there is no commonly accepted solution to these questions, although some answers have been offered (see, for example, King).

MOLECULAR EVOLUTION IN THE TEST TUBE

The phenomenal growth of molecular information in biology has sparked various attempts to demonstrate evolutionary processes in the laboratory. Since this approach usually demands considerable biochemical analyses as well as rigid control over genetic and environmental conditions, and rapid generation times, most of these studies have been performed with microbial organisms. Among the common techniques is to subject a strain of bacteria to a new carbon source (e.g., xylitol) or nitrogen source (e.g., butyramide) which the cells are not able to metabolize properly. When such cells are simultaneously exposed to a mutagenic agent, mutations increase in frequency, and some adaptive mutations may then arise. Natural selection then proceeds by permitting the survival of those bacterial strains with improved metabolic efficiency. Often, evolutionary changes of this kind occur through the adaptation of enzymes that were initially "inefficient" on the new substrate since they were primarily used for other purposes. A variety of such adaptational changes have been noted in experiments by Wu, Betz, and others (see also review of Clarke):

1. Synthesis of the inefficient enzyme may become constitutive through a regulatory mutation, thereby increasing the amount of this enzyme in the presence of the new substrate.
2. A regulatory mutation may enable synthesis of the inefficient enzyme to become inducible by the new substrate.
3. Mutations may occur in the structural gene for the enzyme itself enabling the former inefficient enzyme to metabolize the new substrate more efficiently.
4. Mutations may occur enabling the substrate to enter the cell more easily.

A striking demonstration of enzymatic evolution is shown in an experiment by Campbell and co-workers on an *E. coli* protein. Instead of trying to adapt bacteria to an unusual artificial medium, they exposed a strain carrying a deletion of the β galactosidase *z* gene (pp. 675ff.) to lactose medium. In the absence of β galactosidase, hydrolysis of lactose does not occur. Such colonies of bacteria can be recognized by the fact that they develop a red color on an "indicator" medium in which lactose is present, in contrast to the white color of lactose-utilizing colonies.

Campbell and co-workers found that within one month of growth on a lactose-containing medium, the deficient z^{del} strain gave rise to white colonies that could utilize lactose, although inefficiently. Further growth and selection among these new lactose-utilizing cells then gave rise to a more efficient strain. When exposed to lactose medium unsupplemented with other sugars, the final selected strain of bacterial cells, called *Ebg*,* could form colonies as rapidly as could wild-type *E. coli*.

Various tests then showed that the *Ebg* strain had evolved a lactose-hydrolyzing enzyme completely different from β galactosidase. This new enzyme had a greater molecular weight, different immunological properties, and was sensitive to different ionic conditions. It's genetic locus on the *E. coli* map (at the 59 minute position near the *tol C* locus) was also different from that of the *lac* locus (at the 10 minute position, see Fig. 19–8) which would account for the fact that its synthesis was not regulated in accord with other enzymes in the *lac* operon. Nevertheless, as recently shown by Hall and Hartl, the appearance of this enzyme in some strains of *Ebg* can be regulated by the presence of lactose, although the regulation mechanism is still unknown. These experiments therefore demonstrate that a protein, with only vague affinities for a particular function, can assume that function with remarkable efficiency by a stepwise evolutionary process of mutation and selection.

On the nucleotide level, Mills, Kramer, and Spiegelman have narrowed test-tube evolutionary experiments down to the smallest self-replicating molecule known so far. They began with a Qβ RNA viral molecule about 3600 nucleotides long that could replicate itself *in vitro* upon the addition of replicase enzyme and various ribonucleotide triphosphates (see p. 105). Selection for rapid replication was then practiced by successive transfers of only the earliest replicating molecules to new cultures. Under these conditions successful Qβ molecules need only retain those sequences that enable them to be recognized by the replicase enzyme provided in the culture. That is, they no longer have the need for genes that formerly coded for what are now unnecessary proteins, i.e., coat protein and replicase. Selection was then even further intensified by placing fitness advantages on molecules that could replicate when only a single such molecule is present in an entire culture. This single-stranded molecule, or "plus" strand, must rapidly attract a replicase enzyme to form a complement or "minus" strand which then forms a new plus strand, etc.

By these selective mechanisms the Spiegelman group has achieved what is at present, the ultimate parasite; a self-replicating RNA molecule called Midivariant-1 (or MDV-1) that is only 218 nucleotides long! The nucleotide sequences of the plus and minus strands of MDV-1 are shown in Fig. 35–9, along with the probable secondary structure.

*Abbreviation for "Evolved β galactosidase." The product of the fifth selection cycle was called *Ebg*-5.

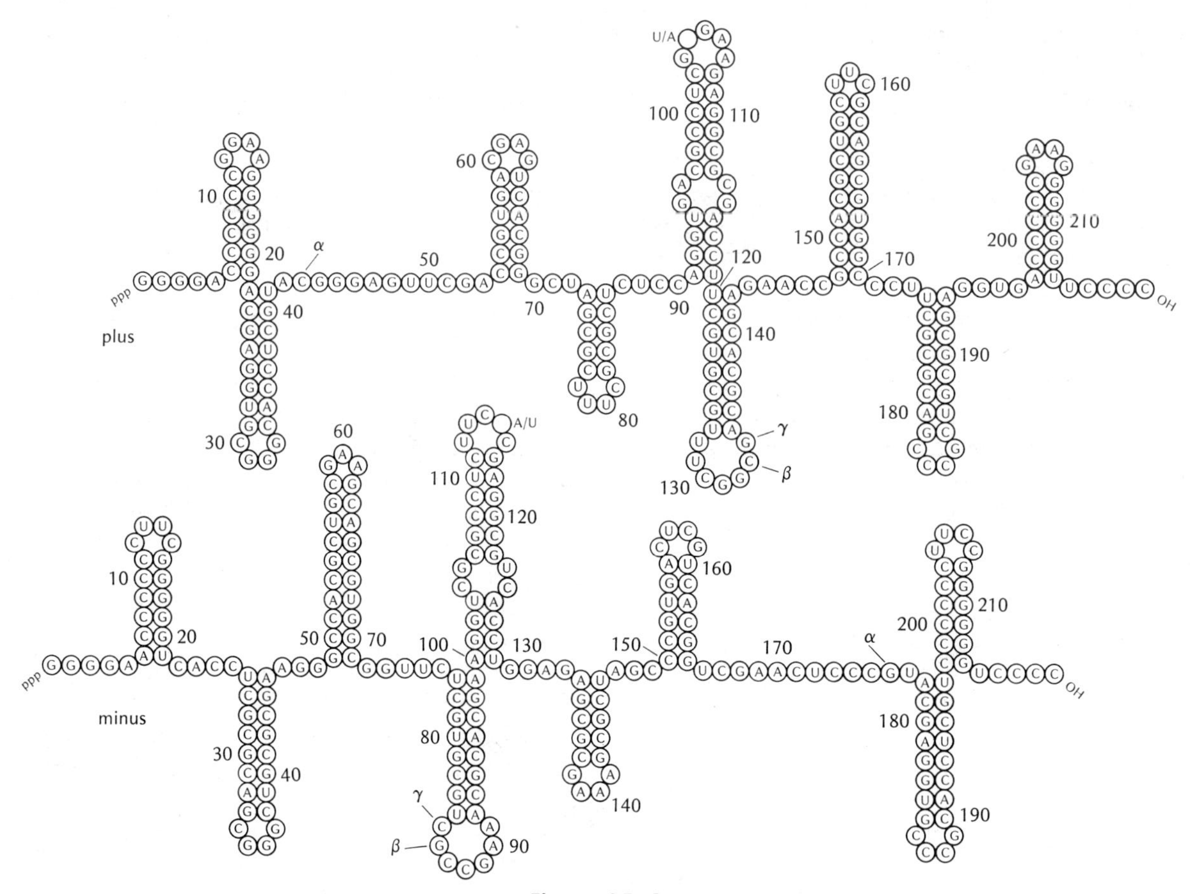

Figure 35–9

Ribonucleotide sequences of the plus and minus strands of MDV-1. The 5′ ends of the strands are on the left, the 3′ ends are on the right. The α, β, and γ designations refer to nucleotide substitutions discovered by Kramer et al. that arose in a strain selected for resistance to ethidium bromide. In the plus strand these substitutions are $\alpha(\#43) = \text{C} \rightarrow \text{U}$, $\beta(\#133) = \text{C} \rightarrow \text{U}$, and $\gamma(\#134) = \text{G} \rightarrow \text{A}$. Complementary substitutions occur of course in the minus strand, $\alpha(\#176) = \text{G} \rightarrow \text{A}$, $\beta(\#186) = \text{G} \rightarrow \text{A}$, $\gamma(\#85) = \text{C} \rightarrow \text{U}$. It is probable that these mutations cause a loss of binding sites for ethidium bromide since the mutant RNA binds less of this compound than does normal MDV-1. (Modified after Mills, Kramer, and Spiegelman.)

Note that although the primary nucleotide sequence of the minus strand is different from the plus strand, the phenotype of both strands are complementary and identical in many respects. It is presumably some feature of this hairpin-looped phenotype which enables the replicase enzyme to recognize the molecule. This system therefore offers the opportunity to investigate the specific nucleotide sequences involved in replicase recognition, and enables a precise determination of those nucleotide changes that occur as the molecule evolves further. For example, selection can now be practiced on this molecule for a variety of adaptations including resistance to inhibitory agents such as ethidium bromide (Kramer et al.) or the ability to replicate in the presence of limited quantities of one of the ribonucleotide triphosphates.

EVOLUTION OF GENETIC SYSTEMS

From a genetic point of view perhaps the most interesting evolutionary events are those that have taken place in the systems of coding and transferring the genetic material itself. As all other traits, genetic systems have also evolved, and trace their ancestry back to that "Aquatic Garden of Eden" out of which all life began. Experiments by Oro, and by Fox and Harada, have shown that the bases, adenine and uracil, may be spontaneously synthesized from compounds that were probably present in the early history of the earth. Ponnamperuma and others have extended these experiments to show the formation of the purine-sugar combinations, adenosine and deoxyadenosine, under presumed primitive conditions. Heating nucleosides such as these with inorganic phosphate can then produce the nucleotides of RNA and DNA.

At present only speculations can exist as to whether RNA or DNA served as the initial genetic material. One point of view is that the present-day sequence, DNA → RNA → protein, also represents the evolutionary sequence. Another possibility is that the first genetic material was RNA, since RNA now serves as the only connection between genes and proteins. In the past, RNA might have served as both a self-duplicating and protein-synthesizing system. Some support for this view exists in the fact that the self-duplication of RNA is certainly possible, and is still preserved in many RNA viruses. The replacement of RNA as genetic material by DNA might then have occurred because of at least two advantages: (1) DNA is more stable, lacking a hydroxyl group at the 2′ position of the sugar; and (2) since RNA had evolved the protein-synthesizing system, the enzymes used for this purpose would not act upon DNA, and DNA could, therefore, restrict itself to the production of templates.

Whatever the mechanism, it is clear that at some point in evolutionary history the two processes of replication and function, embodied, respectively, in nucleic acid and protein, evolved a combined relationship which gave to its bearers an immense selective advantage. Adaptive functions could now be passed on to future generations by the translation of self-replicating nucleotide sequences into functional protein sequences.

The evolution of a gene-protein system then led directly to the evolution of a genetic code. As pointed out by Sonneborn and by Woese, it seems likely that the first proteins made from a particular genic template were not always perfectly alike. Considerable "errors" must have existed in the initial imperfect translation mechanisms, so that many ambiguities could have occurred, e.g., the assignment of different amino acids to the same codon. The fact that such errors can take place is seen presently in the effects caused by streptomycin (p. 670), changes in magnesium-ion concentration, changes in alkalinity, and other translational-error-producing factors. Because of this variation the early proteins produced by a particular nucleotide sequence would probably have been "statistically" alike rather than exactly alike. That being the case, changes in the assignments of particular amino acids to particular codons would not have produced the novel and drastic effects such changes generally cause today. In other words, changed codon assignments in primitive genetic systems might produce proteins that were not markedly different from the normally varied proteins produced by an imperfect translation mechanism. Under such circumstances the genetic code would have an opportunity to evolve and achieve a state in which the fewest translational errors are produced.

The present coding dictionary (Table 28–3) has a number of features which indicate that evolution has probably proceeded in an error-reducing direction. First, note that amino acids having more than one codon have "synonymous" codons that are generally identical for the first two nucleotides (positions I and II) and differ only in the third nucleotide (position III). Since *in vitro* experiments show that position III in many codons is the one most easily subject to translational error, such a coding system would serve to help prevent amino acid substitutions.* Second, we can follow Woese's

*See Crick's wobble hypothesis, pp. 659–660

classification and divide the amino acids into two broad groups: the "functional" amino acids (tyrosine, histidine, lysine, glutamic acid, tryptophan, etc.) involved in establishing enzymatic activity for proteins, and the "nonfunctional" amino acids (phenylalanine, leucine, isoleucine, valine, alanine, threonine, etc.). Note now that translational errors at the next-most-error-prone codon nucleotide (position I) will generally cause the substitution of an amino acid from the same "group." Thus UUU (phenylalanine) may be misread as CUU, AUU, and GUU but still produce an amino acid in the nonfunctional group. Also, the finding that codons bearing pyrimidines (U, C) at the II position are more error-prone during translation than codons bearing purines (G, A) at this position may help account for the fact that the codons of the functional amino acid group are mostly of the latter type, and the nonfunctional group (in which errors are not as important) generally have codons which are of the former type.

Crick suggests that a few amino acids occupied at first most of the codons in the genetic dictionary. New amino acids were then added, taking their codon designations from related amino acids already present. Presumably, at some point the code was "frozen," and all organisms that were derived from this primitive ancestor shared this common code. The fact that only one universal code now exists may therefore indicate that only the bearers of this particular code successfully survived the early evolutionary period. Although speculations such as these are highly conjectural, they serve the important purpose of helping us to look at the genetic code from the same evolutionary point of view from which we look at other attributes of life.

With the evolution of genetic material and perfection of the genetic code, further selective advantages must have accrued to organisms with improved mechanisms of organization and transmission of genetic material. Perhaps the most significant of such developments was the evolution of the chromosome. In one structure many genes could now be arranged according to their most efficient relationships and transmitted as a unit. In lower organisms the clustering of genes with related function (e.g., the histidine loci, p. 623; or

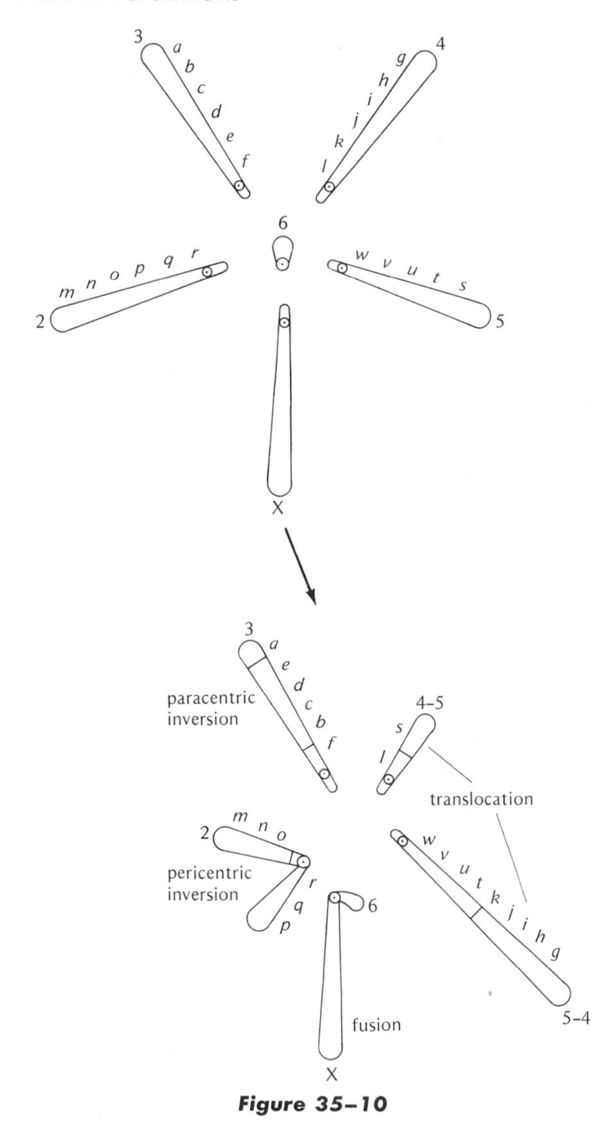

Figure 35–10

Top: primitive Drosophila *metaphase chromosome configuration. Bottom: some of the chromosome changes that occur in different* Drosophila *species during their evolution. (After Stone, 1962.)*

T4 chromosome, Fig. 26–22) indicates that chromosome organization probably serves an essential role in survival. Furthermore, the fact that all chromosomally mapped strains of *E. coli* and *Salmonella* have the same gene sequence shows that considerable selection pressure must exist for particular gene relationships to be maintained in the face of all the many possible chromosome changes.

In higher organisms organization on the level of the chromosome itself has been replaced, at least partially, by other mechanisms of gene control (see Chapters 29 and 30). Numerous higher organisms, therefore, show wide variation in their chromosome systems (Chapters 21 and 22), some of which can be traced through large portions of their evolutionary history. In the *Drosophila* genus, for instance, the basic chromosome constitution of five long arms and a dot, shown in Fig. 35–10, has evolved by inversions, translocations, and fusions into a multitude of different configurations. Wasserman, for example, has traced chromosome evolution throughout a large portion of species in the *repleta* group of this genus (Fig. 35–11) by using salivary chromosome banding patterns, as has Stalker for the *melanica* group and Carson and co-workers for the Hawaiian *Drosophilidae*.

The evolution of genetic systems involves also the development of mitotic and meiotic mechanisms as well as the development of sexual and asexual reproduction in its numerous forms. Some aspects of these topics have been mentioned in previous chapters (see Chapters 2 and 12) and are further discussed in the books of Lewis and John, Darlington, Stebbins, and White.

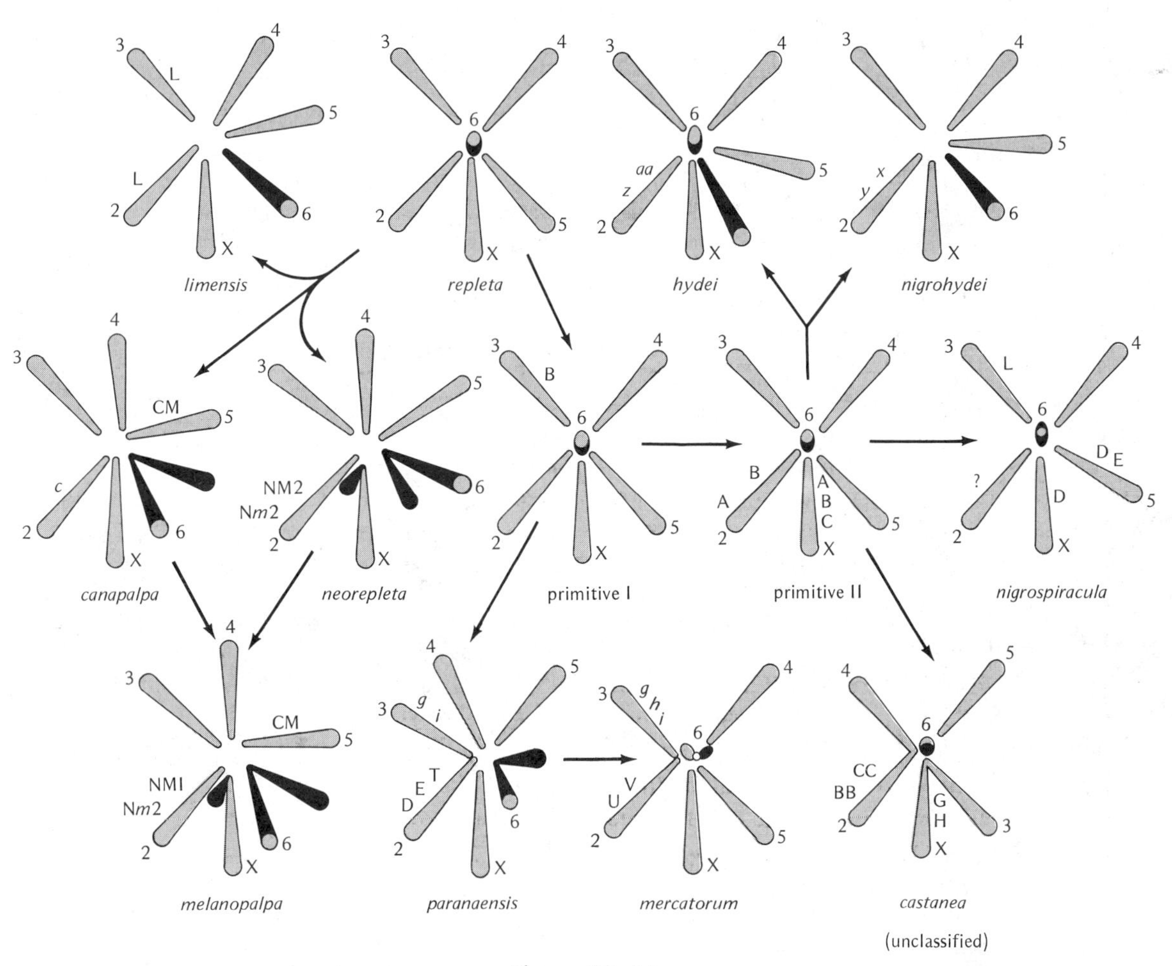

Figure 35–11

Possible paths of chromosomal evolution in some species of the repleta group of Drosophila. (After Stone, 1955; after Wasserman.)

REFERENCES

Anderson, E., 1949. *Introgressive Hybridization.* John Wiley, New York.

Betz, J. L., P. R. Brown, M. J. Smyth, and P. H. Clarke, 1974. Evolution in action. *Nature*, **247**, 261–264.

Boyd, W. C., 1964. Modern ideas on race, in the light of our knowledge of blood groups and other characters with known mode of inheritance. In *Taxonomic Biochemistry and Serology*, C. A. Leone (ed.). Ronald Press, New York, pp. 119–169.

Campbell, J. H., J. A. Lengyel, and J. Langridge, 1973. Evolution of a second gene for β-galactosidase in *Escherichia coli*. *Proc. Nat. Acad. Sci.*, **70**, 1841–1845.

Carson, H. L., D. E. Hardy, H. T. Spieth, and W. S. Stone, 1970. The evolutionary biology of the Hawaiian *Drosophilidae*. In *Essays in Evolution and Genetics in Honor of Theodosius Dobzhansky*, M. K. Hecht and W. C. Steere (eds.). Appleton-Century-Crofts, New York, pp. 437–543.

Cavalli-Sforza, L. L., 1966. Population structure and human evolution. *Proc. Roy. Soc. Lond.,* (*B*), **164,** 362–379.

Clarke, P. H., 1974. The evolution of enzymes for the utilization of novel substrates. In *Evolution in the Microbial World*, M. J. Carlile and J. J. Skehel (eds.). Cambridge Univ. Press, Cambridge, pp. 183–217.

Clausen, J., D. D. Keck, and W. M. Hiesey, 1948. Experimental studies on the nature of species. III. Environmental responses of climatic races of *Achillea*. *Carneg. Inst. Wash. Publ. No. 581*, 1–129.

Coyne, J. A., and B. Grant, 1972. Disruptive selection on I-maze activity in *Drosophila melanogaster*. *Genetics*, **71,** 185–188.

Crick, F. H. C., 1968. The origin of the genetic code. *Jour. Mol. Biol.*, **38,** 367–379.

Darlington, C. D., 1958. *The Evolution of Genetic Systems*, 2nd ed. Basic Books, New York.

Darwin, C., 1859. *The Origin of Species.* John Murray, London. (The 6th edition of this classical work has been reprinted many times.)

Darwin, C., and A. R. Wallace, 1859. On the tendency of species to form varieties; and on the perpetuation of varieties and species by natural means of selection. *Jour. Linn. Soc.*, **3,** 45–62. (Reprinted in the collection of Gabriel and Fogel; see References, Chapter 1.)

Dayhoff, M. O. (ed.), 1973. *Atlas of Protein Sequence and Structure*, Vol. 5, Suppl. 1. National Biomed. Res. Found., Washington, D.C.

Dobzhansky, Th., 1944. Chromosomal races in *Drosophila pseudoobscura* and *D. persimilis*. *Carneg. Inst. Wash. Publ. No. 554*, Washington, D.C., pp. 47–144.

———, 1951. *Genetics and the Origin of Species,* 3rd ed. Columbia Univ. Press, New York.

———, 1958. Genetics of natural populations. XXVII. The genetic changes in populations of *Drosophila pseudoobscura* in the American Southwest. *Evolution*, **12,** 385–401. (Reprinted in Spiess's collection; see References, Chapter 31.)

———, 1970. *Genetics of the Evolutionary Process.* Columbia Univ. Press, New York.

Dobzhansky, Th., L. Ehrman, O. Pavlovsky, and B. Spassky, 1964. The superspecies *Drosophila paulistorum*. *Proc. Nat. Acad. Sci.*, **51.** 3–9.

Ehrman, L., 1960. The genetics of hybrid sterility in *Drosophila paulistorum*. *Evolution*, **14,** 212–223.

Fisher, R. A., E. B. Ford, and J. Huxley, 1939. Taste-testing the anthropoid apes. *Nature*, **144,** 750.

Fitch, W. M., 1972. Does the fixation of neutral mutations form a significant part of observed evolution in proteins? *Brookhaven Symp. Biol.*, **23,** 186–215.

Ford, E. B., 1975. *Ecological Genetics,* 4th ed. Chapman and Hall, London.

Fox, S. W., and K. Dose, 1972. *Molecular Evolution and the Origin of Life*. W. H. Freeman, San Francisco.

Fox, S. W., and K. Harada, 1961. Synthesis of uracil under conditions of a thermal model of prebiological chemistry. *Science*, **133,** 1923–1924.

Frazier, W. A., R. H. Angeletti, and R. A. Bradshaw, 1972. Nerve growth factor and insulin. *Science*, **176,** 482–488.

Grant, V., 1963. *The Origin of Adaptations.* Columbia Univ. Press, New York.

———, 1965. Evidence for the selective origin of incompatibility barriers in the leafy-stemmed *Gilias*. *Proc. Nat. Acad. Sci.*, **54,** 1567–1571.

———, 1971. *Plant Speciation.* Columbia Univ. Press, New York.

Gummerson, K. S., and R. Williamson, 1974. Sequence divergence of mammalian globin messenger RNA. *Nature*, **247,** 265–267.

Hall, B. G., and D. L. Hartl, 1974. Regulation of newly evolved enzymes. I. Selection of a novel lactase regulated by lactose in *Escherichia coli*. *Genetics*, **76,** 391–400.

HEGEMAN, G. D., and S. L. ROSENBERG, 1970. The evolution of bacterial enzyme systems. *Ann. Rev. Microbiol.*, **24,** 429–462.

HILL, R. L., K. BREW, T. C. VANAMAN, I. P. TRAYER, and P. MATTOCK, 1969. The structure, function, and evolution of α-lactalbumin *Brookhaven Symp. Biol.*, **21,** 139–152.

HOROWITZ, N. H., 1945. On the evolution of biochemical syntheses. *Proc. Nat. Acad. Sci.*, **31,** 153–157. (Reprinted in the collection of Gabriel and Fogel; see References, Chapter 1.)

———, 1965. The evolution of biochemical syntheses—retrospect and prospect. In *Evolving Genes and Proteins,* V. Bryson and H. J. Vogel (eds.). Academic Press, New York, pp. 15–23.

HUXLEY, J., 1943. *Evolution: The Modern Synthesis.* Harper & Row, New York.

JUKES, T. H., 1966. *Molecules and Evolution*. Columbia Univ. Press, New York.

———, 1972. Comparison of polypeptide sequences. In *Proceedings of the Sixth Berkeley Symposium on Mathematical Statistics and Probability*, Vol. 5. Univ. of Calif. Press, Berkeley, pp. 101–127.

JUKES, T. H., and R. HOLMQUIST, 1972. Evolutionary clock: Nonconstancy of rate in different species. *Science*, **177,** 530–532.

KETTLEWELL, H. B. D., 1961. The phenomenon of industrial melanism in *Lepidoptera*. *Ann. Rev. Entomol.*, **6,** 245–262.

———, 1973. *The Evolution of Melanism.* Clarendon Press, Oxford.

KIMURA, M., and T. OHTA, 1972. Population genetics, molecular biometry, and evolution. In *Proceedings of the Sixth Berkeley Symposium on Mathematical Statistics and Probability*, Vol. 5. Univ. of Calif. Press, Berkeley, pp. 43–68.

KING, J. L., 1972. The role of mutation in evolution. In *Proceedings of the Sixth Berkeley Symposium on Mathematical Statistics and Probability*, Vol. 5. Univ. of Calif. Press, Berkeley, pp. 69–100.

KOOPMAN, K. F., 1950. Natural selection for reproductive isolation between *Drosophila pseudoobscura* and *D. persimilis*. *Evolution*, **4,** 135–148.

KRAMER, F. R., D. R. MILLS, P. E. COLE, T. NISHIHARA, and S. SPIEGELMAN, 1974. Evolution *in vitro*: Sequence and phenotype of a mutant RNA resistant to ethidium bromide. *Jour. Mol. Biol.*, **89,** 719–736.

LANGLEY, C. H., and W. M. FITCH, 1974. An examination of the constancy of the rate of molecular evolution. *Jour. Mol. Evol.*, **3,** 161–177.

LATIMER, H., 1958. A study of the breeding barrier between *Gilia australis* and *G. splendens.* Ph.D. Thesis, Claremont Univ. College, Claremont, Calif. (see Grant, 1963.)

LAVEN, H., 1959. Speciation by cytoplasmic isolation in the *Culex pipiens* complex. *Cold Sp. Harb. Symp.*, **24,** 166–173.

LEWIS, K. R., and B. JOHN, 1963. *Chromosome Marker*. Churchill, London.

LEWONTIN, R. C., 1972. The apportionment of human diversity. *Evol. Biol.*, **6,** 381–398.

MANGELSDORF, P. C., 1974. *Corn, Its Origin, Evolution and Improvement*. Harvard Univ. Press, Cambridge, Mass.

MANGELSDORF, P. C., and R. G. REEVES, 1959. The origin of corn. *Botanical Museum Leaflets, Harvard University*, **18,** 329–400.

MATHER, K., 1955. Polymorphism as an outcome of disruptive selection. *Evolution*, **9,** 52–61.

———, 1973. *Genetical Structure of Populations.* Chapman and Hall, London.

MAYR, E., 1963. *Animal Species and Evolution*. Harvard Univ. Press, Cambridge, Mass.

MCCARTHY, B. J., and M. N. FARQHAR, 1972. The rate of change of DNA in evolution. *Brookhaven Symp. Biol.*, **23,** 1–41.

MILLER, S. L., and L. E. ORGEL, 1974. *The Origins of Life on the Earth*. Prentice-Hall, Englewood-Cliffs, N.J.

MILLER, S. L., and H. C. UREY, 1959. Organic compound synthesis on the primitive earth. *Science*, **130,** 245–251.

MILLS, D. R., F. R. KRAMER, and S. SPIEGELMAN, 1973. Complete nucleotide sequence of a replicating RNA molecule. *Science*, **180,** 916–927.

MOORE, J. A., 1949. Patterns of evolution in the genus *Rana*. In *Genetics, Paleontology, and Evolution,* G. L. Jepsen, E. Mayr, and G. G. Simpson. (eds.). Princeton Univ. Press, Princeton, N.J., pp. 315–355.

ORO, J., 1961. Mechanism of synthesis of adenine from hydrogen cyanide under possible primitive earth conditions. *Nature*, **191,** 1193–1194.

PATTERSON, J. T., and W. S. STONE, 1952. *Evolution in the Genus Drosophila.* Macmillan, Inc., New York.

PONNAMPERUMA, C., and R. MACK, 1965. Nucleotide synthesis under possible primitive earth conditions. *Science*, **148,** 1221–1223.

PONNAMPERUMA, C., R. MARINER, and C. SAGAN, 1963.

Formation of adenosine by ultraviolet irradiation of a solution of adenine and ribose. *Nature*, **198**, 1199–1200.

RENSCH, B., 1960. *Evolution Above the Species Level.* Columbia Univ. Press, New York.

SCHARLOO, W., 1971. Reproductive isolation by disruptive selection: Did it occur? *Amer. Naturalist,* **105,** 83–86.

SHEPPARD, P. M., 1961. Some contributions to population genetics resulting from the study of the *Lepidoptera. Adv. in Genet.*, **10,** 165–216.

SIMPSON, G. G., 1953. *The Major Features of Evolution.* Columbia Univ. Press, New York.

———, 1964. Organisms and molecules in evolution. *Science*, **146,** 1535–1538.

SOANS, A. B., D. PIMENTEL, and J. S. SOANS, 1974. Evolution of reproductive isolation in allopatric and sympatric populations. *Amer. Nat.*, **108,** 117–124.

SONNEBORN, T. M., 1957. Breeding systems, reproductive methods, and species problems in Protozoa. In *The Species Problem*, E. Mayr (ed.). Amer. Assoc. Adv. Sci., Washington, D.C., pp. 155–324.

———, 1965. Degeneracy of the genetic code: Extent, nature and genetic implications. In *Evolving Genes and Proteins*, V. Bryson and H. J. Vogel (eds). Academic Press, New York, pp. 377–397.

STALKER, H. D., 1966. The phylogenetic relationship of the species in the *Drosophila melanica* group. *Genetics*, **53,** 327–342.

STEBBINS, G. L., JR., 1950. *Variation and Evolution in Plants.* Columbia Univ. Press, New York.

———, 1960. The comparative evolution of genetic systems. In *Evolution After Darwin*, Vol. I, The Evolution of Life, S. Tax (ed.). Univ. of Chicago Press, pp. 197–226.

———, 1971. *Chromosome Evolution in Higher Plants.* Arnold, London.

STERN, C., 1973. *Principles of Human Genetics.* W. H. Freeman, San Francisco.

STONE, W. S., 1955. Genetic and chromosomal variability in *Drosophila. Cold Sp. Harb. Symp.*, **20,** 256–269.

———, 1962. The dominance of natural selection and the reality of superspecies (species groups) in the evolution of *Drosophila. Univ. Texas Publ.*, 6205, 507–537.

THODAY, J. M., 1972. Disruptive selection. *Proc. Roy. Soc. Lond.,* (*B*), **182,** 109–143.

THODAY, J. M., and J. B. GIBSON, 1962. Isolation by disruptive selection. *Nature*, **193,** 1164–1166.

WASSERMAN M., 1960. Cytological and phylogenetic relationships in the *repleta* group of the genus *Drosophila. Proc. Nat. Acad. Sci.*, **46,** 842–859.

WHITE, M. J. D., 1973. *Animal Cytology and Evolution,* 3rd ed. Cambridge Univ. Press, Cambridge.

WOESE, C. R., 1965. On the evolution of the genetic code. *Proc. Nat. Acad. Sci.*, **54,** 1546–1552.

———, 1967. *The Genetic Code: The Molecular Basis for Gene Expression.* Harper & Row, New York.

WU, T. T., E. C. C. LIN, and S. TANAKA, 1968. Mutants of *Aerobacter aerogenes* capable of utilizing xylitol as a novel carbon. *Jour. Bact.* **96,** 447–456.

YCAS, M., 1969. *The Biological Code.* North-Holland Publ. Co., Amsterdam.

ZUCKERKANDL, E., 1965. The evolution of hemoglobin. *Sci. American*, **212,** May issue, pp. 110–118. (Reprinted in the collection of Srb, Owen, and Edgar; see References, Chapter 1.)

ZUCKERKANDL, E., and L. PAULING, 1965. Evolutionary divergence and convergence in proteins. In *Evolving Genes and Proteins*, V. Bryson and H. J. Vogel (eds.). Academic Press, New York, pp. 97–166.

ZUCKERKANDL, E., and W. A. SCHROEDER, 1961. Amino acid composition of the polypeptide chains of *Gorilla* hemoglobin. *Nature*, **192,** 984–985.

36
PROSPECTS FOR THE CONTROL OF HUMAN EVOLUTION

At the apex of our interest in evolution stands an interest in the fate of our own species. In which direction is mankind evolving? Are man's biological endowments satisfactory for his needs? What are the prospects for the control of man's evolution? Our knowledge of genetics obviously offers us the opportunity to answer some aspects of these questions. However, before making this attempt, let us first consider the unique features of *Homo sapiens*.

Man is a distinctive animal. Physically he is large-brained, relatively hairless, standing upright with long legs, short toes, prehensile hands, and opposable thumbs. More distinctive than his physical appearance, however, is man's intelligence, which provides him with his uniquely flexible adaptive behavior; that is, man "learns" from environmental experiences by incorporating such experiences into his behavior. He does this in an essentially Lamarckian fashion, by consciously "acquiring" behavioral responses that answer the needs of specific situations to which he is exposed. This mode of behavior is significantly different from that of other organisms, which are primarily forced to rely for survival upon the "instinctive" responses built into their nervous systems. Man, by contrast, can be raised in different environments and learn to obtain his food, defend himself, provide shelter, and perform numerous cultural tasks in many specialized ways. Most important, man can acquire and transmit such information through language between individuals and between generations.

In short, man has two uniquely different hereditary systems. One is the system that transfers biological information from parent to offspring in the form of genes and chromosomes. The other is the system that transfers cultural information from speaker to listener, from writer to reader, from viewer to spectator, and forms our cultural heritage.

CULTURAL AND BIOLOGICAL EVOLUTION

The changes provided by cultural heredity over the last 10,000 years have been most impressive. We have moved from bands of hunters and fishermen mainly concerned with obtaining food, to complex urban societies in which such concerns occupy relatively little time for many of us. Instead of hunting and primitive food gathering, an increased proportion of our efforts now concern various cultural and technological tasks that could never have been foreseen 10,000 years ago, or even just a generation ago. In some areas one can at present hardly predict from one year to the next where the changes will come from and of what type they will be. This remarkable rapidity in cultural and technological change, at least in the field of science, shows even further promise of increase if we consider the increased proportion of scientists that now exist and are now being trained. Price has provided a widely quoted estimate that, of all scientists that have ever lived, more than 90 percent are alive today!

In contrast to this rapid improvement in cultural and technological heredity, the improvement in man's genetic or biological heredity during this 10,000-year period has been undetectable. In regard to man's most distinguished possession, his brain, there has been no change in size in *Homo sapiens* in the last one hundred thousand years and there is no clear indication that there has been any qualitative change as well. That is, our ancestors of many years ago, given our training, may well have provided the same range and distribution of mentality as we have today. Why this difference in speed between cultural and biological evolution?

By way of oversimplifying, although not too seriously, this contrast can be ascribed to differences between two distinct types of evolution; the mode of inheritance of acquired characters utilized by cultural evolution, and the mode of inheritance through natural selection utilized by biological evolution. The Lamarckian mode of cultural evolution is an extension of the method by which man "learns." It depends upon the presence of conscious agents, that is, humans with brains, that are able to modify inherited cultural information in a direction of greater adaptiveness or utility. Transmission occurs from mind to mind rather than through DNA. Men pick up information from their ancestors and contemporaries, change it purposely to provide improved utility for themselves and their offspring, and pass it on. An adaptive or useful social character is thus acquired, so to speak, through the purposeful modification by man of his culture and technology in what he considers an advantageous direction. The speed with which such modification takes place and the consequent speed of cultural change are limited by numerous factors, but, theoretically, in the long run, primarily by man's inventiveness. Furthermore, the generation time for cultural evolution may be as rapid as communication methods can make it. An improvement can now be proposed in one part of the world and modified in another part with more than airplane speed.

In striking contrast to the speed of cultural evolution is the slow progress of natural selection. The reasons for this are apparent from previous discussions. As far as we know, there are no cellular particles that are sufficiently intelligent to detect or determine the direction of biological evolution, and then change themselves accordingly. Organic evolution, as we have seen, occurs through a process of selection acting upon random genetic changes. According to this view, there are chance differences that arise in genes or combinations of genes (mutation and recombination) which produce a variety of effects on their carriers. These genetic differences furnish an array of genotypes of which the environment chooses for survival those that are reproductively most successful. Genetic evolution is slow since it must await fortuitous accidental genetic changes in DNA before it can proceed, and each change may take a considerable number of generations before it can be incorporated into the population.

This disparity in speed between cultural and biological evolution has emphasized many of our biological inadequacies. Let us consider a few of these.

BIOLOGICAL LIMITATIONS

Because of our advanced technology we are becoming to a large degree sedentary in occupation, but our intestine and appetites do not follow accordingly. Many who live in surplus societies, such as the United States, tend to put on weight and suffer from the accompanying ills. The pains and problems of childbirth are probably a consequence of man's erect posture and are undoubtedly aggravated by the lack of proper physical exertion. The stress of many aspects of social living, ambition, and competition finds much of the human race biologically unprepared, and we suffer from anxiety, ulcers, heart disease, and other socially aggravated illnesses (called "the ulcer belt syndrome" by Comfort). Pollutions of various kinds caused by sewage, tobacco, automobiles, and industry lead to a variety of modern diseases ranging from induced cancer to emphysema and silicosis.

Perhaps one of the most important contrasts between what we would like to be and what we are lies in the difference between biological and cultural maturity. Biologically our efficiency begins to fall soon after we reach the reproductive age of 20 to 30 years. Our cultural efficiency, however, in the sense of contributions that we can make, just begins at that time or somewhat later. Our cultural development is thus limited by our biological decline; that is, our biological heritage stresses reproductive success and hardens the arteries afterward, while our cultural development asks for continued plasticity and longevity.

Postreproductive longevity, unfortunately, is a trait that tends to remain low in many organisms that reach reproductive maturity relatively early in their potential life-span. In the years before civilization, only about half the human population passed the age of 20 and probably not more than 1 out of 10 lived beyond 40. These low longevity values extended into the period of the early Greeks and even into modern periods among primitive people. Life expectancy remained between 20 to 30 years until the Middle Ages, then rose somewhat, and has risen very sharply among Europeans and Americans in the last century, from a life expectancy of about 40 years in 1850 to the present 70 years.

These statistics are important because they indicate that we now have among us an age group, those 40 to 50 years and older, upon whose biological attributes natural selection has never directly operated. In other words, the adaptive traits of these individuals are those they had possessed in the years prior to their reproductive periods, and their post-reproductive fitness is no longer reflected in their relative reproductive success. For example, an individual who has produced three children, but who, at the age of 50 develops cancer or other diseases with genetic components is by this fact no less reproductively successful than an individual of the same age who has produced three children but does not suffer from such diseases. One may, of course, argue that children with healthy grandparents are more fit than children with ill or absent grandparents since they get more attention and care. In social situations, however, such caretaking functions can be easily assigned to other individuals, and it is unlikely that grandparental attention adds to reproductive fitness.*

Healthy old age is, therefore, a trait which only relatively few genetic variants might be expected to attain. At present, 2 percent of the population reaches 90 years and only about 1 per 1000 individuals reaches 100. There is, however, little promise that, even with considerable medical progress, average life expectancy can be raised beyond 80 to 85 years.

A further contrast between past and present biological requirements is in fertility. "Clutch size," a term used to describe the number of offspring born to a nesting pair of birds, is certainly as adaptive a factor in man as it is in birds. That is, the ability of a mature human female under primitive conditions to produce eight or nine offspring during her reproductive period, of which an average of two survive, is of selective value in ensuring that the population will continue in the

* Glass recently (and Muller, earlier) pointed out that senescence and death may have evolutionary value to populations, since such means help to ensure the turnover and replacement of older genotypes by new genotypes that may be better adapted.

face of high infant mortality. However, in many countries, even those that are technologically backward, infant mortality rates have markedly decreased. In countries such as Chile, the rate of infant mortality is now about 10 percent of all births, or half what it was 45 years ago, and in countries such as the United States, infant mortality is about one quarter of what it is in Chile. This overall imbalance between human fecundity and survival rate has led to an exponential growth of the human population, which is now *doubling* at the rate of once every 30 to 40 years, or doubling at a rate that is almost *7000 times* as fast as in primitive paleolithic societies.

This has led to a veritable "population explosion," which, if unchecked, would produce a world population of 50 billion people in only five generations. When we consider that agricultural production is barely able to keep up with its present requirements, the problem is indeed serious, especially if we add the widespread demand among poor people for an increased standard of living. Statistics by the United Nations show no per capita increase in food production since World War II. Many perennial famine areas still suffer from food shortages, and in some areas, such as India, yield per acre is still low, and small agricultural increases have only come from putting more land into cultivation. The increase in agricultural acreage is, however, a limited solution because of the limitation in usable land. In the United States today each person has about 6 acres of agriculturally usable land to supply his needs and also to provide surpluses to be used by some of the agriculturally deficient countries. In England usable acreage falls to one-tenth this value and in Japan to one-thirtieth. Considerable amounts of usable land are still available in some newly developing countries, such as in Africa, and some agricultural gains can be made by using new high-yield varieties, but even these increases will be unsatisfactory if population growth continues.

Fortunately a reduction in human fecundity can be instituted through presently known birth-control methods without the need for evolving such a reduction biologically. The most serious problem that lies ahead in this respect is that of educating people and institutions to utilize and promote such methods. The same approach toward other biological inadequacies, however, such as low longevity and the "ulcer belt syndrome," cannot be undertaken as easily. For these traits there is a need for some sort of controlled genetic change that would help put man in biological harmony with his surroundings. The question that arises is, therefore, whether we can impose a direction upon our biological evolution, and, through our own conscious mediation, approximate the speedier method of inheritance of acquired characters. If this were possible, what kind of genetic change would be desirable?

DELETERIOUS GENES

Let us consider briefly possible changes in the frequencies of genes that have obviously deleterious effects. The number of children that can be classified as born markedly defective, either physically or mentally, can be conservatively estimated at about 20 to 25 in a thousand births, and the mortality rate ascribed to such congenital malformations in the United States is about 15 percent of all infant deaths. Many other defects, although not immediately noted at birth, become apparent during the childhood years and are more widely prevalent than may be imagined. In various studies, about 30 percent or more of hospital admissions for children, and 50 percent of all childhood deaths are ascribed to birth defects or to complications that may have been caused by such defects. Although not all birth defects are genetically produced, the proportion of genetically caused defectives among them is undoubtedly high. If we include other genetic defects that appear later in life, such as muscular dystrophy, diabetes, and so on, this value can probably be doubled quite easily. Now, if in addition we also include less obvious defects which nevertheless have strong genetic components, such as impaired resistance to stress and infection, and various other physical and psychological weaknesses, the effect of deleterious genes must touch at least a majority of our population.

The ubiquity of deleterious genes with lethal effect has been dramatically demonstrated by studies made of the offspring of cousin marriages by Morton, Crow, Muller, and others. These studies have utilized the techniques of detecting and partitioning the genetic load caused by inbreeding (see pp. 814–815) and have shown that outwardly normal individuals in our society carry a genetic load equivalent to that of approximately one to eight deleterious lethal genes ("lethal equivalents") that, if homozygous, would cause early death.

Now, two important questions we can ask are: (1) What accounts for the high prevalence of these deleterious genes, and (2) what, if anything, can we do to get rid of them?

The reasons for their high frequency are not yet fully agreed upon, although there is little question that they all arise originally through mutation. One opinion, held by the late Theodosius Dobzhansky and others, is that such genes, although deleterious in homozygous condition, may offer considerable advantage to their heterozygous carriers by producing some sort of hybrid vigor. According to this theory, a gene will be maintained in the population although the homozygote produced by this gene is relatively inferior in fitness (see p. 766). Another school, formerly headed by the late Herman Muller, believes that such genes produce no advantage of any kind, and that their frequency is now high because the usual effect of natural selection has been artificially reduced. According to Muller, genotypes that were formerly defective and would have been eliminated under more primitive conditions are now kept alive by medical techniques and enabled to pass on their defective genes to their offspring. As we know, a decrease in the selection coefficient against a particular gene causes an increase in the equilibrium frequency of the gene ($q = \sqrt{u/s}$ for a recessive gene, $p = u/s$ for a dominant gene; pp. 770–771). Thus if deleterious genes are not eliminated by selection they will gradually increase in frequency in accord with their mutation rate. Since the mutation rate is usually low, the frequency of any particular gene will increase rather slowly, but since there are many possible deleterious genes, the genetic load will increase significantly.

It is obvious that according to Muller's theory we will be unable to reach any biological harmony until most such genes are removed. If they are not removed and continue to increase in frequency, Muller held out the eventual prospect that the human race will end up with two types of individuals: one kind will be so genetically crippled that they can hardly move, and the other kind will be somewhat less crippled but spend all their time taking care of the first kind. This specter is made even gloomier if we attempt to consider means by which such genes can be eliminated. Since all of us are probably carriers of at least a few deleterious recessive genes, most of which we do not know about, there is little prospect in eliminating them short of mass sterilization.

Serious as this argument may be for deleterious genes that produce severe handicaps, it is undoubtedly exaggerated for genes whose deleterious effects can be treated relatively inexpensively. Nearsightedness, for example, is a trait whose frequency has most likely increased in recent periods yet can nevertheless be corrected quite simply by an optometrist. Furthermore, the fact that natural selection no longer operates to eliminate many genotypes is not necessarily an undesirable feature of modern life. Few individuals would argue today that fire and clothes should be abolished because they are artificial devices that circumvent natural selection by permitting nonfurry genotypes to survive in cold climates. It would also be difficult for us to hearken back to "the good old" prevaccination, presanitation days of smallpox, diphtheria, typhus, cholera, and plague.

In spite of medical and cultural progress, however, the effect of numerous deleterious genes cannot be easily treated, and although Muller may have been overly alarmed about their increase in frequency, we have become more aware of their widespread existence in recent years. Many geneticists have, therefore, turned to exploring the possibility of controlling deleterious effects by artificially changing gene frequencies.

EUGENICS

In its modern form, suggestions for improving human genetic material have come under the name *eugenics*, a term proposed by Francis Galton before the turn of the century.* Galton, concerned with the heredity of quantitative characters such as intelligence, became aware, after reading Darwin, that the evolution of a trait through natural selection could be substituted by the evolution of a trait through social selection. The approach of many early eugenicists, however, reflected their own personal biases as to which characteristics were desirable and which undesirable.† This was true of many English and American eugenicists and reached its culmination in the "racial health" movement in Germany during the 1930s. The Germans promulgated "eugenic" laws establishing themselves as the "master race" destined to be served by "inferior" subject races. Intermarriage between a German "Aryan" and a "non-Aryan" was forbidden, and millions of people were killed because of their membership in "inferior" races or groups. The particular social and political causes that fostered this genocide are not the subject of this book, although it is important to recognize that myths about the "degeneration" of race and intelligence because of mixture with "inferior" types are continually being perpetuated.

In the United States, one of the true racial "melting pots" of the world, there is no evidence for the biological superiority of any particular race in respect to intelligence. Blacks, long considered to be at the bottom of the racial pecking-order, show as wide a range of intelligence as do whites. According to Pettigrew and others, racial differences in IQ examinations have been demonstrated to be remarkably plastic, influenced by such factors as prenatal diet, early cultural surroundings, and even the color of the interviewer in the IQ exam (see also Chapter 10). Predictions that intelligence will steadily decline because of the higher reproductive rate of the lower "unintelligent" social classes and their consequent increase in frequency are also contradicted by facts. In a Scottish survey that covered almost 90 percent of all 11-year-old children in 1932 and again in 1947, no decrease in IQ was found. On the contrary, these studies showed a significant increase in average intelligence during this interval.

The fears expressed by various writers that the abolition of privileged classes in society will lead to "hybridization" and thus to the abolition and "dilution" of superior genotypes are therefore hardly scientifically based. The evidence at present is that high intelligence is not the exclusive genetic property of a particular social class, but rather that its expression can easily be masked in any group by deficiencies in diet, lack of cultural stimulation, and absence of opportunity. As environmental conditions improve, average intelligence scores may also be expected to improve, although genetic differences between individuals will still remain. We may, in fact, predict that equality of economic and educational opportunity for all classes will enable each individual to more nearly achieve his true potential. Society will be the benefactor in producing more creative and inspired individuals such as Leonardo da Vinci, Voltaire, and Newton,

---

*One of the first proposals suggesting that man could be improved through selective breeding was made by Plato in his dialogue *The Republic*. In the ideal philosopher-state described by Plato, only the most physically and mentally fit individuals were to be mated and their offspring raised by the state. Inferior types were to be prevented from mating or their offspring destroyed. Since family relations were absent in *The Republic*, determination of superior and inferior types could be accomplished impartially, and the governing class was selected only "from the most superior." However, the ancient notion of superiority and inferiority was much different from that of Plato. A conquering people was considered "superior," a subjugated people was considered "inferior." In ancient Sparta, for example, some measure of selective breeding seems to have been practised with the purpose of raising a "superior" military ruling class to hold in subjugation the "inferior" servant classes. The fact that cultural disparities usually existed between conquerors and subjects reinforced such notions but did not seriously reflect whether any essential biological differences were responsible for these cultural differences. Were the Spartans biologically "superior" to the Helots, to the Corinthians, Athenians, and so on?

†C. B. Davenport, a New Englander, exemplified this approach in the United States using New Englanders as the standard of comparison for all American nationalities. According to Davenport, presumed social characteristics (e.g., Italian "violence," Jewish "mercantilism," Irish "alcoholism," etc.) had identifiable genetic components. Davenport even described nervousness, cheerfulness, and piety as possible inherited family traits.

who may otherwise die anonymously among the dispossessed sections of our society. As Dobzhansky has stated, there is little to lament in "the passing of social organizations that used the many as a manured soil in which to grow a few graceful flowers of refined culture."

Stripped from racism and provincial prejudice, eugenics may be considered as a serious attempt to diminish human suffering and improve the human gene pool. It has been subdivided into two aspects: (1) negative eugenics, the attempt to decrease the frequency of harmful genes; and (2) positive eugenics, the attempt to increase the frequency of beneficial genes. Negative eugenics involves social discouragement of reproduction by genotypes that are most obviously deleterious. For example, it would be foolish and self-destructive to encourage hemophiliacs, who are being preserved by blood transfusions, to reproduce. Similarly, where known, female carriers of the hemophilia gene should be made aware of their genetic problem and encouraged not to pass it on. These eugenic programs will suffice to control suffering from a number of deleterious genes, although they will not eliminate them, and many of these educational measures are already in practice today. Many deleterious genes that are present in high frequency, however, such as diabetes, and others where the carriers are not known, such as infantile amaurotic idiocy, cannot be controlled in this way (see the very inefficient elimination of recessives under selection, pp. 745 and 761–763).

It might, therefore, seem somewhat more encouraging to place emphasis on positive eugenics—increasing the frequency of beneficial traits rather than merely decreasing the frequency of deleterious genes. Unfortunately many characteristics we consider desirable, such as high intelligence, esthetic sensitivity, good physical health and longevity, are not caused by single genes that are easily identified, but by complexes of many genes acting together in appropriate environments.

In other organisms in which the creation of beneficial gene complexes has been attempted, the methods involve various complicated selection schemes based on selection of families and testing progeny under controlled environmental conditions. As discussed by Lerner and others, the results of these experiments have, in general, improved certain complex characters by some degree but have usually caused the deterioration of others. One characteristic that usually suffers most in such experiments is that overall quality called "fitness"; many highly selected lines end up physically debilitated and sterile. Muller, Crow, and others, however, have pointed to the likelihood that traits such as high intelligence and esthetic sensitivity have not been stringently selected for in the past, and considerable genetic variability for these traits probably exists. Thus, were selection to be instituted for these traits, the population might well respond rapidly without an accompanying fall in fitness. The means of selection themselves, however, assume paramount importance in man. Eugenic measures dictating who is to mate with whom would be intolerable, even presuming that environmental controls can be placed on human activity for selection progress to be evaluated.

As a first approach toward a more acceptable method of positive eugenics than selective mating, Muller and others have proposed the utilization of sperm banks containing the preserved frozen sperm of outstanding creative individuals. According to this method, called *germinal choice* or *eutelegenesis*, married women volunteers would choose to be artificially inseminated by males that were long dead but had highly desirable characteristics. Possible acceptance of this method has some precedence in the fact that between 5000 and 10,000 babies are presently born in the United States on the basis of sperm donors. The cause for most of these donor fertilizations lies in the sterility of the husband, although in some cases genetic incompatibilities between husband and wife (e.g., Rh factor) or genetic defects in the husband (e.g., hemophilia) are responsible. Muller proposed to extend these donor fertilizations by educating couples to desire a highly superior genetic endowment for their children, and by then demonstrating the increased proportion of genetically gifted children that will presumably be produced by this method.

Other proposed eugenic methods involve direct manipulation of human DNA by "genetic surgery" or "genetic engineering." Most or all of these methods may sound like science fiction, but considering the speed with which genetics has been developing, they should at least be mentioned and briefly evaluated.

The first type of genetic surgery is the process of transformation, which, as we know, involves taking the raw DNA from one organism and by cellular insertion allowing it to recombine with the DNA of another organism (Chapter 19). Hopefully, by this process one out of a thousand or one out of a million cells become transformed. Unfortunately the probability is still low that we will come to know which particular sequences of nucleotides determine the many complex characteristics we would like to have. Even were such knowledge possible, as in some procaryotes (see Fig. 28–5), the difficulty of extracting from human chromosomes only the desired sequences to use as donor DNA, coupled with the relatively low frequency of transformation, would make such a technique difficult, if not impossible, to evaluate. Similar criticisms hold true for a possible second mode of genetic surgery, DNA transfer by means of viral particles, or transduction.

A third mode of genetic surgery involves the use of chemical mutagenic agents that would affect specific nucleotides or sections of DNA more than others (see Chapter 24). It might be remotely possible to subject a specific portion of DNA to chemical mutagens in the hope that it will be changed in a desirable direction. The obstacle again, however, is the problem of isolating the desired section of DNA and, in some as yet unknown way, confining the action of the mutagen to only the desired nucleotide or nucleotides.

Another possible eugenic method is parthenogenesis—to induce females bearing desirable genetic constitutions to lay diploid eggs which do not have to be fertilized. Such eggs would more truly reflect the constitutions of their mothers than fertilized eggs, and thereby permit the replication of desirable maternal genotypes. Other proposals include the transplantation of diploid nuclei from desirable genotypes into unfertilized eggs, and the conversion of somatic tissues into embryonic tissues. By these methods "clones" of individuals could be established, all bearing the same genotype. Even if such genetic uniformity is desirable, the techniques necessary to implement these proposals remain to be perfected.*

Although the outlook for positive eugenics is not bright at the present time, the technical obstacles associated with it are probably not insuperable (see the review of Widdus and Ault). Some small successes have so far been achieved, such as the experiments reported by Merril and others in which tissue culture cells derived from galactosemic patients are believed to incorporate a missing enzyme (p. 665). Also, the mRNA of α and β hemoglobin chains have now been isolated and used to produce DNA specific for human hemoglobin by means of reverse transcriptase (p. 106). The use of restriction endonucleases has also opened a large new field enabling the "cutting" and "splicing" of DNA (pp. 425ff.).

Perhaps some form of eutelegenesis or genetic engineering can be developed which would be acceptable and productive. If it is to come, eugenic progress will depend primarily upon the recognition that man's personal happiness is strongly limited by his biological heritage, and that along with man's cultural evolution his development must also be considered in terms of genetic biological evolution. Like other creatures, man evolves, but, unlike other creatures, man knows that he evolves. The control of this biological evolution is therefore the change from reproductive success caused by natural selection to reproductive success caused by

*As a more immediate attack on genetic problems, Lederberg has suggested the development of "medical engineering," or *euphenics*—to affect the phenotype of man by intervening as soon as possible into his genetically defective development. Transplantation of tissues and organs, development of artificial organs, industrial synthesis of specific hormones and enzymes are all suggested devices that can be used to help prolong man's life. Such methods are continually being invented and used by medicine, and are exemplified in recent successful kidney transplants, and in the treatment of diabetes and phenylketonuria. It is also possible that the insertion of specific eucaryotic DNA into procaryotic plasmids (p. 428) may become one method of cheaply producing eucaryotic hormones and enzymes, or of "infecting" patients lacking such proteins with procaryotic molecules that can produce them.

human choice. Daring as it sounds, it is no more daring than the method by which many cultural advances have been and will be made.

Perhaps the most pertinent question we can ask of eugenics is its goal. Even if we assign to eugenics the most moral of motives—that of the good of mankind—it still remains to be determined whether this "good" is known. Can we choose the direction of human evolution with the certainty that this direction leads to what is best for our descendants? Shall we populate the world with the weak or the strong? With the sensitive or the insensitive? Fortunately the answer to this question does not have to be an unequivocal choice of either one type or the other. Rather, we may choose many different genotypes, among whom factors such as intelligence and longevity will undoubtedly rank highly in value.

If there is any serious obstacle to man's biological and cultural progress, it lies in the present disparity between man's control over tremendously powerful energy sources and the absence of knowledgeable control over many important social, economic, and political problems. Under these circumstances the aggressive emotions by which man has traditionally protected himself against others have become, when acted out on a broad national or international scale, genocidal and suicidal.

One aspect of the future is therefore clear. The promise that genetics holds for the benefit of mankind cannot bear fruit without the social development that is able to utilize this knowledge. Freedom from racial prejudice, from economic and political exploitation, and above all from self-destruction, are all prerequisites for the beginnings of freedom from biological limitations. If we can free ourselves of our presently anachronistic social systems and prejudices, our future is bright. Reflecting the vision of numerous authors such as Dobzhansky, Muller, Teilhard de Chardin, and others, Platt has eloquently stated a grand expectation:

"Now for several hundred years the great evolutionary hormones of knowledge and technology have been pressing us, almost without our understanding it, into power and prosperity and communication and interaction, and into increasing tolerance and vision and choice and planning—pressing us, whether we like it or not, into a single coordinated humankind. . . . In a short time we will move, if we survive the strain, to a wealthy and powerful and coordinated world society reaching across the solar system, a society that might find out how to keep itself alive and evolving for thousands or millions or billions of years, a time as long as all of evolution past. It is a tremendous prospect. . . . It is a quantum jump. It is a new state of matter. The act of saving us, if it succeeds, will make us participants in the most incredible event in evolution. It is the step to Man."

REFERENCES

Clow, C. L., F. C. Fraser, C. Laberge, and C. R. Scriver, 1973. On the application of knowledge to the patient with genetic disease. *Progr. Med. Genet.*, **9**, 159–213.

Comfort, A., 1963. Longevity of man and his tissues. In *Man and His Future*, G. Wolstenholme (ed.). Little, Brown, Boston, pp. 217–229.

Crow, J. F., 1961. Mechanisms and trends in human evolution. *Daedalus*, **90**, 416–431.

Davis, B. D., 1970. Prospects for genetic intervention in man. *Science*, **170**, 1279–1283.

Davis, K., 1964. Population. *Proc. XVI Intern. Congr. Zool.*, **9**, 3–14.

Davenport, C. B., 1911. *Heredity in Relation to Eugenics*. Holt, Rinehart and Winston, New York.

Dobzhansky, Th., 1962. *Mankind Evolving*. Yale Univ. Press, New Haven, Conn.

Dunn, L. C., 1962. Cross currents in the history of human genetics, *Amer. Jour. Hum. Genet.*, **14**, 1–13. (Reprinted in Boyer's collection; see References, Chapter 1.)

Ebling, F. J., and G. W. Heath, (eds.), 1972. *The Future of Man*. Academic Press, London.

Glass, B., 1965. *Science and Ethical Values*. Univ. of N. Carolina Press, Chapel Hill.

Haller, M. H., 1963. *Eugenics: Hereditarian Attitudes*

in American Thought. Rutgers Univ. Press, N.J.

Hilton, B., D. Callahan, M. Harris, P. Condliffe, and B. Berkley (eds.), 1972. *Ethical Issues in Human Genetics: Genetic Counseling and the Use of Genetic Knowledge.* Plenum Press, New York.

Jones, A., and W. F. Bodmer, 1974. *Our Future Inheritance: Choice or Chance?* Oxford Univ. Press, Oxford.

Lederberg, J., 1963. Biological future of man. In *Man and His Future,* G. Wolstenholme (ed.), Little, Brown, Boston, pp. 263–273.

Maynard Smith, J., 1965. Eugenics and utopia. *Daedalus,* **94,** 487–505. (Reprinted in Morris's collection; see References, Chapter 31.)

Medawar, P. B., 1959. *The Future of Man.* Basic Books, New York.

Morton, N. E., J. F. Crow, and H. J. Muller, 1956. An estimate of the mutational damage in man from data on consanguineous marriages. *Proc. Nat. Acad. Sci.,* **42,** 855–863. (Reprinted in Spiess's collection; see References, Chapter 31.)

Muller, H. J., 1963. Genetic progress by voluntary conducted germinal choice. In *Man and His Future,* G. Wolstenholme (ed.). Little, Brown, Boston, pp. 247–262.

Pettigrew, T. F., 1964. Race, mental illness and intelligence: A social psychological view. *Eugenics Quart.,* **11,** 189–215.

Platt, J. R., 1965. The step to man. *Science,* **149,** 607–613.

Price, D. J. da S., 1963. *Little Science, Big Science.* Columbia Univ. Press, New York.

Scottish Council for Research in Education, 1949. *The Trend of Scottish Intelligence.* Univ. of London Press, London.

Teilhard de Chardin, P., 1955. *Le Phenomene Humain.* Du Seuil, Paris. (Translated into English and reprinted 1959 under the title *The Phenomenon of Man,* Harper & Row, New York.)

Widdus, R., and C. R. Ault, 1974. Progress in research related to genetic engineering and life synthesis. *Intern. Rev. Cytol.,* **38,** 7–66.

AUTHOR INDEX

Numbers in *italics* indicate pages listing complete reference citations.

SUBJECT INDEX

A page number in **boldface** indicates mention of the subject in a figure or figure legend.